Ribozymes

Ribozymes

Volume 1

Edited by

Sabine Müller
Benoît Masquida
Wade Winkler

WILEY-VCH

Editors

Sabine Müller
University Greifswald
Institut für Biochemie
Felix-Hausdorff-Str. 4
17489 Greifswald
Germany

Benoît Masquida
CNRS – Université de Strasbourg
UMR 7156
Génétique Mollaire Génomique
Microbiologie
4 allée Konrad Roentgen
67084 Strasbourg
France

Wade Winkler
The University of Maryland
Cell Biology & Molecular Genetics
3112 Biosciences Bldg.
MD
United States

Cover
Courtesy of Dr. Benoît Masquida

Library of Congress Card No.:
applied for

British Library Cataloguing-in-Publication Data
A catalogue record for this book is available from the British Library.

Bibliographic information published by the Deutsche Nationalbibliothek
The Deutsche Nationalbibliothek lists this publication in the Deutsche Nationalbibliografie; detailed bibliographic data are available on the Internet at <http://dnb.d-nb.de>.

Print ISBN: 978-3-527-34454-3
ePDF ISBN: 978-3-527-81455-8
ePub ISBN: 978-3-527-81453-4
oBook ISBN: 978-3-527-81452-7

Typesetting Straive, Chennai, India
Printing and Binding CPI Group (UK) Ltd, Croydon, CR0 4YY

Printed on acid-free paper

C083571_240621

Contents

Preface

Ribozymes, a neologism that appeared in 1982 in a paper of Tom Cech, triggered the blossom of a field of research on which the RNA community gathered its countless efforts over years. Thus, after the discovery of the first naturally occurring catalytic RNAs, more than 30 years ago, research in the field of ribozymes and RNA catalysis has made tremendous progress. In the 1990s, many of the catalytic RNAs known today were first identified in nature. In parallel to these discoveries, the powerful SELEX method for discovery of novel nucleic acids was invented, which allowed the development of artificial ribozymes with rather diverse functionalities. Over the years, investigation into the structure and mechanism of ribozymes led to a deep understanding of their catalytic strategies. Today, ribozymes are understood to an extent that it is possible to utilize rational design and molecular engineering to construct catalytic RNAs with pre-defined function. Yet, there is still much to be learned. Improvements in high-throughput bioinformatics approaches are still fostering the discovery of new ribozymes and novel genomic locations of known motifs in highly diverse genetic contexts for all branches of life. In spite of an apparent loss of attraction due to the discovery of the RNAi and CRISPR/cas mechanisms, which at times seemed more appealing, ribozymes still inspire activity of many research groups. Even after three decades of research following discovery of the first catalytic RNA, ribozyme research has not lost the intriguing and highly motivating flair of the first days. There are still many questions to be addressed and much is waiting to be discovered.

This book aims to survey what we have learned over the past 35 years about ribozymes and nucleic acid catalysis and to present the today state of the art. It musters over 30 chapters demonstrating this activity. From the study of artificial ribozymes developed by Darwinian selection to natural ribozymes including the large translation and splicing machineries, via ribozyme engineering, and their biochemical and biophysical studies, this book takes the reader through a thrilling journey. An enormous knowledge has been accumulated on ribozyme structure, function and mechanism. However, the wealth of data also makes apparent the missing parts, which encompass what are the cellular processes regulated by ribozymes and how this is done. For instance, only little is known about the role of ribozymes identified in the human genome, and the biological roles of the new ribozymes discovered in bacteria have just started to be studied. Ribozymes deserve a continuous look from researchers, and state of the art investigation methods need to be coupled with classical ones to unravel the still unknown.

We wish to thank all participating authors for the tremendous work. Their efforts and high-quality contributions have made this book come true. We hope we have succeeded in providing a source of information on mechanistic and structural aspects of nucleic acid catalysis, on tools and methods for characterization, engineering and application of ribozymes, as well as on the key questions, strategies and challenges in ribozyme research today.

December 2019

Sabine Müller
Benoît Masquida
Wade Winkler

Foreword

The study of ribozymes, RNA-based catalysts, and subsequently nucleic acid-based catalysis has been immensely instrumental in the push of efforts toward a deeper understanding of RNA structure and function. The field attracted many scientists trained in chemistry, biology, or computer science. In 1993, a few years after the Noble Prize in Chemistry (https://www.nobelprize.org/prizes/lists/all-nobel-prizes-in-chemistry) was awarded to Tom Cech [1] and Sidney Altman [2] for the discovery of ribozymes, the RNA Society (www.rnasociety.org) was founded and two years later the *RNA Journal* (https://rnajournal.cshlp.org) started. Ever since, the annual RNA meetings have gathered more than 1000 scientists from all over the world. Many new techniques and approaches were developed accelerating the pace of discoveries on RNA. For many years, the meetings started with a session on "RNA Catalysis." However, with the avalanche of new data, new RNAs, and new biology, the session on catalysis session has dwindled. Billions of years ago, RNA did start the chemistry of the game of life, but was overwhelmed by the initiated evolutionary processes.

The central group of the chemical and biological feats achieved by RNA is the ribose hydroxyl O2′, at the same time key actor and Achilles's heel. And a DNA phosphodiester bond is 10^4–10^5 less prone to cleavage than a RNA phosphodiester bond [3]. As Paracelsus wrote several centuries ago: "All things *are* poison *and* nothing *is* without poison; *only the* dose makes *a* thing not *a* poison".

The catalytic power of the 2′-hydroxyl group must be strongly controlled and amplified only at specific location in the RNA sequence. Catalysis generally is initiated by the formation of an anionic O2′ oxygen. In any biological system, catalysis requires accessibility, local molecular dynamics and solvent molecules (water, ions, or small ligands). A water molecule (or a hydroxide ion), or a solvated divalent ion, or an amine group from a ligand can thus capture the proton from the 2′-hydroxyl group. Afterwards, depending on the available mobility of the nucleotide, the anionic O2′ oxygen can attack its own 3′-phosphate group or another phosphodiester bond. Other types of chemical reactions have also been achieved using ribozymes as described in several chapters of this book.

Accessibility and local dynamics depend on the RNA sequence and the ensuing RNA architecture. The distinctive mark of nucleic acids is the formation of pairs between the bases of the nucleotides with the complementary Watson–Crick pairs the most frequent ones. However, such pairs form only regular helices and any complex fold or assemblies of helices require linking segments that engage in some type of non-Watson–Crick base pairs. Depending on the complexity and compactness of the RNA fold, not many bases will remain unpaired with a large number of degrees of freedom. Such regions are particularly prone to phosphodiester cleavages through hydrolysis or metal ion attack. Further, it was noticed a long time ago that, in single-stranded regions, dinucleotide steps with sequence pyrimidine-adenosine (UpA and CpA) were particularly sensitive to cleavage [4, 5]. These early observations were later thoroughly studied [6, 7]. Such cleavages are regularly observed in control lanes during gel electrophoresis. The precise molecular mechanism for these spontaneous cleavages in YpA sequences, however, has remained elusive. Recently, a surprising interpretation, based on mass spectrometry data and implying the *syn* conformation of the A and its protonation at N3 position, has been put forward [8].

In each ribozyme, cleavage occurs at a very precise dinucleotide location and generally with weak sequence dependence. The more extensively studied are the endonucleolytic, ribozymes (see Chapters 1 and 2). New data and results confirm the involvement of unexpected chemistry with anionic guanosine residues acting as a general base for capturing the O2′-hydroxyl proton [9, 10]. Interestingly, a proposal has been forwarded in which the tautomeric enol form of guanosine in which N1(G) carries an in-plane electron doublet would capture the proton from the O2′-hydroxyl group [11, 12]. Interestingly, such tautomers have been observed in functional ribosome crystals and related to the occurrence of miscoding at the first and second codon positions following the formation of tautomeric GoU pairs [13, 14]. The tautomeric ratio of G is around 1 in 10^4, a value close to the average miscoding error in bacteria [15]. The tautomeric forms trapped in a constrained environment (stacking, H-bond between the amino group of G and an anionic phosphate oxygen, minor groove contacts, etc.) are stabilized as observed in several crystal structures [12, 16]. The frequent occurrence of stacked Gs forming H-bonds through their Watson–Crick edge to phosphate anionic oxygens has been analyzed and described thoroughly [17]. It has also been shown that the tautomeric form coexists with the anionic form [18].

Overall, this book is timely and unique in its breadth of content. The book describes the great diversity of nucleic acid base catalysis beautifully. The reader is conveyed to a chemical journey extending from biologically functional ribozymes (like RNaseP or group I and II introns) to engineered and designed ribozymes as well as DNAzymes. For example, we now have three distinct structural environments with a natural 2′–5′ phosphodiester linkage: the group II ribozyme [19, 20], the spliceosome [21], and the lariat-capping ribozyme [22]. Despite the clear structural similarities between the active sites of group II ribozyme and of the spliceosome, the striking observation is how the 2′–5′ linkage promotes a highly constrained environment with nested non-Watson–Crick pairs and sugar-phosphate contacts.

Given those three-dimensional visions, one can only wonder at the structural and molecular fitness of nucleic acids, especially RNA. Here I have selected some topics that I enjoy particularly, and I apologize for not discussing many other aspects that are covered in this book. Indeed, some chapters extend to biotechnology and to ribozymes used in diagnostics and therapeutics. Finally, the major tools and techniques used for the analysis of nucleic acid structures are presented up to date. In this regard, I cannot resist citing Sydney Brenner [23, 24], Nobel Prize 2002 in Physiology or Medicine: "Progress in science depends on new techniques, new discoveries and new ideas, probably in that order."

Eric Westhof
Architecture et Réactivité de l'ARN
Université de Strasbourg
Institut de biologie moléculaire et cellulaire du CNRS
2 allée Konrad Roentgen, 67084 Strasbourg, France
August 2019

References

1 Cech, T.R., Zaug, A.J., and Grabowski, P.J. (1981). In vitro splicing of the ribosomal RNA precursor of *Tetrahymena*: involvement of a guanosine nucleotide in the excision of the intervening sequence. *Cell* 27 (3 Pt 2): 487–496.
2 Guerrier-Takada, C., Gardiner, K., Marsh, T. et al. (1983). The RNA moiety of ribonuclease P is the catalytic subunit of the enzyme. *Cell* 35 (3 Pt 2): 849–857.
3 Thompson, J.E., Kutateladze, T.G., Schuster, M.C. et al. (1995). Limits to catalysis by ribonuclease A. *Bioorg. Chem.* 23 (4): 471–481.
4 Cannistraro, V.J., Subbarao, M.N., and Kennell, D. (1986). Specific endonucleolytic cleavage sites for decay of *Escherichia coli* mRNA. *J. Mol. Biol.* 192 (2): 257–274.
5 Kierzek, R. (1992). Hydrolysis of oligoribonucleotides: influence of sequence and length. *Nucleic Acids Res.* 20 (19): 5073–5077.
6 Kaukinen, U., Lyytikainen, S., Mikkola, S., and Lonnberg, H. (2002). The reactivity of phosphodiester bonds within linear single-stranded oligoribonucleotides is strongly dependent on the base sequence. *Nucleic Acids Res.* 30 (2): 468–474.
7 Soukup, G.A. and Breaker, R.R. (1999). Relationship between internucleotide linkage geometry and the stability of RNA. *RNA* 5 (10): 1308–1325.
8 Fuchs, E., Falschlunger, C., Micura, R., and Breuker, K. (2019). The effect of adenine protonation on RNA phosphodiester backbone bond cleavage elucidated by deaza-nucleobase modifications and mass spectrometry. *Nucleic Acids Res.* 47(14): 7223–7234.
9 Wilson, T.J., Liu, Y., Li, N.S. et al. (2019). Comparison of the structures and mechanisms of the pistol and hammerhead ribozymes. *J. Am. Chem. Soc.* 141 (19): 7865–7875.

10 Bevilacqua, P.C. (2003). Mechanistic considerations for general acid-base catalysis by RNA: revisiting the mechanism of the hairpin ribozyme. *Biochemistry* 42 (8): 2259–2265.

11 Pinard, R., Hampel, K.J., Heckman, J.E. et al. (2001). Functional involvement of G8 in the hairpin ribozyme cleavage mechanism. *EMBO J.* 20 (22): 6434–6442.

12 Singh, V., Fedeles, B.I., and Essigmann, J.M. (2015). Role of tautomerism in RNA biochemistry. *RNA* 21 (1): 1–13.

13 Demeshkina, N., Jenner, L., Westhof, E. et al. (2012). A new understanding of the decoding principle on the ribosome. *Nature* 484 (7393): 256–259.

14 Rozov, A., Wolff, P., Grosjean, H. et al. (2018). Tautomeric G*U pairs within the molecular ribosomal grip and fidelity of decoding in bacteria. *Nucleic Acids Res.* 46 (14): 7425–7435.

15 Parker, J. (1989). Errors and alternatives in reading the universal genetic code. *Microbiol. Rev.* 53 (3): 273–298.

16 Westhof, E., Yusupov, M., and Yusupova, G. (2014). Recognition of Watson–Crick base pairs: constraints and limits due to geometric selection and tautomerism. *F1000Prime Rep.* 6: 19.

17 Zirbel, C.L., Sponer, J.E., Sponer, J. et al. (2009). Classification and energetics of the base-phosphate interactions in RNA. *Nucleic Acids Res.* 37 (15): 4898–4918.

18 Kimsey, I. J., Petzold, K., Sathyamoorthy, B., Stein, Z. W., and Al-Hashimi, H. M. (2015) Visualizing transient Watson-Crick-like mispairs in DNA and RNA duplexes. *Nature* 519, 315–320.

19 Toor, N., Keating, K.S., Taylor, S.D., and Pyle, A.M. (2008). Crystal structure of a self-spliced group II intron. *Science* 320 (5872): 77–82.

20 Costa, M., Walbott, H., Monachello, D. et al. (2016). Crystal structures of a group II intron lariat primed for reverse splicing. *Science* 354 (6316).

21 Wilkinson, M.E., Fica, S.M., Galej, W.P. et al. (2017). Postcatalytic spliceosome structure reveals mechanism of 3′-splice site selection. *Science* 358 (6368): 1283–1288.

22 Meyer, M., Nielsen, H., Olieric, V. et al. (2014). Speciation of a group I intron into a lariat capping ribozyme. *Proc. Natl. Acad. Sci. U.S.A.* 111 (21): 7659–7664.

23 Brenner, S. (2002). Life sentences: detective rummage investigates. *Genome Biol.* 3 (9): comment1013.1011-1013.1012.

24 Robertson, M. (1980). Biology in the 1980s, plus or minus a decade. *Nature* 285 (5764): 358–359.

Part I

Nucleic Acid Catalysis: Principles, Strategies and Biological Function

1

The Chemical Principles of RNA Catalysis

Timothy J. Wilson and David M. J. Lilley

The University of Dundee, Cancer Research UK Nucleic Acid Structure Research Group, MSI/WTB Complex, Dow Street, Dundee DD1 5EH, UK

1.1 RNA Catalysis

Ribozymes are enzymes that are made of RNA rather than protein. Their function is to accelerate the rates of chemical reactions. This chapter discusses the chemical principles of catalysis as applied to biological macromolecules. Except for the peptidyl transferase reaction of the ribosome, the known natural ribozymes all carry out phosphoryl transfer reactions, either transesterification (including nucleotidyl transfer) or hydrolysis. We shall therefore focus on these reactions principally. However, RNA can bind small molecules with great selectivity, and indeed the riboswitches exploit this ability in many ways to control gene expression. One, the glmS (glucosamine-6-phosphate riboswitch) ribozyme, uses a bound molecule as a coenzyme, and it is not impossible that other ribozymes that use coenzymes in a wider range of chemistry remain to be discovered. Ribozymes that have been selected in the laboratory demonstrate that a wider range of chemistry can be supported by RNA catalysis [1–3].

One of the most powerful tools that we can use to study the ribozyme mechanism is X-ray crystallography. Having a knowledge of the three-dimensional structure is invaluable. Yet this is not enough and may even be misleading. This may happen for several possible reasons. First, the RNA sequence may have been reduced too far, removing key elements required for proper structure and function. This occurred with the hammerhead ribozyme [4], where the removal of critical elements that formed a tertiary interaction led to a remodeling of the active center [5]. The active species is rarely studied. The structure of the ribozyme has often been modified to prevent activity, and the possible consequences of the modification need to be considered. Second, crystal contacts may induce structural changes, as found in the twister ribozyme where the nucleotide 5′ to the scissile phosphate was pulled out of the active site by interaction with guanine in a symmetry-related ribozyme molecule leading to loss of an in-line geometry [6]. Lastly, of course, a crystal structure can never capture the transition state because, by definition, it is fleeting. The best we

Ribozymes, First Edition. Edited by Sabine Müller, Benoît Masquida, and Wade Winkler.

can do is to find a related high-energy intermediate if it exists or try to model it with a transition state analog as was done for the hairpin [7] and hammerhead [8] ribozymes. But we always need to apply chemical insight when looking at crystal structures of ribozymes, and frequently we must extrapolate what we find to deduce the events occurring in the transition state. Ultimately only kinetic measurements probe transition state properties. It is thus the combination of structural analysis, kinetic measurements, and atomic mutagenesis that allow us to approach an understanding of the catalytic chemical mechanism.

1.2 Rates of Chemical Reactions and Transition State Theory

The possible trajectory of a chemical reaction is described by a potential energy surface plotting the free energy at each point on the reaction landscape. The reaction will take the path of lowest free energy between the starting material and product, and the highest point along that path is the transition state, generally described as a saddle point on the reaction trajectory. In transition state theory [9–11], we calculate the rate of reaction as the concentration of the activated complex (i.e. the species at the point of highest free energy, or transition state) multiplied by the rate of passage out of that state that will be related to bond vibrational frequencies. An equilibrium between the ground state and transition state is postulated, leading to an expression for the rate (k) as

$$k = \kappa \cdot \left({}^{k_B T}/_{h} \right) \exp \left({}^{-\Delta G^{\ddagger}}/_{RT} \right) \tag{1.1}$$

where k_B is Boltzmann's constant, h is Planck's constant, R is the gas constant, and T is the absolute temperature. κ is a factor that measures the probability of the transition state proceeding to form the product. The key parameter here is $\Delta G^{\ddagger}$, which is the free energy of activation that must be supplied to promote the substrate to the transition state. The population of the activated complex thus governs the rate of reaction that will be determined by the energy barrier according to statistical mechanics. Parenthetically, we note that it is mechanistically valuable to be able to relate chemical rates to equilibrium properties, and that linear free energy relationships are very important in the analysis of catalysis.

Transition state theory shows that chemical catalysis is a question of reducing the energetic barrier to the formation of the activated complex, i.e. stabilizing the transition state to lower its free energy. This can occur in a variety of ways, including electrostatic interactions, stabilization by hydrogen bonding, and formation of covalent complexes. We shall discuss this further for the phosphoryl transfer reactions that occur in the ribozymes.

The activation free energy $\Delta G^{\ddagger}$ can, of course, be parsed into enthalpy and entropies of activation. The latter can be particularly important in bimolecular reactions, which involve the loss of translational and rotational entropy in the formation of the activated complex. Chemical catalysis occurs in two forms: homogeneous catalysis, where reactants and catalyst are in the same phase, and heterogeneous catalysis, where typically the reaction occurs on some surface. In a sense, macromolecular catalysis is somewhat intermediate. When two reactants bind to a macromolecular catalyst, the loss of translational and rotational freedom is partially "paid for" in advance, and the effective concentration of the reactants is increased. Thus in peptidyl transferase reactions, the peptidyl- and aminoacyl-transfer RNAs (tRNAs) are bound to the ribosome and oriented ready to condense to form a new peptide bond. In the group, I intron ribozyme first stage reaction exogenous guanidine reacts when it is bound to the ribozyme. Activation entropy can also appear in more subtle ways, as the reorganization of solvent during the reaction, for example.

1.3 Phosphoryl Transfer Reactions in the Ribozymes

All the known natural ribozymes apart from ribosomal peptidyl transferase catalyze phosphoryl transfer, so we shall focus here on phosphoryl transfer reactions [12]. Figure 1.1 shows the reactions catalyzed by the nucleolytic ribozymes, and the group II and I self-splicing intron ribozymes. These are all transesterification reactions that involve the nucleophilic attack of a ribose hydroxyl group on the phosphodiester linkage of RNA. For the nucleolytic and group II ribozymes, the nucleophile is a 2′-hydroxyl, on the adjacent or a remote ribose, respectively. For the first stage of the group I ribozymes, the nucleophile is the 3′-hydroxyl of a guanosine molecule. In the case of RNaseP, the nucleophile is water, and the group II ribozyme can also undergo a hydrolytic reaction.

The phosphorus atom in a phosphodiester is tetrahedral, with four sp^3 orbitals bonded to two bridging and two non-bridging oxygen atoms. The $pK_a = 1$, so there is a negative charge; the bonds to the non-bridging O atoms have partial double bond character by pπ–dπ interaction, and the charge is delocalized. Nucleophilic attack of the oxygen of ROH or HOH requires the participation of the P d orbital, forming a phosphorane intermediate with sp^3d character that is close to the transition state (Figure 1.2a). The phosphorane is trigonal bipyramidal, with three equatorial O atoms and the nucleophilic and leaving group O atoms in the apical positions [14] (Figure 1.2b). Therefore to generate the phosphorane, the nucleophile must attack in-line with the P and the leaving group O.

The cleavage reaction for the nucleolytic ribozymes is shown in Figure 1.3. The products of the reaction are a cyclic 2′,3′-phosphate, and a 5′-hydroxyl. For the hammerhead and hairpin ribozymes, it has been demonstrated that the reaction proceeds with inversion of chirality at the phosphorus atom [15, 16]. This is consistent with the S_N2 mechanism shown, proceeding via the phosphorane. Kinetic isotope effects measured for the formation of cyclic 2′,3′-phosphate from uridine 3′-*m*′-nitrobenzyl phosphate [17] were indicative of a phosphorane-type transition state. Using a series of oligonucleotides containing a nucleoside analog with 2′-C-β-branched substituents with fluoro substitution having a range of pK_a values, Piccirilli and coworkers [18] measured the Brønsted parameter β_{nuc}. This gives a measure of the change in charge on the nucleophile approaching the transition state. Ye et al. obtained $\beta_{nuc} = 0.75 \pm 0.15$. The corresponding Brønsted

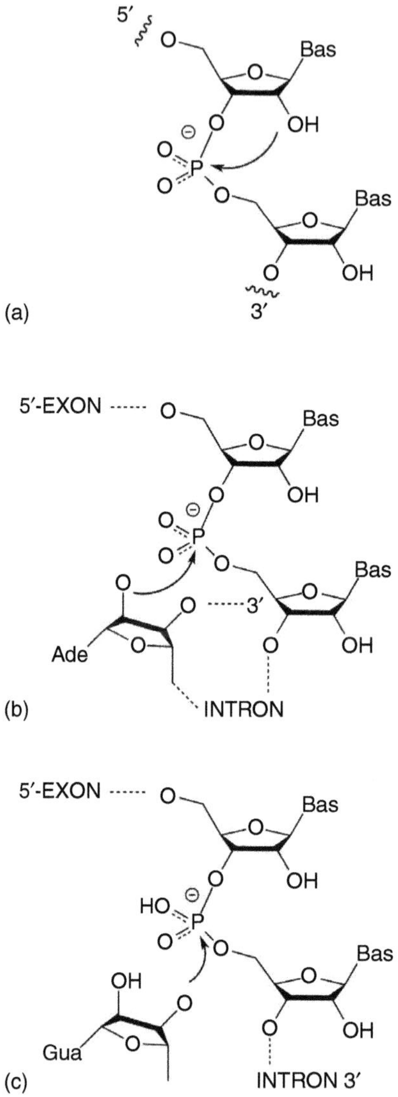

Figure 1.1 Three ribozyme-catalyzed transesterification reactions. In the nucleolytic ribozymes (a), the O2′ attacks the adjacent 3′-P with departure of the O5′. In the group II intron ribozyme (b), there is a similar reaction except that the O2′ nucleophile is remote within the intron. The nucleophile in the first reaction of the group I intron ribozyme (c) is the O3′ of an exogenous guanosine molecule.

parameter for the leaving group β_{lg} was measured by Lönnberg and coworkers [19] using a series of model uridine 3′-phosphate monoalkyl esters. They obtained $\beta_{lg} = -1.28 \pm 0.02$. These values are consistent with a concerted mechanism for the cleavage reaction, with significant development of charge on the oxygen atoms of the nucleophile and leaving groups.

1.4 Catalysis of Phosphoryl Transfer

We shall now consider how the cleavage reaction for the nucleolytic ribozymes could be catalyzed. These ribozymes achieve about a million-fold enhancement in rate

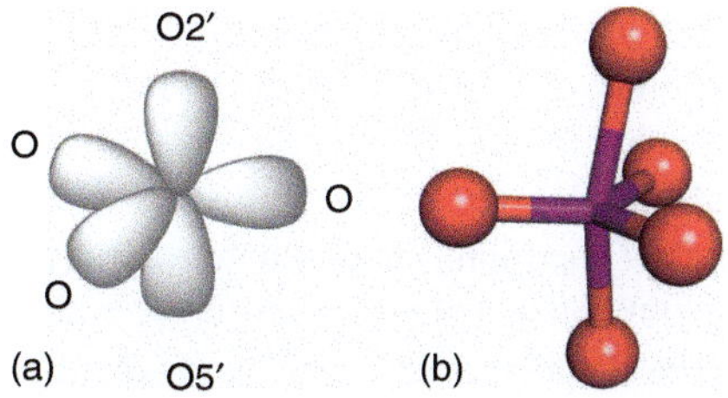

Figure 1.2 The geometry of a phosphorane. (a) A representation of phosphorus orbital hybridization in a phosphorane. The phosphorus atom is sp^3d hybridized and has trigonal bipyramidal geometry. (b) The structure of a vanadate transition state analog taken from the crystal structure of the hairpin ribozyme [13]. The vanadate mimics the conformation of the penta-coordinate phosphorane that is close to the transition state for the ribozyme transesterification reaction. The nucleophile and leaving group are in the apical positions, and three oxygen atoms lie in the central plane.

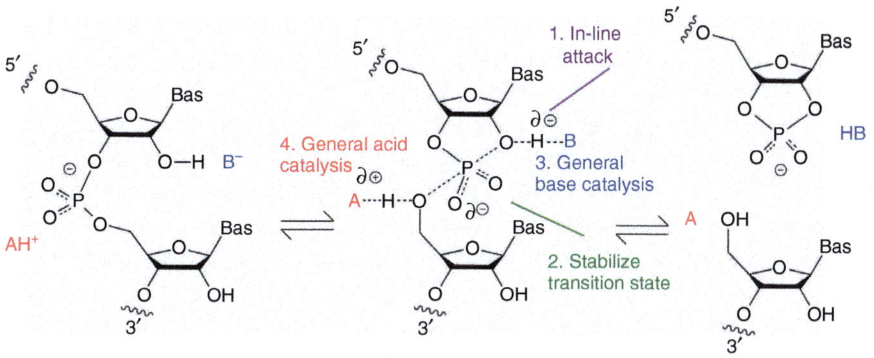

Figure 1.3 The chemical mechanism of the nucleolytic ribozymes and possible catalytic strategies. Cleavage proceeds left to right, and ligation right to left. The transition state is approximated by the central structure, with a pentavalent, trigonal-bipyramidal phosphorane structure. Four potential catalytic strategies are indicated. (1) is the in-line trajectory of attack, (2) stabilization of the transition state structurally or electrostatically, (3) deprotonation of the nucleophile, and (4) protonation of the oxyanion leaving group. By the principle of microscopic reversibility, this reaction scheme is symmetrical. For example, B^- acts as a general base to deprotonate the O2′ nucleophile in the cleavage reaction, and thus, BH acts as a general acid to protonate the O2′ leaving group in the ligation reaction, where O5′ is now the nucleophile.

over the reaction in a flexible dinucleotide, and a number of catalytic strategies may contribute to this; several authors have listed the potential contributions [20–22]. While this is a useful guide to our chemical thinking, in the end, everything comes down to stabilization of the transition state, i.e. lowering the activation barrier $\Delta G^{\ddagger}$ in Eq. (1.1). All the contributions are interconnected, and catalysis is multifactorial. With that caveat, we can consider four major contributions to catalysis of the phosphoryl transfer reaction (Figure 1.3).

1. *Alignment of nucleophilic attack.* As noted above, the nucleophile O, P, and leaving group O are aligned in the phosphorane, and the nucleophile must attack with

a trajectory that is co-linear with the P and the leaving O. A flexible dinucleotide will randomly sample the in-line geometry, but ribozymes are larger and structured. The required orbital overlap is not a sharp function of orientation, so some deviation from a perfect line of attack will be tolerated. Pre-alignment by the structure of a ribozyme may contribute a part of the total rate enhancement, but this is likely to be significantly less than 10^2. Of course, the RNA structure must not get trapped in a conformation that is markedly out of alignment – indeed, this is the basis of in-line probing of RNA structure [20].

2. *Structural and/or electrostatic stabilization of the transition state.* Two aspects of the phosphorane distinguish it from the substrates and products, i.e. a change in the conformation and a redistribution of electric charge. In principle, if the RNA is, to some degree, complementary to the structure of the transition state, it may stabilize it by the formation of interactions not found in the ground state. We have noted an example in the twister ribozyme, where a stereospecific hydrogen bonding interaction between a guanine N2 and a non-bridging O atom of the scissile P leads to a 100-fold difference in cleavage rate [22]. The phosphorane is formally dianionic, and the juxtaposition of positive charge could stabilize the charge redistribution. In principle, this might be that of a metal ion or nucleobase, or due to proton transfer as discussed in the following section 1.5.
3. *Deprotonation of the nucleophile.* The value of the Brønsted parameter β_{nuc} [18] indicates that there is a significant accumulation of positive charge on the O2′ nucleophile, and this could be reduced by the action of a general base to remove the proton. Moreover, the resulting alkoxide ion is a much stronger nucleophile than the hydroxyl group.
4. *Protonation of the leaving group.* Similarly, the large negative value of β_{lg} [19] shows that negative charge accumulates on the O5′ leaving group. Protonation of O5′ by a general acid will therefore result in a superior leaving group.

Catalytic strategies 3 and 4 collectively constitute general acid–base catalysis and are generally coordinated. They shall be discussed further in the following section 1.5.

1.5 General Acid–Base Catalysis in Nucleolytic Ribozymes

In specific acid–base catalysis, proton transfer occurs with water (including OH^- and H_3O^+), while general acid–base catalysis involves different (often) organic proton donors and acceptors that are weak acids and bases. In the current context, this means components of the RNA, i.e. nucleobases, 2′-OH groups, or bound hydrated metal ions. It could also involve a bound small molecule acting as a coenzyme, and *glm*S provides one example of this [23–25]. Jenks [26] has drawn attention to the important role of proton transfer in enzymes. In the nucleolytic ribozymes, general acid–base catalysis provides the largest contribution to the catalytic rate enhancement [27]. The most common general base and acid in the nucleolytic ribozymes are nucleobases, and especially that of guanine acting as a general base in the cleavage reaction, but as we shall see, other functionalities can also play a role.

The mechanism of the general acid–base-catalyzed cleavage reaction is shown in Figure 1.3. The general base deprotonates the O2′ nucleophile, and the general acid protonates the 5′-oxyanion leaving group. Some ribozymes catalyze the reverse ligation reaction. In that case by the principle of microscopic reversibility, the general acid and base exchange roles, and their required state of protonation. The catalytic power of macromolecular catalysts employing general acid-base catalysis will be limited by two aspects: the fraction of catalyst that is in an appropriate state of protonation to be active and the reactivity of the general acid and base.

1.5.1 The Fraction of Active Catalyst, and the pH Dependence of Reaction Rates

In the cleavage reaction depicted in Figure 1.3, the general acid must have a proton to donate, and the general base must be able to accept a proton. In other words, to be active, the ribozyme must have a protonated general acid and a deprotonated general base. The observed rate of cleavage (k_{obs}) will be lower than the rate when the ribozyme is fully active (k_{cleave}) according to:

$$k_{obs} = k_{cleave} \cdot f_A \cdot f_B \tag{1.2}$$

where f_A is the fraction of protonated acid and f_B is the fraction of deprotonated base. f_A and f_B will be a function of pH, according to the pK_a values of the general acid and base. pH is arguably the most powerful experimental tool we have in probing general acid–base catalytic mechanisms. For example, the general base can exist either as B^- or BH (e.g. for guanine; the corresponding species for adenine would be B and BH^+) at high and low pH values, respectively. The fraction of the two forms will be given by

$$[B^-]/[BH] = 10^{(pH-pK_a)} \tag{1.3}$$

When the base is operating at a pH that is the same as its pK_a, then half of the molecules will be in the required unprotonated form. If the base has a pK_a = 10, and the reaction is carried out at pH = 7, then most molecules will be protonated (inactive as a base), and only one molecule in 1000 will be in the active unprotonated form. However, we shall see later that this is compensated to some degree by a higher reactivity of the base due to its high pK_a. Similarly, if this moiety is acting as a general acid, then 999 molecules out of 1000 will be in the protonated form that can act as a general acid. But these species will be reluctant to donate that proton, i.e. they will be relatively unreactive.

In the cleavage reactions of the hairpin and Varkud satellite (VS) ribozymes, the general acid and base are the nucleobases of adenine and guanine, respectively (see Chapter 3). The apparent pK_a values (i.e. the values measured from the pH dependence of cleavage rate) in the context of the VS ribozyme reaction have been measured at 5.2 and 8.4, respectively [28]. Neither is close to physiological pH, and we would expect that only one ribozyme molecule in 1000 would be active. Bevilacqua [29] carried out a general analysis for the case of a ribozyme using a general acid and base (actually formulated for the hairpin ribozyme, but it applies generally), deriving a partition function to calculate f_A and f_B for given values of pK_a:

$$f_A = \left\{1+10^{\left(pK_a^B-pH\right)}\right\} / \left\{1+10^{\left(pK_a^B-pH\right)}+10^{\left(pK_a^B-pK_a^A\right)}+10^{\left(pH-pK_a^A\right)}\right\} \quad (1.4)$$

$$f_B = \left\{1+10^{\left(pH-pK_a^A\right)}\right\} / \left\{1+10^{\left(pK_a^B-pH\right)}+10^{\left(pK_a^B-pK_a^A\right)}+10^{\left(pH-pK_a^A\right)}\right\} \quad (1.5)$$

where the pK_a values of the general acid and base are written pK_a^A and pK_a^B, respectively. Figure 1.4a shows f_A and f_B plotted as a function of pH for the case of pK_a values of 5.2 and 8.4 (simulating the case of the VS ribozyme). At the low pH end, $f_A = 1$ (fully protonated), and then falls in a log-linear manner as the pH rises. Over the same range, f_B rises until it reaches a plateau (fully deprotonated) toward the high pH end. Equation (1.2) shows that the pH dependence of the reaction will reflect the product $f_A \cdot f_B$, that is also plotted in Figure 1.4a. $f_A \cdot f_B$ rises at the low pH end in the regime where $f_A = 1$, but f_B is increasing steadily. It then begins to tail off as f_A begins to fall while f_B is still rising. Eventually, f_B begins to saturate (f_B approaches 1) while f_A is steadily falling in a log-linear fashion. The net result is a bell-shaped curve. It is important to note that the initial rise at low pH is due to deprotonation of the general base, and the fall at high pH is due to general acid deprotonation. However, the reducing rise in $f_A \cdot f_B$ approaching the peak is due to the deprotonation of the acid, not the base.

The shape of this curve fits the experimental data for the VS ribozyme very well (see Chapter 3) [28]. Note that because the general acid has a low pK_a value, the general base has a high pK_a value, and over three units separate the two, the maximum value of $f_A \cdot f_B$ is only 6×10^{-4}. In protein enzymes, the imidazole side chain of histidine is frequently used in general acid–base catalysis. The pK_a of imidazole is normally close to neutrality, so when two histidine residues are used for general acid–base catalysis, a considerably higher value of $f_A \cdot f_B$ is achieved.

The pK_a values for the adenine and guanine nucleobases in the hairpin ribozyme are further apart than those in the VS ribozyme, mostly because the pK_a of guanine in the hairpin ribozyme is higher than that in the VS ribozyme. We have computed an $f_A \cdot f_B$ simulation for pK_a values of 6.0 and 10.0 that corresponds to the hairpin ribozyme in Figure 1.4b. Because the pK_a values are now separated by four units, a distinct plateau forms, where the deprotonation of acid and base compensate in $f_A \cdot f_B$ to create the flat top. Moreover, the eventual fall of $f_A \cdot f_B$ does not occur until a high pH is reached at which measurement of cleavage rate is not possible (i.e. the fall lies in the shaded region of the plot). The overall shape is therefore masked in the experimental profile of rate versus pH, which can therefore be mistaken for a single ionization. This has led to persistent serious errors of interpretation in past studies.

Returning to the case of nucleobases of pK_a values of 5.2 and 8.4, we may ask what pH profile will result for the reverse (ligation) reaction, where the nucleobase of low pK_a acts as a general base in deprotonated form, and that of high pK_a acts are general acid in its protonated form. This is simulated in Figure 1.4c. Now both f_A and $f_B = 1$ over much of the range, and the maximum value of $f_A \cdot f_B = 0.9$. Yet the shape of

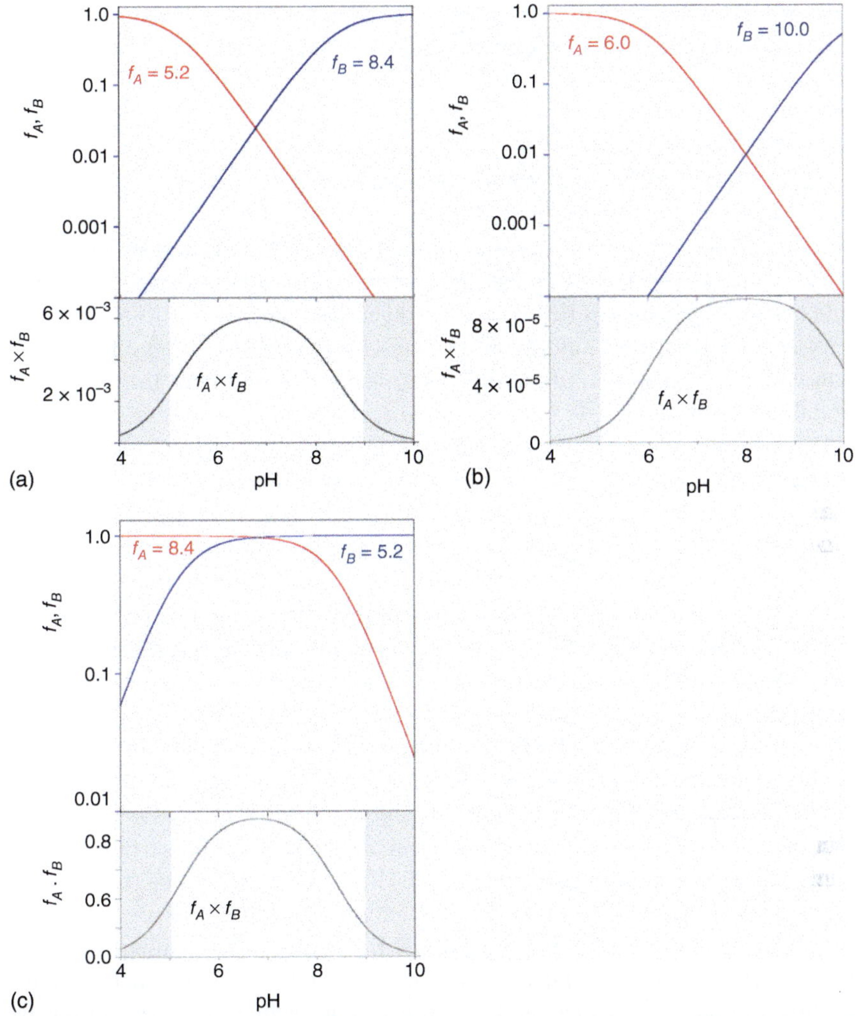

Figure 1.4 Simulations of the pH dependence of general acid–base catalyzed ribozyme reactions. For the three cases shown, the fraction of protonated acid (f_A) and deprotonated base (f_B) have been calculated using Eqs. (1.4, 1.5) and plotted as a function of pH over the range 4–10 (upper plots). The product ($f_A \cdot f_B$) is also plotted as a function of pH (lower plots) with the experimentally inaccessible regions between pH 4–5 and 9–10 shown grayed out. This curve simulates the pH dependence of the reaction rate. These plots have been generated for three cases. (a) $f_A = 5.2$ and $f_B = 8.4$. This generates a bell-shaped curve of $f_A \cdot f_B$ versus pH. This is close to the pH dependence of the VS ribozyme. (b) $f_A = 6.0$ and $f_B = 10.0$. The higher pK_a of the base shifts the reduction in $f_A \cdot f_B$ at higher pH to an inaccessible value, so that the curve appears to reach a stable plateau. This represents the situation with the hairpin ribozyme. (c), $f_A = 8.4$ and $f_B = 5.2$. The pK_a values for the acid and base have been exchanged with respect to part (a). Although the absolute value of $f_A \cdot f_B$ is now higher; the shape of the curve is identical to that in part (a). Thus the pH dependence alone cannot give an assignment of the acid and base; this is known as the principle of kinetic ambiguity.

$f_A \cdot f_B$ versus pH is identical to that where the general base and acid have high and low pK_a values, respectively. This is termed the principle of kinetic ambiguity, and as a consequence, the pH dependence of the reaction cannot reveal which nucleobase is acting as general base and which is the general acid. Other approaches must be used to determine this – see Chapter 3.

Nucleobases participating in proton transfer may also respond to the electrostatic influence of nearby nucleotides that can add complexity to the shape of the pH dependence of reaction rate. Bevilacqua and coworker [30] have recently derived partition functions that take into account cooperative interactions between general acid and base and the effect of titration of nearby nucleotides.

Nucleobase substitution or atomic mutagenesis is generally carried out in conjunction with rate versus pH measurements to investigate the roles for particular functionalities in ribozymes. To take an example, when it was suspected that G630 in the VS ribozyme was acting as the general base in the cleavage reaction, it was substituted by diaminopurine (i.e. replacing O6 by an amine) with a significantly lower pK_a of around 5 [28]. This should shift the curve of f_B versus pH strongly left (i.e. to lower pH) in our simulations, and so the peak of $f_A \cdot f_B$ moves close to pH = 5. The experimental curve followed this closely [28], confirming that the higher pK_a was indeed due to G630. Note, however, that this experiment by itself does not allow us to determine whether G630 is acting as general acid or general base. Similar experiments were performed to probe the role of G33 as a general base in the twister ribozyme [6].

Atomic mutagenesis, coupled with rate versus pH measurements, can provide compelling evidence for the involvement of a given nucleobase. In the twister ribozyme, the adenine (A1) immediately 3′ to the scissile phosphate was suspected to be acting as the general acid in the cleavage reaction, so was subjected to atomic mutagenesis. Replacement of adenine N7 by CH raises its pK_a by about 1.5 units, and it was found that this substitution at A1 raised the rate of cleavage by the twister ribozyme fivefold at pH = 8.5 [22]. While there are a number of ways mutagenesis can lower the catalytic power of a ribozyme, it is much harder to explain how the rate can be increased. But this can be easily rationalized in terms of the fraction of active ribozyme. The A1N7C substitution displaces the curve of f_A versus pH toward that of f_B, raising the peak value of $f_A \cdot f_B$. In this way, the fraction of active catalyst is higher than for the unmodified ribozyme, resulting in a faster-observed rate.

The mechanism of the twister ribozyme illustrates a factor that is normally disregarded. A purine nucleobase has three ring-nitrogen atoms that can be protonated: N3, N7, and N1, in order of decreasing acidity. In other ribozymes, it is N1 that participates in proton transfer, but nucleotide A1 of the twister ribozyme acts as a general acid by donating a proton from the highly acidic N3 [22]. The shape of the rate versus pH curve is determined by the macroscopic pK_a of the nucleobase. However, the relative extent of protonation of the three nitrogen atoms is determined by their respective microscopic pK_as. For adenosine the extent of protonation is N3 0.7%, N7 3.2%, N1 96.1% [31]. Equations (1.2)–(1.5) above assume a single site of protonation, and this is reasonable when N1 participates in proton transfer. However, to explain the results of atomic mutagenesis for the twister ribozyme, it is necessary to modify Eq. (1.2):

$$k_{obs} = k_{cleave} \cdot f_A \cdot f_B \cdot f_{N3H} \tag{1.6}$$

where f_A and f_B are the fractions of protonated acid and base, respectively, and the extra factor f_{N3H} is the fraction of protonation that occurs specifically at N3 [22].

1.5.2 The Reactivity of General Acids and Bases

We have seen that the nucleobases typically have pK_a values that are removed significantly far from neutrality. Clearly, a disadvantage compared to a protein enzyme using histidine for general acid–base catalysis. Yet the reactivity of these bases is also a function of pK_a, and this works in the opposite sense to the population of the active form, as discussed above. In general base catalysis the rate of ribozyme cleavage is related to the pK_a of the general base by the Brønsted equation:

$$\log k_{cleave} = B - \beta \cdot pK_a \tag{1.7}$$

where β is the extent of proton transfer in the transition state, and B is a constant specific to the reaction studied. An analogous linear free energy relation can be written for acidity. The activity of an hepatitis delta virus (HDV) ribozyme mutated in the key cytosine nucleotide that acts as the general acid was found to be restored by bases such as imidazole within the solution [32]. A value of $\beta = 0.5$ was calculated by measuring the reaction rate as a function of the pK_a of the exogenous base [33, 34]. Thus the intrinsic rate for a base of pK_a = 10 will be 32 times higher than one of pK_a = 7. This higher reactivity partially compensates for the small population of unprotonated base that can act in general base catalysis at physiological pH. In the HDV ribozyme experiment, the exogenous base is acting as an acid, so the opposite relationship applies. The intrinsic rate for an acid of pK_a = 10 will be 32 times lower than one of pK_a = 7, and this is partially offset by the higher population of protonated acid at physiological pH.

1.6 pK_a Shifting of General Acids and Bases in Nucleolytic Ribozymes

The generally accepted pK_a values for the adenine and guanine nucleobases are 4.2 and 9.5, respectively. If they remained at these values in a ribozyme such as the VS then only one molecule in about 10^5 would be in the correct state of protonation to catalyze the cleavage reaction. This fraction could become more favorable if the pK_a values were shifted closer to neutrality, and for the VS ribozyme, apparent values of 5.2 and 8.4 were measured experimentally [28] for the cleavage reaction. Higher values of the lower apparent pK_a have been measured for the hairpin ribozyme as 6.3 [35] and 6.9 for the twister ribozyme [6]. In the context of the electronegative environment of RNA, it is relatively easy to explain raised pK_a values. The most remarkable of these is the twister ribozyme. The adenine (A1) that acts as the general acid is held in position by a number of hydrogen bonds, two of which involve both protons of N6 that are donated to non-bridging O atoms of successive phosphate groups, that carry a negative charge. It is this unusual environment that raises the pK_a of the adenine virtually three units. This is necessary because, in the catalytic mechanism, it is the highly acidic N3, not the usual N1, that donates a proton to the O5′ oxyanion leaving group (see Chapter 3) [22].

It is harder to see how the pK_a of guanine can be lowered in a ribozyme, and in general, this is less variable. The apparent pK_a of G630 in the VS ribozyme has an unusually low value of 8.4, but this was measured in a high Mg^{2+} concentration. It is quite likely that there is an ion binding site close to G630 [36], and the positively charged ion lowers its apparent pK_a.

1.7 Catalytic Roles of Metal Ions in Ribozymes

Being anionic polyelectrolytes, RNA molecules are associated with many metal ions, some of which may participate directly in catalysis. Both monovalent and divalent metal ions can interact with RNA, although it is the divalent ions that are more likely to bind specifically and function in catalysis. In solution, metal ions are hydrated, with an inner shell of water molecules that are tightly bound. A magnesium ion normally has a first coordination sphere of six water molecules arranged in an octahedral geometry (Figure 1.5). The Mg^{2+}–O distance is 2.1 Å, and according to ligand

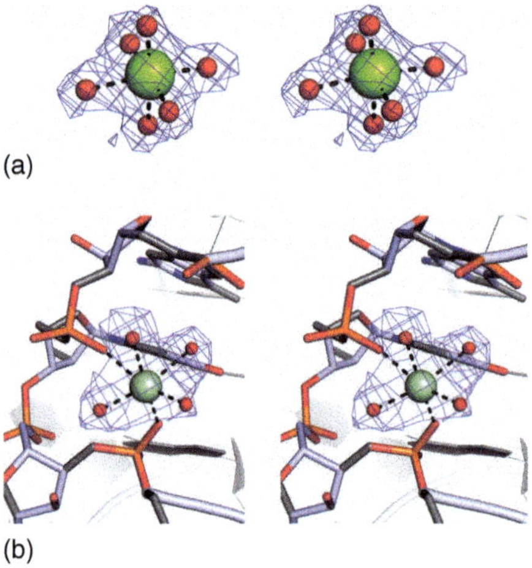

Figure 1.5 Hexaquo-magnesium ions. The Mg^{2+} ion has an inner hydration sphere comprising six tightly bound water molecules with octahedral symmetry. The water molecules may form hydrogen bonds to the RNA or be replaced by RNA ligands such as non-bridging phosphate O atoms. These images are taken from ions bound to ribozyme structures, and the electron density clearly reveals the positions of the hydrating water molecules. These images are shown as parallel-eye stereoscopic pairs. (a) A Mg^{2+} ion with all six inner-sphere water molecules. (b) A Mg^{2+} ion bound to a tight turn in the backbone of the twister ribozyme [6]. This ion has exchanged two inner-sphere water molecules for phosphate non-bridging O atoms, retaining four water molecules of hydration.

field theory, the bond has significant covalent character. These water molecules are not readily displaced, and the great majority of ions associated with RNA retain a full hydration sphere. These are not specifically bound at one location, and exhibit fast exchange. This is termed outer-sphere binding or "atmospheric" binding. Some water molecules within the first hydration sphere may be hydrogen-bonded to acceptors on the RNA, and thus can be located in crystal structures. Monovalent ions virtually always bind as outer-sphere complexes. Most divalent ions such as Mg^{2+} will also be bound in an outer sphere fashion. However, inner-sphere water molecules can sometimes be substituted by one or more RNA ligands (often a non-bridging phosphate O) if a suitable binding pocket can form. In that case, we call this inner-sphere binding, and the complex is in slow exchange. We observed seven hydrated Mg^{2+} ions bound to the twister sister (TS) ribozyme structure [37], all with octahedral symmetry. Two were outer-sphere complexes (retaining full coordination spheres of water molecules), three had exchanged a single inner-sphere water molecule for RNA ligands, and two had exchanged two water molecules. All the RNA ligands were non-bridging phosphate oxygen atoms apart from one where a cytosine O2 was directly bonded to the metal.

Small ions near the top of the periodic table have weakly polarizable orbitals and are classed as "hard" ions. Larger ions with more electrons are more polarizable and are classed as "soft" ions. In general, hard ions bind preferentially to similarly hard anions, so that bound Mg^{2+} ions will normally be found attached to oxygen ligands. By contrast, Mn^{2+} or Cd^{2+} ions bind more avidly to soft atoms like sulfur, and this can be the basis of a way to investigate the roles of metal ions in the transition states of ribozyme reactions.

In general, metal ions are indispensable to the folding of RNA, to lower the electrostatic repulsion between the phosphate groups. In most cases, this can be achieved by monovalent ions, albeit in higher concentration (2 or 3 logs typically) than required for divalent ions, so outer-sphere binding is generally sufficient to achieve folding into the active conformation. However, we sometimes observe that very tight turns in the backbone of RNA may be bridged by an Mg^{2+} ion as an inner-sphere complex (Figure 1.5).

Site-specifically bound metal ions can directly participate in the chemistry of catalysis in a number of different ways:

- They can bind to the reactants and organize and stabilize the structure of the transition state.
- They can stabilize developing negative charge in the transition state electrostatically. An atmosphere of outer-sphere metal ions could also achieve some stabilization of the transition state and, likely, this occurs quite generally in the nucleolytic ribozymes.
- Metal ions can act as Lewis acids, binding directly to reactants to activate them.
- Lastly, hydrated metal ions can act in general acid–base catalysis. Ions like Mg^{2+} are weakly acidic, whereby one of the inner-sphere water molecules can lose a proton, i.e.

$$[Mg(H_2O)_6]^{2+} + H_2O \rightleftharpoons [Mg(H_2O)_5OH]^{+} + H_3O^{+}$$

with $pK_a = 11.4$. There is good evidence in the HDV ribozyme that a bound metal ion acts as a general base to deprotonate the nucleophilic O2′ [38], the TS ribozyme probably acts in a similar manner [37] (see Chapter 3), and we have recently obtained evidence that a bound metal ion acts as a general acid to protonate the O5′ leaving group of the pistol ribozyme [55].

There are a number of tests that can point to the direct involvement of metal ions in ribozyme chemistry. Ribozymes such as the hairpin or twister are active in high concentrations of monovalent ions, at a rate that is 1/10th of that in Mg^{2+} ions [39]. However, when the metal ion is required to participate directly in the reaction, that factor increases to $\sim 10^{-5}$, e.g. as found in the TS ribozyme [37]. The requirement for inner-sphere coordination can be tested by replacing Mg^{2+} with $Co^{3+}(NH_3)_6$ ions. The two ions are structurally similar, but the ammine ligands of the latter exchange extremely slowly. Thus if inner-sphere ligand exchange is required, $Co^{3+}(NH_3)_6$ cannot replace Mg^{2+} with retention of activity. For example, the HDV ribozyme is essentially inactive in the presence of $Co^{3+}(NH_3)_6$ ions [38].

In contrast to the nucleolytic ribozymes, the larger ribozymes, such as the self-splicing introns and RNaseP, seem to have rejected general acid–base catalysis, in favor of acting as metalloenzymes. Many protein enzymes, including nucleases and polymerases, carry out phosphoryl transfer reactions using two metal ions to activate the nucleophile, position components, and stabilize the transition state [40]. Over a number of years, Herschlag, Piccirilli, and their coworkers explored the role of metal ions in the catalytic mechanism of the group I intron ribozyme using a combination of atomic mutagenesis and careful reaction kinetics [41–46]. Contacts between the metal ions and the transition state were studied by synthesis of sulfur- or amino-substituted substrates (stereo-selectively where relevant), and then looking for restoration of activity using softer metal ions such as Mn^{2+} or Cd^{2+}. This is often termed metal ion rescue. These experiments indicated that three metal ions are bound to the transition state of the group I intron ribozyme (Figure 1.6a). These are:

- *Metal ion A*. Bound to the 3′ and *pro*S non-bridging O atoms of the scissile phosphate. This probably functions by stabilizing the transition state as a negative charge develops on the leaving group.
- *Metal ion B*. Bound to the O3′ of the guanosine. This interaction would be expected to activate the nucleophile, possibly acting as a Lewis acid.
- *Metal ion C*. Bound to the O2′ of the guanosine plus the *pro*S non-bridging O of the scissile phosphate. Here the role is less obvious, but the metal ion could serve to position the reacting groups, as well as providing further electrostatic stabilization.

Two metal ions were observed bound in the active site of the *Azoarcus* group I intron ribozyme poised at the second stage [47, 48] (Figure 1.6b). One was bound to the 3′ and *pro*S non-bridging O atoms of the scissile phosphate, i.e. as deduced for metal ion A. The other was bound like a combination of metal ions B and C, interacting with O2′ and O3′ of the guanosine as well as the *pro*S non-bridging O of the scissile. The difference between the conclusions from structural and functional studies could reflect disorder within the crystal, or perhaps a difference between

(a)

(b)

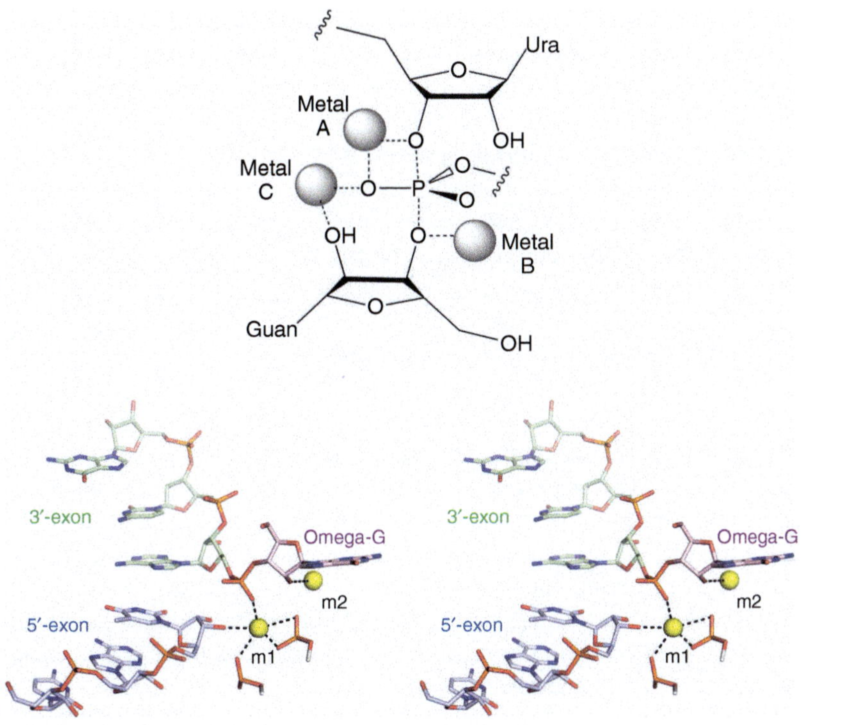

Figure 1.6 The positions of Mg^{2+} ions bound in the transition state of the group I intron ribozyme. (a) The locations of divalent metal ions deduced from systematic analysis of the kinetics of ribozymes with atomic substitutions (especially S for O) coupled with the exchange of Mg^{2+} for softer metal ions [41–46]. (b) Parallel-eye stereoscopic view of the crystal structure of the *Azoarcus* group I intron ribozyme trapped prior to the second step of the splicing reaction [47]. The 3′-OH of the 5′-exon is poised to make a nucleophilic attack on the adjacent phosphodiester linkage in the 3′-exon, leaving the terminal guanine nucleotide (Ω-G) bound to the ribozyme. Two metal ions (m1 and m2) are coordinated in the active center. m1 is directly bound to the O3′ nucleophile.

a necessarily ground-state structure and kinetic measurements that can probe the transition state.

1.8 The Choice Between General Acid–Base Catalysis and the Use of Metal Ions

The majority of ribozymes use either general acid–base catalysis or act as metalloenzymes. The same division of catalytic mechanism is found in protein enzymes performing phosphoryl transfer reaction. RNaseA uses histidine side chains to remove and donate protons [49, 50] while a typical restriction enzyme uses Mg^{2+} ions. All nucleolytic ribozymes seem to use general acid–base catalysis, where nucleobases frequently adopt the role taken by histidine in RNaseA (Chapter 3). By contrast, RNaseP and the self-splicing introns have generally evolved metal ion-based catalytic mechanisms. The reason for the distinction is not clear. An active center in which metal ions play the key catalytic role may be more amenable to remodeling in between two-stage reactions like those of the group I and II introns. It is perhaps easier to evolve an RNA that uses bound metal ions, given its polyanionic nature. No ribozyme derived by in vitro selection has proved to use nucleobases, and the close similarity of the active sites of the hairpin and VS ribozymes (Chapter 3) suggests there could be relatively few ways to use nucleobases to catalyze phosphoryl transfer reactions.

1.9 The Limitations to RNA Catalysis

By comparison with proteins, the catalytic resources of RNA are very limited. These are just four rather similar heterocyclic nucleobases, 2′-hydroxyl groups, and hydrated metal ions. This is perhaps reflected in the catalytic rate enhancements achieved and the limited range of reactions that are catalyzed. Part of the limitation comes from the pK_a values of the nucleobases; for example, only a small fraction of the VS ribozyme carrying out cleavage is active at a given time. If due allowance is made for that, then the catalytic efficiency becomes comparable to RNaseA [51]. Rate could well be a limitation, and the majority of ribozymes that exist in contemporary cells do not undergo multiple turnovers.

The lack of a wide array of chemical resources may be a greater limitation on the range of chemistry catalyzed. In the RNA world hypothesis, it is necessary that RNA would catalyze a much greater range of chemistry. We can speculate that the chemical repertoire of RNA might be expanded if small molecules could bind and act as coenzymes. RNA is very good at binding small molecule ligands with great specificity, exemplified by the great range of ligands for the riboswitches [52]. Riboswitches have been identified that bind a number of coenzymes, including thiamine pyrophosphate (TPP), flavin mononucleotide (FMN), *S*-adenosylmethionine (SAM), S-adenosylhomocysteine (SAH), tetrahydrofolate (THF), Ado-cobalamine. GlmS (Chapter 3) provides a precedent for a ribozyme using a coenzyme, where bound glucosamine-6-phosphate is the probable general acid in the cleavage reaction. The most abundant group of riboswitches are those that bind TPP [53]. TPP is a very versatile coenzyme, involved in the formation and breakage of carbon–carbon bonds, e.g. in transketolase. As we have discussed previously [54], we should consider the possibility that a ribozyme might bind TPP as a coenzyme, to catalyze a new range of metabolic interconversions. The discovery of such novel ribozymes would be exciting indeed!

Acknowledgment

Work on RNA catalysis in Dundee is funded by Cancer Research UK under program Grant A18604.

References

1 Seelig, B. and Jäschke, A. (1999). A small catalytic RNA motif with Diels–Alderase activity. *Chem. Biol.* 6 (3): 167–176.

2 Sengle, G., Eisenfuh, R.A., Arora, P.S. et al. (2001). Novel RNA catalysts for the Michael reaction. *Chem. Biol.* 8 (5): 459–473.

3 Tsukiji, S., Pattnaik, S.B., and Suga, H. (2003). An alcohol dehydrogenase ribozyme. *Nat. Struct. Biol.* 10 (9): 713–717.

4 Khvorova, A., Lescoute, A., Westhof, E., and Jayasena, S.D. (2003). Sequence elements outside the hammerhead ribozyme catalytic core enable intracellular activity. *Nat. Struct. Biol.* 10 (9): 1–5.

5 Martick, M. and Scott, W.G. (2006). Tertiary contacts distant from the active site prime a ribozyme for catalysis. *Cell* 126 (2): 309–320.

6 Liu, Y., Wilson, T.J., McPhee, S.A., and Lilley, D.M. (2014). Crystal structure and mechanistic investigation of the twister ribozyme. *Nat. Chem. Biol.* 10 (9): 739–744.

7 Rupert, P.B., Massey, A.P., Sigurdsson, S.T., and Ferré-D, Amaré, A.R. (2002). Transition state stabilization by a catalytic RNA. *Science* 298 (5597): 1421–1424.

8 Mir, A. and Golden, B.L. (2016). Two active site divalent Ions in the crystal structure of the hammerhead ribozyme bound to a transition state analogue. *Biochemistry* 55 (4): 633–636.

9 Evans, M.G. and Polyani, M. (1935). Some applications of the transition state method to the calculation of reaction velocities, especially in solution. *Trans. Faraday Soc.* 31: 875–894.

10 Eyring, H. (1935). The activated complex in chemical reactions. *J. Chem. Phys.* 3: 107–114.

11 Pelzer, H. and Wigner, E. (1935). Über die geschwindigkeitskon stante von austauschreaktionen *Z. Phys. Chem.* B15: 445.

12 Oivanen, M., Kuusela, S., and Lonnberg, H. (1998). Kinetics and mechanisms for the cleavage and isomerization of the phosphodiester bonds of RNA by Brönsted acids and bases. *Chem. Rev.* 98 (3): 961–990.

13 Rupert, P.B. and Ferré-D'Amaré, A.R. (2001). Crystal structure of a hairpin ribozyme-inhibitor complex with implications for catalysis. *Nature* 410: 780–786.

14 Westheimer, F.H. (1968). Pseudo-rotation in the hydrolysis of phosphate esters. *Acc. Chem. Res.* 1: 70–78.

15 van Tol, H., Buzayan, J.M., Feldstein, P.A. et al. (1990). Two autolytic processing reactions of a satellite RNA proceed with inversion of configuration. *Nucleic Acids Res.* 18 (8): 1971–1975.

16 Slim, G. and Gait, M.J. (1991). Configurationally defined phosphorothioate-containing oligoribonucleotides in the study of the mechanism of cleavage of hammerhead ribozymes. *Nucleic Acids Res.* 19 (6): 1183–1188.

17 Gerratana, B., Sowa, G.A., and Cleland, C.W. (2000). Characterization of the transition-state structures and mechanisms for the isomerization and cleavage reactions uridine 3′-m-nitrobenzyl phosphate. *J. Am. Chem. Soc.* 122 (51): 12615–12621.

18 Ye, J.D., Li, N.S., Dai, Q., and Piccirilli, J.A. (2007). The mechanism of RNA strand scission: an experimental measure of the Brønsted coefficient, beta nuc. *Angew. Chem.* 46 (20): 3714–3717.

19 Kosonen, M., Youseti-Salakdeh, E., Strömberg, R., and Lönnberg, H. (1997). Mutual isomerization of uridine 2′- and 3′-alkylphosphatesand cleavage to a 2′,3′-cyclic phosphate: the effect of the alkyl group on the hydronium and hydroxide-ion-catalyzed reactions. *J. Chem. Soc. Perkin Trans.* 2: 2661–2666.

20 Soukup, G.A. and Breaker, R.R. (1999). Relationship between internucleotide linkage geometry and the stability of RNA. *RNA* 5 (10): 1308–1325.

21 Koo, S., Novak, T., and Piccirilli, J.A. (2008). Catalytic mechanism of the HDV ribozyme. In: *Ribozymes and RNA Catalysis* (eds. D.M.J. Lilley and F. Eckstein). Cambridge: Royal Society of Chemistry.

22 Wilson, T.J., Liu, Y., Domnick, C. Kath-Schorr, S. and D. M. J. Lilley, D. M. J. (2016). The novel chemical mechanism of the twister ribozyme. *J. Am. Chem. Soc.* 138 (19): 6151–6162.

23 McCarthy, T.J. Plog, M. A., Floy, S. A., Jansen, J. A., Soukup, J. K. and Soukup, G. A. (2005). Ligand requirements for glmS ribozyme self-cleavage. *Chem. Biol.* 12 (11): 1221–1226.

24 Klein, D.J. and Ferré-D'Amaré, A.R. (2006). Structural basis of glmS ribozyme activation by glucosamine-6-phosphate. *Science* 313 (5794): 1752–1756.

25 Cochrane, J.C., Lipchock, S.V., Smith, K.D., and Strobel, S.A. (2009). Structural and chemical basis for glucosamine 6-phosphate binding and activation of the glmS ribozyme. *Biochemistry* 48 (15): 3239–3246.

26 Jenks, W.P. (1987). *Catalysis in Chemistry and Enzymology*. New York: Dover Publications Inc.

27 Li, Y. and Breaker, R.R. (1999). Kinetics of RNA degradation by specific base catalysis of transesterification involving the 2′-hydroxyl group. *J. Am. Chem. Soc.* 121: 5364–5372.

28 Wilson, T.J., McLeod, A.C., and Lilley, D.M.J. (2007). A guanine nucleobase important for catalysis by the VS ribozyme. *EMBO J.* 26 (10): 2489–2500.

29 Bevilacqua, P.C. (2003). Mechanistic considerations for general acid–base catalysis by RNA: revisiting the mechanism of the hairpin ribozyme. *Biochemistry* 42 (8): 2259–2265.

30 Frankel, E.A. and Bevilacqua, P.C. (2018). Complexity in pH-dependent ribozyme kinetics: dark pK_a shifts and wavy rate-pH profiles. *Biochemistry* 57 (5): 483–488.

31 Kapinos, L.E., Operschall, B.P., Larsen, E., and Sigel, H. (2011). Understanding the acid–base properties of adenosine: the intrinsic basicities of N1, N3 and N7. *Chemistry* 17 (29): 8156–8164.

32 Perrotta, A.T., Shih, I., and Been, M.D. (1999). Imidazole rescue of a cytosine mutation in a self-cleaving ribozyme. *Science* 286 (5437): 123–126.

33 Shih, I.H. and Been, M.D. (2001). Involvement of a cytosine side chain in proton transfer in the rate-determining step of ribozyme self-cleavage. *Proc. Natl. Acad. Sci. U.S.A.* 98 (4): 1489–1494.

34 Nakano, S., Proctor, D.J., and Bevilacqua, P.C. (2001). Mechanistic characterization of the HDV genomic ribozyme: assessing the catalytic and structural contributions of divalent metal ions within a multichannel reaction mechanism. *Biochemistry* 40 (40): 12022–12038.

35 Nahas, M.K. et al. (2004). Observation of internal cleavage and ligation reactions of a ribozyme. *Nat. Struct. Mol. Biol.* 11 (11): 1107–1113.

36 Zamel, R. and Collins, R.A. (2002). Rearrangement of substrate secondary structure facilitates binding to the *Neurospora* VS ribozyme. *J. Mol. Biol.* 324 (5): 903–915.

37 Liu, Y., Wilson, T.J., and Lilley, D.M.J. (2017). The structure of a nucleolytic ribozyme that employs a catalytic metal ion. *Nat. Chem. Biol.* 13: 508–513.

38 Nakano, S., Chadalavada, D.M., and Bevilacqua, P.C. (2000). General acid–base catalysis in the mechanism of a hepatitis delta virus ribozyme. *Science* 287: 1493–1497.

39 Murray, J.B., Seyhan, A.A., Walter, N.G. et al. (1998). The hammerhead, hairpin and VS ribozymes are catalytically proficient in monovalent cations alone. *Chem. Biol.* 5: 587–595.

40 Steitz, T.A. and Steitz, J.A. (1993). A general 2-metal-ion mechanism for catalytic RNA. *Proc. Natl. Acad. Sci. U.S.A.* 90 (14): 6498–6502.

41 Shan, S.O., Yoshida, A., Sun, S.G. et al. (1999). Three metal ions at the active site of the *Tetrahymena* group I ribozyme. *Proc. Natl. Acad. Sci. U.S.A.* 96 (22): 12299–12304.

42 Shan, S., Kravchuk, A.V., Piccirilli, J.A., and Herschlag, D. (2001). Defining the catalytic metal ion interactions in the *Tetrahymena* ribozyme reaction. *Biochemistry* 40 (17): 5161–5171.

43 Forconi, M., Lee, J., Lee, J.K. et al. (2008). Functional identification of ligands for a catalytic metal ion in group I introns. *Biochemistry* 47 (26): 6883–6894.

44 Forconi, M., Sengupta, R.N., Piccirilli, J.A., and Herschlag, D. (2010). A rearrangement of the guanosine-binding site establishes an extended network of functional interactions in the Tetrahymena group I ribozyme active site. *Biochemistry* 49 (12): 2753–2762.

45 Sengupta, R.N., Herschlag, D., and Piccirilli, J.A. (2012). Thermodynamic evidence for negative charge stabilization by a catalytic metal ion within an RNA active site. *ACS Chem. Biol.* 7 (2): 294–299.

46 Sengupta, R.N. et al. (2016). An active site rearrangement within the *Tetrahymena* group I ribozyme releases nonproductive interactions and allows formation of catalytic interactions. *RNA* 22 (1): 32–48.

47 Adams, P.L., Stahley, M.R., Wang, J., and Strobel, S.A. (2004). Crystal structure of a self-splicing group I intron with both exons. *Nature* 430: 45–50.

48 Stahley, M.R. and Strobel, S.A. (2005). Structural evidence for a two-metal-ion mechanism of group I intron splicing. *Science* 309 (5740): 1587–1590.

49 Thompson, J.E. and Raines, R.T. (1994). Value of general acid–base catalysis to ribonuclease A. *J. Am. Chem. Soc.* 116: 5467–5468.

50 Raines, R.T. (1998). Ribonuclease A. *Chem. Rev.* 98 (3): 1045–1066.

51 Wilson, T.J. et al. (2010). Nucleobase-mediated general acid–base catalysis in the Varkud satellite ribozyme. *Proc. Natl. Acad. Sci. U.S.A.* 107: 11751–11756.

52 Breaker, R.R. (2011). Prospects for riboswitch discovery and analysis. *Mol. Cell* 43 (6): 867–879.

53 Winkler, W., Nahvi, A., and Breaker, R.R. (2002). Thiamine derivatives bind messenger RNAs directly to regulate bacterial gene expression. *Nature* 419 (6910): 952–956.

54 Wilson, T.J. and Lilley, D.M.J. (2015). RNA catalysis – is that it? *RNA* 21 (4): 534–537.

55 T. J. Wilson, Y. Liu N. S. Li, Q. Dai, J. A. Piccirilli and D. M. J. Lilley (2019). Comparison of the structures and mechanisms of the pistol and hammerhead ribozymes. *J. Amer. Chem. Soc* 141, 7865–7875.

2

Biological Roles of Self-Cleaving Ribozymes

Christina E. Weinberg

Leipzig University, Institute for Biochemistry, Brüderstraße 34, 04103 Leipzig, Germany

2.1 Introduction

After more than 30 years of ribozyme research, we know of nine self-cleaving ribozyme classes today. Each of these classes shows a unique secondary and tertiary structure that enables a site-specific intramolecular cleavage of a phosphodiester linkage. Much research effort in the past three decades has focused on solving the three-dimensional structures, deciphering the precise catalytic mechanisms and defining other biochemical characteristics of these catalytic RNAs. In all known examples, the scission of the RNA backbone occurs through an internal phosphoester-transfer reaction where the 2′ oxygen of a ribose attacks the adjacent 3′ phosphate to result in one product with a 2′,3′-cyclic phosphate and another product exhibiting a 5′ hydroxyl at its terminus. All self-cleaving ribozymes use acid–base catalysis to support cleavage and often employ divalent metal ions, usually magnesium, to stabilize an active ribozyme structure.

In the past, self-cleaving ribozyme discovery has been mostly serendipitous. However, computational methods enabled the targeted discovery of novel self-cleaving ribozyme classes and more examples of known classes in a variety of organisms. Computational searches based on the secondary structure information of known self-cleaving ribozymes identified more representatives of the same classes throughout the tree of life, for example, for hepatitis δ virus (HDV) [1] and hammerhead ribozymes [2–5]. Furthermore, experimental analysis of a computational method to discover any type of RNA structure led to the finding that two of these RNA structures represented additional self-cleaving ribozyme classes [6, 7]. One of them is found in the 5′ untranslated region (UTR) of bacterial mRNA genes and represents the only known natural metabolite-responsive self-cleaving ribozyme [8]. The second was the twister self-cleaving ribozyme, whose discovery led to the observation that self-cleaving RNAs in bacteria are often located in close proximity to each other and to certain types of genes [7]. This facilitated the targeted discovery of the latest three novel self-cleaving ribozyme classes: twister sister, hatchet, and pistol [9].

These computational approaches were complemented by experimental methods. An in vitro selection, which used transcripts from the fragmented human genome

Ribozymes, First Edition. Edited by Sabine Müller, Benoît Masquida, and Wade Winkler.

as a selection pool, isolated several ribozymes, one of which is a conserved mammalian sequence with an HDV-like secondary structure, which resides in an intron of the CPEB3 gene [10]. Furthermore, experimental approaches shed light on the evolutionary history of ribozymes, when it was found that the catalytic core of the hammerhead ribozyme commonly arises in in vitro selection experiments from synthetic pools. Thus, this ribozyme might have evolved multiple times independently. These investigations suggest that RNAs with the ability to cleave the phosphodiester

Table 2.1 Summary of known self-cleaving ribozyme classes and their abundance, distribution, and biological functions.

Ribozyme class and subtype		Number of examples	Distribution	Biological functions
Hammerhead	Type I	190915[a)]	Bacteria, archea, plants, metazoans	For replication of viroids, viroid-like RNAs [13]
	Type II	155	Bacteria, archea	In PLEs [14], retrozymes [15], SINEs [16], and other unknown functions
	Type III	2723	Bacteria, archea, plants	
Hairpin		8	Viroids	Viroid replication [13]
Hepatitis δ virus (HDV)-like		8715[b)]	With helper virus in humans, higher eukaryotes, insects, bacteria	HDV RNA replication [13] R2 elements [17], LINE element L1Tc [18], and other unknown functions
Varkud satellite (VS)		1	*Neurospora* (fungus)	Satellite RNA replication [19]
glmS		313 (150 env) [20]	Bacteria, mostly Gram-positive	Metabolite-responsive regulation of gene expression [8]
Twister	P1	1155 [7]	Bacteria, plants, fungi, and metazoans, such as insects, fish, and flatworm [7]	Unknown
	P3	9 [7]		
	P5	331 [7]		
Twister sister		5 (242 env) [9]	Bacteria [9]	Unknown
Hatchet		1 (302 env) [9]	Metagenomic (presumed bacterial) [9]	Unknown
Pistol		35 (875 env) [9]	Bacteria [9]	Unknown

Self-cleaving ribozyme classes including subtypes (where applicable) are listed in order of discovery. Number of self-cleaving ribozyme examples for each class are given according to Rfam [21], unless another reference is listed. Rfam numbers include environmental sequences (env).

a) This number includes hammerhead type I, HH9, and HH10 Rfam families.

b) This number includes Rfam families for HDV, CPEB, HDV-*F. prausnitzii*, drz- agam-1, and drz-agam2-2.

linkage of nucleic acids can evolve fairly easily, and there are likely other ribozymes in nature capable of catalyzing self-scission awaiting discovery [11, 12].

Many of the early discoveries of self-cleaving ribozymes arose when scientists experimentally investigated a biological system that happened to use a ribozyme. However, a new wave of discoveries has been possible without reference to a biological function. This poses the challenge of determining the roles of these catalytic RNAs in nature. One issue connected to this task is the drastic increase in the mere number of known self-cleaving ribozyme examples, which lies already in the thousands. Another challenge is that the function of genes nearby self-cleaving ribozymes examples is frequently unknown, which limits their use to gather hints at the overall connection with self-cleaving ribozyme biology.

In this chapter, the currently known functions of self-cleaving ribozymes in biology are summarized (see Table 2.1 for an overview). These functions include a role in replication of viroids and in other circular RNAs such as HDV and Varkud satellite (VS) RNAs (Table 2.1). Additionally, the involvement of some self-cleaving ribozymes in retrotransposition in a variety of different transposable elements is discussed. Although the exact ribozyme function is well understood in some cases, in many other instances, including ribozymes not present in transposable elements, their precise purpose remains a mystery. In these cases, further studies are necessary to determine a detailed mechanistic contribution of these self-cleaving ribozymes to the biology of the organism. Lastly, the only known biological function of a self-cleaving ribozyme class in bacteria is described.

2.2 Use of Self-cleaving Ribozymes for Replication

Self-cleaving ribozymes play an important role in viroids, viroid-like satellite RNAs, and HDV RNAs. These three subviral entities share several common features such as small size, circular RNA structure, pathogenicity, and usage of some form of rolling circle mechanism (explained below) to replicate [22–25]. Their circular genome protects them from exonuclease-mediated degradation, and their strong self-complementarity renders endoribonucleases, which often target single-stranded RNA, ineffective. To support their replication, some viroids, viroid-like RNAs, and HDV RNAs contain self-cleaving ribozymes.

2.2.1 Viroids

Viroids, which do not code for any protein, can infect plants without the presence of a helper virus [13]. They rely entirely on the transcription and processing machinery of their hosts to replicate and, in addition, use their own RNA elements to achieve propagation [26]. Viroids can be divided into two distinct families. The family of Avsunviroidae is named after the *avocado sun blotch viroid* (also abbreviated as ASBVd), and the family of Pospiviroidae is named after the *potato spindle tuber viroid* (abbreviated as PSTVd).

In the family of Avsunviroidae, several integral steps of viroid replication are carried out by ribozymes. Hammerhead ribozymes have been found in both sense and antisense strands, and they play a key role in viroid replication through their site-specific cleavage of oligomeric viroid RNA transcripts into monomers (Figure 2.1a) [25, 27]. The very first self-cleaving ribozymes to be discovered were hammerhead ribozymes reported in ASBVd and a viroid-like satellite RNA (explained in the next section) [27, 28]. Later, additional examples were found in *peach latent mosaic viroid* (PLMVd) [29], *chrysanthemum chlorotic mottle viroid* (CChMVd) [30], eggplant latent viroid [31], and others [32, 33]. Self-cleaving ribozymes have been used in the field of synthetic biology for the construction of RNA-based elements to regulate gene expression [34, 35]. Recently, CChMVd hammerhead ribozymes exhibited particularly robust cleavage in mammalian and bacterial cell culture, making them promising candidates for the design of artificial ribozyme constructs to regulate gene expression in vivo [36].

Viroids in the family of Avsunviroidae replicate their circular ssRNA genomes through a "symmetric" pathway (Figure 2.1a). The process starts with a circular ssRNA referred to as the plus (+) strand. A DNA-dependent RNA polymerase provided by the host transcribes the circular (+) strand into an oligomeric transcript, referred to as the (−) strand. Because the (+) strand is circular, the RNA polymerase creates multiple concatemeric copies, and in this sense the resulting linear transcript is oligomeric or multimeric. Despite their natural DNA dependency, these host polymerases can be forced to accept an RNA template [37–39]. The newly generated (−) strand folds and cleaves, catalyzed by the hammerhead ribozymes, into monomeric units (Figure 2.1a) [13]. Cleavage likely occurs during transcription. Host enzymes (see following paragraph) circularize each unit, and the same process is repeated to obtain a (+) strand circular genome (Figure 2.1a). The more abundant strand in a cell is, by convention, called the (+) strand [13].

Hammerhead motifs are usually stabilized by tertiary interactions between loops that flank the ribozyme core, which leads to robust self-cleavage rates [40–42]. However, ASBVd hammerhead ribozyme examples have been discovered that consist of only the minimal ribozyme core. This core structure lacks additional sequence necessary to facilitate tertiary interactions, making these ribozyme examples thermodynamically unstable and very slow to cleave. Their extremely poor activity as monomeric ribozymes suggests that self-cleavage most likely occurs through dimeric hammerhead structures transiently formed by the oligomeric RNA transcripts during replication [13, 43]. Once the oligomeric transcripts are cleaved by the ribozyme, they have 5′-hydroxyl and 2′,3′ cyclic phosphodiester termini typical of nucleolytic ribozyme products. These RNA ends can be ligated by plant tRNA ligases (Figure 2.1a) [44–46], and the resulting 2′-phosphate can be removed by an enzyme with 2′-phosphotransferase activity [47]. This ligation results once more in a circular RNA.

Pospiviroidae replicate in the nucleus of infected cells by an asymmetric rolling circle mechanism that is entirely carried out by proteins (Figure 2.1b top). This pathway starts with the repeated transcription of the circular (+) strand RNA by host nuclear-encoded polymerase (NEP). The resulting oligomeric (−) strand contains

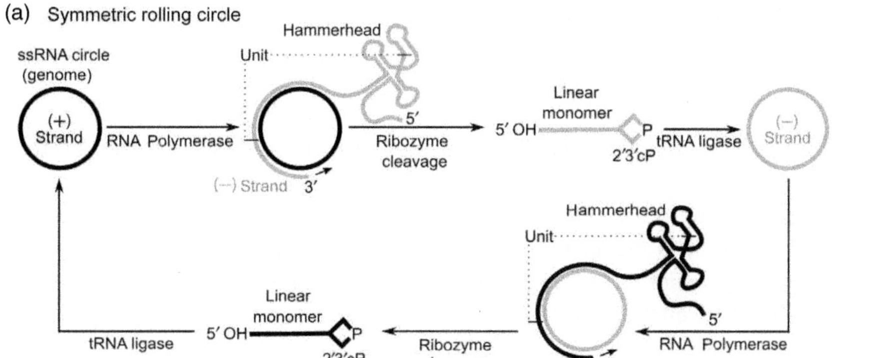

Figure 2.1 Rolling circle replication mechanisms in viroids and viroid-like satellite RNAs. (a) Symmetric rolling circle mechanism in viroids of the Avsunviroidae family and viroid-like RNAs. Circular (+) strand RNA is transcribed by DNA-dependent RNA polymerase into oligomeric (−) strand RNAs. Dotted lines indicate cleavage sites that define a single unit within the oligomeric RNA. Units are separated by hammerhead ribozyme cleavage and circularized by host enzymes, such as tRNA ligase. The circular (−) strand RNA is used for a second round of amplification, yielding the (+) strand genome. In viroid-like RNAs, a hairpin instead of a hammerhead ribozyme could catalyze the cleavage of (+) strand oligomeric transcript as well as the ligation of the (+) strand linear monomer into a circular RNA. (b) Asymmetric rolling circle mechanism in viroids of the Pospiviroidae family (top) and viroid-like RNAs (bottom). In Pospiviroidae, the nuclear-encoded polymerase (NEP) generates an oligomeric transcript, the (−) strand. Dotted lines indicate cleavage sites that define a single unit within the oligomeric RNA. The linear (−) strand is used as template for transcription of oligomeric (+) strands. CCRs fold and direct RNase III cleavage to generate unit-length products, which are circularized by DNA ligase 1. This process is similar for viroid-like RNAs (bottom), except that instead of CCRs, hammerhead ribozymes, which are encoded in the (+) strand, cleave the oligomeric transcript into linear monomers. These unit-length transcripts are circularized either by ribozyme-mediated or enzymatic ligation.

many consecutive antisense copies of the (+) strand. The (−) strand oligomers serve as templates for the synthesis of oligomeric (+) strands that are then cleaved into monomers and ligated into circular (+) strand RNAs (Figure 2.1b top) [13]. However, Pospiviroidae do not contain self-cleaving ribozymes, but harbor several conserved sequence motifs within their rodlike secondary structure. One of these structures, the so-called central conserved region (CCR), is used by a class III RNase enzyme as the cleavage site to separate oligomeric (+) strand transcripts into monomers (Figure 2.1b top). The resulting monomers contain 5′-phosphate and 3′-hydroxyl termini that are covalently connected, remarkably, by nuclear DNA ligase 1, which the viroid coopts to circularize RNA substrates (Figure 2.1b top) [48].

2.2.2 Viroid-like Satellite RNAs

Viroid-like RNAs, which do not encode any proteins, are similar to viroids in that they use ribozymes in their rolling circle replication. However, in addition to hammerhead ribozymes [28, 49], some viroid-like RNAs also employ hairpin ribozymes for the cleavage of oligomeric transcripts into monomers [50]. Viroid-like RNAs are called satellite RNAs because they are dependent on a co-infection with a helper RNA virus [13] that supplies, at least in part, the RNA polymerase complex and a coat protein that supports satellite RNA transmission [33]. After ribozyme-mediated co-transcriptional cleavage of viroid-like RNA oligomeric transcripts, the resulting monomers are circularized either by a cytoplasmic isoform of the tRNA ligase [44] or, in contrast to viroids, by RNA self-ligation catalyzed by the ribozyme. The high RNA ligase activity of the hairpin ribozyme suggests that it is able to circularize monomeric rolling circle intermediates efficiently and without the help of protein enzymes [50, 51].

Depending on whether the asymmetric or symmetric rolling circle pathway is used by satellite RNAs, they contain a self-cleaving ribozyme on their (−) strand. In satellite RNAs that use asymmetric rolling circle replication (meaning there is no circular RNA intermediate), hammerhead ribozymes are only found in the (+) strand (Figure 2.1b bottom). Those satellite RNAs replicating via a symmetric rolling circle mechanism contain a ribozyme on both strands (+ and −). These ribozymes could be either a hammerhead ribozyme or a hammerhead on one strand and a hairpin ribozyme on the other [50, 52]. In the satellite RNA of the cereal yellow dwarf virus-RPV RNA, the encoded hammerhead ribozyme is only active in oligomeric transcripts, but not in single hammerhead copies present in monomeric transcripts. In unit-length RNAs, the hammerhead ribozyme adopts a pseudoknot structure that prevents catalytic activity. Only in oligomeric transcripts can two hammerhead ribozymes come together, fold into a structure lacking the inhibitory pseudoknot, and become active [13, 53].

Recently, a connection between viroids and viroid-like RNAs containing hammerhead ribozymes to retrozymes was suggested. Possibly these circular plant pathogens may have emerged de novo from the population of abundant retrozyme circular RNAs present in plant transcriptomes [54].

2.2.3 Hepatitis δ Virus RNA

A subviral entity closely related to viroids is the HDV RNA. It has been found in higher eukaryotes, including humans, and has a genome size of ~1700 nt [55]. Similarly to viroid-like satellite RNAs, HDV RNA depends on a helper virus, hepatitis B virus, for transmission [33]. In contrast to its viroid and viroid-like relatives, HDV RNA encodes a protein called the delta antigen (δAg) [55]. As with some other RNA viruses, the δAg coding sequence is found in the antigenomic polarity of HDV [56]. This means that the δAg mRNA is actually transcribed as a linear molecule from the (+) strand RNA and becomes equipped with 5′-cap and 3′-polyA tails to facilitate translation initiation through the conventional ribosome-scanning mechanism [57]. Two δAg isoforms, small (S) and large (L) δAg, are produced from the same open reading frame (ORF). The S-δAg, which contains a nuclear localization signal (NLS) and an RNA-binding motif that imports HDV RNA into the nucleus, supports viral replication. The large isoform of the delta antigen (L-δAg), which is 19 amino acids longer than S-δAg, is produced through RNA editing of the same δAg mRNA in later stages of the viral infection, and it is required for viral particle assembly [58].

The genomic region comprising the δAg is classified as one of two distinct domains of HDV RNA. The other domain corresponds to a ~350 nt viroid-like sequence that folds into a rodlike secondary structure [56, 59, 60] and contains the ribozymes necessary for replication in its terminal region. HDV RNA replicates autonomously in the nucleus through a symmetric rolling circle mechanism catalyzed by host enzymes such as a DNA-dependent RNA polymerase accompanied by co-transcriptional self-cleavage of HDV ribozymes [61–63]. HDV ribozyme cleavage generates RNA fragments with a 2′,3′-cyclic phosphate and a 5′-hydroxyl group that allow circularization by another host-specific enzyme, likely similar to tRNA ligases in plants [64].

2.2.4 *Neurospora* Varkud Satellite RNAs Replicate Using a DNA Intermediate

Natural isolates of the fungus *Neurospora crassa* contain in their mitochondria an 881 nt DNA plasmid. The transcription product of this Varkud plasmid is an oligomeric RNA that harbors a self-cleaving ribozyme involved in the processing of intermediates during rolling circle replication [19, 65–67]. Interestingly, the VS ribozyme performs cleavage and ligation reactions, which are both necessary for the replication of the satellite RNA, as a dimer [68, 69]. In *Neurospora,* catalytic RNAs are part of a rolling circle mechanism that includes a DNA stage. In this process, an oligomeric RNA is transcribed from the VS plasmid DNA template by the *Neurospora* mitochondrial RNA polymerase using a promoter located immediately upstream of the ribozyme. The ribozymes cleave the transcript and then ligate these monomeric RNAs into circular RNA intermediates, which are subsequently used as a template for reverse transcription. This process is carried out by the reverse transcriptase (RT) encoded on the Varkud plasmid to yield full-length (−) strand cDNAs. After displacement or degradation of the RNA template, synthesis of the (+) strand DNA and ligation to generate a closed circular DNA presumably occur [67].

2.3 Self-cleaving Ribozymes as Part of Transposable Elements

Transposons, which are a type of mobile genetic element, can move themselves or copies of themselves to different locations within the genome. Two classes of transposable elements (TEs) are distinguished based on the mechanism of transposition. The first class includes all those elements that transpose by a so-called "copy and paste" mechanism. Transposons of this class, referred to as retrotransposons, are first copied from the genomic locus by transcription into RNA. This RNA intermediate is then reverse-transcribed into cDNA and inserted back into the genome at a new position. Retrotransposons usually encode proteins such as reverse transcriptases (RTs) or endonucleases (ENs) that facilitate the insertion into the host genome [70].

There are several retrotransposon subclasses: (i) Transposable elements with long terminal repeats (LTRs) that encode their own RT to convert the RNA intermediate into DNA, (ii) long interspersed nuclear elements (LINEs), which do not contain LTRs, but still harbor an RT gene, (iii) short interspersed nuclear elements (SINEs), which neither have LTRs nor code for an RT, and (iv) *Penelope*-like elements (PLEs), which cannot be grouped as either of these previous classes, as they code for RT and EN, but do not carry typical LTRs at their ends, but rather the so-called *Penelope* LTRs (PLTRs). Within the class of retrotransposons, it is common to classify transposable elements based on whether or not they contain LTRs [70].

The second class of transposable elements includes those that transpose by a "cut and paste" mechanism. This mechanism is used by DNA transposons and does not include any RNA intermediates, but instead relies on several transposase enzymes [70]. As these elements do not contain self-cleaving ribozymes, they are not discussed further here.

Finally, transposable elements of both classes, retrotransposons and DNA transposons, can be grouped into autonomous and nonautonomous TEs. Autonomous means that these transposons are self-sufficient because they bring all necessary enzymatic or ribozymatic equipment needed for transposition. Nonautonomous transposons, however, rely on autonomous transposable elements also present in the organism and hijack their transposition machinery [70].

In the following sections, examples of self-cleaving ribozymes found in diverse subclasses of retrotransposons are discussed, including a site-specific non-LTR retrotransposon called R2, PLEs, SINEs in *Schistosoma*, and retrozymes.

2.3.1 R2 Elements: Non-LTR Retrotransposons that Use HDV-like Ribozymes for Retrotransposition

R2 elements were first discovered in ribosomal DNA (rDNA) loci of *Drosophila melanogaster* in the early 1980s, at which time non-LTR retrotransposons had not been discovered [71, 72]. Soon these elements were found in many other insect species, such as the silk moth *Bombyx mori* [73, 74], as well as arthropods, and other taxa of animals including nematodes, birds, and tunicates. In all these examples, R2 elements are exclusively found at a specific site in 28S rRNA genes – the "R" referring to this rRNA gene association (Figure 2.2a,b). In fact, R2 elements are only able to integrate into the genome at this specific location. This site-specific transposition made it possible to investigate this element closely, making it one of the best-understood non-LTR retrotransposons [17].

Located 74 bp downstream of the R2 element in the 28S rDNA locus, sometimes the R1 element can be found (Figure 2.2a) [71, 72]. Sequence analysis of both elements revealed structures similar to those of LINEs in mammals [75, 76] and in *Drosophila* [77], hinting at a possible role as transposable element [78, 79].

The full-length R2 element is derived from co-transcription with 28S rRNA (Figure 2.2c) [80]. The transcript contains a conserved stretch of nucleotides in its 5′ UTR that can be folded into a double-pseudoknotted structure with five base-paired regions, conforming to the HDV ribozyme structure originally discovered in HDV [62, 81, 82]. The HDV-like ribozyme precedes an ORF that codes for the R2 protein (Figure 2.2c). This multi-domain protein harbors DNA-binding domains, a reverse transcriptase domain with an N-terminal RNA-binding domain and a C-terminal region with a putative thumb and zinc-binding domain, followed by an endonuclease domain (Figure 2.2d). Acting in concert, these RNA and protein components mediate retrotransposition, presumably with the aid of additional host factors.

After the host RNA polymerase I transcribes the full-length R2 transcript starting from the common promotor of the 28S rRNA, the RNA intermediate is cleaved at its 5′ end by the site-specific scission of the HDV-like ribozyme that resides in the first ~184 nt of the R2 element (Figure 2.2c) [81]. This explains why previous experiments never identified a promotor at the 5′ end of the R2 elements [83], but instead transcripts of extensively 5′-truncated elements were observed [81, 84]. The 5′ ends of these transcripts correspond to ribozyme cleavage sites rather than to transcription start sites. The HDV-like ribozyme not only frees the transposable element from the 28S rRNA co-transcript but likely also serves as an internal ribosomal entry site (IRES) to promote translation of the R2 protein (Figure 2.2e) [85–87]. Furthermore, it was shown that the 5′ and 3′ UTRs of the R2 RNA intermediate serve as binding sites for the R2 protein in the integration process (Figure 2.2f) [88].

In the R2 protein, the endonuclease domain at the C-terminal end of the R2 ORF is responsible for an initial nick in the DNA bottom strand (Figure 2.2d,f). This domain contains an active site similar to that of type IIS restriction enzymes, instead of an apurinic endonuclease (APE) domain usually found in other non-LTR retrotransposons [89]. Characteristic for this type of restriction endonuclease is that the catalytic and DNA-binding domains are separate. Thus, this enzyme binds the DNA a short distance from the cleavage site. Indeed, R2 protein footprint analysis showed that DNA contacts of the R2 protein were made upstream and downstream of the insertion site [90, 91], which corresponds to the current model of R2 integration.

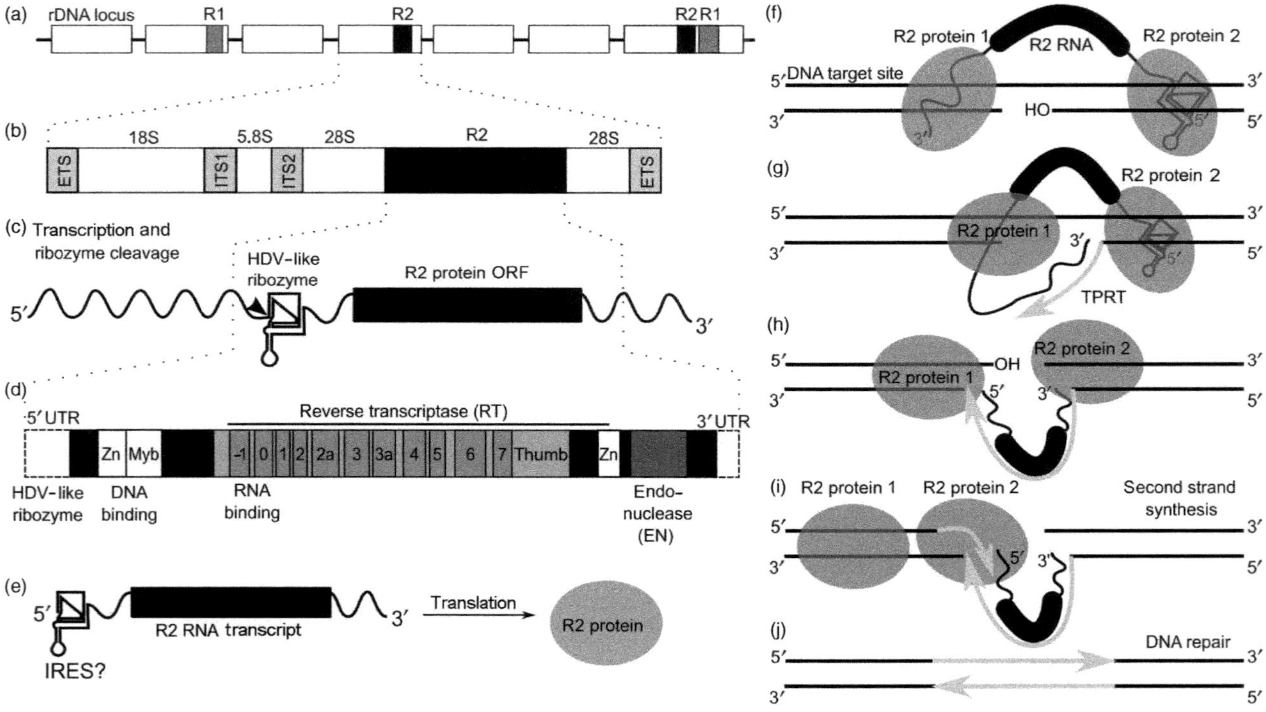

Figure 2.2 The R2 element in *Bombyx mori* and its transposition mechanism. (a) Genomic rDNA locus comprising multiple rDNA units. Each rDNA unit (white box) contains 18S, 5.8S, and 28S genes. Some rDNA units contain either R2 (black box), R1 (dark gray box), or both. (b) Organization of one rDNA unit with an R2 element inserted into a specific site within the 28S rDNA is shown. The external transcribed spacers (ETS) and internal transcribed spacers (ITS1 and 2) found in the precursor rRNA are depicted as light gray boxes. (c) Transcription of the R2 element in *B. mori* yields an RNA consisting of the HDV-like ribozyme in its 5′ UTR, the open reading frame (ORF) for the R2 protein and a 3′ UTR. (d) An expanded view of the R2 ORF highlights the protein domains. The R2 protein consists from its N-terminus to C-terminus of the following domains: C_2H_2 zinc-finger and Myb-like nucleic acid-binding motifs and the reverse transcriptase (RT) domain including two additional conserved motifs designated −1 and 0 and motifs 1–7, followed by a potential thumb domain, a potential zinc-binding domain that could play a role in DNA binding, and finally the endonuclease domain (EN). Dotted lines designate untranslated regions. (e) Translation of the ORF generates R2 protein. Translation initiation likely occurs through IRES-like structure of the 5′ UTR. R2 proteins can bind the 5′ and 3′ ends of the R2 element RNA. (f) The RNA/protein complexes bind the target site for insertion into the 28S gene. The R2 protein that is bound to the 3′ end of the R2 transcript (R2 protein 1) cleaves the DNA bottom strand. (g) The resulting 3′-hydroxyl from the DNA nick is used to prime reverse transcription of the R2 transcript (TPRT). Light gray arrow represents cDNA. (h) When the reverse transcription reaches the 5′ end of the R2 transcript, the R2 transcript is removed from the R2 protein 2. This R2 protein becomes active and cleaves the DNA top strand, which is now used to initiate second-strand synthesis. (i) During second-strand synthesis, the RNA is removed from the cDNA either by RNase H or displacement. (j) After complete synthesis of the entire R2 element, the nicks are sealed by host DNA repair mechanisms.

R2 endonuclease cleaves one strand of the DNA target strand site-specifically. The resulting 3′ DNA end is used to prime reverse transcription of the R2 RNA (Figure 2.2f). This process, which is common to many non-LTR retrotransposons but was first discovered in the R2 element, was termed target-primed reverse transcription (TPRT) (Figure 2.2g) [92]. TPRT is most efficient at the 28S gene insertion site with RNA containing the 3′ UTR of the R2 element as template, so R2 elements integrate at this specific part of the 28S gene [17, 93]. The R2 RT can also prime, although less efficiently, the reverse transcription of any RNA using the 3′ end of any other RNA or single-stranded DNA even if there is no complementarity between the template and the primer [17, 93]. While the exact 3′ end of the R2 element does not appear critical for the TPRT mechanism [94, 95], the exact 5′ end is important, as first suggested by in vivo integration results [96, 97].

Target-primed reverse transcription generates a cDNA copy of the R2 element that is covalently attached to the DNA bottom strand (Figure 2.2g). The RT domain in charge of facilitating this process of TPRT consists of the typical motifs 1–7 of an reverse transcriptase and two conserved sequence motifs (termed 0 and −1) at its N-terminus. These domains are involved in RNA binding, and their similarity to other RTs suggests a close evolutionary relationship of R2 retrotransposons to other transposable elements [98, 99]. R2 RT shows a higher processivity than most other RTs, which is beneficial, as early RT dissociation from the template causes truncated (dead) copies of R2 elements [100, 101].

Integral to the understanding of R2 integration was the finding that the R2 protein can bind the 5′ and 3′ ends of the R2 transcript [88]. The bound 5′ RNA end is about 300 nt in length and contains a pseudoknot that is conserved across silk moths [102, 103]. If the R2 protein binds to the 5′ end of the R2 RNA intermediate, it is directed to bind downstream of the insertion site (Figure 2.2f, referred to as R2 protein 2). However, if the 3′ end of the R2 RNA is bound, the R2 protein interacts with the target DNA upstream of the insertion site (Figure 2.2f, referred to as R2 protein 1). This R2 protein upstream of the insertion site initiates the retrotransposition process by cleaving the bottom DNA strand, and the resulting 3′-hydroxyl is used for the first-strand synthesis of the R2 element by TPRT (as mentioned above). As the R2 RNA is used for first-strand synthesis, it is removed from the downstream R2 protein, which is bound to the 5′ end of the RNA (Figure 2.2g,h). This activates the downstream R2 protein to cleave the top DNA strand (Figure 2.2h) and to subsequently use the generated 3′-hydroxyl group to synthesize the second strand (Figure 2.2i). The R2 RT neither harbors a conserved RNase H domain nor shows detectable RNase H activity in vitro. Therefore, it is likely that second-strand synthesis by R2 RT allows direct displacement of the annealed R2 RNA [93, 101], instead of RNase H-mediated degradation of the RNA strand in the RNA:DNA hybrid (Figure 2.2i). Initiation of second-strand synthesis has not yet been documented in vitro [91, 92, 101, 104], and it is possible that this step is carried out by a host polymerase in vivo [17]. After second-strand synthesis is complete, the DNA repair system of the host seals remaining nicks in the DNA (Figure 2.2j).

Variations in the priming of the second-strand synthesis have been observed. These variations can lead to R2 elements with diverse 5′ ends. Some R2 elements have an exact 28S/R2 boundary, but other R2 elements include additional insertions of unspecific sequences. This diversity arises due to differing locations of the HDV-like ribozyme cleavage site in different R2 elements. If the HDV-like ribozyme cleaves the R2 RNA so that a small portion of the 28S rRNA remains, second-strand synthesis is facilitated because a heteroduplex between the cDNA and the DNA insertion site can form. This stabilizes the integration intermediate, and second-strand synthesis commences generating exact 5′ ends [97]. In *Drosophila simulans*, for example, ribozyme cleavage occurs in a GC-rich region of the 28S rRNA either 13 or 28 nucleotides upstream of the insertion site, leaving a small portion of the 28S rRNA to create a heteroduplex [82, 105]. If, in a second scenario, the HDV-like ribozyme cleavage site lies right at the 28S/R2 junction, so that there are no 28S sequences as part of the cDNA, the R2 element must use so-called microhomology regions to facilitate second-strand synthesis. This is possible because the R2 RT adds up to five non-templated nucleotides to the 3′ end of the cDNA. This serendipitously generates microhomologies to prime second-strand DNA synthesis and gives rise to sequence variation at the 5′ junctions of different integrated copies of R2 [96, 104].

A side product of ribozyme-mediated processing of the 28S co-transcript is the propagation of several nonautonomous parasites of R2, so-called SIDEs (short internally deleted elements). SIDEs are R2 elements that have lost their ORF but have retained the ribozyme and 3′ UTR. While the 3′ UTR enables binding by the R2 integration machinery, the HDV-like ribozyme catalyzes the co-transcriptional cleavage from the 28S transcript. These two features allow SIDEs to efficiently transpose in hosts that contain R2 elements [106].

2.3.2 HDV-like Ribozymes in Other Non-LTR Retrotransposon Lineages

Non-LTR retrotransposons in rDNA (such as the R2 element) and in telomeres (such as SART retrotransposons) insert sequence-specifically into the host. However, there have been some examples of HDV-like ribozymes discovered that are found in association with other retroelements that insert with little specificity.

In a study that used secondary structure information instead of mere sequence homology alignments, HDV-like ribozymes were discovered to be widespread in nature [1]. Several of the HDV-like ribozymes discovered in this search immediately suggested a biological role through their genomic context or gene association. For example, HDV-like ribozymes in *Strongylocentrotus purpuratus* and *Anopheles gambiae* were found within or near genes coding for RT-like proteins. Also, HDV-like ribozymes in nematodes were found in hundreds of intergenic copies that were always located between conserved downstream sequences and different upstream sequences. These ribozymes could be part of retrotransposons. Other ribozyme examples could play roles in gene regulation, as the HDV-like ribozyme discovered in *Faecalibacterium prausnitzii* might suggest [1]. This HDV-like ribozyme is always found upstream of phosphoglucosamine mutase (GlmM), and there are recent suggestions for an involvement in metabolite-dependent gene regulation [107].

Identification of HDV-like ribozymes in different retrotransposon types and in different species suggests a widespread use of these catalytic RNAs in retrotransposition steps such as 5′ processing, translation initiation, and potentially TPRT [87]. If elements are transcribed as part of a longer transcript, ribozyme cleavage liberates the retrotransposon from the co-transcribed flanking sequences. This might be the case for so-called Baggins retrotransposons and retrotransposon-like elements (RTEs), which contain HDV-like ribozymes, but do not insert site-specifically. These transposable elements frequently map to introns or are found immediately downstream of LTR retrotransposons, which ensures their expression through co-transcription with these other genetic elements. In such a scenario, it appears likely that the presence of HDV-like ribozymes enables 5′ processing [87]. Additional examples of hammerhead and HDV-like ribozymes were described in *Schistosoma*, ticks, and lamprey, where they play a role in processing of SINEs [108] and LINEs [109]. HDV-like ribozymes were also found near LTR retrotransposons in the antisense direction to the LTR ORF, which makes it unlikely that they function in 5′ processing. Their function in these occurrences is unknown [87].

Furthermore, it was shown that HDV-like ribozymes support translation initiation in the absence of a 5′ cap and 3′ polyA required for translation in eukaryotes possibly through their close resemblance to IRESs. Several HDV-like ribozyme examples were tested in in vitro or in vivo translation assays and revealed a translation efficiency similar to or higher than that of the Hepatitis C virus (HCV) IRES positive control [87]. Therefore, it seems likely that the HDV-like ribozyme acts similarly to an IRES by presumably binding the translation machinery and enabling translation initiation.

The genomic distribution of another type of retrotransposon found in *Trypanosoma cruzi* is considered random, although some specificity for insertions has been reported. These retrotransposons are called L1Tc and have been found in isolated loci as tandem repeats and associated with genomic regions rich in repetitive DNA sequences [110]. The L1Tc is a LINE element, which means it encodes its transposition machinery (Figure 2.3a). In addition, L1Tc contains an HDV-like ribozyme in its 5′ UTR, which was shown to be active in co-transcriptional cleavage assays. The ribozyme can trigger the release of the element from long polycistronic transcripts, which are typical transcript forms for trypanosomes [111]. The L1Tc bears an internal promotor, designated Pr77, which preserves the autonomous character of the element [112]. This promotor overlaps with the first 77 nucleotides of the L1Tc element (Figure 2.3a). Thus, the 5′ region of L1Tc can serve as promotor on the DNA level [112] and at the RNA level resembles an HDV-like ribozyme that cleaves and generates transcripts with a homogenous 5′ end (Figure 2.3a) [111]. The promotor-derived transcripts are translated despite the absence of a capped leader structure at the L1Tc RNA's 5′-end region. This suggests a cap-independent translation mechanism, presumably through the use of an IRES [111] as explained above.

2.3.3 *Penelope*-like Elements (PLEs) Contain Hammerhead Ribozymes

Besides LTR and non-LTR retrotransposons, a distinct third class has been described: PLEs [114, 115]. These transposable elements were first discovered to be functionally active in the fly *Drosophila virilis* where they cause hybrid dysgenesis. Hybrid

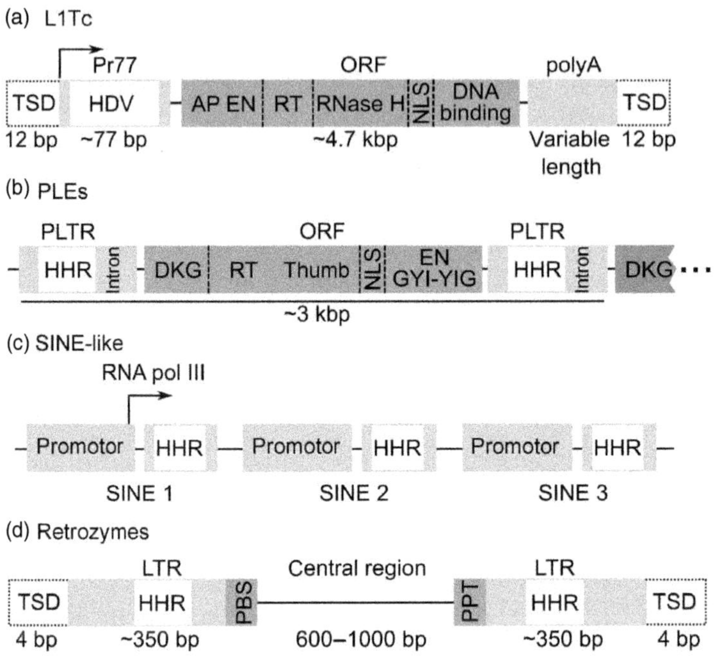

Figure 2.3 Self-cleaving ribozymes as parts of transposable elements. (a) Schematic representation of the simplified composition of L1Tc retrotransposons from *Trypanosoma cruzi*. The element is flanked by target site duplications (TSD) of usually 12 bp, and it encodes a protein with an apurinic/apyrimidinic endonuclease (AP EN) domain, a reverse transcriptase (RT) and RNase H domain, and a DNA-binding domain. The first 77 nt of L1Tc harbor an HDV ribozyme (HDV) and on the DNA level correspond to an internal promoter (Pr77) that generates abundant and translatable transcripts [110–112]. (b) Schematic representation of the composition of *Penelope*-like elements (PLEs). PLEs occur as tandem or multi-copy repeats in which an ORF is flanked by *Penelope* long terminal repeats (PLTRs). The ORF codes for a protein with RT and EN domains. The EN belongs to the GIY-YIG class of endonucleases. The PLTRs contain a hammerhead ribozyme (HHR) [14] and have been shown to also contain an intron in some PLEs [98]. (c) Schematic representation of small interspersed nuclear element-like retrotransposons in *Schistosoma*. These elements are often found in repetitive sequences and consist of a promotor followed by a hammerhead ribozyme (HHR). All promotors could initiate transcription. However, if the promotor of the element labeled SINE1 enables RNA polymerase III-mediated transcription, a long transcript with several copies of the element designated SINE 1–3 is produced. If ribozyme cleavage then occurs in SINE 1, the resulting SINE 2 and 3 could be produced from their new genomic location after transposition by using the promotor of SINE 2 for transcription initiation. However, if ribozyme cleavage occurred at all ribozyme cleavage sites in each SINE, single SINE copies are generated. This ceases further transposition of the element after integration into the genome, because single SINE copies contain the promotor in their 3′ end; thus transcription of the element cannot be assured [108, 113]. (d) Schematic representation of the composition of retrozymes. Retrozymes are flanked by target site duplications (TSDs) and LTRs, which contain hammerhead ribozymes (HHR). The central region does not contain an ORF and is flanked by primer binding site (PBS) and polypurine tract (PPT) elements needed for priming of DNA synthesis from the RNA element [15].

dysgenesis describes a phenomenon in which unrelated, different kinds of transposable elements are mobilized upon crossing a female fly from the wild with a male fly from an established laboratory strain. Such a cross results in a high level of gonadal sterility in the F_1 generation, chromosomal nondisjunction and recombination, and the occurrence of multiple phenotypic mutations [114].

PLEs have an average length of about 3 kb (Figure 2.3b). They consist of direct LTR-like elements called *Penelope*-LTRs (PLTRs), which encode a hammerhead ribozyme [14] and flank a region coding for reverse transcriptase (RT) and endonuclease (EN) protein domains [115]. The hammerhead ribozyme secondary structure consists of three stems (I, II, and III) around a catalytic core that contains 15 highly conserved nucleotides that are essential for ribozyme cleavage. Three permuted forms of this catalytic RNA have been found, named according to which stem connects the ribozyme to the rest of the transcript, as type I, type II, or type III [5]. PLEs are often found in a tandem arrangement as they tend to insert into or adjacent to preexisting PLE copies. It is hypothesized that a tandem arrangement is needed for a functional element, because it would allow the 3′ sequence of one element to represent the upstream sequence of the following PLE. If the 3′ sequence of the upstream element contains a promotor, a transcriptionally active unit would be formed. A similar mechanism was observed for *HeT-A* elements in *Drosophila* [116]. Indeed, it was shown that the PLTR from the upstream element contains the start site of the *Penelope* transcript and gives rise to the 5′ UTR of the PLE [98].

PLEs have been identified in different phyla in eukaryotes. They are massively widespread in metazoan genomes, including distant invertebrates, from cnidarians to chordates including arthropods such as wasps, moths, and termites [14, 117]. PLEs were also found in vertebrates such as fish and different reptiles such as turtles and snakes, but not in mammalian genomes [14]. A PLE distribution including the genomes of fungi and many animals, protists, and plants [118] corresponds to the occurrence of hammerhead ribozyme variants with a particularly short stem III conserved in the PLTRs [14, 117]. This short stem III renders these hammerhead ribozymes thermodynamically unstable, similar to minimal hammerhead ribozyme examples found in ASBVd (see Section 2.2.1). This thermodynamic instability promotes a cleavage as a hammerhead dimer instead of a monomer. In addition to the numerous hammerhead ribozyme examples with a short stem III, fungi, protists, and plants also contain PLEs with canonical type-I hammerhead ribozymes in their PLTRs with typical hammerhead ribozyme helices I, II, and III, as well as the conserved nucleotides involved in tertiary interactions [14]. Interestingly, hammerhead ribozymes that are part of PLEs are highly abundant in some termite gut metagenomes with up to 58 ribozyme examples per megabase of sequence observed [117]. This represents a higher density of hammerhead ribozymes in termites than in all other previously studied metazoans [14]. This observation suggests that PLEs are especially prolific in these termite species; however a genome sequence of these termites is needed to provide a definitive answer.

A tentative hypothesis for the role of hammerhead ribozymes in the PLEs' mode of transposition seems plausible, where the hammerhead ribozymes would self-cleave the transcript of the retroelement [14]. This would generate ligation-compatible

5′ and 3′ RNA ends, resulting in circular RNAs that could be the template for oligomeric retrotranscription and genomic insertion through the participation of the RT and EN activities encoded by the retrotransposon [119]. This would also supply an alternate explanation for the typical head-to-tail tandem arrangement of genomic PLEs, which require multimeric copies to become competent for further retrotranspositions [115]. Furthermore, the tandem arrangement of PLEs appears to be necessary for efficient ribozyme cleavage. The hammerhead ribozymes found in PLEs are mostly type I with a palindromic, often extremely short loop in stem III. As such short stem III structures have been previously hypothesized to be thermodynamically unstable [43], it appears likely that these RNAs do not cleave as monomers, but dimers. This has been proven experimentally for select examples [14, 43, 117]. Therefore, a tandem architecture might enable hammerhead ribozymes to form dimers that allow the ribozyme to cleave. In vitro self-cleavage rates determined for PLE hammerhead ribozymes with a short stem III have been low, even as dimers. In fact, these ribozymes cleave so slowly in vitro that the biological relevance of this cleavage is questionable. Therefore, it is plausible that these ribozymes rely in addition on other factors in vivo to increase cleavage speed. These factors could be proteins such as RNA chaperones or other cofactors [14].

Another distinguishing feature of PLE retrotransposons is the fact that several representatives of these elements from different organisms contain introns. For example, in multiple genomic copies of the *Penelope* element from *D. virilis,* several short (50–70 bp) introns are present [98]. Inherent to the transposition of retroelements is the reverse transcription of an RNA intermediate into cDNA, which is then inserted into the genome. These RNA intermediates are subject to splicing. In fact, the splicing of artificially introduced introns into prospective retroelements is used to prove a transposition pathway relying on RNA [120, 121]. When RT-PCR experiments on total RNA showed that introns in PLEs can be correctly spliced [98], the question arose on how introns can still be preserved in a retrotransposon. One possibility is that PLEs actually transpose through a DNA-based mechanism, in which the endonuclease could catalyze important steps of transposition. But it is likewise possible that an RNA-based pathway for transposition is enabled, if, for example, the unspliced RNA represents a preferred template for reverse transcription and transposition [115].

PLEs are autonomous retrotransposons as they contain an ORF coding for an RT and EN, connected by a "linker" segment of variable length (Figure 2.3b). The RT domain consists of seven motifs usually found in RTs. The N-terminus of the RT displays a conserved extension of ∼190 amino acids containing a highly conserved DKG motif (consisting of the amino acids aspartic acid, lysine, glycine) (consisting of the amino acids aspartic acid, lysine, glycine). Other than the DKG motif, little or no protein sequence homology is observed. The function of the DKG motif is unknown, but protease or nucleic acid-binding functions are discussed as possibilities. Such activities have been found immediately upstream of the RTs in other LTR and non-LTR retrotransposons [122]. At its C-terminus, the RT is followed by a conserved sequence that could function as a thumb domain. Thumb domains ensure RT processivity and primer template interaction. Although this C-terminal extension corresponds in its length and position to other thumb domains, it does not show sequence similarity to known ones. However, telomerase thumb domains with no sequence similarity to those of retroviral RTs also function as thumb domain [115, 123, 124]. Therefore the conserved domain at the C-terminus of the RT in PLEs could likely function as a thumb domain as well.

A linker segment of variable length connects the thumb domain to the EN domain. In most PLE examples, the linker contains a bipartite NLS, in which two clusters of basic amino acids are separated by a 9–12 amino acid linker. The PLE EN belongs to a protein family containing the catalytic module typical of GIY-YIG endonucleases and not to the restriction enzyme-like or apurinic/apyrimidinic-like ENs typical for non-LTR retrotransposons [118]. These GIY-YIG endonucleases occur only in bacterial/organellar group I introns and transpose through DNA. In PLEs, the GIY-YIG module contains a stretch of highly conserved amino acids that are part of a CCHH motif, which overlaps with the endonuclease [125].

The PLE structure is most consistent with a TPRT-like mechanism of transposition. In TPRT, the endonuclease supplied by the retrotransposon cleaves chromosomal DNA to generate a primer for reverse transcription. This is supported by the frequent observation of 5′ truncations and variable-length target site duplications (TSDs) [118]. Apart from PLEs, a TPRT mechanism is typical of non-LTR retrotransposons such as the R2 element, telomerases, and group II introns [126].

PLEs show phylogenetic connections to bacterial self-splicing introns and are believed to predate telomerases and most eukaryotic retrotransposons [127]. It has been previously shown through phylogenetic reconstruction analysis that RTs found in PLEs form a sister clade to telomerase reverse transcriptases (TERTs) [98]. TERTs are specialized nonmobile RTs responsible for adding telomeric repeats to the ends of linear chromosomes in most eukaryotes. Hammerhead ribozymes with short stem III were previously observed that are speculated to associate with PLEs. These hammerhead ribozymes often occur next to RT genes. Since these RTs are most similar to known PLE-RTs and TERTs, it is likely that those unusual hammerhead ribozymes are indeed present in PLEs [14, 117].

2.3.4 Hammerhead Ribozymes Associated with Repetitive Elements in *Schistosoma mansoni*

Certain SINE-like retrotransposons, which are another type of non-autonomous retrotransposon, harbor a self-cleaving ribozyme. In different *Schistosoma* species, active hammerhead ribozymes have been found as part of SINE-like retrotransposons [108, 113]. RNA polymerase III transcribes these elements, which are often found as part of repetitive sequences (Figure 2.3c). It has been shown in vitro that the hammerhead ribozyme liberates SINE copies from multimeric transcripts by cleavage in cis and that it is also capable of cleavage in trans [108]. Following ribozyme-mediated cleavage, the RNA is reverse-transcribed and inserted into the genome. However, because the polymerase III promotor is located only at the 3′ end of cleaved multimeric transcripts, single copies of the SINE represent dead ends of transposition (Figure 2.3c) [108]. Rather the element needs to insert into other genomic locations with at least two adjacent SINE elements to ensure transcription (Figure 2.3c). That way the 3′ promotor of the second element can be used for transcription initiation of the following elements. These repetitive SINEs represent one example of "selfish elements" along with PLEs, R2 elements, and other transposons. Selfish elements are defined as repeated sequences deemed not useful to the host, but are merely maintained because they have discovered sequence-specific replication and amplification strategies [108].

2.3.5 Retrozymes: A New Class of Plant Retrotransposons that Contains Hammerhead Ribozymes

In 2016, a novel class of transposable elements in plants has been described that also harbors self-cleaving RNAs and thus represents another example of ribozyme function [15]. These elements were called "retrozymes".

Retrozymes range in size from about 1 to 1.5 kb and are found in plants, mostly in eudicots, with some examples in ferns, monocots, and algae [15]. These elements are delimited by 4 bp TSDs and LTRs of about 350 bp, each of which harbors a hammerhead ribozyme (Figure 2.3d). These ribozyme-containing repeats flank a variable central region of about 600–1000 bp that does not seem to encode a protein. Therefore, these transposable elements are classified as nonautonomous retrotransposons because they need the proteins from autonomous selfish elements for their successful propagation and genome insertion (see below).

While the overall retrozyme sequences are highly heterogeneous and share almost no sequence homology between species (except the hammerhead ribozyme motif), there are two small conserved domains typical of Ty3/gypsy LTR retrotransposons [15]. These two conserved domains, the primer binding site (PBS) and a polypurine tract (PPT), flank the central region of the retrozyme (Figure 2.3d). During the mobilization of LTR retrotransposons, these sequences are required to prime DNA synthesis from the linear RNA transcript.

This general composition is similar to previously described nonautonomous retrotransposons, so-called terminal-repeat retrotransposons in miniature (TRIMs) and small long terminal repeat retrotransposons (SMARTs) [128, 129]. Like retrozymes, these elements depend on autonomous LTR retrotransposons also present in the genome.

The retrozyme life cycle shares features of the propagation of viroids and virus satellites, and there is also evidence that retrozymes may have evolved from PLEs [54]. Retrozymes are transcribed from their DNA into linear RNA molecules that are able to cleave themselves. When the ribozymes cleave themselves, a linear RNA molecule is produced with a 2′,3′-cP and a 5′ hydroxyl group. Therefore, the ribozyme could support a ligation of these ends through its own activity or just provide the site of action for a protein ligase present in plants, for example, tRNA ligases in *Arabidopsis thaliana* [44] and *Solanum melongena* [13]. These enzymes are able to connect RNAs with exactly those RNA ends resulting from ribozyme cleavage. Either mechanism would yield a circular, covalently closed RNA molecule that could be amplified through rolling circle amplification (Figure 2.1) and integrated into the host genome with the help of proteins provided by autonomous retrotransposons also present in the plant, such as Ty3/gypsy.

2.4 Hammerhead Ribozymes with Suggested Roles in mRNA Biogenesis

There are some self-cleaving ribozyme examples that have been suggested to be involved in mRNA biogenesis due to their genomic location. Hammerhead ribozymes in amniotes [130, 131] and HDV RNA in CPEB3 genes in mammals are found in an intron [10]. These associations suggest possible ribozyme function in pre-mRNA processing and alternative splicing, although experimental proof is lacking. In some mammalian C-type lectin (Clec2) and Clec2-like genes, an unusual type-III hammerhead ribozymes has been described, which consists of two separate regions that together comprise the hammerhead core. In the genome, hundreds of nucleotides lie between the hammerhead regions. When both fragments come together in vitro, the ribozyme is activated and cleaves. The ribozyme cleavage site lies between the translation termination and polyadenylation signal within the 3′ UTR. Thus upon cleavage, the polyadenylation signal from the 3′ end of the mRNA is removed, which leads to a reduction in protein expression in vivo [132]. The discovery and investigations of this unusual ribozyme example have fueled a new search for trans-cleaving ribozymes [133].

2.5 The *glmS* Ribozyme Regulates Glucosamine-6-phosphate Levels in Bacteria

In contrast to the previously discussed roles of self-cleaving ribozymes in eukaryotes and the recent hypothesis of a metabolite-responsive HDV-like ribozyme in *F. prausnitzii* [107], the only well-established bacterial example of a self-cleaving ribozyme for which the biological role has been deciphered is the *glmS* ribozyme.

The *glmS* ribozyme, which is located in the 5′ UTR of the *glmS* mRNA, cleaves itself upon binding glucosamine-6-phosphate (GlcN6P) and regulates the expression of the *glmS* gene, which produces GlcN6P, in a negative feedback loop [8]. Because of its metabolite-responsive gene regulation, this RNA not only is a self-cleaving ribozyme but also qualifies as a riboswitch. Riboswitches are often employed by bacteria to regulate gene expression in response to metabolites or ions. The genes regulated are usually involved in the biosynthesis or homeostasis of the molecule detected [134]. The *glmS* gene encodes the enzyme glutamine:fructose-6-phosphate amidotransferase, which generates GlcN6P from fructose-6-phosphate and glutamine. This reaction is the initial step in the pathway for the production of UDP-*N*-acetyl-glucosamine (UDP-GlcNAc), an essential precursor of cell wall biosynthesis.

The *glmS* ribozyme regulates gene expression in many Gram-positive bacteria and some Gram-negative bacteria [6, 20]. Binding of GlcN6P promotes RNA cleavage via internal phosphoester transfer with rate enhancements of greater than 6 orders of magnitude [135, 136]. Other molecules such as glucosamine (GlcN) or glucose-6-phosphate (Glc6P) that lack the phosphate or amino group are not able to support ribozyme self-cleavage as efficiently. Extensive biochemical and crystallographic studies have revealed that the *glmS* riboswitch uses its ligand as a cofactor to promote RNA self-cleavage rather than as allosteric effector [8, 137, 138] and that the overall ribozyme structure only changes minimally between ligand-bound and apo states. This indicates a preformed binding pocket and a rigid structure of the active site in which the 2′-oxygen nucleophile and leaving group are already aligned before the ligand is bound [139].

In *Bacillus subtilis*, the mechanism by which self-cleavage enables control over gene expression has been deciphered [140]. Rather than affecting translation initiation or transcript elongation, mRNA stability is reduced by ribozyme action. Cleavage of the *glmS* mRNA generates RNA ends typical for nucleolytic ribozymes: a 5′ cleavage product with a 2′,3′-cyclic phosphate and a 3′ cleavage product with a 5′-hydroxyl group [8, 137, 141]. Cleaved *glmS* ribozyme mRNAs with a 5′ hydroxyl were shown to be specifically targeted by RNase J1 and rapidly degraded [140]. This leads to a decrease of *glmS* mRNA and subsequently reduced GlmS protein product.

Today there are many ribozyme representatives known in bacteria. Their sheer abundance and high conservation suggest important functions [7, 9]. The discovery that in bacteria self-cleaving ribozymes are often found near each other or certain types of genes [7] provides new opportunities to explore this finding. As there is so little known about self-cleaving ribozymes in bacteria, apart from *glmS*, there is a need to further explore this nearly completely uncharted research area.

2.6 The Biological Roles of Many Ribozymes Are Unknown

Taking into account newly discovered classes and the recent discoveries of an expanded distribution of long-known self-cleaving ribozyme classes [1, 5, 7, 9], what we know about the biology of self-cleaving RNAs surely represents merely the tip of the iceberg. The twister self-cleaving ribozyme class, for example, is extremely widespread, and still we know nothing about its function [7]. Particularly with the rise of ribozyme motifs found in bacteria, it becomes apparent that we understand even less about the roles of self-cleaving RNAs in this domain, as the *glmS* ribozyme is the only bacterial representative whose biological function is known. Therefore, future investigations should focus on elucidating the importance of self-cleaving ribozymes for the biology of bacteria.

Understanding the biological role of self-cleaving ribozymes and detailed elucidation of their employed mechanisms often enables a variety of new research topics. For example, *glmS* ribozymes and other RNA-based regulatory systems represent promising targets for antibacterial drugs [142]. The interference of a synthetic analog of GlcN6P that causes the misregulation of *glmS* riboswitch-mediated gene expression can thus have an impact on bacterial cell survival [143–146]. Such antibacterial compounds employ a new mode of action distinct from known antibiotics classes, which makes them especially valuable in fighting bacterial strains with antibiotic resistance.

In synthetic biology, self-cleaving RNAs of different classes have been applied to regulate gene expression in bacteria and eukaryotes, for example, as aptazymes, which are artificial RNAs that cleave themselves upon sensing a ligand [34, 35].

The discovery of more examples of known self-cleaving ribozyme classes and the discovery of entirely new classes could provide more clues that could lead to deciphering the biological roles of self-cleaving ribozymes. For example, when several HDV-like ribozymes were discovered by a computational search, the genomic context immediately suggested the biological roles of some representatives [1]. However, the exact biological significance of some self-cleaving ribozymes in eukaryotes has not yet been determined, despite locating some of them to distinct transposable elements. In addition, there are many self-cleaving ribozymes in eukaryotes that are not located near known transposable elements and whose biological significance is unclear (Table 2.1).

Generally, the biological roles of self-cleaving ribozymes have been hard to decipher as they are often found near genes of unknown function or are predicted in environmental sequences or organisms whose cultivation and genetic manipulation are not yet possible. However, the broad distribution, high abundance, and strict conservation of self-cleaving ribozymes emphasize their importance in biology.

2.7 Conclusion

More than 30 years of self-cleaving ribozyme research has shed light on their three-dimensional structures, allowed an exploration of their biochemical characteristics, and enabled an in-depth analysis of their precise catalytic mechanisms. Researchers investigated self-cleaving ribozyme distribution along the tree of life and learned how to turn serendipitous ribozyme discovery into targeted searches. The investigation of self-cleaving ribozyme function beyond mere structural or biochemical description is starting to unravel more and more biological roles of these catalytic RNAs. Most of these roles so far have been deciphered for eukaryotic examples, where they are often connected to mobile genetic elements. These examples, as intriguing as they are, represent isolated instances. This highlights the need for further exploration of the biology connected to these catalytic RNAs. Thus, the elucidation of the biological roles of self-cleaving ribozymes, especially in bacteria, is one of the most underexplored areas of RNA research today.

Acknowledgments

I would like to thank Zasha Weinberg for discussions about the biological roles of self-cleaving ribozymes and critical reading of this chapter. This work was supported by the "Fonds der chemischen Industrie im Verband der Chemischen Industrie e.V." (Grant# 661601).

References

1 Webb, C.-H.T., Riccitelli, N.J., Ruminski, D.J., and Lupták, A. (2009). Widespread occurrence of self-cleaving ribozymes. *Science* 326 (5955): 953.

2 de la Peña, M. and Garcìa-Robles, I. (2010). Ubiquitous presence of the hammerhead ribozyme motif along the tree of life. *RNA* 16 (10): 1943–1950.

3 Perreault, J., Weinberg, Z., Roth, A. et al. (2011). Identification of hammerhead ribozymes in all domains of life reveals novel structural variations. *PLoS Comput. Biol.* 7 (5): e1002031.

4 Seehafer, C., Kalweit, A., Steger, G. et al. (2011). From alpaca to zebrafish: hammerhead ribozymes wherever you look. *RNA* 17 (1): 21–26.

5 Hammann, C., Luptak, A., Perreault, J., and de la Peña, M. (2012). The ubiquitous hammerhead ribozyme. *RNA* 18 (5): 871–885.

6 Barrick, J.E., Corbino, K.A., Winkler, W.C. et al. (2004). New RNA motifs suggest an expanded scope for riboswitches in bacterial genetic control. *Proc. Natl. Acad. Sci. U.S.A.* 101 (17): 6421–6426.

7 Roth, A., Weinberg, Z., Chen, A.G.Y. et al. (2014). A widespread self-cleaving ribozyme class is revealed by bioinformatics. *Nat. Chem. Biol.* 10 (1): 56–60.

8 Winkler, W.C., Nahvi, A., Roth, A. et al. (2004). Control of gene expression by a natural metabolite-responsive ribozyme. *Nature* 428 (6980): 281–286.

9 Weinberg, Z., Kim, P.B., Chen, T.H. et al. (2015). New classes of self-cleaving ribozymes revealed by comparative genomics analysis. *Nat. Chem. Biol.* 11 (8): 606–610.

10 Salehi-Ashtiani, K., Luptak, A., Litovchick, A., and Szostak, J.W. (2006). A genome-wide search for ribozymes reveals an HDV-like sequence in the human CPEB3 gene. *Science* 313 (5794): 1788–1792.

11 Salehi-Ashtiani, K. and Szostak, J.W. (2001). In vitro evolution suggests multiple origins for the hammerhead ribozyme. *Nature* 414 (6859): 82–84.

12 Tang, J. and Breaker, R.R. (2000). Structural diversity of self-cleaving ribozymes. *Proc. Natl. Acad. Sci. U.S.A.* 97 (11): 5784–5789.

13 Flores, R., Grubb, D., Elleuch, A. et al. (2011). Rolling-circle replication of viroids, viroid-like satellite RNAs and hepatitis delta virus: variations on a theme. *RNA Biol.* 8 (2): 200–206.

14 Cervera, A. and de la Peña, M. (2014). Eukaryotic Penelope-like retroelements encode hammerhead ribozyme motifs. *Mol. Biol. Evol.* 31 (11): 2941–2947.

15 Cervera, A., Urbina, D., and La de Peña, M. (2016). Retrozymes are a unique family of non-autonomous retrotransposons with hammerhead ribozymes that propagate in plants through circular RNAs. *Genome Biol.* 17 (1): 135.

16 Spotila, L.D., Hirai, H., Rekosh, D.M., and Lo Verde, P.T. (1989). A retroposon-like short repetitive DNA element in the genome of the human blood fluke *Schistosoma mansoni*. *Chromosoma* 97 (6): 421–428.

17 Eickbush, T.H. and Eickbush, D.G. (2015). Integration, regulation, and long-term stability of R2 retrotransposons. *Microbiol. Spectr.* 3 (2).

18 Macías, F., Afonso-Lehmann, R., López, M.C. et al. (2018). Biology of *Trypanosoma cruzi* retrotransposons: from an enzymatic to a structural point of view. *Curr. Genom.* 19 (2): 110–118.

19 Collins, R.A. (2002). The *Neurospora* Varkud satellite ribozyme. *Biochem. Soc. Trans.* 30 (Pt 6): 1122–1126.

20 McCown, P.J., Roth, A., and Breaker, R.R. (2011). An expanded collection and refined consensus model of glmS ribozymes. *RNA* 17 (4): 728–736.

21 Kalvari, I., Argasinska, J., Quinones-Olvera, N. et al. (2018). Rfam 13.0: shifting to a genome-centric resource for non-coding RNA families. *Nucleic Acids Res.* 46 (D1): D335–D342.

22 Branch, A.D. and Robertson, H.D. (1984). A replication cycle for viroids and other small infectious RNAs. *Science* 223 (4635): 450–455.

23 Hutchins, C.J., Keese, P., Visvader, J.E. et al. (1985). Comparison of multimeric plus and minus forms of viroids and virusoids. *Plant Mol. Biol.* 4 (5): 293–304.

24 Sharmeen, L., Kuo, M.Y., Dinter-Gottlieb, G., and Taylor, J. (1988). Antigenomic RNA of human hepatitis delta virus can undergo self-cleavage. *J. Virol.* 62 (8): 2674–2679.

25 Daròs, J.A., Marcos, J.F., Hernandez, C., and Flores, R. (1994). Replication of avocado sunblotch viroid – evidence for a symmetrical pathway with 2 rolling circles and hammerhead ribozyme processing. *Proc. Natl. Acad. Sci. U.S.A.* 91 (26): 12813–12817.

26 Flores, R., Minoia, S., Carbonell, A. et al. (2015). Viroids, the simplest RNA replicons: how they manipulate their hosts for being propagated and how their hosts react for containing the infection. *Virus Res.* 209: 136–145.

27 Hutchins, C.J., Rathjen, P.D., Forster, A.C., and Symons, R.H. (1986). Self-cleavage of plus and minus RNA transcripts of avocado sunblotch viroid. *Nucleic Acids Res.* 14 (9): 3627–3640.

28 Prody, G.A., Bakos, J.T., Buzayan, J.M. et al. (1986). Autolytic processing of dimeric plant virus satellite RNA. *Science* 231 (4745): 1577–1580.

29 Hernández, C. and Flores, R. (1992). Plus and minus RNAs of peach latent mosaic viroid self-cleave in vitro via hammerhead structures. *Proc. Natl. Acad. Sci. U.S.A.* 89 (9): 3711–3715.

30 Navarro, B. and Flores, R. (1997). Chrysanthemum chlorotic mottle viroid: unusual structural properties of a subgroup of self-cleaving viroids with hammerhead ribozymes. *Proc. Natl. Acad. Sci. U.S.A.* 94 (21): 11262–11267.

31 Fadda, Z., Daròs, J.A., Fagoaga, C. et al. (2003). Eggplant latent viroid, the candidate type species for a new genus within the family Avsunviroidae (hammerhead viroids). *J. Virol.* 77 (11): 6528–6532.

32 Wu, Q., Wang, Y., Cao, M. et al. (2012). Homology-independent discovery of replicating pathogenic circular RNAs by deep sequencing and a new computational algorithm. *Proc. Natl. Acad. Sci. U.S.A.* 109 (10): 3938–3943.

33 Flores, R., Gago-Zachert, S., Serra, P. et al. (2014). Viroids: survivors from the RNA world? *Annu. Rev. Microbiol.* 68: 395–414.

34 Vinkenborg, J.L., Karnowski, N., and Famulok, M. (2011). Aptamers for allosteric regulation. *Nat. Chem. Biol.* 7 (8): 519–527.

35 Felletti, M. and Hartig, J.S. (2017). Ligand-dependent ribozymes. *Wiley Interdiscip. Rev. RNA* 8 (2).

36 Wurmthaler, L.A., Klauser, B., and Hartig, J.S. (2018). Highly motif- and organism-dependent effects of naturally occurring hammerhead ribozyme sequences on gene expression. *RNA Biol.* 15 (2): 231–241.

37 Mühlbach, H.P. and Sänger, H.L. (1979). Viroid replication is inhibited by alpha-amanitin. *Nature* 278 (5700): 185–188.

38 Schindler, I. and Mühlbach, H.P. (1992). Involvement of nuclear DNA-dependent RNA-polymerases in potato spindle tuber viroid replication – a reevaluation. *Plant Sci.* 84 (2): 221–229.

39 Navarro, J.A., Vera, A., and Flores, R. (2000). A chloroplastic RNA polymerase resistant to tagetitoxin is involved in replication of *Avocado sunblotch* viroid. *Virology* 268 (1): 218–225.

40 Khvorova, A., Lescoute, A., Westhof, E., and Jayasena, S.D. (2003). Sequence elements outside the hammerhead ribozyme catalytic core enable intracellular activity. *Nat. Struct. Biol.* 10 (9): 708–712.

41 de la Peña, M., Gago, S., and Flores, R. (2003). Peripheral regions of natural hammerhead ribozymes greatly increase their self-cleavage activity. *EMBO J.* 22 (20): 5561–5570.

42 Martick, M. and Scott, W.G. (2006). Tertiary contacts distant from the active site prime a ribozyme for catalysis. *Cell* 126 (2): 309–320.

43 Forster, A.C., Davies, C., Sheldon, C.C. et al. (1988). Self-cleaving viroid and newt RNAs may only be active as dimers. *Nature* 334 (6179): 265–267.

44 Englert, M. and Beier, H. (2005). Plant tRNA ligases are multifunctional enzymes that have diverged in sequence and substrate specificity from RNA ligases of other phylogenetic origins. *Nucleic Acids Res.* 33 (1): 388–399.

45 Englert, M., Latz, A., Becker, D. et al. (2007). Plant pre-tRNA splicing enzymes are targeted to multiple cellular compartments. *Biochimie* 89 (11): 1351–1365.

46 Nohales, M.-Á., Molina-Serrano, D., Flores, R., and Daròs, J.A. (2012). Involvement of the chloroplastic isoform of tRNA ligase in the replication of viroids belonging to the family Avsunviroidae. *J. Virol.* 86 (15): 8269–8276.

47 Steiger, M.A., Kierzek, R., Turner, D.H., and Phizicky, E.M. (2001). Substrate recognition by a yeast 2′-phosphotransferase involved in tRNA splicing and by its *Escherichia coli* homolog. *Biochemistry* 40 (46): 14098–14105.

48 Nohales, M.-Á., Flores, R., and Daròs, J.A. (2012). Viroid RNA redirects host DNA ligase 1 to act as an RNA ligase. *Proc. Natl. Acad. Sci. U.S.A.* 109 (34): 13805–13810.

49 Forster, A.C. and Symons, R.H. (1987). Self-cleavage of plus and minus RNAs of a virusoid and a structural model for the active sites. *Cell* 49 (2): 211–220.

50 Buzayan, J.M., Gerlach, W.L., and Bruening, G. (1986). Nonenzymatic cleavage and ligation of RNAs complementary to a plant-virus satellite RNA. *Nature* 323 (6086): 349–353.

51 Fedor, M.J. (2000). Structure and function of the hairpin ribozyme. *J. Mol. Biol.* 297 (2): 269–291.

52 Flores, R., Hernandez, C., de La Pena, M. et al. (2001). Hammerhead ribozyme structure and function in plant RNA replication. *Ribonucleases* 341: 540–552.

53 Song, S.I., Silver, S.L., Aulik, M.A. et al. (1999). Satellite cereal yellow dwarf virus-RPV (satRPV) RNA requires a DouXble hammerhead for self-cleavage and an alternative structure for replication. *J. Mol. Biol.* 293 (4): 781–793.

54 de La Peña, M. and Cervera, A. (2017). Circular RNAs with hammerhead ribozymes encoded in eukaryotic genomes: the enemy at home. *RNA Biol.* 14 (8): 985–991.

55 Makino, S., Chang, M.F., Shieh, C.K. et al. (1987). Molecular cloning and sequencing of a human hepatitis delta (delta) virus RNA. *Nature* 329 (6137): 343–346.

56 Wang, K.S., Choo, Q.L., Weiner, A.J. et al. (1986). Structure, sequence and expression of the hepatitis delta (delta) viral genome. *Nature* 323 (6088): 508–514.

57 Taylor, J. and Pelchat, M. (2010). Origin of hepatitis delta virus. *Future Microbiol.* 5 (3): 393–402.

58 Taylor, J.M. (2006). Structure and replication of hepatitis delta virus RNA. *Curr. Top. Microbiol. Immunol.* 307: 1–23.

59 Rizzetto, M., Hoyer, B., Canese, M.G. et al. (1980). delta Agent: association of delta antigen with hepatitis B surface antigen and RNA in serum of delta-infected chimpanzees. *Proc. Natl. Acad. Sci. U.S.A.* 77 (10): 6124–6128.

60 Chen, P.J., Kalpana, G., Goldberg, J. et al. (1986). Structure and replication of the genome of the hepatitis delta virus. *Proc. Natl. Acad. Sci. U.S.A.* 83 (22): 8774–8778.

61 Kuo, M.Y., Sharmeen, L., Dinter-Gottlieb, G., and Taylor, J. (1988). Characterization of self-cleaving RNA sequences on the genome and antigenome of human hepatitis delta virus. *J. Virol.* 62 (12): 4439–4444.

62 Ferre-D'Amare, A.R., Zhou, K.H., and Doudna, J.A. (1998). Crystal structure of a hepatitis delta virus ribozyme. *Nature* 395 (6702): 567–574.

63 Chadalavada, D.M., Cerrone-Szakal, A.L., and Bevilacqua, P.C. (2007). Wild-type is the optimal sequence of the HDV ribozyme under cotranscriptional conditions. *RNA* 13 (12): 2189–2201.

64 Reid, C.E. and Lazinski, D.W. (2000). A host-specific function is required for ligation of a wide variety of ribozyme-processed RNAs. *Proc. Natl. Acad. Sci. U.S.A.* 97 (1): 424–429.

65 Saville, B.J. and Collins, R.A. (1990). A site-specific self-cleavage reaction performed by a novel RNA in *Neurospora* mitochondria. *Cell* 61 (4): 685–696.

66 Saville, B.J. and Collins, R.A. (1991). RNA-mediated ligation of self-cleavage products of a *Neurospora* mitochondrial plasmid transcript. *Proc. Natl. Acad. Sci. U.S.A.* 88 (19): 8826–8830.

67 Kennell, J.C., Saville, B.J., Mohr, S. et al. (1995). The VS catalytic RNA replicates by reverse transcription as a satellite of a retroplasmid. *Genes Dev.* 9 (3): 294–303.

68 Ouellet, J., Byrne, M., and Lilley, D.M.J. (2009). Formation of an active site in trans by interaction of two complete Varkud satellite ribozymes. *RNA* 15 (10): 1822–1826.

69 Suslov, N.B., DasGupta, S., Huang, H. et al. (2015). Crystal structure of the Varkud satellite ribozyme. *Nat. Chem. Biol.* 11 (11): 840–846.

70 Craig, N.L., Craigie, R., Gellert, M., and Lambowitz, A.M. (eds.) (2002). *Mobile DNA II*. American Society of Microbiology.

71 Dawid, I.B. and Rebbert, M.L. (1981). Nucleotide-sequences at the boundaries between gene and insertion regions in the rDNA of *Drosophila melanogaster*. *Nucleic Acids Res.* 9 (19): 5011–5020.

72 Roiha, H., Miller, J.R., Woods, L.C., and Glover, D.M. (1981). Arrangements and rearrangements of sequences flanking the 2 types of rDNA insertion in *Drosophila melanogaster*. *Nature* 290 (5809): 749–753.

73 Fujiwara, H., Ogura, T., Takada, N. et al. (1984). Introns and their flanking sequences in *Bombyx mori* rDNA. *Nucleic Acids Res.* 12 (17): 6861–6869.

74 Eickbush, T.H. and Robins, B. (1985). *Bombyx mori* 28S ribosomal genes contain insertion elements similar to the Type I and II elements of *Drosophila melanogaster*. *EMBO J.* 4 (9): 2281–2285.

75 Loeb, D.D., Padgett, R.W., Hardies, S.C. et al. (1986). The sequence of a large L1Md element reveals a tandemly repeated 5′ end and several features found in retrotransposons. *Mol. Cell. Biol.* 6 (1): 168–182.

76 Hattori, M., Kuhara, S., Takenaka, O., and Sakaki, Y. (1986). L1 family of repetitive DNA sequences in primates may be derived from a sequence encoding a reverse transcriptase-related protein. *Nature* 321 (6070): 625–628.

77 Fawcett, D.H., Lister, C.K., Kellet, E., and Finnegan, D.J. (1986). Transposable elements controlling I-R hybrid dysgenesis in *Drosophila melanogaster* are similar to mammalian LINEs. *Cell* 47 (6): 1007–1015.

78 Burke, W.D., Calalang, C.C., and Eickbush, T.H. (1987). The site-specific ribosomal insertion element type-II of Bombyx mori (R2Bm) contains the coding sequence for a reverse transcriptase-like enzyme. *Mol. Cell. Biol.* 7 (6): 2221–2230.

79 Xiong, Y. and Eickbush, T.H. (1988). The site-specific ribosomal DNA insertion element R1Bm belongs to a class of non-long-terminal-repeat retrotransposons. *Mol. Cell. Biol.* 8 (1): 114–123.

80 Eickbush, D.G., Ye, J., Zhang, X. et al. (2008). Epigenetic regulation of retrotransposons within the nucleolus of *Drosophila*. *Mol. Cell. Biol.* 28 (20): 6452–6461.

81 Eickbush, D.G. and Eickbush, T.H. (2010). R2 retrotransposons encode a self-cleaving ribozyme for processing from an rRNA cotranscript. *Mol. Cell. Biol.* 30 (13): 3142–3150.

82 Eickbush, D.G., Burke, W.D., and Eickbush, T.H. (2013). Evolution of the R2 retrotransposon ribozyme and its self-cleavage site. *PLoS One* 8 (9): e66441.

83 George, J.A. and Eickbush, T.H. (1999). Conserved features at the 5′ end of *Drosophila* R2 retrotransposable elements: implications for transcription and translation. *Insect Mol. Biol.* 8 (1): 3–10.

84 Eickbush, D.G. and Eickbush, T.H. (2003). Transcription of endogenous and exogenous R2 elements in the rRNA gene locus of *Drosophila melanogaster*. *Mol. Cell. Biol.* 23 (11): 3825–3836.

85 Kieft, J.S. (2008). Viral IRES RNA structures and ribosome interactions. *Trends Biochem. Sci.* 33 (6): 274–283.

86 Berry, K.E., Waghray, S., and Doudna, J.A. (2010). The HCV IRES pseudoknot positions the initiation codon on the 40S ribosomal subunit. *RNA* 16 (8): 1559–1569.

87 Ruminski, D.J., Webb, C.H., Riccitelli, N.J., and Luptak, A. (2011). Processing and translation initiation of non-long terminal repeat retrotransposons by hepatitis delta virus (HDV)-like self-cleaving ribozymes. *J. Biol. Chem.* 286 (48): 41286–41295.

88 Christensen, S.M., Ye, J., and Eickbush, T.H. (2006). RNA from the 5′ end of the R2 retrotransposon controls R2 protein binding to and cleavage of its DNA target site. *Proc. Natl. Acad. Sci. U.S.A.* 103 (47): 17602–17607.

89 Yang, J., Malik, H.S., and Eickbush, T.H. (1999). Identification of the endonuclease domain encoded by R2 and other site-specific, non-long terminal repeat retrotransposable elements. *Proc. Natl. Acad. Sci. U.S.A.* 96 (14): 7847–7852.

90 Christensen, S. and Eickbush, T.H. (2004). Footprint of the retrotransposon R2Bm protein on its target site before and after cleavage. *J. Mol. Biol.* 336 (5): 1035–1045.

91 Christensen, S.M. and Eickbush, T.H. (2005). R2 target-primed reverse transcription: ordered cleavage and polymerization steps by protein subunits asymmetrically bound to the target DNA. *Mol. Cell. Biol.* 25 (15): 6617–6628.

92 Luan, D.D., Korman, M.H., Jakubczak, J.L., and Eickbush, T.H. (1993). Reverse transcription of R2Bm RNA is primed by a nick at the chromosomal target site – a mechanism for non-LTR retrotransposition. *Cell* 72 (4): 595–605.

93 Bibillo, A. and Eickbush, T.H. (2002). The reverse transcriptase of the R2 non-LTR retrotransposon: continuous synthesis of cDNA on non-continuous RNA templates. *J. Mol. Biol.* 316 (3): 459–473.

94 Luan, D.D. and Eickbush, T.H. (1995). RNA template requirements for target DNA-primed reverse transcription by the R2 retrotransposable element. *Mol. Cell. Biol.* 15 (7): 3882–3891.

95 Luan, D.D. and Eickbush, T.H. (1996). Downstream 28S gene sequences on the RNA template affect the choice of primer and the accuracy of initiation by the R2 reverse transcriptase. *Mol. Cell. Biol.* 16 (9): 4726–4734.

96 Eickbush, D.G., Luan, D.D., and Eickbush, T.H. (2000). Integration of *Bombyx mori* R2 sequences into the 28S ribosomal RNA genes of *Drosophila melanogaster*. *Mol. Cell. Biol.* 20 (1): 213–223.

97 Fujimoto, H., Hirukawa, Y., Tani, H. et al. (2004). Integration of the 5′ end of the retrotransposon, R2Bm, can be complemented by homologous recombination. *Nucleic Acids Res.* 32 (4): 1555–1565.

98 Arkhipova, I.R., Pyatkov, K.I., Meselson, M., and Evgen'ev, M.B. (2003). Retroelements containing introns in diverse invertebrate taxa. *Nat. Genet.* 33 (2): 123–124.

99 Blocker, F.J.H., Mohr, G., Conlan, L.H. et al. (2005). Domain structure and three-dimensional model of a group II intron-encoded reverse transcriptase. *RNA* 11 (1): 14–28.

100 Bibillo, A. and Eickbush, T.H. (2002). High processivity of the reverse transcriptase from a non-long terminal repeat retrotransposon. *J. Biol. Chem.* 277 (38): 34836–34845.

101 Kurzynska-Kokorniak, A., Jamburuthugoda, V.K., Bibillo, A., and Eickbush, T.H. (2007). DNA-directed DNA polymerase and strand displacement activity of the reverse transcriptase encoded by the R2 retrotransposon. *J. Mol. Biol.* 374 (2): 322–333.

102 Kierzek, E., Kierzek, R., Moss, W.N. et al. (2008). Isoenergetic penta- and hexanucleotide microarray probing and chemical mapping provide a secondary structure model for an RNA element orchestrating R2 retrotransposon protein function. *Nucleic Acids Res.* 36 (6): 1770–1782.

103 Kierzek, E., Christensen, S.M., Eickbush, T.H. et al. (2009). Secondary structures for 5′ regions of R2 retrotransposon RNAs reveal a novel conserved pseudoknot and regions that evolve under different constraints. *J. Mol. Biol.* 390 (3): 428–442.

104 Bibillo, A. and Eickbush, T.H. (2004). End-to-end template jumping by the reverse transcriptase encoded by the R2 retrotransposon. *J. Biol. Chem.* 279 (15): 14945–14953.

105 Stage, D.E. and Eickbush, T.H. (2009). Origin of nascent lineages and the mechanisms used to prime second-strand DNA synthesis in the R1 and R2 retrotransposons of *Drosophila*. *Genome Biol.* 10 (5).

106 Eickbush, D.G. and Eickbush, T.H. (2012). R2 and R2/R1 hybrid non-autonomous retrotransposons derived by internal deletions of full-length elements. *Mob. DNA* 3 (1): 10.

107 Passalacqua, L.F.M., Jimenez, R.M., Fong, J.Y., and Lupták, A. (2017). Allosteric modulation of the *Faecalibacterium prausnitzii* hepatitis delta virus-like ribozyme by glucosamine 6-phosphate: the substrate of the adjacent gene product. *Biochemistry* 56 (45): 6006–6014.

108 Ferbeyre, G., Smith, J.M., and Cedergren, R. (1998). Schistosome satellite DNA encodes active hammerhead ribozymes. *Mol. Cell. Biol.* 18 (7): 3880–3888.

109 Tay, W.T., Behere, G.T., Batterham, P., and Heckel, D.G. (2010). Generation of microsatellite repeat families by RTE retrotransposons in lepidopteran genomes. *BMC Evol. Biol.* 10: 144.

110 Bringaud, F., Bartholomeu, D.C., Blandin, G. et al. (2006). The *Trypanosoma cruzi* L1Tc and NARTc non-LTR retrotransposons show relative site specificity for insertion. *Mol. Biol. Evol.* 23 (2): 411–420.

111 Sanchez-Luque, F.J., Lopez, M.C., Macias, F. et al. (2011). Identification of an hepatitis delta virus-like ribozyme at the mRNA 5′-end of the L1Tc retrotransposon from *Trypanosoma cruzi*. *Nucleic Acids Res.* 39 (18): 8065–8077.

112 Heras, S.R., Lopez, M.C., Olivares, M., and Thomas, M.C. (2007). The L1Tc non-LTR retrotransposon of *Trypanosoma cruzi* contains an internal RNA-pol II-dependent promoter that strongly activates gene transcription and generates unspliced transcripts. *Nucleic Acids Res.* 35 (7): 2199–2214.

113 Laha, T., McManus, D.P., Loukas, A., and Brindley, P.J. (2000). Sj alpha elements, short interspersed element-like retroposons bearing a hammerhead ribozyme motif from the genome of the oriental blood fluke *Schistosoma japonicum. BBA-Gene Struct. Expr.* 1492 (2–3): 477–482.

114 Evgen'ev, M.B., Zelentsova, H., Shostak, N. et al. (1997). Penelope, a new family of transposable elements and its possible role in hybrid dysgenesis in *Drosophila virilis. Proc. Natl. Acad. Sci. U.S.A.* 94 (1): 196–201.

115 Evgen'ev, M.B. and Arkhipova, I.R. (2005). Penelope-like elements – a new class of retroelements: distribution, function and possible evolutionary significance. *Cytogenet. Genome Res.* 110 (1–4): 510–521.

116 Danilevskaya, O.N., Arkhipova, I.R., Traverse, K.L., and Pardue, M.L. (1997). Promoting in tandem: the promoter for telomere transposon HeT-A and implications for the evolution of retroviral LTRs. *Cell* 88 (5): 647–655.

117 Lünse, C.E., Weinberg, Z., and Breaker, R.R. (2016). Numerous small hammerhead ribozyme variants associated with Penelope-like retrotransposons cleave RNA as dimers. *RNA Biol.* 14 (11): 1499–1507.

118 Arkhipova, I.R. (2006). Distribution and phylogeny of Penelope-like elements in eukaryotes. *Syst. Biol.* 55 (6): 875–885.

119 Pyatkov, K.I., Arkhipova, I.R., Malkova, N.V. et al. (2004). Reverse transcriptase and endonuclease activities encoded by Penelope-like retroelements. *Proc. Natl. Acad. Sci. U.S.A.* 101 (41): 14719–14724.

120 Boeke, J.D., Garfinkel, D.J., Styles, C.A., and Fink, G.R. (1985). Ty elements transpose through an RNA intermediate. *Cell* 40 (3): 491–500.

121 Nakayashiki, H., Kiyotomi, K., Tosa, Y., and Mayama, S. (1999). Transposition of the retrotransposon MAGGY in heterologous species of filamentous fungi. *Genetics* 153 (2): 693–703.

122 Arkhipova, I.R., Lyubomirskaya, N.V., and Ilyin, Y.V. (1995). *Drosophila Retrotransposons*. Georgetown, TX; Heidelberg: Landes Co.; Springer-Verlag.

123 Peng, Y., Mian, I.S., and Lue, N.F. (2001). Analysis of telomerase processivity: mechanistic similarity to HIV-1 reverse transcriptase and role in telomere maintenance. *Mol. Cell* 7 (6): 1201–1211.

124 Hossain, S., Singh, S., and Lue, N.F. (2002). Functional analysis of the C-terminal extension of telomerase reverse transcriptase. A putative "thumb" domain. *J. Biol. Chem.* 277 (39): 36174–36180.

125 Kowalski, J.C., Belfort, M., Stapleton, M.A. et al. (1999). Configuration of the catalytic GIY-YIG domain of intron endonuclease I-Tevl: coincidence of computational and molecular findings. *Nucleic Acids Res.* 27 (10): 2115–2125.

126 Zimmerly, S., Guo, H., Perlman, P.S., and Lambowitz, A.M. (1995). Group II intron mobility occurs by target DNA-primed reverse transcription. *Cell* 82 (4): 545–554.

127 Gladyshev, E.A. and Arkhipova, I.R. (2007). Telomere-associated endonuclease-deficient Penelope-like retroelements in diverse eukaryotes. *Proc. Natl. Acad. Sci. U.S.A.* 104 (22): 9352–9357.

128 Witte, C.P., Le, Q.H., Bureau, T., and Kumar, A. (2001). Terminal-repeat retrotransposons in miniature (TRIM) are involved in restructuring plant genomes. *Proc. Natl. Acad. Sci. U.S.A.* 98 (24): 13778–13783.

129 Gao, D., Chen, J., Chen, M. et al. (2012). A highly conserved, small LTR retrotransposon that preferentially targets genes in grass genomes. *PLoS One* 7 (2): e32010.

130 Garcia-Robles, I., Sanchez-Navarro, J., and de La Pena, M. (2012). Intronic hammerhead ribozymes in mRNA biogenesis. *Biol. Chem.* 393 (11): 1317–1326.

131 de la Peña, M. and Garcìa-Robles, I. (2010). Intronic hammerhead ribozymes are ultraconserved in the human genome. *EMBO Rep.* 11 (9): 711–716.

132 Martick, M., Horan, L.H., Noller, H.F., and Scott, W.G. (2008). A discontinuous hammerhead ribozyme embedded in a mammalian messenger RNA. *Nature* 454 (7206): 899–902.

133 Webb, C.-H.T. and Lupták, A. (2018). Kinetic parameters of trans scission by extended HDV-like ribozymes and the prospect for the discovery of genomic trans-cleaving RNAs. *Biochemistry* 57 (9): 1440–1450.

134 McCown, P.J., Corbino, K.A., Stav, S. et al. (2017). Riboswitch diversity and distribution. *RNA* 23 (7): 995–1011.

135 McCarthy, T.J., Plog, M.A., Floy, S.A. et al. (2005). Ligand requirements for glmS ribozyme self-cleavage. *Chem. Biol.* 12 (11): 1221–1226.

136 Wilkinson, S.R. and Been, M.D. (2005). A pseudoknot in the 3′ non-core region of the glmS ribozyme enhances self-cleavage activity. *RNA* 11 (12): 1788–1794.

137 Klein, D.J. and Ferre-D'Amare, A.R. (2006). Structural basis of glmS ribozyme activation by glucosamine-6-phosphate. *Science* 313 (5794): 1752–1756.

138 Bingaman, J.L., Zhang, S., Stevens, D.R. et al. (2017). The GlcN6P cofactor plays multiple catalytic roles in the glmS ribozyme. *Nat. Chem. Biol.* 13 (4): 439–445.

139 Cochrane, J.C., Lipchock, S.V., and Strobel, S.A. (2007). Structural investigation of the glmS ribozyme bound to its catalytic cofactor. *Chem. Biol.* 14 (1): 97–105.

140 Collins, J.A., Irnov, I., Baker, S., and Winkler, W.C. (2007). Mechanism of mRNA destabilization by the glmS ribozyme. *Genes Dev.* 21 (24): 3356–3368.

141 Scott, W.G., Martick, M., and Chi, Y.-I. (2009). Structure and function of regulatory RNA elements: ribozymes that regulate gene expression. *Biochim. Biophys. Acta* 1789 (9–10): 634–641.

142 Blount, K.F. and Breaker, R.R. (2006). Riboswitches as antibacterial drug targets. *Nat. Biotechnol.* 24 (12): 1558–1564.

143 Lim, J., Grove, B.C., Roth, A., and Breaker, R.R. (2006). Characteristics of ligand recognition by a glmS self-cleaving ribozyme. *Angew. Chem. Int. Ed. Engl.* 45 (40): 6689–6693.

144 Mayer, G. and Famulok, M. (2006). High-throughput-compatible assay for glmS riboswitch metabolite dependence. *ChemBioChem* 7 (4): 602–604.

145 Lünse, C.E., Schmidt, M.S., Wittmann, V., and Mayer, G. (2011). Carba-sugars activate the glmS-riboswitch of *Staphylococcus aureus*. *ACS Chem. Biol.* 6 (7): 675–678.

146 Schüller, A., Matzner, D., Lünse, C.E. et al. (2017). Activation of the glmS ribozyme confers bacterial growth inhibition. *ChemBioChem* 18 (5): 435–440.

Part II

Naturally Occurring Ribozymes

3

Chemical Mechanisms of the Nucleolytic Ribozymes

Timothy J. Wilson and David M. J. Lilley

School of Life Sciences, The University of Dundee, Cancer Research UK Nucleic Acid Structure Research Group, MSI/WTB Complex, Dow Street, Dundee DD1 5EH, UK

3.1 The Nucleolytic Ribozymes

The nucleolytic ribozymes are a group of relatively small RNA species (most are <100 nucleotides) that catalyze site-specific self-cleavage reactions. They catalyze a transesterification reaction in which the O2′ attacks the adjacent phosphodiester linkage by an S_N2 mechanism [1, 2], with departure of the O5′ and formation of a cyclic 2′,3′-phosphate (Figure 3.1). Observed rates of cleavage of 1–10 min^{-1} are typical, representing a $\geq 10^5$-fold rate enhancement above background for a dinucleotide. Some of the nucleolytic ribozymes will carry out the reverse ligation reaction [3–8]. Repeated cycles of cleavage and ligation reactions have been observed in a ribozyme by single-molecule experiments [9].

The first nucleolytic ribozymes to be discovered carried out processing of replication intermediates [10]. These included the hammerhead [11, 12], hairpin [3, 13, 14], and Varkud satellite (VS) [15] ribozymes. The hepatitis delta virus (HDV) ribozymes were discovered in the complementary single-stranded genomes of the human pathogen HDV [16–18]. Like the hammerhead and hairpin ribozymes, the HDV ribozyme cleaves single-stranded RNA replicated by a rolling circle process. Over a decade went by before the next distinct nucleolytic ribozyme was discovered. Remarkably, this was a riboswitch that bound glucosamine-6-phosphate (Glc6P) [19]. While the majority of riboswitches alter their structure on binding a ligand to control gene expression either at the transcriptional or translational level, the *glmS* riboswitch turned out to be a ribozyme that cleaved mRNA in response to the binding of ligand. Another decade passed before the next nucleolytic ribozymes came along, but then four came at once. A new bioinformatic pipeline in the Breaker Lab [20] identified the twister, twister sister (TS), pistol and hatchet ribozymes [21]. Thus, altogether, we now have nine nucleolytic ribozymes for structural and mechanistic study, and this has provided the opportunity to make comparisons and perceive some subgroupings.

Ribozymes, First Edition. Edited by Sabine Müller, Benoît Masquida, and Wade Winkler.

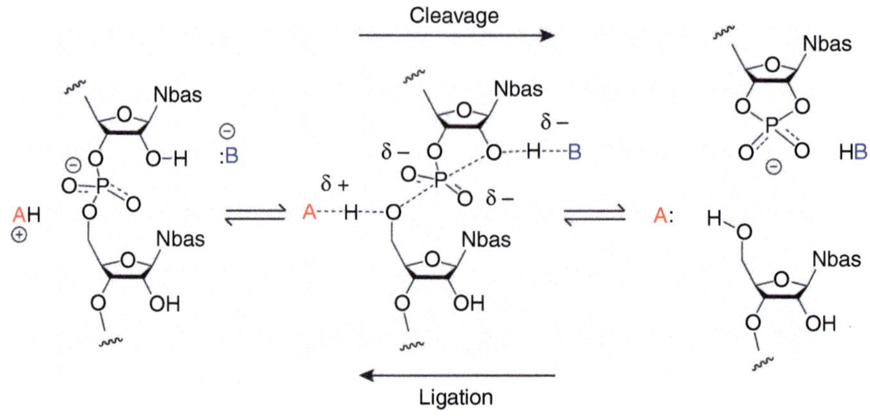

Figure 3.1 The chemical mechanism of the nucleolytic ribozymes catalyzed by general acid–base catalysis. The cleavage reaction is left to right and the ligation reaction right to left. The central phosphorane structure approximates as the transition state for the reaction. In the cleavage reaction, B^- acts as a general base to deprotonate the O2′ nucleophile, while AH^+ protonates the 5′-oxyanion leaving group. A number of examples are presented in Chapter 1.

3.2 Some Nucleolytic Ribozymes Are Widespread

Some of the nucleolytic ribozymes are widespread in genomes. The first hammerhead ribozymes were discovered in the plant viroids, although one was found in amphibia [22]. More recently, they have been found in plant [23] and mammalian [24, 25] genomes, and in fact quite widely in nature [26]. HDV-related ribozymes have also been found to be encoded in the human genome [27] and more widely [28]. The twister ribozyme is also found widely in bacteria, fungi, plants, and animals [29], and *in vitro* ribozyme activity was demonstrated in several cases. On the other hand, some nucleolytic ribozymes are not commonly found. The VS ribozyme was isolated from a wild strain of *Neurospora*, and has never been found again.

3.3 Secondary Structures of Nucleolytic Ribozymes – Junctions and Pseudoknots

The secondary structures of eight of the nucleolytic ribozymes are depicted in Figure 3.2. Clearly both helical junctions and pseudoknots can generate autonomously folding, small functional RNA species. The hammerhead, hairpin, and VS ribozymes are organized by helical junctions. The hammerhead ribozyme has a three-way junction that creates the active site. The hairpin ribozyme is based on a four-way junction that directs two loop-containing helical arms so that the loops interact to create the active center. The dynamics of this junction are critical to the function of the ribozyme [30]. The VS ribozyme has a number of three-way junctions, the folding of which generates the active structure [31, 32]. The active site is again created by a loop–loop interaction, which can occur both in cis, and

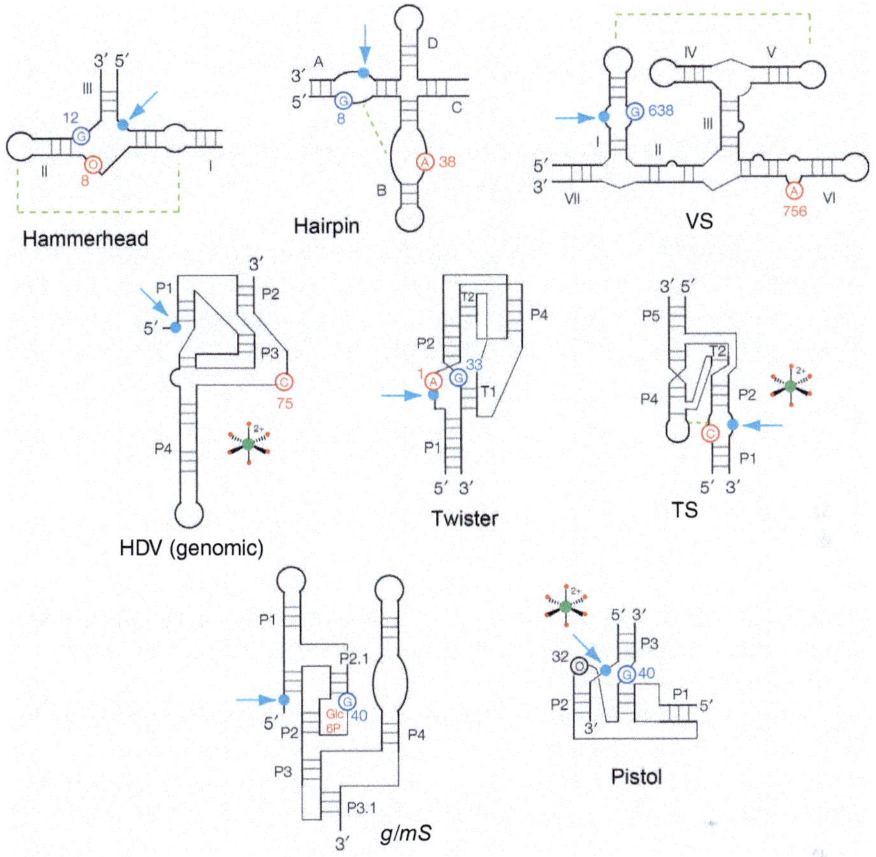

Figure 3.2 Secondary structure depictions of the nucleolytic ribozymes with the scissile phosphates (cyan) arrowed, and the general acid (red) and base (blue) shown. Broken green lines indicate important tertiary contacts. Three of the ribozymes act with the direct involvement of metal ions shown by green spheres with octahedral coordination of red water molecules of hydration. Base pairing is indicated by the gray lines, but the numbers shown do not represent the actual numbers of base pairs. The style labeling of helices varies between ribozymes, reflecting those used historically. The more recent ribozymes all label helices as P (pairing) or T (tertiary), while a number of older ribozymes (hammerhead, VS ribozymes) use Roman numerals. The hairpin ribozyme names the four helices forming the four-way junction A through D. These names are used throughout this chapter.

in trans between two VS ribozyme molecules [33]. The remaining ribozymes all contain pseudoknots, with or without additional junctions. The pseudoknots found in the nucleolytic ribozymes are each of different character. The HDV ribozyme has a central H-type pseudoknot that makes additional long-range basepairing interactions with the 5′ and 3′ terminal regions of the RNA. The *glmS* ribozyme has two sequential H-type pseudoknots – these could be disconnected with a single break in the backbone. The twister ribozyme has a double pseudoknot structure in which the strand reverses between the T1 and T2 tertiary contacts. As far as we know, this reversed double-pseudoknot fold is unique to twister. The terminal loop of P4 participates in basepairing to generate both T1 and T2. In some twister ribozymes, there are one or two additional helices, creating a three- or four-way helical junction at the top end of the structure as drawn. The TS ribozyme is based around a three (in some cases four)-way helical junction, but like twister the terminal loop of P4 participates in long-range tertiary interactions in the ribozyme. Finally, the pistol ribozyme contains a single pseudoknot structure together with a three-way helical junction.

It is worth noting that in the early studies of nucleolytic ribozymes the RNA species studied were frequently reduced in size too far in an attempt to obtain the smallest active species [34]. This resulted in some important architectural elements becoming deleted. These included the tertiary contact between helices I and II in the hammerhead ribozyme [35] and the four-way helical junction of the hairpin ribozyme [36].

3.4 Catalytic Players in the Nucleolytic Ribozymes

As we shall see, general acid–base catalysis is a common theme in the nucleolytic ribozymes. The selection of general acid and base is restricted to the four heterocyclic nucleobases and the 2′-hydroxyl group of the ribose. In addition, RNA is a polyanion that binds hydrated metal ions; these can also play a role in catalysis. All the nucleolytic ribozymes employ one or two nucleobases in proton transfer in the transition state, of which the most common is guanine. In the majority of cleavage reactions by these ribozymes, guanine N1 accepts a proton from the O2′ nucleophile. Nucleobases also act as general acids, donating a proton to the O5′ leaving group in the cleavage reaction, and both adenine and cytosine act in this role. Metal ions can act in a variety of ways. Inner-sphere water molecules can act directly in general acid–base catalysis. The charge on the ion can also play a role.

As discussed in Chapter 1, general acid–base catalysis will be most efficient if the acid and base have pK_a values close to neutrality. In protein enzymes, the imidazole side chain of histidine is most commonly used in general acid–base catalysis. The pK_a of imidazole is close to neutrality, so that a significant fraction of the enzyme is in the correct state of protonation for activity. By contrast, the natural pK_a values of the nucleobases (as nucleosides) are either high (guanine, 9.4) or low (adenine, 3.5, cytosine, 4.2). This means that either a small fraction of the catalyst is in the correct state of protonation for activity, although this is offset by high intrinsic reactivity, or alternatively, the catalyst is mostly in the active state but has low reactivity. A key role of the environment of the active center is to shift pK_a values toward neutrality, elevating the active fraction. In the context of the electronegative backbone of RNA, it is relatively easy to conceive how pK_a values can be shifted up, and we shall see that the apparent pK_a of adenine in the twister ribozyme is almost 7 [37]. Similarly, the pK_a of guanine in the hairpin ribozyme is raised to about 10.5, but in the VS ribozyme it is lowered to 8.4 [38], probably due to the close proximity of a positively charged metal ion. The pK_a of a 2′-hydroxyl group is ~13.5, so again this should be lowered if it is to be effective in catalysis. Once again, it is probable that a nearby metal ion would achieve this. This environmental aspect is very significant in raising the observed rates of reaction in these ribozymes.

3.5 The Hairpin and VS Ribozymes: The G Plus A Mechanism

We discuss the hairpin and VS ribozymes together because they share a strong mechanistic similarity. However, their overall structures are completely unrelated. The structural framework of the hairpin ribozyme is a four-way helical junction in which internal loops on adjacent arms interact to create the active center [36] (Figure 3.2). By contrast, the VS ribozyme is based upon two key three-way helical junctions [31, 32, 39]. Its active site is also generated by an interaction between two internal loops, although it was shown that this can occur between two VS molecules in trans [33]. Despite the difference in overall architecture of the hairpin and VS RNA, the common feature is that the scissile phosphates and active-site components are contained within two interacting internal loops. As we shall see below, this similarity extends much deeper than this.

The rates of cleavage or ligation of both ribozymes are strongly pH dependent, consistent with proton transfer occurring in the transition state. The observed rate (k_{obs}) will be related to the intrinsic rate for ribozyme in its required state of protonation (i.e. having a deprotonated base and protonated acid, giving the maximum possible rate k_{max}) by

$$k_{obs} = k_{max} \cdot f_a \cdot f_b \tag{3.1}$$

where f_a and f_b are the fractions of protonated acid and unprotonated base, respectively (see Chapter 1). The pH versus rate profile for the VS ribozyme is a classic bell shape, fitting pK_a values of 5.2 and 8.4 (Figure 3.3a). Both pK_a values are lower than would be expected if a metal ion acted in general acid–base catalysis.

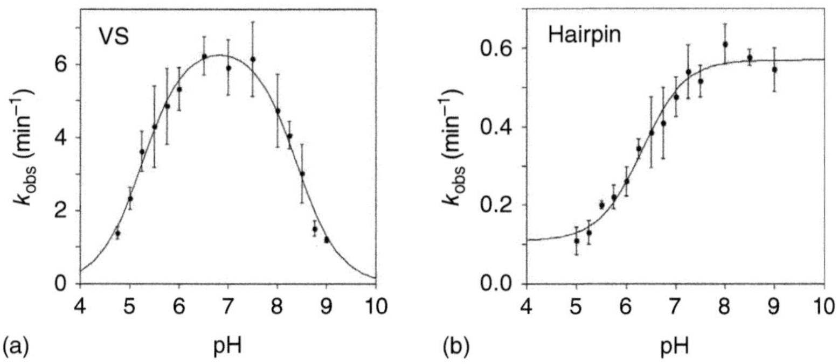

Figure 3.3 Experimental rate-pH profiles for the VS and hairpin ribozymes. (a) The pH dependence of the cleavage reaction for the VS ribozyme, in the presence of 200 mM Mg^{2+} ions [38]. (b) The pH dependence of the cleavage reaction for the hairpin ribozyme in its four-way junction form in the presence of 10mM Mg^{2+} ions [9]. Source: (a) Wilson et al. [38].

Earlier experiments in which individual nucleotide substitutions were made in the VS ribozyme had implicated G638 (in the substrate loop that includes the scissile phosphate) [38] and A756 (in the A730 loop that interacts with the substrate loop) [40–44]. The pH dependence of reaction rates varied with substitution of these two nucleotides [38], consistent with their direct roles in general acid–base catalysis. These experiments alone cannot specifically assign the general acid and base, but this was achieved by 5′-phosphorothiolate substitution that raises the leaving group strength so that protonation is no longer required. It was observed that while the cleavage rate of an A756G mutant VS ribozyme was impaired 10^4-fold, full activity was restored by 5′-phosphorothiolate substitution at the scissile phosphate [45]. This demonstrates that A756 is the general acid. The pH dependence of the A756G ribozyme (where only titration of the general base should now affect the reaction rate) was fully consistent with the assignment of G638 as the general base. This assignment was also suggested on the basis of changes in nucleobase pK_a with ionic environment [46]. Thus, we can conclude that VS ribozyme cleavage reaction is catalyzed by G638 in its deprotonated state acting as the general base to remove the nucleophilic O2′ proton, and A756 in its protonated state acting as the general acid to protonate the O5′ leaving group (Figure 3.4a). The pH dependence of a VS G638DAP ribozyme cleavage reaction indicated that proton transfer contributes at least 10^2 to 10^3 fold to the catalytic power [38].

The crystallographic structure of the VS ribozyme turned out to be a hard nut to crack, and was only determined [47] more than two decades after its discovery

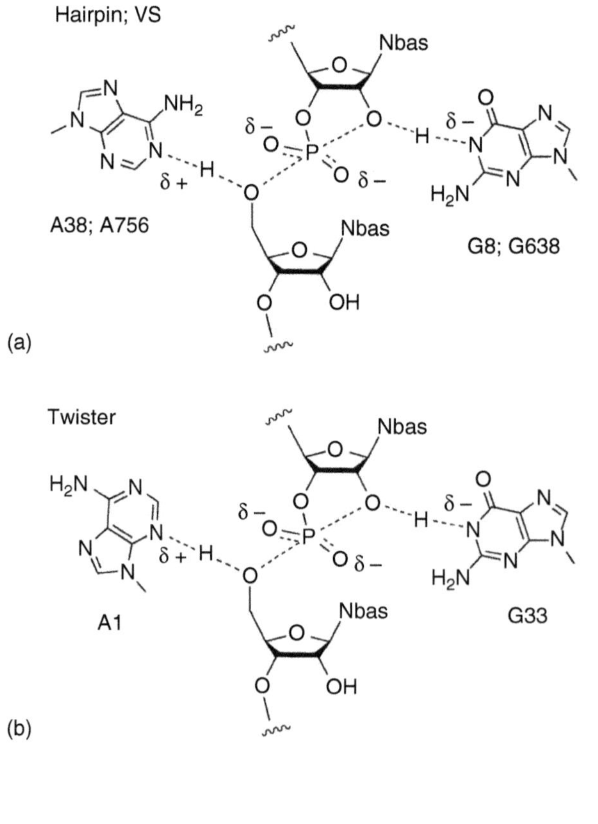

Figure 3.4 Proposed mechanisms for three ribozymes using guanine and adenine nucleobases as general base and acid, respectively, in their cleavage reactions. (a) The hairpin and VS ribozyme mechanisms. Guanine N1 and adenine N1 are the general base and acid, respectively. (b) The mechanism of the twister ribozyme. This differs from the hairpin and VS ribozyme mechanisms in that the proton is donated by the more acidic adenine N3.

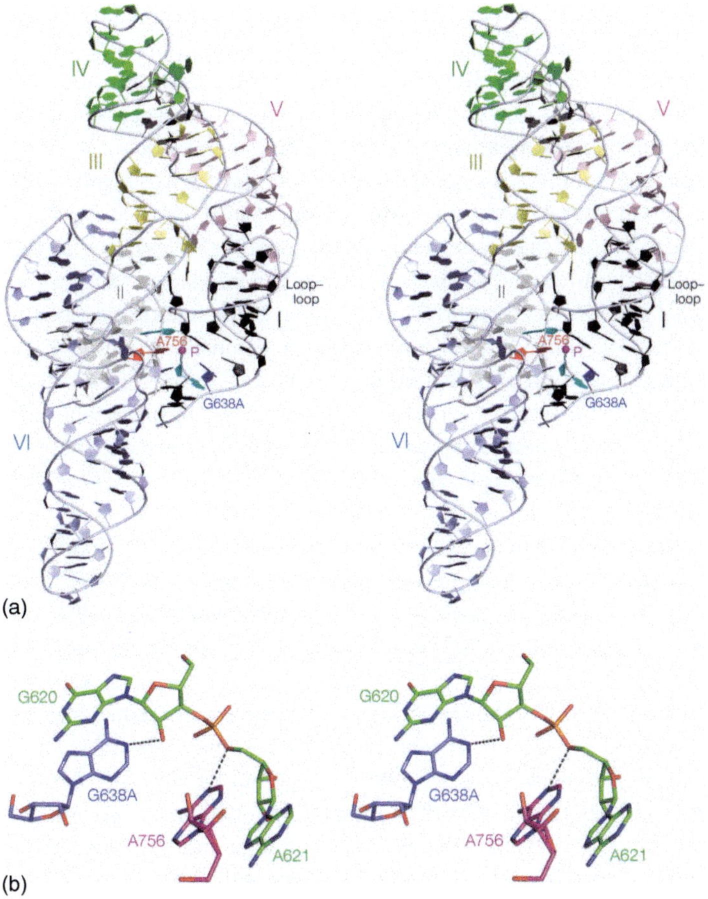

Figure 3.5 The structure of the VS ribozyme in dimeric form. In this and subsequent molecular graphics images, stereoscopic images are presented as parallel-eye pairs. This representation was generated from PDB ID: 4R4V at 3.07 Å resolution [47]. The ribozyme crystallized was a G638A mutant used to prevent self-cleavage occurring during crystallization. (a) A functional ribozyme unit, comprising the substrate stem-loop I from one monomer and the remaining parts of the ribozyme from another. The position of the scissile phosphate analog is indicated by the red sphere. (b) The active center of the ribozyme. The key catalytic participants are the nucleobases of G638 (replaced by adenine in this structure) and A756, the general base and acid, respectively, in the cleavage reaction. Note that in this structure G638A and A756 are contributed by different monomer ribozymes [33].

(Figure 3.5a). All the mechanistic studies were performed in the absence of a crystal structure. Yet when the structure was finally solved, it was found that G638 and A756 were indeed adjacent to the O2′ and O5′ atoms, respectively (Figure 3.5b), consistent with their deduced roles in catalysis.

At first sight, the pH dependence of the cleavage and ligation rates for the hairpin ribozyme appears rather different from that of the VS ribozyme (Figure 3.3b).

At low pH the rates rise, achieve a plateau around neutrality, and remain there up to pH 9 [9, 48, 49]. No reduction in activity was found at the high pH end, i.e. no bell-shaped pH dependence like that of the VS ribozyme. This has led to considerable confusion in the literature, for which we have made an in-depth critical assessment [50]. However, if the ribozyme mechanism were based on general acid–base catalysis equivalent to that of the VS, but one of the catalytic players had a pK_a ≥10, then simulations (see Chapter 1) show that the fall in reaction rates at high pH will begin at a pH higher than that experimentally accessible (~pH 9, due to the instability of the RNA structure), and so the lowering of the rate is not observed in practice [50]. A critical guanine nucleotide (G8) was suggested by substitution experiments [48, 51–53], and bell-shaped pH dependence of reaction rates with G8DAP (DAP = 2,6-diaminopurine) [48, 54], imidazole [55], and 8-azaguanine [56] substitutions lent further support. Thomas and Perrin [57] showed that a 2′-bromoacetamide affinity label replacing the O2′ nucleophile was transferred to the G8 nucleobase.

X-ray crystal structures of the hairpin ribozyme have been solved for both the four-way junction form [58] (Figure 3.6a) and a more minimalist version lacking the junction [59–61]. A crystal structure in which the scissile phosphate was replaced by vanadate to mimic the pentacoordinate phosphorane transition state [62] provided the greatest insight into the catalytic mechanism. It showed that G8 N1 was hydrogen bonded to the nucleophilic O2′ [62] (Figure 3.6b), thus well positioned to act as a general base in the cleavage reaction. In addition, an adenine (A38) N1 was positioned to hydrogen bond to the O5′ (the expected leaving group) when protonated, consistent with a role as general acid. It had been shown that ablation of the A38 nucleobase lowered cleavage activity by 10^4-fold [63], and NAIM experiments supported an important role for A38 in ligation [64]. Wedekind and coworkers [65] found that substitution by N1-deazaadenine did not alter the structure of the ribozyme, yet led to no detectable activity. Using a transition state analogue, the Wedekind group also measured a microscopic pK_a = 6.27 for N1 of A38 by Raman crystallography, in excellent agreement with kinetic data, and showed that the elevated pK_a required hydrogen bonding between A38 N6 and the *pro*R nonbridging oxygen of the scissile phosphate [66].

Phosphorothiolate experiments confirmed that G8 was the general base and A38 the general acid in the hairpin ribozyme cleavage reaction [54]. Thus, the rate of cleavage of an A38P hairpin ribozyme (P = purine) was restored 3.6×10^3-fold by 5′-phosphorothiolate substitution, and the remaining pH dependence was consistent with that of G8 alone. These phosphorothiolate substitution experiments are critical and provide additional information beyond assigning the general acid. If the adenine nucleotides of the hairpin or VS ribozymes were acting to provide electrostatic stabilization of the transition state [63], then activity should not be restored to a mutant in this position by the 5′-phosphorothiolate substitution.

Thus, we find that the hairpin and VS ribozymes share closely similar mechanisms of catalysis. The cleavage reaction mechanism is based to a significant extent on general acid–base catalysis in which N1 of a deprotonated guanine (G8 or G638) acts as general base to remove a proton from the O2′ nucleophile, and the N1 of a protonated adenine (A38 or A756) acts as general acid to transfer a proton onto the

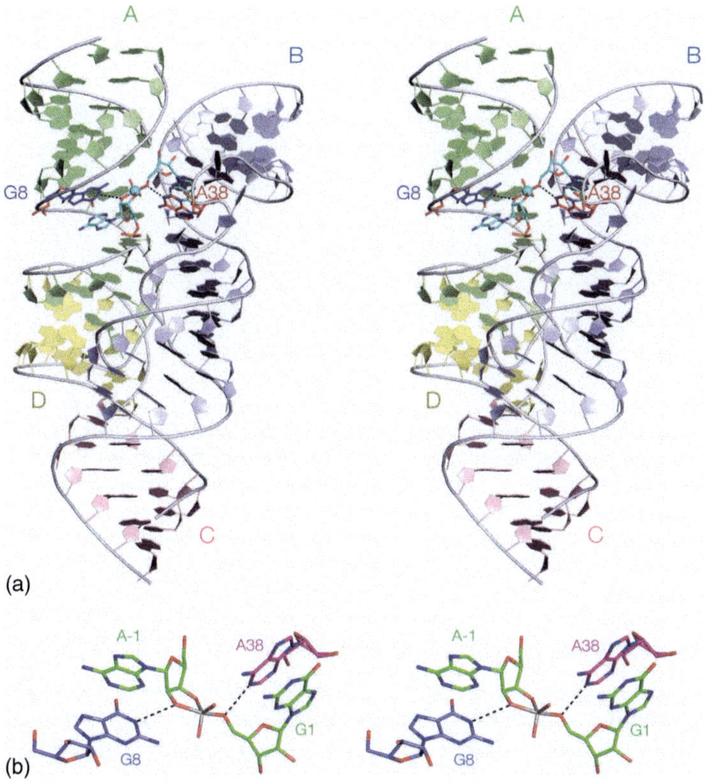

Figure 3.6 The structure and active center of the hairpin ribozyme in its natural four-way junction form. This representation was generated from PDB ID: 1M5O at 2.2 Å resolution [58]. (a) The complete ribozyme, with the four helices differentiated by color. An antiparallel four-way junction acts as a framework that juxtaposes the internal loops of the A and B helices to create the active center of the ribozyme. This shows an approximately side view chosen to show the interaction between the loops of the A and B helices. The nucleotides flanking the cleavage site are shown cyan with the scissile phosphate drawn as a sphere. G8 (blue) and A38 (red) are indicated. (b) The active site, with a vanadate replacing the scissile phosphate as a transition state analog. The geometry of the vanadium dictates that this structure approximates to that of a phosphorane with inline attack. G8 N1 and A38 N1 are hydrogen bonded to the O2′ and O5′, respectively, shown by broken lines.

O5′-leaving group (Figure 3.4a). By the principle of microscopic reversibility, in the reverse (ligation) reaction, the roles of the active site nucleobases will be reversed, with a neutral adenine acting as the general base to remove the O5′ proton, and guanine in its neutral state acting as the general acid to protonate the O2′ leaving group.

Although the global structures of the hairpin and VS ribozymes are completely unrelated, the topology of the active sites is closely similar (Figure 3.7). Both are formed by the interaction of two internal loops, and if they are drawn to show the same strand polarity it can be seen that all the key participants in the reaction are in equivalent positions. Thus, the guanine nucleotides (G8 and G638) are in one loop on

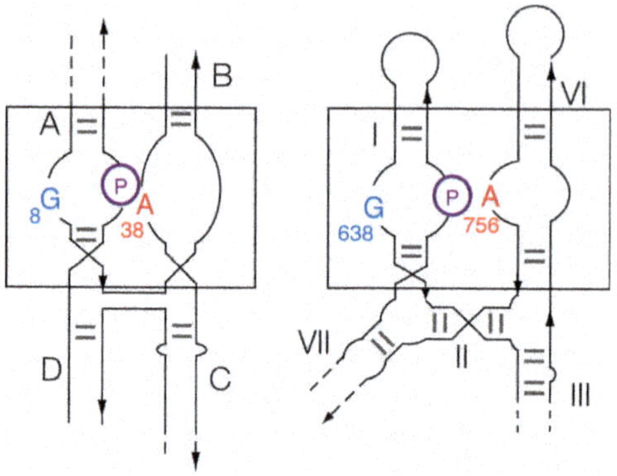

Figure 3.7 Schematic of the structures of the hairpin and VS ribozyme drawn to illustrate the topological equivalence of the two active sites. The scissile phosphate (P circled) and the catalytic adenine and guanine nucleobases are highlighted for both ribozymes. While the overall secondary structures for the two ribozymes are unrelated, they can be arranged to bring the interacting internal loops side by side (boxed in both ribozymes). These are drawn so that the strand polarities are the same for both. The relative positioning of the key functional elements then occupy the equivalent positions with respect to the corresponding loops.

the strand opposite to the scissile phosphate, while the adenine nucleotides (A38 and A756) are in the other loop. The crystal structures of both ribozymes [47, 62] show that the guanine and adenine nucleobases are stacked on those flanking the scissile phosphates on the 5′ and 3′ sides, respectively (Figures 3.5b and 3.6b). This splays apart the flanking nucleotides, helping achieve an inline geometry that is required for S_N2 attack of the nucleophile, and positions the general base and acid to act in proton transfer.

The main limitation to reaction rates catalyzed by the nucleolytic ribozymes arises from the pK_a values of the participating nucleotides. Adenine and guanine as free nucleotides have $pK_a = 3.5$ and 9.5, respectively, and if they remain at these values then only one ribozyme molecule in a million would be in the correct state of protonation to catalyze the reaction. However, these ribozymes have evolved a structure with a local environment that tends to shift the pK_a values closer to neutrality. The electronegative environment of RNA will tend to raise the pK_a of adenine, and we see that it is 6.0 in the hairpin and 5.2 in the VS ribozymes. In the following section (3.6), we shall see that it is even higher in the twister ribozyme [67]. The pK_a of G638 in the VS ribozyme is lowered to 8.4, likely due to proximity to a bound Mg^{2+} ion. These values would bring the fraction of active ribozyme up to 1 in 10^4, a 100-fold improvement on the active fraction expected if the pK_a values were unshifted.

3.6 The Twister Ribozyme: A G Plus A Variant

The twister ribozyme [29] arguably has the most completely understood catalytic mechanism of all the nucleolytic ribozymes at the current time, and makes a

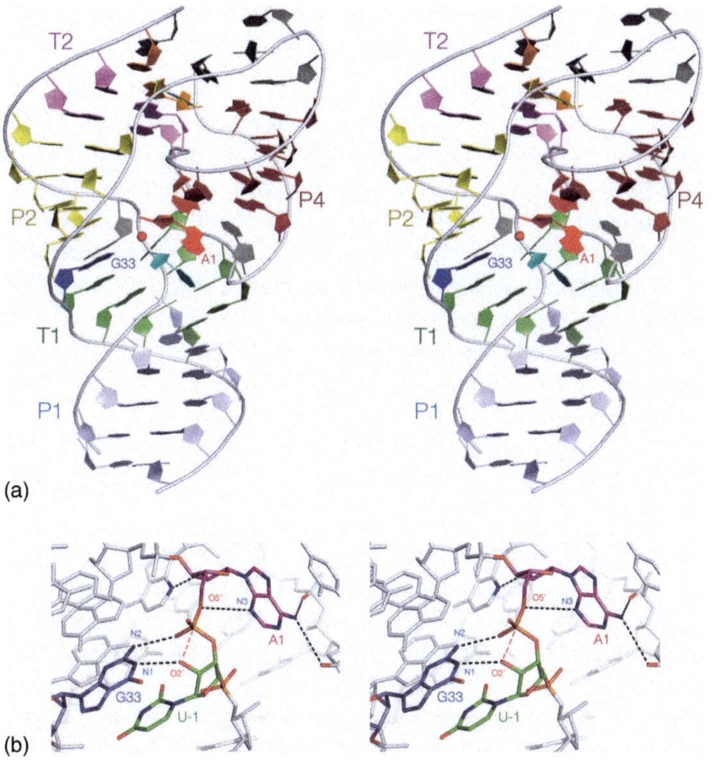

Figure 3.8 The structure of the twister ribozyme. This representation was generated from PDB ID: 4OJI at 2.34 Å resolution [37]. (a) The overall structure of the reversed double-pseudoknot structure. Helices are differentiated by color. T1 and T2 are the two long-range tertiary interactions. The position of the scissile phosphate is indicated by the red sphere. (b) Close view of the catalytic center of the ribozyme. Note that in this structure U-1 has been rotated around to lie under G33. This places the O2′ of U-1 in position for inline nucleophilic attack, shown by the broken red line. Broken black lines indicate hydrogen bonds.

valuable comparison with the hairpin and VS ribozymes. We solved the crystal structure of the twister ribozyme at 2.3-Å resolution [37] (Figure 3.8a), showing that it folds into a double, inverted pseudoknot structure that was confirmed in two further studies [68, 69]. As so often is the case, it was clear that the structure of the ribozyme in the crystals could not correspond exactly to the active conformation. In our structure, the nucleobases flanking the scissile phosphate were directed in the same direction, rather than splayed apart as required by the inline trajectory of S_N2 attack. The U-1 5′ to the scissile phosphate was pulled out of the active center by hydrogen bonding to a guanine of a crystallographically symmetry-related molecule. However, we found that a simple, unhindered rotation of U-1 in the major groove to bring it under G33 would generate a geometry well-aligned for inline attack [37] (Figure 3.8b), confirmed by a molecular dynamic analysis [70]. It seems highly probable that this represents close to the active conformation of the ribozyme.

The rate of cleavage of the twister ribozyme exhibits a bell-shaped dependence upon pH [37] indicative of proton transfer in the transition state. Fitting the data to the double-ionization model indicated the participation of groups with pK_a values of 6.9 and 9.5. The upper value corresponds to a guanine, assigned to G33 by substitution experiments. In the remodeled structure, G33 is perfectly positioned to abstract the proton from the O2′ nucleophile, accepting it at N1 (Figure 3.8b). In addition, G33 N2 is hydrogen bonded to the *pro*R O of the scissile phosphate. We have measured a 100-fold stereospecific effect of substituting sulfur for this oxygen that depended on the presence of the G33 exocyclic amine (i.e. N2) [67]. It seems probable that this interaction helps stabilize the transition state.

G33 lies in the same internal loop as the scissile phosphate, on its opposite strand. This conforms to the same topology as the hairpin and VS ribozymes, but what then is the group titrating at pK_a = 6.9? The most likely candidate would be an adenine, although the high pK_a indicates that this must have a rather unusual environment. Most of the conserved adenine nucleotides of the twister ribozyme can be excluded from a catalytic role either because of pH-insensitivity of cleavage rate to substitution, or because of structural implausibility. This only leaves A1, the nucleotide immediately 3′ to the scissile phosphate. A1 is stacked at the base of the P4 helix, fixed in place by three hydrogen bonds. Of particular significance, the two protons of A1N6 are donated to nonbridging O atoms of successive phosphate groups in the T1 helix. The charge on these phosphate groups will raise the pK_a of A1. The positioning of A1 contributes to the inline geometry, and directs its N3 toward the O5′ of the scissile phosphate. It is sterically not possible to orient the usual N1 toward the leaving group, and deaza substitutions in A1 showed that A1 N1C has elevated activity, whereas the activity of A1 N3C was virtually undetectable [67]. N3 has a microscopic pK_a of 1.5, and so would be expected to represent about 1% of protonated adenine. But to compensate for this, the special environment of A1 raises its macroscopic pK_a almost to neutrality. For this case, the observed rate of catalysis k_{obs} will be related to the intrinsic rate for fully A1-protonated ribozyme (k_{max}) by

$$k_{obs} = k_{max} \cdot f_a \cdot f_b \cdot f_{N3H} \tag{3.2}$$

where f_a and f_b are the fractions of protonated acid and base, respectively, and the extra factor f_{N3H} is the fraction of protonation that occurs specifically at N3. Thus although f_{N3H} is approximately 0.01, f_a will be ~100-fold higher than its value in the VS ribozyme, where the observed low pK_a is 5.2. Thus, these factors essentially cancel to give similar observed rates of cleavage.

In summary, we see that the twister ribozyme exploits all the potential catalytic strategies that we discussed in Chapter 1.

1. The structure is easily remodeled into a conformation that is well aligned for S_N2 attack of the O2′ on the scissile phosphate. This will likely make a modest contribution to the catalytic rate enhancement.
2. The transition state is stabilized by accepting a hydrogen bond from G33 N2.
3. G33 N1 removes a proton from the O2′ to make it a stronger nucleophile.
4. A1 N3 donates a proton to the oxyanion leaving group.

Three and four constitute concerted general acid–base catalysis and probably contribute at least a 1000-fold to the overall catalytic power of the ribozyme. We have seen that the special environment of A1 makes this possible, and thus the whole structure of the ribozyme has evolved to generate this local environment in the active center.

We should mention that an alternative mechanism has been proposed for the twister ribozyme. For an impartial perspective on this, the reader is directed to an excellent review by Breaker [71].

The mechanism of the twister ribozyme is a special case of the G+A mechanism employed by the hairpin and hammerhead ribozymes (compared in Figure 3.4). All three ribozymes use the nucleobases of guanine as a general base and adenine as general acid, but the twister ribozyme is unique in that the adenine lies immediately 3′ to the scissile phosphate, and uses its N3 to transfer a proton.

3.7 The Hammerhead Ribozyme: A 2′-Hydroxyl as a Catalytic Participant

The hammerhead ribozyme was the first nucleolytic ribozyme to be discovered [11], and the first to yield to crystallographic study [72, 73]. The hammerhead ribozyme is based upon a central three-way helical junction (Figure 3.2), and folding in metal ions involves coaxial stacking of two helices (II and III) and a remodeling of the junction [74, 75]. However, the initial crystal structures were not in agreement with mechanistic studies [76], and it was later shown that this version had been cut down too far, removing sequences outside the core that could make additional tertiary contacts necessary for full activity [35]. The extended form folded under physiological conditions [77], and exhibited faster cleavage rates [78]. Martick and Scott [79] solved the structure of the extended ribozyme, revealing the details of a tertiary interaction between the terminal loop of stem II and an internal receptor within stem I (Figure 3.9a). The tertiary contact resulted in a substantial alteration of the structure of the core of the ribozyme to a structure in much better agreement with mechanistic and substitution studies [80].

The rate of cleavage by the hammerhead ribozyme rises across the whole accessible range of pH [81], suggesting that there are no participating groups of low pK_a. Simultaneous substitution of G8 and G12 by 2,6-diaminopurine led to bell-shaped pH dependence [82], pointing to a role for these nucleotides in catalysis. Examination of the crystal structure of the extended form of the hammerhead ribozyme [79] showed that the nucleobase of G12 is positioned with its N1 directed toward the nucleophilic O2′ (Figure 3.9b), thus poised to act as general base in a manner similar to the hairpin, VS, and twister ribozymes. This would account for the rising rate of cleavage with pH for the unmodified ribozyme. It is also supported by chemical modification data in which a substrate modified by 2′-bromoacetamide alkylated G12 N1 [83]. By contrast, the nucleobase of G8 is stacked on the nucleotide 3′ to the scissile phosphate, but makes a Watson–Crick base pair to C3 and is not in a position to participate in catalysis. Moreover, there is no other nucleobase in a position that

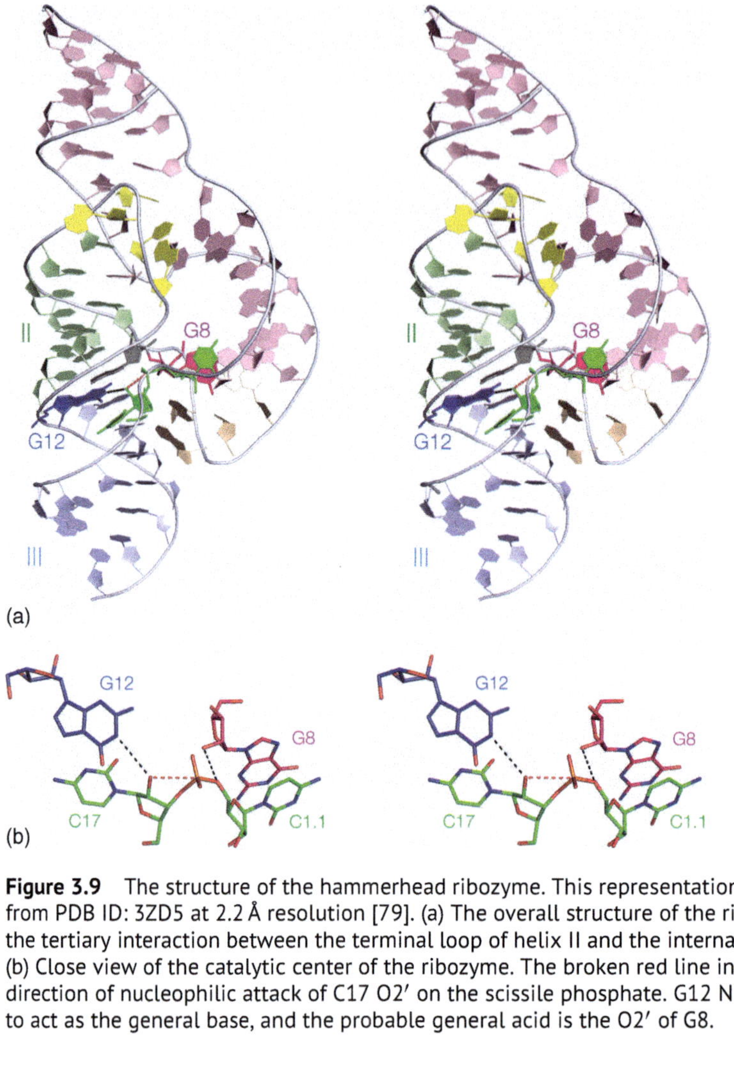

Figure 3.9 The structure of the hammerhead ribozyme. This representation was generated from PDB ID: 3ZD5 at 2.2 Å resolution [79]. (a) The overall structure of the ribozyme. Note the tertiary interaction between the terminal loop of helix II and the internal loop in helix I. (b) Close view of the catalytic center of the ribozyme. The broken red line indicates the direction of nucleophilic attack of C17 O2′ on the scissile phosphate. G12 N1 is in position to act as the general base, and the probable general acid is the O2′ of G8.

is consistent with general acid catalysis. However, the O2′ of G8 is located adjacent to the O5′ of the scissile phosphate (Figure 3.9b), well positioned to donate a proton to the oxyanion-leaving group. This is consistent with a marked alteration in the rate-pH profile for an atomic mutant where the O2′ of G8 was replaced by NH_2. While the cleavage rate of the unmodified hammerhead rises log-linearly between pH 5 and 9, that of the G8 2′NH_2 ribozyme reaches a plateau at pH 6 (consistent with the pK_a of 6.2 for a 2′NH_2 group [84]), and falls again beyond pH = 9 [85]. The normal pK_a of a 2′-hydroxyl group is 13.5. If it remains at this value in the ribozyme the great majority will have a proton to donate, but will be very unlikely to donate it to another group. While it is easy to see how groups (such as adenine) can have raised pK_a values in the electronegative context of RNA, it is harder to see how pK_a values can be lowered. We suggested that the low pK_a of G638 in the VS ribozyme was probably due to close proximity to a metal ion [38], and the same is likely to be true for the G8 O2′ in the hammerhead. Activity in monovalent metal ions and Co(III) hexammine [81, 86, 87] indicates that there is no direct role for metal ions in catalysis, but a secondary effect is quite possible. Although no ion was found bound to the G8 O2′ in the crystal structure, one was bound close by in the study by Martick and Scott [79], and quantum mechanics/molecular mechanics (QM/MM) calculations suggested that it could move closer to lower the pK_a of the G8 O2′ in the transition state [88]. This hypothesis is supported by the observation that 5′-phosphorothiolate substitution restores the loss of cleavage activity that resulted from either the absence of divalent cations or the presence of a deoxyribose at position 8 [83], and by the observation that the cleavage rate exhibits a log-linear relationship as a function of the metal ion pK_a [83]. Thus, the hammerhead is another example of a ribozyme evolving a fold that tunes the chemical environment required for catalysis.

In summary, the current evidence supports general acid–base catalysis in the hammerhead ribozyme in which G12 N1 acts as the general base and G8 O2′ functions as the general acid (Figure 3.10a).

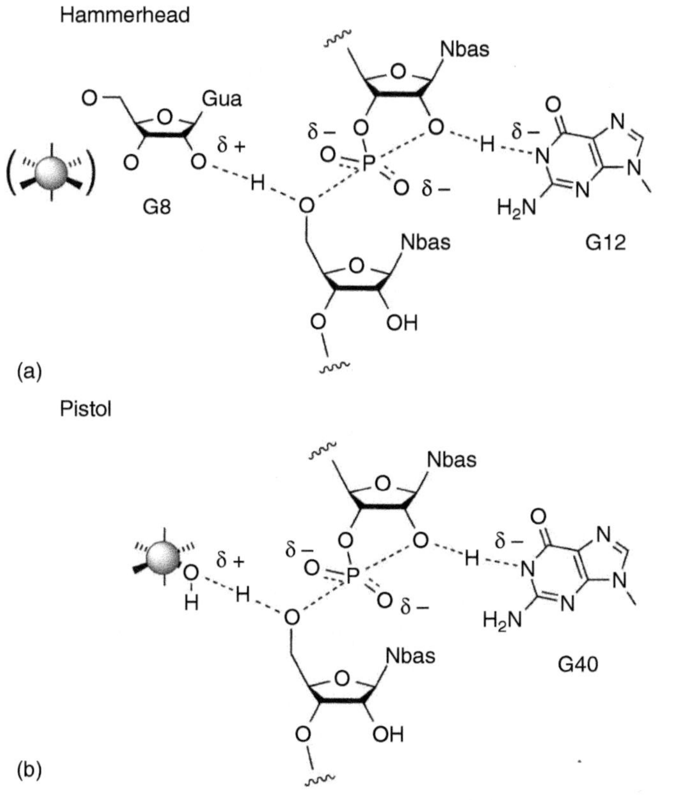

Figure 3.10 Proposed mechanisms for the hammerhead and pistol ribozymes. Both ribozymes use guanine N1 as the general base, but differ in the general acid. (a) The hammerhead ribozyme mechanism. The general acid is the 2′-hydroxyl group of G8, whose pK_a is probably lowered by interaction with a metal ion. (b) The pistol ribozyme mechanism. In this case, the general acid is a metal ion-bound water molecule. The ion is probably located in part by interaction with the O2′ of G32.

3.8 The Hepatitis Delta Virus Ribozyme: A Direct Role for a Metal Ion

In contrast to larger ribozymes such as the self-splicing introns and ribonuclease P, there is no evidence for the direct participation of hydrated metal ions in the catalytic mechanisms of the nucleolytic ribozymes discussed up to this point. The hammerhead, hairpin, and VS ribozymes have been shown to retain some activity at pH = 8 in the presence of 4 M Li^+ as the only metal ion [86], and the hairpin ribozyme is fully active in Co(III) hexammine ions [89–91]. In marked contrast, the HDV ribozyme is essentially inactive in both 4 M Li^+ and in Co(III) hexammine ions [92]. This makes a sharp distinction between the HDV ribozyme and the others we have discussed to this point, and points to a direct role for a metal in the catalytic mechanism. However, the rate of cleavage by the HDV ribozyme is pH dependent [92], again indicating proton transfer in the transition state, consistent with general acid–base catalysis.

The HDV ribozyme exists in genomic and antigenomic forms, corresponding to the (+) and (−) strands of the viral satellite RNA [93, 94]. These have closely similar secondary structures, based upon a double pseudoknot fold (Figure 3.2). Doudna and coworkers [95, 96] have solved crystal structures of the genomic form of the ribozyme, confirming the overall double pseudoknot structure (Figure 3.11a). Unlike the ribozymes discussed so far, there were no guanine nucleobases closer than 9 Å from the scissile phosphate, and none was correctly oriented to act in general acid–base catalysis. There was however a cytosine nucleobase (C75 – corresponding to C76 in the antigenomic form) oriented so that its N3 was pointing toward the scissile phosphate at a distance of 3.7 Å (Figure 3.11b). An antigenomic HDV C76U ribozyme was essentially inactive, but activity was significantly restored by addition of exogenous imidazole in the medium [98]. On the basis of the crystal structures, it was suggested that C75 would act as general base to catalyze the cleavage reaction [95]. If the cytosine nucleobase participates in general acid–base catalysis, then what is the other participant?

Bevilacqua and coworkers [92] showed that the rate of cleavage of the genomic HDV ribozyme increased log-linearly with pH, and with unit gradient, up to pH = 6.5, and remained at that plateau at higher pH values. The data fitted a $pK_a = 6.1$ that could conceivably correspond to a cytosine in an electronegative environment. The other participant would have to have a high pK_a, and this could be a hydrated metal ion as discussed above. The plateau would then correspond to full deprotonation of cytosine if it were acting as a general base as suggested. Alternatively, the plateau could be due to compensatory simultaneous deprotonation of cytosine acting as the general acid and a metal hydroxide acting as the general base, so that f_A and f_B cancel each other out over that range of pH as discussed previously. Clarification came when Nakano et al. [92] repeated the pH titration of reaction rate in the absence of divalent cations, with 1 M Na^+ ions to ensure ribozyme folding. Thereupon the pH dependence inverted, with a cleavage rate that was maximal at low pH and declined linearly as the pH was increased, corresponding to a $pK_a = 5.7$. Under these conditions, water takes the role of the

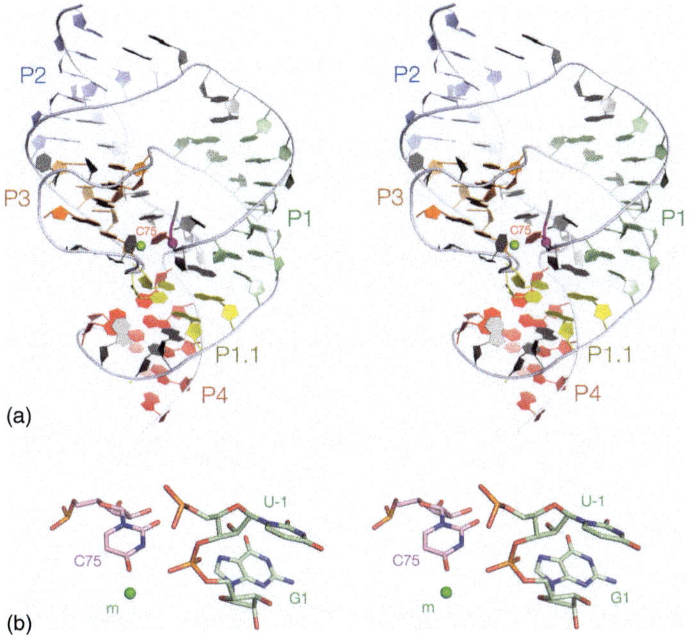

Figure 3.11 The structure of the HDV ribozyme. (a) The overall structure of the ribozyme, generated from PDB ID: 1SJ3 at 2.2 Å resolution [96]. The magenta sphere denotes the scissile phosphate in the center of the pseudoknot structure. (b) Close view of the catalytic center of the ribozyme, showing the proximity of C75 and a metal ion (m) to the scissile phosphate. This image was drawn using PDB ID: 1VC7 at 2.4 Å resolution. The conformation is not inline, and the positions do not fully reflect the established roles of the cytosine nucleobase [97] and hydrated metal ion [92]. It is likely both will be repositioned in the active geometry to some degree.

hydrated metal ion and its constant concentration means it is not a factor in the kinetics, isolating the titration of C75. Since the rate fell with the increase in pH, it indicates that C75 acts as the general acid. Lupták et al. [99] subsequently measured the pK_a of C75 in the context of the genomic HDV ribozyme by ^{13}C NMR as 5.4, in reasonable agreement with the value reported by the kinetic measurements of the cleavage reaction. Reaction rate as a function of Mg^{2+} concentration fitted a Hill coefficient of unity, consistent with the binding of a single ion participating in the reaction. A definitive assignment of the general acid and base was made by Das and Piccirilli [97] using phosphorothiolate modification. They found that 5′-phosphorothiolate substitution resulted in almost full restoration of activity of the antigenomic HDV C76U ribozyme. Furthermore, the pH dependence of this modified ribozyme then only reflected the deprotonation of the base, because the deprotonation of the general acid is no longer kinetically significant. They further showed that C76 N3C (3-deazacytosine) gave undetectable levels of activity, but cleavage was restored by 5′-phosphorothiolate substitution. In the genomic HDV ribozyme product complex C75 N3 was 2.7 Å from the O5′ atom [95] (Figure 3.11b), so the cytosine nucleobase can be well positioned to act as general acid.

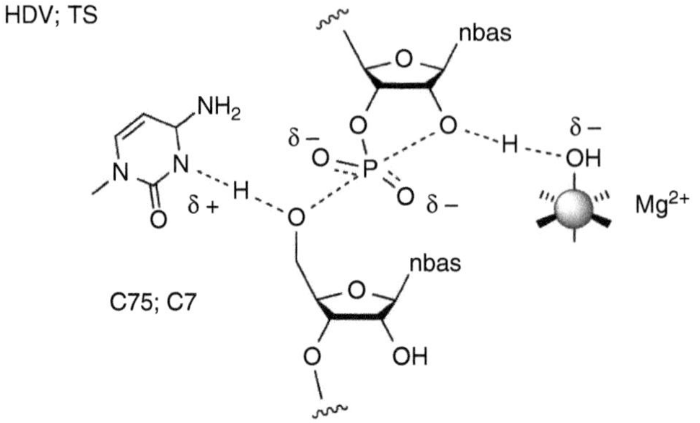

Figure 3.12 Proposed mechanisms for the HDV and TS ribozymes. These ribozymes probably have the same mechanism of general acid–base catalysis, in which the general base is a metal ion-bound hydroxide ion, and the general acid is cytosine N3. The mechanism for the HDV ribozyme is well defined by current data, but that for the TS ribozyme cannot be regarded as proven at present.

Taken together, the best evidence suggests that HDV uses general acid–base catalysis mediated by a metal hydroxide as the general base and cytosine N3 as the general acid (Figure 3.12). This is in marked contrast to the other nucleolytic ribozymes discussed thus far.

3.9 The Twister Sister (TS) Ribozyme: Another Metallo-Ribozyme

When the TS ribozyme was identified by bioinformatic analysis by Breaker and coworkers [21], it was noted that the secondary structure appeared to be quite similar to that of the twister ribozyme and so named twister sister (Figure 3.2). However, the position of strand cleavage relative to the overall fold is not the same, and as we shall see the two ribozymes are mechanistically quite different. So we shall refer to the new ribozyme as TS.

We solved the structure of a form of the ribozyme with a three-way helical junction (termed TS-4 by Weinberg et al. [21]) at 2.0-Å resolution [100]. The helices form two coaxially stacked groups (P1–P2–T2 and P4–P5) organized by the helical junction, and linked by two tertiary contacts (Figure 3.13a). The structure was found to include a number of bound Mg^{2+} ions, five of which had exchanged one or two inner-sphere water molecules with RNA oxygen atoms, one of which we shall return to shortly. The loop containing the scissile phosphate is quasi-helical, coaxial with and located between P1 and P2, with extensive hydrogen bonding. The structure has recently been confirmed by Ren and coworkers [101].

Weinberg et al. [21] showed that the rate of cleavage of the TS-1 form of the ribozyme (based on a four-way helical junction) increases with pH up to neutrality, remaining at a plateau at higher pH values. Once again, this indicated proton transfer in the cleavage reaction. They also showed that the rate increased

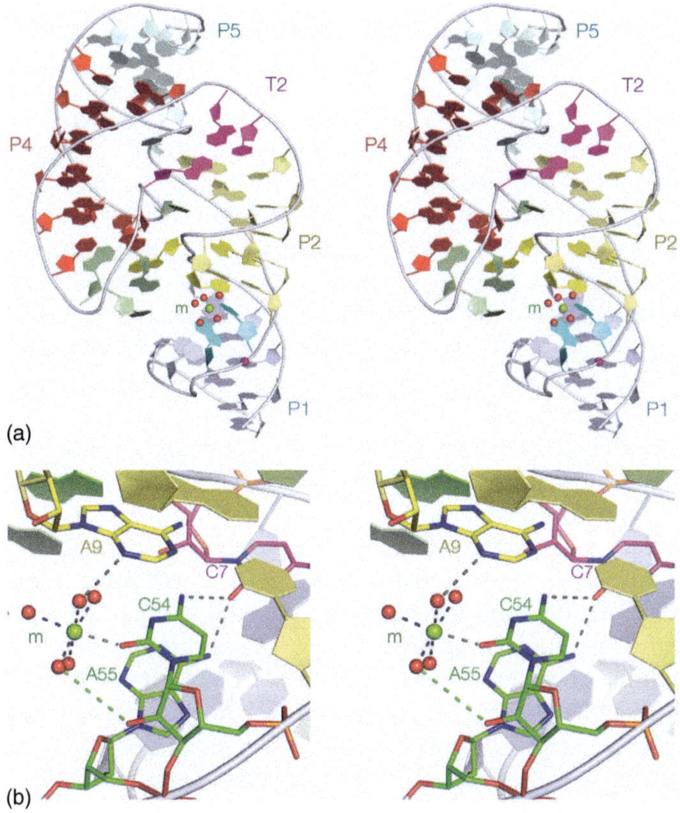

Figure 3.13 The structure of the TS ribozyme. This representation was generated from PDB ID: 5T5A at 2.0 Å resolution [100]. (a) The overall structure of the ribozyme. The structure is a pseudoknot, and T2 is a long-range tertiary interaction. The red sphere denotes the scissile phosphate, and an important hydrated metal ion is labeled m. (b) Close view of the catalytic center of the ribozyme. C54 and A55 flank the scissile phosphate. A metal ion (m) is directly bonded to C54 O2, and an inner-sphere water molecule is hydrogen bonded to A9 N3. Another inner-sphere water molecule is in position to act as the general base, shown by the connection (broken green line) to the modeled C54 O2′ nucleophile. Substitution of C7 results in a large reduction in cleavage rate, and it is likely this nucleobase is directly involved in catalysis. It is probable that this structure undergoes significant remodeling into the active geometry.

log-linearly with Mg^{2+} ion concentration. We found that the activity for the TS-4 ribozyme was extremely low in either 1 M Li^+ or 1 mM Co(III) hexammine ions ($k_{obs} = 5.7 \times 10^{-5}$ and $<1 \times 10^{-5}$, respectively) [100], the absence of divalent cations having an effect similar in magnitude to that observed for HDV. This indicates the direct participation of a metal ion in the TS ribozyme reaction. In addition, we found that the rate of cleavage in the presence of divalent metal ions was log-linearly dependent on the pK_a of the metal (i.e. the ease of deprotonation of an inner-sphere water molecule) [100]. This strongly indicates that a hydrated metal ion is involved in proton transfer in the cleavage reaction. Thus in contrast to the twister ribozyme, TS is clearly a metallo-ribozyme.

Our crystal structure of the TS-4 ribozyme contains a candidate catalytic metal ion [100]. This is directly bound to C54 immediately 3′ to the scissile phosphate, whose O2 has displaced an inner-sphere water of hydration (Figure 3.13b). Another inner-sphere water molecule is H-bonded to N2 of C54. A further inner-sphere water molecule is close to the O2′ of C54, i.e. the reaction nucleophile. So, in this structure, the hydrated metal ion is well placed to act as the general base. However, it should be noted that the nucleobases of U54 and A55 flanking the scissile phosphate are oriented in similar directions, not in the "splayed-apart" conformation required to generate inline nucleophilic attack. So, this structure must represent a prereactive conformation that will become remodeled to some degree approaching the transition state. In principle, the metal ion may relocate in this process.

If a hydrated metal ion acts as a potential general base, what might act as a general acid? Extensive mutagenesis suggests that modification of C7 most consistently results in the largest reduction of cleavage activity (Y. Liu, T.J. Wilson, and D. M. J. Lilley, unpublished data). In our crystal structure C7 is 12 Å from the scissile phosphate, and part of the extensively H-bonded network of the L1 loop (Figure 3.13b). But if the active center of the ribozyme is remodeled to create the required inline geometry, this might also reposition C7 to allow it to participate directly in catalysis. We have considered the possibility that the angle between the axes of the two groups of stacked helices at the junction changes to make a more intimate interaction between the terminal loop of P4 (that makes all the tertiary interactions) and the L1 loop containing the scissile phosphate. However, mutagenesis of the nucleotides in P4 has eliminated them as participants in proton transfer [100]. Thus, the key catalytic players of the TS ribozyme are most probably a bound hydrated metal ion and the nucleobase of C7 (Figure 3.12), but this is not known with certainty at the present time. We are still some way from a full understanding of the catalytic mechanism of this ribozyme.

3.10 The Pistol Ribozyme: A Metal Ion as the General Acid

The pistol ribozyme is another of the new group of ribozymes discovered in the last five years by Breaker and coworkers [21, 29], along with twister and TS. Like all the nucleolytic ribozymes, the reaction involves nucleophilic attack of the O2′ on its adjacent phosphorus atom (Figure 3.1), and the catalyzed reaction is relatively fast at 10 min^{-1} [102]. The consensus secondary structure indicated that the RNA would fold as a pseudoknot, and this has been demonstrated in three independent crystal structures of pistol ribozymes from three strains [85, 103, 104] (Figure 3.14a). The structure comprises four double-helical sections, one of which is the pseudoknot formed by the loop of P1. Interestingly, when the core of the ribozyme comprising the pseudoknot and helices P2 and P3 are compared with the hammerhead ribozyme there is a striking similarity, with helices II, I, and III in equivalent positions. Once

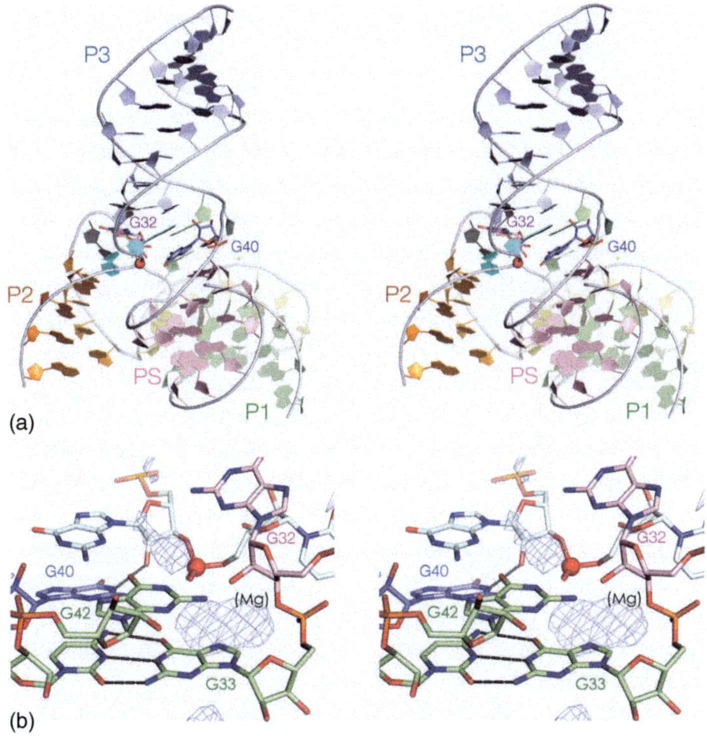

Figure 3.14 The structure of the pistol ribozyme. This representation was generated from our crystal structure determined at 2.8 Å resolution [85]. The red sphere is the scissile phosphate. (a) The overall structure of the ribozyme. The structure is a pseudoknot, where stem PS is formed by the terminal loop of helix P1. The strand that connects the P1 and P2 helices (colored yellow, in the rear) interacts with the minor groove of the P1 helix. (b) Close view of the catalytic center of the ribozyme. G40 N1 is hydrogen bonded to the O2′ nucleophile and acts as the general base in the cleavage reaction. The $\mathbf{F}_o - \mathbf{F}_c$ electron density map shows an elongated section that most probably corresponds to a bound Mg^{2+} ion. This is surrounded by atoms whose substitution leads to a large reduction in the rate of cleavage, including G32 O2′, G33 N7, and the *pro*R oxygen of the scissile phosphate. An inner-sphere water molecule hydrating the metal ion is the probable general acid in the cleavage reaction.

key functional nucleotides are identified in the pistol ribozyme (discussed below), we see that this similarity becomes even more striking.

The rate of cleavage rises with pH in a log-linear manner up to pH = 8.75, and then falls at higher pH values. Substitution of G40I shifted the pH maximum down by ~0.5 pH units [85], strongly implicating G40 in proton transfer in the transition state. Its position in the crystal structures with its N1 2.5 Å from the O2′ nucleophile suggests that it could serve as general base in the cleavage reaction. Interestingly, the position of G40 is equivalent to that of G12 in the hammerhead, where it serves as general base. Thus the general base is assigned, but the identity of a general acid is less obvious. Examination of the structures suggests that nucleobases are unlikely to act as the general acid since the nearest is ≥5 Å from the O5′. Taking the analogy with the hammerhead ribozyme a step further, we considered the possibility that the O2′ of G32 (or A32; ribozymes with A or G at position 32 are equally active) might donate a proton to the O5′-leaving group. Its position in the structure is equivalent to G8 in the hammerhead, but the O2′ is 5.1 Å from the O5′. Moreover, while an A32 O2′NH_2 substitution led to a 100-fold reduction of cleavage rate, the pH profile was essentially unaltered from the natural sequence [85]. This stands in marked contrast to the hammerhead ribozyme, where the equivalent substitution resulted in a wide plateau in the pH-rate profile (see above). Thus the two ribozymes are mechanistically distinct in this respect, and the O2′ does not act directly as the general acid in the pistol ribozyme. So if G32 O2′ is not the general acid in that ribozyme, what is?

The rate of cleavage in Co(III) hexammine is lowered three logs, strongly indicating a role for metal ions in catalysis. In our crystal structure of the pistol ribozyme, we noted that there was an elongated section of unassigned electron density within a cavity adjacent to the scissile phosphate (Figure 3.14b), and Ren et al. have made a similar observation [103]. The most likely occupant would be a hydrated metal ion, and perhaps this might be playing an important role in the catalysis. Atomic mutagenesis of several groups lining this cavity resulted in a reduction in cleavage rate by three logs, including a stereoselective effect of O to S substitution at the *pro*R oxygen of the scissile phosphate. Perhaps most convincingly, the cleavage rate exhibited a log-linear relationship as a function of the metal ion pK_a over 4 orders of magnitude [85]. This shows that reaction rate depends directly on the acidity of the metal bound water, consistent with the hydrated metal ion acting as the general acid in the cleavage reaction.

Thus, it seems that the pistol ribozyme uses both a familiar and a novel strategy. It uses guanine N1 as the general base, in common with many ribozymes, and it is also a metalloenzyme (Figure 3.10b). But, significantly, it is the first nucleolytic ribozyme known to use a hydrated metal ion as general acid to protonate the oxyanion-leaving group in the cleavage reaction. Of course, if the pistol ribozyme also catalyzes the reverse ligation reaction (possible, but not known at this time), then the metal ion will function as general base deprotonating O5′ nucleophile in a similar manner to the HDV ribozyme in cleavage.

3.11 The *glmS* Ribozyme: Participation of a Coenzyme

We have left discussion of the *glmS* ribozyme until last because it uniquely uses a coenzyme to catalyze its cleavage reaction. It is moreover the only ribozyme that is simultaneously a riboswitch, used to control gene expression in bacteria by a ligand-dependent degradation of mRNA. It was discovered in the Breaker lab as a repeated motif in the 5′-UTR of a number of genes in Gram-positive bacteria. When carrying out inline probing experiments on the RNA, Winkler et al. [19] found a site of strong self-cleavage, creating 5′-OH and cyclic 2′3′-termini. The activity required the presence of Glc6P, with half-maximal activity at ~200 μM. The rate of cleavage was enhanced 5 orders of magnitude in the presence of the Glc6P ligand. The observed rate of cleavage was found to depend on reaction pH, but since the binding of the ligand is pH dependent this is difficult to interpret with confidence. The ribozyme was active in the presence of a number of divalent metal ions [105]. It was also active in high concentrations of monovalent metal ions, as well as Co(III) hexammine, so the direct involvement of a metal ion is unlikely.

In principle, the activation of the *glmS* ribozyme might be a result of Glc6P acting as an allosteric effector or a coenzyme. Yet the $\sim10^5$-fold activation would be very high for allostery, and evidence rapidly accumulated that Glc6P was directly participating in the reaction chemistry. Soukup and coworkers [106] found that the phosphate of the ligand was dispensable, and even serinol could promote activity. So long as the amine was present, the framework of Glc6P was not essential. Furthermore, chemical footprinting and crosslinking experiments indicated that the structure of the *glmS* ribozyme was unchanged on binding ligand. This was ultimately confirmed by X-ray crystallography [107, 108]. Structures were obtained for a product complex of the *Bacillus subtilis glmS* ribozyme [107] and a complex of Glc6P bound to intact *glmS* ribozyme from *Bacillus anthracis* [109] (Figure 3.15a). The ribozyme contains two sequential pseudoknot structures. A short helix (P2.1) is packed in between two coaxial stacks of helices, and linked by a pseudoknot at one end and a junction at the other. It includes an internal loop that contains the ligand binding site and the scissile phosphate. The nucleobases of the nucleotides flanking the scissile phosphate are splayed apart, generating a good inline trajectory for attack.

Both the sugar and the phosphate of the Glc6P are H-bonded to the loop, and the glucopyranoside ring stacks on a guanine nucleobase. The core of the ribozyme is preorganized in the absence of ligand, excluding allostery as a mechanism. However, the amine of the Glc6P is 2.9 Å from the O5′-leaving group, well positioned to act as general acid (Figure 3.15b). The pK_a of the amine is 8.2, and should therefore be quite effective in this role. If Glc6P acts as the general acid, can a general base be identified? G40 N1 (in the numbering system for *B. subtilis*; the corresponding nucleotide in *B. anthracis* is G33) is 3.2 Å from O2′ nucleophile, consistent with a role as general base. A *glmS* G40A mutant ribozyme was severely impaired, cleaving at 5×10^{-5} min^{-1} [110]. Thus, the probable mechanism of the *glmS* ribozyme is general acid–base catalysis, with a guanine N1 as general base and the amine of the Glc6P ligand acting as general acid.

3.12 A Classification of the Nucleolytic Ribozymes Based on Catalytic Mechanism

In the last few years, the amount of structural and mechanistic data on nucleolytic ribozymes has increased substantially. The comparison between the information derived from eight different ribozymes creates an opportunity to find common themes and contrasts. All these ribozymes use general acid–base catalysis to provide a major contribution to the catalytic rate enhancement. Despite the relative paucity of potential catalytic groups in RNA, there is a spread of elements that participate directly in the reactions among the different ribozymes, and flexibility in the way they are employed. These are summarized in Table 3.1.

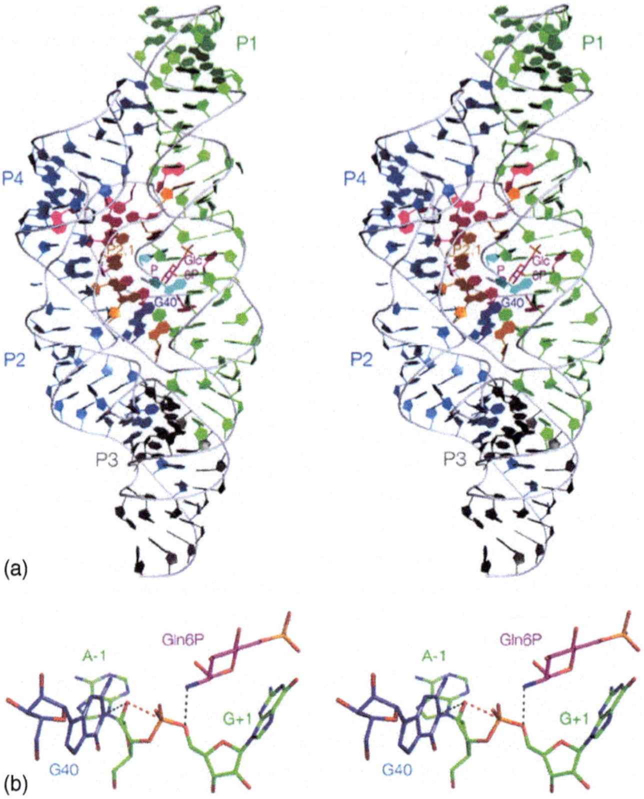

Figure 3.15 The structure of the *glmS* ribozyme bound to glucose-6-phosphate. This representation was generated from PDB ID: 2HO7 at 2.9 Å resolution [107]. (a) The overall structure of the ribozyme. The glucose-6-phosphate (Glc-6P) is shown in stick form, colored magenta. The scissile phosphate is obscured in this view, but is denoted by P and flanked by A-1 and G+1 colored cyan. (b) Close view of the catalytic center of the ribozyme. The A-1 O2′ is positioned for inline nucleophilic attack (broken red line). G40 N1 is well placed to act as general base. An amino group has been modeled onto Glc6P to generate glucosamine-6-phosphate; this amine is positioned to act as general acid to protonate the G+1 O5′-leaving group.

We have seen that ring nitrogen atoms of nucleobases are the most frequent elements used in general acid–base catalysis. The commonest strategy is a guanine nucleobase acting as general base in the cleavage reaction of 6/8 ribozymes, with N1 accepting a proton from the O2′ nucleophile. Guanine may be favored because its N2 can bond to a nonbridging oxygen of the scissile phosphate, helping to position the general base and stabilize the transition state. This has been shown to occur for the hairpin and twister ribozymes and may occur in others. In addition, cooperativity between the negative charge of deprotonated guanine and the positive charge of a protonated general acid such as adenine can shift both pK_a values toward neutrality and raise the rate of cleavage [111].

The nature of the general acid is rather more variable, with nucleobases, hydrated metal ions, 2′-hydroxyl groups, and even exogenous small molecules playing this

Table 3.1 Catalytic participants in the cleavage reactions of the nucleolytic ribozymes. This provides a classification of the ribozymes.

Ribozyme	base	acid	notes
Hairpin	G N1	A N1	
VS	G N1	A N1	
Twister	G N1	A N3	acid adjacent to scissile P
Hammerhead	G N1	O2′	O2′ activated by M^{2+}
Pistol	G N1	M^{2+} H_2O	
HDV	M^{2+} OH^-	C N3	
TS	M^{2+} OH^-	C N3 ?	
glmS	G N1	Glc6P amine	use of a coenzyme

role. The hairpin and VS ribozymes use adenine N1, and this in conjunction with guanine N1 as general base clearly places them as a subgroup within the nucleolytic ribozymes (Figure 3.4a), as a result of convergent evolution. Twister also uses adenine as the general acid, but the proton is donated by the more acidic N3 (Figure 3.4b), and a special environment raises the pK_a abnormally high to compensate for the lower population of the protonated form that is required. Twister is also unusual in that the participating nucleotide is immediately 3′ to the scissile phosphate. HDV and probably TS ribozymes use a metal hydroxide as general base (Figure 3.12). There is good evidence that for HDV the general acid is a cytosine nucleobase. This is probably also true for the TS ribozymes, but this has not been definitely established. But we tentatively class HDV and TS ribozymes as another subgroup in the table of nucleolytic ribozymes.

The hammerhead and pistol ribozymes share some common elements of structure and mechanism. Both use guanine N1 as general base, and both have an important role for a 2′-hydroxyl group in general acid catalysis. However, the functional role for the O2′ is different for the two ribozymes. There is good evidence that the 2′OH of G8 in the hammerhead is the general acid (Figure 3.10a), and a metal ion most probably lowers its pK_a to act in this role. By contrast, the roles are likely to be approximately reversed in the pistol ribozyme, with the A32 O2′ helping to position a hydrated metal ion that is the general acid in the cleavage reaction (Figure 3.10b). It is possible that the hammerhead and pistol ribozymes evolved from a common ancestor, so that in contrast to the hairpin and VS ribozyme group, the hammerhead and pistol ribozymes may have arisen from divergent evolution.

It will not have escaped the notice of attentive readers that we have not yet discussed the remaining nucleolytic ribozyme that emerged from the bioinformatic

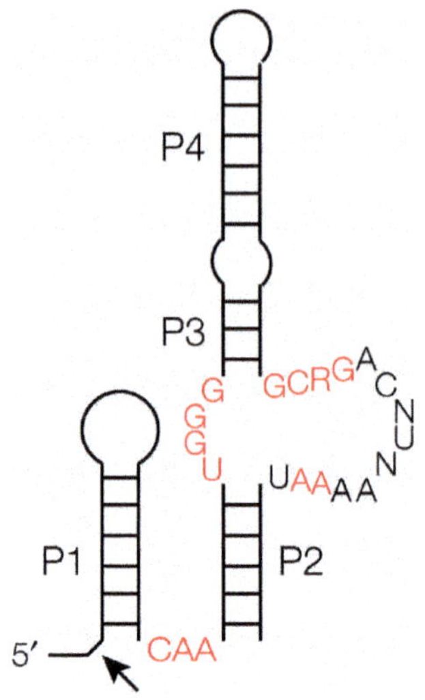

Figure 3.16 The sequence and proposed secondary structure of the hatchet ribozyme. This is a consensus sequence and secondary structure adapted from Breaker and coworkers, derived from the analysis of 210 hatchet sequences. Nucleotides colored red are >97% conserved. The position of ribozyme cleavage is arrowed. Source: Adapted from Li et al. [112].

analysis of Weinberg et al. [21], i.e. the hatchet ribozyme[1]. We have left this to one side because at the present time there is no crystallographic structure. The secondary structure and conserved sequence elements are shown in Figure 3.16. The biochemical properties of hatchet RNA were investigated by Li et al. [112], and it now makes an interesting exercise to see if those data can suggest how the new ribozyme might fit into any of the categories we have discussed. Can we use the insight gained from analysis of the eight nucleolytic ribozymes to indicate how the hatchet ribozyme might function? The conserved nucleotides (shown in red in Figure 3.16) include all four nucleotides, so any nucleobase involvement is possible, and we are not aware of any mutational exercise performed on the ribozyme. The pH dependence of cleavage rate exhibited a log-linear increase from low pH to neutrality, thereafter remaining flat to higher pH. This is equivalent to the simulation we presented in Figure 1.3b and corresponds to one pK_a of 7.5 and another of a high undetermined value. This is similar to the situation with the hairpin ribozyme where the high pK_a value is that of guanine, but the data of Li et al. [112] suggest an alternative explanation in this case. They noted a strong inhibition of cleavage on addition of hexammine Co(III) ions, even in the presence of near-saturating concentrations of Mg^{2+} ions. Furthermore, no reaction was observed in high concentrations of monovalent cations alone. This suggests the direct involvement of a hydrated metal ion in the catalysis. The lower pK_a likely corresponds to adenine or cytosine, and thus it is possible that hatchet is another HDV-like metalloenzyme ribozyme where the metal ion acts as general base in the cleavage reaction. This will however require further investigation, hopefully including the 3D structure in due course.

Lastly, the *glmS* ribozyme is clearly in a unique class in using a small-molecule coenzyme in catalysis. Once again the general base in a guanine N1, but the general

1 A structure was subsequently obtained for the hatchet ribozyme by Ren and coworkers [113]. This broadly confirms the overall secondary structure. However unfortunately the active center was apparently structured incorrectly, so it is not possible to draw clear mechanistic conclusions.

acid is the amine of the ligand Glc6P. This is the only example of a ribozyme using a coenzyme that has been discovered to date. The precedent for small-molecule involvement was set by the group I intron ribozyme that uses guanosine in the first stage of its splicing reaction; however, that is a reactant, not a coenzyme. Nevertheless, the riboswitches demonstrate the remarkable ability of RNA to bind ligands with great specificity, and we wonder if there are other ribozymes using coenzymes to perform different chemical reactions [114], or at the very least if this was not a common feature of RNA-based catalysis during the early evolution of life.

The catalytic resources of RNA are very limited, yet the nucleolytic ribozymes appear to use these to the maximum. They exploit these resources to accelerate reactions to a remarkable extent that is comparable to that of protein enzymes in some cases.

Acknowledgments

We thank our coworkers and collaborators for their contributions to the studies on the structure and mechanisms of nucleolytic ribozymes, including Yijin Liu, Stephanie Kath-Schorr, Joe Piccirilli (University of Chicago), and Darrin York (Rutgers University). We acknowledge the financial support of Cancer Research UK to our research.

References

1 van Tol, H., Buzayan, J.M., Feldstein, P.A. et al. (1990). Two autolytic processing reactions of a satellite RNA proceed with inversion of configuration. *Nucleic Acids Res.* 18 (8): 1971–1975.

2 Slim, G. and Gait, M.J. (1991). Configurationally defined phosphorothioate-containing oligoribonucleotides in the study of the mechanism of cleavage of hammerhead ribozymes. *Nucleic Acids Res.* 19 (6): 1183–1188.

3 Buzayan, J.M., Gerlach, W.L., and Bruening, G. (1986). Non-enzymatic cleavage and ligation of RNAs complementary to a plant virus satellite RNA. *Nature* 323: 349–353.

4 Saville, B.J. and Collins, R.A. (1991). RNA-mediated ligation of self-cleavage products of a *Neurospora* mitochondrial plasmid transcript. *Proc. Natl. Acad. Sci. U.S.A.* 88: 8826–8830.

5 Komatsu, Y., Koizumi, M., Sekiguchi, A., and Ohtsuka, E. (1993). Cross-ligation and exchange reactions catalyzed by hairpin ribozymes. *Nucleic Acids Res.* 21 (2): 185–190.

6 Jones, F.D., Ryder, S.P., and Strobel, S.A. (2001). An efficient ligation reaction promoted by a Varkud Satellite ribozyme with extended 5′- and 3′-termini. *Nucleic Acids Res.* 29 (24): 5115–5120.

7 McLeod, A.C. and Lilley, D.M.J. (2004). Efficient, pH-dependent RNA ligation by the VS ribozyme in *trans*. *Biochemistry* 43 (4): 1118–1125.

8 Canny, M.D., Jucker, F.M., and Pardi, A. (2007). Efficient ligation of the *Schistosoma* hammerhead ribozyme. *Biochemistry* 46 (12): 3826–3834.

9 Nahas, M.K., Wilson, T.J., Hohng, S. et al. (2004). Observation of internal cleavage and ligation reactions of a ribozyme. *Nat. Struct. Mol. Biol.* 11 (11): 1107–1113.

10 Prody, G.A., Bakos, J.T., Buzayan, J.M. et al. (1986). Autolytic processing of dimeric plant virus satellite RNA. *Science* 231 (4745): 1577–1580.

11 Hutchins, C.J., Rathjen, P.D., Forster, A.C., and Symons, R.H. (1986). Self-cleavage of plus and minus RNA transcripts of avocado sunblotch viroid. *Nucleic Acids Res.* 14: 3627–3640.

12 Forster, A.C. and Symons, R.H. (1987). Self-cleavage of plus and minus RNAs of a virusoid and a structural model for the active sites. *Cell* 49: 211–220.

13 Hazeloff, J.P. and Gerlach, W.L. (1988). Simple RNA enzymes with new and highly specific endoribonuclease activities. *Nature* 334: 585–591.

14 Hampel, A. and Tritz, R. (1989). RNA catalytic properties of the minimum (−)sTRSV sequence. *Biochemistry* 28: 4929–4933.

15 Saville, B.J. and Collins, R.A. (1990). A site-specific self-cleavage reaction performed by a novel RNA in *Neurospora* mitochondria. *Cell* 61: 685–696.

16 Kuo, M.Y., Sharmeen, L., Dinter-Gottlieb, G., and Taylor, J. (1988). Characterization of self-cleaving RNA sequences on the genome and antigenome of human hepatitis delta virus. *J. Virol.* 62 (12): 4439–4444.

17 Perrotta, A.T. and Been, M.D. (1990). The self-cleaving domain from the genomic RNA of hepatitis delta virus: sequence requirements and the effects of denaturant. *Nucleic Acids Res.* 18 (23): 6821–6827.

18 Perrotta, A.T. and Been, M.D. (1991). A pseudoknot-like structure required for efficient self-cleavage of hepatitis delta virus RNA. *Nature* 350: 434–436.

19 Winkler, W.C., Nahvi, A., Roth, A. et al. (2004). Control of gene expression by a natural metabolite-responsive ribozyme. *Nature* 428 (6980): 281–286.

20 Weinberg, Z., Wang, J.X., Bogue, J. et al. (2010). Comparative genomics reveals 104 candidate structured RNAs from bacteria, archaea, and their metagenomes. *Genome Biol.* 11 (3): R31.

21 Weinberg, Z., Kim, P.B., Chen, T.H. et al. (2015). New classes of self-cleaving ribozymes revealed by comparative genomics analysis. *Nat. Chem. Biol.* 11 (8): 606–610.

22 Epstein, L.M. and Gall, J.G. (1987). Self-cleaving transcripts of satellite DNA from the newt. *Cell* 48: 535–543.

23 Przybilski, R., Graf, S., Lescoute, A. et al. (2005). Functional hammerhead ribozymes naturally encoded in the genome of *Arabidopsis thaliana*. *Plant Cell* 17 (7): 1877–1885.

24 Martick, M., Horan, L.H., Noller, H.F., and Scott, W.G. (2008). A discontinuous hammerhead ribozyme embedded in a mammalian messenger RNA. *Nature* 454 (7206): 899–902.

25 de la Pena, M. and Garcia-Robles, I. (2010). Intronic hammerhead ribozymes are ultraconserved in the human genome. *EMBO Rep.* 11 (9): 711–716.

26 de la Pena, M. and Garcia-Robles, I. (2010). Ubiquitous presence of the hammerhead ribozyme motif along the tree of life. *RNA*, 16: 1943–1950.

27 Salehi-Ashtiani, K., Luptak, A., Litovchick, A., and Szostak, J.W. (2006). A genomewide search for ribozymes reveals an HDV-like sequence in the human CPEB3 gene. *Science* 313 (5794): 1788–1792.

28 Webb, C.H., Riccitelli, N.J., Ruminski, D.J., and Luptak, A. (2009). Widespread occurrence of self-cleaving ribozymes. *Science* 326 (5955): 953.

29 Roth, A., Weinberg, Z., Chen, A.G. et al. (2014). A widespread self-cleaving ribozyme class is revealed by bioinformatics. *Nat. Chem. Biol.* 10 (1): 56–60.

30 Tan, E., Wilson, T.J., Nahas, M.K. et al. (2003). A four-way junction accelerates hairpin ribozyme folding via a discrete intermediate. *Proc. Natl. Acad. Sci. U.S.A.* 100 (16): 9308–9313.

31 Lafontaine, D.A., Norman, D.G., and Lilley, D.M. (2001). Structure, folding and activity of the VS ribozyme: importance of the 2-3-6 helical junction. *EMBO J.* 20 (6): 1415–1424.

32 Lafontaine, D.A., Norman, D.G., and Lilley, D.M.J. (2002). The global structure of the VS ribozyme. *EMBO J.* 21: 2461–2471.

33 Ouellet, J., Byrne, M., and Lilley, D.M.J. (2009). Formation of an active site in trans by interaction of two complete Varkud Satellite ribozymes. *RNA* 15: 1822–1826.

34 Lilley, D.M.J. (2003). Ribozymes – a snip too far? *Nat. Struct. Biol.* 10 (9): 672–673.

35 Khvorova, A., Lescoute, A., Westhof, E., and Jayasena, S.D. (2003). Sequence elements outside the hammerhead ribozyme catalytic core enable intracellular activity. *Nat. Struct. Biol.* 10 (9): 1–5.

36 Murchie, A.I.H., Thomson, J.B., Walter, F., and Lilley, D.M.J. (1998). Folding of the hairpin ribozyme in its natural conformation achieves close physical proximity of the loops. *Mol. Cell* 1: 873–881.

37 Liu, Y., Wilson, T.J., McPhee, S.A., and Lilley, D.M. (2014). Crystal structure and mechanistic investigation of the twister ribozyme. *Nat. Chem. Biol.* 10 (9): 739–744.

38 Wilson, T.J., McLeod, A.C., and Lilley, D.M.J. (2007). A guanine nucleobase important for catalysis by the VS ribozyme. *EMBO J.* 26 (10): 2489–2500.

39 Lipfert, J., Ouellet, J., Norman, D.G. et al. (2008). The complete VS ribozyme in solution studied by small-angle X-ray scattering. *Structure* 16: 1357–1367.

40 Lafontaine, D.A., Wilson, T.J., Norman, D.G., and Lilley, D.M.J. (2001). The A730 loop is an important component of the active site of the VS ribozyme. *J. Mol. Biol.* 312 (4): 663–674.

41 Lafontaine, D.A., Wilson, T.J., Zhao, Z.-Y., and Lilley, D.M.J. (2002). Functional group requirements in the probable active site of the VS ribozyme. *J. Mol. Biol.* 323: 23–34.

42 Hiley, S.L., Sood, V.D., Fan, J., and Collins, R.A. (2002). 4-thio-U cross-linking identifies the active site of the VS ribozyme. *EMBO J.* 21 (17): 4691–4698.

43 Zhao, Z., McLeod, A., Harusawa, S. et al. (2005). Nucleobase participation in ribozyme catalysis. *J. Am. Chem. Soc.* 127 (14): 5026–5027.

44 Suydam, I.T., Levandoski, S.D., and Strobel, S.A. (2010). Catalytic importance of a protonated adenosine in the hairpin ribozyme active site. *Biochemistry* 49 (17): 3723–3732.

45 Wilson, T.J., Li, N.-S., Lu, J. et al. (2010). Nucleobase-mediated general acid–base catalysis in the Varkud satellite ribozyme. *Proc. Natl. Acad. Sci. U.S.A.* 107: 11751–11756.

46 Smith, M.D., Mehdizadeh, R., Olive, J.E., and Collins, R.A. (2008). The ionic environment determines ribozyme cleavage rate by modulation of nucleobase pK_a. *RNA* 14 (9): 1942–1849.

47 Suslov, N.B., DasGupta, S., Huang, H. et al. (2015). Crystal structure of the Varkud satellite ribozyme. *Nat. Chem. Biol.* 11 (11): 840–846.

48 Pinard, R., Hampel, K.J., Heckman, J.E. et al. (2001). Functional involvement of G8 in the hairpin ribozyme cleavage mechanism. *EMBO J.* 20 (22): 6434–6442.

49 Kuzmin, Y.I., Da Costa, C.P., and Fedor, M.J. (2004). Role of an active site guanine in hairpin ribozyme catalysis probed by exogenous nucleobase rescue. *J. Mol. Biol.* 340 (2): 233–251.

50 Wilson, T.J. and Lilley, D.M. (2011). Do the hairpin and VS ribozymes share a common catalytic mechanism based on general acid–base catalysis? A critical assessment of available experimental data. *RNA* 17 (2): 213–221.

51 Grasby, J.A., Mersmann, K., Singh, M., and Gait, M.J. (1995). Purine functional groups in essential residues of the hairpin ribozyme required for catalytic cleavage of RNA. *Biochemistry* 34 (12): 4068–4076.

52 Shippy, R., Siwkowski, A., and Hampel, A. (1998). Mutational analysis of loops 1 and 5 of the hairpin ribozyme. *Biochemistry* 37 (2): 564–570.

53 Wilson, T.J., Zhao, Z.-Y., Maxwell, K. et al. (2001). Importance of specific nucleotides in the folding of the natural form of the hairpin ribozyme. *Biochemistry* 40: 2291–2302.

54 Kath-Schorr, S., Wilson, T.J., Li, N.S. et al. (2012). General acid–base catalysis mediated by nucleobases in the hairpin ribozyme. *J. Am. Chem. Soc.* 134 (40): 16717–16724.

55 Wilson, T.J., Ouellet, J., Zhao, Z.Y. et al. (2006). Nucleobase catalysis in the hairpin ribozyme. *RNA* 12 (6): 980–987.

56 Liu, L., Cottrell, J.W., Scott, L.G., and Fedor, M.J. (2009). Direct measurement of the ionization state of an essential guanine in the hairpin ribozyme. *Nat. Chem. Biol.* 5 (5): 351–357.

57 Thomas, J.M. and Perrin, D.M. (2006). Active site labeling of G8 in the hairpin ribozyme: implications for structure and mechanism. *J. Am. Chem. Soc.* 128 (51): 16540–16545.

58 Rupert, P.B. and Ferré-D'Amaré, A.R. (2001). Crystal structure of a hairpin ribozyme-inhibitor complex with implications for catalysis. *Nature* 410: 780–786.

59 Alam, S., Grum-Tokars, V., Krucinska, J. et al. (2005). Conformational heterogeneity at position U37 of an all-RNA hairpin ribozyme with implications for metal binding and the catalytic structure of the S-turn. *Biochemistry* 44 (44): 14396–14408.

60 Salter, J., Krucinska, J., Alam, S. et al. (2006). Water in the active site of an all-RNA hairpin ribozyme and effects of Gua8 base variants on the geometry of phosphoryl transfer. *Biochemistry* 45 (3): 686–700.

61 Spitale, R.C., Volpini, R., Mungillo, M.V. et al. (2009). Single-atom imino substitutions at A9 and A10 reveal distinct effects on the fold and function of the hairpin ribozyme catalytic core. *Biochemistry* 48 (33): 7777–7779.

62 Rupert, P.B., Massey, A.P., Sigurdsson, S.T., and Ferré-D'Amaré, A.R. (2002). Transition state stabilization by a catalytic RNA. *Science* 298 (5597): 1421–1424.

63 Kuzmin, Y.I., Da Costa, C.P., Cottrell, J.W., and Fedor, M.J. (2005). Role of an active site adenine in hairpin ribozyme catalysis. *J. Mol. Biol.* 349 (5): 989–1010.

64 Ryder, S.P., Oyelere, A.K., Padilla, J.L. et al. (2001). Investigation of adenosine base ionization in the hairpin ribozyme by nucleotide analog interference mapping. *RNA* 7 (10): 1454–1463.

65 Spitale, R.C., Volpini, R., Heller, M.G. et al. (2009). Identification of an imino group indispensable for cleavage by a small ribozyme. *J. Am. Chem. Soc.* 131 (17): 6093–6095.

66 Liberman, J.A., Guo, M., Jenkins, J.L. et al. (2012). A transition-state interaction shifts nucleobase ionization toward neutrality to facilitate small ribozyme catalysis. *J. Am. Chem. Soc.* 134 (41): 16933–16936.

67 Wilson, T.J., Liu, Y., Domnick, C. et al. (2016). The novel chemical mechanism of the twister ribozyme. *J. Am. Chem. Soc.* 138 (19): 6151–6162.

68 Eiler, D., Wang, J., and Steitz, T.A. (2014). Structural basis for the fast self-cleavage reaction catalyzed by the twister ribozyme. *Proc. Natl. Acad. Sci. U.S.A.* 111 (36): 13028–13033.

69 Ren, A., Kosutic, M., Rajashankar, K.R. et al. (2014). In-line alignment and Mg^{2+} coordination at the cleavage site of the *env22* twister ribozyme. *Nat. Commun.* 5: 5534.

70 Gaines, C.S. and York, D.M. (2016). Ribozyme catalysis with a twist: active state of the twister ribozyme in solution predicted from molecular simulation. *J. Am. Chem. Soc.* 138: 3058–3065.

71 Breaker, R.R. (2017). Mechanistic debris generated by twister ribozymes. *ACS Chem. Biol.* 12 (4): 886–891.

72 Pley, H.W., Flaherty, K.M., and McKay, D.B. (1994). Three-dimensional structure of a hammerhead ribozyme. *Nature* 372: 68–74.

73 Scott, W.G., Finch, J.T., and Klug, A. (1995). The crystal structure of an all-RNA hammerhead ribozyme: a proposed mechanism for RNA catalytic cleavage. *Cell* 81 (7): 991–1002.

74 Bassi, G., Møllegaard, N.E., Murchie, A.I.H. et al. (1995). Ionic interactions and the global conformations of the hammerhead ribozyme. *Nat. Struct. Biol.* 2 (1): 45–55.

75 Bassi, G.S., Murchie, A.I.H., Walter, F. et al. (1997). Ion-induced folding of the hammerhead ribozyme: a fluorescence resonance energy transfer study. *EMBO J.* 16 (24): 7481–7489.

76 Peracchi, A., Beigelman, L., Scott, E.C. et al. (1997). Involvement of a specific metal ion in the transition of the hammerhead ribozyme to its catalytic conformation. *J. Biol. Chem.* 272 (43): 26822–26826.

77 Penedo, J.C., Wilson, T.J., Jayasena, S.D. et al. (2004). Folding of the natural hammerhead ribozyme is enhanced by interaction of auxiliary elements. *RNA* 10 (5): 880–888.

78 Canny, M.D., Jucker, F.M., Kellogg, E. et al. (2004). Fast cleavage kinetics of a natural hammerhead ribozyme. *J. Am. Chem. Soc.* 126 (35): 10848–10849.

79 Martick, M. and Scott, W.G. (2006). Tertiary contacts distant from the active site prime a ribozyme for catalysis. *Cell* 126 (2): 309–320.

80 Nelson, J.A. and Uhlenbeck, O.C. (2008). Hammerhead redux: does the new structure fit the old biochemical data? *RNA* 14 (4): 605–615.

81 Curtis, E.A. and Bartel, D.P. (2001). The hammerhead cleavage reaction in monovalent cations. *RNA* 7 (4): 546–552.

82 Han, J. and Burke, J.M. (2005). Model for general acid–base catalysis by the hammerhead ribozyme: pH-activity relationships of G8 and G12 variants at the putative active site. *Biochemistry* 44 (21): 7864–7870.

83 Thomas, J.M. and Perrin, D.M. (2009). Probing general acid catalysis in the hammerhead ribozyme. *J. Am. Chem. Soc.* 131 (3): 1135–1143.

84 Aurup, H., Tuschl, T., Benseler, F. et al. (1994). Oligonucleotide duplexes containing 2′-amino-2′-deoxycytidines: thermal stability and chemical reactivity. *Nucleic Acids Res.* 22 (1): 20–24.

85 Wilson, T.J., Liu, Y., Li, N.S., Dai, Q., Piccirilli, J.A., and Lilley, D.M.J. (2019). Comparison of the structures and mechanisms of the pistol and hammerhead ribozymes. *J. Amer. Chem. Soc.* 141: 7865–7875.

86 Murray, J.B., Seyhan, A.A., Walter, N.G. et al. (1998). The hammerhead, hairpin and VS ribozymes are catalytically proficient in monovalent cations alone. *Chem. Biol.* 5: 587–595.

87 O'Rear, J.L., Wang, S., Feig, A.L. et al. (2001). Comparison of the hammerhead cleavage reactions stimulated by monovalent and divalent cations. *RNA* 7 (4): 537–545.

88 Lee, T.S., Silva Lopez, C., Giambasu, G.M. et al. (2008). Role of Mg^{2+} in hammerhead ribozyme catalysis from molecular simulation. *J. Am. Chem. Soc.* 130 (10): 3053–3064.

89 Hampel, A. and Cowan, J.A. (1997). A unique mechanism for RNA catalysis: the role of metal cofactors in hairpin ribozyme cleavage. *Chem. Biol.* 4: 513–517.

90 Hegg, L.A. and Fedor, M.J. (1995). Kinetics and thermodynamics of intermolecular catalysis by hairpin ribozymes. *Biochemistry* 34 (48): 15813–15828.

91 Young, K.J., Gill, F., and Grasby, J.A. (1997). Metal ions play a passive role in the hairpin ribozyme catalysed reaction. *Nucleic Acids Res.* 25 (19): 3760–3766.

92 Nakano, S., Chadalavada, D.M., and Bevilacqua, P.C. (2000). General acid–base catalysis in the mechanism of a hepatitis delta virus ribozyme. *Science* 287: 1493–1497.

93 Branch, A.D. and Robertson, H.D. (1991). Efficient trans cleavage and a common structural motif for the ribozymes of the human hepatitis d agent. *Proc. Natl. Acad. Sci. U.S.A.* 88: 10163–10167.

94 Rosenstein, S.P. and Been, M.D. (1991). Evidence that genomic and antigenomic RNA self-cleaving elements from hepatitis delta virus have similar secondary structures. *Nucleic Acids Res.* 19 (19): 5409–5416.

95 Ferré-d'Amaré, A.R., Zhou, K., and Doudna, J.A. (1998). Crystal structure of a hepatitis delta virus ribozyme. *Nature* 395: 567–574.

96 Ke, A., Zhou, K., Ding, F. et al. (2004). A conformational switch controls hepatitis delta virus ribozyme catalysis. *Nature* 429 (6988): 201–205.

97 Das, S.R. and Piccirilli, J.A. (2005). General acid catalysis by the hepatitis delta virus ribozyme. *Nat. Chem. Biol.* 1 (1): 45–52.

98 Perrotta, A.T., Shih, I., and Been, M.D. (1999). Imidazole rescue of a cytosine mutation in a self-cleaving ribozyme. *Science* 286 (5437): 123–126.

99 Luptak, A., Ferre-D'Amare, A.R., Zhou, K. et al. (2001). Direct pK_a measurement of the active-site cytosine in a genomic hepatitis delta virus ribozyme. *J. Am. Chem. Soc.* 123 (35): 8447–8452.

100 Liu, Y., Wilson, T.J., and Lilley, D.M.J. (2017). The structure of a nucleolytic ribozyme that employs a catalytic metal ion. *Nat. Chem. Biol.* 13: 508–513.

101 Zheng, L., Mairhofer, E., Teplova, M. et al. (2017). Structure-based insights into self-cleavage by a four-way junctional twister-sister ribozyme. *Nat. Commun.* 8 (1): 1180.

102 Harris, K.A., Lunse, C.E., Li, S. et al. (2015). Biochemical analysis of pistol self-cleaving ribozymes. *RNA* 21 (11): 1852–1858.

103 Ren, A., Vusurovic, N., Gebetsberger, J. et al. (2016). Pistol ribozyme adopts a pseudoknot fold facilitating site-specific in-line cleavage. *Nat. Chem. Biol.* 12 (9): 702–708.

104 Nguyen, L.A., Wang, J., and Steitz, T.A. (2017). Crystal structure of Pistol, a class of self-cleaving ribozyme. *Proc. Natl. Acad. Sci. U.S.A.* 114 (5): 1021–1026.

105 Roth, A., Nahvi, A., Lee, M. et al. (2006). Characteristics of the *glmS* ribozyme suggest only structural roles for divalent metal ions. *RNA* 12 (4): 607–619.

106 McCarthy, T.J., Plog, M.A., Floy, S.A. et al. (2005). Ligand requirements for *glmS* ribozyme self-cleavage. *Chem. Biol.* 12 (11): 1221–1226.

107 Klein, D.J. and Ferré-D'Amaré, A.R. (2006). Structural basis of *glmS* ribozyme activation by glucosamine-6-phosphate. *Science* 313 (5794): 1752–1756.

108 Cochrane, J.C., Lipchock, S.V., Smith, K.D., and Strobel, S.A. (2009). Structural and chemical basis for glucosamine 6-phosphate binding and activation of the *glmS* ribozyme. *Biochemistry* 48 (15): 3239–3246.

109 Cochrane, J.C., Lipchock, S.V., and Strobel, S.A. (2007). Structural investigation of the *GlmS* ribozyme bound to its catalytic cofactor. *Chem. Biol.* 14: 97–105.

110 Klein, D.J., Been, M.D., and Ferré-D'Amaré, A.R. (2007). Essential role of an active-site guanine in *glmS* ribozyme catalysis. *J. Am. Chem. Soc.* 129 (48): 14858–14859.

111 Frankel, E.A. and Bevilacqua, P.C. (2018). Complexity in pH-dependent ribozyme kinetics: dark pK_a shifts and wavy rate–pH profiles. *Biochemistry* 57 (5): 483–488.

112 Li, S., Lunse, C.E., Harris, K.A., and Breaker, R.R. (2015). Biochemical analysis of hatchet self-cleaving ribozymes. *RNA* 21 (11): 1845–1851.

113 Zheng, L., Falschlunger, C., Huang, K. et al. (2020). *Proc. Natl. Acad. Sci. U.S.A.* 116 (22): 10783–10791.

114 Wilson, T.J. and Lilley, D.M.J. (2015). RNA catalysis--is that it? *RNA* 21 (4): 534–537.

4

The *glmS* Ribozyme and Its Multifunctional Coenzyme Glucosamine-6-phosphate

Juliane Soukup

Creighton University, Department of Chemistry, 2500 California Plaza, Omaha, NE 68178, USA

4.1 Introduction

In nature, catalysts are used to expedite biochemical reactions and achieve specificity. Proteins were assumed to be the only biological catalysts until catalytic nucleic acid molecules – ribozymes – were discovered and found to fold to form active sites that support chemical reactions [1–5]. Ribozymes are noncoding RNAs that promote chemical transformations with rate enhancements similar to those of protein enzymes. Another essential feature in nature is the ability of biomolecules to specifically bind ligands and/or small molecules in order to control gene expression, which is fulfilled by a different group of noncoding RNAs. Riboswitches are novel RNA elements that control the metabolic state of microorganisms by directly binding metabolites/ligands and regulating gene expression [6–8]. Both ribozymes and riboswitches provide novel targets and tools for use as RNA therapeutics. Ribozymes have been developed to target defined RNA sequences for cleavage of viral and disease-causing RNAs, and riboswitches, due to the fact that they control expression of essential metabolic genes in bacteria, are novel targets for antibiotic development.

4.2 Ribozymes

With the initial discovery of catalytic RNAs over 35 years ago [1–3] came a long-standing interest in and dedication to investigating the structure and function of these RNA molecules. One specific area of interest has focused on the mechanisms of RNA catalysis, which has resulted in a vast knowledge of RNA catalysts that are responsible for peptide bond formation in the ribosome [9, 10], RNA splicing in the spliceosome [11], RNA self-splicing by group I [12] and group II introns [13, 14], and RNA processing of transfer RNAs (tRNAs) by RNase P [15]. In addition, a large number of ribozymes are nucleolytic, self-cleaving RNAs, including the hammerhead, hairpin, hepatitis delta virus (HDV), Varkud satellite

Ribozymes, First Edition. Edited by Sabine Müller, Benoît Masquida, and Wade Winkler.

(VS), *glmS*, twister, twister sister (TS), pistol, and hatchet ribozymes (reviewed in [16]). Ribozymes perform a surprising number of reactions even though RNA and its building blocks have similar heterocycles and very limited diversity with regard to functional groups. In addition, the functional groups found on nucleotide bases, as well as the sugar-phosphate backbone, do not have pK_as near neutrality (the imino protons of A and C have pK_as near 4, and the imino protons of G and U have pK_as near 9), and so, it is assumed that ribozymes may not be able to utilize certain catalytic mechanisms of action, such as proton transfer reactions [17, 18]. Due to the intrinsic functional group limitations of RNA, a number of ribozymes actually employ help from small-molecule ligands. Examples of small-molecule ligands include metal ions (utilized by a number of ribozymes), an exogenous guanine nucleoside/nucleotide or an internal adenine nucleotide (utilized by group I and group II introns, respectively), and cyclic di-guanosyl-5′-monophosphate (di-GMP) along with guanosine-5′-triphosphate (GTP; employed by the c-di-GMP class II riboswitch with a downstream group I intron) [19–23]. In contrast to the examples given, the ONLY ribozyme that utilizes a small molecule/metabolite as a *coenzyme* for catalysis is the *glmS* self-cleaving ribozyme.

4.3 Riboswitches

Riboswitches are noncoding elements within messenger RNAs (mRNAs) that directly bind to cellular metabolites and modulate gene expression [6–8]. Riboswitches provide a novel mode of RNA-mediated genetic regulation that requires no auxiliary protein factors. RNA serves as the sole molecular switch that modulates transcription, translation, or RNA processing through conformational changes prompted by direct interaction with a specific cellular metabolite [24–26]. In order to exert control over gene expression, riboswitches must couple the task of ligand recognition with that of modulating a requisite aspect of gene expression. Consequently, riboswitches are generally composed of two interdependent domains that include a natural ligand binding or aptamer domain [27] and an "expression platform" [24–26] whose precise conformation impacts gene expression. These highly structured domains reside in untranslated regions of mostly prokaryotic mRNAs and serve as metabolite-sensitive switches. A stunning array of riboswitches have been characterized that respond to a variety of small molecules. To date, nearly 40 different classes of riboswitches have been discovered that recognize metabolites such as coenzyme B_{12}, flavin mononucleotide (FMN), thiamine pyrophosphate (TPP), *S*-adenosylmethionine (SAM), lysine, glycine, guanine, adenine, Mg^{2+}, fluoride, and glucosamine-6-phosphate (GlcN6P), to name a few (reviewed in [6, 24–26, 28–30]). All the riboswitches identified in prokaryotes have been found in the 5′-untranslated regions (5′-UTRs) of metabolic genes. The single eukaryotic riboswitch identified to date (that binds TPP) has been found in introns within 5′-UTRs and 3′-UTRs [31, 32]. Riboswitches are typically found in the mRNAs of genes that comprise the biosynthetic pathway responsible for producing the cognate metabolite. Therefore, they afford an elegant mechanism for feedback regulation of biosynthetic pathways that allows organisms to respond to the metabolic state of the cell.

In observing the vast array of metabolites recognized by riboswitch RNAs, it is of interest that many of these small molecules/metal ions are derived from RNA components [29]. In contrast, there are just a few riboswitch classes that selectively bind to metabolites that are not directly derived from RNA components. These riboswitch RNAs bind to and control the expression of genes related to the metabolism of amino acids (lysine, glycine, and glutamine), ions (Mg^{2+}, Mn^{2+}, F^-, and Ni^{2+}/Co^{2+}), and genes regulating key metabolic steps in the synthesis of GlcN6P, azaaromatic ligands, and guanidine [29].

4.4 The *glmS* Riboswitch/Ribozyme

The *glmS* riboswitch stands out among the known classes of riboswitches as the only member with intrinsic catalytic capability and one of just a few that binds a metabolite that is not directly derived from RNA components. Similarly, the *glmS* ribozyme stands out among the known classes of natural RNA catalysts as the only ribozyme requiring a coenzyme for activity [33, 34]. The discovery of this RNA by [33] was quite exciting, and studies on the structure and function of the *glmS* riboswitch/ribozyme continue today. This review will highlight the details of this small, self-cleaving, uniquely coenzyme-dependent ribozyme.

The *glmS* riboswitch was first discovered in *Bacillus subtilis*. To classify as a riboswitch, a noncoding RNA must exhibit the following characteristics: (i) the RNA must specifically bind a metabolite, (ii) the RNA must undergo a conformational change induced by metabolite binding, and (iii) the RNA must modulate gene expression in a metabolite-dependent manner [24–26]. In the presence of the amine-containing sugar GlcN6P, the *glmS* riboswitch/ribozyme undergoes self-cleavage, which sets this riboswitch apart from other riboswitch classes that influence gene expression largely through transcription termination or inhibition of translation initiation. The criterion for influencing gene expression is fulfilled by *glmS* riboswitch self-cleavage, after which the associated mRNA is further targeted for degradation [35]. Another indicator of riboswitch function is the modularity of riboswitch elements. The *glmS* riboswitch/ribozyme has been shown to control expression of β-galactosidase production when placed upstream of the reporter gene open reading frame [33].

The *glmS* riboswitch/ribozyme requires a coenzyme for activity [33, 34]. The GlcN6P coenzyme plays a direct role in the catalytic mechanism of RNA self-cleavage [34], whereas other RNA catalysts may require exogenous compounds only as co-substrates or allosteric effectors. A notable example is the group I intron that utilizes exogenous guanosine as a substrate but not a catalyst per se of the splicing mechanism [20]. Both biochemical and biophysical methods have been used to investigate interactions between the *glmS* ribozyme and GlcN6P. These studies have revealed the essential components of the coenzyme and the RNA, and they provide evidence for the mechanism of action (see discussions below).

4.5 Biological Function of the *glmS* Ribozyme

The catalytic *glmS* riboswitch resides in the 5′-UTR of the *glmS* mRNA of numerous Gram-positive bacteria including the antibiotic-resistant superbug *Staphylococcus aureus* (specifically methicillin-resistant *Staphylococcus aureus* [MRSA]) and the opportunistic bacterium *Enterococcus faecalis*, which has intrinsic antibiotic resistance and a remarkable capacity for developing antibiotic resistance [36, 37]. In total, comparative sequence analyses have identified 2126 *glmS* ribozyme representatives, with newer members identified in *Clostridium difficile* [29, 30, 38–41]. Recently, *C. difficile* has surpassed MRSA as the most common hospital-acquired infection [36]. The Centers for Disease Control and Prevention has listed "*C. difficile*" as an "urgent" antimicrobial threat. In addition to those bacteria that have developed antibiotic resistance, a number of other Gram-positive members also pose a threat to human health, such as *Bacillus anthracis*, a bioterror threat; the foodborne pathogens *Bacillus cereus* and *Listeria monocytogenes*; and the causative agent of tetanus disease *Clostridium tetani*.

The *glmS* mRNA is responsible for producing the GlmS enzyme, fructose-6-phosphate amidotransferase (also called GlcN6P synthase), which catalyzes a reaction that produces GlcN6P [42]. The production of this amino sugar is the first committed step in a metabolic pathway that ultimately leads to synthesis of the bacterial cell wall [42]. The *glmS* riboswitch/ribozyme controls the expression of the GlmS enzyme in Gram-positive bacteria, whereas, in Gram-negative bacteria, as well as eukaryotes, GlcN6P synthase is allosterically controlled by the reaction end product, making the *glmS* riboswitch/ribozyme in Gram-positive bacteria a unique control mechanism that is essential to bacterial cell viability [42, 43]. It is because of this feature that there is great interest in understanding the structure and function of the *glmS* riboswitch/ribozyme RNA, as it may be an excellent target for antibiotic drug development.

The *glmS* riboswitch/ribozyme selectively recognizes GlcN6P with a 1 : 1 stoichiometry and is 10^6- to 10^7-fold activated by this metabolic product of the GlmS enzyme [33, 34]. The binding of coenzyme to the *glmS* RNA, with an apparent K_m of ~200 μM, initiates an internal phosphoester transfer reaction that results in self-cleavage and inactivation of the mRNA. RNA inactivation has been shown to involve RNase J1, a homolog of the *Escherichia coli* RNase E protein [35]. Therefore, the *glmS* ribozyme facilitates feedback inhibition of its metabolic pathway. In addition, mutation of the *glmS* ribozyme sequence, resulting in the loss of self-cleavage, causes a loss of sporulation by *B. subtilis* [35].

The role of GlcN6P as a coenzyme for *glmS* ribozyme self-cleavage was first shown biochemically [34] and then biophysically [44–49]. In addition to kinetic analyses and crystallographic studies with GlcN6P and closely related analogs, other compounds (including carbasugars, fluoro-carbasugars, and the phosphatase-inert compounds 6-deoxy-6-phosphonomethyl analog and the 6-*O*-malonyl ether) have been shown to stimulate self-cleavage albeit to a lesser degree [34, 50–55]. These results lead to the interesting question of how various ligands may affect *glmS* self-cleavage in vivo, as there are many different sugars varying in concentration within the cell. To investigate riboswitch function in a biological environment, Fedor and coworker utilized a unique biological system that they had previously developed to study hairpin ribozyme folding in *Saccharomyces cerevisiae* [56]. This work demonstrated that levels of hexose metabolites in yeast can modulate function of the *glmS* ribozyme through either inhibition or activation of self-cleavage [56]. Therefore, the *glmS* riboswitch must integrate chemical signals from a number of sources to fine-tune gene expression to the metabolic needs of the cell. In order to develop drugs that might act as antibiotics by targeting the *glmS* ribozyme, a thorough analysis of all structural requirements of the RNA and ligand must be examined and appreciated.

4.6 *glmS* Ribozyme Structure and Function – Initial Biochemical Analyses

The *glmS* ribozyme contains four paired regions (P1–P4) that fold to form a catalytic core (Figure 4.1) [33]. This ~150 nt RNA is dependent on a doubly pseudoknotted core of ~75 nt that contains both the cleavage site and GlcN6P binding site [44–49]. The nested pseudoknots are a defining feature of the three-dimensional structure of the *glmS* ribozyme, with the P2.1 and P2.2 pseudoknots composing the catalytic core. P1 and P2 contain phylogenetically conserved base pairs that are essential for the formation of the catalytic core, whereas the P3–P4 domains are not required for self-cleavage. However, a region 3′ to the core segments was shown to enhance catalytic activity through additional pseudoknot formation [57]. In addition, the "catalytic core" of the *glmS* ribozyme was investigated by mutational analyses of key nucleotides and examination of highly conserved nucleotide identities across bacterial species. The *glmS* ribozyme was shown to contain an intricate, multistrand core that utilizes pseudoknot formation to correctly fold the RNA for subsequent interaction with its coenzyme [58].

Prior to crystallization of the *glmS* ribozyme, structural data was obtained using nucleotide analog interference mapping (NAIM) and nucleotide analog interference suppression (NAIS) studies (utilizing the *B. cereus* sequence). NAIM and NAIS revealed backbone and nucleobase functional groups essential for ligand-dependent ribozyme function. Using NAIM, the biochemical contribution of a single chemical group within a nucleic acid can be defined using a nucleotide analog that modifies the atom(s) of interest [59]. To provide a general means of detection, the analogs are chemically tagged with a 5′-phosphorothioate, a derivative that replaces the 5′ nonbridging R_p oxygen atom with a sulfur. The phosphorothioate linkage serves as a chemical marker because it is selectively cleaved by iodine [60]. After analog incorporation, functional RNAs are distinguished from nonfunctional by their ability to undergo self-cleavage. If a particular chemical group is essential for function, then the activity will be impaired for the subset of RNA variants that are modified at the essential positions. This result is considered nucleotide analog interference.

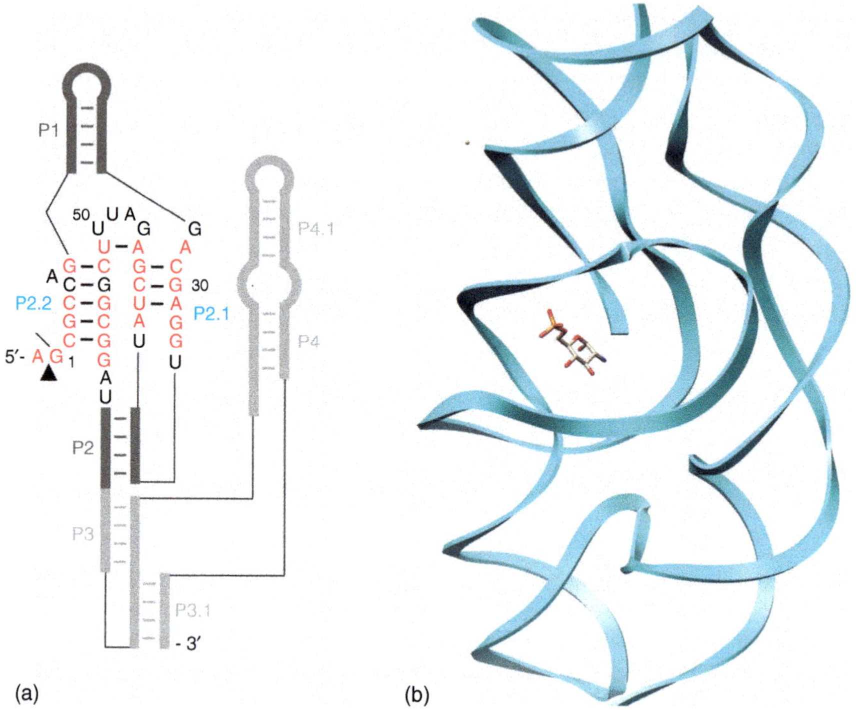

Figure 4.1 Structure of the *glmS* ribozyme. (a) Secondary structure. The core sequence (P2.1 and P2.2) of the *Bacillus cereus glmS* ribozyme is shown with highly conserved nucleotides in red (nucleotide identity 90% conserved) and black (nucleotide identity 75% conserved), while requisite structural elements (P1 and P2) and peripheral structural elements (P3 and P4) are schematically shown in dark gray and light gray, respectively. The arrowhead denotes the site of self-cleavage. (b) Tertiary structure. The ribbon model depicts the structure of the *T. tengcongensis glmS* ribozyme (turquoise) bound to glucosamine-6-phosphate and a substrate RNA with a 2′,5′-phosphodiester linkage determined by X-ray crystallography (PDB ID: 3B4C rendered using chimera, http://www.cgl.ucsf.edu/chimera) [46]. The coenzyme GlcN6P is shown as a stick drawing. Nucleotide identities for the 90% conserved nucleotides in the catalytic core are exactly the same for the *B. cereus* and *T. tengcongensis glmS* ribozymes.

NAIM assays were performed on both the P1–P3 and P1–P4 *glmS* ribozymes in order to determine if the peripheral P4 domain had any effect on essential functional group interactions. In addition, NAIM was performed with the natural ligand GlcN6P and allowed identification of necessary functional groups and potential sites of ligand and metal ion interactions. Experiments performed with phosphorothioate analogs and the thiophilic divalent metal ion manganese identified metal ion binding sites that clustered within the catalytic core. A strong site of interference occurred at residue C2, near the cleavage site of the *glmS* ribozyme, independent of the presence of the P4 domain. However, for the P2.1 region, strong phosphorothioate effects were observed for C29, G30, and A31 (*B. cereus* numbering) for the truncated P1–P3 ribozyme, while weak effects occurred at A28, G30, and A31 for the full-length P1–P4 ribozyme. Additionally, the P1–P3 ribozyme demonstrated weak phosphorothioate effects in P2.2 at G3 and G7, while the P1–P4 ribozyme exhibited strong interference at G3. In both P1–P3 and P1–P4 ribozymes, similar but not identical patterns of phosphorothioate effects reside within the P2.1 and P2.2 segments and helped identify two metal ion binding sites not directly associated with coenzyme binding [61].

Further NAIM studies identified essential 2′-hydroxyl (2′-OH) and nucleobase functionalities. In the presence of Mg^{2+} and GlcN6P, with either the P1–P3 or P1–P4 ribozymes, identical strong 2′-deoxy interferences at A58 and U59 were observed. In P2.1, however, strong 2′-deoxy effects were observed at G27 and A28 for the P1–P3 ribozyme but at C29 and G30 for the P1–P4 ribozyme. NAIM using purine analogs and the full-length P1–P4 ribozyme resulted in interference patterns indicative of contacts to the exocyclic amine of A31, G30, and G57 and to the N7 imine of A28 and G30, whereas steric restraints were implicated for A31 and A58 [61].

Additional studies were performed using a unique variation of the NAIS technique. Previously, NAIS has been employed to identify tertiary interactions within other ribozymes, where analysis is based on the principle that if a tertiary interaction is disrupted by site-specific deletion or alteration of one functional group in an interacting pair, then no additional energetic penalty will result from deletion or alteration of the second functional group. For *glmS* studies, the typical NAIS methodology was modified by combining the use of a ligand analog (serving as a site-specific modification) with interference mapping of the full-length *glmS* RNA to identify sites at which nucleotide analog incorporation that previously caused interference with cognate ligand were tolerated with a ligand analog. NAIS experiments were performed using the ligand analog glucosamine (GlcN), which lacks the 5′-phosphate functionality of GlcN6P but has been previously demonstrated to stimulate P1–P4 *glmS* ribozyme self-cleavage activity at a reduced rate. Thus, GlcN-dependent suppressions of interferences in the *glmS* ribozyme reveal contacts between the phosphate moiety of GlcN6P and functional groups within the RNA catalyst.

Within P2.2, GlcN-dependent suppression of phosphorothioate interference was observed at G3, but not at C2. For the P2.1 segment, suppression of phosphorothioate interference was observed at A28 and G30, but not at A31 [61]. Therefore, NAIS data suggest that three *glmS* ribozyme phosphate oxygens compose a portion of the ligand binding site responsible for recognition of the phosphate moiety of GlcN6P. Furthermore, NAIS performed in the presence of Mg^{2+} and GlcN for the P1–P4 ribozyme demonstrated a site of 2′-deoxy interference suppression at G30, essential imine (A28) and amine (A31) groups, and steric boundaries within the phosphate recognition site (identified by NAIS at A58 and A31 using diaminopurine as nucleotide analog). The composition of the ligand binding site indicates that GlcN6P binds adjacent to the cleavage site within a catalytic core composed of interactions among the highly conserved sequence segments within P2.1 and P2.2 [61].

NAIM and NAIS data aided in visualizing how the ligand was recognized by the RNA and introduced a possible role for ligand in the mechanism of self-cleavage as the ligand binding site was found to lie close to the cleavage site [61]. To further investigate the ribozyme active site, hydroxyl radical footprinting and cross-linking analyses in the absence and presence of GlcN6P were performed. These results showed little structural change in the RNA in response to the ligand, indicating preorganization of the ligand binding pocket [62]. Fluorescence resonance energy transfer (FRET) analyses and footprinting with terbium(III) and RNase V1 also indicated no significant changes in secondary and tertiary structure upon ligand binding [48]. In contrast, NAIM analyses of full-length P1–P4 versus truncated P1–P3 *glmS* RNA identified small changes in essential nucleotide identities within the catalytic core. P4 is thought to stabilize the catalytic core as its presence enhances catalytic activity and mutations that disrupt its tertiary structure can be compensated for by increasing Mg^{2+} concentration [57]. Biochemical analyses of the *glmS* ribozyme indicate that the RNA is structurally adaptable, dependent on the presence of the P3 and P4 domains. In addition, the catalytic core is relatively plastic despite being essential to ligand binding and catalysis [61].

Another important aspect of *glmS* structure and function involves its folding pathway. Ribozymes are very prone to misfolding, and the initial process of RNA folding does seem to be a rate-limiting step for the *glmS* ribozyme as cleavage initiated with Mg^{2+} and GlcN6P is slow ($3\,min^{-1}$) compared with self-cleavage initiated by the addition of GlcN6P to *glmS* RNA pre-folded in the presence of Mg^{2+} ($72\,min^{-1}$) [63]. More recently, *glmS* ribozyme production was optimized to obtain homogeneous, natively folded RNA for detailed kinetic analyses [64]. This *glmS* ribozyme was allowed to pre-fold, which resulted in a monophasic, fast self-cleavage reaction ($k_{obs} = 80\,min^{-1}$) that went to completion (90%), and the RNA was monomeric and natively folded, according to the small-angle X-ray scattering (SAXS) radius of gyration determined, as well as SAXS profile overlays with crystal structure data [64]. To further understand folding of the *glmS* ribozyme, optical force microscopy, using optical tweezers, was performed to measure folding dynamics and cleavage rates over a range of forces and ligand concentrations. These studies utilized a core *glmS* ribozyme (just 66 nucleotides in length), and results indicate that folding of a specific structural element, the P2.2 duplex, controls active site formation and catalysis [65]. This folded state, however, is only weakly stable, regardless of coenzyme concentration. This is in stark contrast to what is observed for nearly all other riboswitches, where the binding energy of the ligand is converted into additional RNA folding stability, which favors the ligand-bound configuration.

These biochemical analyses revealed a great deal about the structure and function of the *glmS* ribozyme prior to crystallography studies. Furthermore, the fact that they were performed in solution may better indicate physiologically important features of ligand binding and catalysis by the *glmS* RNA. Further biochemical and computational studies will be cited below in the context of describing more intricately the roles of metal ions, nucleobases, and the coenzyme GlcN6P in the active site of the *glmS* ribozyme.

4.7 *glmS* Ribozyme Structure and Function – Initial Crystallographic Analysis

Crystal structures of the *glmS* ribozyme have provided atomic level detail of this coenzyme-dependent RNA catalyst. One of the first crystal structures solved was that of the *glmS* pre-cleaved ribozyme in the presence or absence of the competitive inhibitor glucose-6-phosphate (Glc6P). This structure revealed a rigid RNA preformed in a coenzyme binding competent state, similar to that proposed from biochemical assays [44, 47, 62]. The RNA exhibited a double pseudoknot structure that is nearly identical to that of the post-cleavage state structure or to a structure determined with the natural ligand GlcN6P that included a 2′-*O*-methyl modification at the cleavage site preventing catalysis [44–47, 49]. When crystallized with GlcN6P, interactions of the *glmS* RNA with the ligand were found to involve hydrated Mg^{2+} coordinated to the ligand phosphate [45]. Distinct among the binding interactions seen for other riboswitches and their ligands, the *glmS* coenzyme binding site is open and solvent accessible [44]. Crystallographic data has aided in defining contacts between RNA functional groups and the ligand sugar ring [44–47, 49]. As additional crystal structures have been solved, more insight has been gained into the active site and coenzyme binding site. Helix P2.2 and the G1 nucleotide within the active *glmS* ribozyme have been shown to be key RNA features necessary for coenzyme binding and subsequent catalysis [47].

In addition, X-ray crystallography and kinetic experiments have identified a site-specific mutation within the catalytic core that results in a loss of catalysis but not coenzyme binding, indicating that a conserved guanine (G33 in *B. cereus* and G40 in *Thermoanaerobacter tengcongensis*) within the active site plays a role in the catalytic mechanism [46, 49]. Furthermore, crystallographic studies with the competitive inhibitor Glc6P and the nonnatural sugar mannosamine-6-phosphate (MaN6P) revealed similar binding interactions with the RNA as observed for the natural ligand. Glc6P cannot support *glmS* self-cleavage, but MaN6P does at a rate about sevenfold slower than GlcN6P [34, 49]. However, the binding affinity for MaN6P is weak compared with GlcN6P, which is likely an important indication of how the ribozyme selects against ligands that are structurally and chemically similar to GlcN6P [49]. Further crystallographic studies will be cited below in the context of describing more intricately the roles of metal ions, nucleobases, and the coenzyme GlcN6P in the active site of the *glmS* ribozyme.

4.8 Metal Ion Usage by the *glmS* Ribozyme

Metal ions play key roles in RNA structure and function. This is especially true for catalytically active ribozymes. While all ribozymes require metal ions for folding, few utilize essential metal ions in their catalytic mechanisms. Therefore, many ribozymes retain catalytic activity (sometimes at a reduced rate), as long as the folding requirements are fulfilled by monovalent, divalent, or multivalent metal ions. Although physiologically divalent metal ions are probably most relevant to *glmS* structure and function, self-cleavage activity is supported by an exchange-inert Cobalt(III) complex and by molar concentrations of monovalent ions [66, 67]. Cobalt hexamine is a structural mimic of hexahydrated Mg^{2+} and is inert because exchange of inner-sphere amines and protons is slow relative to the self-cleavage reaction, so interaction of cobalt hexamine with RNA can only occur through electrostatic and outer-sphere interactions [68, 69]. *glmS* self-cleavage is supported by cobalt hexamine (at ~1 mM) at a rate only slightly slower than that obtained in the presence of saturating Mg^{2+}, indicating that the ribozyme does not require inner-sphere divalent metal ion coordination for catalysis [66].

Monovalent metal ions can also support *glmS* self-cleavage; however, molar concentrations are required [66, 67]. This provides additional evidence that inner-sphere metal ion coordination is not essential. With saturating GlcN6P (10 mM) and monovalent metal ion at 1 M, the *glmS* ribozyme exhibits greater self-cleavage activity in the presence of Li^+ followed by $NH_4{}^+$ and Na^+. Under these conditions, nominal self-cleavage activity was observed in the presence of K^+, Rb^+, and Cs^+ [67]. In consideration of the physicochemical properties of metal ions, correlations of ribozyme activity are apparent with ionic radius and absolute hardness. Interestingly, Li^+ and Mg^{2+} have near-identical properties that provide for maximal performance of the *glmS* ribozyme (ionic radius of $Li^+ = 0.59$Å and $Mg^{2+} = 0.72$ Å and absolute hardness of $Li^+ = 35.12\eta$ and $Mg^{2+} = 47.59\eta$) [67]. Otherwise, ribozyme reactivity with metal ions appears to correlate with absolute hardness to the point that ionic radius exceeds approximately 1.3–1.5 Å.

Although *glmS* ribozyme self-cleavage can still occur in the presence of monovalent metal ions or exchange-inert cobalt(III) complexes, the rates of these reactions are lower. This suggests that not all roles of divalent metal ions may be compensated for by other positively charged ions. Furthermore, although metal ions do not directly promote *glmS* catalysis [66], it has been shown that metal ion identity and the varying physicochemical properties of metal ions (monovalent or divalent) have an impact on *glmS* ribozyme self-cleavage [67]. Specifically, metal ion identity influenced the overall apparent pK_a of ribozyme self-cleavage, and metal ion binding largely reflected phosphate oxygen affinity [67]. These results suggest that metal ions likely play indirect roles in supporting the mechanism of catalysis. Further studies focused on the rapid ligand binding chemical catalysis step observed for pre-folded *glmS* RNA were in agreement with other published data, indicating a key role for divalent metal ions in coordinating the phosphate of GlcN6P in the ligand binding pocket [34, 44–47, 49, 61, 67]. In addition, GlcN did not promote rapid ligand binding and catalysis in the presence of divalent cations. Therefore, it can be assumed that folding of the ribozyme is relatively independent of cation identity but that rapid initiation of catalysis upon the addition of ligand requires a more precise metal ion presence in order to properly coordinate the ligand phosphate functionality [70].

A recent study looked more generally at the role of Mg^{2+} ions in the active site of the *glmS* ribozyme–GlcN6P coenzyme complex [71]. Computational methods, including classical molecular dynamics and mixed quantum mechanical/molecular mechanical (QM/MM) approaches, were employed to investigate the effects of Mg^{2+} ions in the active site of the *glmS* ribozyme. Although no crystal structures of the *glmS* holoribozyme (with coenzyme bound) or the wild-type aporibozyme (without the coenzyme) have been solved with a Mg^{2+} ion near the cleavage site, these authors pursued this possibility [71]. The simulations utilized the *T. tengcongensis glmS* ribozyme crystal structure in the pre-cleavage state (a deoxy substitution at the A-1 position) with the GlcN6P coenzyme (PDB ID: 2Z75) [47]. Since no Mg^{2+}

ions were resolved in the active site of this crystal structure, in the simulations, one was manually added to the active site at slightly different positions. The presence of a Mg^{2+} ion near the scissile phosphate oxygen atoms resulted in electrostatic repulsion of the coenzyme, disruption of hydrogen bonding interactions, and blocked nucleophilic attack. Positioning of a Mg^{2+} ion near the scissile phosphate was therefore considered anticatalytic. Further experiments placed a Mg^{2+} ion near the Hoogsteen face of G40 (G33 in the *B. cereus* ribozyme). These computations indicated a potentially catalytically favorable result as a Mg^{2+} ion at this position helped to stabilize negative charge and shift the pK_a of the N1 of G40 [71].

Further work by Bevilacqua and coworkers utilized ribozyme kinetic assays and sulfur substitution at the nonbridging oxygen atoms at the cleavage site to investigate thio effects and metal ion rescue in both the *glmS* holoenzyme and apoenzyme. Work on the holoenzyme revealed large thio effects, especially for the R_p thio substrates [64]. Further calculations suggest that the larger thio effect for the R_p thio substrate could be because this phosphate oxygen hydrogen-bonds to the coenzyme GlcN6P. In addition, the large thio effects for the holoenzyme are also observed in 3 M K^+ and were not rescued by the thiophilic metal ion Cd^{2+}, indicating that the holoenzyme does not utilize a divalent metal ion in its active site [64]. For the apoenzyme, in contrast to the holoenzyme, strong metal ion rescue was observed, indicating an important metal ion interaction in the active site of the apoenzyme [64]. These results suggest a model in which a divalent metal ion substitutes for the missing GlcN6P coenzyme in the apoenzyme and fulfills roles that the coenzyme performs in the holoenzyme. The authors propose that the roles of the coenzyme include the release of the nucleophile and charge stabilization of the nonbridging oxygens during self-cleavage [64].

4.9 In Vitro Selected *glmS* Catalyst Loses Coenzyme Dependence

Directly related to the work just described on the role of metal ions in *glmS* ribozyme self-cleavage, in vitro selection was used to select a GlcN6P coenzyme-independent *glmS* ribozyme. Only nucleotides near the active site of the wild-type *T. tengcongensis glmS* ribozyme were mutagenized, and after 10 rounds of selection, three mutations were found in all the sequences resembling the wild-type *glmS* RNA – U49A, U51A, and G65A (this selected RNA is termed *glmS*^{AAA}) [72, 73]. The *glmS*^{AAA} ribozyme is active in Mg^{2+}, with no additional reactivity observed when GlcN6P is present. In addition to reactivity in Mg^{2+}, the *glmS*^{AAA} ribozyme is equally active in Mn^{2+} and ~40-fold more active in Ca^{2+} (although the reactivity of *glmS*^{AAA} in Ca^{2+} is ~30-fold slower than the GlcN6P-dependent activity of the wild-type *glmS* ribozyme) [72]. In contrast to the wild-type *glmS* ribozyme, *glmS*^{AAA} is inactive in molar concentrations of most monovalent metal ions and in cobalt hexamine, the latter indicating a requirement for inner-sphere cation coordination [72]. Subsequent work by the Ferré-D'Amaré group involved further selections to evolve a Ca^{2+}-specific ribozyme that is >10 000 times more active in Ca^{2+} than Mg^{2+}, inactive in 100 mM Mg^{2+}, and is not responsive to GlcN6P [74].

4.10 Essential Coenzyme GlcN6P Functional Groups

Since the initial discovery of the *glmS* ribozyme, much effort has been applied to understanding the role of the natural ligand GlcN6P. One of the first key questions that had to be addressed was whether GlcN6P played a role as an allosteric effector or as a coenzyme for *glmS* self-cleavage. The role of metabolites/ligands that bind riboswitch RNAs is predominantly as allosteric effectors, where their job is to change the shape of the riboswitch RNA in order to affect gene expression. Early key experiments on the *glmS* ribozyme indicated that its ligand, GlcN6P, acts as a coenzyme for *glmS* self-cleavage, as the ligand is integral to catalysis [34]. The fact that the *glmS* RNA does not undergo a dramatic folding change in the presence of the ligand indicated that GlcN6P was not acting as an allosteric effector [44–48, 62]. As support for the role of GlcN6P as a coenzyme, it has been demonstrated that *glmS* self-cleavage both requires and correlates with the acid dissociation constant (pK_a) of the amine functionality of GlcN6P and related compounds [34]. In addition, ligand analogs that lack the amine functionality cannot support *glmS* self-cleavage. One such ligand, Glc6P, acts as a competitive inhibitor at high concentration, binding to the *glmS* RNA but failing to support self-cleavage [34, 51]. These initial results illustrated the necessity of the ligand amine functionality to self-cleavage and an expanded capacity for biological RNA catalysis through the use of a coenzyme [34]. In vitro selection attempted to identify *glmS* ribozyme variants with an expanded capacity to recognize other ligand analogs. Although variants that could support self-cleavage at a reduced rate were identified, all required the natural ligand, further supporting an essential role for GlcN6P as a coenzyme [75].

In determining that the ligand amine functionality was essential for *glmS* ribozyme activity, a variety of ligand analogs indicated that other functional groups aid in enhancing coenzyme binding and catalysis. Kinetic analyses with ligand analogs indicated the necessity of at least one hydroxyl group adjacent to the amine (Figure 4.2) [34, 44–46, 49–51, 54]. It appeared that the 4-hydroxyl (4-OH) group served as a hydrogen bond donor and that the 1-OH group may be involved in molecular recognition by RNA [34, 44–47, 49–51]. Other ligand analogs that open the sugar ring cannot support *glmS* self-cleavage. Interestingly, the *glmS* ribozyme from *S. aureus* underwent self-cleavage in the presence of a carbasugar analog of GlcN6P with activity similar to that of the natural ligand, indicating that the ring oxygen does not substantially contribute to ligand binding or RNA self-cleavage [50, 55]. Finally, the *glmS* ribozyme from *B. anthracis* binds with selectivity to the α-anomer of GlcN6P, and binding is pH-dependent [76]. These results and previous work [34, 50, 51, 76] indicate the importance of stereospecificity and protonation state in ligand binding to the RNA. Atomic resolution of ligand–RNA interactions from crystallographic studies provides additional insight into the multiple hydrogen bonding interactions within the *glmS* ribozyme–coenzyme active site, which

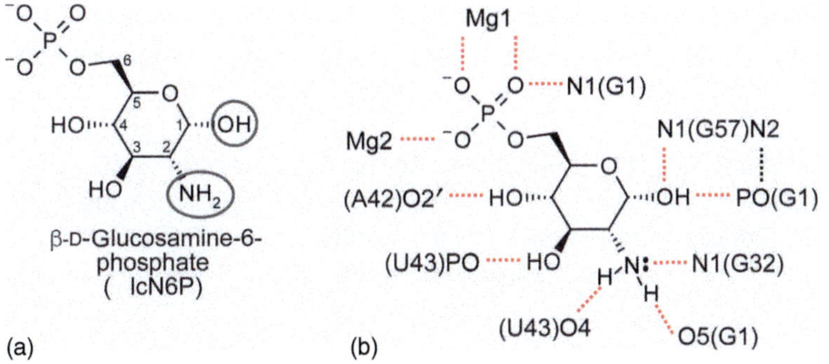

Figure 4.2 GlcN6P coenzyme and recognition by the *glmS* ribozyme. (a) Structure of GlcN6P with the requisite hydroxyl and amine groups indicated (gray circle). (b) GlcN6P recognition by the *glmS* ribozyme. Depicted are contacts (red dotted lines) to each of two divalent metal ions (Mg1 and Mg2) and to nucleotide functional groups within the core of the *glmS* ribozyme.

involve functional groups on the RNA (G1(5′-O), G1(pro-R_p), A42(2′-O), U43(O4), U43(pro-R_p), G57(N2), and G57(N1)) and on the coenzyme (N2 and O1 positions) [44–47, 49].

An important feature of the natural ligand GlcN6P is the phosphate group, which increases the pK_a of the amine functionality in solution. Although GlcN can stimulate ribozyme self-cleavage, the reaction rate and apparent affinity is substantially reduced, reflecting the contribution of phosphate interactions with the RNA and metal ions upon binding [34]. Furthermore, the solution pK_a for GlcN is essentially equal to the apparent pK_a for *glmS* ribozyme self-cleavage with GlcN or GlcN6P [34]. These data suggest that the ligand phosphate influence on pK_a is quenched upon interaction with the ribozyme and substantially shifted toward neutrality. In addition, glucosamine-6-sulfate (GlcN6S) can support ribozyme self-cleavage to the same extent as GlcN6P, albeit at concentrations at least 100-fold greater [33]. Finally, a series of nine GlcN6P analogs with phosphatase-inert linkages were synthesized, and two of these phosphonate mimics (6-deoxy-6-phosphonomethyl analog and the 6-*O*-malonyl ether) were able to support *glmS* ribozyme self-cleavage at rates of ~1 min^{-1} [54]. These results provide strong support for the notion that the phosphate is critical to GlcN6P positioning in the riboswitch "active site." This is seen in the phosphonate series, in which only the 6-deoxy-6-phosphonomethyl analog possessing a single methylene (CH_2) unit in place of the bridging phosphate oxygen is effective. Deletion of this methylene or insertion of an additional methylene all but abolishes this *glmS* self-cleavage activity [54]. In addition to this apparent positioning constraint, a dianionic end group also appears to be advantageous. This can be seen in the carboxylate mimic series, wherein a monocarboxylate is nearly inactive but the dicarboxylate (6-*O*-malonyl ether) supports *glmS* self-cleavage [54]. These studies indicate that the phosphate group plays a role in recognition of the ligand by the RNA and affects protonation state of the coenzyme relevant to catalysis [34, 44–46, 49–51, 61, 67, 76–81].

Further synthesis and functional analysis of ligand analogs will elucidate the intricacies of ligand binding and the role of the coenzyme in *glmS* self-cleavage. Although many specific interactions occur between the *glmS* RNA and GlcN6P allowing for discrimination against closely related compounds, there appears to remain some flexibility for the design of novel ligand analogs that either support or inhibit *glmS* self-cleavage. Such compounds might work as agonists or antagonists to perturb expression of the GlmS enzyme and amino sugar metabolism and therefore may be promising leads for future antibiotic drug development.

4.11 Mechanism of *glmS* Ribozyme Self-Cleavage

4.11.1 Importance of Coenzyme GlcN6P

The role of GlcN6P as a coenzyme has been described and intensely studied; however, its exact function in the mechanism of self-cleavage is still under investigation. Ribozymes can use multiple catalytic strategies to accomplish efficient and fast rates of self-cleavage. The Breaker Lab has described four catalytic strategies: in-line nucleophilic attack, deprotonation of the 2′-OH nucleophile, protonation of the 5′-OH atom, and stabilization of charge on the nonbridging oxygen atoms [82, 83]. Recently, the Bevilacqua Lab added two more strategies to this list: acidification of the 2′-OH and the release of the 2′-OH nucleophile from inhibitory interactions [64]. In considering these six strategies, the *glmS* ribozyme could utilize a number of these with the GlcN6P coenzyme intricately involved.

The *glmS* ribozyme self-cleavage reaction proceeds using an acid–base mechanism. A number of hypotheses regarding the identity of the general acid and base functional groups for *glmS* self-cleavage have come from different biochemical and biophysical analyses. The pH dependence of *glmS* self-cleavage implies that GlcN6P acts as the general acid and/or base for the reaction. Specifically, it has been proposed that the coenzyme amine acts as a general acid, protonating the 5′ oxygen leaving group of G1 during *glmS* ribozyme self-cleavage [44, 45]. Molecular dynamics simulations have also predicted that GlcN6P acts as a general acid based on results that upon binding of GlcN6P to *glmS* ribozyme, the pK_a of the amine functionality decreased by 2.2 pH units from the solution pK_a of 8.2 due to the active site environment [78]. In contrast, the pK_a of the amine group slightly increased to about 8.4 upon binding to a G40A inactive mutant of the *glmS* ribozyme (*T. tengcongensis* ribozyme numbering corresponding to G33 in the *B. cereus* ribozyme) [78]. The hypothesis that the coenzyme amine acts as a general acid predicts that the N1 of G40/G33 functions as the general base. Although there is some support for this hypothesis, there is much conflicting evidence.

Subsequent to early work on the catalytic role(s) played by the coenzyme GlcN6P, recent studies have proposed even more responsibilities to the coenzyme. Work described above that centered on identifying the role of metal ions in the *glmS* active site also focused on realizing and understanding the large number of catalytic roles portrayed by GlcN6P. Experiments utilized both the *glmS* holoenzyme and apoenzyme RNAs and measured thio effects and metal ion rescue [64]. Large stereospecific normal thio effects and a lack of metal ion rescue in the holoribozyme posit that nucleobases and the coenzyme play direct chemical roles in catalysis and aid in aligning the active site for self-cleavage. In contrast, experiments with the aporibozyme revealed large stereospecific inverse thio effects, suggesting that GlcN6P disrupts an inhibitory interaction involving the nucleophile. The aporibozyme also exhibited strong metal ion rescue pointing to a role for the coenzyme in electrostatic stabilization. Simulations utilizing classical and QM/MM calculations support the thio and metal ion effects observed experimentally. Taking all this data into account, the authors propose that the coenzyme GlcN6P plays multiple catalytic roles in the *glmS* holoenzyme, including (i) donating a proton to the leaving group as the general acid, (ii) aligning the active site, (iii) disrupting an inhibitory hydrogen bond involving the nucleophile (so activating the 2′-OH), and (iv) stabilizing the charge development during the self-cleavage reaction [64].

In thinking about the added potential roles for the coenzyme, the Bevilacqua group continued their thio effect experiments; however, they utilized *glmS* ribozyme variants (both *glmS* holoribozyme and aporibozyme constructs that contained a G57A mutation) and a GlcN6P analog (1-deoxy-GlcN6P). Both the N2 of G57 and the O1 of the coenzyme are known to be involved in critical hydrogen bonding interactions in the *glmS* ribozyme–coenzyme active site, so using the G57A mutant removes one of these hydrogen bonding partners, while the 1-deoxy-GlcN6P removes another. Thio effects were investigated with a two-piece *glmS* ribozyme construct in order to use different substrates (oxo at the scissile phosphate, R_p thio, S_p thio, or a dithio substrate), different ribozymes (G57A holoenzyme, wild-type holoenzyme, or G57A apoenzyme), and different coenzymes (GlcN6P or 1-deoxy-GlcN6P) [84]. Because previous results indicate that the *glmS* holoribozyme does not bind divalent metal ions at the nonbridging oxygen atoms of the scissile phosphate, any thio effects observed in these assays should be due to disruption of hydrogen binding within the active site. Kinetic assays with different combinations of substrate, ribozyme, and coenzyme resulted in the loss of differing numbers of hydrogen bond donors, which allowed for investigation of how the 2′-OH nucleophile is activated [84]. The authors proposed that the wild-type (oxo) holoribozyme is activated by competitive hydrogen bonding for the *pro-*R_p nonbridging oxygen between three hydrogen bond donors: GlcN6P(O1), GlcN6P(N2), and G57(N2). The presence of simultaneous hydrogen bonds between the *pro-*R_p oxygen and any two of these three donors efficiently prohibits hydrogen bond formation with the 2′-OH nucleophile, releasing it for proton abstraction by the general base. It seems that this excessive network of hydrogen bonds is essential for potent activation of the 2′-OH nucleophile, as any of the single variants tested leads to strong inhibition of the oxo substrate [84]. In experiments with the G57A+1-deoxy-GlcN6P double variant, where two of the three potential hydrogen bond donors at the *pro-*R_p oxygen are missing, an inverse thio effect is observed at the *pro-*R_p position. In this construct, one *pro-*R_p oxygen lone pair is left to hydrogen bond with the 2′-OH nucleophile. This work reported the first inverse thio effect for the *glmS* ribozyme in the presence of a coenzyme, although the G57A mutant *glmS* RNA was utilized [84]. Deletion of all three potential hydrogen bond donors involving the *pro-*R_p oxygen in the G57A aporibozyme also leads to an inverse thio effect at the *pro-*R_p position due to the release of both *pro-*R_p oxygen lone pairs. This allows for the inhibitory 2′-OH nucleophile to interact with the *pro-*R_p nonbridging oxygen and supports the notion that the *glmS* ribozyme uses an overdetermined set of competing hydrogen bond donors in its active site to ensure potent activation of the 2′-OH nucleophile and regulation of catalysis by the coenzyme [84]. This work provides a great example of the varied feats performed by the coenzyme GlcN6P.

4.11.2 pH-Reactivity Profiles

A mechanistic interpretation of the pH-reactivity profile for the *glmS* ribozyme has been previously proffered and provides insight into how the catalyst might function [85, 86]. Similar types of analyses have been performed on other catalytic RNAs, such as the HDV ribozyme [85–87]. Briefly, the *glmS* ribozyme exhibits a pH-reactivity profile with an apparent pK_a above which a slope of zero may reflect the inverse relationship of the catalyst's general acid and general base contributions (Figure 4.3). Considering all the work that has been done collecting pH-reactivity profiles of the *glmS* ribozyme, similar results indicate an apparent reaction pK_a of 6.9–7.8 relative to the pK_a of 8.2 for GlcN6P in solution [34, 48, 76, 79]. In considering these results and the hypothesis that GlcN6P functions primarily as a general acid catalyst for *glmS* ribozyme self-cleavage following its solution pK_a of 8.2, there exists a requirement for a relatively strong general base catalyst with a pK_a of ~10 to achieve the observed kinetic profile (Figure 4.3a).

An alternative interpretation of the *glmS* ribozyme pH-reactivity profile proposes that GlcN6P functions as the coenzyme by playing roles as both a general base and acid catalyst (Figure 4.3b). In this model, the kinetic profile of the ribozyme reflects the contributions of GlcN6P operating at its liganded pK_a [34]. This model is supported by studies that indicate that the *glmS* ribozyme can tune the catalytically critical pK_a of its coenzyme amine group. Raman crystallography directly measured the pK_a of GlcN6P coenzyme binding to RNA, where the pK_a of the amine was lowered to 7.26 and the pK_a of the coenzyme phosphate was raised to 6.35 from its solution pK_a of 5.98 [79]. Therefore, the *glmS* ribozyme appears to aid coenzyme function by tuning the pK_a of the amine group to allow for optimal acid–base catalysis.

In addition to the fact that the *glmS* ribozyme exhibits an apparent reaction pK_a that simply and precisely corresponds to the pK_a of the coenzyme's amine functionality, the pH-reactivity profile provides further information about mechanism [34]. The shape of the pH reactivity curve suggests that the coenzyme acts first as a general base and then as a general acid (Figure 4.3b). In other words, maximal activity is provided by deprotonated coenzyme, whose function as a general base leads to the formation of protonated coenzyme that can then protonate the 5′ leaving group. Although it has been argued that the apparent role of GlcN6P as a general acid catalyst prevents the coenzyme from acting as a general base catalyst [45], the rationale

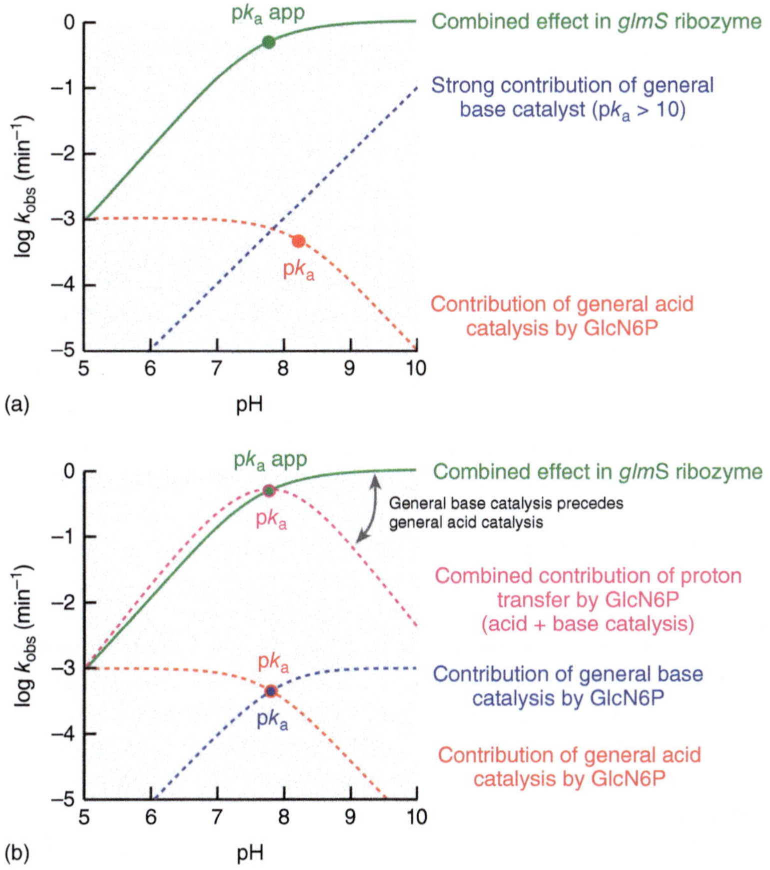

Figure 4.3 Interpretation of *glmS* ribozyme pH-reactivity profile. (a) By analogy to the acid–base chemistry of the HDV ribozyme, an approximation of the actual pH-reactivity profile for the *glmS* ribozyme (green) would require that the contribution of GlcN6P functioning solely as a general acid catalyst following its solution pK_a (red) be combined with the contribution of a general base catalyst with a pK_a of ~10 (blue). (b) An alternative interpretation of the *glmS* ribozyme pH-reactivity profile. An approximation of the actual pH-reactivity profile for the *glmS* ribozyme (green) might represent the combined contributions of both general base and general acid catalysis by GlcN6P (magenta), where if general base catalysis precedes and yields general acid catalysis, then the reactivity profile is not diminished above the apparent pK_a of GlcN6P.

presented here makes clear the necessity that both general base and acid catalysis are somehow inherently interdependent.

A recent study further supports this hypothesis [81]. GlcN6P analogs were assayed for their ability to support self-cleavage, and a strong correlation was found between the pH dependence of self-cleavage and the intrinsic acidity of the GlcN6P analogs. The analogs with low binding affinity exhibited rate enhancements that were proportional to their intrinsic acidity. This linear free energy relationship between coenzyme efficiency and acid dissociation constant supports a mechanism wherein the coenzyme is directly involved as a general acid–base catalyst. In this same study, the ligand analogs differed in their Brønsted acid–base strengths, and therefore, a Brønsted plot measured the sensitivity of the self-cleavage reaction to the acid–base strength of the catalyst. A high value for the Brønsted coefficient was reported ($\beta \sim 0.7$), indicating that the transition state involves a significant amount of proton transfer [81].

4.11.3 Role of an Active Site Guanine

While it is clear that the coenzyme GlcN6P plays a multitude of catalytic roles in *glmS* self-cleavage and there is good evidence that a Mg^{2+} ion near the active site helps stabilize negative charge and shift the pK_a of an active site guanine [71], it is still unclear what role an active site guanine may play. Crystallographic studies proposed that the N1 of G33 (G33 in the *B. cereus* ribozyme and G40 in *T. tengcongensis*) functions as the general base due to its position adjacent to the 2′-OH of A-1 at the scissile phosphodiester linkage [44, 45], and the pK_a of this nucleobase functionality is consistent with this model. However, a number of observations contradict this relatively simple interpretation. Primarily, the model predicts that such an independently functioning general base catalyst would support measurable activity in the absence of GlcN6P as a general acid catalyst at neutral or basic pH. However, the ribozyme is inactive in the absence of coenzyme [34]. In addition, the apparent pK_a of the ribozyme reaction does not reflect a pH-averaged value consistent with that of the proposed general acid and general base catalysts. The model additionally predicts that mutation of G33 to adenosine with a nucleobase pK_a of 3.9 should support catalysis at near-neutral pH, where the functional form of both the general acid and general base would be predominant. However, the G33A mutation has been demonstrated to completely inactivate the *glmS* ribozyme [45–47]. The *glmS* ribozyme therefore defies simple mechanistic explanation based on independently functioning general acid and base catalysts.

In order to investigate the role of the active site guanine, QM/MM free energy simulations and pK_a calculations were performed [88]. These analyses differ from other widely used methods, such as molecular dynamics simulations using classical force fields or traditional QM/MM geometry optimizations, as the widely used methods have limitations on the data they can provide. Utilizing QM/MM free energy simulations that combine umbrella sampling and a finite temperature string method, the authors of this study were able to propose a role for the active site guanine. Their calculations suggest that an external base deprotonates either G40/G33(N1) or possibly the A-1(O2′) [88]. In one of the proposed models presented, the self-cleavage reaction is initiated by deprotonation of G40/G33(N1). G40/G33(N1), rather than the A-1(O2′), is deprotonated because the pK_a of A-1(O2′) is much higher than that of G40/G33(N1) in the initial state, probably due to the hydrogen bonding interactions between A-1(O2′) and the *pro-*R_p oxygen. The self-cleavage reaction proceeds with nucleophilic attack by A-1(O2′) on the scissile phosphate, proton transfer from A-1(O2′) to G40/G33(N1), and ultimately, cleavage of the RNA backbone between A-1 and G1 is complete. This model is supported by the QM/MM free energy simulations as well as experimental data [88].

Following from the work discussed above, it is proposed that an "external base" deprotonates the N1 of G40/G33. If we assume that the coenzyme is the external base, the question remains how the general base catalysis initiated by GlcN6P at a site distal to the 2′-OH of A-1 can ultimately activate a better-positioned general base catalyst such as G40/G33. Closer examination of the proximity of functional groups within the active site of the *glmS* ribozyme reveals a plausible mechanism of proton transfer between the coenzyme's amine functionality and the N1 of G33, as the ultimate general base catalyst. While a proton relay was originally proposed to involve bound water molecules [44], which are not observed in other crystal forms of the *glmS* ribozyme [45–47], it has subsequently been proposed [61, 86] that active site nucleotide functional groups, proven to be important for catalysis by NAIM, support a scheme for proton transfer. The coenzyme's amine group is specifically modeled to initiate a proton relay through the intervening N1 of G32 (Figure 4.2b), which contacts a nonbridging phosphate oxygen at the scissile phosphodiester linkage. In this way, the N1 of G33 and the coenzyme's amine are simultaneously and necessarily activated to respectively serve as the ultimate general base for deprotonation of the 2′ OH of A-1 and the general acid for protonation of the 5′ oxygen leaving group of G1. Further support comes from studies that examined the pH dependence of self-cleavage by wild-type and mutant *glmS* ribozymes and reported that the apparent pK_a values did not support a role for G33 or any other active site guanine in general base catalysis [80]. Furthermore, using pH-fluorescence profiles with ribozymes containing a fluorescent guanosine analog, 8-azaguanosine, at position 33, the authors observed that the pH-dependent step in catalysis did not involve G33 deprotonation [80]. These results are further strengthened by molecular dynamics experiments that determined that G40/G33 plays a structural stabilization role within the active site of the *glmS* ribozyme but not a direct chemical role [77]. This points to an alternative role for G40/G33, specifically for the protonated N1 position, in proton transfer and/or transition state stabilization [80].

One final supporting data point is that the nucleobase identity at G32 is strictly conserved in *glmS* ribozymes [29, 30, 41, 75], and preliminary results indicate that mutation at this site results in substantial loss of self-cleavage activity (J. Soukup, unpublished). Therefore, the proposed comprehensive model appropriately predicts that any perturbation in the chain of events in the proton relay is equally detrimental to ribozyme activity (i.e. general base and acid catalysis are inherently interdependent). Therefore, the proposed mechanism is consistent with the entirety of available biochemical and biophysical data.

4.12 Potential for Antibiotic Development Affecting *glmS* Ribozyme/Riboswitch Function

Biochemical and biophysical analyses have provided considerable information regarding the unique coenzyme-dependent catalysis of the *glmS* ribozyme. Coenzyme functional groups required for binding and the coenzyme amine functionality and pK_a perturbation are all essential aspects that impact optimal *glmS* ribozyme self-cleavage activity. Although the requirements for efficient ligand binding and catalysis are many, there appear to remain opportunities for development of compounds that can support ribozyme self-cleavage and riboswitch function. In bacteria, the riboswitch must be able to discriminate against metabolites related to its natural ligand in order to tightly regulate metabolic pathways. However, it has been demonstrated that ligand analogs can substitute for the natural metabolite, affording the opportunity to artificially modulate *glmS* ribozyme/riboswitch activity and metabolic gene expression, and ultimately inhibit bacterial growth [89–91]. Therefore, continued investigation of both *glmS* ribozyme and riboswitch structure and function may identify metabolite analogs that can act as prospective antimicrobial agents.

Acknowledgments

This publication was made possible by grants from the National Institute for General Medical Science (NIGMS) (5P20GM103427 and R15M083641), a component of the National Institutes of Health (NIH) and its contents are the sole responsibility of the authors and do not necessarily represent the official views of NIGMS or NIH.

References

1 Cech, T.R., Zaug, A.J., and Grabowski, P.J. (1981). *In vitro* splicing of the ribosomal RNA precursor of *Tetrahymena*: involvement of a guanosine nucleotide in the excision of the intervening sequence. *Cell* 27 (3 Pt 2): 487–496.
2 Kruger, K., Grabowski, P.J., Zaug, A.J. et al. (1982). Self-splicing RNA: autoexcision and autocyclization of the ribosomal RNA intervening sequence of *Tetrahymena*. *Cell* 31: 147–157.
3 Guerrier-Takada, C., Gardiner, K., Marsh, T. et al. (1983). The RNA moiety of ribonuclease P is the catalytic subunit of the enzyme. *Cell* 35: 849–857.
4 Breaker, R.R. (1999). Catalytic DNA: in training and seeking employment. *Nat. Biotechnol.* 17: 422–423.
5 Li, Y. and Breaker, R.R. (1999). Deoxyribozymes: new players in the ancient game of biocatalysis. *Curr. Opin. Struct. Biol.* 9: 315–323.
6 Roth, A. and Breaker, R.R. (2009). The structural and functional diversity of metabolite-binding riboswitches. *Annu. Rev. Biochem.* 78: 305–334.
7 Dambach, M.D. and Winkler, W.C. (2009). Expanding roles for metabolite-sensing regulatory RNAs. *Curr. Opin. Microbiol.* 12: 161–169.
8 Henkin, T.M. (2008). Riboswitch RNAs: using RNA to sense cellular metabolism. *Genes Dev.* 22: 3383–3390.
9 Moore, P.B. and Steitz, T.A. (2003). The structural basis of large ribosomal subunit function. *Annu. Rev. Biochem.* 72: 813–850.
10 Nissen, P., Hansen, J., Ban, N. et al. (2000). The structural basis of ribosome activity in peptide bond synthesis. *Science* 289: 920–930.

11 Newman, A. (2001). Molecular biology: RNA enzymes for RNA splicing. *Nature* 413: 695–696.

12 Doudna, J.A. and Cech, T.R. (2002). The chemical repertoire of natural ribozymes. *Nature* 418: 222–228.

13 Lehmann, K. and Schmidt, U. (2003). Group II introns: structure and catalytic versatility of large natural ribozymes. *Crit. Rev. Biochem. Mol. Biol.* 38: 249–303.

14 Pyle, A.M. and Lambowitz, A.M. (2006). Group II introns: ribozymes that splice RNA and invade DNA. In: *The RNA World* (eds. R.F. Gesteland, T.R. Cech and J.F. Atkins), 469–506. Cold Spring Harbor, New York: Cold Spring Harbor Laboratory Press.

15 Frank, D.N. and Pace, N.R. (1998). Ribonuclease P: unity and diversity in a tRNA processing ribozyme. *Annu. Rev. Biochem.* 67: 153–180.

16 Lilley, D.M.J. (2003). The origins of RNA catalysis in ribozymes. *Trends Biochem. Sci.* 28: 495–501.

17 Bevilacqua, P.C., Brown, T.S., Nakano, S., and Yajima, R. (2004). Catalytic roles for proton transfer and protonation in ribozymes. *Biopolymers* 73: 90–109.

18 Jencks, W.P. (1969). *Catalysis in Chemistry and Enzymology*. New York: Dover Publications.

19 Steitz, T.A. and Steitz, J.A. (1993). A general two-metal-ion mechanism for catalytic RNA. *Proc. Natl. Acad. Sci. U.S.A.* 90: 6498–6502.

20 Cech, T.R. (1990). Self-splicing of group I introns. *Annu. Rev. Biochem.* 59: 543–568.

21 Toor, N., Keating, K.S., Taylor, S.D., and Pyle, A.M. (2008). Crystal structure of a self-spliced group II intron. *Science* 320: 77–82.

22 Toor, N., Keating, K.S., and Pyle, A.M. (2009). Structural insights into RNA splicing. *Curr. Opin. Struct. Biol.* 19: 260–266.

23 Lee, E.R., Baker, J.L., Weinberg, Z. et al. (2010). An allosteric self-splicing ribozyme triggered by a bacterial second messenger. *Science* 329: 845–848.

24 Nudler, E. and Mironov, A.S. (2004). The riboswitch control of bacterial metabolism. *Trends Biochem. Sci.* 29: 11–17.

25 Winkler, W.C. and Breaker, R.R. (2005). Regulation of bacterial gene expression by riboswitches. *Annu. Rev. Microbiol.* 59: 487–517.

26 Winkler, W.C. and Breaker, R.R. (2003). Genetic control by metabolite-binding riboswitches. *ChemBioChem* 4: 1024–1032.

27 Gold, L., Polisky, B., Uhlenbeck, O., and Yarus, M. (1995). Diversity of oligonucleotide functions. *Annu. Rev. Biochem.* 64: 763–797.

28 Breaker, R.R. (2009). Riboswitches: from ancient gene-control systems to modern drug targets. *Future Microbiol.* 4: 771–773.

29 McCown, P.J., Corbino, K.A., Stav, S. et al. (2017). Riboswitch diversity and distribution. *RNA* 23 (7): 995–1011.

30 McCown, P.J., Roth, A., and Breaker, R.R. (2011). An expanded collection and refined consensus model of *glmS* ribozymes. *RNA* 17 (4): 728–736.

31 Sudarsan, N., Barrick, J.E., and Breaker, R.R. (2003). Metabolite-binding RNA domains are present in the genes of eukaryotes. *RNA* 9: 644–647.

32 Kubodera, T., Watanabe, M., Yoshiuchi, K. et al. (2003). Thiamine-regulated gene expression of *Aspergillus oryzae thiA* requires splicing of the intron containing a riboswitch-like domain in the 5′-UTR. *FEBS Lett.* 555: 516–520.

33 Winkler, W.C., Nahvi, A., Roth, A. et al. (2004). Control of gene expression by a natural metabolite-responsive ribozyme. *Nature* 428 (6980): 281–286.

34 McCarthy, T.J., Plog, M.A., Floy, S.A. et al. (2006). Ligand requirements for *glmS* ribozyme self-cleavage. *Chem. Biol.* 12 (11): 1221–1226.

35 Collins, J.A., Irnov, I., Baker, S., and Winkler, W.C. (2007). Mechanism of mRNA destabilization by the *glmS* ribozyme. *Genes Dev.* 21 (24): 3356–3368.

36 Miller, B.A., Chen, L.F., Sexton, D.J., and Anderson, D.J. (2011). Comparison of the burdens of hospital onset, healthcare facility associated *Clostridium difficile* infection and of healthcare-associated infection due to methicillin-resistant *Staphylococcus aureus* in community hospitals. *Infect. Control Hosp. Epidemiol.* 32: 387–390.

37 Paulsen, I.T., Banerjei, L., Myers, G.S. et al. (2003). Role of mobile DNA in the evolution of vancomycin-resistant *Enterococcus faecalis*. *Science* 299 (5615): 2071–2074.

38 McCown, P.J., Winkler, W.C., and Breaker, R.R. (2012). Mechanism and distribution of *glmS* ribozymes. *Methods Mol. Biol.* 848: 113–129.

39 Griffiths-Jones, S., Moxon, S., Marshall, M. et al. (2005). Rfam: annotating noncoding RNAs in complete genomes. *Nucleic Acids Res.* 33 (database issue): D121–D124.

40 Griffiths-Jones, S., Bateman, A., Marshall, M. et al. (2003). Rfam: an RNA family database. *Nucleic Acids Res.* 31 (1): 439–441.

41 Barrick, J.E., Corbino, K.A., Winkler, W.C. et al. (2004). New RNA motifs suggest an expanded scope for riboswitches in bacterial genetic control. *Proc. Natl. Acad. Sci. U.S.A.* 101 (17): 6421–6426.

42 Milewski, S. (2002). Glucosamine-6-phosphate synthase: the multi-facets enzyme. *Biochim. Biophys. Acta* 1597 (2): 173–192.

43 Badet-Denisot, M.-A., Rene, L., and Badet, B. (1993). Mechanistic investigations on glucosamine-6-phosphate synthase. *Bull. Soc. Chim. Fr.* 130: 249–255.

44 Klein, D.J. and Ferré-D'Amaré, A.R. (2006). Structural basis of *glmS* ribozyme activation by glucosamine-6-phosphate. *Science* 313 (5794): 1752–1756.

45 Cochrane, J.C., Lipchock, S.V., and Strobel, S.A. (2007). Structural investigation of the *glmS* ribozyme bound to its catalytic cofactor. *Chem. Biol.* 14 (1): 97–105.

46 Klein, D.J., Been, M.D., and Ferré-D'Amaré, A.R. (2007). Essential role of an active-site guanine in *glmS* ribozyme catalysis. *J. Am. Chem. Soc.* 129 (48): 14858–14859.

47 Klein, D.J., Wilkinson, S.R., Been, M.D., and Ferré-D'Amaré, A.R. (2007). Requirement of helix P2.2 and nucleotide G1 for positioning the cleavage site and cofactor of the *glmS* ribozyme. *J. Mol. Biol.* 373 (1): 178–189.

48 Tinsley, R.A., Furchak, J.R., and Walter, N.G. (2007). Trans-acting *glmS* catalytic riboswitch: locked and loaded. *RNA* 13 (4): 468–477.

49 Cochrane, J.C., Lipchock, S.V., Smith, K.D., and Strobel, S.A. (2009). Structural and chemical basis for glucosamine-6-phosphate binding and activation of the *glmS* ribozyme. *Biochemistry* 48 (15): 3239–3246.

50 Lünse, C.E., Schmidt, M.S., Wittmann, V., and Mayer, G. (2011). Carba-sugars activate the *glmS*-riboswitch of *Staphylococcus aureus*. *ACS Chem. Biol.* 6 (7): 675–678.

51 Lim, J., Grove, B.C., Roth, A., and Breaker, R.R. (2006). Characteristics of ligand recognition by a *glmS* self-cleaving ribozyme. *Angew. Chem. Int. Ed.* 45 (40): 6689–6693.

52 Cochrane, J.C. and Strobel, S.A. (2008). Riboswitch effectors as protein enzyme cofactors. *RNA* 14 (6): 993–1002.

53 Wang, G.N., Lau, P.S., Li, Y.F., and Ye, X.S. (2012). Synthesis and evaluation of glucosamine-6-phosphate analogues as activators of *glmS* riboswitch. *Tetrahedron* 68: 9405–9412.

54 Fei, X., Holmes, T., Diddle, J. et al. (2014). Phosphatase-inert glucosamine 6-phosphate mimics serve as actuators of the *glmS* riboswitch. *ACS Chem. Biol.* 9: 2875–2882.

55 Matzner, D., Schüller, A., Seitz, T. et al. (2017). Fluoro-carba-sugars are glycomimetic activators of the *glmS* ribozyme. *Chemistry* 23: 12604–12612.

56 Watson, P.Y. and Fedor, M.J. (2011). The *glmS* riboswitch integrates signals from activating and inhibitory metabolites *in vivo*. *Nat. Struct. Mol. Biol.* 18 (3): 359–363.

57 Wilkinson, S.R. and Been, M.D. (2005). A pseudoknot in the 3′ non-core region of the *glmS* ribozyme enhances self-cleavage activity. *RNA* 11 (12): 1788–1794.

58 Soukup, G.A. (2006). Core requirements for *glmS* ribozyme self-cleavage reveal a putative pseudoknot structure. *Nucleic Acids Res.* 34 (3): 968–975.

59 Strobel, S. and Shetty, K. (1997). Defining the chemical groups essential for *Tetrahymena* group I intron function by nucleotide analog interference mapping. *Proc. Natl. Acad. Sci. U.S.A.* 94: 2903–2908.

60 Gish, G. and Eckstein, F. (1988). DNA and RNA sequence determination based on phosphorothioate chemistry. *Science* 240: 1520–1522.

61 Jansen, J.A., McCarthy, T.J., Soukup, G.A., and Soukup, J.K. (2006). Backbone and nucleobase contacts to glucosamine-6-phosphate in the *glmS* ribozyme. *Nat. Struct. Mol. Biol.* 13 (6): 517–523.

62 Hampel, K.J. and Tinsley, M.M. (2006). Evidence for preorganization of the *glmS* ribozyme ligand binding pocket. *Biochemistry* 45 (25): 7861–7871.

63 Brooks, K.M. and Hampel, K.J. (2009). A rate-limiting conformational step in the catalytic pathway of the *glmS* ribozyme. *Biochemistry* 48 (24): 5669–5678.

64 Bingaman, J.L., Zhang, S., Stevens, D.R. et al. (2017). The GlcN6P cofactor plays multiple catalytic roles in the *glmS* ribozyme. *Nat. Chem. Biol.* 13: 439–445.

65 Savinov, A. and Block, S.M. (2018). Self-cleavage of the *glmS* ribozyme core is controlled by a fragile folding element. *Proc. Natl. Acad. Sci. U.S.A.* 115: 11976–11981.

66 Roth, A., Nahvi, A., Lee, M. et al. (2006). Characteristics of the *glmS* ribozyme suggest only structural roles for divalent metal ions. *RNA* 12 (4): 607–619.

67 Klawuhn, K., Jansen, J.A., Souchek, J. et al. (2010). Analysis of metal ion dependence in *glmS* ribozyme self-cleavage and coenzyme binding. *ChemBioChem* 11 (18): 2567–2571.

68 Jou, R. and Cowan, J.A. (1991). Ribonuclease H activation by inert transition-metal complexes. Mechanistic probes for metallocofactors: insights on the metallobiochemistry of divalent magnesium ion. *J. Am. Chem. Soc.* 113 (17): 6685–6686.

69 Cowan, J.A. (1993). Metallobiochemistry of RNA. $Co(NH_3)_6^{3+}$ as a probe for Mg^{2+}(*aq*) binding sites. *J. Inorg. Biochem.* 49 (3): 171–175.

70 Brooks, K.M. and Hampel, K.J. (2011). Rapid steps in the *glmS* ribozyme catalytic pathway: cation and ligand requirements. *Biochemistry* 50 (13): 2424–2433.

71 Zhang, S., Stevens, D.R., Goyal, P. et al. (2016). Assessing the potential effects of active site Mg^{2+} ions in the *glmS* ribozyme–cofactor complex. *J. Phys. Chem. Lett.* 7: 3984–3988.

72 Lau, M.W. and Ferré-D'Amaré, A.R. (2013). An *in vitro* evolved *glmS* ribozyme has the wild-type fold but loses coenzyme dependence. *Nat. Chem. Biol.* 9: 805–810.

73 Lau, M.W. and Ferré-D'Amaré, A.R. (2016). *In vitro* evolution of coenzyme-independent variants from the *glmS* ribozyme structural scaffold. *Methods* 106: 76–81.

74 Lau, M.W., Trachman, R.J. 3rd, and Ferré-D'Amaré, A.R. (2017). A divalent cation-dependent variant of the *glmS* ribozyme with stringent Ca^{2+} selectivity co-opts a preexisting nonspecific metal ion-binding site. *RNA* 23: 355–364.

75 Link, K.H., Guo, L., and Breaker, R.R. (2006). Examination of the structural and functional versatility of *glmS* ribozymes by using *in vitro* selection. *Nucleic Acids Res.* 34 (17): 4968–4975.

76 Davis, J.H., Dunican, B.F., and Strobel, S.A. (2011). *glmS* Riboswitch binding to the glucosamine-6-phosphate α-anomer shifts the pK_a toward neutrality. *Biochemistry* 50 (33): 7236–7242.

77 Banás, P., Walter, N.G., Sponer, J., and Otyepka, M. (2010). Protonation states of the key active site residues and structural dynamics of the *glmS* riboswitch as revealed by molecular dynamics. *J. Phys. Chem. B* 114 (26): 8701–8712.

78 Xin, Y. and Hamelberg, D. (2010). Deciphering the role of glucosamine-6-phosphate in the riboswitch action of *glmS* ribozyme. *RNA* 16 (12): 2455–2463.

79 Gong, B., Klein, D.J., Ferré-D'Amaré, A.R., and Carey, P.R. (2011). The *glmS* ribozyme tunes the catalytically critical pK_a of its coenzyme glucosamine-6-phosphate. *J. Am. Chem. Soc.* 133 (36): 14188–14191.

80 Viladoms, J., Scott, L.G., and Fedor, M.J. (2011). An active-site guanine participates in *glmS* ribozyme catalysis in its protonated state. *J. Am. Chem. Soc.* 133 (45): 18388–18396.

81 Viladoms, J. and Fedor, M.J. (2012). The *glmS* ribozyme cofactor is a general acid-base catalyst. *J. Am. Chem. Soc.* 134 (46): 19043–19049.

82 Breaker, R.R., Emilsson, G.M., Lazarev, D. et al. (2003). A common speed limit for RNA-cleaving ribozymes and deoxyribozymes. *RNA* 9: 949–957.

83 Emilsson, G.M., Nakamura, S., Roth, A., and Breaker, R.R. (2003). Ribozyme speed limits. *RNA* 9: 907–918.

84 Bingaman, J.L., Gonzalez, I.Y., Wang, B., and Bevilacqua, P.C. (2017). Activation of the *glmS* ribozyme nucleophile via overdetermined hydrogen bonding. *Biochemistry* 56: 4313–4317.

85 Soukup, G.A. and Soukup, J.K. (2009). Structure and mechanism of the *glmS* ribozyme. In: *Non-Protein Coding RNAs – Springer Series in Biophysics* (eds. N. Walter, S.A. Woodson and R.T. Batey), 129–143. Berlin, Germany: Springer.

86 Soukup, J.K. (2013). The structural and functional uniqueness of the *glmS* ribozyme. In: *Catalytic RNA, Vol 120, Progress in Molecular Biology and Translational Science* (ed. G.A. Soukup), 173–194. London, UK: Academic Press.

87 Bevilacqua, P.C. (2003). Mechanistic considerations for general acid-base catalysis by RNA: revisiting the mechanism for the hairpin ribozyme. *Biochemistry* 42 (8): 2259–2265.

88 Zhang, S., Ganguly, A., Goyal, P. et al. (2015). Role of the active site guanine in the *glmS* ribozyme self-cleavage mechanism: quantum mechanical/molecular mechanical free energy simulations. *J. Am. Chem. Soc.* 137: 784–798.

89 Sudarsan, N., Wickiser, J.K., Nakamura, S. et al. (2003). An mRNA structure in bacteria that controls gene expression by binding lysine. *Genes Dev.* 17 (21): 2688–2697.

90 Sudarsan, N., Cohen-Chalamish, S., Nakamura, S. et al. (2005). Thiamine pyrophosphate riboswitches are targets for the antimicrobial compound pyrithiamine. *Chem. Biol.* 12 (12): 1325–1335.

91 Schüller, A., Matzner, D., Lünse, C.E. et al. (2017). Activation of the *glmS* ribozyme confers bacterial growth inhibition. *ChemBioChem* 18: 435–440.

5

The Lariat Capping Ribozyme

Henrik Nielsen[1], Nicolai Krogh[1], Benoît Masquida[2], and Steinar Daae Johansen[3]

[1] *University of Copenhagen, Department of Cellular and Molecular Medicine, 3 Blegdamsvej, Copenhagen, Denmark*
[2] *CNRS-Université de Strasbourg, UMR 7156, Génétique Moléculaire Génomique Microbiologie, 21 rue René Descartes, Strasbourg, France*
[3] *NORD University, Genomics group, 11 Universitetsalléen, Bodø, Norway*

5.1 Introduction

The lariat capping ribozyme (LCrz) constitutes an independent class of ribozymes. LCrz is the only known natural ribozyme that specifically modifies the 5′end of an RNA species. It shows structural resemblance to the group I intron splicing ribozymes (GrIrz), and a plausible model for its evolutionary origin can be put forward. The function of LCrz in mRNA capping represents a remarkable example of molecular adaptation. This chapter provides an update of a previous book chapter and reviews on LCrz [1–5] and includes biochemical and structural aspects, discovery of new variants, and a high-resolution structure from X-ray crystallography. Furthermore, a section on application of LCrz as a research tool is included.

5.1.1 The Basics

The LCrz is an approximately 180 nt ribozyme that catalyzes formation of an unusual mRNA cap in which the first and third nucleotides are joined by a 2′, 5′phosphodiester bond – in other words, a 3 nt lariat at the 5′end of the mRNA. It is also referred to as "the branching ribozyme," because the reaction leading to the formation of the lariat cap (LC) involves a nucleophilic attack of the 2′-hydroxyl (2′-OH) of an internal residue at a nearby phosphodiester bond, yielding a branched nucleotide. As a result, the RNA strand in which the LCrz is embedded is cleaved, leaving a 5′fragment with a 3′OH and a 3′fragment with the 3 nt lariat at its 5′end (Figure 5.1a).

LCrzs have striking structural resemblance to group I intron splicing ribozymes (GrIrz) and have only been found in complex "twin-ribozyme" group I introns inserted into nuclear rDNA. Twin-ribozyme introns harbor a GrIrz, the LCrz, and a

Ribozymes, First Edition. Edited by Sabine Müller, Benoît Masquida, and Wade Winkler.

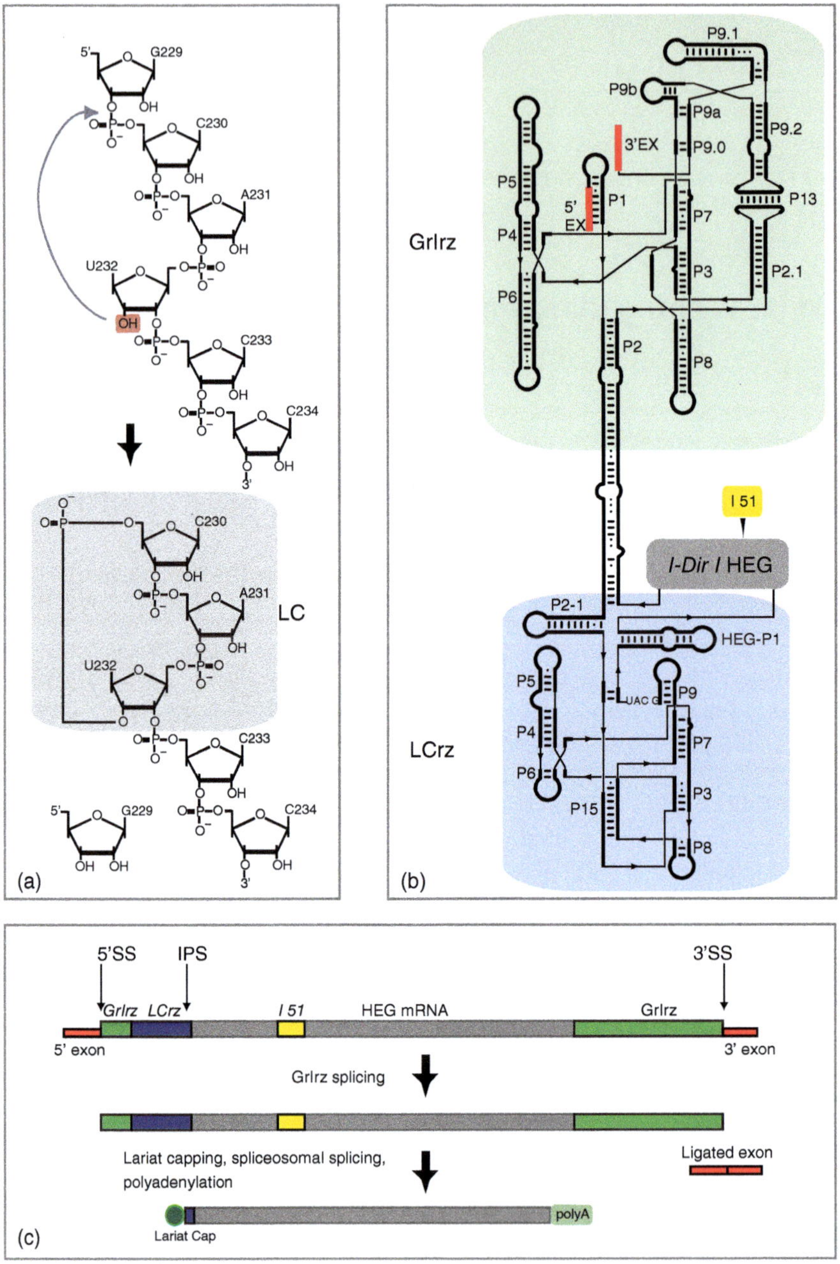

Figure 5.1 (a) Branching reaction catalyzed by LCrz. The 2′ OH of the internal residue U232 (in the *Didymium* version of LCrz) makes a nucleophilic attack at the phosphodiester at C230. This results in cleavage at the internal intron processing site (IPS) liberating a 5′ fragment with a 3′ OH and a 3′ fragment with a 5′ lariat in which the first and third nucleotides of the sequence 5′ CAU are linked by a 2′,5′phoshodiester bond. In the natural system, the lariat caps the intron-encoded homing endonuclease mRNA. (b) Base-pairing diagram of the twin-ribozyme intron from *Didymium iridis*. The lariat capping ribozyme and the homing endonuclease gene are inserted into P2 of the group I splicing ribozyme. The three functional units are recognized as separate domains in the structure. (c) Expression of the homing endonuclease mRNA occurs by GrIrz self-splicing followed by lariat capping by LCrz, removal of a spliceosomal intron (I51), and polyadenylation.

gene encoding a homing endonuclease (HEG; Figure 5.1b). The GrIrz is responsible for catalyzing the steps in intron splicing and circularization, whereas the LCrz is specialized to facilitate the expression of the intron-encoded protein by mRNA capping. The homing endonuclease (HE) facilitates intron mobility by catalyzing DNA cleavage of intron-less alleles [6, 7]. However, the expression of the protein from within an intron in a gene transcribed by RNA polymerase (RNAP) I does not provide the mRNA with the usual features that promote export and translation. It is believed that lariat capping functionally substitutes for the m^7G cap. Furthermore, the presence of a spliceosomal intron within the HEG and conventional polyA signals constitutes a remarkable adaptation that brings the HEG mRNA onto the normal RNA pol II expression pathway, resulting in polysome-associated HEG mRNA (Figure 5.1c) [8]. So far the distribution of LCrz is limited to eukaryotic microorganisms. These comprise the myxomycete *Didymium iridis*, several species of the amoeboflagellate *Naegleria*, and a few *Allovahlkampfia* species. Notably, these unicellular organisms have in common that they harbor nuclear group I introns and express their proteins from capped mRNAs.

5.1.2 A Brief Account of the Discovery of the Lariat Capping Ribozyme

The initial discovery of the LCrz was made during characterization of a large group I intron in the small subunit (SSU) ribosomal RNA (rRNA) of the myxomycete *D. iridis* [9, 10]. Using splicing assays of a number of deletion constructs, it was found that two ribozyme domains were located within the intron and that both could be folded into group I-like ribozyme structures. Accordingly, they were named GIR1 (group I-like ribozyme 1) and GIR2. GIR2 was shown to be a conventional splicing ribozyme belonging to the IE subgroup, and GIR1 was shown to be a cleavage ribozyme acting at an intron internal site, named IPS (intron processing site). Surprisingly, primer extension analyses revealed two stop signals three nucleotides apart that were labeled IPS1 and IPS2, and for a long time it was believed that cleavage occurred by two consecutive hydrolytic cleavage events [11]. Subsequently, a similar (albeit not identical) organization of a group I intron was found in several species of the amoeboflagellate *Naegleria* [12]. After the basic processing scheme was mapped, the GIR1 ribozyme from *Naegleria* was biochemically characterized and optimized through SELEX experiments in the Cech lab as a prelude to crystallization [13, 14]. As described later, the delimitation of the *Naegleria* ribozyme had the unfortunate consequence that the ribozyme was described and selected as a hydrolytic cleavage ribozyme that had little bearing on the in vivo ribozyme. In characterization of the *Didymium* ribozyme, more emphasis was made on defining borders of the ribozyme that would preserve its true catalytic activity [15]. This resulted in the realization that one of the primer extension stops was a stop at a branched nucleotide rather than a runoff product. This observation implied only one cleavage site (IPS; former IPS1) and provided the key to unravel the reaction mechanism [15], which was later supported by structural studies using X-ray crystallography [16]. In the course of elucidating the true catalytic activity and function of the ribozyme, it was renamed "the lariat capping ribozyme."

5.1.3 Readers Guide to Nomenclature

The description of the LCrz in this chapter requires a certain amount of specific nomenclature as well as nomenclature from group I intron research. Group I introns in rRNA are named according to the system described by Johansen and Haugen [17]. The name of the host species and the insertion site within rRNA (using *Escherichia coli* numbering) are reflected in the descriptor. Thus, *Dir*.S956-1 means the first group I intron described at position 956 in the SSU rRNA from *D. iridis*. The LCrz is found inserted in different subgroups of group I introns (IC1 or IE), and here, the classification follows the system of Michel and Westhof [18] based on differences in the structure of peripheral elements. Whenever the description of LCrz is specific for an individual variant of the ribozyme, the species name is incorporated into the name, e.g. *Dir*LCrz for the *D. iridis* ribozyme. Conversely, the generic "LCrz" is used for description of general features. Numbering of nucleotides in LCrz is according to their position within the twin-ribozyme context. However, when studies of length variants of the ribozyme are described out of the natural context, the cleavage site (IPS) is used as a fix point to describe how much was included. The minimal variant of *Dir*LCrz begins at pos. 73 within the twin-ribozyme intron and ends at pos. 251. This variant includes 157 nt upstream and 22 nt downstream of the IPS, respectively. It is thus described as *Dir*LCrz-157.22.

5.1.4 The Species Involved

Only one *D. iridis* isolate (Panama 2; Pan2) has been reported to harbor a mobile twin-ribozyme intron with a functional LCrz among the *Didymium* myxomycetes [6, 9, 10], and extensive searches for nuclear group I introns in other myxomycetes have not turned up new examples [19]. *D. iridis* preys on microorganisms and dead organic matter on the forest floor. The complex life cycle comprises haploid amoebae, flagellates, cysts, diploid amoebae, and a syncytial plasmodium that can differentiate into sporangia with haploid spores [20, 21]. During vegetative growth, the organisms undergo frequent cycles of encystment and excystment depending on humidity. This affects the expression of rRNA, including the activity of the LCrz [22].

The second find of LCrz was in several species of the amoeboflagellate *Naegleria* [12]. *Nae*LCrz resides in a different subgroup (IC1) of group I introns inserted into a different site (S516) compared with that of *Dir*LCrz (Table 5.1). In contrast to the *Didymium* intron that likely was recently acquired by horizontal transfer [4], the *Naegleria* intron appears to have been gained early and inherited in a strictly vertical fashion [23]. The intron is found in 29/78 of the known isolates suggesting frequent losses [4, 24]. Only haploid life forms have been observed in *Naegleria*, indicating that transfer of the intron by homing is infrequent or non-existing. According to the Goddard and Burt model [25], HEG-containing introns undergo cycles of gain (by homing) and loss (by sequence drift and deletion) of the intron. We infer that homing in *Naegleria* must occur in a setting different from sexual reorganization or the selection pressure on the HE must be for an alternative function of the protein.

The most recent finding was that of an LCrz in *Allovahlkampfia* [26]. This variant was identified by database searches. The sequence was derived from biological

Table 5.1 Comparison of currently known LCrz.

	***Dir*LCrz**	***Nae*LCrz**	***Asp*LCrz**
Group I intron			
Host intron, type	IE	IC1	IC1
Insertion site	S956	S516	S516
Inheritance	Horizontal	Vertical	—
LCrz			
Insertion of LCrz	P2	P6	P6
Full size (estimated)	approx. 180 nt	197–216 nt	approx. 200 nt
Optimal size (in vitro)	157.22	180.18 (*Npr*LCrz)	169.28
Branching rate (maximum)	0.085 min^{-1}	0.3 min^{-1}	0.08 min^{-1}

material that was no longer available, and thus, *Asp*LCrz was synthesized based on sequence information only [24]. The assignment to the *Allovahlkampfia* genus was based on the flanking SSU rRNA sequence that placed the sequence as a sister species to the acrasid slime molds, clearly distinct from the free-living *Naegleria* amoeboflagellates [26]. Both the splicing ribozyme and the HEG have many substitutions compared with the *Naegleria* consensus, and the LCrz contains 33 positions that deviate from the corresponding invariant positions in *Nae*LCrz. Similar to observations in *Naegleria* [19], the twin-ribozyme intron is optional among *Allovahlkampfia* species and strains [27]. The sequence-based structures of the known LCrz variants suggest that there are two main types only: the "*Didymium* type" and the "*Naegleria* type" comprising all species of *Naegleria* and *Allovahlkampfia*.

5.2 Reactions Catalyzed by LCrz

Most natural ribozymes are found as an integral part of larger RNA molecules that are impractical to study. A first step is to identify the relevant borders that allow the ribozyme to be studied in isolation. From inspection of the secondary structure diagram of the twin-ribozyme intron from *D. iridis* depicted in Figure 5.1b, this appears less complex since LCrz is inserted at the tip of the P2 helix and clearly constitutes a separate structural domain. However, this is deceptive. The key thing to consider is instead that LCrz is a cleavage ribozyme located within the pre-rRNA transcript. Untimely cleavage of the rRNA precursor would be detrimental to the host. Thus, LCrz is unlikely to fold directly into its active conformation represented by the secondary structure diagram. Rather, it folds into an inactive conformation that becomes activated only when splicing of the intron has occurred. This coordination of LCrz activity with other processing events in the twin-ribozyme intron is dealt with in Section 5.4. In this section, the focus is on the activities of the isolated LCrz.

5.2.1 The Branching Reaction

The branching (or lariat capping) reaction is the natural reaction of LCrz and the only observed reaction in cellular RNA. The reaction is a transesterification at the IPS and results in cleavage of the RNA, leaving a 5′ fragment with a 3′ OH and a 3′ fragment with a tiny lariat at the 5′ end. The lariat is made by joining of the first and third nucleotides by a 2′, 5′ phosphodiester bond. In *Dir*LCrz, the branching reaction is initiated by a nucleophilic attack of the 2′ OH of U232 at the phosphate of C230 (Figure 5.1a). This was originally demonstrated in a trans-cleavage experiment using a core ribozyme that was truncated at A222 in L9 (7 nt upstream of the IPS; Figure 5.2) and a substrate carrying the 7 missing nucleotides upstream of IPS followed by 22 nt downstream of the IPS (i.e. 7.22). Using individual deoxynucleotide substitutions of the critical nucleotides C230–C233, it was shown that only substitution of the 2′ OH of U232 completely prevented the branching reaction. The lariat structure was deduced by classical enzymatic degradation and thin-layer chromatography [15]. This view of the reaction mechanism was later supported by X-ray crystallography [16]. The minimal version of the *Didymium* ribozyme that carry out the branching reaction in vitro is *Dir*LCrz-157.22. The in vitro branching rate for this variant is 0.085 min^{-1} (Table 5.1), which is only 1 order of magnitude less than the cleavage rates of most optimized minimal cleavage ribozymes (e.g. 0.2–0.5 min^{-1} for the hairpin [29], 1.0 min^{-1} for the Varkud satellite [30], and 0.5–2.0 min^{-1} for the hammerhead ribozyme [31]). The *Naegleria*-type ribozymes generally conform to what has been studied for *Dir*LCrz in terms of core structure and cleavage kinetics, but detailed analyses of the mechanism and structure of the active site have not been performed.

5.2.2 Ligation and Hydrolysis

In vitro experiments have revealed two additional reactions catalyzed by LCrz, reversal of the branching reaction (here referred to as ligation) and cleavage at the IPS by hydrolysis. There is no evidence that these reactions occur in vivo. Ligation is very efficient in vitro. In *Dir*LCrz constructs that include more than 166 nt upstream of the IPS, ligation completely masks the branching reaction. In this situation, the reaction instead appears as hydrolytic cleavage at IPS. The reason for this is that the ribozyme undergoes repeated cycles of branching and ligation with occasional cleavages by hydrolysis that accumulate because the hydrolytic cleavage cannot be reversed. The hydrolytic cleavage reaction also occurs independently of branching in constructs where the nucleophilic 2′ OH of U232 is not properly presented at the

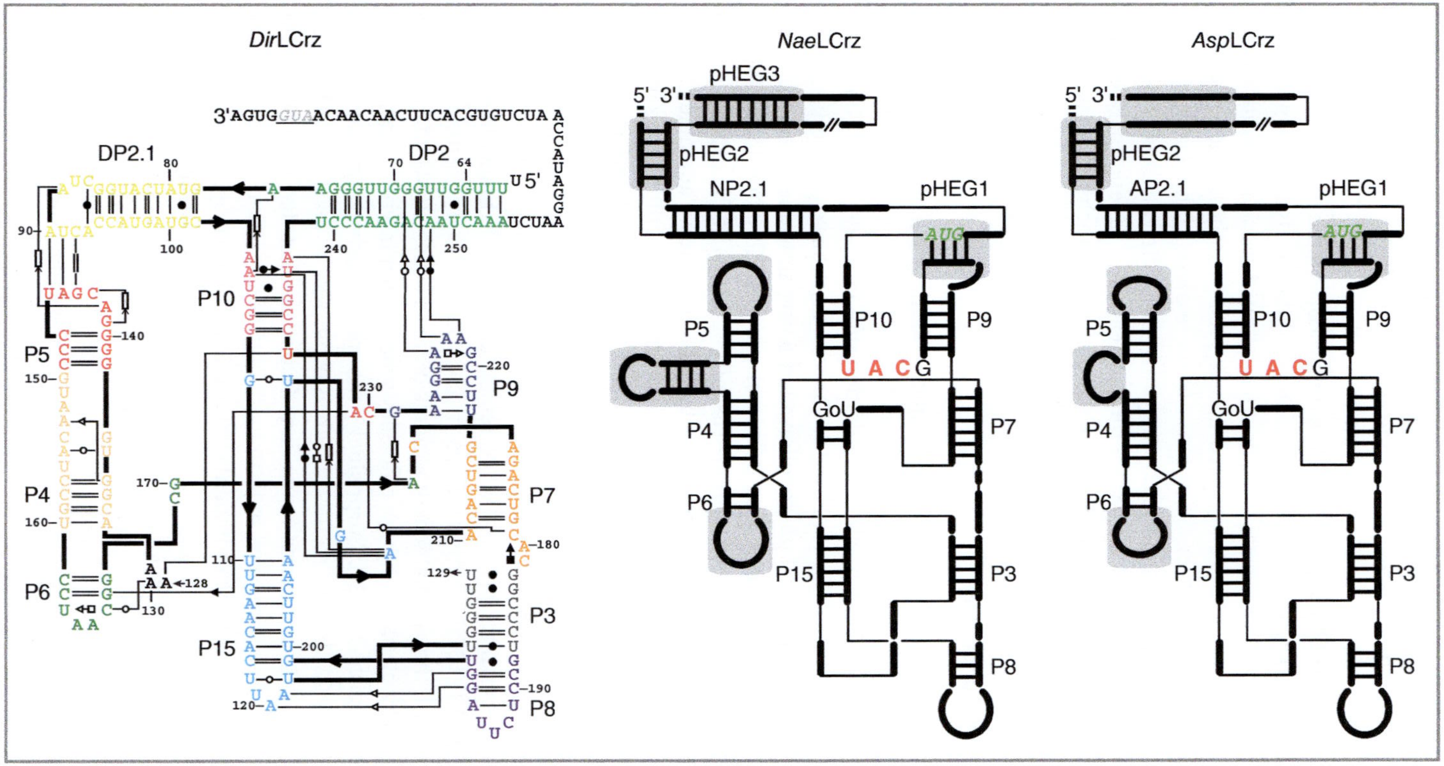

Figure 5.2 Secondary structure diagrams of the three known LCrz. *Dir*LCrz is shown in details with base-pairing symbols according to the geometric classification system [28] and color of stems similar to 3D structures shown in Figure 5.3. The structure is adapted from the circularly permuted form used in X-ray crystallography. The initiation codon in the downstream HEG encoding ORF is in green lettering and underlined. *Nae*LCrz and *Asp*LCrz are shown as schematics to emphasize the similarities and the different organization of the flanking elements compared with *Dir*LCrz. The lariat capping sequence 5′-CAU and the HEG initiation codon 5′-AUG are in red and green lettering, respectively.

active site. Hydrolysis is much slower than branching and is particularly observed in constructs that are shortened too much downstream of the IPS and in many mutants.

The three different types of reaction can be experimentally separated and studied individually. The branching reaction is isolated from ligation by addition of 2 M urea to the reaction [15, 32]. The slight denaturation may assist in the release of products or inhibit their reassociation such that ligation is inhibited. Hydrolysis is negligible at these conditions. Conversely, ligation can be studied at acidic conditions (e.g. pH = 5.5) at which branching is inhibited. Finally, the hydrolysis reaction can by studied in 3′ shortened constructs or mutants. In the analysis of the reaction, cleavage by branching is easily distinguished from hydrolysis by primer extension analysis that yield stops three nucleotides apart for the two reactions. Kinetic analyses have yielded estimated rates for *Dir*LCrz-166.22 of 0.085 min^{-1} (branching), 1 min^{-1} (ligation), and 0.01 min^{-1} (hydrolysis), respectively [16].

5.2.3 Reaction Conditions

Kinetic cleavage analysis of LCrz usually involves a folding step for 10 min at 45 °C in 25 mM $MgCl_2$ and 1 M KCl at acidic conditions (10 mM acetate, pH = 5.5), followed by initiation of the reaction by addition of a cleavage buffer containing the same salt and HEPES-KOH at pH = 7.5. Variants that cleave predominantly by branching display a rapid initial phase followed by a very slow phase and a fraction (usually <10%) of the molecules that do not cleave, probably due to misfolding that is not resolved within the time frame (usually 120 min) of the experiment. In contrast, variants that cleave predominantly by hydrolysis cleave at a much slower rate and frequently go to completion. For natural variants, vast differences in cleavage rate among 11 representatives of *Nae*LCrz probably reflect differences in vitro only [24]. The most efficient of these variants, *Npr*LCrz-191.28, shows pronounced cleavage by branching during in vitro transcription, i.e. at conditions that are much closer to physiological conditions than the standard conditions used for cleavage assays. A shorter variant, *Npr*LCrz-191.18, retains considerable branching activity, but in *Npr*LCrz-191.15, the activity drops to 1/40. Interestingly, ligation has not been observed with the optimal length variants of this ribozyme [24]. Many length variants of *Dir*LCrz cleave with high efficiency when tested in vivo in various host organisms (unpublished, [33]). Surprisingly, the model variant for branching in vitro (*Dir*LCrz-157.22) is very inefficient when expressed in yeast [33]. However, *Dir*LCrz-166.22 catalyzes efficient branching, despite the fact that branching with this variant is completely masked in vitro by the ligation reaction [15, 32]. The in vitro reaction conditions were thoroughly studied in early work by Jabri et al. using *Nan*LCrz-178.19 [13]. Unfortunately, this variant cleaves exclusively by hydrolysis only, and subsequent optimizations by SELEX experiments [14] may have reflected a more loose organization of the active site. Systematic studies of *Dir*LCrz reaction conditions have not been performed, but conditions appear in general to be the same as found by Jabri et al. [13]. However, the mutations that relieved the *Nan*LCrz of high salt requirement for hydrolytic cleavage did not have a similar effect in the *Dir*LCrz context [4].

5.3 The Structure of the LCrz Core

The sources of structural information of LCrz comprise phylogenetic analyses, chemical and enzymatic structure probing, extensive mutation studies, and, in the case of *Dir*LCrz, small-angle X-ray scattering and X-ray crystallography. LCrz can be divided into a core that carries out branching in isolation and regulatory domains or sequence elements that are integrated into the flanking regions communicating with the core.

5.3.1 The Detailed Structure of *Dir*LCrz

The similarity of LCrz to GrIrz was noted from its discovery. GrIrz can be represented (Figure 5.2) as a bundle of three helical stacks, the P4–P6 "scaffolding domain," the P3–P9 "catalytic domain," and the P1–P2 "substrate domain." Specifically, LCrz resembles a GrIrz at the second step of splicing where the substrate domain consists of a P10–P1–P2 stack. In LCrz, the distinctive difference is the replacement of P2 with the pseudoknot P15, which can be viewed as a pairing between the 5′ strand of a P2 and the 3′ strand of P8 and J8/7. The original model was supported by mutational studies and structure probing [11, 34].

An important step forward was revision of the original model of LCrz through molecular modeling inspired by GrIrz crystal structures, including that of the *Azoarcus* $tRNA^{Ile}$ intron at the second step of splicing [35, 36] to which *Dir*LCrz is mostly similar [37]. Several important structural changes were made to the model and confirmed by mutational analyses (Figure 5.2). First, it was realized that P15 should be extended by three base pairs with a GoU pair at the top. Compared with GrIrz, this could be derived from a P1 having lost its 5′ exon such that the residues from the internal guide sequence could base pair with J8/7. Viewed in this context, two A residues that in GrIrz are involved in positioning of the substrate domain are now participating directly in this domain through base pairing. Second, in the GoU pair, U207 was demonstrated as the pairing partner of G109, rather than the originally proposed U232. Third, a critical role of J5/4, and particularly A153 was demonstrated. In the revised model, residues that are critical for the branching reaction lie within a pocket formed at the interface of P10, P15, P7, and J5/4. All the distinctive features of LCrz concentrate in this pocket and result in a topology very different from that observed in GrIrz crystal structure models. The structure of the essential J8/7 segment of GrIrz is dramatically changed and has been partly replaced by residues belonging to the LCrz lariat fold J9/10. Other key features are the detachment of the nucleophile from the GoU pair at the catalytic site and a structural alteration of the GoU pair receptor. In contrast, the G-binding pocket in P7 was conserved from GrIrz.

These and other observations constituted the starting point for crystallizing *Dir*LCrz [16]. The wild-type unreacted form was solved to 3.85 Å resolution, and an inactive form that was circularly permutated over the scissile bond was solved at 2.5 Å. The two structures show similar folds and were confirmed to have very similar shapes in solution by small-angle X-ray scattering experiments [16]. The structures confirmed the highly constrained double pseudoknotted (P15–P3–P7) core. On top of the three principal domains, the structure provided details on the *Didymium*-specific regulatory domain P2–P2.1. These included the organization of the three-way junction that connects the domain to P10 and nucleates the folding of the active site by acting as a receptor for A209. The motion of the loop containing A209 is allowed through the conformation of the GoU base pair, which adopts a trans conformation, instead of a cis conformation in GrIrz, leading A209 in the adequate direction. Two important tertiary interactions position P2.1 over L5 and P2 over L9, respectively. The former forms a kissing complex with three perfectly matched Watson–Crick base pairs, while the latter results from the interaction of three adenines from L9 with the shallow groove of P2. These interactions tether the core and forms an overall characteristic signet ring shape of LCrz in which the P2.1 and P4–P6 domains align perpendicular to the core (Figure 5.3a). The position of the P4–P6 element is one of several examples of how a structural element has been conserved from GrIrz at the secondary structure level yet has adopted a completely new structure and function.

In the circularly permutated form of the ribozyme, the nucleophile U232 (O2′) is positioned 5.1 Å from the C230 (O5′) where cleavage takes place. G229, which is homologous to ωG from GrIrz, is not bound to its binding site in P7. This is similar to what is observed in the post-cleavage form of GrIrz [38] and suggests that the LCrz structure represents a post-cleavage state. The structure of the wild-type form of the ribozyme lacked densities for C230 and A231. Thus, the details of the active site including the position of putative catalytic metal ions are still not known. However, the present evidence strongly suggests that the active site pocket is built from a combination of residual GrIrz elements and *Dir*LCrz specific elements (Figure 5.3b). The former comprise the G-binding site, GrIrz-derived elements that have been redefined and gained a new function, e.g. the GoU pair and its receptor. The latter comprise the three-way junction in the regulatory domain and the lariat fold sequence, 5′ CAU. Together, the structural work leaves a picture of a highly structured, preformed core scaffold in which the active site can only assemble through contributions from peripheral elements that are subject to conformational switching.

5.3.2 Structure of the *Naegleria*-type LCrz

Although *Didymium* and *Naegleria* are distantly related genera, the catalytic core organization of the two types of LCrz appears highly similar, including the three domain structure and the diagnostic P15 pseudoknot [1] (Figure 5.2). The most prominent differences are *Naegleria*-specific insertions and extensions in J4/5 of the P4–P6 domain and a distinctly different structure of the putative regulatory domain J2/10 [11, 34]. The structure diagram is supported by analysis of sequence variation among many isolates from *Naegleria* [1, 4, 12, 24], as well as early work involving structure probing and characterization of the reaction in *Naegleria andersoni* [13]. Tang et al. [24] compared sequence variation between the LCrz variants from *Naegleria* that differed the most in sequence and cleavage rate by a

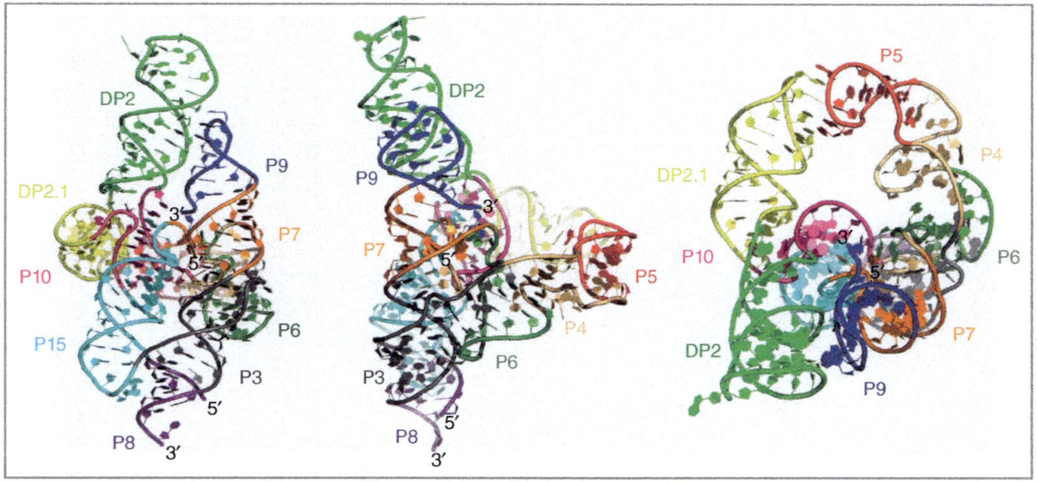

(a)

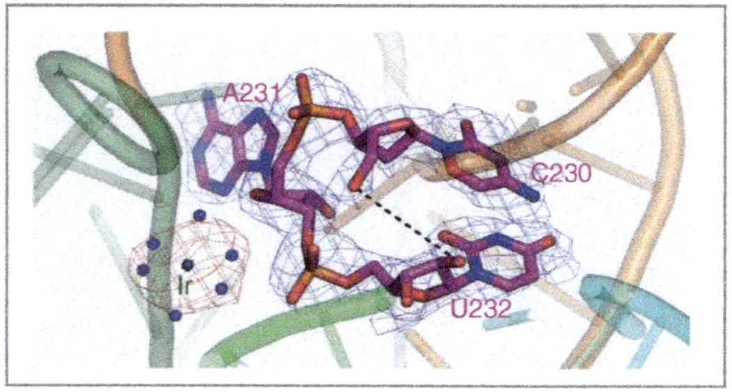

(b)

Figure 5.3 (a) Three different views of the structure of *Dir*LCrz. In the leftmost view, *Dir*LCrz is represented according to the orientation used for group I intron ribozymes with P9 in the upper right corner and P15 (corresponding to P1 and P2 in group I intron ribozymes) in the lower left. In the middle view, the ribozyme is rotated by 90° clockwise on the vertical axis. In the view to the right, the first view is rotated by 90° on a horizontal axis to show the signet ring shape resulting from the orientation of P4–P6 and DP2–DP2.1 domains. Discontinuities in the backbone are due to the absence of some residues in electron density maps. (b) Close-up of the lariat fold in the structure model derived from X-ray crystallography of a circularly permuted form of *Dir*LCrz at 3.85 Å resolutions. The lariat fold is essentially preformed through folding of the core and has a highly constrained backbone with bases pointing outward to make specific interactions. The structure likely represents a post-cleavage situation with a distance of 5.1 Å from the nucleophilic O2′ group of U232 to the 5′ OH group (the phosphate is lacking in the construct). The presence of an iridium hexamine ion in the vicinity of the lariat fold could indicate the involvement of a metal ion site in catalysis.

domain-swapping experiment. The experiment suggested that the vast differences in J5/4 were of little importance in contrast to differences in L9 and the sequences immediately upstream of P10 (see Section 5.4).

The *Asp*LCrz (Figure 5.2) belongs to the *Naegleria*-type of LCrz, but due to the evolutionary distance between *Allovahlkampfia* and *Naegleria*, the ribozyme has 33 positions that deviate from invariant positions in the *Naegleria* ribozymes [26]. Most of the differences are located in loop regions or found as compensatory changes in stems, including four changes in the characteristic P15 pseudoknot. Thus, the *Asp*LCrz sequence lends strong support to the structure diagram of the *Naegleria*-type LCrz. The most interesting differences are located immediately upstream of P10 in the junction to the splicing ribozyme. This junction is considerably shorter than in *Nae*LCrz and forms a conserved interaction with sequences downstream of the ribozyme, overlapping the HE mRNA (see Section 5.4). Thus, the core of the ribozyme appears highly conserved among distantly related species, with considerable variation in flanking, regulatory sequences.

5.4 Communication Between LCrz and Flanking Elements

Twin-ribozyme intron biology implies that LCrz is not simply a constitutive capping ribozyme, but involved in regulation as well. Three pathways have been described that differ in the relative order of execution of the GrIrz and LCrz ribozyme activities with dramatic consequences for the expression of rRNA as well as the HE mRNA (Figure 5.4a): (i) During normal growth, GrIrz acts first to splice the twin-ribozyme intron and thus to ligate rRNA exons. Subsequently, the LCrz caps the mRNA that becomes exported to the cytoplasm where it is found in the polysome fraction and presumably translated [8]. (ii) Cellular stress induces the circularization pathway of GrIrz processing, in which the ribozyme catalyzes 3′ splice site hydrolysis followed by formation of a full-length intron circle without ligation of rRNA exons [39]. In this pathway, the LCrz remains inactive. (iii) During starvation-induced encystment, which is a frequent event in *Didymium*, the pre-rRNA is cleaved by LCrz, and the GrIrz remains inactive [22]. This is believed to facilitate rapid production of the HE mRNA during excystment prior to transcription of new pre-rRNA and constitutes an argument for a function of the HE protein in pre-rRNA processing in addition to the well-established function in homing of the intron. Together these observations demonstrate that a mechanism for on/off switching of both the GrIrz and LCrz must exist.

5.4.1 Group I Ribozyme Switching

Conformational switching in RNA involves competition between at least two alternative structures. The competition can be solely due to differences in thermodynamic stability. In molecular biology, however, conformational switching is mostly linked to other processes. Switching can depend on cellular factors,

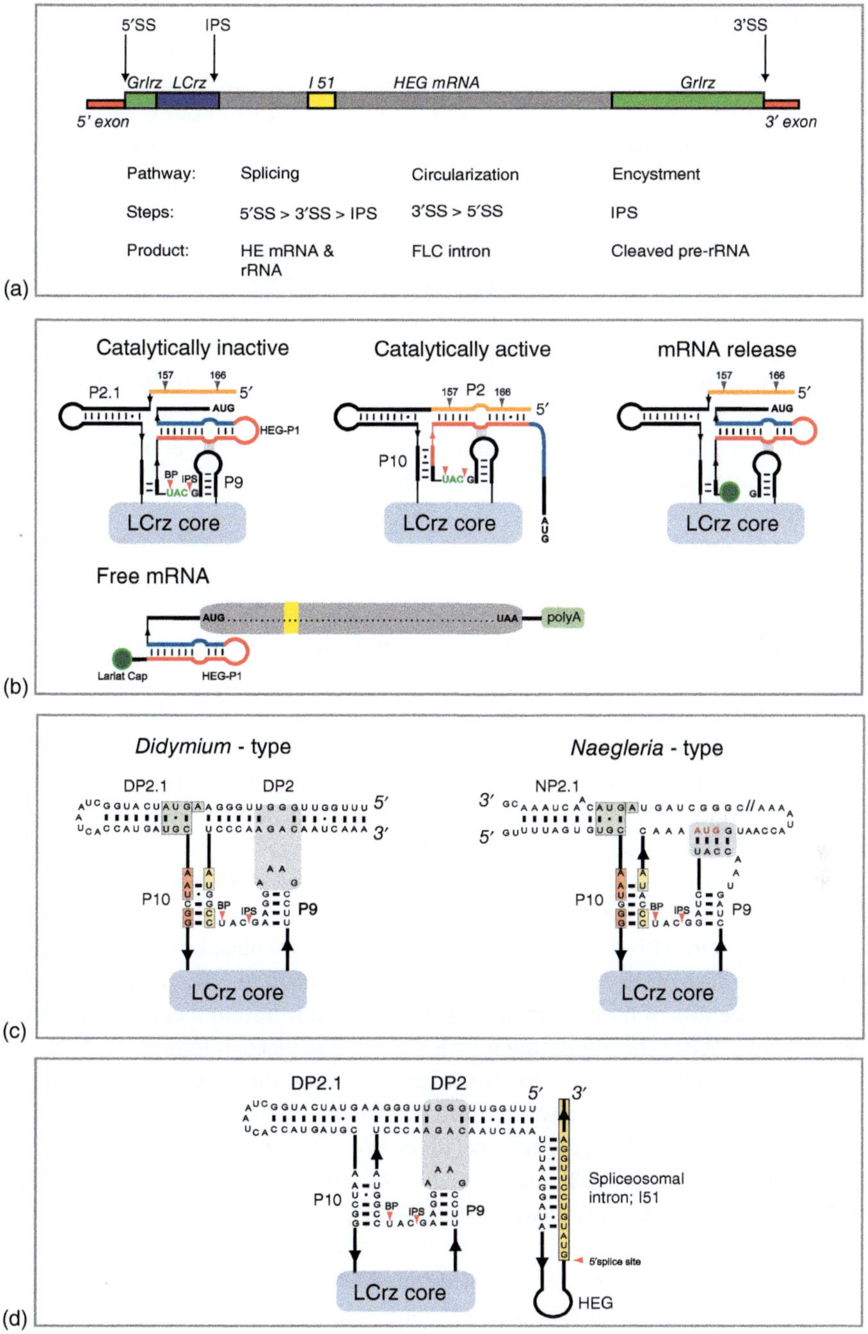

Figure 5.4 (a) The three processing pathways that have been characterized for the *Dir*.S956-1 intron. See main text for further explanation. (b) Diagram showing the conformational switching involving the mutually exclusive P10 and HEG P1 structures. HEG P1 forms upon transcription, thus preventing P10 formation and LCrz branching. After GrIrz splicing, P10 forms and lariat capping occurs. Post-cleavage HEG P1 formation is instrumental in the release of the mRNA from LCrz. At later stages, HEG P1 is found as a 5′ UTR element in the HE mRNA. (c) Outline of the regulatory P2.1 element in the two main types of LCrz. (d) Outline of the potential base-pair interaction between the 5′ part of I51 and the catalytically active form of *Dir*LCrz.

e.g. ligands that bind and stabilize one of the alternative structures, or it can be self-induced [40]. Frequently, switching is an integrated part of a time course of events. RNA switching is particularly relevant in the present case since processing of the intron is linked to rDNA transcription and pre-rRNA processing. In the case of the GrIrz, the switch between the splicing and circularization pathways is mostly understood in terms of competition for docking into the single guanosine binding site in P7 between the exogenous guanosine cofactor (exoG) and the intron terminal guanosine (ωG). ExoG binding promotes the splicing pathway and ωG binding promotes the circularization pathway [39, 41]. The switch between the pathways is not fully understood. If transcription and pre-rRNA processing rates influence the switch, this could constitute signals that would allow the intron to form full-length circular RNA during cellular stress. These circles are "genomic" in the sense that they contain all intronic sequence information and have been hypothesized to be involved in mobility. In this way, cellular stress could trigger an escape route of the intron. Interestingly, both pathways are shut down during starvation-induced encystment.

5.4.2 LC Ribozyme Switching

For *Dir*LCrz, the off switching appears to be better understood and relies on transcriptional order. Nucleotides from the 3′ strand of P10 can fold into an alternative local hairpin (HEG P1) that is co-transcriptionally favored (Figure 5.4b) and may be stabilized by acting as a receptor for L9 [42]. This has been demonstrated in vitro by structure probing [32, 42]. The formation of HEG P1 precludes the formation of the active site and leaves LCrz inactive during pre-rRNA transcription, at least in a window that allows completion of the 1.4 kb twin-ribozyme intron. This is important because LCrz has the ability to cleave nascent transcripts as shown in yeast with constructs that lack the ability to form HEG P1 [33]. Group I intron splicing is an early event in pre-rRNA processing. After the intron is spliced out, the HEG P1 is unfolded, the three-way junction P2.1–P2–P10 is formed, and the LCrz becomes activated. The details of how splicing triggers this conformational change are currently unknown. After the branching reaction has taken place, refolding of HEG P1 occurs in order to release the HE mRNA, essentially "pulling" the lariat cap out of the active site [16, 42]. Evidence comes from in vitro experiments demonstrating the inhibitory effect of HEG P1 on the reversal of the branching reaction in [32]. HEG P1 is presumably found in the 5′UTR of the HE mRNA and may play a role in enhancing translation of the unusual, lariat-capped mRNA [8].

Considering the critical role of HEG P1 in regulating *Dir*LCrz, it is interesting that this element is absent in the *Naegleria*-type LCrz. This could be related to the different structural context (insertion in P6 rather than P2 of GrIrz) or reflects an independent origin of the LCrz or a different path of the adaptation process. From early on, it was noted that a base-pairing interaction between 4 nt of L9 and the first nucleotides in the open reading frame of the HEG was conserved among *Nae*LCrz [12, 23, 24]. However, it was only recently realized that sequences upstream of P10 form two conserved base-pairing interactions in both *Nae*LCrz and *Asp*LCrz with sequences further downstream within the HEG ORF [26]. These pairings are proximal to the L9/HEG ORF pairings and set up a three-way junction with P10 that resembles the situation in *Dir*LCrz (Figure 5.4c).

Structural modeling of *Nae*LCrz suggests that L5 can form a tertiary interaction with the most upstream of the two stems to tether the core similarly to the situation in *Dir*LCrz. The interactions are conserved in *Asp*LCrz, and all interactions are consistent with mutational studies, including domain-swapping experiments. Thus, it appears that a regulatory domain comprising P2 and P2.1 is in fact conserved among all known examples of LCrz, albeit with some structural variation. As a consequence, these stems are referred to with a species-specific name, e.g. DP2.1 for *Didymium* P2.1 (Figure 5.4c). It remains to be explored if AP2.1 and NP2.1 participate in conformational switch regulation of the branching reaction and fold back onto the catalytic core, as seen for the DP2.1 [16, 32] and how these structural differences in general impact the biology of the twin-ribozyme intron in *Naegleria*.

5.4.3 A Role of Spliceosomal Intron I51 in *Dir*LCrz Regulation?

The HEG of *Dir*.S956-1 harbors a small spliceosomal intron (I51) (Figure 5.1b,c). The presence of spliceosomal introns in eukaryotic nuclear rDNA are highly unusual, but similar introns with cognate 5′ splice site, 3′ splice site, and branch site features have been reported in several nuclear group I intron HEGs [7, 8, 43, 44]. A closer inspection of I51 revealed a 10 nt sequence complementary to the flanking sequence of the active form of *Dir*LCrz (Figure 5.4d). A strong base-pairing interaction like this would probably lock *Dir*LCrz into the mode of lariat capping: a result apparently consistent with the observation that the lariat capping reaction dominates in vivo. This potential interaction, however, remains to be tested and further assessed in *Didymium*.

5.5 Reflections on the Evolutionary Aspect of LCrz

Phylogenetic analysis is useful to understand the evolution of a particular gene in a biological setting. In the case of LCrz, this is complicated by the fact that both group I introns and HE are mobile elements. Thus, the relevance of the host organism in such analyses is unclear although the host contributes the transcriptional apparatus for expression of the intron and "host factors" for at least some processing steps. LCrz has so far only been found in the setting of twin-ribozyme introns and appears to have coevolved with both the GrIrz and the HEG [4, 5, 41]. The *Naegleria*-type LCrz appears to have been acquired early, perhaps prior to the separation of the *Naegleria* and *Allovahlkampfia* lineages based on comparison of the sequences of all three elements of their twin-ribozyme introns [27, 45]. Within the *Naegleria* genus, it is inherited in a strictly vertical fashion [4, 23]. In contrast, the *Didymium* LCrz has not been found in other myxomycetes and thus appears to be acquired recently [4]. Interestingly, a sister intron to *Dir*.S956-1 has been found in the myxomycete *Diderma* at the exact same rDNA position as the *Didymium* intron and with an almost identical GrIrz and a very similar HEG inserted into the P2 segment. However, the *Diderma* intron lacks LCrz and may represent a preexisting "receptor" for LCrz insertion. An interesting possibility is that the two main examples of LCrz originated independently and that more sporadic occurrences will be found among eukaryotic microorganisms.

5.5.1 A Model for the Emergence of LCrz

Many myxomycetes have very high numbers of group I introns inserted into their rDNA, e.g. isolates of *Fuligo* and *Diderma* have 12 and 20 or more rDNA introns, respectively [5, 41], and HEGs are relatively frequent (about 1/10 are HEG-containing introns). The feeding habits of myxomycetes involve preying on other microorganisms and may facilitate the horizontal transfer of mobile genomic intron elements. In this scenario, the resemblance between *Dir*LCrz and the eubacterial IC3 group I introns may not be all that surprising. We speculate that a bacterial group I intron invaded another group I intron upstream of an HEG. Because the intron was inserted inside a preexisting intron, the selection pressure for splicing of the most recent intron was relieved, and the new insertion left to degenerate or adapt to the new setting. At the sequence level, one can envisage different paths that would lead to the topological change required for transformation into an LCrz. A single transposition event of the 5′ GUGUUC stretch from the 3′ strand of P15 of the wild-type LCrz to A120 in J15/3 changes the topology and the base-pairing scheme compared with the *Azoarcus* $tRNA^{Ile}$ intron (see Figure S4 in Ref. [37]). Alternatively, sequence drift resulting in alternative base pairing within the core could lead to formation of the double-pseudoknotted core of LCrz that in turn would allow for shortening of the principal domains as seen in LCrz compared with the *Azoarcus* intron (Figure 5.5a). An intriguing observation is that the sequence of the lariat fold, 5′ CAU, is identical to the anticodon sequence of $tRNA^{Ile}$ located immediately upstream of the *Azoarcus* intron. If not a coincidence, the presence of this sequence joined to ωG would imply an unusual mode of insertion of the intron.

The structure and catalytic activity of LCrz adds to the picture of the GrIrz scaffold as being a highly versatile framework in RNA biology. The ribozyme can perform splicing and hydrolysis reactions in both cis and trans [46], act as a ligase [47], and work as an allosteric ribozyme regulated by the second-messenger c-di-GMP [48]. It can assemble itself from activated pieces [49] and form cooperative networks that sustain self-replication [50]. LCrz extends the list to comprise lariat formation and may be the most radical example. Here, key structural elements of the scaffold (e.g. J8/7) have been redefined, and a 2′ OH is activated as the nucleophile instead of a 3′ OH.

5.5.2 An Evolutionary Path to Spliceosomal Splicing?

It is widely accepted that spliceosomal splicing originated from self-splicing group II intron ribozymes (reviewed in [51]). This is based on the similarities in reaction, including the lariat formation in the first step, as well as structural similarities. Given

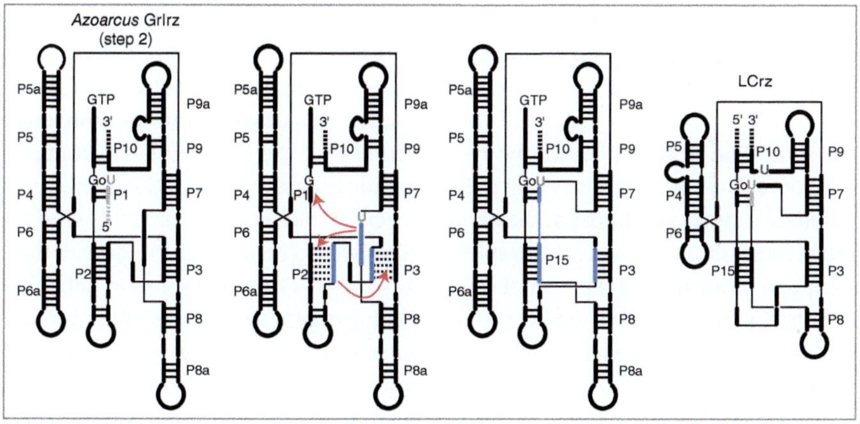

(a)

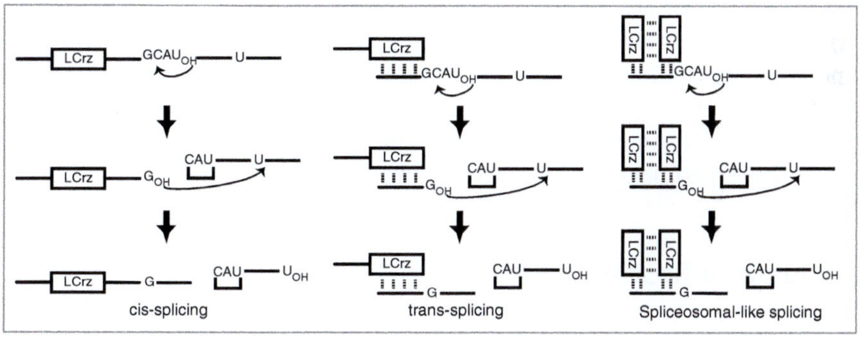

(b)

Figure 5.5 (a) Model for the evolution of a GrIrz into LCrz. The evolution is assumed to take place in a setting where a GrIrz has been inserted into a preexisting GrIrz upstream of a HEG. Thus, the ribozyme is functionally redundant and can undergo sequence drift. (i) The GrIrz, here represented by the *Azoarcus* tRNA intron ribozyme, adopts alternative pairings involving J8/7, P2, and P3 at a stage corresponding to prior to the second catalytic step. (ii) The absence of the 5′ exon favors alternatively folded species by drifting of neighboring sequence stretches, altering the overall secondary structure. (iii) The LC fold is selected due to the appearance of a branching reaction, allowing the formation of the lariat and conferring an increased half-life to the downstream homing endonuclease mRNA. (iv) The gain in energetic stabilization due to the presence of the new pseudoknot P3–P15 allows for a shortening of some peripheral elements, leading to the final version of LCrz that is always shorter than GrIrz. (b) Diagram showing a hypothetical path from LCrz to spliceosomal-like splicing. The first panel shows the development of cis-splicing by re-recruitment of the lariat-capped RNA from the first step and attack mediated by G229, homologous to the ωG. The middle panel shows the transformation from cis- to trans-splicing, and the third panel shows distribution of the catalytic RNA into pieces, similar to snRNA's in spliceosomal splicing.

that GrIrz is ancient and that the emergence of LCrz demonstrates that the GrIrz scaffold can give rise to a branching reaction, it is reasonable to speculate on an evolutionary path from GrIrz to spliceosomal-like splicing. Such a path is outlined in Figure 5.5b. The transformation of GrIrz to LCrz was described in the previous section 5.5.1. The next step would be re-recruitment of the released lariat-capped RNA out of register in such a way that the guanosine bound in P7 (G229 in *Dir*LCrz) could make an attack of a phosphodiester in a reaction that would be formally equivalent to reversal of the second step of splicing. This would result in the release of the lariat-capped "intron" and ligation of flanking "exons." Thus, the reaction appears very similar to group II intron/spliceosomal splicing except for the size of the lariat. The reaction in cis could evolve into a reaction in trans relatively easy as exemplified by the work on standard splicing activities of GrIrz from both *Tetrahymena* and *Azoarcus* [52, 53]. Perhaps related to this, the recruitment of substrates in trans could relieve the branching reaction from the tight constraints of placement of the nucleophile in relation to the cleavage site and thus extend the size of the lariat. Finally, the ribozyme could undergo evolution to be coded in pieces. Assembly of GrIrz scaffolds from pieces has been described [49], and group II intron ribozymes can be complemented in trans [54]. Thus, the described path leads to spliceosomal-like splicing. It would be conceptually interesting to demonstrate the feasibility of such a path by in vitro evolution experiments.

5.6 LCrz as a Research Tool

The use of ribozymes in combination with in vitro transcription provides a flexible and inexpensive approach to production of large quantities of RNA molecules with well-defined ends. As an example, providing the primary transcript with an upstream hammerhead ribozyme and a downstream HDV ribozyme is a convenient way to release an RNA of interest with 5′ OH and a 3′ cyclic phosphate ends, respectively, e.g. for X-ray crystallography [55]. A further advantage in this example is that the constraints on the sequence of the RNA produced are minimal. LCrz can similarly be used as part of an in vitro transcript upstream of the sequence of interest to release an RNA with a lariat cap at its 5′ end. The sequence constraints for such a construct have not been worked out in detail, but within the lariat cap sequence 5′ CAU, only the A is absolutely required (unpublished observations). Downstream of this sequence, the following 19 nt are normally included in constructs, but much of this sequence is likely to be dispensable. The lariat cap will in practical terms constitute a blocked 5′ end. This could, for instance, be useful in the assembly of large RNA modules in RNA origami or in the synthesis of therapeutic RNAs as a protection against 5′ exonucleases.

Many ribozymes have been shown to be transferable to non-native organisms and have been used as experimental tools to study RNA processing and gene expression in bacterial, fungal, and mammalian systems [56, 57]. LCrz is active in vitro in the absence of protein cofactors and hence likely to be transferable. We have demonstrated efficient lariat capping in organisms as different as *E. coli*, yeast, and human cell lines (unpublished, [33]). Importantly, these observations rule out an absolute requirement for native cofactors for LCrz in a cellular environment. LCrz has the additional advantage that the lariat cap is resistant to cellular 5′–3′ degradation activities when expressed ectopically. Specifically, the lariat cap is not a substrate for cellular debranching enzymes [15], and the particular structure of the lariat cap [16], including the absence of an exposed 5′ end, makes it unlikely to be a substrate for decapping enzymes and 5′–3′ exonucleases.

We have used lariat capping of a model transcript encoding green fluorescent protein (GFP) in yeast with the dual purpose of studying the fate of lariat-capped transcripts and the cap dependency of individual steps in gene expression [33]. The m^7GpppN cap structure of eukaryotic mRNA 5′ ends plays a critical role in many aspects of gene expression, including mRNA splicing, polyadenylation, export, translation initiation, and overall stability [58–60]. Given this plethora of functions, it is experimentally attractive to dissect the role of the cap in distinct cellular processes by manipulating its structure, preferably on individual mRNA. However, present strategies to this end have shortcomings. Genetic interference with cap synthesis affects all mRNA species and inhibits cell growth, making it difficult to distinguish direct from indirect effects caused by the mutation. As an alternative, switching the promoter to redirect mRNA transcription from RNAPII to RNAPIII is a possibility, but along with changed cap status, RNAPIII transcription also affects pre-mRNA splicing and 3′ end formation, restricting its applicability. Finally, releasing mRNA from primary transcripts through the action of small cleavage ribozymes [61, 62], RNase P, or RNase III [63] all leaves ends that are substrates for cellular exonucleases, as does transcription in vivo by T7 RNAP [64]. In contrast, we proposed that transplantation of LCrz could be used to create 5′ end stable transcripts in which the m^7G cap was missing, exposing its roles in mRNA metabolic processes.

The LCrz variant used (*Dir*LCrz-166.22) had essential parts of the inactive conformation deleted to promote direct folding into the active conformation. Consistent with this, we observed by analysis of chromatin-associated transcripts that a significant fraction was cleaved during transcription. Other experiments showed that complete cleavage was attainable and that the extent of 5′ end processing by the LCrz depended on the choice of promoter. The transcripts were 3′ end-processed at the expected site. An effect of lariat capping on the efficiency of polyadenylation was suggested by analysis in strains lacking one or both of the major deadenylases, Ccr4p and Pan2p. However, efficient polyadenylation followed by deadenylation by another deadenylase could not be ruled out. In any case, the extent of polyadenylation was compatible with nuclar stability, and the transcript was eventually exported to the cytoplasm where it was found as a lariat-capped and oligoadenylated transcript evenly distributed in the cytoplasm as evidenced by *in situ* hybridization. The GFP-mRNA was very poorly translated ($<1\%$ of control m^7G capped GFP-mRNA), and translation was not rescued by insertion of a templated polyA tail, pointing to the lack of an m^7G cap as the cause of poor translation. The lariat-capped GFP-mRNA had 3-fold increased half-life, probably resulting from a block of the decapping step in mRNA turnover. Instead, the lariat-capped mRNA was likely turned over by 3′–5′ degradation by the cytoplasmic exosome. The fate of the lariat cap itself is unknown.

The study pointed to two steps in gene expression where lariat capping may be used as a tool to unravel cap dependency: the initial polyadenylation of the transcript and the recruitment of the cytoplasmic exosome for 3′–5′ degradation.

The feasibility of expressing relatively stable lariat-capped transcripts in the yeast cytoplasm that are poorly translated paves the way for at least two new types of experiments (Figure 5.6). Ongoing work demonstrates that the GFP encoding sequence can be replaced with many other genes of interest and that the level of expression of the lariat-capped RNA can be controlled using tetracycline-regulated promoters. Such transcripts can be designed to compete with endogenous transcripts for factors (by "sponging") and thus to study the role of these factors in, e.g. localization or translation of the mRNA with minimal perturbation of the cell. In another line of studies, we have inserted yeast IRES elements into lariat-capped GFP transcripts and showed considerable rescue of GFP translation. Such constructs allow for dissection of the function of IRES elements and provide an alternative to previous approaches for IRES characterization. Here, one strategy relied on insertion of a stable hairpin at the 5′ end of the transcript [65]. The m^7G cap was present, but the hairpin interfered with its functions. Thus, the m^7G cap and the IRES were not rigorously uncoupled. Another strategy was to use transfection with mRNA with a nonfunctional ApppG cap [66]. These transcripts were made in vitro and thus escaped the normal mRNP assembly pathway. The lariat capping approach has the advantage of generating transcripts that are otherwise processed normally while providing the transcript with a cap that is virtually inert in stimulating translation. Finally, lariat-capped GFP mRNA may be used for screening of genomic libraries for sequences with IRES function using cell sorting based on fluorescence.

5.7 Conclusions and Unsolved Problems

The LCrz is the hallmark of complex twin-ribozyme introns. At this stage, its primary role in promoting expression of the HE mRNA through capping has been established. The basics of the ribozyme reaction and the underlying structural aspects have been described, and an additional role in timing of the capping reaction and release of the mRNA in relation to intron splicing has been outlined. Even so, many questions remain. Considering the established function of the HE in intron transfer during sexual reorganization, it is not intuitive why the mRNA is produced at relatively high levels in growing cells. The mRNA is found on polysomes and likely being translated. Is there a second function of the protein, e.g. in promoting processing steps of the intron RNA, including the LCrz? A thorough analysis of the reaction conditions of the lariat capping reaction has not been conducted, and it is noteworthy that most (but not all) variants require high salt and magnesium concentrations and elevated temperatures for optimal cleavage. Perhaps could the HE protein relieve the LCrz of the nonphysiological conditions? The twin-ribozyme introns can undergo three different processing pathways, and RNA conformational switching appears to play key roles in regulation of these. Here, our understanding of these switching mechanisms is still very limited, but at present, the system appears to be one of the

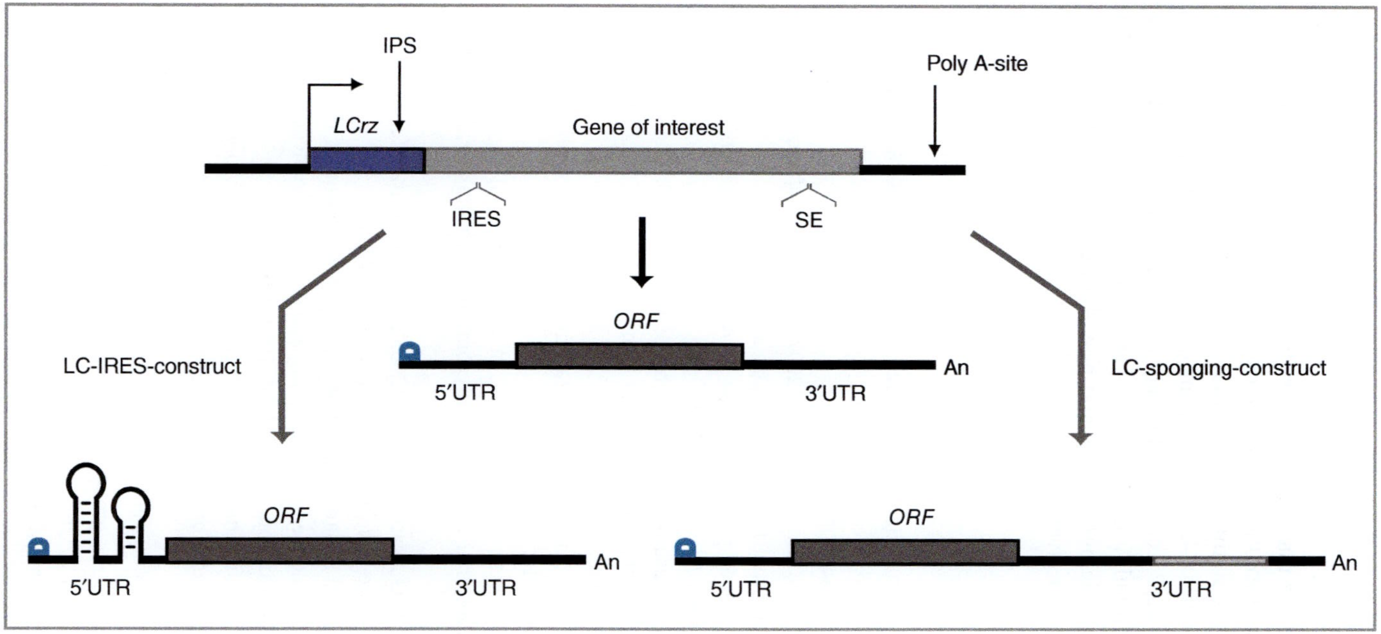

Figure 5.6 Lariat capping as a tool. Lariat capping constructs are made as fusion constructs with an upstream lariat capping ribozyme in front of the gene of interest. A polyA site or a templated polyA tail is placed further downstream, depending on the application. The ribozyme cleaves the nascent transcript at the IPS and releases a lariat-capped mRNA. In the current constructs this adds 22 nt of non-native sequence to the 5′end of the transcript. Two different applications are sketched. To the left, an IRES is inserted to drive translation in the otherwise very lowly translated lariat-capped mRNA. This can be used to characterize IRES elements or to screen for IRES activity. To the right, a sequence-similar construct is used for competition with endogenous mRNA for mRNA-binding factors that can be RNA (e.g. miRNA) or proteins. By adding, deleting, or mutating such elements (here referred to as sponging elements [SE]) on the construct, the interactions can be dissected and the consequences on the cellular phenotype scored.

most complex (almost) "all-RNA" functional entities. At the structural level, only a single stage in the catalytic cycle has been captured by X-ray crystallography, and many aspects, e.g. the role of metal ions in catalysis, are still unsolved. Finally, the LCrz appears to be of particular interest from an evolutionary perspective because the close relatedness to GrIrz may reveal clues to how RNA molecules evolve. Is it possible to recapitulate the evolution of a branching activity from a GrIrz by in vitro evolution experiments? And in the larger picture, is it possible to take this one step further and evolve spliceosomal-like splicing, one of the key molecular processes in human cells, in the test tube? Certainly, the twin-ribozyme introns found in myxomycetes and other eukaryotic microorganisms are an example of how complex even the humblest genetic element can be and how their study can contribute to our appreciation of molecular diversity.

References

1 Johansen, S., Einvik, C., and Nielsen, H. (2002). DiGIR1 and NaGIR1: naturally occurring group I-like ribozymes with unique core organization and evolved biological role. *Biochimie* 84: 905–912.
2 Johansen, S.D., Haugen, P., and Nielsen, H. (2007). Expression of protein-coding genes embedded in ribosomal DNA. *Biol. Chem.* 388: 679–686.
3 Nielsen, H. and Johansen, S.D. (2007). A new RNA branching activity: the GIR1 ribozyme. *Blood Cells Mol. Dis.* 38: 102–109.
4 Nielsen, H., Beckert, B., Masquida, B., and Johansen, S.D. (2008). The GIR1 Branching Ribozyme. In: *Ribozymes and RNA catalysis* (eds. D.M.J. Lilley and F. Eckstein). RSC Publishing. Cambridge. Ch. 12: 230–253.
5 Hedberg, A. and Johansen, S.D. (2013). Nuclear group I introns in self-splicing and beyond. *Mob. DNA* 4: 17.
6 Johansen, S., Elde, M., Vader, A. et al. (1997). In vivo mobility of a group I twintron in nuclear ribosomal DNA of the myxomycete *Didymium iridis*. *Mol. Microbiol.* 24: 737–745.
7 Haugen, P., Wikmark, O.G., Vader, A. et al. (2005). The recent transfer of a homing endonuclease gene. *Nucleic Acids Res.* 33: 2734–2741.
8 Vader, A., Nielsen, H., and Johansen, S. (1999). In vivo expression of the nucleolar group I intron-encoded I-DirI homing endonuclease involves the removal of a spliceosomal intron. *EMBO J.* 18: 1003–1013.
9 Decatur, W.A., Einvik, C., Johansen, S., and Vogt, V.M. (1995). Two group I ribozymes with different functions in a nuclear rDNA intron. *EMBO J.* 14: 4558–4568.
10 Johansen, S. and Vogt, V.M. (1994). An intron in the nuclear ribosomal DNA of *Didymium iridis* codes for a group I ribozyme and a novel ribozyme that cooperate in self-splicing. *Cell* 76: 725–734.
11 Einvik, C., Nielsen, H., Westhof, E. et al. (1998). Group I-like ribozymes with a novel core organization perform obligate sequential hydrolytic cleavages at two processing sites. *RNA* 4: 530–541.

12 Einvik, C., Decatur, W.A., Embley, T.M. et al. (1997). *Naegleria* nucleolar introns contain two group I ribozymes with different functions in RNA splicing and processing. *RNA* 3: 710–720.

13 Jabri, E., Aigner, S., and Cech, T.R. (1997). Kinetic and secondary structure analysis of *Naegleria andersoni* GIR1, a group I ribozyme whose putative biological function is site-specific hydrolysis. *Biochemistry* 36: 16345–16354.

14 Jabri, E. and Cech, T.R. (1998). In vitro selection of the *Naegleria* GIR1 ribozyme identifies three base changes that dramatically improve activity. *RNA* 4: 1481–1492.

15 Nielsen, H., Westhof, E., and Johansen, S. (2005). An mRNA is capped by a 2′,5′ lariat catalyzed by a group I-like ribozyme. *Science* 309: 1584–1587.

16 Meyer, M., Nielsen, H., Olieric, V. et al. (2014). Speciation of a group I intron into a lariat capping ribozyme. *Proc. Natl. Acad. Sci. USA* 111: 7659–7664.

17 Johansen, S. and Haugen, P. (2001). A new nomenclature of group I introns in ribosomal DNA. *RNA* 7: 935–936.

18 Michel, F. and Westhof, E. (1990). Modelling of the three-dimensional architecture of group I catalytic introns based on comparative sequence analysis. *J. Mol. Biol.* 216: 585–610.

19 Nandipati, S.C., Haugli, K., Coucheron, D.H. et al. (2012). Polyphyletic origin of the genus *Physarum* (Physarales, Myxomycetes) revealed by nuclear rDNA mini-chromosome analysis and group I intron synapomorphy. *BMC Evol. Biol.* 12: 166.

20 Dee, J. (1997). Genes and development in *Physarum*. *Trends Genet.* 3: 208–213.

21 Stephenson, S.L., Fiore-Donno, A.M., and Schnittler, M. (2011). Myxomycetes in soil. *Soil Biol. Biochem.* 43: 2237–2242.

22 Vader, A., Johansen, S., and Nielsen, H. (2002). The group I-like ribozyme DiGIR1 mediates alternative processing of pre-rRNA transcripts in *Didymium iridis*. *Eur. J. Biochem.* 269: 5804–5812.

23 Wikmark, O.G., Einvik, C., De Jonckheere, J.F., and Johansen, S.D. (2006). Short-term sequence evolution and vertical inheritance of the *Naegleria* twin-ribozyme group I intron. *BMC Evol. Biol.* 6: 39.

24 Tang, Y., Nielsen, H., Birgisdottir, A.B., and Johansen, S. (2011). A natural fast-cleaving branching ribozyme from the amoeboflagellate *Naegleria pringsheimi*. *RNA Biol.* 8: 997–1004.

25 Goddard, M.R. and Burt, A. (1999). Recurrent invasion and extinction of a selfish gene. *Proc. Natl. Acad. Sci. USA* 96: 13880–13885.

26 Tang, Y., Nielsen, H., Masquida, B. et al. (2014). Molecular characterization of a new member of the lariat capping twin-ribozyme introns. *Mob. DNA* 5: 25.

27 Geisen, S., Bonkowski, M., Zhang, J., and De Jonckheere, J.F. (2015). Heterogeneity in the genus *Allovahlkampfia* and the description of the new genus Parafumarolamoeba (Vahlkampfiidae; Heterolobosea). *Eur. J. Protistol.* 51: 335–349.

28 Leontis, N.B., Stombaugh, J., and Westhof, E. (2002). The non-Watson–Crick base pairs and their associated isostericity matrices. *Nucleic Acids Res.* 30: 3497–3531.

29 Fedor, M.J. (2000). Structure and function of the hairpin ribozyme. *J. Mol. Biol.* 297: 269–291.

30 Lafontaine, D.A., Wilson, T.J., Norman, D.G., and Lilley, D.M. (2001). The A730 loop is an important component of the active site of the VS ribozyme. *J. Mol. Biol.* 312: 663–674.

31 Clouet-d'Orval, B. and Uhlenbeck, O.C. (1997). Hammerhead ribozymes with a faster cleavage rate. *Biochemistry* 36: 9087–9092.

32 Nielsen, H., Einvik, C., Lentz, T.E. et al. (2009). A conformational switch in the DiGIR1 ribozyme involved in release and folding of the downstream I-DirI mRNA. *RNA* 15: 958–967.

33 Krogh, N., Pietschmann, M., Schmid, M. et al. (2017). Lariat capping as a tool to manipulate the 5′ end of individual yeast mRNA species in vivo. *RNA* 23: 683–695.

34 Einvik, C., Nielsen, H., Nour, R., and Johansen, S. (2000). Flanking sequences with an essential role in hydrolysis of a self-cleaving group I-like ribozyme. *Nucleic Acids Res.* 28: 2194–2200.

35 Adams, P.L., Stahley, M.R., Gill, M.L. et al. (2004). Crystal structure of a group I intron splicing intermediate. *RNA* 10: 1867–1887.

36 Adams, P.L., Stahley, M.R., Kosek, A.B. et al. (2004). Crystal structure of a self-splicing group I intron with both exons. *Nature* 430: 45–50.

37 Beckert, B., Nielsen, H., Einvik, C. et al. (2008). Molecular modelling of the GIR1 branching ribozyme gives new insight into evolution of structurally related ribozymes. *EMBO J.* 27: 667–678.

38 Lipchock, S.V. and Strobel, S.A. (2008). A relaxed active site after exon ligation by the group I intron. *Proc. Natl. Acad. Sci. USA* 105: 5699–5704.

39 Nielsen, H., Fiskaa, T., Birgisdottir, A.B. et al. (2003). The ability to form full-length intron RNA circles is a general property of nuclear group I introns. *RNA* 9: 1464–1475.

40 Nagel, J.H. and Pleij, C.W. (2002). Self-induced structural switches in RNA. *Biochimie* 84: 913–923.

41 Nielsen, H. and Johansen, S.D. (2009). Group I introns: moving in new directions. *RNA Biol.* 6: 375–383.

42 Birgisdottir, A.B., Nielsen, H., Beckert, B. et al. (2011). Intermolecular interaction between a branching ribozyme and associated homing endonuclease mRNA. *Biol. Chem.* 392: 491–499.

43 Haugen, P., Reeb, V., Lutzoni, F., and Bhattacharya, D. (2004). The evolution of homing endonuclease genes and group I introns in nuclear rDNA. *Mol. Biol. Evol.* 21: 129–140.

44 Johansen, S.D., Vader, A., Sjottem, E., and Nielsen, H. (2006). In vivo expression of a group I intron HEG from the antisense strand of Didymium ribosomal DNA. *RNA Biol.* 3: 157–162.

45 Walochnik, J. and Mulec, J. (2009). Free-living amoebae in carbonate precipitating microhabitats of karst caves and a new vahlkampfiid amoeba, *Allovahlkampfia* spelaea gen. nov., sp. nov. *Acta Protozool.* 48: 25–33.

46 Inoue, T., Sullivan, F.X., and Cech, T.R. (1985). Intermolecular exon ligation of the rRNA precursor of *Tetrahymena*: oligonucleotides can function as 5′ exons. *Cell* 43: 431–437.

47 Vicens, Q. and Cech, T.R. (2009). A natural ribozyme with 3′,5′ RNA ligase activity. *Nat. Chem. Biol.* 5: 97–99.

48 Lee, E.R., Baker, J.L., Weinberg, Z. et al. (2010). An allosteric self-splicing ribozyme triggered by a bacterial second messenger. *Science* 329: 845–848.

49 Draper, W.E., Hayden, E.J., and Lehman, N. (2008). Mechanisms of covalent self-assembly of the *Azoarcus* ribozyme from four fragment oligonucleotides. *Nucleic Acids Res.* 36: 520–531.

50 Vaidya, N., Manapat, M.L., Chen, I.A. et al. (2012). Spontaneous network formation among cooperative RNA replicators. *Nature* 491: 72–77.

51 Novikova, O. and Belfort, M. (2017). Mobile group II introns as ancestral eukaryotic elements. *Trends Genet.* 33: 773–783.

52 Amini, Z.N. and Muller, U.F. (2015). Increased efficiency of evolved group I intron spliceozymes by decreased side product formation. *RNA* 21: 1480–1489.

53 Dolan, G.F. and Muller, U.F. (2014). Trans-splicing with the group I intron ribozyme from *Azoarcus*. *RNA* 20: 202–213.

54 Hetzer, M., Wurzer, G., Schweyen, R.J., and Mueller, M.W. (1997). Trans-activation of group II intron splicing by nuclear U5 snRNA. *Nature* 386: 417–420.

55 Meyer, M. and Masquida, B. (2016). Polyacrylamide gel electrophoresis for purification of large amounts of RNA. *Methods Mol. Biol.* 1320: 59–65.

56 Sullenger, B.A. and Gilboa, E. (2002). Emerging clinical applications of RNA. *Nature* 418: 252–258.

57 Long, M.B., Jones, J.P. 3rd, Sullenger, B.A., and Byun, J. (2003). Ribozyme-mediated revision of RNA and DNA. *J. Clin. Invest.* 112: 312–318.

58 Topisirovic, I., Svitkin, Y.V., Sonenberg, N., and Shatkin, A.J. (2011). Cap and cap-binding proteins in the control of gene expression. *Wiley Interdiscip. Rev. RNA* 2: 277–298.

59 Gonatopoulos-Pournatzis, T. and Cowling, V.H. (2014). Cap-binding complex (CBC). *Biochem. J.* 457: 231–242.

60 Ramanathan, A., Robb, G.B., and Chan, S.H. (2016). mRNA capping: biological functions and applications. *Nucleic Acids Res.* 44: 7511–7526.

61 Dower, K., Kuperwasser, N., Merrikh, H., and Rosbash, M. (2004). A synthetic A tail rescues yeast nuclear accumulation of a ribozyme-terminated transcript. *RNA* 10: 1888–1899.

62 Meaux, S. and Van Hoof, A. (2006). Yeast transcripts cleaved by an internal ribozyme provide new insight into the role of the cap and poly(A) tail in translation and mRNA decay. *RNA* 12: 1323–1337.

63 Meaux, S., Lavoie, M., Gagnon, J. et al. (2011). Reporter mRNAs cleaved by Rnt1p are exported and degraded in the cytoplasm. *Nucleic Acids Res.* 39: 9357–9367.

64 Dower, K. and Rosbash, M. (2002). T7 RNA polymerase-directed transcripts are processed in yeast and link 3′ end formation to mRNA nuclear export. *RNA* 8: 686–697.

65 Reineke, L.C. and Merrick, W.C. (2009). Characterization of the functional role of nucleotides within the URE2 IRES element and the requirements for eIF2A-mediated repression. *RNA* 15: 2264–2277.

66 Gilbert, W.V., Zhou, K., Butler, T.K., and Doudna, J.A. (2007). Cap-independent translation is required for starvation-induced differentiation in yeast. *Science* 317: 1224–1227.

6

Self-Splicing Group II Introns

Isabel Chillón and Marco Marcia

European Molecular Biology Laboratory, Grenoble Outstation, 71 Avenue des Martyrs, 38042 Grenoble, France

6.1 Introduction

Group II introns (GIIi) are self-splicing ribozymes that ensure efficient maturation of coding transcripts [1, 2] and/or act as retrotransposable elements [3], thus having a profound impact on gene expression and evolution [4] and the potential to be used as biotechnological tools [5]. After reviewing the milestones that brought us from identifying GIIi to determining their high resolution structures, in this chapter, we will discuss the functional, evolutionary, and biological roles of these ribozymes, zoom into GIIi at the molecular level to describe their intricate secondary and tertiary structure, and finally summarize the impact of GIIi research on understanding more complex splicing machineries, like the spliceosome, and on developing new tools to study other emerging large non-coding RNAs.

6.2 Milestones in the Characterization of Group II Introns

Inadvertently observed by electron microscopy in the late 1970s [6], GIIi were identified and named as such in the early 1980s, when a new class of self-splicing RNA with conserved secondary structures was discovered in the mitochondria of the yeast *Saccharomyces cerevisiae* (Figure 6.1) [7]. In the following years, GIIi were also found in the chloroplast genomes of plants [8], in bacteria [9], in archaea [10], and in the mitochondrial genome of an annelid worm [11]. Soon after their identification, studies on their catalytic mechanism revealed that GIIi self-splice via two consecutive transesterification reactions, like group I introns [12, 13], but differently from the latter, GIIi predominantly produce a lariat, not circular, product, forming a 2′–5′ phosphodiester bond at their branch site [12–14]. Lariat formation and structural similarity with small nuclear RNAs (snRNAs) suggested that GIIi represent the evolutionary ancestors of spliceosomal snRNAs [15]. Since GIIi are sub-megadalton

Ribozymes, First Edition. Edited by Sabine Müller, Benoît Masquida, and Wade Winkler.

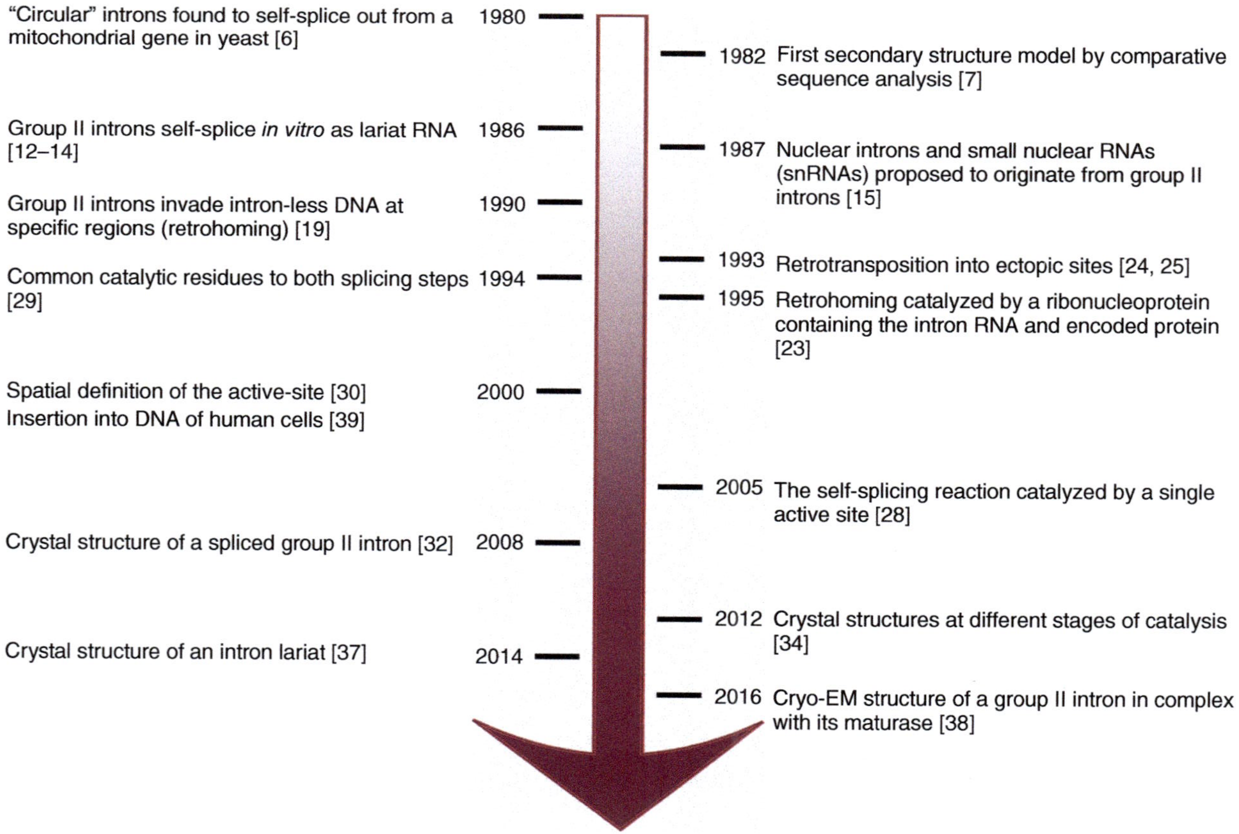

Figure 6.1 Milestones in the research of GIIi. Schematic representation of selected breakthroughs in the study of GIIi. References to the cited works are indicated in brackets.

enzymes that can be produced and tested in vitro, they became a model system to dissect the splicing mechanism relevant to eukaryotic nuclear splicing. In addition, the reversible nature of the GIIi transesterification reaction soon proved it feasible to test reverse splicing in vitro and thus explore the mechanism of intron mobility [16–18]. GIIi mobility is a process that was first investigated in 1990, when the introns ail and ai2 from *S. cerevisiae* were observed to invade an intron-less strain in a process called homing [19]. In the same years, it became evident that within many GIIi sequences resides an open reading frame encoding the so-called intron-encoded protein (IEP) [20], an evolutionary and functional intron partner that acts as a maturase (or chaperone) [21] and favors intron mobility via its reverse transcriptase (RT) and endonuclease (En) domains [22]. GIIi and their cognate IEPs were proven to form a functional ribonucleoprotein required for retrohoming into intron-less targets [23] and for retrotransposition into ectopic genomic sites [24, 25].

Along with biochemical and enzymatic investigations, much of the initial understanding about GIIi catalysis came from evolutionary and sequence conservation analyses. Bioinformatics efforts, coupled with biochemical studies, identified a key set of absolutely essential catalytic residues [26, 27] that form a single active site for both steps of splicing [28] and scaffolding elements that form tertiary interactions conferring structural integrity to the GIIi [28–31].

Finally, in 2008, almost 30 years after GIIi discovery, the first three-dimensional (3D) structure of the *Oceanobacillus iheyensis* group IIC intron (*O.i.*I1) was obtained [32], soon followed by a gallery of other structures capturing different catalytic stages [33, 34], different folding conformations [35], the lariat state of an intron chimera [36], and the architecture of the other two intron classes, group IIB [37] and group IIA, the latter in complex with its IEP [38]. These structures are now opening up unexplored possibilities to rationally engineer or target GIIi with small ligands for potential biotechnological and medical applications [39].

6.3 Evolutionary Conservation and Biological Role

Spreading of GIIi across all domains of life results from the exceptional ability of these ribozymes to act as mobile genetic elements and from the fact that GIIi have been conserved throughout evolution. Introns are acquired by vertical inheritance from a common ancestor and by horizontal transfer across species. Horizontal transfer can proceed via retrohoming, by which GIIi insert themselves into their native genomic loci, or via ectopic transposition, by which GIIi target ectopic sites [40, 41].

6.3.1 Phylogenetic Classifications

Phylogenetic classification of GIIi is based on properties of the RNA, of its cognate IEP, or of both partners (Figures 6.2 and 6.3). First, based on RNA sequence and structural conservation, GIIi are classified into three main groups: IIA, IIB, and IIC, mainly differing in how they recognize exon junctions via exon binding site (EBS)–intron binding site (IBS) interactions (Figure 6.2) [51]. In class IIA, two EBS

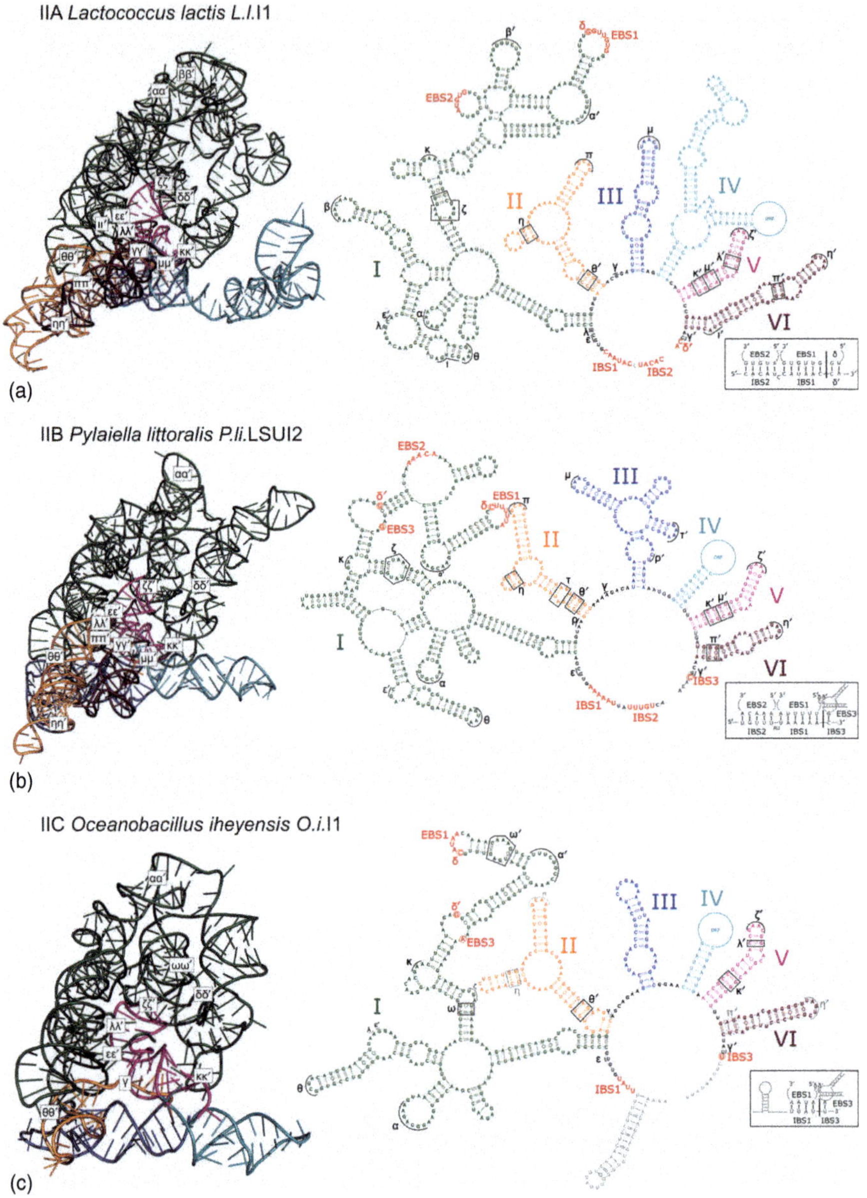

Figure 6.2 Secondary and tertiary structures of GIIi. Schematic secondary RNA structures depicting three GIIi representatives of the main classes IIA (a), IIB (b), and IIC (c). The sequences correspond to the wild-type introns as provided in the bacterial GIIi database [42] but omitting part of domain 4 and the ORF. Domains are color coded. Sequences involved in base-pairing IBS–EBS interactions are indicated in red. Sequences participating in tertiary long-range interactions are boxed and named by Greek letters. Tertiary interactions in gray color in the *O.i.*I1 intron represents potential long-range contacts deduced by comparison with *P.li.*LSUI2 but absent in the crystal structure. The inset represents the special configuration of the RNA motifs involved in intron–exon contacts during exon ligation. The 3D structures represented correspond to PDB files 5G2X (*L.l.*I1), 4R0D (*P.l.*LSUI2), and 4FAQ (*O.i.*I1).

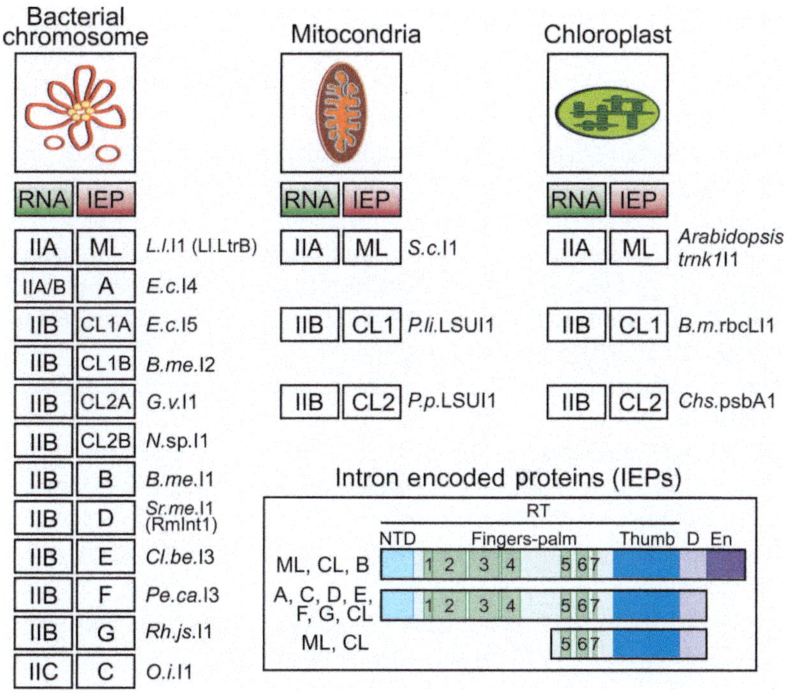

Figure 6.3 Evolutionary classification. The indicated classes are a compilation of different phylogenetic analyses performed in organellar and/or bacterial genomes [20, 42–50]. Representative examples are indicated at the right of each category. The inset represents the typical domain structure of the IEP of every major class of proteins. Abbreviations are as follows: *B.m.*, *Bryopsis maxima*; *B.me*, *Bacillus megaterium*; *Chs.*, *Chlamydomonas* sp.; *Cl.be.*, *Clostridium beijerinckii*; *E.c.*, *Escherichia coli*; *G.v.*, *Gloeobacter violaceus*; *L.l.*, *Lactococcus lactis*; *N.sp.*, *Nostoc* sp.; *O.i.*, *Oceanobacillus iheyensis*; *P.li.*, *Pylaiella littoralis*; *P.p.*, *Porphyra purpurea*; *Pe.ca.*, *Pelobacter carbinolicus*; *Rh.js.*, *Rhodococcus jostii*; *S.c.*, *Saccharomyces cerevisiae*; *Sr.me.*, *Sinorhizobium meliloti*.

sequences (EBS1 and EBS2) bind to two corresponding IBS sequences (IBS1 and IBS2) upstream of the 5′-splice junction, while the 3′-exon binds the intron through a 1- or 2-nucleotide-long sequence (δ′) complementary to a sequence (δ) adjacent to EBS1 (Figure 6.2a). Class IIB differs from IIA in that the δ′ nucleotide is located in an internal loop of domain 1 (in red in Figure 6.2b). The interaction with the 3′-exon is then established through an EBS3 sequence located in the same internal loop as δ′, such that EBS3 and the 3′-exon (IBS3) are reciprocally connected by tertiary interactions (Figure 6.2b) [52]. Finally, in class IIC, the exon recognition occurs as in class IIB, but the EBS2–IBS2 interaction is missing and in part substituted by a stem loop of a rho-independent transcription terminator (Figure 6.2c) [53].

Second, based on IEP sequence and structural conservation, GIIi are grouped into nine main categories (chloroplast-like [CL], mitochondrial-like [ML], and bacterial class A-G) (Figure 6.3). IEPs usually contain three conserved domains: (i) the RT domain, homologous to the spliceosomal Prp8 rather than telomerase or viral RTs [54] and further divided into the N-terminal domain (NTD), the fingers–palm catalytic domain, and the thumb maturase domain; (ii) the DNA binding domain (DBD); and (iii) the DNA En domain (Figure 6.3, inset) [38]. CL IEPs (found in chloroplast and mitochondria of fungi, algae, and plants and in certain bacteria and archaea), ML IEPs (found in mitochondria, chloroplasts, and bacteria), and bacterial class B IEPs all possess full-length RT domains. Instead, bacterial A, C, D, E, F, and G IEPs and archaeal D and E IEPs lack the En domain (Figure 6.3) [55, 56]; thus, these introns require coupling retrohoming to replication (see Section 6.3.3).

Third, based on the coevolution of introns and cognate IEPs [43], a mixed classification was also proposed, according to which class IIA introns coupled to mitochondrial ML IEP belong to the so-called IIA1 category, IIB introns coupled to class D IEP belong to the IIB3 category, and so on [44].

Despite these classification efforts, the lack of an extensive primary sequence conservation in both the RNA and the IEP and the difficulties in correlating sequence alignments made classifications controversial. For instance, introns associated to the bacterial IEP class F, which were originally identified as a conserved group [57], were successively divided into several monophyletic groups based on further analyses of newly identified open reading frames [45]. Currently, group F comprises members of the original class F and other previously unclassified introns. A new group G has also been described, which is phylogenetically related to F and comprises only previously unclassified introns but no intron of the original class F [45, 58].

6.3.2 Differentiation and Evolutionarily Acquired Properties

Despite deriving from a common ancestor, GIIi have developed species-specific mechanisms and properties during evolution, such that organellar introns are clearly distinct from bacterial introns.

First, organellar introns frequently interrupt essential genes, thus requiring efficient splicing that is often assisted by host factors [59], such as ATP-dependent DEAD-box helicases like Mss116 in yeast or CYT-19 in *Neurospora* [60]. However, such introns often lack or possess truncated, not functional IEPs and have thus lost mobility during evolution. For example, intron 5 of *S. cerevisiae oxi3* gene coding for the cytochrome oxidase subunit 1 (ai5γ) lacks any ORF [7], whereas intron 4 of mitochondrial NADH dehydrogenase subunit 1 (nad1I4) of various angiosperms does harbor an IEP ORF (*mat*R), but this protein possesses degenerate RT motifs (except for a well-conserved thumb domain) and no En domain (Figure 6.3, inset) [61]. Second, some organellar introns contain fragmented ribozyme moieties and undergo trans-splicing, which frequently requires the formation of host protein complexes [62]. Other degenerated introns, as those present in wheat mitochondria, follow unusual pathways, i.e. the formation of circles containing non-encoded nucleotides to ensure proper splicing [63].

Differently from organellar introns, bacterial GIIi generally possess full-length ribozymes that undergo cis-splicing and functional IEPs that promote efficient retrotransposition into empty genomic DNA targets [64, 65]. Moreover, bacterial introns frequently interrupt nonessential genes, many of which are also mobile elements. As a result, bacterial GIIi are subjected to a different type of evolutionary pressure to promote splicing, and they are considered true retroelements [66, 67]. A particularly curious example is the bacterial intron RmInt1 (*Sr.me.*I1), a symbiont of the nitrogen-fixing plant *Sinorhizobium meliloti* that interrupts the transposase of a high-copy-number insertion sequence from the IS630 family [68]. Production of such transposase would be deleterious to the host, so to prevent transposase expression, RmInt1 has evolved as a deliberately inefficient, low-fidelity splicing ribozyme. For instance, RmInt1 produces <1% correctly spliced exons [66], partly due to competing IBS1 sites in 3′-exon that cause the formation of unproductive side products during splicing [69, 70] and partly due to a run of cytidine residues at its 3′-splice site that reduces splicing fidelity and causes the formation of circular products with extra exonic residues [71, 72].

6.3.3 Spreading and Survival in the Host Genome

GIIi evolutionary survival and spreading across species depends on the efficiency of retrohoming and ectopic transposition, two processes that are influenced by the structural architecture of the GIIi RNA and/or IEP and by the genetic environment of the host [73].

Retrohoming occurs via target DNA-primed reverse transcription (TPRT), which is a three-step mechanism [74]. In the first step, the excised lariat intron forms a complex with its IEP, binds DNA at any accessible site [75], and then searches for features that facilitate the formation of specific interactions with its DNA target sequence (homing site) [76], which ultimately lead to dsDNA unwinding. Unwinding allows the RNA to establish base pairings with the target DNA. GIIi IEPs establish additional contacts with the distal region of both exons [77], but these recognition requirements seem to be less stringent for introns with no En domain or lacking the IEP [78, 79]. Target recognition interactions are similar but not identical to those described during splicing. For example, the IBS2–EBS2 interaction and distal interactions at 5′ and 3′-exon regions are dispensable for efficient splicing of RmInt1 but strictly needed for retrohoming [78, 80]. Once the intron has reverse spliced into its homing site, a free 3′-hydroxyl (OH) end needs to be generated for reverse transcription to start. Depending on the origin of the 3′-OH end, retrohoming follows two different pathways (Figure 6.4). In the En-dependent and replication-independent retrohoming pathway (Figure 6.4a), i.e. typical of Ll.LtrB from *Lactococcus lactis* (*L.l.*I1) [81], the En domain cleaves the opposite strand (antisense) of the DNA target site, and the resulting 3′-OH end is then used by the RT domain to produce a cDNA complementary to the intron (Table 6.1). In contrast, in the En-independent and replication-dependent retrohoming pathway, i.e. typical of RmInt1 or of retrohoming Ll.LtrB En^- mutants, the free 3′-OH end is generated by a nascent strand synthetized by DNA polymerase III (PolIII) (Figure 6.4b). RmInt1, which reverse splices into transient ssDNA because its IEP lacks the En and most DBD domains [74], predominantly uses the lagging strand (nascent Okazaki fragments or RNA primers synthesized by the primase) [84], while Ll.LtrB En^- mutants, which can insert themselves into dsDNA, use the leading strand from an approaching replication fork [88]. Other mechanisms to initiate reverse transcription have also been described

Figure 6.4 GIIi retrohoming pathways. (a) Schematic showing retrohoming in introns coding an IEP with an endonuclease domain (i.e. Ll.LtrB). The sense strand, containing the DNA target site, is represented by a black arrow; the antisense strand is represented by a gray arrow. The reverse transcriptase (RT) and endonuclease (En) domains of the IEP are indicated. The host factor RNaseH is represented as an H letter in a green circle. (b) Schematic showing retrohoming in introns coding an IEP lacking the endonuclease domain or with an En domain containing impairing mutations (i.e. RmInt1 and Ll.LtrB YRT mutant). Both parental DNA strands are represented by black arrows. The leading strand synthesized by the DNA polymerase III is indicated by a blue arrow. In the lagging strand, the RNA primers synthesized by the primase are displayed as violet arrows; the Okazaki fragments, which are synthesized from the RNA primers, are indicated as light purple arrows. The products of the reaction, after replication is complete, are indicated at the bottom: the black arrows represent the original parental strains; the blue arrow, the leading strand; and the light purple arrow, the lagging strand. In all cases, the exons surrounding the intron and the empty DNA target sites are represented by boxes with red (E1) and green (E2) bright colors in the sense strand and by boxes with red (E1) and green (E2) pale colors in the antisense strand. Source: (a) Adapted from Cousineau et al. [81].

(Table 6.1), i.e. yeast aI2 intron retrohoming requires double-strand break repair (DSBR) recombination with the intron donor site [82]. The third and final steps during intron retrohoming depend on host protein activities. Tens of retrohoming factors have been identified in *Escherichia coli* [89]: RNA degradation factors, such as RNaseH1 or PolI acting as 5′–3′-exonuclease; second-strand cDNA synthesis factors, dependent on Pol III and replication start enzymes; and nick ligation factors, i.e. DNA ligase A. Some of these factors seem to have different roles during retrohoming in introns lacking the En domain. For instance, ExoIII (3′–5′ dsDNA exonuclease) inhibits the retrohoming of Ll.LtrB in *E. coli*, but it stimulates that of RmInt1 in *Si. meliloti*, probably to accommodate retrohoming into the replication process [90].

Different from retrohoming, retrotransposition [86] occurs by direct reverse splicing into dsDNA or ssDNA at ectopic sites similar, but not identical to the homing site, and its mechanism depends on the genomic environment [91]. For instance, Ll.LtrB retrotransposes predominantly in an En-independent manner into ssDNA in its native host *L. lactis* but in an En-dependent manner into dsDNA in the heterologous host *E. coli* [87]. Environmental signals can also stimulate retrotransposition: amino acid and glucose starvation promote retrotransposition through the small molecule alarmones ppGpp and cAMP, respectively [92]. Besides favoring genetic diversity, retrotransposition is also a hazard, which the host organism counteracts via a battery of mechanisms, i.e. deploying proteins, such as the RNA decay complex RNaseE, to limit GIIi spread in response to particular physiological conditions [93], or gradually inactivating functional mobile introns by accumulation of mutation, fragmentation, and structural loss at the 3′-splice junction and in the IEP [40, 67, 94, 95]. GIIi, especially bacterial ones, counterbalance such tight control from the host, i.e. exceptionally using non-cognate alternative IEPs after horizontal transfer, as for the *Sr.md*.I1 IEP from *Sinorhizobium medicae* that supports RmInt1 homing in *Si. meliloti* [65], or evolving to protect the host, as for the interruption of mobile elements by RmInt1 (see Section 6.3.2).

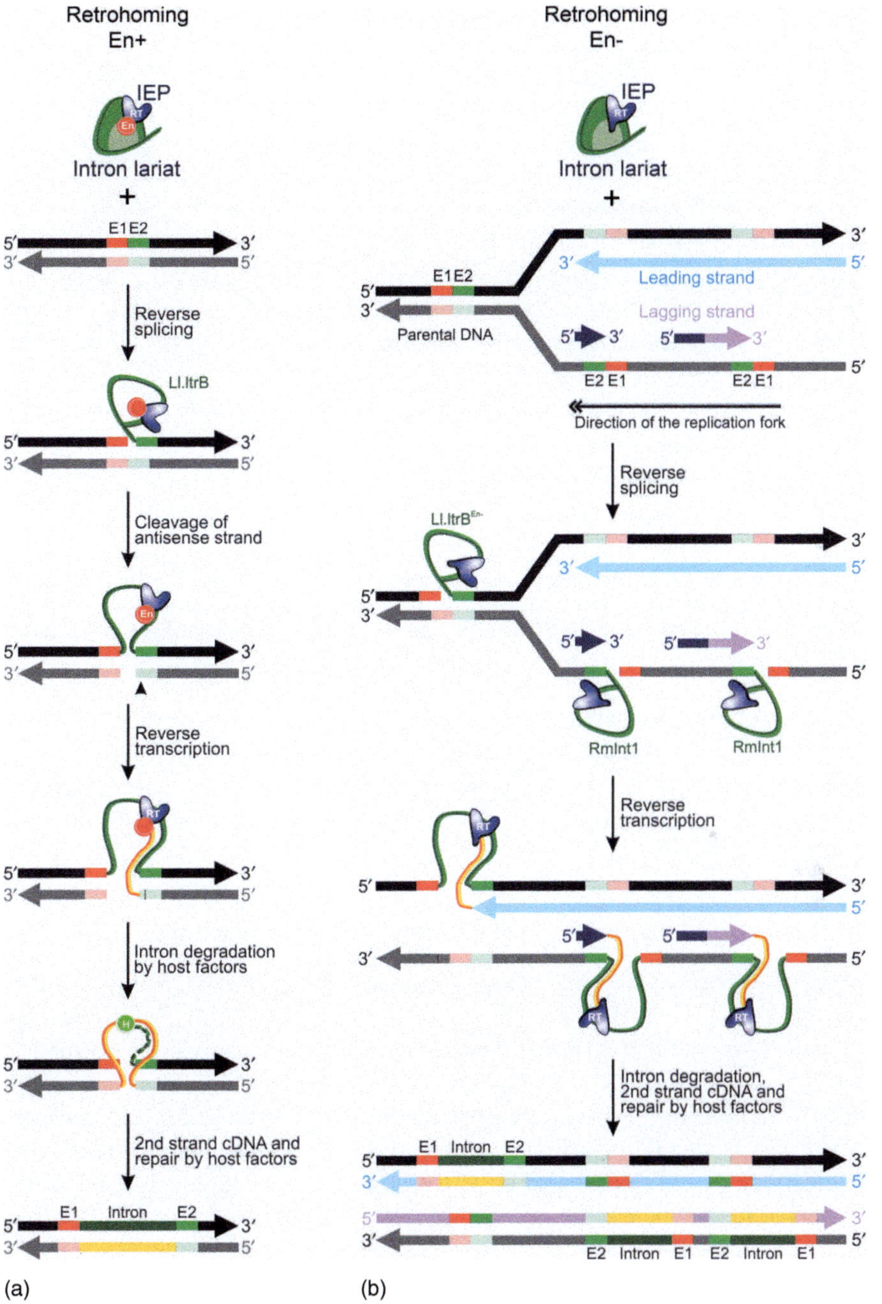
Retrohoming
En+
IEP
RT
En
Intron lariat
+
5′
E1 E2
3′
3′
5′
Reverse
splicing
Ll.ltrB
Cleavage of
antisense strand
Reverse
transcription
Intron degradation
by host factors
2nd strand cDNA and
repair by host factors
E1
Intron
E2
(a)
Retrohoming
En-
Leading strand
Lagging strand
Parental DNA
E2 E1
Direction of the replication fork
Ll.ltrB[En-]
RmInt1
Intron degradation,
2nd strand cDNA and
repair by host factors
(b)

Table 6.1 Characteristics of retrohoming and retrotransposition pathways in GIIi.

	Characteristics[a)]	Example	References
	Retrohoming		
Yeast	Homologous recombination		
	Endonuclease dependent	*S.c.*aI2	[82]
Bacteria	No homologous recombination		
	Endonuclease dependent – coupled to replication in leading strand	Ll.LtrB	[83]
	Endonuclease independent – coupled to replication into lagging strand	RmInt1	[84]
	Endonuclease independent – nickase dependent	Ll.LtrB	[85]
	Retrotransposition		
All	No homologous recombination		
	Endonuclease independent – coupled to replication into lagging strand	Ll.LtrB	[86]
	Endonuclease dependent – into dsDNA	Ll.LtrB	[87]

a) Characteristics are not exhaustive. They refer to most of the introns described in each category.

6.4 Structural Architecture

GIIi owe their ability to catalyze the splicing reaction to the fact that they adopt a well-ordered 3D structure, which is conserved across different GIIi classes (Figure 6.2) [32, 34, 37, 38].

6.4.1 Secondary Structure and Long-Range Tertiary Interactions

GIIi possess a conserved arrangement of six secondary structural domains (D1–D6, Figure 6.2) [7].

D1 is the largest and 5′-terminal domain and is further subdivided into subdomains (D1a, D1b, D1c, D1d1, D1d2, and D1[i/ii]). D1 controls intron folding by forming a rigid five-way junction and two flexible internal loops in D1c and D1d (see Section 6.4.2). Folded D1 is further stabilized by a kissing loop interaction between D1b and D1d2 (α–α′) common to all GIIi classes and by additional interactions idiosyncratic to specific classes, i.e. kissing loop (β–β′) between D1c and D1d2 in class IIA and some IIB introns and a ribose zipper between D1d1 and D1d2 (ω–ω′) in class IIC introns. D1 also provides specific interactions for the downstream domains (i.e. two tetraloop:tetraloop-receptor interactions, one between D1c and D2 [θ–θ′] and another between D1d1 and D5 [ζ–ζ′]; interaction between D1c1 and the internal bulge of D6 [ι–ι′, [96]]) and for the flanking exons (EBS1 and EBS2 in D1d2 of classes IIA and IIB, and EBS3 or δ in D1d1; see also Section 6.3.1). Finally, D1 contributes to stabilize the active site by two main interaction network: the so-called Z-anchor, which connects D1c to nucleotides near the 5′-splice junction

(ε–ε') and near the two-nucleotide bulge in D5 (λ–λ'), and the κ loop, in which D1d1 interacts with nucleotides of the D5 catalytic triad (κ–κ'). Due to this intricate network of interactions, D1 serves as a functional control element contributing to the efficiency and fidelity of the splicing reaction [35].

D2, D3, and D4 form peripheral, but not less important motifs. D2, which rarely harbors an IEP ORF, interacts with D1c (see above) and D6 (η–η'), helping orient the 3′-splice junction in the second step of splicing, and it connects to D3 via the highly conserved catalytic junction J2/3, which likely mediates structural rearrangements during splicing [34]. D3 allosterically contributes to shaping the active site via its terminal loop 3a, which connects to D5 nucleotides adjacent to the catalytic triad (μ–μ') [97]. By contrast, D4 points out of the intron core, and in most introns, it harbors the IEP ORF [57].

D5 and D6 are the key functional domains. D5 is the most conserved GIIi region [98] and presents (i) the so-called catalytic triad that forms a triple helix together with J2/3, (ii) a terminal loop that interacts with D1 (see above), and (iii) the so-called 2-nucleotide bulge that bends the D5 stem to interact both with the triple helix and with motifs adjacent to the 5′- and 3′-splice junctions. D6, which is less conserved and less structurally characterized, harbors the bulged nucleotide (mostly an adenosine) that forms the lariat branch site, it is connected to the 3′-splice junction, it establishes key interactions for correct positioning during catalysis (i.e. η–η' and ι–ι'; see above), and it is anchored into the active site by interactions between the penultimate nucleotide and J2/3 (γ–γ') and between EBS3 and IBS3. Finally, some GIIi possess a structured extension at the 3′-end, called domain 7 (D7), which may improve the efficiency of the second step of splicing [99, 100].

6.4.2 Folding

GIIi folding is a tightly controlled process in which the mature active site is assembled when the splice junctions are properly aligned to ensure splicing fidelity and occurs in a rapid, energetically inexpensive manner to maximize splicing efficiency. Current studies on ai5γ and on *O.i.*I1 have shown that these introns evolved a partially hierarchical, apparently two-step folding pathway, different from that of other ribozymes like RNase P and group I introns [101–104]. According to a "first comes, first folds" paradigm [35], D1 folds first (within seconds in vitro) guided by the formation of the κ motif in D1d1 [103], in a concerted, cooperative, and synchronized manner, avoiding off-pathway kinetic traps [105]. In vivo, the IEP or other cellular chaperones facilitate this process [106]. Intra-D1 motifs and interactions – such as the Z-anchor, T-loop, κ motif, α–α', ω–ω', and EBS1–IBS1 – are formed at this stage, as confirmed by a crystal structure of isolated D1 [35]. As soon as D1 is properly folded, downstream domains are immediately assembled [104, 107]. Two flexible regions in D1c and D1d1 ensure conformational sampling to search for interaction partners in D2–D6 and become stabilized upon the formation of the lock-and-key recognition θ–θ' and κ–κ' interactions and of the ζ–ζ' and λ–λ' interactions, which form by induced fit [35]. Final assembly of the active site structure and the trigger of the first catalytic event likely depend on a conformational switch in the J2/3 region, which may rotate between an inactive (toggled) and an active (triple helix) state along the splicing cycle [34].

6.4.3 Stabilization by Solvent and IEP

Besides intramolecular long-range interactions, the GIIi structure is stabilized by solvent molecules and the IEP. Ions play a dual role in a metallo-ribozyme like the GIIi [17, 101]. A heteronuclear four metal ion center in the intron core is essential to promote catalysis (see Section 6.4.4). However, ions also bind key structural regions, such as secondary structure elements, like GA mismatches or GU wobble pairs, or tertiary structure motifs, such as the S-turn (D3) or the T-loop (D1a), or nucleotides involved in long-range interactions, such as tetraloop-receptor interactions, the Z-anchor, and the ω–ω′ ribose zipper [108]. Notably, more than 30% of site-specific ions present in the mature GIIi structure are already bound to the D1 folding intermediate [35, 108]. Among such site-bound ions, most correspond to mono- and divalent metals, but some correspond to the anionic side chains of acidic residues of the IEP [108]. Detailed insights into IEP recruitment and positioning on the GIIi ribozyme can now be obtained from the cryo-EM structure of Ll.LtrB [38]. The N-terminal RT, fingers–palm and thumb domains, and the C-terminal DBD and En domains of the LtrA IEP bind D1 and D4 of the Ll.LtrB RNA [109]. Additionally, LtrA interacts with exonic segments of the Ll.LtrB RNA. Such intron–IEP interactions facilitate splicing. Equivalent contacts of the IEP with the target DNA are likely also important for recognition of the cognate site during retrotransposition (see also Section 6.3.3).

6.4.4 Active Site and Reaction Mechanism

GIIi self-splicing proceeds via a series of consecutive transesterification reactions, which can now be described at medium to high resolution (2.7 to ~4.0 Å) on the basis of crystal structures. In the first step of splicing, a water molecule or the 2′-OH of the branch site attacks the 5′-splice junction, separating the 5′-exon from an intron/3′-exon linear (in case of hydrolysis, [110]) or lariat (in case of branching, [13]) intermediate. In the precatalytic state (PDB ID: 4FAQ), the catalytic triad (C358–G359–C360 in *O.i.*I1), the two-nucleotide bulge (A376–C377), and the J2/3 junction (A287–G288–C289) adopt a triple helix conformation in the active site (Figure 6.5). Such motifs bind a four metal ion cluster, formed by two divalent (M1 and M2) and by two monovalent ions (K1 and K2). Coordinated by M1 and M2, the scissile phosphate presents its pro-R oxygen atom to the two metals. Additionally, M1 coordinates the 3′-OH of the last nucleotide of the exon, while M2 coordinates the reaction nucleophile. Molecular dynamics simulations suggest that the bulk water accepts the nucleophile's proton during its attack on the scissile phosphate [111]. K1 stabilizes the triple helix conformation, whereas K2 stabilizes the 5′-end of the intron inducing a kink of ~50° at the scissile position (Figure 6.5) [112]. Upon hydrolysis (PDB ID: 4FAR), the scissile phosphate rotates toward K2, while the 5′-exon and the triple helix do not displace significantly.

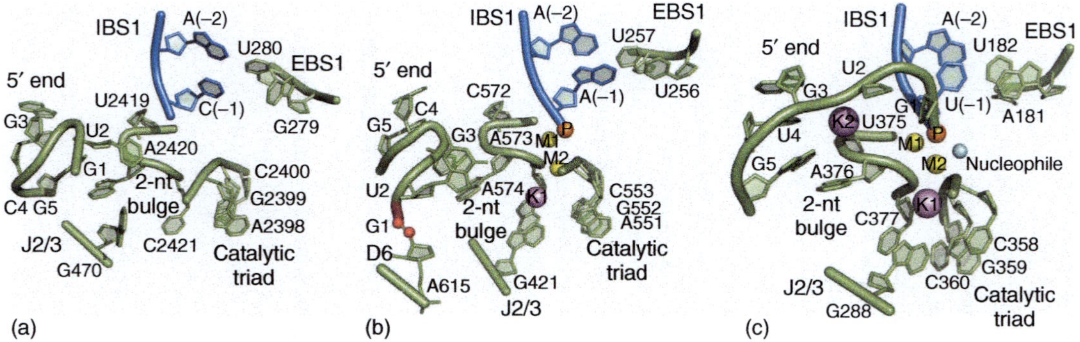

Figure 6.5 Active site representations of three group II introns of know 3D structure. (a) Ll.LtrB in the postcatalytic state (PDB ID: 5G2X). Ions are not modeled due to the low structural resolution (3.8 Å). (b) *P.li.*LSUI1 in the postcatalytic lariat form (PDB ID: 4R0D). Lariat bond is depicted in red. (c) *O.i.*I1 in the precatalytic state (PDB ID: 4FAQ). The intron backbone is depicted as a cartoon representation in green. The intron (IBS1 site) is depicted in blue. When present, the scissile phosphate is represented as an orange sphere, M1 and M2 catalytic ions as yellow spheres, K1 and K2 ions as purple spheres, and the reaction nucleophile as a cyan sphere.

At this stage of the reaction, the 5′-exon is already optimally placed to act as the nucleophile of the second step of splicing, but the 5′-end of the intron needs to be released from the active site to make space for the 3′-splice junction. There is no precise structural description of how the 3′-splice junction is recruited, because a structure of the pre-second step of splicing is missing. However, the 5′-end of the intron could represent a steric constrain that impedes the access of the 3′-splice junction to the catalytic core, particularly in the lariat conformation, when it is bound at the branch site in D6 just few nucleotides away from the last intron nucleotide. Based on enzymatic and structural data (PDB ID: 4FAU and 4FAX), it was therefore proposed that release of first-step products and recruitment of second-step reactants depend on a conformational change of the J2/3 junction and two-nucleotide bulge motifs that transiently breaks the triple helix conformation [34].

After such rearrangement, the active site triple helix is formed again along with the intact heteronuclear metal ion cluster M1-M2-K1-K2 remaining bound to the fully spliced postcatalytic intron and is thus likely to contribute to retrotransposition (PDB ID: 4E8M, 4E8K, and 4FAW; [34]).

Interestingly, the GIIi active site is very similar to that of many other divalent metal ion-dependent DNA/RNA processing enzymes, including other splicing machineries like the spliceosome, nucleases like Cas9, and DNA/RNA polymerases [34, 113]. All these enzymes adopt a similar active site geometry, with the two divalent ions (M1 and M2) chelated by acidic groups of strictly conserved catalytic residues, the phosphodiester substrate bound at specific binding sites [114], and the two positively charged residues (K1 and K2) located in the second coordination shell of M1/M2, possibly reflecting common needs of DNA and RNA processing enzymes to ensure the fidelity, specificity, and efficiency of their reaction [113].

6.5 Lessons and Tools from Group II Intron Research

Besides providing molecular insights into the splicing reaction, evolutionary, functional, and structural work on the GIIi also facilitated the understanding of other splicing machineries, stimulated the study of other large RNAs, and guided the design of biotechnological tools to exploit the GIIi genetic potential.

6.5.1 Analogies to Other Splicing Machineries

GIIi are evolutionarily, functionally, and chemically related to the eukaryotic spliceosome [1]. Both machineries form lariat products and possess a nearly identical active site dependent on metal ions for catalysis [115]. The strictly conserved AGC catalytic triad and two-nucleotide bulge in spliceosomal U6 snRNA [116] are analogous to the GIIi triad and bulge in D5 [117]. In both systems, these conserved triads form triple helices with other RNA components of the active site, the J2/3 junction in GIIi and the ACAGAGA box in spliceosomal U6 snRNA [28]. Mechanistically, the two systems are also analogous: both undergo a two-step reaction with the same stereochemistry [115], which involves rearrangements between the splicing steps that facilitate the exchange of first- and second-step reactants and products in the active site [118]. Finally, the GIIi IEP possesses an analogous molecular architecture and thus likely a common evolutionary origin, with spliceosomal subunit Prp8 [119], and both proteins establish analogous contacts with the RNA cores of GIIi and the spliceosomal RNA, respectively [120].

6.5.2 Lessons to Study Other Large Non-coding RNAs

Besides important lessons for the biology and mechanism of splicing, GIIi research also stimulated the development of experimental approaches to study other large RNA molecules. GIIi are among the largest RNAs that have been crystallized so far. Crystallization and in general structural biology techniques have strict requirements in terms of purity and homogeneity of the targets. Methods that have been developed over the years to produce, purify, and probe the conformational stability and secondary architecture of the GIIi have recently been used successfully to gain initial insights into a novel class of transcripts, long non-coding RNAs, which participate in epigenetic regulation of development, response to stress, and adaptation to the environment [121]. The non-denaturing purification methods required to crystallize GIIi [32] have so far been proven to be an ideal method to produce many long non-coding RNAs [122, 123]. Crystallization strategies developed for the GIIi also revealed suitable for other ribozymes and large RNAs [35, 124, 125].

6.5.3 Biotechnological Applications of GIIi

Both the GIIi RNA and IEP can be engineered into biotechnological tools, i.e. for gene editing in bacteria or for the development of highly processive RTs [5, 126]. Because of their ability to recognize and modify target DNA sites with high efficiency, GIIi pioneered the use of RNA-guided site-specific genome modification [127]. Engineering of the EBS–IBS interactions aided by computer algorithms [128] can retarget GIIi and induce the resulting “targetrons” to integrate into desired genomic locations [129], i.e. to inactivate a specific gene in bacteria or to introduce a selectable marker or gene of interest embedded in GIIi D4 [130–132]. These systems are less prone to off-target integrations than other gene editing tools (i.e. one in 10^6 for Ll.LtrB, compared with one in 10^4 for CRISPR-Cas9) [87, 133] but face two bottlenecks, particularly in eukaryotic cells. First, the formation of the intron–IEP complex in the eukaryotic nucleus is challenging [134, 135]. Second, nonsense-mediated decay of GIIi, translational repression of IEPs, and the dependency of GIIi folding on high magnesium concentrations are difficult to overcome [136]. Thus, implementation of targetrons in eukaryotes was so far successful only for Ll.LtrB and RmInt1, for which retrohoming into plasmid DNA targets could be obtained, but at low efficiency [39, 137]. Moreover, the GIIi IEPs can be engineered for molecular biology. Their RT domains typically have high fidelity and processivity to ensure efficient splicing and mobility. Thus, they are particularly suitable, i.e. for creation of RNA-seq libraries [138]. For instance, the *Thermosynechococcus elongatus* thermostable group II intron reverse transcriptase (TGIRT) has higher thermal stability (T_m = 81 versus 57 °C), processivity (up to

~700 versus ~200 nt long), and fidelity (two- to fourfold higher) than retroviral RTs [139], while *Eubacterium rectale* RT (marathon RT) can synthesize up to 9.5 kb cDNA with comparable fidelity to TGIRT [140].

6.6 Perspectives and Open Questions

Despite the tremendous achievements in GIIi research, questions still remain open on the mechanism of catalysis, IEP association, and biotechnological potential of this important ribozyme. Regarding the mechanism of catalysis, the state preceding the second step of splicing can currently only be derived from the structure of a postcatalytic group IIB intron (PDB ID: 4R0D), while the geometry of the branch site before the first step of splicing can only be inferred from the precatalytic structure of an intron that splices through hydrolysis (PDB ID: 4FAQ). The dynamics that lead to intron rearrangements between the two steps of splicing can only be deduced from intron structures obtained in the presence of nonphysiological ions (PDB ID: 4FAU, 4FAX). Molecular dynamics simulations coupled to fluorescence spectroscopy of the active site derivatives, novel time-resolved approaches in crystallography, or electron microscopy may possibly provide clues about the energy, timing, and concatenation of events required during such crucial catalytic transitions.

The oligomeric state of the IEP remains controversial. Current structures reveal the presence of a single intron-bound monomer for Ll.LtrB, but such mode of interaction leaves exposed a predicted protein binding surface on GIIi D1 [38]. Clarifying the mode of intron–IEP recognition may have profound effects in our understanding of intron folding, stability, and retrotransposition.

Finally, the optimization of targetrons for gene editing in eukaryotes remains a challenge, which – if overcome – would boost genome engineering thanks to the high efficiency, fidelity, and rare off-target effects of GIIi.

In summary, the GIIi retroelements, which has kept researchers busy for the last 30 years, still has great potential for future investigation and may lead to unexpected, beneficial discoveries in the years to come.

Acknowledgments

We acknowledge Dr. Olga Fedorova (Yale University, New Haven, USA) and Dr. María Dolores Molina-Sánchez and Dr. Rafael Nisa-Martínez (Estación Experimental del Zaidín, Consejo Superior de Investigaciones Científicas [EEZ-CSIC], Granada, Spain) for critical reading of the manuscript.

References

1 Cech, T.R. (1986). The generality of self-splicing RNA: relationship to nuclear mRNA splicing. *Cell* 44: 207–210.

2 Pyle, A.M. and Lambowitz, A.M. (2006). Group II introns: ribozymes that splice RNA and invade DNA. In: *The RNA World*, 3e (eds. R.F. Gesteland, T.R. Cech and J.F. Atkins), 469–505. Cold Spring Harbor: Cold Spring Harbor Press.
3 Dai, L. and Zimmerly, S. (2002). The dispersal of five group II introns among natural populations of *Escherichia coli*. *RNA* 8: 1294–1307.
4 Mattick, J.S. (1994). Introns: evolution and function. *Curr. Opin. Genet. Dev.* 4: 823–831.
5 Enyeart, P.J., Mohr, G., Ellington, A.D., and Lambowitz, A.M. (2014). Biotechnological applications of mobile group II introns and their reverse transcriptases: gene targeting, RNA-seq, and non-coding RNA analysis. *Mob. DNA* 5: 2–20.
6 Halbreich, A., Pajot, P., Foucher, M. et al. (1980). A pathway of cytochrome *b* mRNA processing in yeast mitochondria: specific splicing steps and an intron-derived circular RNA. *Cell* 19: 321–329.
7 Michel, F., Jacquier, A., and Dujon, B. (1982). Comparison of fungal mitochondrial introns reveals extensive homologies in RNA secondary structure. *Biochimie* 64: 867–881.
8 Michel, F., Umesono, K., and Ozeki, H. (1989). Comparative and functional anatomy of group II catalytic introns – a review. *Gene* 82: 5–30.
9 Ferat, J.L. and Michel, F. (1993). Group II self-splicing introns in bacteria. *Nature* 364: 358–361.
10 Dai, L. and Zimmerly, S. (2003). ORF-less and reverse-transcriptase-encoding group II introns in archaebacteria, with a pattern of homing into related group II intron ORFs. *RNA* 9: 14–19.
11 Valles, Y., Halanych, K.M., and Boore, J.L. (2008). Group II introns break new boundaries: presence in a bilaterian's genome. *PLoS One* 3: e1488.
12 van der Veen, R., Arnberg, A.C., van der Horst, G. et al. (1986). Excised group II introns in yeast mitochondria are lariats and can be formed by self-splicing *in vitro*. *Cell* 44: 225–234.
13 Peebles, C.L., Perlman, P.S., Mecklenburg, K.L. et al. (1986). A self-splicing RNA excises an intron lariat. *Cell* 44: 213–223.
14 Schmelzer, C. and Schweyen, R.J. (1986). Self-splicing of group II introns *in vitro*: mapping of the branch point and mutational inhibition of lariat formation. *Cell* 46: 557–565.
15 Sharp, P.A. (1987). Splicing of messenger RNA precursors. *Science* 235: 766–771.
16 Augustin, S., Muller, M.W., and Schweyen, R.J. (1990). Reverse self-splicing of group II intron RNAs *in vitro*. *Nature* 343: 383–386.
17 Podar, M., Perlman, P.S., and Padgett, R.A. (1995). Stereochemical selectivity of group II intron splicing, reverse splicing, and hydrolysis reactions. *Mol. Cell. Biol.* 15: 4466–4478.
18 Chin, K. and Pyle, A.M. (1995). Branch-point attack in group II introns is a highly reversible transesterification, providing a potential proofreading mechanism for 5′-splice site selection. *RNA* 1: 391–406.
19 Meunier, B., Tian, G., Macadre, C. et al. (1990). Group II introns transpose in yeast mitochondria. In: *Structure Function and Biogenesis of Energy Transfer*

Systems: Proceedings of the International Symposium on Structure, Function, and Biogenesis of Energy Transfer Systems, Bari, Italy (9–11 July 1989), 1e (eds. E. Quagliariello, S. Papa, F. Palmieri and C. Saccone), 169–174. Amsterdam: Elsevier.

20 Toor, N., Hausner, G., and Zimmerly, S. (2001). Coevolution of group II intron RNA structures with their intron-encoded reverse transcriptases. *RNA* 7: 1142–1152.

21 Lambowitz, A.M. and Perlman, P.S. (1990). Involvement of aminoacyl-tRNA synthetases and other proteins in group I and group II intron splicing. *Trends Biochem. Sci.* 15: 440–444.

22 Kennell, J.C., Moran, J.V., Perlman, P.S. et al. (1993). Reverse transcriptase activity associated with maturase-encoding group II introns in yeast mitochondria. *Cell* 73: 133–146.

23 Zimmerly, S., Guo, H., Eskes, R. et al. (1995). A group II intron RNA is a catalytic component of a DNA endonuclease involved in intron mobility. *Cell* 83: 529–538.

24 Mueller, M.W., Allmaier, M., Eskes, R., and Schweyen, R.J. (1993). Transposition of group II intron aI1 in yeast and invasion of mitochondrial genes at new locations. *Nature* 366: 174–176.

25 Sellem, C.H., Lecellier, G., and Belcour, L. (1993). Transposition of a group II intron. *Nature* 366: 176–178.

26 Michels, W.J. Jr., and Pyle, A.M. (1995). Conversion of a group II intron into a new multiple-turnover ribozyme that selectively cleaves oligonucleotides: elucidation of reaction mechanism and structure/function relationships. *Biochemistry* 34: 2965–2977.

27 Koch, J.L., Boulanger, S.C., Dib-Hajj, S.D. et al. (1992). Group II introns deleted for multiple substructures retain self-splicing activity. *Mol. Cell. Biol.* 12: 1950–1958.

28 de Lencastre, A., Hamill, S., and Pyle, A.M. (2005). A single active-site region for a group II intron. *Nat. Struct. Mol. Biol.* 12: 626–627.

29 Chanfreau, G. and Jacquier, A. (1994). Catalytic site components common to both splicing steps of a group II intron. *Science* 266: 1383–1387.

30 Boudvillain, M., de Lencastre, A., and Pyle, A.M. (2000). A tertiary interaction that links active-site domains to the 5′ splice site of a group II intron. *Nature* 406: 315–318.

31 Boudvillain, M. and Pyle, A.M. (1998). Defining functional groups, core structural features and inter-domain tertiary contacts essential for group II intron self-splicing: a NAIM analysis. *EMBO J.* 17: 7091–7104.

32 Toor, N., Keating, K.S., Taylor, S.D., and Pyle, A.M. (2008). Crystal structure of a self-spliced group II intron. *Science* 320: 77–82.

33 Marcia, M., Somarowthu, S., and Pyle, A.M. (2013). Now on display: a gallery of group II intron structures at different stages of catalysis. *Mobile DNA* 4: 14–26.

34 Marcia, M. and Pyle, A.M. (2012). Visualizing group II intron catalysis through the stages of splicing. *Cell* 151: 497–507.

35 Zhao, C., Rajashankar, K.R., Marcia, M., and Pyle, A.M. (2015). Crystal structure of group II intron domain 1 reveals a template for RNA assembly. *Nat. Chem. Biol.* 11: 967–972.

36 Costa, M., Walbott, H., Monachello, D. et al. (2016). Crystal structures of a group II intron lariat primed for reverse splicing. *Science* 354: aaf9258.

37 Robart, A.R., Chan, R.T., Peters, J.K. et al. (2014). Crystal structure of a eukaryotic group II intron lariat. *Nature* 514: 193–197.

38 Qu, G., Kaushal, P.S., Wang, J. et al. (2016). Structure of a group II intron in complex with its reverse transcriptase. *Nat. Struct. Mol. Biol.* 23: 549–557.

39 Guo, H., Karberg, M., Long, M. et al. (2000). Group II introns designed to insert into therapeutically relevant DNA target sites in human cells. *Science* 289: 452–457.

40 Fernandez-Lopez, M., Munoz-Adelantado, E., Gillis, M. et al. (2005). Dispersal and evolution of the *Sinorhizobium meliloti* group II RmInt1 intron in bacteria that interact with plants. *Mol. Biol. Evol.* 22: 1518–1528.

41 LaRoche-Johnston, F., Monat, C., and Cousineau, B. (2016). Recent horizontal transfer, functional adaptation and dissemination of a bacterial group II intron. *BMC Evol. Biol.* 16: 223–234.

42 Candales, M.A., Duong, A., Hood, K.S. et al. (2012). Database for bacterial group II introns. *Nucleic Acids Res.* 40: D187–D190.

43 Simon, D.M., Kelchner, S.A., and Zimmerly, S. (2009). A broadscale phylogenetic analysis of group II intron RNAs and intron-encoded reverse transcriptases. *Mol. Biol. Evol.* 26: 2795–2808.

44 Toro, N. (2003). Bacteria and archaea group II introns: additional mobile genetic elements in the environment. *Environ. Microbiol.* 5: 143–151.

45 Toro, N. and Martinez-Abarca, F. (2013). Comprehensive phylogenetic analysis of bacterial group II intron-encoded ORFs lacking the DNA endonuclease domain reveals new varieties. *PLoS One* 8: e55102.

46 Zimmerly, S., Hausner, G., and Wu, X. (2001). Phylogenetic relationships among group II intron ORFs. *Nucleic Acids Res.* 29: 1238–1250.

47 Brouard, J.S., Turmel, M., Otis, C., and Lemieux, C. (2016). Proliferation of group II introns in the chloroplast genome of the green alga *Oedocladium carolinianum* (Chlorophyceae). *Peer J.* 4: e2627.

48 Hausner, G., Olson, R., Simon, D. et al. (2006). Origin and evolution of the chloroplast trnK (matK) intron: a model for evolution of group II intron RNA structures. *Mol. Biol. Evol.* 23: 380–391.

49 Odom, O.W., Shenkenberg, D.L., Garcia, J.A., and Herrin, D.L. (2004). A horizontally acquired group II intron in the chloroplast psbA gene of a psychrophilic *Chlamydomonas*: in vitro self-splicing and genetic evidence for maturase activity. *RNA* 10: 1097–1107.

50 Toro, N., Molina-Sanchez, M.D., and Fernandez-Lopez, M. (2002). Identification and characterization of bacterial class E group II introns. *Gene* 299: 245–250.

51 Michel, F., Costa, M., and Westhof, E. (2009). The ribozyme core of group II introns: a structure in want of partners. *Trends Biochem. Sci.* 34: 189–199.

52 Costa, M., Michel, F., and Westhof, E. (2000). A three-dimensional perspective on exon binding by a group II self-splicing intron. *EMBO J.* 19: 5007–5018.

53 Toor, N., Robart, A.R., Christianson, J., and Zimmerly, S. (2006). Self-splicing of a group IIC intron: 5′ exon recognition and alternative 5′ splicing events implicate the stem-loop motif of a transcriptional terminator. *Nucleic Acids Res.* 34: 6461–6471.

54 Zhao, C. and Pyle, A.M. (2016). Crystal structures of a group II intron maturase reveal a missing link in spliceosome evolution. *Nat. Struct. Mol. Biol.* 23: 558–565.

55 Zimmerly, S. and Semper, C. (2015). Evolution of group II introns. *Mobile DNA* 6: 7–25.

56 Nagy, V., Pirakitikulr, N., Zhou, K.I. et al. (2013). Predicted group II intron lineages E and F comprise catalytically active ribozymes. *RNA* 19: 1266–1278.

57 Simon, D.M., Clarke, N.A., McNeil, B.A. et al. (2008). Group II introns in eubacteria and archaea: ORF-less introns and new varieties. *RNA* 14: 1704–1713.

58 Toro, N., Martinez-Abarca, F., Molina-Sanchez, M.D. et al. (2018). Contribution of mobile group II introns to *Sinorhizobium meliloti* genome evolution. *Front. Microbiol.* 9: 627.

59 Solem, A., Zingler, N., Pyle, A.M., and Li-Pook-Than, J. (2009). Group II introns and their protein collaborators. In: *Non-protein Coding RNAs*, 1e (eds. N.G. Walter, S.A. Woodson and R.T. Batey), 167–182. Berlin: Springer.

60 Cao, W., Coman, M.M., Ding, S. et al. (2011). Mechanism of Mss116 ATPase reveals functional diversity of DEAD-box proteins. *J. Mol. Biol.* 409: 399–414.

61 Wahleithner, J.A., MacFarlane, J.L., and Wolstenholme, D.R. (1990). A sequence encoding a maturase-related protein in a group II intron of a plant mitochondrial nad1 gene. *Proc. Natl. Acad. Sci. U.S.A.* 87: 548–552.

62 Reifschneider, O., Marx, C., Jacobs, J. et al. (2016). A ribonucleoprotein supercomplex involved in trans-splicing of organelle group II introns. *J. Biol. Chem.* 291: 23330–23342.

63 Massel, K., Silke, J.R., and Bonen, L. (2016). Multiple splicing pathways of group II trans-splicing introns in wheat mitochondria. *Mitochondrion* 28: 23–32.

64 Nisa-Martinez, R., Jimenez-Zurdo, J.I., Martinez-Abarca, F. et al. (2007). Dispersion of the RmInt1 group II intron in the *Sinorhizobium meliloti* genome upon acquisition by conjugative transfer. *Nucleic Acids Res.* 35: 214–222.

65 Martinez-Rodriguez, L., Garcia-Rodriguez, F.M., Molina-Sanchez, M.D. et al. (2014). Insights into the strategies used by related group II introns to adapt successfully for the colonisation of a bacterial genome. *RNA Biol.* 11: 1061–1071.

66 Chillon, I., Martinez-Abarca, F., and Toro, N. (2011). Splicing of the *Sinorhizobium meliloti* RmInt1 group II intron provides evidence of retroelement behavior. *Nucleic Acids Res.* 39: 1095–1104.

67 Dai, L. and Zimmerly, S. (2002). Compilation and analysis of group II intron insertions in bacterial genomes: evidence for retroelement behavior. *Nucleic Acids Res.* 30: 1091–1102.

68 Martinez-Abarca, F., Zekri, S., and Toro, N. (1998). Characterization and splicing *in vivo* of a *Sinorhizobium meliloti* group II intron associated with particular insertion sequences of the IS630-Tc1/IS3 retroposon superfamily. *Mol. Microbiol.* 28: 1295–1306.

69 Costa, M., Michel, F., Molina-Sanchez, M.D. et al. (2006). An alternative intron–exon pairing scheme implied by unexpected *in vitro* activities of group II intron RmInt1 from *Sinorhizobium meliloti*. *Biochimie* 88: 711–717.

70 Costa, M., Michel, F., and Toro, N. (2006). Potential for alternative intron–exon pairings in group II intron RmInt1 from *Sinorhizobium meliloti* and its relatives. *RNA* 12: 338–341.

71 Chillon, I., Molina-Sanchez, M.D., Fedorova, O. et al. (2014). *In vitro* characterization of the splicing efficiency and fidelity of the RmInt1 group II intron as a means of controlling the dispersion of its host mobile element. *RNA* 20: 2000–2010.

72 Molina-Sanchez, M.D., Martinez-Abarca, F., and Toro, N. (2006). Excision of the *Sinorhizobium meliloti* group II intron RmInt1 as circles *in vivo*. *J. Biol. Chem.* 281: 28737–28744.

73 Novikova, O. and Belfort, M. (2017). Mobile group II introns as ancestral eukaryotic elements. *Trends Genet.* 33: 773–783.

74 Lambowitz, A.M. and Belfort, M. (2015). Mobile bacterial group II introns at the crux of eukaryotic evolution. *Microbiol. Spectr.* 3. 10.1128/microbiolspec.MDNA3-0050-2014.

75 Aizawa, Y., Xiang, Q., Lambowitz, A.M., and Pyle, A.M. (2003). The pathway for DNA recognition and RNA integration by a group II intron retrotransposon. *Mol. Cell* 11: 795–805.

76 Guo, H., Zimmerly, S., Perlman, P.S., and Lambowitz, A.M. (1997). Group II intron endonucleases use both RNA and protein subunits for recognition of specific sequences in double-stranded DNA. *EMBO J.* 16: 6835–6848.

77 Mohr, G., Smith, D., Belfort, M., and Lambowitz, A.M. (2000). Rules for DNA target-site recognition by a lactococcal group II intron enable retargeting of the intron to specific DNA sequences. *Genes Dev.* 14: 559–573.

78 Jimenez-Zurdo, J.I., Garcia-Rodriguez, F.M., Barrientos-Duran, A., and Toro, N. (2003). DNA target site requirements for homing *in vivo* of a bacterial group II intron encoding a protein lacking the DNA endonuclease domain. *J. Mol. Biol.* 326: 413–423.

79 Xiang, Q., Qin, P.Z., Michels, W.J. et al. (1998). Sequence specificity of a group II intron ribozyme: multiple mechanisms for promoting unusually high discrimination against mismatched targets. *Biochemistry* 37: 3839–3849.

80 Barrientos-Duran, A., Chillon, I., Martinez-Abarca, F., and Toro, N. (2011). Exon sequence requirements for excision *in vivo* of the bacterial group II intron RmInt1. *BMC Mol. Biol.* 12: 24–31.

81 Cousineau, B., Smith, D., Lawrence-Cavanagh, S. et al. (1998). Retrohoming of a bacterial group II intron: mobility via complete reverse splicing, independent of homologous DNA recombination. *Cell* 94: 451–462.

82 Eskes, R., Liu, L., Ma, H. et al. (2000). Multiple homing pathways used by yeast mitochondrial group II introns. *Mol. Cell. Biol.* 20: 8432–8446.

83 Zimmerly, S., Guo, H., Perlman, P.S., and Lambowitz, A.M. (1995). Group II intron mobility occurs by target DNA-primed reverse transcription. *Cell* 82: 545–554.

84 Martinez-Abarca, F., Barrientos-Duran, A., Fernandez-Lopez, M., and Toro, N. (2004). The RmInt1 group II intron has two different retrohoming pathways for mobility using predominantly the nascent lagging strand at DNA replication forks for priming. *Nucleic Acids Res.* 32: 2880–2888.

85 Novikova, O., Smith, D., Hahn, I. et al. (2014). Interaction between conjugative and retrotransposable elements in horizontal gene transfer. *PLoS Genet.* 10: e1004853.

86 Cousineau, B., Lawrence, S., Smith, D., and Belfort, M. (2000). Retrotransposition of a bacterial group II intron. *Nature* 404: 1018–1021.

87 Coros, C.J., Landthaler, M., Piazza, C.L. et al. (2005). Retrotransposition strategies of the *Lactococcus lactis* Ll.LtrB group II intron are dictated by host identity and cellular environment. *Mol. Microbiol.* 56: 509–524.

88 Zhong, J. and Lambowitz, A.M. (2003). Group II intron mobility using nascent strands at DNA replication forks to prime reverse transcription. *EMBO J.* 22: 4555–4565.

89 Smith, D., Zhong, J., Matsuura, M. et al. (2005). Recruitment of host functions suggests a repair pathway for late steps in group II intron retrohoming. *Genes Dev.* 19: 2477–2487.

90 Nisa-Martinez, R., Molina-Sanchez, M.D., and Toro, N. (2016). Host factors influencing the retrohoming pathway of group II intron RmInt1, which has an intron-encoded protein naturally devoid of endonuclease activity. *PLoS One* 11: e0162275.

91 Dickson, L., Huang, H.R., Liu, L. et al. (2001). Retrotransposition of a yeast group II intron occurs by reverse splicing directly into ectopic DNA sites. *Proc. Natl. Acad. Sci. U.S.A.* 98: 13207–13212.

92 Coros, C.J., Piazza, C.L., Chalamcharla, V.R. et al. (2009). Global regulators orchestrate group II intron retromobility. *Mol. Cell* 34: 250–256.

93 Coros, C.J., Piazza, C.L., Chalamcharla, V.R., and Belfort, M. (2008). A mutant screen reveals RNase E as a silencer of group II intron retromobility in *Escherichia coli*. *RNA* 14: 2634–2644.

94 Molina-Sanchez, M.D. and Toro, N. (2015). Inactivation of group II intron RmInt1 in the *Sinorhizobium meliloti* genome. *Sci. Rep.* 5: 12036.

95 Toro, N., Martinez-Rodriguez, L., and Martinez-Abarca, F. (2014). Insights into the history of a bacterial group II intron remnant from the genomes of the nitrogen-fixing symbionts *Sinorhizobium meliloti* and *Sinorhizobium medicae*. *Heredity* 113: 306–315.

96 Li, C.F., Costa, M., and Michel, F. (2011). Linking the branchpoint helix to a newly found receptor allows lariat formation by a group II intron. *EMBO J.* 30: 3040–3051.

97 Toor, N., Keating, K.S., Fedorova, O. et al. (2010). Tertiary architecture of the *Oceanobacillus iheyensis* group II intron. *RNA* 16: 57–69.

98 Peebles, C.L., Zhang, M., Perlman, P.S., and Franzen, J.S. (1995). Catalytically critical nucleotide in domain 5 of a group II intron. *Proc. Natl. Acad. Sci. U.S.A.* 92: 4422–4426.

99 Stabell, F.B., Tourasse, N.J., and Kolsto, A.B. (2009). A conserved 3′ extension in unusual group II introns is important for efficient second-step splicing. *Nucleic Acids Res.* 37: 3202–3214.

100 Stabell, F.B., Tourasse, N.J., Ravnum, S., and Kolsto, A.B. (2007). Group II intron in *Bacillus cereus* has an unusual 3′ extension and splices 56 nucleotides downstream of the predicted site. *Nucleic Acids Res.* 35: 1612–1623.

101 Swisher, J.F., Su, L.J., Brenowitz, M. et al. (2002). Productive folding to the native state by a group II intron ribozyme. *J. Mol. Biol.* 315: 297–310.

102 Fedorova, O., Waldsich, C., and Pyle, A.M. (2007). Group II intron folding under near-physiological conditions: collapsing to the near-native state. *J. Mol. Biol.* 366: 1099–1114.

103 Waldsich, C. and Pyle, A.M. (2007). A folding control element for tertiary collapse of a group II intron ribozyme. *Nat. Struct. Mol. Biol.* 14: 37–44.

104 Su, L.J., Waldsich, C., and Pyle, A.M. (2005). An obligate intermediate along the slow folding pathway of a group II intron ribozyme. *Nucleic Acids Res.* 33: 6674–6687.

105 Pyle, A.M., Fedorova, O., and Waldsich, C. (2007). Folding of group II introns: a model system for large, multidomain RNAs? *Trends Biochem. Sci.* 32: 138–145.

106 Fedorova, O., Solem, A., and Pyle, A.M. (2010). Protein-facilitated folding of group II intron ribozymes. *J. Mol. Biol.* 397: 799–813.

107 Fedorova, O. and Zingler, N. (2007). Group II introns: structure, folding and splicing mechanism. *Biol. Chem.* 388: 665–678.

108 Marcia, M. and Pyle, A.M. (2014). Principles of ion recognition in RNA: insights from the group II intron structures. *RNA* 20: 516–527.

109 Watanabe, K. and Lambowitz, A.M. (2004). High-affinity binding site for a group II intron-encoded reverse transcriptase/maturase within a stem-loop structure in the intron RNA. *RNA* 10: 1433–1443.

110 Podar, M., Chu, V.T., Pyle, A.M., and Perlman, P.S. (1998). Group II intron splicing *in vivo* by first-step hydrolysis. *Nature* 391: 915–918.

111 Casalino, L., Palermo, G., Rothlisberger, U., and Magistrato, A. (2016). Who activates the nucleophile in ribozyme catalysis? An answer from the splicing mechanism of group II introns. *J. Am. Chem. Soc.* 138: 10374–10377.

112 Chan, R.T., Robart, A.R., Rajashankar, K.R. et al. (2012). Crystal structure of a group II intron in the pre-catalytic state. *Nat. Struct. Mol. Biol.* 19: 555–557.

113 Genna, V., Colombo, M., De Vivo, M., and Marcia, M. (2018). Second-shell basic residues expand the two-metal-ion architecture of DNA and RNA processing enzymes. *Structure* 26: 40–50.

114 Steitz, T.A. and Steitz, J.A. (1993). A general two-metal-ion mechanism for catalytic RNA. *Proc. Natl. Acad. Sci. U.S.A.* 90: 6498–6502.

115 Fica, S.M., Tuttle, N., Novak, T. et al. (2013). RNA catalyses nuclear pre-mRNA splicing. *Nature* 503: 229–234.

116 Madhani, H.D. and Guthrie, C. (1992). A novel base-pairing interaction between U2 and U6 snRNAs suggests a mechanism for the catalytic activation of the spliceosome. *Cell* 71: 803–817.

117 Keating, K.S., Toor, N., Perlman, P.S., and Pyle, A.M. (2010). A structural analysis of the group II intron active site and implications for the spliceosome. *RNA* 16: 1–9.

118 Konarska, M.M., Vilardell, J., and Query, C.C. (2006). Repositioning of the reaction intermediate within the catalytic center of the spliceosome. *Mol. Cell* 21: 543–553.

119 Galej, W.P., Oubridge, C., Newman, A.J., and Nagai, K. (2013). Crystal structure of Prp8 reveals active site cavity of the spliceosome. *Nature* 493: 638–643.

120 Galej, W.P., Toor, N., Newman, A.J., and Nagai, K. (2018). Molecular mechanism and evolution of nuclear pre-mRNA and group II intron splicing: insights from cryo-electron microscopy structures. *Chem. Rev.* 118: 4156–4176.

121 Pyle, A.M. (2014). Looking at lncRNAs with the ribozyme toolkit. *Mol. Cell* 56: 13–17.

122 Chillon, I., Marcia, M., Legiewicz, M. et al. (2015). Native purification and analysis of long RNAs. *Methods Enzymol.* 558: 3–37.

123 Somarowthu, S., Legiewicz, M., Chillon, I. et al. (2015). HOTAIR forms an intricate and modular secondary structure. *Mol. Cell* 58: 353–361.

124 Marcia, M. (2016). Using molecular replacement phasing to study the structure and function of RNA. *Methods Mol. Biol.* 1320: 233–257.

125 Marcia, M., Humphris-Narayanan, E., Keating, K.S. et al. (2013). Solving nucleic acid structures by molecular replacement: examples from group II intron studies. *Acta Crystallogr., Sect. D: Biol. Crystallogr.* 69: 2174–2185.

126 Wei, M.Q., Mengesha, A., Good, D., and Anne, J. (2008). Bacterial targeted tumour therapy-dawn of a new era. *Cancer Lett.* 259: 16–27.

127 Doudna, J.A. and Charpentier, E. (2014). Genome editing. The new frontier of genome engineering with CRISPR-Cas9. *Science* 346: 1258096.

128 Zhuang, F.L., Karberg, M., Perutka, J., and Lambowitz, A.M. (2009). EcI5, a group IIB intron with high retrohoming frequency: DNA target site recognition and use in gene targeting. *RNA* 15: 432–449.

129 Zhong, J., Karberg, M., and Lambowitz, A.M. (2003). Targeted and random bacterial gene disruption using a group II intron (targetron) vector containing a retrotransposition-activated selectable marker. *Nucleic Acids Res.* 31: 1656–1664.

130 Heap, J.T., Pennington, O.J., Cartman, S.T. et al. (2007). The ClosTron: a universal gene knock-out system for the genus *Clostridium*. *J. Microbiol. Methods* 70: 452–464.

131 Garcia-Rodriguez, F.M., Hernandez-Gutierrez, T., Diaz-Prado, V., and Toro, N. (2014). Use of the computer-retargeted group II intron RmInt1 of *Sinorhizobium meliloti* for gene targeting. *RNA Biol.* 11: 391–401.

132 Karberg, M., Guo, H.T., Zhong, J. et al. (2001). Group II introns as controllable gene targeting vectors for genetic manipulation of bacteria. *Nat. Biotechnol.* 19: 1162–1167.

133 Kim, D., Lim, K., Kim, S.T. et al. (2017). Genome-wide target specificities of CRISPR RNA-guided programmable deaminases. *Nat. Biotechnol.* 35: 475–480.

134 Chalamcharla, V.R., Curcio, M.J., and Belfort, M. (2010). Nuclear expression of a group II intron is consistent with spliceosomal intron ancestry. *Genes Dev.* 24: 827–836.

135 Reinoso-Colacio, M., Garcia-Rodriguez, F.M., Garcia-Canadas, M. et al. (2015). Localization of a bacterial group II intron-encoded protein in human cells. *Sci. Rep.* 5: 12716.

136 Mastroianni, M., Watanabe, K., White, T.B. et al. (2008). Group II intron-based gene targeting reactions in eukaryotes. *PLoS One* 3: e3121.

137 Garcia-Rodriguez, F.M., Barrientos-Duran, A., Diaz-Prado, V. et al. (2011). Use of RmInt1, a group IIB intron lacking the intron-encoded protein endonuclease domain, in gene targeting. *Appl. Environ. Microbiol.* 77: 854–861.

138 Carrell, S.T., Tang, Z., Mohr, S. et al. (2018). Detection of expanded RNA repeats using thermostable group II intron reverse transcriptase. *Nucleic Acids Res.* 46: e1.

139 Mohr, S., Ghanem, E., Smith, W. et al. (2013). Thermostable group II intron reverse transcriptase fusion proteins and their use in cDNA synthesis and next-generation RNA sequencing. *RNA* 19: 958–970.

140 Zhao, C., Liu, F., and Pyle, A.M. (2018). An ultraprocessive, accurate reverse transcriptase encoded by a metazoan group II intron. *RNA* 24: 183–195.

141 Manigrasso J, Chillon I, Genna V, Vidossich P, Somarowthu S, Pyle AM, De Vivo M and Marcia M (2020). Visualizing group II intron dynamics between the first and second steps of splicing. *Nat Commun 11*, 2837. doi: 10.1038/s41467-020-16741-4.

7

The Spliceosome: an RNA–Protein Ribozyme Derived From Ancient Mobile Genetic Elements

Erin L. Garside[1], Oliver A. Kent[2], and Andrew M. MacMillan[2]

[1] *University of Alberta, Department of Biochemistry, Edmonton, AB T6G 2H7, Canada*
[2] *University Health Network, Princess Margaret Cancer Centre, 101 College Street, Toronto, ON M5G 1L7, Canada*

7.1 Discovery of Introns and Splicing

In 1977, efforts to map the adenovirus genome – specifically the correspondence between viral mRNAs and their coding DNA – showed the lack of strict complementarity between the two [1, 2]. This discovery of a split gene structure where coding "exon" sequences were separated by noncoding intervening or "intron" sequences was quickly confirmed to be common to eukaryotic precursor messenger RNAs (pre-mRNAs) although the number of introns present in an organism varies from tens of thousands to several dozen. The observation of a discontinuity between the DNA template and the mature mRNA in eukaryotes suggested that the initially transcribed pre-mRNA was modified to yield the mature product, a process referred to as splicing [3].

Intervening sequences have since been characterized in a wide variety of contexts including precursor ribosomal RNAs. The pre-rRNA of the ciliate *Tetrahymena* contains an intervening sequence belonging to the larger class of group I introns. In 1981, Cech et al. reported that in the course of their studies of ribosomal RNA transcription, they had observed splicing of the *Tetrahymena* pre-rRNA consistent with the splicing activity being intrinsic to the intron itself as an autocatalytic activity [4]. At about the same time, the RNA component of the RNase P enzyme was shown to possess the catalytic activity mediating the processing of pre-tRNAs [5]. These two discoveries overturned the idea that protein enzymes were the sole biocatalysts – a dogma that had developed since the discovery of the catalytic activity of the protein urease.

Intron excision in eukaryotic pre-mRNAs takes place through two sequential transesterification reactions. In the first step, the 2′ hydroxyl of a conserved adenosine within the intron carries out the attack at the 5′ splice site, generating the free 5′ exon and an intron-3′ exon lariat intermediate. This is followed by attack of the 5′ exon at the 3′ splice site to yield the ligated exon and a branched lariat intron product (Figure 7.1) [6]. In the autocatalytic group I reaction, the 3′ hydroxyl

Ribozymes, First Edition. Edited by Sabine Müller, Benoît Masquida, and Wade Winkler.

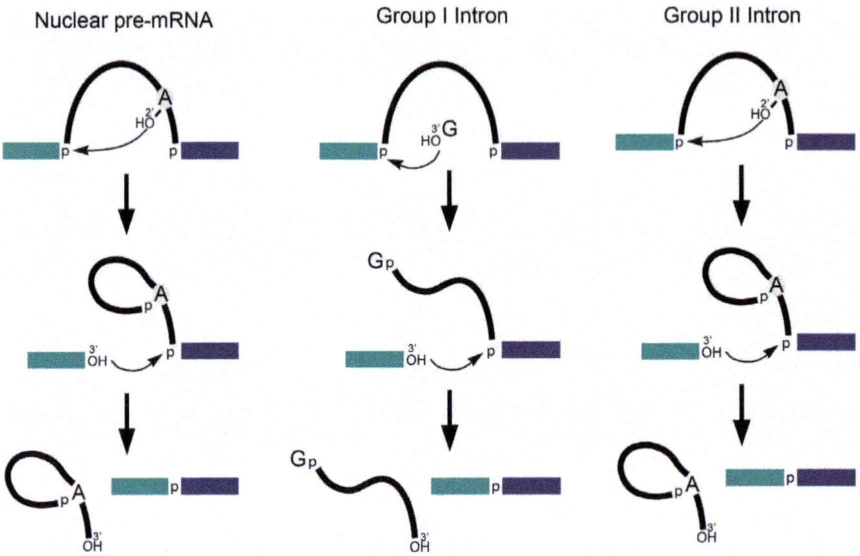

Figure 7.1 Chemical mechanisms of splicing. Splicing of nuclear pre-mRNA (left), group I (middle), group II (right) introns. Source: Adapted from Cech [6].

of an exogenous guanosine cofactor attacks the 5′ splice site, yielding the free 5′ exon and linear 3′ product with guanosine ligated at the 5′ end of the intron. The attack of the 5′ exon at the 3′ splice site produces the ligated exon product and free intron (Figure 7.1) [7]. A third class of long intervening sequences distinct from those contained within nuclear pre-mRNAs and the group I RNAs are the group II introns, encoded in the mitochondria and chloroplasts of plants and fungi and are also widely distributed in bacteria [8]. As with the group I introns, splicing of the precursor group II RNA is an autocatalytic event in which the RNA component, ranging in size from ~90 to thousands of nucleotides, of a ribonucleoprotein particle (RNP) is the essential and sufficient basis of splicing mediated by two conserved domains, V and VI, found within the intron. Intriguingly, the mechanism of group II splicing is directly analogous to that of pre-mRNA splicing involving initial attack of an adenosine within the intron at the 5′ splice site followed by displacement at the 3′ splice site to produce the ligated product (Figure 7.1c) [9]. The mechanistic similarity immediately raises the question about the relationship between group II intron and pre-mRNA splicing.

7.2 snRNPs and the Spliceosome

Characterization of nuclear pre-mRNA splicing in both yeast and metazoans showed that the splicing process was dependent on a large RNP assembly, referred to as the spliceosome [10, 11], involving five small nuclear ribonucleoprotein particles (snRNPs): U1, U2, U4, U5, and U6, each containing the eponymous small nuclear RNA (snRNA) as well as associated proteins. These spliceosomal snRNAs represent a subclass of uridine-rich RNAs that had been identified and partially characterized in the early 1970s although at that time no cellular role for these RNAs was known [12]. One challenge in identifying spliceosomal snRNAs in diverse species was the lack of conserved sequence. For example, the Ul snRNA from budding yeast is much larger than U1 snRNA in humans (900 versus 152 nucleotides). There is however a remarkable conservation of secondary snRNA structure from yeast to humans with short sequences of primary structure identity.

Pre-mRNA splicing is directed by conserved sequences with the introns: sequences at the 5′ and 3′ splice sites, the branch sequence containing the first step adenosine nucleophile, and in metazoans, a pyrimidine-rich sequence separating the branch and 3′ splice site sequences (Figure 7.2a). Conserved features of the spliceosomal snRNAs include a complementarity within both the U1 and U6 snRNAs to the consensus 5′ splice site and complementarity between the U2 snRNA and the branch point sequence (BPS). An extensive U4 and U6 snRNA base pairing conserved across species is mutually exclusive with a highly conserved U2–U6 snRNA interaction (Figure 7.2b). A consequence of the U2–U6 snRNA interaction is the formation of a U6 structure, the internal stem loop (ISL) shown to be critical for splicing. This combination of U2 and U6 snRNA base pairings with the pre-RNA along with the U2–U6 pairing serves to juxtapose all of the reactive elements of the intron (Figure 7.2c).

7.3 The Spliceosomal Cycle

Analysis by non-denaturing gel electrophoresis of pre-mRNAs incubated in splicing extracts revealed the sequential formation of large complexes – E (early in humans) or CC (commitment complex in yeast), A, B, and C – on the RNA. This was ultimately determined to represent the stepwise association of snRNPs and non-snRNP splicing factors with the pre-mRNA substrate to assemble the active spliceosome; further genetic, biochemical, and structural analyses have elaborated this pathway with evidence for additional distinct spliceosomal complexes (Figure 7.3).

In the assembly pathway, initial recognition of pre-mRNA introns involves base pairing of the Ul RNA with the 5′ splice site along with association of protein factors, SF1 and U2AF in metazoans, with the branch region, pyrimidine tract and 3′ splice site to form the E complex. All of these interactions are transient and presumably relate to overall regulation of splicing accuracy and efficiency. The U1 snRNA–splice site interaction is in some cases dispensable. Depletion of U1 snRNP from metazoan splicing extracts can be compensated for by the presence of SR proteins a set of factors required for constitutive splicing that also regulate alternative splicing events [14]. In addition, U1 snRNP is entirely absent in the minimal spliceosome of the red alga *Cyanidioschyzon merolae*, an organism containing only 27 introns [15].

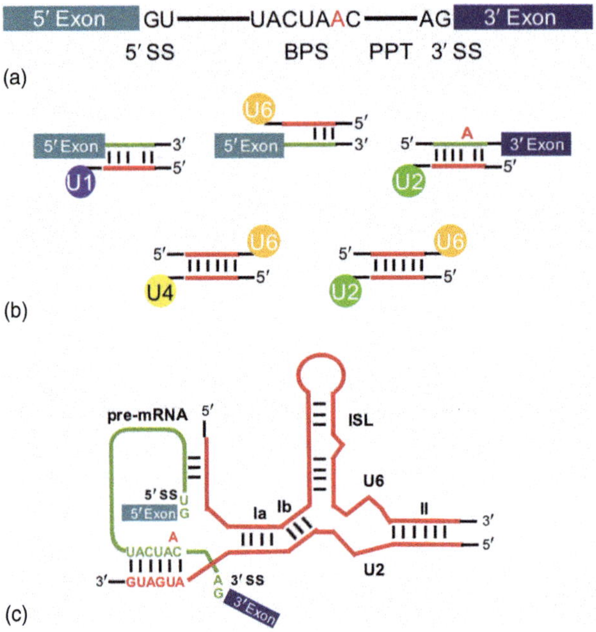

Figure 7.2 Recognition of pre-mRNA introns. (a) Conserved intron sequences at the 5′ and 3′ splice sites and at the branch point sequence (BPS) (branch A in red). The polypyrimidine tract in metazoans separates the BPS and 3′ splice site (b). (Top) Conserved complementarity between the U1 and U6 snRNAs and 5′ splice site and the U2 snRNA and BPS. (Bottom) Mutually exclusive pairings between U4 and U6 snRNA and between U2 and U6 snRNA. (c) U6 snRNA–5′ splice site interaction combined with U2 snRNA–BPS interaction along with U2 snRNA–U6 snRNA association juxtaposes reactive elements of the intron.

The initial protein-based recognition of the 3′ splice site region is replaced by the recruitment of the U2 snRNP to the intron, an association mediated by base pairing of U2 snRNA with the branch sequence, resulting in the formation of the A complex. This base pairing is incomplete, resulting in the bulging of the branch adenosine thereby selected as the nucleophile for the first step of splicing. Displacement of U1 snRNP from the 5′ splice site is concomitant with the association of the U4/U6–U5 tri-snRNP mediated by U6 snRNA-splice site interaction, yielding the B complex. Within the tri-snRNP, there is an extensive base pairing between the U4 and U6 snRNAs. Unwinding of the U4/U6 base pairing allows the pairing of the U2 and U6 snRNAs with each other, an association that was shown to be required for the two transesterification steps of splicing. Activation of the B complex to the B^{act} complex leads to the first transesterification followed by rearrangement to the C complex. The second step occurs in the subsequent C* complex. Further rearrangement to post-catalytic P and intron lariat spliceosome (ILS) complexes result in the release of the ligated mRNA and recycling of spliceosomal factors (Figure 7.3; reviewed in [13]).

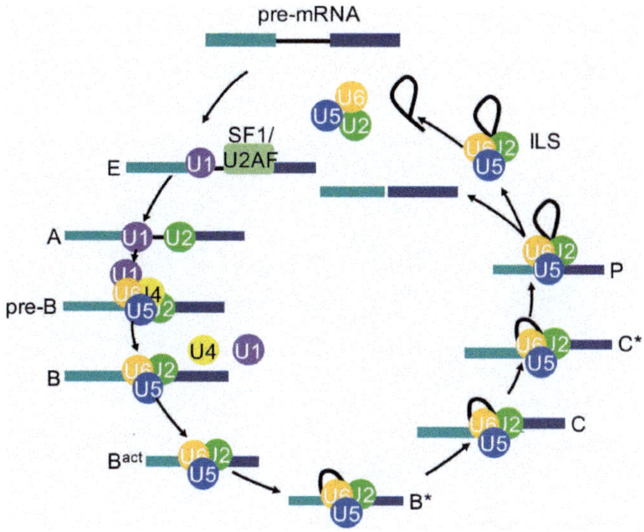

Figure 7.3 Spliceosome assembly pathway. Source: Adapted from Will and Luhrmann [13].

7.4 Chemistry of Splicing

The mechanisms of group I, group II, and nuclear pre-mRNA splicing are distinct but share intriguing commonalities. The first two classes of splicing are autocatalytic where a folded intron structure is both necessary and sufficient for promotion of the two transesterifications, while pre-mRNA splicing requires the formation of the spliceosome (a large RNP complex) on the RNA substrate. All three reactions were found to require the presence of divalent magnesium suggesting that the mechanism of the splicing reactions shares similarities with known metalloenzymes.

The discovery that the chemistry of splicing was conserved between the group II and nuclear introns led to the speculation, given the snRNA content of the spliceosome, that pre-mRNA splicing within the spliceosomal RNP was an RNA-catalyzed reaction. The direct association of U1 snRNA with the 5′ splice site also suggested a possible direct role of snRNA in the chemistry of the splicing reaction [16]. The subsequent discovery of interactions between U6 snRNA and the 5′ splice site, the role of U2 snRNA in selecting the branch nucleophile, and the close proximation of U5 snRNA to the splice sites were also consistent with an intimate role if not requirement for the snRNAs in splicing catalysis. Further to these findings, the observation of diverse classes of group II introns where the recruitment of two or three RNAs was required for assembly of the catalytic core resembled the requirement of *in trans* snRNAs for the spliceosomal reaction and reinforced the possible evolutionary relationship between autocatalytic group II and nuclear pre-mRNA splicing (Figure 7.4) [17].

Protein enzymes are involved in a diverse range of phosphoryl transfer reactions. Steitz and Steitz proposed a general two metal mechanism for enzymatic RNA

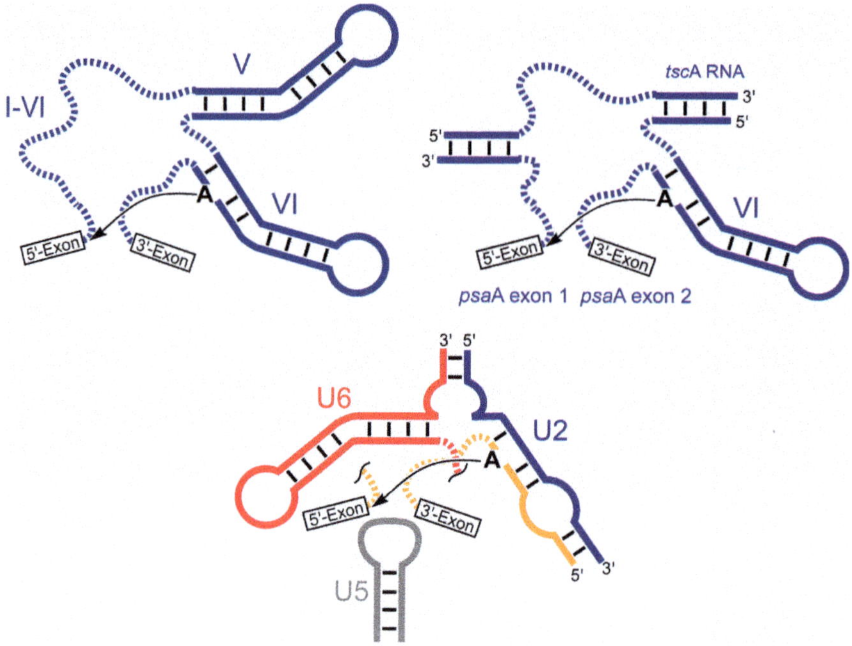

Figure 7.4 Comparison of group II and spliceosomal snRNA structures. (Top left) Participation of domains V and VI of a folded group II intron in splicing. (Top right) Three component assembly of a group II intron through association of chloroplast *psaA* introns mediated by the tscA RNA. (Bottom) Association of U6, U2, and U5 snRNAs at the heart of the pre-mRNA splicing machinery. Source: Adapted from Sharp [17].

cleavage based on the mechanism of the 3′,5′ exonuclease of DNA polymerase I in *Escherichia coli* [18]. This reaction yields a 5′ phosphate and 3′ hydroxyl without amino acid side chains participating directly in the chemistry of the reaction. Two metal ions 3.8 Å apart coordinated by three Asp and one Glu residue form inner-sphere contacts with the scissile phosphate and water nucleophile: one metal activates the nucleophile, while the second stabilizes the transition state and the 3′ oxyanion leaving group (Figure 7.5a). In the hydrolysis reaction catalyzed by *E. coli* alkaline phosphatase, two Zn^{2+} ions, coordinated by His and Asp side chains held 3.9 Å apart, are responsible for stabilizing the transition state in each of the two steps. A number of other enzymes catalyzing phosphoryl transfer, including the RNase H (RH) domain of HIV reverse transcriptase (RT), and the P1 nuclease, feature an active site with two metal ions positioned 3.9 Å apart. This suggests a common mechanism of phosphoryl transfer, conserved between evolutionarily unrelated proteins, involving two catalytic metal ions bound at the enzyme active site. In this model, the metal ions promote the activation of an incoming nucleophile by assisting in its deprotonation, stabilize the phosphoryl intermediate/transition state, and assist in the stabilization of the leaving group. This model of catalysis is perfectly compatible with a structured RNA active site positioning two catalytic metals in the appropriate position and was therefore proposed for the group I, group II, and spliceosomal reactions (Figure 7.5b).

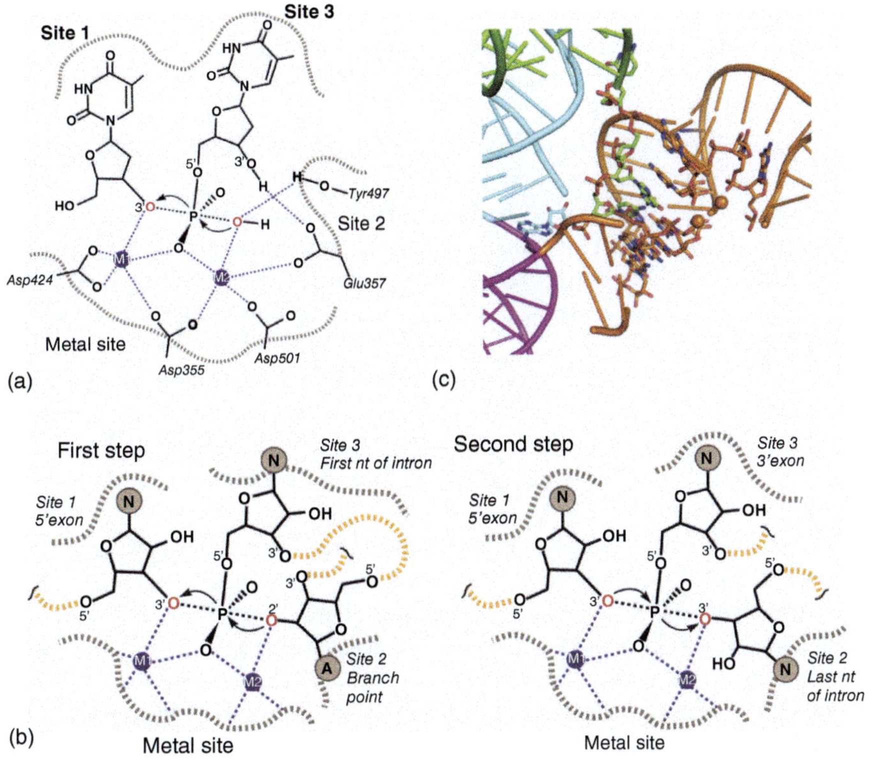

Figure 7.5 Two metal ion hypothesis. (a) Two metals at the active site of *E. coli* DNA polymerase I. (b) Proposed active site of the group II intron and spliceosome. (c) High-resolution X-ray structural analysis of the *Oceanobacillus iheyensis* Group II intron (PDB ID: 3BWP).

The two-metal hypothesis was ultimately supported by high-resolution X-ray structures of the group I [19] and group II [20] introns (Figure 7.5c). In the crystal structure of the group II intron from *Oceanobacillus iheyensis*, two metals 3.9 Å apart were observed at the proposed active site. The first has two inner-sphere ligands, phosphate oxygens within domain V, while the second is coordinated by phosphate oxygens from what has been referred to as the catalytic triad [20].

In the absence of high-resolution structural analysis, biochemical experiments involving chemically modified pre-mRNAs or snRNAs proved to be a powerful means to investigate spliceosomal catalysis. A key experimental approach has been the substitution of sulfur for oxygen in the phosphodiester linkages at non-bridging or bridging positions within the pre-mRNA substrate or snRNA. Impairment of reaction by a sulfur substitution can represent evidence of a required oxygen coordination to a metal ion such as Mg^{2+} because of the weak affinity of sulfur for magnesium. This requirement can be definitively demonstrated in an experiment where thiophilic Mn^{2+} or Cd^{2+} rescues the catalytic activity of the sulfur substitution.

Placement of phosphorothioates at both the 5′ and 3′ splice sites, within pre-mRNA substrates, allowed the determination of the stereochemistry of the cleavage reactions [21]. The observation of an inversion of stereochemistry in both steps suggested that each consisted of a single in-line SN_2-like displacement reaction analogous to other phosphoryl transfer reactions. Furthermore, the fact that only one of two possible sulfur substitutions, in the non-bridging pro-Sp position, at both splice sites allowed the reaction to proceed suggested that the other pro-Rp oxygen could be involved in metal coordination.

Substitution of the 3′ hydroxyl with sulfur in the scissile phosphate was also used to test the metal requirement for splicing [22]. Here, replacement of the Mg^{2+} cofactor with thiophilic Mn^{2+} allowed an otherwise impeded reaction to proceed in a metal rescue demonstrating stabilization of the leaving group by direct metal coordination.

Extensive genetic and biochemical studies established the importance of both internal U6 snRNA and U2–U6 snRNA structures in the catalysis of both steps of splicing [23]. Specifically, a conserved bulge in the U6 ISL as well as the presence of an AGC "catalytic triad" mirrored structures in domain V at the active site of the group II intron. Reconstitution of yeast U6 snRNP with U6 snRNA site-specifically modified with phosphorothioates allowed the testing of metal coordination in rescue experiments [24]. This study established a network of snRNA–metal interactions consistent with two RNA bound metals at the heart of the spliceosome at a single active site. Specifically, this work suggested that specific non-bridging oxygens on the *Saccharomyces cerevisiae* G78 and U80 of U6 snRNA coordinate catalytic metals during branching, while U59 and U80 coordinate metals for exon ligation.

In the spliceosome, the fact that phosphorothioate substitution impeded splicing, extended by metal rescue, and compared with the high-resolution X-ray analysis of the group II intron active site strongly supports a more detailed model of snRNA and substrate interaction with metal at the active site (Figure 7.6).

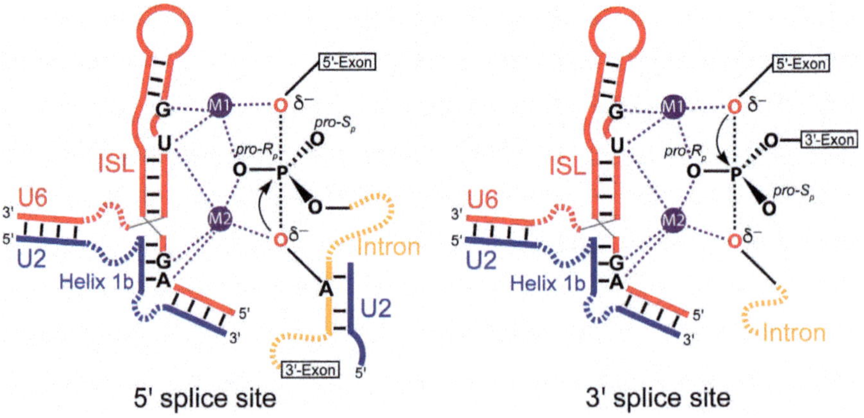

Figure 7.6 Model for the spliceosome active site. Inhibition of splicing by phosphorthioate substitution combined with metal rescue experiments and comparison to high-resolution X-ray structural analysis of the group II intron provide the basis of a model for the spliceosome active site at the time of the first (left) and second (right) transesterifications.

7.5 Spliceosome Structural Analysis

Spliceosomal structural studies have included X-ray and NMR analyses of protein and RNA components, RNP complexes, and cryo-EM studies of larger assemblies [25–27]. The heterogeneity of spliceosomes resulting from the complex structural rearrangements involved in spliceosome assembly, activation, and catalysis long suggested that single-particle analysis as available through cryo-electron microscopy would be the most fruitful approach in terms of a detailed structural study of the spliceosome. This has turned out to be the case, but it has been only recently with advances in data acquisition technologies that high-resolution structure determination has been possible.

Multiple structures of pre-spliceosomal and spliceosomal complexes have now been described. These include pre-catalytic, activated, and post-first and second step spliceosomes. These structures have validated conclusions based on decades of extensive genetic and biochemical experiments while providing a medium- to high-resolution structural model of the spliceosome active site. They reveal a sequential assembly of an RNA structure with two RNA-bound metals, consistent with genetic and biochemical studies, positioned to catalyze the two splicing steps. However, they also reveal the intimate role of proteins in stabilizing and regulating the formation and activity of the RNA core of the splicing machinery.

7.6 Spliceosome Structures

7.6.1 Pre-spliceosome: Tri-snRNP

In the pre-spliceosomal U4/U5/U6 tri-snRNP, as visualized by cryo-EM, the U4 and U6 snRNAs are base-paired, and this association persists in the spliceosomal B complex. U4 and U6 snRNAs form two stems separated by a U4 stem loop (Figure 7.7a,b); the proposed catalytic residues of U6 snRNA base-pair with U4 snRNA [28, 29]. Upstream of this duplex, U6 snRNA forms a 5′ stem loop structure distinct from the ISL that will form at the active site of the spliceosome; the 5′ end of the U6 snRNA ACAGA sequence that binds the 5′ splice site forms the tail end of the 5′ hairpin (Figure 7.7b). The nucleotides 5′ to this in U6 base-pair to the hairpin structure with the top of stem loop 1 of U5 snRNA, which will ultimately base-pair with the 3′ end of the 5′ exon. The sequestering of residues involved in catalysis suggests that the U4/U6 structure may therefore be regarded as a locking mechanism that prevents premature formation of the spliceosomal active site in advance of earlier proofreading steps.

A prominent feature of the tri-snRNP is PRP8, the largest protein in the spliceosome and a component of the U5 snRNP [30]. A variety of analyses established a link between PRP8 and the active site of the spliceosome. In photochemical cross-linking experiments, PRP8 was shown to directly interact with the 5′ splice site, the 3′ splice site, and the branch region [31]. In yeast genetic experiments, defects in splicing caused by splice site mutations are rescued by a large number of PRP8 suppressor

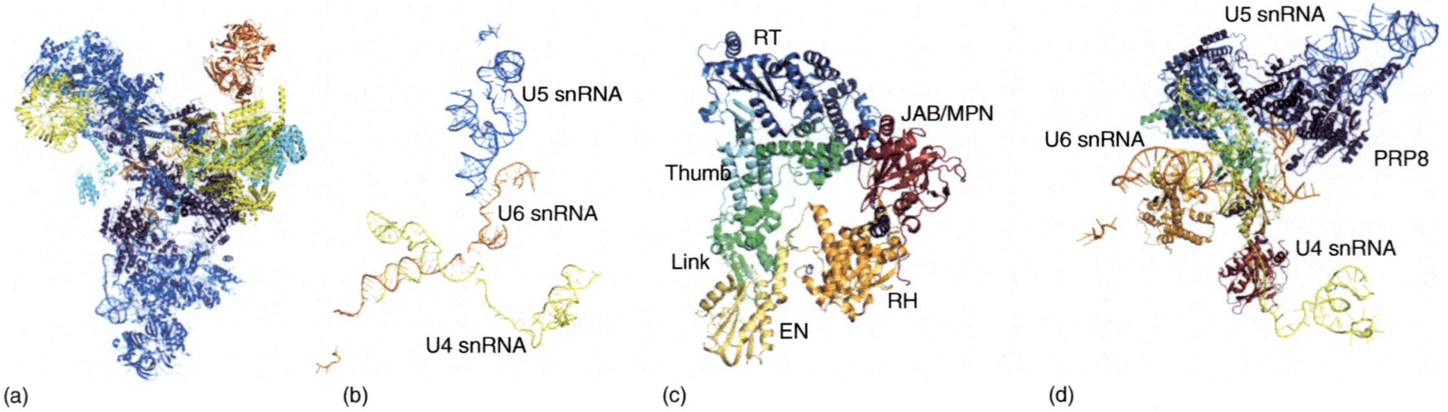

Figure 7.7 The pre-spliceosome: tri-snRNP structure. (a) Cryo-EM structure of the *S. cerevisiae* tri-snRNP (PDB: 5GAN). (b) RNA core of the *S. cerevisiae* tri-snRNP. U6 snRNA (orange) is extensively base-paired with U4 snRNA (yellow). Loop 1 of U5 snRNA (blue) interacts with U6 snRNA. (c) High-resolution X-ray structure of PRP8 showing reverse transcriptase-like (RT blue), linker region (green), endonuclease-like (EN; yellow), RNase H-like (RH; orange), and Jab/MPN (red) domains (PDB ID: 4143). (d) The RNA core of the tri-snRNP is bound by PRP8. Colors are consistent for Figures 7.7-7.12. Source: Adapted from Nguyen et al. [28].

alleles [30]. The clustering of these alleles alongside the cross-linking data raised the provocative notion that the active site of the spliceosome could contain both protein and RNA components. The original X-ray structural analysis of a large portion of PRP8 in a complex with the U5 snRNP assembly factor Aar2 confirmed the predicted presence of endonuclease (EN)-like and RT-like domains along with the previously characterized RNase H domain and the Jab/MPN domain that regulates the activity of Brr2, the ATPase responsible for the unwinding of the U4/U6 duplex [32] (Figure 7.7c).

The RT active site within PRP8 is mutated such that metal binding is precluded; although the RH and EN domains retain canonical metal binding sites, mutagenesis abrogating these sites does not affect yeast viability suggesting that these domains are structural in function. In the tri-snRNP, PRP8 scaffolds the spliceosomal RNA structure within the tri-snRNP with the U5/U6 snRNA interaction sandwiched between the N-terminal (NT) domain and the PRP8 core and the single-stranded 5′ tail of U6 snRNA passing through the cavity between these domains (Figure 7.7d).

7.6.2 Pre-spliceosome: A Complex

The structure of the *S. cerevisiae* A complex containing both the U1 and U2 snRNPs, associated with each other through two interfaces, and bound to pre-mRNA has been modeled based on cryo-EM data with a highest resolution from 4.0 (U1 snRNP) to 4.9 Å (U2 snRNP; Figure 7.8a). The extended U2 snRNP structure includes the SF3B particle, a highly conserved multi-protein complex.

At the 5′ end of the U1 snRNA, ten nucleotides become ordered through base pairing with the 5′ splice site with the duplex stabilized by interaction with two proteins Yhc1 and Luc7 through backbone (Yhc1, Luc7) and minor groove (Luc7) contacts. This includes zinc finger-mediated backbone contacts to the intron (+1 to +4 positions; Yhc1) and snRNA (U5-C8; Luc7). Yhc1 additionally stabilizes the duplex through contacts to the intron backbone at positions +5 and +6 (Figure 7.8b).

Within the U2 snRNP, the BPS UACUAAC is duplexed with the GUAGUA of U2 snRNA, and the branch A is bulged from the duplex, the basis of its selection as the first step nucleophile. In the A complex, this branch duplex is associated with SF3B, at a distance of ~130 Å from the 5′ splice site, with the branch adenosine nestled in the HEAT repeats of Hsh155 (SF3B1 in humans), the largest component of the SF3B complex (Figure 7.8c). Biochemical characterization of the human spliceosome suggests an early recognition and sequestering of the bulged duplex by SF3B6, a highly conserved protein that is absent in yeast SF3B, which presumably transitions to an SF3B1–branch duplex association [33].

7.6.3 B Complex

The structure of the *S. cerevisiae* B complex including the U2 and tri-snRNP has been captured with a highest resolution of 3 Å and represents a pre-spliceosome following the dissociation of U1 snRNP but before the Brr2-mediated unwinding of the U4/U6 snRNA duplex (Figure 7.9a) [28]. In the B complex, the disposition of U4/U6 and U5

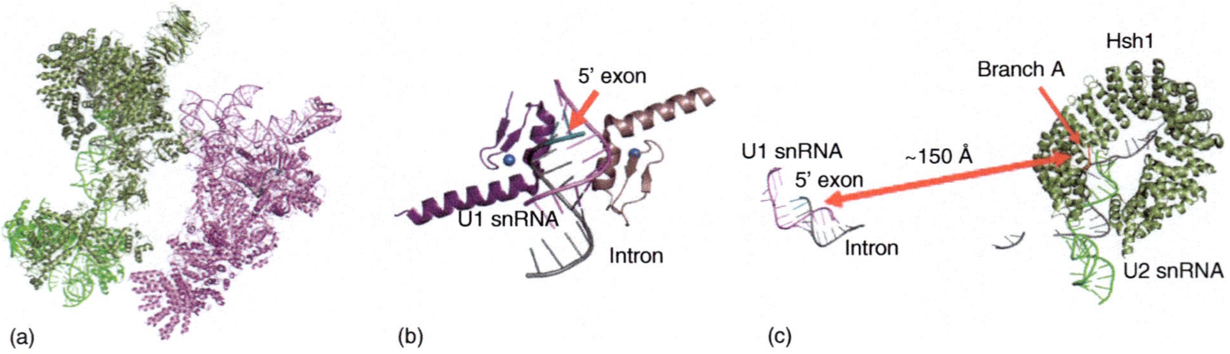

Figure 7.8 The pre-spliceosome: A complex structure. (a) The A complex is composed of U1 snRNP (violet) and U2 snRNP (green) interacting with the intron 5′ splice site and branch region, respectively. (b) Yhc1 (purple) and Luc7 (mauve) stabilize the 5′ splice site/U1 snRNP duplex with a zinc finger and an α-helix. (c) The branch A and 5′ splice site are separated by ~130 Å and the SF3b protein Hsh155. The branch A is bulged out of the branch point sequence/U2 snRNA duplex, selecting it as the nucleophile for the first step of splicing.

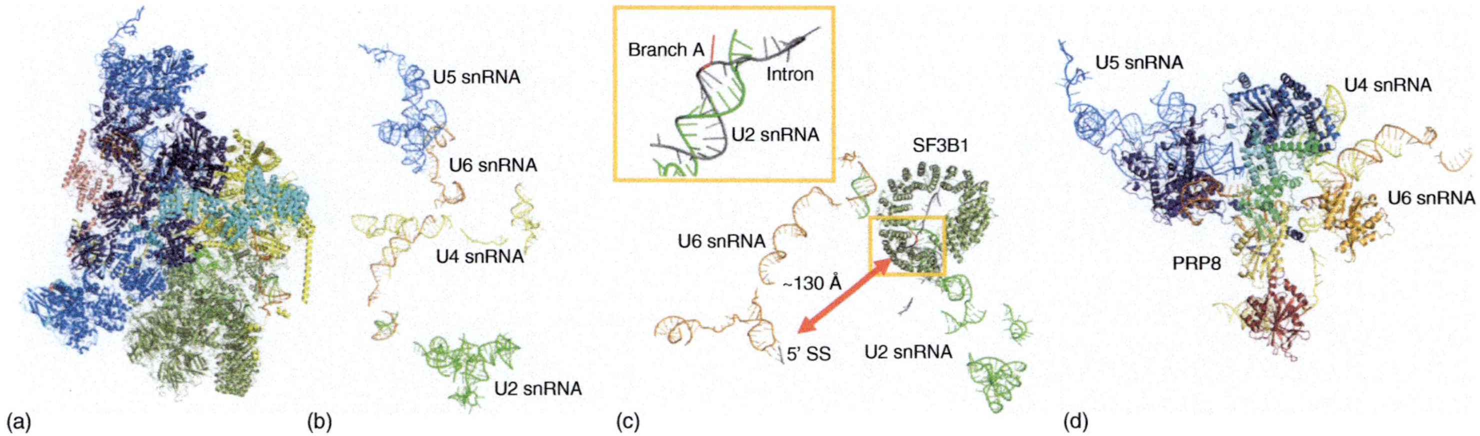

Figure 7.9 B complex structure. (a) Cryo-EM structure of the *S. cerevisiae* B complex (PDB: 5NRL). (b) snRNA organization in the B complex, with U2 snRNA in green. (c) (Inset) The BPS forms an imperfect duplex with U2 snRNA to bulge out the branch A (red) and select it as the nucleophile for the first step of splicing. (Right) The branch duplex is sequestered by SF3B1 and separated by ~130 Å from the 5′ splice site. (d) PRP8 scaffolds the RNA core.

snRNAs is very similar to that in the tri-snRNP. The 3′ end of U6 snRNA base-pairs with the 5′ end of U2 snRNA to form U2/U6 helix II (Figure 7.9b). Binding of the tri-snRNP to U2 snRNP is stabilized by interactions between the spliceosomal proteins Prp3 and Hsh155 on one end and Brr2 and Cus1 on the other.

While the exons are not visible in this structure, three nucleotides just downstream of the 5′ splice site are observed base-pairing with U6 snRNA. The two reactive sites for the first step of splicing, the 5′ splice site and branch duplex, continue to be separated physically as the spliceosome assembles presumably to prevent premature splicing using incorrect splice sites: the branch duplex remains associated with Hsh155 ~130 Å from the most visible 5′ nucleotide of the pre-mRNA (Figure 7.9c).

In this complex, the single-stranded region of U6 snRNA is sandwiched between the N-terminal and core domains of PRP8. The exon-binding loop of U5 snRNA lies alongside the N-terminal domain and contacts U6 snRNA. The U4 stem and U4/U6 duplex lie on the opposite side of PRP8, along the RT domain. The U5 stem loop 1 is positioned between the two halves of the N-terminal domain (Figure 7.9d).

In the B complex, U6 snRNA is bound to the 5′ splice site, with base-pairing interactions distinct from those seen in the activated spliceosome. The 3′ end of the final 5′ splice site-binding sequence on U6 snRNA (ACAGAGA) is in a stem loop, stabilized by PRP8, and shortly downstream it is bound to U4 snRNA, preventing it from binding the 5′ splice site. Instead, the 5′ splice site is base-paired with the loop of the U6 snRNA stem while a β hairpin in the linker between the RH and Jab/MPN domains lies alongside the stem loop.

7.6.4 Activated B Complex

Major structural rearrangements occur during the transition from the B to activated B (B^{act}) complex, the latter structure determined with a core 3 Å resolution by cryo-electron microscopy [34] (Figure 7.10a). Here, the U4 snRNP is displaced due to the Brr2-mediated unwinding of the U4/U6 snRNA duplex to release U4 snRNA

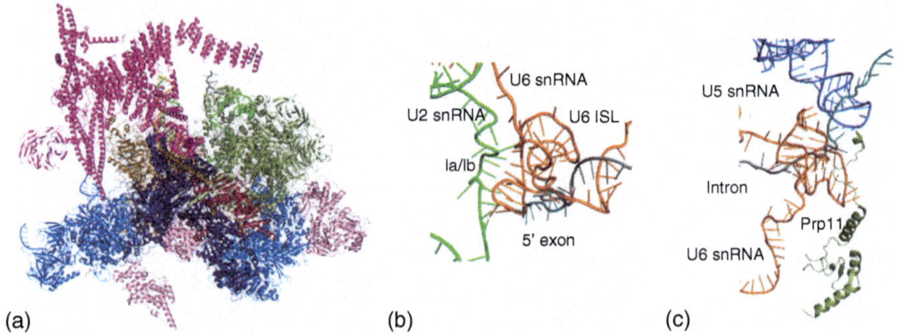

Figure 7.10 Activation of the spliceosome: B^{act} complex structure. (a) Cryo-EM structure of the *S. cerevisiae* B^{act} complex (PDB: 5GM6). (b) RNA at the core of the spliceosome. U2 and U6 snRNA base pairs form helices Ia and Ib; U6 snRNA forms the ISL. The 5′ splice site (gray) base pairs with U6 snRNA. (c) Prp11 blocks association of the branch duplex with the active site.

and associated proteins, which allows formation of the U6 ISL and the U2 and U6 snRNA base pairing in U2/U6 helix Ia and Ib (Figure 7.10b). A three-helix junction forms and includes the U2 snRNA/BPS duplex, the duplex between U6 snRNA and the 5′ splice site, and the U6 ISL representing the catalytic center of the spliceosome (Figure 7.10b). However, an N-terminal region of the SF3a protein Prp11 lies across the active site blocking the first step (Figure 7.10c).

7.6.5 C and C* Complexes

The first step of splicing occurs in what is referred to as the B* complex that leads to a transition to the C complex (Figure 7.11a). As the C complex is formed upon branching of the intron, it hints at the arrangement of the B* complex, of which there is no structure. The cryo-EM structure of the C complex, solved with a resolution of 2.9 Å at the center of the spliceosome, was captured directly after the first step of splicing, as evidenced by the presence of the lariat intron and first step protein factors [35].

SF3B is not present in the B* or C complexes, allowing for the rearrangement of the spliceosome into an active conformation and freeing the branch duplex for participation in the first step of splicing. Dissociation of Prp11 along with the SF3a complex opens up space alongside the U6 ISL for the U2 snRNA/BPS duplex (Figure 7.11b). The removal of Prp11 also allows the protein Cefl to stabilize the active site created by U2 and U6 snRNAs. The removal of Hsh155 frees the branch adenosine and allows the RH domain of PRP8 to swing up alongside the branch duplex. The U2 snRNA/BPS duplex rotates around U2 snRNA to bring the branch A into proximity with the 5′ splice site, which remains in place. The adenine base of the branch A is hydrogen-bonded to the uridine two bases 5′ to it, which directs the 2′ hydroxyl toward the 5′ splice site. The branched 5′ splice site product of the first step is visible in the structure (Figure 7.11c).

The U5 and U6 snRNAs do not move between the B^{act} and C complexes: U5 snRNA remains bound to the 5′ exon, and U6 snRNA maintains the U6 ISL, its interactions with the 5′ end of the intron, and its base pairing with U2 snRNA. U2 snRNA adopts the catalytically competent IIc stem. Modeled into this structure are two metal ions coordinated by the backbones of U80 and the catalytic triad (A59, G60, C61; Figure 7.11d) of U6 snRNA consistent with previous genetic and biochemical work.

The catalytic core is stabilized by numerous proteins. U2 and U5 snRNAs remain nestled against PRP8, with the U6 ISL in the NT domain (Figure 7.11e). Yju2 binds between the tail of U6 snRNAs, following helices Ia and Ib, and the U2/BPS duplex; its N-terminus pokes into the RNA core. Cwc25 stabilizes the U2/BPS duplex. Isyl lies between the BPS/U2 snRNA duplex and the helix formed by U6 snRNA and the 5′ splice site, while Cefl stabilizes the U6 ISL (Figure 7.11f).

The C* complex is very similar to the C complex and was solved by cryo-EM to a resolution of 4 Å (Figure 7.12a) [36]. The structure shows the rearrangement of first step products and the positioning of the second step substrates for catalysis (Figure 7.12b). The U2/U6 snRNA catalytic core remains in the same conformation

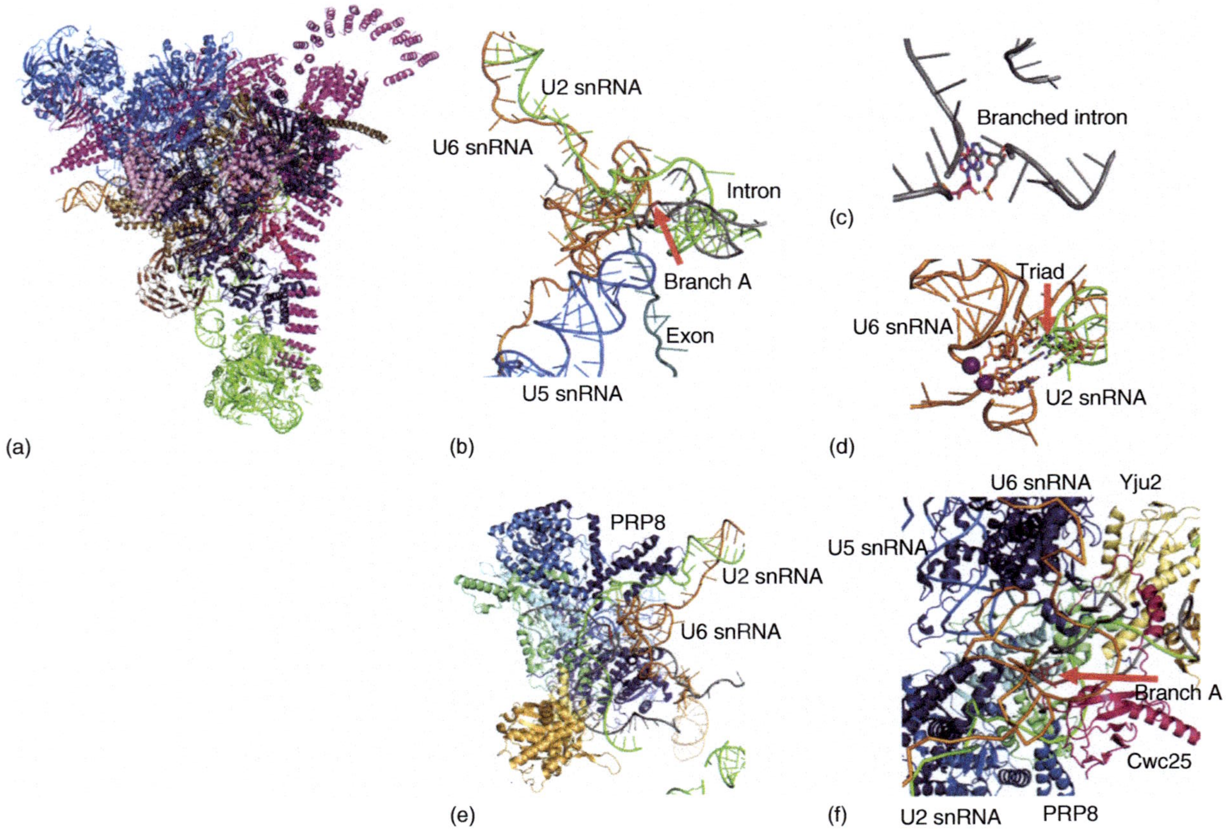

Figure 7.11 C complex structure. (a) Cryo-EM structure of the *S. cerevisiae* C complex (PDB: 5GMK). (b) RNA organization shows rearrangement to bring the branch duplex into the active site. (c) Branched intron. (d) Positioning of catalytic metals and catalytic triad at the active site. (e) PRP8 binds the RNA core. (f) First step factors, Yju2 and Cwc25, are bound to the spliceosome, suggesting this complex was captured directly after branching.

during the transition from the C to C* complex. A rotation around A30 of U2 snRNA moves the branch A and 5′ end of the intron out of the active site by 15–20 Å. The BPS/U2 snRNA duplex, and the rest of U2 snRNA, flips up to sit against the RH domain of PRP8. Prp17 binds alongside the distal part of the intron/U2 helix (Figure 7.12c). The β hairpin in the RH domain of PRP8 inserts into the minor groove of the BPS/U2 snRNA duplex, with the tip extending toward the U6 ISL (Figure 7.12d). The base pairing between U6 snRNA and the 5′ end of the intron is maintained and moves with the branch A out of the core of the spliceosome.

In this structure, the 5′ exon remains base-paired to loop 1 of U5 snRNA. Its free 3′ hydroxyl is positioned to attack the 3′ splice site, which has moved into the active site. The 3′ splice site and exon do not appear to be base-paired, but are held in place by PRP8 (Figure 7.12e). The 3′ end of the intron is bound between the RH and linker, and the 3′ exon is bound between the RT and linker.

7.6.6 P Complex

The P complex arises from exon ligation in the C* complex and has been structurally characterized to a resolution of 3.7 Å by cryo-EM (Figure 7.13a) [37]. The RNA core and protein organization of the post-catalytic spliceosome is very similar to that of the C* spliceosome. The exons have been ligated but are still held in place by the U5 snRNA (5′ exon) and PRP8 (3′ exon). The U2 snRNA duplexed to the BPS is still separated from the U6 snRNA by the hairpin in the RH domain of PRP8.

In this structure, the last base of the intron interacts with the first base of the intron while stacking on U6 snRNA A51. The branch A interacts with the penultimate nucleotide of the intron. These interactions project the 3′ hydroxyl of the 3′ splice site into the active site, suggesting that this is the position of the intron during exon ligation (Figure 7.13b).

The 3′ splice site and 3′ exon are separated by an α-helix finger from the linker region of PRP8. In their uncleaved state, this interaction would open the scissile phosphate to expose it for nucleophilic attack (Figure 7.13c). The β finger of the RH domain interacts with the branch duplex on the other side of the RNA to stabilize recognition of the 3′ splice site.

7.6.7 Intron Lariat Spliceosome Complex

The post-catalytic ILS complex from *Schizosaccharomyces pombe* was the first spliceosome structure to be solved by cryo-EM. The resolution of this structure ranges from 2.9 Å at the core to 8 Å at the periphery, with an average resolution of 3.6 Å [38].

The RNA core of the spliceosome is highly similar to that of the *S. cerevisiae* P complex. Consistent with an ILS, the mRNA is not present in this structure. This structure represented the first high-resolution model of the U2 and U6 snRNA catalytic structures. The intron lariat with the branch adenosine of the lariat rotated slightly so it no longer points into the active site is still bound to the spliceosome with the 5′ splice site base-paired to U6 snRNA and the branch sequence duplexed with U2 snRNA.

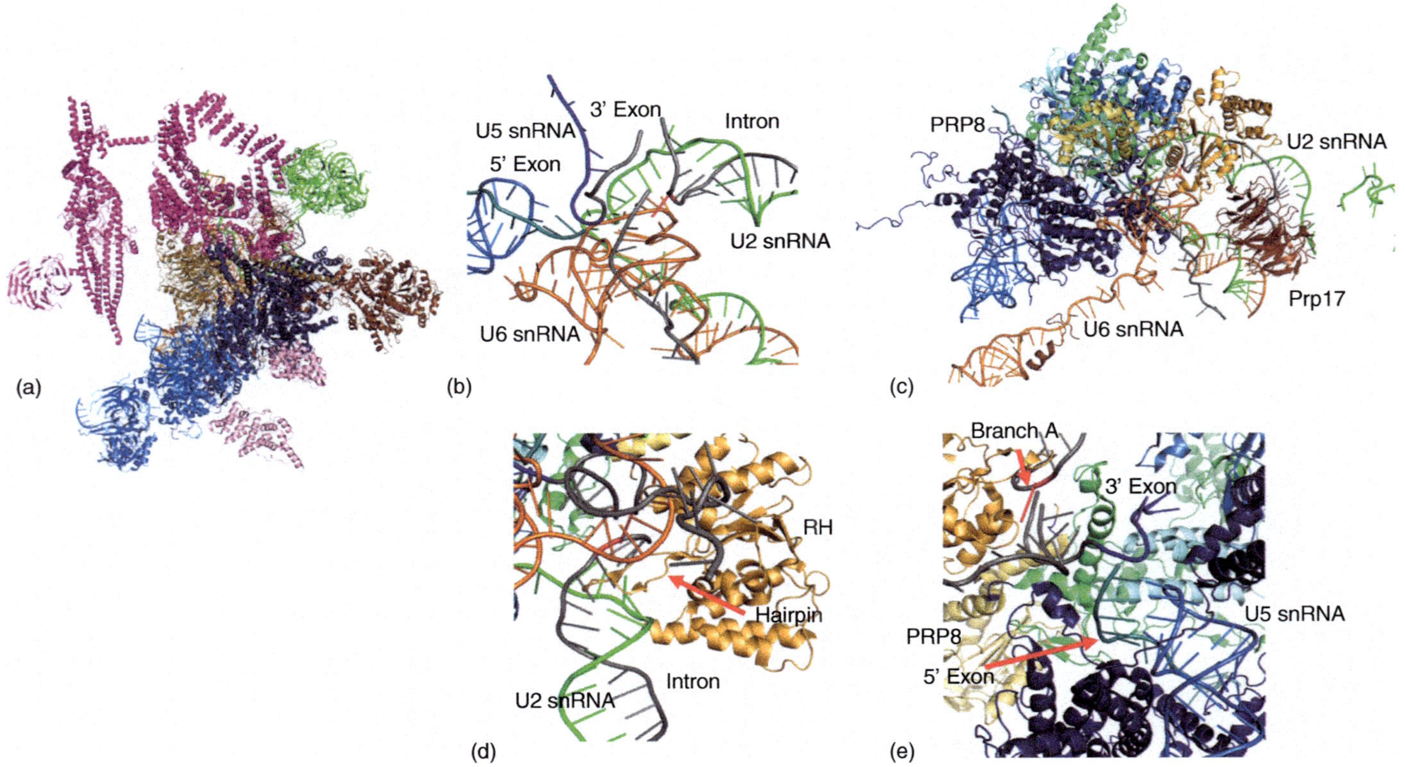

Figure 7.12 C* complex structure. (a) Cryo-EM structure of the *S. cerevisiae* C* complex (PDB: 5WSG). (b) RNA core of the spliceosome following rearrangement of first step products and positioning of the second step substrates for catalysis. The 5′ exon is base-paired to U5 snRNA, while the 5′ SS is base-paired with U6 snRNA. The branch A has rotated out of the active site to make way for the 3′ exon. (c) PRP8 and Prp17 stabilize the RNA core. (d) The β hairpin of PRP8's RH inserts into the minor groove of the BPS/U2 snRNA duplex. (e) The 3′ exon is bound by the linker of PRP8.

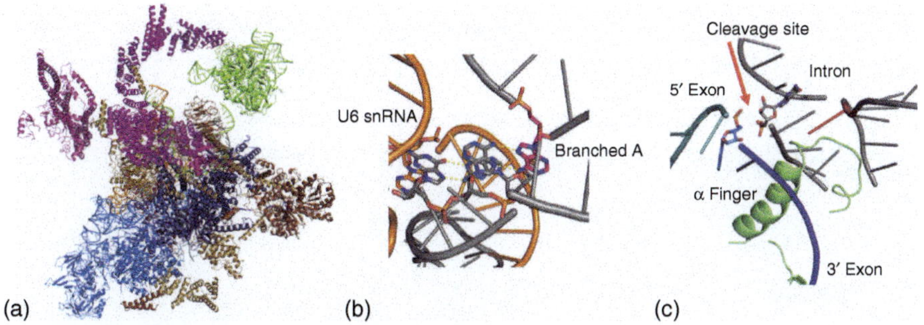

Figure 7.13 Post-catalytic spliceosome: P complex structure. (a) Cryo-EM structure of the *S. cerevisiae* P complex (PDB: 6EXN). (b) The branch A is held in place by noncanonical base pairing with the A at the 3′ splice site. That A stacks on the final G of the intron, which itself interacts with the first G of the 5′ splice site. (c) The 3′ splice site wraps around an α-helix in the linker of PRP8 to expose the scissile phosphate for exon ligation.

7.7 Insights from Spliceosome Disassembly

Following exon ligation, disassembly of the spliceosome is an active process and enables recycling of the spliceosomal components. The U6 snRNP protein Prp24 is involved in U6 snRNA rearrangement, specifically the transition between the U2/U6 catalytic RNA core of the spliceosome and free U6 snRNA in the pathway to regenerate the U4/U6 snRNA base pairing of the tri-snRNP structure. The X-ray structure of the four RNA recognition motifs (RRMs) of Prp24 bound to a portion of U6 snRNA revealed a complex interlocking topology with the RNA threaded through a central cavity in the protein [39]. The U6 ACAGA sequence required for 5′ splice site recognition is sequestered via multiple interactions with Prp24. The nucleotides of the ISL including the AGC triad responsible for positioning of the catalytic metals are not contacted by protein, but their inactive duplex conformation is a function of the overall RNA–protein complex structure.

7.8 Conservation of Spliceosomal and Group II Active Sites

Group II introns are packaged as RNP complexes where the protein component assists promoting the autocatalytic RNA splicing activity. The cryo-EM structure of endogenously spliced *Lactococcus lactis* group II intron was determined in a complex with the intron-encoded LtrA maturase [40].

In addition to the maturase activity, LtrA also has RT and EN activities. Despite low sequence conservation, the structures of the maturase and PRP8 RT are highly similar (Figure 7.14a). The X-ray structure features the domain VI lariat, domain V with the conserved catalytic triad in association with the EBS1 exon-binding sequence and the thumb portion of the RT domain. There is a remarkable correspondence between this assembly and the catalytic core of the spliceosome

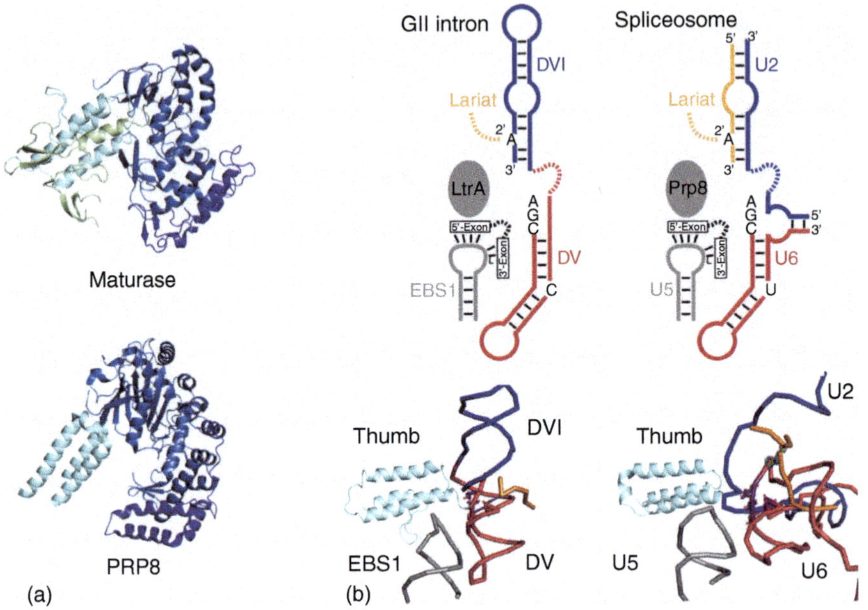

Figure 7.14 Comparison of group II and spliceosomal RNP. (a) Detail of structures highlighting the RT domains of group II maturase (PDB: 5G2X) and PRP8 (PDB: 4I43) with the thumb region in cyan. (b) Schematic of spliced group II and spliceosomal active sites (top). Cryo-EM structures of the maturase–Group II intron and the spliceosome active site (bottom).

including the positioning of the U5 snRNA and the thumb portion of the PRP8 RT domain (Figure 7.14b). This conservation of active sites provided additional support to the origin of spliceosomal introns as group II transposable elements.

7.9 Summary and Perspectives

The activity of the spliceosomal ribozyme is very tightly regulated at multiple stages to ensure fidelity of splicing but also to allow modulation of splice site selection in alternative splicing. This regulation is achieved by sequential recognition of splice sites by snRNAs and protein factors as well as the complex set of snRNA structural rearrangements that occur during the spliceosomal cycle. A common feature of these rearrangements is the sequestering of reactive components of the spliceosome within inactive snRNA conformations by virtue of pre-catalytic snRNA–substrate association or through transient interaction with protein factors.

The observations of the catalytic and autocatalytic properties of RNA, conserved chemical mechanisms, and essential, conserved snRNA–pre-mRNA interactions long suggested both that spliceosome was a ribozyme and that this complex RNP evolved from an ancient group II-like intron. A subset of group II introns are mobile elements capable of propagating themselves within the host genome; these introns contain open reading frames including those encoding for EN and RT activities involved in the insertion of the intron into the host DNA. RH domains are also found in eukaryotic retroviruses. The observation of these domains in PRP8 strongly supports the relationship between invading genetic elements and especially self-splicing group II introns with pre-mRNA splicing in a progenitor eukaryote. This link has been strengthened by the high-resolution structures of both group II introns and spliceosomes that have confirmed the conservation of their cores both with respect to catalytic RNA structures and their protein association [41].

The remarkable propagation of the group II intron through the eukaryotic genome and the transition of the autocatalytic machinery into a complex RNP ribozyme greatly increased the complexity of gene regulation in eukaryotes. The modulation of splicing by *trans* factors enabled the differential patterns of splicing observed in a temporal and tissue-specific fashion that are the hallmarks of complex eukaryotic organisms.

References

1 Berget, S.M., Moore, C., and Sharp, P.A. (1977). Spliced segments at the 5′ terminus of adenovirus 2 late mRNA. *Proc. Natl. Acad. Sci. USA* 74 (8): 3171–3175.

2 Chow, L.T., Gelinas, R.E., Broker, T.R., and Roberts, R.J. (1977). An amazing sequence arrangement at the 5′ ends of adenovirus 2 messenger RNA. *Cell* 12 (1): 1–8.

3 Burge, C.B., Tuschl, T., and Sharp, P.A. (1999). Splicing of precursors to mRNAs by the spliceosome. In: *The RNA World II* (eds. R.F. Gesteland, T.R. Cech and J.F. Atkins), 525–560. Cold Spring Harbor: Cold Spring Harbor Laboratory Press.

4 Cech, T.R., Zaug, A.J., and Grabowski, P.J. (1981). In vitro splicing of the ribosomal RNA precursor of Tetrahymena: involvement of a guanosine nucleotide in the excision of the intervening sequence. *Cell* 27 (3 Pt 2): 487–496.

5 Guerrier-Takada, C., Gardiner, K., Marsh, T. et al. (1983). The RNA moiety of ribonuclease P is the catalytic subunit of the enzyme. *Cell* 35 (3 Pt 2): 849–857.

6 Cech, T.R. (1990). Self-splicing of group I introns. *Annu. Rev. Biochem.* 59: 543–568.

7 Padgett, R.A., Konarska, M.M., Grabowski, P.J. et al. (1984). Lariat RNA's as intermediates and products in the splicing of messenger RNA precursors. *Science* 225 (4665): 898–903.

8 Lambowitz, A.M. and Zimmerley, S. (2011). Group II introns: mobile ribozymes that invade DNA. In: *RNA Worlds* (eds. J.F. Atkins, R.F. Gesteland and T.R. Cech), 525–560. Cold Spring Harbor: Cold Spring Harbor Laboratory Press.

9 Peebles, C.L., Perlman, P.S., Mecklenburg, K.L. et al. (1986). A self-splicing RNA excises an intron lariat. *Cell* 44 (2): 213–223.

10 Brody, E. and Abelson, J. (1985). The "spliceosome": yeast pre-messenger RNA associates with a 40S complex in a splicing-dependent reaction. *Science* 228 (4702): 963–967.

11 Grabowski, P.J., Seiler, S.R., and Sharp, P.A. (1985). A multicomponent complex is involved in the splicing of messenger RNA precursors. *Cell* 42 (1): 345–353.

12 Weinberg, R.A. and Penman, S. (1968). Small molecular weight monodisperse nuclear RNA. *J. Mol. Biol.* 38 (3): 289–304.
13 Will, C.L. and Luhrmann, R. (2011). Spliceosome structure and function. *Cold Spring Harb. Perspect. Biol.* 3 (7): a003707.
14 Crispino, J.D., Blencowe, B.J., and Sharp, P.A. (1994). Complementation by SR proteins of pre-mRNA splicing reactions depleted of U1 snRNP. *Science* 265 (5180): 1866–1869.
15 Stark, M.R., Dunn, E.A., Dunn, W.S. et al. (2015). Dramatically reduced spliceosome in *Cyanidioschyzon merolae*. *Proc. Natl. Acad. Sci. USA* 112 (11): E1191–E1200.
16 Lerner, M.R., Boyle, J.A., Mount, S.M. et al. (1980). Are snRNPs involved in splicing? *Nature* 283 (5743): 220–224.
17 Sharp, P.A. (1991). Five easy pieces. *Science* 254 (5032): 663.
18 Steitz, T.A. and Steitz, J.A. (1993). A general two-metal-ion mechanism for catalytic RNA. *Proc. Natl. Acad. Sci. USA* 90 (14): 6498–6502.
19 Stahley, M.R. and Strobel, S.A. (2005). Structural evidence for a two-metal-ion mechanism of group I intron splicing. *Science* 309 (5740): 1587–1590.
20 Toor, N., Keating, K.S., Taylor, S.D., and Pyle, A.M. (2008). Crystal structure of a self-spliced group II intron. *Science* 320 (5872): 77–82.
21 Moore, M.J. and Sharp, P.A. (1993). Evidence for two active sites in the spliceosome provided by stereochemistry of pre-mRNA splicing. *Nature* 365 (6444): 364–368.
22 Sontheimer, E.J., Sun, S., and Piccirilli, J.A. (1997). Metal ion catalysis during splicing of premessenger RNA. *Nature* 388 (6644): 801–805.
23 Madhani, H.D. and Guthrie, C. (1994). Dynamic RNA–RNA interactions in the spliceosome. *Annu. Rev. Genet.* 28: 1–26.
24 Fica, S.M., Tuttle, N., Novak, T. et al. (2013). RNA catalyses nuclear pre-mRNA splicing. *Nature* 503 (7475): 229–234.
25 Ritchie, D.B., Schellenberg, M.J., and MacMillan, A.M. Spliceosome structure: piece by piece. *Biochem. Biophys. Acta* 1789 (9-10): 624–633.
26 Fica, S.M. and Nagai, K. (2017). Cryo-electron microscopy snapshots of the spliceosome: structural insights into a dynamic ribonucleoprotein machine. *Nat. Struct. Mol. Biol.* 24 (10): 791–799.
27 Shi, Y. (2017). Mechanistic insights into precursor messenger RNA splicing by the spliceosome. *Nat. Rev. Mol. Cell Biol.* 18 (11): 655–670.
28 Plaschka, C., Lin, P.C., and Nagai, K. (2017). Structure of a pre-catalytic spliceosome. *Nature* 546 (7660): 617–621.
29 Nguyen, T.H.D., Galej, W.P., Bai, X.C. et al. (2016). Cryo-EM structure of the yeast U4/U6.U5 tri-snRNP at 3.7 Å resolution. *Nature* 530 (7590): 298–302.
30 Grainger, R.J. and Beggs, J.D. (2005). Prp8 protein: at the heart of the spliceosome. *RNA* 11 (5): 533–557.
31 Teigelkamp, S., Newman, A.J., and Beggs, J.D. (1995). Extensive interactions of PRP8 protein with the 5′ and 3′ splice sites during splicing suggest a role in stabilization of exon alignment by U5 snRNA. *EMBO J.* 14 (11): 2602–2612.

32 Galej, W.P., Oubridge, C., Newman, A.J., and Nagai, K. (2013). Crystal structure of Prp8 reveals active site cavity of the spliceosome. *Nature* 493 (7434): 638–643.

33 Schellenberg, M.J., Dul, E.L., and MacMillan, A.M. (2011). Structural model of the p14/SF3b155·branch duplex complex. *RNA* 17 (1): 155–165.

34 Yan, C., Wan, R., Bai, R. et al. (2016). Structure of a yeast activated spliceosome at 3.5 Å resolution. *Science* 353 (6302): 904–911.

35 Wan, R., Yan, C., Bai, R. et al. (2016). Structure of a yeast catalytic step I spliceosome at 3.4 Å resolution. *Science* 353 (6302): 895–904.

36 Yan, C., Wan, R., Bai, R. et al. (2017). Structure of a yeast step II catalytically activated spliceosome. *Science* 355 (6321): 149–155.

37 Wilkinson, M.E., Fica, S.M., Galej, W.P. et al. (2017). Postcatalytic spliceosome structure reveals mechanism of 3′-splice site selection. *Science* 358 (6368): 1283–1288.

38 Yan, C., Hang, J., Wan, R. et al. (2015). Structure of a yeast spliceosome at 3.6 Å resolution. *Science* 349 (6253): 1182–1191.

39 Montemayor, E.J., Curran, E.C., Liao, H.H. et al. (2014). Core structure of the U6 small nuclear ribonucleoprotein at 1.7 Å resolution. *Nat. Struct. Mol. Biol.* 21 (6): 544–551.

40 Qu, G., Kaushal, P.S., Wang, J. et al. (2016). Structure of a group II intron in complex with its reverse transcriptase. *Nat. Struct. Mol. Biol.* 23 (6): 549–557.

41 Zhao, C. and Pyle, A.M. (2016). Crystal structures of a group II intron maturase reveal a missing link in spliceosome evolution. *Nat. Struct. Mol. Biol.* 23 (6): 558–565.

8

The Ribosome and Protein Synthesis

Paul Huter, Michael Graf, and Daniel N. Wilson

University of Hamburg, Institute for Biochemistry and Molecular Biology, 20146 Hamburg, Germany

8.1 Central Dogma of Molecular Biology

Conservation of the genome, its transfer, and faithful implementation of the information stored within are fundamentally important steps in every cell. About 50–60 years ago, the central dogma of molecular biology gained prominence for establishing the sequential occurrence of these vital events and their interconnectedness [1, 2]. The classical view describes a consecutive order of events, in which deoxyribonucleic acid (DNA) can either replicate itself to maintain the genomic integrity (replication) or transfer its information into ribonucleic acid (RNA) molecules (transcription), which in turn serve as templates for the synthesis of proteins (translation). Together, replication, transcription, and translation form the three founding pillars of the dogma of molecular biology. This model is valid to this day; however, extensive studies over the past decades have broadened our understanding of the molecular mechanisms behind it. Both replication and transcription require the recruitment of macromolecular machines. While replication of the genome requires the action of DNA polymerases, transcription of information from DNA to RNA is mediated by RNA polymerases. RNA molecules are a heterogeneous population that fulfill various roles in a cell and can be divided into two major classes, namely, noncoding RNA (ncRNA) and messenger RNA (mRNA). Recent studies have shown that ncRNAs pursue different functions such as catalysis of chemical reactions (e.g. ribosomal RNA [rRNA]), serving as adaptor molecules (e.g. transfer RNA [tRNA]) or structural scaffolds (e.g. rRNA) and regulating gene expression (e.g. micro RNA [miRNA]) [3, 4]. On the other hand, mRNAs contain the information of genes, which can be decoded and translated into proteins. This process is called translation and is mediated by ribosomes. In contrast to the other two molecular machines, the ribosome consists predominantly of rRNA, and the structural and mechanistic themes of the core components are conserved among the three phylogenetic kingdoms of life. However, there are certain differences between, and as well as within, each kingdom concerning size, regulation, and composition, to name but a few [5, 6]. Here we focus exclusively on the prokaryotic

Ribozymes, First Edition. Edited by Sabine Müller, Benoît Masquida, and Wade Winkler.

ribosomal machinery of *Escherichia coli*. Unless mentioned otherwise, all ribosomes within this chapter refer to the *E. coli* ribosome.

8.2 Structure of the *E. coli* Ribosome

The ribosome is a multifunctional cellular complex with an approximate molecular mass of 2.3 MDa. It consists of two separate subunits, namely, the large ribosomal subunit (LSU) (50S) and the small ribosomal subunit (SSU) (30S), which together form a complete 70S ribosome. Both of the subunits are made out of rRNA and proteins. The SSU is composed of the 16S rRNA and 21 ribosomal proteins (rProteins), forming the typical morphological shape of the SSU, which can be divided into head, neck, and body. The body itself can be further segmented into shoulder, platform, foot, and spur (also known as toe) (Figure 8.1a). The LSU includes the 23S rRNA, 5S rRNA, and 33 rProteins. It forms a rounded base with three protuberances called the L1 stalk, the central protuberance, and the L7/L12 stalk (Figure 8.1b). Both subunits join by forming intermolecular bridges, providing an intersubunit space, which is important for the accommodation of tRNAs (Figure 8.1c). The ribosome possesses three binding sites for tRNAs, namely, the A-site (acceptor site), the P-site (peptidyl site), and the E-site (exit site). All three tRNAs associate with the ribosome by forming contacts with the SSU and LSU (Figure 8.1d–f). Each of the subunits harbors a crucial functional core. The SSU comprises the decoding center (DC), where the correct tRNA is selected according to the information stored within the mRNA (Figure 8.1d,f). The LSU contains the peptidyl transferase center (PTC), which links amino acids (aa) to a nascent polypeptide chain (NC) that passes through the ribosomal exit tunnel (Figure 8.1e,f). High-resolution structures revealed that both functional cores are made of rRNA with the closest rProteins being too far-off to directly participate in their respective enzymatic reaction, leading to the suggestion that the ribosome is a ribozyme [9]. This is supported by a vast amount of biochemical data, revealing that translation can occur in the absence of translation factors as well as many rProteins [10–14]. While mutations in rProteins are widely tolerated, manipulation of highly conserved rRNA residues, in contrast, results in inactivation of ribosomes [10, 15, 16]. Furthermore, it was shown that small in vitro evolved RNA molecules are capable of catalyzing the main chemical reactions of peptide synthesis [17]. Although evidence for a functional ribosome that solely consists of rRNA is lacking, the structural and biochemical insights point toward the direction that the ribosome is a classical ribozyme [18, 19]. The role of rProteins, on the other hand, is to improve efficiency and accuracy of translation, binding of tRNAs and translation factors and folding the rRNA in its functional state [20].

8.3 Translation Cycle

Translation is the most energy-consuming pathway in a growing *E. coli* cell. Approximately 50% of the energy in form of ATP and guanosine-5′-triphosphate (GTP)

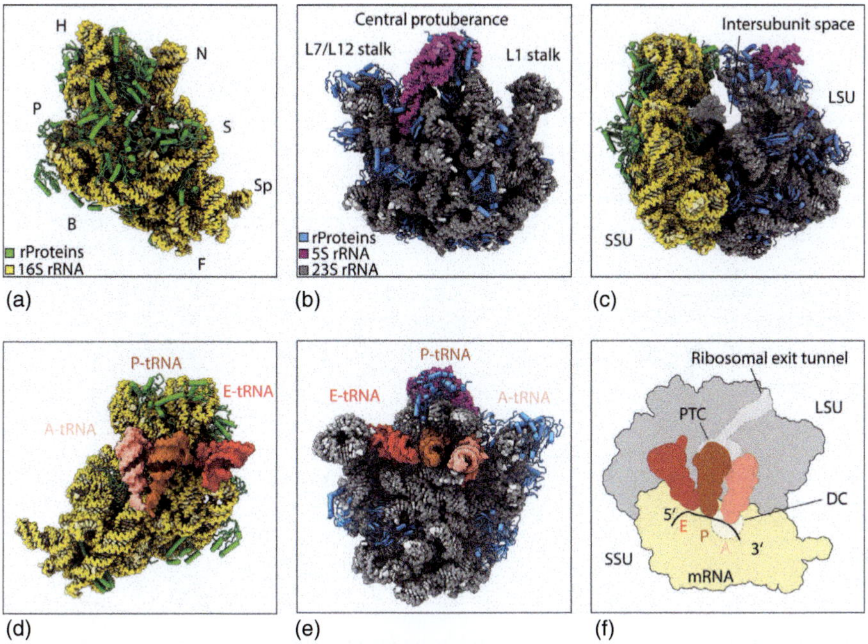

Figure 8.1 Structural overview of the bacterial ribosome. (a) View on the SSU from the solvent side. The 16S rRNA (yellow) and rProteins (green) of the SSU are shown. The major subdivision of the SSU is indicated: H, head; N, neck; B, body; P, platform; S, shoulder; F, foot; and SP, spur (also known as toe) [7]. (b) View on the LSU from the solvent side. The 23S rRNA (gray), 5S rRNA (magenta), and rProteins (blue) are indicated [8]. (c) Overview of a vacant 70S ribosome with the intersubunit space indicated. (d) View on the SSU from the interface with A-, P-, and E-tRNAs (salmon, brown, red). (e) View on the LSU from the interface with A-, P-, and E-tRNAs bound (red, brown, salmon) (PDB: 5AFI, 1VY4). (f) Schematic representation of the 70S ribosome showing the LSU (gray); SSU (yellow); A-, P-, and E-tRNAs (red, brown, salmon); and mRNA (black). The side of the decoding center (DC) on the SSU, the peptidyl transferase center (PTC), and the ribosomal exit tunnel on the LSU are depicted as dashed line.

is consumed during protein synthesis [21, 22]. Due to the enormous energy costs, translation is a tightly regulated and monitored process as errors during protein synthesis would have devastating effects. Therefore, ribosomes interact not only with mRNAs or tRNAs but also with many different protein factors like elongation factors or rescuing factors, which ensure fast and accurate translation. The order in which ribosomes interact with these different factors is dictated by the four steps of translation. These steps are (i) initiation; (ii) elongation, which can be further subdivided into decoding, peptide bond formation, and translocation; (iii) termination; and (iv) recycling (Figure 8.2). It is worth noting that the last step, recycling, and the first step, initiation, are connected, as the dissociation of both subunits from each other allows them to participate in another round of initiation. Hence, translation should be imagined as a circular process, and the order of events is often described as the translation cycle. Each single step and the corresponding sub-steps will be described in more detail in the following sections 8.3.1–8.3.4.

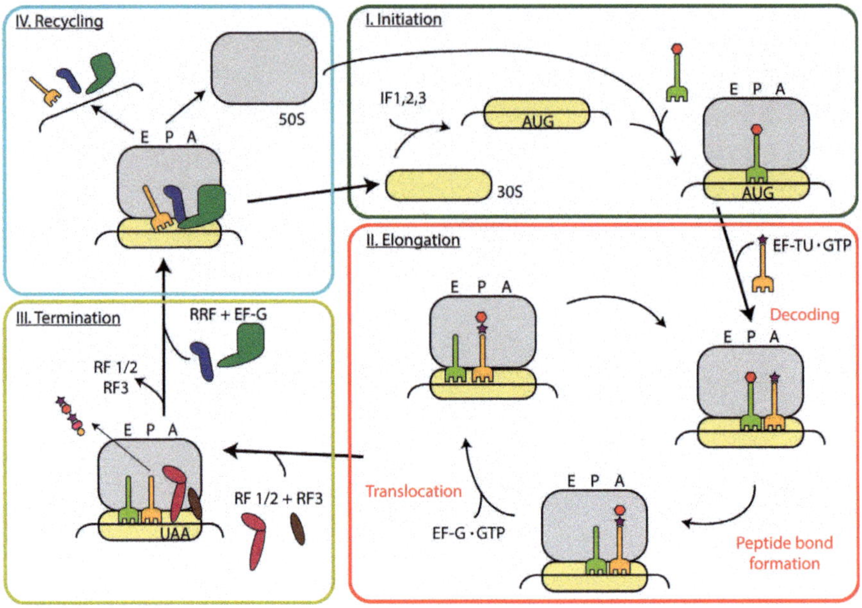

Figure 8.2 The translation cycle. During initiation (green), the SSU incorporates the mRNA and interacts with initiation factors (IFs) 1, 2, and 3 followed by the binding of the LSU and positioning of the initiator tRNA by IF2 to form the 70S initiation complex (70SIC). This step is followed by elongation (red), during which tRNAs are delivered to the A-site of the ribosome. After establishing a correct interaction between the anticodon of the tRNA and the codon on the mRNA (decoding), the ribosome catalyzes the transfer of the peptide from the P-site tRNA on to the A-site tRNA (peptide bond formation), prolonging the NC by an additional aa. Subsequently, elongation factor G (EF-G) binds and translocates the tRNAs from the A-site to the P-site and the P-site to the E-site. This leads to an empty A-site, allowing another round of elongation until the appearance of a stop codon in the A-site of the ribosome (beige, termination). Stop codons are recognized by either RF1 or RF2 (release factor), which hydrolyze the ester bond between the tRNA and the NC, allowing the polypeptide chain to be released. RF3 mediates the release of the RF1/2 and allows EF-G and ribosomal recycling factor (RRF) to split the ribosome (cyan, recycling). After dissociation of the subunits from each other, they can participate in another round of translation.

8.3.1 Initiation

Initiation of translation in bacteria starts either during transcription of mRNA or primarily on full-length mRNA transcripts [23]. It involves the formation of the 30S preinitiation complex (30PIC), during which the mRNA, the initiator tRNA ($tRNA_i$), and the initiation factors (IFs) 1, 2, and 3 bind to the SSU. After establishing the correct interaction between the anticodon of the $tRNA_i$ and the start codon on the mRNA, the 30PIC is converted into a functional 30S initiation complex (30SIC), which is capable of recruiting the LSU. The newly formed 70S initiation complex (70SIC) allows peptide bond formation between the $tRNA_i$ and the incoming tRNA in the A-site. Although having a limited set of factors involved, initiation is a slow process. The assembly of ribosomes on the mRNA can last for several seconds, whereas elongation repeats itself 20 times per second [24, 25]. Therefore, initiation is the rate-limiting step of translation [26, 27].

Formation of the 30PIC

After dissociation of the two subunits during recycling, IF3 binds the 30S subunit either before or after release of the mRNA and tRNA [28]. IF3 binds at the platform of the small subunit and adopts an open conformation [28, 29]. The N-terminal domain (NTD) is located close to the binding site of the $tRNA_i$, whereas the C-terminal domain (CTD) interacts with loop 790 of h24 (rRNA helix of the small subunit) (Figure 8.3a) [28, 30]. Binding of IF3 prevents reassociation of the LSU before binding of the remaining IFs, mRNA, and $tRNA_i$ [33, 34]. Subsequently, IF2 binds and forms a short-lived complex together with IF3 and the SSU, which

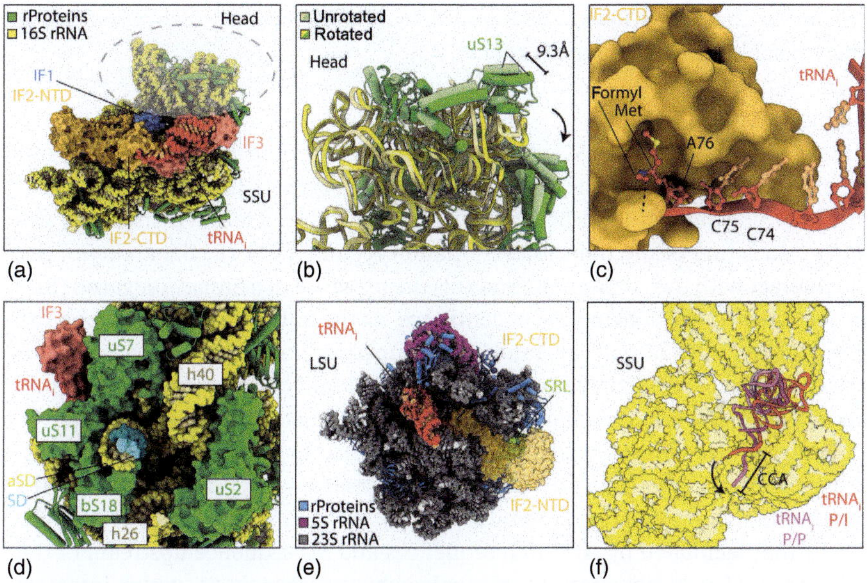

Figure 8.3 Main events during formation of the 70S initiation complex. (a) Structure of the 30PIC containing IF1 (blue), IF2 (gold), IF3 (salmon), and $tRNA_i$ (red). The ribosomal head is indicated with dashed lines (PDB: 5MLV) [30]. (b) Close-up view on the ribosomal head and comparison between an unrotated (PDB: 5AFI) (khaki, 16S rRNA; light green, rProteins) and rotated conformation in the presence of IFs (PDB: 5MLV) (yellow, 16S rRNA; green, rProteins) of the SSU. Indicated is the clockwise rotation of the head toward the platform. The position of uS13 indicates a shift of 9.3 Å [30, 31]. (c) Close-up view on the interaction between IF2-CTD, CCA end of $tRNA_i$, and the formylmethionine. The additional interactions formed by the formyl moiety (dashed lines) leads to the specificity of IF2 for $tRNA_i$ (PDB: 5MLV) [30]. (d) View on the aSD/SD helix (SD sequence, cyan) on the SSU with the surrounding proteins (green) and helices (yellow), forming the mRNA entrance (PDB: 5MLV) [30]. (e) Conformation of IF2-GDPNP relative to the LSU after formation of the 70SIC before GTP hydrolysis. After formation of the 70SIC, the G-domain of IF2 comes in close contact with the sarcin–ricin loop (SRL) (green) (PDB: 3JCJ) [32]. (f) View on the SSU after GTP hydrolysis. As a consequence of the hydrolysis of IF2 bound GTP, the $tRNA_i$ is rearranged from its P/I (red, PDB: 5LMV) into the P/P (purple, 5AFI) state, reorienting the CCA end into a conformation that allows peptide bond formation on the LSU.

becomes stabilized by the binding of IF1. After the initial binding of the IFs on the SSU, IF1 and IF3 synergistically induce a rotation of the head toward the platform of the SSU, reminiscent of the conformation of a rotated ribosome (Figure 8.3b) [35, 36]. In this conformation, IF1 and IF2 interact with several important regions of the SSU as well as with the $tRNA_i$. IF1 binds to the ribosomal A-site in vicinity to h44, the 530 loop, and rProtein S12 and stabilizes IF2 and IF3 on the small subunit (Figure 8.3a) [30]. IF2 consists of three major parts, namely, the N-terminal part, the central "G-domain," and the C-terminal part. The NTD of IF2 contacts IF1 and S12 and is thought to be in charge for the initial binding of IF2 to the small subunit [35]. The G-domain contains the structural elements for binding and hydrolysis of GTP [37, 38]. The C-terminal part of IF2 recruits the $tRNA_i$ [30, 39, 40]. The $tRNA_i$ is aminoacylated with methionine, whose αNH_2 group gets formylated by formyl methyl transferase (FMT). The modification of the αNH_2 group leads to a specificity for the $tRNA_i$ toward IF2 and discrimination against elongation factor thermo unstable (EF-Tu) (Figure 8.3c) [41, 42].

Recruitment of the mRNA

Binding of the mRNA to the SSU can happen any time during 30PIC formation and is therefore independent of the composition of the complex [28, 43]. The recruitment rather depends on inherited features of the mRNA itself, which are (i) the secondary structure of the translation initiation region (TIR), (ii) the ability of the TIR to interact with the rProtein S1, and (iii) the presence of a Shine–Dalgarno (SD) sequence [43–46]. The docking site or "entrance" of the mRNA is located around the platform of the SSU (h26, h28, h40) and is made up of several positively charged rProteins that are able to interact with differently folded mRNAs (Figure 8.3d) [47–50]. It was shown that weak secondary structures favor translation initiation; however also highly structured mRNAs are able to be recruited to the docking site [51–53]. This suggests that, although all mRNAs have the same docking site, the strength of folding influences the details of recruitment [28]. Recruitment of the mRNA to the SSU is further facilitated by A/U-rich sequences and SD sequence upstream of the start codon. A/U-rich sequences are recognized by the rProtein S1, which interacts with the mRNA upstream of the SD sequence and therefore facilitates docking and unfolding of the mRNA [50, 54–56]. The SD sequence can be found around 7–10 bases upstream of the start codon and has the consensus sequence AGGAGG. The SD sequence assists in placing the start codon of the mRNA into the P-site of the ribosome by forming interactions with the anti-SD sequence located in the 3′-end of the 16S rRNA, with the SD/anti-SD helix forming at the interface of the ribosomal head and back of the platform (Figure 8.3d) [45, 49, 57–59]. It is noteworthy that not all mRNAs contain all the abovementioned features. In this context, there are three different kinds of mRNAs, namely, those that have an SD sequence upstream of the start codon (SD-containing mRNAs), those that do not have an SD sequence upstream of the start codon (non-SD mRNAs) and are dependent on the bacterial species, and those that directly start with AUG (leaderless mRNAs). The abundance of these types can change significantly [58, 60, 61]. SD-containing mRNAs are the most favorable mRNAs for translation, as they utilize every feature needed for fast initiation. On the other hand, non-SD mRNAs do not contain an SD sequence in proximity to the start codon. It remains elusive how ribosomes, programmed with these mRNAs, are able to faithfully position the start codon at the P-site, despite forming a stable complex with the SSU [61, 62]. Last but not least, leaderless mRNAs lack the 5′-end upstream of start codon. It was shown that they mainly associate with 70S rather than SSU [63]. Moreover, they do not require the interaction of S1 and other proteins that form the initial docking site of the mRNA [64, 65].

Formation of the 30SIC

The main event during transition to the 30SIC is the formation of an interaction between the anticodon of the $tRNA_i$ and the start codon of the mRNA. It requires the correct positioning of a start codon into the ribosomal P-site. This process is facilitated by the formation of an SD/anti-SD pairing, but does not necessarily rely on this interaction, as non-SD mRNA and leaderless mRNA are still able to initiate translation [61, 62, 66, 67]. Presumably, the SD/anti-SD helix helps to increase the concentration of start codon triplets near the P-site, whereas its recognition is achieved kinetically by IFs [66, 68]. In bacteria, all start codon triplets share a U at the second position. AUG, GUG, and UUG are the most common ones with AUG being the most favored [69]. Recognition of the correct start codon and tRNA allows placement of the $tRNA_i$ in its peptidyl/initiation (P/I) state by forming contacts between the anticodon stem loop (ASL) and stop codon [32, 35, 36, 47]. Thereby, the 30S·IF1·IF2·$tRNA_i$ complex becomes further stabilized, whereas IF3 becomes destabilized [28, 61]. IF3 interferes with the binding of the LSU as it occupies a space important for the formation of an intersubunit bridge. Destabilization of IF3 might work as the trigger for LSU association [33, 35, 41, 62].

Formation of the 70SIC

The LSU binds the SSU in its rotated state with all IFs and the $tRNA_i$ attached to it [47, 62]. First of all, IF3 dissociates from the complex before GTP hydrolysis [70]. Next, IF2 interacts with the sarcin–ricin loop (SRL) inducing the hydrolysis of GTP bound to IF2, causing several structural rearrangements (Figure 8.3e) [71, 72]. Firstly, IF2 undergoes a conformational change that loosens its contacts with both subunits as well as with the $tRNA_i$ [70, 73]. Secondly, IF1 leaves the complex [62]. Subsequently, the transition of the SSU from a rotated state to an unrotated state occurs, and thereby the $tRNA_i$ becomes rearranged in its P/P position (Figure 8.3f) [32, 35, 73, 74]. Lastly, IF2 leaves the ribosome, committing the 70SIC for elongation [73, 74].

8.3.2 Elongation

During elongation, the ribosome has to translate the genetic code into amino acid sequences. It starts with the dissociation of IF2 from the 70SIC, leaving an empty A-site behind. The vacant A-site is recognized by a ternary complex, which consists of EF-Tu, aminoacyl tRNA (aa-tRNA), and GTP. Based on the complementarity between the anticodon of the tRNA and the codon of the mRNA, the ternary complex either gets rejected or is allowed to accommodate in the ribosomal A-site (decoding). Once the correct tRNA is selected, the ribosome transfers the growing peptide chain to the aa-tRNA in the A-site and therefore elongates the NC by one additional amino acid (peptide bond formation). Subsequently, elongation factor G (EF-G) binds to the ribosome and translocates the mRNA in 3′-direction by one codon, causing the now deacylated tRNA to move to the E-site and the A-site tRNA to the P-site (translocation). Hence, the A-site is vacant, allowing the next ternary complex to enter. The three main steps of elongation continue until the ribosome encounters a stop codon at the A-site, marking the end of elongation and initiating termination. In the following section these three steps of elongation will be described in more detail.

Decoding

Throughout elongation, the ribosome discriminates between correct and incorrect ternary complexes, dictated by the mRNA codon in the ribosomal A-site (Figure 8.4a). These ternary complexes contain EF-Tu, GTP, and one of 50 different aa-tRNAs matching one or more of the 61 canonical codons on the mRNA [78]. The basis for the discrimination is the base complementarity of the ASL of the tRNA and the codon on the mRNA. Taking the "wobble hypothesis" into consideration that states that the first two positions of the codon create the coding specificity (explanation follows later), three possibilities arise to describe the complementarity between the mRNA and the tRNA [79]. A tRNA can be either cognate (no mismatch between the first and second position), near-cognate (one mismatch between the first and second position), or non-cognate (no match between the first and second position) [80]. However, studies showed that base pairing alone is not sufficient to account for the low average misincorporation rate of 3×10^{-3} [81, 82]. Moreover, antibiotics and mutations affecting the ribosome can induce miscoding, demonstrating that the ribosome augments the obtained fidelity rates [83]. The kinetic proofreading model describes a two-step mechanism during accommodation of the A-site tRNA, in which the ribosome discriminates between cognate and near-cognate tRNAs [84, 85]. Under physiological conditions it was shown that non-cognate tRNAs are not accepted by the ribosome and therefore are excluded from the discussions [86]. Initial selection begins with binding of the ternary complex to the A-site in a codon-independent manner, followed by codon recognition (Figure 8.4b,c). Dependent on the complementarity of the anticodon/codon helix, the ternary complex either is rejected or undergoes a conformational change triggering EF-Tu-bound GTP hydrolysis (Figure 8.4d,e) [87]. EF-Tu rearranges in the GDP-bound conformation and dissociates from the tRNA [86, 88, 89]. Release of EF-Tu from the ternary complex marks the second proofreading step. Dependent on the nature of the anticodon/codon helix, the tRNA is rejected or accommodates within the ribosomal A-site followed by peptide bond formation (Figure 8.4f) [86]. The first molecular details on decoding were based on X-ray structures of isolated 30S subunit crystals programmed with either cognate or near-cognate ASLs [80, 90]. Binding of the ASL to the mRNA results in a double helical structure (anticodon/codon helix). This induces universally conserved nucleotides A1492 and A1493 to flip out of the internal loop of helix 44 and interact with the minor groove of the anticodon/codon helix utilizing A-minor motifs [80, 90]. These motifs consist of consecutive adenines

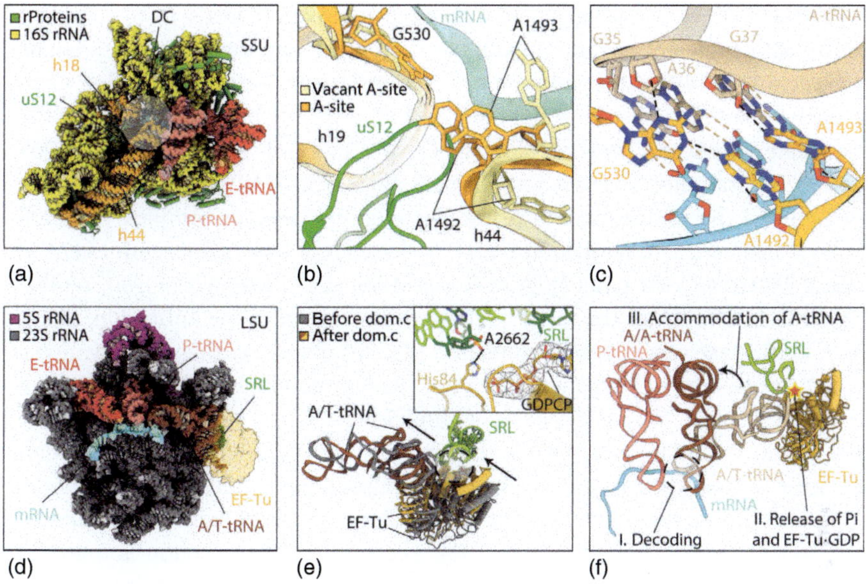

Figure 8.4 Decoding and accommodation of the A-site tRNA. (a) View on the SSU with a vacant A-site. The decoding center is indicated by dashed lines with contributing rRNA helices shown in orange. mRNA is shown in cyan, and P-tRNA and E-tRNA are shown in salmon and red, respectively (4WZD). (b) Close-up view on the decoding center. G530 of h18 and A1492 and A1493 of h44 adapt different positions in the presence (orange) or absence (yellow) of A-site tRNA. uS12 (green) reaches toward the decoding center but is too distant to directly contribute to the process (PDB: 4WZD, 4V87). (c) View on the anticodon/codon helices (cyan, mRNA; tan, A-tRNA) with hydrogen bond interactions depicted as dashed lines in tan. The flipped-out positions of A1492 and A1493 as well as G530 in its *anti*-conformation contribute additional interactions (black dashed lines), increasing the stability of the anticodon/codon helix (PDB: 4WZD, 4V87). (d) View on the LSU in the presence of the ternary complex in the A-site. A/T tRNA (brown), EF-Tu (gold), and the sarcin–ricin loop (green) are shown (PDB: 5UYK). (e) Position of the ternary complex before (gray) and after (brown colors) domain closure. Inset: After domain closure, the catalytic His84 of EF-Tu is interacting with the backbone of A2662 of the 23S rRNA and coordinates the nucleophilic attack of a water molecule on the γ-phosphate of GTP (here GDPCP) (PDB: 5UYK, 5UYQ). (f) GTP hydrolysis and release of inorganic phosphate induces structural rearrangements within EF-Tu and releases EF-Tu from the ribosome. The A-site tRNA shifts from the A/T state toward the A/A state (PDB: 5UYK, 4WZD). Source: (b, c) Adapted from Demeshkina et al. [75]; Rozov et al. [76]; (d, e) Adapted from Loveland et al. [77].

that insert into the minor groove of RNA helices and thus contribute to their stability [91] (Figure 8.4b,c). Additionally, G530 switches from the *syn*-conformation to the *anti*-conformation. The structural rearrangements allow A1493 to interact with the first base pair, whereas A1492 acts together with G530 to monitor the second position. Interactions with the third position is less stringent as it involves fewer hydrogen bonds, providing a structural explanation [80, 90] for the wobble hypothesis (Figure 8.4c) [79]. The energy derived from a cognate interaction prevents the tRNA to dissociate from the ribosome and induces domain closure, a movement of the head toward the shoulder of the 30S, triggering hydrolysis of EF-Tu-bound

GTP (Figure 8.4d,e). However, a near-cognate interaction leads to a disturbed geometry within the anticodon/codon helix, disrupting important hydrogen bonds between A1492, A1493, G530, and the ASL/codon interaction and displacing them from each other. The resulting energy penalty leads to the dissociation of the near-cognate tRNA and prevents domain closure [80, 90]. X-ray analysis of 70S ribosomes programmed with cognate or near-cognate tRNAs and longer mRNAs further refined the understanding of the decoding process. In complex with 70S ribosomes, near-cognate anticodon/codon helices adopt base pairs reminiscent of canonical Watson–Crick pairs (Figure 8.4c). Moreover, nucleotides G530, A1492, and A1493 seem to adopt the same conformation relative to the anticodon/codon helix as in the case of a cognate interaction [75, 76]. Based on these observations, decoding errors derive from mismatched anticodon/codon helices that mimic a Watson–Crick shape. The energy costs of mimicking a Watson–Crick-like geometry requires a keto–enol tautomerization, pushing the equilibrium toward dissociation of the near-cognate ternary complex. This is also a bona fide explanation why non-cognate tRNAs never surpass the initial selection checkpoint as the energy penalty would be even higher [75, 76, 92]. Recognition of the codon is followed by domain closure, which is identical for cognate and near-cognate interactions mimicking Watson–Crick geometry (Figure 8.4e) [75, 77]. Recent cryo-electron microscopy (cryo-EM) structures explain the necessity of the Watson–Crick geometry for domain closure. Near-cognate interaction with a G:U wobble base pair fails to stabilize G530 to an extend that is favorable for domain closure and therefore keeps the ribosome in an open conformation [77]. During domain closure, the ternary complexes get pushed toward SRL of the LSU. This allows the catalytic important His84 of EF-Tu to interact with A2662 of the SRL, hydrolyzing GTP by coordinating a water molecule for nucleophilic attack on the γ-phosphate of GTP (Figure 8.4e) [77, 87]. Hydrolysis and release of the inorganic phosphate induce conformational changes within EF-Tu and eventually its release from the ribosome. Conformational changes in EF-Tu leads the aa-tRNA to shift from its EF-Tu-bound state (A/T state) to the A/A state, a movement that includes the accommodation of the aminoacyl end of the tRNA into the PTC (Figure 8.4f) [87, 93, 94]. tRNA accommodation is fast for cognate tRNAs but slow for near-cognate tRNAs, therefore providing an evidence for the second proofreading step during tRNA selection [86, 88]. Structural insights that explain the different accommodation rates are lacking; however misalignment of the anticodon/codon helix as well as steric restrains imposed by the ribosome are likely [78, 95].

Peptide Bond Formation

Formation of a peptide bond between the aa-tRNA and peptidyl-tRNA is the main chemical reaction during protein synthesis. It involves the aminolysis of the ester bond that links the NC to the ribose of the P-site tRNA and transfer of the chain on to the A-site tRNA, adding an additional residue [96]. This happens in a stepwise manner. After the accommodation of the aa-tRNA, a nucleophilic attack of the α-amine of the aa-tRNA on to the carbonyl carbon of the peptidyl-tRNA occurs. This leads to the formation of a transition state, the rate-limiting step of peptide bond formation [96, 97]. In comparison, the transition state decomposes fast on the ribosome while it accumulates in solution. Hence, the ribosome catalyzes this reaction by 10^5- to 10^7-fold faster compared with reactions in solution, highlighting the significance of the translation apparatus [98]. Fast breakdown of the transition state results in a deacylated tRNA in the P-site and a peptidyl-tRNA in the A-site [97]. Structural and biochemical studies point out that the reaction on the ribosome is RNA driven [99, 100]. Hence it is not surprising that the PTC, the catalytic center of peptide bond formation, is composed of a cluster of universally conserved rRNA nucleotides located within the central loop of domain V of the 23S rRNA (Figure 8.5a) [104, 105]. These residues form a tight cavity that serves as a platform for the accommodation of the CCA ends of the A-site and P-site tRNAs, bringing them in close contact and shielding them from the close surrounding (Figure 8.5b,c). This is in line with experiments showing that the LSU alone is capable of peptide bond formation [106].

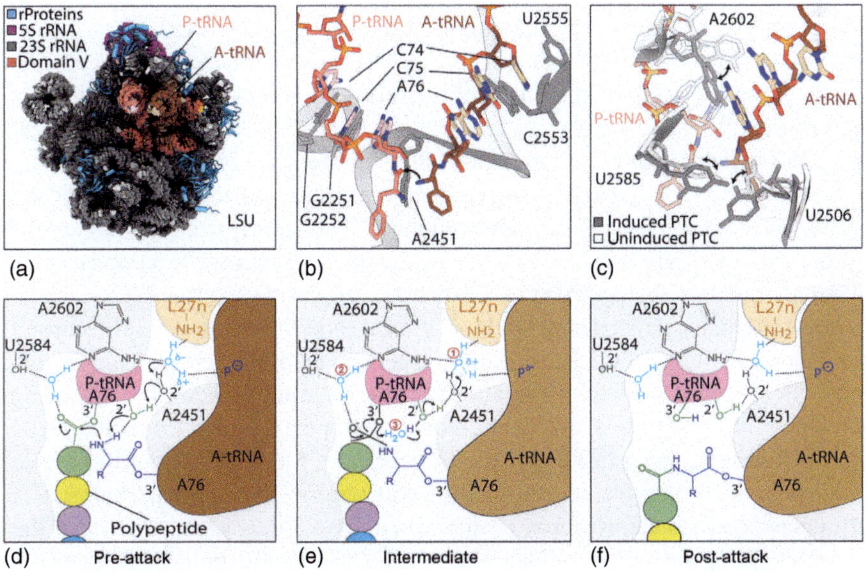

Figure 8.5 Conformation of residues involved in the proton wire of the PTC. (a) The catalytic center (orange) of peptide bond formation is composed of rRNA residues of domain V of the 23S rRNA (gray). A- and P-tRNA are shown in brown and salmon, respectively. H68 and H69 of the 23S rRNA were removed for reasons of visibility of the PTC (PDB: 4V5D). (b, c) Accommodation of A-site tRNA leads to an induced fit within the PTC. Residues U2555 and C2553 of 23S rRNA interact with C74 and C75 of the A-site tRNA. Stabilization of the CCA end allows the amino acid moiety of the A-site tRNA to interact with residue A2451. Accompanied by accommodation of the A-site tRNA are several shifts within residues of the PTC, with U2506, U2585, and A2602 showing the most severe movements. U2506 moves into a position that allows accommodation of the A-site tRNA, while residues U2585 and A2602 adapt positions that expose the peptidyl ester for nucleophilic attack. (PDB: 4V5D, 4V51) [101, 102]. (d–f) Model of the proton wire mechanism in the presence of three trapped water molecules (blue) inside the PTC. Source: (a) Adapted from Voorhees et al. [101]; (e, f) Adapted from Polikanov et al. [103].

Crystal structures of *Haloarcula marismortui* 50S subunits provided a first detailed picture. In the absence of A-site tRNA, the PTC adopts the uninduced state. C74 and C75 of the P-site tRNA form Watson–Crick base pairing with G2251 and G2252 holding the acceptor end in place (Figure 8.5b). A76 is stacking on to A2451 and is in hydrogen bond distant to A2450 [101, 103, 105, 107]. Nucleotides A2062, A2451, and U2585 form a pocket around the ester group of the peptidyl-tRNA to sterically exclude water that would otherwise hydrolyze the ester bond (Figure 8.5c) [108, 109]. G2583 forms a G:U wobble pair together with U2506 blocking the A-site pocket (Figure 8.5b). Binding of the A-site tRNA causes an induced fit that is a conformational change in the PTC (referred as the induced state). A-tRNA nucleotides C74 stacks with U2555, C75 base pairs with G2253, and A76 forms an A-minor motif with G2583 (Figure 8.5b) [107–110]. The latter interaction leads to disruption of the G2583–U2506 wobble base pair, shifting U2506 away from the A-site pocket. Hence, the CCA end becomes stabilized, resulting in the positioning of the amino acid moiety that allows formation of hydrogen bonds with the N3 and 2′-OH of A2451 and the 2′-OH of A76 (Figure 8.5b). Accompanied by the A-tRNA accommodation is a shift of A2602 and U2585, thereby exposing the peptidyl-tRNA ester for a nucleophilic attack (Figure 8.5c) [108, 111]. Thus, the PTC mainly functions as an entropic trap, leading to a significant reduction of the reaction entropy [98, 108, 109]. Most enzymes contribute to chemical reactions in two ways. The first contribution is to arrange the substrates in close proximity, allowing them to react. The second step involves the functional groups of the enzyme that influence the chemistry of the reaction. The contribution of both steps can differ significantly from one enzyme to the other. An important question is whether the peptidyl transfer reaction is solely dependent on substrate alignment or if functional groups within the PTC exist that influence the chemical reactivity [99]. Biochemical experiments as well as the crystal structures of *H. marismortui* 50S suggested a general acid/base mechanism involving the N3 of A2451. However, mutations of this adenine as well as neighboring nucleotides did not lead to a significant reduction in peptide bond formation [110–114]. Furthermore, using full-length tRNA substrates for kinetic experiments, no pH dependency for peptide bond formation was observed [115, 116]. Therefore, a substrate-assisted mechanism was proposed with the 2′-OH group of A76 of the P-site tRNA, being the main candidate [117–119]. The α-amino group of the aa-tRNA forms an extensive network of hydrogen bonds with the N3 of A2451 and the 2′-OH group of A76 [108, 109]. As it was demonstrated that the former is neglectable, a substrate-assisted mechanism was suggested, in which a proton is transferred from the nucleophile on the 2′-OH group to the 3′-OH group of A76 [108, 109, 117]. Indeed, deletion of the 2′-OH of the A76 showed a decrease of activity by at least 100-fold [119–122]. Hence, a proton-shuttling mechanism was proposed, either as a six- or eight-membered mechanism, in which two or three protons together with a water molecule coordinate the attack of the nucleophile in a fully concerted manner [97, 108, 123, 124]. Recent high-resolution structures of pre-attack and post-catalysis states 70S ribosomes of *Thermus thermophilus* provided further molecular details [101, 103]. Owing to the higher resolution, the position of three water molecules (W1–3) inside the PTC could be identified. These water molecules are coordinated by A2451, U2584, and A2602, the N-terminal part of L27 as well as with the A76 of both tRNAs (Figure 8.5d). Based on these findings, a proton wire mechanism was suggested in which residues of the PTC and both tRNAs obtain a catalytic role by activating the water molecules [103]. A tight network of hydrogen bonds between the N6 of A2602, the phosphate oxygen of A76 of the A-site tRNA, the 2′-OH of A2451, and the N-terminus of L27 coordinates W1 to the attacking amine and shields W1 from exchange with the bulk solvent (*pre-attack*) (Figure 8.5d). This guarantees a geometry that allows the concerted attack by the α-amine of the aa-tRNA onto the carbonyl carbon of the peptidyl-tRNA via transfer of the proton from the nucleophile via A76 2′-OH of the P-site tRNA to the A2451 2′-OH to W1 [97, 103]. The result is a tetrahedral intermediate state, which is stabilized by the donation of a proton to the negatively charged ester carbonyl carbon by W2 (Figure 8.5e). The ribosome facilitates the fast breakdown of this state to prevent premature termination by hydrolysis [96, 97, 103]. On the basis of the 70S structure, a transfer of a proton from the positively charged W1 to W3 to form a H_3O^+ ion is suggested that catalyzes the hydrolysis of the intermediate state into their respective products (Figure 8.5f) [103]. Thus, the ribosome provides a prearranged proton-transfer network that contributes to the reaction besides being an entropic trap [98, 103]. Another question raised by this study is the contribution of L27 to peptide bond formation. In bacterial ribosomes, L27 represents the closest rProtein to the PTC. From crystal and cryo-EM structures, it was shown that, upon accommodation of the A-site, the last three N-terminal residues become ordered and protrude into the PTC, which is not the case for complexes that are not competent for peptide bond formation [101, 103, 125]. Thus, L27 stabilizes residues of the 23S rRNA and both tRNAs upon accommodation of the A-site tRNA [101, 103]. Additionally L27 is suggested to coordinate the W1 molecule and, together with the A-tRNA A76 and the 5′phosphate oxygen of A2451, is the source responsible for the deprotonation of the α-amine (Figure 8.5f) [103]. In contrast to structural studies, biochemical studies reveal a contradictory picture. Whereas one study demonstrated that deletion of the three N-terminal residues leads to a decreased rate of translation [126], other studies reported that ribosomes lacking L27 have no defect in peptide bond formation [127]. Further biochemical and kinetic experiments are needed to further address the contribution of L27 and therefore address the question as to whether the ribosome is really a ribozyme.

Translocation

Following peptide bond formation, the ribosome is occupied with a deacylated tRNA in the P-site and a peptidyl-tRNA in the A-site (*PRE-state*). In order to allow translation to continue, the ribosome has to translocate the bound tRNAs together with their associated mRNA codons from the P-site to the E-site and the A-site to the P-site. As a result, the ribosome provides a vacant A-site programmed with the next codon of the mRNA (*POST-state*). Translocation happens in a stepwise manner, with the acceptor ends of tRNAs moving first with respect to the LSU (A/P, P/E) followed by a movement of the tRNA with respect to the SSU (P/P, E/E) [93]. The driving force behind translocation is large conformational changes of the SSU that coordinate the movement of mRNA and tRNAs to their respective place [128]. The ribosome possesses the capability to perform these movements on its own. However, in the absence of factors, this process is bidirectional, meaning the ribosome translocates backward and forward. It requires the assistance of EF-G and GTP that drives translocation into a unidirectional process by acting like a pawl and therefore preventing back-translocation [129–131]. Researchers in recent years revealed the underlying mechanisms and could dissect the events of translocation in a sequential order. Following peptide bond formation the CCA ends of the tRNAs spontaneously move from the A-site to the P site and P-site to the E-site on the LSU (A/P and P/E hybrid states) (Figure 8.6a) [135, 136]. The driving force for these movements is the deacylated state of the P-site tRNA, which has a high affinity for the E-site [137, 138]. Accommodation of the CCA end of the deacylated tRNA into the E-site results in a vacant LSU P-site, which, on the other hand, has a strong affinity for peptidyl-tRNAs [139, 140]. Coupled to the movement of the acceptor stem of the deacylated tRNA on the LSU is a counterclockwise (CCW) rotation of the SSU relative to the LSU (~3–10°), often referred to as "ratcheting" (Figure 8.6b) [128, 132, 141, 142]. In the absence of EF-G, kinetic studies observed a back-and-forth fluctuation between the classic non-rotated state and the rotated hybrid state. Interestingly, EF-G can bind both states [93, 135, 143]. The binding of EF-G, however, accelerates the formation of the rotated hybrid state and subsequently drives the reaction toward the formation of the POST-complex (P/P, E/E) (Figure 8.6b) [144, 145]. Rotation of the small subunit moves domain I (dI) of EF-G in close proximity to the SRL. domain I (dI) is structurally similar to other translational GTPases (traGTPases), comprising the important G subdomain required for the hydrolysis of GTP [146]. Hence, the close proximity to the SRL triggers GTP hydrolysis and thereby the release of inorganic phosphate [147–150]. Akin for EF-Tu during decoding, the energy released by hydrolysis does not have an influence on translocation per se, but is stored to dissociate EF-G after translocation. This is exemplified by the fact that in presence of non-hydrolyzable GTP analogs, a single round of translocation can occur [150, 151]. Binding of EF-G triggers an additional independent movement of the SSU termed the head swivel [152]. During head swiveling, the head of the SSU subunits rotates clockwise (CW) toward the platform (Figure 8.6b). Thereby, it frees the path for the movement of tRNAs on the SSU, which is usually blocked by the head domain [153]. Translocation of the mRNA is passive and dependent on the movement of A- and P-tRNAs [154]. However, translocation of mRNA cannot be observed in the presence of only a deacylated tRNA in the P-site [154] and requires at least an ASL in the A-site [154]. Simultaneously with the CW movement of the head, the body and platform start to inverse their movement in a CW direction [152, 155]. The exact time point of this reverse rotation is not known. Cryo-EM and crystallographic studies were able to capture an intermediate state, with body–platform being back-rotated to 3.5°, while the head swivel reaches its maximum rotation [155–157]. The tRNAs in the complex are configured in an ap/P and pe/E state (Figure 8.6d). These states are characterized by accommodated acceptor stems on the LSU as well as the ASL on the SSU in their respective tRNA binding sites. However, on the SSU some tRNA elements still

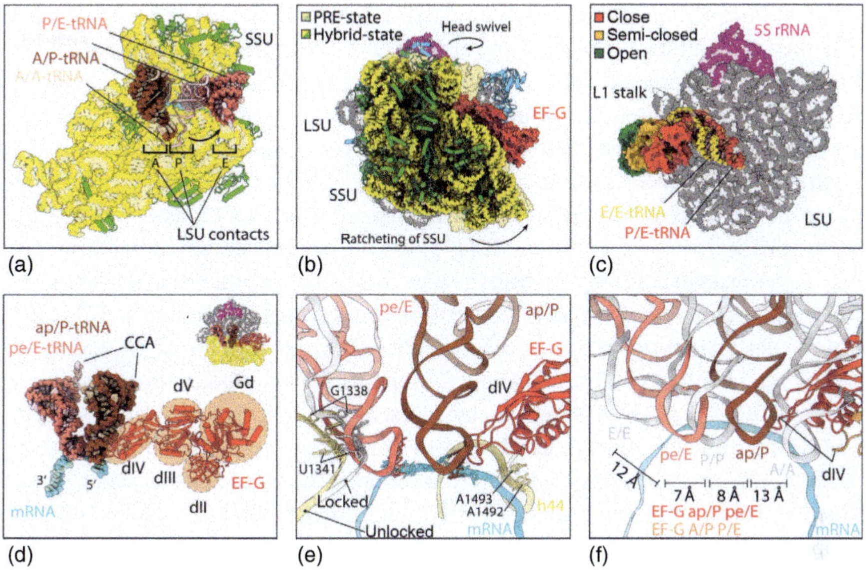

Figure 8.6 Structural rearrangements of the ribosome and EF-G during translocation. (a) Transition of tRNAs relative to the SSU from the A/A (tan) to A/P (brown) and P/P (pink) to P/E (salmon) states during the initial ratcheting of the ribosome in the absence of EF-G (PDB: 4V7C). (b) Comparison between the PRE (khaki, rRNA; light green, SSU proteins) and hybrid state (yellow, rRNA; green, SSU proteins) relative to the LSU (gray, 23S rRNA; magenta, 5S rRNA; blue, LSU proteins) is indicated. Binding of EF-G (red) triggers an additional movement of the head of the SSU called the "head swivel" that is a clockwise rotation of the head toward the platform of the SSU (PDB: 4V5D, 4V7D) [101, 132]. (c) Conformation of the L1 stalk shown in the open (green), semi-closed (orange), and closed (red) positions. The L1 stalk facilitates the movement of the tRNA from a P/E (red) to E/E (yellow) state. Dissociation of the E-tRNA from the ribosomes moves the L1 stalk back into its open position (PDB: 4V9D, 4V5D, 4V9D). (d) View from the top of the ribosome (see inlet) with mRNA (cyan), ap/P tRNA (brown), pE/E tRNA (salmon), and EF-G (red) being shown. Further indicated is the arrangement of domains of EF-G with Gd being the "G-domain" (PDB: 4V7B). (e) Close-up view on the mRNA path. Domain IV (dIV) of EF-G disrupts important interactions of A1492 and A1493 of h44 (yellow), destabilizing the anticodon/codon helix and allowing the A/P tRNA to transit into the ap/P (brown) conformation. EF-G induced head swiveling induces a structural rearrangement from a locked state (gray) to an unlocked state (yellow) of h29 and neighboring residues that allows accommodation of pe/E (salmon) tRNA (PDB: 4V7B). (f) Comparison of different tRNA states (gray, A/A + P/P + E/E; brown, ap/P; salmon, pe/E) and their movement along the mRNA (cyan) path (distances between the different states are indicated). Two different states of dIV of EF-G are shown (tan, EF-G in complex with A/P + P/E tRNA; red, EF-G in complex with ap/P + pe/E tRNA), showing the extension of dIV during translocation (PDB: 4V5D, 4V7D, 4V7B) [101, 132, 133]. Source: (a) Adapted from Brilot et al. [132]; (d, e) Adapted from Gao et al. [134].

interact with residues of the A- and P-sites on the SSU head domain [156–158]. This might represent a late-stage intermediate showing that reverse rotation of the SSU body and platform is initiated before head swiveling reaches its maximum. Followed by positioning of the ASL and mRNA in the SSU is a back swivel of the head and further reverse rotation of the SSU body and platform transforming the ribosome in its classic non-rotated state with a vacant A-site (POST-state; P/P, E/E) [152]. Coupled to the movements of the SSU and tRNAs is the conformation of the L1 stalk of the LSU that compromises parts of helices 76, 77, and 78 of 23S rRNA, as well as the L1 protein. Dependent on the rotation of body and platform, the L1 stalk was visualized in three different conformations. In the non-rotated classic state, the L1 configuration can be described as outward or open, directed away from the ribosomal core. In the rotated hybrid state, the L1 stalks undergoes a 45–60 Å movement, interacting with the elbow of P/E tRNA [143, 153, 154]. Upon completion of translocation, the L1 protein is still contacting the now fully accommodated E-tRNA but has shifted by 20 Å compared with the open conformation (Figure 8.6c) [143]. Dissociation of the deacylated tRNA moves the L1 stalk back to its open position. Based on these observed conformations, it is likely that the L1 stalk facilitates the movement of the E-site tRNA at different stages as suggested by [159]. Not only the ribosome but also EF-G undergoes conformational changes throughout translocation. Structural studies with EF-G in solution and bound to different states of translocation reveal a transition from a free conformation in solution into an extended conformation on the ribosome [132–134, 158, 160]. Notably, one X-ray structure observed the compacted form of EF-G on the ribosome. However, the observed compact form might have been the result of fusing EF-G to L9 of the neighboring ribosome and/or the usage of the antibiotic dityromycin trapping EF-G and therefore may not represent a physiological state [161]. The extension of EF-G is driven by the progressive movement of domain IV, resulting in docking of this domain in the A-site of the SSU as observed in non-rotated post-translocation complexes [132, 134, 158]. Structures of intermediate states showed that domain IV contacts the ASL of the peptidyl-tRNA still bound to the SSU A-site as well as important residues of the DC (Figure 8.6e,f) [132–134, 158]. These contacts might form barriers that need to be destabilized by EF-G for fast translocation. The fully accommodated domain IV in the POST-state might, on the other hand, work as a barrier or pawl that prevents back-translocation of the peptidyl-tRNA from the P-site to the A-site [162].

8.3.3 Termination

The presence of one of three stop codons within the A-site terminates protein synthesis by releasing the NC from the ribosome. These three stop codons are encoded as UAG, UAA, and UGA. In contrast to canonical codons, stop codons are recognized by class I release factors that mediate the hydrolysis of the ester bond of the peptidyl-tRNA. Release factor 1 (RF1) recognizes UAG and UAA codons, whereas release factor 2 (RF2) is specific for UGA and UAA. After release of the NC, the class II release factor RF3 binds the ribosome and dissociates RF1/2 from the ribosome. RF1/2 share highly conserved regions and consist of four domains with domains 2, 3, and 4 of the factors overlapping with the canonical binding site of A-site tRNA [163]. For a long time, it remained elusive how these decoding factors are capable of discriminating between the different stop codons or if they indirectly recognize stop codons through interactions with the ribosome. Swapping of conserved domains between RF1 and RF2 revealed the presence of a crucial tripeptide motif, namely, P(A/V)T in RF1 and SPF in RF2, located in a loop of domain 2 [164]. Exchanging these motifs between both RFs changes the specificity toward the stop codon, suggesting that the tripeptide motif efficiently deciphers stop codons, in an anticodon-like manner [164, 165]. Four high-resolution crystal structures of RF1/2 bound to both their respective stop codons explain the molecular mechanism behind deciphering stop codons [166–169]. P(A/V)T/SPF motives are located in loops that are directed toward the decoding site, interacting with the stop codon. The stop codon itself adopts an unusual conformation with the first two bases stacking and the third base being sandwiched between G530 and residues of the release factors. Surprisingly, only a single amino acid in both motifs is in direct contact with the second position of the respective stop codon, namely, the T186 for RF1 and the S206 for RF2 [166–169]. This is in agreement with studies from Ito et al., showing a prerequisite of those two aa in overexpressed RF1/2 mutants in ΔRF1 or ΔRF2 strains [164]. The acceptance of an A and a G for RF2 might be due to the potential of serine to interact with A and G at this position. U1 position of the stop codon is recognized by backbone elements of the decoding factors that interact with N3 of uridine and explains the restriction to U at this position [166–169]. Due to the backbone interaction, mutations failed to confirm this interaction. However, by introducing noncanonical RNA residues at the first position, Erlacher and coworkers were able to show that this interactions relies on the exocyclic group of uridine, explaining its exclusiveness at this position [170]. RF1 monitors the third position via interactions of Thr194 and Q181, whereas RF2 interaction depends on T194 [166–169]. Mutational studies show that exchange of aa adjacent to the tripeptide motifs can change the specificity of RF1/2 [171–173]. These findings, however, should not question the importance of the tripeptide motives, but rather highlight an elaborate network of interactions in which the P(A/V)T/SPF motif is a prerequisite (Figure 8.7a,b). Using metal ion fluorescence resonance energy transfer, Trappl and Joseph could show that upon recognition of the stop codon, RF1 opens from a closed to an open extended conformation on the ribosome. By contrast, this induced fit does not happen in the presence of a sense codon [177]. Structures of the isolated decoding factors reveal a tight packing of domains 2 and 3 against each other, whereas bound to the ribosome, domain 3 escapes this packing and is orientated toward the PTC [166–169, 174]. Opening of domain 3 requires a rearranged state of a switch loop within RF1/2, connecting domains 3 and 4. This rearranged state of the switch loop is stabilized by residues of the decoding site that adopt an alternative conformation upon stop codon recognition by RFs (Figure 8.7c) [166, 168]. Thus, recognition of the stop codon is coupled to the opening of RFs on the ribosome (Figure 8.7d). Sequence alignments between all kingdoms showed the abundance of a GGQ motif in all RFs [178]. In bacteria this motif is found within the tip of domain 3, placing it next to A76 of the peptidyl-tRNA upon RF opening. Mutations affecting the first and second glycines abolished hydrolysis, whereas mutations of glutamine were tolerated, indicating a direct involvement of the motif for hydrolysis [178, 179]. This came as a surprise as the glutamine is post-translationally methylated in vivo, leading to enhanced activity for NC release [180, 181]. From structural analysis it came apparent that the side chain of Gln

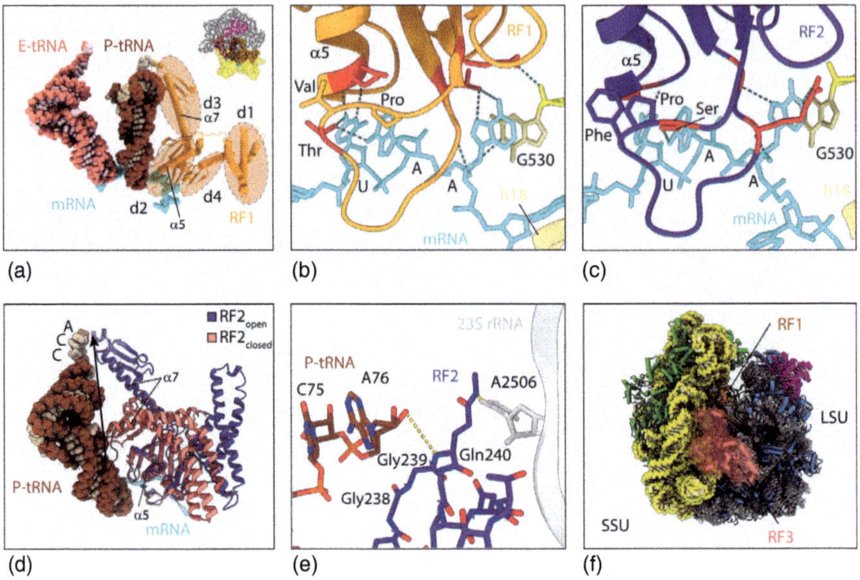

Figure 8.7 Termination of translation in the presence of a stop codon. (a) View from the top of the ribosome (see inlet) with mRNA (cyan), P-tRNA (brown), E-tRNA (salmon), and RF1 (orange). Further indicated is the domain arrangement of RF1 (PDB: 4V63). (b) Interaction of the PVT motif of RF1 (orange) with the UAA stop codon of the mRNA (cyan). Residues of RF1 that are directly involved in decoding the stop codon are marked in red (PDB: 4V63). (c) Same as (b) but in the presence of RF2 and the corresponding SPF motif (PDB: 4V67). (d) After decoding of the stop codon, RF2 transits from a closed conformation (salmon) to an open conformation (blue) placing the tip of domain 3 toward the PTC. This involves a movement of 60 Å of domain 3 compared with the closed RF2 conformation (PDB: 4V67, 1GQE) [167, 174]. (e) Backbone interaction between the catalytic important Gln240 of RF2 (blue) and A76 of P-tRNA (brown) is shown (PDB: 6C5L). (f) Structure of RF3 (salmon) bound to the ribosome in the presence of RF1 (orange) (PDB: 6GXM). Source: (a, b) Adapted from Laurberg et al. [166]; (c) Adapted from Korostelev et al. [167]; (e) Adapted from Zeng and Jin [175]; (f) Adapted from Graf et al. [176].

of the GGQ motif in RF1 is orientated away from A76 and the catalytic center. The backbone NH, however, is in hydrogen bonding distance to the 3′-OH of A76. Mutation of Gln to Pro in RF1 eliminates the NH backbone interaction [182] and thereby abolishes the activity of RF1 (Figure 8.7e) [167]. Recent structures, which used RFs carrying the methylated glutamine, showed a tighter packing of the glutamine against its neighboring residues, thereby additionally stabilizing the backbone of the glutamine [175, 183]. The two glycines, on the other hand, are not directly involved in catalysis but are important for the conformation and integrity of the loop [166, 167]. Likewise, during peptide bond formation, the presence of an A-site substrate induces a similar conformational change within the PTC [179]. In the presence of RF1/2, this includes residues U2506 that overlaps with the binding site of RFs and U2585 that moves away from the ester bond, allowing its hydrolysis by a water molecule [103]. This in turn leads to the following situation: accommodation of domain 3 and conformational changes within the PTC allow the activation of an attacking water molecule. The backbone of Gln interacts via its NH group with the 3′-OH of A76, whereas its side chain is shielding the PTC from other nucleophiles larger than water (Figure 8.7e) [166–169, 184]. Similar to peptide bond formation, this reaction proceeds through a tetrahedral intermediate [166–174, 177, 178, 185]. Breakdown of this state results in deacylated tRNA and free NC. In contrast to peptide bond formation, hydrolysis of the NC is less well understood. One model suggests that 2′-OH of A76 acts as a protein shuttle by accepting a proton from the nucleophilic water and subsequently transferring it to the leaving group on the 3′-OH that is stabilized by the main chain amide of the glutamine [98, 108, 119, 185, 186]. Other models suggest a stepwise proton transfer with only one proton moving at the same time [123]. After hydrolysis the class II release factor RF3 binds the ribosome and stimulates the release of RF1/2 [187, 188]. Like EF-Tu and EF-G, RF3 is a traGTPase and binds the ribosome preferentially in complex with GTP (Figure 8.7f) [189–191]. The binding site of RF3 overlaps with their position on the ribosome [192–195]. From crystal structures it is evident that in the absence of RF3, but in the presence of RF1/2, the ribosome adopts a non-rotated state [166–169]. By contrast, it was shown that in the presence of RF3, but in the absence of RF1/2, the ribosome is in a rotated state [176, 192, 193, 195]. Recent biophysical and structural studies further refined the understanding for this process. Association of RF3-GTP with 70S ribosomes facilitates the transition of post-hydrolysis complexes containing RF1/2 from a non-rotated state to a rotated state [176] and thereby removes the decoding factors [189, 196, 197]. Thereby, RF3 moves as a rigid body along with the SSU relative to the LSU, which leads to interactions of catalytic important G-domain of RF3 with the SRL of the LSU. Thus, RF3 fulfills a dual role. First, RF3 mediates subunit rotation, which leads to the ejection of RF1/2, and second, RF3 facilitates its own dissociation via contacts with the SRL, triggering the hydrolysis of GTP to GDP and Pi. Interestingly, RF3 mutants that are able to bind GTP but are deficient in hydrolysis are also able to induce this transformation, suggesting that the presence of the hydrolyzable analog is needed but not its hydrolysis [189, 198]. It rather seems that hydrolysis is needed to reset ribosomes back to a non-rotated state by dissociating RF3 from the ribosome [189, 191, 198]. Last but not least it should be noted that cells lacking RF3 show no growth defect [199]. Even more, RF3 is only found in a subset of bacteria, suggesting that RF3 is not part of a conserved mechanism but an auxiliary factor fine-tuning the RNA machinery [200].

8.3.4 Recycling

Dissociation of RF1/2 results in a post-termination complex (PoTC) programmed with a deacylated tRNA in a hybrid P/E state with mRNA. To return the ribosomal subunits to the pool of translating ribosome, the ribosome has to be disassembled into its individual components. Recycling is mediated by three factors, namely, the ribosome recycling factor (RRF), EF-G, and IF3. RRF binds the ribosome in its rotated state (Figure 8.8a) [197, 201, 202]. Binding of RRF to the ribosomal A-site prevents reassociation of RF1/2, as evident from single-molecule Förster

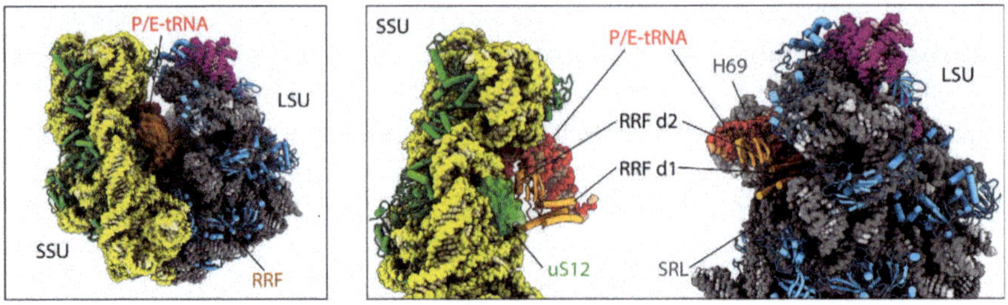

Figure 8.8 Post-termination 70S ribosome in complex with RRF and P/E-tRNA. (a) Front view on the 70S ribosome programmed with RRF (orange) and P/E-tRNA (red). (b) Interactions of RRF and P/E-tRNA with respect to the SSU and LSU. The domain arrangement of RRF is indicated as d1 or d2.

resonance energy transfer (smFRET) experiments. While binding of RF1/2 converts the ribosome to a non-rotated state, increasing concentrations of RRF stabilize the rotated state. This is also in agreement with structural studies, showing that binding of RRF to a physiological PoTC was only observed in the rotated state [153, 201, 203]. Notably, there was also a structure of RRF bound to a non-rotated ribosome. However, in this crystal structure, the ASL of a P-tRNA was used, thereby allowing accommodation of RRF as it would otherwise clash with the body of the tRNA [204]. RRF consists of two domains connected by a linker and binds the A- and P-site cleft of the LSU. Domain II is contacting S12 and intersubunit bridge B2a on the SSU, while domain I forms several contacts with 23S rRNA on the LSU (Figure 8.8b) [201, 203, 205, 206]. Binding of RRF alone is not sufficient to split subunits but requires EF-G to do so [207–209]. Reconstructions of time-resolved cryo-EM showed a conformational change of RRF upon binding of EF-G that directs domain II toward domain I bringing it into closer contact with the inter-subunit bridge B2a [201, 210]. This structural rearrangement together with domain IV of EF-G is thought to split B2a and thereby facilitates subunit dissociation [153, 201, 210–212]. Experimentally it was shown that additional to the presence of RRF and EF-G, recycling requires the hydrolysis of GTP bound to EF-G to split the 70S ribosome into subunits [202, 213]. However, it remains elusive if hydrolysis or resulting Pi release triggers the abovementioned structural changes or if they are required to dissociate EF-G from the ribosome. Upon subunit dissociation, RRF and EF-G leave the ribosome, and IF3 binds the ribosome, preventing rebinding of the LSU [34, 202, 208, 214]. Furthermore, it was shown that binding of IF3 accelerates the departure of P-site tRNA and thereby indirectly causes the dissociation of the mRNA [34, 202]. Thus, IF3 facilitates the departure of tRNA and mRNA but also connects recycling with translation initiation by preventing reassociation of the LSU. Noteworthy, the above-described pathway of recycling might just be one possible way to split subunits. There are many possibilities described in the literature how this might be achieved. For example, a recent paper suggests that in the presence of RRF and EF-G, the mRNA is first to leave the 70S, followed by the tRNA and delayed subunit dissociation. However, in the presence of an upstream

SD sequence, mRNA and tRNA dissociate after 70S splitting [215]. Interestingly, mRNAs used for recycling experiments mostly contain a start codon followed by a stop codon with an upstream SD sequence, pushing the experimental outcome toward prior splitting followed by tRNA/mRNA dissociation. On the other hand, the absence of an SD sequence would provide a pool of vacant 70S that can be used to initiate translation [63, 215, 216]. Another model suggests that EF-G translocates RRF into the P-site and thereby dissociates tRNA and hence indirectly the mRNA [214]. This would again provide a vacant 70S pool. Another explanation for the discrepancy in experimental outcomes is the presence of different Mg^{2+} concentrations. The lower the Mg^{2+} concentration is, the more likely it is to observe a 70S splitting event. In summary, there might be several pathways to return ribosomes to the translation pool.

References

1 Crick, F.H. (1958). On protein synthesis. *Symp. Soc. Exp. Biol.* 12: 138–163.

2 Crick, F. (1970). Central dogma of molecular biology. *Nature* 227 (5258): 561–563.

3 Huttenhofer, A., Schattner, P., and Polacek, N. (2005). Non-coding RNAs: hope or hype? *Trends Genet.* 21 (5): 289–297.

4 Dogini, D.B., Pascoal, V.D., Avansini, S.H. et al. (2014). The new world of RNAs. *Genet. Mol. Biol.* 37 (1 Suppl): 285–293.

5 Melnikov, S., Ben-Shem, A., Garreau de Loubresse, N. et al. (2012). One core, two shells: bacterial and eukaryotic ribosomes. *Nat. Struct. Mol. Biol.* 19 (6): 560–567.

6 Anger, A.M., Armache, J.P., Berninghausen, O. et al. (2013). Structures of the human and *Drosophila* 80S ribosome. *Nature* 497 (7447): 80–85.

7 Schluenzen, F., Tocilj, A., Zarivach, R. et al. (2000). Structure of functionally activated small ribosomal subunit at 3.3 A resolution. *Cell* 102: 615–623.

8 Yusupov, M.M., Yusupova, G.Z., Baucom, A. et al. (2001). Crystal structure of the ribosome at 5.5 A resolution. *Science* 292 (5518): 883–896.

9 Schmeing, T.M. and Ramakrishnan, V. (2009). What recent ribosome structures have revealed about the mechanism of translation. *Nature* 461 (7268): 1234–1242.

10 Dabbs, E.R. (1986). Mutant studies on the prokaryotic ribosome. In: *Structure, Function and Genetics of Ribosomes* (eds. B. Hardesty and G. Kramer), 733–748. Springer-Verlag.

11 Gavrilova, L.P., Kostiashkina, O.E., Koteliansky, V.E. et al. (1976). Factor-free non-enzymic and factor-dependent systems of translation of polyU by *E. coli* ribosomes. *J. Mol. Biol.* 101: 537–552.

12 Leder, P. and Nirenberg, M. (1964). RNA codewords and protein synthesis, II. Nucleotide sequence of a valine RNA codeword. *Proc. Natl. Acad. Sci. U.S.A.* 52: 420–427.

13 Lill, R., Robertson, J.M., and Wintermeyer, W. (1986). Affinities of tRNA binding sites of ribosomes from *Escherichia coli*. *Biochemistry* 25 (11): 3245–3255.
14 Traub, P. and Nomura, M. (1969). Structure and function of the *Escherichia coli* ribosome. Mechanism of assembly of 30S ribosomes studied *in vitro*. *J. Mol. Biol.* 40: 391–413.
15 Triman, K.L., Peister, A., and Goel, R.A. (1998). Expanded versions of the 16S and 23S ribosomal RNA mutation databases (16SMDBexp and 23SMDBexp). *Nucleic Acids Res.* 26 (1): 280–284.
16 Sato, N.S., Hirabayashi, N., Agmon, I. et al. (2006). Comprehensive genetic selection revealed essential bases in the peptidyl-transferase center. *Proc. Natl. Acad. Sci. U.S.A.* 103 (42): 15386–15391.
17 Zhang, B. and Cech, T.R. (1997). Peptide bond formation by in vitro selected ribozymes. *Nature* 390 (6655): 96–100.
18 Noller, H.F. (2012). Evolution of protein synthesis from an RNA world. *Cold Spring Harbor Perspect. Biol.* 4 (4): a003681.
19 Noller, H.F., Hoffarth, V., and Zimniak, L. (1992). Unusual resistance of peptidyltransferase to protein extraction procedures. *Science* 256: 1416–1419.
20 Wilson, D.N. and Nierhaus, K.H. (2005). Ribosomal proteins in the spotlight. *Crit. Rev. Biochem. Mol. Biol.* 40 (5): 243–267.
21 Buttgereit, F. and Brand, M.D. (1995). A hierarchy of ATP-consuming processes in mammalian cells. *Biochem. J.* 312 (Pt 1): 163–167.
22 Russell, J.B. and Cook, G.M. (1995). Energetics of bacterial growth: balance of anabolic and catabolic reactions. *Microbiol. Rev.* 59 (1): 48–62.
23 Passalacqua, K.D., Varadarajan, A., Ondov, B.D. et al. (2009). Structure and complexity of a bacterial transcriptome. *J. Bacteriol.* 191 (10): 3203–3211.
24 Fluitt, A., Pienaar, E., and Viljoen, H. (2007). Ribosome kinetics and aa-tRNA competition determine rate and fidelity of peptide synthesis. *Comput. Biol. Chem.* 31 (5-6): 335–346.
25 Young, R. and Bremer, H. (1976). Polypeptide-chain-elongation rate in *Escherichia coli* B/r as a function of growth rate. *Biochem. J.* 160 (2): 185–194.
26 Gualerzi, C.O. and Pon, C.L. (1990). Initiation of messenger-RNA translation in prokaryotes. *Biochemistry* 29 (25): 5881–5889.
27 Laursen, B.S., Sorensen, H.P., Mortensen, K.K., and Sperling-Petersen, H.U. (2005). Initiation of protein synthesis in bacteria. *Microbiol. Mol. Biol. Rev.* 69 (1): 101–123.
28 Milon, P. and Rodnina, M.V. (2012). Kinetic control of translation initiation in bacteria. *Crit. Rev. Biochem. Mol. Biol.* 47 (4): 334–348.
29 Carter, A.P., Clemons, W.M. Jr., Brodersen, D.E. et al. (2001). Crystal structure of an initiation factor bound to the 30S ribosomal subunit. *Science* 291 (5503): 498–501.
30 Hussain, T., Llacer, J.L., Wimberly, B.T. et al. (2016). Large-scale movements of IF3 and tRNA during bacterial translation initiation. *Cell* 167 (1): 133–144.e13.

31 Fischer, N., Neumann, P., Konevega, A.L. et al. (2015). Structure of the *E. coli* ribosome-EF-Tu complex at <3 A resolution by C-corrected cryo-EM. *Nature* 520 (7548): 567–570.
32 Sprink, T., Ramrath, D.J., Yamamoto, H. et al. (2016). Structures of ribosome-bound initiation factor 2 reveal the mechanism of subunit association. *Sci. Adv.* 2 (3): e1501502.
33 Dallas, A. and Noller, H.F. (2001). Interaction of translation initiation factor 3 with the 30S ribosomal subunit. *Mol. Cell* 8 (4): 855–864.
34 Karimi, R., Pavlov, M., Buckingham, R., and Ehrenberg, M. (1999). Novel roles for classical factors at the interface between translation termination and initiation. *Mol. Cell* 3: 601–609.
35 Julian, P., Milon, P., Agirrezabala, X. et al. (2011). The Cryo-EM structure of a complete 30S translation initiation complex from *Escherichia coli*. *PLoS Biol.* 9 (7): e1001095.
36 Simonetti, A., Marzi, S., Myasnikov, A.G. et al. (2008). Structure of the 30S translation initiation complex. *Nature* 455 (7211): 416–420.
37 Gualerzi, C.O., Severini, M., Spurio, R. et al. (1991). Molecular dissection of translation initiation factor-IF2 – evidence for two structural and functional domains. *J. Biol. Chem.* 266: 16356–16362.
38 Wienk, H., Tishchenko, E., Belardinelli, R. et al. (2012). Structural dynamics of bacterial translation initiation factor IF2. *J. Biol. Chem.* 287 (14): 10922–10932.
39 Caban, K., Pavlov, M., Ehrenberg, M., and Gonzalez, R.L. Jr., (2017). A conformational switch in initiation factor 2 controls the fidelity of translation initiation in bacteria. *Nat. Commun.* 8 (1): 1475.
40 Guenneugues, M., Caserta, E., Brandi, L. et al. (2000). Mapping the fMet-tRNA^{fMet} binding site of initiation factor IF2. *EMBO J.* 19 (19): 5233–5240.
41 Antoun, A., Pavlov, M.Y., Lovmar, M., and Ehrenberg, M. (2006). How initiation factors tune the rate of initiation of protein synthesis in bacteria. *EMBO J.* 25 (11): 2539–2550.
42 Boelens, R. and Gualerzi, C.O. (2002). Structure and function of bacterial initiation factors. *Curr. Protein Pept. Sci.* 3: 107–119.
43 Studer, S. and Joseph, S. (2006). Unfolding of mRNA secondary structure by the bacterial translation initiation complex. *Mol. Cell* 22 (1): 105–115.
44 Shine, J. and Dalgarno, L. (1974). The 3′-terminal sequence of *E. coli* 16S rRNA: complementarity to nonsense triplets and ribosome binding sites. *Proc. Natl. Acad. Sci. U.S.A.* 71 (4): 1342–1346.
45 Skorski, P., Leroy, P., Fayet, O. et al. (2006). The highly efficient translation initiation region from the *Escherichia coli* rpsA gene lacks a Shine–Dalgarno element. *J. Bacteriol.* 188 (17): 6277–6285.
46 Steitz, J.A. and Jakes, K. (1975). How ribosomes select initiator regions in mRNA: Base pair formation between the 3′ terminus of 16S rRNA and the mRNA during initiation of protein synthesis in *E. coli*. *Proc. Natl. Acad. Sci. U.S.A.* 72 (12): 4734–4738.
47 Allen, G., Zavialov, A., Gursky, R. et al. (2005). The cryo-EM structure of a translation initiation complex from *Escherichia coli*. *Cell* 121 (5): 703–712.

48 Jenner, L., Romby, P., Rees, B. et al. (2005). Translational operator of mRNA on the ribosome: how repressor proteins exclude ribosome binding. *Science* 308 (5718): 120–123.

49 Kaminishi, T., Wilson, D.N., Takemoto, C. et al. (2007). A snapshot of the 30S ribosomal subunit capturing mRNA via the Shine–Dalgarno interaction. *Structure* 15 (3): 289–297.

50 Marzi, S., Myasnikov, A.G., Serganov, A. et al. (2007). Structured mRNAs regulate translation initiation by binding to the platform of the ribosome. *Cell* 130 (6): 1019–1031.

51 Allert, M., Cox, J.C., and Hellinga, H.W. (2010). Multifactorial determinants of protein expression in prokaryotic open reading frames. *J. Mol. Biol.* 402 (5): 905–918.

52 Kudla, G., Murray, A.W., Tollervey, D., and Plotkin, J.B. (2009). Coding-sequence determinants of gene expression in *Escherichia coli*. *Science* 324 (5924): 255–258.

53 Nakamoto, T. (2006). A unified view of the initiation of protein synthesis. *Biochem. Biophys. Res. Commun.* 341 (3): 675–678.

54 Demo, G., Rasouly, A., Vasilyev, N. et al. (2017). Structure of RNA polymerase bound to ribosomal 30S subunit. *elife* 6: e28560.

55 Komarova, A.V., Tchufistova, L.S., Supina, E.V., and Boni, I.V. (2002). Protein S1 counteracts the inhibitory effect of the extended Shine–Dalgarno sequence on translation. *RNA* 8 (9): 1137–1147.

56 Sengupta, J., Agrawal, R., and Frank, J. (2001). Visualization of protein S1 within the 30S ribosomal subunit and its interaction with messenger RNA. *Proc. Natl. Acad. Sci. U.S.A.* 98 (21): 11991–11996.

57 Jacob, W.F., Santer, M., and Dahlberg, A.E. (1987). A single base change in the Shine–Dalgarno region of 16S rRNA of *Escherichia coli* affects translation of many proteins. *Proc. Natl. Acad. Sci. U.S.A.* 84 (14): 4757–4761.

58 Ma, J., Campbell, A., and Karlin, S. (2002). Correlations between Shine–Dalgarno sequences and gene features such as predicted expression levels and operon structures. *J. Bacteriol.* 184 (20): 5733–5745.

59 Yusupova, G., Jenner, L., Rees, B. et al. (2006). Structural basis for messenger RNA movement on the ribosome. *Nature* 444 (7117): 391–394.

60 Chang, B., Halgamuge, S., and Tang, S.L. (2006). Analysis of SD sequences in completed microbial genomes: non-SD-led genes are as common as SD-led genes. *Gene* 373: 90–99.

61 Scharff, L.B., Childs, L., Walther, D., and Bock, R. (2011). Local absence of secondary structure permits translation of mRNAs that lack ribosome-binding sites. *PLoS Genet.* 7 (6): e1002155.

62 Milon, P., Konevega, A.L., Gualerzi, C.O., and Rodnina, M.V. (2008). Kinetic checkpoint at a late step in translation initiation. *Mol. Cell* 30 (6): 712–720.

63 Grill, S., Lodei, P., Gualerzi, C.O., and Bläsi, U. (2000). Selective stimulation of translation of leaderless mRNA by initiation factor 2: evolutionary implications for translation. *EMBO J.* 19: 4101–4110.

64 Kaberdina, A.C., Szaflarski, W., Nierhaus, K.H., and Moll, I. (2009). An unexpected type of ribosomes induced by kasugamycin: a look into ancestral times of protein synthesis? *Mol. Cell* 33 (2): 227–236.
65 Moll, I., Grill, S., Grundling, A., and Blasi, U. (2002). Effects of ribosomal proteins S1, S2 and the DeaD/CsdA DEAD-box helicase on translation of leaderless and canonical mRNAs in *Escherichia coli*. *Mol. Microbiol.* 44 (5): 1387–1396.
66 Calogero, R.A., Pon, C.L., Canonaco, M.A., and Gualerzi, C.O. (1988). Selection of the mRNA translation initiation region by *Escherichia coli* ribosomes. *Proc. Natl. Acad. Sci. U.S.A.* 85: 6427–6431.
67 Van Etten, W.J. and Janssen, G.R. (1998). An AUG initiation codon, not codon-anticodon complementarity, is required for the translation of unleadered mRNA in *Escherichia coli*. *Mol. Microbiol.* 27 (5): 987–1001.
68 Canonaco, M.A., Gualerzi, C.O., and Pon, C.L. (1989). Alternative occupancy of a dual ribosomal binding site by mRNA affected by translation initiation factors. *Eur. J. Biochem.* 182: 501–506.
69 Sussman, J.K., Simons, E.L., and Simons, R.W. (1996). *Escherichia coli* translation initiation factor 3 discriminates the initiation codon *in vivo*. *Mol. Microbiol.* 21 (2): 347–360.
70 Goyal, A., Belardinelli, R., Maracci, C. et al. (2015). Directional transition from initiation to elongation in bacterial translation. *Nucleic Acids Res.* 43 (22): 10700–10712.
71 Qin, H., Grigoriadou, C., and Cooperman, B.S. (2009). Interaction of IF2 with the ribosomal GTPase-associated center during 70S initiation complex formation. *Biochemistry* 48 (22): 4699–4706.
72 La Teana, A., Gualerzi, C.O., and Dahlberg, A.E. (2001). Initiation factor IF 2 binds to the alpha-sarcin loop and helix 89 of *Escherichia coli* 23S ribosomal RNA. *RNA* 7 (8): 1173–1179.
73 Myasnikov, A., Marzi, S., Simonetti, A. et al. (2005). Conformational transition of initiation factor 2 from the GTP- to GDP-bound state visualized on the ribosome. *Nat. Struct. Mol. Biol.* 12 (12): 1145–1149.
74 Marshall, R.A., Aitken, C.E., and Puglisi, J.D. (2009). GTP hydrolysis by IF2 guides progression of the ribosome into elongation. *Mol. Cell* 35 (1): 37–47.
75 Demeshkina, N., Jenner, L., Westhof, E. et al. (2012). A new understanding of the decoding principle on the ribosome. *Nature* 484 (7393): 256–259.
76 Rozov, A., Demeshkina, N., Westhof, E. et al. (2015). Structural insights into the translational infidelity mechanism. *Nat. Commun.* 6: 7251.
77 Loveland, A.B., Demo, G., Grigorieff, N., and Korostelev, A.A. (2017). Ensemble cryo-EM elucidates the mechanism of translation fidelity. *Nature* 546 (7656): 113–117.
78 Rodnina, M.V. and Wintermeyer, W. (2001). Fidelity of aminoacyl-tRNA selection on the ribosome: kinetic and structural mechanisms. *Annu. Rev. Biochem.* 70: 415–435.
79 Crick, F.H.C. (1966). Codon-anticodon pairing: the wobble hypothesis. *J. Mol. Biol.* 19: 548–555.

80 Ogle, J.M., Brodersen, D.E., Clemons, W.M. Jr., et al. (2001). Recognition of cognate transfer RNA by the 30S ribosomal subunit. *Science* 292 (5518): 897–902.

81 Loftfield, R.B. (1963). The frequency of errors in protein biosynthesis. *Biochem. J.* 89: 82–92.

82 Eisinger, J., Feuer, B., and Yamane, T. (1971). Codon-anticodon binding in tRNAphe. *Nat. New Biol.* 231 (21): 126–128.

83 Ogle, J.M. and Ramakrishnan, V. (2005). Structural insights into translational fidelity. *Annu. Rev. Biochem.* 74: 129–177.

84 Rodnina, M.V., Fricke, R., and Wintermeyer, W. (1994). Transient conformational states of aminoacyl-tRNA during ribosome binding catalyzed by elongation factor Tu. *Biochemistry* 33 (40): 12267–12275.

85 Rodnina, M.V., Fricke, R., Kuhn, L., and Wintermeyer, W. (1995). Codon-dependent conformational change of elongation factor Tu preceding GTP hydrolysis on the ribosome. *EMBO J.* 14: 2613–2619.

86 Pape, T., Wintermeyer, W., and Rodnina, M. (1999). Induced fit in initial selection and proofreading of aminoacyl-tRNA on the ribosome. *EMBO J.* 18 (13): 3800–3807.

87 Voorhees, R.M., Schmeing, T.M., Kelley, A.C., and Ramakrishnan, V. (2010). The mechanism for activation of GTP hydrolysis on the ribosome. *Science* 330 (6005): 835–838.

88 Pape, T., Wintermeyer, W., and Rodnina, M.V. (1998). Complete kinetic mechanism of elongation factor Tu-dependent binding of aminoacyl-tRNA to the A site of the *E. coli* ribosome. *EMBO J.* 17 (24): 7490–7497.

89 Rodnina, M.V., Pape, T., Fricke, R. et al. (1996). Initial binding of the elongation factor Tu-center-dot-GTP-center-dot-aminoacyl-tRNA complex preceding codon recognition on the ribosome. *J. Biol. Chem.* 271 (2): 646–652.

90 Ogle, J.M., Murphy, F.V., Tarry, M.J., and Ramakrishnan, V. (2002). Selection of tRNA by the ribosome requires a transition from an open to a closed form. *Cell* 111 (5): 721–732.

91 Lescoute, A. and Westhof, E. (2006). The A-minor motifs in the decoding recognition process. *Biochimie* 88 (8): 993–999.

92 Rozov, A., Demeshkina, N., Westhof, E. et al. (2016). New structural insights into translational miscoding. *Trends Biochem. Sci.* 41 (9): 798–814.

93 Blanchard, S.C., Kim, H.D., Gonzalez, R.L. Jr., et al. (2004). tRNA dynamics on the ribosome during translation. *Proc. Natl. Acad. Sci. U.S.A.* 101 (35): 12893–12898.

94 Schmeing, T.M., Voorhees, R.M., Kelley, A.C. et al. (2009). The crystal structure of the ribosome bound to EF-Tu and aminoacyl-tRNA. *Science* 326 (5953): 688–694.

95 Fischer, N., Neumann, P., Bock, L.V. et al. (2016). The pathway to GTPase activation of elongation factor SelB on the ribosome. *Nature* 540 (7631): 80–85.

96 Satterthwait, A.C. and Jencks, W.P. (1974). The mechanism of the aminolysis of acetate esters. *J. Am. Chem. Soc.* 96 (22): 7018–7031.

97 Hiller, D.A., Singh, V., Zhong, M., and Strobel, S.A. (2011). A two-step chemical mechanism for ribosome-catalysed peptide bond formation. *Nature* 476 (7359): 236–239.

98 Sievers, A., Beringer, M., Rodnina, M.V., and Wolfenden, R. (2004). The ribosome as an entropy trap. *Proc. Natl. Acad. Sci. U.S.A.* 101 (21): 7897–7901.

99 Polacek, N. and Mankin, A.S. (2005). The ribosomal peptidyl transferase center: structure, function, evolution, inhibition. *Crit. Rev. Biochem. Mol. Biol.* 40 (5): 285–311.

100 Rodnina, M.V., Beringer, M., and Wintermeyer, W. (2007). How ribosomes make peptide bonds. *Trends Biochem. Sci.* 32 (1): 20–26.

101 Voorhees, R.M., Weixlbaumer, A., Loakes, D. et al. (2009). Insights into substrate stabilization from snapshots of the peptidyl transferase center of the intact 70S ribosome. *Nat. Struct. Mol. Biol.* 16 (5): 528–533.

102 Selmer, M., Dunham, C., Murphy, F.t. et al. (2006). Structure of the 70S ribosome complexed with mRNA and tRNA. *Science* 313 (5795): 1935–1942.

103 Polikanov, Y.S., Steitz, T.A., and Innis, C.A. (2014). A proton wire to couple aminoacyl-tRNA accommodation and peptide-bond formation on the ribosome. *Nat. Struct. Mol. Biol.* 21 (9): 787–793.

104 Ban, N., Nissen, P., Hansen, J. et al. (2000). The complete atomic structure of the large ribosomal subunit at 2.4 Å resolution. *Science* 289: 905–920.

105 Nissen, P., Hansen, J., Ban, N. et al. (2000). The structural basis of ribosome activity in peptide bond synthesis. *Science* 289 (5481): 920–930.

106 Wohlgemuth, I., Beringer, M., and Rodnina, M.V. (2006). Rapid peptide bond formation on isolated 50S ribosomal subunits. *EMBO Rep.* 7 (7): 699–703.

107 Kim, D. and Green, R. (1999). Base-pairing between 23S rRNA and tRNA in the ribosomal A site. *Mol. Cell* 4: 859–864.

108 Schmeing, T.M., Huang, K.S., Kitchen, D.E. et al. (2005). Structural insights into the roles of water and the 2′ hydroxyl of the P site tRNA in the peptidyl transferase reaction. *Mol. Cell* 20 (3): 437–448.

109 Schmeing, T.M., Huang, K.S., Strobel, S.A., and Steitz, T.A. (2005). An induced-fit mechanism to promote peptide bond formation and exclude hydrolysis of peptidyl-tRNA. *Nature* 438 (7067): 520–524.

110 Nissen, P., Ippolito, J.A., Ban, N. et al. (2001). RNA tertiary interactions in the large ribosomal subunit: the A-minor motif. *Proc. Natl. Acad. Sci. U.S.A.* 98 (9): 4899–4903.

111 Hansen, J.L., Schmeing, T.M., Moore, P.B., and Steitz, T.A. (2002). Structural insights into peptide bond formation. *Proc. Natl. Acad. Sci. U.S.A.* 99 (18): 11670–11675.

112 Muth, G.W., Ortoleva-Donnelly, L., and Strobel, S.A. (2000). A single adenosine with a neutral pKa in the ribosomal peptidyl transferase center. *Science* 289 (5481): 947–950.

113 Polacek, N., Gaynor, M., Yassin, A., and Mankin, A.S. (2001). Ribosomal peptidyl transferase can withstand mutations at the putative catalytic nucleotide. *Nature* 411 (6836): 498–501.

114 Thompson, J., Kim, D.F., O'Connor, M. et al. (2001). Analysis of mutations at residues A2451 and G2447 of 23S rRNA in the peptidyltransferase active site of the 50S ribosomal subunit. *Proc. Natl. Acad. Sci. U.S.A.* 98 (16): 9002–9007.

115 Beringer, M., Bruell, C., Xiong, L. et al. (2005). Essential mechanisms in the catalysis of peptide bond formation on the ribosome. *J. Biol. Chem.* 280 (43): 36065–36072.

116 Youngman, E.M., Brunelle, J.L., Kochaniak, A.B., and Green, R. (2004). The active site of the ribosome is composed of two layers of conserved nucleotides with distinct roles in peptide bond formation and peptide release. *Cell* 117 (5): 589–599.

117 Dorner, S., Panuschka, C., Schmid, W., and Barta, A. (2003). Mononucleotide derivatives as ribosomal P-site substrates reveal an important contribution of the 2′-OH to activity. *Nucleic Acids Res.* 31 (22): 6536–6542.

118 Erlacher, M.D., Lang, K., Wotzel, B. et al. (2006). Efficient ribosomal peptidyl transfer critically relies on the presence of the ribose 2′-OH at A2451 of 23S rRNA. *J. Am. Chem. Soc.* 128 (13): 4453–4459.

119 Weinger, J.S., Parnell, K.M., Dorner, S. et al. (2004). Substrate-assisted catalysis of peptide bond formation by the ribosome. *Nat. Struct. Mol. Biol.* 11 (11): 1101–1106.

120 Aqvist, J., Lind, C., Sund, J., and Wallin, G. (2012). Bridging the gap between ribosome structure and biochemistry by mechanistic computations. *Curr. Opin. Struct. Biol.* 22 (6): 815–823.

121 Huang, Y. and Sprinzl, M. (2011). Peptide bond formation on the ribosome: the role of the 2′-OH group on the terminal adenosine of peptidyl-tRNA and of the length of nascent peptide chain. *Angew. Chem. Int. Ed.* 50 (32): 7287–7289.

122 Zaher, H.S., Shaw, J.J., Strobel, S.A., and Green, R. (2011). The 2′-OH group of the peptidyl-tRNA stabilizes an active conformation of the ribosomal PTC. *EMBO J.* 30 (12): 2445–2453.

123 Kuhlenkoetter, S., Wintermeyer, W., and Rodnina, M.V. (2011). Different substrate-dependent transition states in the active site of the ribosome. *Nature* 476 (7360): 351–354.

124 Wallin, G. and Aqvist, J. (2010). The transition state for peptide bond formation reveals the ribosome as a water trap. *Proc. Natl. Acad. Sci. U.S.A.* 107 (5): 1888–1893.

125 Huter, P., Arenz, S., Bock, L.V. et al. (2017). Structural basis for polyproline-mediated ribosome stalling and rescue by the translation elongation factor EF-P. *Mol. Cell* 68 (3): 515–527. e6.

126 Maguire, B.A., Beniaminov, A.D., Ramu, H. et al. (2005). A protein component at the heart of an RNA machine: the importance of protein l27 for the function of the bacterial ribosome. *Mol. Cell* 20 (3): 427–435.

127 Maracci, C., Wohlgemuth, I., and Rodnina, M.V. (2015). Activities of the peptidyl transferase center of ribosomes lacking protein L27. *RNA* 21 (12): 2047–2052.

128 Frank, J. and Agrawal, R.K. (2000). A ratchet-like inter-subunit reorganization of the ribosome during translocation. *Nature* 406: 318–322.

129 Shoji, S., Walker, S.E., and Fredrick, K. (2006). Reverse translocation of tRNA in the ribosome. *Mol. Cell* 24 (6): 931–942.

130 Konevega, A.L., Fischer, N., Semenkov, Y.P. et al. (2007). Spontaneous reverse movement of mRNA-bound tRNA through the ribosome. *Nat. Struct. Mol. Biol.* 14 (4): 318–324.

131 Liu, G., Song, G., Zhang, D. et al. (2014). EF-G catalyzes tRNA translocation by disrupting interactions between decoding center and codon-anticodon duplex. *Nat. Struct. Mol. Biol.* 21 (9): 817–824.

132 Brilot, A.F., Korostelev, A.A., Ermolenko, D.N., and Grigorieff, N. (2013). Structure of the ribosome with elongation factor G trapped in the pretranslocation state. *Proc. Natl. Acad. Sci. U.S.A.* 110 (52): 20994–20999.

133 Ramrath, D.J., Lancaster, L., Sprink, T. et al. (2013). Visualization of two transfer RNAs trapped in transit during elongation factor G-mediated translocation. *Proc. Natl. Acad. Sci. U.S.A.* 110 (52): 20964–20969.

134 Gao, Y.G., Selmer, M., Dunham, C.M. et al. (2009). The structure of the ribosome with elongation factor G trapped in the posttranslocational state. *Science* 326 (5953): 694–699.

135 Munro, J.B., Altman, R.B., O'Connor, N., and Blanchard, S.C. (2007). Identification of two distinct hybrid state intermediates on the ribosome. *Mol. Cell* 25 (4): 505–517.

136 Moazed, D. and Noller, H.F. (1989). Intermediate states in the movement of transfer RNA in the ribosome. *Nature* 342: 142–148.

137 Rheinberger, H.-J. and Nierhaus, K.H. (1983). Testing an alternative model for the ribosomal peptide elongation cycle. *Proc. Natl. Acad. Sci. U.S.A.* 80: 4213–4217.

138 Schmeing, T.M., Moore, P.B., and Steitz, T.A. (2003). Structures of deacylated tRNA mimics bound to the E site of the large ribosomal subunit. *RNA* 9 (11): 1345–1352.

139 Semenkov, Y., Shapkina, T., Makhno, V., and Kirillov, S. (1992). Puromycin reaction for the A-site-bound peptidyl-transfer RNA. *FEBS Lett.* 296 (2): 207–210.

140 Sharma, D., Southworth, D.R., and Green, R. (2004). EF-G-independent reactivity of a pre-translocation-state ribosome complex with the aminoacyl tRNA substrate puromycin supports an intermediate (hybrid) state of tRNA binding. *RNA* 10 (1): 102–113.

141 Agirrezabala, X., Lei, J., Brunelle, J.L. et al. (2008). Visualization of the hybrid state of tRNA binding promoted by spontaneous ratcheting of the ribosome. *Mol. Cell* 32 (2): 190–197.

142 Julian, P., Konevega, A.L., Scheres, S.H. et al. (2008). Structure of ratcheted ribosomes with tRNAs in hybrid states. *Proc. Natl. Acad. Sci. U.S.A.* 105 (44): 16924–16927.

143 Cornish, P.V., Ermolenko, D.N., Noller, H.F., and Ha, T. (2008). Spontaneous intersubunit rotation in single ribosomes. *Mol. Cell* 30 (5): 578–588.

144 Chen, C., Stevens, B., Kaur, J. et al. (2011). Single-molecule fluorescence measurements of ribosomal translocation dynamics. *Mol. Cell* 42 (3): 367–377.

145 Belardinelli, R., Sharma, H., Caliskan, N. et al. (2016). Choreography of molecular movements during ribosome progression along mRNA. *Nat. Struct. Mol. Biol.* 23 (4): 342–348.

146 Maracci, C. and Rodnina, M.V. (2016). Review: translational GTPases. *Biopolymers* 105 (8): 463–475.

147 Belitsina, N.V., Glukhova, M.A., and Spirin, A.S. (1975). Translocation in ribosomes by attachment-detachment of elongation factor G without GTP cleavage: evidence from a column-bound ribosome system. *FEBS Lett.* 54 (1): 35–38.

148 Koch, M., Flur, S., Kreutz, C. et al. (2015). Role of a ribosomal RNA phosphate oxygen during the EF-G-triggered GTP hydrolysis. *Proc. Natl. Acad. Sci. U.S.A.* 112 (20): E2561–E2568.

149 Moazed, D., Robertson, J.M., and Noller, H.F. (1988). Interaction of elongation factors EF-G and EF-Tu with a conserved loop in 23S RNA. *Nature* 334: 362–364.

150 Rodnina, M.V., Savelsbergh, A., Katunin, V.I., and Wintermeyer, W. (1997). Hydrolysis of GTP by elongation factor G drives tRNA movement on the ribosome. *Nature* 385 (6611): 37–41.

151 Pan, D., Kirillov, S.V., and Cooperman, B.S. (2007). Kinetically competent intermediates in the translocation step of protein synthesis. *Mol. Cell* 25 (4): 519–529.

152 Guo, Z. and Noller, H.F. (2012). Rotation of the head of the 30S ribosomal subunit during mRNA translocation. *Proc. Natl. Acad. Sci. U.S.A.* 109 (50): 20391–20394.

153 Dunkle, J.A., Wang, L., Feldman, M.B. et al. (2011). Structures of the bacterial ribosome in classical and hybrid states of tRNA binding. *Science* 332 (6032): 981–984.

154 Fei, J., Bronson, J.E., Hofman, J.M. et al. (2009). Allosteric collaboration between elongation factor G and the ribosomal L1 stalk directs tRNA movements during translation. *Proc. Natl. Acad. Sci. U.S.A.* 106 (37): 15702–15707.

155 Joseph, S. and Noller, H.F. (1998). EF-G-catalyzed translocation of anticodon stem-loop analogs of transfer RNA in the ribosome. *EMBO J.* 17 (12): 3478–3483.

156 Ratje, A.H., Loerke, J., Mikolajka, A. et al. (2010). Head swivel on the ribosome facilitates translocation by means of intra-subunit tRNA hybrid sites. *Nature* 468 (7324): 713–716.

157 Zhou, J., Lancaster, L., Donohue, J.P., and Noller, H.F. (2013). Crystal structures of EF-G-ribosome complexes trapped in intermediate states of translocation. *Science* 340 (6140): 1236086.

158 Zhou, J., Lancaster, L., Donohue, J.P., and Noller, H.F. (2014). How the ribosome hands the A-site tRNA to the P site during EF-G-catalyzed translocation. *Science* 345 (6201): 1188–1191.

159 Bock, L.V., Blau, C., Schroder, G.F. et al. (2013). Energy barriers and driving forces in tRNA translocation through the ribosome. *Nat. Struct. Mol. Biol.* 20 (12): 1390–1396.

160 Czworkowski, J., Wang, J., Seitz, T.A., and Moore, P.B. (1994). The crystal structure of elongation factor G complexed with GDP, at 2.7 Å resolution. *EMBO J.* 13: 3661–3668.

161 Lin, J., Gagnon, M.G., Bulkley, D., and Steitz, T.A. (2015). Conformational changes of elongation factor G on the ribosome during tRNA translocation. *Cell* 160 (1-2): 219–227.

162 Ling, C. and Ermolenko, D.N. (2016). Structural insights into ribosome translocation. *Wiley Interdiscip. Rev. RNA* 7 (5): 620–636.

163 Zhou, J., Korostelev, A., Lancaster, L., and Noller, H.F. (2012). Crystal structures of 70S ribosomes bound to release factors RF1, RF2 and RF3. *Curr. Opin. Struct. Biol.* 22 (6): 733–742.

164 Ito, K., Uno, M., and Nakamura, Y. (2000). A tripeptide 'anticodon' deciphers stop codons in messenger RNA. *Nature* 403 (6770): 680–684.

165 Nakamura, Y., Ito, K., and Ehrenberg, M. (2000). Mimicry grasps reality in translation termination. *Cell* 101 (4): 349–352.

166 Laurberg, M., Asahara, H., Korostelev, A. et al. (2008). Structural basis for translation termination on the 70S ribosome. *Nature* 454 (7206): 852–857.

167 Korostelev, A., Asahara, H., Lancaster, L. et al. (2008). Crystal structure of a translation termination complex formed with release factor RF2. *Proc. Natl. Acad. Sci. U.S.A.* 105 (50): 19684–19689.

168 Korostelev, A., Zhu, J., Asahara, H., and Noller, H.F. (2010). Recognition of the amber UAG stop codon by release factor RF1. *EMBO J.* 29 (15): 2577–2585.

169 Weixlbaumer, A., Jin, H., Neubauer, C. et al. (2008). Insights into translational termination from the structure of RF2 bound to the ribosome. *Science* 322 (5903): 953–956.

170 Hoernes, T.P., Clementi, N., Juen, M.A. et al. (2018). Atomic mutagenesis of stop codon nucleotides reveals the chemical prerequisites for release factor-mediated peptide release. *Proc. Natl. Acad. Sci. U.S.A.* 115 (3): E382–E389.

171 Ito, K., Uno, M., and Nakamura, Y. (1998). Single amino acid substitution in prokaryote polypeptide release factor 2 permits it to terminate translation at all three stop codons. *Proc. Natl. Acad. Sci. U.S.A.* 95 (14): 8165–8169.

172 Korkmaz, G. and Sanyal, S. (2017). R213I mutation in release factor 2 (RF2) is one step forward for engineering an omnipotent release factor in bacteria *Escherichia coli*. *J. Biol. Chem.* 292 (36): 15134–15142.

173 Young, D.J., Edgar, C.D., Poole, E.S., and Tate, W.P. (2010). The codon specificity of eubacterial release factors is determined by the sequence and size of the recognition loop. *RNA* 16 (8): 1623–1633.

174 Vestergaard, B., Van, L., Andersen, G. et al. (2001). Bacterial polypeptide release factor RF2 is structurally distinct from eukaryotic eRF1. *Mol. Cell* 8 (6): 1375–1382.

175 Zeng, F. and Jin, H. (2016). Peptide release promoted by methylated RF2 and ArfA in nonstop translation is achieved by an induced-fit mechanism. *RNA* 22 (1): 49–60.

176 Graf, M., Huter, P., Maracci, C. et al. (2018). Visualization of translation termination intermediates trapped by the Apidaecin 137 peptide during RF3-mediated recycling of RF1. *Nat. Commun.* 9 (1): 3053.

177 Trappl, K. and Joseph, S. (2016). Ribosome induces a closed to open conformational change in release factor 1. *J. Mol. Biol.* 428 (6): 1333–1344.

178 Frolova, L., Tsivkovskii, R., Sivolobova, G. et al. (1999). Mutations in the highly conserved GGQ motif of class 1 polypeptide release factors abolish the ability of human eRF1 to trigger peptidyl-tRNA hydrolysis. *RNA* 5: 1014–1020.

179 Shaw, J.J. and Green, R. (2007). Two distinct components of release factor function uncovered by nucleophile partitioning analysis. *Mol. Cell* 28 (3): 458–467.

180 Dincbas-Renqvist, V., Engstrom, A., Mora, L. et al. (2000). A post-translational modification in the GGQ motif of RF2 from *Escherichia coli* stimulates termination of translation. *EMBO J.* 19 (24): 6900–6907.

181 Heurgue-Hamard, V., Champ, S., Engstrom, A. et al. (2002). The hemK gene in *Escherichia coli* encodes the *N*(5)-glutamine methyltransferase that modifies peptide release factors. *EMBO J.* 21 (4): 769–778.

182 Santos, N., Zhu, J., Donohue, J.P. et al. (2013). Crystal structure of the 70S ribosome bound with the Q253P mutant form of release factor RF2. *Structure* 21 (7): 1258–1263.

183 Pierson, W.E., Hoffer, E.D., Keedy, H.E. et al. (2016). Uniformity of peptide release is maintained by methylation of release factors. *Cell Rep.* 17 (1): 11–18.

184 Jin, H., Kelley, A.C., Loakes, D., and Ramakrishnan, V. (2010). Structure of the 70S ribosome bound to release factor 2 and a substrate analog provides insights into catalysis of peptide release. *Proc. Natl. Acad. Sci. U.S.A.* 107 (19): 8593–8598.

185 Trobro, S. and Aqvist, J. (2009). Mechanism of the translation termination reaction on the ribosome. *Biochemistry* 48 (47): 11296–11303.

186 Shaw, J.J., Trobro, S., He, S.L. et al. (2012). A Role for the 2′ OH of peptidyl-tRNA substrate in peptide release on the ribosome revealed through RF-mediated rescue. *Chem. Biol.* 19 (8): 983–993.

187 Freistroffer, D.V., Pavlov, M.Y., MacDougall, J. et al. (1997). Release factor RF3 in *E. coli* accelerates the dissociation of release factors RF1 and RF2 from the ribosome in a GTP-dependent manner. *EMBO J.* 16 (13): 4126–4133.

188 Goldstein, J.L. and Caskey, C.T. (1970). Peptide chain termination: effect of protein S on ribosomal binding of release factors. *Proc. Natl. Acad. Sci. U.S.A.* 67 (2): 537–543.

189 Adio, S., Sharma, H., Senyushkina, T. et al. (2018). Dynamics of ribosomes and release factors during translation termination in *E. coli*. *elife* 7: e34252.

190 Koutmou, K.S., McDonald, M.E., Brunelle, J.L., and Green, R. (2014). RF3:GTP promotes rapid dissociation of the class 1 termination factor. *RNA* 20 (5): 609–620.

191 Peske, F., Kuhlenkoetter, S., Rodnina, M.V., and Wintermeyer, W. (2014). Timing of GTP binding and hydrolysis by translation termination factor RF3. *Nucleic Acids Res.* 42 (3): 1812–1820.

192 Gao, H., Zhou, Z., Rawat, U. et al. (2007). RF3 induces ribosomal conformational changes responsible for dissociation of class I release factors. *Cell* 129 (5): 929–941.

193 Jin, H., Kelley, A.C., and Ramakrishnan, V. (2011). Crystal structure of the hybrid state of ribosome in complex with the guanosine triphosphatase release factor 3. *Proc. Natl. Acad. Sci. U.S.A.* 108 (38): 15798–15803.

194 Pallesen, J., Hashem, Y., Korkmaz, G. et al. (2013). Cryo-EM visualization of the ribosome in termination complex with apo-RF3 and RF1. *elife* 2: e00411.

195 Zhou, J., Lancaster, L., Trakhanov, S., and Noller, H.F. (2012). Crystal structure of release factor RF3 trapped in the GTP state on a rotated conformation of the ribosome. *RNA* 18 (2): 230–240.

196 Ermolenko, D.N., Majumdar, Z.K., Hickerson, R.P. et al. (2007). Observation of intersubunit movement of the ribosome in solution using FRET. *J. Mol. Biol.* 370 (3): 530–540.

197 Sternberg, S.H., Fei, J., Prywes, N. et al. (2009). Translation factors direct intrinsic ribosome dynamics during translation termination and ribosome recycling. *Nat. Struct. Mol. Biol.* 16 (8): 861–868.

198 Shi, X. and Joseph, S. (2016). Mechanism of translation termination: RF1 dissociation follows dissociation of RF3 from the ribosome. *Biochemistry* 55 (45): 6344–6354.

199 Grentzmann, G., Brechemierbaey, D., Heurgue, V. et al. (1994). Localization and characterization of the gene encoding release factor RF3 in *Escherichia coli*. *Proc. Natl. Acad. Sci. U.S.A.* 91 (13): 5848–5852.

200 Margus, T., Remm, M., and Tenson, T. (2007). Phylogenetic distribution of translational GTPases in bacteria. *BMC Genomics* 8: 15.

201 Fu, Z., Kaledhonkar, S., Borg, A. et al. (2016). Key intermediates in ribosome recycling visualized by time-resolved cryoelectron microscopy. *Structure* 24 (12): 2092–2101.

202 Prabhakar, A., Capece, M.C., Petrov, A. et al. (2017). Post-termination ribosome intermediate acts as the gateway to ribosome recycling. *Cell Rep.* 20 (1): 161–172.

203 Gao, N., Zavialov, A.V., Li, W. et al. (2005). Mechanism for the disassembly of the posttermination complex inferred from cryo-EM studies. *Mol. Cell* 18 (6): 663–674.

204 Weixlbaumer, A., Petry, S., Dunham, C.M. et al. (2007). Crystal structure of the ribosome recycling factor bound to the ribosome. *Nat. Struct. Mol. Biol.* 14 (8): 733–737.

205 Agrawal, R., Sharma, M., Kiel, M. et al. (2004). Visualization of ribosome-recycling factor on the *Escherichia coli* 70S ribosome: functional implications. *Proc. Natl. Acad. Sci. U.S.A.* 101 (24): 8900–8905.

206 Barat, C., Datta, P.P., Raj, V.S. et al. (2007). Progression of the ribosome recycling factor through the ribosome dissociates the two ribosomal subunits. *Mol. Cell* 27 (2): 250–261.

207 Hirokawa, G., Demeshkina, N., Iwakura, N. et al. (2006). The ribosome-recycling step: consensus or controversy? *Trends Biochem. Sci.* 31 (3): 143–149.

208 Peske, F., Rodnina, M., and Wintermeyer, W. (2005). Sequence of steps in ribosome recycling as defined by kinetic analysis. *Mol. Cell* 18 (4): 403–412.

209 Zavialov, A.V., Hauryliuk, V.V., and Ehrenberg, M. (2005). Splitting of the posttermination ribosome into subunits by the concerted action of RRF and EF-G. *Mol. Cell* 18 (6): 675–686.

210 Zhang, D., Yan, K., Zhang, Y. et al. (2015). New insights into the enzymatic role of EF-G in ribosome recycling. *Nucleic Acids Res.* 43 (21): 10525–10533.

211 Pai, R.D., Zhang, W., Schuwirth, B.S. et al. (2008). Structural Insights into ribosome recycling factor interactions with the 70S ribosome. *J. Mol. Biol.* 376 (5): 1334–1347.

212 Yokoyama, T., Shaikh, T.R., Iwakura, N. et al. (2012). Structural insights into initial and intermediate steps of the ribosome-recycling process. *EMBO J.* 31 (7): 1836–1846.

213 Borg, A., Pavlov, M., and Ehrenberg, M. (2016). Complete kinetic mechanism for recycling of the bacterial ribosome. *RNA* 22 (1): 10–21.

214 Hirokawa, G., Nijman, R., Raj, V. et al. (2005). The role of ribosome recycling factor in dissociation of 70S ribosomes into subunits. *RNA* 11 (8): 1317–1128.

215 Chen, Y., Kaji, A., Kaji, H., and Cooperman, B.S. (2017). The kinetic mechanism of bacterial ribosome recycling. *Nucleic Acids Res.* 45 (17): 10168–10177.

216 Qin, B., Yamamoto, H., Ueda, T. et al. (2016). The termination phase in protein synthesis is not obligatorily followed by the RRF/EF-G-dependent recycling phase. *J. Mol. Biol.* 428 (18): 3577–3587.

9

The RNase P Ribozyme

Markus Gößringer, Isabell Schencking, and Roland Karl Hartmann

Institut für Pharmazeutische Chemie, Philipps-Universität Marburg, Faculty of Pharmacy, Institute of Pharmaceutical Chemistry, Marbacher Weg 6, D-35037 Marburg, Germany

9.1 Introduction

Transfer RNAs are generally transcribed as precursor molecules with 5′- and 3′-terminal extensions. RNase P generates the mature tRNA 5′-ends in all domains of life, while the 3′-end trimming is, in most organisms, accomplished by endonucleolytic RNase Z cleavage and CCA addition catalyzed by ATP(CTP):tRNA nucleotidyl transferase [1–3]. Since its discovery around 1980, primarily by the laboratories of Sidney Altman and Norman R. Pace, RNase P was long thought to uniformly consist of a catalytic RNA (ribozyme) subunit and one or more protein cofactors (Figure 9.1, top). This view was consolidated with the finding that even the human nuclear RNase P RNA (H1 RNA), associated with 10 protein cofactors, retains very weak but specific RNA-alone activity in vitro [14]. Yet, in 2008, Rossmanith and coworkers reported the human mitochondrial RNase P to be a complex of three different proteins but lacking any RNA component [9]. One of those protein subunits (MRPP3) was identified as the endonuclease [9] and was later termed PRORP for *pro*teinaceous or *p*rotein-*o*nly *R*Nase *P* [10]. This type of enzyme is not found in Bacteria and Archaea but is widespread in most branches of the eukaryotic phylogenetic tree [15], where it seems to act as a single polypeptide in most cases. Recently, another type of protein-only RNase P was discovered in the hyperthermophilic bacterium *Aquifex aeolicus* and related bacteria of the family Aquificaceae, which likely acquired the corresponding gene by horizontal gene transfer from an archaeon [12]. Whereas the genes encoding the RNA-based RNase P were lost from the Aquificaceae, a few bacteria from other branches of the phylogenetic tree were identified to encode both types of RNase P, the original ribozyme type and a homolog of *Aquifex* RNase P (HARP). Whether the RNase P replacement in the Aquificaceae represents a caprice of evolution or confers specific advantages to these bacteria remains to be elucidated. The structural diversity of RNase P enzymes (summarized in Figure 9.1, including RNase MRP related to eukaryotic nuclear-cytoplasmic RNase P and involved in precursor rRNA processing)

Ribozymes, First Edition. Edited by Sabine Müller, Benoît Masquida, and Wade Winkler.

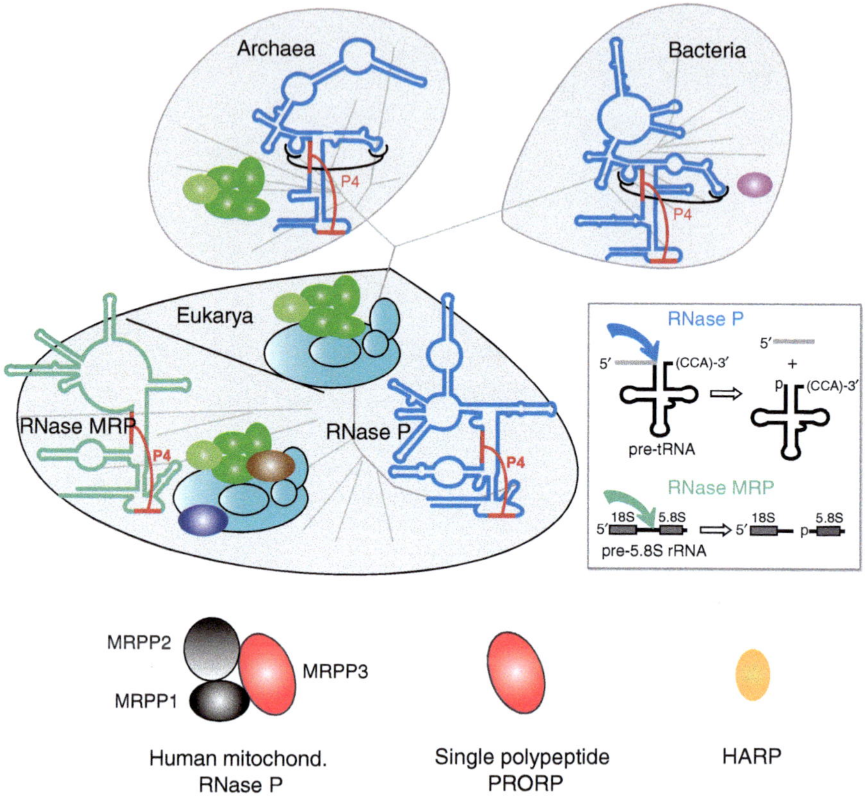

Figure 9.1 Family tree of RNA-based RNase P enzymes (top) and sketches of protein-only RNases P (bottom). RNA-based RNases P consist of an RNA subunit (blue, helix P4 in red) and a single protein subunit in Bacteria (purple oval), five proteins in Archaea (green ovals; the one in lighter green indicates the ribosomal k-turn binding protein L7Ae that was proposed to be homologous to POP3 in yeast and RPP38 in human [4]) and a total of nine or ten proteins in Eukarya (green and light-blue ovals). The RNA subunits share conserved nucleotides and a conserved core structure in and around helix P4 in the catalytic domain [5, 6], which indicates a common origin of all RNase P RNAs. Five of the eukaryal RNase P proteins are homologs of the archaeal protein subunits (green ovals). There is also a sibling of RNase P, RNase MRP, involved in precursor 5.8S rRNA processing (box on the right). A similar core structure of RNase P and MRP RNAs [5] indicates that they are paralogs originating from gene duplication. RNase MRP shares its protein subunits with RNA-based RNase P except for RPR2 (= RPP21), but has two MRP-specific proteins (SNM1 and RMP1, dark blue, and brown ovals) in the yeast system [7]. The sketches for eukaryal RNase P and MRP are based on the composition of the yeast enzymes; the protein ovals are not drawn to scale, yet the largest oval illustrates the largest protein subunit (POP1). The composition of human RNase MRP (1 RNA and 7–10 protein cofactors) is still under debate, and there may be different RNase MRP subpopulations owing to a more dynamic association of some of the protein subunits [8]. In human mitochondria, RNase P is a trimeric protein complex (MRPP1 to 3, black gray and red ovals) devoid of RNA; endonucleolytic activity resides in MRPP3 (red oval) [9]. PRORPs (*pro*tein-only *R*Nase *P*) (red oval) are Eukarya-specific homologs of MRPP3 and function as monomeric protein-only RNase P enzymes [10]. Another type of protein-only RNase P, a small polypeptide of 23 kDa belonging to the PIN_5 domain subfamily of endonucleases [11], was recently discovered in the bacterial family Aquificaceae; homologs of this enzyme, termed HARPs (Homolog of *Aquifex* RNase P) (yellow oval), were found bioinformatically in some other Bacteria and many Archaea, but not in Eukarya [12]. Source: Adapted from Willkomm and Hartmann [13].

makes them a unique model system to compare the molecular strategies utilized by "almost RNA alone," more complex ribonucleoprotein (RNP) as well as protein-only activities for catalyzing the same tRNA-specific endoribonuclease reaction.

In this chapter, we focus on (i) the state of knowledge on the structure and function of bacterial RNase P, (ii) recent efforts to develop antibiotics against bacterial RNase P, and (iii) concepts to utilize RNase P as a tool in gene inactivation.

9.2 Bacterial RNase P

In the vast majority of bacteria, RNase P is a RNP particle composed of a catalytic RNA subunit (P RNA, ~340–400 nt/115–140 kDa, gene *rnpB*) and a small basic protein subunit (P protein, ~120 aa/13 kDa, gene *rnpA*). The protein contributes not more than 10% to the molar mass of the holoenzyme complex. In vitro, the RNA subunit can catalyze the endonuclease reaction in the absence of the protein cofactor, which yet requires increased ionic strength – elevated Mg^{2+} concentrations in particular – for efficient substrate cleavage [16]. Thus, the bacterial RNA-based RNase P is a unique natural ribozyme that has the capability to efficiently catalyze multiple turnover reactions.

9.2.1 P RNA Structure and Evolution

Bacterial P RNAs are highly structured; about two-thirds of their nucleotides are involved in primarily Watson–Crick base-pairing interactions. By phylogenetic-comparative sequence analysis of several hundred bacterial P RNA sequences, the secondary structure and some of the RNA's tertiary interactions could be established several years before crystallographic data became available. Only a subset of helices is absolutely essential for catalytic function. At the same relative position, but with some length variations, these helices build the core structure of all bacterial P RNAs. In contrast, the peripheral helices that are attached to the conserved core structure show variable occurrence in different bacteria and can be individually deleted with only moderate effects on catalytic function [17–21]. Based on the different peripheral structural elements, bacterial P RNAs can be classified into A-type (ancestral) and B-type (*Bacillus*) architectures [22–24]. Sometimes, also a C-type (Chloroflexi) is mentioned in the literature to highlight P RNAs that have features that are intermediate between A- and B-type architectures (exemplified by P RNA from *Thermomicrobium roseum*; [23, 24]). Most bacterial P RNAs are A-type structures, such as *Escherichia coli* P RNA, the major model system in this group. Low G + C Gram-positive bacteria (e.g. *Bacillus subtilis*) mainly contain B-type structures. Helices P6, P13, P14, P16, P17, and P18 are characteristic for A-type and P5.1, P10.1, P15.1, and P15.2 for B-type structures [23] (Figure 9.2). The presence of P13, P14, and P18 is variable among A-type RNAs, and P19 has a patchy occurrence in general (see [31] for a study on the functional role of P19). Bacterial P RNA is composed of two independently folding domains, the "catalytic domain" (C-domain) and "specificity domain" (S-domain) [32, 33] (Figure 9.2). The C-domain includes ~2/3 of all P RNA nucleotides and contains all elements of

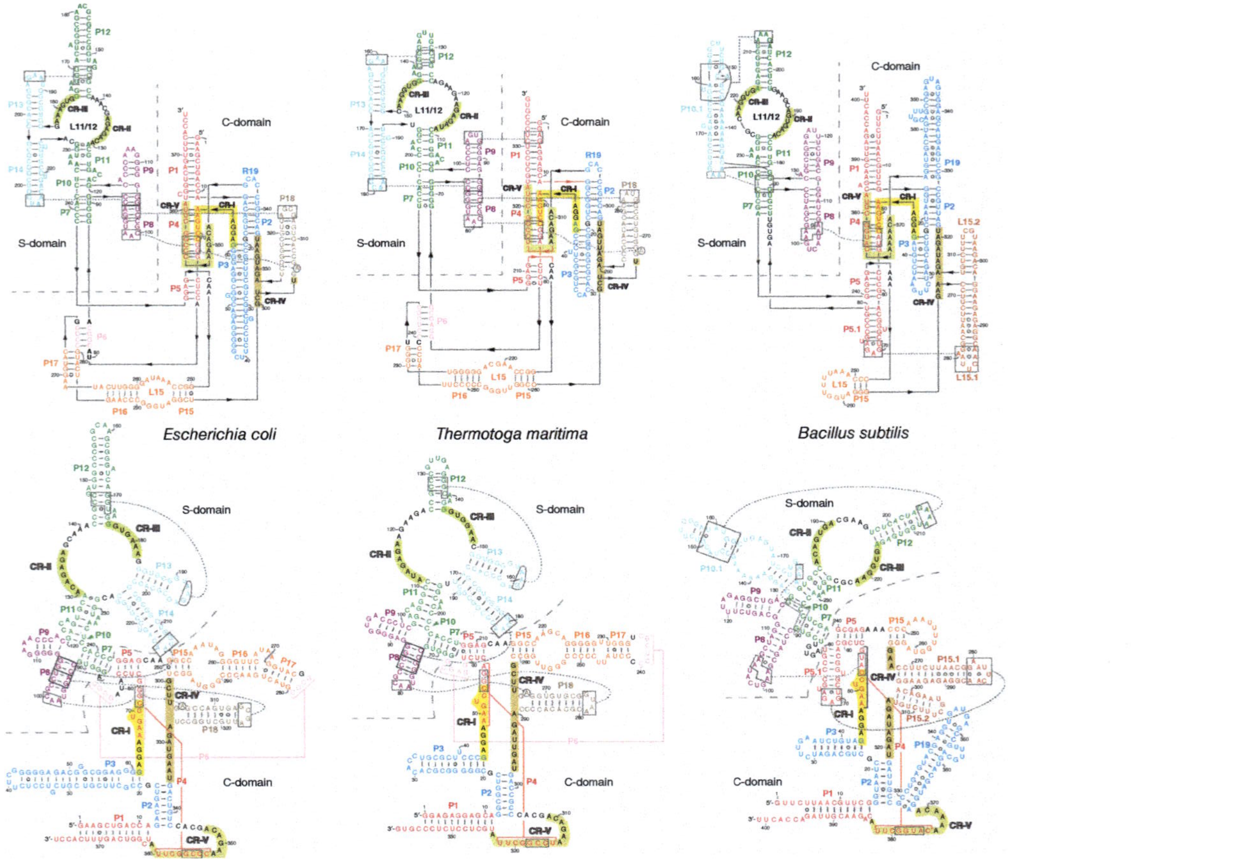

Figure 9.2 Secondary structures of (a) *E. coli* (type A), (b) *T. maritima* (type A), and (c) *B. subtilis* (type B) RNase P RNAs in the presentation by [25, 26] (drawings on the left) or [22] (drawings on the right). The dashed lines separate the respective catalytic (C-) and specificity (S-) domains. The conserved regions CR-I, II, III, IV, and V are highlighted. R19 in the *E. coli* and *T. maritima* structures at the top indicates that P RNAs lacking P19 have a 4-nt linker between P2 and CR-V [25]. Long-range tertiary interactions are depicted by boxes connected by dotted lines. L14:P8, L18:P8 in *E. coli* and *T. maritima* P RNAs and L9:P1 in *T. maritima* P RNA are tetraloop/tetraloop receptor interactions; the features of the L14:P8 interaction are illustrated and explained in the context of Figure 9.7. In the case of the L9:P1 interaction in *T. maritima* P RNA, A96 forms an A-minor type I interaction with bp G4:C331 and G95 a "G-minor" interaction with bp A3:U332, with H bonds of the G95 2′-hydroxyl (2′-OH) to the O2′ of U332, of the 2′-OH of U332 to the N3 of G95, and of the 2-amino group of G95 to the O2 of U332. Tertiary interactions for *B. subtilis* P RNA are depicted based on the crystal structure of *G. stearothermophilus* P RNA (L8:P4, L5.1:L15.1; [30]).

the active site. The isolated C-domain can interact with the P protein and retains catalytic activity, even in the absence of the P protein [34–36]. The S-domain comprises the remaining ~1/3 of the RNA subunit and contributes to substrate specificity and affinity through interacting with the corner of L-shaped tRNA structures (Figures 9.3a and 9.4). Several phylogenetically variable structural elements

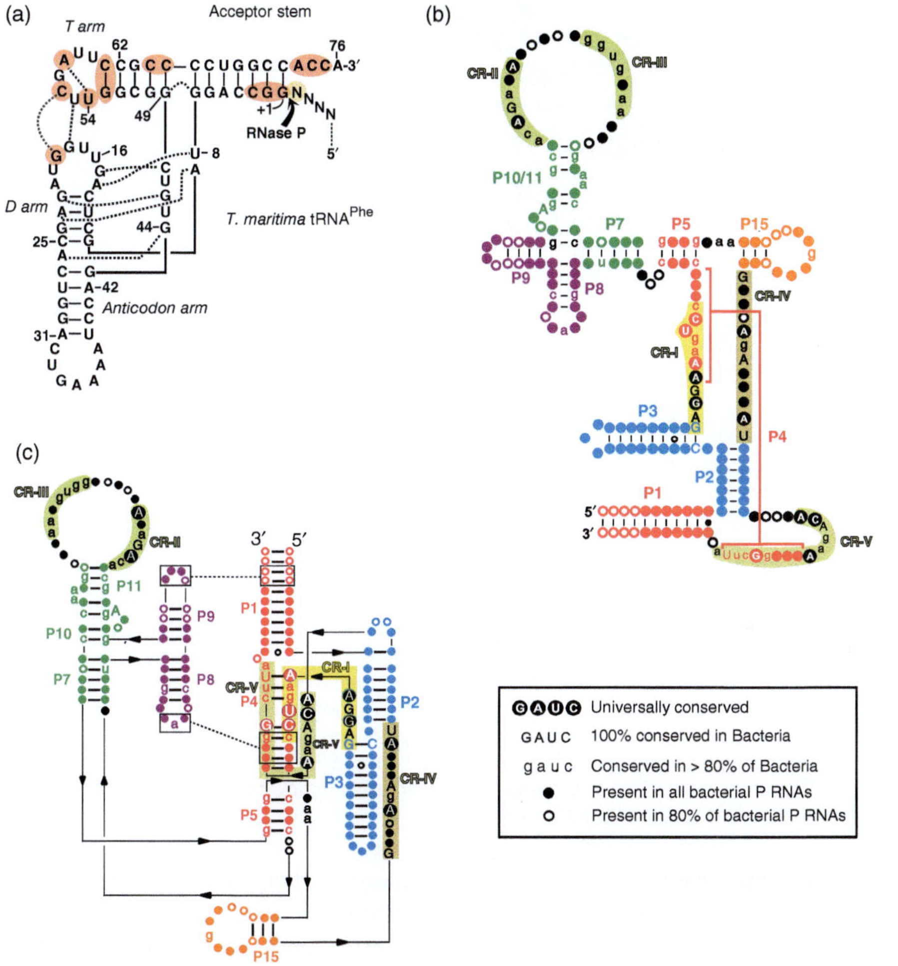

Figure 9.3 (a) Secondary structure of *T. maritima* tRNA^{Phe}, including tertiary interactions (indicated by dotted lines) as inferred from the crystal structure of yeast tRNA^{Phe}. A derivative of *T. maritima* tRNA^{Phe} with an extended anticodon arm was used for co-crystallization with the *T. maritima* RNase P holoenzyme [27]; nucleotides interacting with P RNA are salmon-colored, nucleotide −1 not addressed by the crystal structure is highlighted in yellow to indicate the key role of its 2′-OH for catalytic metal ion binding (see text for details). (b) Bacterial minimum consensus structure, according to [6, 17], except that the nucleotide immediately upstream of the "ggug" motif of CR-III is not indicated as conserved. The conserved regions CR-I, II, III, IV, and V are highlighted as in Figure 9.2. (c), as in panel (b), but shown in the secondary structure presentation of [26].
Source: (a) Adapted from Heide et al. [37]

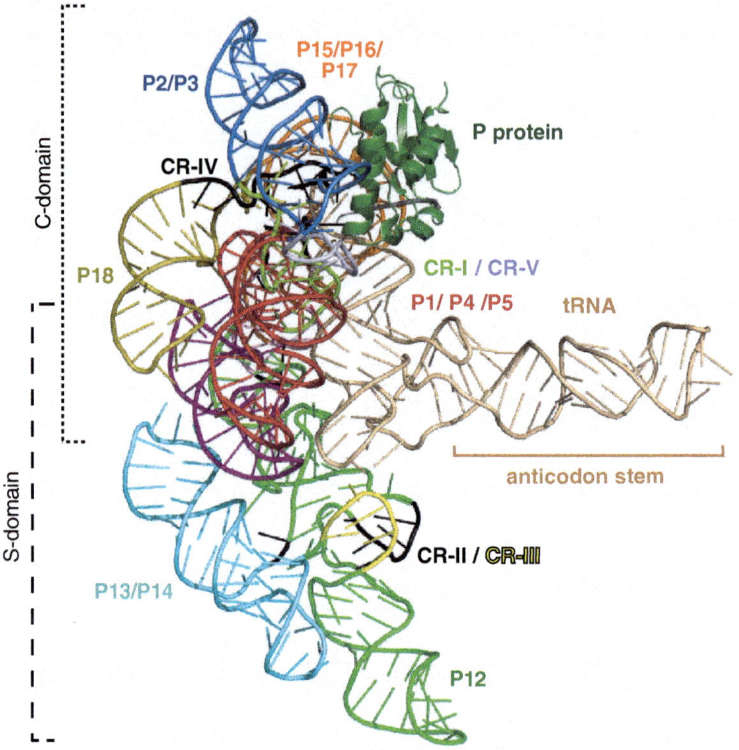

Figure 9.4 Tertiary structure of the *T. maritima* RNase P holoenzyme in complex with tRNA [27]. The P protein (dark green) at the top is shown in the ribbon presentation; the *T. maritima* tRNA^{Phe} with engineered anticodon arm extension is shown in sand color. The anticodon arm of the bound tRNA extends away from the complex, and only the 3′-ACCA end of the acceptor stem and the tRNA corner, where D and T arm interact, make specific contacts (see Figures 9.3a and 9.9) with the slightly bended RNA subunit. The P RNA architecture consists of two layers of helices (P13/P14 and P18 are piled onto the core helices). The areas of the C- and S-domains are marked by brackets on the left. P8/P9 (colored in violet) is located in the background.

located on the surface of the RNA's core structure participate in short-range and long-range docking interactions (tetraloop/tetraloop receptor contacts, A-minor interactions, ribose zippers, dinucleotide platforms, and others; for details, see legend to Figures 9.2 and 9.7), thus functioning as struts to stabilize the RNA's global structure. The set of core helices is also found, although to different extents, in P RNA subunits from Archaea and Eukarya [5, 6, 38], demonstrating that all P RNAs have a common origin. However, archaeal and eukaryal P RNAs lack several of the stabilizing peripheral elements and tertiary contacts. Instead, the protein content of archaeal (5 protein subunits) and eukaryal (9–10 protein subunits) RNase P enzymes is increased to 40% and 70%, respectively. Here, RNA-protein and protein–protein interactions compensate for the lack of the RNA struts [39, 40], as verified and specified by the recent cryo-EM structures of archaeal and eukaryal RNase P as well as RNase MRP [1–4]. Sequence analysis of P RNA sequences indicates that less than 20% of the nucleotide positions are conserved in base identity. The universally conserved nucleotides are concentrated in five regions (conserved regions CR-I to CR-V) that occur at the same relative positions in all P RNAs: CR-I, CR-IV, and CR-V are located in the C-domain and CR-II and CR-III in the S-domain (Figures 9.3b,c and 9.4). Several nucleotides in CR-I and CR-V are part of the most conserved P RNA helix P4 [45].

9.2.2 The Single Protein Subunit

The bacterial P protein, which is not related to any of the archaeal/eukaryal protein subunits, is essential for bacterial RNase P function in vivo. Its overall topology is αßßßαßα, with an unusual left-handed ßαß crossover, and it shares structural homology with the C-terminal domain of ribosomal protein S5 and the domain IV of elongation factor G [46]. Two RNA-binding regions were identified: (i) a large central cleft formed by the N-terminal α-helix and the four-stranded central ß-sheet and (ii) a highly basic sequence stretch (17–18 residues: Lys-X_{4-5}-Ala-X_2–Arg–Asn-X_2–[Lys or Arg] –Arg-X_2–[Arg or Lys]) located in α-helix 2 and the preceding loop within the left-handed (ßαß) crossover structure that is described as the RNR motif. Apart from the conserved RNR consensus motif, bacterial P proteins show a low degree of sequence conservation (~20%–30% identity), but the 3D structures of the structurally analyzed P proteins from *B. subtilis* [46], *Staphylococcus aureus* [47], and *Thermotoga maritima* [48] are very similar. In accordance, complementation studies demonstrated that P proteins from different bacterial species are functionally interchangeable [16, 49, 50].

9.2.3 P RNAs – Architectural Principles, Variations, Idiosyncrasies

Crystallographic analyses disclosed many structural details, largely consistent with or contextualizing available biochemical data. Crystal structures were obtained for full-length P RNAs from *T. maritima* (type A) [51] and *Geobacillus stearothermophilus* (type B) [30] and for isolated S-domains of *Thermus thermophilus* (type A) [29] and *B. subtilis* (type B) P RNAs [28]. The structure of the *T. maritima* holoenzyme in complex with a mature tRNA (solved at 3.8 Å resolution [27]) sharpened the insight into the binding site of the P protein, substrate interaction sites, active site architecture, metal ion binding sites, and the global organization of the tertiary complex. The structural data for the P RNAs of *T. maritima* and *G. stearothermophilus* show high similarity in the catalytic core region (Figure 9.5). Several coaxial helical stacks (e.g. P2/P3, P4/P5, and P8/P9) are fixed and oriented by several long-range docking interactions that were principally identified earlier based on comparative sequence analyses [22, 25, 61, 62]: this includes the L8–P4 (type A and B), L18–P8 (type A), L9–P1 (type A), L14–P8 (type A), L13–P12 (type A), L12–P10.1 (type B), and L5.1–L15.1 (type B) contacts (Figure 9.2). The L5.1/L15.1 contact is unique to B-type ribozymes and plays a key role in stabilizing their tertiary structures [30, 63]. Among the interdomain contacts in A-type RNAs (L8–P4, L18–P8, and L9–P1), only the L8–P4 contact is implemented in all P RNAs. Disruption of the L9–P1 contact in *E. coli* P RNA caused no functional defect, indicating that this interaction

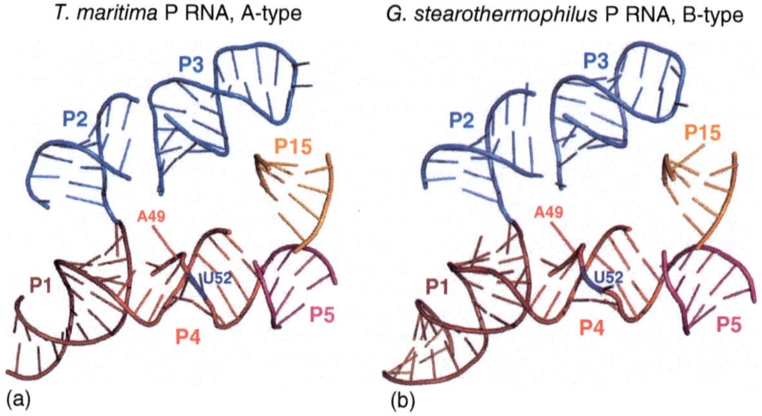

Figure 9.5 Arrangement of core structural elements in the crystal structures of *T. maritima* (A-type) and *G. stearothermophilus* (B-type) P RNAs (PDB entry 3q1r, [27], and PDB entry 2a64, [30]), illustrating their very similar core architectures; note that the universally conserved U52 (U69 in *E. coli*, Figure 9.2), which interacts with Me1 (Figure 9.11b) in the *T. maritima* holoenzyme, is oriented differently in the two structures.

either contributes little to structural stabilization or may not form at all in *E. coli* and closely related P RNAs. However, the contact is operative in addition to L18–P8 in thermostable P RNAs as those from *T. thermophilus* [64] and *T. maritima* [27]. Some A-type P RNAs lack P18 and thus cannot form the L18–P8 contact, but seem to have compensated this by reinforcement of other interactions or possibly by evolving new contacts [18, 65]. The catalytic core regions of both P RNA types are compactly folded and have a remarkably flat surface. In the case of A-type P RNA, the flat surface is formed by two layers of helices lying on top of each other. The larger layer includes the conserved core elements while the smaller layer consists of the stabilizing, peripheral helices P13/14, and P18 (visible in Figure 9.4). The P8–P9 stack is at the interface between the C-domain and the more distal parts of the S-domain (P13/P14 and P7/P10/P11/P12) (Figure 9.6). In crystals of A-type P RNA, the two domains are not coplanar, resulting in a concave overall shape, which is less evident in the *G. stearothermophilus* B-type structure [30]. The more concave shape of A-type P RNA is better predisposed to substrate binding [27]. The particularly flat and expanded shape of the B-type structure may represent a conformer that requires binding of the precursor tRNA (pre-tRNA) substrate to transition to the more concave, catalytically active fold. The loop/loop receptor interactions L14–P8 and L13–P12 stabilize the S-domain fold in A-type structures. B-type P RNA lacking P13 and P14 instead utilize the tetraloop/tetraloop receptor interaction L12-P10.1 and two A-minor interactions (between two adenines in J10.1/11 and the two G:C base pairs at the coaxial stacking interface of P7 and P10; see Figure 9.7) for this purpose. The conserved regions (CR-I, II, III, IV, and V) are clustered in two distinct areas in the overall tertiary structure. Regions CR-I and V (overlapping the universally conserved P4 pseudoknot structure), as well as CR-IV, are part of the active site in the C-domain (Figures 9.4, 9.6, and 9.8). P4, engaged in multiple metal ion interactions [66–68], including the catalytic ones (see below), is in close vicinity of the interface to the protein subunit (Figure 9.8). The protein:RNA interface is located on top of CR-IV/CR-V and further includes P2, P3, and P15; the conserved RNR motif of the protein is in the immediate proximity of CR-IV, P2, and P4 (Figure 9.8a). Though the secondary and tertiary structures of the S-domain of A- and B-type P RNAs are different, they present the two tRNA interaction modules defined by CR-II and CR-III in an equal manner (Figure 9.7). Each of two elements folds into a T loop structure consisting of a pentanucleotide motif composed of a 3-nt-loop termed U-turn (nt 2–4) flanked by a non-Watson–Crick base pair (bases 1 and 5) and followed by a sharp turn or bulged base(s) at its 3′-terminus [69]. The two T loops interact with each other (Figures 9.7 and 9.9d) to bring together the J11/12 and J12/13 strands to form one rigid structural module that is positioned for substrate interaction within the structural framework of the S-domain (see next paragraph). The crystal structure of the *T. maritima* ternary complex (A-type P RNA, P protein, tRNA) suggests that no large conformational changes are necessary for holoenzyme formation and substrate binding, as inferred from the observation that the structures of P RNA and P protein alone are nearly identical to those in the ternary complex. Small local conformational changes include reordering of a short loop at the N-terminus of the protein due to interactions with the P RNA and of the tRNA 3′-end to interact with the L15 loop of the P RNA. The protein subunit is docked at the C-domain's surface rather than being embedded in the RNA structure (Figure 9.8a) and is localized away from the interface between S- and C-domain (Figure 9.4). This is consistent with the absence of major protein-driven conformational changes. Nevertheless, biochemical studies reported measurable effects on the fold of both subunits, RNA and protein, during holoenzyme assembly [70–72].

9.3 Substrate Interaction

In the *T. maritima* ternary complex, the anticodon stem/loop of the bound tRNA points away from the holoenzyme (Figure 9.4), and only the coaxially stacked acceptor stem/T domain engages in intermolecular contacts with the P RNA [73]. The acceptor stem/T domain docks onto the shallow surface that is formed by the concave shape of the P RNA (Figure 9.4). Remarkably, the overall contact area between the two molecules is relatively small. Most RNA–RNA interactions inferred from the crystal structure agree well with data from previous genetic and biochemical studies (e.g. [74–78]). One end of the acceptor stem/T domain, the corner of the L-shaped tRNA (where T and D loop interact), contacts the S-domain of the P RNA. On the opposite site, the 3′-ACCA appendix of the acceptor stem separates from the 5′-end and enters a ~20 Å wide round arch that consists of elements P15/P16/P17 and which is held together by the P6 pseudoknot (Figures 9.6 and 9.9a).

Recognition of the tRNA body by RNase P RNA is based on shape complementarity and specific stacking and H bonding interactions (Figure 9.9). A major contact is formed by intermolecular Watson–Crick base pairs between the 3′-NCCA end (N = any base; discriminator base at position 73) of the tRNA and complementary

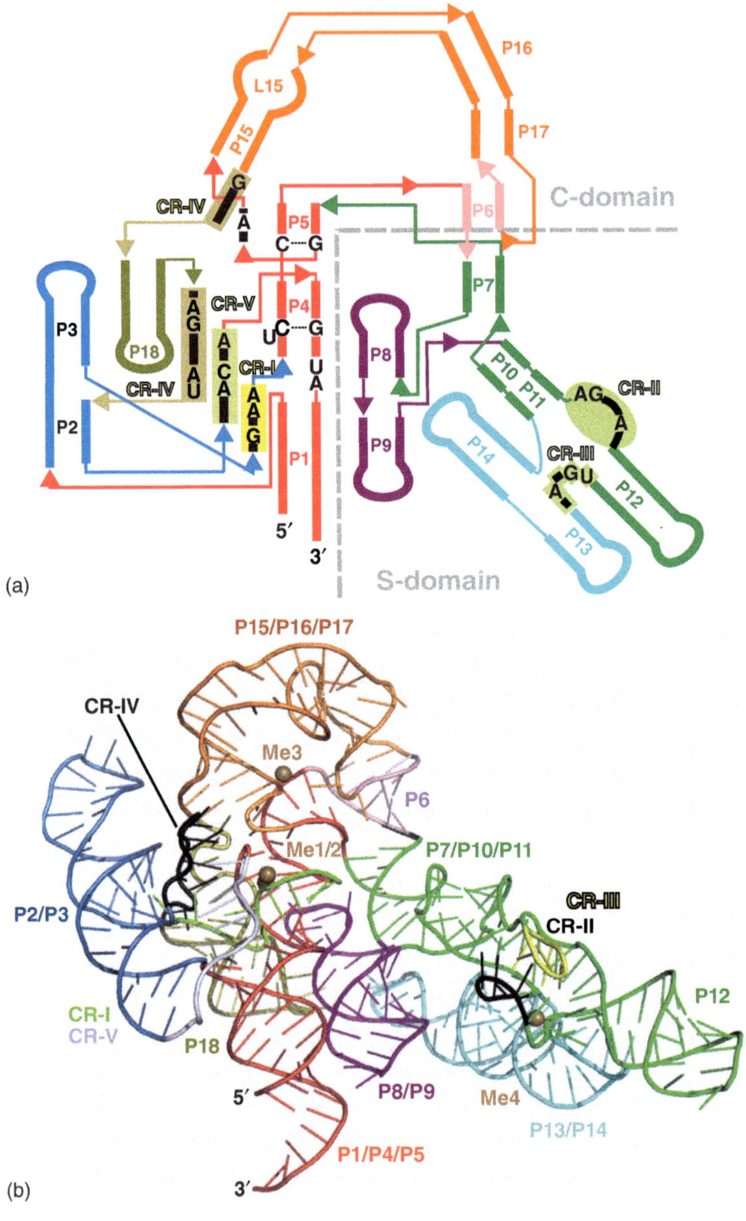

Figure 9.6 Correlation of *T. maritima* P RNA secondary (a) and tertiary (b) structures. (a) The dashed gray line separates C- and S-domains. Arrows indicate the 5′–3′ direction. Coaxially stacked helices are illustrated in the same color and the five regions with the highest sequence conservation (CR-I to CR-V) cluster in two distinct areas in the C- and S-domain, respectively (highlighted in yellow/lime green/ocher tones as in Figure 9.2); note that CR-I and CR-V overlap with helix P4. Highly conserved base identities are indicated as black letters. (b) Structural elements are colored as in panel (a), and the CR regions are marked by distinct colors. The two active site metal ions (Me1 & Me2) and two structurally important metal ions (Me3 & Me4) are shown as sand-colored spheres. The orientation of the molecule visualizes the two CR clusters, the position of P8/P9 sandwiched between the C-domain and the more distal parts of the S-domain, and the P15/16/17/6 round arch into which the tRNA 3′-NCCA end is threaded with support from Me3. Source: (b) Adapted from Reiter et al. [27]

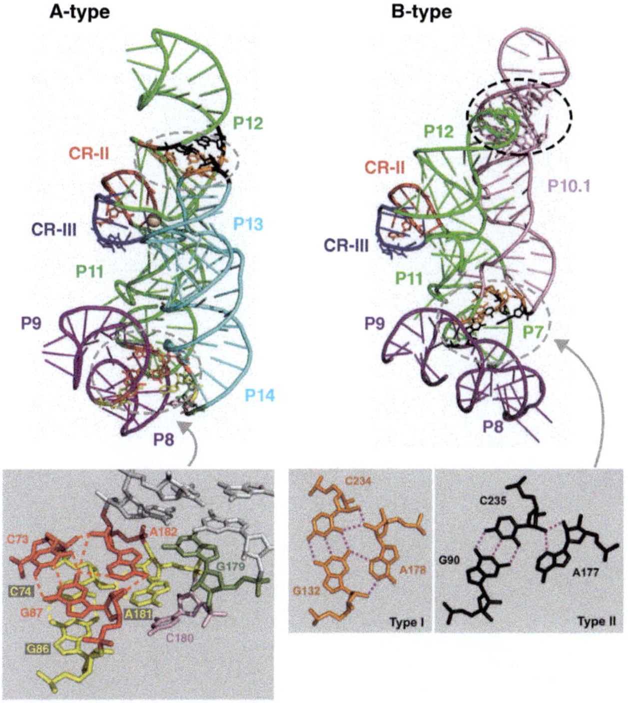

Figure 9.7 Top: tertiary structures of the S-domains from bacterial A-type (*T. maritima*, PDB entries 3q1q and 3q1r [27]) and B-type P RNAs (*B. subtilis*, PDB entry 1nbs [28]). In each S-domain the two T loops that are formed by CR-II (red) and CR-III (blue) constitute a rigid structural module that positions two unstacked nucleotides for substrate interaction in a very similar manner. Metal ion Me4 (sand-colored in the structure on the left), located close to the backbone of A140/A141/G142 at the base of P12, A149/C150 in J12/13 and G152 at the base of P13 stabilizes the universally conserved dual T loop module (see also Figure 9.6b). The P13/P14 stem is unique to A-type and the P10.1 stem unique to B-type P RNAs. There are two intradomain tertiary interactions in type A RNA: a tetraloop/tetraloop receptor contact between L14 and P8, and a similar loop/loop receptor interaction between L13 and P12 also involving two A-minor interactions and H bonding between the 2′-OH groups of G161 and C139 (encircled by dashed gray lines) [27, 29]. In rigid tetraloop/tetraloop receptor interactions, two unpaired purines insert into the minor groove of a helix. Bottom: In the L14:P8 example of *T. maritima* P RNA (bottom left panel), A182 forms a type I A-minor interaction with the bp C73:G87 (all three nucleotides in red) and A181 a type II A-minor interaction with bp C74:G86 (all three nucleotides in yellow); this involves H bonding of the ribose 2′-OH groups of A182/C73 and A181/C74, which represents the minimal version of a ribose zipper; in addition, this kind of interaction involves stacking of the second to the fourth base of the tetraloop (C180, A181, A182). In type B RNA, a tetraloop/tetraloop receptor interaction occurs between the L12 tetraloop and the receptor in P10.1 (dashed black circle). In this type of interaction, A148 and C149 of *B. subtilis* P RNA form a so-called AC platform (also indicated in Figure 9.2), a side-by-side configuration or pseudo base pair within the same helix strand to open the minor groove for interaction with the L12 tetraloop; in detail, A206 stacks on C149, while A207/U147 and A208/C146 form a ribose zipper, and bases A206, A207, and A208 form ~5 additional H bonds to bases and riboses in this platform region. In addition, B-type RNAs form two A-minor contacts (gray dashed circle) from A177 in J10.1/11 to bp G90:C235 (A-minor typ II interaction) and from A178 to bp G132:C234 (A-minor type I) (see bottom cartoons on the right). G90:C235 and G132:C234 are the adjacent closing base pairs of the stacked helices P7 and P10 (see Figure 9.2).

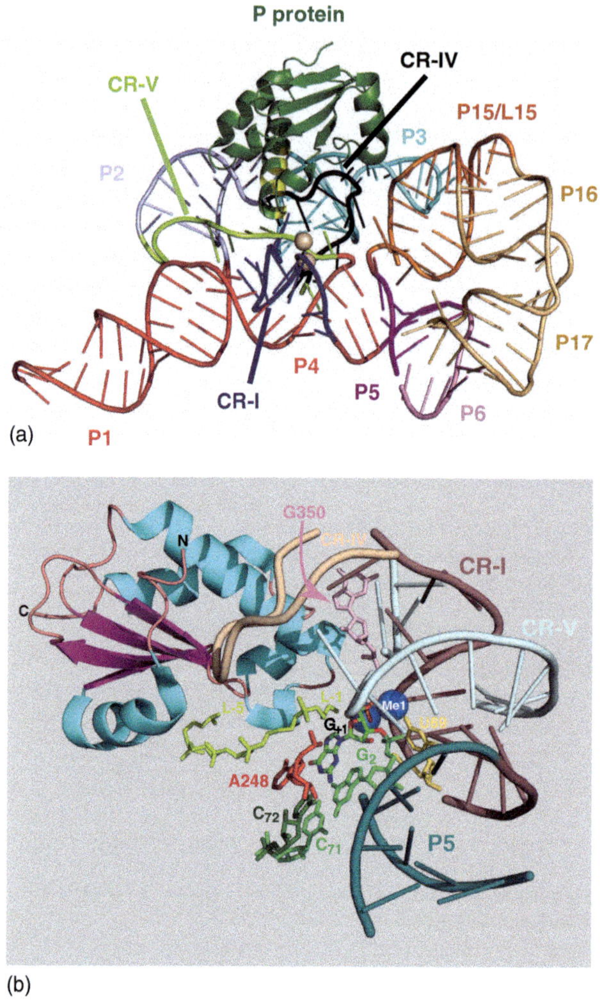

Figure 9.8 (a) Interface between RNA and protein subunit in the post-cleavage state structure of the *T. maritima* RNase P holoenzyme ([27]; PDB entry 3q1r). The protein subunit is in close proximity to the conserved regions CR-I (blue), CR-IV (black), CR-V (lime green) and helices P2, P3, P4, P15. The highly conserved RNR motif (shown in yellow) is located at the site where several of these elements intersect. The two active site metal ions Me1 and Me2 are shown as sand-colored spheres. (b) Illustration of the proximity of the P protein to the active site and the 5′-leader embedded in the central groove of the P protein (positioned in the back) in the post-cleavage state structure of the *T. maritima* RNase P holoenzyme:tRNA complex with soaked-in 5′-leader fragment ([27]; PDB entry 3q1r). CR-I/P4 (brown, 5′-AGGAAAGUCCGGA) and CR-V/P4 (pale cyan, 5′-ACAGAAUCCGGCUU), part of CR-IV (sand-colored, 5′-AGAUUGAU), and P5 (teal) are shown as cartoons; Me1 (in the foreground) and Me2 (in the background) are shown in blue and the backbone of nucleotides −1 to −5 (L − 1 to −5) of the 5′-leader in light green; U69 (*E. coli* numbering, gold-colored), G350 (pink) and A248 (red) are highlighted for comparability with Figure 9.11; tRNA nucleotides G_{+1} (in element color code), G_2:C_{71} (bright green) and C_{72} (dark green) illustrate the end of the acceptor stem.

nucleotides in the C-domain. In the A-type *T. maritima* holoenzyme crystallized with a tRNA carrying an A residue at position 73, three Watson–Crick base pairs between the 3′-ACCA terminus and J16/15 were observed (A_{73}:U256, C_{74}:G255, C_{75}:G254) (Figure 9.9a; [27]). The two base pairs involving C_{74} and C_{75} are the core of this interaction and form independent of base identity at position 73 [75, 79–81]. Biochemical data have also demonstrated Watson–Crick base pairing of tRNA residues C_{74} and C_{75} to two conserved G residues in loop L15 of B-type P RNAs [82, 83]. For the base-pairing interactions in the *T. maritima* holoenzyme, the two RNA strands of the internal loop L15 have to accommodate a ribose zipper

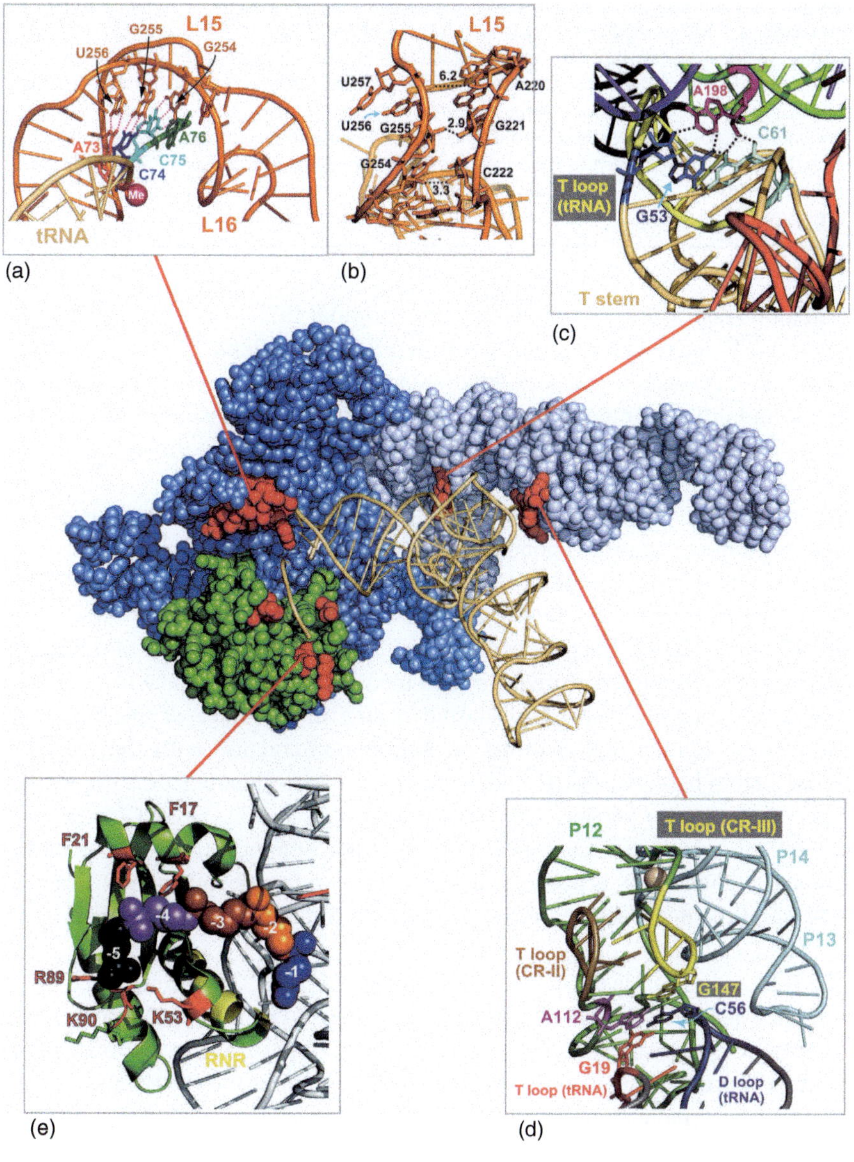

Figure 9.9 Interactions between the *T. maritima* RNase P holoenzyme and (pre-)tRNA. P RNA (C-domain in blue and S-domain in light blue) and P protein (in green) are shown as space-filling presentation; the tRNA and soaked-in 5′-leader (both sand-colored) are shown in the cartoon presentation (backbone wire, bases as sticks). Major substrate recognition elements (red) of the holoenzyme are shown in detail in boxes a to e. (a) Intermolecular base-pairing between the 3′-ACC end of $tRNA^{Phe}$ (A_{73}, C_{74} and C_{75}) and complementary nucleotides in the L15 loop of the C-domain (U256, G255 and G254). This base-pairing interaction is stabilized by metal ion Me3 (magenta sphere) and by (b) the two strands of the L15 loop folding into a ribose zipper conformation. (c) A highly conserved unpaired adenosine (A198) in P11 of the S-domain inserts into the T domain of the tRNA (at G_{53}:C_{61}, see Figure 9.3a) to form a minor groove nucleoside triple interaction (A-minor interaction). (d) Each of the highly conserved regions CR-II (J11/12) and CR-III (J12/13) in the S-domain fold into pentaloop motifs that interact with each other to form two interleaved T loops. The interleaving T loops expose two unstacked nucleotides (A112 in CR-II and G147 in CR-III) that stack on top of the tRNA tertiary base pair G_{19}:C_{56} at the corner of the tRNA structure where D and T loop interact (see also Figure 9.3a). Me4 is shown as a sand-colored sphere. (e) Trace of the soaked-in 5′-leader mimic representing nt −1 to −5 of pre-tRNA; only the backbone density was resolved in the crystals (shown here in the space-filling mode); residues −3 to −5 seem to bind into the central groove of the P protein. P protein residues proposed to be involved in binding of leader nt −4 and −5 (F17, F21, K53, R89, and K90) are shown in red, and residues of the conserved RNR motif are highlighted in yellow.

conformation (Figure 9.9b). Additionally, the terminal tRNA base A_{76} may form a *cis* Hoogsteen/Watson–Crick base pair with G253 in *T. maritima* P RNA. This interaction would be consistent with a c7-deaza interference at A_{76} observed in pre-tRNA processing and tRNA binding assays [84]. A stabilizing metal ion was crystallographically localized close to this interaction site (Figure 9.9a, corresponding to Me3 in Figure 9.6b).

Furthermore, acceptor stem nucleotides G_{+1}, G_2, C_3, and C_{72} (Figure 9.3a) are poised to form minor groove H bonding contacts with nucleotides in P4, J5/15, and possibly P5. The 2-amino group of G_{+1} can donate an H bond to the OP1 of C212 (J5/15), which could explain why bacterial RNases P prefer a G at position +1 [85] (Table 9.1; Figure 9.10a). Biochemical findings that are consistent with the contacts inferred from the crystal structure have been incorporated into Table 9.1.

At the T arm, a conserved unstacked adenosine in the bulge of the P11 stem (A198 in *T. maritima* P RNA) forms an A-minor interaction with the last and conserved T stem base pair G_{53}:C_{61} adjacent to the T loop (Figure 9.9c). The neighboring conserved *trans* Watson–Crick/Hoogsteen base pair U_{54}:A_{58} (T_{54}:m^1A_{58} in natural tRNAs), which stacks on the G_{53}:C_{61} base pair, enhances the interaction: the 2′-hydroxyl (2′-OH) of U_{54} is poised to donate an H bond to the OP1 of A197 (in line with [77]) and the 6-amino group of the conserved A_{58} donates a second hydrogen bond to the OP2 of C199 (Table 9.1). The stack of C_{61} and A_{58} extends to G_{18} and G_{57} (the D loop nucleotide G_{18} intercalates between A_{58} and G_{57}; Figure 9.10c).

Another major contact involves the universally conserved tertiary Watson–Crick base pair G_{19}:C_{56} that connects the T and D loop in all tRNAs. The two unstacked bases interact with two unstacked bases in the S-domain of the P RNA (A112 and G147) to form intermolecular G_{19}-A112 and C_{56}-G147 stacks. In the S-domain, A112 and G147 are exposed by the two interleaving T loops that are formed by CR-II and

Table 9.1 Correlation of potential hydrogen bonding contacts at the P RNA:tRNA binding interface (based on the *Thermotoga maritima* RNase P holoenzyme:tRNA complex [27]), PDB entries 3q1r and 3q1q) with biochemical data.

Distances in the crystal structure of the *T. maritima* RNase P holoenzyme:tRNA complex (3q1q.pdb)	**Consistent with modification interference and mutational data**[a]
tRNA acceptor stem	
G_1[b] O2′ 3.2Å U315 (P4) OP1[c]	A 2′-deoxy modification at G_1 reduced the efficiency of cleavage ofhairpin model substrates by *E. coli* P RNA [54, 196]
C212 O2′ 3.7Å U315 (P4) OP2	For C353 of *E. coli* P RNA, which corresponds to U315 of *T. maritima* P RNA, an *R*p-phosphorothioate modification was shown to interfere with tRNA binding to P RNA [63]; for C247 of *E. coli* P RNA (C212 in *T. maritima*), 2′-deoxy modification also interfered with tRNA binding to P RNA [197]
Alternatively:	
C212 O2′ 3.2Å A213 (J5/15) OP2	This alternative H bond would also be consistent with biochemical data, as *R*p-phosphorothioate modification at A248 (A213 in *T. maritima*) also interfered with tRNA binding to *E. coli* P RNA [63]; this contact may indirectly stabilize the interaction with the tRNA
G_1 O2′ 4.8Å C212 (J5/15) OP2	No specific biochemical evidence
G_1N2 2.7Å C212 (J5/15) OP1	Binding and cleavage of a hairpin model substrate by *E. coli* P RNA was impaired when G_1 is replaced with inosine [196]
Alternatively:	
c212 OP1 3.0Å A213 (J5/15) O2′	No specific biochemical evidence
C_{72} O2′ 3.9Å A213 (J5/15) O2′	No specific biochemical evidence
G_2 O2′ 5.3Å G211 (P5) OP1	No specific biochemical evidence
C_3 O2′ 4.2Å G211 (P5) OP2	2′-deoxy modification at N interfered with cleavage by *E. coli* P RNA [198]
tRNA T stem	
C_{64} OP1 3.1Å C53 (PA) O2′	No specific biochemical evidence
C_{64} O2′ 3.5Å U321 (P4) O2′	2′-deoxy modification at U359/360 (U321/U322 in *T. maritima*) interfered with tRNA binding to *E. coli* P RNA [197]
C_{65} O2′ 5.3Å U322 (P4) O2′	
C_{65} O3′ 3.9Å U322 (P4) O2′	
C_{65} OP2 3.4Å G51 (P4) O2′	No specific biochemical evidence
C_{65} OP1 3.9Å G51 (P4) N2	No specific biochemical evidence

(*continued*)

Table 9.1 (Continued)

G53:C_{61} / A198 (A-minor type I, Figures 9.9c and 9.10c)	2,6-diaminopurine and 2-aminopurine modification at A233 (A198 in *T. maritima*) interfered with pre-tRNA binding to *E. coli* P RNA [129]; A230 to G mutation (A198 in *T. maritima*) and 2 -deoxy modification at C_{61}, individually reduced pre-tRNA affinity for *B. subtilis* P RNA, but in the presence of 2 -deoxy-C_{61}, the A230G mutation had no additional effect [70, 73]; 2′-deoxy modification at G_{53} interfered with cleavage by *E. coli* P RNA [198]
tRNA T loop	
G_{19}:C_{56} / A112, G147 stacking (Figures 9.9d and 9.10c)	Changing the G_{19}:C_{56} base pair to C_{19}:G_{56} reduced pre-tRNA affinity to *B. subtilis* P RNA ~ 20-fold [59]; *R*p-phosphorothioate modification at A130 (A112 in *T. maritima*) interfered with tRNA binding to *E. coli* P RNA [63]
U_{54} O2′ 2.9Å A197 (P11) OP1 A_{58} N6 2.7Å C199 (P11) OP2	2′-deoxy modification at U_{54}, interfered with cleavage by *E. coli* P RNA [198] and reduced pre-tRNA affinity for *B. subtilis* P RNA [73]; nucleoside analog incorporation at A232 and A234 of *E. coli* P RNA (A197/C199 in *T. maritima*) impaired pre-tRNA binding [129]

a) The biochemical data were obtained for *E. coli* or *B. subtilis* P RNA; thus the observed effects pertain to nucleotides corresponding to those given in the left column for *T. maritima* P RNA (for P RNA numbering, see also Figure 9.2).
b) For tRNA nucleotides, the numbers are given as subscripts.
c) The structural P RNA element is given in parentheses after the nucleotide; OP1 corresponds to the pro-*S*p oxygen, OP2 to the pro-*R*p oxygen.

CR-III (Figures 9.7 and 9.9d). Despite the differences in S-domain architecture of A- and B-type P RNAs, the two unstacked bases are presented on the RNA surface in a similar geometry in both P RNA types (Figure 9.7) [29]. The interactions just discussed (Figures 9.9c, d and 9.10c) emphasize the important role of the S-domain for tRNA recognition and binding. Finally, the crystal structure indicates minor groove contacts between T stem nucleotides 64 and 65 and helix P4 (Table 9.1; Figure 9.10b). This may further include an H bond from the 2-amino group of G51 (P4) to the OP1 of tRNA nucleotide 65; the contact could explain the conservation of G51 (G68 in *E. coli*) whose phosphate is in the coordination sphere of the catalytic metal ion Me2 (Figure 9.11a,b).

Through interaction with 5′-leader nucleotides, the P protein also contributes to substrate binding, and furthermore to catalytic efficiency without direct participation of amino acid residues in catalysis. Based on numerous biochemical studies, the major roles of the protein cofactor can be summarized as follows: the P protein, through binding to the 5′-leader of pre-tRNAs, largely increases enzyme-substrate affinity under physiological salt conditions, under which mature tRNA binds with low affinity to RNase P [86–88]. This prevents a severe deceleration of multi-turnover reactions by rate-limiting product release, as seen in P RNA-alone reactions in vitro at elevated salt concentrations [89, 90]. It further prevents product inhibition in vivo where the total concentration of mature tRNA (~0.2–0.5 mM per

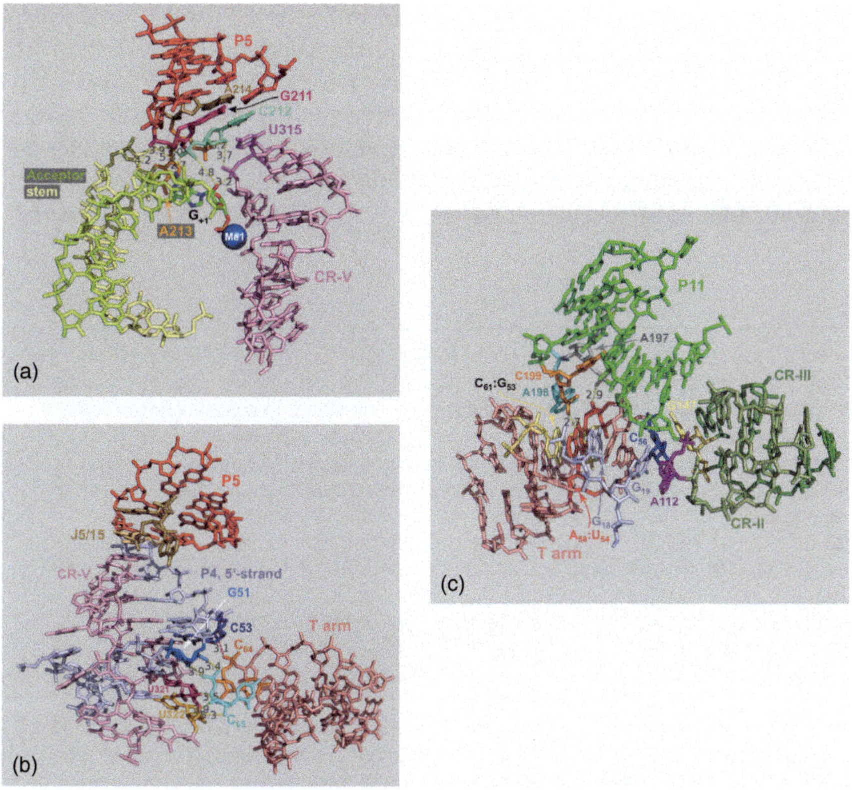

Figure 9.10 (a) Interactions between the *T. maritima* RNase P holoenzyme and the tRNA acceptor stem (primarily minor groove contacts), based on the crystal structure (PDB entry 3q1q). G_{+1} is shown in element color code, following nucleotides of the 5′-strand in lime green, and the 3′-strand in yellow. (b) Contacts of tRNA T stem nucleotides 64 and 65 (see Figure 9.3a) to nucleotides of helix P4. (c) Overview of the A198/G_{53}:C_{61} interaction and other D/T loop contacts with the P RNA S-domain. The numbering of P RNA nucleotides is as in Figure 9.2 for *T. maritima* P RNA and tRNA nucleotides are numbered as in Figure 9.3a. For details, see text and Table 9.1.

E. coli cell [91, 92] largely exceeds that of pre-tRNA at limiting RNase P concentrations (100–1100 RNase P molecules per *E. coli* cell depending on the growth medium [93]). In addition, the P protein, through binding to the 5′-leader and by positioning the pre-tRNA, indirectly enhances the affinity of functionally important magnesium ions in enzyme–substrate (E–S) complexes to permit catalysis at physiological magnesium concentrations [94, 95]. Other functions of the P protein include induction of local conformational changes in the P RNA [71, 72], equalization of pre-tRNA affinities [96], and mediating RNase P dimer formation (specific for *B. subtilis*) [97]. The crystal structure of the *T. maritima* ternary complex revealed that the 5′-end of the bound tRNA is positioned in direct proximity to the highly conserved P RNA regions CR-I/CR-V/P4 and close to the P protein, but there is clearly no direct protein contact to the mature 5′-end of the tRNA (Figure 9.8b). To investigate interactions with the substrate 5′-leader, the crystals were soaked with a 7-nucleotide leader sequence (positions N_{-1} to N_{-5} included in the 3q1r PDB entry refined to 4.2 Å resolution). Although RNA–protein interactions could not be resolved with atomic precision, some conclusions could be drawn in combination with the results from biochemical studies. As established by Fierke and coworkers [86, 88, 98], the P protein interacts extensively with the conformationally extended pre-tRNA leader that binds into its central groove (Figure 9.8b) that is lined with highly conserved amino acid residues (Figure 9.9e). Pre-tRNA with a single 5′-leader nucleotide

Figure 9.11 (a), Left: Transition state model for site-specific phosphodiester hydrolysis of pre-tRNA substrates catalyzed by bacterial (RNase) P RNA, based on previous mechanistic studies and the crystal structure of the *T. maritima* holoenzyme in complex with tRNA [27, 52–56]. An activated water molecule (in magenta) acts as the nucleophile in an S_N2 in-line displacement mechanism. The two non-bridging oxygens and the 5′-oxygen of the scissile phosphodiester are in one plane, whereas the nucleophile and the 3′-oxyanion leaving group occupy the apical positions in the trigonal-bipyramidal transition state. Atoms of active site residues involved in the coordination of the metal ions, Me1 and Me2 (A67, G68, and U69) are numbered according to *E. coli* P RNA and correspond to A49, G51, and U52 of *T. maritima* P RNA (see Figure 9.2). U69 is the universally conserved bulged nucleotide of helix P4. The 2′-OH at nucleotide -1 of pre-tRNA (ribose shown in the assumed C2′-*endo* conformation [57]), not addressed by the post-cleavage state crystal structure [27], was shown in several previous studies to affect the binding and utilization of (catalytic) Mg^{2+} [54, 57–60]) and may interact with Me2 via water molecules of its hydration shell; the model considers evidence that the 2′-OH (in sky blue) accepts an H bond during substrate ground state binding [56] and may further support donation of a proton from a Me2-associated water molecule to the 3′-oxyanion leaving group [56]. Right: Nucleotides in the immediate vicinity of the assumed catalytic metal ions Me1 and Me2 in the crystal structure of the *T. maritima* RNase P:tRNA post-cleavage state complex [27], shown in roughly the same orientation as in the transition state model on the left; G_{+1} and the backbone trace of the leader (at 4.2 Å resolution) are shown in the element color code (C atoms in green, oxygens in red, nitrogen in blue and phosphorus in orange), and individual atoms are indicated for the ribose at nt −1; the O3′ of U_{-1} is at a distance of 3.4 Å to Me2; this structural information was obtained by soaking a 5′-leader fragment into the crystals [27]; for further details, see below. (b) Nucleotides in the immediate vicinity of Me1 and Me2 in the post-cleavage state structure [27]. Residues are numbered according to the numbering system for *E. coli* P RNA (Figure 9.2). L − 1 to −5: backbone trace of the 5′-leader nucleotides −1 to −5 (in element color code). The first two base pairs of the tRNA acceptor stem are shown in green; G350 (*E. coli* numbering) is highlighted (in pink) to distinguish it from A351, and the universally conserved U69 is gold-colored; the 5′-phosphate of G_{+1} is shown in element color code; the O4 of U69, the pro-*Sp* oxygen of A67, and the pro-*Rp* the oxygen of G68 are depicted in red. A248, N332 (G332 in *E. coli*, U294 in *T. maritima*) and N333 (A333 in *E. coli*, G295 in *T. maritima*; see Figure 9.2) are shown here because they are, together with A352, positioned close to the leader nucleotides −1 to −3. (c), Top row: Stacking of A248 (*E. coli* numbering, A213 in *T. maritima* P RNA; see Figure 9.2) on top of tRNA nucleotide C_{72} (distance ~4 Å) in the *T. maritima* holoenzyme:tRNA:5′-leader complex (PDB entry 3q1r), shown in two different orientations; CR-I/P4: nucleotides 5′-AGGAAAGUCCGGA; CR-V: 5′-ACAGAAUCCGGCUU; CR-IV: 5′-GCUU−AGAUUGAU (see Figure 9.2). Bottom row: Comparable orientations of A248 (A256 in *G. stearothermophilus*, A246 in *B. subtilis* P RNA, see Figure 9.2) in the crystal structure of P RNA from *G. stearothermophilus* (PDB entry 2a64 [30]). CR-IV includes the nucleotides as highlighted in Figure 9.2 for *B. subtilis* P RNA; CR-I/P4 and CR-V/P4 include the nucleotides that are equivalent to those specified above for *T. maritima* P RNA.

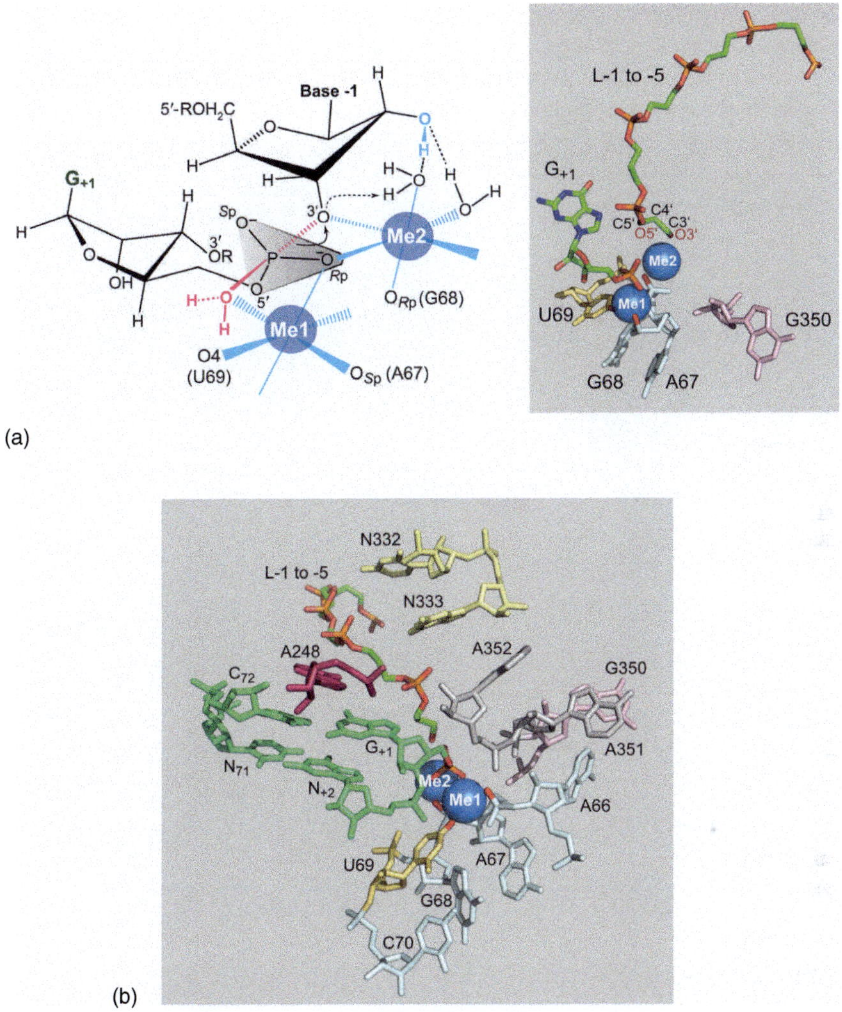

binds with low affinity to bacterial RNase P, even with lower affinity than mature tRNA, and this pertains to P RNA alone and the holoenzyme [86]. From the results of several studies, one can conclude that the 5′-leader:P protein interaction extends from $N_{-2/-3}$ to at least N_{-5} and possibly up to $N_{-7/-8}$ with intrinsic sequence preferences that can accelerate the kinetics of steps preceding cleavage chemistry [86, 96, 99–101]. P RNA nucleotides in the direct vicinity of the leader nucleotides −1 to −3 are A248, A352, and nucleotides 332 and 333 (*E. coli* numbering; see Figure 9.11b). Mutation of P protein residues in its central groove that are assumed to interact with nucleotides at position N_{-4} and N_{-5} (F17, F21, R89, and K90; Figure 9.9e) in the *T. maritima* holoenzyme have a more severe effect on catalytic activity than mutation of amino acids that are in proximity to nucleotides at N_{-1} or N_{-2} [102], in line with N_{-4} and N_{-5} being at the heart of the interaction. For *B. subtilis* RNase P, there is biochemical evidence that residues F20 and Y34 of the *B. subtilis* P protein mediate

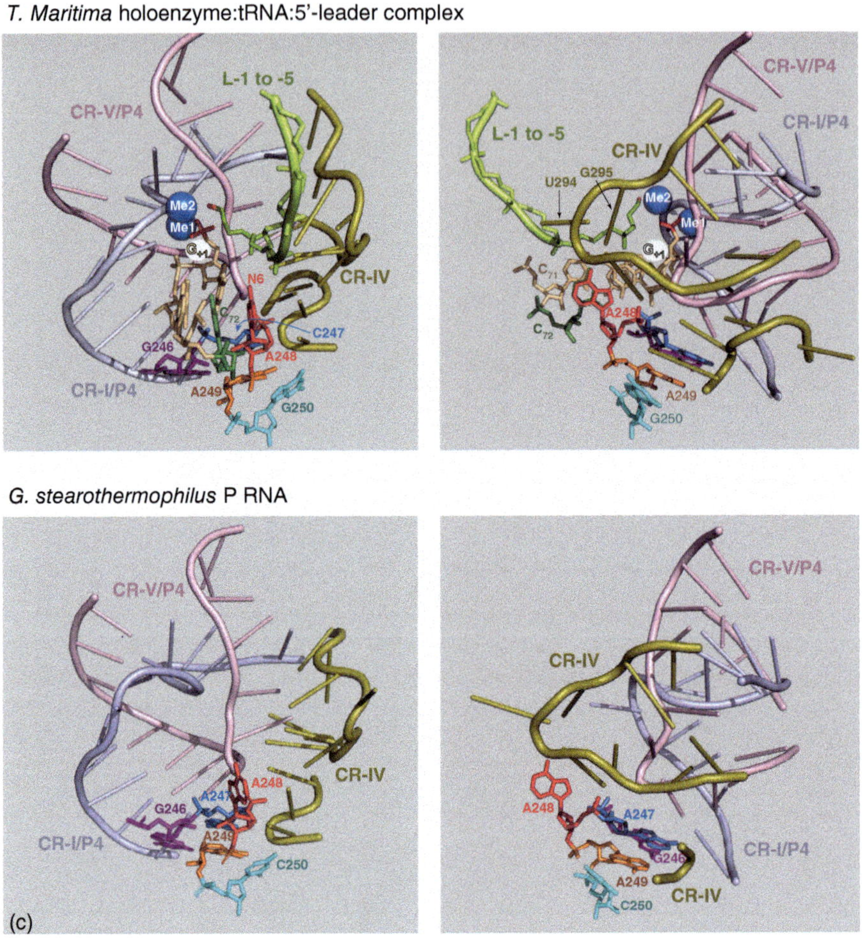

Figure 9.11 *(Continued)*

preferential binding of leaders with an A residue at position −4. *E. coli* RNase P has an overall lower selectivity for nt −4 identity, preferring an A or C residue [99]. This difference is assignable to a Y34 to L34 substitution in the *E. coli* P protein. These findings indicated differences in 5′-leader sequence specificity of RNases P in different bacteria, consistent with the observation that base prevalence at individual leader positions across the cellular pool of pre-tRNAs indeed varies among different bacteria [99]. Harris and coworkers have developed a methodology where they subject a pool of pre-tRNAs with randomized 5′-leader nucleotides to in vitro processing by RNase P. Aliquots are withdrawn at different time points. The amounts of unreacted individual pre-tRNAs are quantified by high-throughput sequencing (RNA-Seq) analysis [103] to infer the cleavage kinetics of all sequence variants relative to a reference substrate under competitive multiple turnover conditions [100,

101]. The authors observed that the P protein of *E. coli* RNase P selects pre-tRNA leader variants in a competitive setup based on nucleotide identity at positions −2 and −3, and this selection occurs at the level of substrate association. This simultaneously mitigates the effect of nt −1 (optimally U) as the primary contributor to catalytic specificity (as measured under single turnover conditions). Also, the coupling of neighboring base identities was particularly pronounced at nt −3 (at the interface to the P protein) and −2 (at the interface to P RNA) [101]. In a related study, the authors further concluded that the identity of the leader nucleotide at −2, and to a lesser extent at −1, determines if the P protein can accelerate cleavage for certain leader sequence variants over others. This is the case e.g. with leaders containing $A_{-2}C_{-1}$ but not with those containing $C_{-2}C_{-1}$. Thus, the identity primarily of nucleotide −2, which is in the contact sphere of the P RNA subunit, determines to which extent sequence-specific binding of leader nucleotides −3 to −6 by the P protein can be energetically coupled to catalysis by the P RNA, and thereby coaxed into faster cleavage kinetics under conditions where an excess amount of different pre-tRNAs competes for binding to a limiting amount of RNase P [100].

In summary, several interaction modules of bacterial RNase P contribute to overall substrate affinity and positioning of substrate and cleavage site. The absence of single interaction modules does not abolish substrate recognition, and this abundancy explains the processing of non-tRNA substrates by bacterial RNase P [1]. Substrate recognition determinants are concentrated in the 5′-leader and 3′-trailer proximal to the cleavage site and in the D/T arm interaction region at the corner of the tRNA L-shape. In the holoenzyme, the RNA subunit mainly contributes to overall substrate recognition and positioning. The protein subunit does not contact any region of the mature tRNA but binds the 5′-leader of the pre-tRNA with certain sequence preferences [99–101], thereby enhancing overall substrate affinity and specificity. Moreover, the protein increases catalytic efficiency by indirectly increasing the affinity of binding sites for functionally important Mg^{2+} ions in E–S complexes, by committing the enzyme to preferential binding of pre-tRNA over mature tRNA, and through favoring a conformational step preceding cleavage chemistry (see Section 9.4). Finally, the bacterial holoenzyme acts more efficiently on non-tRNA substrates than the RNA subunit alone [1]. Therefore, the protein subunit also contributes to broadening the substrate spectrum of the bacterial RNase P enzyme.

9.4 RNA-based Metal Ion Catalysis

9.4.1 The Two-metal Ion Mechanism

RNase P enzymes are metallonucleases. Bacterial P RNA, and RNA-based RNase P variants in general, promote efficient catalysis with Mg^{2+} or Mn^{2+} (albeit at different concentration optima) as the sole divalent metal ion. Both metal ions, but Mg^{2+} more than Mn^{2+}, are proficient in supporting tRNA and P RNA folding, (pre-)tRNA binding to P RNA and catalysis [104–106]. Ca^{2+} ions support catalysis only poorly but are proficient with respect to the other functions [105, 107, 108]. Zn^{2+} ions are defective in promoting (pre-)tRNA binding and RNA folding, but can weakly support catalysis by *E. coli* P RNA in the presence of ammonium salts [109] or when combined with Sr^{2+}, Ba^{2+}, or $[Co(NH_3)_6]^{3+}$ [94, 106]. Sr^{2+} or Ba^{2+} are inefficient in catalysis [104, 106, 109] and only promote weak cleavage by *E. coli* P RNA under specialized conditions (at pH 9.5 and in the presence of 10% ethanol [110]). Pb^{2+}-induced hydrolysis patterns of *E. coli* P RNA obtained in the presence of different divalent metal ions suggested that *E. coli* P RNA adopts a very similar conformation in the presence of Mg^{2+}, Mn^{2+}, Ca^{2+}, Sr^{2+}, Ba^{2+}, whereas transition metals, such as Zn^{2+}, Cd^{2+}, Co^{2+}, Cu^{2+}, and Ni^{2+}, induce changes of the native P RNA conformation [108]. Particularly Ca^{2+}, which essentially prevents catalysis at e.g. pH 6, has been an accepted replacement for Mg^{2+} to mediate proper P RNA folding and high affinity (pre-)tRNA binding [53, 56, 87, 105, 107, 111], and this also pertains to Sr^{2+}, Ba^{2+} and, with restriction, to $[Co(NH_3)_6]^{3+}$ [94, 106, 108, 109]. It should also be mentioned that Ca^{2+}, Sr^{2+}, Zn^{2+}, or $[Co(NH_3)_6]^{3+}$ were shown to compete with catalytic Mg^{2+} ions for binding to P RNA, resulting in inhibition effects [53, 94, 104].

The catalytic mechanism of endonucleolytic phosphodiester hydrolysis by RNA-based RNase P was mainly studied in the RNA-alone setup in the absence of the protein subunit. This simplified model system was based on the original finding that omission of the bacterial protein cofactor (termed P protein, RnpA, or C5 in *E. coli*) can be compensated for in vitro by elevated Mg^{2+} concentrations [16]. The validity of this setup was confirmed in follow-up studies, providing no evidence that functional groups of the protein subunit may be directly involved in catalysis [27, 90, 94, 112]. In 1993, a two-metal ion mechanism, known from protein-catalyzed phosphoryl transfer reactions such as those carried out by the 3′,5′-exonuclease of *E. coli* DNA polymerase I or alkaline phosphatase, was proposed to also apply to ribozyme-catalyzed hydrolytic phosphoryl-transfer reactions that yield 5′-phosphate and 3′-OH termini, which includes RNase P and group I and II self-splicing introns [113]. In this type of mechanism, the two metal ions simultaneously coordinate one of the non-bridging oxygens at the scissile phosphodiester, which neutralizes the developing negative charge in the transition state. One of the two metal ions further activates an H_2O molecule as the nucleophile, while the second metal ion acts as a Lewis acid to facilitate the leaving of the 3′-oxyanion (Figure 9.11a) [113]. To address this model experimentally, cleavage of pre-tRNAs with a single *Rp*- or *Sp*-phosphorothioate modification at the canonical cleavage site were analyzed in reactions catalyzed by *E. coli* and *B. subtilis* P RNAs [52, 53], representing the two major model systems for type A and B architectures of bacterial RNase P. In both cases, the *Rp*-phosphorothioate modification decreased the rate constant for Mg^{2+}-dependent catalysis by at least three orders of magnitude. However, cleavage could be rescued up to ~100-fold by adding (or replacing Mg^{2+} with) thiophilic Cd^{2+} or Mn^{2+} ions, indicating that inner-sphere metal ion coordination of the (pro-)*Rp*-oxygen occurs in the transition state of phosphodiester hydrolysis [52, 53, 114]. Analyses of Cd^{2+} cooperativity for *Rp*-phosphorothioate-modified pre-tRNA cleavage by *E. coli* and *B. subtilis* P RNAs, and competition of catalytically inactive Ca^{2+} ions with Mg^{2+} for the catalytic metal ion binding sites of *B. subtilis*

P RNA, were consistent with two metal ions being essential in the cleavage mechanism [52, 53]. The inhibitory effect of an *R*p-phosphorothioate modification at the cleavage site and the rescue of cleavage in the presence of Cd^{2+} and Mn^{2+} was also seen in the *E. coli* RNase P holoenzyme reaction [112], in line with the notion that the reaction is RNA-catalyzed. Efforts to verify the proposed stabilization of the 3′-oxyanion leaving group by direct coordination to one of the catalytic metal ions using pre-tRNA with a single 3′-*S*-phosphorothiolate inter-nucleotide linkage at the cleavage site were unsuccessful. Cleavage at the modified linkage was completely blocked and could not be rescued by Mn^{2+} or Cd^{2+}; a possible explanation for this finding is that the bulkier sulfur atom at this position excludes the catalytic metal ion(s) from the active site [115]. The concerted mechanism of phosphodiester hydrolysis illustrated in Figure 9.11a, which involves activation of a nucleophilic water molecule by metal coordination, was confirmed by measurement of solvent nucleophile isotope effects [55].

The essential involvement of two metal ions in catalysis by RNase P RNA is consistent with the X-ray structural analysis of the *T. maritima* RNase P holoenzyme in complex with tRNA and a 5′-leader fragment soaked into the crystals [27]. A previous study analyzing the Mg^{2+}-dependence of catalysis by *E. coli* P RNA suggested that at least three metal ions are required for optimal activity [116]. This may have included contributions of Mg^{2+} ions to RNA folding and stabilization of the pre-tRNA:P RNA complex in addition to promoting catalysis itself.

The bacterial RNase P (RNA) reaction shows a linear dependence of log k_{obs} over pH with a slope of ~1 between pH ~5.0 and 6.5 under single turnover conditions [52, 53]. This is thought to reflect the ionization of the Mg^{2+}-bound nucleophilic water molecule (Figure 9.11a), which causes cleavage chemistry to be rate-limiting in this pH range. The flattening of the log k_{obs} over the pH curve at higher pH can be explained by a change in the rate-limiting step rather than reporting the pK_a for deprotonation of the metal-bound water nucleophile. This is consistent with the observation that replacement of the 2′-OH at pre-tRNA nt −1 with a 2′-deoxy or 2′-amino modification, which substantially decreases the rate of the chemical step, extends log k_{obs} over pH linearity to at least pH 8 [54, 77]. The rate-limiting step for cleavage of all-ribose pre-tRNA above pH 6.5 is thought to be a conformational change of E–S complexes that is supported by the P protein [96, 117]. In a fluorescence titration and stopped-flow kinetic study analyzing binding of 5′-fluorescence-labeled pre-tRNA (with varying 5′-leader length) to the *B. subtilis* RNase P holoenzyme, this rate-limiting step above pH 6.5 was experimentally inferred to be a pH-independent isomerization step of initial E–S complexes [111]. In the derived mechanistic model, RNase P binds pre-tRNA in a first bimolecular collision step near the diffusion limit independent of 5′-leader length, mainly recognizing the tRNA moiety in this initial complex. A second unimolecular conformational step, optimally favored in the presence of at least 4-nt long leaders, increases the overall substrate affinity, and this is achieved by a decrease in the rate constant k_{-2} for the reversal of the conformational step (see Scheme 9.1). This isomerization step, which is coupled to interactions of the

5′-leader and the P protein, is proposed to reposition essential functional groups, including catalytic metal ions for efficient catalysis [111].

$$\mathrm{E + S} \underset{k_{-1}}{\overset{k_1}{\rightleftharpoons}} \mathrm{E{\bullet}S} \underset{k_{-2}}{\overset{k_2}{\rightleftharpoons}} \mathrm{E{\bullet}S^*} \xrightarrow{k_{chem}} \mathrm{E{\bullet}L{\bullet}P} \xrightarrow{k_4} \mathrm{E + L + P}$$

Scheme 9.1 Rate constants for individual steps of pre-tRNA processing by bacterial RNase P (RNA). E•S*, E•S complexes after the conformational rearrangement preceding cleavage chemistry defined by k_{chem}; L, cleaved 5′-leader; P, product = mature tRNA. The last two steps are essentially irreversible.

9.4.2 Architecture of the Active Site

Based on the X-ray structures of the *T. maritima* type A RNase P holoenzyme:tRNA complexes with and without 5′-leader mimic (PDB entries 3q1r and 3q1q), two metal ions (Me1 and Me2 in Figure 9.11b) were interpreted as the catalytic ones [27]. The two are sandwiched between the 5′-end of tRNA, the CR-V region, and the 5′-strand of helix P4, including the universally conserved bulged uridine (U69 in *E. coli*, U52 in *G. stearothermophilus* and *T. maritima*). The binding site for Me1 appears to be largely preorganized in bacterial P RNAs, as this metal ion is already present in the *T. maritima* RNase P holoenzyme:tRNA complex without 5′-leader mimic. Moreover, Me1 structurally corresponds to the metal ion binding site M6 in the *G. stearothermophilus* B-type P RNA structure, the only site at which specific Mg^{2+} binding was observed to be clearly favorable in this crystallographic analysis [118]. The second metal ion site (Me2) could only be located after soaking the 5′-leader mimics into crystals of the *T. maritima* RNase P:tRNA complex in the presence of Sm^{3+}. This is interesting given NMR spectroscopic data that identified a metal ion binding site near the ribose of nt −1 in pre-tRNA-derived model substrates, which critically depended on the presence of the 2′-OH at nt −1, suggesting the possibility that Me2 is pre-bound to the pre-tRNA leader nt −1 and brought to the active site upon complexation with the enzyme [57]. The contribution of the 2′-OH at nt −1 to binding of Me2 could be tested by soaking a 5′-leader fragment with a DNA residue at position −1 into crystals of the *T. maritima* RNase P:tRNA complex and testing for Me2 occupancy relative to the all-RNA leader fragment. The distance of the two metal ions (Me1 = Eu^{3+}, Me2 = Sm^{3+}) was inferred as ~4.8 Å in the crystal structure [27]. Pace and coworkers initially disfavored Me1 (M6) as catalytic metal ion based on their *G. stearothermophilus* P RNA structure, arguing that this site is at some distance (12 Å) from the center of catalysis as inferred from biochemical (crosslinking) studies [118]. However, they later reconciled the catalytic center model by Reiter et al. [27] with their own structural analyses on free *G. stearothermophilus* P RNA by invoking that B-type P RNAs with their rather flat shape are not predisposed to bind the pre-tRNA substrate and thus have to undergo a conformational rearrangement to bind a substrate upon initial encounter [119] firmly.

Biochemical data corroborate the described model of the structure and function of the catalytic center, harboring the binding sites for metal ions Me1 and Me2. For example, (i) *R*p- and *S*p-phosphorothioate modifications at A67 and G68 (*E. coli* numbering, Figure 9.2) decreased the cleavage rate constant by 2–4 orders of magnitude [68, 120], (ii) deletion or mutation of the universally conserved U69 decreased the cleavage rate constant ~50- to 80-fold [121], (iii) nucleotide analog interference mapping (NAIM) showed that *R*p-phosphorothioates at A67, G68, and A352 reduced the *cis*-cleavage rate of a P RNA:tRNA conjugate ~200-fold [66], and (iv) were also among the few modifications that caused the strongest interferences with tRNA binding (positioning) [67]; finally, mutation of G350 was found to cause a catalytic defect, and *E. coli* cells expressing a G350C mutant P RNA showed a growth defect under conditions of magnesium depletion relative to cells expressing wild-type P RNA [122].

In retrospect, it is interesting to look at the NAIM data gained in the bacterial RNase P system. Harris and Pace [66], selecting for self-cleavage of a covalent conjugate of the substrate and *E. coli* P RNA, observed only four *R*p-phosphorothioate interferences: at A67, G68, U69 (weaker), and A352, which are nucleotides in the immediate vicinity or coordination sphere (A67, G68, U69) of Me1/2 (Figure 9.11). This, by the way, strongly supports the notion that the catalytic site consists of Me1/2 and its immediate environment. Our lab, using an assay selecting for binding of mature tRNA to *E. coli* P RNA, identified the same sites of *R*p-phosphorothioate interferences plus additional ones [67]. The interference effects at A67, G68, and A352 were also among the strong ones [84]. This indicates that a tRNA can only bind tightly to P RNA when the acceptor stem end is correctly positioned in the active site, that is, in a register in which the E–S complex is poised to proceed to catalysis.

In addition to metal ions Me1 and Me2, two other metal ion binding sites (Me3 and Me4) were localized in the *T. maritima* holoenzyme:tRNA complex upon crystal soaking with Sm^{3+} and Eu^{3+} [27]. Site Me3 is at the 3′-NCCA binding site in L15 (Figures 9.6b and 9.9a), where it stabilizes intermolecular base pairing between the ribozyme and (pre-)tRNA [75, 79, 83, 123]. Me3 is coordinated through phosphate oxygen contacts to tRNA nt A_{73} and C_{74} and P RNA nt A249 and G250 (*E. coli* numbering; [27]; Figure 9.9a). This is in line with Pb^{2+}-induced hydrolysis at A249 [124] and phosphorothioate interferences found at A249/G250 in an assay selecting for high-affinity tRNA binding to *E. coli* P RNA [67, 84, 125]. Metal ion Me4 is located close to the backbone of A140/A141/G142 at the base of P12, A149/C150 in J12/13 and G152 at the base of P13 (Figure 9.2, *T. maritima* P RNA), contributing to the stabilization of the universally conserved T loop region (CR-II and CR-III; Figure 9.6b) of P RNA involved in recognition of the corner of L-shaped tRNA molecules (Figure 9.9d).

9.4.3 The "A248/nt −1" Interaction

The holoenzyme–tRNA complex structure with a soaked 5′-leader mimic, although representing a major leap toward understanding the structural basis of RNase P function, could not clarify how pre-tRNA substrates are aligned in the active site at atomic resolution. Local conformational changes of the E–S complex are likely to occur toward traversing the transition state of phosphodiester hydrolysis. We still do not have a complete picture as to which functional groups of the ribozyme are in contact with the nt −1 immediately upstream of the canonical cleavage site in E–S complexes. Short-range crosslinking studies, using mature tRNA with a single s^6G substitution at position +1 [126] or pre-tRNA with a single s^4U substitution at position −1 [78], identified the highly conserved residue A248 (*E. coli* numbering, A213 in *T. maritima* P RNA) in the J5/15 interhelical junction as a P RNA nucleotide in direct vicinity to the scissile phosphodiester. A biochemical follow-up study addressed the possibility of an interaction between A248 and nt −1 of pre-tRNA by mutational analysis of A248 and pre-tRNA nt −1 combined with 2′-deoxy modifications at nt −1 [127]. This study indeed revealed contact between adenine 248 and the base at nt −1 (optimally U_{-1} in *E. coli* and many other bacteria), which was shown to cooperate with the contact involving the 2′-OH of nt −1 in selecting the canonical cleavage site [127]. The observed, quite complex and interdependent effects of nt 248 and −1 mutations and 2′-deoxy modification at nt −1 on substrate affinity and cleavage site selection led Harris and coworkers to propose *cis* Watson–Crick/Watson–Crick pairing between the bases of nt −1 and A248 as the most plausible structural interpretation of their results [127]. However, this has been questioned in other studies [36, 85, 128], e.g. based on the finding that a 3-methyl-U substitution at −1, sterically blocking the Watson–Crick interface, resulted in no apparent change in substrate affinity and cleavage rate [85]. Kirsebom and coworkers considered an alternative model according to which the N6 of A248 forms a hydrogen bond to the O2 of a pyrimidine at nt −1 and the N7 of A248 accepting an H bond from the 2′-OH at nt −1 [128]. In the *T. maritima* post-cleavage state holoenzyme:tRNA structure, the N6 of A213 (A248 in *E. coli*) is close to (within ~5 Å) the phosphate oxygen OP1 at nt −1 as well as to the ribose backbone and phosphate oxygen OP1 at nt −2 of the soaked-in 5′-leader oligomer [27]. Interestingly, A213 stacks on top of C_{72} at the end of the acceptor stem and exposes its Hoogsteen edge, which may pair with the pre-tRNA base at nt −1 (Figure 9.11c, top). Isosteric *trans*-Hoogsteen/Watson–Crick pairs of A with U, A and C, but not with G, are conceivable [129], which is in line with the observation that a G at position −1 is less favorable for cleavage efficiency [100] and for selection of the canonical cleavage site (between nt −1 and +1) by bacterial RNase P [130, 131]; G is also underrepresented in *E. coli* and *B. subtilis* pre-tRNAs at position −1 [99]. It is further possible that the 6-amino group of A248 may form a bifurcated H bond interaction, one to the base at nt −1 and the second to the phosphate of nt −1 or −2, or the ribose 2′-OH at nt −2. A contact to the 2′-OH at nt −2 would be in line with this substituent being important for cleavage efficiency [58], and a contact to the phosphate of nt −2 would be consistent with a phosphorothioate interference effect at nt −2 on pre-tRNA processing [132]. Interference of N6-methyl and N7-deaza modifications at A248 with substrate binding to *E. coli* P RNA would be consistent with H bonds involving its 6-amino group and N7 at the Hoogsteen edge of A248 [133]. Surprisingly, A249, but not A248, was reported to give rise to N6-methyl and N7-deaza modification interference in self-cleavage assays of covalent P RNA–tRNA conjugates [134]. This discrepancy may be attributed to the quite non-native conditions of the self-cleavage assay (3 M NaCl, 1.25 mM Mg^{2+}, 50 mM PIPES, pH 5.5, assay temperature 50 °C [134]), which may have masked A248 interference. In summary, A248 helps to align the canonical scissile phosphodiester in the active site. The holoenzyme crystal structure indicates stacking of A248 (A213, according to *T. maritima* P RNA numbering) on top of the N_{72} residue (Figure 9.11c), which may be the primary task of residue A248 to support proper positioning of incoming substrates. The "A248/nt −1" interaction was further shown to communicate with other contacts such as the 3′-NCCA interaction [60, 85, 127, 128]. In the crystal structure of *G. stearothermophilus* P RNA, A248 (A256 in *G. stearothermophilus*; A246 in *B. subtilis* P RNA, see Figure 9.2) is located essentially in the same position as in the *T. maritima* holoenzyme:tRNA complex (Figure 9.11c; cf. top and bottom panels), suggesting that this adenine is pre-positioned to stack on the N_{72} residue of the incoming (pre-)tRNA. This concomitantly supports the unstacking of the well-stacked tRNA residues N_{72} and N_{73} [135], thus facilitating threading of the 3′-NCCA end into the P15/16/17/6 round arch. It is also expected that nt −1 traverses distinct conformational states toward reaching the post-cleavage state, during which A248 may change its interaction sphere. As numerous studies have carved out the crucial role of the 2′-OH at nt −1 for cleavage site selection, efficient catalysis, and catalytic metal ion binding [54, 56–60, 127], it remains a major goal of future studies to understand the molecular contact spheres of the ribose at nt −1 during the catalytic cycle.

9.4.4 Specific RNase P Cleavage by the P15 Module

The L15 loop of bacterial P RNAs base pairs with the pre-tRNA NCCA 3′-end and binds several Mg^{2+} ions [136], of which one Mg^{2+} binding site (Me3) could be localized in the *T. maritima* holoenzyme:tRNA structure [27]. Kirsebom and coworkers demonstrated that isolated L15 loop mimics (the loop flanked by short stems on both sides), as short as 31 nt in length, promote very weak but specific pre-tRNA cleavage in the presence of elevated Mg^{2+} concentrations (160 mM), yielding 5′-phosphates and 3′-OH termini as products [136]. As such modules lack the assumed active center that coordinates catalytic metal ions Me1 and Me2 (Figure 9.11a,b), the question arises whether the active site of the RNase P ribozyme can be narrowed down to the immediate environment of Me1/Me2. One may consider the possibility that several regions of the ribozyme (and the protein cofactor) cooperate in substrate positioning to expose the cleavage site and in attracting multiple Mg^{2+} ions to the vicinity of the cleavage site. Two metal ions may be recruited, possibly one from the set of Mg^{2+} ions binding in the vicinity of the cleavage site (e.g. in the P4 or P15 region) and one from solution or pre-bound to the pre-tRNA substrate to promote cleavage. According to current knowledge, catalysis is expected to be most efficient when metal ions at sites Me1 and Me2 in the P4/CR-V region are utilized, but other pairs of Mg^{2+} ions may promote catalysis with lower efficiency as well. However, this requires Mg^{2+} concentrations at least two orders of magnitude higher than in vivo (~1–2 mM freely available Mg^{2+} in the bacterial cytoplasm [137]). Thus, while explaining some of the in vitro data, such a possibility is unlikely to be relevant to function in vivo.

9.5 RNase P as an Antibiotic Target

Bacterial RNase P fulfills two central conditions of a promising drug target. First, its single RNA (*rnpB*) and protein (*rnpA*) subunits belong to the ~7% of bacterial genes that are indispensable for cell viability [138–142]. Second, human nuclear-cytoplasmic and mitochondrial RNase P enzymes structurally differ from their bacterial counterparts in terms of (i) number, identity, and function of protein subunits, (ii) loss of RNA-alone cleavage activity under in vivo conditions (nuclear-cytoplasmic RNase P) or the absence of an RNA component (mitochondrial RNase P), and (iii) the mode of substrate recognition [142–144]. Thus, the prospect of being able to inhibit bacterial RNase P without affecting human RNase P enzymes makes bacterial RNase P attractive as a novel antibacterial target. Since the P RNA and P protein subunit of the bacterial RNase P holoenzyme are both essential for tRNA processing in vivo, the two types of macromolecules can be addressed as targets for enzyme inhibition. Here we briefly summarize previous efforts (more comprehensively discussed in [141]) and then focus on more recent activities in this research area.

9.5.1 P RNA as a Target

One strategy is the inhibition of the catalytic RNA via antisense oligonucleotides (ASOs). The L15 loop, which interacts with tRNA 3′-NCCA ends, has proven to be a particularly accessible target site in bacterial A-type [145–148] as well as B-type P RNAs [63]. Beyond sterically inhibiting base pairing between the L15 loop and the substrate 3′-NCCA end, such ASOs were shown to perturb folding and metal ion coordination in the catalytic core of *E. coli* (A-type) P RNA [148]. An RNA 14-mer as well as iso-sequential LNA (locked nucleic acid) and PNA (peptide nucleic acid) variants targeting the L15 loop caused 50% inhibition of *E. coli* P RNA activity in vitro at concentrations in the lower nanomolar range [147, 148]. The PNA 14-mer coupled to an invasive peptide to improve cellular uptake, was then demonstrated to inhibit the growth of highly permeable *E. coli* AS19 cells at a concentration of 10 μM [148]. In addition, the growth of the strain in the presence of the PNA-peptide conjugate could be rescued by expressing *B. subtilis* P RNA that lacks complementarity to the PNA oligomer, demonstrating sequence-specific inhibition of *E. coli* P RNA. More efficient uptake of ASO-peptide conjugates by clinical bacterial isolates is a major hurdle for such approaches. Recently, conjugates of LNA/DNA mixmer ASOs coupled to the invasive peptide $(KFF)_3K$ and directed against the mRNA of the cell division protein FtsZ were shown to inhibit the growth of pathogenic *S. aureus* strains and to increase the survival rate of *S. aureus*-infected mice [149]. Shorter ASOs (<10 nt), with a higher density of LNA modifications or all-LNA, may give rise to ASO-invasive peptide conjugates with improved bacterial uptake and efficiency.

Still, a disadvantage of ASO-peptide conjugates is their large size and costly synthesis compared with small-molecule antibiotics.

Aminoglycosides are antibacterial agents that primarily target the bacterial ribosome and its decoding center associated with the 30S ribosomal subunit [150]. The basic structure is an aminocyclitol linked to one or more amino sugars by glycosidic linkage(s) (Figure 9.12a). They are polar molecules with a varying number of amino groups (partially) protonated at physiological pH. Among different natural aminoglycosides, neomycin B (NeoB) inhibited *E. coli* RNase P most efficiently (50% inhibition at 35 μM in the RNA-alone and 60 μM in the holoenzyme reaction [152]). Inhibition of both holoenzyme and RNA-alone reaction indicated the binding of aminoglycosides to the P RNA. Increasing the Mg^{2+} concentration decreased the inhibitory effect of NeoB on P RNA processing and the antibiotic protected prominent Pb^{2+} hydrolysis sites in the P RNA. The authors concluded that aminoglycosides electrostatically compete with Mg^{2+} ions for functionally important divalent metal ion binding sites in the RNA [152]. Later, amino groups of NeoB were derivatized with guanidinium, arginyl, or lysyl groups to enhance their inhibitory efficacy [141, 153, 154]. The best-performing hexa-arginine derivative of NeoB, NeoR6 (Figure 9.12a), inhibited *E. coli* RNase P at a 500-fold lower concentration than the parental NeoB [153]. At least 10-fold higher concentrations of NeoR6 were required to achieve 50% inhibition of partially purified human nuclear-cytoplasmic RNase P relative to *E. coli* RNase P [153], suggesting some selectivity for bacterial RNase P enzymes. Arginine derivatives were more effective than guanidinium and lysine derivatives, indicating that side-chain length, flexibility, and chemical properties of the end group determine the inhibitory potency [154]. Like the parental aminoglycosides, the arginine derivatives were inferred to compete with Mg^{2+} for binding to the catalytic RNA [154]. Critical issues are the (i) low specificity of interactions between RNA molecules and aminoglycoside–arginine conjugate, (ii) bacterial uptake efficiency in view of their cationic nature and increased size, and (iii) toxicity. Specificity is clearly an issue, considering that neomycin B and kanamycin B were observed to bind tRNA molecules in the same concentration range where they inhibit bacterial RNase P [155].

Phenothiazine derivatives (Figure 9.12b) have been found to bind to different RNAs, such as the HIV trans-activation response element RNA, the *E. coli* ribosomal A site, coxsackievirus B3 loop D RNA, and the loop B of the poliovirus IRES [156]. Kirsebom and coworkers examined this group of compounds as inhibitors of three bacterial RNase P ribozymes from *E. coli* (A-type), *Mycoplasma hypopneumoniae* (B-type), and *Mycobacterium leprae* (A-type) [157]. An inhibitory effect on the RNA-alone reaction was observed in the μM range (~10–400 μM) for all tested phenothiazine derivatives. Stronger inhibition effects were observed with aromatic amine derivatives (methylene blue, toluidine blue O, thionine acid, azure A and B, and pyronine G). The three antipsychotic phenothiazines with documented antibacterial activities (thioridazine, trifluoperazine, and chlorpromazine) were less inhibitory; the more distantly related tricyclic antidepressants imipramine

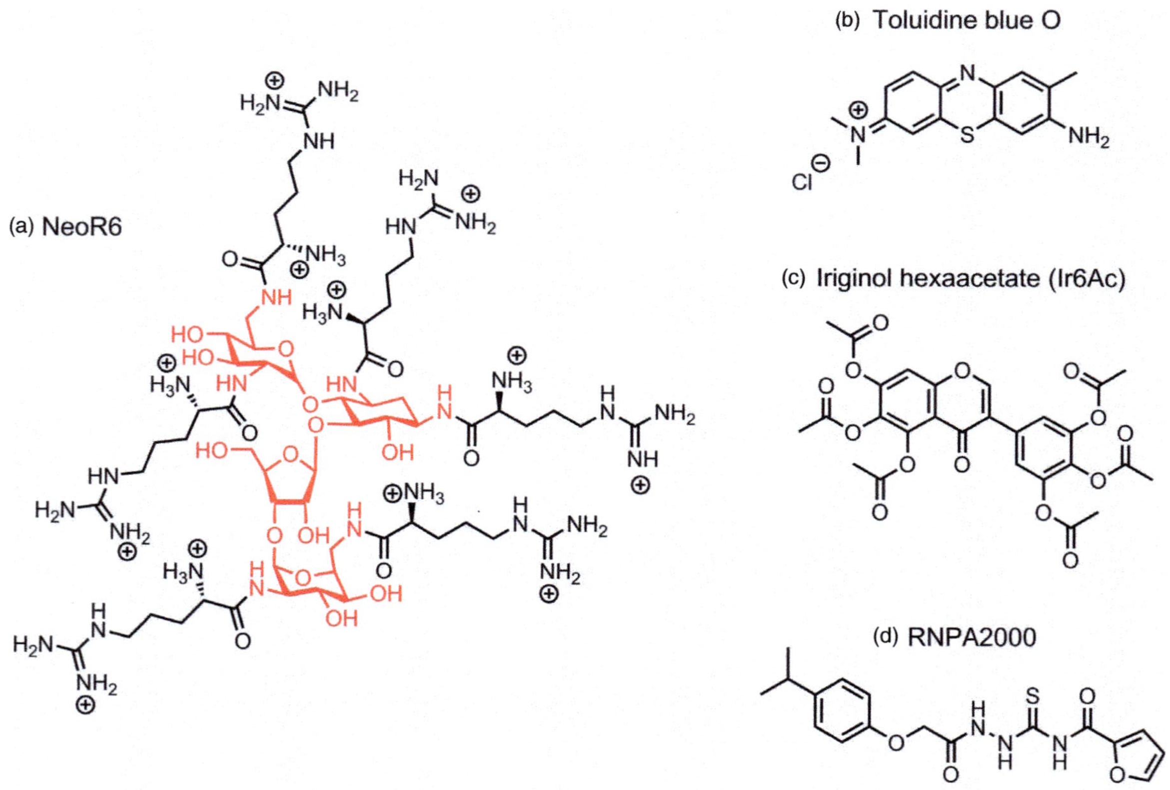

Figure 9.12 Structures of reported bacterial RNase P inhibitors. (a) In NeoR6, the six amino groups of the Neomycin B (NeoB) core (red) were extended by arginine moieties. The pK_a values of the amino groups (at the alpha C-atoms of the arginine moieties) are estimated to be ~9. (b) Resonance structure of Toluidine blue O as an example for a P RNA inhibitor with phenothiazine core. (c) Iriginol is a natural isoflavone that was esterified with acetic acid to obtain iriginol hexaacetate (Ir6Ac). (d) The thiosemicarbazide derivative RNAP2000 was reported to inhibit bacterial RNase P by interacting with its protein subunit.

and amitriptyline (dibenzoazepines), which lack the phenothiazine scaffold and contain a central 7-membered nonaromatic ring, were inactive [157].

In contrast to aminoglycosides, inhibition by the phenothiazine toluidine blue O was independent of pH between 6 and 9 and showed only a minor dependence on salt concentration. Further experiments were performed to investigate the mode of action of phenothiazine derivatives. Varying the preincubation order of substrate, inhibitor, and ribozyme revealed an enhancement of the inhibitory effect when inhibitor (toluidine blue O or thionine acid) and substrate were preincubated before the addition of the ribozyme, indicating that phenothiazine binding to the pre-tRNA makes a major contribution to inhibition of the cleavage reaction. This observation was supported by suppression of Pb^{2+}-induced hydrolysis in the D loop of yeast $tRNA^{Phe}$ in the presence of various phenothiazine derivatives. Cleavage inhibition assays and Pb^{2+}-probing experiments led to the conclusion that phenothiazines have multiple binding sites in functionally important regions of the S- and C-domain of bacterial P RNAs [157]. In conclusion, phenothiazines can be regarded as a scaffold with the potential to be further optimized for inhibition of bacterial RNase P. The quite promiscuous interaction with different RNAs needs to be changed towards a higher binding specificity for bacterial P RNA.

9.5.2 The Bacterial RNase P Holoenzyme as Target

Fierke and coworkers pursued a high-throughput screening approach to identify novel P RNA inhibitors [155]. This was preceded by the development of a fluorescence polarization/anisotropy (FP/FA) assay to measure pre-tRNA cleavage, which is more suitable for high-throughput analyses in microtiter plate formats compared with conventional, gel-based RNase P assays employing ^{32}P-labeled pre-tRNAs. The authors showed a very similar kinetic performance of *B. subtilis* RNase P in the FP/FA real-time assay relative to gel-based assays using 5′-fluorescein- or 5′-^{32}P-labeled pre-tRNA. NeoB inhibited *B. subtilis* RNase P in the FP/FA real-time assay at a similar concentration (50% inhibition at 23 μM) as observed for *E. coli* RNase P (60 μM; [152]). They also showed aminoglycoside binding to the pre-tRNA substrate in the FA assay, where the concentration dependence of binding indicated that this interaction contributes to the inhibitory effect on the RNase P processing reaction. After further optimization of the end-point fluorescence polarization high-throughput assay [155], a library of ~2900 compounds was screened in 384-well microplates. Only one compound, iriginol hexaacetate (Ir6Ac; Figure 9.12c), was identified and confirmed in multiple screens as an inhibitor of *B. subtilis* RNase P, with an IC_{50} of ~0.8 μM in multiple turnover reactions [155]. Ir6Ac had no influence on the FA signal of 5′-fluorescein–pre-tRNA, providing evidence against Ir6Ac binding to the pre-tRNA substrate as well. The effects of Ir6Ac on the steady-state kinetic parameters K_m and k_{cat} pointed to a mixed inhibition model (inhibitor binds to free enzyme and the enzyme–substrate complex, with different affinities). Although Ir6Ac appears as a more specific inhibitor of bacterial RNase P than aminoglycoside derivatives and phenothiazines, the compound is unlikely to display activity in vivo, as its esterified acetyl groups are sensitive to hydrolysis. Future studies will have to clarify whether Ir6Ac is a suitable lead scaffold for the development of metabolically more stable derivatives [155].

9.5.3 P Protein as a Target

Apart from targeting the catalytic P RNA subunit, there have been attempts to address the protein subunit (RnpA) of bacterial RNase P by small molecule inhibitors. Scientists at SmithKline Beecham Pharmaceuticals reported an NMR structure of the *S. aureus* RnpA protein [47] driven by the motivation to explore bacterial RNase P as an antibiotic target. More recently, an inhibitor termed RNPA2000 (a thiosemicarbazide derivative; Figure 9.12d) was reported to inhibit *S. aureus* RNase P by interacting with its protein subunit [151]. The compound was described to have bactericidal activity, not to cause human cell cytotoxicity after 24 h of incubation at a 16-fold higher concentration (256 μg/ml) than the minimal inhibitory concentration (MIC, 16 μg/ml) against *S. aureus* [151]. RNPA2000 has been used as a starting point for the development of more potent inhibitors. In structure–activity studies of RNPA2000, the furan moiety was already replaced, as it has the potential to cause toxic effects upon metabolization [158]. However, the antimicrobial activities of these derivatives were not increased.

Finally, it has still to be demonstrated for RNPA2000 and derivatives, as well as for Ir6Ac, that these compounds do not act as aggregators [159, 160] of the bacterial RNase P protein instead of being specific enzyme inhibitors.

9.6 Application of RNase P as a Tool in Gene Inactivation

9.6.1 The Guide Sequence (GS) Concept

The hallmark of the GS concept is the formation of a bipartite, tRNA-like RNase P substrate through annealing of an RNA oligonucleotide (termed external guide sequence, EGS [161]) to a target RNA (Figure 9.13a [169]). The target RNA is then specifically cleaved by endogenous RNase P activities, an approach that resembles antisense-based inactivation of target RNAs through the recruitment of endogenous RNase H activities [170]. In the *E. coli* system, the concept was inspired by the initial finding that *E. coli* RNase P can process quite efficiently minimal, tRNA-derived hairpin substrates mimicking coaxially stacked acceptor stem and T arm [171]. Recognition of minimal substrates was also shown for other bacterial RNase P enzymes, including the A-type enzyme of *T. thermophilus* [172] and the B-type enzyme of *B. subtilis* [132]. Forster and Altman [161], who introduced the EGS concept, demonstrated that bipartite substrates composed of EGS and target RNA must at least mimic a 7–8 bp long acceptor stem with a dangling 3′-ACCA terminus (later also shown to pertain to type B RNase P from *B. subtilis* [132]). Still, more efficient processing by *E. coli* RNase P was seen with hybrids that base pair over ~12 nt (or up to 19 nt), thus mimicking coaxially stacked acceptor and T stems (Figure 9.13b). The 3′-ACCA terminus is crucial for anchoring the bipartite substrate to the active site of bacterial RNase P via base pairing to the L15 loop of its catalytic RNA subunit (Figure 9.9a). The approach was tested in *E. coli* using plasmid-encoded EGS transcripts targeting regions predicted to be accessible in the mRNA for chloramphenicol acetyl transferase (CAT). EGS expression increased the chloramphenicol sensitivity of the bacteria [164], indicating that EGS and target RNA form complexes in vivo that are recognized as a substrate by endogenous RNase P. Likewise, in a follow-up study, targeting the essential *gyrA* (gyrase A) and *rnpA* (RNase P protein subunit) mRNAs by EGS transcripts strongly reduced the viability of *E. coli* bacteria. Inhibition was dependent on EGS:target RNA complementarity (with tolerance for 3 mismatches in a stretch of 15 bp), and applying more than one EGS simultaneously resulted in additive inhibition effects [165]. In the aforementioned proof-of-concept studies, plasmid-borne EGS expression cassettes were transformed into *E. coli* cells. For antimicrobial applications, Tolmasky and coworkers explored exogenous delivery of EGS agents into bacterial cells, using the *aac(6′)-Ib* resistance gene as a model target that is expressed in clinical isolates of *Acinetobacter baumannii* [173]. The *aac(6′)-Ib* gene expresses an aminoglycoside acetyltransferase (AAC) that inactivates aminoglycoside antibiotics such as amikacin by acetylation of specific amino groups. An EGS forming 13 bp with the *aac(6′)-Ib* mRNA target, identified as being most efficient within a set of tested EGS variants [173], served as a template for the design of iso-sequential LNA/DNA mixmers that were analyzed for their capacity to elicit cleavage of *aac(6′)-Ib* mRNA by endogenous *E. coli* RNase P. The authors found EGS gapmers with 3 to 5 consecutive LNA residues at their 5′-end and the 3′-terminal CCA triplet as LNA to be very efficient. Such modified EGS oligonucleotides increased the stability against degradation in *E. coli* lysates, as expected, and were able to decrease amikacin resistance in the highly permeable *E. coli* strain AS19 [174]. The findings indicated that stabilizing the EGS 3′-CCA base-pairing interaction with the L15 loop of bacterial RNase P is crucial for target RNA cleavage by endogenous RNase P. In a follow-up study, the same group coupled an LNA/DNA gapmer (LNA residues at the 5′-end, 8 DNA residues in the center and the 3′-ACCA as LNA) directed against *aac(6′)-Ib* mRNA to a cell-penetrating peptide [175]. The conjugate was cleaved by *E. coli* RNase P in vitro and impaired growth of *A. baumannii* in the presence of amikacin relative to a control conjugate with a scrambled sequence. Yet, in both studies just mentioned, it remains a formal possibility that mRNA inactivation of such LNA/DNA EGS gapmers in bacterial cells might have included contributions from RNase H-mediated mRNA degradation.

9.6.2 EGS Technology in Eukaryotic Cells

For the application of the GS concept to gene inactivation in human and other eukaryotic cells, two strategies have been pursued. The first approach, basically identical to the EGS approach described above for bacterial cells, involves the formation of bipartite, tRNA-like structures that are cleaved by, e.g. human nuclear-cytoplasmic RNase P (Figure 9.13a). In the second approach, the GS is tethered to

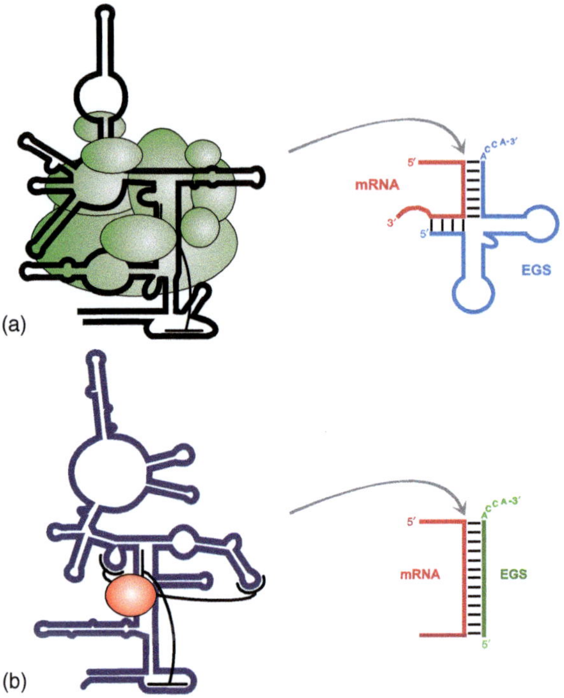

Figure 9.13 Degradation of target mRNAs by external guide sequence (EGS) approaches. (a) A precursor tRNA-like structure is formed by Watson–Crick base pairing between an accessible region of a target mRNA (red) and an EGS RNA (sky blue). This recruits endogenous e.g. human nuclear-cytoplasmic RNase P that endonucleolytically cleaves the target RNA at the site indicated by the gray arrow. The EGS contributes about ~3/4 of the original tRNA structure from the 3′-strand of the D stem to the 3′-end, while the target RNA provides the 5′-leader, the 5′-strand of the acceptor stem, and the 5′-part of the D stem/loop [162]. (b) The same approach, as illustrated in panel (a), but tailored to bacterial systems. Here, more simplified EGS designs can be used, as bacterial RNase P is able to cleave quite efficiently target RNA:EGS complexes that base-pair over ~12–19 nt base pairs, mimicking coaxially stacked acceptor and T stems. The 3′-ACCA terminus is crucial for anchoring the bipartite substrate to the active site of bacterial RNase P via base-pairing to the L15 loop of its catalytic RNA subunit [163–165]. (c), The M1–GS concept, where the guide sequence is tethered to the 3′-end of *E. coli* RNase P RNA, historically termed M1 RNA [163, 166, 167]. The single-stranded 4-nt 3′-overhang of mature M1 RNA is extended by a linker (in orange) of 24–50 nt [163], followed by (optimally) 13–16 nt complementary to the target RNA and terminating with a dangling 3′-ACCA tail that can be extended by a few additional nucleotides [163]. M1–GS conjugates are substantially more active at physiological Mg^{2+} concentrations than the isolated M1 RNA ribozyme acting *in trans* [167]. A variant with enhanced efficiency carrying two base exchanges in the P14 region of M1 RNA (G224A, G225A; see also Figure 9.2) was identified by in vitro selection [168]; there was evidence that the mutations enhance a substrate contact between the P14 region of the catalytic RNA and the target RNA immediately downstream of the GS:target duplex (indicated by the light blue circles connected by the arrow). Applications of the concept to human cells (via expression of an M1–GS conjugate from a Pol II or III promoter) were shown to further benefit from a stimulating effect of human nuclear-cytoplasmic RNase P on M1–GS activity (see text).

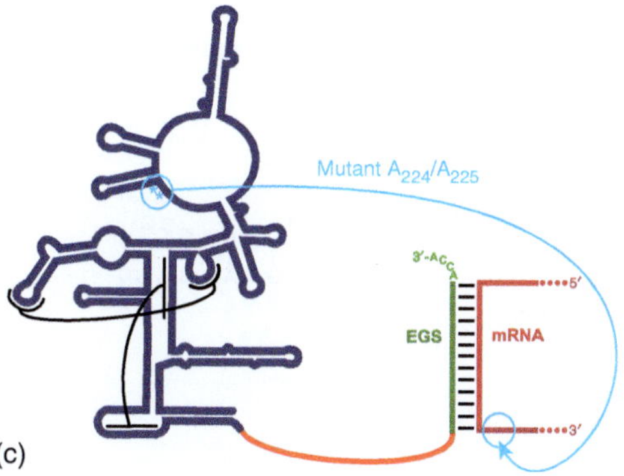

(c)

Figure 9.13 *(Continued)*

the 3′-end of *E. coli* P RNA, such that target RNAs are cleaved *in cis* by the hooked up catalytic RNA upon annealing of the GS to the target RNA (Figure 9.13c).

9.6.3 EGS Oligonucleotides and Recruitment of Human Nuclear-Cytoplasmic RNase P

With respect to the first approach, it became rapidly clear that minimal substrates for bacterial RNase P of the type shown in Figure 9.13b are not cleavable by human nuclear-cytoplasmic RNase P. Altman and coworkers when analyzing the structural requirements for recognition of bipartite tRNA-like substrates by human RNase P, observed cleavage when the EGS contributed ~3/4 of the original tRNA structure from the D loop to the 3′-end, while the target RNA provided the 5′-leader, the 5′-strand of the acceptor stem and the 5′-part of the D stem/loop [176] (Figure 9.13a). Expression of this type of EGS (termed EGS^{CAT}; Figure 9.14a) enabled a ~twofold down-regulation of an ectopically expressed CAT gene in a human lung cancer cell line [162]. Considering that human RNase P cleaved EGS^{CAT}:CAT mRNA hybrids less efficiently than natural tRNA precursors, Yuan and Altman [176] employed in vitro selection to identify more efficient EGS variants. For this purpose, they fused the CAT mRNA part to EGS^{CAT} to generate a unimolecular tRNA-like structure and randomized four nucleotides in the T loop and five in the variable loop region. Additional random mutations over the entire RNA length were introduced by error-prone PCR. After eight rounds selecting for rapid cleavage by human RNase P at increasing stringency, improved variants were identified. The derived EGS (EGS 9; Figure 9.14a) of the best performing clone mediated target RNA cleavage 30 times faster than the original EGS^{CAT}. Kinetic analyses indicated that target affinity and V_{max} were increased for EGS 9 relative to EGS^{CAT}, which correlated with anticodon stem disruption (inferred from structure probing) and the presence of T and variable loop sequences not present in natural tRNAs; in addition, while cleavage mediated by EGS^{CAT} had a Mg^{2+} optimum of 25 mM, this tended to be reduced to lower concentrations for the "faster-cleaving" EGS variants such as EGS 9 [176]. In a recent study, Liu and coworkers [178] expressed EGS transcripts from

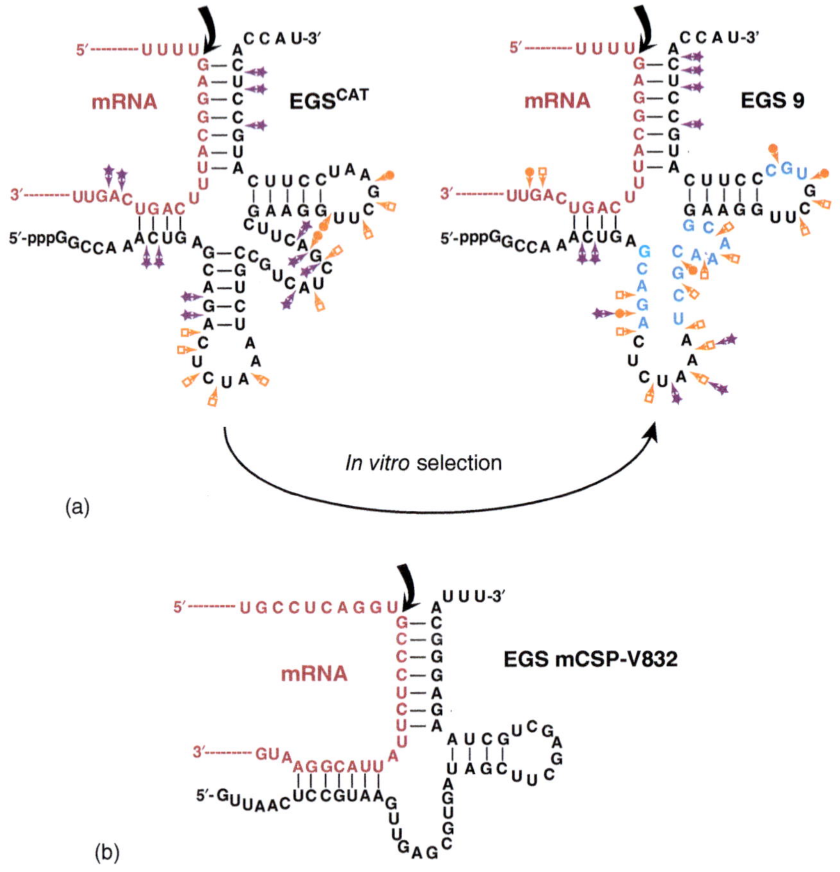

Figure 9.14 EGS variants to induce target RNA cleavage by human nuclear-cytoplasmic RNase P. (a) Left: original EGS^{CAT} directed to an accessible region of mRNA coding for chloramphenicol acetyltransferase (CAT) and derived (with sequence changes) from *E. coli* tRNA^{Tyr}. Right: the most active derivative variant EGS 9 identified by in vitro selection [176], with the regions changed (relative to EGS^{CAT}) on the sequence and/or structural level indicated by blue nucleotides. Both pre-tRNA-like molecules were subjected to in vitro probing by RNases T1 (orange arrows with filled circle), T2 (orange arrows with open squares) and cobra venom nuclease (purple arrows with asterisks). RNase T1 cleaves 3′ of single-stranded G residues, RNase T2 3′ of single-stranded residues with a slight preference for A residues and cobra venom nuclease (RNase V1) cleaves phosphodiester linkages in paired or stacked regions [177]. The altered pattern of RNase susceptibility of EGS 9 versus EGS^{CAT} was interpreted as indicating the formation of new tertiary interactions in EGS 9 that enhance the rate of cleavage by human RNase P [176]. (b) EGS mCSP-V832 is an efficient antiviral EGS used in a recent study [178].

retroviral vectors to target the shared mRNA region of murine cytomegalovirus (MCMV) capsid scaffolding protein (mCSP) and serine proteinase assemblin. The EGS mCSP-V832 (Figure 9.14b), identified by in vitro selection [179] and binding with high affinity to its target, reduced mCSP expression by ~90%, inhibited viral growth by 8000-fold in MCMV-infected cells, and prolonged survival of infected mice [178]. The same strategy was used to inhibit herpes simplex virus 1 gene expression and viral growth [180]. In another variation of the concept, the same group [181] used an attenuated *Salmonella* strain harboring a plasmid encoding an EGS directed against the pre-genomic RNA of Hepatitis B virus (HBV) under control of a eukaryotic promoter. Oral administration of such *Salmonella* bacteria delivered the EGS-expressing plasmid to the livers of HBV-infected mice, resulting in a ~95% decrease of HBV gene expression and a ~200 000-fold reduction in viral DNA levels in the livers and sera of the mice.

It was shown that deletion of the tRNA anticodon arm has little effect on cleavage efficiency by human nuclear-cytoplasmic RNase P, whereas deletion of the D arm decreased efficiency to different extents depending on the specific pre-tRNA substrate [182, 183]. Studies on size-reduced substrates for frog and human nuclear-cytoplasmic RNase P enzymes further revealed that minimal RNAs consisting of acceptor stem and T arm are processed with good efficiencies if the 5′-strands of acceptor and T stems are separated by a bulge of optimally 5–9 nt in length [183, 184]. George and coworkers [185] at Innovir laboratories explored this type of substrate for application of the EGS concept (summarized in Figure 9.15a). The target RNA strand of the model substrate (Figure 9.15a) was designed based on in vitro selected small model substrates for human RNase P [183]; here [185], target RNA and EGS were physically separated in the T loop. The following features were found to be beneficial to cleavage efficiency: (i) a G:C bp as the first bp of the acceptor stem, (ii) a closing G:C or C:G bp in the T stem, (iii) a T loop-derived UUCR (U-turn) motif at the 3′-end of the target sequence, and (iv) 3′-extension of the UUCR motif by 7 A residues (or possibly by any other sequence). All functional EGS variants formed more than 50% G:C base pairs with the target RNA (7–9 out of 12), and extension of acceptor or T stem by not more than 1 bp at the junction of the two stems was permissible. A 3′-NCCA overhang was dispensable, and 5–15 nt in the bulge connecting acceptor and T stems were compatible with good cleavage efficiency.

In another study [187], the same laboratory verified that deletion of anticodon arm and variable loop in EGS:target RNA complexes are compatible with good cleavage efficiency (the same rate constant of 1 min^{-1} was measured for such bimolecular complexes and a continuous pre-tRNA substrate). Acceptor and T stem extensions were critical, which is in line with the mechanistic ruler model according to which eukaryotic nuclear-cytoplasmic RNase P enzymes measure the combined lengths of acceptor and T stems (12 bp) for selection of the cleavage site [183, 188]. D stem extensions were more permissible, but 4–5 bp were found to be optimal [187]. Single nucleotide deletions in the T loop reduced cleavage activity to different extents in a position-dependent manner. Using this type of EGS:target RNA complex (Figure 9.15b), George and coworkers tested a series of chemically modified EGS variants, aiming at an increase of their half-life in biological fluids (human serum), an important issue for therapeutic applications employing exogenous EGS delivery [186]. Stabilization of EGS oligonucleotides while retaining good cleavage efficiency was achieved by the following modification pattern: (i) uniform 2′-*O*-methyl modification of the EGS except for the T loop, (ii) addition of a 3′–3′ inverted 2′-deoxythymidine (iT) residue to the 3′-end, (iii) avoidance of ribopyrimidines in the T loop and (iv) specific modification

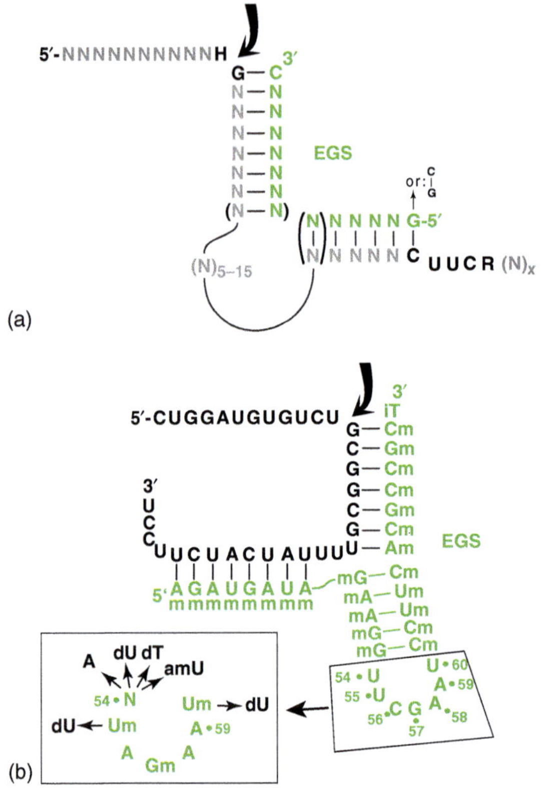

Figure 9.15 EGS:target RNA complexes for cleavage by human nuclear-cytoplasmic RNase P. (a) Minimal EGS 12- or 13-mers [185]; the acceptor or T stem could be extended by 1 G:C bp without compromising cleavage efficiency (indicated by parentheses). All active EGS bound the target RNA with at least 7 G:C base pairs in the two stems. A T loop-like UUCR (R = A or G) sequence downstream of the T stem was found to be optimal, as was the presence of additional nucleotides beyond UUCR [indicated as $(N)_x$]. (b) EGS:target RNA complexes mimicking tRNA molecules that lack an anticodon arm [186]. The target RNA is a fragment derived from the pregenomic RNA of Hepatitis B virus. The EGS was chemically modified to increase nuclease resistance without compromising its induction of target RNA cleavage by human nuclear-cytoplasmic RNase P. This goal was largely achieved by introducing 2′-*O*-methyl modifications (Nm) at all positions except for the T loop (boxed region), by adding a 3′–3′ inverted 2′-deoxythymidine (iT) residue to the EGS 3′-end, by avoiding ribopyrimidines in the T loop combined with specific T loop modifications (box at the left bottom; dN = 2′-deoxy, amU = 2′-amino-2′-deoxyuridine). Rate constants for cleavage were very similar for the parental all-RNA EGS and such heavily modified isosequential EGS variants, about 0.5 min^{-1} in the experimental setup used for reactions catalyzed by human nuclear-cytoplasmic RNase P in vitro [182]. For the experiments in [185], the specific T loop modification pattern (see box on the left) was dU_{54} Um_{55} A_{56} Gm_{57} A_{58} A_{59} Um_{60}.

patterns – primarily 2′-*O*-methyl (Nm) and 2′-deoxy substitutions (dN) – in the T loop (summarized in Figure 9.15b). The EGS design of Figure 9.15b (with dU at position 54 and Um at 55/60; 7 bp each in the hybrid D and acceptor stem), tested for knockdown of two model targets (PKC-α and bcl-xL mRNAs) in T24 bladder carcinoma cells [189], turned out to be very efficient and specific based on Western and Northern blot analyses. A control EGS with reversed T loop sequence (while preserving the individual modifications) was entirely inactive, indicating that the knockdown effect was indeed attributable to cleavage by endogenous RNase P.

9.6.4 The M1–GS Approach

In this conceptual variant, the guide sequence is tethered to the 3′-end of *E. coli* RNase P RNA, traditionally termed M1 RNA [163, 166, 167] (Figure 9.13c). The natural single-stranded 4-nt 3′-overhang of mature M1 RNA is extended by a linker of 24–50 nt [163], followed by (optimally) 13–16 nucleotides complementary to the target RNA and terminating with a dangling 3′-ACCA tail that can be extended by not more than 1–3 nucleotides without loss of efficiency [167]. The concept gained momentum owing to the surprising finding that such M1–GS conjugates are substantially more active at physiological Mg^{2+} concentrations than the *trans*-acting M1 RNA ribozyme [167]. The authors assigned this to a higher Mg^{2+} requirement for substrate binding *in trans* in the absence of the RNase P protein. The following aspects may contribute to explaining this observation.

First, docking of a hybrid helix consisting of guide sequence and target RNA to the active site of M1 RNA does not involve the S-domain that interacts with the corner of L-shaped tRNAs (Figure 9.9; [36, 190]). This may relax the constraints for the Mg^{2+}-dependent folding and conformational alignment of the RNA's C- and S-domains, which is normally required for productive interaction with the substrate. Second, the frequency of (transient) substrate-binding events to the active site of M1 RNA is expected to be increased in covalently linked M1–GS conjugates, and some of these events are productive. Still, physiological conditions are suboptimal for the catalytic performance of M1–GS constructs, considering that the tested M1–GS RNA [167] performed cleavage most efficiently at 62 °C, 100 mM monovalent salt, and 60–100 mM Mg^{2+}. Fusing the guide sequence to a bacterial RNase P RNA C-domain only maybe a conceptual variation of the approach.

Furthermore, Liu and Altman [167] observed ≥30-fold stimulation of target RNA cleavage by M1–GS conjugates in the presence of the C5 protein and ≥5-fold stimulation by partially purified preparations of human nuclear-cytoplasmic RNase P. As this enhancement was not seen with human cell fractions devoid of RNase P activity and since human RNase P is unable to cleave purely helical GS:target RNA complexes, the stimulating effect was assigned to endogenous proteins present in preparations of human nuclear-cytoplasmic RNase P [167].

Liu and coworkers applied the M1-GS approach to viral targets. They moved the concept forward by identifying more active M1–GS RNA variants through in vitro selection from RNA pools randomized in the M1 RNA moiety. Variants with G to A exchanges at positions 224 and 225 at the base of the P14 hairpin in the S-domain of M1 RNA (Figure 9.2) cleaved the GS:target RNA substrate with at least tenfold enhanced efficiency [168], which was mainly attributable to 50 times higher affinity of the mutant M1–GS RNA to the target RNA compared with the wild-type M1–GS. Extension of the hybridizing target sequence by a 3′-tail not complementary to the GS was necessary for this enhancement, and a 4-thio-U residue immediately downstream of the target sequence (at the position of the lower light blue circle in Figure 9.13c) showed enhanced UV crosslinking to M1 RNA nucleotides 224 and 225 when the latter were mutated to A residues. These results indicated a substrate contact between the P14 region of the catalytic RNA and the target RNA immediately downstream of the GS:target duplex (Figure 9.13c), which is as yet poorly understood, considering that the active site of the catalytic RNA is far away from P14 (Figure 9.6b). Irrespectively, the aforementioned ribozyme variants, expressed from retroviral constructs under the control of the U6 promoter, were able to suppress the levels of thymidine kinase mRNA and protein by 95–99% in herpes simplex virus 1-infected mammalian cells [168]. Similar inhibition effects were obtained for other viral targets [191]. The same group used an attenuated *Salmonella* strain to orally deliver antiviral M1–GS ribozymes (encoded on a plasmid with U6 promoter) to mice. The M1–GS RNA, directed against the overlapping mRNA region of two MCMV proteins, M80.5 and protease, was detectable in the spleen and liver of treated mice. Oral treatment of MCMV-infected mice with such *Salmonella*-based vectors resulted in reduced viral gene expression, decreased viral titers, and improved survival in comparison to untreated animals or mice that received *Salmonella* bacteria-harboring plasmids coding for M1–GS ribozymes with altered targeting sequence or inactive ribozyme moiety [192]. The same approach was successfully applied to suppress viral replication in differentiated human THP-1 cells (a model system for macrophages) infected with human cytomegalovirus [193].

9.6.5 Outlook

In the EGS approach, the target RNA is recognized by Watson–Crick base pairing as in all antisense applications. However, recognition by RNase P additionally requires a specific tertiary structure of the EGS:target complex, providing a higher order of specificity and thus predicting less off-target effects, as previously argued by Ma et al. [186]. Indeed, the extremely promising follow-up study by Ma et al. [189] confirmed the specificity as well as the efficiency of this approach. The EGS approach may turn out to be superior to LNA gapmer/RNase H approaches in terms of specificity, as RNase H cleaves RNA:DNA hybrids in regions as short as 4–7 bp. This predicts cleavage of off-target RNAs whose interaction with LNA gapmers involves not more than a few mismatches while preserving a stretch of 4–7 consecutive RNA:DNA base pairs [189]. There is evidence that human non-mitochondrial RNase P is located in the nucleus and cytoplasm (reviewed in [194]), offering the perspective of target RNA cleavage in both compartments. There is also a natural example of an "alienated" RNase P function in the nucleus: human nuclear RNase P is recruited to the nuclear-retained, proto-oncogenic long noncoding RNA (lncRNA) MALAT1 to remove a 3′-terminal tRNA structure from the MALAT1 transcript; this processing event then induces the formation of a triple helix that stabilizes the mature MALAT1 lncRNA against degradation, thus obviating the need for stabilization of the Pol II transcript by 3′-polyadenylation [195, 196].

In future studies, LNA and other modifications of EGS oligonucleotides need to be explored and may further increase the efficiency of this approach. Considering the success story of the 18-meric antisense oligonucleotide Nusinersen (Spinraza™) in the treatment of spinal muscular atrophy (SMA; [197]), intrathecal administration of (LNA-modified) EGS oligonucleotides may be a promising strategy to address neuronal cells. In addition, beyond nanoparticles to deliver EGS oligonucleotides, gymnotic ("naked") delivery of phosphorothioate- and LNA-modified EGS oligonucleotides could be an option to reach cellular targets [198] (reviewed in [199]). Nanoparticle-based or gymnotic EGS delivery approaches may have advantages over gene therapeutic routes utilizing retroviral vectors or attenuated *Salmonella* (see above) for the delivery of EGS expression cassettes. In conclusion, exogenous EGS delivery may well turn out to be a robust contextual alternative to other antisense strategies such as those employing siRNAs and LNA gapmers.

References

1 Hartmann, R.K., Gößringer, M., Späth, B. et al. (2009). The making of tRNAs and more – RNase P and tRNase Z. *Prog. Mol. Biol. Transl. Sci.* 85: 319–368.

2 Ma, M., Li de la Sierra-Gallay, I., Lazar, N. et al. (2017). The crystal structure of Trz1, the long form RNase Z from yeast. *Nucleic Acids Res.* 45 (10): 6209–6216.

3 Wellner, K., Betat, H., and Mörl, M. (2018). A tRNA's fate is decided at its 3′ end: collaborative actions of CCA-adding enzyme and RNases involved in tRNA processing and degradation. *Biochim. Biophys. Acta* 1861 (4): 433–441.

4 Cho, I., Lai, L.B., Susanti, D. et al. (2010). Ribosomal protein L7Ae is a subunit of archaeal RNase P. *Proc. Natl. Acad. Sci. U.S.A.* 107 (33): 14573–14578.

5 Forster, A.C. and Altman, S. (1990). Similar cage-shaped structures for the RNA components of all ribonuclease P and ribonuclease MRP enzymes. *Cell* 62 (3): 407–409.

6 Marquez, S.M., Harris, J.K., Kelley, S.T. et al. (2005). Structural implications of novel diversity in eukaryal RNase P RNA. *RNA* 11 (5): 739–751.

7 Walker, S.C. and Engelke, D.R. (2006). Ribonuclease P: the evolution of an ancient RNA enzyme. *Crit. Rev. Biochem. Mol. Biol.* 41 (2): 77–102.

8 Welting, T.J., Kikkert, B.J., van Venrooij, W.J., and Pruijn, G.J. (2006). Differential association of protein subunits with the human RNase MRP and RNase P complexes. *RNA* 12 (7): 1373–1382.

9 Holzmann, J., Frank, P., Löffler, E. et al. (2008). RNase P without RNA: identification and functional reconstitution of the human mitochondrial tRNA processing enzyme. *Cell* 135 (3): 462–474.

10 Gobert, A., Gutmann, B., Taschner, A. et al. (2010). A single *Arabidopsis* organellar protein has RNase P activity. *Nat. Struct. Mol. Biol.* 17 (6): 740–744.

11 Matelska, D., Steczkiewicz, K., and Ginalski, K. (2017). Comprehensive classification of the PIN domain-like superfamily. *Nucleic Acids Res.* 45 (12): 6995–7020.

12 Nickel, A.I., Wäber, N.B., Gößringer, M. et al. (2017). Minimal and RNA-free RNase P in *Aquifex aeolicus*. *Proc. Natl. Acad. Sci. U.S.A.* 114 (42): 11121–11126.

13 Willkomm, D.K. and Hartmann, R.K. (2007). An important piece of the RNase P jigsaw solved. *Trends Biochem. Sci.* 32 (6): 247–250.

14 Kikovska, E., Svärd, S.G., and Kirsebom, L.A. (2007). Eukaryotic RNase P RNA mediates cleavage in the absence of protein. *Proc. Natl. Acad. Sci. U.S.A.* 104 (7): 2062–2067.

15 Lechner, M., Rossmanith, W., Hartmann, R.K. et al. (2015). Distribution of ribonucleoprotein and protein-only RNase P in eukarya. *Mol. Biol. Evol.* 32 (12): 3186–3193.

16 Guerrier-Takada, C., Gardiner, K., Marsh, T. et al. (1983). The RNA moiety of Ribonuclease P is the catalytic subunit of the enzyme. *Cell* 35 (3): 849–857.

17 Siegel, R.W., Banta, A.B., Haas, E.S. et al. (1996). Mycoplasma fermentans simplifies our view of the catalytic core of ribonuclease P RNA. *RNA* 2 (5): 452–462.

18 Haas, E.S., Brown, J.W., Pitulle, C., and Pace, N.R. (1994). Further perspective on the catalytic core and secondary structure of ribonuclease P RNA. *Proc. Natl. Acad. Sci. U.S.A.* 91 (7): 2527–2531.

19 Darr, S.C., Zito, K., Smith, D., and Pace, N.R. (1992). Contributions of phylogenetically variable structural elements to the function of the ribozyme ribonuclease P. *Biochemistry* 31 (2): 328–333.

20 Waugh, D.S., Green, C.J., and Pace, N.R. (1989). The design and catalytic properties of a simplified ribonuclease P RNA. *Science* 244 (4912): 1569–1571.

21 Schlegl, J., Hardt, W.D., Erdmann, V.A., and Hartmann, R.K. (1994). Contribution of structural elements to *Thermus thermophilus* ribonuclease P RNA function. *EMBO J.* 13 (20): 4863–4869.

22 Haas, E.S., Banta, A.B., Harris, J.K. et al. (1996). Structure and evolution of ribonuclease P RNA in Gram-positive bacteria. *Nucleic Acids Res.* 24 (23): 4775–4782.

23 Haas, E.S. and Brown, J.W. (1998). Evolutionary variation in bacterial RNase P RNAs. *Nucleic Acids Res.* 26 (18): 4093–4099.

24 Brown, J.W. (1999). The ribonuclease P database. *Nucleic Acids Res.* 27 (1): 314.

25 Massire, C., Jaeger, L., and Westhof, E. (1998). Derivation of the three-dimensional architecture of bacterial ribonuclease P RNAs from comparative sequence analysis. *J. Mol. Biol.* 279 (4): 773–793.

26 Tsai, H.Y., Masquida, B., Biswas, R. et al. (2003). Molecular modeling of the three-dimensional structure of the bacterial RNase P holoenzyme. *J. Mol. Biol.* 325 (4): 661–675.

27 Reiter, N.J., Osterman, A., Torres-Larios, A. et al. (2010). Structure of a bacterial ribonuclease P holoenzyme in complex with tRNA. *Nature* 468 (7325): 784–789.

28 Krasilnikov, A.S., Yang, X., Pan, T., and Mondragón, A. (2003). Crystal structure of the specificity domain of ribonuclease P. *Nature* 421 (6924): 760–764.

29 Krasilnikov, A.S., Xiao, Y., Pan, T., and Mondragón, A. (2004). Basis for structural diversity in homologous RNAs. *Science* 306 (5693): 104–107.

30 Kazantsev, A.V., Krivenko, A.A., Harrington, D.J. et al. (2005). Crystal structure of a bacterial ribonuclease P RNA. *Proc. Natl. Acad. Sci. U.S.A.* 102 (38): 13392–13397.

31 Stamatopoulou, V., Toumpeki, C., Vourekas, A. et al. (2014). On the role of the appended P19 element in type A RNAs of bacterial RNase P. *Biochemistry* 53 (11): 1810–1817.

32 Pan, T. (1995). Higher order folding and domain analysis of the ribozyme from *Bacillus subtilis* ribonuclease P. *Biochemistry* 34 (3): 902–909.

33 Loria, A. and Pan, T. (1996). Domain structure of the ribozyme from eubacterial ribonuclease P. *RNA* 2 (6): 551–563.

34 Green, C.J., Rivera-León, R., and Vold, B.S. (1996). The catalytic core of RNase P. *Nucleic Acids Res.* 24 (8): 1497–1503.

35 Loria, A. and Pan, T. (2001). Modular construction for function of a ribonucleoprotein enzyme: the catalytic domain of *Bacillus subtilis* RNase P complexed with *B. subtilis* RNase P protein. *Nucleic Acids Res.* 29 (9): 1892–1897.

36 Wu, S., Kikovska, E., Lindell, M., and Kirsebom, L.A. (2012). Cleavage mediated by the catalytic domain of bacterial RNase P RNA. *J. Mol. Biol.* 422 (2): 204–214.

37 Heide, C., Busch, S., Feltens, R., and Hartmann, R.K. (2001). Distinct modes of mature and precursor tRNA binding to *Escherichia coli* RNase P RNA revealed by NAIM analyses. *RNA* 7 (4): 553–564.

38 Harris, J.K., Haas, E.S., Williams, D. et al. (2001). New insight into RNase P RNA structure from comparative analysis of the archaeal RNA. *RNA* 7 (2): 220–232.

39 Li, D., Willkomm, D.K., and Hartmann, R.K. (2009). Minor changes largely restore catalytic activity of archaeal RNase P RNA from *Methanothermobacter thermoautotrophicus. Nucleic Acids Res.* 37 (1): 231–242.

40 Sinapah, S., Wu, S., Chen, Y. et al. (2011). Cleavage of model substrates by archaeal RNase P: role of protein cofactors in cleavage-site selection. *Nucleic Acids Res.* 39 (3): 1105–1116.

41 Wan, F., Wang, Q., Tan, J. et al. (2019). Cryo-electron microscopy structure of an archaeal ribonuclease P holoenzyme. *Nat. Commun.* 10 (1): 1–13.

42 Wu, J., Niu, S., Tan, M., Huang, C. et al. (2018). Cryo-EM Structure of the Human Ribonuclease P Holoenzyme. *Cell* 175 (5): 1393–1404.e11.

43 Lan, P., Tan, M., Zhang, Y., Niu, S. et al. (2018). Structural insight into precursor tRNA processing by yeast ribonuclease P. *Science* 362 (6415).

44 Perederina, A., Li, D., Lee, H. et al. (2020). Cryo-EM structure of catalytic ribonucleoprotein complex RNase MRP. *Nat. Commun.* 11 (1): 1–10.

45 Hartmann, E. and Hartmann, R.K. (2003). The enigma of ribonuclease P evolution. *Trends Genet.* 19 (10): 561–569.

46 Stams, T., Niranjanakumari, S., Fierke, C.A., and Christianson, D.W. (1998). Ribonuclease P protein structure: evolutionary origins in the translational apparatus. *Science* 280 (5364): 752–755.

47 Spitzfaden, C., Nicholson, N., Jones, J.J. et al. (2000). The structure of ribonuclease P protein from *Staphylococcus aureus* reveals a unique binding site for single-stranded RNA. *J. Mol. Biol.* 295 (1): 105–115.

48 Kazantsev, A.V., Krivenko, A.A., Harrington, D.J. et al. (2003). High-resolution structure of RNase P protein from *Thermotoga maritima*. *Proc. Natl. Acad. Sci. U.S.A.* 100 (13): 7497–7502.

49 Waugh, D.S. and Pace, N.R. (1990). Complementation of an RNase P RNA (*rnpB*) gene deletion in *Escherichia coli* by homologous genes from distantly related eubacteria. *J. Bacteriol.* 172 (11): 6316–6322.

50 Gößringer, M. and Hartmann, R.K. (2007). Function of heterologous and truncated RNase P proteins in *Bacillus subtilis*. *Mol. Microbiol.* 66 (3): 801–813.

51 Torres-Larios, A., Swinger, K.K., Krasilnikov, A.S. et al. (2005). Crystal structure of the RNA component of bacterial ribonuclease P. *Nature* 437 (7058): 584–587.

52 Warnecke, J.M., Fürste, J.P., Hardt, W.D. et al. (1996). Ribonuclease P (RNase P) RNA is converted to a Cd(2+)-ribozyme by a single Rp-phosphorothioate modification in the precursor tRNA at the RNase P cleavage site. *Proc. Natl. Acad. Sci. U.S.A.* 93 (17): 8924–8928.

53 Warnecke, J.M., Held, R., Busch, S., and Hartmann, R.K. (1999). Role of metal ions in the hydrolysis reaction catalyzed by RNase P RNA from *Bacillus subtilis*. *J. Mol. Biol.* 290 (2): 433–445.

54 Persson, T., Cuzic, S., and Hartmann, R.K. (2003). Catalysis by RNase P RNA: unique features and unprecedented active site plasticity. *J. Biol. Chem.* 278 (44): 43394–43401.

55 Cassano, A.G., Anderson, V.E., and Harris, M.E. (2004). Analysis of solvent nucleophile isotope effects: evidence for concerted mechanisms and nucleophilic activation by metal coordination in nonenzymatic and ribozyme-catalyzed phosphodiester hydrolysis. *Biochemistry* 43 (32): 10547–10559.

56 Cuzic-Feltens, S., Weber, M.H., and Hartmann, R.K. (2009). Investigation of catalysis by bacterial RNase P via LNA and other modifications at the scissile phosphodiester. *Nucleic Acids Res.* 37 (22): 7638–7653.

57 Zuleeg, T., Hartmann, R.K., Kreutzer, R., and Limmer, S. (2001). NMR spectroscopic evidence for Mn^{2+}(Mg^{2+}) binding to a precursor-tRNA microhelix near the potential RNase P cleavage site. *J. Mol. Biol.* 305 (2): 181–189.

58 Perreault, J.P. and Altman, S. (1992). Important 2′-hydroxyl groups in model substrates for M1 RNA, the catalytic RNA subunit of RNase P from *Escherichia coli*. *J. Mol. Biol.* 226 (2): 399–409.

59 Brännvall, M., Kikovska, E., and Kirsebom, L.A. (2004). Cross talk between the +73/294 interaction and the cleavage site in RNase P RNA mediated cleavage. *Nucleic Acids Res.* 32 (18): 5418–5429.

60 Zahler, N.H., Sun, L., Christian, E.L., and Harris, M.E. (2005). The pre-tRNA nucleotide base and 2′-hydroxyl at N(−1) contribute to fidelity in tRNA processing by RNase P. *J. Mol. Biol.* 345 (5): 969–985.

61 Brown, J.W., Nolan, J.M., Haas, E.S. et al. (1996). Comparative analysis of ribonuclease P RNA using gene sequences from natural microbial populations reveals tertiary structural elements. *Proc. Natl. Acad. Sci. U.S.A.* 93 (7): 3001–3006.

62 Massire, C., Jaeger, L., and Westhof, E. (1997). Phylogenetic evidence for a new tertiary interaction in bacterial RNase P RNAs. *RNA* 3 (6): 553–556.

63 Walczyk, D., Willkomm, D.K., and Hartmann, R.K. (2016). Bacterial type B RNase P: functional characterization of the L5.1-L15.1 tertiary contact and antisense inhibition. *RNA* 22 (11): 1699–1709.

64 Marszalkowski, M., Willkomm, D.K., and Hartmann, R.K. (2008). Structural basis of a ribozyme's thermostability: P1-L9 interdomain interaction in RNase P RNA. *RNA* 14 (1): 127–133.

65 Marszalkowski, M., Teune, J.H., Steger, G. et al. (2006). Thermostable RNase P RNAs lacking P18 identified in the Aquificales. *RNA* 12 (11): 1915–1921.

66 Harris, M.E. and Pace, N.R. (1995). Identification of phosphates involved in catalysis by the ribozyme RNase P RNA. *RNA* 1 (2): 210–218.

67 Hardt, W.D., Warnecke, J.M., Erdmann, V.A., and Hartmann, R.K. (1995). Rp-phosphorothioate modifications in RNase P RNA that interfere with tRNA binding. *EMBO J.* 14 (12): 2935–2944.

68 Christian, E.L., Kaye, N.M., and Harris, M.E. (2000). Helix P4 is a divalent metal ion binding site in the conserved core of the ribonuclease P ribozyme. *RNA* 6 (4): 511–519.

69 Krasilnikov, A.S. and Mondragón, A. (2003). On the occurrence of the T-loop RNA folding motif in large RNA molecules. *RNA* 9 (6): 640–643.

70 Buck, A.H., Dalby, A.B., Poole, A.W. et al. (2005). Protein activation of a ribozyme: the role of bacterial RNase P protein. *EMBO J.* 24 (19): 3360–3368.

71 Buck, A.H., Kazantsev, A.V., Dalby, A.B., and Pace, N.R. (2005). Structural perspective on the activation of RNase P RNA by protein. *Nat. Struct. Mol. Biol.* 12 (11): 958–964.

72 Guo, X., Campbell, F.E., Sun, L. et al. (2006). RNA-dependent folding and stabilization of C5 protein during assembly of the *E. coli* RNase P holoenzyme. *J. Mol. Biol.* 360 (1): 190–203.

73 Hernandez-Cid, A., Aguirre-Sampieri, S., Diaz-Vilchis, A., and Torres-Larios, A. (2012). Ribonucleases P/MRP and the expanding ribonucleoprotein world. *IUBMB Life* 64 (6): 521–528.

74 Pan, T., Loria, A., and Zhong, K. (1995). Probing of tertiary interactions in RNA: 2′-hydroxyl-base contacts between the RNase P RNA and pre-tRNA. *Proc. Natl. Acad. Sci. U.S.A.* 92 (26): 12510–12514.

75 Kirsebom, L.A. and Svärd, S.G. (1994). Base pairing between *Escherichia coli* RNase P RNA and its substrate. *EMBO J.* 13 (20): 4870–4876.

76 LaGrandeur, T.E., Hüttenhofer, A., Noller, H.F., and Pace, N.R. (1994). Phylogenetic comparative chemical footprint analysis of the interaction between ribonuclease P RNA and tRNA. *EMBO J.* 13 (17): 3945–3952.

77 Loria, A. and Pan, T. (1997). Recognition of the T stem-loop of a pre-tRNA substrate by the ribozyme from *Bacillus subtilis* ribonuclease P. *Biochemistry* 36 (21): 6317–6325.

78 Kufel, J. and Kirsebom, L.A. (1996). Different cleavage sites are aligned differently in the active site of M1 RNA, the catalytic subunit of *Escherichia coli* RNase P. *Proc. Natl. Acad. Sci. U.S.A.* 93 (12): 6085–6090.

79 Busch, S., Kirsebom, L.A., Notbohm, H., and Hartmann, R.K. (2000). Differential role of the intermolecular base-pairs G292-C(75) and G293-C(74) in the reaction catalyzed by *Escherichia coli* RNase P RNA. *J. Mol. Biol.* 299 (4): 941–951.

80 Wegscheid, B. and Hartmann, R.K. (2006). The precursor tRNA 3′-CCA interaction with *Escherichia coli* RNase P RNA is essential for catalysis by RNase P in vivo. *RNA* 12 (12): 2135–2148.

81 Hardt, W.D., Schlegel, J., Erdmann, V.A., and Hartmann, R.K. (1995). Kinetics and thermodynamics of the RNase P RNA cleavage reaction: analysis of tRNA 3′-end variants. *J. Mol. Biol.* 247 (2): 161–172.

82 Svärd, S.G., Kagardt, U., and Kirsebom, L.A. (1996). Phylogenetic comparative mutational analysis of the base-pairing between RNase P RNA and its substrate. *RNA* 2 (5): 463–472.

83 Wegscheid, B. and Hartmann, R.K. (2007). *In vivo* and *in vitro* investigation of bacterial type B RNase P interaction with tRNA 3′-CCA. *Nucleic Acids Res.* 35 (6): 2060–2073.

84 Heide, C., Feltens, R., and Hartmann, R.K. (2001). Purine N7 groups that are crucial to the interaction of *Escherichia coli* RNase P RNA with tRNA. *RNA* 7 (7): 958–968.

85 Kirsebom, L.A. (2007). RNase P RNA mediated cleavage: substrate recognition and catalysis. *Biochimie* 89 (10): 1183–1194.

86 Crary, S.M., Niranjanakumari, S., and Fierke, C.A. (1998). The protein component of *Bacillus subtilis* ribonuclease P increases catalytic efficiency by enhancing interactions with the 5′ leader sequence of pre-tRNAAsp. *Biochemistry* 37 (26): 9409–9416.

87 Kurz, J.C., Niranjanakumari, S., and Fierke, C.A. (1998). Protein component of *Bacillus subtilis* RNase P specifically enhances the affinity for precursor-tRNAAsp. *Biochemistry* 37 (8): 2393–2400.

88 Niranjanakumari, S., Stams, T., Crary, S.M. et al. (1998). Protein component of the ribozyme ribonuclease P alters substrate recognition by directly contacting precursor tRNA. *Proc. Natl. Acad. Sci. U.S.A.* 95 (26): 15212–15217.

89 Reich, C., Olsen, G.J., Pace, B., and Pace, N.R. (1988). Role of the protein moiety of ribonuclease P, a ribonucleoprotein enzyme. *Science* 239 (4836): 178–181.

90 Tallsjö, A. and Kirsebom, L.A. (1993). Product release is a rate-limiting step during cleavage by the catalytic RNA subunit of *Escherichia coli* RNase P. *Nucleic Acids Res.* 21 (1): 51–57.

91 Goodsell, D.S. (1991). Inside a living cell. *Trends Biochem. Sci.* 16 (6): 203–206.

92 Dong, H., Nilsson, L., and Kurland, C.G. (1996). Co-variation of tRNA abundance and codon usage in *Escherichia coli* at different growth rates. *J. Mol. Biol.* 260 (5): 649–663.

93 Dong, H., Kirsebom, L.A., and Nilsson, L. (1996). Growth rate regulation of 4.5 S RNA and M1 RNA the catalytic subunit of *Escherichia coli* RNase P. *J. Mol. Biol.* 261 (3): 303–308.

94 Kurz, J.C. and Fierke, C.A. (2002). The affinity of magnesium binding sites in the *Bacillus subtilis* RNase P x pre-tRNA complex is enhanced by the protein subunit. *Biochemistry* 41 (30): 9545–9558.

95 Sun, L. and Harris, M.E. (2007). Evidence that binding of C5 protein to P RNA enhances ribozyme catalysis by influencing active site metal ion affinity. *RNA* 13 (9): 1505–1515.

96 Sun, L., Campbell, F.E., Zahler, N.H., and Harris, M.E. (2006). Evidence that substrate-specific effects of C5 protein lead to uniformity in binding and catalysis by RNase P. *EMBO J.* 25 (17): 3998–4007.

97 Fang, X.W., Yang, X.J., Littrell, K. et al. (2001). The *Bacillus subtilis* RNase P holoenzyme contains two RNase P RNA and two RNase P protein subunits. *RNA* 7 (2): 233–241.

98 Rueda, D., Hsieh, J., Day-Storms, J.J. et al. (2005). The 5′ leader of precursor tRNAAsp bound to the *Bacillus subtilis* RNase P holoenzyme has an extended conformation. *Biochemistry* 44 (49): 16130–16139.

99 Koutmou, K.S., Zahler, N.H., Kurz, J.C. et al. (2010). Protein-precursor tRNA contact leads to sequence-specific recognition of 5′ leaders by bacterial ribonuclease P. *J. Mol. Biol.* 396 (1): 195–208.

100 Niland, C.N., Anderson, D.R., Jankowsky, E., and Harris, M.E. (2017). The contribution of the C5 protein subunit of *Escherichia coli* ribonuclease P to specificity for precursor tRNA is modulated by proximal 5′ leader sequences. *RNA* 23 (10): 1502–1511.

101 Niland, C.N., Zhao, J., Lin, H.C. et al. (2016). Determination of the specificity landscape for ribonuclease P processing of precursor tRNA 5′ leader sequences. *ACS Chem. Biol.* 11 (8): 2285–2292.

102 Reiter, N.J., Osterman, A.K., and Mondragón, A. (2012). The bacterial ribonuclease P holoenzyme requires specific, conserved residues for efficient catalysis and substrate positioning. *Nucleic Acids Res.* 40 (20): 10384–10393.

103 Guenther, U.P., Yandek, L.E., Niland, C.N. et al. (2013). Hidden specificity in an apparently nonspecific RNA-binding protein. *Nature* 502 (7471): 385–388.

104 Guerrier-Takada, C., Haydock, K., Allen, L., and Altman, S. (1986). Metal ion requirements and other aspects of the reaction catalyzed by M1 RNA, the RNA subunit of ribonuclease P from *Escherichia coli*. *Biochemistry* 25 (7): 1509–1515.

105 Smith, D., Burgin, A.B., Haas, E.S., and Pace, N.R. (1992). Influence of metal ions on the ribonuclease P reaction. Distinguishing substrate binding from catalysis. *J. Biol. Chem.* 267 (4): 2429–2436.

106 Brännvall, M. and Kirsebom, L.A. (2001). Metal ion cooperativity in ribozyme cleavage of RNA. *Proc. Natl. Acad. Sci. U.S.A.* 98 (23): 12943–12947.

107 Hardt, W.D., Schlegl, J., Erdmann, V.A., and Hartmann, R.K. (1993). Gel retardation analysis of *E. coli* M1 RNA-tRNA complexes. *Nucleic Acids Res.* 21 (15): 3521–3527.

108 Brännvall, M., Mikkelsen, N.E., and Kirsebom, L.A. (2001). Monitoring the structure of *Escherichia coli* RNase P RNA in the presence of various divalent metal ions. *Nucleic Acids Res.* 29 (7): 1426–1432.

109 Cuzic, S. and Hartmann, R.K. (2005). Studies on *Escherichia coli* RNase P RNA with Zn^{2+} as the catalytic cofactor. *Nucleic Acids Res.* 33 (8): 2464–2474.

110 Kazakov, S. and Altman, S. (1991). Site-specific cleavage by metal ion cofactors and inhibitors of M1 RNA, the catalytic subunit of RNase P from *Escherichia coli*. *Proc. Natl. Acad. Sci. U.S.A.* 88 (20): 9193–9197.

111 Hsieh, J. and Fierke, C.A. (2009). Conformational change in the *Bacillus subtilis* RNase P holoenzyme–pre-tRNA complex enhances substrate affinity and limits cleavage rate. *RNA* 15 (8): 1565–1577.

112 Warnecke, J.M., Green, C.J., and Hartmann, R.K. (1997). Role of metal ions in the cleavage mechanism by the *E. coli* RNase P holoenzyme. *Nucleosides Nucleotides Nucleic Acids* 16 (5-6): 721–725.

113 Steitz, T.A. and Steitz, J.A. (1993). A general two-metal-ion mechanism for catalytic RNA. *Proc. Natl. Acad. Sci. U.S.A.* 90 (14): 6498–6502.

114 Chen, Y., Li, X., and Gegenheimer, P. (1997). Ribonuclease P catalysis requires Mg^{2+} coordinated to the *pro-*R_P oxygen of the scissile bond. *Biochemistry* 36 (9): 2425–2438.

115 Warnecke, J.M., Sontheimer, E.J., Piccirilli, J.A., and Hartmann, R.K. (2000). Active site constraints in the hydrolysis reaction catalyzed by bacterial RNase P: analysis of precursor tRNAs with a single 3′-S-phosphorothiolate internucleotide linkage. *Nucleic Acids Res.* 28 (3): 720–727.

116 Smith, D. and Pace, N.R. (1993). Multiple magnesium ions in the ribonuclease P reaction mechanism. *Biochemistry* 32 (20): 5273–5281.

117 Loria, A. and Pan, T. (1999). The cleavage step of ribonuclease P catalysis is determined by ribozyme–substrate interactions both distal and proximal to the cleavage site. *Biochemistry* 38 (27): 8612–8620.

118 Kazantsev, A.V., Krivenko, A.A., and Pace, N.R. (2009). Mapping metal-binding sites in the catalytic domain of bacterial RNase P RNA. *RNA* 15 (2): 266–276.

119 Kazantsev, A.V., Rambo, R.P., Karimpour, S. et al. (2011). Solution structure of RNase P RNA. *RNA* 17 (6): 1159–1171.

120 Crary, S.M., Kurz, J.C., and Fierke, C.A. (2002). Specific phosphorothioate substitutions probe the active site of *Bacillus subtilis* ribonuclease P. *RNA* 8 (7): 933–947.

121 Kaye, N.M., Zahler, N.H., Christian, E.L., and Harris, M.E. (2002). Conservation of helical structure contributes to functional metal ion interactions in the catalytic domain of ribonuclease P RNA. *J. Mol. Biol.* 324 (3): 429–442.

122 Rasmussen, T.A. and Nolan, J.M. (2002). G350 of *Escherichia coli* RNase P RNA contributes to Mg^{2+} binding near the active site of the enzyme. *Gene* 294 (1-2): 177–185.

123 Oh, B.K., Frank, D.N., and Pace, N.R. (1998). Participation of the 3′-CCA of tRNA in the binding of catalytic Mg^{2+} ions by ribonuclease P. *Biochemistry* 37 (20): 7277–7283.

124 Ciesiolka, J., Hardt, W.D., Schlegl, J. et al. (1994). Lead-ion-induced cleavage of RNase P RNA. *Eur. J. Biochem.* 219 (1-2): 49–56.

125 Heide, C., Pfeiffer, T., Nolan, J.M., and Hartmann, R.K. (1999). Guanosine 2-NH_2 groups of *Escherichia coli* RNase P RNA involved in intramolecular tertiary contacts and direct interactions with tRNA. *RNA* 5 (1): 102–116.

126 Christian, E.L., McPheeters, D.S., and Harris, M.E. (1998). Identification of individual nucleotides in the bacterial ribonuclease P ribozyme adjacent to the pre-tRNA cleavage site by short-range photo-cross-linking. *Biochemistry* 37 (50): 17618–17328.

127 Zahler, N.H., Christian, E.L., and Harris, M.E. (2003). Recognition of the 5′ leader of pre-tRNA substrates by the active site of ribonuclease P. *RNA* 9 (6): 734–745.

128 Wu, S., Chen, Y., Mao, G. et al. (2014). Transition-state stabilization in *Escherichia coli* ribonuclease P RNA-mediated cleavage of model substrates. *Nucleic Acids Res.* 42 (1): 631–642.

129 Leontis, N.B., Stombaugh, J., and Westhof, E. (2002). The non-Watson-Crick base pairs and their associated isostericity matrices. *Nucleic Acids Res.* 30 (16): 3497–3531.

130 Kirsebom, L.A. and Svärd, S.G. (1992). The kinetics and specificity of cleavage by RNase P is mainly dependent on the structure of the amino acid acceptor stem. *Nucleic Acids Res.* 20 (3): 425–432.

131 Brännvall, M., Pettersson, B.M.F., and Kirsebom, L.A. (2003). Importance of the +73/294 interaction in *Escherichia coli* RNase P RNA substrate complexes for cleavage and metal ion coordination. *J. Mol. Biol.* 325 (4): 697–709.

132 Hansen, A., Pfeiffer, T., Zuleeg, T. et al. (2001). Exploring the minimal substrate requirements for trans-cleavage by RNase P holoenzymes from *Escherichia coli* and *Bacillus subtilis*. *Mol. Microbiol.* 41 (1): 131–143.

133 Siew, D., Zahler, N.H., Cassano, A.G. et al. (1999). Identification of adenosine functional groups involved in substrate binding by the ribonuclease P ribozyme. *Biochemistry* 38 (6): 1873–1883.

134 Kaye, N.M., Christian, E.L., and Harris, M.E. (2002). NAIM and site-specific functional group modification analysis of RNase P RNA: magnesium dependent structure within the conserved P1-P4 multihelix junction contributes to catalysis. *Biochemistry* 41 (14): 4533–4545.

135 Zuleeg, T., Hansen, A., Pfeiffer, T. et al. (2001). Correlation between processing efficiency for ribonuclease P minimal substrate and conformation of the nucleotide -1 at the cleavage position. *Biochemistry* 40 (11): 3363–3369.

136 Kikovska, E., Wu, S., Mao, G., and Kirsebom, L.A. (2012). Cleavage mediated by the P15 domain of bacterial RNase P RNA. *Nucleic Acids Res.* 40 (5): 2224–2233.

137 Alatossava, T., Jutte, H., Kuhn, A., and Kellenberger, E. (1985). Manipulation of intracellular magnesium content in polymyxin B nonapeptide-sensitized *Escherichia coli* by ionophore A23187. *J. Bacteriol.* 162 (1): 413–419.

138 Kobayashi, K., Ehrlich, S.D., Albertini, A. et al. (2003). Essential *Bacillus subtilis* genes. *Proc. Natl. Acad. Sci. U.S.A.* 100 (8): 4678–4683.

139 Gößringer, M., Kretschmer-Kazemi Far, R., and Hartmann, R.K. (2006). Analysis of RNase P protein (*rnpA*) expression in *Bacillus subtilis* utilizing strains with suppressible *rnpA* expression. *J. Bacteriol.* 188 (19): 6816–6823.

140 Wegscheid, B., Condon, C., and Hartmann, R.K. (2006). Type A and B RNase P RNAs are interchangeable in vivo despite substantial biophysical differences. *EMBO Rep.* 7 (4): 411–417.

141 Willkomm, D.K., Pfeffer, P., Reuter, K. et al. (2010). RNase P as a Drug Target. In: *Ribonuclease P* (eds. F. Liu and S. Altman), 235–256. New York: Springer.

142 Mondragón, A. (2013). Structural studies of RNase P. *Annu. Rev. Biophys.* 42: 537–557.

143 Brillante, N., Gößringer, M., Lindenhofer, D. et al. (2016). Substrate recognition and cleavage-site selection by a single-subunit protein-only RNase P. *Nucleic Acids Res.* 44 (5): 2323–2336.

144 Klemm, B.P., Wu, N., Chen, Y. et al. (2016). The diversity of ribonuclease P: protein and RNA catalysts with analogous biological functions. *Biomolecules* 6 (2): 27.

145 Childs, J.L., Poole, A.W., and Turner, D.H. (2003). Inhibition of *Escherichia coli* RNase P by oligonucleotide directed misfolding of RNA. *RNA* 9 (12): 1437–1445.

146 Willkomm, D.K., Gruegelsiepe, H., Goudinakis, O. et al. (2003). Evaluation of bacterial RNase P RNA as a drug target. *ChemBioChem* 4 (10): 1041–1048.

147 Gruegelsiepe, H., Willkomm, D.K., Goudinakis, O., and Hartmann, R.K. (2003). Antisense inhibition of *Escherichia coli* RNase P RNA: mechanistic aspects. *ChemBioChem* 4 (10): 1049–1056.

148 Gruegelsiepe, H., Brandt, O., and Hartmann, R.K. (2006). Antisense inhibition of RNase P: mechanistic aspects and application to live bacteria. *J. Biol. Chem.* 281 (41): 30613–30620.

149 Meng, J., Da, F., Ma, X. et al. (2015). Antisense growth inhibition of methicillin-resistant *Staphylococcus aureus* by locked nucleic acid conjugated with cell-penetrating peptide as a novel FtsZ inhibitor. *Antimicrob. Agents Chemother.* 59 (2): 914–922.

150 Silva, J. and Carvalho, I. (2007). New insights into aminoglycoside antibiotics and derivatives. *CMC* 14 (10): 1101–1119.

151 Eidem, T.M., Lounsbury, N., Emery, J.F. et al. (2015). Small-molecule inhibitors of *Staphylococcus aureus* RnpA-mediated RNA turnover and tRNA processing. *Antimicrob. Agents Chemother.* 59 (4): 2016–2028.

152 Mikkelsen, N.E., Brännvall, M., Virtanen, A., and Kirsebom, L.A. (1999). Inhibition of RNase P RNA cleavage by aminoglycosides. *Proc. Natl. Acad. Sci. U.S.A.* 96 (11): 6155–6160.

153 Eubank, T.D., Biswas, R., Jovanovic, M. et al. (2002). Inhibition of bacterial RNase P by aminoglycoside-arginine conjugates. *FEBS Lett.* 511 (1-3): 107–112.

154 Kawamoto, S.A., Sudhahar, C.G., Hatfield, C.L. et al. (2008). Studies on the mechanism of inhibition of bacterial ribonuclease P by aminoglycoside derivatives. *Nucleic Acids Res.* 36 (2): 697–704.

155 Liu, X., Chen, Y., and Fierke, C.A. (2014). A real-time fluorescence polarization activity assay to screen for inhibitors of bacterial ribonuclease P. *Nucleic Acids Res.* 42 (20): e159.

156 Mayer, M. and James, T.L. (2004). NMR-based characterization of phenothiazines as a RNA binding scaffold. *J. Am. Chem. Soc.* 126 (13): 4453–4460.

157 Wu, S., Mao, G., and Kirsebom, L.A. (2016). Inhibition of bacterial RNase P RNA by phenothiazine derivatives. *Biomolecules* 6 (3): E38.

158 Lounsbury, N., Eidem, T., Colquhoun, J. et al. (2018). Novel inhibitors of *Staphylococcus aureus* RnpA that synergize with mupirocin. *Bioorg. Med. Chem. Lett.* 28 (6): 1127–1131.

159 McGovern, S.L., Caselli, E., Grigorieff, N., and Shoichet, B.K. (2002). A common mechanism underlying promiscuous inhibitors from virtual and high-throughput screening. *J. Med. Chem.* 45 (8): 1712–1722.

160 Viviani, L.G., Piccirillo, E., Cheffer, A. et al. (2018). Be aware of aggregators in the search for potential human ecto-5′-nucleotidase inhibitors. *Molecules* 23 (8): E1876.

161 Forster, A.C. and Altman, S. (1990). External guide sequences for an RNA enzyme. *Science* 249 (4970): 783–786.

162 Yuan, Y., Hwang, E.S., and Altman, S. (1992). Targeted cleavage of mRNA by human RNase P. *Proc. Natl. Acad. Sci. U.S.A.* 89 (17): 8006–8010.

163 Liu, F. and Altmann, S. (1996). Requirements for cleavage by a modified RNase P of a small model substrate. *Nucleic Acids Res.* 24 (14): 2690–2696.

164 Guerrier-Takada, C., Salavati, R., and Altman, S. (1997). Phenotypic conversion of drug-resistant bacteria to drug sensitivity. *Proc. Natl. Acad. Sci. U.S.A.* 94 (16): 8468–8472.

165 McKinney, J., Guerrier-Takada, C., Wesolowski, D., and Altman, S. (2001). Inhibition of *Escherichia coli* viability by external guide sequences complementary to two essential genes. *Proc. Natl. Acad. Sci. U.S.A.* 98 (12): 6605–6610.

166 Guerrier-Takada, C., Li, Y., and Altman, S. (1995). Artificial regulation of gene expression in *Escherichia coli* by RNase P. *Proc. Natl. Acad. Sci. U.S.A.* 92 (24): 11115–11119.

167 Liu, F. and Altman, S. (1995). Inhibition of viral gene expression by the catalytic RNA subunit of RNase P from *Escherichia coli*. *Genes Dev.* 9 (4): 471–480.

168 Kilani, A.F., Trang, P., Jo, S. et al. (2000). RNase P ribozymes selected in vitro to cleave a viral mRNA effectively inhibit its expression in cell culture. *J. Biol. Chem.* 275 (14): 10611–10622.

169 Gopalan, V., Vioque, A., and Altman, S. (2002). RNase P: variations and uses. *J. Biol. Chem.* 277 (9): 6759–6762.

170 Grünweller, A. and Hartmann, R.K. (2007). Locked nucleic acid oligonucleotides: the next generation of antisense agents? *BioDrugs* 21 (4): 235–243.

171 McClain, W.H., Guerrier-Takada, C., and Altman, S. (1987). Model substrates for an RNA enzyme. *Science* 238 (4826): 527–530.

172 Schlegl, J., Fürste, J.P., Bald, R. et al. (1992). Cleavage efficiencies of model substrates for ribonuclease P from *Escherichia coli* and *Thermus thermophilus. Nucleic Acids Res.* 20 (22): 5963–5970.

173 Soler Bistué, A.J., Ha, H., Sarno, R. et al. (2007). External guide sequences targeting the aac(6′)-Ib mRNA induce inhibition of amikacin resistance. *Antimicrob. Agents Chemother.* 51 (6): 1918–1925.

174 Soler Bistué, A.J., Martín, F.A., Vozza, N. et al. (2009). Inhibition of aac(6′)-Ib-mediated amikacin resistance by nuclease-resistant external guide sequences in bacteria. *Proc. Natl. Acad. Sci. U.S.A.* 106 (32): 13230–13235.

175 Jackson, A., Jani, S., Sala, C.D. et al. (2016). Assessment of configurations and chemistries of bridged nucleic acids-containing oligomers as external guide sequences: a methodology for inhibition of expression of antibiotic resistance genes. *Biol. Methods Protoc.* 1 (1): bpw001.

176 Yuan, Y. and Altman, S. (1994). Selection of guide sequences that direct efficient cleavage of mRNA by human ribonuclease P. *Science* 263 (5151): 1269–1273.

177 Helfer, A.C., Romilly, C., Chevalier, C. et al. (2014). Probing RNA structure in vitro with enzymes and chemicals. In: *Handbook of RNA Biochemistry*, 2e, vol. 1 (eds. R.K. Hartmann, A. Bindereif, A. Schön and E. Westhof), 205–230. Weinheim: Wiley-VCH.

178 Li, W., Sheng, J., Xu, M. et al. (2017). Inhibition of murine cytomegalovirus infection in animals by RNase P-associated external guide sequences. *Mol. Ther. Nucleic Acids* 9: 322–332.

179 Zhou, T., Kim, J., Kilani, A.F. et al. (2002). *In vitro* selection of external guide sequences for directing RNase P-mediated inhibition of viral gene expression. *J. Biol. Chem.* 277 (33): 30112–30120.

180 Liu, J., Shao, L., Trang, P. et al. (2016). Inhibition of herpes simplex virus 1 gene expression and replication by RNase P-associated external guide sequences. *Sci. Rep.* 6: 27068.

181 Xia, C., Chen, Y.C., Gong, H. et al. (2013). Inhibition of hepatitis B virus gene expression and replication by ribonuclease P. *Mol. Ther.* 21 (5): 995–1003.

182 Hardt, W.D., Schlegl, J., Erdmann, V.A., and Hartmann, R.K. (1993). Role of the D arm and the anticodon arm in tRNA recognition by eubacterial and eukaryotic RNase P enzymes. *Biochemistry* 32 (48): 13046–13053.

183 Yuan, Y. and Altman, S. (1995). Substrate recognition by human RNase P: identification of small, model substrates for the enzyme. *EMBO J.* 14 (1): 159–168.

184 Carrara, G., Calandra, P., Fruscoloni, P., and Tocchini-Valentini, G.P. (1995). Two helices plus a linker: a small model substrate for eukaryotic RNase P. *Proc. Natl. Acad. Sci. U.S.A.* 92 (7): 2627–2631.

185 Werner, M., Rosa, E., Nordstrom, J.L. et al. (1998). Short oligonucleotides as external guide sequences for site-specific cleavage of RNA molecules with human RNase P. *RNA* 4 (7): 847–855.

186 Ma, M.Y., Jacob-Samuel, B., Dignam, J.C. et al. (1998). Nuclease-resistant external guide sequence-induced cleavage of target RNA by human ribonuclease P. *Antisense Nucleic Acid Drug Dev.* 8 (5): 415–426.

187 Werner, M., Rosa, E., Al Emran, O. et al. (1999). Targeted cleavage of RNA molecules by human RNase P using minimized external guide sequences. *Antisense Nucleic Acid Drug Dev.* 9 (1): 81–88.

188 Carrara, G., Calandra, P., Fruscoloni, P. et al. (1989). Site selection by *Xenopus laevis* RNAase P. *Cell* 58 (1): 37–45.

189 Ma, M., Benimetskaya, L., Lebedeva, I. et al. (2000). Intracellular mRNA cleavage induced through activation of RNase P by nuclease-resistant external guide sequences. *Nat. Biotechnol.* 18 (1): 58–61.

190 Wu, S., Chen, Y., Lindell, M. et al. (2011). Functional coupling between a distal interaction and the cleavage site in bacterial RNase-P-RNA-mediated cleavage. *J. Mol. Biol.* 411 (2): 384–396.

191 Bai, Y., Trang, P., Li, H. et al. (2008). Effective inhibition in animals of viral pathogenesis by a ribozyme derived from RNase P catalytic RNA. *Proc. Natl. Acad. Sci. U.S.A.* 105 (31): 10919–10924.

192 Bai, Y., Gong, H., Li, H. et al. (2011). Oral delivery of RNase P ribozymes by *Salmonella* inhibits viral infection in mice. *Proc. Natl. Acad. Sci. U.S.A.* 108 (8): 3222–3227.

193 Bai, Y., Li, H., Vu, G.P. et al. (2010). *Salmonella*-mediated delivery of RNase P-based ribozymes for inhibition of viral gene expression and replication in human cells. *Proc. Natl. Acad. Sci. U.S.A.* 107 (16): 7269–7274.

194 Hartmann, R.K., Krupp, G., and Hardt, W.D. (1995). Towards a new concept of gene inactivation: specific RNA cleavage by endogenous ribonuclease P. *Biotechnol. Annu. Rev.* 1: 215–265.

195 Zong, X., Nakagawa, S., Freier, S.M. et al. (2016). Natural antisense RNA promotes 3′ end processing and maturation of MALAT1 lncRNA. *Nucleic Acids Res.* 44 (6): 2898–2908.

196 Wilusz, J.E. (2016). Long noncoding RNAs: re-writing dogmas of RNA processing and stability. *Biochim. Biophys. Acta* 1859 (1): 128–138.

197 Rigo, F., Hua, Y., Krainer, A.R., and Bennett, C.F. (2012). Antisense-based therapy for the treatment of spinal muscular atrophy. *J. Cell Biol.* 199 (1): 21–25.

198 Stein, C.A., Hansen, J.B., Lai, J. et al. (2010). Efficient gene silencing by delivery of locked nucleic acid antisense oligonucleotides, unassisted by transfection reagents. *Nucleic Acids Res.* 38 (1): e3.

199 Grünweller, A. and Hartmann, R.K. (2016). Chemical modification of nucleic acids as a key technology for the development of RNA-based therapeutics. *Pharmazie* 71 (1): 8–16.

10

Ribozyme Discovery in Bacteria

Adam Roth[1,2] *and Ronald Breaker*[1,2,3]

[1] *Howard Hughes Medical Institute, New Haven, Connecticut 06520-8103, United States*
[2] *Department of Molecular, Cellular and Developmental Biology, Yale University, New Haven, Connecticut 06520-8103, USA*
[3] *Yale University, Department of Molecular Biophysics and Biochemistry, New Haven, Connecticut 06520-8103, United States*

10.1 Introduction

One of the defining features of modern metabolism is the division of labor among biopolymers. However, the metabolism we observe today likely originated from an era, nearly four billion years ago, in which a single polymer predominated [1]. That polymer, RNA, can template its reproduction from monomers or fragments, thereby enabling the continuous transfer of information. RNA can also serve as a catalyst by adopting stable structures that accelerate specific chemical transformations. This duality of function equips RNA with two key attributes necessary for a self-propagating molecular system at or very near to life's origin. In fact, the first molecule that could have been regarded as alive might very well have been RNA or an evolutionary forerunner quite similar in structure [2].

Vestiges of this ancient era, which is often referred to as the RNA world, abound in contemporary organisms. RNA polynucleotides participate in DNA replication as primers as well as templates, they are involved in all phases of gene control, and they guide the destruction of invading nucleic acids and determine chemical modification sites on target RNAs [3]. As monomers and dimers, RNAs serve as essential signaling molecules that choreograph physiological and developmental responses [4], and most coenzymes and prosthetic groups employed by protein enzymes contain RNA nucleotide or nucleotide-like moieties [5]. However, RNA catalysts, or ribozymes, are the remnants that most powerfully evoke the prowess of the RNA world, and none more so than the ribosome. A large multi-subunit apparatus heavily studded with proteins but a ribozyme at its core, the ribosome is ubiquitous in biology. That the sequences of nearly all proteins in modern organisms are encoded in mRNAs, decoded by tRNA adaptors, and synthesized by what is essentially an RNA machine points to the ancestral ribosome as an invention of the RNA world. Additionally, the mere existence during this era of such an elaborate RNA-based device implies

Ribozymes, First Edition. Edited by Sabine Müller, Benoît Masquida, and Wade Winkler.

an extensive attendant intermediary metabolism [6], one that is likely to have been teeming with enzymes made from RNA.

10.2 Protein Takeover

The proto-ribosome was quite possibly the pièce de résistance of the RNA world, yet it was already stirring the beginnings of a major coup. For all their proficiency, most of the enzymes of the RNA world appear to have been deposed – and resoundingly so. Biocatalysis in contemporary metabolism is almost solely dominated by protein enzymes, and the reasons for this seem clear. In contrast to the quartet of chemically similar nucleobases used by ribozymes, the palette of 20 side chains available to protein enzymes comprises different polarities, charges, functions, and structures and so possesses tremendous versatility in substrate recognition and reaction chemistry. And unlike RNA, which must contend with a negatively charged backbone that can be an impediment to efficient folding, proteins benefit from a peptide backbone whose polar moieties interact with one another to stabilize common turn or loop structures such as α-helices and β-sheets. Together, the diversity of amino acid side chains and the intrinsic structural propensity of the peptide backbone to stabilize structural motifs greatly facilitate folding and provide access to specialized environments like the hydrophobic interiors of membrane bilayers. Moreover, the comparatively small size of amino acids in comparison with nucleotides facilitates the fine-scale spatial organization of protein–enzyme binding pockets, thereby affording greater dexterity in the recognition and transformation of substrates. Proteins have the advantage compared to RNA not only in performance but also in a resource economy. Because the average mass of an amino acid is about one-third that of a nucleotide, far fewer resources are consumed in building a protein enzyme than in producing an RNA of comparable length, which would provide an enormous competitive growth advantage.

10.3 Ribozymes as Evolutionary Holdouts

Viewed in comparison with ribozymes, the advantages of protein enzymes seem overwhelming. Nonetheless, modern organisms contain a smattering of holdouts, RNA catalysts that have somehow managed to escape having their roles subsumed by proteins. Some of these extant ribozyme classes, like the ribosome and ribonuclease P (RNase P), are ubiquitous or nearly so in contemporary biology, pervading all domains of life. Although these ribozymes critically depend on affiliated polypeptide subunits for their functions in vivo, the reactions of these ribonucleoprotein complexes are directly mediated by or involve RNA [7, 8]. In the case of the ribosome, the maintenance of its catalytic core as RNA throughout the course of evolution might very well be linked to the tremendous variety of molecules with which it must interact. This high degree of complexity, in which the reaction products are essential and there are stringent demands placed upon the catalytic RNA to operate in concert with myriad subunits and substrates, makes a wholesale evolutionary transition from ribozyme to protein enzyme fraught. In essence, the existence of too many moving parts (enzymes and substrates) has precluded the replacement of what could be viewed as an anachronistic enzyme.

An analogous argument can be made to explain the retention of a catalytic RNA subunit in RNase P, the family of endoribonucleases involved in processing tRNA precursors, as well as a variety of other RNA substrates. In this case, however, the network of functions that appears to have locked in the RNA subunit might not be directly related to the primary role of RNase P as an endonuclease. Rather, RNase P and various subassemblies thereof have been shown to participate in processes such as transcriptional control and DNA repair [9], and it has been suggested that the simultaneous fulfillment of these roles has demanded the maintenance of the RNA subunit because it serves as a critical organizing center of the ribonucleoprotein complex [10]. The existence in some eukaryotic organelles of distinct protein-only RNase P enzymes, which process the same RNA substrates as their ribonucleoprotein counterparts, supports the contention that neofunctionalization of RNA-containing RNase P beyond its purely catalytic role could indeed be a factor in the preservation of the RNA component [11].

In contrast to RNase P and the ribosome, which act on their substrates in trans, all other known natural ribozymes are self-processing. Such autocatalytic RNAs have specialized roles in niche genetic contexts that might provide a measure of insulation from the evolutionary forces that have all but obliterated RNA catalysts. For example, as mobile genetic elements, group I and group II self-splicing introns exploit auto-excision as a key to their survival [12]. Because auto-excision from a transcript minimizes adverse effects on the host organism, this capability provides access to a greater range of insertion sites. For group II introns especially, this activity is an essential part of its life cycle, as the reverse reaction drives the propagation of the parasitic DNA that encodes it. Ribozymes of the self-cleaving variety, such as those of the hammerhead and hepatitis delta virus (HDV) families [13], offer more examples of RNA catalysts that might have persisted because they perform roles for which they seem ideally suited. These diminutive motifs often reside in viroids and satellite RNAs, which are small, circular single-stranded RNAs that act as infectious agents. Their site-specific cleavage activities are critical for the life cycles of the pathogenic subviral RNAs, as they convert extended linear products resulting from rolling circle replication into unit-length progeny. In playing these roles, ribozymes offer distinct advantages over protein enzymes: they occupy little space, and because they are self-contained and bear no reliance on exogenous protein factors, replication delays are minimized.

10.4 The Role of Serendipity in Early Ribozyme Discoveries

The proposal that biological catalysts could be formed from RNA polymers long predated the identification of natural examples. Struck by the ability of tRNA to adopt a compact globular structure reminiscent of folded proteins, Woese, Crick, and Orgel proposed the elegant hypothesis that life could have originated with RNA if templating activity could exist simultaneously with catalytic activity [14–16]. However, even though such primordial catalytic roles for RNA had been envisioned, it was not at all a foregone conclusion that such RNAs were still in existence. There was no expectation that any ribozymes would have persisted for billions of years in the face of relentless selection pressure for polypeptides. Serendipity had a major role, then, in some of the earliest ribozyme discoveries. Investigations of the splicing mechanisms of an intervening sequence (IVS) occurring in a *Tetrahymena* rRNA precursor led to the first realization that contemporary RNA structures could be imbued with catalytic properties. Efforts to prepare the pre-rRNA for use in assays that might identify splicing factors were short-circuited by excision of the IVS even in the absence of nuclear extract [17]. Subsequent studies demonstrated that this IVS, termed the group I intron, does not require host protein factors for splicing to occur [18]. Several years later, another class of IVSs, referred to as group II introns, which had already been recognized as bearing conserved secondary structures and sequence motifs distinct from those of the group I [19], were also revealed to possess self-splicing activity [20, 21].

Similarly, the first cohort of self-cleaving ribozymes was uncovered while researchers were investigating the life cycles of viroids and satellite RNAs. These small RNA circles were observed to undergo autocatalytic processing in vitro [22–27], with the hammerhead and hairpin ribozymes subsequently emerging as the motifs responsible for these cleavage activities [28–31]. Another distinct class of ribozyme takes its name from the human pathogen in which it was first identified: the HDV. This virus contains a single-stranded circular RNA genome, in which the HDV ribozyme processes multimeric transcripts into monomers via rolling circle replication, analogously to the hammerhead and hairpin motifs in viroids and satellite RNAs [32, 33]. A fourth self-cleaving RNA, the Varkud satellite (VS) ribozyme, is encoded on small plasmids in the mitochondria of certain strains of *Neurospora* [34]. Site-specific processing in vitro was detected in transcripts derived from these plasmids, implying a role for the VS ribozyme in the reproductive cycle of its host satellite element.

In each of the above cases, the detection of cleavage events in circular RNAs attracted the attention of investigators, and because the assays were generally conducted in the absence of protein factors, it was immediately apparent that autocatalytic RNAs might be responsible. In other instances, however, where the ribozyme in question is part of a ribonucleoprotein, the function of the holoenzyme was understood long before an RNA component was appreciated as the catalytic subunit. Regarding the ribosome, for example, views on whether protein or RNA was directly responsible for peptide bond formation evolved gradually over a span of decades. Ribosomal proteins initially received much of the attention as presumed agents of peptidyl transfer, but the striking phylogenetic conservation of ribosomal RNA structure was one of the early observations that emphasized the importance of the nucleic acid component [35]. Nucleotide mapping experiments provided biochemical evidence that tRNAs interacted with highly conserved residues in rRNA [36], and the active participation of the RNA subunits was suggested even more strongly when ribosome preparations subjected to vigorous deproteinization protocols retained peptidyl transferase activity [37]. Ultimately, an atomic resolution structure of the ribosome that was devoid of proteins in the active site established definitively that peptide bond formation is RNA-catalyzed [38].

Similarly, the realization that a catalytic RNA subunit is responsible for the cleavage activity of RNase P occurred well after the initial discovery of the endoribonuclease [39]. Investigations of the roles of the protein and RNA subunits of bacterial RNase P revealed that the RNA component was essential for processing tRNA precursors [40, 41], but only subsequently was it demonstrated that the RNA alone could catalyze the cleavage reaction in vitro [7].

10.5 Ribozymes Emerge from Structured Noncoding RNA Searches

Although serendipity ushered in the era of RNA enzymes, this manner of discovery cannot support sustained efforts to search for additional classes of catalytic RNA. In recent years, though, the hunt for ribozymes and ribozyme candidates has benefited from searches for entirely different types of RNAs. Computational tools initially developed to identify cis-regulatory and other structured noncoding RNAs (ncRNAs) have, not surprisingly, also proved adept in uncovering novel ribozyme families. One type of cis-regulatory RNA that these computational methods were originally designed to find is riboswitches. Existing predominantly in bacteria, riboswitches are structured RNAs usually residing in 5′ untranslated regions (5′ UTRs) that directly bind small-molecule ligands such as metabolites or ions [42]. Riboswitches often exhibit a bipartite structure in which the conserved ligand-binding domain, or aptamer, overlaps another domain, referred to as the expression platform, that directly affects gene expression. As a result of this overlapping structure, riboswitches generally adopt one of two mutually exclusive conformations governed by the occupancy state of the aptamer. In this manner, riboswitches modulate the expression of associated genes by responding directly to the cellular concentrations of their cognate ligands.

Approximately 40 classes of riboswitches have been identified that recognize ligands including coenzymes, amino acids, purine derivatives, second-messenger signaling molecules, and monatomic ions [43]. From an evolutionary perspective, such ligands are effectively immutable, and this results in severe constraints for the RNA structures that grasp them. The creation of a highly selective binding pocket for such a target molecule requires the precise spatial positioning of a constellation of functional groups. This degree of organization, in turn, demands an RNA characterized by highly conserved arrangements of nucleotide sequences and secondary structure elements – precisely the same features that are shared by ribozymes to form active sites utilizing conserved reaction mechanisms. In principle, then, scouring sequence databases for motifs like riboswitches could potentially identify new classes of ribozymes.

Computational pipelines that were initially deployed to search for metabolite-binding RNAs relied on comparative sequence alignments drawn primarily from bacterial genomes. Historically, bacteria have revealed themselves to be an especially rich source for these motifs. Moreover, although RNAs of this type are abundant in the bacterial domain, they are more sparsely distributed elsewhere, with only a few riboswitch classes having been detected in archaea and eukaryotes [43–46]. There are further benefits, however, that are derived from focusing structured ncRNA searches on the bacterial domain. For example, certain features of bacterial genomes offer intrinsic advantages for computational search strategies. Noncoding RNAs, by definition, reside almost exclusively in intergenic regions (IGRs), which in bacteria represent only a minor fraction of the total genome. As a result, restricting interrogated sequences to only these regions streamlines the computational load by reducing unnecessary churn. Also, the sheer number of microbial organisms whose genomes have been sequenced creates vast, ever-expanding databases brimming with taxonomic diversity. Besides simply maximizing the opportunity for structured ncRNA discovery, the size and depth of such repositories allow meaningful conclusions to be drawn about individual motifs when patterns of sequence and structure conservation exist in evolutionarily diverged groups.

Many of the early bioinformatics-based searches for bacterial cis-regulatory RNAs sought conserved sequences and structural elements in IGRs associated with homologous or related genes [47–50]. Because riboswitches specific for a given metabolite often tend to associate with a narrow set of genes that relate directly to that ligand, applying comparative sequence alignment methods in this focused manner was expected to increase the likelihood of identifying novel classes of such RNAs. One motif identified by this approach was initially observed to reside exclusively in the 5′ UTRs of *glmS* genes in certain Gram-positive organisms (Figure 10.1) [47, 52]. The *glmS* gene encodes glutamine-fructose-6-phosphate aminotransferase, which uses glutamine as the nitrogen donor in converting fructose 6-phosphate to glucosamine 6-phosphate (GlcN6P). As the first step in hexosamine biosynthesis, this reaction is essential for the production of peptidoglycan. Although the motif associated with this gene exhibited all the hallmarks of a classical metabolite-binding RNA, its behavior was markedly different. The cognate ligand, GlcN6P, modulated gene expression not by inducing a conformational change that could be relayed to a distinct expression platform, but rather by participating as a coenzyme in site-specific self-cleavage of the structured RNA [52, 53]. The *glmS* RNA may therefore be classified as a riboswitch, despite its uniqueness in effecting gene control through its ribozyme activity. And the *glmS* motif stands apart not only from other riboswitches but also from other ribozymes. Although the group I intron relies on exogenous guanosine to provide the nucleophile in the first step of splicing, only the *glmS* ribozyme harnesses a metabolite as a cofactor (rather than as a substrate) to promote self-processing.

The highly conserved core of the *glmS* RNA establishes a GlcN6P-binding pocket proximal to the active site, with less conserved peripheral elements contributing to the overall stability of the ribozyme [54–57]. Atomic resolution structures reveal that the amine group of docked GlcN6P is potentially poised to neutralize the developing negative charge on the 5′ phosphate oxygen at the cleavage site [54, 55]. Biochemical analyses have provided additional support for the involvement of the ligand as a general acid [58–60], and recent studies have posited even more extensive participation

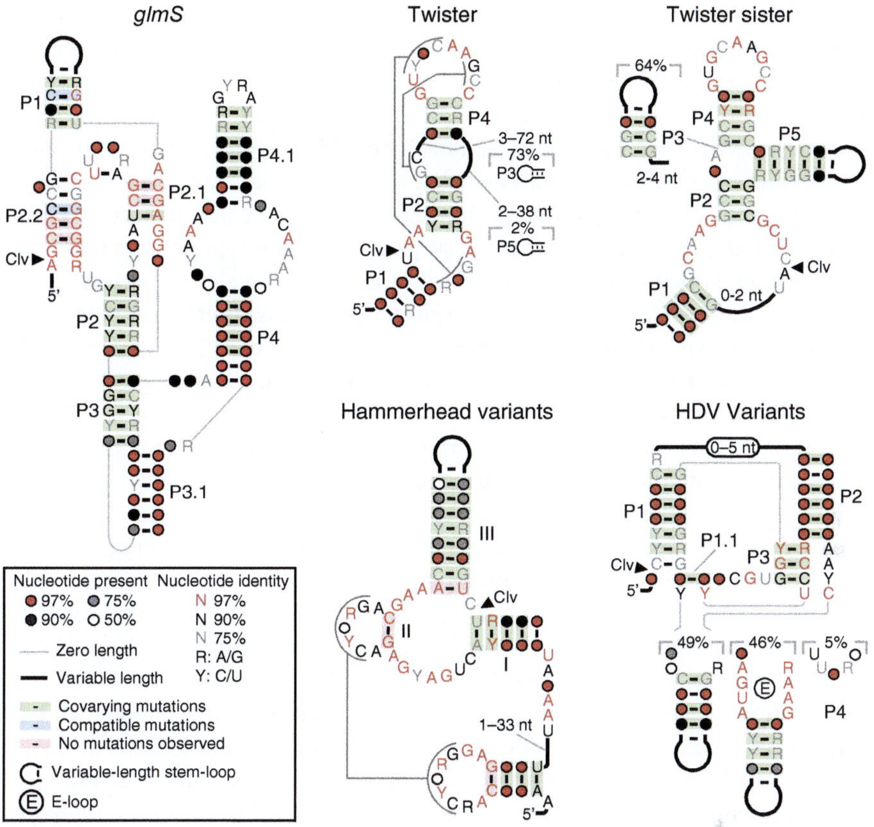

Figure 10.1 Consensus structures for *glmS*, twister and twister sister ribozymes, and for hammerhead and HDV ribozyme variants. For the twister and hammerhead RNAs, sequences within internal and terminal loops that form pseudoknots are indicated by connectors. The percentages of ribozyme representatives containing optional stem-loops at certain positions are indicated. Optional stem-loops and other modular elements have also been observed in *glmS* ribozymes [51], but such occurrences are quite rare (<1%), and so are not depicted in the secondary structure model. Arrowheads denote the cleavage sites (Clv).

of GlcN6P in the cleavage mechanism [61]. Unlike the ligand-binding sites of most riboswitch aptamers, which are flexible when unoccupied, the GlcN6P-docking site of the *glmS* ribozyme is relatively rigid [62, 63], reflecting the key contributions of the ligand to catalysis rather than structure stabilization.

The cleavage site of the *glmS* ribozyme occurs near the 5′ boundary of the motif (Figure 10.1) and therefore far upstream of the start codon of the *glmS* open reading frame (ORF). The reaction mechanism of *glmS* RNA, like that of the other small self-cleaving ribozymes, proceeds via transesterification, generating fragments with 2′,3′-cyclic phosphate and 5′-hydroxyl termini [52]. In *Bacillus subtilis*, the 5′-hydroxyl end group generated by this reaction renders the large cleavage product containing the ORF susceptible to degradation by the exoribonuclease RNase J1, resulting in downregulation of the *glmS* gene [64].

The *glmS* RNA motif emerged from a search strategy that evaluated IGRs based not only on the presence of conserved structural features but also on their associations with related genes. Even when the phylogenetic distribution of this ribozyme was subsequently expanded, the affiliation of *glmS* RNA with its eponymous gene was nearly always preserved [51]. In rare cases where genes other than *glmS* occurred downstream, the enzymes encoded by these outliers nonetheless resided in the same metabolic neighborhood, having a substrate or product molecule in common with the GlmS enzyme. It is not surprising that this type of computational search method, which takes gene context into account, revealed an autocatalytic RNA serving as a regulatory element. The *glmS* ribozyme is, after all, a riboswitch. But most self-cleaving ribozymes do not have primary functions in genetic control, and thus are not known to associate strongly with highly related genes. In an attempt to uncover structured ncRNAs lacking such genetic associations, a related search strategy was undertaken that removed this factor from consideration.

By applying a computational pipeline in which IGRs were initially clustered using only homology searches and without regard to gene context [65, 66], an RNA motif termed twister was identified within a restricted bacterial dataset (Figure 10.1) [67]. An expanded search for additional homologs of this RNA was subsequently conducted on more comprehensive databases that included bacterial, archaeal, and eukaryal genome sequences as well as metagenome sequences. Beyond the relatively few bacterial examples that were scattered throughout the class Clostridia, larger numbers of twister RNAs were uncovered in metagenome sequences and in genomes from a variety of eukaryotic organisms including fungus, insects, fish, and plants. Although many genomes were observed to contain multiple occurrences of twister representatives, the genome of the parasitic blood fluke *Schistosoma mansoni* was particularly striking in this regard, harboring more than 1000 examples of the twister motif.

The twister RNA consensus structure is defined by three short stems conjoined by conserved internal and terminal loops, with two pseudoknots contributing to the tertiary structure (Figure 10.1) [67]. For the majority of twister representatives, the 5′ and 3′ termini reside in the P1 stem. But the motif is characterized by additional stems that occur sporadically, and the termini can reside alternately in these optional stems. Although such circular permutation is rare among structured ncRNAs and has never been detected in riboswitch aptamers, one motif – the hammerhead ribozyme – is frequently subject to this type of variation. As this feature of a known self-cleaving RNA was shared by the twister motif, it suggested that twister RNAs could be serving analogous biochemical roles. Subsequent experiments using twister representatives from various organisms confirmed that this motif represented a novel class of self-cleaving RNAs, capable of rate enhancements as high as 10^{10}-fold [67].

Though they represent structurally distinct motifs, the commonalities between hammerhead and twister RNAs extend beyond just their functions as self-cleaving ribozymes and tolerances for circular permutation. An analysis of the genes residing in the immediate vicinities of twister and hammerhead sequences revealed that these two ribozyme classes often display markedly similar genetic contexts. Genes encoding certain protein domains occur disproportionately within several kilobases of either twister or hammerhead sequences. In addition, these ribozymes often reside near one another. Twister RNAs, for example, frequently inhabit genetic neighborhoods that contain other twister representatives or members of the hammerhead class. The recurring genetic associations displayed by these distinct classes of self-cleaving RNAs suggest that they could function interchangeably in their biological roles. Perhaps more importantly, at least from a discovery standpoint, these associations immediately suggested a method to streamline the hunt for additional self-cleaving ribozymes.

10.6 Ribozymes Beget Ribozymes

Particularly when they are rare, ncRNA motifs are difficult to extract from DNA sequence databases of enormous size. In principle, a comprehensive search of all available IGRs could identify sparsely occurring motifs. But a search of this type would necessitate combing through tens of billions of base pairs of DNA sequence, creating an extreme computational load. Moreover, because such a strategy would likely generate a large surfeit of false positives, efforts to identify high-quality ncRNA candidates amid the noise would be unmanageable. For these reasons, the parceling of databases into restricted subsets can greatly enhance computational search efficiencies [68].

Therefore, the tendencies of hammerhead and twister RNAs to occupy similar genetic contexts provided an opportunity. If these patterns of gene association applied to self-cleaving RNA classes more generally, then they could be exploited to winnow IGRs potentially enriched for ribozyme sequences. To apply this method, a number of genetic elements were selected that resided preferentially near hammerhead and twister RNAs [69]. These elements included ORFs corresponding to 16 conserved protein domains as well as all of the permuted forms of the hammerhead and twister ribozymes themselves. IGRs occurring within several kilobases of each of these ORFs and ribozymes were used to populate the restricted database, and an established computational pipeline [66, 70] was then used to identify structured noncoding motifs from this focused pool of sequences [69].

One of the RNAs to emerge from this search was the twister sister motif, so named because the general organization of its secondary structure loosely resembles that of the twister ribozyme and because highly conserved nucleotide sequences within a terminal loop are shared by the two consensus structures (Figure 10.1). Yet the twister sister motif also possesses conserved nucleotides elsewhere in its consensus structure that are unique, and it lacks two pseudoknots that are defining features of the twister ribozyme. In vitro assays established that, like their structural siblings – twister RNAs – representatives of the twister sister class also possess self-cleavage activity. The cleavage site resides in a domain of the motif – the internal loop conjoining the P1 and P2 stems – that is analogous to the domain containing the cleavage site in twister RNAs. Interestingly, though, the specific site of cleavage for twister sister RNAs occurs on the opposite strand of that analogous internal loop (Figure 10.1). Considering the different positions of the scissile phosphates in twister and twister sister motifs, these may be regarded as distinct ribozyme classes. Comparisons of the atomic resolution structures of these RNAs lend further support to this conclusion, as the active-site architectures and nucleotides contributing to rate enhancement also appear to differ between these ribozymes [71–75]. Nonetheless, and despite their different phylogenetic distributions, it is difficult to rule out the evolutionary descent of twister and twister sister RNAs from a single ancestral ribozyme.

Employing a biased pool approach to seek new families of RNA catalysts also resulted in the identification of unusual structural variants of known ribozyme classes [69]. Multiple examples of hammerhead ribozymes with atypical stem II elements comprising a lone base pair were uncovered in bacterial metagenomes, for example (Figure 10.1). Only one isolated example of a hammerhead RNA conforming to this rare architecture had been previously reported [76]. The presumably weak stem II structure in these newly identified examples is likely to be stabilized by a kissing-loop interaction with a conserved hairpin loop residing immediately 5′ to the hammerhead ribozyme core domain. Hundreds of HDV ribozyme variants harboring an E-loop motif within the P4 stem were also identified in bacterial metagenomes (Figure 10.1) [69], vastly increasing the number of representatives of this HDV ribozyme isoform, for which the only example had previously been described [77]. Many additional examples of this variant subtype were identified in eukaryotes using not only homology searches [69] but also a comparative genomics pipeline in fungi [78].

This search strategy, initially inspired by the common genetic associations of twister and hammerhead ribozymes, was an effective one, uncovering on its first pass a new ribozyme class as well as structural variants of two known classes. This appeared to validate the original premise that the immediate vicinities of certain genetic elements might be enriched for different types of self-cleaving RNAs. Accordingly, the second iteration of this filtered search method was applied, using updated databases containing more recently deposited bacterial DNA sequences and selecting IGRs near an expanded number of ribozyme-associated genes [69]. Two additional self-cleaving RNAs, the pistol and hatchet motifs, were derived from this broader search (Figure 10.2). Based on the uniqueness of their conserved sequences and secondary structure features, these motifs define distinct new ribozyme classes. Approximately two dozen examples of the pistol motif were found in various *Firmicutes* species, typically at a frequency of a single copy per genome. But many hundreds more were identified in metagenome sequences, and this heavily skewed representation in environmental sequence databases was also observed for hatchet and twister sister RNAs [69, 79, 80]. These findings underscore the potential wealth of catalytic ncRNAs possessed by bacterial and archaeal organisms that have not yet been cultured in isolation.

Not every structured ncRNA motif emerging from these biased searches was shown to have self-cleavage activity, however. In fact, of the 39 motifs that were identified using this targeted strategy – referred to as RNAs Associated with Genes Associated with Twister and Hammerheads (RAGATHs) – only the five described

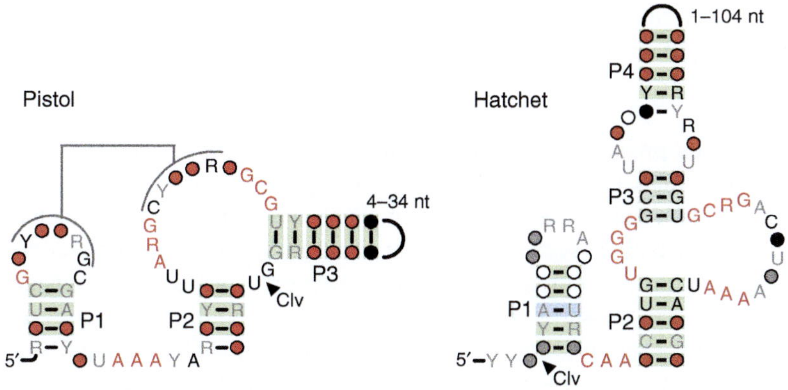

Figure 10.2 Consensus structures for pistol and hatchet ribozymes. Annotations are as described in Figure 10.1.

in this section (twister sister, pistol, hatchet, and the hammerhead and HDV variants) were confirmed as ribozymes. The remaining 34 RAGATHs range in complexity from groupings of relatively simple stem elements to larger and far more elaborate constructions, but for none of these was self-processing activity detected in vitro [68, 69]. These additional motifs, like the five ribozymes identified alongside them, were derived predominantly from environmental sequences, and their roles are unknown. Many of these RNAs display levels of structural complexity and nucleotide conservation that are on par with or even exceed those of known self-cleaving ribozymes. Therefore, some of these RAGATH motifs may be potentially autocatalytic but could require certain factors, conditions, or structural domains that were not provided in preliminary assays of their activities. A small subset of RAGATHs often resides in positions immediately adjacent to known self-cleaving RNAs, raising the possibility that they could function as modular domains, perhaps by engaging in cross talk with the associated ribozymes to regulate their cleavage activities in response to various signals. Whatever functions are fulfilled by this large collection of RAGATHs, the biased search used to find them seems to have uncovered a genetic niche that relies heavily on RNA structures. Many of the genes whose surrounding IGRs provided the fodder for these ncRNA searches bear annotations that indicate relationships to bacteriophage Mu [67, 69]. Thus, whether or not a given RAGATH is catalytic, one explanation for their collective prevalence in these contexts might be that Mu-like phages delegate a substantial fraction of functions such as RNA processing and gene regulation to ncRNA motifs.

10.7 Ribozyme Dispersal Driven by Association with Selfish Elements

The identification of the new catalytic RNA motifs described above has been accompanied by a tremendous expansion in the number of representatives of previously identified ribozyme classes, revealing self-cleaving RNAs to be far more pervasive than originally thought. Using homology- and structure-based search methods, investigators have uncovered multitudes of hammerhead and HDV RNAs in all domains of life [76, 81–84]. However, merely knowing that such motifs are likely to undergo self-processing does little to illuminate their biological roles.

In bacteria, the roles of some self-cleaving ribozymes might pertain to the bacteriophage-related sequences with which they are commonly associated [67, 69, 76, 81]. For example, because transcription of prophage genes might occasionally be initiated from host promoters, the presence of ribozymes near prophage termini could enforce transcript boundaries, helping to eliminate extraneous sequences that might interfere with the replication of the phage. Self-cleaving RNAs could also be utilized simply for processing of polycistronic transcripts, perhaps to generate small trans-acting regulatory RNAs or to differentially modulate the stabilities of various cleaved fragments. In support of this, twister ribozymes placed in the 3′ UTRs of bacterial reporter genes have been demonstrated to exert substantial effects on corresponding expression levels [85]. Arrangements in which two or more ribozymes are interspersed among neighboring ORFs have prompted speculation that religation events following ribozyme-mediated cleavage could generate products such as circular RNAs and reordered transcripts, representing a possible mechanism of alternative splicing in bacteria [76, 86].

In metazoans too, the pervasiveness of self-cleaving ribozymes might stem in large part from their associations with mobile genetic elements. In a variety of organisms, including newts and cave crickets [87, 88], hammerhead ribozymes are encoded by satellite DNAs, which are tandemly repeated sequences that can be dispersed uniformly throughout genomes or restricted to heterochromatic regions of chromosomes. Other genetic elements with similar distributions also appear to commonly exploit self-cleaving RNAs, as a diverse collection of retrotransposons have been found to harbor hammerhead and HDV ribozymes [89–94]. The action of these ribozymes is required to produce appropriately processed transcripts that serve as key intermediates in the life cycles of these retroelements. For example, a feature of the R2 retrotransposon family, which is widely distributed among metazoans [95], is the occurrence of HDV ribozymes near the 5′ termini. Because these retroelements lack internal promoters, they are usually co-transcribed with host genes, and so ribozyme-mediated self-cleavage allows the separation of the R2 transcripts from upstream sequences [91, 93]. It has also been reported that, following strand scission, the doubly pseudoknotted HDV RNA promotes the translation of the downstream ORF, suggesting that this structure has been co-opted to behave in the manner of an internal ribosome entry site [93].

10.8 Domesticated Ribozymes

The prolific dispersal mechanisms of some ribozyme-containing retroelements have resulted in inordinate numbers of autocatalytic RNA sequences in certain genomes. For example, there are thousands of copies of hammerhead ribozyme sequences associated with the Smα family of retroelements in *S. mansoni* [92, 96], and it seems likely that the ~1000 representatives of the twister motif in this same organism [67] could be due to a similar type of association. The genome-wide distribution of ribozyme sequences on this scale ostensibly creates opportunities for exaptation events, in which the self-cleavage activities of ribozymes become unyoked from their selfish elements and are repurposed to function in service to the host organism. Several examples that are putatively derived from such exaptation events have been reported. An HDV-like ribozyme that is highly conserved in mammalian genomes resides in an intron of the single-copy gene *CPEB3*, which is involved in regulating mRNA polyadenylation [97]. Presumably, tuning of the ribozyme cleavage activity could be a factor in determining the level of *CPEB3* gene expression. Conserved hammerhead ribozyme sequences are also observed to have taken up positions in introns and 3′ UTRs of numerous metazoans [76, 98, 99], offering additional support that retroelement-derived self-cleaving RNAs are occasionally scavenged by their hosts and harnessed to influence transcript maturation pathways.

Self-cleaving RNAs are not the only ribozymes whose activities can be subsumed for the benefit of the host. In the bacterium *Azotobacter vinelandii,* a group II intron resides in the *groEL* gene, which encodes a chaperonin whose expression is critical during such environmental stresses as heat shock. Despite inhabiting a mesophilic host, this self-splicing ribozyme exhibits remarkable thermostability, attaining its optimum splicing activity at 65 °C [100]. Because the intron interrupts the *groEL* stop codon, the integrity of the ORF is preserved regardless of whether splicing occurs, and so it is not immediately apparent what differences there might be in protein production from spliced versus unspliced transcripts. Nonetheless, one explanation for the association of a thermotolerant group II intron with a heat shock gene is that temperature-modulated splicing activity could contribute to the regulation of GroEL expression at the posttranscriptional level. Thus, it appears that one fortuitous insertion event might be sufficient to nudge a self-splicing intron from selfishness toward altruism.

Another domestication event of this sort could conceivably have arisen from the integration of a group I intron between sequences corresponding to a riboswitch aptamer and its target gene. Occurring in *Clostridium difficile*, this juxtaposition of a self-splicing RNA with a cyclic-di-GMP-binding domain has resulted in the formation of an allosteric ribozyme, in which the choice between alternative splicing pathways is influenced by the occupancy of the aptamer domain [101]. When the aptamer exists in the bound conformation, one of two competing structural elements within the intron is preferentially stabilized, promoting utilization of a 5′ splice site that generates a fully spliced mRNA. The conjoined exons constituting this spliced product create an ideal ribosome binding site that is optimally positioned in relation to the start codon of the target gene. In the absence of cyclic di-GMP, however, an alternative structure adopted by the aptamer domain elicits concomitant changes in the organization of the group I intron, resulting in the use of an alternative 5′ splice site. This leads to an abortive pathway ultimately yielding a trimmed mRNA that is unlikely to be translated because its 5′ UTR is too short to accommodate a ribosome binding site. Efficient translation of the target gene is also likely to be precluded in unspliced precursor RNAs, due to the absence of a suitably positioned ribosome binding site and to an start codon sequestered by base pairing. Thus, expression of the downstream gene is likely to occur only when cyclic di-GMP accumulates to sufficient levels [102].

10.9 New Ribozymes from Old

New catalytic RNAs can also be derived from existing ribozyme classes in ways that do not depend on the integration of distinct functional modules such as riboswitch aptamers. Sometimes the reshuffling of existing subdomains within a ribozyme class is all that is required. A striking example of this is provided by the identification of a bacterial large ncRNA class, termed CP (circularly permuted) group II RNA, that displays extensive similarities to group II self-splicing introns [103]. The The CP group II RNA appears to contain all of the individual substructures that normally comprise a group II intron, but the order in which they occur is atypical. Domains that reside near the 3′ end of group II ribozymes occupy the 5′ end of CP group II RNAs; accordingly, the order in which the 5′ and 3′ splice sites occur in the corresponding precursor molecules is reversed. As might be expected from the structural similarities of the individual subdomains, the phosphoester transfer chemistry mediated by CP group II RNA in vitro is identical to that of the group II intron. However, because the CP group II ribozyme represents a circularly permuted version of the group II RNA, its reaction products differ markedly from those of canonical group II ribozymes. Instead of the conjoined exons generated by group II RNAs, catalysis by the CP group II ribozyme results in a small circular RNA product. And in place of the lariat structure ultimately assumed by the canonical intron, the reacted form of CP group II RNA adopts a branched T-shaped structure. No functions have yet been ascribed to these unusual reaction products, although it is presumed that they provide some benefit either to the host organism or to an associated selfish element.

Perhaps a more extreme example of the evolution of new function through reorganization of an existing catalytic core is provided by the lariat-capping (LC) ribozyme [104]. This RNA has been identified in certain eukaryotic microorganisms, where it resides immediately upstream of an ORF encoding a homing endonuclease. Although the LC ribozyme appears to have descended from a group I intron, its three-dimensional structure has diverged substantially [105], and its reaction chemistry and biological role are distinct. Unlike the canonical group I intron, which promotes self-splicing via a two-step process, the LC ribozyme catalyzes a one-step branching reaction in which the 2′-hydroxyl group of a residue near the beginning of the ORF attacks a phosphodiester linkage just two nucleotides upstream, resulting in 5′ capping of the mRNA with a three-nucleotide loop and simultaneous release of the upstream LC ribozyme. This tiny lariat structure is likely to extend the half-life of the mRNA by shielding it from 5′ exonucleases, serving in lieu of a 7-methylguanosine cap.

10.10 Will New ncRNAs Broaden the Scope of RNA Catalysis?

Ribozymes are oddities in modern metabolism. Just 15 classes have been discovered, and their catalytic repertoire is generally quite restricted. Whether they emerged during the RNA world era or more recently in evolution, most ribozymes that have persisted to the present day seem to occupy a niche involving phosphoester transfer reactions catalyzed in cis. It is notable, however, that two of the extant ribozymes most compellingly rooted in the RNA world – RNase P and the ribosome – act on their substrates in trans. It is clear from these two examples that ancient RNA world enzymes would not have been limited merely to cleaving, splicing, or otherwise processing themselves.

The catalytic moieties of RNase P and the ribosome, like those of the self-splicing introns and their derivatives, are large RNAs that adopt elaborate, highly conserved secondary and tertiary structures. Such ornate architecture, then, might be a requisite feature for ribozymes that manage intricate catalytic tasks (Figure 10.3). Moreover, there are vanishingly few conserved bacterial ncRNAs of comparable size and structural complexity that have known functions *not* relating to catalysis. Thus, the recent identifications from computational searches of a number of large, highly structured ncRNAs in bacterial genomes and metagenomes [68, 106–108] immediately suggest a tantalizing possibility. Simply by virtue of their sizes and complex secondary structures, these RNAs may reasonably be considered as candidates for ribozymes that catalyze complex reactions (Figure 10.3). The genetic contexts and phylogenetic distributions of many of these ncRNAs do not appear to be consistent with roles in self-cleavage or self-splicing, and for those that have been tested in vitro, no self-processing activities have yet been detected. This leaves open the possibility – or perhaps even establishes the likelihood – that some of these newly discovered RNAs could be enzymes that recognize their substrates in trans. Such hypothetical substrates could be macromolecules, akin to those of RNase P or the ribosome, but small-molecule metabolites should also be considered as possible reactants. Certainly there is ample precedent for the exquisite recognition of metabolites by RNA, set by dozens of riboswitch classes and by the reliance of the group I intron on exogenous guanosine.

Just as contemporary cellular biochemistry depends critically on myriad anabolic and catabolic pathways, a vigorous RNA world metabolism would have required the extensive manipulation of small molecules. It is possible that some of the large complex ncRNAs awaiting characterization are fulfilling just such roles, perhaps even in the very same way that they have been since before the appearance of encoded protein synthesis. Many of these recently reported ncRNAs occupy far more circumscribed phylogenetic niches than the large ribozymes that are currently known, perhaps reflecting their gradual but inexorable slide toward extinction. Whatever their roles, some of these extraordinary molecules should afford a rare glimpse into the ancient era of the RNA world.

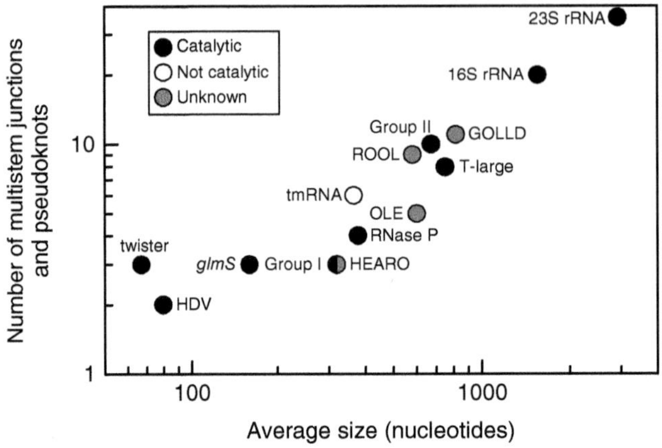

Figure 10.3 Sizes and structural complexities of ribozymes and other structured ncRNAs occurring in bacteria. The number of multistem junctions and pseudoknots occurring within a given ncRNA is used here as an index of structural complexity. This plot includes four examples of ncRNAs whose functions are unknown but whose sizes and structural complexities equal those of known large ribozymes. Because they function as components of the same holoenzyme, 23S and 16S rRNAs are both denoted as catalytic. Transfer-messenger RNA (tmRNA) is one of the only large (>200 nucleotides) structured ncRNAs occurring in bacteria whose function is known not to involve catalysis. RNA classes that occupy roughly the same position in the plot are represented as a bisected circle.

Acknowledgments

RNA research in the Breaker Laboratory is supported by the Howard Hughes Medical Institute and by National Institutes of Health grant P01GM022778.

References

1 Joyce, G.F. (2002). The antiquity of RNA-based evolution. *Nature* 418 (6894): 214–221.
2 Hud, N.V., Cafferty, B.J., Krishnamurthy, R. and Williams, L.D. (2013). The origin of RNA and "my grandfather's axe". *Chem. Biol.* 20 (4): 466–474.
3 Cech, T.R. and Steitz, J.A. (2014). The noncoding RNA revolution-trashing old rules to forge new ones. *Cell* 157 (1): 77–94.
4 Nelson, J.W. and Breaker, R.R. (2017). The lost language of the RNA World. *Sci. Signal.* 10 (483).
5 White, H.B. III., (1976). Coenzymes as fossils of an earlier metabolic state. *J. Mol. Evol.* 7 (2): 101–104.
6 Benner, S.A., Ellington, A.D., and Tauer, A. (1989). Modern metabolism as a palimpsest of the RNA world. *Proc. Natl. Acad. Sci. U.S.A.* 86 (18): 7054–7058.
7 Guerrier-Takada, C., Gardiner, K., Marsh, T. et al. (1983). The RNA moiety of ribonuclease P is the catalytic subunit of the enzyme. *Cell* 35 (3 Pt 2): 849–857.
8 Nissen, P., Hansen, J., Ban, N. et al. (2000). The structural basis of ribosome activity in peptide bond synthesis. *Science* 289 (5481): 920–930.

9 Jarrous, N. (2017). Roles of RNase P and its subunits. *Trends Genet.* 33 (9): 594–603.

10 Gopalan, V., Jarrous, N., and Krasilnikov, A.S. (2018). Chance and necessity in the evolution of RNase P. *RNA* 24 (1): 1–5.

11 Howard, M.J., Liu, X., Lim, W.H. et al. (2013). RNase P enzymes: divergent scaffolds for a conserved biological reaction. *RNA Biol.* 10 (6): 909–914.

12 Lambowitz, A.M. and Belfort, M. (1993). Introns as mobile genetic elements. *Annu. Rev. Biochem.* 62: 587–622.

13 Jimenez, R.M., Polanco, J.A., and Luptak, A. (2015). Chemistry and biology of self-cleaving ribozymes. *Trends Biochem. Sci.* 40 (11): 648–661.

14 Crick, F.H. (1968). The origin of the genetic code. *J. Mol. Biol.* 38 (3): 367–379.

15 Orgel, L.E. (1968). Evolution of the genetic apparatus. *J. Mol. Biol.* 38 (3): 381–393.

16 Woese, C.R. (1967). The evolution of the genetic code. In: *The Genetic Code: the Molecular Basis for Genetic Expression*, 179–195. New York: Harper and Row.

17 Cech, T.R., Zaug, A.J., and Grabowski, P.J. (1981). In vitro splicing of the ribosomal RNA precursor of Tetrahymena: involvement of a guanosine nucleotide in the excision of the intervening sequence. *Cell* 27 (3 Pt 2): 487–496.

18 Kruger, K., Grabowski, P.J., Zaug, A.J. et al. (1982). Self-splicing RNA: autoexcision and autocyclization of the ribosomal RNA intervening sequence of Tetrahymena. *Cell* 31 (1): 147–157.

19 Michel, F. and Dujon, B. (1983). Conservation of RNA secondary structures in two intron families including mitochondrial-, chloroplast- and nuclear-encoded members. *EMBO J.* 2 (1): 33–38.

20 Peebles, C.L., Perlman, P.S., Mecklenburg, K.L. et al. (1986). A self-splicing RNA excises an intron lariat. *Cell* 44 (2): 213–223.

21 van der Veen, R., Arnberg, A.C., van der Horst, G. et al. (1986). Excised group II introns in yeast mitochondria are lariats and can be formed by self-splicing in vitro. *Cell* 44 (2): 225–234.

22 Buzayan, J.M., Gerlach, W.L., and Bruening, G. (1986). Non-enzymatic cleavage and ligation of RNAs complementary to a plant virus satellite RNA. *Nature* 323: 349–353.

23 Feldstein, P.A., Buzayan, J.M., and Bruening, G. (1989). Two sequences participating in the autolytic processing of satellite tobacco ringspot virus complementary RNA. *Gene* 82 (1): 53–61.

24 Haseloff, J. and Gerlach, W.L. (1989). Sequences required for self-catalysed cleavage of the satellite RNA of tobacco ringspot virus. *Gene* 82 (1): 43–52.

25 Hutchins, C.J., Rathjen, P.D., Forster, A.C. and Symons, R.H. (1986). Self-cleavage of plus and minus RNA transcripts of avocado sunblotch viroid. *Nucleic Acids Res.* 14 (9): 3627–3640.

26 Kaper, J.M., Tousignant, M.E., and Steger, G. (1988). Nucleotide sequence predicts circularity and self-cleavage of 300-ribonucleotide satellite of arabis mosaic virus. *Biochem. Biophys. Res. Commun.* 154 (1): 318–325.

27 Prody, G.A., Bakos, J.T., Buzayan, J.M. et al. (1986). Autolytic processing of dimeric plant virus satellite RNA. *Science* 231 (4745): 1577–1580.

28 Forster, A.C. and Symons, R.H. (1987). Self-cleavage of virusoid RNA is performed by the proposed 55-nucleotide active site. *Cell* 50 (1): 9–16.
29 Forster, A.C. and Symons, R.H. (1987). Self-cleavage of plus and minus RNAs of a virusoid and a structural model for the active sites. *Cell* 49 (2): 211–220.
30 Hampel, A. and Tritz, R. (1989). RNA catalytic properties of the minimum (−)sTRSV sequence. *Biochemistry* 28 (12): 4929–4933.
31 Rubino, L., Tousignant, M.E., Steger, G. and Kaper, J.M. (1990). Nucleotide sequence and structural analysis of two satellite RNAs associated with chicory yellow mottle virus. *J. Gen. Virol.* 71 (Pt 9): 1897–1903.
32 Wu, H.N. and Lai, M.M. (1989). Reversible cleavage and ligation of hepatitis delta virus RNA. *Science* 243 (4891): 652–654.
33 Wu, H.N., Lin, Y.J., Lin, F.P. et al. (1989). Human hepatitis delta virus RNA subfragments contain an autocleavage activity. *Proc. Natl. Acad. Sci. U.S.A.* 86 (6): 1831–1835.
34 Saville, B.J. and Collins, R.A. (1990). A site-specific self-cleavage reaction performed by a novel RNA in *Neurospora* mitochondria. *Cell* 61 (4): 685–696.
35 Noller, H.F. (1984). Structure of ribosomal RNA. *Annu. Rev. Biochem.* 53: 119–162.
36 Moazed, D. and Noller, H.F. (1989). Interaction of tRNA with 23S rRNA in the ribosomal A, P, and E sites. *Cell* 57 (4): 585–597.
37 Noller, H.F., Hoffarth, V., and Zimniak, L. (1992). Unusual resistance of peptidyl transferase to protein extraction procedures. *Science* 256 (5062): 1416–1419.
38 Ban, N., Nissen, P., Hansen, J. et al. (2000). The complete atomic structure of the large ribosomal subunit at 2.4 A resolution. *Science* 289 (5481): 905–920.
39 Robertson, H.D., Altman, S., and Smith, J.D. (1972). Purification and properties of a specific *Escherichia coli* ribonuclease which cleaves a tyrosine transfer ribonucleic acid presursor. *J. Biol. Chem.* 247 (16): 5243–5251.
40 Kole, R., Baer, M.F., Stark, B.C. and Altman, S. (1980). *E. coli* RNAase P has a required RNA component. *Cell* 19 (4): 881–887.
41 Stark, B.C., Kole, R., Bowman, E.J. and Altman, S. (1978). Ribonuclease P: an enzyme with an essential RNA component. *Proc. Natl. Acad. Sci. U.S.A.* 75 (8): 3717–3721.
42 Breaker, R.R. (2012). Riboswitches and the RNA world. *Cold Spring Harbor Perspect. Biol.* 4 (2): a003566.
43 McCown, P.J., Corbino, K.A., Stav, S. et al. (2017). Riboswitch diversity and distribution. *RNA* 23 (7): 995–1011.
44 Baker, J.L., Sudarsan, N., Weinberg, Z. et al. (2012). Widespread genetic switches and toxicity resistance proteins for fluoride. *Science* 335 (6065): 233–235.
45 Sudarsan, N., Barrick, J.E., and Breaker, R.R. (2003). Metabolite-binding RNA domains are present in the genes of eukaryotes. *RNA* 9 (6): 644–647.
46 Winkler, W., Nahvi, A., and Breaker, R.R. (2002). Thiamine derivatives bind messenger RNAs directly to regulate bacterial gene expression. *Nature* 419 (6910): 952–956.

47 Barrick, J.E., Corbino, K.A., Winkler, W.C. et al. (2004). New RNA motifs suggest an expanded scope for riboswitches in bacterial genetic control. *Proc. Natl. Acad. Sci. U.S.A.* 101 (17): 6421–6426.

48 Corbino, K.A., Barrick, J.E., Lim, J. et al. (2005). Evidence for a second class of *S*-adenosylmethionine riboswitches and other regulatory RNA motifs in alpha-proteobacteria. *Genome Biol.* 6 (8): R70.

49 Weinberg, Z., Barrick, J.E., Yao, Z. et al. (2007). Identification of 22 candidate structured RNAs in bacteria using the CMfinder comparative genomics pipeline. *Nucleic Acids Res.* 35 (14): 4809–4819.

50 Yao, Z., Barrick, J., Weinberg, Z. et al. (2007). A computational pipeline for high- throughput discovery of cis-regulatory noncoding RNA in prokaryotes. *PLoS Comput. Biol.* 3 (7): e126.

51 McCown, P.J., Roth, A. and Breaker, R.R. (2011). An expanded collection and refined consensus model of glmS ribozymes. *RNA* 17 (4): 728–736.

52 Winkler, W.C., Nahvi, A., Roth, A. et al. (2004). Control of gene expression by a natural metabolite-responsive ribozyme. *Nature* 428 (6980): 281–286.

53 McCarthy, T.J., Plog, M.A., Floy, S.A. et al. (2005). Ligand requirements for glmS ribozyme self-cleavage. *Chem. Biol.* 12 (11): 1221–1226.

54 Cochrane, J.C., Lipchock, S.V., and Strobel, S.A. (2007). Structural investigation of the GlmS ribozyme bound to Its catalytic cofactor. *Chem. Biol.* 14 (1): 97–105.

55 Klein, D.J. and Ferre-D'Amare, A.R. (2006). Structural basis of glmS ribozyme activation by glucosamine-6-phosphate. *Science* 313 (5794): 1752–1756.

56 Roth, A., Nahvi, A., Lee, M. et al. (2006). Characteristics of the glmS ribozyme suggest only structural roles for divalent metal ions. *RNA* 12 (4): 607–619.

57 Wilkinson, S.R. and Been, M.D. (2005). A pseudoknot in the 3′ non-core region of the glmS ribozyme enhances self-cleavage activity. *RNA* 11 (12): 1788–1794.

58 Davis, J.H., Dunican, B.F., and Strobel, S.A. (2011). glmS Riboswitch binding to the glucosamine-6-phosphate alpha-anomer shifts the pK_a toward neutrality. *Biochemistry* 50 (33): 7236–7242.

59 Viladoms, J., Scott, L.G., and Fedor, M.J. (2011). An active-site guanine participates in glmS ribozyme catalysis in its protonated state. *J. Am. Chem. Soc.* 133 (45): 18388–18396.

60 Xin, Y. and Hamelberg, D. (2010). Deciphering the role of glucosamine-6-phosphate in the riboswitch action of glmS ribozyme. *RNA* 16 (12): 2455–2463.

61 Bingaman, J.L., Zhang, S., Stevens, D.R. et al. (2017). The GlcN6P cofactor plays multiple catalytic roles in the glmS ribozyme. *Nat. Chem. Biol.* 13 (4): 439–445.

62 Cochrane, J.C., Lipchock, S.V., Smith, K.D. and Strobel, S.A. (2009). Structural and chemical basis for glucosamine 6-phosphate binding and activation of the glmS ribozyme. *Biochemistry* 48 (15): 3239–3246.

63 Hampel, K.J. and Tinsley, M.M. (2006). Evidence for preorganization of the glmS ribozyme ligand binding pocket. *Biochemistry* 45 (25): 7861–7871.

64 Collins, J.A., Irnov, I., Baker, S., Winkler, W.C. (2007). Mechanism of mRNA destabilization by the glmS ribozyme. *Genes Dev.* 21 (24): 3356–3368.

65 Tseng, H.H., Weinberg, Z., Gore, J. et al. (2009). Finding non-coding RNAs through genome-scale clustering. *J. Bioinf. Comput. Biol.* 7 (2): 373–388.

66 Weinberg, Z., Wang, J.X., Bogue, J. et al. (2010). Comparative genomics reveals 104 candidate structured RNAs from bacteria, archaea, and their metagenomes. *Genome Biol.* 11 (3): R31.

67 Roth, A., Weinberg, Z., Chen, A.G. et al. (2014). A widespread self-cleaving ribozyme class is revealed by bioinformatics. *Nat. Chem. Biol.* 10 (1): 56–60.

68 Weinberg, Z., Lünse, C.E., Corbino, K.A. et al. (2017). Detection of 224 candidate structured RNAs by comparative analysis of specific subsets of intergenic regions. *Nucleic Acids Res.* 45 (18): 10811–10823.

69 Weinberg, Z., Kim, P.B., Chen, T.H. et al. (2015). New classes of self-cleaving ribozymes revealed by comparative genomics analysis. *Nat. Chem. Biol.* 11 (8): 606–610.

70 Yao, Z., Weinberg, Z., and Ruzzo, W.L. (2006). CMfinder – a covariance model based RNA motif finding algorithm. *Bioinformatics* 22 (4): 445–452.

71 Eiler, D., Wang, J., and Steitz, T.A. (2014). Structural basis for the fast self-cleavage reaction catalyzed by the twister ribozyme. *Proc. Natl. Acad. Sci. U.S.A.* 111 (36): 13028–13033.

72 Liu, Y., Wilson, T.J., and Lilley, D.M.J. (2017). The structure of a nucleolytic ribozyme that employs a catalytic metal ion. *Nat. Chem. Biol.* 13 (5): 508–513.

73 Liu, Y., Wilson, T.J., McPhee, S.A. and Lilley, D.M. (2014). Crystal structure and mechanistic investigation of the twister ribozyme. *Nat. Chem. Biol.* 10 (9): 739–744.

74 Ren, A., Kosutic, M., Rajashankar, K.R. et al. (2014). In-line alignment and Mg(2)(+) coordination at the cleavage site of the env22 twister ribozyme. *Nat. Commun.* 5: 5534.

75 Zheng, L., Mairhofer, E., Teplova, M. et al. (2017). Structure-based insights into self-cleavage by a four-way junctional twister-sister ribozyme. *Nat. Commun.* 8 (1): 1180.

76 Perreault, J., Weinberg, Z., Roth, A. et al. (2011). Identification of hammerhead ribozymes in all domains of life reveals novel structural variations. *PLoS Comput. Biol.* 7 (5): e1002031.

77 Sanchez-Luque, F.J., Lopez, M.C., Macias, F. et al. (2011). Identification of an hepatitis delta virus-like ribozyme at the mRNA 5′-end of the L1Tc retrotransposon from *Trypanosoma cruzi*. *Nucleic Acids Res.* 39 (18): 8065–8077.

78 Li, S. and Breaker, R.R. (2017). Identification of 15 candidate structured noncoding RNA motifs in fungi by comparative genomics. *BMC Genomics* 18 (1): 785.

79 Harris, K.A., Lünse, C.E., Li, S. et al. (2015). Biochemical analysis of pistol self-cleaving ribozymes. *RNA* 21 (11): 1852–1858.

80 Li, S., Lünse, C.E., Harris, K.A. and Breaker, R.R. (2015). Biochemical analysis of hatchet self-cleaving ribozymes. *RNA* 21 (11): 1845–1851.

81 de la Pena, M. and Garcia-Robles, I. (2010). Ubiquitous presence of the hammerhead ribozyme motif along the tree of life. *RNA* 16 (10): 1943–1950.

82 Jimenez, R.M., Delwart, E., and Luptak, A. (2011). Structure-based search reveals hammerhead ribozymes in the human microbiome. *J. Biol. Chem.* 286 (10): 7737–7743.

83 Seehafer, C., Kalweit, A., Steger, G. et al. (2011). From alpaca to zebrafish: hammerhead ribozymes wherever you look. *RNA* 17 (1): 21–26.

84 Webb, C.H., Riccitelli, N.J., Ruminski, D.J. and Luptak, A. (2009). Widespread occurrence of self-cleaving ribozymes. *Science* 326 (5955): 953.

85 Felletti, M., Bieber, A., and Hartig, J.S. (2017). The 3′-untranslated region of mRNAs as a site for ribozyme cleavage-dependent processing and control in bacteria. *RNA Biol.* 14 (11): 1522–1533.

86 Hammann, C., Luptak, A., Perreault, J. and de la Pena, M. (2012). The ubiquitous hammerhead ribozyme. *RNA* 18 (5): 871–885.

87 Epstein, L.M. and Gall, J.G. (1987). Self-cleaving transcripts of satellite DNA from the newt. *Cell* 48 (3): 535–543.

88 Rojas, A.A., Vazquez-Tello, A., Ferbeyre, G. et al. (2000). Hammerhead-mediated processing of satellite pDo500 family transcripts from *Dolichopoda* cave crickets. *Nucleic Acids Res.* 28 (20): 4037–4043.

89 Cervera, A. and De la Pena, M. (2014). Eukaryotic penelope-like retroelements encode hammerhead ribozyme motifs. *Mol. Biol. Evol.* 31 (11): 2941–2947.

90 Cervera, A., Urbina, D., and de la Pena, M. (2016). Retrozymes are a unique family of non-autonomous retrotransposons with hammerhead ribozymes that propagate in plants through circular RNAs. *Genome Biol.* 17 (1): 135.

91 Eickbush, D.G. and Eickbush, T.H. (2010). R2 retrotransposons encode a self-cleaving ribozyme for processing from an rRNA cotranscript. *Mol. Cell. Biol.* 30 (13): 3142–3150.

92 Ferbeyre, G., Smith, J.M., and Cedergren, R. (1998). Schistosome satellite DNA encodes active hammerhead ribozymes. *Mol. Cell. Biol.* 18 (7): 3880–3888.

93 Ruminski, D.J., Webb, C.H., Riccitelli, N.J. and Luptak, A. (2011). Processing and translation initiation of non-long terminal repeat retrotransposons by hepatitis delta virus (HDV)-like self-cleaving ribozymes. *J. Biol. Chem.* 286 (48): 41286–41295.

94 Sanchez-Luque, F.J., Lopez, M.C., Carreira, P.E. et al. (2014). The wide expansion of hepatitis delta virus-like ribozymes throughout trypanosomatid genomes is linked to the spreading of L1Tc/ingi clade mobile elements. *BMC Genomics* 15: 340.

95 Luchetti, A. and Mantovani, B. (2013). Non-LTR R2 element evolutionary patterns: phylogenetic incongruences, rapid radiation and the maintenance of multiple lineages. *PLoS One* 8 (2): e57076.

96 Spotila, L.D., Hirai, H., Rekosh, D.M. and Lo Verde, P.T. (1989). A retroposon-like short repetitive DNA element in the genome of the human blood fluke, *Schistosoma mansoni*. *Chromosoma* 97 (6): 421–428.

97 Salehi-Ashtiani, K., Luptak, A., Litovchick, A. and Szostak, J.W. (2006). A genomewide search for ribozymes reveals an HDV-like sequence in the human CPEB3 gene. *Science* 313 (5794): 1788–1792.

98 de la Pena, M. and Garcia-Robles, I. (2010). Intronic hammerhead ribozymes are ultraconserved in the human genome. *EMBO Rep.* 11 (9): 711–716.

99 Martick, M., Horan, L.H., Noller, H.F. and Scott, W.G. (2008). A discontinuous hammerhead ribozyme embedded in a mammalian messenger RNA. *Nature* 454 (7206): 899–902.

100 Adamidi, C., Fedorova, O., and Pyle, A.M. (2003). A group II intron inserted into a bacterial heat-shock operon shows autocatalytic activity and unusual thermostability. *Biochemistry* 42 (12): 3409–3418.

101 Lee, E.R., Baker, J.L., Weinberg, Z. et al. (2010). An allosteric self-splicing ribozyme triggered by a bacterial second messenger. *Science* 329 (5993): 845–848.

102 Chen, A.G., Sudarsan, N., and Breaker, R.R. (2011). Mechanism for gene control by a natural allosteric group I ribozyme. *RNA* 17 (11): 1967–1972.

103 Roth, A., Weinberg, Z., Vanderschuren, K. et al. (2021). A naturally occurring circularly permuted group II intron in bacteria.

104 Nielsen, H., Westhof, E., and Johansen, S. (2005). An mRNA is capped by a 2′,5′ lariat catalyzed by a group I-like ribozyme. *Science* 309 (5740): 1584–1587.

105 Meyer, M., Nielsen, H., Olieric, V. et al. (2014). Speciation of a group I intron into a lariat capping ribozyme. *Proc. Natl. Acad. Sci. U.S.A.* 111 (21): 7659–7664.

106 Harris, K.A. and Breaker, R.R. (2018). Large noncoding RNAs in bacteria. *Microbiol. Spectr.* 6 (4).

107 Puerta-Fernandez, E., Barrick, J.E., Roth, A. and Breaker, R.R. (2006). Identification of a large noncoding RNA in extremophilic eubacteria. *Proc. Natl. Acad. Sci. U.S.A.* 103 (51): 19490–19495.

108 Weinberg, Z., Perreault, J., Meyer, M.M. and Breaker, R.R. (2009). Exceptional structured noncoding RNAs revealed by bacterial metagenome analysis. *Nature* 462 (7273): 656–659.

11

Small Self-Cleaving Ribozymes in the Genomes of Vertebrates

Marcos de la Peña

IBMCP (CSIC-UPV), C/Ingeniero Fausto Elio s/n, 46022, Valencia, Spain

11.1 The Family of Small Self-Cleaving Ribozymes in Eukaryotic Genomes: From Retrotransposition to Domestication

Among all known catalytic RNAs or ribozymes, there is an enigmatic group of small (~50–200 nt) self-cleaving RNAs, which catalyze an intramolecular reaction of 2′-hydroxyl transesterification in a sequence-specific way. This group of nucleolytic ribozymes can carry out RNA self-cleavage at rates only a few times slower than their protein counterparts [1]. At least nine classes of naturally occurring small ribozymes have been described so far, which in chronological order of their discovery are hammerhead ([2, 3], hairpin [4], human hepatitis D [5], Varkud satellite [6], GlmS [7], twister [8], twister sister, hatchet, and pistol [9] ribozymes.

The hammerhead, the hairpin, and the human hepatitis D ribozymes were the first examples of small self-cleaving catalytic RNAs reported at the end of the 1980s. A common trait of these three motifs was their occurrence in the small (~200–1600 nt) circular RNA (circRNA) genomes of a group of atypical subviral agents able to infect either plants (plant virus satellite RNAs and viroids for the hammerhead and the hairpin ribozymes) or humans (the human hepatitis D virus for the hepatitis D ribozyme). It was not until the last decade that we became aware of the widespread occurrence of several of these small ribozymes among prokaryotic and eukaryotic DNA genomes [10]. Regarding eukaryotic genomes, small catalytic RNAs can be found in diverse plants and animals, including our own genome, and carrying out biological functions that we are just starting to learn. The most recent data point to the existence of a tight connection between diverse genomic small ribozymes and retrotransposable elements, either autonomous or nonautonomous [11–14]. These retrotransposon-associated self-cleaving ribozymes usually appear as conserved motifs in the boundaries of the retroelement (either at the 5′ or at both 5′- and 3′-ends), which indicates that RNA self-cleaving activity would be required in a processing step of the mobile element transcripts. Nevertheless, this gene regulatory landscape changes in the case of the genomes of higher vertebrates

Ribozymes, First Edition. Edited by Sabine Müller, Benoît Masquida, and Wade Winkler.

such as amniotes (reptiles, birds, and mammals), which seem to have lost these particular families of ribozyme-containing retrotransposons during their evolution. As described in this chapter, while retrotransposons with small ribozymes have efficiently colonized the genomes of lower vertebrates (fishes and amphibians), the genomes of amniotes have only retained a few highly conserved self-cleaving motifs, which have been evolutionarily selected to play new biological roles as a result of functional domestication.

11.2 The Widespread Case of the Hammerhead Ribozyme: From Bacteria to Vertebrate Genomes

The hammerhead ribozyme (HHR) is the most prevalent small ribozyme in prokaryotic and eukaryotic genomes. This catalytic RNA has a simple structure, composed of three double helixes (named I–III) surrounding a catalytic core of 15 highly conserved nucleotides (Figure 11.1a). Its secondary structure resembles a hammerhead-like shape, although the real 3D structure corresponds with a γ-shaped motif (Figure 11.1b). This architecture is further stabilized through tertiary interactions between the helical stems and loops of helixes I and II, which are required for the highest efficiency of self-cleavage in vivo under conditions of low magnesium concentration [16–18]. There are three possible circularly permuted topologies for the HHR, named types I, II, and III, depending on the open-ended helix (Figure 11.1c). The first HHRs were reported from a group of simple infectious agents of plants, such as plant virus satellite RNAs and viroids, which are exclusively composed of a small, noncoding circRNA genome [2, 3]. Since then, the discovery of newer examples of viroid-like circRNAs with ribozymes in plants has offered a collection of more than 20 different HHR motifs, mostly of the type III architecture [19]. But soon, after the first reports of viroidal HHRs, examples of type I HHRs were described in the repetitive regions (satellite DNA) of the genomes of newts and salamanders [20, 21]. Similar type I HHRs were also reported in the satellite DNA of some trematode [22] and cave cricket [23] genomes, and examples of type III HHRs were later described in the genome of the *Arabidopsis thaliana* plant [24]. Since 2010, different labs have discovered a widespread occurrence of the HHR motif in all domains of life. Different bioinformatic approaches revealed the presence of *bona fide* HHRs in many prokaryotic and eukaryotic genomes [25–28], including our own genome [29], confirming that the HHR is a ubiquitous catalytic RNA in all life kingdoms. Interestingly, whereas prokaryotic and bacteriophage genomes showed the presence of all types of HHRs (type I, II, and III), metazoan and plant genomes almost exclusively contain type I and type III HHR motifs, respectively (Figure 11.1c) [15, 30, 31]. In the case of plants, we have found that diverse genomes of angiosperms encode type III HHRs located within short tandem repeats. These repeats express as abundant small circRNAs, very similar to viroidal RNAs with ribozymes, which are in fact the RNA intermediates of a novel family of nonautonomous retroelements called *retrozymes*, for *retro*elements with HHR*s* [12]. In the genomes of metazoans from cnidarians to lower vertebrates, such as fishes

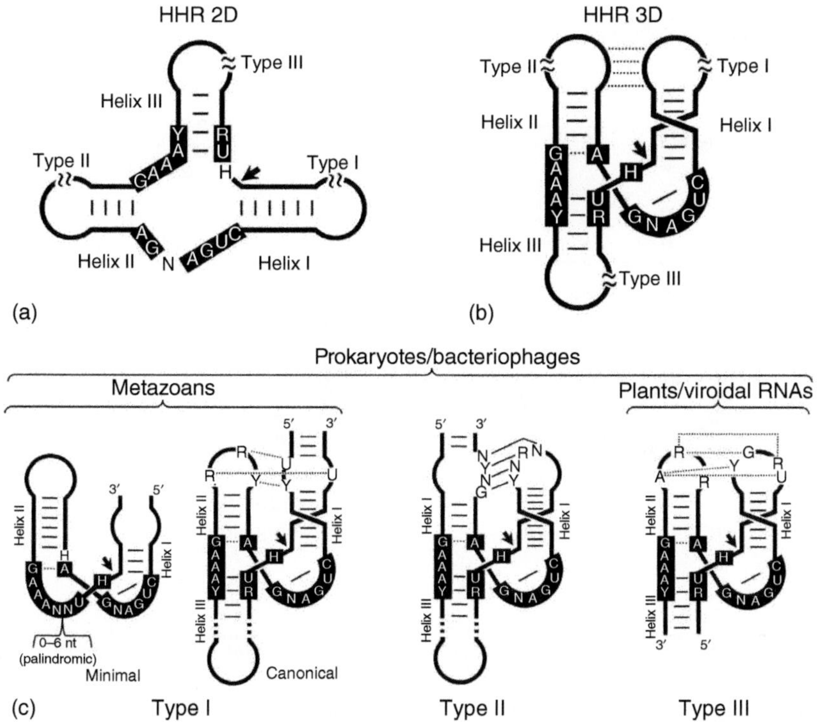

Figure 11.1 Diagrams of the hammerhead ribozyme drawn in (a) the classical secondary or (b) three-dimensional structure representations. (c) Diagrams of each of the three possible topologies (types I, II, and III) of naturally occurring hammerhead ribozymes and their typical distribution among DNA and RNA genomes. Dotted and continuous lines refer to non-canonical and Watson–Crick base pairs, respectively. Characteristic tertiary interactions for each hammerhead type are depicted. Black boxes indicate the conserved nucleotides at the catalytic core. Source: Adapted from Garcia-Robles et al. [15].

and amphibians, we find type I HHR motifs similar to those originally described in the repetitive DNA of newts. These HHRs have been detected in the genomes of many other newts and salamanders, the lamprey fish *Petromyzon marinus*, and the frog *Xenopus tropicalis* and usually as part of a repetitive satellite DNA (Figure 11.2) [27–29, 31]. A typical trait of the HHRs in lower vertebrates resides in their unstable and palindromic helix III (0–2 bp stem), which avoids extensive in vitro self-cleavage as single motifs but not as dimeric motifs through the adoption of an elongated, narrow helix III [3, 32, 33]. We can assume that satellite DNA with HHRs in metazoans could be regarded as another family of nonautonomous retrotransposons with ribozymes or *retrozymes* similar to those described in plant genomes [12, 30].

Moreover, lower vertebrates also contain numerous examples of another variant of type I HHR, which is characterized by the lack of any nucleotide at the helix III, as well as any recognizable tertiary interaction between helixes I and II (Figures 11.1c and 11.2). These minimal HHRs are frequently associated with the ancient family

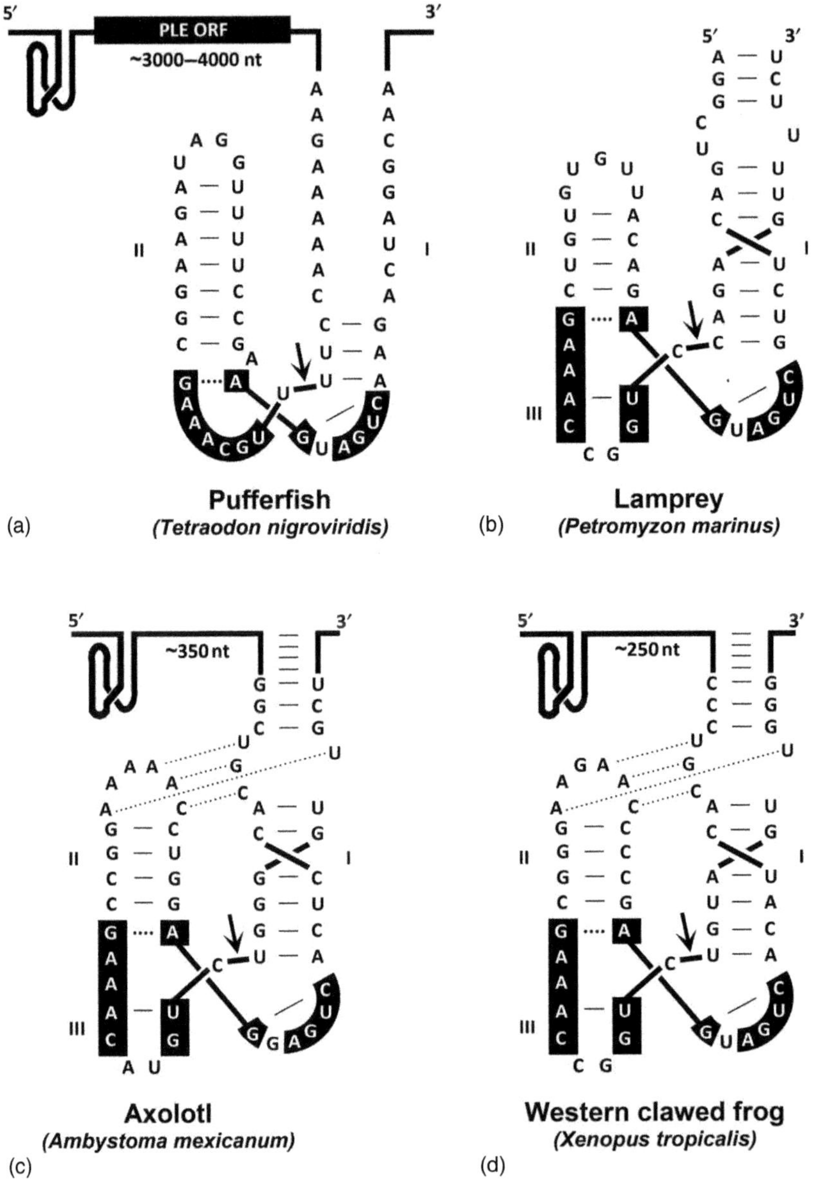

Figure 11.2 (a) Representative example of a minimal type I hammerhead present in the 5' and 3' terminal repeats of a Penelope-like element in the green pufferfish genome. (b) Representative example of a type I hammerhead ribozyme present in the genomes of lampreys. Examples of type I hammerhead ribozymes present in the repetitive satellite DNA of salamanders (c) and frogs (d). The characteristic tertiary loop 1-loop 2 interactions detected in the amphibian but not in the fish motifs, are indicated with dotted lines.

of Penelope-like retrotransposable elements (PLEs) [11, 34]. Although type I HHRs from either autonomous (PLEs) or nonautonomous (retrozymes) retrotransposons do not share any sequence homology, their structural similarities and coexistence in metazoan genomes suggest a relationship among them (Figure 11.2) [11, 25, 29].

As mentioned before, only a few and highly conserved examples of genomic HHRs seem to have subsisted in higher vertebrates such as amniotes. This small set of HHRs in the genomes of amniotes would derive from those motifs present in lower metazoans after different events of domestication or exaptation during the evolution of the genomes of vertebrates.

11.2.1 The Discontinuous HHR in Mammals

The so-called discontinuous hammerhead ribozyme (dis-HHR) was the first example of an HHR in a mammalian genome [35]. A bioinformatic analysis revealed the presence of this atypical HHR motif in the 3′-untranslated regions (UTRs) of C-type lectin-like family 2 (*CLEC2*) genes of some mammals. Although the ribozyme is formally a type III HHR (Figure 11.3), its unusual size and architecture, as well as the kind of tertiary loop–loop interactions, strongly suggest that dis-HHRs would be derived from one or maybe two of the typical type I HHRs present in lower metazoans [15, 25]. As shown in Figure 11.3, mammalian

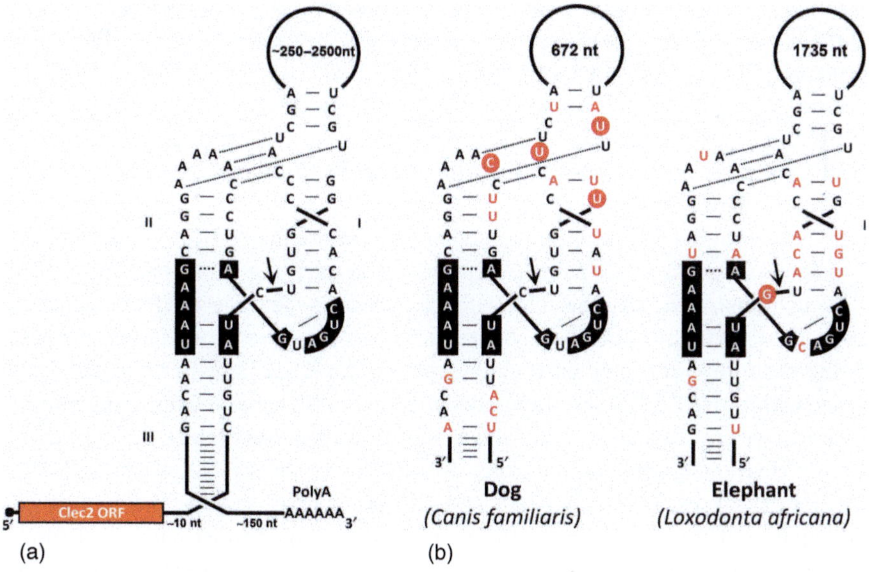

Figure 11.3 (a) Schematic representation of the discontinuous hammerhead ribozyme present at the 3′-UTR of the CLEC2 gene of some mammals (sequence corresponds to the mouse motif). Sizes of helixes I and flanking regions of the mammalian ribozymes are indicated. (b) Examples of deleterious ribozyme variants present in the dog and the elephant genomes. Mutations expected to either affect or abolish the self-cleaving capabilities of the ribozyme with respect to the mouse motif are shown in red and within circles, respectively.

dis-HHRs show a very long helix I, which splits the ribozyme into two parts separated from ~150 to more than 2000 nt, depending on the species. Moreover, the helix III of the dis-HHRs are also unusually large, with ~30 consecutive bp, instead of the 4–6 bp found in classical type III HHR motifs of plants and viroid-like RNAs. But the trait that allows the conclusion that the dis-HHR derives from a type I HHR motif is the set of tertiary interactions between loops 1 and 2, which totally match with the ones reported for most type I HHRs [19, 35].

The dis-HHR motif can be found in the genomes of a variety of mammals and in the 3′-UTR of *CLEC2* genes, such as *CLEC2d* and *CLEC2e* genes. C-type lectins, in general, represent a complex superfamily of mammalian lectins that comprises several different groups with different domain structures. The particular case of the CLEC2 family seems to result from multiple gene duplication episodes during the evolution of mammals, resulting in a different number of copies depending on the organism, with several putative isoforms for each copy [36, 37]. In this regard, dis-HHRs only occur in one or a few of the CLEC2 copies present in a given genome. Comparative analysis of the dis-HHR from diverse mammals shows that the helix III of the ribozyme starts a few nucleotides after the stop codon of the CLEC2 open reading frame and finishes about 100–200 nt before the polyadenylation site of the gene. The dis-HHRs seem to follow a patchy distribution among mammals, with a common presence in rodents and diverse ungulates but absence in primates or most carnivores. The conserved presence of a dis-HHR in a CLEC2 gene of the platypus genome indicates that this ribozyme already existed in the genomes of the mammalian ancestors (Figure 11.4). Consequently, the absence of the dis-HHRs in diverse mammalian genomes would be likely due to a genomic loss during the evolution of these animals. It can be highlighted that diverse mammals seem to retain remnants of the dis-HHR motif, which harbor mutations that are known to abolish the self-cleaving activity of the ribozyme indicating a process of fossilization of the motif (Figure 11.3B).

Functional studies regarding the biological role of the dis-HHR in mammals are still lacking. Original work using a reporter gene carrying this ribozyme at the 3′-UTR demonstrated that the dis-HHR catalyzed cis-cleavage of the messenger RNA (mRNA), which induced an 80% reduction of the gene expression compared with the same reporter gene carrying a mutated dis-HHR [35]. In vivo, dis-HHR cleavage within the 3′-UTR should remove the polyA tail, generally inducing proper nuclear export, circularization, and, consequently, translation of the CLEC2 mRNA. Thus, self-cleavage is not expected to be very efficient in order to allow some CLEC2 expression, and most likely, dis-HHR should have a more specific mechanism in the regulation of the CLEC2 gene. In this regard, and in contrast with most small and compact self-cleaving ribozymes described so far, mammalian dis-HHRs show a quite complex structure where the substrate and enzyme segments are located hundreds of nucleotides away from each other, which potentially suggests multiple folding issues during mRNA transcription. Moreover, these large sequence intervals seem to have increased during mammal evolution, especially if we take the platypus dis-HHR (~150 nt separation) as the most ancient version of this ribozyme. Consequently, we can deduce that this discontinuous format could have been preserved to

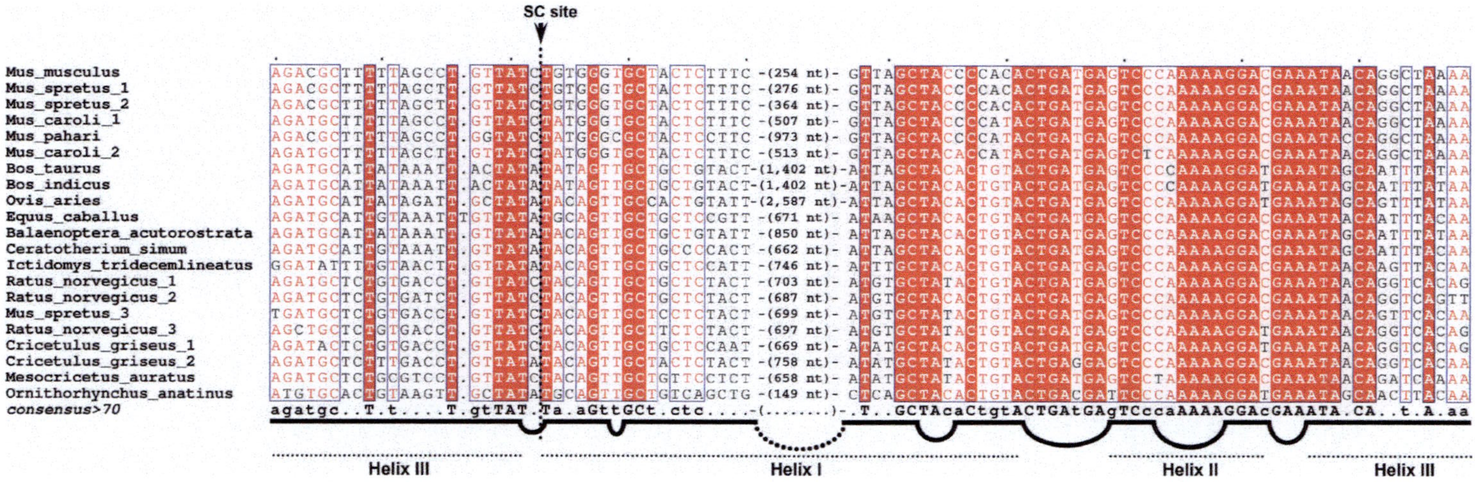

Figure 11.4 DNA sequence alignment of the discontinuous hammerhead motifs from rodents, ungulates, and a monotreme. Only the ribozyme sequences able to adopt the *bona fide* hammerhead motif were used for the alignment. Nucleotide sequences corresponding to helix I extensions were not included in the alignment, but their sizes are indicated. Consensus sequence of the ribozyme alignment is shown at the bottom.

avoid the efficient and fast self-cleavage activity of the classical HHR. Nevertheless, more sophisticated roles in the control of CLEC2 gene expression can be envisaged for the dis-HHR, and due to its particular placement at the 3′-UTR of the gene, posttranscriptional regulation, such as the control of the mRNA decay, microRNA (miRNA) binding, or alternative polyadenylation sites among others, cannot be ruled out [38, 39].

11.2.2 Intronic HHRs in Amniotes

In 2009, following the discovery of the dis-HHRs in Scott's lab, we performed a series of bioinformatic analyses through iterative homology searches using small portions of the HHR motif. The goal of this strategy was to uncover the presence of HHR motifs in diverse genomes by further analysis of the regions surrounding the matches retrieved from homology searches. This simple approach revealed an unexpectedly high number of positive HHR hits in prokaryotic genomes and, most notably, in repetitive DNA regions of eukaryotic genomes [25, 31]. Moreover, these searches also resulted in very few examples of exceptionally conserved type I HHRs in the genomes of all mammals and other amniotes, such as reptiles and birds.

A Highly Conserved HHR in the Intron of an Anti-oncogene

A first HHR motif called HH9 ribozyme was bioinformatically detected in the eighth and large intron (15 kb) of the reversion-inducing cysteine-rich protein with Kazal motifs (RECK) gene present in the chromosome 9 of the human genome. RECK is a key membrane glycoprotein present in all metazoans, which was originally characterized as a tumor suppressor [40]. This factor is known to inhibit the metalloproteinases involved in remodeling the extracellular matrix, a vital step during embryogenesis and angiogenesis [41]. The whole architecture of the intronic RECK HH9 resembles to the classical type I HHR motifs present in the genomes of many lower metazoans [25, 29], including the predicted tertiary interactions between loops 1 and 2 (Figure 11.5A) [18]. The helix III of HH9, however, is different from the ones found in most metazoan HHRs: whereas HH9 helix III has 4 bp capped by a tri-loop, type I HHRs in lower metazoans have palindromic sequences adopting less stable stems (1–2 bp), which usually results in low self-cleaving activity as a monomeric motif [29]. However, HH9 ribozyme shows very fast self-cleavage under low Mg^{2+} concentration, with an in vitro k_{obs} of 2.4 min^{-1} for the human motif, which is in the same order as the fastest natural HHRs described in the literature. On the other hand, HH9 motifs show a quite long and stable helix I, with at least 6 bp on either side of the loop 1. Thus, under physiological conditions, the unimolecular reaction is expected to achieve a dynamic equilibrium of self-cleavage/self-ligation with slow product dissociation. Schistosome type I HHR is known to be as efficient at ligation as it is at cleavage [42], so intronic HH9 could follow in vivo a more complex behavior. Moreover, previous analysis revealed two expressed sequence tag (EST) sequences isolated from *Bos taurus* neural tissues that mapped to the HH9 ribozyme [29]. Both sequences were chimeric molecules resulting from ligation events of two different RNAs; the 5′-side of both ESTs corresponded to the small

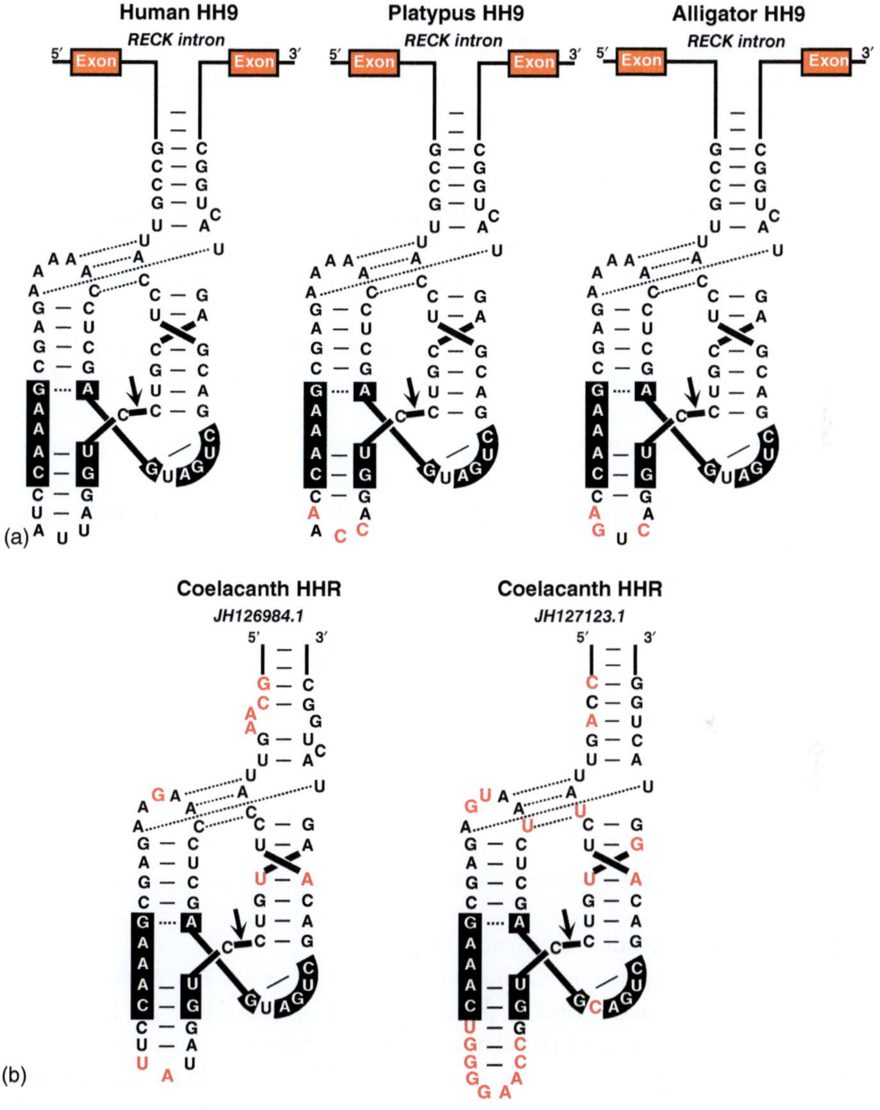

Figure 11.5 (a) Schematic representation of the intronic HH9 hammerhead ribozymes present in the RECK gene of the human, the platypus, and the alligator *Alligator mississippiensis* genomes. Nucleotide changes with respect to the human motif are shown in red. (b) Examples of HH9-like hammerhead ribozymes present in the genome of the coelacanth *Latimeria chalumnae*. Nucleotide differences with the human HH9 motif are shown in red."

nuclear RNAs (snRNAs) U5 or U6, respectively, whereas the 3′-side corresponded to the 3′ fragment resulting from the HH9 self-cleaved intron. These EST sequences suggest that not only the intron harboring HH9 would self-cleave in vivo but also the resulting 3′ product of the self-cleavage could interfere with the splicing machinery. While these chimeric RNA molecules may offer some hints about the mechanism of action of intronic HHRs (see below), no other examples of similar ESTs have been detected so far in the databases.

Comparative genomic analysis of the HH9 motifs among mammals shows a really significant conservation for this ribozyme in the long introns of the *RECK* orthologues of all mammalian genomes so far sequenced (sequence identity >95% for 62 nt). Moreover, this sequence conservation can be extended to the *RECK* orthologues of all birds and even some reptiles such as turtles and alligators, but not lizards (Figure 11.6). The conserved presence of the intronic HH9 in the *RECK* orthologues of most amniotes but not in other tetrapods such as amphibians indicates that the RECK ribozyme would have been functionally fixed during the origin of the earliest amniotic organisms (∼300 Mya). However, the original source of the HH9 motif would be much older. Previous sequence analyses [15] revealed the presence of hundreds of HHRs in the coelacanth genome (*Latimeria chalumnae*) showing high sequence identity with the intronic HH9 motifs of amniotes, with only a few differences restrained to the helix III/loop 3 region of the HHR motif (Figure 11.5). These coelacanth HHRs are usually found as isolated motifs dispersed along the fish genome, with some examples located within annotated gene introns, although none of them correspond to the *RECK* gene. These parallelisms suggest that intronic HHRs in the coelacanth and in amniotes could perform similar biological roles (see below). Although further studies are required, the numerous and widespread HHRs in the coelacanth genome may connect the origin of these particular ribozymes to mobile genetic elements. As the most feasible landscape, some of these particular coelacanth HHRs would have survived and domesticated in the tetrapod genomes that predated amniotes, whereas amphibians would have only retained the HHRs associated to mobile retroelements (Figure 11.2).

A Second Intronic HHR Motif in the Human Genome

A second HHR motif present in the human genome is called HH10 (Figure 11.7A). This type I ribozyme was found in the second and largest intron (11 kb) of the CCDC186 (also known as C10orf118) gene at chromosome 10, which codes for the cutaneous T-cell lymphoma (CTCL) tumor antigen L14-2. This is an uncharacterized protein factor expressed in all tissue types, which has been linked to T-cell lymphomas and other types of cancer [43]. The intronic HH10 motif is conserved in the same CCDC186 locus of many mammals, with the exception of most rodents and lagomorphs, among others. Helixes I and II of the HH10 ribozyme are very similar to the ones in HH9, including loop–loop interactions, with the exception of a G-C to A-U covariation at helix I (Figure 11.5B). Interestingly, this particular covariation is also present in most coelacanth HHRs, which can be regarded as another indication of the evolutionary relationship between the fish and Amniota HHRs. However, the helix III of HH10 is much longer (usually from more than 40 nt

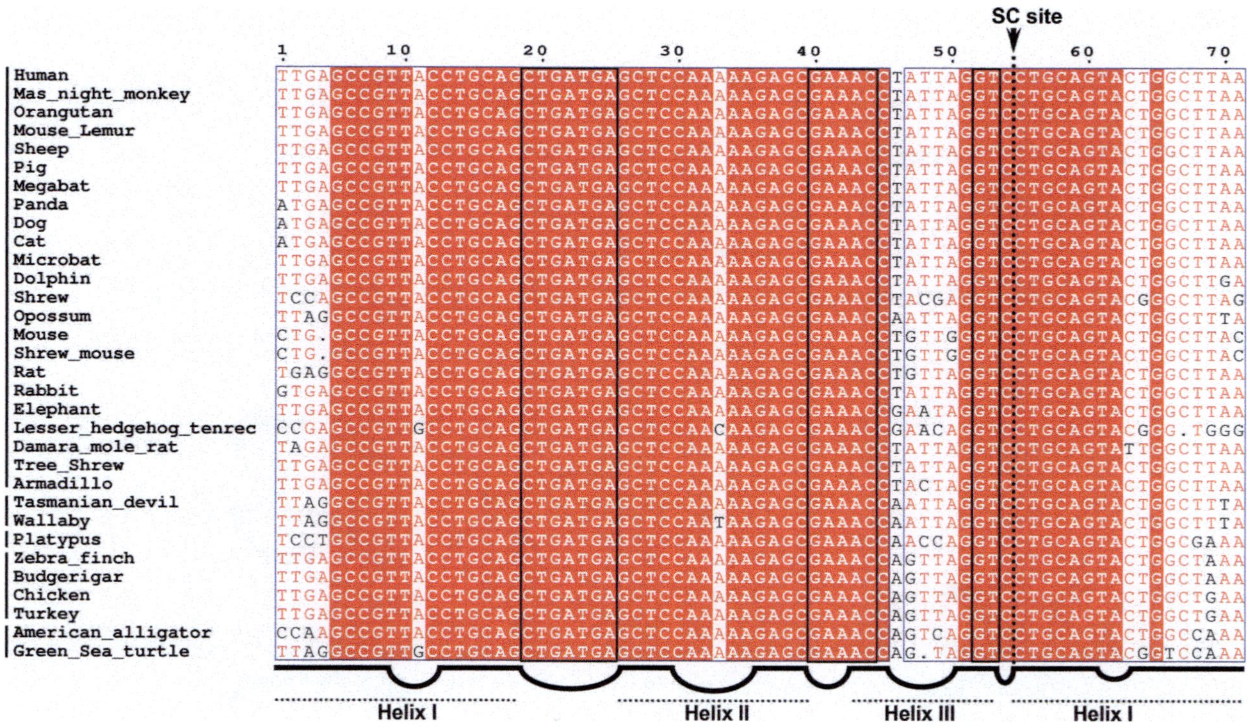

Figure 11.6 DNA sequence alignment of representative HH9 hammerhead motifs conserved in the intron of the RECK gene of all analyzed mammals, birds, alligators, and turtles. Regions corresponding to the three helixes of the ribozyme are indicated at the bottom, with stems and loops as straight and curved lines, respectively.

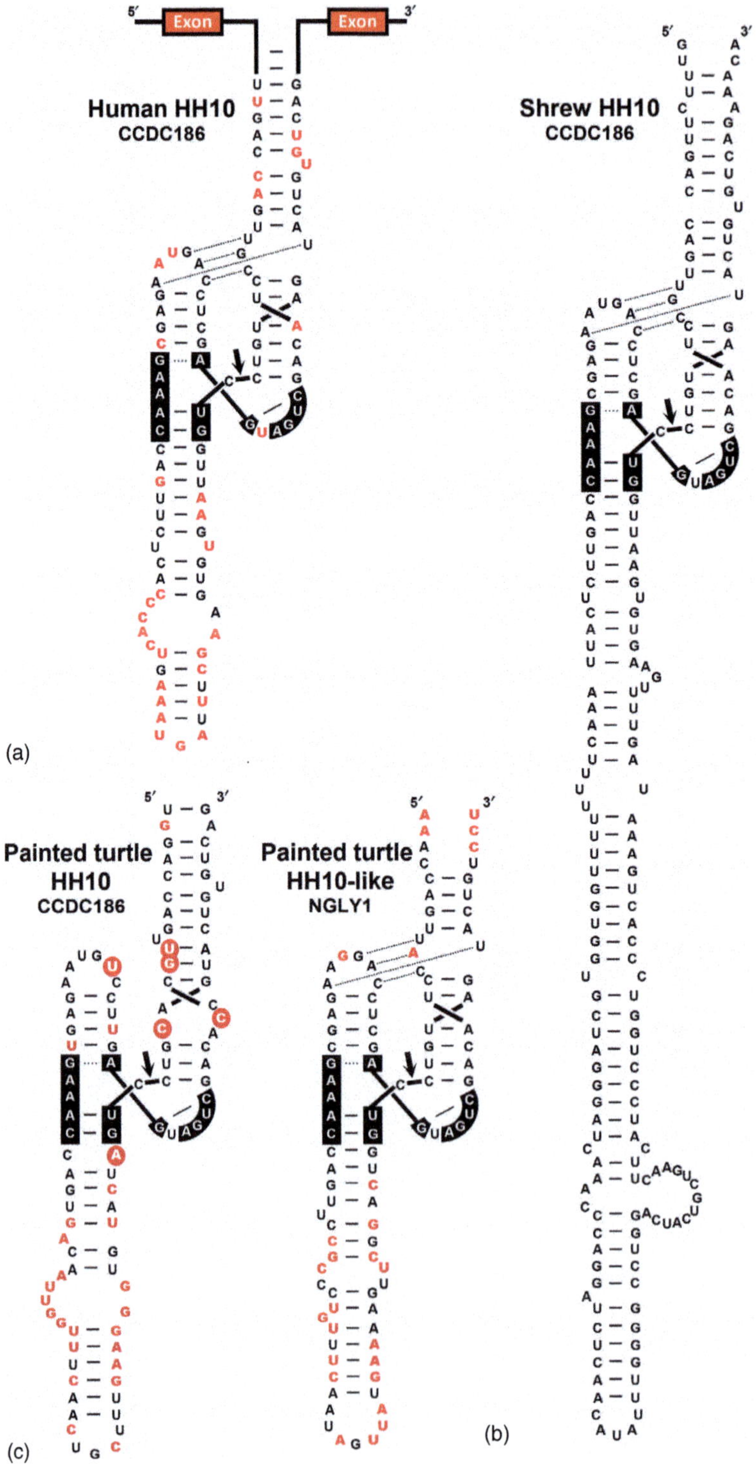
Exon
Exon
5′
3′
Human HH10
CCDC186
Shrew HH10
CCDC186
(a)
Painted turtle
HH10
CCDC186
Painted turtle
HH10-like
NGLY1
(b)
(c)

Figure 11.7 (a) Schematic representation of the intronic HH10 hammerhead ribozyme present in the human CCDC186 gene. Positions showing nucleotide heterogeneity in either the stems (compensatory) or the loops of the HH10 orthologues of most mammals are shown in red. (b) A singular intronic HH10 motif showing a much larger helix III found in the CCDC186 gene of the common shrew (*Sorex araneus*). (c) Two examples of intronic HH10-like motifs found in the CCDC186 and NGLY1 genes of the painted turtle (*Chrysemys picta*). Nucleotide changes with respect to the human HH10 motif are shown in red. Deleterious changes expected to deeply affect or abolish self-cleaving activity are shown within circles. Source: Garcia-Robles et al. [15]. © 2010 De Gruyter.

up to a maximum of 130 nt detected in the HH10 motif of the common shrew *Sorex araneus*) than the ones of HH9 (11 nt) or coelacanth HHRs (11–19 nt) (Figure 11.7). Despite their atypically long helix III, the HH10 motif efficiently self-cleaves during in vitro transcription [27].

HH10 ribozyme not only is absent in the CCDC186 orthologues of diverse mammals but also shows lower sequence conservation than HH9, which altogether indicates a more relaxed selection pressure and a less essential role of the HH10 motif. However, HH10 occurs in the orthologous CCDC186 intron of distant amniotes such as platypus, wallaby, or even the painted turtle, although in this latter case as an inactive relic (Figure 11.7C). This widespread occurrence in amniotes suggests that HH10 in CCDC186, like the HH9 motif in *RECK*, is also a very ancient innovation. In this regard, active HH10-like motifs are present in the first and large (~12 kb) intron of the *N*-glycanase 1 (NGLY1) gene of some reptiles and birds, but not mammals (Figure 11.7C) [15]. Altogether, these conserved HHRs in the introns of different genes of mammals and other amniotes could be regarded as a family of self-cleaving motifs, most likely involved in a new mechanism for the biogenesis of the mRNA (see Section 11.3).

11.3 Other Intronic HHRs in Amniotes: Small Catalytic RNAs in Search of a Function

The high sequence similarity between intronic HHRs in amniotes suggests a common origin for all of them. Previous homology searches revealed more examples of conserved intronic HHRs in the genomes of amniotes, most notably in sauropsids (birds and reptiles), with a high resemblance to the mammalian HH9 [15, 29]. One HH9-like ribozyme was detected within the first intron (9 kb) of the gene encoding the dystrobrevin beta protein (DTNB) (Figure 11.8a). DTNB is expressed in brain tissues that interacts with the dystrophin factor to form the core of the transmembrane dystrophin-glycoprotein complex involved in neuronal signaling [44, 45]. Mammalian orthologues of DTNB have lost the extra first exon/intron present in sauropsids, and consequently, they lack this particular DTNB-HHR. Another example of HH9-like preserved in sauropsids can be found in the eighth and large (28 kb) intron of the transient receptor potential cation channel subfamily M member 3 (TRPM3) gene (Figure 11.8b), which encodes a calcium channel involved in the glucose homeostasis [46]. Orthologous TRPM3 genes in mammals do not seem

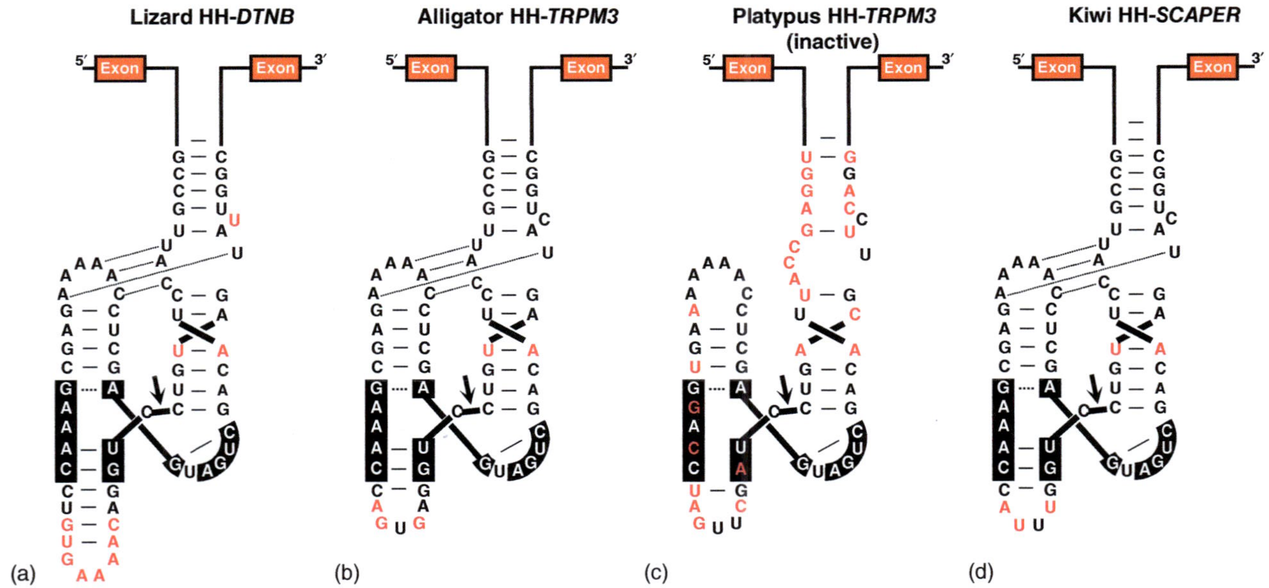

Figure 11.8 (a) Schematic representation of the intronic hammerhead ribozyme conserved in the DTNB gene of reptiles and birds, but not mammals. Nucleotide changes with respect to the human HH10 motif are shown in red. Active (b) and inactive (c) examples of the hammerhead ribozyme motif conserved in the intron of the TRPM3 gene of most sauropsids and platypus, respectively. (d) An intronic hammerhead ribozyme conserved in the SCAPER gene of some birds, such as the common kiwi (*Apteryx australis*).

to contain any intronic HHR with the exception of platypus, where the presence of a catalytically inactive HHR variant (Figure 11.8c) indicates that the TRPM3-HHR would have been lost during the evolution of mammals. A latter example of an intronic HHR can be found preserved in the seventh intron (7 kb) of the gene coding for the S phase cyclin A-associated protein in the endoplasmic reticulum (SCAPER) of some avian genomes, such as the kiwi bird (Figure 11.8d). HHR remnants can be also found in the same SCAPER intron of other bird genomes, harboring different mutations that may reduce or abolish the self-cleaving capabilities of the ribozyme.

Interestingly, the whole sequence of these sauropsid HHRs is very similar to the HH9 motif with the exception of the G-C to A-U covariation at helix I typically found at the HH10 and most coelacanth HHR motifs. These particularities highlight both the importance of the primary sequence of these ribozymes and their evolutionary constraints once they become fixed in a particular locus (Figure 11.8).

Altogether, this collection of intronic HHRs in diverse genes of amniotes would define a whole family of highly conserved self-cleaving motifs, which are expected to perform similar but yet unknown biological roles. At the molecular level, the *bona fide* HHRs present in the genomes of amniotes should efficiently catalyze the intron cleavage during in vivo transcription of the gene, disrupting the continuity of the pre-mRNA. Due to the fast self-cleavage reaction catalyzed by these HHRs and the size of the intron sequence encompassed by the self-cleavage site and the 3′ site of splicing (5–15 kb), RNA cleavage is expected to be completed before intron transcription is finished and, consequently, spliced out. Nevertheless, intronic fragments could remain stably connected after cleavage through the noncovalent base pairing of the helix I or even self-ligated again in internal equilibrium. Both options may depend on not only the stability of helix I but also the cellular conditions during the process of transcription and splicing, like temperature, helicases, hnRNPs, RNA chaperones, or any other protein factors involved during pre-mRNA transcription and splicing. Consequently, a significant fraction of the introns with HHRs can be fully transcribed and spliced out before it is cleaved and the products are detached.

Previous data mining revealed two ESTs from cow fetal and brain tissues, which mapped to the precise 5′-end of the self-cleaved RECK intron fused to U5 or U6 snRNAs [29]. These two ESTs support that intronic HHRs would be active in vivo and, although an artefactual origin cannot be discarded, also indicate that these ribozymes may be likely involved in the splicing process. In this regard, previous analysis revealed that one of the isoforms described for both RECK and CCDC186 genes results from a similar alternative splicing event concerning a cassette exon that can be alternatively spliced in or out [15]. For both genes, these precise cassette exons precede the Iarge introns carrying either the HH9 or HH10 ribozymes (Figure 11.9). In the absence of any other experimental support, these observations suggest a role of the intronic HHRs in the control of alternative splicing. On the other hand, previous work has revealed that some intronic regulatory elements involved in alternative splicing have to be covalently attached to the exons, and the presence of artificial HHRs in the intron disrupts the behavior of such regulatory elements during splicing [48, 49].

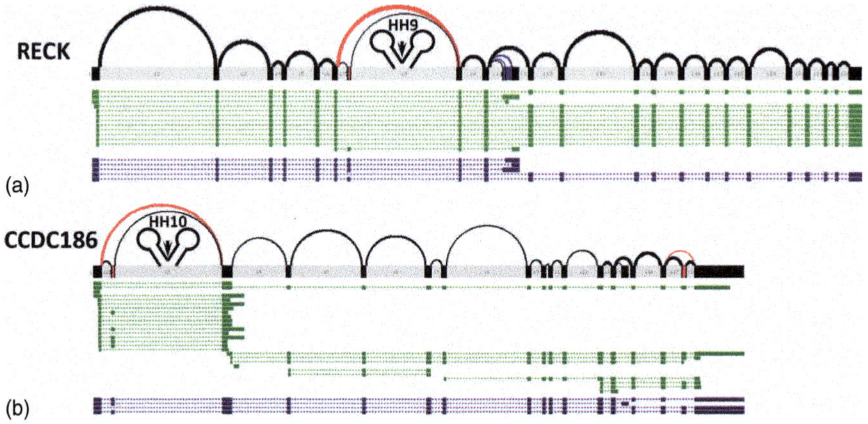

Figure 11.9 Splicing graphs of the human RECK (a) and CCDC186 (b) genes obtained through the Alternative Splicing Gallery web [47]. Introns and exons are shown in gray and black, respectively, with the exception of cassette exons (red) and competing 3′ splice sites (blue). Representative ESTs (green) and ensembl transcript assemblies (purple) are shown under each gene. Symbols for HH9 and HH10 ribozymes are drawn on their corresponding introns. Source: Garcia-Robles et al. [15]. © 2010 De Gruyter.

Regarding the origin of the HHR motifs in the genomes of amniotes, all data indicate that both discontinuous and intronic HHRs preserved in the genomes of reptiles, birds, and mammals would come after a domestication or exaptation process of the motifs present in retroelements of the ancestors of amniotes, such as amphibians and, most notably, coelacanth fishes. This phenomenon of exaptation is a recurrent trend in the evolution of mammals and amniotes, in general, that has been already described for many other traits of diverse transposable elements [50, 51], and the case of the HHR would be one more in this list.

11.4 The Family of the Hepatitis D Virus Ribozymes

The human Hepatitis D virus (HDV, or also called δ-agent) is a circRNA satellite of 1679 nt that depends on the hepatitis B virus for its propagation [52]. Unlike plant viral RNA satellites and viroids with HHRs, HDV does not encode its own polymerase but encodes for two isoforms of a protein factor (the hepatitis D antigen, or HDAg), as well as a self-cleaving ribozyme motif in each polarity of the RNA, the so-called HDV ribozymes [5, 53], which process the multimeric copies of the viral RNA genome during its rolling circle replication. The structure of the HDV ribozyme shows five helixes forming two coaxial stacks and connected as a nested double pseudoknot [54] (Figure 11.10). Since its discovery, the HDV ribozyme was thought to be an exclusive ribozyme of the infectious HDV circRNA, but in 2006, a new HDV-like ribozyme motif was reported in the human genome [55]. After this finding, deeper bioinformatic analysis revealed the widespread presence of HDV-like ribozymes in the genomes of organisms from all branches of life [56].

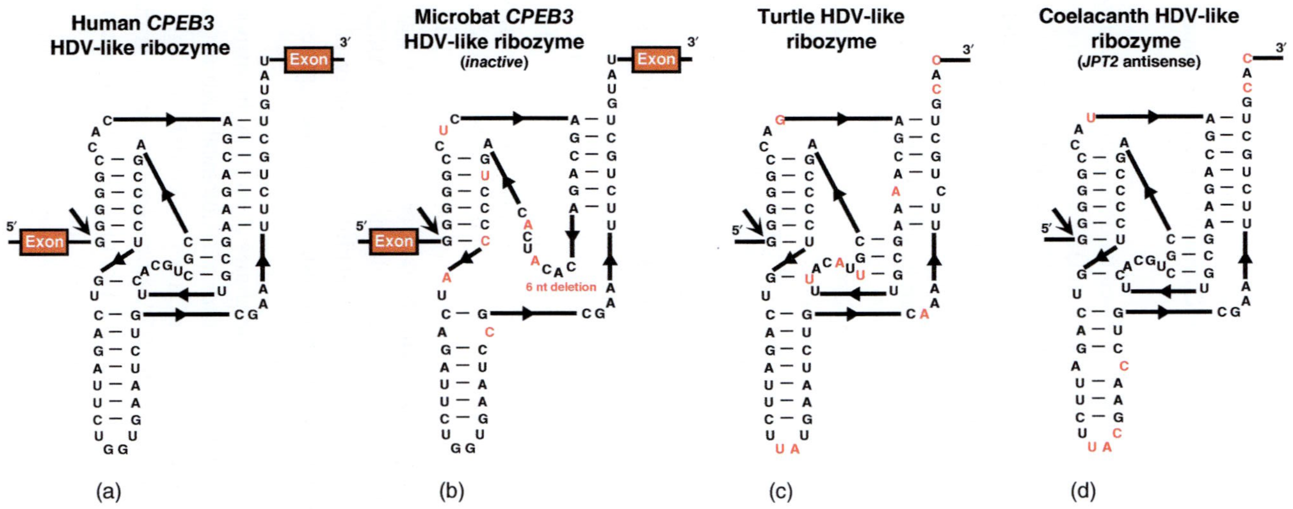

Figure 11.10 Schematic representations of (a) the human and (b) the microbat (*Myotis lucifugus*) CPEB3 HDV-like ribozymes. Very similar HDV-like ribozymes can be found in the genomes of (c) turtles (intergenic motif) and (d) coelacanths (intronic motif in the antisense strand of the JPT2 gene). Nucleotide changes with respect to the human CPEB3 ribozyme are shown in red. Source: Adapted from Ruminski et al. [14].

Most of these new ribozymes were then found to map at the 5′-UTR of autonomous long interspersed nuclear element (LINE) retrotransposons [13, 14], suggesting that human HDV-like ribozyme could be regarded as another example of functional domestication of a particular retrotransposon trait.

11.4.1 An Intronic HDV-Like Ribozyme Conserved in the Genome of Mammals

A genomic HDV-like ribozyme was discovered using in vitro selection experiments from a library of the human genome [55]. This HDV-like motif mapped at the third intron (46 kb, with the ribozyme placed at 12 kb from the next exon) of the *CPEB3* gene encoding the cytoplasmic polyadenylation element-binding protein 3, a factor involved in the regulation of mRNA polyadenylation in the cytoplasm of neurons and oocytes, among others [57]. Thus, this motif is another example of an intronic self-cleaving ribozyme that offers many parallelisms with the family of intronic HHRs, which could shed some light on the biological roles of these intriguing motifs. Moreover, intronic the CPEB3 HDV-like ribozyme appears in the same locus of many mammals, including some marsupials (Figure 11.11), although it does not seem to be preserved in either monotremes (platypus and echidnas) or sauropsids, indicating that this particular innovation was fixed in marsupial and placental mammals about 160 million years ago. The level of sequence conservation of the mammalian *CPEB3* HDV-like ribozyme is lower than the observed one for intronic HHRs (Figure 11.11). Actually, some species such as microbats (Figures 11.10b and 11.11) contain inactive variants of the HDV-like motif showing diverse truncations and point mutations, which indicates that the motif would be following a molecular fossilization process in these particular species. As in the case of the intronic HHRs, examples of HDV-like ribozymes similar to the human motif can be found in the genomes of other amniotes (reptiles and birds) and coelacanths such as *L. chalumnae* [15] or *Latimeria menadoensis* [14] (Figure 11.10). However, these HDV ribozymes in sauropsids and fishes are not intronic but intergenic, and, in some cases, related to retrotransposable elements [14]. Altogether, these observations indicate that, similarly to the evolutionary history of the intronic HHRs, the intronic CPEB3 ribozyme in mammals could be regarded as another exaptation event of a self-cleaving motif.

The human CPEB3 HDV-like ribozyme shows a very low in vitro self-cleavage activity, with a k_{obs} of only ~0.02 min^{-1} at nonphysiological high magnesium concentration [55]. This observed rate constant is two to 3 orders of magnitude slower than previously reported values for the HDV ribozyme [58] or the human intronic HHR [29], suggesting that the human HDV-like motif is either a slow self-cleaving ribozyme or its activity is somehow inhibited in these experiments. More detailed biochemical analyses of the human HDV-like ribozyme indicated that, in vitro, it follows a quite complex kinetic profile, with a mixed population of RNAs showing either slow or slightly faster cleavage rates [59]. Co-transcriptional splicing kinetics together with truncation experiments of the 5′ flanking sequence indicated that the human CPEB3 ribozyme itself would be intrinsically much faster reacting than

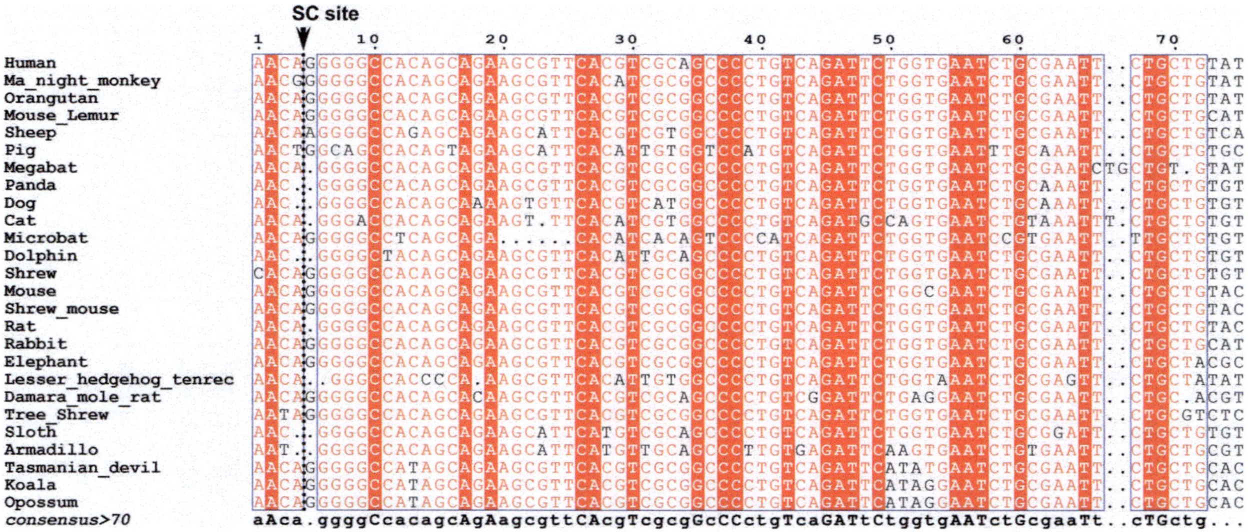

Figure 11.11 DNA sequence alignment of the intronic HDV-like motifs conserved in the *CPEB3* gene of diverse mammals. Deletions and other sequence changes found in diverse mammals are expected to deeply affect or abolish the self-cleaving capabilities of the ribozyme motif. Source: Adapted from Ruminski et al. [14].

expected, which suggest that, in vivo, the intronic sequences flanking the motif and specific protein factors might be involved in the regulation of this self-cleaving motif.

As described in Section 11.2 for the mammalian HHRs, the biological role and molecular mechanism of the intronic CPEB3 ribozyme is still under research. However, a previous study described an intriguing relationship between the intronic HDV-like ribozyme and human memory performance [60]. That work described the existence in humans of a single-nucleotide polymorphism at position 36 of the CPEB3 ribozyme (U to C, SNP rs11186856), which induces a twofold increase in the rate of self-cleavage. It was found that individuals homozygous for this faster ribozyme variant performed poorer in episodic memory tests, particularly when the presented material had emotional valence [60]. Although a connection between a complex trait such as human episodic memory with a simple intronic self-cleaving motif is not evident, it is known that CPEB factors have been related to long-lasting forms of synaptic plasticity and memory formation in model organisms [61], and thus, the connection of the CPEB3 ribozyme to memory performance would not be surprising. Assuming that catalysis of CPEB3 ribozyme is tightly regulated in vivo, we expect that depending on the tissue or specific cellular conditions, the pre-mRNA may follow a particular pathway (i.e. alternative splicing and regulation of the gene expression) induced by the intron self-cleavage. The exclusive and conserved occurrence of intronic ribozymes, such as not only HDV-like but also HHR motifs, in the most complex vertebrates would represent an innovation that has allowed the appearance of exclusive and complex traits in highly evolved organisms, although further investigation will be required to understand the precise molecular mechanisms involved.

In summary, the landscape for the single HDV-like ribozyme in mammals replicates the observed one for the group of intronic HHR ribozymes. It would seem that coelacanth genomes contain a few dozens of transposon-related HDV-like ribozymes, and only a few of them have subsisted in the genomes of amniotes. However, while HDV-like ribozymes in sauropsids are located in intergenic regions and do not point to any specific function, mammals would have domesticated a single motif in the *CPEB3* gene as an innovation that offered an adaptive advantage to these organisms.

11.5 Other Small Self-Cleaving Ribozymes Hidden in the Genomes of Vertebrates?

Since its discovery at the end of the 1980s, a group of small self-cleaving RNAs have been regarded as a group of weird motifs mostly involved in the mechanism of rolling circle replication of a group of weird infectious agents with circRNA genomes. Now, we know that small self-cleaving ribozymes are frequent in a myriad of DNA genomes and mostly are involved in the biology of diverse genetic mobile elements. The striking similarities between self-cleaving motifs of lower metazoans (i.e. invertebrates and lower vertebrates, such as fishes) and the few and highly conserved ribozymes found in higher metazoans (i.e. amniotes) suggest that small ribozymes have followed a domestication process from a primary role in retrotransposition to new biological tasks [50, 62]. While retrotransposons with self-cleaving motifs would be absent or unrecognized in the genomes of amniotes, their small ribozymes have been preserved in diverse noncoding regions of pre-mRNAs, adding an extra level of complexity in these organisms through molecular mechanisms that will be dissected.

However, only two examples (the HHR and the HDV-like ribozymes) out of the nine classes of small self-cleaving RNAs so far described [10] have been found conserved in the genomes of higher vertebrates. In 2004, a totally novel small ribozyme was reported in the human genome, the so-called β-globin co-transcriptional cleavage ribozyme (CotC ribozyme) [63]. In that work, the human β-globin gene was proposed to contain an RNA self-cleaving activity located downstream of the protein-coding sequence and polyA site of the gene. The CoTc motif seems to be quite conserved in primates, but not in other mammals or vertebrates. However, since this initial report, there has been no independent confirmation of these findings. Moreover, subsequent analysis failed to show any ribozyme activity for the CoTc motif [64], and consequently, the autocatalytic and biological significance of this RNA remains to be validated. In any case, this example not only reflects the difficulties for finding new autocatalytic activities but also indicates that the presence of novel self-cleaving RNA motifs of unknown sequence and architecture in the genomes of vertebrates is highly feasible, most especially if we take into account the recent discovery of new small ribozymes, notably in prokaryotes [10]. Moreover, not only RNA self-cleavage but also other simple biochemical and regulatory functions performed by RNA motifs [65] may be hidden in the genomes of the most complex organisms and should be carefully searched in the future.

References

1 Cochrane, J.C. and Strobel, S.A. (2008). Catalytic strategies of self-cleaving ribozymes. *Acc. Chem. Res.* 41: 1027–1035.

2 Hutchins, C.J., Rathjen, P.D., Forster, A.C., and Symons, R.H. (1986). Self-cleavage of plus and minus RNA transcripts of avocado sunblotch viroid. *Nucleic Acids Res.* 14: 3627–3640.

3 Prody, G.A., Bakos, J.T., Buzayan, J.M. et al. (1986). Autolytic processing of dimeric plant virus satellite RNA. *Science* 231: 1577–1580.

4 Buzayan, J.M., Gerlach, W.L., and Bruening, G. (1986). Non-enzymatic cleavage and ligation of RNAs complementary to a plant virus satellite RNA. *Nature* 323: 349–353.

5 Kuo, M.Y., Sharmeen, L., Dinter-Gottlieb, G., and Taylor, J. (1988). Characterization of self-cleaving RNA sequences on the genome and antigenome of human hepatitis delta virus. *J. Virol.* 62: 4439–4444.

6 Saville, B.J. and Collins, R.A. (1990). A site-specific self-cleavage reaction performed by a novel RNA in *Neurospora* mitochondria. *Cell* 61: 685–696.

7 Winkler, W.C., Nahvi, A., Roth, A. et al. (2004). Control of gene expression by a natural metabolite-responsive ribozyme. *Nature* 428: 281–286.

8 Roth, A., Weinberg, Z., Chen, A.G. et al. (2014). A widespread self-cleaving ribozyme class is revealed by bioinformatics. *Nat. Chem. Biol.* 10: 56–60.

9 Weinberg, Z., Kim, P.B., Chen, T.H. et al. (2015). New classes of self-cleaving ribozymes revealed by comparative genomics analysis. *Nat. Chem. Biol.* 11: 606–610.

10 Jimenez, R.M., Polanco, J.A., and Luptak, A. (2015). Chemistry and biology of self-cleaving ribozymes. *Trends Biochem. Sci* 40: 648–661.

11 Cervera, A. and De la Peña, M. (2014). Eukaryotic penelope-like retroelements encode hammerhead ribozyme motifs. *Mol. Biol. Evol.* 31: 2941–2947.

12 Cervera, A., Urbina, D., and de la Peña, M. (2016). Retrozymes are a unique family of non-autonomous retrotransposons with hammerhead ribozymes that propagate in plants through circular RNAs. *Genome Biol.* 17: 135.

13 Eickbush, D.G. and Eickbush, T.H. (2010). R2 retrotransposons encode a self-cleaving ribozyme for processing from an rRNA cotranscript. *Mol. Cell. Biol.* 30: 3142–3150.

14 Ruminski, D.J., Webb, C.H., Riccitelli, N.J., and Luptak, A. (2011). Processing and translation initiation of non-long terminal repeat retrotransposons by hepatitis delta virus (HDV)-like self-cleaving ribozymes. *J. Biol. Chem.* 286: 41286–41295.

15 Garcia-Robles, I., Sanchez-Navarro, J., and De la Peña, M. (2012). Intronic hammerhead ribozymes in mRNA biogenesis. *Biol. Chem.* 393: 1317–1326.

16 De la Peña, M., Gago, S., and Flores, R. (2003). Peripheral regions of natural hammerhead ribozymes greatly increase their self-cleavage activity. *EMBO J.* 22: 5561–5570.

17 Khvorova, A., Lescoute, A., Westhof, E., and Jayasena, S.D. (2003). Sequence elements outside the hammerhead ribozyme catalytic core enable intracellular activity. *Nat. Struct. Biol.* 10: 708–712.

18 Martick, M. and Scott, W.G. (2006). Tertiary contacts distant from the active site prime a ribozyme for catalysis. *Cell* 126: 309–320.

19 De la Peña, M., Garcia-Robles, I., and Cervera, A. (2017). The hammerhead ribozyme: a long history for a short RNA. *Molecules* 22 (1): 78.

20 Epstein, L.M. and Gall, J.G. (1987). Self-cleaving transcripts of satellite DNA from the newt. *Cell* 48: 535–543.

21 Zhang, Y. and Epstein, L.M. (1996). Cloning and characterization of extended hammerheads from a diverse set of caudate amphibians. *Gene* 172: 183–190.

22 Ferbeyre, G., Smith, J.M., and Cedergren, R. (1998). Schistosome satellite DNA encodes active hammerhead ribozymes. *Mol. Cell. Biol.* 18: 3880–3888.

23 Rojas, A.A., Vazquez-Tello, A., Ferbeyre, G. et al. (2000). Hammerhead-mediated processing of satellite pDo500 family transcripts from *Dolichopoda* cave crickets. *Nucleic Acids Res.* 28: 4037–4043.

24 Przybilski, R., Graf, S., Lescoute, A. et al. (2005). Functional hammerhead ribozymes naturally encoded in the genome of *Arabidopsis thaliana*. *Plant Cell* 17: 1877–1885.

25 De la Peña, M. and Garcia-Robles, I. (2010). Ubiquitous presence of the hammerhead ribozyme motif along the tree of life. *RNA* 16: 1943–1950.

26 Jimenez, R.M., Delwart, E., and Luptak, A. (2011). Structure-based search reveals hammerhead ribozymes in the human microbiome. *J. Biol. Chem.* 286: 7737–7743.

27 Perreault, J., Weinberg, Z., Roth, A. et al. (2011). Identification of hammerhead ribozymes in all domains of life reveals novel structural variations. *PLoS Comput. Biol.* 7: e1002031.

28 Seehafer, C., Kalweit, A., Steger, G. et al. (2011). From alpaca to zebrafish: hammerhead ribozymes wherever you look. *RNA* 17: 21–26.

29 De la Peña, M. and Garcia-Robles, I. (2010). Intronic hammerhead ribozymes are ultraconserved in the human genome. *EMBO Rep.* 11: 711–716.

30 De la Peña, M. and Cervera, A. (2017). Circular RNAs with hammerhead ribozymes encoded in eukaryotic genomes: the enemy at home. *RNA Biol.* 14: 985–991.

31 Hammann, C., Luptak, A., Perreault, J., and de la Peña, M. (2012). The ubiquitous hammerhead ribozyme. *RNA* 18: 871–885.

32 Forster, A.C., Davies, C., Sheldon, C.C. et al. (1988). Self-cleaving viroid and newt RNAs may only be active as dimers. *Nature* 334: 265–267.

33 Garrett, T.A., Pabon-Pena, L.M., Gokaldas, N., and Epstein, L.M. (1996). Novel requirements in peripheral structures of the extended satellite 2 hammerhead. *RNA* 2: 699–706.

34 Lunse, C.E., Weinberg, Z., and Breaker, R.R. (2017). Numerous small hammerhead ribozyme variants associated with Penelope-like retrotransposons cleave RNA as dimers. *RNA Biol.* 14: 1499–1507.

35 Martick, M., Horan, L.H., Noller, H.F., and Scott, W.G. (2008). A discontinuous hammerhead ribozyme embedded in a mammalian messenger RNA. *Nature* 454: 899–902.

36 Hao, L., Klein, J., and Nei, M. (2006). Heterogeneous but conserved natural killer receptor gene complexes in four major orders of mammals. *Proc. Natl. Acad. Sci. U.S.A.* 103: 3192–3197.

37 Zhou, H., Kartsogiannis, V., Quinn, J.M. et al. (2002). Osteoclast inhibitory lectin, a family of new osteoclast inhibitors. *J. Biol. Chem.* 277: 48808–48815.

38 Scott, W.G., Horan, L.H., and Martick, M. (2013). The hammerhead ribozyme: structure, catalysis, and gene regulation. *Prog. Mol. Biol. Transl. Sci.* 120: 1–23.

39 Scott, W.G., Martick, M., and Chi, Y.I. (2009). Structure and function of regulatory RNA elements: ribozymes that regulate gene expression. *Biochim. Biophys. Acta* 1789: 634–641.

40 Takahashi, C., Sheng, Z., Horan, T.P. et al. (1998). Regulation of matrix metalloproteinase-9 and inhibition of tumor invasion by the membrane-anchored glycoprotein RECK. *Proc. Natl. Acad. Sci. U.S.A.* 95: 13221–13226.

41 Oh, J., Takahashi, R., Kondo, S. et al. (2001). The membrane-anchored MMP inhibitor RECK is a key regulator of extracellular matrix integrity and angiogenesis. *Cell* 107: 789–800.

42 Canny, M.D., Jucker, F.M., and Pardi, A. (2007). Efficient ligation of the *Schistosoma* hammerhead ribozyme. *Biochemistry* 46: 3826–3834.

43 Hartmann, T.B., Thiel, D., Dummer, R. et al. (2004). SEREX identification of new tumour-associated antigens in cutaneous T-cell lymphoma. *Br. J. Dermatol.* 150: 252–258.

44 Grady, R.M., Wozniak, D.F., Ohlemiller, K.K., and Sanes, J.R. (2006). Cerebellar synaptic defects and abnormal motor behavior in mice lacking alpha- and beta-dystrobrevin. *J. Neurosci.* 26: 2841–2851.

45 Peters, M.F., O'Brien, K.F., Sadoulet-Puccio, H.M. et al. (1997). beta-Dystrobrevin, a new member of the dystrophin family. Identification, cloning, and protein associations. *J. Biol. Chem.* 272: 31561–31569.

46 Wagner, T.F., Loch, S., Lambert, S. et al. (2008). Transient receptor potential M3 channels are ionotropic steroid receptors in pancreatic beta cells. *Nat. Cell Biol.* 10: 1421–1430.

47 Leipzig, J., Pevzner, P., and Heber, S. (2004). The Alternative Splicing Gallery (ASG): bridging the gap between genome and transcriptome. *Nucleic Acids Res.* 32: 3977–3983.

48 Gromak, N., Talotti, G., Proudfoot, N.J., and Pagani, F. (2008). Modulating alternative splicing by cotranscriptional cleavage of nascent intronic RNA. *RNA* 14: 359–366.

49 Pastor, T., Dal Mas, A., Talotti, G. et al. (2011). Intron cleavage affects processing of alternatively spliced transcripts. *RNA* 17: 1604–1613.

50 Okada, N., Sasaki, T., Shimogori, T., and Nishihara, H. (2010). Emergence of mammals by emergency: exaptation. *Genes Cells* 15: 801–812.

51 Platt, R.N. 2nd, Vandewege, M.W., and Ray, D.A. (2018). Mammalian transposable elements and their impacts on genome evolution. *Chromosome Res.* 26: 25–43.

52 Makino, S., Chang, M.F., Shieh, C.K. et al. (1987). Molecular cloning and sequencing of a human hepatitis delta (delta) virus RNA. *Nature* 329: 343–346.

53 Sharmeen, L., Kuo, M.Y., Dinter-Gottlieb, G., and Taylor, J. (1988). Antigenomic RNA of human hepatitis delta virus can undergo self-cleavage. *J. Virol.* 62: 2674–2679.

54 Ferre-D'Amare, A.R., Zhou, K., and Doudna, J.A. (1998). Crystal structure of a hepatitis delta virus ribozyme. *Nature* 395: 567–574.

55 Salehi-Ashtiani, K., Luptak, A., Litovchick, A., and Szostak, J.W. (2006). A genomewide search for ribozymes reveals an HDV-like sequence in the human CPEB3 gene. *Science* 313: 1788–1792.

56 Webb, C.H., Riccitelli, N.J., Ruminski, D.J., and Luptak, A. (2009). Widespread occurrence of self-cleaving ribozymes. *Science* 326: 953.

57 Huang, Y.S. and Richter, J.D. (2004). Regulation of local mRNA translation. *Curr. Opin. Cell Biol.* 16: 308–313.

58 Chadalavada, D.M., Knudsen, S.M., Nakano, S., and Bevilacqua, P.C. (2000). A role for upstream RNA structure in facilitating the catalytic fold of the genomic hepatitis delta virus ribozyme. *J. Mol. Biol.* 301: 349–367.

59 Chadalavada, D.M., Gratton, E.A., and Bevilacqua, P.C. (2010). The human HDV-like CPEB3 ribozyme is intrinsically fast-reacting. *Biochemistry* 49: 5321–5330.

60 Vogler, C., Spalek, K., Aerni, A. et al. (2009). CPEB3 is associated with human episodic memory. *Front. Behav. Neurosci.* 3: 4.

61 Richter, J.D. (2007). CPEB: a life in translation. *Trends Biochem. Sci* 32: 279–285.

62 Bejerano, G., Pheasant, M., Makunin, I. et al. (2004). Ultraconserved elements in the human genome. *Science* 304: 1321–1325.

63 Teixeira, A., Tahiri-Alaoui, A., West, S. et al. (2004). Autocatalytic RNA cleavage in the human beta-globin pre-mRNA promotes transcription termination. *Nature* 432: 526–530.

64 Dye, M.J., Gromak, N., and Proudfoot, N.J. (2006). Exon tethering in transcription by RNA polymerase II. *Mol. Cell* 21: 849–859.

65 Vu, M.M., Jameson, N.E., Masuda, S.J. et al. (2012). Convergent evolution of adenosine aptamers spanning bacterial, human, and random sequences revealed by structure-based bioinformatics and genomic SELEX. *Chem. Biol.* 19: 1247–1254.

Part III

Engineered Ribozymes

12

Phosphoryl Transfer Ribozymes

Razvan Cojocaru and Peter J. Unrau

Simon Fraser University, Department of Molecular Biology and Biochemistry, 8888 University Drive, Burnaby, BC V5A 1S6, Canada

12.1 Introduction

Phosphoryl transfer enzymes play fundamental roles in metabolism. Over the last two decades, a set of phosphoryl transfer ribozymes have been isolated that provide insight into how RNA rather than protein might have been able to sustain many of these centrally important reactions. One function of phosphoryl transfer reactions in modern metabolism is to maintain the balance of energy inside cells by interconverting between mono-, di-, and triphosphate nucleotides (catalyzed by mono- and diphosphate kinases) [1]. The high chemical potential energy stored in the phosphate bonds of di- and triphosphate nucleotides is used by multiple cellular functions to drive otherwise thermodynamically unfavorable reactions [2]. Catalytically, phosphoryl transfer reactions and the complex often multistep chemistry of such reactions make it unclear how well RNA could have promoted similar chemistry early in the evolution of life. Surprisingly, RNA is remarkably good at such chemistry, as this chapter will discuss in detail.

Due to the importance of phosphorylation in current biology, nucleic acid-catalyzed phosphoryl transfer is of special interest from the perspective of a hypothetical RNA world [3–5]. This theory, which posits that RNA catalysis predated protein catalysis early in evolution, is based on several key observations. (i) Three catalytically critical and complex cellular machines are all at their core ribozymes (RNA catalysts). The ribosome, a large RNA–protein complex essential for protein synthesis in all domains of life, mediates peptide bond formation via its ribosomal RNA [6–8]. RNase P, which is found in all three domains of life and is largely responsible for transfer RNA (tRNA) maturation, is an RNA catalyst [9–11]. Recently, the eukaryotic spliceosome, which is thought to have evolved from, and that shares common catalytic mechanisms with the group II self-splicing intron, is also at its core an RNA catalyst [12]. (ii) Metabolically hundreds of protein enzymes use RNA cofactors [13]. This includes not only the activated nucleotides adenosine-5′-triphosphate (ATP) (177 enzymatic reactions), and

Ribozymes, First Edition. Edited by Sabine Müller, Benoît Masquida, and Wade Winkler.

guanosine-5′-triphosphate (GTP) (27 reactions), but also the cap-like redox cofactors such as nicotinamide adenine dinucleotide (NAD) (446 reactions), flavin adenine dinucleotide (FAD) (41 reactions) (numbers from the KEGG Database [14–16]). (iii) Finally, the metabolic organization of nucleotide synthesis suggests that RNA synthesis could have predated DNA synthesis. Together this evidence suggests that RNA catalysis may have preceded the genetic polymer of DNA and the protein catalyzed metabolism of today in an earlier RNA dominated period of evolution.

In Sections 12.2–12.6 that follow, we introduce the different mechanisms by which ribozyme phosphoryl transfer reactions occur and compare them with their protein equivalents. We take a bottom-up approach, putting ribozyme-mediated phosphoryl transfer reactions into the context of nucleotide and nucleoside metabolism, the activation of RNA oligomers, and the assembly and replication of RNA molecules. Our progression begins with kinase ribozymes, which can self-phosphorylate and self-thiophosphorylate their 5′-termini or 2′-internal hydroxyls, followed by self-activation of ribozymes using cyclic trimetaphosphate (TMP). Next, we discuss glycosidic bond-forming ribozymes using 5-phosphoribosyl 1(α)-pyrophosphate (PRPP) as a substrate and the phosphoryl transfer chemistry required to produce activated substrates that are critical for nucleotide synthesis. We then compare and contrast ribozyme-catalyzed and protein-catalyzed capping and ligation reactions, highlighting the multistep mechanism needed for a ribozyme to activate an RNA substrate to carry out either capping or RNA ligation. We conclude this chapter by describing the evolution of ligase ribozymes into versatile polymerase ribozymes capable of bidirectional polymerization (3′ to 5′ or 5′ to 3′) and the recent discovery of ribozyme-catalyzed reverse transcriptase activity.

12.2 Kinase Ribozymes

One major set of phosphoryl transfer reactions is catalyzed by kinase enzymes. Kinases carry out phosphorylation by transferring phosphate groups from high energy donor molecules, such as ATP (pppA), to specific substrates. With 518 protein kinase genes scattered across the human genome, nucleoside monophosphate kinases (NMPKs) and nucleoside diphosphate kinases (NDPKs) are two critical enzyme families that are essential for the synthesis and homeostasis of nucleotides. This dynamic balance is required to maintain cellular metabolic processes and to both synthesize and repair RNA and DNA polymers central to information transfer and storage, respectively [1]. NMPKs and nucleotide kinases (NKs) catalyze the transfer of a terminal γ-phosphoryl group from a nucleoside triphosphate onto an attacking nucleophile in a single step.

Similarly, polynucleotide nucleotide kinase (PNK) type enzymes phosphorylate the termini of RNA or DNA strands. This then allows ligases to join nucleic acid strands together [17], which can play critical roles in either a viral infection or the repair of DNA damage. The attack of either a hydroxyl group (Eq. (12.1)) or a monophosphate (Eq. (12.2)) is achieved by the precise positioning of the relevant nucleophile inside the enzyme active site. In contrast, the divalent metal ion-dependent NDPK enzymes catalyze the phosphorylation of nucleoside diphosphates to nucleoside triphosphates via a two-step mechanism involving a phospho-enzyme intermediate (Eq. (12.3)). This process is intrinsically highly symmetrical as the product of the reaction differs only from the substrates by the identity of N^1 and N^2 (Eqs. (12.3) and (12.4)) [18, 19].

$$\mathbf{p}\mathrm{ppN}^1 + 5'\text{-HO-RNA} \overset{\text{PNK}}{\longleftrightarrow} \mathrm{ppN}^1 + \mathbf{p}\mathrm{RNA} \tag{12.1}$$

$$\mathbf{p}\mathrm{ppN}^1 + \mathrm{pN}^2 \overset{\text{NMPK}}{\longleftrightarrow} \mathrm{ppN}^1 + \mathbf{p}\mathrm{pN}^2 \tag{12.2}$$

$$\mathbf{p}\mathrm{ppN}^1 + \mathrm{NDPK} \longleftrightarrow \mathrm{ppN}^1 + \mathrm{NDPK}\mathbf{p} \tag{12.3}$$

$$\mathrm{NDPK}\mathbf{p} + \mathrm{ppN}^2 \longleftrightarrow \mathrm{NDPK} + \mathbf{p}\mathrm{ppN}^2 \tag{12.4}$$

While NDPKs do not show high specificity toward types of nucleoside bases, NMPKs do, with human tissues containing a thymidylate kinase (dTMPK), a uridylate–cytidylate kinase (UMP–CMPK), five isozymes of adenylate kinase (AK) named AK1–5, and several guanylate kinases (GUKs) [20]. This suggests that the two-step mechanism and the intrinsic symmetry of the NDPK type enzymes might be responsible for their lack of specificity.

Given the role of protein kinases, how would have kinase ribozymes been utilized in an early RNA world? Would RNA have been versatile enough in its catalytic abilities to carry out RNA activation through phosphorylation? One early indication for the validity of RNA catalyzed kinase reactions came from studies of the *Tetrahymena* group I intron. A shortened form of the intron acts as a phosphotransferase ribozyme, catalyzing the transfer of a 3′-phosphate from one oligonucleotide to the 3′-terminal hydroxyl of another [21]. To that end, using in vitro selection from deep random sequences and subsequent laboratory mediated evolution, kinase ribozymes have been isolated that catalyze self-thiophosphorylation and self-phosphorylation of 5′-terminal or 2′-internal hydroxyls (Figure 12.1a,b) [22–26]. All of these kinase ribozymes underwent a similar selection process, where RNA libraries were incubated with either ATPγS or GTPγS, and self-thiophosphorylating kinases were isolated by taking advantage of the chemical specificity conferred by the sulfur modification. Despite being selected for thiophosphoryl transfer, these isolated kinase ribozymes were also able to catalyze phosphoryl transfer reactions, albeit at reduced rates.

The first approach toward selecting kinase ribozymes the deep random sequence used as a starting point a pre-existing (in vitro selected) RNA aptamer domain that binds ATP [27], surrounded by regions of the random sequence [25]. This approach resulted in seven classes of kinase ribozymes, with five acting as 5′-kinases and two as internal 2′-hydroxyl kinases. An engineered variant of one of these ribozymes, Kin.46, was also capable of catalyzing the transfer of γ-thiophosphate from ATPγS to the 5′-hydroxyl of an exogenous RNA oligomer held in the enzyme's active site through Watson–Crick base pairing [28]. The catalyzed reaction has a similar mechanism to the reaction catalyzed by human polynucleotide nucleotide kinase (hPNK), performing multiple turnover chemistry with a k_{cat} of up to 0.17 min^{-1}.

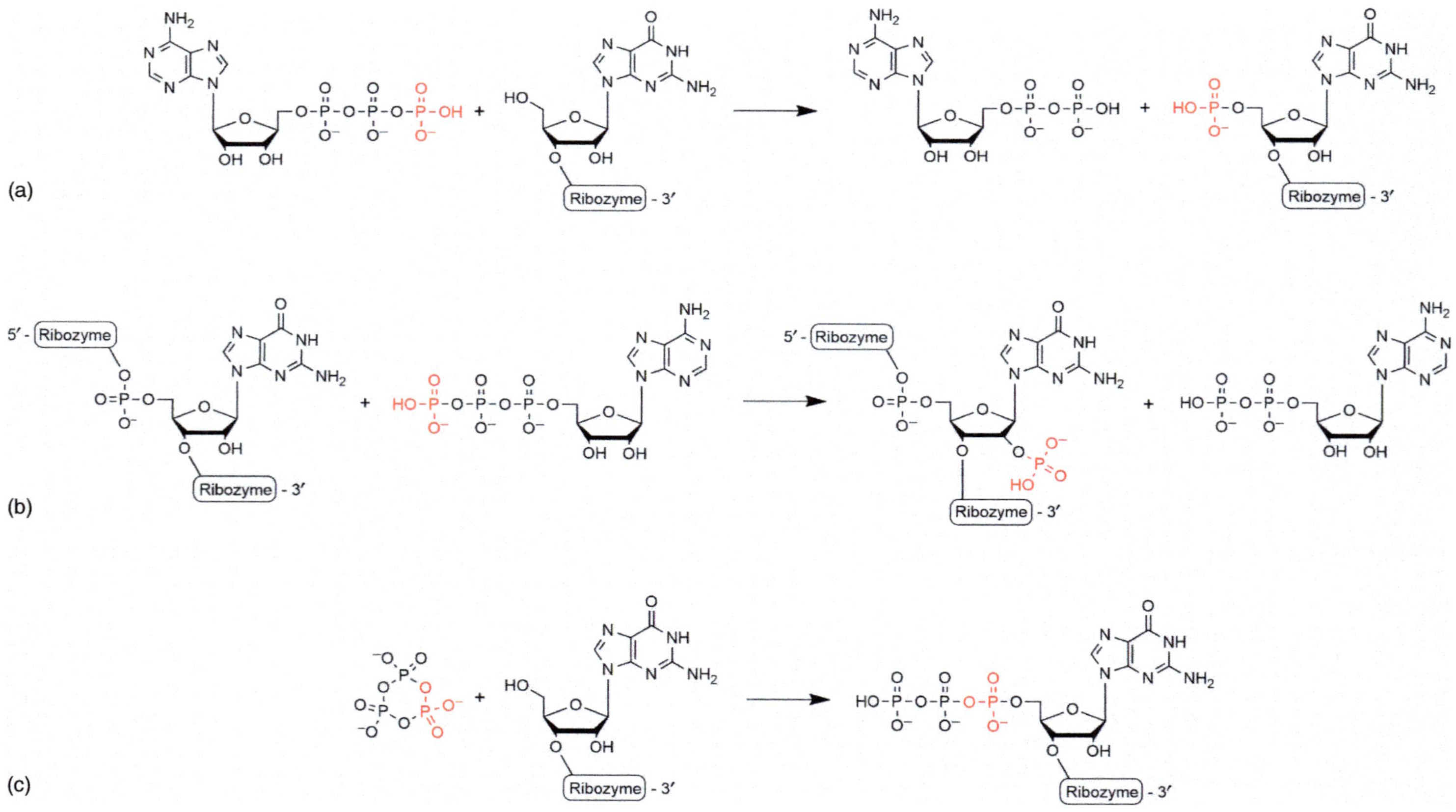

Figure 12.1 Kinase ribozymes have been characterized that catalyze phosphoryl transfer from either an adenosine or guanosine triphosphate to the (a) terminal 5′-hydroxyl or (b) internal 2′-hydroxyl of an RNA molecule. (c) Triphosphorylation ribozymes catalyze the triphosphorylation of RNA 5′-hydroxyl groups with trimetaphosphate (TMP).

ATP and GTPγS were also substrates for Kin.46; however, the k_{cat} was ~100-fold and ~650-fold lower than with ATPγS, respectively, showing high specificity for the donor with which it was selected [28].

Similarly, using an aminoacylase parent ribozyme as a starting point [29], another 23 distinct kinase ribozymes were isolated with 5′-terminal or 2′-internal phosphorylation activity. Using GTPγS as the donor molecule, these ribozymes had catalytic rates ranging from 8×10^{-6} to 6×10^{-4} min^{-1}. The evolution of kinase ribozymes continued with the selection of the ribozyme 2PTmin3.2. Although showing a 5–10-fold lower catalytic rate than the ribozymes selected by Szostak in 1993, this 2′-internal hydroxyl kinase is the first demonstration of a trans-acting ribozyme able to phosphorylate the 2′-hydroxyl of an unpaired exogenous substrate [26]. A special feature of this ribozyme is its ability to execute multiple cycles of dethiophosphorylation and rethiophosphorylation driven by ATPγS, allowing it to act as primitive energy transducing molecular motor [30].

Divalent magnesium cations are critically important in many essential biological ribozymes [31–33] and help stabilize negative charges during the nucleophilic attack and increase the electrophilicity of phosphates. The kinase ribozymes discussed so far are all pH-independent between pH 5 and 9 and are divalent metal ion-dependent, with magnesium being the main metal ion cofactor. One of the last selected kinase ribozymes, and the most divergent in behavior compared to its predecessors, is K28(1–77)C [22, 23, 34]. This ribozyme can perform thiophosphorylation onto itself at two widely separated residues. However, as opposed to the previously described Mg^{2+}-dependent ribozymes, in addition to Mg^{2+}, K28(1–77)C requires a tightly bound Cu^{2+}–GTPγS complex in its active site at a high pH of 8.5 for optimal catalytic activity [23, 34]. The diverse metal ion behavior observed among the selected kinase ribozymes therefore highlights the capability of RNA to evolve complex chemical strategies in an RNA world.

While phosphorylation of a nucleotide hydroxyl is essential in modern metabolism for the subsequent activation of a nucleotide or polymer to a diphosphate or triphosphate, alternative ribozyme-based activation mechanisms have been explored using cyclic TMP as an energy source (Figure 12.1c). Nucleoside triphosphates can be generated directly from nucleosides and TMP at a pH of 12 [35]. TMP is one of the most reactive polyphosphorylating reagents among the polyphosphates [36] and can be generated through the self-reaction of the termini of linear polyphosphates [37]. While the rapid hydrolysis of TMP at high pH [38] is problematic, a ribozyme that facilitates the triphosphate activation of RNA at neutral pH could be useful in an RNA world. Using in vitro selection, over a dozen ribozyme families were selected that are able to triphosphorylate their 5′-hydroxyl termini using TMP [39]. Kinetic analysis of eight individuals showed catalytic rate constants ranging between 0.01 and 0.03 min^{-1} under selective conditions (50 mM TMP, 100 mM $MgCl_2$, pH 8.3). One of the ribozymes, TPR1, with a minimum functional motif of 96 nucleotides, was capable of triphosphorylating RNAs both in *cis* and *trans*. This ribozyme has a k_{cat} of 0.16 min^{-1} and K_M for TMP of 30 mM in optimal buffer conditions (100 mM TMP, 400 mM $MgCl_2$, pH 8.1), making the catalytic rate ~10^7-fold faster than the uncatalyzed reaction in the absence of magnesium.

In summary, in vitro selected kinase ribozymes demonstrate that in an early RNA world, RNA catalysis could have been versatile enough to catalyze the phosphorylation and direct activation of a variety of RNA oligomers. However, kinase ribozymes are currently only able to phosphorylate 5′-terminal hydroxyls or internal 2′ hydroxyls and are unable to catalyze reactions on free nucleosides. The selection of such kinase ribozymes would mimic reactions highly likely to have been important in an early RNA world and it would serve as an important tool in the evolution of an artificial RNA metabolism in the laboratory. The diversity of functional sequences found in kinase ribozyme selections to date indicates the potential in this area for future research. Likewise, the selections that are able to produce mono- and diphosphate kinase ribozymes with mechanisms similar to that found in modern metabolism will likely appear. In contrast to existing monophosphate kinase ribozymes that phosphorylate in a single step, it remains to be determined if diphosphate kinase ribozymes will require a two-step mechanism (Eqs. (12.3) and (12.4)) to overcome the difficult catalytic challenge of charge stabilization in these reactions. While such selections remain to be performed, it appears that RNA will be up to this challenge. As we will see later in this chapter, a number of ribozymes promote two-step chemistry that closely resembles that needed for diphosphate kinase chemistry (see Eqs. (12.3) and (12.4) and Sections 12.4 and 12.5).

12.3 Glycosidic Bond Forming Ribozymes

The de novo and salvage pathways that lead to the synthesis of nucleotides use PRPP as a substrate. While the phosphorylation of purine and pyrimidine mononucleotides ultimately allows the production of RNA and DNA polymers in modern metabolism and is of fundamental importance in core metabolism, reactions involving PRPP do not directly involve phosphoryl transfer type enzymes. Rather, they formally belong to the pentosyltransferase family of enzymes. Nevertheless, we discuss artificially selected glycosidic bond-forming ribozymes as they have important ramifications to how ribozyme-mediated phosphoryl transferase type reaction pathways might have evolved in an RNA world. Such phosphoryl transferase reactions would have been essential in an early RNA world, but have yet to be demonstrated by laboratory RNA selections.

In order to efficiently produce the β-D-ribofuranose sugar found in RNA nucleotides, metabolism must first control three abundant cyclic forms of ribose: α-D-ribopyranose (59% abundance), β-D-ribopyranose (20% abundance), and β-D-ribofuranose (20% abundance), together with small but extremely reactive amounts of the linear aldehyde form of this sugar [40]. Modern metabolism achieves this control by two key phosphoryl-transferase reactions that focus on ribose chemistry toward nucleotide synthesis. Ribokinase takes ribose (R) to ribose-5-phosphate (pR or R5P, Eq. (12.5)).

$$\mathbf{ppp}\mathrm{A} + \mathrm{R} \rightarrow \mathbf{p}\mathrm{R} + \mathrm{ppA} \qquad (12.5)$$

Synthesis of R5P precludes the formation of α-D-ribopyranose and β-D-ribopyranose, which constitutes nearly 80% of the possible ribopyranose isomers. Converting R5P to PRPP via ribose-phosphate diphosphokinase produces the activated substrate PRPP (Eq. (12.6)) that has the correct initial stereochemistry for the synthesis of β nucleotides.

$$\mathbf{ppp}\mathrm{A} + \mathrm{pR} \rightarrow \mathrm{pR}\mathbf{pp} + \mathrm{pA} \tag{12.6}$$

PRPP, which is locked into the α-D-ribofuranose form, simultaneously precludes the ring-opening of the ribose sugar and prevents the formation of the otherwise very reactive linear aldehyde sugar form of ribose. Thus, just as performed by modern metabolism, it appears likely that RNA enzymes must have required these two phosphoryl transfer reactions early in evolution in order to produce nucleotides. Based on the previously discussed 5′-kinase ribozymes (see Section 12.2), there seems to be little doubt that, early in evolution, a ribokinase ribozyme could have existed. While more challenging, the recent discovery of riboswitches that recognize PRPP as a ligand [41], suggests that ribose-phosphate diphosphokinases could have existed as well. Interestingly, in vitro selection experiments for purine nucleotide synthesis ribozyme also provide tangential support for the early existence of a PRPP type synthase ribozyme.

Nucleotide synthesis in an RNA world context has been explored in the lab by the in vitro selection of two PRPP dependent nucleotide synthesis ribozymes [42, 43]. In core metabolism, PRPP plays a central role in pyrimidine synthesis either via the de novo orotidine synthesis or by the salvage nucleotide synthesis pathways. Likewise, de novo purine synthesis or salvage nucleotide synthesis utilizes PRPP. Both ultimately result in the synthesis of β nucleotides. Such chemistry ranges from highly concerted to nearly fully dissociative, as judged by a range of kinetic isotope effect experiments [44]. Selection for a 4-thiouridine nucleotide synthase ribozyme (Figure 12.2a), using PRPP tethered to the 3′-terminus of a high diversity RNA pool, yielded pyrimidine nucleotide synthase ribozymes [43] that were later shown to be highly dissociative [45]. This indicates that the ribozyme mediates chemistry, where the pyrophosphate is nearly fully displaced prior to the final attack of the N1 atom of 4-thiouracil. This pyrimidine chemistry, which is thermodynamically less favorable than purine synthesis, was complemented by the selection of more efficient 6-thioguanosine nucleotide synthase ribozymes, again using PRPP tethered to the 3′ termini of a high diversity RNA pool (Figure 12.2b) [42]. Both ribozyme families showed very high substrate specificity and a marked preference for either their 4-thiouracil or 6-thioguanine substrates, strongly indicating that RNA has no intrinsic difficulty in recognizing small ribozyme substrates.

As just argued, PRPP serves a dual purpose in modern metabolism. First, it activates a single furanose sugar isomer for nucleotide synthesis, and second, it suppresses the formation of other furanose sugar isomers. This, however, presents an evolutionary conundrum given the unstable nature of PRPP itself. Such an activated sugar could not have easily accumulated in a prebiotic world to useful concentrations. Thus, while PRPP might have been available prebiotically in small amounts [46], it seems unlikely that this molecule could have been a core feature

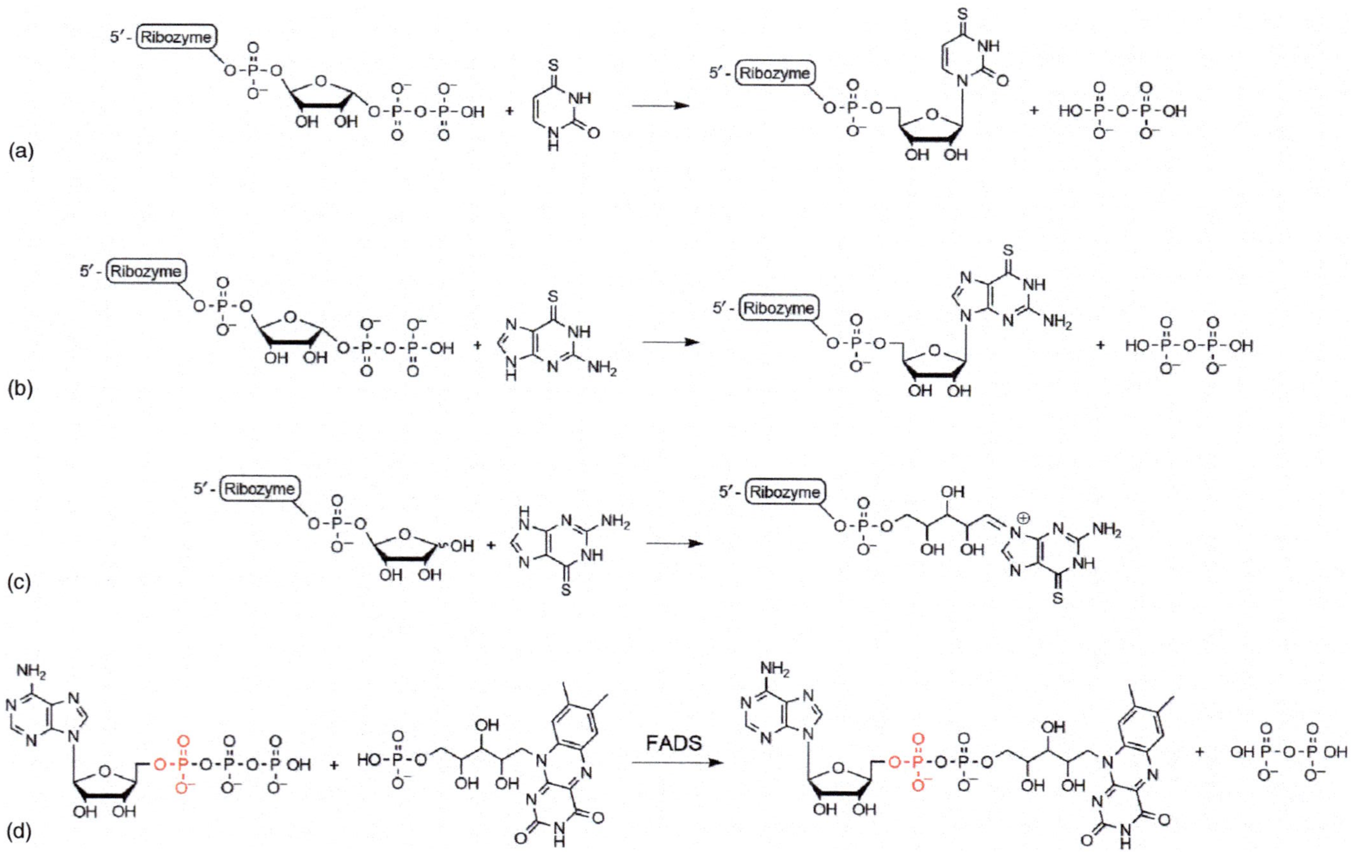

Figure 12.2 Glycosidic bond-forming ribozymes catalyze the synthesis of (a) 4-thiouracil monophosphate (^{4S}UMP) and (b) 6-thioguanosine monophosphate (^{6S}GMP) from RNA tethered 5-phosphoribosyl 1(α)-pyrophosphate (PRPP) and 4-thiouracil (^{4S}Ura) or 6-thioguanine (^{6S}Gua) respectively. (c) A promiscuous in vitro selected ribozyme uses aldehyde chemistry to react a nucleobase with R5P to produce a carbon linker similar to that found in flavin mononucleotide. (d) Flavin adenine dinucleotide synthetase (FADS) attaches an adenine nucleotide to a flavin mononucleotide that contains a flexible five-carbon ribitol linker.

of an RNA world until after the evolution of an RNA metabolism robust enough to allow the reliable harnessing of metabolic energy in polyphosphates, such as ATP, had occurred. Conversely, selective pressure early in evolution would have been strongly driven by life's exponentially increasing demand for both nucleotides and a reliable metabolic energy currency. The exponentially rapid consumption of complex prebiotically synthesized nucleotides [47, 48] and other prebiotic materials that enabled the transition from abiotic chemistry to organized biological processes strongly implies that the metabolic synthesis of activated compounds such as PRPP would have become essential very early in the evolution of an RNA type world.

But how could the multistep metabolic pathways linking ribose to nucleotide synthesis have most naturally emerged early in evolution, when selection pressure would have demanded the almost immediate emergence of such apparently complex catalysts [49]? As the earliest waves of biology strip-mined prebiotic sources of complex metabolites (such as nucleotides), might it have been possible that RNA intrinsically possesses the promiscuity required to rapidly establish and build a complex RNA core metabolism capable of utilizing simpler and simpler prebiotic compounds? Specifically, could it have been that the critical PRPP type phosphoryltransfer type reactions (Eq. (12.6)) and PRPP dependent nucleotide synthesis (Figure 12.2) reactions so essential for modern metabolism are derived from promiscuous ribozyme mediated chemistries?

A third in vitro selection for nucleotide synthesis suggests that such ideas may have validity. A selection for nucleotide synthesis using tethered R5P rather than PRPP found that the linear aldehyde form of R5P could be made to react with 6-thioguanine [50] easily. While the linear ribose chain product of this reaction was not completely characterized, the C–N linkage between sugar and nucleobase (Figure 12.2c) bears a marked resemblance to the linear linkage found between flavin and ribose in flavin mononucleotide (Figure 12.2d). This pseudo-nucleotide is being used to synthesize FAD in modern metabolism. Thus reactive ribose aldehyde chemistry can easily be imagined to lead to the covalent tethering of redox nucleobase cofactors to RNA catalysts early in an RNA world (see Section 12.4 for additional strategies) where redox catalytic diversity was presumably quite limited in the absence of redox activating cofactors.

Quite unexpectedly, simply replacing R5P with PRPP caused this R5P-dependent 6-thioguanine tethering ribozyme to now function as an effective nucleotide synthesis ribozyme [50].

As this promiscuous ribozyme supports both the chemistry of the linear aldehyde form of ribose found in R5P and of the cyclic furanose form of ribose found in PRPP, the following evolutionary scenario might have been possible in an early RNA world: metabolism would have required redox chemistry, just as modern metabolism does today. As demonstrated by the artificially selected 6-thioguanine tethering ribozyme, promiscuous ribozymes that are able to tether redox cofactors to RNA using R5P could have easily evolved in an early RNA metabolism. Such ribozymes might have played a key role in the early emergence of critical flavin-dependent biochemistry that today is essential for a broad range of carbon–carbon, carbon–nitrogen, and carbon–oxygen chemistry. Under these assumptions, the evolution of a single new ribozyme activity early in evolution that was able to phosphorylate R5P to PRPP (Eq. (12.6)) would have rapidly resulted in the evolution of ribozyme-mediated nucleotide synthesis. This process would have been driven by the consumption of the prebiotic supplies of nucleotides by a growing RNA metabolism. This potential link between the tethering of redox factors and the emergence of modern PRPP dependent nucleotide synthesis early in metabolism is substantially untested, and much experimental work remains to fully explore this conjecture. Nevertheless, RNA has proven to be surprisingly versatile at mediating the types of chemistry that are directly relevant to the emergence of centrally important metabolic reactions found in modern life, and ribozyme research has given tantalizing clues as to how the earliest steps in the emergence of an RNA-based metabolism could have occurred.

12.4 Capping Ribozymes

The enzymatic capping of messenger RNAs (mRNAs) is a multistep process in modern eukaryotic metabolism that plays a key role in allowing the cell to test mRNA quality prior to translation [51, 52]. Capping in prokaryotic organisms occurs relatively rarely and is mechanistically distinct from eukaryotic capping. It involves the incorporation of metabolic dinucleotide cofactors into the 5′-terminus of certain mRNAs in a fashion that is speculated to alter mRNA lifetimes as nutrient conditions change [53–56]. Throughout the kingdoms of life, it is striking that metabolism requires many dinucleotide 5′–5′ cap analogs, such as NAD and FAD, that are essential protein enzyme redox cofactors. This suggests that capping in an RNA world would have been an important way to diversify the functionality of RNA catalysts early in evolution.

Two independent in vitro selections have found RNA capping ribozymes [57, 58]. Neither selection was initially focused on the discovery of capping activity. Nevertheless, careful characterization indicated that capping is a straightforward phosphoryl transfer reaction for RNA to mediate. This was surprising as it might have been reasonably expected that stabilizing the charges found on each of the substrates required for capping would have been difficult to achieve. In the first selection, the Yarus laboratory was searching for a mixed carboxylate–phosphate anhydride of the sort normally produced by aminoacyl-tRNA synthetases [57, 59]. Instead, they discovered a quite universal capping ribozyme, Iso6. Similarly, the Unrau laboratory was trying to select for an RNA polymerase using a high diversity random pool annealed to a poly(A) template and incubated with ^{4S}UTP. Instead of the desired polymerization activity, they also found a versatile capping ribozyme, called 6.17. Characterization of this ribozyme leads to the discovery that both ribozymes had very broad substrate requirements [58]. Initially activated with a 5′ triphosphate, both ribozymes could be capped with a range of nucleotides differing in either phosphate number (Eqs. (12.7) and (12.8), Figure 12.3) or base type. As both ribozymes had distinct selective histories, secondary structures, and metal ion requirements, it was perhaps not surprising that their apparent k_{cat} values differed by approximately 60-fold. Curiously, these two ribozymes shared many enzymatic

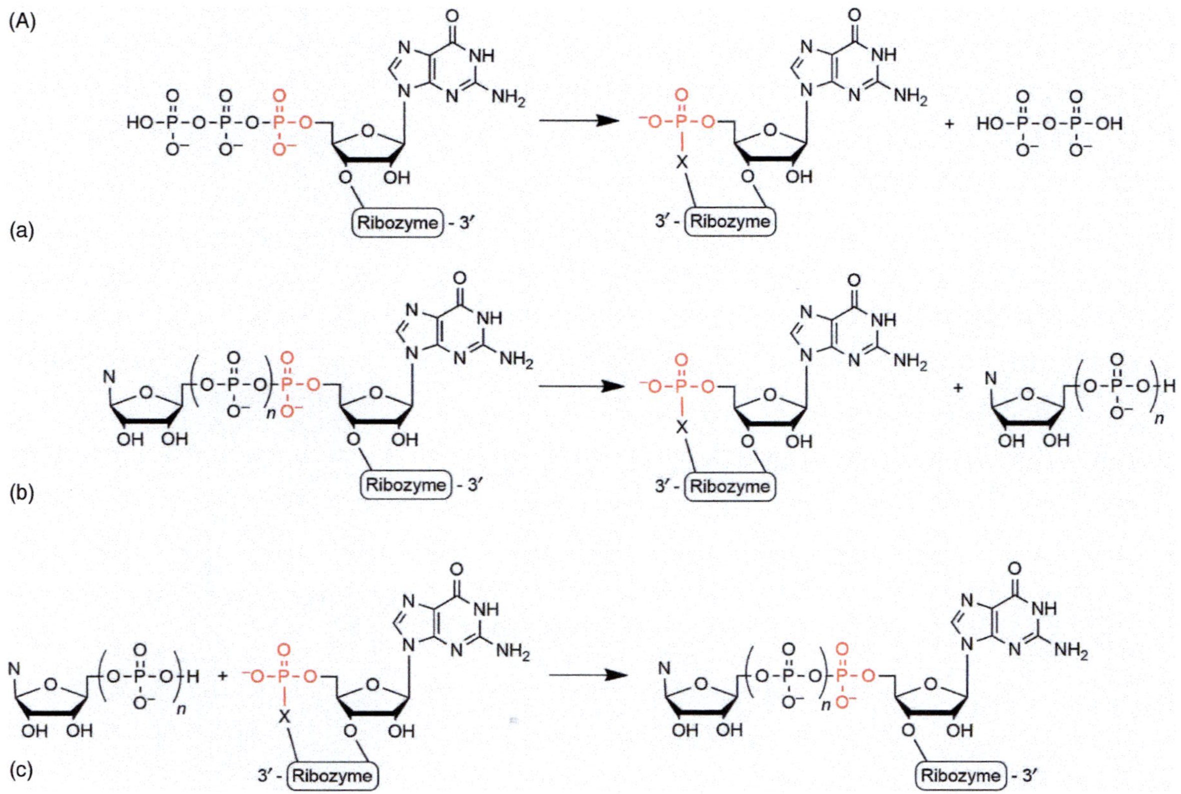

Figure 12.3 Ribozyme mediated self-capping. (A) RNA capping ribozymes catalyze the transfer of RNA oligomers or nucleotides (N = A, U, G, C) with varying numbers of phosphates (n = 1, 2, 3, 4) to the 5′ α-phosphate of an RNA sequence to form an RNA cap (N(5′)p_npRNA cap) through 5′–5′ phosphate–phosphate linkages. This pathway involves an intermediate where the 5′ α-phosphate of an uncapped (a) or capped (b) RNA molecule forms a covalent bond with a 3′ downstream Nitrogen or Oxygen (X). The intermediate is then attacked by RNA oligomers or nucleotides to form an RNA cap or exchange an existing cap (c). (B) The protein-mediated capping proceeds in reverse fashion to that of the ribozyme capping mechanism. Here, the enzyme (E) attacks a reactive 5′ α-phosphate, but this time on the incoming cap nucleotide so as to form a covalent N(5′)pE nucleotide intermediate (a), which in turn is attacked by a ppRNA to form a phosphate–phosphate linkage (b).

(B)

(a)

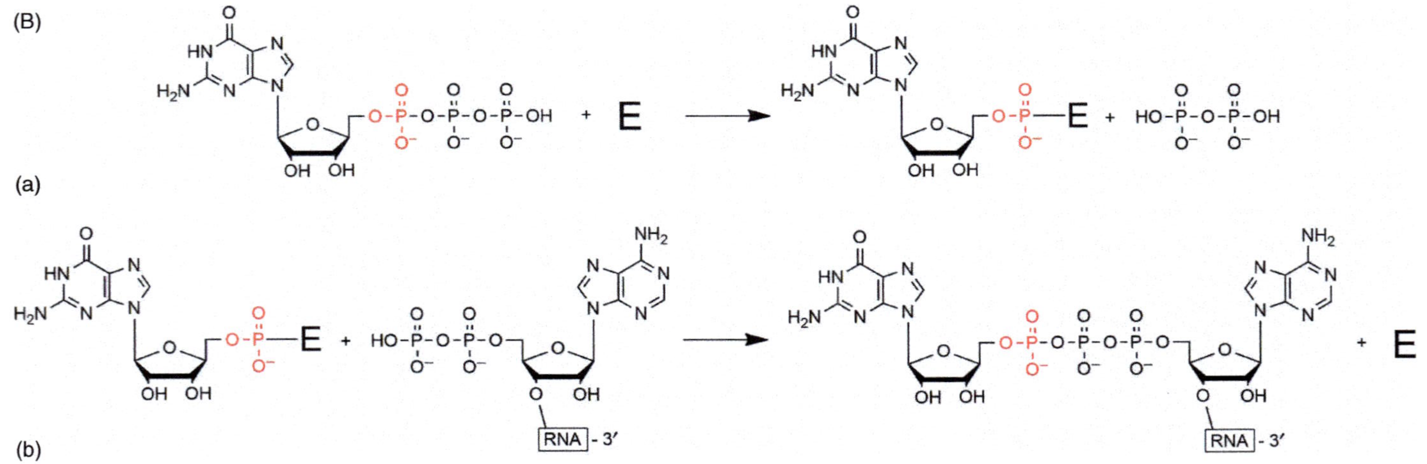

(b)

Figure 12.3 (*Continued*)

similarities apart from this difference in rate. First, both ribozymes become faster at low pH. This is in contrast to ligase ribozymes, which become faster at high pH as the attacking hydroxyl in these ligases must be deprotonated prior to the nucleophilic attack (see Section 12.5). Second, as capping substrates changed in phosphate number (nucleotide monophosphate [NMP], nucleotide diphosphate [NDP], nucleotide triphosphates [NTPs]), the two ribozymes changed catalytic rates in a highly correlated fashion. Both ribozymes had nearly the same relative speeds using NTP and NDP substrates, but both become about 20-fold slower when NMP was used to generate a cap ($n = 1$, Eq. (12.8)). Third, both ribozymes activated the α-phosphate of the triphosphate originally found on the ribozyme (Figure 12.3, Eq. (12.7)), which was followed in a second step by nucleotide capping (Eq. (12.8)). Fourth, both ribozymes could exchange cap structures implying a high level of symmetry in the reaction mechanism (Eq. (12.9)).

$$\text{pp}\mathbf{p}\text{RNA} \rightarrow \mathbf{p} < \text{RNA} + \text{PP}_i \tag{12.7}$$

$$\text{p}_n\text{N} + \mathbf{p} < \text{RNA} \rightarrow \text{N}(5')\text{p}_n\mathbf{p}\text{RNA} \tag{12.8}$$

$$\text{p}_m\text{N}' + \text{N}(5')\text{p}_n\mathbf{p}\text{RNA} \rightarrow \text{N}'(5')\text{p}_m\mathbf{p}\text{RNA} + \text{p}_n\text{N} \tag{12.9}$$

where $n, m = 1, 2, 3, 4$, and N, N′ = A, U, G, and C, < indicates a lariat type activated 5′ phosphate.

These commonalities suggested similarity in mechanism between the two distinct ribozyme isolates, as indicated in Eqs. (12.7)–(12.9) and Figure 12.3.

By synthesizing Rp and Sp α-capped variants of the 6.17 capping ribozyme [60], it was demonstrated that cap exchange retained stereochemistry rather than inverting it. While the covalent intermediate implied by this retaining mechanism was not found, this experiment indicated that both ribozymes performed capping using a two-step mechanism. If initially activated with a triphosphate, both ribozymes would, in their first step, attack the 5′ α-terminal phosphate found on the ribozyme to form a lariat structure (Figure 12.3A(a)), displacing pyrophosphate and inverting the stereochemistry of the reaction center. If activated with a cap, these ribozymes would first displace the nucleotide found immediately proximal to the α-phosphate, while again forming a lariat (Figure 12.3A(b)). While the nucleophile responsible for forming this lariat has yet to be identified, it is likely to involve an amine from a base or a 2′ hydroxyl of a ribose sugar found in either ribozyme's catalytic core. In the second step, the incoming capping substrate now attacks this intermediate, reversing the stereochemistry of the first step to a retaining configuration (Figure 12.3A(c)), thus explaining the symmetry observed with the cap exchange in this ribozyme.

Such a two-step mechanism would also explain the extensive commonalities between the two ribozymes and is fundamentally identical to that of protein-based eukaryotic RNA capping [61]. Both ribozymes appear to only recognize the phosphates immediately proximal to the site of lariat formation. This, together with the two-step mechanism, explains the otherwise perplexing finding that forming a cap with an NMP is less efficient than with NDP or NTP for both ribozymes.

Protein-mediated capping activates the 5′ α-phosphate of the incoming GTP nucleotide, effectively reversing the order of this otherwise equivalent chemistry. First, the protein catalyst attacks the GTP α-phosphate with a lysine, displacing pyrophosphate to form a covalent intermediate, which is directly analogous to the ribozyme lariat structure (Compare Figure 12.3A(a),A(b),B(a)). In the second step, the capping enzyme allows an mRNA, which possesses a terminal 5′ diphosphate, to attack this reactive intermediate to form a final GpppRNA triphosphate cap (Figure 12.3B(b),A(c), $n = 2$). From an evolutionary perspective, this protein strategy has several obvious regulatory advantages. Namely, premature hydrolysis of the GpE intermediate does not inactivate capping as a fresh GTP can still react. Likewise, using ppRNA in the second step allows capping to be maximally regulated by biological processes. Otherwise, RNA-mediated capping by a ribozyme is identical to that of eukaryotic capping and NDPK dependent phosphoryl transfer enzymes from a mechanistic perspective. While recently studied ribozymes have no mechanism to break the symmetry between their initial and final states (allowing them to continuously exchange caps until hydrolysis occurs [60]), further in vitro selection could be used to evolve capping and NDPK-like ribozymes. Capping with redox factors such a nicotinamide mononucleotide (NMN) would have allowed an early RNA world to rapidly evolve redox-dependent metabolic pathways using RNA as a catalyst. More interestingly, the future demonstration of NDPK-like ribozymes would have direct ramifications for the role of RNA in the early emergence of a nucleotide-based energy currency. Such chemistry would be fundamentally similar to that used by NDPK enzymes found in modern metabolism (Eqs. (12.3) and (12.4)).

12.5 Ligase Ribozymes

RNA ligase enzymes are universal in all domains of biology and are able to catalyze the ligation of RNA molecules through phosphodiester bond formation. These reactions play a critical role in RNA repair, splicing, and editing pathways [62–64]. The protein ligases, such as the two branches of bacteriophage T4 ligases, RNA ligase 1 (Rnl1), and RNA ligase 2 (Rnl2) [65], are capable of ligating 5′-phosphorylated RNA (5′-pRNA) termini with a terminal 3′-hydroxyl RNA via a three-step mechanism. The first two steps are chemically analogous to RNA capping (Figure 12.4a,b). The enzymes initially form an activated adenylate intermediate (A(5′)p-E) via the attack of an active site lysine residue on the α-phosphate of pppA, displacing pyrophosphate (PP_i) (Figure 12.4a). Similarly, DNA ligases can also use the nucleotide cap NAD^+ as a substrate, as opposed to pppA, releasing in this case, NMN [66]. The first step confers specificity toward purine nucleotides by burying the purine base in a hydrophobic pocket of conserved residues and exposing the α-phosphate of the nucleotide on the enzyme surface [61]. In the second step, the adenylate is transferred onto the 5′ α-phosphate of the 5′-pRNA to form an A(5′)ppRNA cap (Figure 12.4b). In the third and final step, the enzyme coordinates the nucleophilic attack of a 3′ hydroxyl of an opposing RNA strand to form a final phosphodiester linkage with the A(5′)ppRNA capped RNA, displacing pA (Figure 12.4c). This

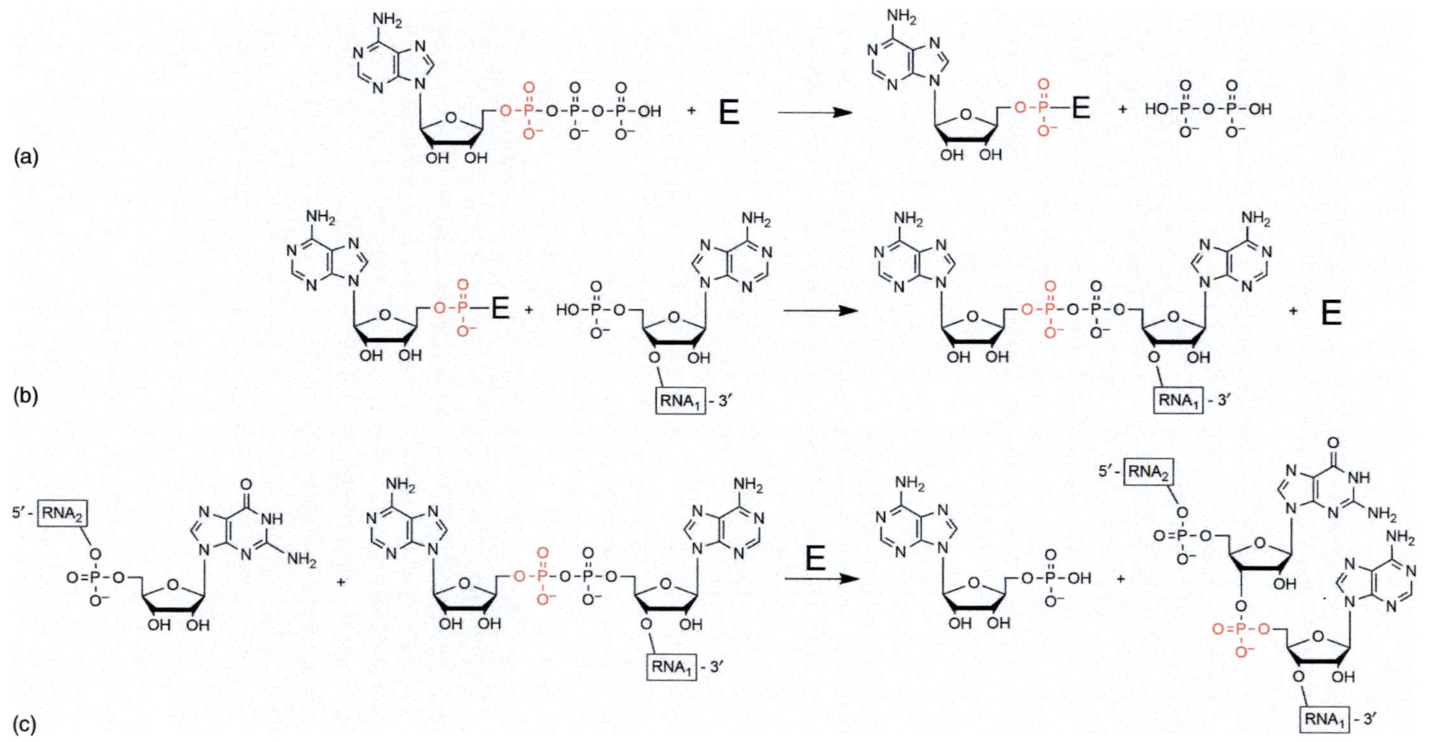

Figure 12.4 Protein mediated RNA ligation proceeds via a three-step mechanism, with the first two steps directly analogous to capping. (a) The enzyme (E) attacks the reactive 5′ α-phosphate on an incoming ATP, forming a covalent A(5′)pE nucleotide intermediate and releasing pyrophosphate. (b) In turn, this activated phosphate is attacked by a 5′-pRNA$_1$ to form an AppRNA$_1$ cap structure. (c) The enzyme then coordinates the attack of a 3′ hydroxyl of an upstream RNA$_2$ strand to AppRNA$_1$ to finally form the phosphodiester linkage while displacing pA.

mechanism has the ultimate advantage of activating an otherwise quiescent 5′ phosphate to enable phosphodiester bond formation. As just discussed (see Section 12.4), ribozymes are capable of two-step capping chemistry, but it remains to be seen if the more complex three-step chemistry performed by protein RNA ligases can be achieved in the laboratory via the evolution of a ribozyme that combines all three steps of this process.

In contrast to protein catalyzed ligation, all ribozyme ligases found to date proceed via a single-step mechanism. Two such naturally occurring ribozymes with ligase activity are found in the hairpin and hammerhead families. Both these ribozymes (see Chapters 3 and 16) catalyze reversible sequence-specific cleavage of the RNA backbone, where the nucleophilic attack of the 2′-hydroxyl onto an adjacent 3′-phosphodiester results in the formation of a 2′–3′-cyclic phosphate and a free 5′-hydroxyl terminus [67–69]. Favoring ligation over cleavage, hairpin ribozymes derived from the (−) strand of *Tobacco ringspot virus* satellite RNA show ligation yields of up to 57% (Figure 12.5a) [70]. Conversely, although cleavage is preferred by hammerhead ribozymes, they are capable of efficient ligation, as observed in organisms such as *Schistosoma*, which show up to 23% of ligated products when starting with a fully cleaved substrate (Figure 12.5a) [71]. The interconversion between cleaved cyclic phosphate and a ligated phosphodiester is nearly thermodynamically neutral. While exploited by viruses to great effect, the susceptibility of the cyclic phosphate to hydrolysis and its intrinsic reversibility likely does not make for an effective ligation mechanism, where the robust preservation of more complex genetic information is required.

A major requirement of an RNA world is the ability of RNA molecules to sustain life through RNA replication. Of particular interest is an RNA replication system that can either polymerize sequential nucleotides onto a growing strand in a template-dependent manner or assemble RNA oligomers via RNA catalyzed ligation reactions. The first in vitro ribozyme selection aimed to isolate ligase ribozymes that are mechanistically indistinguishable from that of RNA polymerases [72]. This selection used the 5′ triphosphate found on RNA transcripts as activation, resulting in seven isolated ribozyme families that fall into three structural classes, each showing an increase in catalytic rate of 10^3–10^4 over the uncatalyzed reaction. Of these three classes, the class I ligase is unique as it catalyzes the formation of a 3′–5′-phosphodiester bond linkage (Figure 12.5b), while the class II and III ligases catalyzed 2′–5′ linkage formation (Figure 12.5c) [73]. Using further in vitro evolution and engineering approaches on the class I ligase resulted in an improved new ribozyme, 210t, which differs from its parent by only a three base pairs rearrangement in its catalytic core. This new ribozyme achieved a catalytic rate comparable to naturally occurring ribozymes such as the *Tetrahymena* group I intron and RNase P RNA, with a k_{cat} of 360 min^{-1} at pH 9.0 as compared to the parent ribozyme with a k_{cat} of 16 min^{-1} at pH 8.0 [74]. In addition to high catalytic rates, in vitro selected ligase ribozymes are quite versatile. This was demonstrated by Ikawa et al. who synthesized and developed an artificial RNA ligase by combining multiple functioning RNA motifs into *cis*-DSL-1S, a ribozyme that can carry out two successive 3′–5′ nucleotidyl polymerization type reactions [75]. Furthermore, through a

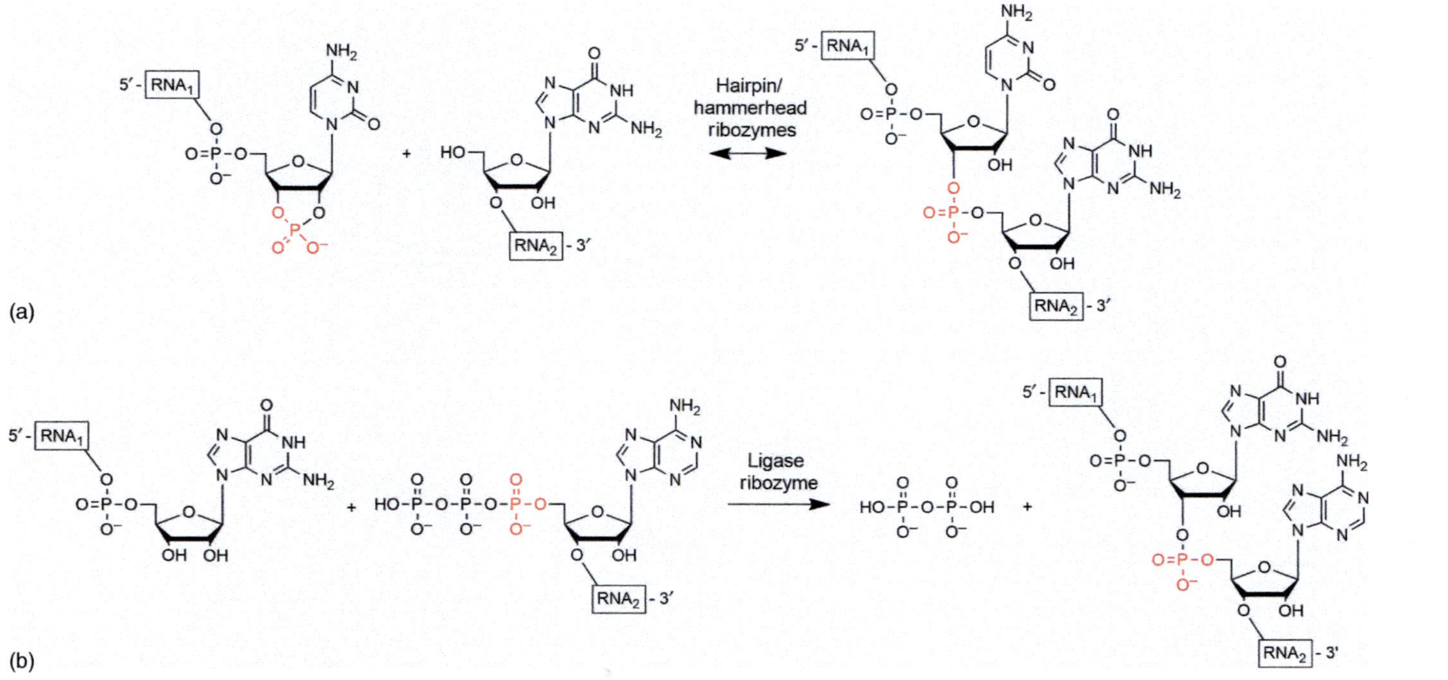

Figure 12.5 Ligase ribozymes catalyze the ligation of an RNA molecule through (a) a 5′ hydroxyl nucleophilic attack on a 2′−3′ cyclic phosphate, as found in the reverse reaction of the hairpin and hammerhead ribozymes, (b) nucleophilic attack of a 3′ hydroxyl or (c) a 2′ hydroxyl on the α-phosphate of a 5′-triphosphate on a downstream RNA molecule and (d) 3′ hydroxyl nucleophilic attack on an AppRNA promoted by ribozymes.

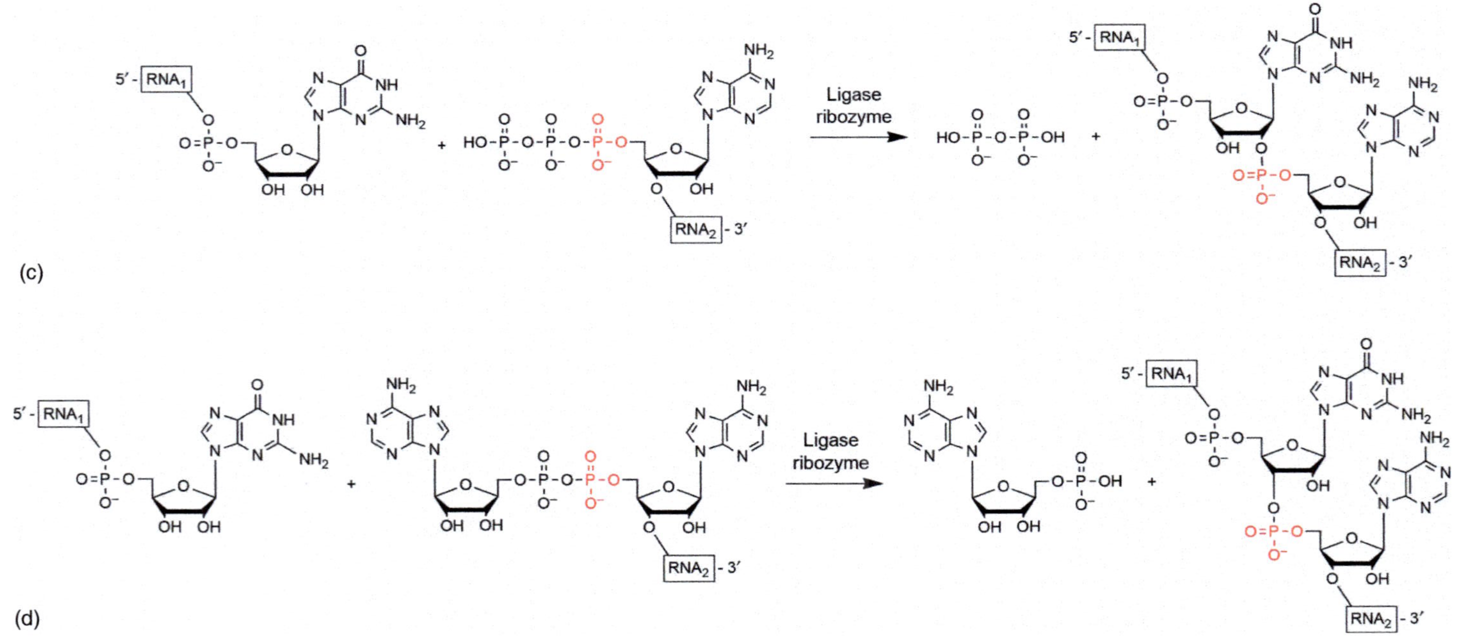

Figure 12.5 *(Continued)*

distinct selection scheme utilizing ATPγS, the ribozyme C06 was selected and was able to catalyze 2′–5′-phosphodiester bond formation with a variety of purine mono-, di-, and trinucleotides [76].

Compared to the modern protein ligases, the ligase ribozymes selected and discussed so far catalyze a single step ligation mechanism as opposed to three steps. As the multistep process fundamentally allows ligation to occur from an unactivated state (i.e. nucleic acid strands initially lacking activation on either the 5′ or 3′ termini), the single-step ligation misses out on the advantages that multistep ligation offers for nucleic acid repair. The question thus rises of whether in the setting of an early RNA world, RNA ribozymes could have evolved to carry out a similar multistep ligation process. As we have seen, phosphorylation is entirely possible (see Section 12.2), and as it stands, all three steps found in protein-mediated ligation reactions can be performed by ribozymes. However, each step is currently performed by a distinct ribozyme. The capping ribozymes described earlier in this chapter (see Section 12.4) can catalyze the first and second steps of the protein RNA ligase mechanism by transferring a nucleotide cap onto the 5′ α-phosphate of an RNA sequence. Furthermore, Hager and Szostak used a pre-existing ATP RNA aptamer domain flanked by random sequences and isolated ligase ribozymes capable of 3′–5′ template-directed ligation of AMP-activated RNA (A(5′)ppRNA) with a catalytic rate enhancement of $\sim 5 \times 10^5$ over the uncatalyzed reaction (Figure 12.5d) [77]. This set of reactions demonstrate that a ribozyme containing all three mechanistic steps of a modern protein ligase could certainly be selected in the future. To that end, there is no reason to believe that RNA could not have easily manipulated and repaired genetic information in an RNA world.

Ligase ribozymes could have also played a key role in early RNA replication through the assembly of smaller, less information-rich oligomers into complex functional RNA molecules. Here the manipulation of genetic information in an RNA world is taken to its logical endpoint, where the construction of complex RNA catalysts could have occurred via the ligation of smaller RNA polymers into larger functional elements. Using the in vitro selected RNA ligase R3C [78] as a starting point, Paul and Joyce developed a self-replicating RNA enzyme [79]. The R3C ribozyme has a simple secondary structure and catalyzes 3′–5′ phosphodiester linkage between two RNA molecules with a catalytic rate of $0.32\,\text{min}^{-1}$. This ribozyme was re-configured to act as a catalyst for its own assembly, joining together two RNA substrates to produce an identical copy of itself in a templated matter. The two-component molecules, A and B, are partially complementary to the ribozyme (R_1), which catalyzes their ligation to form a second identical ribozyme (R_2) (Figure 12.6A). This iteration of the system showed inefficient self-replication with a doubling time of ~17 hours. However, directed evolution guided by kinetic studies [81] has greatly improved the system, achieving doubling times of ~5 minutes with exponential growth rates of $\sim 0.14\,\text{min}^{-1}$ [82]. The R3C ligase was also converted to a cross-catalytic format, where two different ribozymes (R and R′) can synthesize copies of each other from four component molecules (A, A′, B, and B′) (Figure 12.6B) [80]. Both self- and cross-replicating ribozyme systems are capable of indefinite exponential amplification, provided they have an

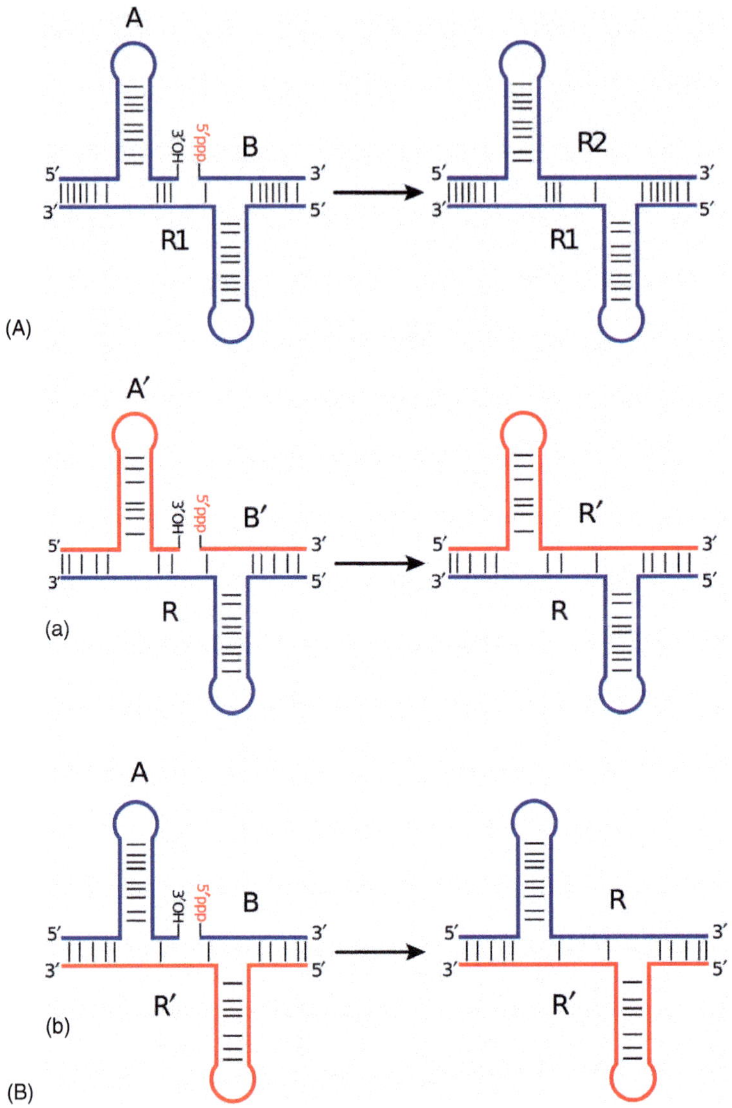

Figure 12.6 Self-replication schematic (A). The ribozyme, R_1, ligates substrates A and B to form an identical ribozyme R_2. Cross-catalytic replication schematic (B). The ribozyme R (blue) ligates substrates A′ and B′ (red) to form ribozyme R′ (red) (a). In a reverse mechanism, the ribozyme R′ (red) ligates substrates A and B (blue) to form ribozyme R (blue) (b). Lines indicate partial hybridization. Source: Adapted from Kim and Joyce [80].

ongoing supply of substrates. It has been proposed that, in an early RNA world, the first enzymatic replication system would have relied on trans-acting ligase ribozymes for the propagation of genetic information by assembling themselves and other catalysts or templates [83], only later to be replaced by RNA-dependent RNA polymerase ribozymes [84].

12.6 Polymerase Ribozymes

Two major strategies have defined attempts to create self-replicating systems out of RNA. The first, as just mentioned, focuses on how complex networks of cooperative molecules can emerge spontaneously in laboratory conditions [79, 80, 82, 83]. These interesting reactions derived fundamentally from exploiting the catalytic core of the *Azoarcus* group I intron ribozyme demonstrate how networks of catalysis could have emerged early in evolution and reinforce how the types of multistep catalytic processes sustained by modern metabolism could have been preceded by RNA. The second strategy explicitly demonstrates how one catalytic activity, namely ligation, could have potentially evolved into the more advanced catalytic function of polymerization. The class I ligase discovered by Bartel and Szostak demonstrated that RNA was capable of the type of chemistry required for RNA polymerization [72]. As modern metabolism uses polymerases for all fundamental forms of information storage and transfer, several laboratories have focused on the development of RNA dependent RNA polymerization ribozymes. Due to the high catalytic efficiency of the class I ligase, efforts have largely focused on converting this ligase ribozyme into a proficient RNA polymerase ribozyme. As another chapter (Chapter 13) will focus on RNA polymerase ribozymes more specifically, we focus here only on how in an early RNA world, such polymerases could have integrated with ribozymes capable of nucleotide synthesis and nucleotide phosphorylation so as to have potentially created the earliest self-sustaining replicative systems.

The class I ligase derived RNA-dependent RNA polymerase ribozymes evolved to date in the laboratory are capable of copying an RNA template by hundreds of nucleotides, albeit on highly repetitive templates [85–88]. They do so by using a two-domain ribozyme [89]. The first domain consists of the class I ligase catalytic core [90], while the second RNA domain was evolved de novo [91] and helps stabilize NTP addition. The resolved 3.0 Å resolution crystal structure of the class I ligase has a resemblance to a tripod [90], while the second RNA domain drapes over the vertex of the ligase core tripod structure [89]. Mechanistically, these RNA-templated RNA polymerase ribozyme families strongly resemble the DNA repair polymerase *Taq* in that they non-processively extend the 3′-terminus of an RNA primer. Most excitingly, these ribozymes have recently been demonstrated to mediate RNA-dependent DNA polymerase activity [92], strongly suggesting that the emergence of DNA genomes could have occurred much earlier in evolution than previously anticipated.

While laboratories have focused on RNA polymerization in the 5′ to 3′ direction (Figure 12.7a), fundamentally 5′ to 3′ and 3′ to 5′ extension are chemically equivalent (Figure 12.7b). While 3′ activated NTPs are untenable due to the presence of reactive 2′ hydroxyls, extension with 5′ activated NTPs can technically proceed in either direction. Interestingly, the Joyce laboratory has observed that when given the opportunity, ribozyme polymerases are indeed capable of extension in either direction [93]. Modern metabolism favors 5′ to 3′ extension presumably because attacking the α phosphate of an incoming nucleotide by a 3′ hydroxyl found on a growing RNA

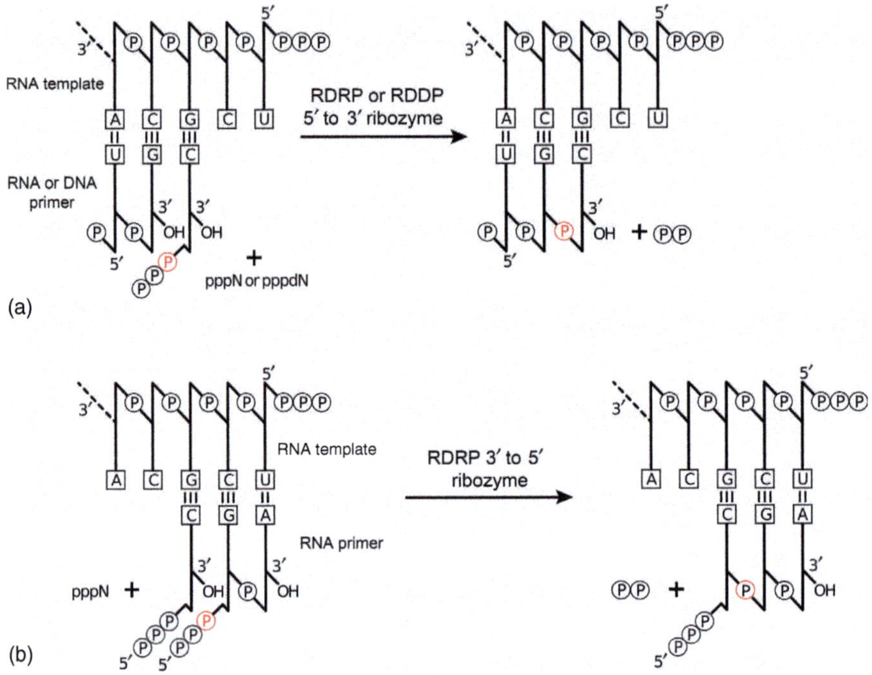

Figure 12.7 Template dependent RNA and DNA polymerization is catalyzed by RNA dependent RNA or DNA polymerase (RDRP or RDDP) Ribozymes. (a) RDRP or RDDP Ribozymes catalyze the formation of a 5′ to 3′-phosphodiester linkage between the 3′ hydroxyl of a template bound RNA primer and the 5′ α-phosphate of either a ribonucleotide or deoxyribonucleotide triphosphate (pppN or pppdN). (b) RDRP Ribozymes catalyzes the formation of a 3′ to 5′-phosphodiester linkage between the 5′ α-phosphate of a template-bound RNA primer and the 3′ hydroxyl of a ribonucleotide triphosphate (pppN). N indicates the possible nucleosides: adenosine (A), cytidine (C), guanosine (G), and uridine (U) in pppN and A, C, G, thymine (T) in pppdN (where dN represents a deoxyribose nucleotide).

strand (Eq. (12.10)) is intrinsically resistant to inadvertent NTP hydrolysis.

$$\mathrm{pp}\mathbf{p}\mathrm{N}_n + \ldots \mathrm{N}_{n-2}\mathrm{pN}_{n-1} \rightarrow \ldots \mathrm{N}_{n-2}\mathrm{pN}_{n-1}\mathbf{p}\mathrm{N}_n + \mathrm{PP_i} \tag{12.10}$$

An active RNA metabolism capable of nucleotide synthesis and phosphorylation would naturally regenerate a nucleoside or nucleotide to an NTP, whereas extension in the 3′–5′ direction (Eq. (12.11)), which is mechanistically equivalent, would be highly sensitive to hydrolysis of the 5′ triphosphate found on the growing strand and would be strongly rejected by evolution:

$$\mathrm{pppN}_n + \mathrm{pp}\mathbf{p}\mathrm{N}_{n-1}\mathrm{pN}_{n-2} \cdots \rightarrow \mathrm{pppN}_n\mathbf{p}\mathrm{N}_{n-1}\mathrm{pN}_{n-2} + \mathrm{PP_i} \tag{12.11}$$

While these polymerase ribozymes are important demonstrations of polymerization activity, they lack many essential features found in protein polymerases that are fully integrated into metabolism. Most importantly, these ribozymes are currently highly unprocessive. Lawrence and Bartel measured in detail the processivity of an RNA polymerase ribozyme derived from the class I ligase [91] and determined that the low processivity was attributed to the weak affinity of the polymerase for the primer-template [94]. Surprisingly, the ribozyme was particularly processive once it achieved accurate primer-template binding and alignment, showing great extension speed and multiple nucleotide additions before dissociation in this specific geometry. Consistent with this, tethering the primer to the polymerase resulted in a ~300-fold increase in polymerization rate [85, 87–89]. This lack of processivity precludes strand displacement, which would be required to replicate RNA genomes into functional RNA genes [95, 96]. Equally important, these ribozymes have no way to specifically initiate or terminate their copying function until recently [97]. Such activities are built into all protein-based DNA-dependent RNA polymerases in modern metabolism, as these functions allow cells to change their transcriptional programs to match changing environmental conditions. Developing improved processivity and mechanisms of initiation and termination similar to those found in protein polymerases, is fundamental to demonstrating the potential of RNA ribozymes to mediate the replication and evolution of an early RNA world metabolism.

12.7 Summary

RNA ribozymes have demonstrated a broad range of phosphoryl transfer reactions that are of direct relevance to the replication and manipulation of RNA in an early RNA world. The potential for a self-consistent RNA metabolism is increasingly supported by in vitro selection of ribozymes in the laboratory that are increasingly capable of metabolically relevant chemistry work in this area.

References

1 Manning, G., Whyte, D.B., Martinez, R. et al. (2002). The protein kinase complement of the human genome. *Science* 298 (5600): 1912–1934.
2 Knowles, J.R. (1980). Enzyme-catalyzed phosphoryl transfer reactions. *Annu. Rev. Biochem.* 49 (1): 877–919.
3 Crick, F.H.C. (1968). The origin of the genetic code. *J. Mol. Biol.* 38 (3): 367–379.
4 Gilbert, W. (1986). The RNA world. *Nature* 319: 618.
5 Orgel, L.E. (1968). Evolution of the genetic apparatus. *J. Mol. Biol.* 38 (3): 381–393.
6 Melnikov, S., Ben-Shem, A., de Loubresse, N.G. et al. (2012). One core, two shells: bacterial and eukaryotic ribosomes. *Nat. Struct. Mol. Biol.* 19 (6): 560–567.
7 Nissen, P., Hansen, J., Ban, N. et al. (2000). The structural basis of ribosome activity in peptide bond synthesis. *Science* 289 (5481): 920–930.
8 Ramakrishnan, V. (2014). The ribosome emerges from a black box. *Cell* 159 (5): 979–984.
9 Guerrier-Takada, C., Gardiner, K., Marsh, T. et al. (1983). The RNA moiety of ribonuclease P is the catalytic subunit of the enzyme. *Cell* 35: 849–857.
10 Klemm, B.P., Wu, N., Chen, Y. et al. (2016). The diversity of ribonuclease P: protein and RNA catalysts with analogous biological functions. *Biomolecules.* 6 (2). https://doi.org/10.3390/biom6020027.

11 Lechner, M., Rossmanith, W., Hartmann, R.K. et al. (2015). Distribution of ribonucleoprotein and protein-only RNase P in Eukarya. *Mol. Biol. Evol.* 32 (12): 3186–3193.

12 Fica, S.M., Tuttle, N., Novak, T. et al. (2013). RNA catalyzes nuclear pre-mRNA splicing. *Nature* 503 (7475): 229–234.

13 White, H.B. (1976). Coenzymes as fossils of an earlier metabolic state. *J. Mol. Evol.* 7 (2): 101–104.

14 Kanehisa, M. and Goto, S. (2000). KEGG: kyoto encyclopedia of genes and genomes. *Nucleic Acids Res.* 28 (1): 27–30.

15 Kanehisa, M., Sato, Y., Kawashima, M. et al. (2016). KEGG as a reference resource for gene and protein annotation. *Nucleic Acids Res.* 44 (D1): D457–D462.

16 Kanehisa, M., Furumichi, M., Tanabe, M. et al. (2017). KEGG: new perspectives on genomes, pathways, diseases and drugs. *Nucleic Acids Res.* 45 (D1): D353–D361.

17 Karimi-Busheri, F., Daly, G., Robins, P. et al. (1999). Molecular characterization of a human DNA kinase. *J. Biol. Chem.* 274 (34): 24187–24194.

18 Berg, J.M., Tymoczko, J.L., and Stryer, L. (2002). Nucleoside monophosphate kinases: catalyzing phosphoryl group exchange between nucleotides without promoting hydrolysis. In: *Biochemistry*, 5e (eds. S. Moran, G.L. Hadler, P. Zimmerman), 389–395. W. H. Freeman and Company.

19 Henzler-Wildman, K.A., Thai, V., Lei, M. et al. (2007). Intrinsic motions along an enzymatic reaction trajectory. *Nature* 450 (7171): 838–844.

20 Van Rompay, A.R., Johansson, M., and Karlsson, A. (2000). Phosphorylation of nucleosides and nucleoside analogs by mammalian nucleoside monophosphate kinases. *Pharmacol. Ther.* 87 (2): 189–198.

21 Zaug, A.J. and Cech, T.R. (1986). The *Tetrahymena* intervening sequence ribonucleic acid enzyme is a phosphotransferase and an acid phosphatase. *Biochemistry* 25 (16): 4478–4482.

22 Biondi, E., Nickens, D.G., Warren, S. et al. (2010). Convergent donor and acceptor substrate utilization among kinase ribozymes. *Nucleic Acids Res.* 38 (19): 6785–6795.

23 Biondi, E., Maxwell, A.W.R., and Burke, D.H. (2012). A small ribozyme with dual-site kinase activity. *Nucleic Acids Res.* 40 (15): 7528–7540.

24 Curtis, E.A. and Bartel, D.P. (2005). New catalytic structures from an existing ribozyme. *Nat. Struct. Mol. Biol.* 12 (11): 994–1000.

25 Lorsch, J.R. and Szostak, J.W. (1994). In vitro evolution of new ribozymes with polynucleotide kinase activity. *Nature* 371 (6492): 31–36.

26 Saran, D., Nickens, D.G., and Burke, D.H. (2005). A trans acting ribozyme that phosphorylates exogenous RNA. *Biochemistry* 44 (45): 15007–15016.

27 Sassanfar, M. and Szostak, J.W. (1993). An RNA motif that binds ATP. *Nature* 364 (6437): 550–553.

28 Lorsch, J.R. and Szostak, J.W. (1995). Kinetic and thermodynamic characterization of the reaction catalyzed by a polynucleotide kinase ribozyme. *Biochemistry* 34 (46): 15315–15327.

29 Illangasekare, M., Sanchez, G., Nickles, T., and Yarus, M. (1995). Aminoacyl-RNA synthesis catalyzed by an RNA. *Science* 267 (5198): 643–647.

30 Saran, D., Held, D.M., and Burke, D.H. (2006). Multiple-turnover thio-ATP hydrolase and phospho-enzyme intermediate formation activities catalyzed by an RNA enzyme. *Nucleic Acids Res.* 34 (11): 3201–3208.

31 Andreini, C., Bertini, I., Cavallaro, G. et al. (2008). Metal ions in biological catalysis: from enzyme databases to general principles. *J. Biol. Inorg. Chem.* 13 (8): 1205–1218.

32 Butcher, S.E. (2011). The spliceosome and its metal ions. *Met. Ions Life Sci.* 9: 235–251.

33 Steitz, T.A. and Steitz, J.A. (1993). A general two-metal-ion mechanism for catalytic RNA. *Proc. Natl. Acad. Sci. U.S.A.* 90 (14): 6498–6502.

34 Biondi, E., Poudyal, R.R., Forgy, J.C. et al. (2013). Lewis acid catalysis of phosphoryl transfer from a copper(II)-NTP complex in a kinase ribozyme. *Nucleic Acids Res.* 41 (5): 3327–3338.

35 Etaix, E. and Orgel, L.E. (1978). Phosphorylation of nucleosides in aqueous solution using trimetaphosphate: formation of nucleoside triphosphates. *J. Carbohydr. Nucleosides Nucleotides* 5: 91–110.

36 Feldmann, W. (1967). Das Trimetaphosphat als Triphosphorylierungsmittel für Alkohole und Kohlenhydrate in wäßriger Lösung. Seine Sonderstellung unter den kondensierten Phosphaten. *Chem. Ber.* 100: 3850–3860.

37 Thilo, E., Schulz, G., and Wichmann, E.-M. (1953). Die Konstitution des Grahamschen und Kurrolschen Salzes. *Z. Anorg. Allg. Chem.* 272: 182–200.

38 Kura, G. (1987). Alkaline hydrolysis of inorganic cyclo-polyphosphates. *Bull. Chem. Soc. Jpn.* 60 (8): 2857–2860.

39 Moretti, J.E. and Müller, U.F. (2014). A ribozyme that triphosphorylates RNA 5′-hydroxyl groups. *Nucleic Acids Res.* 42 (7): 4767–4778.

40 Drew, K.N., Zajicek, J., Bondo, G. et al. (1998). 13C-labeled aldopentoses: detection and quantitation of cyclic and acyclic forms by heteronuclear 1D and 2D NMR spectroscopy. *Carbohydr. Res.* 307 (3): 199–209.

41 Sherlock, M.E., Sudarsan, N., Stav, S., and Breaker, R.R. (2018). Tandem riboswitches form a natural Boolean logic gate to control purine metabolism in bacteria. *eLife* 7.

42 Lau, M.W.L., Cadieux, K.E.C., and Unrau, P.J. (2004). Isolation of fast purine nucleotide synthase ribozymes. *J. Am. Chem. Soc.* 126 (48): 15686–15693.

43 Unrau, P.J. and Bartel, D.P. (1998). RNA-catalysed nucleotide synthesis. *Nature* 395 (6699): 260–263.

44 Berti, P.J. (1999). Determining transition states from kinetic isotope effects. *Methods Enzymol.* 308: 355–397.

45 Unrau, P.J. and Bartel, D.P. (2003). An oxocarbenium-ion intermediate of a ribozyme reaction indicated by kinetic isotope effects. *Proc. Natl. Acad. Sci. U.S.A.* 100 (26): 15393–15397.

46 Akouche, M., Jaber, M., Maurel, M.-C. et al. (2017). Phosphoribosyl pyrophosphate: a molecular vestige of the origin of life on minerals. *Angew. Chem. Int. Ed.* 56 (27): 7920–7923.

47 Powner, M.W. and Sutherland, J.D. (2011). Prebiotic chemistry: a new modus operandi. *Philos. Trans. R. Soc. B Biol. Sci.* 366 (1580): 2870–2877.

48 Stairs, S., Nikmal, A., Bučar, D.-K. et al. (2017). Divergent prebiotic synthesis of pyrimidine and 8-oxo-purine ribonucleotides. *Nat. Commun.* 8.

49 Hayden, E., Lehman, N., and Unrau, P. (2018). Transitions: RNA and ribozymes in the development of life. In: *Astrobiology Handbook* (ed. V.M. Kolb), 379–394. CRC Press; Taylor & Francis Group.

50 Lau, M.W.L. and Unrau, P.J. (2009). A promiscuous ribozyme promotes nucleotide synthesis in addition to ribose chemistry. *Chem. Biol.* 16 (8): 815–825.

51 Hinnebusch, A.G. (2014). The scanning mechanism of eukaryotic translation initiation. *Annu. Rev. Biochem.* 83: 779–812.

52 Jackson, R.J., Hellen, C.U.T., and Pestova, T.V. (2010). The mechanism of eukaryotic translation initiation and principles of its regulation. *Nat. Rev. Mol. Cell Biol.* 11 (2): 113–127.

53 Chen, Y.G., Kowtoniuk, W.E., Agarwal, I. et al. (2009). LC/MS analysis of cellular RNA reveals NAD-linked RNA. *Nat. Chem. Biol.* 5 (12): 879–881.

54 Höfer, K. and Jäschke, A. (2018). Epitranscriptomics: RNA modifications in bacteria and *Archaea*. *Microbiol. Spectr.* 6 (3). https://doi.org/10.1128/microbiolspec.RWR‐0015‐2017.

55 Jäschke, A., Höfer, K., Nübel, G., and Frindert, J. (2016). Cap-like structures in bacterial RNA and epitranscriptomic modification. *Curr. Opin. Microbiol.* 30: 44–49.

56 Luciano, D.J. and Belasco, J.G. (2015). NAD in RNA: unconventional headgear. *Trends Biochem. Sci.* 40 (5): 245–247.

57 Huang, F. and Yarus, M. (1997). 5′-RNA self-capping from guanosine diphosphate. *Biochemistry* 36 (22): 6557–6563.

58 Zaher, H.S., Watkins, R.A., and Unrau, P.J. (2006). Two independently selected capping ribozymes share similar substrate requirements. *RNA* 12 (11): 1949–1958.

59 Huang, F. and Yarus, M. (1997). Versatile 5′ phosphoryl coupling of small and large molecules to an RNA. *Proc. Natl. Acad. Sci. U.S.A.* 94 (17): 8965–8969.

60 Zaher, H.S. and Unrau, P.J. (2006). A general RNA-capping ribozyme retains stereochemistry during cap exchange. *J. Am. Chem. Soc.* 128 (42): 13894–13900.

61 Shuman, S. and Lima, C.D. (2004). The polynucleotide ligase and RNA capping enzyme superfamily of covalent nucleotidyltransferases. *Curr. Opin. Struct. Biol.* 14 (6): 757–764.

62 Abelson, J., Trotta, C.R., and Li, H. (1998). tRNA splicing. *J. Biol. Chem.* 273 (21): 12685–12688.

63 Amitsur, M., Levitz, R., and Kaufmann, G. (1987). Bacteriophage T4 anticodon nuclease, polynucleotide kinase and RNA ligase reprocess the host lysine tRNA. *EMBO J.* 6 (8): 2499–2503.

64 Schnaufer, A., Panigrahi, A.K., Panicucci, B. et al. (2001). An RNA ligase essential for RNA editing and survival of the bloodstream form of *Trypanosoma brucei*. *Science* 291 (5511): 2159–2162.

65 Wang, L.K., Ho, C.K., Pei, Y., and Shuman, S. (2003). Mutational analysis of bacteriophage T4 RNA ligase 1 different functional groups are required for the nucleotidyl transfer and phosphodiester bond formation steps of the ligation reaction. *J. Biol. Chem.* 278 (32): 29454–29462.

66 Pascal, J.M. (2008). DNA and RNA ligases: structural variations and shared mechanisms. *Curr. Opin. Struct. Biol.* 18 (1): 96–105.

67 Feldstein, P.A., Buzayan, J.M., and Bruening, G. (1989). Two sequences participating in the autolytic processing of satellite tobacco ringspot virus complementary RNA. *Gene* 82 (1): 53–61.

68 Hampel, A. and Tritz, R. (1989). RNA catalytic properties of the minimum (-)sTRSV sequence. *Biochemistry* 28 (12): 4929–4933.

69 de la Peña, M., García-Robles, I., and Cervera, A. (2017). The hammerhead ribozyme: a long history for a short RNA. *Molecules* 22 (1): 78.

70 Balke, D., Zieten, I., Strahl, A. et al. (2014). Design and characterization of a twin ribozyme for potential repair of a deletion mutation within the oncogenic CTNNB1-ΔS45 mRNA. *ChemMedChem* 9 (9): 2128–2137.

71 Canny, M.D., Jucker, F.M., and Pardi, A. (2007). Efficient ligation of the *Schistosoma* hammerhead ribozyme. *Biochemistry* 46 (12): 3826–3834.

72 Bartel, D.P. and Szostak, J.W. (1993). Isolation of new ribozymes from a large pool of random sequences [see comment]. *Science* 261 (5127): 1411–1418.

73 Ekland, E.H., Szostak, J.W., and Bartel, D.P. (1995). Structurally complex and highly active RNA ligases derived from random RNA sequences. *Science* 269 (5222): 364–370.

74 Bergman, N.H., Johnston, W.K., and Bartel, D.P. (2000). Kinetic framework for ligation by an efficient RNA ligase ribozyme. *Biochemistry* 39 (11): 3115–3123.

75 Ikawa, Y., Tsuda, K., Matsumura, S., and Inoue, T. (2004). De novo synthesis and development of an RNA enzyme. *Proc. Natl. Acad. Sci. U.S.A.* 101 (38): 13750–13755.

76 Kang, T.J. and Suga, H. (2007). In vitro selection of a 5′-purine ribonucleotide transferase ribozyme. *Nucleic Acids Res.* 35 (12): 4186–4194.

77 Hager, A.J. and Szostak, J.W. (1997). Isolation of novel ribozymes that ligate AMP-activated RNA substrates. *Chem. Biol.* 4 (8): 607–617.

78 Rogers, J. and Joyce, G.F. (2001). The effect of cytidine on the structure and function of an RNA ligase ribozyme. *RNA* 7 (3): 395–404.

79 Paul, N. and Joyce, G.F. (2002). A self-replicating ligase ribozyme. *Proc. Natl. Acad. Sci. U.S.A.* 99 (20): 12733–12740.

80 Kim, D.-E. and Joyce, G.F. (2004). Cross-catalytic replication of an RNA ligase ribozyme. *Chem. Biol.* 11 (11): 1505–1512.

81 Ferretti, A.C. and Joyce, G.F. (2013). Kinetic properties of an RNA enzyme that undergoes self-sustained exponential amplification. *Biochemistry* 52 (7): 1227–1235.

82 Robertson, M.P. and Joyce, G.F. (2014). Highly efficient self-replicating RNA enzymes. *Chem. Biol.* 21 (2): 238–245.

83 Vaidya, N., Manapat, M.L., Chen, I.A. et al. (2012). Spontaneous network formation among cooperative RNA replicators. *Nature* 491 (7422): 72–77.

84 Levy, M. and Ellington, A.D. (2001). The descent of polymerization. *Nat. Struct. Mol. Biol.* 8 (7): 580–582.

85 Attwater, J., Wochner, A., and Holliger, P. (2013). In-ice evolution of RNA polymerase ribozyme activity. *Nat. Chem.* 5 (12): 1011–1018.

86 Attwater, J., Raguram, A., Morgunov, A.S. et al. (2018). Ribozyme-catalysed RNA synthesis using triplet building blocks. *eLife* 7: e35255.

87 Horning, D.P. and Joyce, G.F. (2016). Amplification of RNA by an RNA polymerase ribozyme. *Proc. Natl. Acad. Sci. U.S.A.* 113 (35): 9786–9791.

88 Wochner, A., Attwater, J., Coulson, A., and Holliger, P. (2011). Ribozyme-catalyzed transcription of an active ribozyme. *Science* 332 (6026): 209–212.

89 Wang, Q.S., Cheng, L.K.L., and Unrau, P.J. (2011). Characterization of the B6.61 polymerase ribozyme accessory domain. *RNA* 17 (3): 469–477.

90 Shechner, D.M., Grant, R.A., Bagby, S.C. et al. (2009). Crystal structure of the catalytic core of an RNA-polymerase ribozyme. *Science* 326 (5957): 1271–1275.

91 Johnston, W.K., Unrau, P.J., Lawrence, M.S. et al. (2001). RNA-catalyzed RNA polymerization: accurate and general RNA-templated primer extension. *Science* 292 (5520): 1319–1325.

92 Joyce, G.F. and Samanta, B. (2017). A reverse transcriptase ribozyme. *eLife* 6: e31153.

93 McGinness, K.E., Wright, M.C., and Joyce, G.F. (2002). Continuous in vitro evolution of a ribozyme that catalyzes three successive nucleotidyl addition reactions. *Chem. Biol.* 9 (5): 585–596.

94 Lawrence, M.S. and Bartel, D.P. (2003). Processivity of ribozyme-catalyzed RNA polymerization. *Biochemistry* 42 (29): 8748–8755.

95 Cheng, L.K.L. and Unrau, P.J. (2010). Closing the circle: replicating RNA with RNA. *Cold Spring Harbor Perspect. Biol.* 2 (10): a002204.

96 Joyce, G.F. (2002). The antiquity of RNA-based evolution. *Nature* 418 (6894): 214–221.

97 Cojocaru, R. and Unrau, P.J. (2021). Processive RNA polymerization and promoter recognition in an RNA World. In *Science* [In press].

13

RNA Replication and the RNA Polymerase Ribozyme

Falk Wachowius and Philipp Holliger

MRC Laboratory of Molecular Biology, Francis Crick Avenue, CB2 0QH Cambridge, UK

13.1 Introduction

The idea that a primordial biology based mainly on RNA preceded our current DNA–RNA–protein-based biology, later termed "the RNA world hypothesis" [1], seemed rather far-fetched at the time of its proposal 50 years ago [2, 3]. RNA was recognized as an informational molecule, while catalysis was the realm of proteins and not expected of RNA. However, the perspective changed radically with the landmark discoveries of catalytically active RNAs (ribozymes) such as the group I self-splicing intron (SSI) and RNaseP at the beginning of the 1980s [4, 5]. Even more tantalizing, increasing evidence suggested that the peptidyl transferase center of the ribosome was solely composed of RNA [6, 7], implying that the ribosome, an ancient molecular machine central to all biology, was in fact a ribozyme, i.e. peptide bond formation at the heart of the ribosome is catalyzed by RNA [8]. This led to a paradigm shift by establishing the plausibility of an "RNA world" at the origins of life [1, 9]. A central pillar of such an RNA-centric theory of the origin of life states that in the earliest life forms, both macromolecular catalysis and heredity were based on RNA. One of the attractions of this hypothesis is the simplification from a trimolecular to an unimolecular system, whereby a key transition in the origin of life becomes the emergence of a single RNA molecule capable of "self-replication, mutation and hence evolution toward ever more efficient self-replication" [1]. Over the last 30 years, more and more compelling evidences for an RNA world were brought to light [10, 11]. However, the catalytic potential of RNA seemed limited, as almost all identified ribozymes were catalyzing phosphoester transfer reactions [12, 13], which appeared insufficient to support a prebiotic metabolism.

The onset of RNA in vitro evolution techniques allowed an examination of the principal catalytic potential of RNA and opened up the possibility to recapitulate RNA world reactions in the laboratory. RNA in vitro evolution techniques generated a diverse array of ribozymes, catalyzing essential reactions including C—C bond formation, amide bond formation, or nucleotide synthesis [14, 15], all which would likely have been necessary to support a world dominated by RNA catalysis. The

Ribozymes, First Edition. Edited by Sabine Müller, Benoît Masquida, and Wade Winkler.

cornerstone of the proposed RNA world would have been an RNA-dependent RNA replicase that would be able to replicate its own sequence as well as the genomes of the RNA world. In the "Molecular Biologist's Dream" [16], such a replicase would be the basis for a self-sustained system undergoing Darwinian evolution, fulfilling a minimal definition of life. The minimal requirements such an RNA replicase must fulfill are that the rate of replication, of synthesizing progeny must outpace the rate of degradation of the parent variants, while the fidelity of this copying process needs to be accurate enough to deliver functional variants. More extended requirements include a general replication capacity independent of the template sequence context and a capacity for open-ended Darwinian evolution. In this chapter, we recapitulate the path from the first engineered and selected toward the most advanced RNA polymerase ribozyme (RPR) variants. We discuss structural and biochemical analysis of the RPR, and we emphasize hurdles that still have to overcome in order to reach a self-replicating system based on RNA ribozyme polymerases.

13.2 Nonenzymatic RNA Polymerization

Nonenzymatic RNA monomer polymerization likely preceded enzymatic RNA polymerization both at the origin of life and also in laboratory studies. The four standard RNA nucleobases (A, G, C, U) were likely generated in a prebiotic setting from simpler precursor molecules [17]. Prebiotically plausible synthetic routes generating 2′,3′ cyclic phosphate ribonucleotides have been reconstructed in the laboratory from feedstock molecules as cyanamide, cyanoacetylene, glycolaldehyde, glyceraldehyde, and inorganic phosphate for the pyrimidine nucleobases (C, U) [18] or extended precursors as formamidopyrimidines for the purine nucleobases (A, G) [19]. However, connecting the chemistry of nucleotide synthesis to polymerization remains a problem to be solved [20]. In general, RNA polymerization is entropically disfavored, and therefore chemical activation is needed to drive polymerization. Polymerization of 2′,3′ cyclic phosphate ribonucleotides in aqueous solution showed to be inefficient, yielding only short (<10 nt) RNA strands [21]. This led to the introduction of the highly reactive phosphorimidazolides by Orgel and coworkers (Figure 13.1) that allowed the templated [22, 23] (or untemplated) polymerization catalyzed by Montmorillonite clay [24] of RNA oligonucleotides up to 50 nt. Further developments, e.g. the introduction of a cycle regime and short (3-mer) helper oligonucleotides [25], resulted in the synthesis of a functional part of the hammerhead ribozyme (HHR) by the Szostak group [26], and the recently described activation with 2-amino-imidazole demonstrated more rapid and efficient templated RNA primer extension [27]. This polymerization capacity opens up the principal possibility for nonenzymatic templated RNA replication [28, 29]. However, as this might be sufficient for short RNA sequences, more complex RNA structures would likely pose a more difficult hurdle, restricting the system in its evolutionary capacity. Another problem is posed by the modest fidelity of nonenzymatic RNA copying in the presence of all four ribonucleotides, in particular related to G:U wobble pairing. While the replacement of U with 2-thioU could reduce G:U wobble pairing to a

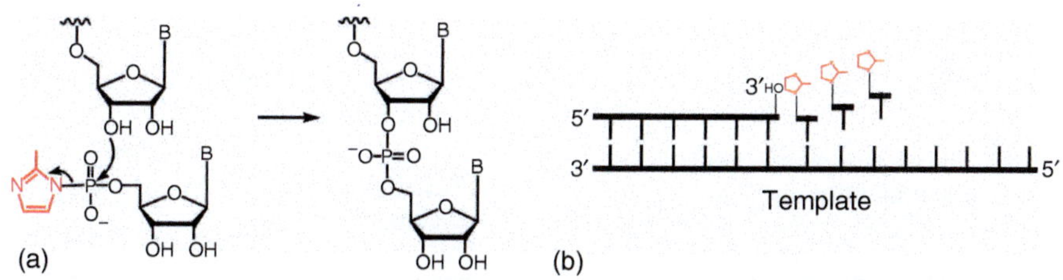

Figure 13.1 Nonenzymatic polymerization of 5′-methylimidazole activated RNA. (a) Nucleophilic attack of the terminal RNA 3′-hydroxyl on the 2-methylimidazole-activated 5′-phosphate of an RNA nucleotide results in the formation of a 3′–5′ RNA phosphodiester linkage. (b) Templated nonenzymatic RNA polymerization. A templated RNA is extended along its 3′ end by 5′-phosphorimidazole-activated RNA nucleotides.

certain extent [30], this also changes the chemical properties of the nucleobase, for example, abrogating the potential of minor groove hydrogen bonding interactions. It is currently unclear to what extent this might affect the phenotypic richness of RNA repertoires in which U is replaced by 2-thio-U, but it would seem likely that these would contain functional ribozymes to a similar extent. Indeed, it could even be argued that the introduction of sulfur might enable a more wide-ranging chemistry, for example, in interaction with heavy metal ions.

How enzymatic (ribozyme)-based RNA replication eventually emerged from pools of nonenzymatically replicating RNA pools is an intriguing question but must have involved gains in efficiency. Another possibility is that using enzymatic catalysis enabled such a system to begin to take advantage of less reactive, kinetically more suitable substrates such as nucleotide di- and triphosphates.

13.3 Enzymatic RNA Polymerization

Enzymatic RNA replication may be either based on a single or cooperative sets of ribozyme molecules. A set of ribozymes with slightly different specificities would resemble a quasispecies and is anticipated to be more robust toward environmental fluctuations compared with a single sequence, but with the drawback of a facile takeover by parasitic sequences. The more prominent understanding of an RNA replicase is based on a defined RNA sequence, capable of self-replication. The (+) strand of this replicase would have had to fulfill a double function by acting as ribozyme and template, with potentially conflicting requirements for both tasks, as replicase function is expected to require a compact folded structure, whereas in general unfolded RNA strands with few secondary structural elements serve as better templates for replication (Figure 13.2). While the (−) strand would only act as template (and therefore does not need to be structured), there is an obvious interdependency with mutations in either strand altering the potential for folding and function, enforcing a balanced coevolution between ribozyme and template function. However, the presence of an enzymatic (or extraneous) mechanism for RNA secondary structure or duplex “unzipping” could relieve ribozyme function

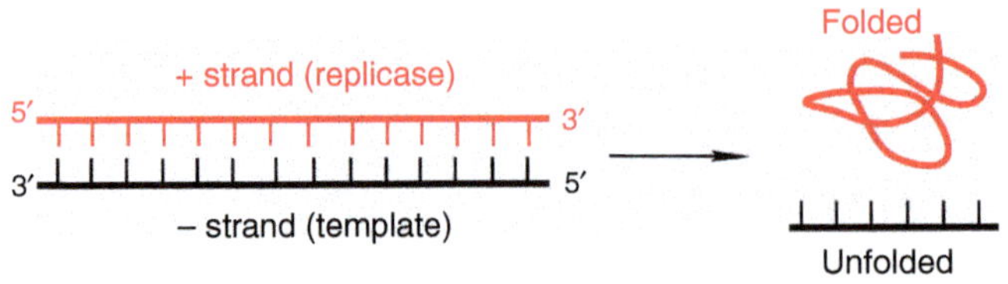

Figure 13.2 Different ideal requirements for RNA template (–) and replicase (+) strands. The replicase (+ strand) should fold spontaneously into stable secondary and tertiary structures, whereas the single-stranded template strand (– strand) should ideally remain unfolded.

from such constraints and decouple replicase sequence evolution from template strand requirements.

Early attempts toward RNA replicase function primarily used oligonucleotides as substrates were based on the modification of existing ribozymes, in particular the *Tetrahymena* group I SSI ribozyme that demonstrated fast (<1 second) phosphodiester transfer catalysis [31]. The reengineering of the self-splicing step of the *Tetrahymena* group I intron allowed single nucleotide additions [32] or the ligation of up to four RNA oligonucleotides on a complementary strand [33]. A minimized variant of the *sunY SSI* ribozyme, consisting of 180 nt, was even able to assemble one of its subunits from five oligonucleotide fragments [34] or (in an optimized sequence context for self-copying) to synthesize its complementary strand from 18 different RNA oligonucleotides each of around 10 nt [35]. The same *sunY SSI* ribozyme is also able to use RNA trinucleotides as substrates [36], but with relatively low fidelity and processivity. Other examples of such self-assembly reactions include the *Azoarcus* group I SSI ribozyme, which can covalently (or non-covalently) self-assemble into active ribozymes from four oligonucleotide fragments [37] and form cooperative replication networks [38], as well as a cross-chiral RNA polymerase [39] that can assemble its own mirror (opposite chirality) ribozyme from 11 oligonucleotides in a templated manner (and vice versa).

The most striking example of RNA self-replication by oligonucleotide assembly is based on two variants of the in vitro selected R3C RNA ligase ribozyme [40]. The two variants were engineered to operate in a cross-catalytic format, so that each variant would assemble the other by templated ligation from two oligonucleotide substrates. In the presence of the four substrate RNAs, addition of either ligase variant initiates exponential amplification (Figure 13.3). However, as replication is based on predefined oligonucleotides, which associate via complementary base pairing, the system is limited in its open-ended evolutionary capacity, as it can only optimize its capacity for cross-ligation. In general, the capacity for emergence of new function through Darwinian evolution from self-assembly cycles involving longer oligonucleotides is hindered by their predefined sequence information and is inversely correlated to oligonucleotide length. Hence fragment assembly-based self-replication systems are restricted to recombination for diversification and can only perform a sparse sampling of the potential RNA sequence space. A replication system based on RNA monomers as substrates in contrast would be able

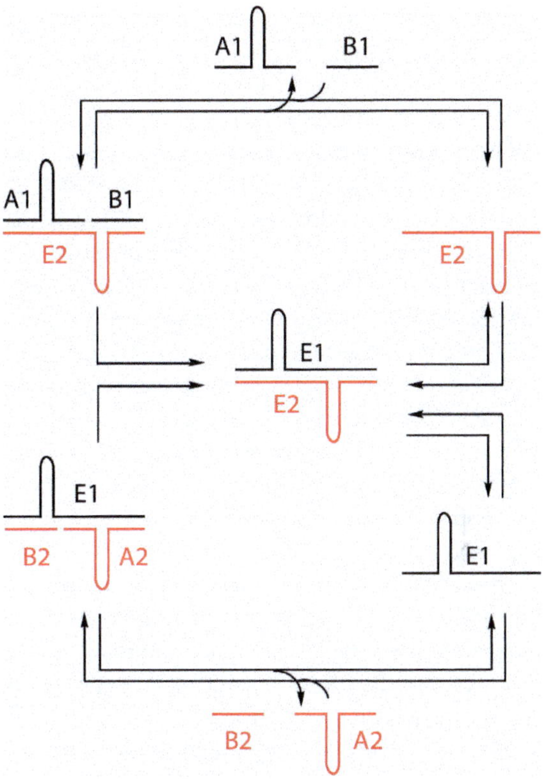

Figure 13.3 Cross-replicating RNA ligases. Ligase E1 catalyzes the ligation of A2 and B2 resulting in active ligase E2, and E2 catalyzes the ligation of active E1 from A1 and B1. Dissociation of E1 and E2 results in copies for another ligation reaction. Source: Lincoln and Joyce [40].

to introduce single-base mutations and therefore explore RNA sequence space more thoroughly. For this reason and because of analogies with current protein RNA polymerases, achieving RNA-catalyzed RNA replication based on mononucleotides has been a key aim in the field.

13.4 Essential Requirements for an RNA Replicator

One obvious paradox is the ab initio generation of an RNA replicase [41], although its emergence in the context of preexisting chemical replication cycles may be conceptually easier to understand. One might ask, if an RNA replicator would spontaneously appear out of a pool of prebiotic oligonucleotide sequences, how long would it survive? Apart from chemical instability, related to the 2′,3′-*cis* diol ribose configuration in RNA, exponential replication of long RNAs in a free diffusing environment readily leads to an emergence of nonfunctional ("parasitic") sequences, i.e. fast replicators based on short and unstructured sequences, which would outcompete the parental replicase. This phenomenon has been well documented and studied in the classic RNA replication experiments by Spiegelman using Qβ replicase [42] that generated the now famous "Spiegelman monsters." Hence, a certain degree of diffusion limiting encapsulation (e.g. mineral pores, micelles, eutectic ice phases, or membranes) [43, 44] would had been already required at the earliest stages of RNA replicase evolution, to contain parasitic sequences, while longer sequences might have been enriched by hydrothermal cycling [45].

RNA self-replication is also constrained by its fidelity and can only be maintained if the error rate remains below a certain level, the so-called error threshold [46, 47], which describes the relationship between replication fidelity and the number of ligation/polymerization reactions, with the permissible per nucleotide error rate inversely proportional to the length of the replicator sequence. Hence a shorter replicator sequence would allow a higher per nucleotide error rate. Although the precise level of the error threshold of RNA (self-) replication may be relaxed for sustained replicase function [48], the RPR (~200 nt) would likely require a c. 0.01 error rate or 99% correct nucleobase incorporations, if one estimates that 80% (160 nt) of its sequence context is essential for ribozyme function, and it would be able to perform approximately 2000 polymerization reactions in its lifetime with at least two (ribozyme and complement) of the possible ten synthesized variants possessing the essential nucleobases ($2/10 = (1-0.01)^{160}$) [49]. In comparison, viral RNA-dependent RNA polymerases exhibit an error rate a magnitude lower of around >0.004 [50] and enable replication of multi-kb RNA viral genomes (e.g. Qβ, ~4 kb) [51]. Fidelity was therefore a strong selection pressure for early RNA replicators as an improvement in fidelity would allow to replicate longer sequences, and such longer sequences could explore more sequence space driving the evolution of RNA replicators with further improved fidelities. Thus over evolutionary time increasingly larger RNA genomes could have been faithfully copied, allowing an increase of the size and complexity of the encoded functional RNAs.

13.4.1 Likelihood of Replicating Sequences in RNA Sequence Space

The minimal size of an RNA replicator is difficult to estimate, but even the smallest ribozymes, e.g. minimized HHR variants, are in the size range of 20–40 nt [52], while robust RNA replicase function, e.g. RNA polymerase function, seems to require RNA sequences of ~150–200 nt [53].

Given the near infinitely small probability of such a long RNA sequence arising by chance, one might ask how densely distributed RNA polymerase/replicase function would be in RNA sequence space. The theoretical RNA sequence space is vast (4^n possible sequences, n = sequence length) and becomes progressively larger with increasing sequence length, but structural and in particular functional space (i.e. the totality of all folded/functional RNA sequences) is much smaller and also depends on environmental conditions. Indeed, the complete coverage of the sequence space of even just a 40 nt long RNA sequence (each possible sequence present at least once) would already require ~1 kg of RNA. Hence, in standard laboratory experiments, only a very small portion of the possible RNA polymerase/replicase ribozyme sequence space can be sampled.

However, it is illustrative to consider the sequence/functional space of shorter functional RNAs, where experimentally accessible random sequence pools can be large enough to completely cover the available sequence space, as was demonstrated for an RNA guanosine-5'-triphosphate (GTP) aptamer selection [54]. Such experiments indicate that sequence space is relatively sparsely populated and that the adaptive landscape is rugged, i.e. that functional "peaks" are isolated and not usually connected through a continuum of RNA sequences. By inference, one would expect the more complex RNA polymerase/replicase ribozymes to be even rarer in sequence space and equally disconnected.

Most in vitro evolution experiments start from 40 to 80 nt long randomized sequence pools, but that may not be sufficient to identify more complex RNA function. Hence, the class I ligase was selected from a 220 nt long randomized pool [55], but as in long randomized pools only a miniscule part of sequence space is covered. It suggests that also more complex function as the class I ligase should be more ubiquitous in extended RNA sequence space. Indeed, the fact that the class I ligase could be isolated from a pool of only 10^{15} molecules, of the theoretical possible 4^{220} possible sequences, suggests that the experimenters were extremely lucky or a rather large number of RNA structures in the sequence space of 220 nt is able to catalyze this specific RNA ligation in some form.

How then could a replicase have emerged, and what biological principles could facilitate the occurrence of more complex RNA function? A general feature of complex (biological) systems is their modularity, i.e. their buildup from simpler structural elements. In the ribosome, the small subunit is responsible for decoding and the large subunit for catalysis. Similarly the RPR consists of a catalytic subunit and a polymerase subunit, which binds the primer/template duplex and is required for in trans polymerization, and also aptazymes and riboswitches are following this modular architecture. These large subunits are themselves built up from simpler elements that can be deduced to a rather small set of simple structural elements, e.g. stem, loop, or hairpin structures, which can be formed from short (10–20 nt) RNA sequences that would be expected with a comparable high frequency in prebiotic oligonucleotide pools.

The frequency of such small structural modules may be identified from bioinformatics analysis, as the main structural features based on Watson–Crick base pairing as secondary structures and tertiary folds can be well predicted for most small RNAs [56]. However, this becomes increasingly challenging for longer RNAs, as most longer RNAs can adopt a range of structures and folds, depending on environmental conditions, making RNA folding and function a highly dynamic process [57]. Indeed, more complex RNA structures/functions appear to be only present in longer random RNA sequence pools [58]. However, their occurrence may not be as rare as expected, as sequence requirements can be relaxed for such small functional modules [48, 59]. Indeed, some sequence regions in the polymerase ribozyme in particular single-stranded regions are more sensitive toward mutations, but the majority of nucleotides in the stem regions could likely be exchanged, as long as Watson–Crick base pairing is maintained, increasing the allowable sequence space and facilitating the generation of stable RNA modules that could combine to generate more complex ribozymes.

How many simple structural elements would be necessary to define a specific function? This also defines the length of necessary random sequence space.

Functional RNAs of comparable low complexity, e.g. the hammerhead motif for RNA cleavage, could be identified multiple times through independent in vitro evolution experiments [60, 61], starting from rather short randomized pools. For even simpler motifs, e.g. the ATP aptamer or the HHR, longer/more complex sequences would not contribute to improved function [58], but more complex RNA function might require a degree of complexity that requires longer RNA sequences. Similarly, an increase in functional activity, e.g. stronger binding of GTP aptamers, required more information or more complex structures [62].

While the likelihood of appearance of such motifs decreases proportionally with increasing length, a buildup from simpler (functional) modules would facilitate the appearance and through recombination increase the evolutionary potential of more complex ribozymes. This hierarchical buildup from shorter sequences and modules could be mimicked for the polymerase ribozyme by the covalent assembly from short RNA sequences (~30 nt) by the hairpin ribozyme [63]. In contrast, a non-covalent assembly from shorter pieces, which yields functional ribozymes in the case of the *Azoarcus* self-splicing group I intron [37], resulted in a severe drop in polymerase activity [63]. More complex function can be also derived through in vitro evolution by selection for new modules on already functional RNA structures, as demonstrated in the generation of the first RPR variant by adding a randomized 78 nt long region to the 3′-end of the class I ligase ribozyme that after 18 selection rounds resulted in the R18 polymerase ribozyme [53].

13.4.2 Reaction Conditions for RNA Replication

The notion of a spontaneous emergence of RNA replication (both enzymatic and nonenzymatic) is complicated by the fact that prebiotic conditions are more likely to be generally hostile to RNA stability and replication, than beneficial, although precise information is lacking. Indeed, the most important physicochemical parameters such as pH, temperature, salinity/minerals, or the presence of small molecule cofactors can only be guessed and likely would vary widely, depending on the specific location and geochemical setting on early earth. Therefore, experiments on both nonenzymatic and enzymatic RNA replications and the evolution of RPR in the laboratory are largely uncoupled from possible early life conditions and explore a "best-case scenario." Nevertheless, useful information on a prebiotically plausible set of boundary conditions and environments can be learned from such experiments.

For example, it is clear that temperature in particular is a critical factor for RNA replication. However, somewhat counterintuitively, cold temperatures, while generally slowing down chemical reactions, can be beneficial to RNA replication due to a phase change in the aqueous solvent. As liquid water is cooled below its freezing point, freezing generates a eutectic phase, where solutes are excluded from the growing ice crystals and become concentrated in an interstitial brine. This results in a strong concentration effect (500–1000 times) of all solutes and generates a quasi-compartment. The eutectic phase of water ice was shown to be a preferential medium for both nonenzymatic [64] and enzymatic RNA polymerization [65]. While proteinaceous polymerases (e.g. T7 RNA polymerase) are not active at subzero temperatures, the polymerase ribozyme even demonstrated increased activity at −7 °C compared with ambient temperatures over longer time spans, also related to the reduced RNA degradation rates at lower temperatures [65]. However, short heat pulses ("snap heating") might help in strand separation, and freeze-thaw cycles have been shown to be beneficial for RNA structure re-equilibration and ligation catalyzed by the hairpin ribozyme [63], demonstrating the likely necessity for both cold and hot temperature scenarios for RNA evolution at the origin of life. Reaction conditions or additives can also profoundly affect ribozyme activity but often may be specifically beneficial for one setting as citrate for fatty acid membrane compartmentalized nonenzymatic RNA replication [66] but might show neutral (or even detrimental) effects on RPR activity under different conditions [67].

13.4.3 The Strand Separation Problem

Dissociation of RNA duplex strands after primer extension on long RNA templates (such as e.g. the RPR sequence itself) is a particularly vexing problem for continuous RNA replication, as RNA duplexes >20 nt are very stable and challenging to dissociate, requiring heating in the presence of high concentrations of denaturants such as urea or a profound shift of pH [68].

That is why complementary RNA oligonucleotides are generally applied in 8M urea polyacrylamide gels to resolve long extended primer/template RNA duplexes [35, 69]. Once the double-stranded RNA dissociates into single strands, there will be competition between the intermolecular reannealing of the synthesized strand to the template and the intramolecular folding of either strand into a functional, 3D structure. In the case of a complex functional RNA as the RPR, which is expected to form extensive intrastrand secondary structure, elements such intrastrand folding might stabilize strand separation in particular in the context of fast heating followed by "snap cooling," which can kinetically freeze certain secondary structures [70]. Alternatively, even single-base modifications as nucleobase methylations at the Watson–Crick site can affect RNA duplex/hairpin equilibria and thereby modulate RNA secondary structure [71]. Therefore, an efficient RNA self-replicator would likely have a sequence context that would avoid the facile formation of a long dsRNA helix by either (+) or (−) strand, implying the formation of stable secondary (and tertiary) structures on either strand. Indeed, sequences that fold into structures with few accessible (Watson–Crick face) sites would be preferred over long single-stranded regions that would easily base-pair with the template strand and would inactivate the polymerase. However, such highly structured RNAs also pose as a block to replication as template secondary structures must be unraveled for replication.

13.5 The Class I Ligase and the First RNA Polymerase Ribozymes

In vitro evolution allows an unbiased exploration of the catalytic ability of RNA beyond its known functionality and to identify new ribozymes that catalyze unique reaction types or adopt different folds to perform already existing reactions, e.g. RNA ligation [72]. The advantage of in vitro selection experiments lies in the open design of sequence space, substrates, and scope of the reaction and their facile modification throughout the experiment. In a pioneering in vitro selection experiment, Bartel and Szostak [55] started from a 220 nt randomized RNA sequence pool, containing ~10^{15} different sequences and selected for templated in cis RNA ligase function by applying the triphosphorylated 5′-end of the pool and an externally added RNA as substrates. The substrates are aligned by Watson–Crick base pairing, while the ribozyme is catalyzing the nucleophilic attack of the 3′-OH on the α-phosphate of the triphosphate with the release of pyrophosphate and the subsequent formation of a phosphodiester bond (Figures 13.4a and 13.5a). From the final selection round, 65 different ribozymes promoted self-ligation, which demonstrated the principal capability to isolate RNA ligase ribozymes by in vitro selection from long sequence libraries. Although biased by the specific selection method and conditions, this also provided an estimate (~10^{-12}) of the likelihood of this specific RNA ligase activity in the applied RNA sequence space. Further reselection and mutagenesis [74, 75] identified sequence similarities between the different isolated ribozymes that allowed their clustering into three different sequence classes (I, II, III). While both class II and III variants only formed 2′–5′ RNA linkages, the class I ligase demonstrated strict regioselectivity for the formation of the canonical 3′–5′ linkages and very fast ligation kinetics with a $k_{cat} = 100\,min^{-1}$ that represents a rate enhancement of 10^9 over the unanalyzed reaction. While impressive and still one of the fastest ribozymes ever isolated, such rates are still significantly slower than proteinaceous enzymes. For example, T7 RNA polymerase is faster by a factor of 10^5. The kinetic analysis of the ligation reaction shows a direct correspondence between pH and reaction rate, pointing toward a single deprotonation event, likely that of the attacking 3′-hydroxyl, as the rate-limiting step [76]. Hence the chemical step is rate limiting at physiological pH, while at higher pH product release becomes the rate-limiting factor, with an anticipated k_{cat} above $1000\,min^{-1}$ at pH 9. Ligation of RNA substrates including mismatches would be expected to show faster dissociation rates and might allow an increase of the catalytic rate at higher pH but would also require higher substrate concentrations (as the K_m would be increased). Interestingly, the class I ligase, *Tetrahymena* group I intron, and RNAseP all show k_{cat} values around $100\,min^{-1}$ at physiological pH (7.5), which might be a general speed limit for RNA-catalyzed phosphotransfer reactions in this pH range. The class I ligase requires ~50 mM Mg^{2+} for optimal ligation rates but is already completely folded at Mg^{2+} concentrations as low as 1 mM [77], with a rather strong specificity for Mg^{2+} over other divalent cations. The substitution of Mg^{2+} for Mn^{2+} leads to reduced ligation rates, while with all other Me^{2+} ions, no substantial ligation activity could be measured, with the exception of Ca^{2+}, which is able to partially replace Mg^{2+}, although the addition of low millimolar concentrations of Mg^{2+} is still required for ligation activity [78].

The 220 nt pool sequence for the ligase selection was generated from three pieces with lengths of 72, 76, and 72 nt. While classes II and III were in the sequence frame of either of the pieces, the class I ligase required all three segments. Even a minimized version of the class I ligase still contained 93 nt, which shows that the

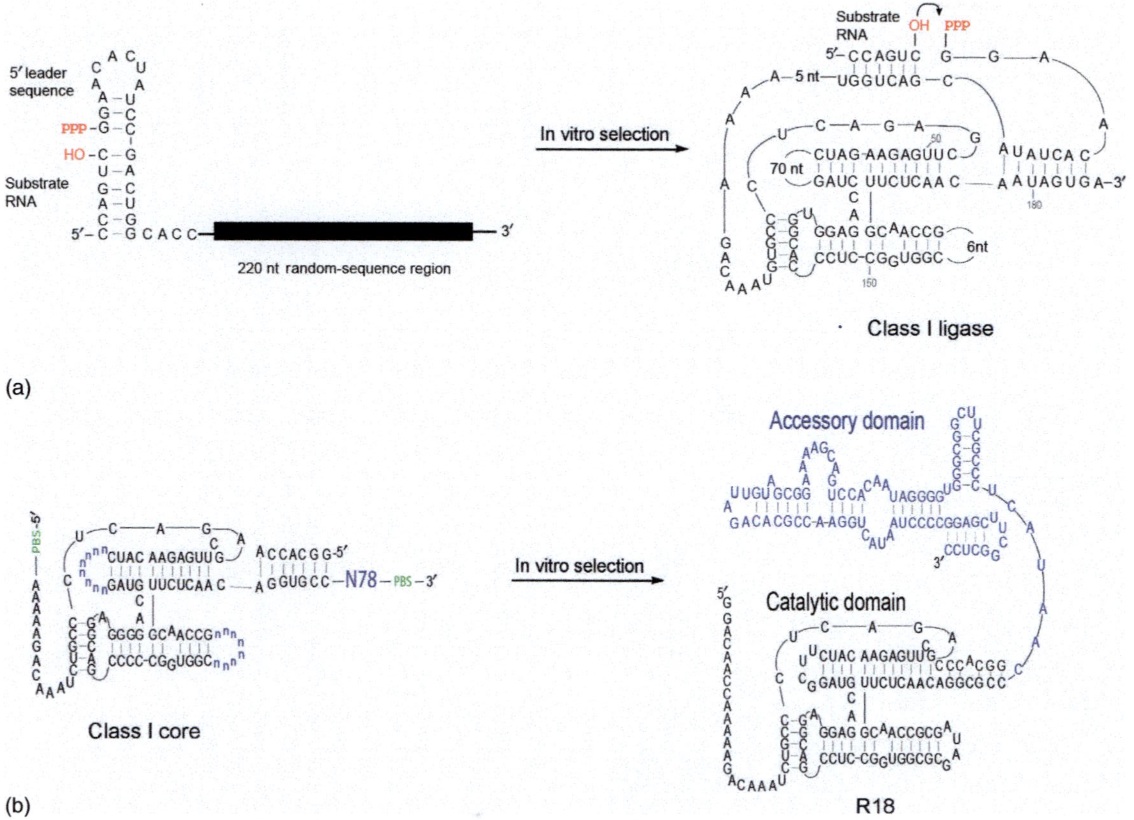

Figure 13.4 Selection of the class I ligase and the R18 RNA polymerase ribozyme. (a) For the class I ligase selection, the substrate RNA was bound via Watson–Crick base pairing to the 5′-leader sequence of the 220 nt random pool RNA sequence, and after in vitro selection and engineering, the class I ligase was generated, able to catalyze the ligation of an RNA substrate to the 5′-triphosphorylated end of the ribozyme. (b) Selection from the class I ligase core sequence including a randomized region (N78) at the 3′-end of the core generated the R18 polymerase ribozyme, able to perform multiple polymerization reactions. PBS, primer binding sites.

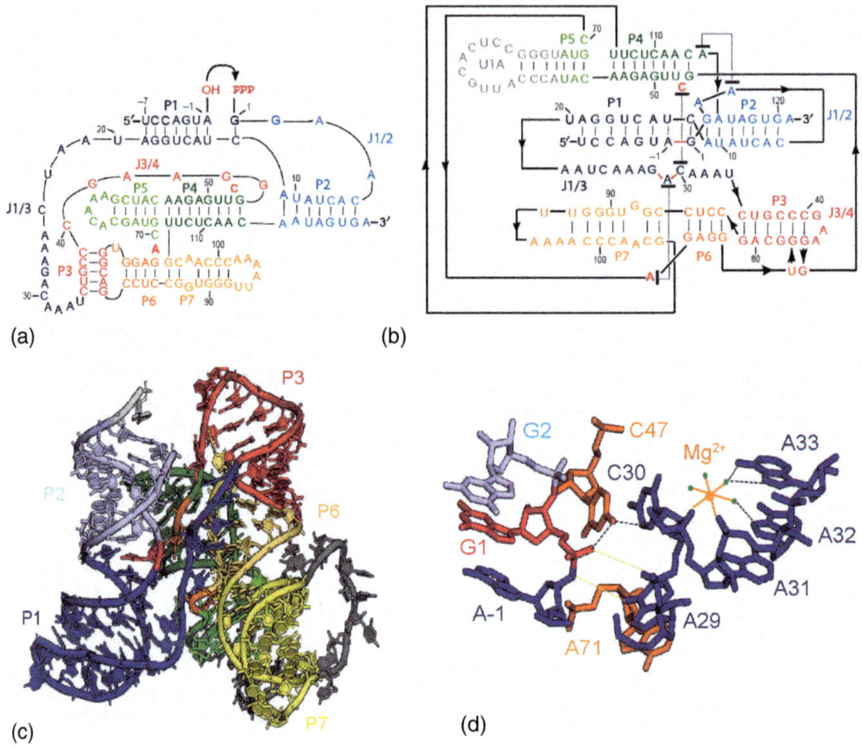

Figure 13.5 Structure of the class I RNA ligase. (a) Secondary structure of the class I ligase bound to its substrate RNA. The black arrow indicates the nucleophilic attack of the substrate RNA 3′-hydroxyl on the 5′-triphosphate α-phosphate of the ribozyme. Different helices and loops are indicated and color coded. (b) Secondary structure of the crystallization construct, color coded as in (a), indicating relative domain orientation and coaxial stacking of the helices. The U1A loop was added to facilitate crystallization. The red dash indicates the A–G ligation junction; the catalytic nucleobases C47 and A71 and the backbone phosphates of the catalytic site are depicted in orange. Stacking interactions are connected with gray lines and terminate in gray bars. (c) Ribbon presentation of the crystallized ligase structure (PDB: 3HHN), with the G (G1) at the ligation junction indicated in red and the catalytic residues C47 and A71 in orange. (d) Representation of the active site. Black dashes indicate hydrogen bonding, and yellow dashes indicate proximity between the A29 and C30 phosphate oxygens and the ligation junction. Source: Adapted from Shechner et al. [73].

level of complexity required could only appear out of long randomized regions. Modifications of the substrates and the template sequence by its hybridization to the 3′ arm of the P2 stem led to multi-turnover catalysis and true intermolecular in trans activity, whereby ligation proceeds as a trimolecular reaction with none of the RNA substrates covalently linked to the ribozyme. Further modification of the template sequence allowed incorporation of a single nucleotide triphosphate and led to the first RPR function, enabling the successive template-dependent polymerization of three nucleotide triphosphates [79], demonstrating the principal capability of RNA-catalyzed RNA polymerization in a manner analogous to protein-catalyzed RNA polymerization. The chemical advantage of applying triphosphates as substrates lies in the kinetic stability of the phosphoanhydride (and specifically the phosphodiester) bond in aqueous environments [80] and the fact that the pyrophosphate leaving group (PPi) provides a driving force for the polymerization reaction, as pyrophosphorolysis (i.e. the reverse reaction) is strongly disfavored. Furthermore, pp_i does not participate in another competing phosphodiester transfer reactions as is the case for nucleoside or oligonucleotide leaving groups generated by SSI catalysis. Interestingly, in these early RPR variants, a pppGGA substrate was still preferred over GTP (pppG), whereby the GA extension interacts in a non-Watson–Crick manner with the ligase core and presumably enhances substrate binding and positioning for nucleophilic attack [81].

In a bold new selection approach directly for RNA polymerase function, Johnston et al. [53] detached the P1 helix from the class I ligase and replaced it with a free primer template duplex, resulting in the loss of the 5′-end of the P2 helix, while the P2 duplex was restored by the addition of an external heptamer sequence, and the J1/3 region became the 5′-end of the ligase core (Figure 13.4b). Subsequently a randomized RNA pool of 78 nt was attached to the 3′-end of the ligase sequence, and after 18 selection rounds for polymerase activity, the R18 ribozyme was identified, which could extend a primer along a template sequence by 14 nt in 24 hours [53], with both the processivity domain and the P2 completing heptamer sequence essential for polymerase activity. However, primer extension by 14 nt was only observed on the most preferential template sequence, with a more general polymerization capacity of 4–8 nt. Also slight changes in the starting primer sequence (− or + one nucleotide) could decrease polymerization by an order of magnitude [82]. One of the reasons for this high sensitivity of the polymerase ribozyme to changes in primer and/or template sequence lies in its weak interaction with the primer/template duplex. As polymerization reactions are performed with reactants in the micromolar concentration range (<10 μM), the K_d for primer/template binding by the polymerase lies in the low millimolar range, depending on sequence context. However, an increase in ribozyme concentration (>15 μM) leads to a decrease of polymerization activity, making primer/template binding the rate-limiting step [82]. Hence, only few polymerization reactions are performed before the polymerase dissociates from the RNA substrate.

A general property of favorable RNA template sequences applied for RNA ribozyme polymerization reactions includes similar sequence contexts, comprising few secondary structure elements. Cytosine-rich templates are principally preferred [83], which might be related to the original selection templates for the class I ligase but is likely also an intrinsic property related to the stronger stacking and base-pairing properties of the guanosine substrates, as a similar trend can be observed in nonenzymatic RNA polymerization [84].

Several strategies were applied to increase the affinity of the polymerase for the primer/template duplex and improve polymerase activity. The attachment of hydrophobic anchors on the polymerase and the primer/template duplex and their colocalization on micelles [85] resulted in a modest 3- to 20-fold increase in product yield, depending on the sequence context. Another strategy was based on the attachment of a cationic arginine cofactor at ten different sites on the ribozyme, using conjugates of arginine with short DNA or RNA oligonucleotides to decrease charge repulsion between the polymerase and the primer/template duplex [86]. However, this strategy led to no improvements in polymerase activity. The installation of G/T-rich DNA oligonucleotide sequences connected to the ribozyme near the active site resulted in improvements in polymerization processivity by an average of 10 nt [87], also depending on sequence context. Another hurdle is posed by the short half-life (~24 hours) of the RNA polymerase under optimal reaction conditions (200 mM Mg^{2+}, pH 8.3), making polymerization a race against ribozyme and template hydrolysis. Indeed, multiple polymerization reactions mainly occur because of the fast catalytic rate of the polymerase that counterbalances RNA degradation and its weak affinity for the primer/template duplex.

13.6 Structural Insight into the Catalytic Core of the RNA Polymerase Ribozyme

For a better understanding of ribozyme polymerase function and to potentially guide rational design strategies, the structure of the catalytic subunit of the polymerase ribozyme, analogous to the class I ligase, was determined by X-ray crystallography [73]. The structure comprises the post-catalytic state while loop 5 was replaced with the U1A-binding loop to aid crystallization by complexation with the U1A protein (Figure 13.5a,b). The overall architecture resembles a tripod, made up from three coaxial stacked stem-loop structures, with two domains formed from regions of the catalytic core domain (P3–P6–P7 and P4–P5) and one by the primer/template duplex (P1–P2), while the legs join at the catalytic center, where the incoming nucleoside triphosphate stacks onto the primer 3′-end and base-pairs with the first nucleotide of the template strand (Figure 13.5c). Unpaired loop regions contribute to and organize the active site by interaction with the core tripod structure by a series of minor groove interactions; in particular the P1 and P6 stems are positioned by two A minor motifs in the J 1/3 joining region, while the P2 helix and the J3/4 loop are on top of the ligase. The active site is formed by a cytosine (C47) and two backbone phosphates (A29, A30) (Figure 13.5d). However, the post-ligation structure lacks pyrophosphate, and no metal ions were detected in close proximity to the active site.

To visualize the catalytic center of the class I ligase in more detail, the structure of a slightly modified ligase variant was determined [88]. As ligation activity of the class I ligase is fast ($k_{cat} = 100\ min^{-1}$), the catalytic cytosine (C47) was mutated to uridine, which diminishes the catalytic rate by a factor of $>10^4$, and Mg^{2+} was substituted for Ca^{2+} and Sr^{2+} that still allow folding into a conformation indistinguishable from the Mg^{2+}-induced conformation [77] but leads to a substantial ($>10^5$) decrease in catalytic rate. This pre-ligation structure reveals an unusual curled conformation of the triphosphate by docking the β- and γ-phosphates above the G1 base that is markedly different from proteinaceous polymerases, wherein the triphosphate is drawn away from the plane of the nucleobase [89]. Also, the NTPs in proteinaceous polymerases are coordinated by Mg^{2+} ions bound to more than one phosphate group, wherein in the class I ligase only the 5′-γ-phosphate appears to coordinate a Mg^{2+}–H_2O cluster. The α-phosphate is positioned in an optimal geometry for in-line attack by the primer 3′-hydroxyl, with an angle between the 3′-hydroxyl, α-phosphate, and oxygen of the leaving group of 176°, very close to the ideal 180°. The backbone phosphates of A29 and C30 both coordinate a catalytic metal ion in the pre-ligation complex that was not observed in the post-ligation structure. Proteinaceous RNA polymerases apply a two metal ion mechanism for catalysis; while metal A activates the 3′-hydroxyl for nucleophilic attack, metal B aids in loss of the pyrophosphate leaving group. Detailed analysis of functional groups of the catalytic nucleobase C47 revealed that the N4 amine and not the N3 imine participates in catalysis, but both the C30 2′-hydroxyl and the C47 N4 amine only function electrostatically by acting as hydrogen bond donors during catalysis [88]. The catalytic mechanism proposes that the A29 and C30 pro-R_p phosphate oxygens coordinate a catalytic Mg^{2+} cofactor that activates the 3′-OH of the substrate RNA for nucleophilic attack, while the N4 of C47 and the 2′-OH of C30 are part of a hydrogen bonding network that stabilizes the transition state and the pyrophosphate leaving group. This catalytic mechanism resembles that of the hepatitis delta virus (HDV) ribozyme that applies a metal ion to deprotonate the nucleophile, and the N3 of the active site cytosine stabilizes the leaving group [90]. However, the class I ligase uses the cytosine N4 amine and does not perform general acid–base catalysis as the HDV.

In the absence of a high-resolution structure of the whole polymerase ribozyme (catalytic and accessory domain), the proposed arrangement of the accessory domain was probed by mutation analysis [91]. This revealed that the accessory domain likely sits on top of the catalytic core and contains a 8 nt purine-rich loop that is positioned between the J 3/4 stem loop of the ligase core and the AL4 triloop of the accessory domain and interacts with the core sequence, likely stabilizing the incoming nucleoside triphosphate (Figure 13.6). In a recent study applying 6-Thio-GTP (6sGTP) as substrate for polymerization during in vitro selection, starting from the R18 polymerase variant [92], a RPR variant including three mutations (A156U, C79U, C113U) was identified, which promoted 200-fold improved incorporation kinetics for 6sGTP, while in particular A156U in the purine-rich loop of the accessory domain showed a strong effect (>50-fold) mapping out the NTP binding site (Figure 13.6).

The polymerase appears to make essential contacts with the primer/template duplex via three defined 2′-hydroxyl groups, with two on the primer (position −2 and −3 from the 3′-terminus) and one on the template sequence (position −3 from the primer 3′-terminus) and three weaker hydrogen bonds at positions +3, +4, and +5 in the single-stranded region of the template. The most severe effect on the catalytic rate by replacement of the 2′-hydroxyl with a 2′-deoxy function was observed for the 2′-position of the primer 3′-end, likely by lowering the pK_a of the neighboring 3′-hydroxyl or by coordinating an essential metal ion [93]. This data generates a low-resolution picture on the interplay between accessory domain and catalytic subunit of the polymerase ribozyme, but only a high-resolution structure of the complete polymerase ribozyme would give a detailed insight into the mechanism of RPR polymerization.

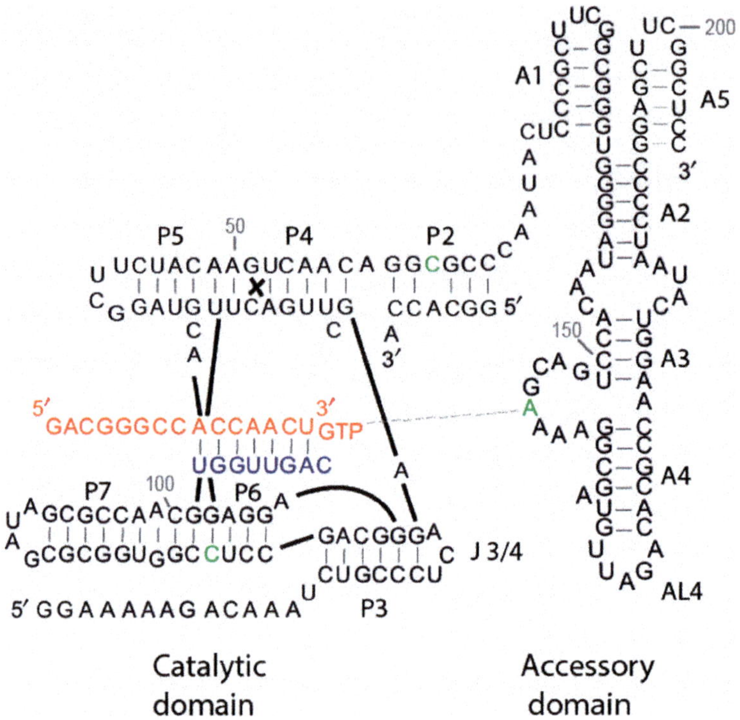

Figure 13.6 Secondary structure representation of the R18 polymerase ribozyme including a primer/template substrate sequence (depicted in orange/blue, respectively) based on Wang et al. [91]. The three nucleotides (A156, C79, C113) likely involved in NTP binding [92] are indicated in green. Source: Adapted from Wang et al. [91].

13.7 Selection for Improved Polymerase Activity I

Further efforts to select for improved RPR from the R18 pool sequences resulted in new polymerase variants [94], but none of these variants showed major improvements in polymerization capacity or sequence generality, leading to the assumption that the R18 polymerase ribozyme might be trapped in a local fitness maximum. In addition, the in cis (intramolecular) selection system, i.e. the covalent linkage of the RNA substrates to the ribozyme pool, resulted in substantial loss of catalytic activity when selected polymerase ribozymes were engineered to operate in trans. A selection method that would allow the coupling of genotype and phenotype and at the same time select for intermolecular (in trans) activity would be highly beneficial. The coupling of genotype and phenotype inside a water-in-oil (w/o) emulsion system was first described for protein selections [95]. Due to the need for compartmentalization in an inert oil phase (allowing maximally 10^9–10^{10} aqueous compartments/ml oil phase), the clonal sequence diversity is reduced by a factor of 10^5 compared with conventional in vitro selections that normally start from a pool diversity of ~10^{15} different sequences. In a "tour de force" w/o selection scheme, performing selections in the liter scale to approach a sequence diversity of ~10^{15} and starting from a mutagenized library based on the R18 polymerase ribozyme, an improved variant (B6.61) was identified that could polymerize up to 20 nt on a specific primer/template sequence [96]. The in-emulsion selection was based on a biotin capture probe, followed by a second selection step, applying the 4-Thio-UTP/AMP gel selection strategy that was applied for the original selection of the polymerase [53]. This selection is based on electrophoresis in a urea-PAGE including a small amount of *N*-acryloyl aminophenylmercuric acetate that reacts covalently with and impedes migration of RNAs containing 4-Thio-U. The B6.61 sequence differs from the wild-type sequence by four nucleotide insertions at the 5′-end and a A170G mutation while the increased polymerization activity is mainly due to an improvement in fidelity (Figure 13.7).

An alternative w/o emulsion selection strategy termed CBT (compartmentalized bead tagging) that links polymerase ribozyme genes to the corresponding ribozyme

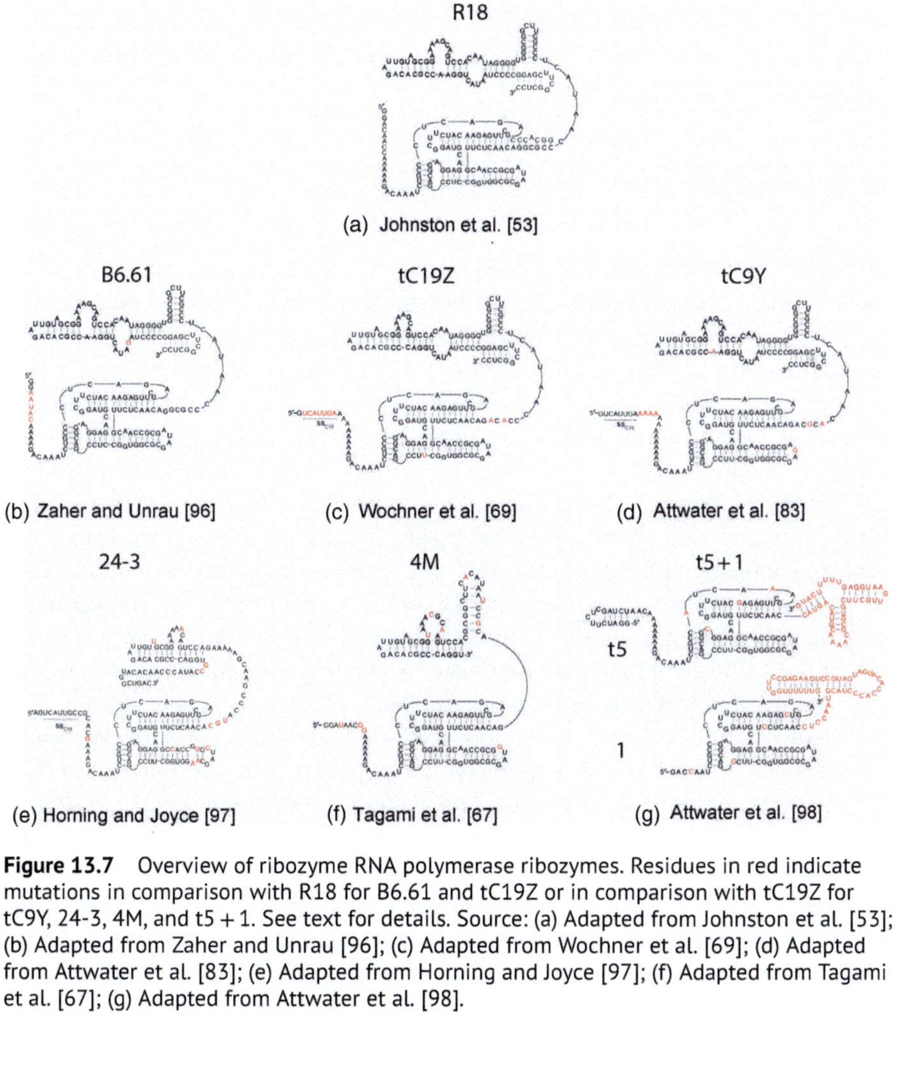

Figure 13.7 Overview of ribozyme RNA polymerase ribozymes. Residues in red indicate mutations in comparison with R18 for B6.61 and tC19Z or in comparison with tC19Z for tC9Y, 24-3, 4M, and t5 + 1. See text for details. Source: (a) Adapted from Johnston et al. [53]; (b) Adapted from Zaher and Unrau [96]; (c) Adapted from Wochner et al. [69]; (d) Adapted from Attwater et al. [83]; (e) Adapted from Horning and Joyce [97]; (f) Adapted from Tagami et al. [67]; (g) Adapted from Attwater et al. [98].

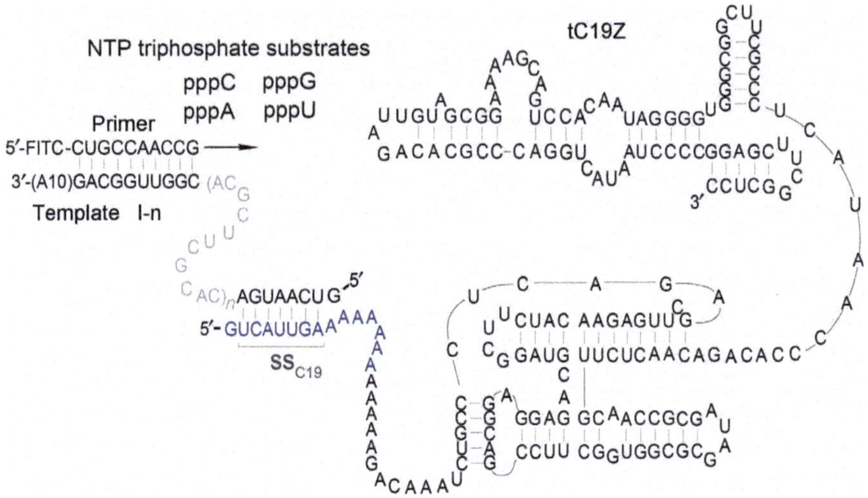

Figure 13.8 Principle of templated RNA polymerization catalyzed by the RNA polymerase ribozyme (tC19Z variant). The template strand (template I-n, here sequence of 11 nucleobases, indicated in gray) is bound at its 5′-end through Watson–Crick base pairing to the 5′-end of the ribozyme via the short *taq* sequence ss_{C19}. The primer binds at the 5′-end of the template strand, and the ribozyme catalyzes the extension of the primer along the template strand using nucleotide triphosphates as substrates.

via microbead display was developed to select for new improved polymerase variants. The starting sequence for selection included a randomized sequence at the 5′-end of the ribozyme, and after three selection rounds, a new improved polymerase ribozyme variant could be identified (c19), in which the 5′-randomized region was transformed into a hairpin structure followed by a 6 nt long single-stranded sequence complementary to the 3′-end of the template sequence [69] (Figures 13.7 and 13.8). In addition, a single point mutation (G93A) in the P2 stem compensated for the missing complementary P2 helix that was not included in the selection experiment. Reengineering of the 5′-end by replacing the hairpin sequence with a A_4 linker sequence generated the (truncated) tc19 polymerase ribozyme that was able to polymerize up to 95 nt on a preferential template sequence via direct tethering of the template sequence to the 5′-end hexanucleotide tag of the polymerase. In a separate selection effort for broader sequence generality, the Z polymerase ribozyme was obtained that included four mutations (C60U, G93A, G95A, and A159C). The combination of the beneficial traits from the tc19 selection, as the 5′-tag sequence and the G93A mutation, and the four mutations from the Z selection resulted in the tc19Z ribozyme, with significantly improved polymerase activity and fidelity that allowed the polymerization of a functional transcript of a minimal HHR (24 nt) [69]. The transfer of the ribozyme extension step of the CBT selection protocol to the eutectic phase of water ice generated the Y polymerase ribozyme variant that included the cold adaptive mutations (G93A, C97A, and U72G) and after addition of the 5′-tag sequence [69] yielded the tC9Y polymerase ribozyme (Figure 13.7) that was able to polymerize up to 206 nt at ambient temperatures (17 °C), which is slightly longer than its own sequence (203 nt) [83]. However, such long-range RNA synthesis was only possible on preferential template sequences such as the unstructured 11-mer repeat sequence (5′-(CACGCUUCGCA)$_n$-3′), and even then the yield of the full length 206 nt product was low (<0.5%).

13.8 Selection for Improved Polymerase Activity II

The currently most advanced polymerase ribozyme variant (24-3) (Figure 13.7) was selected based on a mutagenized library of the tc19Z variant using a sophisticated selection method based on the synthesis of functional RNA aptamers [97]. First, the polymerase was challenged to extend a primer along a template encoding a GTP or cyanocobalamin RNA aptamer, enabling a selection strategy based on binding to the aptamer ligand, followed by a gel-shift selection procedure. Progressively the selection stringency was increased by extending the length of the RNA to be synthesized and/or by decreasing the incubation time. The 24-3 variant included 17 new mutations in comparison with tc19Z both in the accessory domain and the catalytic core sequence, here mainly in the P7 stem. The average rate of primer extension of 1.2 nt min^{-1} is around 100-fold faster compared with the "wild-type" (tC19Z) polymerase variant. However, this increase in polymerization speed comes at the cost of reduced fidelity (92.0% compared with 96.6% for tc19Z). This higher error rate seems to derive mainly from an increased tendency of 24-3 to tolerate G:U wobble pairing. The 24-3 polymerase ribozyme variant was also able to synthesize complex structured RNAs including a phenylalanine transfer RNA (tRNA), by extending a primer sequence by 62 nt, although with a modest final yield (~0.07% after 72 hours). Strikingly, the 24-3 polymerase ribozyme was able to perform "ribo-PCR" by amplifying a specific RNA sequence n a PCR-like fashion. For this purpose, the 24 nt RNA template was composed of two 10 nt primer binding sites, flanking the sequence AGAG that was amplified up to 40 000-fold through repeated thermal cycling, demonstrating a proof of principle for RNA-catalyzed RNA amplification, although further improvements are likely needed to allow a general amplification of complex functional RNAs. The PCR-like amplification capacity opens up the possibility to directly select for functional RNA, based on exponential instead of linear amplification, a key characteristic of Darwinian evolution. Another interesting property of the 24-3 polymerase ribozyme is reverse transcriptase activity, i.e. the synthesis of a DNA strand from an RNA template [99]. The 24-3 variant was able to incorporate all four dNTPs and to synthesize unstructured DNA sequences up to 32 nt, but still requires an RNA template, and hence cannot function as a DNA-templated DNA polymerase. The 24-3 polymerase ribozyme was also applied to label the 3′-end of RNA or DNA by incorporating different functionalized nucleotide analogs guided by a template strand [100] and demonstrated a rather general 3′-labeling capability (UTP analogs showed to be slower and less efficient) by incorporating fluorophore, biotin, azide, or alkyne moieties into model RNA and DNA sequences. However, despite these wide range of applications, the 24-3 variant remains to some extent plagued by the general problems of all RPR variants derived from R18, including a preference for unstructured C-rich templates, inefficient strand displacement, and in addition a reduced fidelity that limits the yield and functionality of the RNAs synthesized by 24-3.

All polymerase ribozyme variants identified so far struggle to copy highly structured RNA sequences, including the polymerase sequence itself. There is therefore a fundamental paradox at the heart of RNA-catalyzed RNA synthesis, whereby RNA needs to fold into a defined and stable structure to encode function (such as RPR activity), but these very same structures prevent such folded sequences from acting as efficient templates for RNA replication. In an attempt to overcome this paradox, a new (in-ice) selection approach was developed, utilizing RNA trinucleotide triphosphates (pppNNN, triplets) instead of NTP monomers as substrates [98]. The library for selection was based on the Z polymerase ribozyme core, in which the accessory domain was replaced with a new random 3′-region of 30 nt. After seven rounds of in vitro selection, a new variant with weak triplet polymerization activity could be identified, and after a further 14 rounds of selection, six new ribozyme families were identified by next-generation sequencing (NGS), each with a unique 3′-region. Interestingly, one variant (1) was dominating the pool, even though it showed no polymerase activity by itself. However, in combination with each of the other triplet polymerase ribozyme variants (t2–t6), it gave a substantial boost to triplet polymerization activity. It seems to exert its effect as a non-catalytic cofactor by forming a 1 : 1 non-covalent heterodimeric holoenzyme complex with the catalytic triplet polymerase ribozyme subunit, whereby cooperation between both subunits leads to both increased activity and processivity, even in the absence of template tethering (as is required for all the mononucleotide polymerase ribozymes). This RNA cofactor (1) bears a unique 5′-hairpin sequence that is essential for its interaction with the catalytically active triplet polymerase ribozyme variants. In particular, one heterodimeric triplet polymerase ribozyme variant (t5 + 1) (Figure 13.7) was able to synthesize through very stable secondary structures with a calculated T_m above 90 °C and could synthesize its own sequence (+ strand) and the corresponding (– strand) in five segments starting from short (8 nt) primer sequences and using mostly triplets as substrates. These t5 + 1 ribozyme-synthesized fragments could be further ligated by t5 + 1 to regenerate an active triplet polymerase ribozyme subunit (t5) ribozyme, demonstrating an emerging capacity for self-replication. However, polymerization/ligation yields were still modest and would require very significant improvements before robust self-replication could take place. In addition, the cofactor sequence is only slightly shorter than the catalytically active ribozyme subunit, resulting in a substantial size increase of the combined polymerase ribozyme, substantially increasing the synthetic burden for self-replication.

Interestingly, the combination of in-ice reaction conditions and the application of trimer substrates not only enabled copying of highly structured RNA templates due to the triplets capacity to open up RNA secondary structures but also unlocked noncanonical RNA synthesis modes such as primer free polymerization and also polymerization in the noncanonical 3′–5′ direction. Despite these unusual properties, the fidelity of triplet polymerization of 97% (per position) on defined template sequences reaches that of the better monomer polymerase variants

[69, 83], presumably due to the incorporation of selection for fidelity (rather than simply RNA synthesis) from the start. The new 3′-domain of the triplet polymerase ribozyme appears responsible for the fidelity increase by readout of minor groove hydrogen bond donor/acceptor patterns, as the introduction of 2-thiouracil into the triplet substrates, which abolishes minor groove H-bonding, resulted in a loss of this fidelity. Nevertheless, compared with proteinaceous RNA polymerases, the fidelity of all RPR is modest, presumably due to only weak geometric discrimination at the NTP incorporation step, indicative of an open catalytic site [101].

Another issue is the comparatively short half-life of RPR in their preferred, high Mg^{2+} reaction buffers. In-ice reaction conditions result in a longer half-life in comparison with ambient temperatures, also related to the requirement of reduced Mg^{2+} concentrations for optimal polymerization activity. However, polymerization kinetics are also reduced in the ice phase; hence alternative reaction conditions that would combine fast polymerization kinetics with increased polymerase ribozyme half-lives would be advantageous. Previous selection efforts to lower the Mg^{2+} dependency of the class I ligase yielded ligase variants that showed improved ligation activity compared with the wild-type (WT) under the applied selection conditions (10 mM Mg^{2+}, pH 6), but with rate constants 2 orders of magnitude lower than the WT variant (k_{cat} = 300 min^{-1}) and as compared with the original reaction conditions (60 mM Mg^{2+}, pH 8), while reduction of the Mg^{2+} concentration to 1 mM led to a further 10-fold drop in rate constants (k_{cat} = 0.3 min^{-1}) [78]. In a similar vein, a new polymerase selection for reduced Mg^{2+} dependency at ambient temperatures was performed, called tethered self-tagging (TST), a selection strategy similar to the previously developed CBT method [69]. TST selection was performed in the presence of different peptides (derived from ribosomal peptides), in particular a lysine decapeptide (K_{10}) and at low Mg^{2+} concentrations (2 mM). This yielded the RPR variant 4M [67] (Figure 13.7), which contains multiple mutations clustered at the 5′-end and in the shortened accessory domain. The addition of the previously identified 5′-hexanucleotide *taq* sequence [69] that base-pairs with the 3′-end of the template strand and was beneficial for long primer extensions generated the tc9-4M polymerase variant. Not only K_{10} but also the structural similar ornithine homodecapeptide (independent of their chirality D or L) resulted in a boost of polymerase activity on a preferential template sequence in the presence of Mg^{2+}-ion concentrations as low as 4 mM, with RPR activity detectable at Mg^{2+} concentrations as low as 1 mM in the presence of the molecular crowder PEG8000. Furthermore, this substantial lowering of the requirement for Mg^{2+} allowed ribozyme-catalyzed polymerization inside phospholipid (1-palmitoyl-2-oleoylphosphatidylcholine [POPC]) vesicles. The remaining absolute requirement of Mg^{2+} ions in the low millimolar concentration range for polymerase activity, even in the presence of K_{10}, points toward a mainly structural role of the positively charged K_{10} decamer, which appears to act by promoting the docking of the primer/template duplex into the polymerase ribozyme, presumably by reducing the charge repulsion. The partial replacement of Mg^{2+} with peptides in the RPR demonstrates a possible scenario for the transition from functional RNA structures toward an RNA/peptide world, with the capability to form more complex structure and function. Simple homopolymeric peptides like K_{10} may be accessible by prebiotic chemistry and therefore could have been present before the onset of coded protein synthesis.

13.9 Conclusion and Outlook

Looking back, it is astonishing how the RPR was developed from its beginnings as an RNA ligase (itself isolated from a random sequence pool) toward the comparatively sophisticated RNA polymerase function, able to perform hundreds of polymerization reactions and to synthesize complex functional RNA. However, the holy grail of the field and a cornerstone of the original RNA world hypothesis, an RNA (self)-replicase, able to continuously replicate its own sequence context and to improve through Darwinian evolution remains elusive. Optimism is nevertheless justified as several of the current crop RPR variants already show many features expected of an RNA replicase, including regioselective formation of RNA 3′–5′ linkages with high speed and accuracy, the unspecific in trans recognition of a primer/template helix, and the template-dependent polymerization of RNA mononucleotides with good fidelity. While we are undoubtedly getting closer, the remaining challenges on the road to robust self-replication remain daunting and might require the integration of non-RNA molecular components (such as cationic peptides or lipids) or noncanonical nucleotide chemistries such as 2-ThioU or diaminopurine, which have shown superior properties in some polymerization settings.

What are possible limits of ribozyme-catalyzed RNA polymerization? All current RPR variants are based on the catalytic core of the class I ligase core that was isolated from a selection for RNA ligation [75]. In general, all polymerase variants were gained through further improvements of the original R18 polymerase sequence, while none of the selections escaped the overall fold of the class I ligase. Therefore, a new selection approach from "scratch" might yield RPR activity in different RNA sequence/structural contexts, and these may be able to catalyze more general RNA polymerization (more efficiently) or at least tell us if the class I ligase fold is the best and most likely RNA structure capable of RNA polymerization, discoverable in the sequence space accessible by in vitro evolution. However, even the anew discovery of the class I ligase-like fold would not exclude the existence of sequences with similar or improved polymerase ribozyme activities. Indeed, different natural ribozymes, e.g. the hammerhead (HHR), hairpin ribozyme (HP), hepatitis delta virus (HDV) and Varkud satellite (VS) ribozyme, are able to catalyze RNA cleavage comprising both different structural folds and catalytic strategies, but in vitro evolution mainly yielded an HHR motif [61], presumably as it represents the simplest solution for RNA cleavage in RNA sequence space.

While the discovery of a self-replicating RNA replicase would open a great many new lines of investigation and alter our perspective on RNA function, it would be unlikely to settle arguments regarding the plausibility of an RNA world or even the appearance of an RNA replicase at the origins of life. Indeed, arguments such as the improbability of the spontaneous appearance of an RPR out of the random pool of short RNA oligonucleotides in a prebiotic setting are hard to counter. Alternative chemistries and/or invocation of a pre-RNA world are unlikely to provide an answer as no rival genetic polymer has so far been identified that could compete with the chemical and structural diversity and plasticity of RNA. Indeed, even though protein sequence space is orders of magnitude larger, RNAs are capable of enormous complexity and sophistication as exemplified in ribozymes such as the group II intron or in RNA protein complexes as the ribosome, rivaling that of protein enzymes. Additionally, RNA with only four different but structurally and chemically similar heterocycles is much more robust toward mutations [48], allowing more rapid adaptive walks through the lower-dimensional RNA sequence space [102]. Another advantage of RNA over proteins as a possible polymer at the origin of life lies in the specific recognition mechanism of complementary sequences by Watson–Crick base pairing, which allows easy intermolecular complex formation and enables heredity and evolution at the molecular level. Hence, despite all the criticisms and its undoubted shortcomings, the RNA world and with it an RNA replicase, likely based on an RPR, remain an attractive and in our opinion currently still the most plausible scenario for the emergence of self-replication and evolution at the origin of life.

References

1 Gilbert, W. (1986). Origin of life: the RNA world. *Nature* 319: 618.

2 Crick, F.H. (1968). The origin of the genetic code. *J. Mol. Biol.* 38: 367–379.

3 Orgel, L.E. (1968). Evolution of the genetic apparatus. *J. Mol. Biol.* 38: 381–393.

4 Kruger, K., Grabowski, P.J., Zaug, A.J. et al. (1982). Self-splicing RNA: autoexcision and autocyclization of the ribosomal RNA intervening sequence of *Tetrahymena*. *Cell* 31: 147–157.

5 Guerrier-Takada, C., Gardiner, K., Marsh, T. et al. (1983). The RNA moiety of ribonuclease P is the catalytic subunit of the enzyme. *Cell* 35: 849–857.

6 Noller, H.F., Hoffarth, V., and Zimniak, L. (1992). Unusual resistance of peptidyl transferase to protein extraction procedures. *Science* 256: 1416–1419.

7 Nissen, P., Hansen, J., Ban, N. et al. (2000). The structural basis of ribosome activity in peptide bond synthesis. *Science* 289: 920–930.

8 Cech, T.R. (2000). Structural biology. The ribosome is a ribozyme. *Science* 289: 878–879.

9 Pressman, A., Blanco, C., and Chen, I.A. (2015). The RNA world as a model system to study the origin of life. *Curr. Biol.* 25: 953–963.

10 Cech, T.R. (2012). The RNA worlds in context. *Cold Spring Harbor Perspect. Biol.* 4: a006742.

11 Cech, T.R. and Steitz, J.A. (2014). The noncoding RNA revolution-trashing old rules to forge new ones. *Cell* 157: 77–94.

12 Wilson, T.J., Liu, Y., and Lilley, D.M. (2016). Ribozymes and the mechanisms that underlie RNA catalysis. *Front. Chem. Sci. Eng.* 10: 178–185.

13 Wachowius, F., Attwater, J., and Holliger, P. (2017). Nucleic acids: function and potential for abiogenesis. *Q. Rev. Biophys.* 50: e4.

14 Wilson, D.S. and Szostak, J.W. (1999). In vitro selection of functional nucleic acids. *Annu. Rev. Biochem.* 68: 611–647.

15 Martin, L.L., Unrau, P.J., and Muller, U.F. (2015). RNA synthesis by in vitro selected ribozymes for recreating an RNA world. *Life* 5: 247–268.

16 Joyce, G.F. and Orgel, L.E. (1993). Prospects for understanding the origin of the RNA world. In: *The RNA World* (eds. R.F. Gesteland and J.F. Atkins), 1–25. Cold Spring Harbor, NY: Cold Spring Harbor Laboratory Press.

17 Sutherland, J.D. (2016). The origin of life – out of the blue. *Angew. Chem. Int. Ed.* 55: 104–121.

18 Powner, M.W., Gerland, B., and Sutherland, J.D. (2009). Synthesis of activated pyrimidine ribonucleotides in prebiotically plausible conditions. *Nature* 459: 239–242.

19 Becker, S., Thoma, I., Deutsch, A. et al. (2016). A high-yielding, strictly regioselective prebiotic purine nucleoside formation pathway. *Science* 352: 833–836.

20 Benner, S.A., Kim, H.J., and Carrigan, M.A. (2012). Asphalt, water, and the prebiotic synthesis of ribose, ribonucleosides, and RNA. *Acc. Chem. Res.* 45: 2025–2034.

21 Verlander, M.S. and Orgel, L.E. (1974). Analysis of high molecular weight material from the polymerization of adenosine cyclic 2′, 3′-phosphate. *J. Mol. Evol.* 3: 115–120.

22 Lohrmann, R. and Orgel, L.E. (1980). Efficient catalysis of polycytidylic acid-directed oligoguanylate formation by Pb^{2+}. *J. Mol. Biol.* 142: 555–567.

23 Inoue, T. and Orgel, L.E. (1982). Oligomerization of (guanosine 5′-phosphor)-2-methylimidazolide on poly(C): an RNA polymerase model. *J. Mol. Biol.* 162: 201–217.

24 Huang, W. and Ferris, J.P. (2006). One-step, regioselective synthesis of up to 50-mers of RNA oligomers by montmorillonite catalysis. *J. Am. Chem. Soc.* 128: 8914–8919.

25 Deck, C., Jauker, M., and Richert, C. (2011). Efficient enzyme-free copying of all four nucleobases templated by immobilized RNA. *Nat. Chem.* 3: 603–608.

26 Prywes, N., Blain, J.C., Del Frate, F., and Szostak, J.W. (2016). Nonenzymatic copying of RNA templates containing all four letters is catalyzed by activated oligonucleotides. *eLife* 5: e17756.

27 Li, L., Prywes, N., Tam, C.P. et al. (2017). Enhanced nonenzymatic RNA copying with 2-aminoimidazole activated nucleotides. *J. Am. Chem. Soc.* 139: 1810–1813.

28 Inoue, T. and Orgel, L.E. (1983). A nonenzymatic RNA polymerase model. *Science* 219: 859–862.

29 Szostak, J.W. (2012). The eightfold path to non-enzymatic RNA replication. *J. Syst. Chem.* 3: 2.

30 Heuberger, B.D., Pal, A., Del Frate, F. et al. (2015). Replacing uridine with 2-thiouridine enhances the rate and fidelity of nonenzymatic RNA primer extension. *J. Am. Chem. Soc.* 137: 2769–2775.

31 Herschlag, D. and Cech, T.R. (1990). Catalysis of RNA cleavage by the *Tetrahymena thermophila* ribozyme. 1. Kinetic description of the reaction of an RNA substrate complementary to the active site. *Biochemistry* 29: 10159–10171.

32 Zaug, A.J. and Cech, T.R. (1986). The intervening sequence RNA of *Tetrahymena* is an enzyme. *Science* 231: 470–475.

33 Doudna, J.A. and Szostak, J.W. (1989). RNA-catalysed synthesis of complementary-strand RNA. *Nature* 339: 519–522.

34 Doudna, J.A., Couture, S., and Szostak, J.W. (1991). A multisubunit ribozyme that is a catalyst of and template for complementary strand RNA synthesis. *Science* 251: 1605–1608.

35 Green, R. and Szostak, J.W. (1992). Selection of a ribozyme that functions as a superior template in a self-copying reaction. *Science* 258: 1910–1915.

36 Doudna, J.A., Usman, N., and Szostak, J.W. (1993). Ribozyme-catalyzed primer extension by trinucleotides: a model for the RNA-catalyzed replication of RNA. *Biochemistry* 32: 2111–2115.

37 Hayden, E.J. and Lehman, N. (2006). Self-assembly of a group I intron from inactive oligonucleotide fragments. *Chem. Biol.* 13: 909–918.

38 Vaidya, N., Manapat, M.L., Chen, I.A. et al. (2012). Spontaneous network formation among cooperative RNA replicators. *Nature* 491: 72–77.

39 Sczepanski, J.T. and Joyce, G.F. (2014). A cross-chiral RNA polymerase ribozyme. *Nature* 515: 440–442.

40 Lincoln, T.A. and Joyce, G.F. (2009). Self-sustained replication of an RNA enzyme. *Science* 323: 1229–1232.

41 Robertson, M.P. and Joyce, G.F. (2012). The origins of the RNA world. *Cold Spring Harbor Perspect. Biol.* 4: a003608.

42 Mills, D.R., Peterson, R.L., and Spiegelman, S. (1967). An extracellular Darwinian experiment with a self-duplicating nucleic acid molecule. *Proc. Natl. Acad. Sci. U.S.A.* 58: 217.

43 Schrum, J.P., Zhu, T.F., and Szostak, J.W. (2010). The origins of cellular life. *Cold Spring Harbor Perspect. Biol.* 2: a002212.

44 Joyce, G.F. and Szostak, J.W. (2018). Protocells and RNA self-replication. *Cold Spring Harbor Perspect. Biol.* 10.

45 Kreysing, M., Keil, L., Lanzmich, S., and Braun, D. (2015). Heat flux across an open pore enables the continuous replication and selection of oligonucleotides towards increasing length. *Nat. Chem.* 7: 203–208.

46 Eigen, M. (1971). Self-organization of matter and evolution of biological macromolecules. *Naturwissenschaften* 58: 465–523.

47 Eigen, M., Mccaskill, J., and Schuster, P. (1988). Molecular quasi-species. *J. Phys. Chem.* 92: 6881–6891.

48 Kun, A., Santos, M., and Szathmary, E. (2005). Real ribozymes suggest a relaxed error threshold. *Nat. Genet.* 37: 1008–1011.

49 Bartel, D.P. (1999). Re-creating an RNA replicase. In: *The RNA World* (eds. R.F. Gesteland, T.R. Cech and J.F. Atkins), 143–161. Cold Spring Harbor, NY: Cold Spring Harbor Laboratory Press.

50 Ward, C.D. and Flanegan, J.B. (1992). Determination of the poliovirus RNA polymerase error frequency at eight sites in the viral genome. *J. Virol.* 66: 3784–3793.

51 Brown, D. and Gold, L. (1996). RNA replication by Q beta replicase: a working model. *Proc. Natl. Acad. Sci. U.S.A.* 93: 11558–11562.

52 McCall, M.J., Hendry, P., Mir, A.A. et al. (2000). Small, efficient hammerhead ribozymes. *Mol. Biotechnol.* 14: 5–17.

53 Johnston, W.K., Unrau, P.J., Lawrence, M.S. et al. (2001). RNA-catalyzed RNA polymerization: accurate and general RNA-templated primer extension. *Science* 292: 1319–1325.

54 Jimenez, J.I., Xulvi-Brunet, R., Campbell, G.W. et al. (2013). Comprehensive experimental fitness landscape and evolutionary network for small RNA. *Proc. Natl. Acad. Sci. U.S.A.* 110: 14984–14989.

55 Bartel, D.P. and Szostak, J.W. (1993). Isolation of new ribozymes from a large pool of random sequences. *Science* 261: 1411–1418.

56 Zuker, M. (2003). Mfold web server for nucleic acid folding and hybridization prediction. *Nucleic Acids Res.* 31: 3406–3415.

57 Mustoe, A.M., Brooks, C.L., and Al-Hashimi, H.M. (2014). Hierarchy of RNA functional dynamics. *Annu. Rev. Biochem.* 83: 441–466.

58 Sabeti, P.C., Unrau, P.J., and Bartel, D.P. (1997). Accessing rare activities from random RNA sequences: the importance of the length of molecules in the starting pool. *Chem. Biol.* 4: 767–774.

59 Kobori, S. and Yokobayashi, Y. (2016). High-throughput mutational analysis of a twister ribozyme. *Angew. Chem. Int. Ed.* 55: 10354–10357.

60 Salehi-Ashtiani, K. and Szostak, J.W. (2001). In vitro evolution suggests multiple origins for the hammerhead ribozyme. *Nature* 414: 82–84.

61 Hammann, C., Luptak, A., Perreault, J., and de la Pena, M. (2012). The ubiquitous hammerhead ribozyme. *RNA* 18: 871–885.

62 Carothers, J.M., Oestreich, S.C., and Szostak, J.W. (2006). Aptamers selected for higher-affinity binding are not more specific for the target ligand. *J. Am. Chem. Soc.* 128: 7929–7937.

63 Mutschler, H., Wochner, A., and Holliger, P. (2015). Freeze-thaw cycles as drivers of complex ribozyme assembly. *Nat. Chem.* 7: 502–508.

64 Monnard, P.A., Kanavarioti, A., and Deamer, D.W. (2003). Eutectic phase polymerization of activated ribonucleotide mixtures yields quasi-equimolar incorporation of purine and pyrimidine nucleobases. *J. Am. Chem. Soc.* 125: 13734–13740.

65 Attwater, J., Wochner, A., Pinheiro, V.B. et al. (2010). Ice as a protocellular medium for RNA replication. *Nat. Commun.* 1: 76.

66 Adamala, K. and Szostak, J.W. (2013). Nonenzymatic template-directed RNA synthesis inside model protocells. *Science* 342: 1098–1100.

67 Tagami, S., Attwater, J., and Holliger, P. (2017). Simple peptides derived from the ribosomal core potentiate RNA polymerase ribozyme function. *Nat. Chem.* 9: 325–332.

68 Mariani, A., Bonfio, C., Johnson, C.M., and Sutherland, J.D. (2018). pH-driven RNA strand separation under prebiotically plausible conditions. *Biochemistry* 57: 6382–6386.

69 Wochner, A., Attwater, J., Coulson, A., and Holliger, P. (2011). Ribozyme-catalyzed transcription of an active ribozyme. *Science* 332: 209–212.

70 Cantor, C.R. and Schimmel, P.R. (1995). *Biophysical Chemistry: The Behavior of Biological Macromolecules*. New York: Freeman.

71 Micura, R., Pils, W., Hobartner, C. et al. (2001). Methylation of the nucleobases in RNA oligonucleotides mediates duplex-hairpin conversion. *Nucleic Acids Res.* 29: 3997–4005.

72 Joyce, G.F. (2007). Forty years of in vitro evolution. *Angew. Chem. Int. Ed.* 46: 6420–6436.

73 Shechner, D.M., Grant, R.A., Bagby, S.C. et al. (2009). Crystal structure of the catalytic core of an RNA-polymerase ribozyme. *Science* 326: 1271–1275.

74 Ekland, E.H. and Bartel, D.P. (1995). The secondary structure and sequence optimization of an RNA ligase ribozyme. *Nucleic Acids Res.* 23: 3231–3238.

75 Ekland, E.H., Szostak, J.W., and Bartel, D.P. (1995). Structurally complex and highly active RNA ligases derived from random RNA sequences. *Science* 269: 364–370.

76 Bergman, N.H., Johnston, W.K., and Bartel, D.P. (2000). Kinetic framework for ligation by an efficient RNA ligase ribozyme. *Biochemistry* 39: 3115–3123.

77 Bergman, N.H., Lau, N.C., Lehnert, V. et al. (2004). The three-dimensional architecture of the class I ligase ribozyme. *RNA* 10: 176–184.

78 Bagby, S.C., Bergman, N.H., Shechner, D.M. et al. (2009). A class I ligase ribozyme with reduced Mg^{2+} dependence: selection, sequence analysis, and identification of functional tertiary interactions. *RNA* 15: 2129–2146.

79 Ekland, E.H. and Bartel, D.P. (1996). RNA-catalysed RNA polymerization using nucleoside triphosphates. *Nature* 382: 373–376.

80 Westheimer, F.H. (1987). Why nature chose phosphates. *Science* 235: 1173–1178.

81 Glasner, M.E., Yen, C.C., Ekland, E.H., and Bartel, D.P. (2000). Recognition of nucleoside triphosphates during RNA-catalyzed primer extension. *Biochemistry* 39: 15556–15562.

82 Lawrence, M.S. and Bartel, D.P. (2003). Processivity of ribozyme-catalyzed RNA polymerization. *Biochemistry* 42: 8748–8755.

83 Attwater, J., Wochner, A., and Holliger, P. (2013). In-ice evolution of RNA polymerase ribozyme activity. *Nat. Chem.* 5: 1011–1018.

84 Wu, T. and Orgel, L.E. (1992). Nonenzymatic template-directed synthesis on hairpin oligonucleotides. 2. Templates containing cytidine and guanosine residues. *J. Am. Chem. Soc.* 114: 5496–5501.

85 Muller, U.F. and Bartel, D.P. (2008). Improved polymerase ribozyme efficiency on hydrophobic assemblies. *RNA* 14: 552–562.

86 Yao, C., Moretti, J.E., Struss, P.E. et al. (2011). Arginine cofactors on the polymerase ribozyme. *PLoS One* 6: e25030.

87 Yao, C. and Muller, U.F. (2011). Polymerase ribozyme efficiency increased by G/T-rich DNA oligonucleotides. *RNA* 17: 1274–1281.

88 Shechner, D.M. and Bartel, D.P. (2011). The structural basis of RNA-catalyzed RNA polymerization. *Nat. Struct. Mol. Biol.* 18: 1036–1042.

89 Castro, C., Smidansky, E.D., Arnold, J.J. et al. (2009). Nucleic acid polymerases use a general acid for nucleotidyl transfer. *Nat. Struct. Mol. Biol.* 16: 212–218.

90 Das, S.R. and Piccirilli, J.A. (2005). General acid catalysis by the hepatitis delta virus ribozyme. *Nat. Chem. Biol.* 1: 45–52.

91 Wang, Q.S., Cheng, L.K., and Unrau, P.J. (2011). Characterization of the B6.61 polymerase ribozyme accessory domain. *RNA* 17: 469–477.

92 Akoopie, A. and Muller, U.F. (2018). The NTP binding site of the polymerase ribozyme. *Nucleic Acids Res.* 46: 10589–10597.

93 Muller, U.F. and Bartel, D.P. (2003). Substrate 2′-hydroxyl groups required for ribozyme-catalyzed polymerization. *Chem. Biol.* 10: 799–806.

94 Lawrence, M.S. and Bartel, D.P. (2005). New ligase-derived RNA polymerase ribozymes. *RNA* 11: 1173–1180.

95 Tawfik, D.S. and Griffiths, A.D. (1998). Man-made cell-like compartments for molecular evolution. *Nat. Biotechnol.* 16: 652–656.

96 Zaher, H.S. and Unrau, P.J. (2007). Selection of an improved RNA polymerase ribozyme with superior extension and fidelity. *RNA* 13: 1017–1026.

97 Horning, D.P. and Joyce, G.F. (2016). Amplification of RNA by an RNA polymerase ribozyme. *Proc. Natl. Acad. Sci. U.S.A.* 113: 9786–9791.

98 Attwater, J., Raguram, A., Morgunov, A.S. et al. (2018). Ribozyme-catalysed RNA synthesis using triplet building blocks. *elife* 7: e35255.

99 Samanta, B. and Joyce, G.F. (2017). A reverse transcriptase ribozyme. *eLife* 6: e31153.

100 Samanta, B., Horning, D.P., and Joyce, G.F. (2018). 3′-End labeling of nucleic acids by a polymerase ribozyme. *Nucleic Acids Res.* 46: e103.

101 Attwater, J., Tagami, S., Kimoto, M. et al. (2013). Chemical fidelity of an RNA polymerase ribozyme. *Chem. Sci.* 4: 2804–2814.

102 Schultes, E.A. and Bartel, D.P. (2000). One sequence, two ribozymes: implications for the emergence of new ribozyme folds. *Science* 289: 448–452.

14

Maintenance of Genetic Information in the First Ribocell

Ádám Kun

Centre for Ecological Research, Institute of Evolution, Karolina út 29, 1113 Budapest, Hungary

14.1 The Ribocell and the Stages of the RNA World

The RNA world [1–3] is an era during the origin of life when RNA played the role of both information storage molecule and enzymes. Ribozymes are the embodiments of this last function. They not only witnessed the origin of life but also made it happen. According to Tibor Gánti [4, 5], absolute life criteria are (i) inherent unity, (ii) inherent stability (homeostasis), the presence of (iii) a metabolism, (iv) an information carrying subsystem, and (v) processes regulated and controlled by a program. As the workhorse of metabolism, ribozymes helped to fulfill criteria (ii) and (iii). Information (criterium iv) are the complementary strands of ribozymes. RNA, in its dual roles as enzymes and information, regulates the processes of the ribocell. Thus, ribozymes were truly at the hearth of what it meant to be alive. The first absolute life criterium is realized by a lipid membrane, and the encapsulation of RNA enzymes and information was the first major evolutionary transition [6–8], which led to the formation of the first living cell. In modern cells, metabolism is run mostly by peptide enzymes, and information is stored in DNA, but RNAs shuttle information between DNA and peptides, and they still have a significant role in the regulation of cellular processes [9–14].

Questions about the origin of life can be roughly divided into structural and dynamical categories. With respect to ribozymes, structural questions mostly relate to what a certain RNA molecule can do, while dynamical questions to the evolution and/or the ecology of the molecules. For example, it is a structural question whether RNA can catalyze all reaction steps of a given pathway, and it is a dynamical question whether these ribozymes can act together in a common environment or whether they can all evolve within one system. Most of the experimental literature on the origin of life focuses on structural questions. Here, we will focus on dynamical questions, especially on the extent of metabolic complexity a ribocell could have achieved and the hurdles it needed to overcome to reach it.

In this chapter, the RNA world is considered to last till DNA takes over as the main information storage molecule (Figure 14.1). It can be argued that after the

Ribozymes, First Edition. Edited by Sabine Müller, Benoît Masquida, and Wade Winkler.

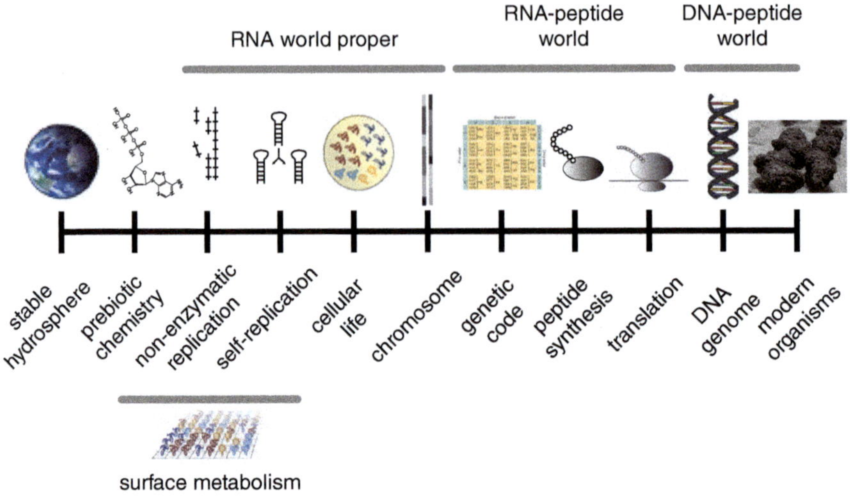

Figure 14.1 Stages of the RNA world. Between the two well-defined endpoints of the origin of life lies the RNA world. Nonenzymatic RNA replication could have led to the evolution of a self-replicating enzyme, which kick-started metabolism. This stage was probably still unfolded on mineral surfaces. Cellular life marks the beginning of life. The error catastrophe was overcome when the chromosome evolved. This led to complex metabolism, which could recruit amino acids and then evolve peptide synthesis, thus leading to the RNA–peptide world stage. With the invention of DNA genome, the RNA world yielded to the now present DNA–peptide world.

invention of translation, it is no longer an RNA world. However, the takeover of peptides cannot happen overnight. Thus, when the first translated peptide rolled out of a ribosome, most enzymatic functions are still fulfilled by ribozymes. The number of ribozymes might have steadily decreased as the number of peptide enzymes increased, but this, by necessity, is a gradual process. Consequently, even at this stage, RNAs act both as an information storage molecule(s) and as enzymes, albeit RNA's role in this last function is withdrawing. The defining feature of the RNA world remains intact. Interestingly, RNA lost most of its information storage role (only RNA viruses store their genetic information in RNA) but retained some of its catalytic functions: the catalytic core of the ribosome, which is found in all living organisms, is made up of RNA [15, 16] (see also Chapter 8). Not as widespread as the ribosome, but there are some natural ribozymes as well, like the RNase P [17] (see also Chapter 9), the group I introns [18] (see also Chapter 5), the group II introns [19] (see also Chapter 6), the hammerhead ribozyme [20], the hairpin ribozyme [21], the hepatitis delta virus and hepatitis delta virus-like ribozymes [22], the *Neurospora* Varkud satellite ribozyme [23], the *glmS* ribozyme [24], and the twister ribozymes [25] (see also Chapters 3 and 4).

14.1.1 Replication of the Genetic Information

Evolution is a powerful mechanism that can produce novelty and considerable increase in complexity, but it requires reproduction, variation, and heredity. Reproduction is the capability of exponential growth achieved via autocatalysis [26, 27], what chemical systems can exhibit [28–32]. However, heredity and variation are characteristics that rarely are featured together in a system. The von Kiedrowski-type replicators [33], for example, exhibit heredity, but the reactants cannot be changed without destroying their ability for autocatalysis. Compositional information [34], on the other hand, has ample variability but lacks heredity [35]. Modular polymers with complementary modules, such as RNA and DNA, have the potential for evolution. Variation in the form of mutations arises during RNA/DNA replication naturally (see below), and via complementarity, information is passed on. The copy of a strand is yielded in two steps: first, the complementary strand is produced, and by copying it, a replica of the original strand forms. However, without a mechanism for copying RNA, it also does not exhibit heredity and thus cannot evolve. For example, RNA molecules of intermediate length can form on clay surfaces [36, 37], but as they form de novo and not as copies of another RNA molecule, there is no heredity. Thus, assuming RNA to be the primordial molecule that led to life also requires self-replication to be the first function on that path.

Self-replication of RNA molecules, i.e. RNA-dependent RNA polymerization, is the key step leading to life. While nonenzymatic replication of RNAs [38] could have bootstrapped the evolution of ribozymes, an RNA-dependent RNA polymerase is a necessity (see also Chapter 13). A fully functional example of this ribozyme is so far elusive. Attempts to modify a ligase ribozyme to act as a template-directed polymerase first yielded an enzyme capable of adding mere six nucleotides [39]. Later, this was improved to add 14 [40] and then up to 20 nucleotides [41], which is still only about 10% of the length of the RNA molecule (Table 14.1). A breakthrough came in 2011 when a ribozyme capable of synthesizing a strand of up to 98 nucleotides was selected [44]. Another derived ribozyme can extend a primer with more than 200 nucleotides [45] but requires a very specific template. Unfortunately, this specific template is not its own sequence or its complementary sequence. Thus, ribozymes are capable of synthetizing long strands based on a template, but their generality is still a question.

14.1.2 On the Metabolic Complexity of Ribocells

Autocatalytic replication of the informational polymer is just one of the functions a living cell needs to exhibit. It should possess a membrane, which is also autocatalytically formed [48]. And it should supply all the building blocks that it cannot take up from the environment via its metabolism. Metabolism is also autocatalytic [49]. A minimal living cell (cf. the chemoton [50]) consists of the three subsystems

Table 14.1 RNA-dependent RNA polymerase ribozymes.

Ribozyme name	Error rate	Length	Processivity	References	Note
b1-233t	1.5×10^{-1}	98	6	[39]	
R18	3.3×10^{-2}	198	14	[40]	
B6.61	n.a.	193	20	[41, 42]	
R18	4.1×10^{-2}	198	23–32	[43]	In ice
R18	4.26×10^{-2}	192		[44]	
tC19	2.68×10^{-2}	198	95	[44]	
Z	n.a.	221	n.a.	[44]	Generality increased
tC19Z	8.8×10^{-3}	198	24	[44]	Able to copy hammerhead minizyme
Y	1.7×10^{-2} 5.2×10^{-2a}	195	30	[45]	
tC9Y	2.3×10^{-2}	202	206 118^{a}	[45]	Specific template
24-3	8×10^{-2}	180	61	[46]	Able to copy functional RNA
	3.4×10^{-2}				
t5 + 1	2.6×10^{-2}	220 + 135	n.a.	[47]	Triplet polymerase
Zcore	8.7×10^{-2}	110	n.a.	[47]	Triplet polymerase
0core	9.5×10^{-2}	112	n.a.	[47]	Triplet polymerase

n.a., data not available.
[a] In ice.

of information storage, metabolism, and encapsulation. Systems having only two of these subsystems are called infrabiological systems (*sensu* Szathmáry [51–53]). There are three infrabiological systems possible: an informational–metabolic system, an encapsulated informational system, and an encapsulated metabolic system. One of these infrabiological systems were the precursor of the first living cell. In contemporary organisms, lipid membranes encapsulate DNA information, and the system is maintained via a metabolism catalyzed by peptide enzymes. While now all subsystems are realized by chemically different entities, their chemical identity is not set into stone. Viruses, for example, are encapsulated informational systems without metabolism. Their DNA or RNA genome is encapsulated by a proteinaceous capsid. Thus, encapsulation can also be done with proteins, not only with lipids. In the RNA world, one molecular species, RNA, fulfills the role of information storage and enzyme. Ribozyme-based replication is an infrabiological system in itself. It has information, as it can be copied, and the information is used to produce the ribozyme that is capable of copying the information. Consequently, the infrabiological system having information and metabolism is especially important for the origin of life.

The informational–metabolic stage probably he on some mineral surface. A prebiotic soup or warm little ponds (cf. Darwin) dilute material too much and foster hydrolysis as opposed to polymerization. Some kind of semi-compartmentalization can solve this problem. For example, negatively charged RNA molecules can be attached to mineral surfaces having positive charge, e.g. pyrite. In such system, the formation of larger organic molecules could be favored [54]. Furthermore, mineral surfaces can catalyze reactions [37, 55] and in an enantioselective manner, which may be responsible for the homochirality we observe today [56, 57]. They can also protect ribozymes from the pervasive UV radiation [58] of the primordial environment.

At the stage of surface metabolism, there could have already been a budding metabolism catalyzed by a small array of ribozymes. Apart from the copying of the genetic information, the assembly of nucleotides from its constituents could have been a function that enhanced the local proliferation of ribozymes. On the other end of the origin of life, in the stage shifting from RNA to DNA information storage, a small array of enzymes is not enough; a complex metabolism run by potentially hundreds if not thousands of genes is required. The metabolic pathways and characteristics common to all living organisms were already in place in the last universal common ancestor (LUCA). This entity is a very late descendant of the first cell, albeit there are some claims that it could still had an RNA genome [59]. LUCA definitely had peptide enzymes; the universality of the genetic code and the similarities of the ribosomes are testimonies of it. Translation, the process that produces polypeptides based on the information stored in a nucleic acid, requires a considerable number of enzymes. The question naturally arises, how to proceed from a system having one or a few enzymatic activities to one having several hundreds.

In this chapter, we first discuss the obstacles in the path of increasing number of ribozymes. Then, the minimal ribozyme diversity of the first ribocell is estimated.

14.2 The Error Thresholds

An evolving system requires heredity, which is achieved by copying nucleic acid polymers via base pairing. A complementary base is ligated at the end of the new polymer, resulting – at the end of the replication process – in a complementary strand. The replication of this strand yields the copy of the original strand. Consequently, by this two-stage process, genetic information can be propagated. However, there are errors during this process resulting in changed copies of the original strand. These errors are the mutations that, on the one hand, provide the variation on which selection can act but, on the other hand, they can destroy a functional ribozyme.

14.2.1 Introducing the Error Threshold

Let us assume that there is a sequence that should be copied as it will fold into an enzyme. We will call this sequence the enzyme sequence (it might also be called the master sequence or the wild-type sequence). It replicates with a replication rate of a_{enzyme}. During replication, mutation can occur, and the resulting sequence is not the enzyme sequence but a parasite. A parasite accepts the catalytic aid of the ribozymes, but it does not contribute to the system as it does not have any enzymatic activity or other useful role. The probability of such mutation is μ. The parasite replicates with a rate of a_{parasite}. There is a uniform, sequence-independent death rate (or washout rate) D. The concentrations of the enzyme (x_{enzyme}) and the parasite (x_{parasite}) are governed by the following two equations:

$$\frac{dx_{\text{enzyme}}}{dt} = a_{\text{enzyme}}(1-\mu)x_{\text{enzyme}} - Dx_{\text{enzyme}} \tag{14.1}$$

$$\frac{dx_{\text{parasite}}}{dt} = a_{\text{enzyme}}\mu x_{\text{enzyme}} + a_{\text{parasite}}x_{\text{parasite}} - Dx_{\text{parasite}} \tag{14.2}$$

This is a simplification of the quasispecies model of Eigen [60], which included all possible mutants of the master sequence. Here, we lump them together as proposed by Maynard Smith [61]. The persistence of the enzyme in equilibrium ($x_{\text{enzyme}} > 0$) requires that

$$(1-\mu)a_{\text{enzyme}} > a_{\text{parasite}} \tag{14.3}$$

Thus, not only has the replication rate of the enzyme be higher than that of the parasite, but also it has to be considerably higher if mutation rate is high. Given the replication rate of the enzyme and the parasite, a critical mutation rate, the error threshold, can be determined. The error threshold is thus the critical mutation rate above which information cannot be maintained despite it having higher replication rate than the parasite. However, while there could be mutants that have lower replication rates than the wild-type sequence, there are a considerable number of them having lower replication rates.

Shorter sequences usually have faster replication rates. The mutation leading to a shorter sequence is called deletion. A deletion can occur, for example, by slippage of the replicase on longer stretches of repetitions [62], which then results in either an insertion or a deletion (together called indels). In contemporary organisms, deletions are more frequent than insertions [63]. Most indels are rather short; 80%+ of them are 1–10 bp [64] or 1–5 bp long [65, 66]. Small indels can already destroy enzymatic activity, but longer deletions are required for considerably faster replicating parasites. Longer deletions, while rarer, are also observable. Thus, there will be shorter and faster mutants competing with the wild-type ribozyme.

We know from the pioneering work of Sol Spiegelman and coworkers [67] that faster replicating, nonfunctional mutations of a functional RNA go to fixation. Starting from the about 3300–3600 nucleotides long RNA genome of the Qβ phage and replicating it, they have arrived – after 75 passages – at an RNA replicating 15 times faster than the original, but being only 550 nucleotides long. This RNA was not a functional phage. In a similar experiment [68], we have replicated a modified *Neurospora* Varkud satellite (VS) ribozyme with the Qβ replicase and after some time transferred a sample to a fresh solution of NTPs and replicase. After a few transfers, the ribozyme could not be detected in the population. The population was dominated by a sequence roughly third of the length of the functional ribozyme and replicating nearly twice as fast.

Consequently, if there is only selection for replication speed, then a shorter mutant of the wild-type enzyme will outcompete it, and the functional RNA (the information) will be lost. Phages can survive the high error rate of their replicases [69–73] and the competition with their faster replicating mutants by selection on function. Only functional virions can infect a new host and replicate in it. The higher-level evolutionary unit, the capsid-encapsulated virus genome, the virion, allows for the apparent replication rate of the functional virus to be higher than that of its shorter mutants. Similarly, when ribozymes are encapsulated into droplets and droplets are selected for further replication based on total enzymatic activity, then the ribozyme can be maintained despite the constant reemergence of the shorter and faster replicating mutants [68]. Thus, a higher-level evolutionary unit is required to satisfy Eq. (14.3). This is assumed in all models of the error threshold either explicitly or implicitly. We also make this assumption here, and I will further elaborate on compartmentalization in the next section.

14.2.2 The Fitness Landscape and Neutrality of Mutations

The mutation rate μ in Eq. (14.3) is the probability of replication resulting in a sequence that is not an enzyme, but a parasite. While mutations change the genotype, they do not necessarily change the phenotype. In the original formulation of the error threshold [60], all mutations led out of the master sequence, i.e. all mutations were considered to result in a different phenotype. Even if we remain true to the original model, not all mutations to a coding DNA sequence (the genotype) result in a change of the amino acid sequence of the coded peptide (the phenotype). The genetic code is degenerate: multiple triplets code for the same amino acid. These synonymous mutations are neutral, and the fitness of the individual bearing the mutated sequence is the same as those harboring the wild-type sequence. As the mutation rate μ is the rate at which the sequence changes to another having lower fitness, the actual mutation rate of the replication process can be higher. Furthermore, even if an amino acid changes, it might not affect the activity and stability of the peptide. Experiments determining the distribution of fitness effects [74] show that there are non-synonymous mutations with neutral effect on fitness. According to extensive mutagenesis of *Salmonella enterica*'s HisA protein, an isomerase in the L-histidine biosynthesis pathway, 2.5% of the non-synonymous mutations are neutral [75]. Furthermore, assaying 64% of all possible single mutants of the antibiotic resistance factor TEM-1 β-lactamase, 320 (32.3%) were found to have the same minimum inhibitory concentration to antibiotics as the wild-type [76]. And 4.8% of the analyzed mutations of ribosomal proteins are neutral [77]. Another study analyzed the E3 ubiquitin ligase activity of 5153 mutants of the RING domain of breast cancer 1 protein (BRCA1) [78]: 90 (1.7%) had a ligase activity score not differing more than 1% of the wild-type's, and 435 (8.4%) were within 5% of the value of the activity score. Thus, not all mutations lead to decrease of fitness.

At the early stages of the origin of life, coded peptides were not yet present, and the faithful replication of ribozymes was the key problem [2]. With regard to RNA, there are mutations that do not affect the secondary structure of it [79–81]. It was estimated that compared with the 4^L different sequences of length L, the number of different structures is 2.35^L [82]. Accordingly, there are considerably more sequences than structures. Usually, a few (1–3) mutations do not change the secondary structure of an RNA. Thus, there is a phenotypic error threshold [83, 84], which is the critical error rate above which the phenotype cannot be maintained despite selection for it.

Neutral mutations are mostly substitutions, i.e. mis-incorporations of a noncanonical base pair into the sequence. While the two base-pair system offers some protection against mutations [85, 86], there are still ample possibilities for base substitutions. Both effects are deeply rooted in the chemistry of the bases. Hydrogen bonds can form between guanine (G) and cytosine (C) and between adenine (A) and uracil (U). In these canonical base pairs, there is always a larger purine derivative (G or A) facing a smaller pyrimidine derivative (C or U). The size difference allows for the easy recognition of A–G and C–U mispairs, and consequently, transversions are rare [87]. Transitions (A $\leftrightarrow$ G and U $\leftrightarrow$ C mutations) are not this easy to catch. In their normal state, the noncanonical base pairs A–C are rarely formed, but the G–U bond is quite strong and plays an important role in RNA secondary structure. There could be base pairs whose donor–acceptor side chains are orthogonal to each other, and noncanonical purine–pyrimidine pairs would be disfavored. Only two such pairs could exist, and it would further lessen mutation probabilities [85, 86]. Resistance to mutations is important, but not the sole determinant of a base suitability, and the prebiotic environment exerted its own selective force on them [88, 89].

Mutations occur mostly for chemical reasons. The current set of two base pairs are such that by tautomerization their hydrogen donor–acceptor characteristic change to mimic that of the other base of the same size. Consequently, the imino form of adenine pairs with cytosine and the imino form of cytosine with adenine. Similarly, the enol forms of guanine or uracil form base pairs with the other base. In these unfavorable states, noncanonical base pairs form, which then could be inherited. Tautomerization is the main mechanism by which base substitutions occur [90]. Spontaneous deamination is another chemical reason via which mutations can arise. Cytosine becomes uracil, adenine hypoxanthine, and guanine xanthine. These last two cannot be found in an RNA or DNA, and thus, error-correcting mechanisms might be able to detect them. However, uracil is natural in RNA (but not in DNA), and the ensuing transition goes undetected. While reactive oxygen species can cause oxidative deamination, which can lead to mutation, the oxygen level at the origin of life was very low, and this source of mutation was probably not important. Nonenzymatic copying of RNA has an error rate of at least 0.01 mutation per base per replication [29].

Mutation rates are often given as mutation per base per replication, but so far, there is no length in our formulation of the error threshold. The mutation rate μ is given as mutation per sequence per replication. The two quantities can be easily exchanged. Let us assume that the enzyme to be replicated has a length of L and the per base mutation rate is denoted by u. Then

$$(1 - \mu) = (1 - u)^L \tag{14.4}$$

We can transform Eq. (14.3) into the more frequently seen form:

$$L < \frac{\ln s}{u} \tag{14.5}$$

where $s = a_{\text{enzyme}}/a_{\text{parasite}}$. If we assume that $\ln s \approx 1$ and $u = 0.01$, then the length of the maintainable enzyme is smaller than 100 nucleotides. While there are ample examples of ribozymes with less than a hundred nucleotides, it is clearly not enough for the genome of a whole ribo-organism. Moreover, even the putative RNA-dependent RNA polymerase ribozymes are longer than 100 nucleotides, they are around 200 nucleotides long (Table 14.1). Longer enzymes could be more accurate, but accuracy is required for a longer enzyme in the first place. Thus, we arrive at the Eigen's paradox: "no large genome without enzymes, and no enzymes without a large genome" [61].

However, as discussed earlier, the formulation of the error threshold in Eq. (14.5) does not take the possibility of neutral mutations into account. A formula for the phenotypic error threshold based on the neutrality of some of the mutations was derived from first principles [84, 91]:

$$L < \frac{-\ln s}{\ln((1-u) + \lambda - (1-u)\lambda)} \tag{14.6}$$

The fraction of neutral mutations (λ) can be estimated by the analysis of RNA secondary structures. The secondary structure of an RNA is a good proxy for its structure [92], and it can be calculated easily [93–95]. Computationally, each position can be mutated and all of the $3L$ sequences differing by only one nucleotide analyzed. An average fraction of neutral mutation can then be obtained. The range of λ for a set of 305 ribozyme sequences is between 9% and 71% (median and mean are 30%) [96]. Alternatively, one can average over a sample of sequences folding into a target structure, as was done for the tRNA^{Phe} structure [97]. The mean λ was found to be 0.2871 ± 0.2489. While the average fraction of neutral mutant seems to be rather similar, there is considerable variation between and within sequences. Mutations in single-stranded regions of the structure change the structure less frequently, than mutations in double-stranded regions. Generalizing from the literature, we have proposed [98] that structural elements in a secondary structure can be classified into four types: neutral structure, connecting structure, forbidden structure and critical structure. Neutral structures can be freely changed, and in some cases, they can even be removed. Connection structures position the critical elements, and as long as the structure is intact, they can fulfill their role. These are the parts of the structure that give it its mutational robustness. Forbidden structures are not found in functional RNAs, as their presence abolishes the function. Critical structures harbor sites that are important for their exact chemical characteristic; often, they are the catalytic site or the substrate-binding sites of the ribozyme. These sites cannot be inferred from the secondary structure alone, only wet-lab experiments can tell us about their existence. Their presence overestimates the neutrality obtained from pure secondary structure studies.

A fitness landscape, a map from genotype to phenotype to fitness, was constructed based on mutagenesis data for the VS ribozyme and the hairpin ribozyme [83, 92]. We were able to show the true extent of the difference between the genotypic and the phenotypic error threshold. The maintainable sequence length was 6–7 times

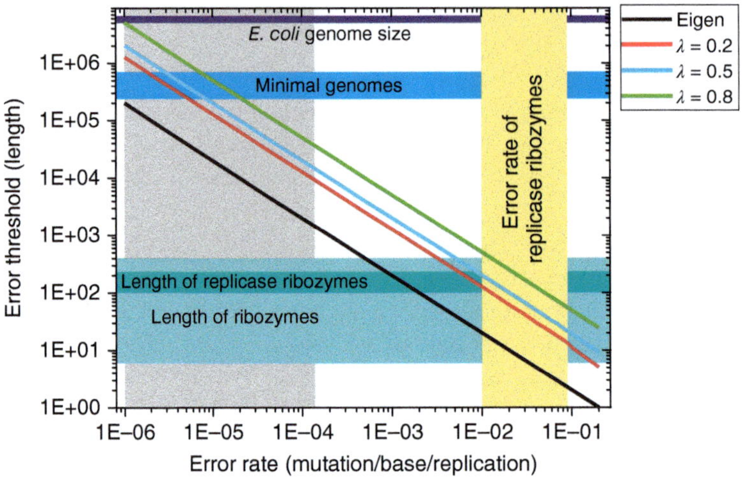

Figure 14.2 The error threshold for the original Eigen's formulation (Eq. (14.6)) (dark gray line) and for various fractions of neutral mutations (Eq. (3.3.7)), when $\ln s = 1$. Populations characterized by the parameter space below the lines are viable, while above it the sequence cannot be maintained. The vertical colored regions represent the error rate of viral RNA replicases (light gray) and the replicase ribozymes (yellow). The horizontal regions represent milestones in length, such as the replicase ribozymes, the minimal present-day genomes of bacteria and *Escherichia coli* (an example of a free-living organism).

as much as previously thought based on Eq. (14.5). This considerable increase in maintainable genome length allows us to replicate known ribozymes [96], but still, a magnitude lower error rate would be needed for the replication of the genome of a minimal ribo-organism, and it is a long shot from the genome size of contemporary living beings (Figure 14.2).

The mutation rate of the RNA-dependent RNA polymerase ribozyme should be quite low. While for self-replication the error rate of 0.1% per base per replication would be low enough, mutation rates for these ribozymes are higher (Table 14.1). Presently, the main problems with these ribozymes are processivity and generality. They can extend a primer by a very limited number of nucleotides, far less than their size. So far, there was little effort to lower the mutation rate of the replicase ribozymes. Theory tells us [99] that if a modest increase in size can increase the fidelity of the enzyme, then genome size and replication fidelity can gradually increase. So far, increase in length increased the fidelity of the replicase, albeit there are considerable variations and very few data points (Figure 14.3).

14.3 Compartmentalization

Splitting information into smaller pieces was seen as the solution for the error threshold by Eigen and Schuster [100]. Owing to their short length, fragments can be replicated by error-prone replicases, even though the replication of the whole genetic information in one piece (in one chromosome) is not feasible due to the

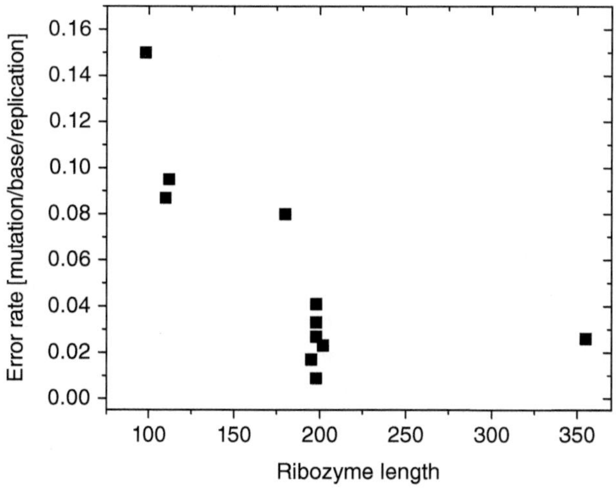

Figure 14.3 Error rate of the replicase ribozymes as function of their length. The 350 bases long ribozyme is a system with a 220 bases long ribozyme and a 135 bases long helper, see Table 14.1.

error threshold. Thus, there will be good copies of fragments, and consequently, no information is lost because of mutations. However, the independently replicating fragments or ribozymes are in competition with each other; hence, the strong replicative coupling in the hypercyclic organization was envisioned. In a hypercycle, each of the members can catalyze the replication of the next member. While short cycles can be stable, they can be destroyed by various parasites, and their evolvability is very limited [101]. So, while Eigen and Schuster have identified the problem, they were not able to give a satisfactory solution to it. There should be other mechanisms to allow the coexistence of independently replicating ribozymes.

A living cell is a prime example of molecular cooperation. Replicators toil for the greater good of the whole. Individually, they would all be better off just accepting the catalytic aid or benefits the other replicators give and not giving anything in return. That is the central problem of the evolution of cooperation: while greater benefits can be reaped if everyone cooperates compared with when no one does, the highest payoff is obtained by exploiting others. The rather bleak message is that rational actors should not cooperate hold when interacting entities meet randomly. However, if the cooperators meet with each other more frequently than with cheaters (parasites), then cooperation could be the evolutionarily favored outcome [102]. There should be some form of population structure or viscosity, so that the benefits of cooperation are reaped by the cooperators and not by the parasites. Encapsulation into a lipid vesicle ensures that the assistance of the molecular cooperators and the harm caused by parasites stay local.

14.3.1 Surface Metabolism and Transient Compartmentalization

Cellular encapsulation is the ultimate form of compartmentalization, but earlier stages could also exist. Mineral surfaces, for example, limit the diffusion of macromolecules, thus creating a viscous population. Small compartments in porous rocks at hydrothermal vents can house a rudimentary metabolism by providing chemical energy [103–106]. It can also offer the higher-level selection required for the coexistence of functional ribozymes with parasites [107]. Not only hard rocks but also ice can be a form of compartmentalization [43]. As RNA is quite labile, some reactions are better carried out in ice [108–110]. Irrespective of the exact nature of the surface, theory suggest that a diverse set of replicators can coexist on it [101, 111–114]. Replicators can locally enhance their own replication as well as the replication of other replicators. While parasites thus can gain enzymatic boost to their replication, localities where they proliferate become less and less conductive to growth. Thus, the parasites' own replication limits their spread. Moreover, the parasites that can actually coexist with the ribozymes are the ones that do not replicate much faster than the ribozymes [115]. The aggressive, fast parasites dominate their own neighborhood quickly, and then without any enzymes around, they die. However, a weaker parasite would allow the enzymes to grow, and thus, the environment remains such that they can still grow. While these parasites, at the moment, are a drain on the resources of the system, they can become useful by evolving an enzymatic function [116]. As there is no stabilizing selection on their function, as they have none, they can freely explore sequence space and might hit upon some useful function. Surfaces, however, cannot sustain an arbitrary diverse metabolism, as not many different types can coexist [117].

Transient compartmentalization would be the next step toward fully cellular life. In such a system, replicators are fully compartmentalized in some stages of their existence, but not all of it. For example, drying lipid vesicles transforms them into lamellar structures releasing their content. Upon wetting the system, the newly forming vesicles take up some material, e.g. nucleic acids from their environment [118]. The drying–wetting cycles are conductive to condensation reactions as during the drying phase, concentration can be relatively high, and the membranes also organize the compounds [119]. Thus, there is a prebiotically plausible way to have transient compartmentalization.

We have investigated the dynamics of a transiently compartmentalized system [68] with the help of in vitro compartmentalization and microfluidics [120, 121]. Tiny (12 pl) aqueous droplets were loaded with the Qβ replicase, NTP, a modified *Neurospora* VS ribozyme and a substrate. The VS ribozyme [23] is a self-cleaving ribozyme that can also be modified to be a trans-acting ribozyme [122]. The droplets can be selected based on the concentration of the cleaved substrate, which is a proxy for the number of functional ribozymes in the droplet. Selected droplets were collected, and their content pooled. New droplets were formed from this pool, and the RNAs inside the droplets were allowed to replicate again. The RNAs had only spent a portion of their life cycle compartmentalized; hence, it is a transient compartmentalization. Even such incomplete compartmentalization allows the ribozymes to coexist with the parasites [68]. Droplets in which parasites proliferate and achieve high concentration will have low product concentration and are not selected. Here, coexistence rests on the number of RNA encapsulated in the droplets at the beginning of the encapsulated phase. If there is one or a few droplets at the beginning, then the ribozyme can coexist with the parasites; otherwise, it cannot.

14.3.2 The Stochastic Corrector Model

Full compartmentalization means that cells stay intact and their content is passed on to the next generation. At the origin of cellular life, we cannot assume the cell to have full control over cell division. It was a stochastic process. Even if we would have a chromosome in the cell containing all the information, due to the stochastic nature of primordial cell division, information could be lost. Just before cell division, there would be two copies of the chromosome in the cell. Each independently would either get to one or the other daughter cell. Half of the time, each daughter cells will have one chromosome, but in the other half of the time, only one of the cells will have chromosomes, and the other will end up empty. Thus, information can be lost due to the stochastic nature of chromosome segregation. This is the assortment load. A way to avoid such a loss of information is to have more copies of the chromosome in the cell. The probability of having zero chromosome in a particular daughter cell after random division is $(1/2)^{\nu_{max}}$, where ν_{max} is the number of chromosomes in the parent cell before division. With increasing number of chromosomes, the probability of ending up with an empty cell is diminishing. However, because of the error threshold, chromosomes could not have evolved before a sufficiently accurate replicase evolved.

The first cell encapsulated independently replicating ribozymes. Let us assume that there was τ different types of ribozymes, each having a function indispensable for the cell. Let us further assume that cells divide when their internal concentration reaches some predefined value, measured by the number of RNAs in the cell (ν_{max}). At this point, the cell divides, and each of the ribozymes assorts randomly to daughter cells. If there are exactly two copies of each ribozyme in the cell before division, then the probability of them assorting evenly to the daughter cells is 0.5^{τ}; thus, the probability of ending up with two viable daughter cells is diminishing as the number of different types increases. When there are more than one type to be maintained, both daughter cells can end up unviable, as one can lack one of the essential genes while the other the opposite. Here again, more copies of the ribozymes (redundancy) help alleviate the assortment load. There is a sharp boundary between the redundancy allowing for a certain number of ribozyme types to coexist and not permitting them to coexist. We call this the second error threshold [123], as there is a critical redundancy below which information is lost and the population is not viable. We have found that at least 100 types can coexist [123], which, as discussed in the next section, would be enough for a minimal ribocell to function. However, two problems remain: (i) internal competition between the independently replicating ribozymes can still destroy the system, and (ii) parasites can still outcompete functional RNAs within a cell. Both problems are exacerbated when the cells are allowed to grow larger and have more RNAs.

Ribozymes are competing for the same resources: the available NTPs and the replicase ribozyme. As the limiting resources are the same, it already limits the number of RNAs that can coexist [124]. Moreover, if there are differences in their growth rates (e.g. differences in their affinities to the replicase), then, due to internal competition, the faster replicating ribozymes can dominate the population. The longer the RNAs can replicate, the lower the frequency of the slower growing ones will be at the time of cell division. This again can lead to loss of information. Similar problem is caused by the appearance of parasites. Parasites, by their quicker replication, will take resources and space from the ribozymes, thereby lowering the effective redundancy in the cell. Upon cell division, the number of ribozymes is less than ν_{max}, increasing the probability that one or more of the required ribozyme types will be lost from one or both of the daughter cells.

The stochastic nature of ribozyme assortment into daughter cells can alleviate some of the problems associated with greater copy number. Just by chance, even from an imbalanced ribozyme distribution, a daughter cell can end up with a favorable internal composition. For example, if parasites assort mostly to one of the daughter cells, then the other will be mostly free of them. Or if most of the faster replicating ribozymes will go to one of the cells, then the internal competition will be lessened in the other for a while. This mechanism was termed the stochastic correction of internal composition, and the model framework described here is the stochastic corrector model [125, 126] (Figure 14.4). It was shown that two types can easily coexist, even if their replication rates are different [125]. Furthermore, in an infinite population, an arbitrary number of types can coexist [127]. All possible internal compositions and all possible divisions are realized, and thus, selection can act on the rare but very beneficial, stochastically generated compositions. In a finite population, only a limited number of genes can coexist [128] at mutation

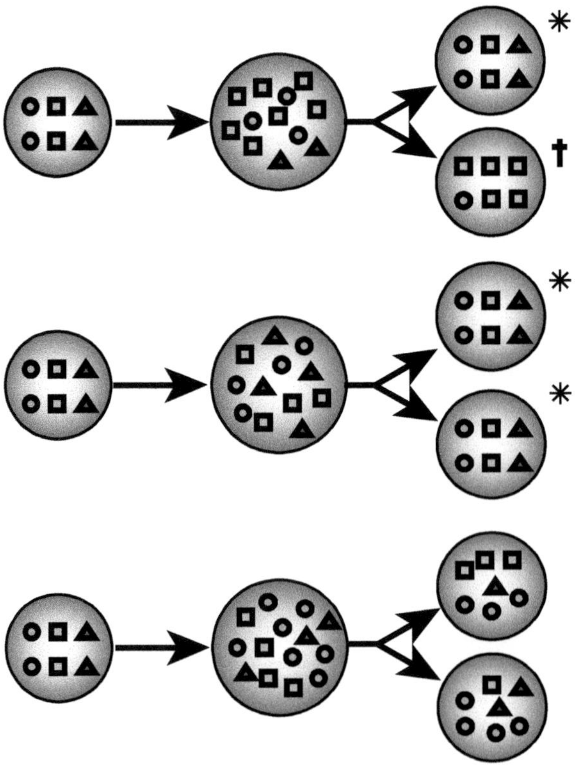

Figure 14.4 The stochastic corrector model. Three essential independently replicating ribozymes are in the ribocells depicted by a circle, a square, and a triangle. Initially, they all have the same concentration, and this uniform distribution is advantageous for the ribocell. The ribozymes multiply, and the internal distribution of the constituents can become uneven. Stochastic division can restore the beneficial distribution (marked with an asterisk); however, some daughter cells can end up with a missing gene, and it becomes unviable (marked with a cross).

rates compared with that of replicase ribozymes. Compositions can also become better by primordial sex, the exchange and mixing of genetic material between cells [129, 130]. While bad compositions can become better, good compositions can lose their good status by primordial sex. Sex can facilitate replicator coexistence to some extent, but it is not a universal remedy for the error catastrophes.

While we do not yet know how much information can stably coexist in a compartmentalized system, it seems that the path to life is a narrow one fraught with dangers. Primordial cells need to navigate between the mythical Scylla and Charybdis of the origin of life [131, 132]: on the one side, too little redundancy increases assortment load, and on the other, internal competition and parasites swamp the cells. In between, we need enough different genes to coexist so as the cells function and serve as the basis for further evolution. It is important to note again that once the fidelity of replication can increase, the information content of the cell can also increase, and gradually, there could be more and more complex systems [99]. Thus, the question arise, how many genes are needed for a minimal ribocell?

14.4 Minimal Gene Content of the First Ribocell

A minimal organism has as few genes as possible. Most research in this field focuses on DNA–peptide organisms, deriving the minimal gene set based on contemporary living bacteria [52, 133–135]. Present-day metabolic pathways might not be the same as ones in a ribocell, albeit some vestiges of the primordial metabolism are still with us [136]. Functionally, present-day metabolism and primordial metabolism need to fulfill the same roles.

Minimal gene sets found in contemporary organisms can be as low as 140 genes: *Tremblaya princeps* has 140 genes [137], *Nasuia deltocephalinicola* has 167 genes [138], *Hodgkinia cicadicola* has 189 genes [139], *Carsonella ruddii* has 213 genes [140], *Zinderia insecticola* has 231 genes [141], and *Sulcia muelleri* has 263 genes [141–145]. However, these symbionts of insects are barely alive in the sense that they lack genes for membrane and cell wall synthesis and lack transporters, most of carbon metabolism [146], and some even lack some genes for DNA replication and translation. Other symbionts and intracellular parasites have around 500–600 genes (*Mycoplasma genitalium*, *Buchnera* sp. [135]). The minimalized, synthetic *Mycoplasma mycoides* JCVI-syn3.0 genome consists of 473 genes [147].

Moya and coworkers [148] have compared eight bacterial genomes to establish the minimal common set of genes found in all of them. This gave an estimate of functional minimal set of genes required for a living cell. Their estimate includes 16 genes for the replication of the genetic material, 106 genes for translation, 15 genes for enzyme folding and modification, 5 genes for cellular processes, and 56 genes for energetic and intermediary metabolism, giving a grand total of 198 genes (the original estimate also included eight poorly characterized genes, which we omit here). Later, they suggested 50 enzymes to be able to fulfill all minimal functionality for the intermediary metabolism [149]. This intermediary metabolism produces energy from glucose via glycolysis; assembles nucleotide triphosphates and deoxynucleotide triphosphates from ribose, nucleobases, and phosphates; forms a lipid species; and produces the required coenzymes.

An RNA organism requires considerably less genes than the above organisms and estimates, as there is no translation. One cannot emphasize enough how many genes are required for translation, which contributes to the difficulty of understanding its evolution [150]. If we subtract the genes for translation and the genes for dNTP production from the minimal 198 genes [148], we arrive at 88 genes. The original set of genes included ones for peptide folding and salvage, and while peptides are not yet present in a ribocell, ribozymes might also require chaperons and salvage pathways. However, as discussed below, the number of genes for cell-level processes is probably underestimated. Compared with comparison-based estimates, the minimalized *M. mycoides* JCVI-syn3.0 includes considerable number of genes for regulation (9), cell division (1), and transport (31) [147]. This is also true for an inferred minimal *Bacillus subtilis* genome [151]. Consequently, the minimal gene content of a ribo-organism might be around 100 genes.

14.4.1 Intermediate Metabolism

The most important enzymatic function of a ribocell is the replication of the genetic material. It requires at least one enzyme, the RNA-dependent RNA polymerase (see Section 14.1.1). However, this one enzyme might not be enough for the replication of RNA. Most of the known replicases require a primer; the synthesis of which replicase and the initiation of the replication process are thus complicated [152]. While there is an example of RNA replicase that requires no primer [47], that system consists of two RNAs, the enzyme and a coenzyme. Thus, even the most elementary process of a cell, the replication of the genetic material, requires two or more functional RNAs.

It has been demonstrated that replicase ribozymes are able to replicate aptamers [46, 47], transfer RNA (tRNA) [46], or the hammerhead ribozyme [44]. Thus, the replication of functional RNA is within the capabilities of the selected replicases. However, none can – at the moment – replicate itself. The problem lies with their processivity, i.e. the number of nucleotides they are able to add to the growing strand (see Table 14.1). In order to achieve full self-replication, the replicase had to be in many pieces, each of them replicated independently, and the functional ribozyme self-assembles, or it is being ligated together by ligase ribozymes [153]. Replicases can self-assemble from their parts, but the successful copying of their parts as well as their complementary sequences have not yet been demonstrated. The replicase, which operates by the addition of triplets [47], can copy some fragments of itself and its complementary strand, but full self-replication is not yet achieved.

Still, there is an often overlooked enzymatic activity that is required for the core function of RNA replication [152, 154]: the unwinding of the resulting double-stranded RNA. Double-stranded RNA is inert in the sense that the ribozyme strand cannot fulfill its catalytic role, nor can the template strand be replicated. Something has to unzip the two strands. Mostly, thermal cycling/gradient [155, 156] or some other oscillatory process [157] is assumed to take care of this conundrum. While short RNAs can be reliably separated in this manner, longer double-stranded

RNAs are still bound too strongly, and the environment conductive for separation is also damaging to RNA strands. Some helicase function is needed in the ribocell. It was suggested that the ancestor of the small subunit of the ribosome had such a function [158]. Most probably, it also adds at least one functional RNA to our list of minimal functions.

Apart from the template-based polymerization of RNA and associated functions, the ribocell needs a constant supply of activated nucleotides (NTPs) [159]. While nucleosides can form from formamide [160–162] and other plausible prebiotic synthesis have also been proposed [163, 164] and they can even be activated to some extent [165], supplies will run out quickly once ribocells begin to consume them. There should be ribozymes that contribute to the formation of activated nucleotides. Activated ribose and a nucleobase can be condensed by a ribozyme [166, 167] to yield nucleosides. Phosphorylation of a single nucleoside has not yet been demonstrated, but 5′-OH of RNA [168] or 3′-OH of a DNA [169] can be triphosphorylated by a ribozyme. Generally, nucleic acid oligomers can be phosphorylated by ribozymes [170–174]. In these ribozymes, the substrate oligomer is bound to the enzyme via base pairing. That is the simplest way of substrate binding, and it also makes the interaction specific. We know that a wide variety of RNA aptamers can bind nucleoside triphosphates [175–179] and catalyze phosphorylation; thus, a ribozyme catalyzing the phosphorylation of nucleosides is conceivable. With that, the replication of the genetic information would be no longer dependent on the exogenous supply of activated nucleotides. However, there should still be a supply of ribose and nucleobases. For both fatty acid and phospholipid membranes, ribose has the best permeability coefficient among aldopentoses and hexoses, and consequently, it can accumulate inside a ribocell [180]. The formose reaction [181] under the right conditions [182] can supply ribose. As for the nucleobases, they were probably supplied by the environment for a long period in the RNA world. Modern biosynthesis of nucleobases are complicated and require a substantial number of enzymes, which suggest that a minimal ribocell did not had the ability to synthesize nucleobases de novo. Nucleotides can spontaneously diffuse through membranes composed of fatty acids [183, 184] and can also diffuse through membranes composed of phospholipids with 12–14 long carbon chains [185]. Consequently, ribocells could rely on exogenous nucleobase supply.

A replicase, a helicase, and some enzyme to produce activated nucleotides are still a long shot from even a minimal metabolism. The minimal intermediate metabolism proposed [149] – apart from the assembly of nucleotides – is able to harness energy, synthetize the membrane constituent, and produce the cofactors (Figure 14.5). Energy (ATP) can be generated via glycolysis. For the investment of two ATPs, the cell gains four ATP from a molecule of glycose. Other sugars can be channeled to this pathway either via the pentose phosphate pathway or by specific isomerases able to convert sugars into sugars in that pathway. The pentose phosphate pathway can then also be employed to produce ribose from other sugars. Whether all of these functions can be catalyzed by ribozymes is still an open empirical question, but as the repertoire of ribozymes seems to be quite diverse [3, 187], we can assume that it is the case.

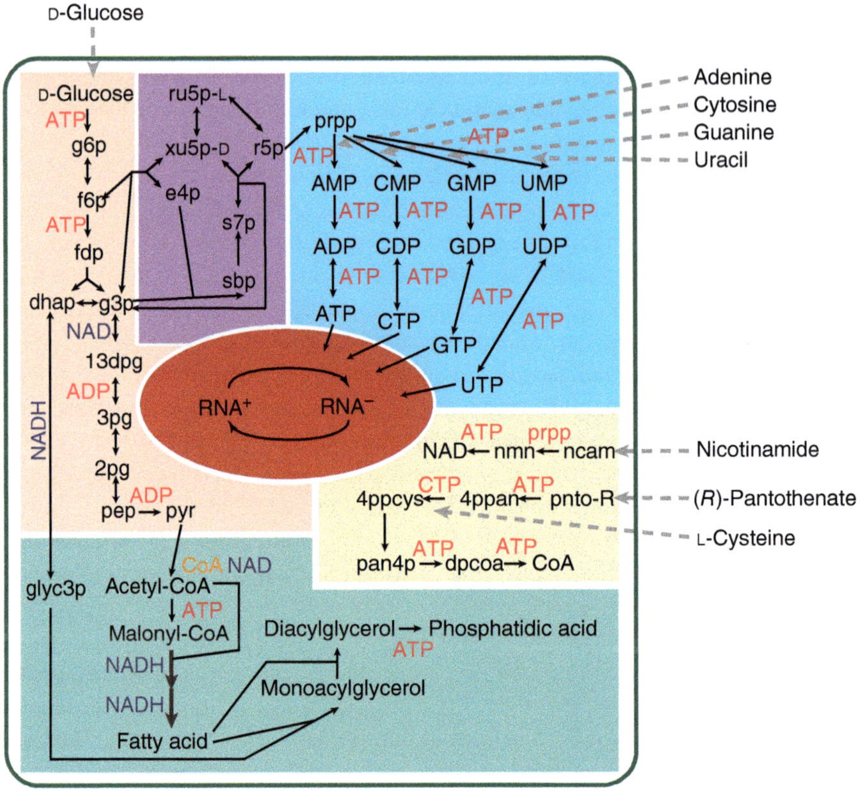

Figure 14.5 A hypothetical minimal metabolism for a ribocell. The replication of the RNA is provided with nucleotides and energy (where ADP is consumed, ATP is produced); apart from CoA and NAD synthesis, a sketch of phospholipid biosynthesis is also shown. Organic compounds to be taken up are depicted outside of the cell membrane. Abbreviations are from the BiGG database [186]. 13dpg, 3-phospho-D-glyceroyl phosphate; 2pg, D-glycerate 2-phosphate; 3pg, 3-phospho-D-glycerate; 4ppan, D-4′-phosphopantothenate; 4ppcys, *N*-((R)-4-phosphopantothenoyl)-L-cysteine; dhap, dihydroxyacetone phosphate; dpcoa, dephospho-CoA; e4p, D-erythrose 4-phosphate; f6p, D-fructose 6-phosphate; fdp, D-fructose 1,6-bisphosphate; g3p, glyceraldehyde 3-phosphate; g6p, D-glucose 6-phosphate; glyc3p, glycerol 3-phosphate; ncam, nicotinamide; nmn, β-nicotinamide D-ribonucleotide; pan4p, pantetheine 4′-phosphate; pep, phosphoenolpyruvate; pnto-*R*, (*R*)-pantothenate; prpp, 5-phospho-α-D-ribose 1-diphosphate; pyr, pyruvate; r5p, α-D-ribose 5-phosphate; ru5p-L, L-ribulose 5-phosphate; s7p, sedoheptulose 7-phosphate; sbp, sedoheptulose-bisphosphatase; xu5p-D: D-xylulose 5-phosphate.

14.4.2 Cell-Level Processes

Too often, when contemplating metabolism, cell-level processes are forgotten. We know more and more about the metabolic networks of organisms [188], and the core of this network is remarkably similar. But, transport, regulation of gene expression, control of cell division, etc. are very diverse.

For example, we have already assumed in the above minimal intermediate metabolism that the ribocell can take up glycose or other sugar sources, nucleobases, the precursors of cofactors, some amino acids, and inorganic materials (like phosphate). This requires some transporters as the membrane should not and cannot be fully permeable to these molecules. The lipid bilayer surrounding the cell not only provides the encapsulation needed for group selection to work but keeps the valuable materials inside the cell. Ions, for example, phosphorylated compounds, can hardly cross the cell membrane [189, 190], so they mostly stay inside or cannot enter the cell. Thus, the same mechanism protecting the cell from losing synthetized compounds also hinders the uptake of materials. Some kind of transport that controls the in- and outflow of material is required. RNA can change the permeability of the membrane [191], and ribozymes can even act as membrane transporters [192] allowing control over the exchange of material with the environment. Whether RNA and the kind of membrane produced/formed by the ribocell can modulate permeability to the extent required is an empirical question. We have argued [150] that the original role for polypeptides could have been the formation of pores and channels. These polypeptides – if they existed – were not translated, just polymerized. Thus, while the full apparatus of translation was not required, a ribozyme – much like the ribosome – capable of amino acid polymerization is needed. In this scenario, the transport or permeability modifying RNA is replaced by a polymerase, so the required number of genes is roughly the same.

One of the absolute life criteria is that processes are regulated and controlled. Genes for regulation and cell-level processes are quite rare when homologous genes in multiple organisms are concerned. These genes are very much environment dependent and do not conserve well. However, they are still extremely important for the functioning of the cell. Their numbers are underestimated in minimal gene content estimates. On the other hand, present-day functional RNAs are directly connected either with translation or with the regulation of gene expression [9–11]; thus, the regulatory role of RNA is well preserved. But it is not so easy to pinpoint what kind of regulation does a rudimentary ribo-organism need.

Enzymatic activity can be controlled in two ways: either the enzyme itself responds to a signal, switching on and off, or the transcription – copying – of the enzyme from the template is affected by a signal. Both can be realized by RNAs. Present-day functional RNAs, like small interfering RNAs and microRNAs, mostly act post-transcriptionally [193], inhibiting translation from messenger RNAs (mRNAs). In an RNA world, such element would affect the product of transcription, the ribozymes themselves. And as such, the ribozymes then would fall into the class of allosterically controllable ribozymes, also referred to as aptazymes. Rationally designed and in vitro evolved aptazymes can respond to temperature change, light, small molecules, or oligonucleotides [194–198]. A smaller set of RNAs can also directly control translation, inhibiting or facilitating the production of enzymes from the chromosome. Furthermore, small molecules can modulate regulatory regions in the chromosome, as in the lac operon. Such regulatory mechanism could have existed in the RNA world as well. As for our estimate of minimal gene content, the information to be stored is longer with the incorporation of regulatory regions of genes or the effector-recognition domains of allosterically controlled ribozymes. Especially for the shorter ribozymes, some increase in length might be possible if it conveys a fitness advantage. We should not forget that the error threshold still looms large on the horizon.

The evolution of chromosome solves the problem stemming from the random distribution of independently replicating genes to daughter cells. At the same time, it also allows very specialized enzymes to evolve [199]. But it also requires an array of enzymes to function [200], as ribozymes need to be transcribed from a large molecule containing all ribozymes ligated together. While cleavage seems to be an easily evolvable function, it needs to be site specific across the chromosome, so it only cleaves between ribozymes, but not within them. This is yet another cell-level function a ribo-organism requires.

A proposed function set will pose a challenge to empirical ribozyme research: ribozymes should be able to catalyze the proposed reactions, and these ribozymes need to be able to work together in one compartment. Not only the maintenance of large set of independent replicators but also working together biochemically is a challenge. Ribozymes are usually metallozymes [201, 202], and altering the prevalent metal ion concentration can change their enzymatic activity [203]. Ribozymes are evolved by themselves, and there is scarce indication that multiple ribozymes can function at the same time. We know that two [204, 205] or three [206] engineered ribozymes can work in concert. The challenge is to have 60–100 ribozymes to work together in a cell. As we are very far from realizing this number, a more accurate estimate would be superfluous.

We are a long way from solving the mystery of the first cell, but more and more of the puzzle pieces are known. The problems, both dynamical and structural, have been identified, and for some, solutions proposed. Here, we have reviewed some of the dynamical problems the first cell needed to overcome via having the right set of ribozymes cooperating with each other.

Acknowledgments

The research was funded by the National Research, Development and Innovation Office (NKFIH) under Grant numbers K119347 and GINOP-2.3.2-15-2016-00057 and by the Volkswagen Stiftung initiative "Leben? – Ein neuer Blick der Naturwissenschaften auf die grundlegenden Prinzipien des Lebens" under project "A unified model of recombination in life." This work was carried out as part of EU COST action CM1304 "Emergence and Evolution of Complex Chemical Systems."

References

1 Yarus, M. (2011). *Life from an RNA World: The Ancestor Within*. Harvard, USA: Harvard University Press.

2 Kun, Á., Szilágyi, A., Könnyű, B. et al. (2015). The dynamics of the RNA world: insights and challenges. *Ann. N.Y. Acad. Sci.* 1341: 75–95.

3 Joyce, G.F. (2002). The antiquity of RNA-based evolution. *Nature* 418 (6894): 214–220.
4 Gánti, T. (2003). *The Principles of Life*. Oxford: Oxford University Press.
5 Gánti, T. (1971). *Az Élet Principiuma*. Budapest: Gondolat.
6 Szathmáry, E. (2015). Toward major evolutionary transitions theory 2.0. *Proc. Natl. Acad. Sci. U.S.A.* 112 (33): 10104–10111.
7 Maynard Smith, J. and Szathmáry, E. (1995). *The Major Transition in Evolution*. Oxford, UK: W.H. Freeman.
8 Szathmáry, E. and Maynard Smith, J. (1995). The major evolutionary transitions. *Nature* 374: 227–232.
9 Meli, M., Albert-Fournier, B., and Maurel, M.C. (2001). Recent findings in the modern RNA world. *Int. Microbiol.* 4 (1): 5–11.
10 Spirin, A.S. (2002). Omnipotent RNA. *FEBS Lett.* 530 (1–3): 4–8.
11 Collins, L.J., Kurland, C.G., Biggs, P., and Penny, D. (2009). The modern RNP world of eukaryotes. *J. Hered.* 100 (5): 597–604.
12 Huang, B. and Zhang, R. (2014). Regulatory non-coding RNAs: revolutionizing the RNA world. *Mol. Biol. Rep.* 41 (6): 3915–3923.
13 Patil, V.S., Zhou, R., and Rana, T.M. (2014). Gene regulation by non-coding RNAs. *Crit. Rev. Biochem. Mol. Biol.* 49 (1): 16–32.
14 Ghildiyal, M. and Zamore, P.D. (2009). Small silencing RNAs: an expanding universe. *Nat. Rev. Genet.* 10 (2): 94–108.
15 Moore, P.B. and Steitz, T.A. (2002). The involvement of RNA in ribosome function. *Nature* 418 (6894): 229–235.
16 Nissen, P., Hansen, J., Ban, N. et al. (2000). The structural basis of ribosome activity in peptide bond synthesis. *Science* 289 (5481): 920–930.
17 Guerrier-Takada, C., Gardiner, K., Marsh, T. et al. (1983). The RNA moiety of ribonuclease P is the catalytic subunit of the enzyme. *Cell* 35 (3): 849–857.
18 Kruger, K., Grabowski, P., Zaug, A.J. et al. (1982). Self-splicing RNA: autoexcision and autocyclization of the ribosomal RNA intervening sequence of *Tetrahymena*. *Cell* 31 (1): 147–157.
19 Peebles, C.L., Perlman, P.S., Mecklenburg, K.L. et al. (1986). A self-splicing RNA excises an intron lariat. *Cell* 44 (2): 213–223.
20 Forster, A.C. and Symons, R.H. (1987). Self-cleavage of plus and minus RNAs of a virusoid and a structural model for the active site. *Cell* 49 (2): 211–220.
21 Hampel, A. and Tritz, R.R. (1989). RNA catalytic properties of the minimum (−)sTRSV sequences. *Biochemistry* 28 (12): 4929–4933.
22 Sharmeen, L., Kuo, M.Y.P., Dinner-Gottlieb, G., and Taylor, J. (1988). Antigenomic RNA of human hepatitis delta viruses can undergo self-cleavage. *J. Virol.* 62 (8): 2674–2679.
23 Saville, B.J. and Collins, R.A. (1990). A site-specific self-cleavage reaction performed by a novel RNA in *Neurospora* mitochondria. *Cell* 61 (4): 685–696.
24 Winkler, W.C., Nahvi, A., Roth, A. et al. (2004). Control of gene expression by a natural metabolite-responsive ribozyme. *Nature* 428 (6980): 281–286.
25 Roth, A., Weinberg, Z., Chen, A.G.Y. et al. (2014). A widespread self-cleaving ribozyme class is revealed by bioinformatics. *Nat. Chem. Biol.* 10 (1): 56–60.

26 Bag, B.G. and von Kiedrowski, G. (1996). Templates, autocatalysis and molecular replication. *Pure Appl. Chem.* 68 (11): 2145.

27 Szathmáry, E. and Gladkih, I. (1989). Sub-exponential growth and coexistence of non-enzymatically replicating templates. *J. Theor. Biol.* 138 (1): 55–58.

28 Zachar, I., Kun, Á., Fernando, C., and Szathmáry, E. (2013). Replicators: from molecules to organisms. In: *Handbook of Collective Robotics: Fundamentals and Challenges* (ed. S. Kernbach), 473–501. Pan Stanford Publishing.

29 Orgel, L.E. (1992). Molecular replication. *Nature* 358 (6383): 203–209.

30 Bissette, A.J. and Fletcher, S.P. (2013). Mechanisms of autocatalysis. *Angew. Chem. Int. Ed.* 52 (49): 12800–12826.

31 Kassianidis, E. and Philp, D. (2006). Design and implementation of a highly selective minimal self-replicating system. *Angew. Chem. Int. Ed.* 45 (38): 6344–6348.

32 Patzke, V. and Von Kiedrowski, G. (2007). Self replicating systems. *ARKIVOC*: 293–310.

33 von Kiedrowski, G. (1986). A self-replicating hexadeoxynucleotide. *Angew. Chem. Int. Ed.* 25 (10): 932–935.

34 Segré, D., Ben-Eli, D., and Lancet, D. (2000). Compositional genomes: prebiotic information transfer in mutually catalytic noncovalent assemblies. *Proc. Natl. Acad. Sci. U.S.A.* 97 (8): 4112–4117.

35 Vasas, V., Szathmáry, E., and Santos, M. (2010). Lack of evolvability in self-sustaining autocatalytic networks constraints metabolism-first scenarios for the origin of life. *Proc. Natl. Acad. Sci. U.S.A.* 107 (4): 1470–1475.

36 Huang, W. and Ferris, J.P. (2003). Synthesis of 35–40 mers of RNA oligomers from unblocked monomers. A simple approach to the RNA world. *Chem. Commun.*: 1458–1459.

37 Ferris, J.P. (2006). Montmorillonite-catalysed formation of RNA oligomers: the possible role of catalysis in the origins of life. *Philos. Trans. R. Soc. London, Ser. B* 361 (1474): 1777–1786.

38 Szostak, J. (2012). The eightfold path to non-enzymatic RNA replication. *J. Syst. Chem.* 3 (1): 2.

39 Ekland, E.H. and Bartel, D.P. (1996). RNA-catalysed RNA polymerization using nucleoside triphosphates. *Nature* 382: 373–376.

40 Johnston, W.K., Unrau, P.J., Lawrence, M.S. et al. (2001). RNA-catalyzed RNA polymerization: accurate and general RNA-templated primer extension. *Science* 292 (5520): 1319–1325.

41 Zaher, H.S. and Unrau, P.J. (2007). Selection of an improved RNA polymerase ribozyme with superior extension and fidelity. *RNA* 13 (7): 1017–1026.

42 Wang, Q.S., Cheng, L.K.L., and Unrau, P.J. (2011). Characterization of the B6.61 polymerase ribozyme accessory domain. *RNA* 17 (3): 469–477.

43 Attwater, J., Wochner, A., Pinheiro, V.B. et al. (2010). Ice as a protocellular medium for RNA replication. *Nat. Commun.* 1: 76.

44 Wochner, A., Attwater, J., Coulson, A., and Holliger, P. (2011). Ribozyme-catalyzed transcription of an active ribozyme. *Science* 332 (6026): 209–212.

45 Attwater, J., Wochner, A., and Holliger, P. (2013). In-ice evolution of RNA polymerase ribozyme activity. *Nat. Chem.* 5: 1011–1018.

46 Horning, D.P. and Joyce, G.F. (2016). Amplification of RNA by an RNA polymerase ribozyme. *Proc. Natl. Acad. Sci. U.S.A.* 113 (35): 9786–9791.

47 Attwater, J., Raguram, A., Morgunov, A.S. et al. (2018). Ribozyme-catalysed RNA synthesis using triplet building blocks. *eLife* 7: e35255.

48 Szathmáry, E. (1999). Chemes, genes, memes: a revised classification of replicators. *Lect. Math. Life Sci.* 26: 1–10.

49 Kun, Á., Papp, B., and Szathmáry, E. (2008). Computational identification of obligatorily autocatalytic replicators embedded in metabolic networks. *Genome Biol.* 9: R51.

50 Gánti, T. (2003). *Chemoton Theory*. New York: Kluwer Academic/Plenum Publishers.

51 Szathmáry, E. (2006). The origin of replicators and reproducers. *Philos. Trans. R. Soc. London, Ser. B* 361 (1474): 1761–1776.

52 Szathmáry, E. (2005). Life: in search of the simplest cell. *Nature* 433: 469–470.

53 Szathmáry, E., Santos, M., and Fernando, C. (2005). Evolutionary potential and requirements for minimal protocells. *Top. Curr. Chem.* 259: 167–211.

54 Wächtershäuser, G. (1998). Origins of life in an iron–sulfur world. In: *The Molecular Origins of Life* (ed. A. Brack), 207–218. Cambridge: Cambridge University Press.

55 Ferris, J.P. (2002). Montmorillonite catalysis of 30–50 mers oligonucleotides: laboratory demonstration of potential steps in the origin of the RNA world. *Origins Life Evol. Biosphere* 32 (4): 311–332.

56 Joshi, P.C., Aldersley, M.F., and Ferris, J.P. (2011). Homochiral selectivity in RNA synthesis: montmorillonite-catalyzed quaternary reactions of D,L-purine with D,L-pyrimidine nucleotides. *Origins Life Evol. Biosphere* 41 (3): 213–236.

57 Hazen, R.M., Filley, T.R., and Goodfriend, G.A. (2001). Selective adsorption of L- and D-amino acids on calcite: implications for biochemical homochirality. *Proc. Natl. Acad. Sci. U.S.A.* 98 (10): 5487–5490.

58 Biondi, E., Branciamore, S., Maurel, M.-C., and Gallori, E. (2007). Montmorillonite protection of an UV-irradiated hairpin ribozyme: evolution of the RNA world in a mineral environment. *BMC Evol. Biol.* 7 (suppl. 2): S2.

59 Poole, A.M. and Logan, D.T. (2005). Modern mRNA proofreading and repair: clues that the last universal common ancestor possessed an RNA genome? *Mol. Biol. Evol.* 22 (6): 1444–1455.

60 Eigen, M. (1971). Selforganization of matter and the evolution of biological macromolecules. *Naturwissenschaften* 10: 465–523.

61 Maynard Smith, J. (1983). Models of evolution. *Proc. R. Soc. London, Ser. B* 219 (1216): 315–325.

62 Levinson, G. and Gutman, G.A. (1987). Slipped-strand mispairing: a major mechanism for DNA sequence evolution. *Mol. Biol. Evol.* 4 (3): 203–221.

63 Gregory, T.R. (2004). Insertion–deletion biases and the evolution of genome size. *Gene* 324: 15–34.

64 Zhang, W., Sun, X., Yuan, H. et al. (2008). The pattern of insertion/deletion polymorphism in *Arabidopsis thaliana*. *Mol. Genet. Genomics* 280 (4): 351–361.

65 Bhangale, T.R., Rieder, M.J., Livingston, R.J., and Nickerson, D.A. (2005). Comprehensive identification and characterization of diallelic insertion–deletion polymorphisms in 330 human candidate genes. *Hum. Mol. Genet.* 14 (1): 59–69.

66 Boschiero, C., Gheyas, A.A., Ralph, H.K. et al. (2015). Detection and characterization of small insertion and deletion genetic variants in modern layer chicken genomes. *BMC Genomics* 16 (1): 562.

67 Mills, D.R., Peterson, R.E., and Spiegelman, S. (1967). An extracellular Darwinian experiment with a self-duplicating nucleic acid molecule. *Proc. Natl. Acad. Sci. U.S.A.* 58: 217–224.

68 Matsumura, S., Kun, Á., Ryckelynck, M. et al. (2016). Transient compartmentalization of RNA replicators prevents extinction due to parasites. *Science* 354 (6317): 1293–1296.

69 Tromas, N. and Elena, S.F. (2010). The rate and spectrum of spontaneous mutations in a plant RNA virus. *Genetics* 185 (3): 983–989.

70 Campagnola, G., McDonald, S., Beaucourt, S. et al. (2015). Structure-function relationships underlying the replication fidelity of viral RNA-dependent RNA polymerases. *J. Virol.* 89 (1): 275–286.

71 Huang, J., Brieba, L.G., and Sousa, R. (2000). Misincorporation by wild-type and mutant T7 RNA polymerases: identification of interactions that reduce misincorporation rates by stabilizing the catalytically incompetent open conformation. *Biochemistry* 39 (38): 11571–11580.

72 Drake, J.W. (1993). Rates of spontaneous mutation among RNA viruses. *Proc. Natl. Acad. Sci. U.S.A.* 90: 4171–4175.

73 Sanjuán, R., Nebot, M.R., Chirico, N. et al. (2010). Viral mutation rates. *J. Virol.* 84 (19): 9733–9748.

74 Eyre-Walker, A. and Keightley, P.D. (2007). The distribution of fitness effects of new mutations. *Nat. Rev. Genet.* 8 (8): 610–618.

75 Lundin, E., Tang, P.-C., Guy, L. et al. (2018). Experimental determination and prediction of the fitness effects of random point mutations in the biosynthetic enzyme HisA. *Mol. Biol. Evol.* 35 (3): 704–718.

76 Jacquier, H., Birgy, A., Le Nagard, H. et al. (2013). Capturing the mutational landscape of the beta-lactamase TEM-1. *Proc. Natl. Acad. Sci. U.S.A.* 110 (32): 13067–13072.

77 Lind, P.A., Berg, O.G., and Andersson, D.I. (2010). Mutational robustness of ribosomal protein genes. *Science* 330 (6005): 825–827.

78 Starita, L.M., Young, D.L., Islam, M. et al. (2015). Massively parallel functional analysis of BRCA1 RING domain variants. *Genetics* 200 (2): 413–422.

79 Huynen, M.A., Stadler, P.F., and Fontana, W. (1996). Smoothness within ruggedness: the role of neutrality in adaptation. *Proc. Natl. Acad. Sci. U.S.A.* 93 (1): 397–401.

80 Huynen, M.A. (1996). Exploring phenotype space through neutral evolution. *J. Mol. Evol.* 43: 165–169.

81 van Nimwegen, E., Crutchfield, J.P., and Huynen, M.A. (1999). Neutral evolution of mutational robustness. *Proc. Natl. Acad. Sci. U.S.A.* 96 (17): 9716–9720.

82 Haslinger, C. and Stadler, P.F. (1999). RNA structure with pseudo-knots: graph-theoretical and combinatorial properties. *Bull. Math. Biol.* 61 (3): 437–467.

83 Kun, Á., Santos, M., and Szathmáry, E. (2005). Real ribozymes suggest a relaxed error threshold. *Nat. Genet.* 37 (9): 1008–1011.

84 Takeuchi, N., Poorthuis, P.H., and Hogeweg, P. (2005). Phenotypic error threshold; additivity and epistasis in RNA evolution. *BMC Evol. Biol.* 5 (1): 9.

85 Szathmáry, E. (1991). Four letters in the genetic alphabet: a frozen evolutionary optimum? *Proc. R. Soc. London, Ser. B* 245: 91–99.

86 Szathmáry, E. (1992). What is the optimum size for the genetic alphabet? *Proc. Natl. Acad. Sci. U.S.A.* 89: 2614–2618.

87 Hershberg, R. and Petrov, D.A. (2010). Evidence that mutation is universally biased towards AT in bacteria. *PLoS Genet.* 6 (9): e1001115.

88 Krishnamurthy, R. (2015). On the emergence of RNA. *Isr. J. Chem.* 55 (8): 837–850.

89 Eschenmoser, A. (1999). Chemical etiology of nucleic acid structure. *Science* 284 (5423): 2118–2124.

90 Fu, L.-Y., Wang, G.-Z., Ma, B.-G., and Zhang, H.-Y. (2011). Exploring the common molecular basis for the universal DNA mutation bias: revival of Löwdin mutation model. *Biochem. Biophys. Res. Commun.* 409 (3): 367–371.

91 Schuster, P. and Stadler, P.F. (1999). Nature and evolution of early replicons. In: *Origin and Evolution of Viruses* (eds. E. Domingo, R.G. Webster and J. Holland), 1–24. New York: Academic Press.

92 Kun, Á., Maurel, M.-C., Santos, M., and Szathmáry, E. (2005). Fitness landscapes, error thresholds, and cofactors in aptamer evolution. In: *The Aptamer Handbook* (ed. S. Klussmann), 54–92. Weinheim: WILEY-VCH Verlag GmbH & Co. KGaA.

93 Hofacker, I.L. (2003). Vienna RNA secondary structure server. *Nucleic Acids Res.* 31: 3429–3431.

94 Hofacker, I.L., Fontana, W., Stadler, P.F. et al. (1994). Fast folding and comparison of RNA secondary structures. *Monatsh. Chem.* 125: 167–188.

95 Lorenz, R., Bernhart, S.H., Höner zu Siederdissen, C. et al. (2011). ViennaRNA Package 2.0. *Algorithms Mol. Biol.* 6 (1): 26.

96 Szilágyi, A., Kun, Á., and Szathmáry, E. (2014). Local neutral networks help maintain inaccurately replicating ribozymes. *PLoS One* 9 (10): e109987.

97 Reidys, C., Forst, C.V., and Schuster, P. (2001). Replication and mutation on neutral networks. *Bull. Math. Biol.* 63: 57–94.

98 Kun, Á. and Szathmáry, E. (2015). Fitness landscapes of functional RNAs. *Life* 5 (3): 1497–1517.

99 Scheuring, I. (2000). Avoiding Catch-22 of early evolution by stepwise increase in copying fidelity. *Selection* 1: 13–23.

100 Eigen, M. and Schuster, P. (1979). *The Hypercycle: A Principle of Natural Self-Organization.* Berlin: Springer-Verlag.

101 Szilágyi, A., Zachar, I., Scheuring, I. et al. (2017). Ecology and evolution in the RNA world: dynamics and stability of prebiotic replicator systems. *Life* 7 (4): 48.
102 Nowak, M.A. (2006). Five rules for the evolution of cooperation. *Science* 314 (5805): 1560–1563.
103 Koonin, E.V. and Martin, W. (2005). On the origin of genomes and cells within inorganic compartments. *Trends Genet.* 21 (12): 647–654.
104 Martin, W. and Russell, M.J. (2007). On the origin of biochemistry at an alkaline hydrothermal vent. *Philos. Trans. R. Soc. London, Ser. B* 362 (1486): 1887–1926.
105 Sleep, N.H., Bird, D.K., and Pope, E.C. (2011). Serpentinite and the dawn of life. *Philos. Trans. R. Soc. London, Ser. B* 366 (1580): 2857–2869.
106 Russell, M.J., Hall, A.J., and Martin, W. (2010). Serpentinization as a source of energy at the origin of life. *Geobiology* 8 (5): 355–371.
107 Branciamore, S., Gallori, E., Szathmáry, E., and Czárán, T. (2009). The origin of life: chemical evolution of a metabolic system in a mineral honeycomb? *J. Mol. Evol.* 69 (5): 458–469.
108 Vlassov, A.V., Johnston, B.H., Landweber, L.F., and Kazakov, S.A. (2004). Ligation activity of fragmented ribozymes in frozen solution: implications for the RNA world. *Nucleic Acids Res.* 32 (9): 2966–2974.
109 Trinks, H., Schröder, W., and Biebricher, C.K. (2005). Ice and the origin of life. *Origins Life Evol. Biosphere* 35 (5): 429–445.
110 Kanavarioti, A., Monnard, P.-A., and Deamer, D.W. (2001). Eutectic phases in ice facilitate nonenzymatic nucleic acid synthesis. *Astrobiology* 1 (3): 271–281.
111 Czárán, T. and Szathmáry, E. (2000). Coexistence of replicators in prebiotic evolution. In: *The Geometry of Ecological Interactions* (eds. U. Dieckmann, R. Law and J.A.J. Metz), 116–134. Cambridge: Cambridge University Press.
112 Czárán, T., Könnyű, B., and Szathmáry, E. (2015). Metabolically coupled replicator systems: overview of an RNA-world model concept of prebiotic evolution on mineral surfaces. *J. Theor. Biol.* 381: 39–54.
113 Takeuchi, N. and Hogeweg, P. (2009). Multilevel selection in models of prebiotic evolution II: a direct comparison of compartmentalization and spatial self-organization. *PLoS Comput. Biol.* 5 (10): e1000542.
114 Hogeweg, P. and Takeuchi, N. (2003). Multilevel selection in models of prebiotic evolution: compartments and spatial self-organization. *Origins Life Evol. Biosphere* 33 (4–5): 375–403.
115 Colizzi, E.S. and Hogeweg, P. (2016). Parasites sustain and enhance RNA-like replicators through spatial self-organisation. *PLoS Comput. Biol.* 12 (4): e1004902.
116 Könnyű, B., Czárán, T., and Szathmáry, E. (2008). Prebiotic replicase evolution in a surface-bound metabolic system: parasites as a source of adaptive evolution. *BMC Evol. Biol.* 8: 267.
117 Könnyű, B. and Czárán, T. (2013). Spatial aspects of prebiotic replicator coexistence and community stability in a surface-bound RNA world model. *BMC Evol. Biol.* 13: 204.

118 Deamer, D.W. and Barchfeld, G.L. (1982). Encapsulation of macromolecules by lipid vesicles under simulated prebiotic conditions. *J. Mol. Evol.* 18 (3): 203–206.

119 Rajamani, S., Vlassov, A., Benner, S. et al. (2008). Lipid-assisted synthesis of RNA-like polymers from mononucleotides. *Origins Life Evol. Biosphere* 38 (1): 57–74.

120 Guo, M.T., Rotem, A., Heyman, J.A., and Weitz, D.A. (2012). Droplet microfluidics for high-throughput biological assays. *Lab Chip* 12 (12): 2146–2155.

121 Griffiths, A.D. and Tawfik, D.S. (2006). Miniaturising the laboratory in emulsion droplets. *Trends Biotechnol.* 24 (9): 395–402.

122 Guo, H.C.T. and Collins, R.A. (1995). Efficient *trans*-cleavage of a stem-loop RNA substrate by a ribozyme derived from *Neurospora* VS RNA. *EMBO J.* 14 (2): 368–376.

123 Hubai, A.G. and Kun, Á. (2016). Maximal gene number maintainable by stochastic correction – the second error threshold. *J. Theor. Biol.* 405: 29–35.

124 Szilágyi, A., Zachar, I., and Szathmáry, E. (2013). Gause's principle and the effect of resource partitioning on the dynamical coexistence of replicating templates. *PLoS Comput. Biol.* 9 (8): e1003193.

125 Szathmáry, E. and Demeter, L. (1987). Group selection of early replicators and the origin of life. *J. Theor. Biol.* 128 (4): 463–486.

126 Grey, D., Hutson, V., and Szathmáry, E. (1995). A re-examination of the stochastic corrector model. *Proc. R. Soc. London, Ser. B* 262 (1363): 29–35.

127 Fontanari, J.F., Santos, M., and Szathmáry, E. (2006). Coexistence and error propagation in pre-biotic vesicle models: a group selection approach. *J. Theor. Biol.* 239 (2): 247–256.

128 Hubai, A.G. and Kun, Á. (2017). The coexistence of independent genes is aided by multilevel selection, but only to a limited extent. *Modelling Biological Evolution 2017: Developing Novel Approaches*, Leicester, UK.

129 Vig-Milkovics, Z., Zachar, I., Kun, Á. et al. (2019). Moderate sex between protocells can balance between a decrease in assortment load and an increase in parasite spread. *J. Theor. Biol.* 462: 304–310.

130 Santos, M., Zintzaras, E., and Szathmáry, E. (2003). Origin of sex revisited. *Origins Life Evol. Biosphere* 33: 405–432.

131 Niesert, U., Harnasch, D., and Bresch, C. (1981). Origin of life between Scylla and Charybdis. *J. Mol. Evol.* 17 (6): 348–353.

132 Mizuuchi, R. and Ichihashi, N. (2018). Sustainable replication and coevolution of cooperative RNAs in an artificial cell-like system. *Nat. Ecol. Evol.* https://doi.org/10.1038/s41559-41018-40650-z.

133 Koonin, E.V. (2000). How many genes can make a cell: the minimal-gene-set concept. *Annu. Rev. Genomics Hum. Genet.* 1 (1): 99–116.

134 Fehér, T., Papp, B., Pál, C., and Pósfai, G. (2007). Systematic genome reductions: theoretical and experimental approaches. *Chem. Rev.* 107 (8): 3498–3513.

135 Islas, S., Becerra, A., Luisi, P.L., and Lazcano, A. (2004). Comparative genomics and the gene complement of a minimal cell. *Origins Life Evol. Biosphere* 34 (1): 243–256.

136 Benner, S.A., Ellington, A.D., and Tauer, A. (1989). Modern metabolism as a palimpsest of the RNA world. *Proc. Natl. Acad. Sci. U.S.A.* 86 (18): 7054–7058.

137 McCutcheon, J.P. and von Dohlen, C.D. (2011). An interdependent metabolic patchwork in the nested symbiosis of mealybugs. *Curr. Biol.* 21 (16): 1366–1372.

138 Bennett, G.M. and Moran, N.A. (2013). Small, smaller, smallest: the origins and evolution of ancient dual symbioses in a phloem-feeding insect. *Genome Biol. Evol.* 5 (9): 1675–1688.

139 McCutcheon, J.P., McDonald, B.R., and Moran, N.A. (2009). Origin of an alternative genetic code in the extremely small and GC-rich genome of a bacterial symbiont. *PLoS Genet.* 5 (7): e1000565.

140 Tamames, J., Gil, R., Latorre, A. et al. (2007). The frontier between cell and organelle: genome analysis of *Candidatus Carsonella ruddii*. *BMC Evol. Biol.* 7 (1): 181.

141 McCutcheon, J.P. and Moran, N.A. (2010). Functional convergence in reduced genomes of bacterial symbionts spanning 200 My of evolution. *Genome Biol. Evol.* 2: 708–718.

142 Chang, H.-H., Cho, S.-T., Canale, M.C. et al. (2015). Complete genome sequence of "*Candidatus Sulcia muelleri*" ML, an obligate nutritional symbiont of maize leafhopper (*Dalbulus maidis*). *Genome Announc.* 3 (1): e01483-01414.

143 McCutcheon, J.P. and Moran, N.A. (2007). Parallel genomic evolution and metabolic interdependence in an ancient symbiosis. *Proc. Natl. Acad. Sci. U.S.A.* 104 (49): 19392–19397.

144 Woyke, T., Tighe, D., Mavromatis, K. et al. (2010). One bacterial cell, one complete genome. *PLoS One* 5 (4): e10314.

145 Wu, D., Daugherty, S.C., Van Aken, S.E. et al. (2006). Metabolic complementarity and genomics of the dual bacterial symbiosis of sharpshooters. *PLoS Biol.* 4 (6): e188.

146 McCutcheon, J.P. and Moran, N.A. (2011). Extreme genome reduction in symbiotic bacteria. *Nat. Rev. Microbiol.* 10: 13–26.

147 Hutchison, C.A., Chuang, R.-Y., Noskov, V.N. et al. (2016). Design and synthesis of a minimal bacterial genome. *Science* 351 (6280): aad6253.

148 Gil, R., Silva, F.J., Peretó, J., and Moya, A. (2004). Determination of the core of a minimal bacterial gene set. *Microbiol. Mol. Biol. Rev.* 68 (3): 518–537.

149 Gabaldón, T., Peretó, J., Montero, F. et al. (2007). Structural analyses of a hypothetical minimal metabolism. *Philos. Trans. R. Soc. London, Ser. B* 362 (1486): 1761–1762.

150 Kun, Á. and Radványi, Á. (2018). The evolution of the genetic code: Impasses and challenges. *Biosystems* 164: 217–225.

151 Reuß, D.R., Commichau, F.M., Gundlach, J. et al. (2016). The blueprint of a minimal cell: *MiniBacillus*. *Microbiol. Mol. Biol. Rev.* 80 (4): 955–987.

152 Cheng, L.K.L. and Unrau, P.J. (2010). Closing the circle: replicating RNA with RNA. *Cold Spring Harbor Perspect. Biol.* 2 (10): a002204.

153 Mutschler, H., Wochner, A., and Holliger, P. (2015). Freeze–thaw cycles as drivers of complex ribozyme assembly. *Nat. Chem.* 7 (6): 502–508.

154 Kováč, L., Nosek, J., and Tomáška, L.u. (2003). An overlooked riddle of life's origins: energy-dependent nucleic acid unzipping. *J. Mol. Evol.* 57 (1): S182–S189.

155 Kreysing, M., Keil, L., Lanzmich, S., and Braun, D. (2015). Heat flux across an open pore enables the continuous replication and selection of oligonucleotides towards increasing length. *Nat. Chem.* 7: 203–208.

156 Krammer, H., Möller, F.M., and Braun, D. (2012). Thermal, autonomous replicator made from transfer RNA. *Phys. Rev. Lett.* 108 (23): 238104.

157 Ball, R. and Brindley, J. (2014). Hydrogen peroxide thermochemical oscillator as driver for primordial RNA replication. *J. R. Soc. Interface* 11 (95): 20131052.

158 Zenkin, N. (2012). Hypothesis: Emergence of translation as a result of RNA helicase evolution. *J. Mol. Evol.* 74 (5): 249–256.

159 Martin, L., Unrau, P., and Müller, U. (2015). RNA synthesis by *in vitro* selected ribozymes for recreating an RNA world. *Life* 5 (1): 247.

160 Saladino, R., Crestini, C., Pino, S. et al. (2012). Formamide and the origin of life. *Phys. Life Rev.* 9 (1): 84–104.

161 Saladino, R., Crestini, C., Costanzo, G. et al. (2001). A possible prebiotic synthesis of purine, adenine, cytosine, and 4(3H)-pyrimidinone from formamide implications for the origin of life. *Bioorg. Med. Chem.* 9 (5): 1249–1253.

162 Pino, S., Sponer, J., Costanzo, G. et al. (2015). From formamide to RNA, the path is tenuous but continuous. *Life* 5 (1): 372.

163 Powner, M.W., Gerland, B., and Sutherland, J.D. (2009). Synthesis of activated pyrimidine ribonucleotides in prebiotically plausible conditions. *Nature* 459: 239–242.

164 Patel, B.H., Percivalle, C., Ritson, D.J. et al. (2015). Common origins of RNA, protein and lipid precursors in a cyanosulfidic protometabolism. *Nat. Chem.* 7 (4): 301–307.

165 Saladino, R., Šponer, J., Šponer, J. et al. (2018). Chemomimesis and molecular Darwinism in action: from abiotic generation of nucleobases to nucleosides and RNA. *Life* 8 (2): 24.

166 Lau, M.W.L., Cadieux, K.E.C., and Unrau, P.J. (2004). Isolation of fast purine nucleotide synthase ribozymes. *J. Am. Chem. Soc.* 126 (48): 15686–15693.

167 Unrau, P.J. and Bartel, D.P. (1998). RNA-catalysed nucleotide synthesis. *Nature* 395 (6699): 260–263.

168 Moretti, J.E. and Müller, U.F. (2014). A ribozyme that triphosphorylates RNA 5′-hydroxyl groups. *Nucleic Acids Res.* 42 (7): 4767–4778.

169 Camden, A.J., Walsh, S.M., Suk, S.H., and Silverman, S.K. (2016). DNA oligonucleotide 3′-phosphorylation by a DNA enzyme. *Biochemistry* 55 (18): 2671–2676.

170 Biondi, E., Maxwell, A.W.R., and Burke, D.H. (2012). A small ribozyme with dual-site kinase activity. *Nucleic Acids Res.* 40 (15): 7528–7540.

171 Li, Y. and Breaker, R.R. (1999). Phosphorylating DNA with DNA. *Proc. Natl. Acad. Sci. U.S.A.* 96 (6): 2746–2751.

172 Curtis, E.A. and Bartel, D.P. (2005). New catalytic structures from an existing ribozyme. *Nat. Struct. Mol. Biol.* 12: 994–1000.

173 Lorsch, J.R. and Szostak, J.W. (1994). *In vitro* evolution of new ribozymes with polynucleotide kinase activity. *Nature* 371: 31–36.

174 Saran, D., Nickens, D.G., and Burke, D.H. (2005). A trans acting ribozyme that phosphorylates exogenous RNA. *Biochemistry* 44 (45): 15007–15016.

175 Jiménez, J.I., Xulvi-Brunet, R., Campbell, G.W. et al. (2013). Comprehensive experimental fitness landscape and evolutionary network for small RNA. *Proc. Natl. Acad. Sci. U.S.A.* 110 (37): 14984–14989.

176 Davis, J.H. and Szostak, J.W. (2002). Isolation of high-affinity GTP aptamers from partially structured RNA libraries. *Proc. Natl. Acad. Sci. U.S.A.* 99 (18): 11616–11621.

177 Sassanfar, M. and Szostak, J.W. (1993). An RNA motif that binds ATP. *Nature* 364 (6437): 550–553.

178 Vu, M.M.K., Jameson, N.E., Masuda, S.J. et al. (2012). Convergent evolution of adenosine aptamers spanning bacterial, human, and random sequences revealed by structure-based bioinformatics and genomic SELEX. *Chem. Biol.* 19 (10): 1247–1254.

179 Curtis, E.A. and Liu, D.R. (2013). Discovery of widespread GTP-binding motifs in genomic DNA and RNA. *Chem. Biol.* 20 (4): 521–532.

180 Sacerdote, M.G. and Szostak, J.W. (2005). Semipermeable lipid bilayers exhibit diastereoselectivity favoring ribose. *Proc. Natl. Acad. Sci. U.S.A.* 102 (17): 6004–6008.

181 Breslow, R. (1959). On the mechanism of the formose reaction. *Tetrahedron Lett.* 1 (21): 22–26.

182 Ricardo, A., Carrigan, M.A., Olcott, A.N., and Benner, S.A. (2004). Borate minerals stabilize ribose. *Science* 303 (5655): 196–196.

183 Mansy, S.S. and Szostak, J.W. (2008). Thermostability of model protocell membranes. *Proc. Natl. Acad. Sci. U.S.A.* 105 (36): 13351–13355.

184 Mansy, S.S. (2010). Membrane transport in primitive cells. *Cold Spring Harbor Perspect. Biol.* 2 (8): a002188.

185 Chakrabarti, A.C., Breaker, R.R., Joyce, G.F., and Deamer, D.W. (1994). Production of RNA by a polymerase protein encapsulated within phospholipid vesicles. *J. Mol. Evol.* 39 (6): 555–559.

186 King, Z.A., Lu, J., Dräger, A. et al. (2016). BiGG models: a platform for integrating, standardizing and sharing genome-scale models. *Nucleic Acids Res.* 44 (D1): D515–D522.

187 Chen, X., Li, N., and Ellington, A.D. (2007). Ribozyme catalysis of metabolism in the RNA World. *Chem. Biodivers.* 4 (4): 633–655.

188 Feist, A.M., Herrgard, M.J., Thiele, I. et al. (2009). Reconstruction of biochemical networks in microbial organisms. *Nat. Rev. Microbiol.* 7 (2): 129–143.

189 Davis, B.D. (1958). On the importance of being ionized. *Arch. Biochem. Biophys.* 78 (2): 497–509.

190 Paula, S., Volkov, A.G., Van Hoek, A.N. et al. (1996). Permeation of protons, potassium ions, and small polar molecules through phospholipid bilayers as a function of membrane thickness. *Biophys. J.* 70 (1): 339–348.

191 Khvorova, A., Kwak, Y.-G., Tamkun, M. et al. (1999). RNAs that bind and change the permeability of phospholipid membranes. *Proc. Natl. Acad. Sci. U.S.A.* 96 (19): 10649–10654.

192 Janas, T., Janas, T., and Yarus, M. (2004). A membrane transporter for tryptophan composed of RNA. *RNA* 10 (10): 1541–1549.

193 Großhans, H. and Filipowicz, W. (2008). The expanding world of small RNAs. *Nature* 451: 414–416.

194 Silverman, S.K. (2003). Rube Goldberg goes (ribo)nuclear? Molecular switches and sensors made from RNA. *RNA* 9 (4): 377–383.

195 Frommer, J., Appel, B., and Müller, S. (2015). Ribozymes that can be regulated by external stimuli. *Curr. Opin. Biotechnol.* 31: 35–41.

196 Kuwabara, T., Warashina, M., and Taira, K. (2000). Allosterically controllable ribozymes with biosensor functions. *Curr. Opin. Chem. Biol.* 4 (6): 669–677.

197 Navani, N.K. and Li, Y. (2006). Nucleic acid aptamers and enzymes as sensors. *Curr. Opin. Chem. Biol.* 10 (3): 272–281.

198 Breaker, R.R. (2002). Engineered allosteric ribozymes as biosensor components. *Curr. Opin. Biotechnol.* 13 (1): 31–39.

199 Szilágyi, A., Kun, Á., and Szathmáry, E. (2012). Early evolution of efficient enzymes and genome organization. *Biol. Direct* 7: 38.

200 Szathmáry, E. and Maynard Smith, J. (1993). The evolution of chromosomes II. Molecular mechanisms. *J. Theor. Biol.* 164 (4): 447–454.

201 Scott, W.G. (2007). Ribozymes. *Curr. Opin. Struct. Biol.* 17 (3): 280–286.

202 Musiari, A., Rowinska-Zyrek, M., Gallo, S., and Sigel, R.K.O. (2014). Metal ions in ribozymes and riboswitches. In: *DNA in Supramolecular Chemistry and Nanotechnology* (eds. E. Stulz and G.H. Clever), 412–433. Wiley.

203 Landweber, L.F. and Pokrovskaya, I.D. (1999). Emergence of a dual-catalytic RNA with metal-specific cleavage and ligase activities: the spandrels of RNA evolution. *Proc. Natl. Acad. Sci. U.S.A.* 96 (1): 173–178.

204 Drude, I., Vauléon, S., and Müller, S. (2007). Twin ribozyme mediated removal of nucleotides from an internal RNA site. *Biochem. Biophys. Res. Commun.* 363 (1): 24–29.

205 Welz, R., Bossmann, K., Klug, C. et al. (2003). Site-directed alteration of RNA sequence mediated by an engineered twin ribozyme. *Angew. Chem. Int. Ed.* 42 (21): 2424–2427.

206 Vaidya, N., Manapat, M.L., Chen, I.A. et al. (2012). Spontaneous network formation among cooperative RNA replicators. *Nature* 491 (7422): 72–77.

15

Ribozyme-Catalyzed RNA Recombination

Benedict A. Smail and Niles Lehman

Science Research & Teaching Center, Portland State University, 1719 SW 10th Ave, Portland, OR 97201, USA

15.1 Introduction

RNA recombination is the process of one or more RNA molecules undergoing a chemical reaction, typically a trans-esterification that produces multiple distinct RNA products. In nature, the most familiar examples of RNA–RNA recombination are the self-splicing group I and group II introns, but it can also be directed by numerous ribozymes, it is prevalent in RNA viruses, and it is required for the maturation of split transfer RNA (tRNA) transcripts of certain Archaea. RNA and DNA recombination catalyzed by protein enzymes is a common feature of all extant biology and is an intrinsic component of genome replication and DNA repair.

An RNA–RNA recombination reaction is distinct from both DNA recombination and RNA ligation. In the former, genetic recombination of two double strands is facilitated by double-strand crossover. (In RNA viruses, homologous RNA recombination can occur by template switching, in which a viral reverse transcriptase detaches from one template strand and reattaches to another [1].) DNA repair in bacteria is mediated by the RecA protein enzyme, which has structural homologues across all kingdoms of life [2]. On the other hand, RNA ligation is the direct joining of the 5′ and 3′ ends of two separate pieces of RNA with a phosphate. RNA ligations have a high energy barrier to reacting and must either be catalyzed or have a high energy leaving group such as a diphosphate or an imidazole. In contrast, true RNA recombination is a catalyzed or uncatalyzed trans-esterification that occurs with little to no change in free energy, is non-homologous, and has a nucleotide, nucleoside, or polynucleotide leaving group [3]. Trans-esterification is the most common type of chemistry catalyzed by all known ribozymes.

Given the universality of recombination in biology, and among other things its utility in biological organisms to avoid Muller's ratchet [3], it seems plausible to conclude that its evolutionary origins are extremely ancient. RNA recombination may predate the earliest known life forms and may possibly have been a key contributing mechanism by which life arose. It is logical, therefore, to wonder whether recombination might have been an intrinsic feature of the RNA world or ancient ribozymes.

Ribozymes, First Edition. Edited by Sabine Müller, Benoît Masquida, and Wade Winkler.

Here we will provide a short review of ribozyme-catalyzed RNA recombination, discussing at length the modified *Azoarcus* group I intron, which has been extensively characterized and has the presently unique ability to self-assemble and catalyze its own formation via recombination. The *Azoarcus* recombinase ribozyme and its ability to form spontaneous networks [4] through cooperative autocatalysis provide an alternative theory to an RNA replicase as a means of prebiotic self-replication, bootstrapping, and diversification in the putative RNA world.

15.2 RNA Recombination Chemistry

A simple physical definition of recombination can be symbolized as A•B + C•D = A•D + C•B, where A, B, C, and D are individual components that are exchanged in the reaction, but whose general characteristics and identity remain unchanged, and the dot (•) represents a covalent bond [3, 5].

In the broadest sense of the word, many chemical reactions are recombinations. Using the simple model above as guide, it can be seen that a condensation is a reaction in which A is some molecule and B is a hydroxyl group attached to A; it follows that C is a hydrogen atom attached to D, which is another molecule. Another technical, example is the reduction of a ketone with sodium borohydride, in which electrons are exchanged for a hydrogen atom and ionic sodium is reduced to metal. Alkene metathesis is an example of a strict recombination reaction in organic chemistry. On the other hand, an addition reaction, such as the addition of a hydrogen halide to an alkene, is not a recombination in any sense because all parts of each reactant become included in the product(s).

It is important to note that an RNA–RNA ligation is essentially a condensation. In this case it is the attack of a ribose hydroxyl on a phosphate group to form a phosphor–ester linkage, with water as a leaving group. While this could be construed to fit the definition of a recombination above, for genetic purposes it is not, because water has no genetic utility as an information-bearing molecule. As this reaction is inherently unfavorable in water, RNA and DNA ligation in nature is typically accomplished by the use of both an enzyme and a cofactor such as ATP or NAD^+. Sequential ligation or polymerization by a ribozyme, which has been a long-term goal of many RNA world proponents, is thus a fundamentally challenging series of reactions that have typically been created with the help of high energy leaving groups such as diphosphates or imidazole.

In contrast to RNA–RNA ligation by condensation, RNA–RNA recombination has the principal virtue of achieving a phosphor–ester linkage with trans-esterification. Instead of a high-energy leaving group, the attack of a ribose hydroxyl on a phosphate requires a nucleoside, nucleotide, or polynucleotide leaving group. The reaction can take place in water and its principle limitation is its reversibility. However, in a prebiotic RNA world without optimized sequences, a greater degree of reversibility may have been useful for sequence space exploration and swapping of information.

15.3 *Azoarcus* Group I Intron

The natural *Azoarcus* group I intron is an intervening sequence in the isoleucine pre-tRNA of the *Azoarcus* purple bacterium, a nitrogen-fixing heterotroph of the phylum beta-proteobacteria [6]. The original group I intron is found adjacent and downstream of the CAU anticodon and excises itself from its tRNA precursor, cyclizes the IVS, and splices the remaining exons together to form the mature tRNA. At just over 200 nucleotides (nt) long, the *Azoarcus* intron is also one of the smallest known natural group I introns, has a high G + C content, and is relatively heat stable, features that have made it attractive candidate for detailed experimentation.

Group I introns are distinct from group II introns according to their mechanism. In group II introns, the nucleophile is the 2′ hydroxyl of an adenine residue and the intermediate is a lariat RNA. This mechanism has recently been discovered in the spliceosome, and phylogenetic analysis suggests that there is an evolutionary relationship between group II introns and the spliceosome [7]. However, the fundamental properties of group I and group II introns are similar. In both cases, the ribozymes are introns that self-cleave, splice together the exons, and circularize the intron. Both group I and group II introns are found in all three domains of life in ribosomal RNA, tRNA, and messenger RNA, and their widespread presence in biologically critical RNAs suggests that they could have been a feature of any early life that utilized RNA.

Group I introns can be modified by removing their 5′ exons and ensuring an endogenous, terminal guanosine nucleotide to facilitate trans-splicing outside of their natural environment. In addition, the 5′ and 3′ ends can be lengthened or shortened to improve splicing or recognition of the substrate. The foundation for this approach first began with the *Tetrahymena* ribozyme, which was transcribed without its 5′ exon and was shown to catalyze the sequential recombination of pentacytidine into polycytidine oligomers up to 30 nt in length [8]. The ribozyme contains an internal template or guide sequence that is required for recognition of external substrates; for the *Tetrahymena* ribozyme, it is a 6-nt GGAGGG sequence.

As with the *Tetrahymena* ribozyme, the *Azoarcus* group I intron can be detached from its exons to create an *Azoarcus* recombinase ribozyme. Following its discovery from phylogenetic analysis of proteobacteria, one early study demonstrated that an *Azoarcus* ribozyme can catalyze the sequential addition of short RNA oligomers by driving the self-splicing reaction partly in reverse [9]. Starting with an *Azoarcus* group I intron that had been modified to include a guanosine at its 3′ terminus and two ligated exons of 10 nt each (termed E1 and E2), John Burke and colleagues showed that the ribozyme could catalyze elongation of the ligated exons by sequentially adding the E2 sequence [9]. This resulted in a range of products from 30 (E1•E2•E2) to 180 nt in length (E1•$[E2]_{17}$). E1 contains the characteristic CAU tag sequence as its last 3 nt in a 10-nt guide sequence; this junction is the target of attack by the endogenous guanosine to produce the ribozyme-substrate covalent intermediate (here we use A to indicate ribozyme: A•E2). The end of E2 also contains a guanosine and the last 4 nt are identical to the last four of the constructed ribozyme; this results in a new E1•E2 attacking the covalent intermediate to form the elongated product.

15.4 Crystal Structure

The crystal structure of the *Azoarcus* ribozyme containing both its exons has been solved to 3.1 Å and provided a comprehensive understanding of the trans-esterification mechanism [10, 11]. The structure shows that the *Azoarcus* group I intron is an obligate metalloenzyme, requiring magnesium ions in the active site to facilitate catalysis. The terminal 3′-OH of the 5′ exon is positioned inline for nucleophilic attack on the scissile phosphate following the CAU tag, which is complementary to the internal guide sequence (IGS) of GUG. The last of these three base pairs is an invariant wobble G–U pair, which is recognized by an adenosine-rich wobble receptor motif. The splice site is held in place by an extensive network of tertiary interactions that help maintain the structure of the intron and retain the 5′ exon prior to ligation to the 3′ exon. Two metals, M1 and M2, which are probably both magnesium, play intrinsic catalytic roles in the active site. M1 coordinates the nucleophilic 3′ oxygen of the 5′ exon (uridine in the CAU tag) to the scissile phosphate, while M2 coordinates the terminal 2′ hydroxyl of the 3′-guanosine and may coordinate a water molecule that bridges to the pro-R_p oxygen of the scissile phosphate. The M1 likely acts as a Lewis acid to activate the 3′ hydroxyl for nucleophilic attack and stabilize the negative charge on the intermediate [10].

15.5 Mechanism

In the laboratory setting, the modified *Azoarcus* ribozyme catalyzes recombination by two distinct biochemical mechanisms [12]. The first of these has been termed the *tF2* mechanism and is considered more primitive (Figure 15.1a). In this single-step mechanism, the ribozyme facilitates attack of a 3′ hydroxyl of one substrate on the overhanging 5′ end of another substrate. The nucleophile is the 3′ hydroxyl of the first substrate, which may be hydrogen bonded to the IGS. It has never been conclusively determined if the 2′ hydroxyl can also participate in this reaction, and if so, whether it is more or less common than the 3′ hydroxyl. As expected from the position of the nucleophile relative to the 5′ end of the second substrate, recombination via the *tF2* mechanism produces a leaving group that is usually only a nucleoside or dimer. Although the equivalent of a ligation has been reported [12], this is extremely unlikely because neither the 5′ nor 3′ end is typically phosphorylated and the additional G nucleotide is more likely the result of polymerase error during nucleotide sequence analysis.

The second method of *Azoarcus* ribozyme catalysis is the more complex, multi-step *R2F2* mechanism (Figure 15.1b). In this mechanism the endogenous, terminal G nucleophile of the ribozyme attacks the phosphor–ester bond just prior

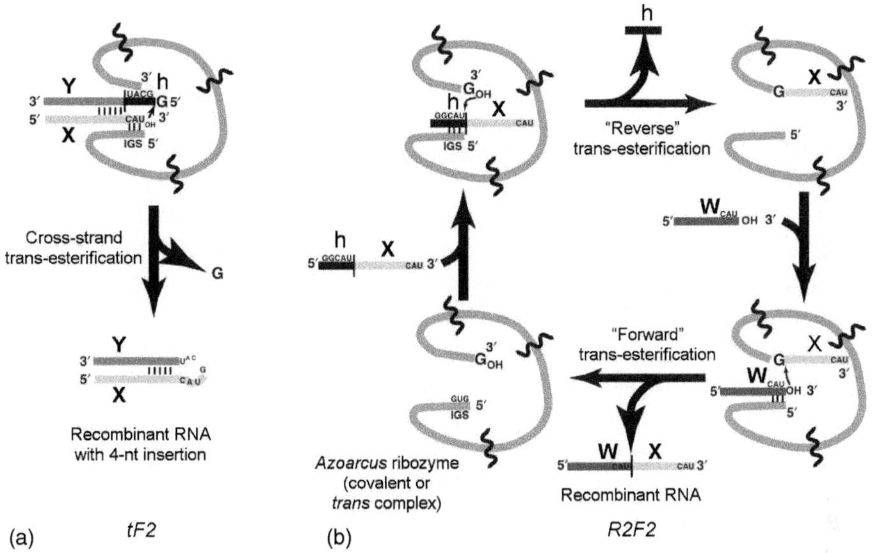

Figure 15.1 Two mechanisms of trans-esterification as catalyzed by the *Azoarcus* ribozyme. (a) The *tF2* mechanism; attack of the 3′-OH of the terminal uridine on the last phosphor–ester bond of the header sequence produces a recombinant molecule with a 4-nt insert. (b) The *R2F2* mechanism; attack of the 3′-OH of the ribozyme's endogenous, terminal guanosine on fragment h**X** releases the header sequence and produces **R•X**, an enzyme–substrate covalent intermediate. Subsequent attack at this juncture by the terminal 3′-OH of fragment **W** releases the ribozyme and restores its endogenous, terminal guanosine, producing the recombinant fragment **W•X**. Source: Draper et al. [12]. © Oxford University.

to the 5′ nucleotides bound to the IGS to produce a covalent intermediate longer than the original ribozyme. The short oligomer leaving group detaches from the guide sequence and is replaced by a second substrate whose 3′ terminal nucleotide attacks the same phosphor–ester bond formed in the previous step, linking the two substrates and releasing the ribozyme as the leaving group with its terminal guanosine remaining in place. By regenerating the guanosine nucleophile, this last step thus allows the ribozyme to turn over and the net reaction effectively splices the two substrates together (minus the short insertion in the first substrate, which is complementary to the IGS).

15.6 Model for Prebiotic Chemistry

A simple utility of the *Azoarcus* ribozyme is to recombine RNA fragments to produce other functional ribozymes, and in this regard it is a model for possible prebiotic recombination of RNA fragments. In 2003, the previous work of Burke and coworkers [9] on the *Azoarcus* ribozyme's ability to polymerize substrates was expanded [13] to modify the ribozyme to use only a 3-nt IGS of GUG. Two substrates were rationally designed to form a hammerhead ribozyme upon recombination at a CAU junction, which was placed in a variable loop region of the hammerhead not critical for self-cleavage. Upon binding to the ribozyme's IGS, the CAU substrate is subject to attack by the endogenous 3′ guanosine of the ribozyme to form a covalent enzyme substrate intermediate. In a second step, a second substrate can bind to the IGS and its terminal U can attack the intermediate at the recently formed phosphor–diester bond to form a functional hammerhead ribozyme that may subsequently undergo self-cleavage in a fully one-pot reaction. This experiment also demonstrated that as a recombinase for medium sized RNAs, the *Azoarcus* ribozyme is faster and more effective than a comparable *Tetrahymena* ribozyme, despite the latter's apparent larger catalytic rate [14]. Nevertheless, the *Tetrahymena* ribozyme has some unique recombinatoric utility; variants of the *Tetrahymena* ribozyme have been shown to repair truncated *lacZ* transcripts in *Escherichia coli* bacteria [15], and mutant human p53 genes [16] through recombination-based reactions.

A critical feature of the modified *Azoarcus* ribozyme that distinguishes it from more experimentally probed *Tetrahymena* ribozyme is the smaller guide sequence. The computational space of a 3-nt guide sequence is 64 possible sequences, but for a 6-nt IGS such as that of the *Tetrahymena* RNA, it is nearly 4100. This implies that the 414-nt *Tetrahymena* ribozyme cannot act as its own substrate, which Cech noted in his early investigations of its catalytic potential [8]. Apart from the fact that the *Tetrahymena* ribozyme does not possess a complement for its own IGS, it is statistically unlikely that a competent matching sequence will be found in any RNA molecule with equal proportions of each base whose sequence length is an order of magnitude smaller than the sequence space of all 6mers. By requiring only 3 nt to recognize the IGS, the *Azoarcus* ribozyme can accept a wide range of substrates as it is considerably more likely to find a 3-nt sequence in a random RNA sequence than a 6-nt one.

The apparent issue with the *Tetrahymena* ribozyme as a recombinase for RNA fragments is that product release from the 6-nt guide sequence can be rate-limiting. This is not a significant problem for the 3-nt guide sequence of the *Azoarcus* ribozyme. As a result, the *Azoarcus* ribozyme is a more efficient recombinase for splicing together short RNA pieces [13]. Indeed, one of the criticisms of RNA polymerase ribozymes as engines of prebiotic assembly and diversity is that their substrates do not easily detach [17], a problem easily overcome by the use of *Azoarcus* or *Azoarcus*-like recombinases that can assemble virtually any RNA construct provided it contains an accessible CAU tag.

An illustrative example of this flexibility is the ability of the *Azoarcus* ribozyme to recombine a full-length (c. 150 nt) class I ligase ribozyme using two smaller substrate molecules modified with the same CAU tag that is complementary to the guide sequence, similar to previous experiments with the hammerhead ribozyme [18]. As with the hammerhead, a variant of the class I ligase can be split into two inert precursors and a CAU sequence inserted in an innocuous part of the ribozyme devoid of significant structure. Using a one-pot reaction containing two substrates that are precursors to the ligase ribozyme and one substrate for the ligase ribozyme, *Azoarcus*-RNA-mediated recombination produces a full-length ligase ribozyme, which subsequently ligates its own substrate [18].

Given the propensity of the *Azoarcus* ribozyme to catalyze recombination of a variety of RNA substrates, it seems logical that it should be able to recombine itself. Although the ability of a full length *Azoarcus* ribozyme to catalyze its own assembly from two or more precursors molecules can be inferred from studies on the hammerhead and the ligase, it is not intuitive that a fragmented ribozyme could spontaneously assemble a catalytic *trans* complex. However an original precedent was established by Doudna and Cech who broke the *Tetrahymena* ribozyme into three distinct pieces and showed that these could spontaneously assemble into a functional ribozyme, primarily through tertiary interactions [19]. Although the main purpose of that research was to study the tertiary structure of the *Tetrahymena* ribozyme, it was immediately clear that large RNA structures might be able to form *trans* complexes with low dissociation constants that could retain activity.

Using the *Tetrahymena* ribozyme as a model for the *Azoarcus* ribozyme, it might be expected that the *Azoarcus* ribozyme could also be fragmented in such a way that the pieces would reassemble into a *trans* complex. Moreover, while the *Tetrahymena* ribozyme cannot effectively act as its own substrate due to its 6-nt guide sequence, the *Azoarcus* ribozyme has far fewer constraints with its 3-nt GUG guide sequence. In fact, the natural *Azoarcus* ribozyme possesses a single, internal CAU triplet sequence. Based on this, one might anticipate that the *Azoarcus* ribozyme could (i) catalyze its own assembly from separate fragments and (ii) be broken into pieces that could reassemble spontaneously into a *trans* complex that would then catalyze its own assembly. This has indeed transpired to be true, as discussed in the next section.

15.7 Spontaneous Self-assembly of *Azoarcus* RNA Fragments

The critical feature necessary for spontaneous self-assembly of the *Azoarcus* ribozyme is the insertion of the CAU tag sequence at a number of junctions within the ribozyme. The ribozyme may then be broken into fragments containing this tag sequence at the 3′ end. Although there is clearly a limit to the number of CAUs that may be inserted, the lack of other specific requirements has made it possible to split the *Azoarcus* ribozyme into several fragments, each with a characteristic CAU. Importantly, the tag sequence must be located in such a place to not significantly disrupt catalytic activity upon assembly. Moreover, the insert must be placed in a location where it will be accessible to the ribozyme. Using these criteria, it has been possible to create four fragments of the *Azoarcus* RNA, **W**, **X**, **Y**, and **Z**, by inserting a CAU tag into three tetraloop regions that are not significantly involved in folding or catalysis. In addition to the tag sequence, a header sequence (h) must be added to serve as the leaving group in the recombination reaction. Incubation of the four fragments results in self-assembly [20] (Figure 15.2a).

In the first step of *Azoarcus* RNA self-assembly, the four fragments **W**, h**X**, h**Y**, and h**Z** form a *trans* complex that is stabilized extensively through tertiary interactions (Figure 15.2b). In the second step, the *trans* complex catalyzes successive

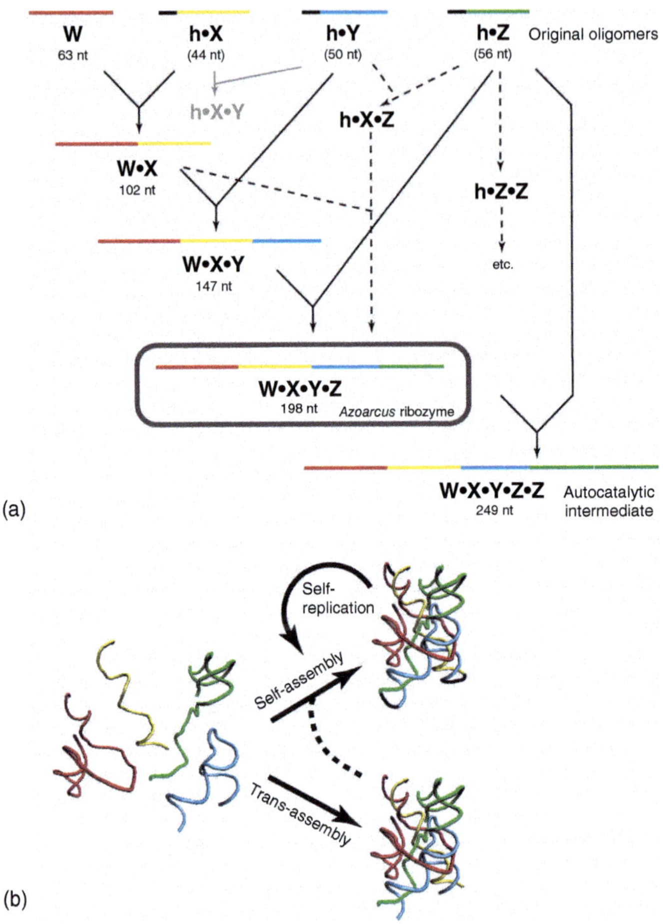

Figure 15.2 Spontaneous self-assembly of the *Azoarcus* ribozyme from four fragments. (a) One possible sequence of recombination reactions leading from the fragments (**W**, h**X**, h**Y**, and h**Z**) to the full-length ribozyme (**W•X•Y•Z**). (b) Involvement of a non-covalent *trans* complex in initiating the recombination reactions. Source: Hayden and Lehman [20]. Reproduced with permission of Elsevier. © 2006.

recombination of the fragments until the full-length ribozyme is formed. The usual first reaction is recombination by the *R2F2* mechanism of **W** and **X** to produce the covalent product **W•X**. In this reaction, the GGCAU 5′ head (h) of **X** is bound to the guide sequence and the ribozyme (R) attacks to produce the covalent complex R•**X**. Then, the CAU 3′ tail of W binds to the guide sequence and the terminal U attacks R•**X** to produce R and **W•X**. By ensuring that the **W**, **X**, and **Y** fragments end in a characteristic CAU, a tail fragment is not necessary.

The typical progression of product formation for the full-length *Azoarcus* ribozyme is **W•X** to **W•X•Y** to **W•X•Y•Z**, although other paths are possible [20]. The recombination of wild-type **W** and **X** to form **W•X** and **W•X•Y** and **Z** to form **W•X•Y•Z** is almost always through the *R2F2* mechanism. However, the recombination of **X** and **Y** or **W•X** and **Y** to **W•X•Y** can follow the *tF2* mechanism.

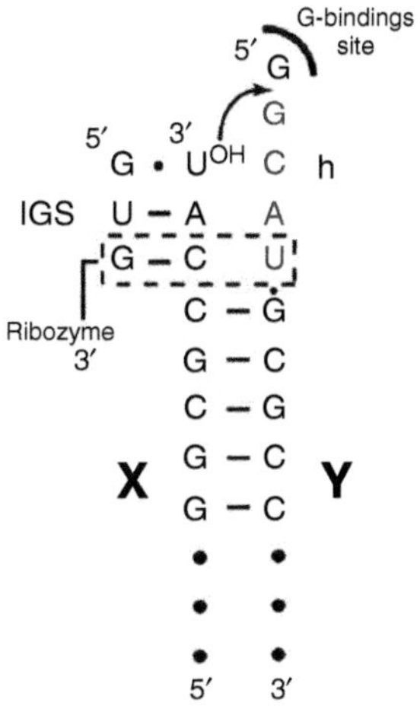

Figure 15.3 Base pairing between two fragments (e.g. **X** and **Y**) that is required for the first-step *tF2* mechanism to operate. Source: Hayden and Lehman [20]. Reproduced with permission of Elsevier. © 2006.

Here, the ribozyme binds a partially double-stranded duplex **X–Y**, where the **X** fragment binds to the GUG guide sequence and the **Y** fragment, including the head, is hydrogen bonded to **X** (Figure 15.3). The terminal U of **X** can attack the GGCAU head of **Y** to join the fragments with a 4-nt insert of GCAU. Slippage in the active site can reduce the insert to CAU but because the head is not phosphorylated, the insert cannot likely be GGCAU; it is unclear how ligation could be operative without a 5′ phosphate group.

Starting with the four fragments of the *Azoarcus* RNA, there are six possible routes to the final covalently contiguous ribozyme. All six require assembly of the *trans* complex to catalyze recombination of the individual components (Figure 15.2b). Once the *trans* complex is present, there are three unique recombination junctions and six unique possible paths to the final product (Table 15.1).

Of these paths, the first (in italic font) is by far the most common route. **X•Y** is not a populated intermediate [20], which is evidence against the third and fourth routes, while **W•X** is the most common intermediate at the second step. Notably, by taking advantage of the internal CAU tag sequence, self-assembly from five fragments (**W**, h**X**, h**Y**, h$\mathbf{Z_1}$, and h$\mathbf{Z_2}$) has recently been achieved [21].

Table 15.1 Possible routes from *Azoarcus* RNA fragments to a full-length covalent ribozyme.

Step 1	Step 2	Step 3	Step 4
W+X+Y+Z	***W•X+Y+Z***	***W•X•Y+Z***	***W•X•Y•Z***
W+X+Y+Z	**W•X+Y+Z**	**W•X+Y•Z**	**W•X•Y•Z**
W+X+Y+Z	**X•Y+W+Z**	**W•X•Y+Z**	**W•X•Y•Z**
W+X+Y+Z	**X•Y+W+Z**	**X•Y•Z+W**	**W•X•Y•Z**
W+X+Y+Z	**Y•Z+W+X**	**X•Y•Z+W**	**W•X•Y•Z**
W+X+Y+Z	**Y•Z+W+X**	**W•X+Y•Z**	**W•X•Y•Z**

15.8 Autocatalysis

The self-assembly of the four *Azoarcus* ribozyme fragments via recombination has been shown to proceed in an autocatalytic fashion. Firstly, the product (**W•X•Y•Z**) may be tested directly for autocatalysis by doping the each of the putative third steps with the full-length product [22]. These correspond to the reactions in Step 3 of Table 15.1: **W•X•Y + Z**, **W•X + Y•Z**, and **W + X•Y•Z**. Doping of a full-length product for each recombination junction produces a linear increase in the initial rate of **W•X•Y•Z** product up to 2 μM, indicating autocatalysis (Figure 15.4). However, the rate of autocatalytic efficiency must be calculated to consider the formation of *trans* complexes in each reaction. **W•X•Y + Z** form *trans* complexes with the highest efficiency so it has the lowest autocatalytic efficiency, whereas **W•X + Y•Z** has the highest autocatalytic efficiency because the **W•X** and **Y•Z** fragments are the least effective at forming a *trans* complex.

Each of the covalent intermediates to the full-length ribozyme has some degree of autocatalysis. Starting with 1 μM of each of the four fragments (**W**, **X**, **Y**, and **Z**), the reaction can be doped with the full length covalently linked ribozyme or one of its intermediates and compared to a reaction without it. In this way, it can be shown that while **W•X** has only limited autocatalytic potential, the **W•X•Y** construct, which contains the bulk of the ribozyme, demonstrates a notable increase in the rate of its own production [22]. The most effective autocatalyst is the full-length ribozyme; thus the total autocatalysis increases as longer molecules are made. The rate of autocatalytic efficiency of *Azoarcus* ribozyme self-assembly compares favorably with other reported autocatalytic systems, such as that of a ligase ribozyme [23].

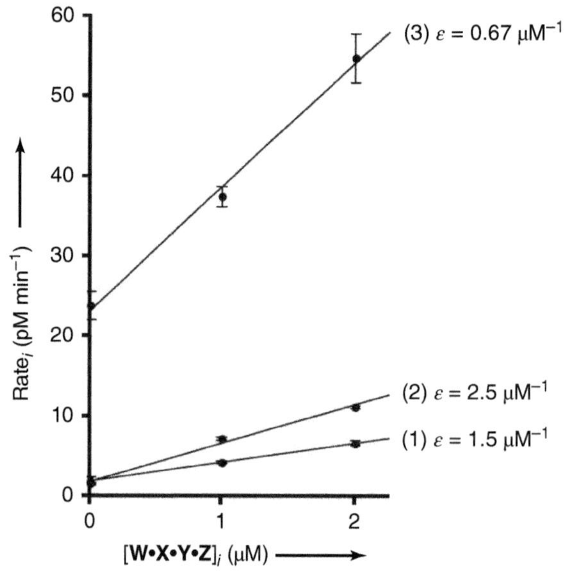

Figure 15.4 Demonstration of autocatalysis during *Azoarcus* ribozyme self-assembly. Reaction (1) = **W + XYZ**; reaction (2) = **WX + YZ**; reaction (3) = **WXY + Z**. Reaction rate plotted as a function of the concentration of doped in final product (**WXYZ**). Linear trends indicative of autocatalysis; ε is the relative autocatalytic efficiency. Source: Hayden et al. [22]. Reproduced with permission of John Wiley & Sons. © 2008.

15.9 Cooperative Self-assembly

The GUG guide sequence and the CAU tag sequence of the *Azoarcus* ribozyme can be modified without significant loss of activity provided that the tag sequence retains some degree of complementarity to the guide sequence. The G–C pair of the guide sequence and tag is considered the most important, and mutations of this pair to a non-Watson–Crick pair drastically reduce the activity of the ribozyme [12]. The G–U wobble is also important for magnesium catalysis in the active site. However, the middle nucleotide of both the guide sequence and tag can be varied without a complete loss of activity. Fourfold variation at each of the middle nucleotides can be used to create 16 distinct **W**-containing genotypes. These genotypes can then be compared for both self-assembly and cross-assembly.

In addition to randomized nucleotides in the center of the guide sequence, each recombination junction of the *Azoarcus* ribozyme can be fragmented to examine the effect of the *Azoarcus* RNA genotypes at that junction. By creating the pairs $_{\mathrm{GMG}}\mathbf{W}_{\mathrm{CNU}}$ and h**XYZ**, $_{\mathrm{GMG}}\mathbf{WX}_{\mathrm{CNU}}$ and h**YZ**, and $_{\mathrm{GMG}}\mathbf{WXY}_{\mathrm{CNU}}$ and h**Z**, where M and N are any nucleotide and GMG is complementary to CNU, there are 48 possible unique *trans* complexes that can be formed, or 16 at each junction. Each of these *trans* complexes will preferentially catalyze assembly of covalent ribozymes according to the guide sequence and tag composition. The higher the degree of complementarity between the guide sequence and the tag, the higher the rate of assembly. For example, *trans* complexes that contain a guide sequence of GAG will preferentially assemble fragments with CUU tags and hence those that contain both the GAG guide sequence and the CUU tag will catalyze their own covalent products. Such ribozymes might be termed *selfish*, for their ability to catalyze themselves. However, if the tag is a very poor match to the guide sequence, such as GUG and CCU, the ribozyme is much less likely to assemble itself, and much more likely to assemble covalent ribozymes containing the CAU tag. The CCU tag is assembled by *trans* complexes with the GGG guide sequence and if those *trans* complexes contained the CAU tag, these ribozymes would catalyze each other's assembly; they might be termed *cooperative*. In this regard, the randomized genotypes of *Azoarcus* RNA have the potential for forming chemical networks via cross-assembly.

When these three sets of *Azoarcus* RNA fragments are incubated, all 48 (16×3) possible covalent ribozymes can arise in solution [4]. Because assembly of the *trans* complex is driven primarily by tertiary interactions, it is expected that there will be roughly equal proportions of each *trans* complex. However, the autocatalytic potential of each *trans* complex is vastly different; *trans* complexes with mismatched middle pairs in the IGS-tag interaction have severely limited potential for self-assembly by autocatalysis.

An example of this evidence for cross-catalysis can be seen with incubation of three subsets of the 48 possible IGS-tag interactions that have poor individual self-assembly rates. The pairs $_{\mathrm{GUG}}\mathbf{W}_{\mathrm{CGU}}$ and h**XYZ**, $_{\mathrm{GAG}}\mathbf{WX}_{\mathrm{CAU}}$ and h**YZ**, and $_{\mathrm{GCG}}\mathbf{WXY}_{\mathrm{CUU}}$ and h**Z** all have poor individual self-assembly rates as a result of the non-Watson–Crick pair in the middle nucleotides, and when they are incubated in isolation, they show a very poor yield. The total yield of covalent ribozyme is increased 125 times when all six fragments are incubated together in solution, indicating that assembly of covalent ribozymes is almost certainly driven by cross-catalysis. This particular subset of *Azoarcus* RNA fragments can thus be termed a *cooperative* set, in which the entire set has substantially more autocatalytic potential than any of the individual pairs in isolation [4].

Which *trans* complexes are the most efficient? Is it the ones that can assemble themselves or those that cooperatively assemble other ribozymes? To assess whether selfish or cooperative ribozymes are more effective, the cooperative set described above can be compared to a selfish set containing the fragment pairs $_{GUG}\mathbf{W}_{CAU}$ and h**XYZ**, $_{GAG}\mathbf{WX}_{CUU}$ and h**YZ**, and $_{GCG}\mathbf{WXY}_{CGU}$ and h**Z**. Each of these subsets self-assembles well in isolation, and the first pair represents the canonical *Azoarcus* ribozyme IGS and tag. In isolation, the net yield of these "selfish" ribozymes exceeds the yield of the cooperative set. However, when both the cooperative set and selfish set described above are incubated together, the growth of covalent ribozymes from the cooperative set exceeds that of the selfish set. This result is seen with similar sets, suggesting that it is not a fluke but a general phenomenon [4]. These results present an interesting contrast to evolutionary theory in which groups of cooperative replicators grow more rapidly than selfish ones, and groups consisting of both selfish and cooperative replicators are eventually dominated by the selfish [24, 25].

The network of *Azoarcus* RNA genotypes that can cooperate to give rise to covalent ribozymes with different characteristics is reminiscent of a hypercycle except that it produces itself with recombination. Although the *Azoarcus* ribozyme network system does not grow exponentially, it does have the ability to buffer informational decay; mutations in the guide sequence and tag can be offset by complementary mutations to allow the network to persist and maintain the essential catalytic ability of the assembled ribozyme. Because cooperative replicators outcompete selfish replicators and a mixture of both expands the catalytic network, the *Azoarcus* ribozyme network implies that molecular ecological succession is a plausible route to complexity and stability [4, 26].

15.10 Game Theoretic Treatment

The cross-assembly of *Azoarcus* RNA fragments by ribozymes of different genotypes (as characterized by changes to the guide sequence and tag sequence) has provided the first known example of game theory being manifested at the level of nucleic acid chemistry [27]. If the different genotypes are made to compete for a shared resource in order to assemble either their own genotypes or other genotypes, the resulting dynamics can be convincingly modeled with evolutionary game theory.

The most straightforward setup for the *Azoarcus* RNA network to manifest game theory is to take 16 genotypes of the $_{GMG}\mathbf{WXY}_{CNU}$ construct and have them compete for the shared resource **Z** (Figure 15.5). The **WXY** construct contains the bulk of the

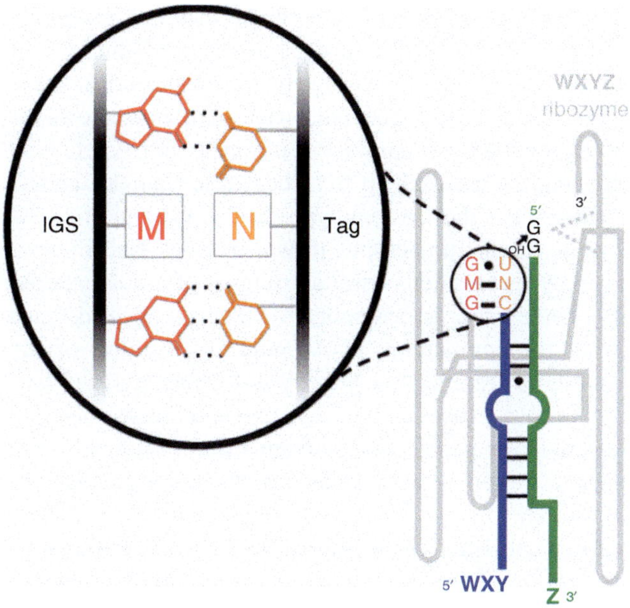

Figure 15.5 Self-assembly of the *Azoarcus* ribozyme from **WXY** and **Z** fragments. Variation in the middle nucleotides of the IGS and tag of the **WXY** fragment (M and N, respectively), leads to 16 possible **WXY** genotypes. Source: Yeates et al. [27]. © 2016 PNAS.

ribozyme, including both the guide sequence and tag and forms *trans* complexes with the highest efficiency. When mixed with **Z** on their own, all 16 genotypes have some ability to self-assemble but the rate of self-assembly is highest when M and N are complementary [27].

In order to create canonical two-player models of evolutionary game theory, such as the Prisoner's Dilemma, the Stag Hunt, and Hawk–Dove scenarios, different genotypes of *Azoarcus* RNA can be pitted against one another in the same pot. If two genotypes are competed against one another, there are 120 possible competitions that can take place among the 16 *Azoarcus* RNA genotypes (16 for the first, multiplied by 15 for the second, divided by 2 to eliminate redundancies). Some of these genotypes would be expected to efficiently self-assemble, while others cross assemble. In all cases, *trans* complexes of both genotypes should form at a roughly similar rate, but the efficiency of self-assembly and cross-assembly will vary according to the genotype.

In order to create a "dominance" scenario, in which one genotype is far more prevalent than another, a representative competition is that of CG versus GA, where the first letter is the nucleotide at M in the guide sequence and the second is for N in the tag. The CG genotype is a Watson–Crick pair and so it assembles itself quite robustly. The GA genotype is not a Watson–Crick pair, hence assembling itself poorly. Furthermore, the two possible cross-assembly pairs CA and GG are weak interactions, so this competition does not have significant cross assembly. As a result, the CG type is the predominant type among the assembled covalent ribozymes.

An example of counter-dominance, which is akin to the classical Prisoner's Dilemma scenario, can be found with the pairs CA and GG. Neither of these pairs are Watson–Crick pairs and neither assembles as well as Watson–Crick pairs. Nevertheless, in isolation, the CA genotype assembles itself with a rate constant three times higher than the GG genotype. In comparison, however, the CA genotype is far more effective at catalyzing the assembly of GG ribozymes, whereas the GG genotype cannot effectively assemble the CA genotype. In this example the GG genotype comes to dominate the competition despite its worse self-assembly rate.

The *Azoarcus* ribozyme genotypes also demonstrate examples of mutual coexistence, driven either by self-assembly or cross-assembly. An example of coexistence by self-assembly results from the competition of AU and UC. Neither of the cross-assembly pairs (AC and UU) are effective, but the A–U Watson–Crick pair and U–C pyrimidine pair both prove to be relatively effective at self-assembly. The result also shows that Watson–Crick pairs alone, though usually sufficient, are not inherently required for effective self-assembly. This result is an example of the classical Stag Hunt game of Biology.

Finally, an example of mutual coexistence by cross-assembly can be found with the inverse of the pairs in the Stag Hunt game: AC and UU. Neither of these pairs efficiently self-assembles, but the cross-assembly pairs are AU and UC, which have moderate self-assembly rates. Incubated together, both genotypes will persist in covalent ribozymes at over 40% each. This last example is equivalent to the Hawk–Dove game of evolutionary dynamics, in which each population is dependent on the other [27, 28].

In theory, the network of *Azoarcus* RNA genotypes can be expanded to include all 120 genotypes and an experimentally unfeasible number of scenarios [29]. A recent paper describes all possible three-player *Azoarcus* RNA networks with mathematical models [30] and suggests that the total number of cooperative or semi-cooperative scenarios exceeds the number of selfish scenarios. One representative scenario that has been performed in vitro is a rock-paper-scissors game with a cyclical arrangement of dominance scenarios. Using the pairs AA, UC, and GU, it can be found that a network of *Azoarcus* ribozymes predictably forms an equilibrium of all three genotypes. In isolation, the AA pair beats UC due to the superiority of the Watson–Crick pair in one of the cross-assembly reactions, resulting in a roughly 70% yield for the AA ribozyme. In the same vein, UC is preferentially assembled over GU and GU is assembled over AA. However, when all three genotypes are incubated together, each genotype comes to represent less than 40% of the total population [30].

15.11 Significance of Game Theoretic Treatments

In an autocatalytic network, or cycle of RNA reactions, sequence alone is not an absolute requirement and networks may be able to persist as long as the integrity of the catalytic unit is maintained. Point mutations need not inactivate entire catalysts, and the *Azoarcus* ribozyme retains its catalytic potential despite what is likely a critical change in the active site; the alteration of either its IGS or tag. Thus, it may have been possible on the primitive Earth to have a network of self-assembling ribozymes with different genotypes. The existence of different genotypes could provide a buffer against informational decay and assist in the recovery of catalytic potential lost by mutations or information swapping. New information could be incorporated by recombination with other strands that drifted into the network; in a large enough network, the cost of incorporating new information would be more easily mitigated. The use of game theory to quantitatively predict the outcome of evolutionary dynamics scenarios in a network of *Azoarcus* ribozymes is the first of its kind in a strictly chemical system.

15.12 Other Recombinase Ribozymes

The *Azoarcus* RNA has not been the only ribozyme used to piece together RNA molecules after the *Tetrahymena* group I intron. The hairpin ribozyme has been shown to ligate together short oligomers, exploiting the innate reactivity of the 2′,3′-cyclic phosphate [31]. While these reactions are technically not recombination, they do lead to a form of self-reproduction that could parallel – or be integrated into – a recombination-based RNA network [32]. The hairpin is a powerful ribozyme for such recombination as it can catalyze both the forward (strand scission) and reverse (ligation) phases of the reaction. This approach has been utilized by Holliger, who also used a variant of the hairpin ribozyme to piece together the 24-3 RNA polymerase ribozyme from RNA oligomers as short as 17 nt [33]. Similar to the reaction catalyzed by *Azoarcus* RNA, the recombination reaction of the hairpin ribozyme required a 3-nt guide sequence to be inserted in each of the substrate strands that were combined to form the polymerase ribozyme. Although the researchers doped each reaction with an increasing amount of substrate to drive the reaction in the forward direction via Le Chatelier's principle, it remains an impressive demonstration of the power of small RNA recombinases to assist in forming large complex RNAs.

Perhaps one of the more intriguing models of recombination from small linear RNAs of relevant prebiotic lengths is the reaction originally developed by Vlassov and coworkers [34], shown in Figure 15.6.

In this reaction, two similar, short RNA strands are brought in close proximity by a template strand, or the splint. The splint, oriented from 3′ to 5′ binds to both

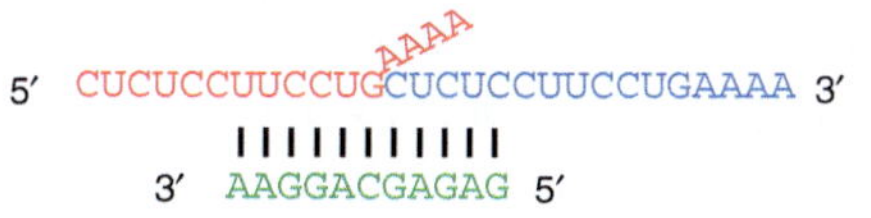

Figure 15.6 Arrangement of three short RNA oligomers in a spontaneous RNA–RNA recombination reaction. The green 11-mer oligomer serves as a splint to catalyze the recombination of two copies of the other 16-mer oligomer to give the 28-mer recombinant product 5′-CUCUCCUUCCUGCUCUCCUUCCUGAAAA-3′ (plus the 4-mer AAAA). Source: Adapted from Lutay et al. [34], Nechaev et al. [35].

strands with a unique region of full Watson–Crick complementarity for each. The poly-A tail of the top strand does not bind to the splint and instead drifts free above the double-stranded complex. In the first step of the recombination reaction, the poly-A tail is spontaneously (and specifically) cleaved, leaving a terminal guanosine with a 2′,3′-cyclic phosphate. The guanosine remains hydrogen-bonded to a cytosine in the splint and in the second step, the cyclic phosphate is attacked by the nearby 5′-hydroxyl of the second strand above the splint, forming either a 2′–5′ or 3′–5′ linkage between the two top strands. Digestion of the product strand, a 28-mer, with ribonuclease T1, which cleaves only the 3′–5′ bond at the 3′ end of a guanosine, has demonstrated that a majority of the linkages formed in the cleavage and ligation reaction are of the 2′–5′ variety [34].

Although the splint in this model is too small to form a significant tertiary structure and does not catalyze a chemical reaction in the traditional sense of an enzyme, it has all the properties of a multiple-turnover ribozyme [36]. It substantially accelerates the cleavage and ligation reaction that occurs at negligible levels in its absence; it may dissociate from the complex upon completion from the reaction, it is substrate-specific according to its hydrogen bonding, and it is neither consumed nor altered in the reaction. Moreover, it is possible for the reaction to happen a second time, producing a 40-nt strand from the product 28-mer. It is possible that structure affects the reaction in a limited way: the 11-nt base-paired region is enough to form an A-form alpha helix, which may promote cleavage of the overhanging tail. The basic features of the reaction imply that nearly every short RNA strand could potentially be a ribozyme if given the right substrates, and while at first glance this seems like a grandiose statement, there is a merit of truth to it in that the general reaction scheme is easily duplicated with wide range of short RNAs [36].

Recombination has also been characterized as a general means of exploring sequence space [5, 37] and maintaining genomic integrity in viruses. In RNA viruses such as coronaviruses, mutations in the lengthy RNA genome are commonplace and homologous recombination may help preserve the most functional parts of the genome [1]. On the other hand, RNA recombination may also facilitate rapid evolution; for example, children immunized with poliovirus vaccine of different serotypes have detectable levels of recombinant viruses eight days after vaccine administration [38].

Although it is a common assumption that homologous recombination is the primary form of recombination among RNA viruses, Chetverin has pointed out that it is difficult to detect non-homologous recombinants and defective interfering RNA because, among other reasons, such mutants are usually not competent [39]. One solution to this problem is to amplify RNA pieces with Qβ replicase in a cell-free system [40], which has demonstrated that viral RNA can recombine in physiological conditions in the presence of magnesium ions and without any ligase or protein factor. In addition, a study of a pyrimidine nucleotide synthase ribozyme found that non-homologous recombination is an efficient way to isolate the catalytic core of a ribozyme, suggesting that non-homologous recombination could be a means of molecular evolution [41]. Thus, it appears that capacity for RNA–RNA recombination is an intrinsic property of RNA molecules.

15.13 Conclusions

The general acceptance that ribozyme catalysis is ancient, as evidenced by the all-RNA active site of the ribosome [42], has made the RNA World hypothesis much more meaningful, and indeed this motive has been one of the principle motivations behind research into ribozymes since their initial discovery. As a result, there is a considerable amount of literature today on artificial ribozymes, including such curiosities as the Diels–Alderase ribozyme, the class I ligase, and aptamers. Nonetheless, despite the fascinating and potential utility of these constructs, such as the use of aptamers to treat health problems, the fact remains that the vast majority of naturally occurring ribozymes catalyze some form of trans-esterification and as such are already equipped with the essential chemical tools to perform recombination. It may be that there was an RNA world that was completely supplanted by protein enzymes, as Gilbert suggested [43], but if one counts natural ribozymes as relics, it seems reasonable to conclude that trans-esterification chemistry, and therefore recombination, had to have played a role in any RNA world that may have existed. The *Azoarcus* ribozyme, with its ability to self-assemble into a *trans* complex that catalyzes its own formation with covalent bonds from inactive oligomers, is an example of RNA structural formation, catalysis, and bootstrapping that could have occurred on the early Earth and been the precursor to an RNA world or a ribonucleoprotein world that preceded life.

Using the evidence and background provided by the recombination abilities of the *Azoarcus* RNA, it has become possible to construct a narrative of the origin of life of complexity by recombination. That narrative runs something like this: In the beginning, there were small RNA oligomers formed by some abiotic or mineral catalyzed processes. These very short RNAs would not have exceeded 15–20 nt in length and would not by themselves be significant catalysts. However, a general process of recombination over time could have increased the size, length, and structural diversity of early RNA pools until there were fragments large enough to assemble into *trans* complexes or transient, catalytic tertiary structures [44]. These molecules could then have been the first enzymes. Some of these molecules would have had recombinase activity that could accelerate the process of forming covalent bonds between trans complexes or even shorter RNAs; shorter versions of the *Azoarcus* recombinase have been selected in vitro in fact [45]. Recombinases that acted only at short, sequence-specific junctions could have recombined a wide variety of substrates, providing a mechanism of generating diversity. Constant turnover from recombination events and catalyzed hydrolysis would have eliminated unstable structures and favored the formation of stable structures, leading to a gradual ecological succession of stable molecules and their component precursors. Finally, some of the larger products of this recombination explosion would have had real and powerful catalytic activity. This could have come in the form of an RNA polymerase ribozyme capable of synthesizing full-length copies of itself or its component precursors (which would recombine into the full-length polymerase ribozyme). Perhaps, it could have been a primitive peptidyl-transferase center that polymerized short amino acid chains; these polypeptides could have become primitive RNA polymerases. Brought in proximity, a primitive peptidyl-transferase ribozyme and a polypeptide RNA polymerase could have formed an irreversible wheel that accelerated diversity, length, and activity until they could finally copy each other and form the ribonucleoprotein world, an obvious precursor to the last universal common ancestor.

References

1 Lai, M.M. (1992). RNA recombination in animal and plant viruses. *Microbiol. Rev.* 56: 61–79.

2 Brendel, V., Brocchieri, L., Sandler, S.J. et al. (1997). Evolutionary comparisons of RecA-like proteins across all major kingdoms of living organisms. *J. Mol. Evol.* 44: 528–541.

3 Lehman, N., Arenas, C.D., White, W.A., and Schmidt, F.J. (2011). Complexity through recombination: from chemistry to biology. *Entropy* 13: 17–37.

4 Vaidya, N., Manapat, M.L., Chen, I.A. et al. (2012). Spontaneous network formation among cooperative RNA replicators. *Nature* 491: 72–77.

5 Pesce, D., Lehman, N., and de Visser, J.A.G.M. (2016). Sex in a test tube: testing the benefits of in vitro recombination. *Philos. Trans. R. Soc. Lond. Ser. B Biol. Sci.* 371.

6 Reinhold-Hurek, B. and Shub, D.A. (1992). Self-splicing introns in tRNA genes of widely divergent bacteria. *Nature* 357: 173–176.

7 Fica, S.M., Tuttle, N., Novak, T. et al. (2013). RNA catalyses nuclear pre-mRNA splicing. *Nature* 503: 229–234.

8 Zaug, A.J. and Cech, T.R. (1986). The intervening sequence RNA of *Tetrahymena* is an enzyme. *Science* 231: 470–475.

9 Chowrira, B.M., Berzal-Herranz, A., and Burke, J.M. (1993). Novel RNA polymerization reaction catalyzed by a group I ribozyme. *EMBO J.* 12: 3599–3605.

10 Adams, P.L., Stahley, M.R., Kosek, A.B. et al. (2004). Crystal structure of a self-splicing group I intron with both exons. *Nature* 430: 45–50.

11 Adams, P.L., Stahley, M.R., Gill, M.L. et al. (2004). Crystal structure of a group I intron splicing intermediate. *RNA* 10: 1867–1887.

12 Draper, W.E., Hayden, E.J., and Lehman, N. (2008). Mechanisms of covalent self-assembly of the *Azoarcus* ribozyme from four fragment oligonucleotides. *Nucleic Acids Res.* 36: 520–531.

13 Riley, C.A. and Lehman, N. (2003). Generalized RNA-directed recombination of RNA. *Chem. Biol.* 10: 1233–1243.

14 Kuo, L.Y., Davidson, L.A., and Pico, S. (1999). Characterization of the *Azoarcus* ribozyme: tight binding to guanosine and substrate by an unusually small group I ribozyme. *Biochim. Biophys. Acta* 1489: 281–292.

15 Sullenger, B.A. and Cech, T.R. (1994). Ribozyme-mediated repair of defective mRNA by targeted, trans-splicing. *Nature* 371: 619–622.

16 Watanabe, T. and Sullenger, B.A. (2000). Induction of wild-type p53 activity in human cancer cells by ribozymes that repair mutant p53 transcripts. *Proc. Natl. Acad. Sci. U.S.A.* 97: 8490–8494.

17 Cheng, L.K.L. and Unrau, P.J. (2010). Closing the circle: replicating RNA with RNA. *Cold Spring Harb. Perspect. Biol.* 2: a002204.

18 Hayden, E.J., Riley, C.A., Burton, A.S., and Lehman, N. (2005). RNA-directed construction of structurally complex and active ligase ribozymes through recombination. *RNA* 11: 1678–1687.

19 Doudna, J.A. and Cech, T.R. (1995). Self-assembly of a group I intron active site from its component tertiary structural domains. *RNA* 1: 36–45.

20 Hayden, E.J. and Lehman, N. (2006). Self-assembly of a group I intron from inactive oligonucleotide fragments. *Chem. Biol.* 13: 909–918.

21 Jayathilaka, T. and Lehman, N. (2018). Spontaneous covalent self-assembly of the *Azoarcus* ribozyme from five fragments. *ChemBioChem* 19: 217–220.

22 Hayden, E.J., von Kiedrowski, G., and Lehman, N. (2008). Systems chemistry on ribozyme self-construction: evidence for anabolic autocatalysis in a recombination network. *Angew. Chem. Int. Ed. Engl.* 47: 8424–8428.

23 Paul, N. and Joyce, G.F. (2002). A self-replicating ligase ribozyme. *Proc. Natl. Acad. Sci. U.S.A.* 99: 12733–12740.

24 Turner, P.E. and Chao, L. (1999). Prisoner's dilemma in an RNA virus. *Nature* 398: 441–443.

25 Kerr, B., Riley, M.A., Feldman, M.W., and Bohannan, B.J.M. (2002). Local dispersalpromotes biodiversity in a real-lifegame of rock-paper-scissors. *Nature* 418: 171–174.

26 Arsène, S., Ameta, S., Lehman, N. et al. (2018). Coupled catabolism and anabolism in autocatalytic RNA sets. *Nucleic Acids Res.* 46 (18): 9660–9666.

27 Yeates, J.A.M., Hilbe, C., Zwick, M. et al. (2016). Dynamics of prebiotic RNA reproduction illuminated by chemical game theory. *Proc. Natl. Acad. Sci. U.S.A.* 113: 5030–5035.

28 Nowak, M.A. and Sigmund, K. (2004). Evolutionary dynamics of biological games. *Science* 303: 793–799.

29 Yeates, J.A.M., Nghe, P., and Lehman, N. (2017). Topological and thermodynamic factors that influence the evolution of small networks of catalytic RNA species. *RNA* 23: 111–121.

30 Mathis, C., Ramprasad, S.N., Walker, S.I., and Lehman, N. (2017). Prebiotic RNA network formation: a taxonomy of molecular cooperation. *Life (Basel)* 7.

31 Gwiazda, S., Salomon, K., Appel, B., and Müller, S. (2012). RNA self-ligation: from oligonucleotides to full length ribozymes. *Biochimie* 94: 1457–1463.

32 Hieronymus, R., Godehard, S.P., Balke, D., and Müller, S. (2016). Hairpin ribozyme mediated RNA recombination. *Chem. Commun. (Camb.)* 52: 4365–4368.

33 Mutschler, H., Wochner, A., and Holliger, P. (2015). Freeze-thaw cycles as drivers of complex ribozyme assembly. *Nat. Chem.* 7: 502–508.

34 Lutay, A.V., Zenkova, M.A., and Vlassov, V.V. (2007). Nonenzymatic recombination of RNA: possible mechanism for the formation of novel sequences. *Chem. Biodivers.* 4: 762–767.

35 Nechaev, S.Y., Lutay, A.V., Vlassov, V.V., and Zenkova, M.A. (2009). Non-enzymatic template-directed recombination of RNAs. *Int. J. Mol. Sci.* 10: 1788–1807.

36 Smail, B.A., Clifton, B.E., Mizuuchi, R., and Lehman, N. (2019). Spontaneous advent of genetic diversity in RNA populations through multiple recombination mechanisms. *RNA* 25 (4): 453–464.

37 Lehman, N. and Unrau, P.J. (2005). Recombination during in vitro evolution. *J. Mol. Evol.* 61: 245–252.

38 Minor, P.D., John, A., Ferguson, M., and Icenogle, J.P. (1986). Antigenic and molecular evolution of the vaccine strain of type 3 poliovirus during the period of excretion by a primary vaccine. *J. Gen. Virol.* 67: 693–706.

39 Chetverin, A.B. (1999). The puzzle of RNA recombination. *FEBS Lett.* 460: 1–5.

40 Chetverina, H.V., Demidenko, A.A., Ugarov, V.I., and Chetverin, A.B. (1999). Spontaneous rearrangements in RNA sequences. *FEBS Lett.* 450: 89–94.

41 Wang, Q.S. and Unrau, P.J. (2005). Ribozyme motif structure mapped using random recombination and selection. *RNA* 11: 404–411.

42 Nissen, P., Hansen, J., Ban, N. et al. (2000). The structural basis of ribosome activity in peptide bond synthesis. *Science* 289: 920–930.

43 Gilbert, W. (1986). Origin of life: the RNA world. *Nature* 319: 618.

44 Blokhuis, A. and Lacoste, D. (2017). Length and sequence relaxation of copolymers under recombination reactions. *J. Chem. Phys.* 147: 094905.

45 Burton, A.S. and Lehman, N. (2010). Enhancing the prebiotic relevance of a set of covalently self-assembling, autorecombining RNAs through in vitro selection. *J. Mol. Evol.* 70: 233–241.

16

Engineering of Hairpin Ribozymes for RNA Processing Reactions

Robert Hieronymus, Jikang Zhu, Bettina Appel, and Sabine Müller

University Greifswald, Institute for Biochemistry, Felix-Hausdorff-Str. 4, 17487 Greifswald, Germany

16.1 Introduction

Over the past 30 years, the structure and the mechanism of ribozymes and other functional RNAs have been extensively studied and nowadays are well understood in many cases. Thus, a large number of RNAs have been developed into specialized tools with potential applications in molecular biology, pharmaceutical and environmental diagnostics, or molecular medicine [1–7]. Ribozymes occurring in nature have been known since the beginning of the 1980s, when the first two catalytic RNAs, the *Tetrahymena* ribozyme and the catalytic RNA subunit of RNase P, were discovered [8, 9]. Other catalytic RNA motifs followed, among those the hammerhead and hairpin ribozymes, two rather small RNA structures derived from plant virus satellite RNAs, where they play essential roles in the processing of replication intermediates. In particular, the ribozyme motifs mediate cleavage of large multimeric RNAs that are produced by rolling-circle replication, to unit length, and assist their circularization, before those are packaged and carried over to other cells [10, 11].

Owing to their small size and rather simple structure, the hammerhead and the hairpin ribozyme are two of the most extensively studied catalytic RNA motifs. A large amount of data on their structure and mechanism have been collected and contributed to understanding these ribozymes to an extent that allows to turn them into useful tools. Also, the rather large sequence variability of both ribozymes is an advantage, making them particularly well suited for engineering into new functional structures. Thus, a number of species were designed for the knockdown of specific RNA substrates by adapting the substrate-binding domain to cleave the specific target sequence [12].

Apart from application in diagnostics and therapy, ribozymes play an important role in the origin of life. The hypothesis of the RNA world puts ribozymes in a central role as catalysts of RNA-based life [13]. Thus, a number of RNA engineering projects in our lab have focused on the engineering of ribozymes for RNA processing reactions. We have particularly worked on the hairpin ribozyme, which owing

Ribozymes, First Edition. Edited by Sabine Müller, Benoît Masquida, and Wade Winkler.

to its characteristic cleavage–ligation nature is particularly well suited for RNA processing involving both reactivities and thus has been engineered into variants that support such essential reactions as RNA splicing and recombination [14].

16.2 The Naturally Occurring Hairpin Ribozyme

The hairpin ribozyme (Figure 16.1) is derived from the negative strand of the satellite RNA of tobacco ringspot virus and mediates the reversible cleavage of a specific phosphodiester bond of a suitable RNA substrate, thus generating products with characteristic ends: a 2′,3′-cyclic phosphate and a free 5′-OH group [15–17].

As mentioned above, the hairpin ribozyme is a structural motif embedded in a viral satellite RNA, where it mediates self-cleavage and likewise self-ligation of intermediates occurring during rolling-circle replication. However, in vitro, the hairpin ribozyme can act as a true multiple turnover catalyst, supporting cleavage or ligation of separate RNA substrates (Figure 16.1) [17]. The cleavage reaction proceeds via in-line attack of the 2′-hydroxyl group on the neighboring phosphate, leading to the departure of the adjacent 5′-oxygen and to the generation of the 2′,3′-cyclic phosphate. The rate enhancement over the non-catalyzed reaction is about 10^7–10^9. Typical cleavage rates are between 0.1 and 1 molecule per minute at physiological salt concentrations, dependent on the particular assay and conditions. The reverse reaction, ligation, proceeds via the same reaction path in opposite direction, and involves nucleophilic attack of the 5′-OH group of one fragment onto the phosphorus of the 2′,3′-cyclic phosphate of the other, leading to ring opening and consequently to end-joining of the two fragments. Ligation is kinetically favored and proceeds about five times faster than cleavage. In contrast to other ribozymes, the hairpin ribozyme is quite efficient in ligation catalysis. This is because the hairpin ribozyme binds its substrates in a rather rigid structure, such that the conformation of the substrate-binding domain before and after cleavage/ligation remains virtually the same. Thus, the entropic cost of ligation is rather low and can be compensated by the favorable enthalpy of the ligation reaction that results from the ring opening of the cyclic phosphate. However, this applies only when substrates remain bound to the ribozyme long enough to allow ligation to proceed. If dissociation of cleavage fragments/ligation substrates is fast, cleavage will be the favored reaction. Hence, the preference for ligation or cleavage can be controlled by the structure of the ribozyme–substrate complex. Ligation is favored if substrates are tightly bound to the ribozyme folded into a stable structure, while cleavage occurs from ribozyme–substrate complexes that are less stable, yet stable enough to fold into a catalytically competent structure [18]. The secondary structure of the minimal catalytic motif shown in Figure 16.1a would favor cleavage, owing to the rather short RNA fragments, which upon cleavage would quickly dissociate. Stabilization of the ribozyme–substrate complex by extending the substrate-binding arms, as will be shown below, would result in three- or even four-way junction structures and thus support ligation. The minimal motif consists of four Watson–Crick base-paired helices separated by two internal loops A and B. Essential nucleotides for activity

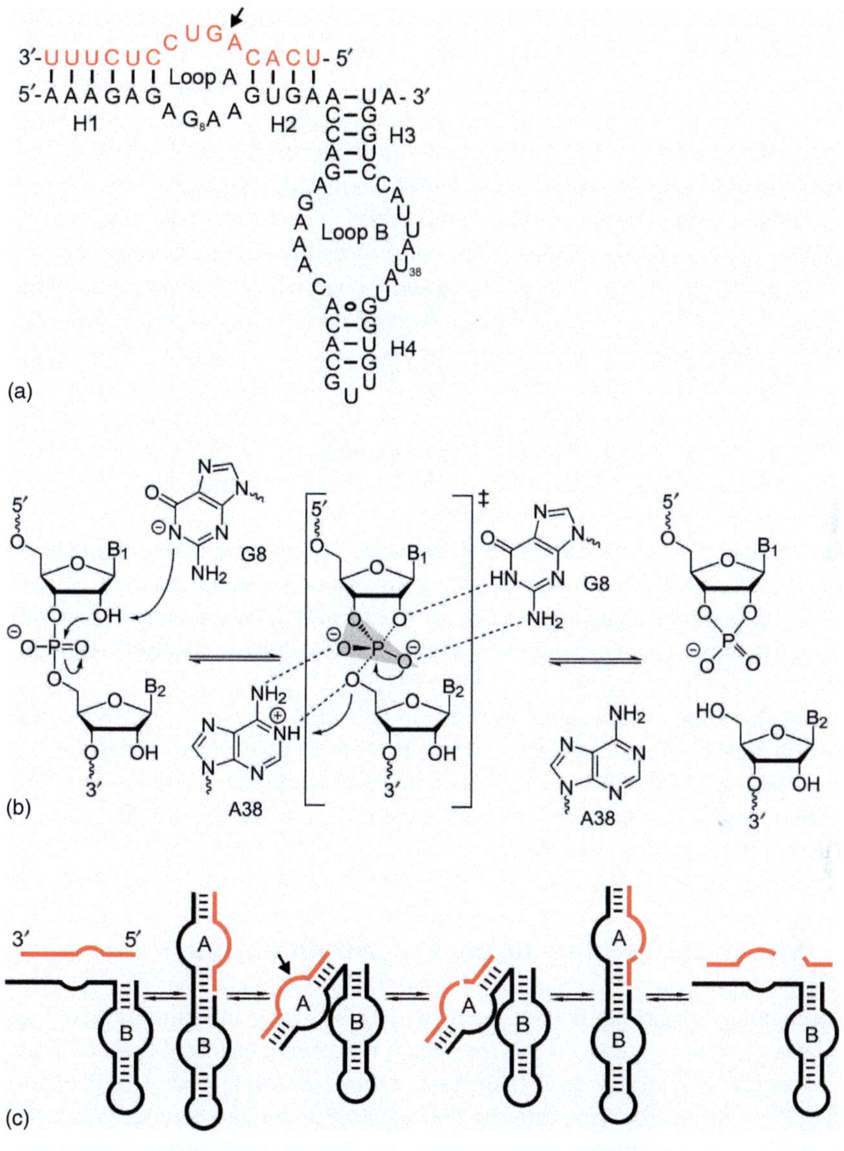

Figure 16.1 (a) Secondary structure of the hairpin ribozyme. The arrow denotes the site of reversible cleavage. (b) Mechanism of reversible phosphodiester cleavage. (c) Conformational dynamics of the ribozyme during one catalytic cycle.

reside within the two loops [19], and linker insertion [20, 21] as well as chemical modification studies [22] of the helix 2–3 junction indicated that helix 2 is not coaxially stacked upon helix 3 in the catalytically competent structure. This allowed for the conclusion that a bend between helix 2 and 3 is required to bring essential elements in loop B proximal to the cleavage site located in loop A. Hence, the hairpin ribozyme has to fold into a docked conformation with close contacts between loops

A and B for catalysis [23]. The docked conformation is stabilized by divalent metal ions and generates the local environment in which catalysis occurs. A Watson–Crick base pair between a loop A guanosine (G + 1) and a loop B cytidine (C25) as well as an interdomain ribose zipper further stabilize the docked state. Moreover, fluorescence resonance energy transfer (FRET) measurements demonstrated that docking is stabilized in four-helix junction constructs compared with minimal hairpin ribozymes [24–26]. Substrate binding and product dissociation cannot take place as long as the complex is locked in the docked state, but proceed from an extended conformation, in which the two ribozyme domains are coaxially stacked upon one another (Figure 16.1c). In order to allow catalysis to proceed in a multiple turnover fashion, the ribozyme iteratively switches between the open (extended) and closed (docked) conformation, thus binding substrates and releasing products [27].

To achieve catalysis, the ribozyme employs RNA functional groups of its nucleotide components. The crystal structure of a precursor form of the RNA and biochemical experiments have suggested that A38 as active-site adenosine and G8 as active-site guanosine adopt the role of a general acid or base, respectively [28–31]. Metal ions or other catalytic cofactors are not used for active-site chemistry, yet are essential to neutralize the Coulomb potential brought about by the negatively charged phosphates in the folded structure. This role is typically served by magnesium ions but can be taken over also by other cations or positively charged molecules, such as spermine and spermidine [32–34].

Taken together, the unique cleavage–ligation behavior of the hairpin ribozymes renders it an ideal candidate for engineering into tools for RNA processing reactions involving RNA cleavage and ligation in a controlled fashion. Structural manipulation of the ribozyme–substrate complex allows to favor the one or the other activity and to use those in the desired order.

16.3 Structural Variants of the Hairpin Ribozyme

Most biochemical experiments have been carried out using the minimal hairpin ribozyme as shown in Figures 16.1a. However, it has been recognized early on that the four-way junction structure, as it appears in the virus satellite RNA, assists the folding of the hairpin ribozyme into the docked state with close contacts between loops A and B. Four-way junction constructs have a slightly stabilized structure compared with the two arm hairpin ribozymes (Figure 16.2), thus being more efficient in catalysis, and most importantly would strongly favor ligation [18, 35]. As shown by Burke and coworkers, three-way junction hairpin ribozymes also support folding in the active structure with the two loops A and B in close proximity [36]. Thus, a number of four-way and three-way junction hairpin ribozymes (Figure 16.2b,c) have been used for structure and function studies [18, 25, 34, 36–38] or for the application of hairpin ribozyme constructs for ligation purposes [14, 39–45]. In addition, reverse-joined and artificially branched hairpin ribozymes [46–50] or constructs with the two domains physically separated [51, 52] (Figure 16.2d–f) have been studied mainly with the aim of demonstrating essential interdomain contacts required for catalysis.

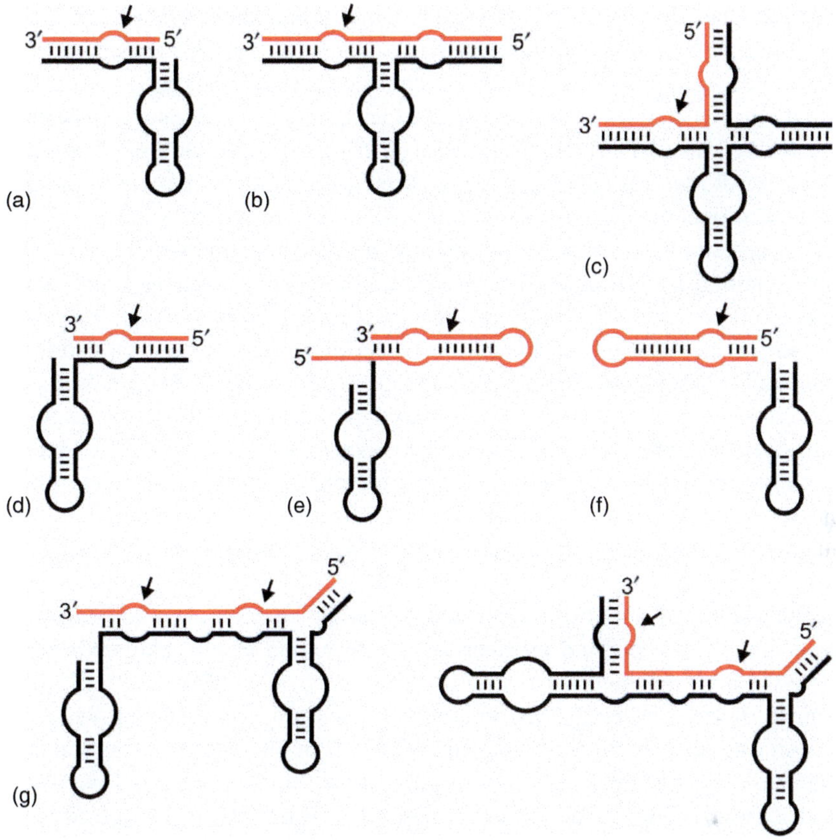

Figure 16.2 Structural variants of the hairpin ribozyme (schematically). (a) Minimal motif, (b) three-way junction, (c) four-way junction, (d) reverse-joined, (e) artificially branched, (f) physically separated domains, and (g) twin ribozymes (tandem duplication and reverse-joined).

A structural variant of the hairpin ribozyme that has been developed and extensively studied in our laboratory is twin ribozymes (Figure 16.2g). Our twin ribozymes are derived from the hairpin ribozyme by tandem duplication [39] or by a combination of a conventional hairpin ribozyme unit with a reverse-joined hairpin ribozyme [48, 50]. Those structures strictly controlled catalyze two chain cleavages and two ligation events and hence mediate the exchange of defined patches of RNA sequence within a suitable RNA substrate [39–41, 43, 53]. Twin ribozymes have been engineered for RNA repair and for demonstration of RNA-mediated RNA recombination as will be discussed in more detail (Section 16.4).

16.4 Hairpin Ribozymes that are Regulated by External Effectors

Allosteric regulation of enzyme activity is a basic principle in nature. Accordingly, a large number of RNA structures have been developed consisting of a catalytic domain connected to a specific aptamer via a communication module [3]. This enables allosteric control of ribozyme activity; depending on the presence of the specific ligand, activity is switched on or off or at least regulated up or down. Hence, the specific ligand acts as an external effector or inhibitor. Not all systems conform to the classical principle of allosteric regulation, which would require binding of the allosteric cofactor to a region outside the catalytic domain, followed by a conformational change in the binding region that propagates to the catalytic domain. Other strategies such as inhibitor control, complementation, targeted ribozyme-attenuated probe strategies, or expansive regulatory strategies were also used [54].

There are several hairpin ribozyme constructs that are regulated by external oligonucleotides [38, 55–58] (Figures 16.3 and 16.4). In a hairpin ribozyme variant obtained from in vitro selection, the short oligonucleotide effector binds to a complementary sequence located in the hairpin loop closing helix 4 (Figure 16.3a). In the absence of the effector, the stem–loop structure of helix 4 is distorted, such that the catalytically essential loop B cannot fold properly. After binding of the oligonucleotide effector, the catalytically competent structure is restored, and activity is switched on [55].

We have developed a hairpin ribozyme derivative that requires an oligonucleotide effector as a structural element for the formation of the catalytic center, thus applying a complementation strategy rather than true allosteric activation [38] (Figure 16.3b). An inactive hairpin ribozyme variant was designed, wherein inactivation was achieved by mutation from C25 to G to interrupt the interdomain G–C Watson–Crick base pair and by the introduction of G–U wobble base pairs and mismatches in helix 4 to destabilize domain B. Addition of the oligonucleotide would restore the structure of the B domain and thus activity of the formerly inactive ribozyme. We have shown that the oligonucleotide effector is able to invade

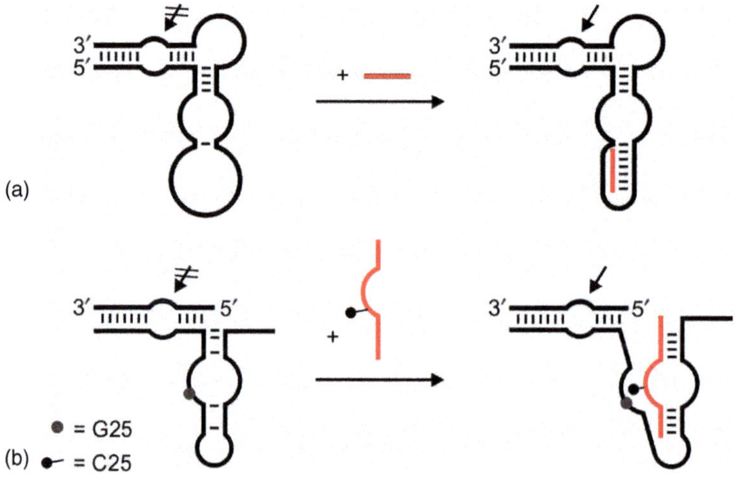

Figure 16.3 Hairpin ribozyme variants that can be activated by external oligonucleotide effectors. Source: (a) Based on Komatsu et al. [55]; (b) Based on Vauléon and Müller [38].

minimal hairpin ribozyme structures as well as three-way junction constructs, allowing both variants to cleave their substrates almost completely (96%). For this, a five- to eightfold excess of the effector oligonucleotide was needed, and the cleavage rate was found being about fivefold lower as compared to the wild-type hairpin ribozyme [38].

Other hairpin ribozyme variants that can be induced or suppressed by external effector oligonucleotides by interfering with the docking process that brings domains A and B in close proximity have been designed by Famulok and coworkers [56]. Those hairpin ribozyme variants are responsive to the *trp* leader messenger RNA (mRNA), which is the RNA sequence that is bound by L-tryptophan-activated trp-RNA-binding attenuation protein. A third domain C was added via a pseudo-three-way junction to allow the formation of alternative stable RNA motifs, such as the internal stem–loop structure in rHP-TRAP (Figure 16.4a) or a pseudo-half-knot in iHP-TRAP (Figure 16.4b). Addition of an oligonucleotide (or an entire *trp* mRNA) that is complementary to the sequence segment of domain C altered the activity of the ribozyme by forcing domains A and B into a stretched conformation (Figure 16.4a). The same RNA effector was used also with an inducible hairpin ribozyme variant, triggering a conformational change of the inactive variant iHP-TRAP, to generate a pseudo-half-knot structure that restores the proper ribozyme fold and thus activates the ribozyme (Figure 16.4b). The maximum rate increase was found at an equimolar ratio of the RNA effector and rHP-TRAP or iHP-TRAP. This strategy is easily adaptable to other sequences and thus can be used for the detection of a specific RNA target. In another series of experiments, two regulatory factors were used to switch ribozyme activity in opposite directions [57]. The key element was a ribozyme variant that harbors an aptamer sequence responsive to flavine mononucleotide (FMN) in

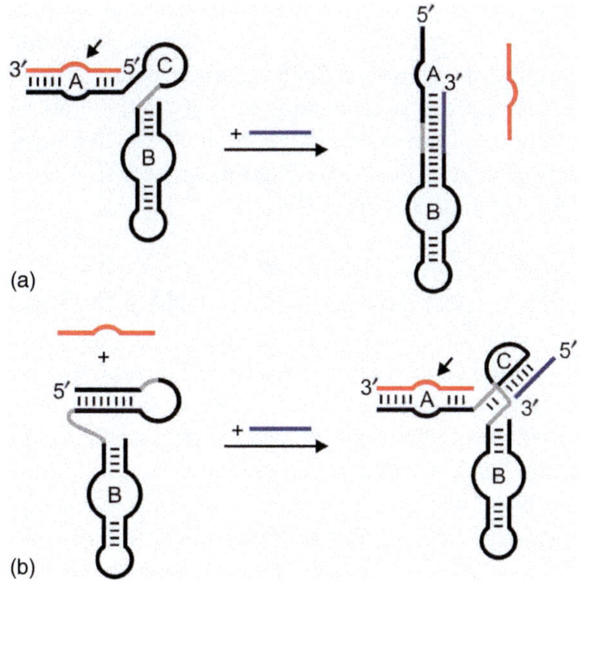

Figure 16.4 Repressible (a) and inducible (b) hairpin ribozyme variants. Red, hairpin ribozyme–substrate; blue, repressor oligonucleotide (a), inducer oligonucleotide (b). Source: Adapted from Najafi-Shoushtari et al. [56].

domain C of the ribozyme structure. Binding of FMN to its aptamer facilitated the docking of domains A and B and thus induced activity. A short oligonucleotide complementary to the aptamer inhibited ribozyme activity by forming an extended conformer that cannot catalyze the cleavage reaction. The addition of FMN to the inactive oligonucleotide–ribozyme complex neutralized the inhibitory effect of the oligonucleotide and hence activates the ribozyme. Similarly, other hairpin ribozyme variants were constructed that contain aptamer sequences in their structure and thus can be regulated by external cofactors [58–62].

FMN-dependent ribozyme variants were also developed in our laboratory [59, 63]. In our constructs, helix 4 of the parent ribozyme was replaced by a communication module connecting the ribozyme to the FMN-specific aptamer that previously was identified by selection in vitro [64] and characterized by nuclear magnetic resonance (NMR) spectroscopy [65] (Figure 16.5). The resulting aptazyme was shown to be responsive to FMN. It readily cleaved its substrate in the presence of FMN, whereas in the absence of the ligand, the activity was considerably reduced. Hence, allosteric activation by FMN was clearly observed, with a 24-fold increase of the rate constant as compared to the reaction in the absence of the allosteric effector [59]. This increase in activity was sufficient to observe the regulation of the aptazyme in dependence on the degree of the FMN oxidation state. FMN in its oxidized state has a planar shape, owing to its conjugated aromatic ring system. Reduction of FMN destroys the aromatic character and induces a change in the molecular geometry from the planar structure to a rooflike bent shape [66, 67] (Figure 16.5). ^{1}H NMR analysis of the FMN–aptamer complex in solution showed that FMN binding to the aptamer originates from hydrophobic stacking of the planar isoalloxazine ring involving the G10–U12–A25 base triple (Figure 16.5b,c) [65]. The specificity is associated with an FMN–adenine base pair within the complex. Hence, binding of FMN is largely stabilized by hydrophobic stacking of the aromatic isoalloxazine ring with the bases of the aptamer. This characteristic feature allowed FMN binding to be controlled by oxidation/reduction of the isoalloxazine ring. Addition of dithionite to the FMN activated ribozyme leads to inactivation, owing to the change of the molecular shape of FMN and loss of its binding capacity [59]. Iterative cycles of oxidation/reduction by dithionite and oxygen allow reversible switching of the activity. Instead of using dithionite/oxygen, activity control can be also achieved by electrochemical reduction/oxidation of FMN [63].

16.5 Twin Ribozymes for RNA Repair and Recombination

As mentioned above, the hairpin ribozyme has a unique cleavage/ligation behavior, based on the fact that the internal equilibrium between cleavage and ligation is shifted toward ligation [35]. This, however, only applies when fragment dissociation is slow, such that fragments remain bound to the ribozyme long enough to allow ligation to take place. On the other hand, if dissociation of cleavage fragments is faster than ligation, cleavage will be the preferred reaction. One of the first projects where we have made use of this characteristic feature was the engineering of twin

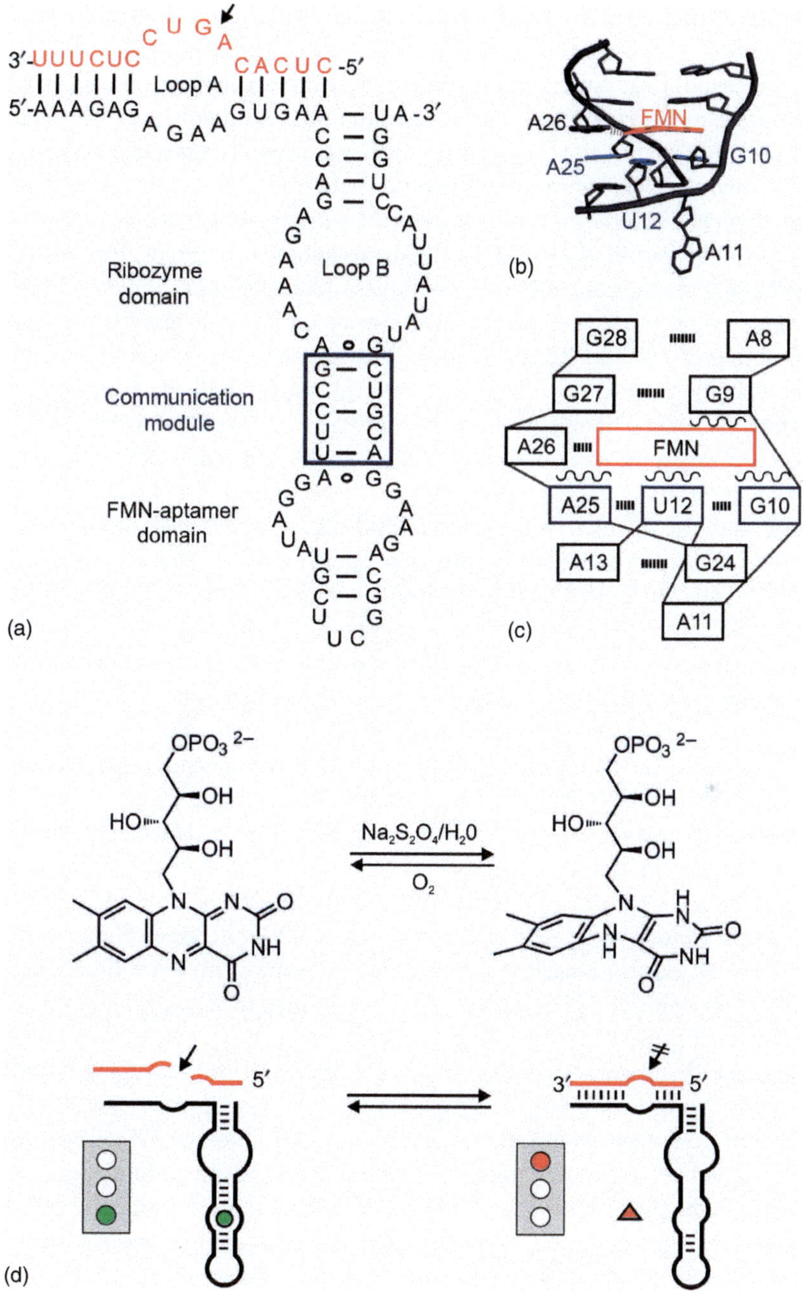

Figure 16.5 FMN-responsive hairpin ribozyme. (a) Secondary structure of the aptazyme. (b) Aptamer structure with bound FMN (PDB DOI: 10.2210/pdb1FMN/pdb). (c) Schematic representation of FMN binding to the aptamer. (d) Change of the molecular shape of FMN in response to the oxidation state. Source: (a, c, d) Adapted from Strohbach et al. [59].

ribozymes [39]. Twin ribozymes are derived from the hairpin ribozyme by tandem duplication, such that upon binding of a suitable substrate, two cleavage/ligation sites are generated (Figure 16.6). Initially, twin ribozymes were engineered for RNA double cleavage [50]. However, the specific cleavage/ligation properties of the hairpin ribozyme mentioned above can be used to remove a predefined sequence patch out of an RNA substrate followed by ligating a separate synthetic "repair" fragment into the gap left behind. This requires catalysis of two cleavage events and two ligations in a strictly controlled fashion. The driving force for the fragment exchange lies in the specific design of the ribozyme–substrate complex. It needs to be engineered in a manner that allows easy and fast dissociation of the cleaved-out fragment versus tight binding of the repair fragment, in order to remove the one fragment and to preferentially ligate the other. This was achieved by destabilization of binding of the fragment to be removed by looped out bases in the substrate or ribozyme part [39, 41] or by mismatches [40], whereas the fragment to be ligated would form a contiguous Watson–Crick duplex with the ribozyme. This way, the fragment to be removed forms fewer base pairs with the ribozyme and hence is less stably bound than the fragment to be ligated, resulting in favorable cleavage of the one and preferred ligation of the other. Thus, dependent on the specific design of the twin ribozyme–substrate complex, fragments of equal lengths have been exchanged [40], or a shorter patch was replaced with a longer one [39, 40, 43, 48, 53] and vice versa [41] (Figure 16.6). Thus, the system mimics the repair of short deletions, insertions, and base replacement mutations with up to 53% yield [40].

Owing to our initial motivation of engineering RNA tools for repair of genetic disorders at the level of RNA, we have developed a twin ribozyme targeting the mutated transcript of the *CTNNB1* gene, encoding β-catenin, a major player of the wnt-signaling pathway [68, 69]. Unfortunately, functional repair has not been achieved, albeit in vitro repair of a truncated version of the full-length transcript was successfully demonstrated [43]. Furthermore, a twin ribozyme was developed for the repair of a mutated version of the mRNA encoding the green fluorescent protein. Repair of the full-length transcript and translation into the functional protein was achieved in vitro [53], though the application of twin ribozymes in cells remained challenging.

Apart from the therapeutic potential, twin ribozymes are interesting models of RNA catalysts that may have played a role in early life. The described process of controlled fragment exchange mimics the process of recombination, which in modern biochemistry is defined as the exchange of genetic material between different organisms that leads to the production of offspring with combinations of traits that differ from those found in either parent. Thus, in the RNA world, twin ribozymes might have supported RNA recombination by the described fragment exchange mechanism and hence contributed to extending the sequence space and the function of RNA molecules in early life forms. However, with a length of about 140 nucleotides, twin ribozymes are rather long and fairly complex species for simple life forms. This led us to think about engineering of simpler hairpin ribozyme descendants for support of RNA recombination, as will be described in Section 16.6.

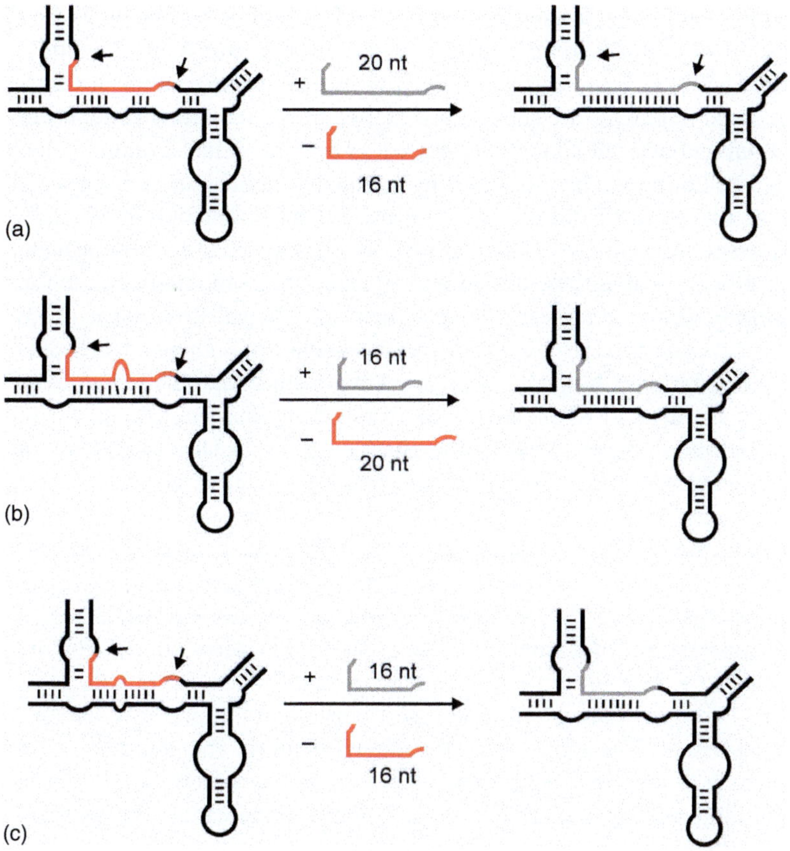

Figure 16.6 Twin ribozyme-mediated RNA recombination. The fragments to be removed form fewer base pairs with the ribozyme than the fragments to be ligated, resulting in favorable cleavage of the one and preferred ligation of the other. Depending on the specific design of the twin ribozyme–substrate complex, a shorter patch is replaced with a longer one (a), a longer patch is replaced with a shorter one (b), or fragments of equal lengths are exchanged (c).

16.6 Hairpin Ribozymes as RNA Recombinases

As we wanted to design a model for early RNA world recombination, the hairpin ribozyme is a superior starting point for engineering: It is relatively small and thus has a higher potential of emergence at the genesis of the RNA world; only 39% of its sequence is conserved, which makes it a pleasing starting material for rational design [16, 19, 70]. Most importantly, it supports RNA cleavage and ligation, the two reactions required for recombination scenarios, to virtually the same extent [27]. The preference for the one or the other reaction is dependent on and hence can be controlled by the structural stability of the ribozyme–substrate complex, thus allowing to direct ribozyme function by structural manipulation.

The concept was to design a hairpin ribozyme variant that is capable of binding two different RNA molecules, both being composed of a non- and a pro-functional region (Figure 16.7) [44]. Upon cleavage of the two individual RNA substrates non- and pro-functional fragments are generated. The two pro-functional fragments are designed to preferentially rebind to the ribozyme, to become ligated to a functional RNA. Thus, in addition to follow reactions by fragment length analysis, successful recombination can be confirmed by employing the recombination product in a functional assay. According to this concept, we decided that a trans-cleaving hammerhead ribozyme should be the desired product of recombination. Like the hairpin ribozyme, the hammerhead ribozyme is a rather small RNA. Apart from the conserved catalytic core, it consists of variable sequence domains, which lets it fit perfectly in a rational design concept, allowing to customize the sequence as needed without significant loss of activity. We inserted the cleavage/ligation site of the wild-type hairpin ribozyme A*GUC (the asterisk marks the cleaved linkage) in the loop region of stem II of the desired hammerhead ribozyme (Figure 16.7). This sequence ensures maximum reaction rates of hairpin ribozyme supported reactions when processing the two RNA substrates. This is important to ensure sufficient supply of cleavage fragments to be ligated to the final recombination product. The binding regions next to the cleavage/ligation site can be arbitrarily adjusted in both the substrate and the ribozyme. Thus, taken into account that, for the hairpin ribozyme, the equilibrium between cleavage and ligation can be controlled by the binding strength between substrate and ribozyme, pro-functional domains of the substrates were designed to bind strongly, whereas the non-functional parts bind weakly to the hairpin ribozyme. As a result, the system is directed to preferential binding and ligation of the two pro-functional fragments obtained from cleavage of the two substrates and rejection of non-functional sequence patches. Preferred binding of the pro-functional fragments over the non-functional ones was achieved by simply increasing the number of base pairs in the ribozyme–substrate complex.

For each consideration we made, we used the secondary structure prediction software RNA structure [71] to calculate the lowest free energy of the particular substrate/product–ribozyme complex. On the basis of these values, we choose the most appropriate complexes for the final system. One of the two substrates was 5′-end labeled with ATTO680 to follow the individual reactions for recombination on a LI-COR 4300 DNA sequencer.

In the initial recombination assays, the final product was obtained with only 2–9% yield. Extensive optimization of the experimental setup and of reaction conditions (temperature, magnesium ion concentration, fragment ribozyme ratios) yields of up to 76% were reached [44]. Successful recombination was confirmed by fragment length analysis and by a functional assay of hammerhead ribozyme (the recombination product) catalyzed cleavage of an externally added RNA substrate [44]. Thus, a rather simple RNA structure, as small as the hairpin ribozyme, is capable of supporting cleavage and cross-ligation of RNA substrates and thus is a superior model of an RNA recombinase ribozyme that may have played an important role in early life forms.

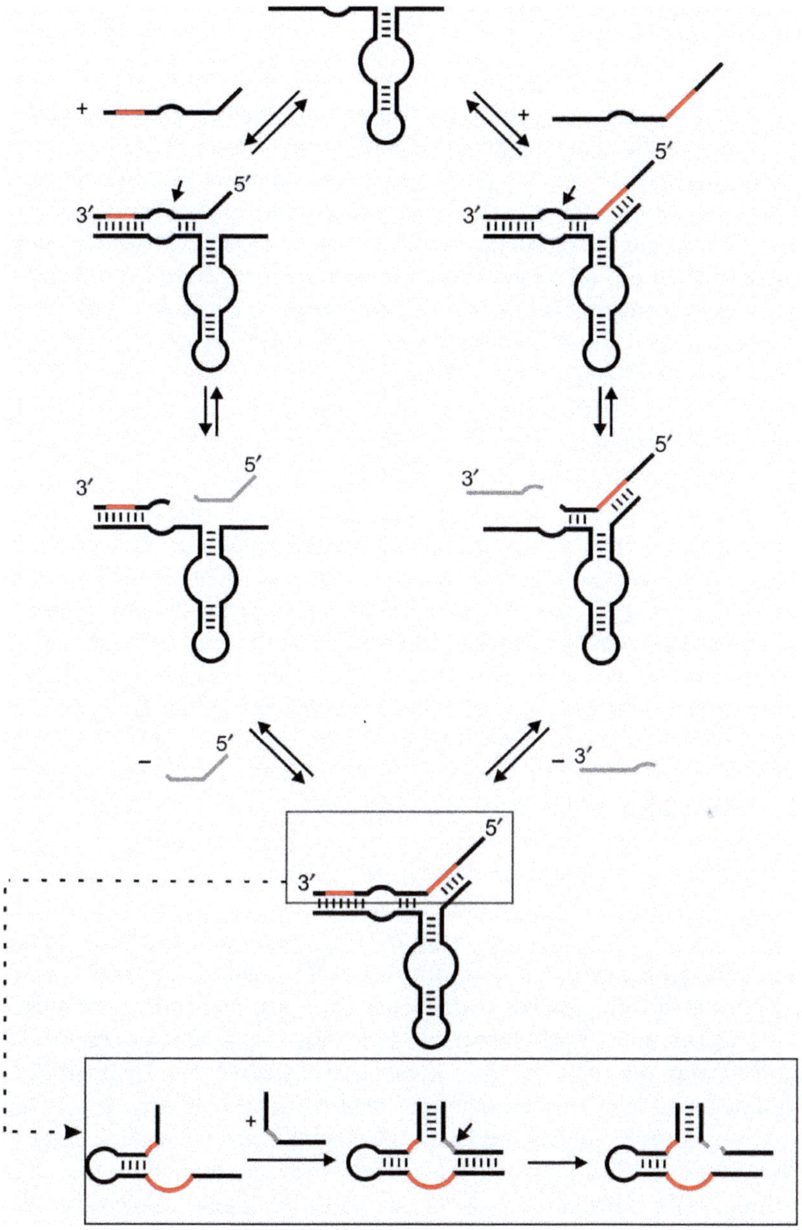

Figure 16.7 Scheme of hairpin ribozyme-mediated recombination as described in [44]. Two substrates, both being composed of a non- (gray) and a pro-functional (red/black) region are bound to and cleaved by the same hairpin ribozyme. The two pro-functional fragments are designed to preferentially rebind to the ribozyme, to become ligated to a functional RNA. The recombination product is a trans-cleaving hammerhead ribozyme that was used for proof of recombination in a functional assay (bottom). The red patches mark the conserved sequences of the hammerhead ribozyme catalytic core. Source: Adapted from Hieronymus et al. [44].

16.7 Self-Splicing Hairpin Ribozymes

Another scenario where a combination of RNA cleavage and ligation is needed in a scheme of subsequent reactions is RNA splicing. Accordingly, we have designed hairpin ribozyme variants that undergo self-processing to form oligomers and circular RNAs (circRNAs) [72–74]. circRNAs appear in all kingdoms of life, are produced from protein-coding and noncoding genes by a process named back-splicing, and are supposed to fulfill various biological functions [75]. In the process of back-splicing, exons are spliced from pre-mRNAs in noncanonical order: In contrast to regular splicing, where an upstream 5′-splice site is linked to a downstream 3′-splice site such that a linear RNA is formed, now, a downstream 5′-splice site is linked to an upstream 3′-splice site to yield a circRNA [76, 77].

We have engineered hairpin ribozyme variants that mimic the process of back-splicing by programming a linear in vitro transcribed RNA to cleave off their 5′- and 3′-termini and to subsequently ligate the remaining RNA stretch in intramolecular fashion to form a circRNA (Figure 16.8). In detail, the linear RNA can form two alternative active conformations (A and B), both favoring self-cleavage, because the resulting fragments are less tightly bound and can easily dissociate. After cleaving off one of the two ends, the ribozyme refolds to adopt the other cleavage-active conformation and to perform the second cleavage. Thus, first, the 5′-end is removed, followed by the 3′-end or vice versa, dependent on the respective folding path (A or B in Figure 16.8). The remaining RNA stretch, still being a hairpin ribozyme structure, carries the characteristic ends (5′-OH and 2′,3′-cyclic phosphate) to undergo ligation. This proceeds in an intramolecular fashion to yield circRNA as predicted. As analyzed by electrophoresis through denaturing polyacrylamide gels, indeed, all expected intermediates and products were observed in the reaction mixture [72, 73]. To our surprise, we also discovered intermolecular ligation resulting in concatemeric versions of the parent RNA [74]. Projected into the RNA world, this demonstrates RNA activity, which for one would have enabled topology changes of RNA species (linear to circular) and for the other may have contributed to increasing the genome size (oligomerization). At some point, small RNA genomes would have transited to larger RNA genomes requiring some kind of ligation reactions that join RNA pieces together. The latter may be particularly interesting when an ensemble of different hairpin ribozyme species with variable sequences would undergo back-splicing followed by oligomerization in mixed composition. This, however, remains to be experimentally tested. The general potential of the hairpin ribozyme to act as an RNA ligase or polymerase in early life has been described also by others. Vlassov et al. demonstrated ligation of a wide variety of RNA substrates by truncated and fragmented derivatives of the hairpin ribozyme at low temperatures [78], and recent work from the Holliger Lab has demonstrated noncanonical 3′–5′ extension of RNA with 2′,3′-cyclic phosphates catalyzed by the hairpin ribozyme [79]. A related system to our self-processing hairpin ribozymes was engineered by Diegelman and Kool, however, with a different purpose: circular RNAs and trans-cleaving hairpin ribozymes should be produced by rolling-circle transcription [80]. To this end, circular DNA oligonucleotides

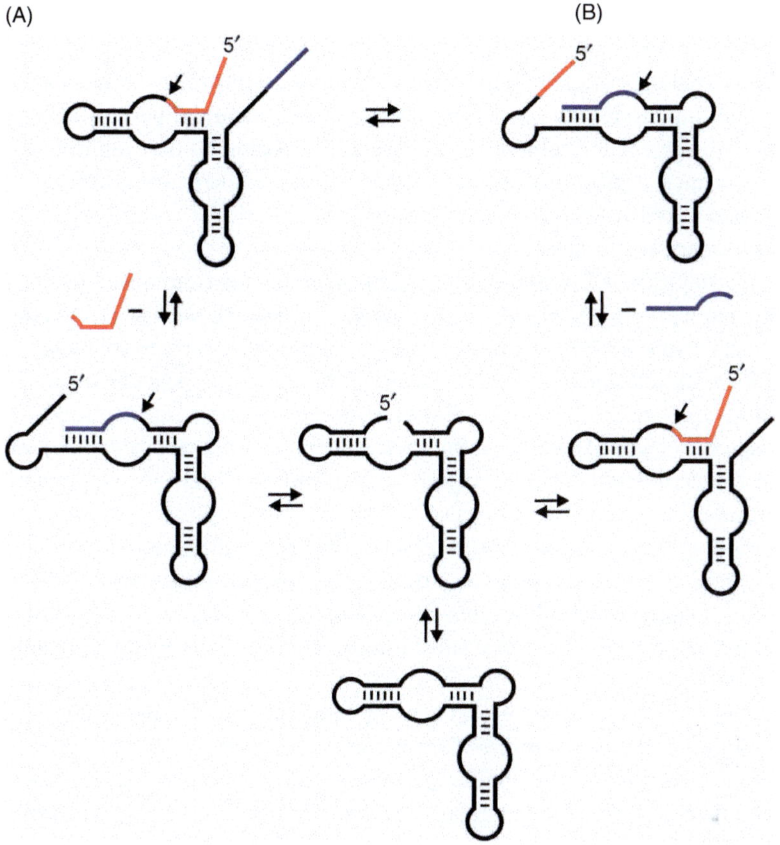

Figure 16.8 Scheme of hairpin ribozyme-mediated RNA circularization as described in [74]. A linear in vitro transcribed RNA can form two alternative active conformations (A and B), both favoring self-cleavage. After cleaving off one of the two ends (red in A; blue in B), the ribozyme refolds to adopt the other cleavage-active conformation and to cleave off the other terminal sequence (blue in A, red in B). The remaining RNA stretch, still being a hairpin ribozyme structure, carries the characteristic ends (5′-OH and 2′,3′-cyclic phosphate, not shown) to undergo intramolecular ligation and become circularized. Source: Adapted from Petkovic et al. [74].

encoding the hairpin ribozyme and ribozyme cleavable sequences were designed and transcribed in vitro by *Escherichia coli* RNA polymerase. This led to the production of long concatemeric RNAs, which subsequently underwent self-processing to unit-length linear and circular RNAs. The circular RNAs were products of self-ligation of the linear RNAs, and the linear RNAs were capable of cleaving appropriate substrates in trans. This result very well agrees with the observation made by us: self-processed hairpin ribozymes can be catalytically active in trans despite the presence of self-binding domains. Regarding the role of circRNA in early life, one may assume that circular species had an advantage over linear ones, reflected in features needed to overcome obstacles in the self-replication of primitive RNAs. End-to-end copying of a linear template requires the definition/identification of a specific initiation site. Otherwise, replication could initiate anywhere making complete replication rather unlikely. CircRNAs may have replicated in a rolling-circle-like process without the need for a defined initiation site. The only requirement for complete replication then would be that replication continues at least once around the circular template. Thus, circRNAs would have a higher chance to survive in error-prone, primitive self-replicating RNA systems and hence an evolutionary advantage.

Since the vast majority of RNA splicing in today's organisms is achieved by the highly regulated and precise removal of introns from pre-mRNAs, we also started to look at hairpin ribozyme-derived structures that would mimic the process of regular splicing. The RNA stretch that mimics the intron is flanked on both sides by a cleavage/ligation site sequence (A*GUC) and a short recognition sequence to exactly position the cleavage site (Figure 16.9). As in the back-splicing model described above, a linear in vitro transcribed RNA was designed to fold into two alternative cleavage favoring conformations. After the first cleavage event, refolding into the alternative conformation occurs to position the second splice site for cleavage. The second cleavage leads to the removal of the intron, and since the two exon-mimicking RNA stretches have the characteristic functional termini (5′-OH and 2′,3′-cyclic phosphate), exons are ligated. The success of the entire process strongly depends on the stability and catalytic competence of the involved starting and intermediate RNA folds. The main driving force for preferential intron removal and exon ligation is the more stable binding of the exons to the ribozyme than of the intron. Initial results indicate successful self-splicing in the described scenario, underscoring the high potential of the hairpin ribozyme to support yet another RNA processing reaction with relevance for early life forms.

16.8 Closing Remarks

Engineering of nucleic acid catalysts with tailored features by the contribution of rational design is a powerful strategy with potential applications. The many examples of ribozyme engineering demonstrate that nowadays, we have understood these systems to a level that allows their design into specific applications. The hairpin ribozyme variants described herein demonstrate the strong relationship between the structure and function of this catalytic RNA. Structural manipulation has resulted in variants with new properties, yet conserving the basic functionalities, including RNA cleavage and ligation. In a number of variants, activity became assessable, as for example, in the FMN-responsive hairpin ribozyme, which consists of the in vitro selected aptamer for FMN appended to the ribozyme structure (Figure 16.5). With regard to the RNA world hypothesis, our efforts on hairpin ribozyme engineering have demonstrated that a rather small RNA can perform a number of RNA processing reactions, including circularization, oligomerization, recombination, and splicing. The hairpin ribozyme can make itself from even shorter fragments, which cooperate to form a functional complex [45]. Therefore, it

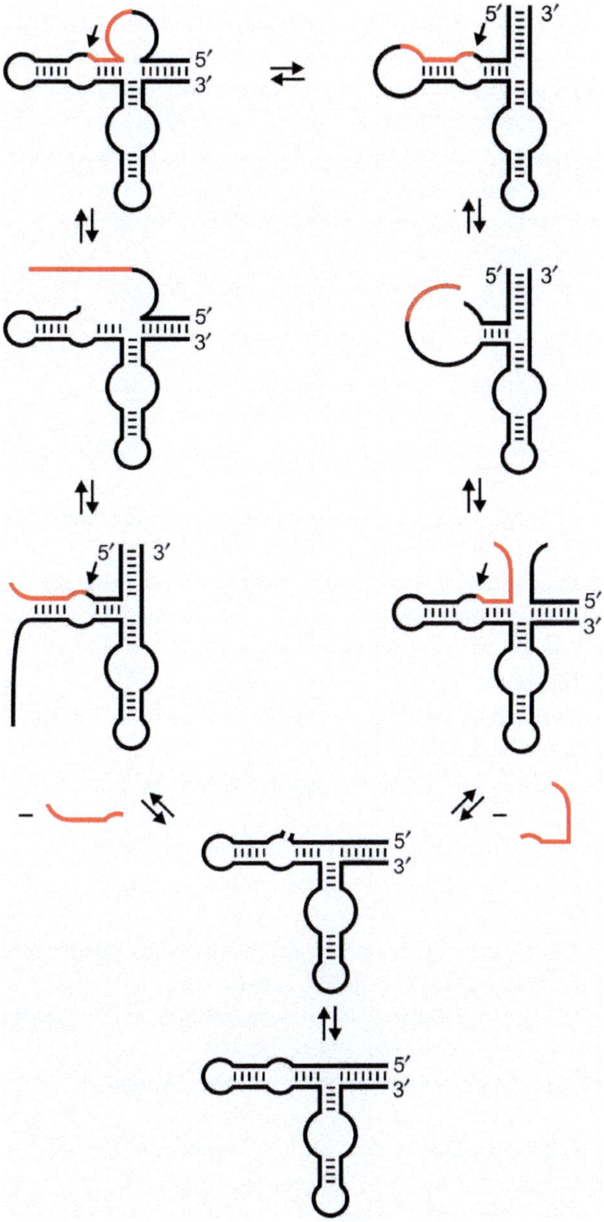

Figure 16.9 Scheme of hairpin ribozyme-mediated self-splicing. A linear in vitro transcribed RNA can fold into two alternative cleavage favoring conformations. After the first cleavage event, refolding into the alternative conformation occurs to position the second splice site for cleavage. The second cleavage leads to removal of the intron (red). The two exon-mimicking RNA stretches form a stable structure that brings the two characteristic functional termini (5′-OH and 2′,3′-cyclic phosphate, not shown) in close proximity and thus become ligated.

is a good model, mimicking the emergence and further development of functional entities starting from short RNA strands to molecules of higher complexity with extended genetic space and functionality. At the example of the hairpin ribozyme, we have demonstrated that by careful structural manipulation, the inherent cleavage and ligation activity can be tuned to support diverse RNA processing pathways, which in the RNA world may have contributed to (i) the emergence of RNAs with catalytic activity, (ii) the extension of genetic space by oligomerization and/or recombination, and (iii) the control of activity by splicing or allosteric regulation.

References

1 Guo, P. (2010). The emerging field of RNA nanotechnology. *Nat. Nanotechnol.* 5: 833.
2 Weigand, J.E. and Suess, B. (2009). Aptamers and riboswitches: perspectives in biotechnology. *Appl. Microbiol. Biotechnol.* 85 (2): 229.
3 Link, K.H. and Breaker, R.R. (2009). Engineering ligand-responsive gene-control elements: lessons learned from natural riboswitches. *Gene Ther.* 16: 1189.
4 Mehlhorn, A., Rahimi, P., and Joseph, Y. (2018). Aptamer-based biosensors for antibiotic detection: a review. *Biosensors* 8 (2).
5 Liu, M., Khan, A., Wang, Z. et al. (2019). Aptasensors for pesticide detection. *Biosens. Bioelectron.* 130: 174–184.
6 Louttit, C., Park, K.S., and Moon, J.J. (2019). Bioinspired nucleic acid structures for immune modulation. *Biomaterials* 217: 119287.
7 Yokobayashi, Y. (2019). Aptamer-based and aptazyme-based riboswitches in mammalian cells. *Curr. Opin. Chem. Biol.* 52: 72–78.
8 Kruger, K., Grabowski, P.J., Zaug, A.J. et al. (1982). Self-splicing RNA: autoexcision and autocyclization of the ribosomal RNA intervening sequence of *Tetrahymena*. *Cell* 31 (1): 147–157.
9 Guerrier-Takada, C. and Altman, S. (1984). Catalytic activity of an RNA molecule prepared by transcription in vitro. *Science* 223 (4633): 285–286.
10 Haseloff, J. and Gerlach, W.L. (1988). Simple RNA enzymes with new and highly specific endoribonuclease activities. *Nature* 334 (6183): 585–591.
11 Symons, R.H. (1997). Plant pathogenic RNAs and RNA catalysis. *Nucleic Acids Res.* 25 (14): 2683–2689.
12 Mulhbacher, J., St-Pierre, P., and Lafontaine, D.A. (2010). Therapeutic applications of ribozymes and riboswitches. *Curr. Opin. Pharmacol.* 10 (5): 551–556.
13 Joyce, G.F. (1996). Building the RNA world ribozymes. *Curr. Biol.* 6 (8): 965–967.
14 Hieronymus, R. and Muller, S. (2019). Engineering of hairpin ribozyme variants for RNA recombination and splicing. *Ann. N.Y. Acad. Sci.* 1447 (1): 135–143.
15 Feldstein, P.A., Buzayan, J.M., and Bruening, G. (1989). Two sequences participating in the autolytic processing of satellite tobacco ringspot virus complementary RNA. *Gene* 82 (1): 53–61.
16 Hampel, A. and Tritz, R. (1989). RNA catalytic properties of the minimum (−)sTRSV sequence. *Biochemistry* 28 (12): 4929–4933.

17 Haseloff, J. and Gerlach, W.L. (1989). Sequences required for self-catalysed cleavage of the satellite RNA of tobacco ringspot virus. *Gene* 82 (1): 43–52.

18 Fedor, M.J. (1999). Tertiary structure stabilization promotes hairpin ribozyme ligation. *Biochemistry* 38 (34): 11040–11050.

19 Berzal-Herranz, A., Joseph, S., Chowrira, B.M. et al. (1993). Essential nucleotide sequences and secondary structure elements of the hairpin ribozyme. *EMBO J.* 12 (6): 2567–2573.

20 Feldstein, P.A. and Bruening, G. (1993). Catalytically active geometry in the reversible circularization of 'mini-monomer' RNAs derived from the complementary strand of tobacco ringspot virus satellite RNA. *Nucleic Acids Res.* 21 (8): 1991–1998.

21 Komatsu, Y., Koizumi, M., Nakamura, H., and Ohtsuka, E. (1994). Loop-size variation to probe a bent structure of a hairpin ribozyme. *J. Am. Chem. Soc.* 116 (9): 3692–3696.

22 Butcher, S.E. and Burke, J.M. (1994). Structure-mapping of the hairpin ribozyme: magnesium-dependent folding and evidence for tertiary interactions within the ribozyme–substrate complex. *J. Mol. Biol.* 244 (1): 52–63.

23 Rupert, P.B. and Ferré-D'Amaré, A.R. (2001). Crystal structure of a hairpin ribozyme–inhibitor complex with implications for catalysis. *Nature* 410 (6830): 780–786.

24 Murchie, A.I.H., Thomson, J.B., Walter, F., and Lilley, D.M.J. (1998). Folding of the hairpin ribozyme in its natural conformation achieves close physical proximity of the loops. *Mol. Cell* 1 (6): 873–881.

25 Walter, N.G., Burke, J.M., and Millar, D.P. (1999). Stability of hairpin ribozyme tertiary structure is governed by the interdomain junction. *Nat. Struct. Biol.* 6 (6): 544–549.

26 Walter, N.G., Chan, P.A., Hampel, K.J. et al. (2001). A base change in the catalytic core of the hairpin ribozyme perturbs function but not domain docking. *Biochemistry* 40 (8): 2580–2587.

27 Zhuang, X., Kim, H., Pereira, M.J. et al. (2002). Correlating structural dynamics and function in single ribozyme molecules. *Science* 296 (5572): 1473–1476.

28 Wilson, T.J., Nahas, M., Araki, L. et al. (2007). RNA folding and the origins of catalytic activity in the hairpin ribozyme. *Blood Cells Mol. Dis.* 38 (1): 8–14.

29 Spitale, R.C., Volpini, R., Heller, M.G. et al. (2009). Identification of an imino group indispensable for cleavage by a small ribozyme. *J. Am. Chem. Soc.* 131 (17): 6093–6095.

30 Suydam, I.T., Levandoski, S.D., and Strobel, S.A. (2010). Catalytic importance of a protonated adenosine in the hairpin ribozyme active site. *Biochemistry* 49 (17): 3723–3732.

31 Mlynsky, V., Banas, P., Hollas, D. et al. (2010). Extensive molecular dynamics simulations showing that canonical G8 and protonated A38H+ forms are most consistent with crystal structures of hairpin ribozyme. *J. Phys. Chem. B* 114 (19): 6642–6652.

32 Earnshaw, D.J. and Gait, M.J. (1998). Hairpin ribozyme cleavage catalyzed by aminoglycoside antibiotics and the polyamine spermine in the absence of metal ions. *Nucleic Acids Res.* 26 (24): 5551–5561.

33 Stolze, K., Holmes, S.C., Earnshaw, D.J. et al. (2001). Novel spermine–amino acid conjugates and basic tripeptides enhance cleavage of the hairpin ribozyme at low magnesium ion concentration. *Bioorg. Med. Chem. Lett.* 11 (23): 3007–3010.

34 Welz, R., Schmidt, C., and Müller, S. (2001). Spermine supports catalysis of hairpin ribozyme variants to differing extents. *Biochem. Biophys. Res. Commun.* 283 (3): 648–654.

35 Nesbitt, S.M., Erlacher, H.A., and Fedor, M.J. (1999). The internal equilibrium of the hairpin ribozyme: temperature, ion and pH effects. *J. Mol. Biol.* 286 (4): 1009–1024.

36 Esteban, J.A., Walter, N.G., Kotzorek, G. et al. (1998). Structural basis for heterogeneous kinetics: reengineering the hairpin ribozyme. *Proc. Natl. Acad. Sci. U.S.A.* 95 (11): 6091–6096.

37 Komatsu, Y., Shirai, M., Yamashita, S., and Ohtsuka, E. (1997). Construction of hairpin ribozymes with a three-way junction. *Bioorg. Med. Chem.* 5 (6): 1063–1069.

38 Vauléon, S. and Müller, S. (2003). External regulation of hairpin ribozyme activity by an oligonucleotide effector. *ChemBioChem* 4 (2–3): 220–224.

39 Welz, R., Bossmann, K., Klug, C. et al. (2003). Site-directed alteration of RNA sequence mediated by an engineered twin ribozyme. *Angew. Chem. Int. Ed.* 42 (21): 2424–2427.

40 Vauléon, S., Ivanov, S.A., Gwiazda, S., and Müller, S. (2005). Site-specific fluorescent and affinity labelling of RNA by using a small engineered twin ribozyme. *ChemBioChem* 6 (12): 2158–2162.

41 Drude, I., Vauléon, S., and Müller, S. (2007). Twin ribozyme mediated removal of nucleotides from an internal RNA site. *Biochem. Biophys. Res. Commun.* 363 (1): 24–29.

42 Drude, I., Strahl, A., Galla, D. et al. (2011). Design of hairpin ribozyme variants with improved activity for poorly processed substrates. *FEBS J.* 278 (4): 622–633.

43 Balke, D., Zieten, I., Strahl, A. et al. (2014). Design and characterization of a twin ribozyme for potential repair of a deletion mutation within the oncogenic CTNNB1-ΔS45 mRNA. *ChemMedChem* 9 (9): 2128–2137.

44 Hieronymus, R., Godehard, S.P., Balke, D., and Muller, S. (2016). Hairpin ribozyme mediated RNA recombination. *Chem. Commun.* 52 (23): 4365–4368.

45 Gwiazda, S., Salomon, K., Appel, B., and Mueller, S. (2012). RNA self-ligation: from oligonucleotides to full length ribozymes. *Biochimie* 94 (7): 1457–1463.

46 Komatsu, Y., Kanzaki, I., Shirai, M., and Ohtsuka, E. (1997). A new type of hairpin ribozyme consisting of three domains. *Biochemistry* 36 (32): 9935–9940.

47 Ivanov, S.A., Volkov, E.M., Oretskaya, T.S., and Müller, S. (2004). Chemical synthesis of an artificially branched hairpin ribozyme variant with RNA cleavage activity. *Tetrahedron* 60 (41): 9273–9281.

48 Ivanov, S.A., Vauléon, S., and Müller, S. (2005). Efficient RNA ligation by reverse-joined hairpin ribozymes and engineering of twin ribozymes consisting of conventional and reverse-joined hairpin ribozyme units. *FEBS J.* 272 (17): 4464–4474.

49 Komatsu, Y., Kanzaki, I., Koizumi, M., and Ohtsuka, E. (1995). Modification of primary structures of hairpin ribozymes for probing active conformations. *J. Mol. Biol.* 252 (3): 296–304.

50 Schmidt, C., Welz, R., and Müller, S. (2000). RNA double cleavage by a hairpin-derived twin ribozyme. *Nucleic Acids Res.* 28 (4): 886–894.

51 Butcher, S.E., Heckman, J.E., and Burke, J.M. (1995). Reconstitution of hairpin ribozyme activity following separation of functional domains. *J. Biol. Chem.* 270 (50): 29648–29651.

52 Shin, C., Choi, J.N., Song, S.I. et al. (1996). The loop B domain is physically separable from the loop A domain in the hairpin ribozyme. *Nucleic Acids Res.* 24 (14): 2685–2689.

53 Balke, D., Becker, A., and Muller, S. (2016). In vitro repair of a defective EGFP transcript and translation into a functional protein. *Org. Biomol. Chem.* 14 (28): 6729–6737.

54 SILVERMAN, S.K. (2003). Rube Goldberg goes (ribo)nuclear? Molecular switches and sensors made from RNA. *RNA* 9 (4): 377–383.

55 Komatsu, Y., Nobuoka, K., Karino-Abe, N. et al. (2002). In vitro selection of hairpin ribozymes activated with short oligonucleotides. *Biochemistry* 41 (29): 9090–9098.

56 Najafi-Shoushtari, S.H., Mayer, G., and Famulok, M. (2004). Sensing complex regulatory networks by conformationally controlled hairpin ribozymes. *Nucleic Acids Res.* 32 (10): 3212–3219.

57 Najafi-Shoushtari, S.H. and Famulok, M. (2005). Competitive regulation of modular allosteric aptazymes by a small molecule and oligonucleotide effector. *RNA* 11 (10): 1514–1520.

58 Najafi-Shoushtari, S.H. and Famulok, M. (2007). DNA aptamer-mediated regulation of the hairpin ribozyme by human α-thrombin. *Blood Cell Mol. Dis.* 38 (1): 19–24.

59 Strohbach, D., Novak, N., and Müller, S. (2006). Redox-active riboswitching: allosteric regulation of ribozyme activity by ligand-shape control. *Angew. Chem. Int. Ed.* 45 (13): 2127–2129.

60 Hall, B., Hesselberth, J.R., and Ellington, A.D. (2007). Computational selection of nucleic acid biosensors via a slip structure model. *Biosens. Bioelectron.* 22 (9): 1939–1947.

61 Meli, M., Vergne, J., and Maurel, M.-C. (2003). In vitro selection of adenine-dependent hairpin ribozymes. *J. Biol. Chem.* 278 (11): 9835–9842.

62 Li, Y.-L., Vergne, J., Torchet, C., and Maurel, M.-C. (2009). In vitro selection of adenine-dependent ribozyme against Tpl2/Cot oncogene. *FEBS J.* 276 (1): 303–314.

63 Strohbach, D., Turcu, F., Schuhmann, W., and Müller, S. (2008). Electrochemically induced modulation of the catalytic activity of a reversible redoxsensitive riboswitch. *Electroanalysis* 20 (9): 935–940.

64 Burgstaller, P. and Famulok, M. (1994). Isolation of RNA aptamers for biological cofactors by in vitro selection. *Angew. Chem. Int. Ed.* 33 (10): 1084–1087.

65 Fan, P., Suri, A.K., Fiala, R. et al. (1996). Molecular recognition in the FMN – RNA aptamer complex. *J. Mol. Biol.* 258 (3): 480–500.

66 Moonen, C.T., Vervoort, J., and Muller, F. (1984). Carbon-13 nuclear magnetic resonance study on the dynamics of the conformation of reduced flavin. *Biochemistry* 23 (21): 4868–4872.

67 Moonen, C.T., Vervoort, J., and Muller, F. (1984). Reinvestigation of the structure of oxidized and reduced flavin: carbon-13 and nitrogen-15 nuclear magnetic resonance study. *Biochemistry* 23 (21): 4859–4867.

68 Korinek, V., Barker, N., Morin, P.J. et al. (1997). Constitutive transcriptional activation by a β-catenin-Tcf complex in $APC^{-/-}$ colon carcinoma. *Science* 275 (5307): 1784–1787.

69 Morin, P.J., Sparks, A.B., Korinek, V. et al. (1997). Activation of β-catenin-Tcf signaling in colon cancer by mutations in β-catenin or APC. *Science* 275 (5307): 1787–1790.

70 Pérez-Ruiz, M., Barroso-delJesus, A., and Berzal-Herranz, A. (1999). Specificity of the hairpin ribozyme: sequence requirements surrounding the cleavage site. *J. Biol. Chem.* 274 (41): 29376–29380.

71 Reuter, J.S. and Mathews, D.H. (2010). RNAstructure: software for RNA secondary structure prediction and analysis. *BMC Bioinf.* 11: 129.

72 Pieper, S., Vauléon, S., and Müller, S. (2007). RNA self-processing towards changed topology and sequence oligomerization. *Biol. Chem.*: 388, 743–746.

73 Petkovic, S. and Muller, S. (2013). RNA self-processing: formation of cyclic species and concatemers from a small engineered RNA. *FEBS Lett.* 587 (15): 2435–2440.

74 Petkovic, S., Badelt, S., Block, S. et al. (2015). Sequence-controlled RNA self-processing: computational design, biochemical analysis, and visualization by AFM. *RNA* 21 (7): 1249–1260.

75 Li, X., Yang, L., and Chen, L.L. (2018). The biogenesis, functions, and challenges of circular RNAs. *Mol. Cell* 71 (3): 428–442.

76 Jeck, W.R. and Sharpless, N.E. (2014). Detecting and characterizing circular RNAs. *Nat. Biotechnol.* 32 (5): 453–461.

77 Petkovic, S. and Muller, S. (2015). RNA circularization strategies in vivo and in vitro. *Nucleic Acids Res.* 43 (4): 2454–2465.

78 Vlassov, A.V., Johnston, B.H., Landweber, L.F., and Kazakov, S.A. (2004). Ligation activity of fragmented ribozymes in frozen solution: implications for the RNA world. *Nucleic Acids Res.* 32 (9): 2966–2974.

79 Mutschler, H. and Holliger, P. (2014). Non-canonical 3′-5′ extension of RNA with prebiotically plausible ribonucleoside 2′,3′-cyclic phosphates. *J. Am. Chem. Soc.* 136 (14): 5193–5196.

80 Diegelman, A.M. and Kool, E.T. (1998). Generation of circular RNAs and trans-cleaving catalytic RNAs by rolling transcription of circular DNA oligonucleotides encoding hairpin ribozymes. *Nucleic Acids Res.* 26 (13): 3235–3241.

17

Engineering of the *Neurospora* Varkud Satellite Ribozyme for Cleavage of Nonnatural Stem-Loop Substrates

Pierre Dagenais, Julie Lacroix-Labonté, Nicolas Girard, and Pascale Legault

Université de Montréal, Pavillon Roger Gaudry, Département de biochimie et de médecine moléculaire, 6128, Succ. centre-ville, Montréal, QC H3C 3J7, Canada

17.1 Introduction

The *Neurospora* Varkud satellite (VS) ribozyme was discovered ~30 years ago as part of an RNA plasmid (881 nucleotides) located in the mitochondria of the *Varkud-1c* strain and other natural isolates of *Neurospora* [1–3]. To this day, the in vivo function of the VS RNA plasmid remains elusive since this RNA does not contain any open reading frame of significant length and it does not confer any specific phenotype to its host [1]. However, the VS ribozyme displays two key enzymatic activities *in vitro* that likely contribute to the replication pathway of the VS RNA [4]: self-cleavage allows processing of multimeric transcripts into linear monomers and self-ligation of monomers allows formation of the predominant circular form [1, 5]. The self-cleavage products were shown to contain 2′,3′-cyclic phosphate and 5′-hydroxyl termini [1] as seen with other members of the small self-cleaving ribozyme family, including the hammerhead, hairpin, hepatitis delta virus (HDV), *glmS*, twister, twister-sister, hatchet, and pistol ribozymes. Over the years, the VS ribozyme has proven to be an excellent model system for understanding fundamental principles of RNA structure, function, and engineering [3, 6–11]. The VS ribozyme has attracted particular interest because of its unique structure and its ability to specifically recognize and cleave a folded stem-loop substrate.

At the time of its discovery, the VS ribozyme was considered a novel RNA not related in sequence to any known catalytic RNA [1]. Thirty years later, it remains a unique functional RNA only found in a few isolates of *Neurospora*. Despite sharing similarities with the hairpin ribozyme in terms of active site topology and the roles of two key nucleobases in the enzymatic mechanism, its secondary and tertiary structures are unique in comparison with other known ribozymes [7, 8, 11]. Furthermore, its mode of substrate recognition, which involves bipartite tertiary recognition of a stem-loop substrate and an activating conformational change, is also unique among known ribozymes. Herein, we review biochemical, biophysical, and structural

Ribozymes, First Edition. Edited by Sabine Müller, Benoît Masquida, and Wade Winkler.

biology studies that have explored the relationship between the VS ribozyme and its substrate as well as engineering studies that have identified derivatives of the VS ribozyme that can cleave nonnatural stem-loop substrates. First, we will describe the relatively simple sequence and secondary structure changes that are compatible with substrate cleavage by the wild-type VS ribozyme and then examine the structural and thermodynamic investigations that have allowed a thorough understanding of substrate recognition and enabled structure-guided engineering studies. We will summarize results from engineering studies aimed at designing VS-derived ribozymes that can cleave nonnatural stem-loop substrates, in which important features considerably differ from the natural substrate. These results will be presented in the context of a detailed mechanism for substrate recognition by the VS ribozyme that recapitulates previous investigations and should help guide future engineering studies.

17.2 Simple Primary and Secondary Structure Changes Compatible with Substrate Cleavage by the VS Ribozyme

17.2.1 Circular Permutations and *trans* Cleavage

In vitro investigations have determined that the minimal contiguous sequence of the VS RNA required for self-cleavage contains 154 nucleotides (nt) [12]. The secondary structure of a ribozyme that contains this minimal self-cleaving region was characterized by site-directed mutagenesis and chemical probing under semi-denaturing conditions [in the absence of magnesium ions (Mg^{2+})] and shown to form six main paired regions organized around two three-way junctions (Figure 17.1) [13]. This secondary structure has provided a standard framework for initial biochemical investigations aimed at better defining the unique functional and structural properties of the VS ribozyme. In this structural context, the substrate is positioned at the extreme 5′-end of the sequence and adopts a stem-loop structure, termed stem-loop I (SLI). Site-directed mutagenesis and chemical probing studies performed under native conditions (in the presence of Mg^{2+}) have provided initial support for magnesium-dependent tertiary folding of the VS ribozyme. This involves a long-range kissing-loop interaction (KLI) in which bases of SLI and stem-loop V (SLV) form a short helix containing three Watson–Crick (WC) base pairs and conversion of the unshifted ground-state conformation of SLI to a shifted conformation (as shown in Figure 17.1; [13, 14]). The presence of Mg^{2+} also leads to phosphodiester bond cleavage between G_{620} and A_{621} in the internal loop of SLI [16]. However, since high concentrations of monovalent salts also support the same cleavage reaction, it was concluded that divalent metal ions do not play an important role in the chemistry of the reaction, but instead are important for folding of the tertiary structure [16, 17].

Several other VS ribozyme sequences have been investigated *in vitro* that deviate from the standard framework in which the SLI substrate is attached to the 5′-end of the ribozyme (Figure 17.2a), illustrating the diverse secondary structure contexts that are compatible with substrate cleavage. Given that the natural multimeric VS

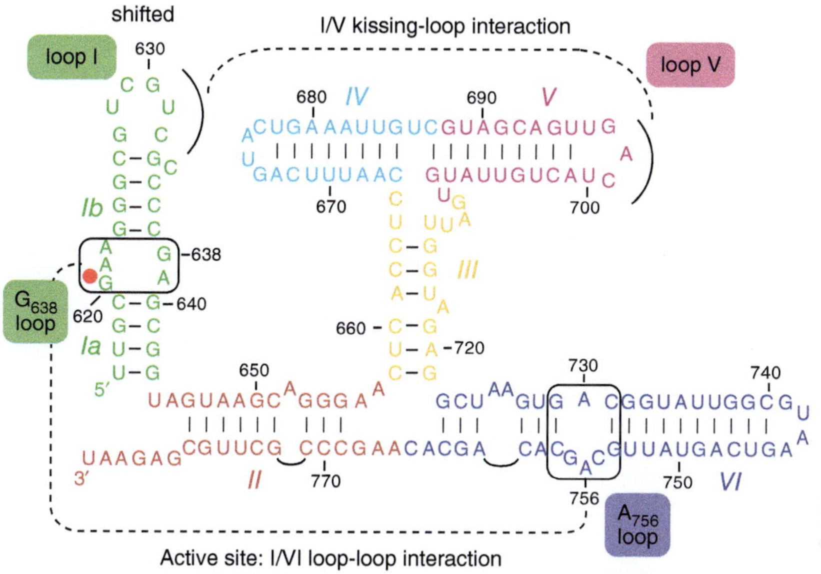

Figure 17.1 Sequence and proposed secondary structure of a minimal VS ribozyme compatible with *cis* cleavage of its 5′-substrate [13, 14]. The six paired domains are shown in different colors. The SLI substrate is represented in its shifted conformation [14]. Recognition of the substrate (green) engages loop I in a KLI with loop V (pink) and the G_{638} internal loop in a loop/loop interaction with the A_{756} internal loop (purple). The cleavage site is marked with a red dot. Source: Girard et al. [15]. Reprinted with permission from Oxford University Press. © 2006.

RNA produced in vivo consists of a series of head-to-tail monomers, it was proposed that a given monomer could prefer a downstream substrate rather than the one immediately upstream. In agreement with this hypothesis, a circular permutation variant in which the SLI substrate is attached to the 3′-end of the ribozyme via a linker sequence (Figure 17.2b; 3′-permutation) instead of being attached directly to its 5′-end (Figure 17.2a; 5′-permutation) was shown to be effective for ligation [19] and allowed for more efficient self-cleavage [20]. Increasing the size of the linker attaching SLI to the ribozyme in the 5′-circular permutation also leads to faster cleavage, suggesting that this linker relieves a steric constraint between the substrate and ribozyme that reduces the cleavage rate of the minimal contiguous self-cleaving ribozyme [21]. Synthesis of a *trans* form of the ribozyme lacking SLI allows cleavage of non-covalently bound substrates with multiple turnover (Figure 17.2c), demonstrating that this *trans* ribozyme is a true enzyme [22]. Given that the VS ribozyme has minimal sequence requirements 5′ of the cleavage site [12], the trans-cleavage activity has been exploited as a biochemical tool to facilitate the purification of *in vitro* transcribed RNAs with a homogeneous 3′-end [23, 24]. For this application, the RNA of interest is synthesized with a VS SLI substrate at its 3′-end and then cleaved by a trans-cleaving ribozyme that is either co-transcribed or independently produced (not shown; [23, 24]). Moreover, *trans* cleavage can also occur under conditions that allow *cis* ribozymes to dimerize, whereby each protomer cleaves the substrate

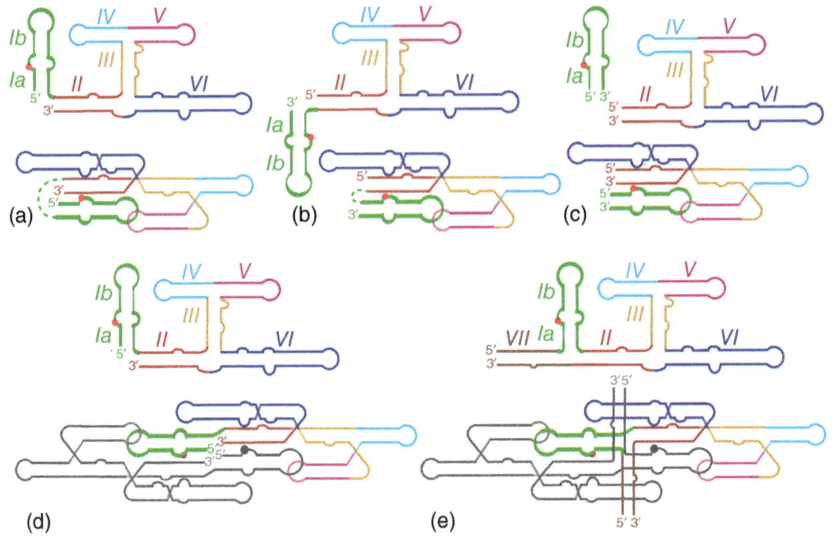

Figure 17.2 Diverse sequence and structural contexts for the SLI substrate that are compatible with cleavage by the VS ribozyme. (a) Minimal contiguous VS ribozyme compatible with *cis* cleavage of the 5′-substrate. (b) Circular permutation variant compatible with *cis* cleavage of the 3′-substrate. (c) VS ribozyme compatible with *trans* cleavage of an isolated SLI substrate that together form a substrate/ribozyme (*S*/*R*) complex. (d) The same sequence as in (a) can also dimerize at high concentration to allow *trans* cleavage of the 5′-substrate. (e) Extended VS ribozyme sequence that positions the SLI substrate as part of a three-way junction. This RNA sequence can form dimers that allow *trans* cleavage of the 5′-substrate. Ribozyme secondary structures in (a) through (e) are shown both as in the original model [13] and according to the orientation of paired regions in the crystal structure [18].

of the partner protomer. This has been observed both for a minimal VS ribozyme sequence (Figure 17.2d; [25, 26]) and for an extended VS ribozyme sequence that positions the SLI substrate as part of a three-way junction (Figure 17.2e; [27]).

17.2.2 The I/V Kissing-Loop Interaction and the Associated Conformational Change in SLI

The natural VS ribozyme has a unique and complex mode of substrate recognition. It first recognizes its substrate through formation of a magnesium-dependent KLI between SLI and SLV, which is associated with a conformational change in SLI (Figure 17.3). Site-directed mutagenesis and chemical modification studies have provided strong support for formation of three WC base pairs at the kissing-loop (KL) interface between bases 630–632 of SLI and 697–699 of SLV (Figure 17.3a; [14]). Single-base substitution of any of these six bases by all other possible bases (A, G, C, or U) causes at least a 20-fold decrease in the cleavage rate, although most often a 1000-fold decrease is observed. The only exception was a fivefold decrease for the $C_{632}U$ mutation; this substitution is minimally disruptive since it converts the C_{632}–G_{697} pair to a U–G pair. Most interestingly, some compensatory mutations that restore pairing for the three WC base pairs at the KLI also restore self-cleavage activity (within fivefold; Figure 17.3a). However, only certain WC compensatory mutations supported activity, indicating that additional functional and/or structural constraints limit the sequence diversity of the two interacting loops in active variants. Mutational studies also lead to the identification of a UNR consensus sequence in both loop I (U_{628}–C_{629}–G_{630}) and loop V (U_{696}–G_{697}–A_{698}) and the prediction of U-turn structures in both loops.

Formation of the KLI is associated with a helix shift that changes the base-pairing register in stem Ib and reorganizes the cleavage-site internal loop so that it becomes cleavable (Figure 17.3b). This conformational switch in SLI was demonstrated to be essential for both the ligation and the cleavage activities [19]. Ligation-based *in vitro* selection studies allowed for the identification of a large pool of SLI sequence variants that could be ligated, from which a subset of 52 sequences were assayed for self-cleavage activity. Covariation analysis of active sequences failed to fit a single secondary structure model. Instead, the selected SLI variants could be divided into two subsets according to the secondary structure of stem Ib, one group forming an unshifted helix and the other one forming a shifted helix (Figure 17.3b). Chemical modification using dimethyl sulfate provided strong evidence that the unshifted conformation of an isolated SLI substrate predominates in the absence of ribozyme both in the absence and presence of Mg^{2+}. The addition of both ribozyme and Mg^{2+} was found necessary to induce the shifted SLI conformation. When combining results

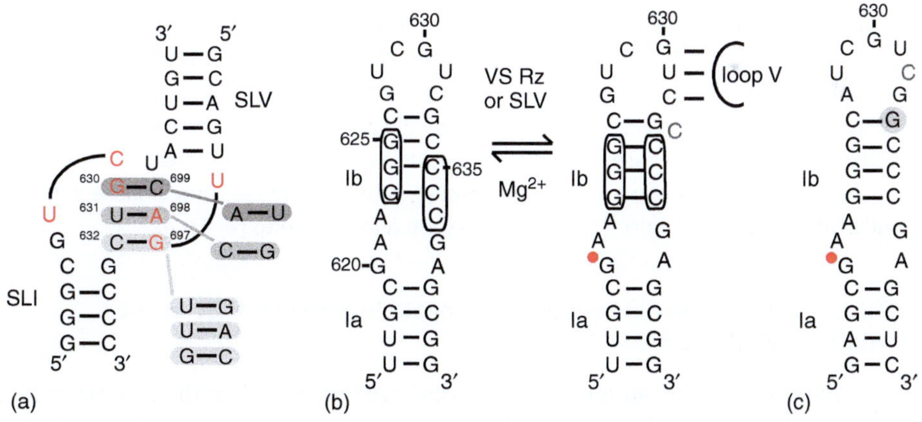

Figure 17.3 Formation of the KLI and associated conformational change. (a) Model of the KLI based on site-directed mutagenesis studies with the proposed three WC base pairs between residues 630–632 of SLI and 697–699 of SLV. Residues forming the typical UNR sequence of U-turn motifs in loop I and loop V are represented with red letters. Base-pair substitutions at the KL interface that are compatible with cleavage activity (within fivefold) are indicated. (b) Conformational change in the SLI substrate from an unshifted to a shifted conformation upon interaction with the VS ribozyme (VS Rz) or an isolated stem-loop V (SLV) in the presence of magnesium ions (Mg^{2+}). (c) Example of a pre-shifted substrate carrying the C634G substitution (shaded in gray), which prevents formation of the unshifted state. Source: (a) Rastogi et al. [14]. (b) Source: Adapted from Bouchard and Legault [24]. (c) Bouchard et al. [28].

from *in vitro* selection with site-directed mutagenesis, it was concluded that variant SLI substrates that are locked in the unshifted conformation are inactive, whereas those that are active are either shiftable, like the wild-type SLI, or pre-shifted, i.e. locked in the shifted conformation (Figure 17.3c). Formation of the shifted conformation reconfigures the cleavage site internal loop from a 6-nt symmetric loop to a 5-nt asymmetric loop, which was later shown to intimately associate with the A_{756} internal loop to form the active site (Figure 17.1).

Follow-up biochemical studies have demonstrated that formation of the KLI is necessary to allow the secondary structure rearrangement in SLI, and that binding of an isolated SLV in the presence of Mg^{2+} is sufficient to allow formation of the KLI and the structural rearrangement in SLI [29]. Detailed thermodynamic investigations with isolated stem-loops (SLI and SLV) have shown that pre-shifted SLI variants have remarkably high affinities for SLV, which are higher than observed for shiftable SLI variants [24]. The helix shift is associated with an energy cost of 3.0 kcal mol^{-1} for SLI substrates with a stem Ia stabilized by four WC base pairs and 1.8 kcal mol^{-1} for SLI substrates in which stem Ia is absent, indicating that a stable stem Ia contributes to stabilizing the unshifted conformation and decreases the stability of the KLI. Thus, these binding studies with shiftable substrates helped rationalize the inhibitory effect of a stable stem Ia on cleavage activity [12, 21, 30–32]. However, given that insertion of a linker sequence between SLI and the rest of the ribozyme also increases cleavage activity [21], disruption of stem Ia in self-cleaving ribozymes increases the cleavage activity by at least two mechanisms: stabilization of the KLI and conversion of the non-pairing 3′-strand into a short activating linker.

Formation of the I/V KLI also facilitates the formation of a tertiary interaction between the G_{638} and the A_{756} internal loops. More specifically, intimate association of these two loops leads to formation of the active site and subsequent substrate cleavage. In the proposed general acid–base cleavage mechanism, the G_{638} nucleobase acts as a general base that extracts a proton from the 2′-OH, making it a better nucleophile to attack the adjacent phosphate, whereas the A_{756} nucleobase acts as the acid that donates a proton to the 5′-oxygen leaving group [8, 33–42].

17.2.3 Summary of SLI Sequences Compatible with Cleavage by the Wild-Type VS Ribozyme

Over the years, several variant SLI sequences have been tested for the cleavage reaction as part of diverse lines of investigation. Taken together, these studies establish the SLI sequence diversity compatible with cleavage activity and allow us to derive some general principles for substrate recognition and cleavage. SLI substrates that can be efficiently cleaved by the wild-type VS ribozyme have been found to contain either shiftable (Figure 17.3b) or pre-shifted stem-loops (Figure 17.3c). Stem Ia plays a regulatory role on the cleavage–ligation equilibrium of shiftable substrates, with longer stems favoring the ligation reaction over the cleavage reaction [20, 36]. Destabilizing stem Ia also increases the cleavage activity of some shiftable substrates, but has very little effect on the cleavage of pre-shifted substrates [12, 21, 30–32]. A wide range of stem Ib sequences are compatible with substrate cleavage, as shown in

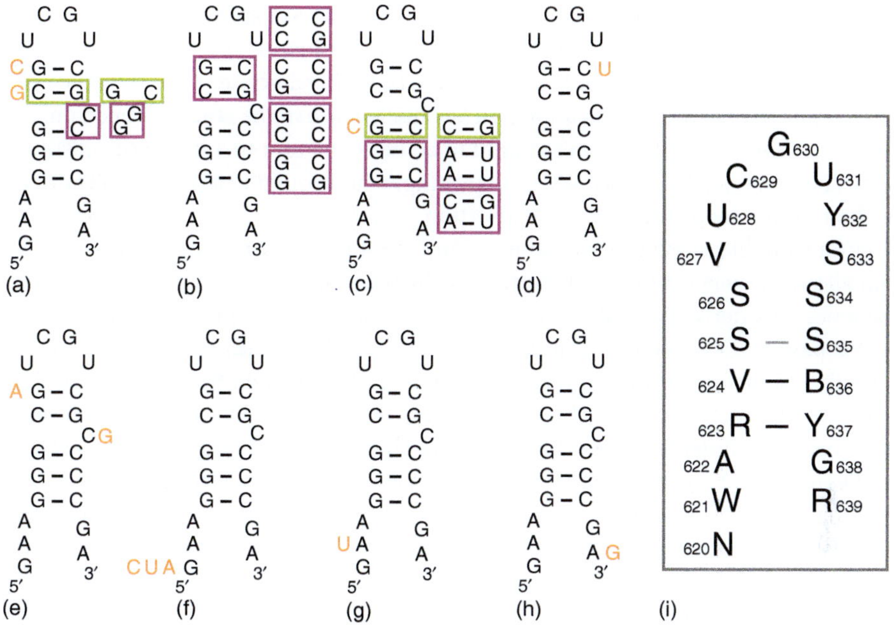

Figure 17.4 SLI substrate variants that can be cleaved by the wild-type VS ribozyme. (a)–(h) The variant sequences are shown in the context of the shifted conformation of the minimal SLI wild-type substrate (residues 621–639). Site-specific substitutions are shown in gold next to the modified residue. Substitutions of groups of residues are boxed and reported on the secondary structure with the same color box. (i) Consensus sequence of the minimal SLI substrate that can be cleaved by the wild-type VS ribozyme based on data presented in (a)–(h) and using nomenclature standards (B is C, G or U; N is A, G, C, or U; R is A or G; S is G or C; V is A, C, or G; W is A or U; Y is U or C). Site-specific and compensatory substitutions provide evidence for obligatory (black bar) or not strictly necessary base pairing (gray bar). Source: (a) Beattie et al. [13]. (b) Adapted from Rastogi et al. [14]. (c) Adapted from Andersen and Collins [19]. (d) Adapted from Rastogi et al. [14]. (e) Bouchard et al. [28]. (f) Guo et al. [12]. (g) Adapted from Hiley et al. [43]. (h) Adapted from Wilson et al. [40].

the context of the shifted state in Figure 17.4a–d. There is limited sequence requirement for residues that were initially presumed to form the two loop closing base pairs (C_{626}–G_{633} and G_{627}–C_{632} base pairs); sequences that allow base pairing are allowed but not essential for cleavage activity (Figure 17.4a–c; [13, 14]). Furthermore, the 623–637 and 624–636 base pairs closing the internal loop were found to be required in stem Ib, with the need for a purine at position 623 and a pyrimidine at 637 (Figure 17.4c; [13, 19]). In contrast, there is not a strong requirement for formation of the 625–635 base pair (Figure 17.4c). In terms of the terminal loop, its sequence is restricted to allow formation of a U-turn structure and the KLI, with the C_{632}U (Figure 17.4d; [14]) and the G_{627}C/G_{627}A (Figure 17.4a,e; [28]) substitutions being permitted. The sequence diversity of the cleavage site internal loop is also fairly limited. Exceptionally, substitution of residue G_{620} located immediately 5′ of the cleavage site with a U or an A does not significantly affect the efficiency of self-cleavage within a minimal ribozyme, whereas an SLI substrate with a C substitution at this position is cleaved more slowly (Figure 17.4f; [12]). In contrast, the substitution of G_{638} for C, A, or U was found to be highly detrimental to the cleavage activity and lead to its identification as a key nucleobase in the reaction mechanism [40]. Residues A_{621} and A_{639} were found to be more tolerant to a few base substitutions (Figure 17.4g,h; [40, 41, 43]). Interestingly, nucleotide insertion is permitted in the internal loop, with a CUAA insertion 3′ of A_{639} resulting in faster cleavage rates for shiftable substrates (not shown; [20]). To summarize the sequence diversity that allows cleavage of the SLI substrate by the wild-type VS ribozyme, a consensus sequence was derived based on previous reports along with the minimal number of WC base pairs needed in stem Ib (Figure 17.4i). This consensus sequence indicates that sequence restriction for cleavage of the SLI substrate by the wild-type VS ribozyme is mostly limited to the terminal loop that forms the KLI and the internal loop that contributes to create the active site. Since ligation-based *in vitro* selection studies identified functional SLI variants that have not been tested yet in the cleavage reaction [19], it is likely that the sequence diversity may be even larger than currently established (Figure 17.4i).

17.3 The Structural Context

Knowledge of the VS ribozyme three-dimensional structure has grown slowly over the years. Early attempts to crystallize the ribozyme for structure determination were unsuccessful, likely due to its dynamic nature and its propensity to form dimers and multimers at high concentrations. However, several low-resolution structural models were reported based on biochemical [31, 44], FRET [45] and SAXS data [46]. In addition, NMR studies of several isolated subdomains have been conducted to characterize several aspects of VS ribozyme function, and eventually an NMR-based model of the substrate/ribozyme (*S*/*R*) complex was determined that describes the open state of the complex (for a recent review: [11]). In parallel, crystallization of a dimeric form of the ribozyme provided the first high-resolution structures of the closed state in which the G_{638} and A_{756} internal loops associate to form the active site [18, 47].

17.3.1 NMR Investigations of the VS Ribozyme

NMR Studies of Kissing-Loop Complexes

Initial NMR investigations were aimed at better understanding formation of the SLI/SLV complex and defining the structures of both the inactive and active conformations of the G_{638} loop [48–50] as well as the structures of the free SLI and SLV loops [49, 51, 52]. In addition, NMR and isothermal titration calorimetry (ITC) studies of several KL complexes using a common SLV sequence and several variant SLI sequences provided critical insights into the KLI-dependent helix shift of the SLI substrate [24, 53]. Thermodynamic investigations by ITC provided evidence that the pre-shifted SLI substrates form more stable complexes (K_d ~0.5 μM) than the shiftable substrates (K_d = 12.5–65 μM) with SLV [24]. 1D imino proton NMR studies revealed that formation of the KLI leads to destabilization of the SLI loop closing base pair both for the shiftable and pre-shifted substrate variants [53]. These results are consistent with the fact that a stable base pair closing loop I is not essential (Figure 17.4i) and support a model in which formation of the KLI destabilizes the loop I closing base pairs and induces a helix shift in the SLI substrate, which leads to a crucial conformational change in the G_{638} loop (Figure 17.5a).

The NMR structure of the SLI/SLV complex determined using a non-cleavable pre-shifted SLI provided the first high-resolution structural and dynamic information of the KLI [53]. First, it confirmed that both loop I and loop V adopt typical U-turn structures in the complex. These U turns allow the bases after the sharp turn to be exposed and available to interact with the other loop. This leads to the formation of four base pairs at the KL interface that stack on one another to create a short A-form-like helix. Three of these base pairs adopt a typical WC/WC configuration as initially predicted from site-directed mutagenesis studies, while the fourth one is a WC/Sugar Edge C–A base pair, which is part of the C_{629}–A_{701}–U_{695} base triple that stacks on stem I (Figure 17.5a). NMR line shape and NOE analyses indicated that several residues at the KL interface, including those that are not stably paired (C_{626}, A_{627}, U_{628}, C_{632}, G_{633}, G_{634} of SLI; U_{695}, U_{696}, and U_{700} of SLV) undergo dynamic exchange on different timescales. Nevertheless, the KL complex forms a stable structure, with extensive stacking interactions at the KLI and between the KLI and the adjacent stems. Comparison of NMR structures of the SLI and SLV terminal loops between their free and bound forms indicate that formation of the KLI induces only minor structural changes in SLV, but involves a disorder-to-order transition in SLI (Figure 17.5a; [11]).

A Divide-and-Conquer Strategy

For the NMR studies, a divide-and-conquer approach was used because structure determination of the entire ribozyme by NMR is extremely challenging given its size, tendency to dimerize and conformational dynamics [54]. The general idea of this approach was to determine high-resolution structures of isolated subdomains and use these structures to derive the structure of a complex formed between the SLI substrate and *trans* ribozyme (*S*/*R* complex). As part of these studies, high-resolution NMR structures of all non-helical subdomains of the *S*/*R* complex were determined [50–52, 54–57]. For each of these domains, precautions were taken to define minimal structural elements that could be investigated in isolation but that were still relevant to the function of the ribozyme. By combining the results of several NMR investigations, a structural schematic could be drawn that summarizes base-pairing and stacking interactions of key structural elements, and a three-dimensional model was built for the *S*/*R* complex [54]. The NMR-based model was built computationally by fragment assembly, superposing overlapping regions of each of the isolated subdomain structures. These studies revealed that the two three-way junctions both adopt a specific topology (family C topology), which clearly defined the orientation of their attached stems. As a result, helical domains II, III, and IV are more or less colinear forming the main backbone of the structure, whereas stem-loops

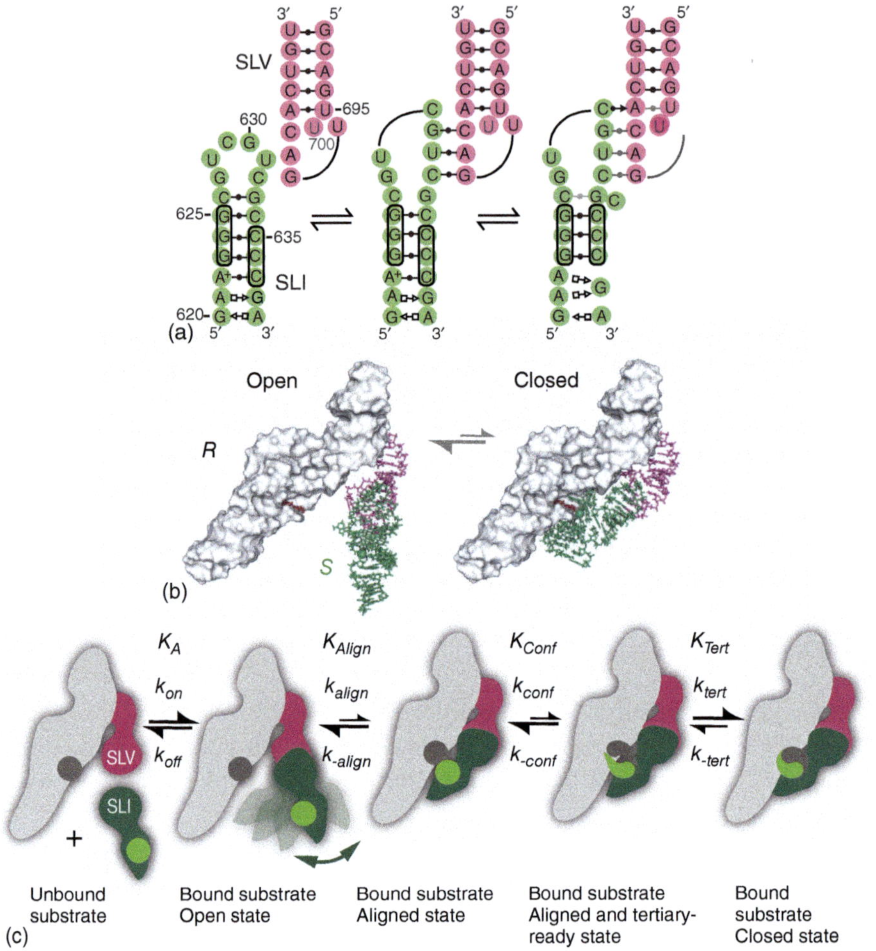

Figure 17.5 Bipartite binding and conformational changes associated with substrate recognition in the VS ribozyme. (a) Proposed model for formation of the KLI between an isolated SLV and a shiftable substrate [53]. In the free state, the SLI substrate adopts a unshifted conformation with a disordered terminal loop, whereas SLV adopts a compact U-turn fold. Based on NMR data, it was hypothesized that the initial formation of the three WC base pairs at the KL interface is associated with a disorder-to-order transition in loop I and destabilization of the closing $C_{626}-G_{633}$ base pair, which triggers the helix shift in SLI. (b) Representation of the open and closed state of the VS ribozyme *S/R* complex. (c) Simple kinetic model for binding of a pre-shifted substrate to the *trans* VS ribozyme (see text). Source: (a) Adapted from Dagenais et al. [11]. (b,c) Girard et al. [15]. Reproduced with permission of John Wiley & Sons. © 2019.

V and VI are oriented in parallel on each side of this backbone like an arm and the opposite leg (Figure 17.2c). In this topological context, SLV is free to form the KLI with SLI, however formation of this interaction positions the SLI cleavage site internal loop rather far away (~27 Å in one of the models) from the A_{756} internal loop of the ribozyme. Thus, the NMR model of the *S/R* complex describes the open state of the ribozyme (Figure 17.5b). This open state represents the ground-state conformation of the *S*/*R* complex that is compatible with the solvent accessibility of active site residues (C_{755} and A_{756}) established from a joint analysis of chemical probing and NMR data [13, 55]. This open conformation is also compatible with ITC investigations that yield similar dissociation constants for the SLI/SLV and SLI/trans ribozyme complexes [13] and with a dynamic model of the VS ribozyme derived from single-molecule fluorescence resonance energy transfer experiments [58].

17.3.2 Crystal Structures of a Dimeric Form of the VS Ribozyme

Crystal structures of a dimeric form of the VS ribozyme that capture its closed state were reported for two active site variants (G_{638}A and A_{756}G) at 3.1-Å resolution [18]. These structures both display an extended ribozyme conformation locked into a domain-swap organization, where the pre-shifted substrate of each protomer associates in *trans* with the ribozyme core of the other protomer (Figure 17.2e). This topological arrangement is consistent with previous *in vitro* evidence for *trans* cleavage of typically *cis*-cleaving ribozymes (Figures 17.1 and 17.2d; [25–27]). It is also in agreement with cleavage studies of a partial concatemeric VS RNA sequence containing one ribozyme core and two copies of SLI; the ribozyme core predominantly cleaves the downstream SLI substrate separated from the core by 0.7 kb rather than the SLI substrate immediately upstream [21]. The crystal structure of the dimer can be simplified to represent a single *S*/*R* complex, in which the SLI substrate of one protomer binds to ribozyme core of the other protomer through both the I/V KLI and the G_{638}/A_{756} loop–loop interaction. This latter interaction typical of the closed state allows formation of the active site, in which the two key nucleobases, G_{638} and A_{756}, are near the scissile phosphate poised to perform their roles as general base and acid in the cleavage mechanism (Figure 17.5b; [8, 33–42]). Only small local atomic repositioning would be required to create the in-line conformation of the reactive groups and position the key nucleotides for the cleavage reaction [11].

17.3.3 Open and Closed States of the *S*/*R* Complex

Although the NMR and crystal structures represent different states of the VS ribozyme, they are in very good agreement with each other in terms of the overall topology of helical domains and detailed structural characteristics of individual subdomains. A complete description of the similarities and small differences between these structures was previously reported to develop an integrated understanding of the structure and dynamics of the VS ribozyme [11]. Of note, the dynamic equilibrium between the open and closed states significantly contributes to substrate recognition as part of the cleavage mechanism (Figure 17.5c). In the open state, the SLI substrate forms the highly stable KLI, but its G_{638} loop does not associate with the A_{756} loop as seen in the closed state. The open state likely predominates within the conformation landscape of the *S*/*R* complex in solution. In addition, structural comparisons between the open and closed states indicate that the G_{638} and A_{756} loops undergo local remodeling upon binding. In summary, high-resolution structural studies have significantly contributed to define substrate recognition as a multistep process (Figure 17.5c) that involves (i) formation of the KLI, which is associated with an activating conformational change for SLI shiftable substrate; (ii) alignment of the G_{638} and A_{756} internal loops; (iii) conformational changes within these two internal loops; and (iv) the actual formation of the tertiary interaction between these loops. Thus, structural studies have provided a wealth of knowledge about substrate recognition that should be extremely valuable for exploring how the VS ribozyme can be engineered to perform novel functions.

17.4 Structure-Guided Engineering Studies

Engineering studies of the VS ribozyme have been conducted to investigate the possibility of adapting the VS ribozyme to cleave alternate RNA stem-loop substrates that differ substantially from the natural substrate. This would be extremely relevant for biotechnological applications, since the RNA stem-loop is a secondary structural element that is commonly found in both cellular and viral RNAs. So far, there have been successful results testing for helix compensation between stem Ib and stem V and for substituting the natural KLI with other KLIs. Moreover, these studies revealed engineering principles about the VS ribozyme that could be applied to the rational design of other functional RNAs. In particular, the results highlight the importance of considering conformational sampling of key structural domains when designing RNAs with novel functions.

17.4.1 Helix-Length Compensation

A helix-length compensation study tested the possibility of designing VS ribozyme derivatives that would cleave an SLI substrate in which the length of stem Ib is different than in the natural substrate [59]. Given the approximate co-linearity of stem I and stem V in initial models of the KL complex, it was predicted that efficient cleavage could be obtained for SLI substrates with either an increase or a decrease in the number of base pairs in stem Ib as long as the change was compensated by subtracting or adding base pairs in stem V. In order to test this hypothesis systematically, variant substrates and ribozymes were prepared with six different helix lengths in both stems Ib and V, and then the cleavage activity of all combinations of substrates and ribozymes was determined (Figure 17.6a,b). To facilitate these investigations, single-turnover kinetic studies of a given ribozyme were performed using multiple substrates in the same cleavage reaction, and the cleavage of the different size substrates was monitored by denaturing gel electrophoresis. It was found that a single ribozyme can generally cleave 2–3 substrates with a similar number of base pairs in stem Ib, revealing some level of substrate promiscuity with respect to helix length (Figure 17.6c). In addition, a certain degree of helix-length compensation was observed between stem Ib and stem V (Figure 17.6c). These results showed that the VS ribozyme could be modified to allow efficient cleavage of SLI substrates with 3–6 base pairs in stem Ib. Thus, this study demonstrated the adaptability of the VS ribozyme architecture and opened the possibility of adapting the VS ribozyme for

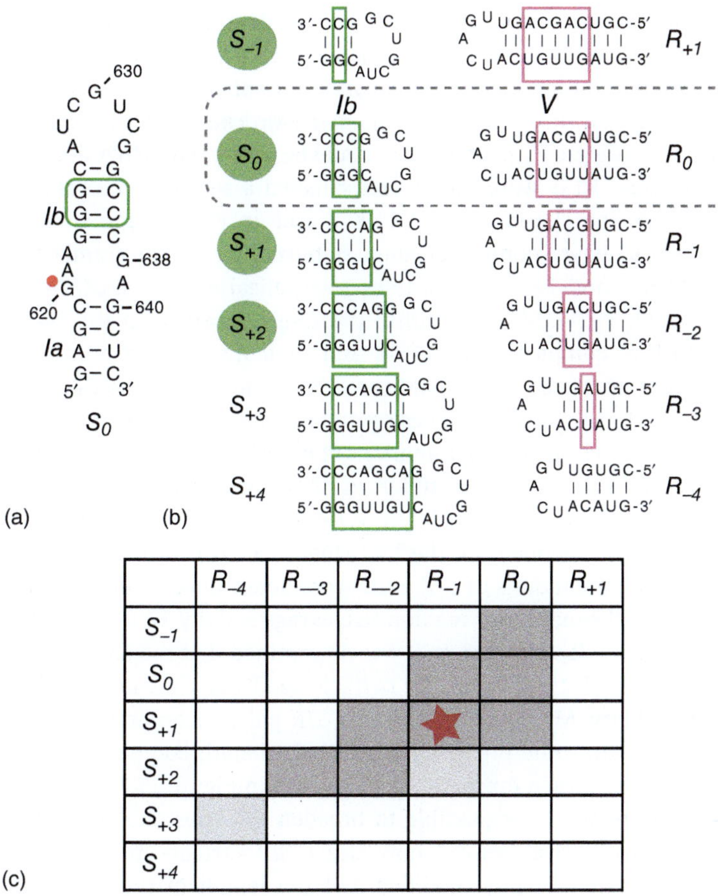

Figure 17.6 Sequence and activity of rationally engineered VS ribozyme variants as part of a helix-length compensation study. (a) Sequence and proposed secondary structure of the S_0 substrate. (b) Sequences of the *S*/*R* variants. Only the modified regions of the substrate (stem-loop Ib) and *trans* ribozyme (stem-loop V) are shown. (c) Grayscale heatmap of k_{cat}/K_M values for the *S*/*R* variants. The most active *S*/*R* pair ($k_{cat}/K_M = 14\,\text{min}^{-1}\,\mu\text{M}^{-1}$) is highlighted by a red star. The dark gray, medium gray, and white shadings correspond to k_{cat}/K_M values that are respectively less than fivefold, between 5- and 50-fold, and more than 50-fold lower compared to the most active *S*/*R* pair. Substrate names shaded with green ovals are those that can be cleaved efficiently by at least one VS ribozyme variant (k_{cat}/K_M less than fivefold of the most active *S*/*R* pair). Source: Lacroix-Labonté et al. [59].

cleavage of other nonnatural substrates, such as SLI substrates with a different terminal loop that could be recognized by a different KLI.

17.4.2 Kissing-Loop Substitutions

Subsequent studies identified derivatives of the VS ribozyme that would allow substrate cleavage using different KLIs [60]. A multistep rational engineering procedure was used, which could be adapted to either other structural domains in the VS ribozyme or other functional RNA. The first step in these studies consists of identifying an RNA module with an important function. The second step is to perform a bioinformatic search of the RCSB protein data bank for this RNA module. Using the online version of FR3D, 113 instances of KLIs were identified within a curated list of non-redundant RNA-containing structures determined by X-ray crystallography, NMR spectroscopy or cryo-EM. The third step was to select from this list of candidate KLIs, those that could be suitable for RNA domain substitution. Two suitable KLIs were found that satisfied a set of structural and thermodynamic criteria: the HIV-1 TAR/TAR* KLI and an rRNA large subunit L88/L22 KLI. The two KLIs are nonhomologous, stabilized by a small number of base pairs (TAR/TAR*: 6 and L88/L22: 4), their interhelical angles (TAR/TAR*: 125°–160° and L88/L22: 137°–154°) are within the range observed in the NMR structure of the SLI/SLV complex (125°–175°), and their dissociation constants (TAR/TAR*: 0.0059 μM and L88/L22: 3.8 μM) are similar to that of the SLI/SLV complex (0.24 μM). The fourth step was to design and prepare the engineered RNA by module substitution using various combinations of helix lengths. For each of the two selected KLIs, several substrates (5–6) and *trans* ribozymes (5–6) were prepared that contain different lengths of stem Ib and stem V, respectively, with their terminal loop sequences modified to reconstitute the KLI (Figure 17.7b). In the last step, the cleavage rate of all combinations of S/R pairs were determined to identify the combination of helix lengths that yields maximum cleavage activity for a given KLI.

Based on single-turnover cleavage assays, it was determined that the natural KLI of the VS ribozyme can be substituted with the two selected KLIs while maintaining substantial cleavage activity. Moreover, the variant S/R pairs incorporating these two KLIs display similar substrate promiscuity and helix-length compensation as the reference S_0/R_0 pair, as shown for the S_{TAR}/R_{TAR} pairs in Figure 17.7a–c. Thus, these studies demonstrate that it is possible to broaden the substrate specificity of the VS ribozyme, since derivatives can now cleave SLI variants in which the natural terminal loop is significantly modified and the length of the adjoining stem is varied to some extent. However, the k_{obs} values of the most active S/R KLI variants determined under identical conditions were reduced by 50-fold (L88/L22 KLI) and 160-fold (TAR/TAR* KLI) compared to the reference complex S_0/R_0. The reduced cleavage rates were not a consequence of impaired folding, since all substrate variants were shown to adopt the expected stem-loop structure by native gel electrophoresis and all ribozyme variants were shown to assume the proper fold by SHAPE analysis. For the most active L22/L88 KLI variant, the reduced cleavage rate could be explained in part by the higher dissociation constant of the KL complex, but this argument is not valid for the most active TAR/TAR* KLI variant. In this case, the KL complex is 40 times more stable than the I/V KL complex, and thus, it was hypothesized that the reduction in catalytic activity could be due to a defect in formation of the active site as a consequence of the intrinsic structure and dynamics of the surrogate KLI.

17.4.3 Role of KLI Dynamics in the Cleavage Reaction

To address the role of dynamics at the KLI in the function of the VS ribozyme, molecular dynamics (MD) studies of three KL complexes (I/V, TAR/TAR*, and L88/L22)

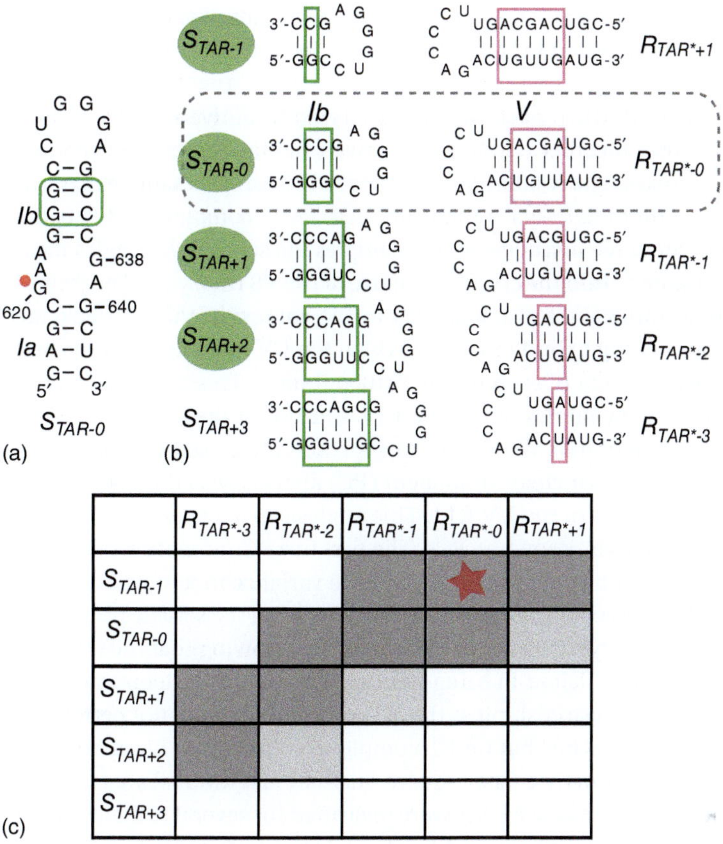

Figure 17.7 Sequence and activity of rationally engineered VS ribozyme variants as part of a KLI substitution study. (a) Sequence and proposed secondary structure of the $S_{TAR^*\text{-}0}$ substrate. (b) Sequences of the S_{TAR}/R_{TAR^*} variants. Only the modified regions of the substrate (stem-loop Ib) and *trans* ribozyme (stem-loop V) are shown. (c) Grayscale heatmap of k_{cat}/K_M values for the $S_{TAR}R_{TAR^*}$ variants. The most active S_{TAR}/R_{TAR^*} pair ($k_{cat}/K_M = 53 \times 10^{-2}$ min^{-1} μM^{-1}) is highlighted by a red star. The dark gray, medium gray and white shadings correspond to k_{cat}/K_M values that are respectively less than fivefold, between 5- and 50-fold, and more than 50-fold lower compared to the most active S_{TAR}/R_{TAR^*} pair. Substrate names shaded with green ovals are those that can be cleaved efficiently by at least one VS ribozyme variant (k_{cat}/K_M less than fivefold of the most active S_{TAR}/R_{TAR^*} pair). Source: (a,b) Lacroix-Labonté et al. [60]. © 2016 Oxford University Press.

were conducted [15]. For simplicity, only the results with the I/V and TAR/TAR* complexes are described here. Temperature replica exchange molecular dynamics (T-REMD) were carried out as an enhanced sampling method, which is well suited to investigate conformational ensembles of RNA populations. The T-REMD simulations consisted in running ~60 MD simulations in parallel at temperatures that were evenly distributed from 300 to 375 K with periodic exchange of the RNA coordinates between neighbor replicates. They were run for 50 ns, with 2-fs steps and frames recorded every 100 steps. The resulting 250 000 frames of the 300-K trajectory were subsequently analyzed to better understand the intrinsic dynamics of the KL complexes.

Each frame of the T-REMD trajectories was analyzed to derive the three Euler angles (α, β, and γ) that together define the relative orientation between the two stems of the KL complexes. The results indicate that the I/V complex samples a much larger region of the Euler angle space than the TAR/TAR* complex, and it is the only KL complex of the three investigated that samples the set of Euler angles measured for the closed state within the crystal structure of the VS ribozyme. The frames of the T-REMD trajectories were also analyzed in the context of the VS ribozyme crystal structure, by creating hybrid models in which the I/V KLI within the crystal structure was replaced by that of each of the 250 000 frames. This analysis demonstrated that the intrinsic dynamics of the I/V KLI, but not of the TAR/TAR* KLI, enables a large subpopulation of MD frames to approach the closed state in which the active site is formed. A principal component (PC) analysis was then performed to characterize the dynamics of the I/V KLI. This analysis extracts nonoverlapping concerted atomic movements, known as PCs. The first five PCs of motion of the I/V KL complex, which account for more than 70% of the variance in atomic positions, defined this KLI as a multi-axial RNA junction that can bring the scissile phosphate close to its position in the crystal structure. These results showing that the intrinsic motion of the I/V KLI is sufficient to bring the G_{638} loop and A_{756} loop into close proximity were particularly astonishing as the MD simulations were not performed in the context of the VS ribozyme, but for KL complexes in isolation. Local hotspots of motion were identified from root-mean square fluctuations (RMSFs) analysis. For the I/V complex, RMSF values above 1.5 Å were measured for several residues in SLI (C_8-G_{12}, G_{16}, C_{17}) and only one residue in SLV (U_{15}). In contrast, for the TAR/TAR* complex, RMSF values are smaller and only a few residues displayed RMSF values above 1.5 Å (C_8 and C_9 of TAR and U_{10} and U_{11} of TAR*), in agreement with more restricted dynamics for this KLI.

In summary, MD investigations of the I/V KL complex have led to a better understanding of the ensemble of RNA conformations that populate the open state and the main modes of motion that lead to formation of the active site (Figure 17.5c). In contrast, more restricted conformational sampling was observed in the MD studies of the TAR/TAR* KL complex, supporting the idea that formation of the active site is hindered in VS ribozyme variants carrying this KL substitution, which would explain the lower cleavage rates. Moreover, the MD studies of the I/V KLI highlight the importance of unpaired residues at the KLI to increase sampling of the conformational space. It was postulated that adding unpaired residues at the TAR/TAR* KLI may provide a useful way to improve cleavage of the TAR-derived substrate by a *trans* ribozyme derivative.

17.4.4 Improving the Cleavage Activity of a Designer Ribozyme

To improve the cleavage activity of the S_{TAR}/R_{TAR*} ribozyme variant, *in vitro* selection studies were conducted. In the starting RNA pool for the selection, 10 residues from the SLV sequence were randomized in order to identify SLV sequences that were optimal for cleavage of the TAR-derived substrate, which contains the HIV-1 TAR RNA sequence in stem Ia and stem-loop Ib, but preserved the natural G_{638} loop necessary for formation of the active site (Figure 17.7a). The active ribozyme sequences from the RNA pool that self-cleaved during the transcription and subsequent incubation period were selected by denaturing gel electrophoresis and amplified by RT-PCR. After six rounds of selection and amplification, the resulting pool of active ribozyme sequences was subjected to next-generation sequencing.

To define a consensus sequence, only the loop V sequences that were enriched in the final pool were considered. These sequences were analyzed to determine the potential number of consecutive standard WC base pairs that could form with loop I. Overall, 96% of the selected sequences have the potential to form at minimum 4 WC base pairs and 64% of them can form a maximum of 4 WC base pairs, indicating that 4 WC base pairs are necessary and sufficient to allow cleavage of the resulting ribozyme variant. Given that only 2% of the ribozyme pool has the potential to form the maximum 6 WC base pairs, as found for the rationally designed $S_{\text{TAR-0}}/R_{\text{TAR*-0}}$ pair (Figure 17.7b), increasing the thermodynamic stability of the KLI may not be beneficial, but rather inhibit the activity. Two main classes of loop V sequences were found that contain the preferred CCCA pattern starting either at position +4 or +5 of the randomized sequence. In both cases, formation of 4–5 WC base pairs at the KLI could leave several residues unpaired in both kissing loops, thereby providing more dynamics at the KLI for formation of the active site and efficient substrate cleavage.

Interestingly, a selected ribozyme sequence ($R_{\text{TAR-S1}}$) was found, which potentially forms four consecutive WC base pairs with S_{TAR} and cleaves the S_{TAR} substrate (Figure 17.8a) with a relatively high single-turnover rate ($k_{\text{obs}} = 5.8\,\text{min}^{-1}$), similar to the rate of the parental S_0/R_0 complex [$k_{\text{obs}} = 12.2\,\text{min}^{-1}$ (16)] (Figure 17.8b). In addition, control variant ribozymes, R_0 and $R_{\text{TAR*-0}}$, which can form either 3 or 6 WC consecutive base pairs, respectively, with S_{TAR}, both cleave this substrate with a much lower rate ($k_{\text{obs}} \leq 0.03\,\text{min}^{-1}$) than $R_{\text{TAR-S1}}$. These results underline the importance of a KLI that is appropriately stable, yet sufficiently dynamic for efficient substrate cleavage by the VS ribozyme. From a T-REMD simulation of the TAR/TAR-S1 KL complex, it was found that the TAR/TAR-S1 KL complex explores a much larger region of the Euler angle space than the original TAR/TAR* KL complex. This is likely due to the increased dynamics of the TAR/TAR-S1 KL complex, which leaves several unpaired residues at the KLI (Figure 17.8c; [15]).

In summary, the *in vitro* selection results further support the substrate binding model in which productive cleavage depends on the formation of a stable KLI that is also sufficiently dynamic to conformationally sample the closed state (Figure 17.5c). The highly stable KLI (high K_A) of the rationally designed $S_{\text{TAR-0}}/R_{\text{TAR*-0}}$ pair may better anchor the substrate in the S/R complex, but its restricted dynamics limits sampling of the closed state ($k_{-\text{align}} \gg k_{\text{align}}$). In contrast, the much less stable KLI (lower K_A) of the S_{TAR}/R_0 pair may allow better sampling of the conformational space, but would not anchor the substrate for a sufficient period of time to allow for formation of the closed state and subsequent cleavage. Thus, the much higher cleavage activity of both the parental S_0/R_0 and selected $S_{\text{TAR}}/R_{\text{TAR-S1}}$ complexes must

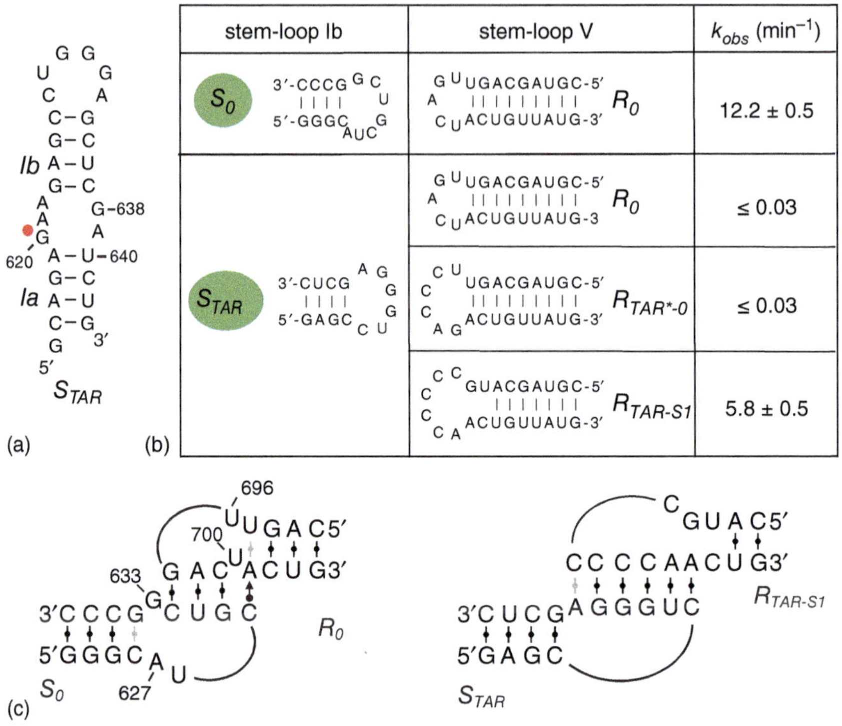

Figure 17.8 Improving the activity of a kissing-loop substitution variant of the VS ribozyme by making the KLI more dynamic. (a) Sequence and proposed secondary structure of the S_{TAR} substrate. (b) Observed cleavage rate (k_{obs}) of selected *S*/*R* pairs. Only the modified region of the substrate (stem-loop Ib) and *trans* ribozyme (stem-loop V) are shown. (c) Structural schematics of the KLI for the S_0/R_0 pair based on the NMR structure of the SLI/SLV complex [53] and of the $S_{TAR}/R_{TAR\text{-}S1}$ pair based on the T-REMD simulation [15]. Gray base pairs symbols indicate more unstable and/or transient base pairs. Source: Bouchard and Legault [53].

result from a KLI that is both appropriately stable and sufficiently dynamic to drive the two key forward steps of substrate recognition leading to cleavage.

17.5 Summary and Future Prospects for VS Ribozyme Engineering

Over the years, investigations of the VS ribozyme have helped define the substrate requirements for cleavage by the wild-type ribozyme, including the position of the substrate within the secondary structure of the ribozyme, its conformation in either a shiftable or pre-shifted state, as well as the diversity of cleavable stem-loop sequences. Moreover, engineering studies have shown that VS ribozyme variants can be used to cleave substrates that contain varying numbers of base pairs in stem Ib or where loop I has been drastically modified. These studies have also contributed in defining a thermodynamic model for substrate recognition by the VS ribozyme as well as establishing general principles that could be applied to other engineering studies with the VS ribozyme or other functional RNAs.

The mechanistic model for substrate binding by the VS ribozyme (Figure 17.5c) should prove useful for future investigations of the VS ribozyme. Using this model, it is likely that derivatives of the VS ribozyme can be designed to cleave substrate variants with many different loop I sequences besides those of the HIV-1 TAR/TAR* KLI and of the ribosomal RNA L22/L88 KLI [15]. Highly functional designer ribozymes must not only be able to form a stable KLI ($K_d \leq \sim 1\,\mu M$) but also be sufficiently dynamic. In this context, the master RNA designer would want to maximize cleavage activity by keeping in mind the possibility of incorporating unpaired residues at the KL interface to enhance the intrinsic dynamics of the KLI. In addition, they could increase the cleavage activity by optimizing the lengths of the two stems connected through the KLI. Although SLV sequences that allow optimal cleavage activity could be identified either by rational engineering or *in vitro* selection, a combination of approaches would most likely yield optimal results. For example, it might be possible to further improve the cleavage activity of the selected $S_{TAR}/R_{TAR\text{-}S1}$ complex by systematically optimizing the lengths of stem Ib and stem V. In addition, the design of optimal KLIs may benefit from computational modeling prior to experimental validation. For example, a T-REMD simulation could be performed, followed by analysis of Euler angles and principle components of motion to help predict the functionality of VS ribozyme variants carrying a new KLI. From a wider perspective, lessons learned from engineering VS ribozyme variants with new KLIs could be adapted to optimize the substitution of other RNA elements within the VS ribozyme or, more generally, to integrate exogenous RNA structural elements into other RNAs to create chimeric RNAs with novel functions.

Given that the stem-loop RNA sequence element is widespread in both cellular and viral RNAs, future engineering studies of the VS ribozyme should also address the possibility of designing variant ribozymes that can cleave alternate substrates with different internal loop sequences. However, this may represent an exceptionally challenging task given that the G_{638} internal loop forms part of the active site and contributes a key catalytic residue in the cleavage mechanism. In addition, such engineering studies may seem unrealistic given that there is still a small subset of ribozyme active sites known at this time. Future studies are needed to expand our knowledge of ribozymes and their active sites as well as to further explore the great potential for engineering RNA molecules with novel functions.

References

1 Saville, B.J. and Collins, R.A. (1990). A site-specific self-cleavage reaction performed by a novel RNA in *Neurospora* mitochondria. *Cell* 61: 685–696.

2 Collins, R.A. and Saville, B.J. (1990). Independent transfer of mitochondrial chromosomes and plasmids during unstable vegetative fusion in *Neurospora*. *Nature* 345: 177–179.

3 Collins, R.A. (2002). The *Neurospora* Varkud satellite ribozyme. *Biochem. Soc. Trans. Rev.* 30: 1122–1126.
4 Kennell, J.C., Saville, B.J., Mohr, S. et al. (1995). The VS catalytic RNA replicates by reverse transcription as a satellite of a retroplasmid. *Genes Dev.* 9: 294–303.
5 Saville, B.L. and Collins, R.A. (1991). RNA-mediated ligation of self-cleavage products of a *Neurospora* mitochondrial plasmid transcript. *Proc. Natl. Acad. Sci. U.S.A.* 88: 8826–8830.
6 Lilley, D.M. (2004). The Varkud satellite ribozyme. *RNA* 10: 151–158.
7 Lilley, D.M.J. (2008). The Hairpin and Varkud Satellite Ribozymes *Ribozymes and RNA Catalysis* (eds. D.M.J. Lilley and F. Eckstein), 66–91. Cambridge: Royal Society of Chemistry.
8 Wilson, T.J. and Lilley, D.M. (2011). Do the hairpin and VS ribozymes share a common catalytic mechanism based on general acid–base catalysis? A critical assessment of available experimental data. *RNA* 17: 213–221.
9 Lilley, D.M. (2011). Catalysis by the nucleolytic ribozymes. *Biochem. Soc. Trans.* 39: 641–646.
10 Wilson, T.J., Liu, Y., and Lilley, D.M.J. (2016). Ribozymes and the mechanisms that underlie RNA catalysis. *Front. Chem. Sci. Eng.* 10: 178–185.
11 Dagenais, P., Girard, N., Bonneau, E., and Legault, P. (2017). Insights into RNA structure and dynamics from recent NMR and X-ray studies of the *Neurospora* Varkud satellite ribozyme. *WIREs RNA* 8: e1421.
12 Guo, H.C.T., De Abreu, D.M., Tillier, E.R.M. et al. (1993). Nucleotide sequence requirements for self-cleavage of *Neurospora* VS RNA. *J. Mol. Biol.* 232: 351–361.
13 Beattie, T.L., Olive, J.E., and Collins, R.A. (1995). A secondary-structure model for the self-cleaving region of *Neurospora* VS RNA. *Proc. Natl. Acad. Sci. U.S.A.* 92: 4686–4690.
14 Rastogi, T., Beattie, T.L., Olive, J.E., and Collins, R.A. (1996). A long-range pseudoknot is required for activity of the *Neurospora* VS ribozyme. *EMBO J.* 15: 2820–2825.
15 Girard, N., Dagenais, P., Lacroix-Labonte, J., and Legault, P. (2019). A multi-axial RNA joint with a large range of motion promotes sampling of an active ribozyme conformation. *Nucleic Acids Res.* 47: 3739–3751.
16 Collins, R.A. and Olive, J.E. (1993). Reaction conditions and kinetics of self-cleavage of a ribozyme derived from *Neurospora* VS RNA. *Biochemistry* 32: 2795–2799.
17 Murray, J.B., Seyhan, A.A., Walter, N.G. et al. (1998). The hammerhead, hairpin and VS ribozymes are catalytically proficient in monovalent cations alone. *Chem. Biol.* 5: 587–595.
18 Suslov, N.B., DasGupta, S., Huang, H. et al. (2015). Crystal structure of the Varkud satellite ribozyme. *Nat. Chem. Biol.* 11: 840–846.
19 Andersen, A. and Collins, R.A. (2000). Rearrangement of a stable RNA secondary structure during VS ribozyme catalysis. *Mol. Cell* 5: 469–478.
20 Zamel, R., Poon, A., Jaikaran, D. et al. (2004). Exceptionally fast self-cleavage by a *Neurospora* Varkud satellite ribozyme. *Proc. Natl. Acad. Sci. U.S.A.* 101: 1467–1472.

21 Poon, A.H., Olive, J.E., McLaren, M., and Collins, R.A. (2006). Identification of separate structural features that affect rate and cation concentration dependence of self-cleavage by the *Neurospora* VS ribozyme. *Biochemistry* 45: 13394–13400.

22 Guo, H.C.T. and Collins, R.A. (1995). Efficient *trans*-cleavage of a stem-loop RNA substrate by a ribozyme derived from *Neurospora* VS RNA. *EMBO J.* 14: 368–376.

23 Ferré-D'Amaré, A.R. and Doudna, J.A. (1996). Use of *cis*- and *trans*-ribozymes to remove 5′ and 3′ heterogeneities from milligrams of *in vitro* transcribed RNA. *Nucleic Acids Res.* 24: 977–978.

24 Bouchard, P. and Legault, P. (2014). A remarkably stable kissing-loop interaction defines substrate recognition by the *Neurospora* VS Ribozyme. *RNA* 20: 1451–1464.

25 Olive, J.E., De Abreu, D.M., Rastogi, T. et al. (1995). Enhancement of *Neurospora* VS ribozyme cleavage by tuberactinomycin antibiotics. *EMBO J.* 14: 3247–3251.

26 Olive, J.E. and Collins, R.A. (1998). Spermine switches a *Neurospora* VS ribozyme from slow *cis* cleavage to fast trans cleavage. *Biochemistry* 37: 6476–6484.

27 Ouellet, J., Byrne, M., and Lilley, D.M. (2009). Formation of an active site in trans by interaction of two complete Varkud Satellite ribozymes. *RNA* 15: 1822–1826.

28 Bouchard, P., Lacroix-Labonté, J., Desjardins, G. et al. (2008). Role of SLV in SLI substrate recognition by the *Neurospora* VS ribozyme. *RNA* 14: 736–748.

29 Andersen, A.A. and Collins, R.A. (2001). Intramolecular secondary structure rearrangement by the kissing interaction of the *Neurospora* VS ribozyme. *Proc. Natl. Acad. Sci. U.S.A.* 98: 7730–7735.

30 Beattie, T.L. and Collins, R.A. (1997). Identification of functional domains in the self-cleaving *Neurospora* VS ribozyme using damage selection. *J. Mol. Biol.* 267: 830–840.

31 Rastogi, T. and Collins, R.A. (1998). Smaller, faster ribozymes reveal the catalytic core of *Neurospora* VS RNA. *J. Mol. Biol.* 277: 215–224.

32 Lafontaine, D.A., Norman, D.G., and Lilley, D.M. (2001). Structure, folding and activity of the VS ribozyme: importance of the 2-3-6 helical junction. *EMBO J.* 20: 1415–1424.

33 Sood, V.D. and Collins, R.A. (2002). Identification of the catalytic subdomain of the VS ribozyme and evidence for remarkable sequence tolerance in the active site loop. *J. Mol. Biol.* 320: 443–454.

34 Lafontaine, D.A., Wilson, T.J., Zhao, Z.-Y., and Lilley, D.M.J. (2002). Functional group requirements in the probable active site of the VS ribozyme. *J. Mol. Biol.* 323: 23–34.

35 Jones, F.D. and Strobel, S.A. (2003). Ionization of a critical adenosine residue in the *Neurospora* Varkud satellite ribozyme active site. *Biochemistry* 42: 4265–4276.

36 McLeod, A.C. and Lilley, D.M. (2004). Efficient, pH-dependent RNA ligation by the VS ribozyme in trans. *Biochemistry* 43: 1118–1125.

37 Zhao, Z.Y., McLeod, A., Harusawa, S. et al. (2005). Nucleobase participation in ribozyme catalysis. *J. Am. Chem. Soc.* 127: 5026–5027.

38 Smith, M.D. and Collins, R.A. (2007). Evidence for proton transfer in the rate-limiting step of a fast-cleaving Varkud satellite ribozyme. *Proc. Natl. Acad. Sci. U.S.A.* 104: 5818–5823.

39 Smith, M.D., Mehdizadeh, R., Olive, J.E., and Collins, R.A. (2008). The ionic environment determines ribozyme cleavage rate by modulation of nucleobase pK_a. *RNA* 14: 1942–1949.

40 Wilson, T.J., McLeod, A.C., and Lilley, D.M. (2007). A guanine nucleobase important for catalysis by the VS ribozyme. *EMBO J.* 26: 2489–2500.

41 Jaikaran, D., Smith, M.D., Mehdizadeh, R. et al. (2008). An important role of G638 in the *cis*-cleavage reaction of the *Neurospora* VS ribozyme revealed by a novel nucleotide analog incorporation method. *RNA* 14: 938–949.

42 Wilson, T.J., Li, N.S., Lu, J. et al. (2010). Nucleobase-mediated general acid-base catalysis in the Varkud satellite ribozyme. *Proc. Natl. Acad. Sci. U.S.A.* 107: 11751–11756.

43 Hiley, S.L., Sood, V.D., Fan, J., and Collins, R.A. (2002). 4-thio-U cross-linking identifies the active site of the VS ribozyme. *EMBO J.* 21: 4691–4698.

44 Hiley, S.L. and Collins, R.A. (2001). Rapid formation of a solvent-inaccessible core in the *Neurospora* Varkud satellite ribozyme. *EMBO J.* 20: 5461–5469.

45 Lafontaine, D.A., Norman, D.G., and Lilley, D.M. (2002). The global structure of the VS ribozyme. *EMBO J.* 21: 2461–2471.

46 Lipfert, J., Ouellet, J., Norman, D.G. et al. (2008). The complete VS ribozyme in solution studied by small-angle X-ray scattering. *Structure* 16: 1357–1367.

47 DasGupta, S., Suslov, N.B., and Piccirilli, J.A. (2017). Structural basis for substrate helix remodeling and cleavage loop activation in the Varkud satellite ribozyme. *J. Am. Chem. Soc.* 139: 9591–9597.

48 Michiels, P.J., Schouten, C.H.J., Hilbers, C.W., and Heus, H.A. (2000). Structure of the ribozyme substrate hairpin of *Neurospora* VS RNA: a close look at the cleavage site. *RNA* 6: 1821–1832.

49 Flinders, J. and Dieckmann, T. (2001). A pH controlled conformational switch in the cleavage site of the VS ribozyme substrate RNA. *J. Mol. Biol.* 308: 665–679.

50 Hoffmann, B., Mitchell, G.T., Gendron, P. et al. (2003). NMR structure of the active conformation of the Varkud satellite ribozyme cleavage site. *Proc. Natl. Acad. Sci. U.S.A.* 100: 7003–7008.

51 Campbell, D.O. and Legault, P. (2005). NMR structure of the Varkud satellite ribozyme stem-loop V RNA and magnesium-ion binding from chemical-shift mapping. *Biochemistry* 44: 4157–4170.

52 Campbell, D.O., Bouchard, P., Desjardins, G., and Legault, P. (2006). NMR structure of Varkud satellite ribozyme stem-loop V in the presence of magnesium ions and localization of metal-binding sites. *Biochemistry* 45: 10591–10605.

53 Bouchard, P. and Legault, P. (2014). Structural insights into substrate recognition by the *Neurospora* Varkud satellite ribozyme: importance of U-turns at the kissing-loop junction. *Biochemistry* 53: 258–269.

54 Bonneau, E., Girard, N., Lemieux, S., and Legault, P. (2015). The NMR structure of the II-III-VI three-way junction from the *Neurospora* VS ribozyme reveals a critical tertiary interaction and provides new insights into the global ribozyme structure. *RNA* 21: 1621–1632.

55 Desjardins, G., Bonneau, E., Girard, N. et al. (2011). NMR structure of the A730 loop of the *Neurospora* VS ribozyme: insights into the formation of the active site. *Nucleic Acids Res.* 39: 4427–4437.

56 Bonneau, E. and Legault, P. (2014). Nuclear magnetic resonance structure of the III-IV-V three-way junction from the Varkud Satellite ribozyme and identification of magnesium-binding sites using paramagnetic relaxation enhancement. *Biochemistry* 53: 6264–6275.

57 Bonneau, E. and Legault, P. (2014). NMR localization of divalent cations at the active site of the *Neurospora* VS ribozyme provides insights into RNA-metal-ion interactions. *Biochemistry* 53: 579–590.

58 Pereira, M.J., Nikolova, E.N., Hiley, S.L. et al. (2008). Single VS ribozyme molecules reveal dynamic and hierarchical folding toward catalysis. *J. Mol. Biol.* 382: 496–509.

59 Lacroix-Labonté, J., Girard, N., Lemieux, S., and Legault, P. (2012). Helix-length compensation studies reveal the adaptability of the VS ribozyme architecture. *Nucleic Acids Res.* 40: 2284–2293.

60 Lacroix-Labonté, J., Girard, N., Dagenais, P., and Legault, P. (2016). Rational engineering of the *Neurospora* VS ribozyme to allow substrate recognition via different kissing-loop interactions. *Nucleic Acids Res.* 44: 6924–6934.

18

Chemical Modifications in Natural and Engineered Ribozymes

Stephanie Kath-Schorr

University of Cologne, Department of Chemistry, Greinstrasse 4, 50939 Köln, Germany

18.1 Introduction

Despite being restricted to four building blocks, ribonucleic acids can fold into a multitude of structures promoting not only genetic information or recognition of binding partners but also its being a catalytically active component, a ribozyme itself. Soon after the discovery of ribozymes in the 1980s [1, 2], the chemical alteration within ribozyme sequences has introduced an exciting new field for fundamental research and potential applications [3–5].

The preparation of ribozymes [6] and deoxyribozymes [7] (catalytic DNA) using nonnatural nucleotide building blocks is attractive because of two reasons: Firstly, for RNA-based catalysts, chemical modifications, primarily at the ribose moiety, add to the overall stability of the nucleic acid polymer. Thus, ribose modifications such as 2′-fluoro and 2′-methoxy groups or locked nucleic acid (LNA) building blocks aim at enabling the (therapeutic) application of ribozymes in a cellular environment or even in whole organisms by providing resistance against degradation in the cell [5]. Secondly, the catalytic power of such engineered ribozymes in vivo (which is usually determined in vitro) is in most cases fairly low with observed cleavage rates within the range of few events per minute. The attachment of nonnatural functional groups to natural nucleobases can expand the scope of catalysis by both ribozymes and deoxyribozymes. This approach serves two main purposes, either the catalytic activity of an existing ribozyme shall be improved or novel ribozymes catalyzing an entirely artificial reaction shall be developed [5, 8, 9]. Importantly, simple incorporation of nonnatural building blocks in existing ribozyme sequences at random positions usually does not result in increased or altered reactivity. Therefore, either de novo development of ribozymes by in vitro selection is required with one or several of the four nucleobases bearing a nonnatural modification. Alternatively, such nonnatural building blocks are incorporated at carefully chosen individual positions within the ribozyme sequence, for example, with the intention to stabilize helical structures [10].

Ribozymes, First Edition. Edited by Sabine Müller, Benoît Masquida, and Wade Winkler.

In this chapter, existing approaches in functionalization and selection of ribozymes using nonnatural nucleotide building blocks are presented. Several examples for chemical modifications introduced into natural ribozyme sequences and their applications are briefly discussed in Section 18.2. Section 18.3 concentrates on the introduction of chemically modified nucleotides during in vitro selection for the selection of novel (deoxy-)ribozymes.

18.2 Chemical Modifications to Study Natural Ribozymes

Solid phase RNA synthesis allows to introduce various chemically modified nucleobases at specific positions within the sequence of a ribozyme in order to prepare modified RNAs with lengths below 100 nucleobases. Longer ribozyme sequences can be prepared by ligation of short synthetic oligonucleotides [11–13]. An alternative approach is the preparation of RNA sequences by in vitro transcription using chemically modified triphosphate building blocks, which results in modification of the entire RNA molecule. T7 RNA polymerase and their engineered descendants can accept diverse unnatural ribonucleoside triphosphates, ranging from 2′-fluoro modified nucleoside triphosphates to larger modifications such as 5-aminoallylcytidine-5′-triphosphate or 5-ethynyluridine-5′-triphosphate [14]. Examples for ribozymes with modified nucleobases are given in the following sections (18.2.1 - 18.2.2).

18.2.1 Modified Nucleotides for Mechanistic and Structural Studies on Ribozymes

Understanding the folding and underlying mechanisms of ribozyme cleavage is essential for studying their occurrence and function as well as exploiting catalytic oligonucleotides for diverse applications. The structure determination of intact self-cleaving ribozymes in the solid state by X-ray diffraction usually requires a chemical modification at the active site preventing cleavage [15–19]. Often, a 2′-methoxy substitution of the ribose unit positioned 5′ at the cleavage site is employed, but other 2′-modifications are also effective [15–19]. Studying the effect of phosphorothioate substitutions at the scissile phosphate allows further insights into the catalytic mechanism of ribozymes [20, 21]. The present review chapter is focused on the mechanistic consequences of chemical modifications more remote to the formed and cleaved phosphoester bonds: For example, chemical probing by mutagenesis such as in nucleotide analogue interference mapping (NAIM) allows mapping of functionally important regions of a ribozyme [22–26]. Additionally, recent advantages in deep sequencing approaches now allow screening of several thousands of mutant ribozymes consisting of only natural nucleotides for self-cleavage and provide an extremely fast access to structural and functional elements of novel ribozymes [27–30].

Studying mechanisms of ribozyme cleavage in solution by introducing nonnatural nucleotides into the ribozyme sequence is an extensively utilized approach by various groups. The exchange of individual nucleobases that are involved in catalysis in sequences of small self-cleaving ribozymes by nonnatural nucleobases shifting their local pK_a values, allowed to decipher their catalytic mechanisms. Comparing the pH-dependent observed cleavage rates of ribozyme constructs with different nucleotide modifications can be employed to decipher a mechanism that is primarily based on general acid–base catalysis (for more information, see Chapter 1) [2, 16, 31–40].

Furthermore, site-specific incorporation of reporter groups allows deciphering the complex folding pathways of ribozymes by spectroscopic approaches [17, 41–50]. Reporter groups are incorporated either directly by solid phase RNA synthesis or post-synthetically in a two-step reaction, e.g. by click chemistry or using (N-Hydroxysuccinimide) NHS-ester chemistry [51]. Careful selection of the modification side enables studying the overall fold of the ribozyme, even on a single-molecule level [52–55]. In most cases, trans-cleaving ribozymes hybridized from several shorter RNA sequences are used. If small self-cleaving ribozymes are prepared from synthetic oligonucleotides, an interesting observation can be made: Frequently, a significant fraction of the RNA does not self-cleave, in particular, if long RNA sequences are prepared by solid phase synthesis. Besides improper folding, chemical side reactions during solid phase synthesis resulting in the removal of the 2′-*O*-protecting group lead to 3′- to 2′-phosphate migration and can increase this effect [56]. Thus, an approach via in vitro transcription is highly desirable to synthesize fully functional ribozymes in vitro. Our group has recently demonstrated that in vitro transcribed glmS and CPEB3 ribozymes with site-specific modifications introduced by an expanded genetic alphabet are fully active [57, 58]. This method is an ideal tool to site-specifically incorporate reporter groups such as fluorophores into ribozyme sequences studying their folding dynamics in solution.

18.2.2 Stabilization of Ribozymes by Chemical Modifications for in Cell Applications

The introduction of chemical modifications into natural ribozyme sequences to achieve resistance against nucleases and improve their therapeutic potential for in cell application has been studied for over two decades [59–61]. Synthetic hammerhead ribozymes have been prepared possessing several 2′-methoxy (**1**), 2′-NH_2 (**2**), 2′-fluoro (**3**), or 2′-*O*-allyl (**4**) or 2′-*C*-allyl (**5**) modifications (Figure 18.1) as well as 3′–3′ linked nucleotides that have been introduced as cap analogs [62]. The serum half-lives of such ribozymes greatly increased, but their catalytic potential was slightly reduced compared to unmodified hammerhead ribozyme sequences [59]. Solid phase synthesis of such modified ribozymes is mandatory as only certain nucleotide positions can be substituted without loss of catalytic activity [60, 63]. Vester and coworkers introduced LNA (**6**) (Figure 18.1) nucleotides into the sequence of hammerhead ribozymes by solid phase synthesis. The LNA-containing hammerhead ribozymes showed improved cleavage activity even with shortened

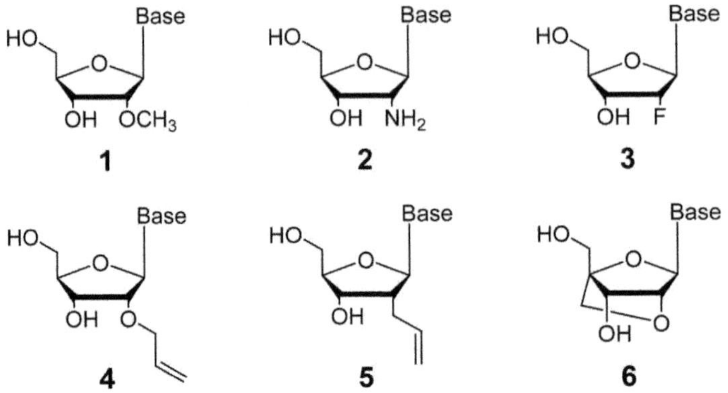

Figure 18.1 Ribose modifications to stabilize ribozymes for in cell applications (2′-methoxy (**1**), 2′-amino (**2**), 2′-fluoro (**3**), 2′-*O*-allyl (**4**), 2′-*C*-allyl (**5**), and LNA (**6**)).

substrate binding arms, which might facilitate access to target sequences in therapeutic applications [64]. Active hammerhead and hairpin ribozymes have further been prepared by intrastrand click ligation bearing a non-natural backbone at the catalytic site providing a novel tool for the assembly of larger RNAs or RNA/DNA hybrid molecules [65].

In contrast to chemical modifications of individual nucleobases, Mirkin and coworkers pursued an alternative approach stabilizing a ribozyme by incorporation into spherical nanoparticles functionalized with oligonucleotides. A synthetic hammerhead construct targeting the human (O-6-Methylguanine-DNA Methyltransferase) MGMT gene [66] was covalently ligated to spherical nucleic acids and transfected into live cells [67]. The authors could demonstrate cleavage of the MGMT messenger RNA (mRNA) and thus knockdown of the MGMT protein by this approach. An interesting approach for light control of RNA folding and function by posttranscriptional RNA labeling was pursued by the Kool group. They showed that the activity of in vitro transcribed hammerhead ribozymes (poly-)acylated at 2′-hydroxy groups with photoactive protecting groups can be restored by light exposure and that the concept is in principle applicable in cell culture [68].

18.3 In Vitro Selection with Chemically Modified Nucleotides: Expanding the Scope of DNA and RNA Catalysis

In vitro selection [69] has been employed to evolve artificial ribozymes with various catalytic functions. In addition to cleavage and ligation of individual phosphodiester bonds and ribozymes catalyzing, for example, RNA polymerization [70, 71], reverse transcription [72], C—O [73–75], C—C [76], C—N [3, 77], or C—S [78] bond formations have been selected [79]. Achieving multiple turnover catalysis in bimolecular reactions with two small molecules by artificial ribozymes is challenging. In principle, this is feasible by selecting for binding to a transition-state analog. However, most selection strategies include tethering one or both small molecule reactants covalently to RNA, either to part of the ribozyme itself or to a short RNA sequence that binds to the ribozyme via base pairing. In rare cases, artificial ribozymes are selected, which can accelerate bimolecular small molecule reactions without covalent attachment of the reactant to the tether as demonstrated for several ribozymes catalyzing Diels–Alder reactions [80–82] and an RNA capping ribozyme [83].

In order to expand the chemical scope of nucleic acid catalysis, advantages of implementing nonnatural nucleotides bearing functional groups that are absent in natural nucleic acids in selection strategies are obvious. In the selection of DNA aptamers, the application of canonical nucleobases bearing artificial side chains has proven to be a valuable approach producing aptamers with low dissociation rate constants (Slow Off-rate Modified Aptamers [SOMAmers]) [84]. Here, even the introduction of two modified bases in one Systematic Evolution of Ligands by EXponential enrichment (SELEX) experiment has been reported [85]. Similarly, DNA enzymes for sequence-specific RNA cleavage have been selected with imidazolyl modified nucleobases, which are essential for catalysis [86–88].

In RNA selection, ribozymes capable of catalyzing the Diels–Alder reaction have both been developed using modified nucleobases (pyridine-modified uridine derivatives [81, 89]) and exclusively natural nucleobases [80, 82]. The introduction of functional groups, which are not naturally occurring in RNA for the selection of ribozymes with novel functions as demonstrated for the Diels–Alderase [81], provides a useful tool to expand the chemical scope of the RNA.

18.3.1 General Aspects for In Vitro Selection Using Unnatural Nucleotides

In vitro selection using modified nucleoside triphosphates as building blocks requires careful consideration of the choice of the chemical modification attached to the nucleobases and the enzymes used in amplifying such modified nucleic acids [14]. First of all, it is essential that base-pairing interactions are not disturbed by the attachment of a chemical functionality to the nucleobase. Likewise, these modified nucleoside triphosphates need to be efficiently incorporated by DNA or/and RNA polymerases with high fidelity at all sequence positions. Sequential incorporation of modified nucleoside triphosphates by polymerases can be restricted and needs to be investigated. Lastly, the modified nucleic acid has to function as new template in the subsequent selection round. Thus, concerted efforts of different groups have resulted in a variety of modified nucleoside triphosphates compatible with enzymatic amplification by standard DNA polymerases [84, 90–96]. Comprehensive reviews about modified nucleosides in SELEX are available [84, 92, 94, 95, 97], and only few selected examples are mentioned in the course of this chapter.

An in-depth study of the incorporation efficiencies of base-modified nucleoside triphosphate analogues testing a wide range of DNA polymerases for polymerase chain reaction (PCR) amplification was reported by Famulok and coworkers [98, 99]. Others specifically investigated the incorporation of protein-like side chains, in particular amino and imidazolyl functionalities, by modified triphosphates in

PCR reactions [100, 101]. Structural evidence for efficient DNA replication by KlenTaq DNA polymerase incorporating C5 modified pyrimidine and C7 modified 7-deazapurine nucleoside triphosphates is provided by the Marx group [102].

The selection of RNA catalysts with modified nucleobases adds additional complexity to the SELEX cycle [103, 104]. First of all, those modifications need to be efficiently introduced during in vitro transcription from an unmodified DNA library [14]. After the selection of catalytically active sequences, reverse transcription of those sequences with high efficiency and fidelity into their corresponding cDNA is required. Consequently, most examples using chemically modified nucleobases in the selection of nucleic acid catalysts have been performed selecting deoxyribozymes, thus avoiding a reverse transcription step in the selection process.

An interesting approach is presented by Liu and coworkers [105]. The authors use a ligase-based polymerization method to assemble nucleic acid polymers with different side chain compositions for aptamer selection. This strategy could in principle be extended to select for catalytically active polymer sequences. An alternative approach that avoid using modified DNA as a template in PCR reactions (which is a critical step in DNA SELEX) previously established in aptamer selection [106–108] was applied in deoxyribozyme selection by the Perrin laboratory [109]. After enrichment of catalytically active sequences in one SELEX cycle, these modified DNA sequences are covalently linked to their unmodified DNA template strands that were initially used in PCR [109].

18.3.2 Selection of Deoxyribozymes with Modified Nucleotides

Recent progress in in vitro selection strategies has led to a variety of DNA-based catalysts [7, 110]. Different deoxyribozymes (DNAzymes) have been developed for target-specific RNA cleavage with the intention to use such deoxyribozymes for cleavage of specific mRNAs in vivo [111]. In contrast to ribozymes, deoxyribozymes can be designed to function at low divalent metal ion concentrations [112] or even in the absence of M^{2+} by using modified nucleotide building blocks for in vitro selection. Protein-like functional groups such as amines, carboxylic acids, guanidinium, and imidazole groups are attached to natural nucleotides and provide important functional groups that are absent in natural nucleic acids. Evolution of such RNA cleaving deoxyribozymes is, besides proof of concept studies, primarily driven by the goal to develop therapeutic tools for site-specific mRNA cleavage in cells.

Barbas and coworkers used imidazolyl groups to extend the chemical functionality of the DNA library selecting for an RNA cleaving deoxyribozyme [86]. Their work evolved a small deoxyribozyme capable of site-specific RNA cleavage at physiological pH with an observed catalytic rate of over 1 min^{-1} [86]. This deoxyribozyme folds into a hairpin structure with three imidazole moieties located close to the cleavage site being essential for activity. Micromolar concentrations of Zn^{2+} are required for catalysis [86]. Perrin et al. simultaneously used amine (lysine analog, ^{Aa}dU, Figure 18.2) and imidazole (histidine analog, ^{Im}dA) modified nucleoside trisphosphates in one selection experiment to select a deoxyribozyme as RNAse A mimic cleaving a single ribophosphodiester linkage in its substrate.

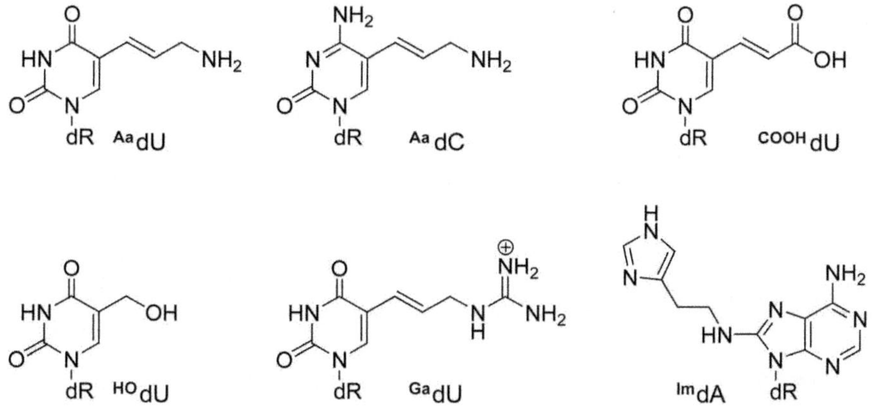

Figure 18.2 Examples for nucleosides with functional groups mimicking amino acid side chains, which have been employed for the selection of novel deoxyribozymes (dR, deoxyribose). Amino (^{Aa}dU, ^{Aa}dC), carboxy (^{COOH}dU), hydroxy-(^{HO}dU), guanidinium (^{Ga}dU), and imidazole (^{Im}dA) groups provide novel chemical properties and are essential for catalysis [113–117]. Source: Adapted from Hollenstein et al. [113].

A general acid–base catalysis mechanism is presumably responsible for its catalytic activity, which was observed even in the absence of divalent metal ions [114, 115, 118]. This is a remarkable example on how protein-like side chains can enhance the functionality of deoxyribozymes. The same laboratory reported a self-cleaving deoxyribozyme possessing tyrosine side chains [119]. A similar objective was pursued by Williams and coworkers. They employed a combination of a C5-imidazolyl-modified dUTP and 3-(aminopropynyl)-7-deaza-dATP to select an RNA cleaving deoxyribozyme [87]. Despite being only moderately active, this DNAzyme accelerated the site-specific cleavage of an all-RNA target in the absence of divalent metal ions. Perrin and coworkers extended their initial approach by incorporating three different chemical modifications, imidazole, ammonium, and guanidinium groups (Figure 18.2, ^{Aa}dC, ^{Ga}dU, ^{Im}dA), achieving efficient RNA cleavage at neutral pH independent of divalent metal ions [113].

A major challenge in the selection of Mg^{2+}-independent RNA cleaving deoxyribozymes is the evolving sequences, which are capable of multiple turnover. Very recently, the Perrin group succeeded and generated a Mg^{2+}-independent deoxyribozyme for RNA cleavage with multiple turnover acting on an all-RNA target sequence [116].

The Silverman laboratory developed the first example of a deoxyribozyme selected from a chemically modified DNA library, which catalyzes a reaction other than phosphodiester cleavage [117]. This deoxyribozyme can promote the hydrolysis of aliphatic amides and requires a single primary amine as modification for catalysis. Amide hydrolysis has been a particularly difficult reaction to achieve by nucleic acid catalysis [4, 120].

The in vitro selection procedure started from a deoxyribozyme pool bearing a randomized region of 40 nucleotides containing either ^{Aa}dU, ^{COOH}dU, or ^{HO}dU

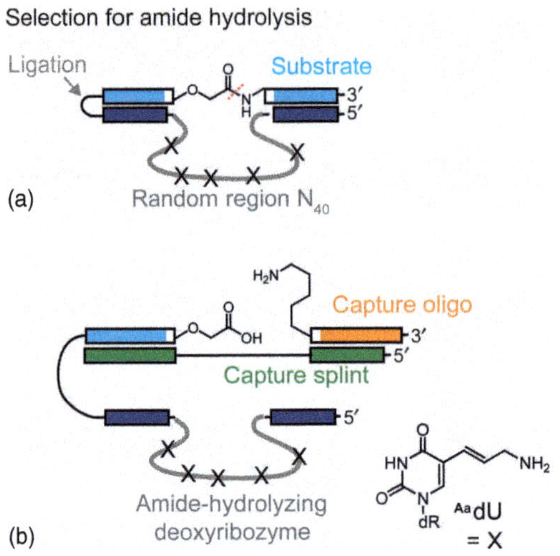

Figure 18.3 In vitro selection strategy for an amide-hydrolyzing deoxyribozyme using nonnatural nucleotides (X) in the random region of the DNA pool by Silverman and coworkers. If DNA catalyzed hydrolysis occurs, the capture oligonucleotide is directed to the hydrolysis product via the capture splint. After covalent attachment of the capture oligonucleotide to the deoxyribozyme sequence, its elongated sequence can be detected and separated from non-cleaving pool sequences by gel electrophoresis. (a) Before catalysis and (b) after catalysis. Source: Adapted from Zhou et al. [117].

(Figure 18.2) [117]. Via splint ligation, a substrate consisting of two oligonucleotides linked by an aliphatic amide bond was attached to the deoxyribozyme pool. Those sequences, which catalyze hydrolysis of the amide bond are captured using two additional oligonucleotides, a capture splint and a capture oligonucleotide bearing a 5′-amino group (Figure 18.3). For each selection round, the PCR amplification of enriched pool sequences was performed using the corresponding modified (Uridine-5'-triphosphate) UTP. Successful catalysts were identified with ^{Aa}dU-modified sequences. Remarkably, only at two positions, the ^{Aa}dU modification is essential for catalysis and the deoxyribozyme even showed reduced activity with only one of these positions being modified by ^{Aa}dU. Selections with ^{HO}dU and ^{COOH}dU gave rise to further catalysts for amide hydrolysis [117]. This example shows that deoxyribozymes modified with nonnatural nucleotides have a remarkable catalytic potential and might lead to the identification of artificial, nucleic acid catalyzed proteases.

18.3.3 Artificial Ribozymes with Nonnatural Nucleobases

The selection of ribozymes employing nonnatural nucleobases is highly complex. Both reactions, in vitro transcription and reverse transcription, are critical steps, and their efficiency and fidelity can vary enormously depending on the nucleobase modification present [121–124]. Moreover, RNA, in contrast to DNA, can adopt a variety of conformations. For straightforward catalytic reactions (such as cleavage or ligation of phosphodiester backbones), selection with natural nucleobases is sufficient. Thus, examples of novel, in vitro selected ribozymes with nonnatural nucleobases are scarce: A ribozyme for carbon–carbon bond formation (Diels–Alderase) has been evolved using pyridyl-modified nucleobases [81], an amide synthase ribozyme has been selected with a 5′-imidazole uridine analogue [125], and an RNA ligase was selected using *N*^{6}-(6-aminohexyl)-adenosine instead of natural adenosine [126]. Joyceand coworkers used 5-bromouridine in the selection of a ribozyme catalyzing phosphoester transfer [127]. In their approach they first replaced uridine nucleobases in the Tetrahymena group I ribozyme by 5-bromouridine. After construction of a library consisting of variants of this 5-bromouridine modified ribozyme, five rounds of in vitro selection led to the identification of ribozymes with strongly increased catalytic activity [127].

18.3.4 Catalysts With Nonnatural Backbones: XNAzymes

To apply chemically modified nucleic acid building blocks in in vitro selection strategies, the efficient enzymatic amplification of those sequences is essential. Natural polymerases might tolerate modifications at the nucleobase and can be engineered to accept a wider range of modifications [85, 88]. In contrast, synthetic nucleic acid polymers (xeno nucleic acids [XNAs]) possess entirely artificial backbones. Such novel nucleic acid backbones provide strongly increased stability against degradation by nucleases or pH-dependent degradation, as well as exceeding hybridization stabilities [91]. Pioneering work from Holliger and coworkers using directed evolution and rational design strategies has led to the development of DNA polymerases capable of synthesizing different XNAs [128, 129]. For the application of XNA in in vitro selection strategies, an efficient reverse transcription, in this case the correct decoding of the XNA and assembly of the corresponding natural DNA polymer, is a further prerequisite and has been achieved [124, 128, 129]. Holliger and coworkers succeeded in the selection of XNAzymes, catalysts based on synthetic nucleic acids bearing an artificial backbone [130]. Four different XNA chemistries were employed for in vitro selection (X-SELEX): arabino nucleic acids (ANAs), 2′-fluoroarabino nucleic acids (FANAs), hexitol nucleic acids (HNAs), and cyclohexene nucleic acids (CeNAs) (Figure 18.4) [130, 131]. In all cases, specialized engineered polymerases were employed to synthesize and reverse transcribe the XNA libraries.

XNAzymes cleaving an internal RNA (based on all four abovementioned artificial backbone chemistries) as well two XNAzymes (based on the FANA backbone), one possessing RNA–RNA ligase activity and another remarkably having XNA–XNA ligase activity, were evolved (Figure 18.4b,c) [130, 131]. Chaput and coworkers demonstrated that in vitro selection of a FANA enzyme can also be achieved using natural polymerases. The XNAzyme site-specifically cleaves RNA with a remarkable rate enhancement comparable to natural ribozymes [132]. Those first examples of in vitro selection of active XNAzymes show the remarkable potential, which shall be exploited for therapeutic applications and material sciences. Furthermore, these experiments can give a glimpse into the early evolution of replicating biomolecules.

18.4 Outlook

With only few examples of evolved artificial ribozymes catalyzing reactions other than cleavage or ligation of phosphodiester bonds, it is impossible to predict the limits of catalysis by RNA and DNA. In particular, requirements for bimolecular catalysis of reactions involving small molecules by ribozymes remain unknown due to the limited range of examples reported. Introducing unnatural nucleotides into the selection process for new ribozymes as presently successfully done in aptamer

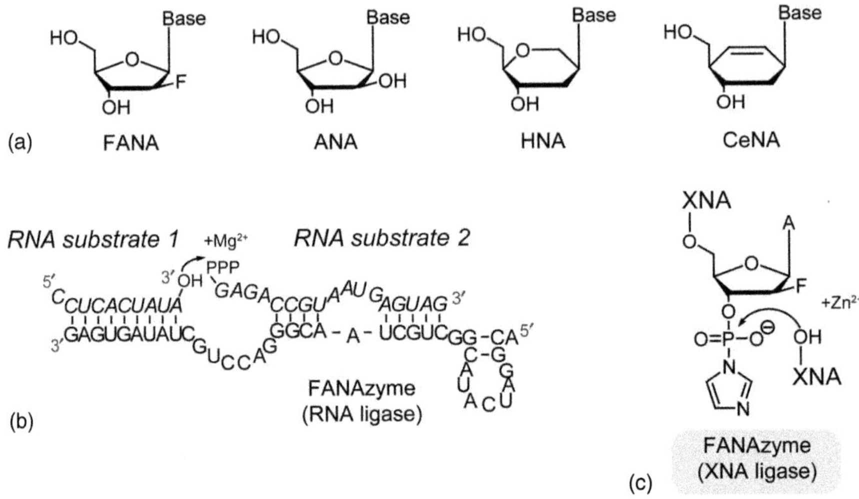

Figure 18.4 XNA and XNA-based catalysts developed by Holliger and coworkers. (a) Nucleic acid building blocks with artificial backbones: arabino nucleic acids (ANA), 2′-fluoroarabino nucleic acids (FANA), hexitol nucleic acids (HNA), and cyclohexene nucleic acids (CeNA). (b) First example of an RNA–RNA ligase XNAzyme based on FANA. (c) An XNAzyme catalyzing the ligation of a 2′-fluoroarabino backbone. Activation of the 3′-phosphate as 3′-phosphorylimidazolide is essential. Source: Adapted from Taylor et al. [130].

selection and in few examples selecting (deoxy-)ribozymes will certainly allow access to a wider chemical range of catalytic transformations.

One step further will be applying an expanded genetic alphabet based on a six-letter code for the selection of RNA and in particular of ribozymes. Hirao and coworkers reported the first promising examples of DNA-based SELEX with unnatural, hydrophobic base pairs, which do not rely on hydrogen bonding for recognition, at specific positions in an otherwise randomized sequence evolving high affinity DNA aptamers [133–135]. The application of such unnatural base pairs in the selection of RNA aptamers and ribozymes has never been demonstrated to date. Besides overcoming the obstacles during SELEX regarding the accessible sequence space for a randomized region with six variables instead of four, direct sequencing of such unnatural base pairs is not yet possible. However, recent progress in effective fourth-generation sequencing techniques will eventually allow the direct sequencing of an expanded genetic alphabet. Being able to use such techniques in a standardized manner for a six-letter genetic alphabet will potentially revolutionize the selection of aptamers and offer an enormous potential to select novel ribozymes.

References

1 Tanner, N.K. (1999). Ribozymes: the characteristics and properties of catalytic RNAs. *FEMS Microbiol. Rev.* 23 (3): 257–275.

2 Lilley, D.M. (2005). Structure, folding and mechanisms of ribozymes. *Curr. Opin. Struct. Biol.* 15 (3): 313–323.
3 Wilson, C. and Szostak, J.W. (1995). In vitro evolution of a self-alkylating ribozyme. *Nature* 374 (6525): 777–782.
4 Dai, X., De Mesmaeker, A., and Joyce, G.F. (1995). Cleavage of an amide bond by a ribozyme. *Science* 267 (5195): 237–240.
5 Silverman, S.K. (2008). Nucleic acid enzymes (ribozymes and deoxyribozymes): in vitro selection and application. In: *Wiley Encyclopedia of Chemical Biology* (ed. T.P. Begley). Wiley https://doi.org/10.1002/9780470048672.wecb406.
6 Wilson, D.S. and Szostak, J.W. (1999). In vitro selection of functional nucleic acids. *Annu. Rev. Biochem.* 68: 611–647.
7 Silverman, S.K. (2016). Catalytic DNA: scope, applications, and biochemistry of deoxyribozymes. *Trends Biochem. Sci.* 41 (7): 595–609.
8 Silverman, S.K. and Baum, D.A. (2009). Use of deoxyribozymes in RNA research. *Methods Enzymol.* 469: 95–117.
9 Silverman, S.K. (2009). Deoxyribozymes: selection design and serendipity in the development of DNA catalysts. *Acc. Chem. Res.* 42 (10): 1521–1531.
10 Looser, V., Langenegger, S.M., Haner, R., and Hartig, J.S. (2007). Pyrene modification leads to increased catalytic activity in minimal hammerhead ribozymes. *Chem. Commun.* 42: 4357–4359.
11 Stark, M.R. and Rader, S.D. (2014). Efficient splinted ligation of synthetic RNA using RNA ligase. *Methods Mol. Biol.* 1126: 137–149.
12 Kershaw, C.J. and O'Keefe, R.T. (2012). Splint ligation of RNA with T4 DNA ligase. *Methods Mol. Biol.* 941: 257–269.
13 Paredes, E., Evans, M., and Das, S.R. (2011). RNA labeling, conjugation and ligation. *Methods* 54 (2): 251–259.
14 Dellafiore, M.A., Montserrat, J.M., and Iribarren, A.M. (2016). Modified nucleoside triphosphates for in vitro selection techniques. *Front. Chem.* 4: 18.
15 Zheng, L., Mairhofer, E., Teplova, M. et al. (2017). Structure-based insights into self-cleavage by a four-way junctional twister-sister ribozyme. *Nat. Commun.* 8 (1): 1180.
16 Gebetsberger, J. and Micura, R. (2017). Unwinding the twister ribozyme: from structure to mechanism. *Wiley Interdiscip. Rev.: RNA* 8 (3): e1402.
17 Dagenais, P., Girard, N., Bonneau, E., and Legault, P. (2017). Insights into RNA structure and dynamics from recent NMR and X-ray studies of the *Neurospora* Varkud satellite ribozyme. *Wiley Interdiscip. Rev.: RNA* 8 (5): e1421.
18 Suslov, N.B., DasGupta, S., Huang, H. et al. (2015). Crystal structure of the Varkud satellite ribozyme. *Nat. Chem. Biol.* 11 (11): 840–846.
19 Ferre-D'Amare, A.R., Zhou, K., and Doudna, J.A. (1998). Crystal structure of a hepatitis delta virus ribozyme. *Nature* 395 (6702): 567–574.
20 Hougland, J.L., Kravchuk, A.V., Herschlag, D., and Piccirilli, J.A. (2005). Functional identification of catalytic metal ion binding sites within RNA. *PLoS Biol.* 3 (9): e277.
21 Bevilacqua, P.C. and Yajima, R. (2006). Nucleobase catalysis in ribozyme mechanism. *Curr. Opin. Chem. Biol.* 10 (5): 455–464.

22 Oyelere, A.K., Kardon, J.R., and Strobel, S.A. (2002). pK_a perturbation in genomic hepatitis delta virus ribozyme catalysis evidenced by nucleotide analogue interference mapping. *Biochemistry* 41 (11): 3667–3675.

23 Ryder, S.P. and Strobel, S.A. (1999). Nucleotide analog interference mapping. *Methods* 18 (1): 38–50.

24 Ryder, S.P. and Strobel, S.A. (1999). Nucleotide analog interference mapping of the hairpin ribozyme: implications for secondary and tertiary structure formation. *J. Mol. Biol.* 291 (2): 295–311.

25 Waldsich, C. (2008). Dissecting RNA folding by nucleotide analog interference mapping (NAIM). *Nat. Protoc.* 3 (5): 811–823.

26 Basu, S., Pazsint, C., and Chowdhury, G. (2004). Analysis of ribozyme structure and function by nucleotide analog interference mapping. *Methods Mol. Biol.* 252: 57–75.

27 Kobori, S. and Yokobayashi, Y. (2018). Analyzing and tuning ribozyme activity by deep sequencing to modulate gene expression level in mammalian cells. *ACS Synth. Biol.* 7 (2): 371–376.

28 Nomura, Y., Chien, H.C., and Yokobayashi, Y. (2017). Direct screening for ribozyme activity in mammalian cells. *Chem. Commun.* 53 (93): 12540–12543.

29 Kobori, S., Takahashi, K., and Yokobayashi, Y. (2017). Deep sequencing analysis of aptazyme variants based on a pistol ribozyme. *ACS Synth. Biol.* 6 (7): 1283–1288.

30 Kobori, S. and Yokobayashi, Y. (2016). High-throughput mutational analysis of a twister ribozyme. *Angew. Chem. Int. Ed.* 55 (35): 10354–10357.

31 Wilson, T.J., Liu, Y., Domnick, C. et al. (2016). The novel chemical mechanism of the twister ribozyme. *J. Am. Chem. Soc.* 138 (19): 6151–6162.

32 Kath-Schorr, S., Wilson, T.J., Li, N.S. et al. (2012). General acid–base catalysis mediated by nucleobases in the hairpin ribozyme. *J. Am. Chem. Soc.* 134 (40): 16717–16724.

33 McCown, P.J., Winkler, W.C., and Breaker, R.R. (2012). Mechanism and distribution of glmS ribozymes. *Methods Mol. Biol.* 848: 113–129.

34 Webb, C.H. and Luptak, A. (2011). HDV-like self-cleaving ribozymes. *RNA Biol.* 8 (5): 719–727.

35 Ferre-D'Amare, A.R. and Scott, W.G. (2010). Small self-cleaving ribozymes. *Cold Spring Harbor Perspect. Biol.* 2 (10): a003574.

36 Strobel, S.A. and Cochrane, J.C. (2007). RNA catalysis: ribozymes, ribosomes, and riboswitches. *Curr. Opin. Chem. Biol.* 11 (6): 636–643.

37 Dubecky, M., Walter, N.G., Sponer, J. et al. (2015). Chemical feasibility of the general acid/base mechanism of glmS ribozyme self-cleavage. *Biopolymers* 103 (10): 550–562.

38 Bevilacqua, P.C. (2003). Mechanistic considerations for general acid–base catalysis by RNA: revisiting the mechanism of the hairpin ribozyme. *Biochemistry* 42 (8): 2259–2265.

39 Nakano, S., Chadalavada, D.M., and Bevilacqua, P.C. (2000). General acid–base catalysis in the mechanism of a hepatitis delta virus ribozyme. *Science* 287 (5457): 1493–1497.

40 Lott, W.B., Pontius, B.W., and von Hippel, P.H. (1998). A two-metal ion mechanism operates in the hammerhead ribozyme-mediated cleavage of an RNA substrate. *Proc. Natl. Acad. Sci. U.S.A.* 95 (2): 542–547.

41 Leamy, K.A., Assmann, S.M., Mathews, D.H., and Bevilacqua, P.C. (2016). Bridging the gap between in vitro and in vivo RNA folding. *Q. Rev. Biophys.* 49: e10.

42 Kobitski, A.Y., Schafer, S., Nierth, A. et al. (2013). Single-molecule FRET studies of RNA folding: a Diels–Alderase ribozyme with photolabile nucleotide modifications. *J. Phys. Chem. B* 117 (42): 12800–12806.

43 Cardo, L., Karunatilaka, K.S., Rueda, D., and Sigel, R.K. (2012). Single molecule FRET characterization of large ribozyme folding. *Methods Mol. Biol.* 848: 227–251.

44 Wilson, T.J., Nahas, M., Araki, L. et al. (2007). RNA folding and the origins of catalytic activity in the hairpin ribozyme. *Blood Cells Mol. Dis.* 38 (1): 8–14.

45 Kobitski, A.Y., Nierth, A., Helm, M. et al. (2007). Mg^{2+}-dependent folding of a Diels–Alderase ribozyme probed by single-molecule FRET analysis. *Nucleic Acids Res.* 35 (6): 2047–2059.

46 Hobartner, C. and Silverman, S.K. (2005). Modulation of RNA tertiary folding by incorporation of caged nucleotides. *Angew. Chem. Int. Ed. Eng.* 44 (44): 7305–7309.

47 Edwards, T.E. and Sigurdsson, S.T. (2005). EPR spectroscopic analysis of U7 hammerhead ribozyme dynamics during metal ion induced folding. *Biochemistry* 44 (38): 12870–12878.

48 Hammann, C., Norman, D.G., and Lilley, D.M. (2001). Dissection of the ion-induced folding of the hammerhead ribozyme using 19F NMR. *Proc. Natl. Acad. Sci. U.S.A.* 98 (10): 5503–5508.

49 Koutmou, K.S., Casiano-Negroni, A., Getz, M.M. et al. (2010). NMR and XAS reveal an inner-sphere metal binding site in the P4 helix of the metallo-ribozyme ribonuclease P. *Proc. Natl. Acad. Sci. U.S.A.* 107 (6): 2479–2484.

50 Furtig, B., Richter, C., Schell, P. et al. (2008). NMR-spectroscopic characterization of phosphodiester bond cleavage catalyzed by the minimal hammerhead ribozyme. *RNA Biol.* 5 (1): 41–48.

51 Kath-Schorr, S. (2015). Cycloadditions for studying nucleic acids. *Top. Curr. Chem.* 374 (1): 1–27.

52 Blanco, M. and Walter, N.G. (2010). Analysis of complex single-molecule FRET time trajectories. *Methods Enzymol.* 472: 153–178.

53 Walter, N.G., Zhuang, X.W., Kim, H. et al. (2003). Correlating structural dynamics and function in single ribozyme molecules. *Biophys. J.* 84 (2): 183a.

54 Walter, N.G. (2003). Probing RNA structural dynamics and function by fluorescence resonance energy transfer (FRET). In: *Current Protocols in Nucleic Acid Chemistry*. Chapter 11, Unit 11 10. RNA Folding Pathways, Editor: Gary D. Glick, Current Protocols in Nucleic Acid Chemistry, Publisher: John Wiley & Sons, Inc, Page Range: 11.10.1–11.10.23

55 Wilson, T.J., Nahas, M., Ha, T., and Lilley, D.M. (2005). Folding and catalysis of the hairpin ribozyme. *Biochem. Soc. Trans.* 33 (Pt 3): 461–465.

56 Reese, C.B. (2005). Oligo- and poly-nucleotides: 50 years of chemical synthesis. *Org. Biomol. Chem.* 3 (21): 3851–3868.

57 Eggert, F., Kulikov, K., Domnick, C. et al. (2017). Illuminated by foreign letters – strategies for site-specific cyclopropene modification of large functional RNAs via in vitro transcription. *Methods* 120: 17–27.

58 Eggert, F. and Kath-Schorr, S. (2016). A cyclopropene-modified nucleotide for site-specific RNA labeling using genetic alphabet expansion transcription. *Chem. Commun.* 52 (45): 7284–7287.

59 Beigelman, L., McSwiggen, J.A., Draper, K.G. et al. (1995). Chemical modification of hammerhead ribozymes. Catalytic activity and nuclease resistance. *J. Biol. Chem.* 270 (43): 25702–25708.

60 Klopffer, A.E. and Engels, J.W. (2003). The effect of universal fluorinated nucleobases on the catalytic activity of ribozymes. *Nucleosides Nucleotides Nucleic Acids* 22 (5-8): 1347–1350.

61 Hendry, P., McCall, M.J., Stewart, T.S., and Lockett, T.J. (2004). Redesigned and chemically-modified hammerhead ribozymes with improved activity and serum stability. *BMC Chem. Biol.* 4 (1): 1.

62 Citti, L. and Rainaldi, G. (2005). Synthetic hammerhead ribozymes as therapeutic tools to control disease genes. *Curr. Gene Ther.* 5 (1): 11–24.

63 Olsen, D.B., Benseler, F., Aurup, H. et al. (1991). Study of a hammerhead ribozyme containing 2′-modified adenosine residues. *Biochemistry* 30 (40): 9735–9741.

64 Christiansen, J.K., Lobedanz, S., Arar, K. et al. (2007). LNA nucleotides improve cleavage efficiency of singular and binary hammerhead ribozymes. *Bioorg. Med. Chem.* 15 (18): 6135–6143.

65 El-Sagheer, A.H. and Brown, T. (2010). New strategy for the synthesis of chemically modified RNA constructs exemplified by hairpin and hammerhead ribozymes. *Proc. Natl. Acad. Sci. U.S.A.* 107 (35): 15329–15334.

66 CITTI, L., ECKSTEIN, F., CAPECCHI, B. et al. (1999). Transient transfection of a synthetic hammerhead ribozyme targeted against human MGMT gene to cells in culture potentiates the genotoxicity of the alkylation damage induced by mitozolomide. *Antisense Nucleic Acid Drug Dev.* 9 (2): 125–133.

67 Rouge, J.L., Sita, T.L., Hao, L. et al. (2015). Ribozyme-spherical nucleic acids. *J. Am. Chem. Soc.* 137 (33): 10528–10531.

68 Velema, W.A., Kietrys, A.M., and Kool, E.T. (2018). RNA control by photoreversible acylation. *J. Am. Chem. Soc.* 140 (10): 3491–3495.

69 Klug, S.J. and Famulok, M. (1994). All you wanted to know about SELEX. *Mol. Biol. Rep.* 20 (2): 97–107.

70 Wochner, A., Attwater, J., Coulson, A., and Holliger, P. (2011). Ribozyme-catalyzed transcription of an active ribozyme. *Science* 332 (6026): 209–212.

71 Horning, D.P. and Joyce, G.F. (2016). Amplification of RNA by an RNA polymerase ribozyme. *Proc. Natl. Acad. Sci. U.S.A.* 113 (35): 9786–9791.

72 Samanta, B. and Joyce, G.F. (2017). A reverse transcriptase ribozyme. *Elife* 6: e31153.
73 Lohse, P.A. and Szostak, J.W. (1996). Ribozyme-catalysed amino-acid transfer reactions. *Nature* 381 (6581): 442–444.
74 Jenne, A. and Famulok, M. (1998). A novel ribozyme with ester transferase activity. *Chem. Biol.* 5 (1): 23–34.
75 Illangasekare, M., Sanchez, G., Nickles, T., and Yarus, M. (1995). Aminoacyl-RNA synthesis catalyzed by an RNA. *Science* 267 (5198): 643–647.
76 Fusz, S., Eisenfuhr, A., Srivatsan, S.G. et al. (2005). A ribozyme for the aldol reaction. *Chem. Biol.* 12 (8): 941–950.
77 Zhang, B. and Cech, T.R. (1997). Peptide bond formation by in vitro selected ribozymes. *Nature* 390 (6655): 96–100.
78 Sengle, G., Eisenfuhr, A., Arora, P.S. et al. (2001). Novel RNA catalysts for the Michael reaction. *Chem. Biol.* 8 (5): 459–473.
79 Silverman, S.K. (2009). Artificial functional nucleic acids: aptamers, ribozymes, and deoxyribozymes identified by in vitro selection. In: *Functional Nucleic Acids for Analytical Applications* (eds. Y. Li and Y. Lu). New York: Springer Science + Business Media, LLC.
80 Seelig, B. and Jäschke, A. (1999). A small catalytic RNA motif with Diels–Alderase activity. *Chem. Biol.* 6 (3): 167–176.
81 Tarasow, T.M., Tarasow, S.L., and Eaton, B.E. (1997). RNA-catalysed carbon–carbon bond formation. *Nature* 389 (6646): 54–57.
82 Agresti, J.J., Kelly, B.T., Jäschke, A., and Griffiths, A.D. (2005). Selection of ribozymes that catalyse multiple-turnover Diels–Alder cycloadditions by using in vitro compartmentalization. *Proc. Natl. Acad. Sci. U.S.A.* 102 (45): 16170–16175.
83 Huang, F., Yang, Z., and Yarus, M. (1998). RNA enzymes with two small-molecule substrates. *Chem. Biol.* 5 (11): 669–678.
84 Rohloff, J.C., Gelinas, A.D., Jarvis, T.C. et al. (2014). Nucleic acid ligands with protein-like side chains: modified aptamers and their use as diagnostic and therapeutic agents. *Mol. Ther. Nucleic Acids* 3: e201.
85 Gawande, B.N., Rohloff, J.C., Carter, J.D. et al. (2017). Selection of DNA aptamers with two modified bases. *Proc. Natl. Acad. Sci. U.S.A.* 114 (11): 2898–2903.
86 Santoro, S.W., Joyce, G.F., Sakthivel, K. et al. (2000). RNA cleavage by a DNA enzyme with extended chemical functionality. *J. Am. Chem. Soc.* 122 (11): 2433–2439.
87 Sidorov, A.V., Grasby, J.A., and Williams, D.M. (2004). Sequence-specific cleavage of RNA in the absence of divalent metal ions by a DNAzyme incorporating imidazolyl and amino functionalities. *Nucleic Acids Res.* 32 (4): 1591–1601.
88 Thomas, J.M., Yoon, J.-K., and Perrin, D.M. (2009). Investigation of the catalytic mechanism of a synthetic DNAzyme with protein-like functionality: an RNaseA mimic? *J. Am. Chem. Soc.* 131 (15): 5648–5658.
89 Tarasow, T.M., Kellogg, E., Holley, B.L. et al. (2004). The effect of mutation on RNA Diels–Alderases. *J. Am. Chem. Soc.* 126 (38): 11843–11851.

90 Lapa, S.A., Chudinov, A.V., and Timofeev, E.N. (2016). The toolbox for modified aptamers. *Mol. Biotechnol.* 58 (2): 79–92.

91 Pinheiro, V.B. and Holliger, P. (2014). Towards XNA nanotechnology: new materials from synthetic genetic polymers. *Trends Biotechnol.* 32 (6): 321–328.

92 Tolle, F. and Mayer, G. (2013). Dressed for success – applying chemistry to modulate aptamer functionality. *Chem. Sci.* 4 (1): 60–67.

93 Lauridsen, L.H., Rothnagel, J.A., and Veedu, R.N. (2012). Enzymatic recognition of 2′-modified ribonucleoside 5′-triphosphates: towards the evolution of versatile aptamers. *ChemBioChem* 13 (1): 19–25.

94 Chen, T. and Romesberg, F.E. (2014). Directed polymerase evolution. *FEBS Lett.* 588 (2): 219–229.

95 Hollenstein, M. (2012). Nucleoside triphosphates – building blocks for the modification of nucleic acids. *Molecules* 17 (11): 13569–13591.

96 Betz, K., Malyshev, D.A., Lavergne, T. et al. (2012). KlenTaq polymerase replicates unnatural base pairs by inducing a Watson–Crick geometry. *Nat. Chem. Biol.* 8 (7): 612–614.

97 Diafa, S. and Hollenstein, M. (2015). Generation of aptamers with an expanded chemical repertoire. *Molecules* 20 (9): 16643–16671.

98 Jager, S. and Famulok, M. (2004). Generation and enzymatic amplification of high-density functionalized DNA double strands. *Angew. Chem. Int. Ed.* 43 (25): 3337–3340.

99 Jager, S., Rasched, G., Kornreich-Leshem, H. et al. (2005). A versatile toolbox for variable DNA functionalization at high density. *J. Am. Chem. Soc.* 127 (43): 15071–15082.

100 Lee, S.E., Sidorov, A., Gourlain, T. et al. (2001). Enhancing the catalytic repertoire of nucleic acids: a systematic study of linker length and rigidity. *Nucleic Acids Res.* 29 (7): 1565–1573.

101 Gourlain, T., Sidorov, A., Mignet, N. et al. (2001). Enhancing the catalytic repertoire of nucleic acids. II. Simultaneous incorporation of amino and imidazolyl functionalities by two modified triphosphates during PCR. *Nucleic Acids Res.* 29 (9): 1898–1905.

102 Bergen, K., Steck, A.L., Strutt, S. et al. (2012). Structures of KlenTaq DNA polymerase caught while incorporating C5-modified pyrimidine and C7-modified 7-deazapurine nucleoside triphosphates. *J. Am. Chem. Soc.* 134 (29): 11840–11843.

103 Vaish, N.K., Fraley, A.W., Szostak, J.W., and McLaughlin, L.W. (2000). Expanding the structural and functional diversity of RNA: analog uridine triphosphates as candidates for in vitro selection of nucleic acids. *Nucleic Acids Res.* 28 (17): 3316–3322.

104 Zinnen, S.P., Domenico, K., Wilson, M. et al. (2002). Selection, design, and characterization of a new potentially therapeutic ribozyme. *RNA* 8 (2): 214–228.

105 Chen, Z., Lichtor, P.A., Berliner, A.P. et al. (2018). Evolution of sequence-defined highly functionalized nucleic acid polymers. *Nat. Chem.* 10 (4): 420–427.

106 Ichida, J.K., Zou, K., Horhota, A. et al. (2005). An in vitro selection system for TNA. *J. Am. Chem. Soc.* 127 (9): 2802–2803.

107 Yu, H., Zhang, S., and Chaput, J.C. (2012). Darwinian evolution of an alternative genetic system provides support for TNA as an RNA progenitor. *Nat. Chem.* 4 (3): 183–187.

108 MacPherson, I.S., Temme, J.S., Habeshian, S. et al. (2011). Multivalent glycocluster design through directed evolution. *Angew. Chem. Int. Ed.* 50 (47): 11238–11242.

109 Renders, M., Miller, E., Hollenstein, M., and Perrin, D. (2015). A method for selecting modified DNAzymes without the use of modified DNA as a template in PCR. *Chem. Commun.* 51 (7): 1360–1362.

110 Schlosser, K. and Li, Y. (2009). Biologically inspired synthetic enzymes made from DNA. *Chem. Biol.* 16 (3): 311–322.

111 Liu, M., Chang, D., and Li, Y. (2017). Discovery and biosensing applications of diverse RNA-cleaving DNAzymes. *Acc. Chem. Res.* 50 (9): 2273–2283.

112 Faulhammer, D. and Famulok, M. (1997). Characterization and divalent metal-ion dependence of in vitro selected deoxyribozymes which cleave DNA/RNA chimeric oligonucleotides. *J. Mol. Biol.* 269 (2): 188–202.

113 Hollenstein, M., Hipolito, C.J., Lam, C.H., and Perrin, D.M. (2013). Toward the combinatorial selection of chemically modified DNAzyme RNase A mimics active against all-RNA substrates. *ACS Comb. Sci.* 15 (4): 174–182.

114 Perrin, D.M., Garestier, T., and Hélène, C. (2001). Bridging the gap between proteins and nucleic acids: a metal-independent RNAseA mimic with two protein-like functionalities. *J. Am. Chem. Soc.* 123 (8): 1556–1563.

115 Lermer, L., Roupioz, Y., Ting, R., and Perrin, D.M. (2002). Toward an RNaseA mimic: a DNAzyme with imidazoles and cationic amines. *J. Am. Chem. Soc.* 124 (34): 9960–9961.

116 Wang, Y., Liu, E., Lam, C.H., and Perrin, D.M. (2018). A densely modified M^{2+}-independent DNAzyme that cleaves RNA efficiently with multiple catalytic turnover. *Chem. Sci.* 9 (7): 1813–1821.

117 Zhou, C., Avins, J.L., Klauser, P.C. et al. (2016). DNA-catalyzed amide hydrolysis. *J. Am. Chem. Soc.* 138 (7): 2106–2109.

118 Ting, R., Thomas, J.M., Lermer, L., and Perrin, D.M. (2004). Substrate specificity and kinetic framework of a DNAzyme with an expanded chemical repertoire: a putative RNaseA mimic that catalyzes RNA hydrolysis independent of a divalent metal cation. *Nucleic Acids Res.* 32 (22): 6660–6672.

119 Lam, C.H., Hipolito, C.J., Hollenstein, M., and Perrin, D.M. (2011). A divalent metal-dependent self-cleaving DNAzyme with a tyrosine side chain. *Org. Biomol. Chem.* 9 (20): 6949–6954.

120 Brandsen, B.M., Hesser, A.R., Castner, M.A. et al. (2013). DNA-catalyzed hydrolysis of esters and aromatic amides. *J. Am. Chem. Soc.* 135 (43): 16014–16017.

121 Vaught, J.D., Dewey, T., and Eaton, B.E. (2004). T7 RNA polymerase transcription with 5-position modified UTP derivatives. *J. Am. Chem. Soc.* 126 (36): 11231–11237.

122 Cozens, C. and Pinheiro, V.B. (2018). XNA synthesis and reverse transcription by engineered thermophilic polymerases. *Curr. Protoc. Chem. Biol.* 10: e47.

123 Chen, X., El-Sagheer, A.H., and Brown, T. (2014). Reverse transcription through a bulky triazole linkage in RNA: implications for RNA sequencing. *Chem. Commun.* 50 (57): 7597–7600.

124 Crouzier, L., Dubois, C., Edwards, S.L. et al. (2012). Efficient reverse transcription using locked nucleic acid nucleotides towards the evolution of nuclease resistant RNA aptamers. *PLoS One* 7 (4): e35990.

125 Wiegand, T.W., Janssen, R.C., and Eaton, B.E. (1997). Selection of RNA amide synthases. *Chem. Biol.* 4 (9): 675–683.

126 Teramoto, N., Imanishi, Y., and Ito, Y. (2000). In vitro selection of a ligase ribozyme carrying alkylamino groups in the side chains. *Bioconjugate Chem.* 11 (6): 744–748.

127 Dai, X. and Joyce, G.F. (2000). In vitro evolution of a ribozyme that contains 5-bromouridine. *Helv. Chim. Acta* 83 (8): 1701–1710.

128 Cozens, C., Pinheiro, V.B., Vaisman, A. et al. (2012). A short adaptive path from DNA to RNA polymerases. *Proc. Natl. Acad. Sci. U.S.A.* 109 (21): 8067–8072.

129 Pinheiro, V.B., Taylor, A.I., Cozens, C. et al. (2012). Synthetic genetic polymers capable of heredity and evolution. *Science* 336 (6079): 341–344.

130 Taylor, A.I., Pinheiro, V.B., Smola, M.J. et al. (2015). Catalysts from synthetic genetic polymers. *Nature* 518 (7539): 427–430.

131 Taylor, A.I. and Holliger, P. (2015). Directed evolution of artificial enzymes (XNAzymes) from diverse repertoires of synthetic genetic polymers. *Nat. Protoc.* 10 (10): 1625–1642.

132 Wang, Y., Ngor, A.K., Nikoomanzar, A., and Chaput, J.C. (2018). Evolution of a general RNA-cleaving FANA enzyme. *Nat. Commun.* 9 (1): 5067.

133 Kimoto, M., Yamashige, R., Matsunaga, K. et al. (2013). Generation of high-affinity DNA aptamers using an expanded genetic alphabet. *Nat. Biotechnol.* 31 (5): 453–457.

134 Kimoto, M., Matsunaga, K., and Hirao, I. (2016). DNA aptamer generation by genetic alphabet expansion SELEX (ExSELEX) using an unnatural base pair system. *Methods Mol. Biol.* 1380: 47–60.

135 Kimoto, M., Nakamura, M., and Hirao, I. (2016). Post-ExSELEX stabilization of an unnatural-base DNA aptamer targeting VEGF165 toward pharmaceutical applications. *Nucleic Acids Res.* 44 (15): 7487–7494.

19

Ribozymes for Regulation of Gene Expression

Julia Stifel and Jörg S. Hartig

University of Konstanz, Department of Chemistry, Room L904, Mailbox A704, Universitätsstr. 10, 78457 Konstanz, Germany

19.1 Introduction

Transcriptional gene switches are widely used and well-established systems to control gene expression. Complementary to these systems that often rely on the expression of additional transcription factors, switches consisting solely of RNA emerged as attractive tools utilizing new regulatory mechanisms [1]. Composed of short RNA sequences, these switches can be integrated into target sequences even if the coding space within a plasmid or genome is limited.

19.2 Conditional Gene Expression Control by Riboswitches

General structures and functions of RNA-based gene regulatory systems resemble the features of naturally occurring riboswitches. Riboswitches as important elements of gene expression regulation in nature were first described in 2002 [2–4]. Among other mechanisms of gene expression control by switching RNA states, riboswitches are the only class to modulate gene expression without the use of intermediate molecules or proteins [5]. Many naturally occurring riboswitches are found in the 5′-untranslated region (5′-UTR) of messenger RNAs (mRNAs) involved in metabolite transport or biosynthesis [6]. Hence, riboswitches are often involved in feedback regulation of the production or utilization of the sensed ligand. The general architecture of a riboswitch can be divided into two major functional modules: a sensor domain and an expression platform (Figure 19.1). Sensor domains are evolutionarily highly conserved aptamer sequences that bind to a specific ligand [7]. Binding of the ligand leads to a conformational change of the RNA and altered gene expression mediated through the so-called expression platform. The most basic engineered riboswitches only consist of an aptamer inserted into different positions of the untranslated region of a gene. Aptamers are RNA sequences that

Ribozymes, First Edition. Edited by Sabine Müller, Benoît Masquida, and Wade Winkler.

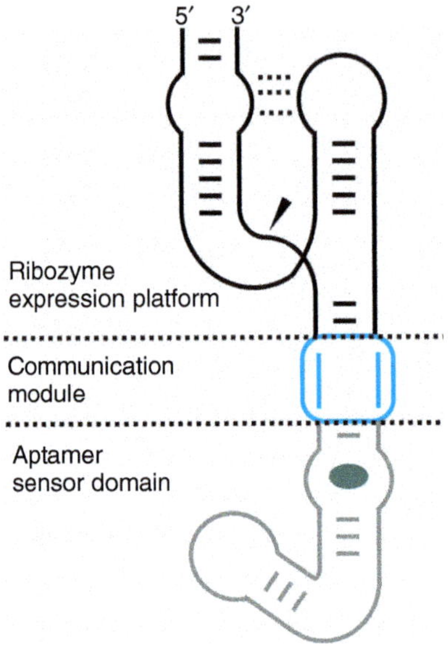

Figure 19.1 *General structure of an aptazyme.* The tripartite construct can be divided into the ribozyme as an expression platform (shown in black), the aptamer domain (shown in grey) and a communication module connecting these two domain (shown in blue). The cleavage site of the ribozyme is indicated by an arrowhead, a ligand bound to the aptamer is represented as a green dot.

are able to bind small molecules with high affinity and specificity [7]. Ligand binding leads to structural changes of the RNA molecule, which can be exploited for regulatory functions [8]. The first attempt to artificially control gene expression with the use of aptamers was carried out by insertion of an aptamer sensing the aminoglycoside antibiotic kanamycin A or tobramycin into the 5′-UTR of a reporter gene [9]. The resulting construct allowed ligand-controlled gene expression in vivo and in vitro presenting the first example of an engineered riboswitch several years before riboswitches were first discovered in nature [3]. Aptamers either are part of naturally occurring riboswitches or can be selected in vitro [7, 10].

19.3 Allosteric Ribozymes as Engineered Riboswitches

In order to create more complex artificial riboswitches, an aptamer domain can be combined with an engineered expression platform [11]. Mostly self-cleaving ribozymes are used as engineered expression platforms that are linked to an aptamer domain to design synthetic riboswitches. In nature, the concept of a self-cleaving ribozyme as an expression platform is found in the *glmS* riboswitch. In the presence of glucosamine-6-phosphate, the mRNA of the ribozyme undergoes self-cleavage, regulating expression levels of the GlmS gene [12]. The combination of an aptamer and a ribozyme domain results in allosteric ribozymes or aptazymes. The potential of ribozymes to act as expression platforms for regulating gene expression was first described by Yen et al. By inserting active and inactive versions of hammerhead ribozymes (HHRs) into different positions in the 5′- and 3′-UTR of a gene, they showed efficient regulatory function of gene expression upon ribozyme cleavage both in mammalian cell culture and in primary cells in mice [13]. In order to fine-tune gene expression, the reaction catalyzed by the ribozyme needs to be controlled in a ligand-responsive manner. Many achievements have been made over the last two decades to engineer ribozymes into ligand-dependent regulators of gene expression. The two major tasks in this development were the development of functional, ligand-responsive aptamer–ribozyme combinations and the implementation of these sequences into target RNAs in order to affect gene expression upon self-cleavage.

Artificial riboswitches based on self-cleaving ribozymes have been developed to function as both allosteric inducers and allosteric inhibitors of gene expression. The important feature distinguishing activating from repressing sequences is the composition of the communication module connecting the aptamer and the ribozyme. Hence, the major challenge in engineering a functional allosteric ribozyme is posed by finding the right connection sequence between the sensory aptamer domain and the ribozyme as an expression platform. First aptazymes arose from a combination of a natural HHR sequence with an aptamer sequence recognizing ATP or theophylline in a modular rational design approach in vitro [14]. By combining these two modules, the self-cleavage reaction of the HHR can be influenced by either the presence or absence of the ligand. Following this modular rational design approach based on Watson–Crick base-pairing rules and thermodynamic stabilities of RNA duplexes, various aptazymes based on HHR, hepatitis delta virus (HDV) ribozymes, or artificial ribozymes were generated in vitro [14–19]. Aptazyme designs based on this modular rational design were supported by computational methods [20, 21].

19.4 In Vitro Selection Methods

In further studies, the concept of modular rational design was combined with in vitro selection techniques [22]. The first example of an in vitro selection approach utilized a pool of RNAs carrying randomized nucleotide sequences in the communication module between an HHR and an aptamer sequence recognizing flavin mononucleotide (FMN). Sequences were separated by polyacrylamide gel electrophoresis to differentiate cleaved (active ribozyme) and uncleaved (inactive ribozyme) sequences in both the absence and presence of FMN. Sequences of the communication modules were analyzed and further characterized. While all activating communication sequences exhibit the original composition of nucleotides, all sequences showing ligand-dependent inhibition of the ribozyme carry a 1–2 nucleotides deletion. Mechanistic studies revealed a helix slippage mechanism wherein localized base-pairing changes within the communication module (Figure 19.2). Binding energy from a ligand–aptamer complex induces a rearrangement that can be described as a "slippage" in the communication module and a displacement of the adjacent G–C base pair needed for a stable folding of the ribozymes catalytic core. Replacement of the FMN aptamer for aptamers with new ligand specificities revealed that the communication modules are not universally applicable but need compatible "matched pairs" of aptamers and communication modules. In vitro selections and mixed approaches for new combinations of aptamers and ribozymes revealed a wide variety of aptazymes active in vitro [22–28].

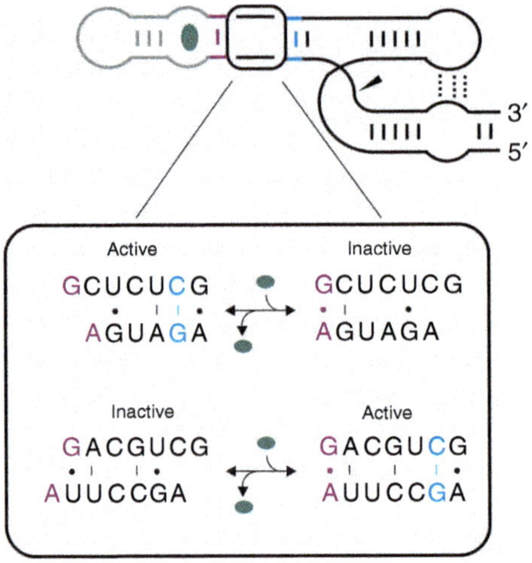

Figure 19.2 *Helix slippage mechanism.* The signal of ligand binding is transmitted from the aptamer domain (shown in grey) to the expression platform (shown in black) through the communication module (shown as a black box). The ligand is represented as a green dot. Ligand binding leads to a stabilisation and pairing of G and A (shown in red) in direct proximity to the ligand binding site. Depending on the architecture of the communication module this slippage either establishes or disrupts the G-C base-pair adjacent to the catalytic core of the ribozyme (shown in blue). By this mechanism the cleavage activity can either be activated or inhibited upon ligand addition.

The first ligand-dependent ribozyme functional in vivo was rationally designed by Ellington and coworkers [29]. By inserting a theophylline-responsive aptamer into the sequence of a group I self-splicing intron, they created various aptazyme variants and investigated both in vitro activation rates and in vivo gene regulatory effects. They observed several discrepancies between the performance of their aptazymes in vitro and their regulatory efficiency in bacteria. This observation was supported by Breaker and coworkers: Aptazyme sequences originating from an in vitro selection did not show effects in regulating a reporter gene in mammalian cell culture [30]. Aptazyme sequences originating from an in vitro screening performed in the Suess Lab were shown to regulate gene expression in yeast but not in mammalian cell culture [31]. Reasons for the inability of highly active aptazymes in vitro to regulate gene expression in a cellular environment might be due to differences in the folding pathways of the RNA sequences [32]. Mechanistic modeling and kinetic RNA folding simulations performed in the Keasling Lab predicted ribozyme folding, catalysis, and RNA half-life as the most important characteristics for functional gene regulatory effects [33]. Predicted gene expression levels precisely matched the experimental results in *Escherichia coli*.

19.5 In Vivo Screening Methods

Alternative strategies to determine aptazyme sequences functional in different organisms are direct in vivo screening and in vivo selection approaches. Here, the ligand-induced (or ligand-repressed) cleavage reaction is directly coupled to the expression of a reporter gene readout, and pools of sequences diversified at crucial positions such as the communication module are sampled for functional control of gene expression. In bacteria, the most effective way to couple the cleavage of small self-cleaving ribozymes to reporter gene expression was an insertion of the aptazyme into the 5′-UTR of green fluorescent protein (GFP). Aptazymes were designed in a way that the ribosomal binding site (RBS) is integrated into one stem of the ribozyme and sequestered by a complementary sequence [34–36] (Figure 19.3). In this conformation, the mRNA is not accessible to the small ribosomal subunit; thus, no translation will occur. As soon as the ribozyme undergoes self-cleavage, the RBS gets released, and translation is initiated. Plasmid libraries carrying a pool

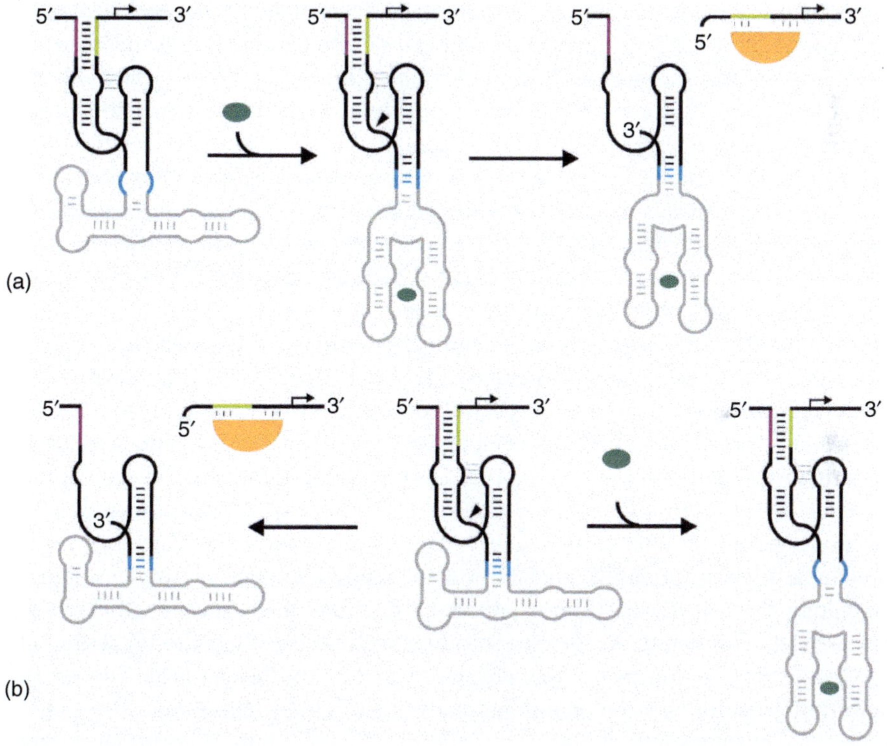

Figure 19.3 Mechanism of ON- and OFF-switches in hammerhead-based aptazymes based on Shine-Dalgarno sequestration in *E. coli*. The hammerhead ribozyme (shown in black) is connected to the ligand binding aptamer (shown in grey) through a communication module (shown in blue). The aptazyme is inserted into the 5′UTR of the gene of interest in a way that the Shine-Dalgarno sequence (shown in green) is included in stem I of the ribozyme and sequestered by a complementary sequence (shown in red) a) ON switch: Binding of the ligand (green dot) activates the hammerhead ribozyme. The active hammerhead ribozyme cleaves (cleavage site indicated by an arrowhead) and sets free the Shine-Dalgarno sequence (green) which is then recognized by the small ribosomal subunit (yellow) and initiates the translation of the gene. b) OFF switch: The active hammerhead ribozyme gets inactivated upon ligand binding, no more cleavage occurs, leading to a repression of gene translation.

of randomized sequences in the communication module can be transformed into *E. coli* with high efficiency. Individual clones are then isolated and cultivated, and the reporter gene levels are easily compared in the absence and presence of the respective ligand [34]. The advantage of this screening method is the simultaneous detection of both ON and OFF switches within the same approach. Being a robust method to identify powerful aptazymes, with increasing library sizes, the isolation of single clones becomes very laborious, demanding for new techniques to handle large libraries. Various methods to screen larger sequence spaces based on cell motility [37] or mortality [38] have been described in literature. However, these methods have their limitations in terms of tolerated expression levels and dynamic ranges of the switches. A promising method to screen large libraries is based on fluorescence-activated cell sorting (FACS). With a fluorescent protein as a reporter gene, clones showing the desired switching performance can be enriched over several rounds of sorting. In the first sort, all clones in the "OFF-state" can be isolated by setting the gate to low fluorescence intensities. Sorted cells will be further cultivated, and gene expression will be induced before the second sorting step to select for the "ON-state." The stringency of selection can be fine-tuned in each sorting step, and desired expression levels can be targeted, which is one of the major advantages of this method. FACS-based library screenings were successfully described for library sizes of 65 000 cells in *E. coli* [39–41]. After enrichment of a population following the predetermined switching behavior, single clones are isolated and further characterized. FACS-based methods are performed in liquid cultures, omitting the highly time-consuming step to plate the libraries on agar dishes and pick single colonies. In eukaryotic systems, self-cleaving aptazymes are mostly regulating gene expression by influencing mRNA stability. Inserted into either 5′- or 3′-UTRs, cleavage of the aptazyme results in a loss of the stabilizing cap structure or poly(A) tail (Figure 19.4). Degradation of the destabilized mRNA consequently leads to a decrease in gene expression [13]. In vivo screening and selection methods in eukaryotic contexts were predominantly developed in yeast, allowing a high throughput. In addition to manual screens of single colonies originating from a diverse library, different methods have been engineered to increase the throughput. An in vivo selection method based on the regulation of the GAL4 transcription factor was developed in the Hartig Lab [42]. The GAL4 transcription factor is a transcriptional activator of *HIS3*, *URA3*, and *lacZ* genes within the chromosomal DNA of the utilized yeast strain. Ligand-dependent switches were enriched by positive and negative selection using *HIS3* and *URA3* as auxotrophy markers, and the switching performance of resulting sequences was subsequently quantified via *lacZ* expression. FACS-based methods similar to the abovementioned in bacteria are equally applicable to *Saccharomyces cerevisiae*. Smolke's group described a FACS-based screening method using a dual-reporter system. In addition to a GFP gene that is regulated by the aptazyme, the plasmid includes a constitutively expressed mCherry gene. The expression of mCherry is used as an internal standard for the normalization of intrinsic or extrinsic variation of gene expression [43]. In further screening methods, they combined a FACS-based enrichment of functional aptazyme sequences with next-generation sequencing

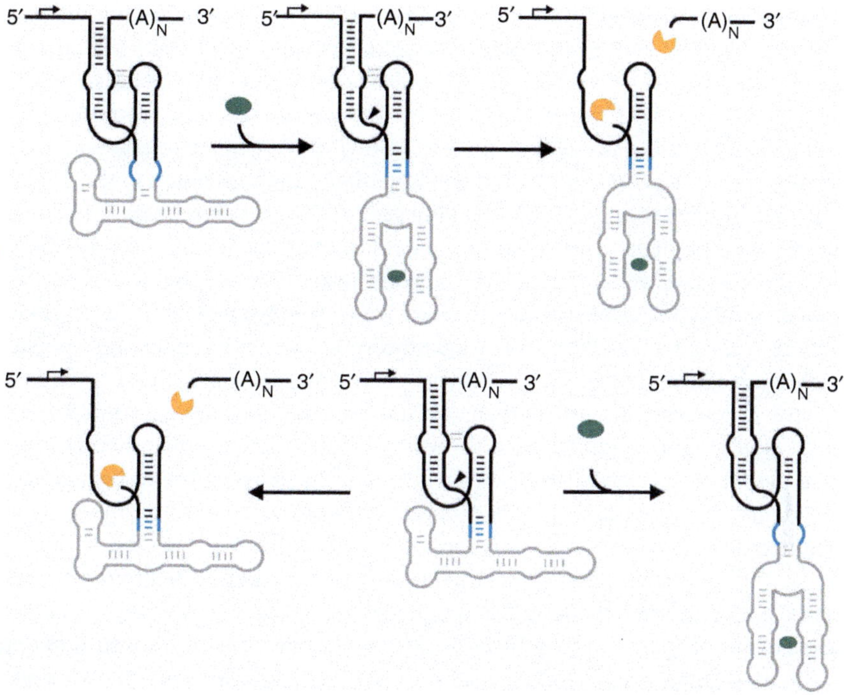

Figure 19.4 Mechanism of ON- and OFF-switches in hammerhead-based aptazymes in the 3′-UTR in *S. cerevisiae*. The hammerhead ribozyme (shown in black) is connected to the ligand binding aptamer (shown in grey) through a communication module (shown in blue). The aptazyme is inserted into the 3′-UTR of a gene of interest. a) OFF-switch: the hammerhead ribozyme is activated upon binding of the ligand (green dot) and undergoes self-cleavage. Cleavage of the polyA-tail ($(A)_n$) leads to degradation of the mRNA and results in repressed expression of the gene. b) ON-switch: ligand binding leads to an inactivation of the hammerhead ribozyme, preventing self-cleavage.

(NGS) [44]. While a lot of improvement was seen in increasing the throughput of in vivo screening methods both in bacteria and yeast, screening in mammalian cell culture still remains a very laborious procedure even for small library sizes [45, 46].

19.6 Rational Design of Allosteric Ribozymes

A rational design approach followed by Beilstein et al. yielded aptazyme sequences functional in HeLa cells [47]. Switches were designed in a way that ligand binding interferes loop–loop interactions between stems I and II of the HHR essential for the ribozyme cleavage reaction in vivo. Zhong et al. characterized a test panel of different communication module sequences to find a correlation between communication module characteristics and gene regulatory function in mammalian cell culture. Based on these correlations, they present a rational design approach to develop communication modules considering hydrogen bonding, base stacking, and the distance to the catalytic core of the ribozyme [48]. Aptazymes containing communication modules obtained by this rational design show high dynamic ranges in mammalian cell culture in response to tetracycline, theophylline, and guanine, thereby outperforming previously developed aptazymes for those ligands. Although the described method yields highly functional switches, the communication module characteristics are only correlated to high basal gene expression, therefore exclusively aiming at aptazymes that downregulate gene expression in the presence of a ligand. Similar strategies of developing genetic ON switches in mammalian cell culture would be highly beneficial in the field of engineering allosteric ribozymes. In addition to the well-studied HHR, mostly used as an expression platform to engineer new types of aptazymes, various other ribozymes have been shown to function as expression platforms in ligand-responsive gene regulators. The small self-cleaving HDV ribozyme was shown to efficiently regulate gene expression in response to guanine in mammalian cell culture [45]. The twister ribozyme was shown to efficiently regulate gene expression in response to theophylline or thiamine pyrophosphate (TPP) in bacteria [49]. The structure of the ribozyme allows the attachment of two aptamers to different sites of the same ribozyme sequence, providing the possibility to create two-input basic Boolean circuits. One of the major obstacles in the further development of allosteric ribozymes is the limited choice of ligands. Although de novo selected aptamers can be obtained from combinatorial nucleic acid libraries for many ligands [50], only a few of them proved functional in cellular systems and therefore useful for in vivo applications [51]. Aptazymes were engineered in bacteria to respond to theophylline, 3-methylxanthine, and TPP [29, 34, 36, 52]. In eukaryotic organisms, available ligands to date are theophylline, guanine, tetracycline, and neomycin [31, 41, 42, 45–47, 53]. For the use of ligand-dependent ribozyme switches in future applications, it will be necessary to expand the repertoire of possible ligands with more favorable properties.

19.7 Applications of Aptazymes for Gene Regulation

Once aptazymes are developed to be active in vivo, they can be employed to regulate gene expression apart from influencing mRNA accessibility or integrity (Figure 19.5). One example is the regulation of transfer RNA (tRNA) activation by attaching a theophylline dependent HHR to the 5′-end of a suppressor tRNA. (Figure 19.5a) The uncleaved sequence disrupts the folding of the typical cloverleaf structure of the tRNA. Only if the ribozyme is cleaved, the tRNA is released and available for processing and aminoacylation. Insertion of an amber stop codon enables regulation of gene expression by ligand-dependent activation of the amber suppressor tRNA. This concept was shown to be functional in vitro [54] as well as in *E. coli* [55] and can be further developed to yield more complex regulatory circuits. By using two different suppressor tRNAs regulated by orthogonal aptazymes, the identity of the incorporated amino acid can be controlled by respective ligand addition [56]. With a combination of mRNA and tRNA regulation, two-input switches following basic Boolean logic can be created [57]. Another example is the regulation of gene expression on the ribosomal RNA (rRNA) level (Figure 19.5b).

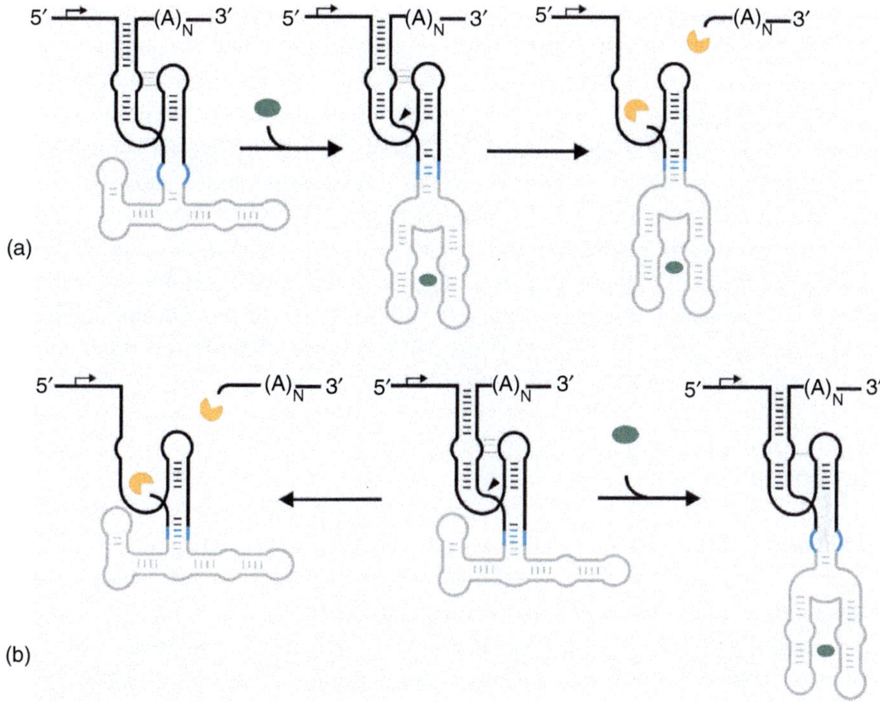

Figure 19.5 *Aptazyme-based regulation of different RNA classes.* a) The aptazyme is connected to a suppressor tRNA. As long as the tRNA is connected to the ribozyme it cannot adopt the correct fold, resulting in degradation. Ligand-induced cleavage leads to liberation of the tRNA that is then processed, charged and can then decode the respective Amber stop codon. b) The aptazyme is inserted into the 16S rRNA. Cleavage of the aptazyme leads to a degradation of the small ribosomal subunit. c) The aptazyme is connected to the 5′ end of a pri-miRNA. Only after ribozyme cleavage the pri-miRNA can be further processed and used for gene knockdown via RNA interference.

Some positions within the 16S rRNA tolerate the insertion of a TPP-dependent hammerhead aptazyme without affecting its activity in translation. The presence of TPP induces cleavage of the ribozyme and consequently cleavage of the 16S rRNA, resulting in a loss of function of the small ribosomal subunit [58]. In eukaryotes, mRNA stability can be regulated not only directly by inserting an aptazyme into the untranslated regions of a target mRNA but also indirectly through RNA interference (Figure 19.5c). This mechanism is implemented by connecting a structural analog of the pri-miRNA sequence with a ligand-inducible aptazyme. Ligand addition allows controlling the liberation of the pri-micro RNA (miRNA) sequence which is further processed by Drosha and Dicer and results in a specific knockdown of a targeted gene [59, 60]. In addition to controlling reporter genes in proof-of-principle studies, various allosteric ribozymes have been used to control effector proteins involved in cellular processes. By inserting a ligand-responsive ribozyme into the 3′-UTR of genes coding for interleukin-2 and interleukin-5, T-cell proliferation can be regulated by ligand addition in mouse and human T-cell lines as well as in mice

[61]. In another application, a ligand-dependent ribozyme was used to regulate p27 or cyclinB1mut as key regulators of the cell cycle. The addition of a ligand allowed arresting the cell cycle in G0/1 or G2/M phase [62]. Owing to the small size of ribozyme-based gene regulators, they can easily be incorporated into viral vectors where only small coding space is available. Allosterically controlled ribozymes were used to control (trans-) gene expression in adeno-associated viral vectors [48, 63], adenoviral oncoviruses [64, 65], measles viruses [64], and alphaviruses [66]. Taken together, ligand-responsive self-cleaving ribozymes possess some distinct advantages that qualify them as artificial genetic switches in many interesting applications. However, the main challenge in the field is the development of novel aptamer sensors that allow for control of gene expression with novel ligands characterized by more favorable properties.

References

1 Chappell, J., Watters, K.E., Takahashi, M.K., and Lucks, J.B. (2015). A renaissance in RNA synthetic biology: new mechanisms, applications and tools for the future. *Curr. Opin. Chem. Biol.* 28 (Supplement C): 47–56.

2 Mironov, A.S., Gusarov, I., Rafikov, R. et al. (2002). Sensing small molecules by nascent RNA: a mechanism to control transcription in bacteria. *Cell* 111 (5): 747–756.

3 Nahvi, A., Sudarsan, N., Ebert, M.S. et al. (2002). Genetic control by a metabolite binding mRNA. *Chem. Biol.* 9 (9): 1043.

4 Winkler, W., Nahvi, A., and Breaker, R.R. (2002). Thiamine derivatives bind messenger RNAs directly to regulate bacterial gene expression. *Nature* 419 (6910): 952–956.

5 Serganov, A. and Nudler, E. (2013). A decade of riboswitches. *Cell* 152 (1-2): 17–24.

6 Mandal, M. and Breaker, R.R. (2004). Gene regulation by riboswitches. *Nat. Rev. Mol. Cell Biol.* 5 (6): 451–463.

7 Ellington, A.D. and Szostak, J.W. (1990). In vitro selection of RNA molecules that bind specific ligands. *Nature* 346 (6287): 818–822.

8 Hermann, T. and Patel, D.J. (2000). Adaptive recognition by nucleic acid aptamers. *Science* 287 (5454): 820–825.

9 Werstuck, G. and Green, M.R. (1998). Controlling gene expression in living cells through small molecule-RNA interactions. *Science* 282 (5387): 296–298.

10 Tuerk, C. and Gold, L. (1990). Systematic evolution of ligands by exponential enrichment: RNA ligands to bacteriophage T4 DNA polymerase. *Science* 249 (4968): 505–510.

11 Desai, S.K. and Gallivan, J.P. (2004). Genetic screens and selections for small molecules based on a synthetic riboswitch that activates protein translation. *J. Am. Chem. Soc.* 126 (41): 13247–13254.

12 Winkler, W.C., Nahvi, A., Roth, A. et al. (2004). Control of gene expression by a natural metabolite-responsive ribozyme. *Nature* 428 (6980): 281–286.

13 Yen, L., Svendsen, J., Lee, J.-S. et al. (2004). Exogenous control of mammalian gene expression through modulation of RNA self-cleavage. *Nature* 431 (7007): 471–476.

14 Tang, J. and Breaker, R.R. (1997). Rational design of allosteric ribozymes. *Chem. Biol.* 4 (6): 453–459.

15 Hartig, J.S., Najafi-Shoushtari, S.H., Grune, I. et al. (2002). Protein-dependent ribozymes report molecular interactions in real time. *Nat. Biotechnol.* 20 (7): 717–722.

16 Beaudoin, J.D. and Perreault, J.P. (2008). Potassium ions modulate a G-quadruplex-ribozyme's activity. *RNA* 14 (6): 1018–1025.

17 Amontov, S. and Jäschke, A. (2006). Controlling the rate of organic reactions: rational design of allosteric Diels–Alderase ribozymes. *Nucleic Acids Res.* 34 (18): 5032–5038.

18 Lam, B.J. and Joyce, G.F. (2009). Autocatalytic aptazymes enable ligand-dependent exponential amplification of RNA. *Nat. Biotechnol.* 27 (3): 288–292.

19 Robertson, M.P. and Ellington, A.D. (2000). Design and optimization of effector-activated ribozyme ligases. *Nucleic Acids Res.* 28 (8): 1751–1759.

20 Penchovsky, R. (2014). Computational design of allosteric ribozymes as molecular biosensors. *Biotechnol. Adv.* 32 (5): 1015–1027.

21 Hall, B., Hesselberth, J.R., and Ellington, A.D. (2007). Computational selection of nucleic acid biosensors via a slip structure model. *Biosens. Bioelectron.* 22 (9): 1939–1947.

22 Soukup, G.A. and Breaker, R.R. (1999). Engineering precision RNA molecular switches. *Proc. Natl. Acad. Sci. U.S.A.* 96 (7): 3584–3589.

23 Soukup, G.A., Emilsson, G.A.M., and Breaker, R.R. (2000). Altering molecular recognition of RNA aptamers by allosteric selection , 11 Edited by D. E. Draper. *J. Mol. Biol.* 298 (4): 623–632.

24 Piganeau, N., Jenne, A., Thuillier, V., and Famulok, M. (2000). An allosteric ribozyme regulated by doxycyline. *Angew. Chem. Int. Ed.* 39 (23): 4369–4373.

25 Gu, H., Furukawa, K., and Breaker, R.R. (2012). Engineered allosteric ribozymes that sense the bacterial second messenger cyclic diguanosyl 5′-monophosphate. *Anal. Chem.* 84 (11): 4935–4941.

26 Najafi-Shoushtari, S.H. and Famulok, M. (2005). Competitive regulation of modular allosteric aptazymes by a small molecule and oligonucleotide effector. *RNA* 11 (10): 1514–1520.

27 Kertsburg, A. and Soukup, G.A. (2002). A versatile communication module for controlling RNA folding and catalysis. *Nucleic Acids Res.* 30 (21): 4599–4606.

28 Helm, M., Petermeier, M., Ge, B. et al. (2005). Allosterically activated Diels–Alder catalysis by a ribozyme. *J. Am. Chem. Soc.* 127 (30): 10492–10493.

29 Thompson, K.M., Syrett, H.A., Knudsen, S.M., and Ellington, A.D. (2002). Group I aptazymes as genetic regulatory switches. *BMC Biotechnol.* 2 (1): 21.

30 Link, K.H., Guo, L., Ames, T.D. et al. (2007). Engineering high-speed allosteric hammerhead ribozymes. *Biol. Chem.* 388 (8): 779.

31 Wittmann, A. and Suess, B. (2011). Selection of tetracycline inducible self-cleaving ribozymes as synthetic devices for gene regulation in yeast. *Mol. BioSyst.* 7 (8): 2419–2427.

32 Kato, Y., Kuwabara, T., Warashina, M. et al. (2001). Relationships between the activities in vitro and in vivo of various kinds of ribozyme and their intracellular localization in mammalian cells. *J. Biol. Chem.* 276 (18): 15378–15385.

33 Carothers, J.M., Goler, J.A., Juminaga, D., and Keasling, J.D. (2011). Model-driven engineering of RNA devices to quantitatively program gene expression. *Science* 334 (6063): 1716–1719.

34 Wieland, M. and Hartig, J.S. (2008). Improved aptazyme design and in vivo screening enable riboswitching in bacteria. *Angew. Chem. Int. Ed.* 47 (14): 2604–2607.

35 Ogawa, A. and Maeda, M. (2007). Aptazyme-based riboswitches as label-free and detector-free sensors for cofactors. *Bioorg. Med. Chem. Lett.* 17 (11): 3156–3160.

36 Ogawa, A. and Maeda, M. (2008). An artificial aptazyme-based riboswitch and its cascading system in *E. coli*. *ChemBioChem* 9 (2): 206–209.

37 Topp, S. and Gallivan, J.P. (2008). Random walks to synthetic riboswitches – a high-throughput selection based on cell motility. *ChemBioChem* 9 (2): 210–213.

38 Sharma, V., Nomura, Y., and Yokobayashi, Y. (2008). Engineering complex riboswitch regulation by dual genetic selection. *J. Am. Chem. Soc.* 130 (48): 16310–16315.

39 Fowler, C.C., Brown, E.D., and Li, Y. (2008). A FACS-based approach to engineering artificial riboswitches. *ChemBioChem* 9 (12): 1906–1911.

40 Lynch, S.A. and Gallivan, J.P. (2009). A flow cytometry-based screen for synthetic riboswitches. *Nucleic Acids Res.* 37 (1): 184–192.

41 Wieland, M., Ausländer, D., and Fussenegger, M. (2012). Engineering of ribozyme-based riboswitches for mammalian cells. *Methods* 56 (3): 351–357.

42 Klauser, B., Atanasov, J., Siewert, L.K., and Hartig, J.S. (2015). Ribozyme-based aminoglycoside switches of gene expression engineered by genetic selection in *S. cerevisiae*. *ACS Synth. Biol.* 4 (5): 516–525.

43 Liang, J.C., Chang, A.L., Kennedy, A.B., and Smolke, C.D. (2012). A high-throughput, quantitative cell-based screen for efficient tailoring of RNA device activity. *Nucleic Acids Res.* 40 (20): e154.

44 Townshend, B., Kennedy, A.B., Xiang, J.S., and Smolke, C.D. (2015). High-throughput cellular RNA device engineering. *Nat. Methods* 12 (10): 989–994.

45 Nomura, Y., Zhou, L., Miu, A., and Yokobayashi, Y. (2013). Controlling mammalian gene expression by allosteric hepatitis delta virus ribozymes. *ACS Synth. Biol.* 2 (12): 684–689.

46 Auslander, S., Ketzer, P., and Hartig, J.S. (2010). A ligand-dependent hammerhead ribozyme switch for controlling mammalian gene expression. *Mol. BioSyst.* 6 (5): 807–814.

47 Beilstein, K., Wittmann, A., Grez, M., and Suess, B. (2015). Conditional control of mammalian gene expression by tetracycline-dependent hammerhead ribozymes. *ACS Synth. Biol.* 4 (5): 526–534.

48 Zhong, G., Wang, H., Bailey, C.C. et al. (2016). Rational design of aptazyme riboswitches for efficient control of gene expression in mammalian cells. *elife* 5: e18858.

49 Felletti, M., Stifel, J., Wurmthaler, L.A. et al. (2016). Twister ribozymes as highly versatile expression platforms for artificial riboswitches. *Nat. Commun.* 7.

50 Mandal, M., Boese, B., Barrick, J.E. et al. (2003). Riboswitches control fundamental biochemical pathways in *Bacillus subtilis* and other bacteria. *Cell* 113 (5): 577–586.

51 Sudarsan, N., Wickiser, J.K., Nakamura, S. et al. (2003). An mRNA structure in bacteria that controls gene expression by binding lysine. *Genes Dev.* 17 (21): 2688–2697.

52 Wieland, M., Benz, A., Klauser, B., and Hartig, J.S. (2009). Artificial ribozyme switches containing natural riboswitch aptamer domains. *Angew. Chem. Int. Ed.* 48 (15): 2715–2718.

53 Win, M.N. and Smolke, C.D. (2007). A modular and extensible RNA-based gene-regulatory platform for engineering cellular function. *Proc. Natl. Acad. Sci. U.S.A.* 104 (36): 14283–14288.

54 Ogawa, A. and Maeda, M. (2008). A novel label-free biosensor using an aptazyme–suppressor-tRNA conjugate and an amber mutated reporter gene. *ChemBioChem* 9 (14): 2204–2208.

55 Berschneider, B., Wieland, M., Rubini, M., and Hartig, J.S. (2009). Small-molecule-dependent regulation of transfer RNA in bacteria. *Angew. Chem. Int. Ed.* 48 (41): 7564–7567.

56 Saragliadis, A. and Hartig, J.S. (2013). Ribozyme-based transfer RNA switches for post-transcriptional control of amino acid identity in protein synthesis. *J. Am. Chem. Soc.* 135 (22): 8222–8226.

57 Klauser, B., Saragliadis, A., Auslander, S. et al. (2012). Post-transcriptional Boolean computation by combining aptazymes controlling mRNA translation initiation and tRNA activation. *Mol. BioSyst.* 8 (9): 2242–2248.

58 Wieland, M., Berschneider, B., Erlacher, M.D., and Hartig, J.S. (2010). Aptazyme-mediated regulation of 16S ribosomal RNA. *Chem. Biol.* 17 (3): 236–242.

59 Kumar, D., An, C.-I., and Yokobayashi, Y. (2009). Conditional RNA interference mediated by allosteric ribozyme. *J. Am. Chem. Soc.* 131 (39): 13906–13907.

60 Bloom, R.J., Winkler, S.M., and Smolke, C.D. (2015). Synthetic feedback control using an RNAi-based gene-regulatory device. *J. Biol. Eng.* 9 (1): 5.

61 Chen, Y.Y., Jensen, M.C., and Smolke, C.D. (2010). Genetic control of mammalian T-cell proliferation with synthetic RNA regulatory systems. *Proc. Natl. Acad. Sci. U.S.A.* 107 (19): 8531–8536.

62 Wei, K.Y. and Smolke, C.D. (2015). Engineering dynamic cell cycle control with synthetic small molecule-responsive RNA devices. *J. Biol. Eng.* 9 (1): 21.

63 Strobel, B., Klauser, B., Hartig, J.S. et al. (2015). Riboswitch-mediated attenuation of transgene cytotoxicity increases adeno-associated virus vector yields in HEK-293 cells. *Mol. Ther.* 23 (10): 1582–1591.

64 Ketzer, P., Kaufmann, J.K., Engelhardt, S. et al. (2014). Artificial riboswitches for gene expression and replication control of DNA and RNA viruses. *Proc. Natl. Acad. Sci. U.S.A.* 111 (5): E554–E562.

65 Ketzer, P., Haas, S.F., Engelhardt, S. et al. (2012). Synthetic riboswitches for external regulation of genes transferred by replication-deficient and oncolytic adenoviruses. *Nucleic Acids Res.* 40 (21): e167.

66 Bell, C.L., Yu, D., Smolke, C.D. et al. (2015). Control of alphavirus-based gene expression using engineered riboswitches. *Virology* 483: 302–311.

20

Development of Flexizyme Aminoacylation Ribozymes and Their Applications

Takayuki Katoh, Yuki Goto, Toby Passioura, and Hiroaki Suga

The University of Tokyo, Graduate School of Science, Department of Chemistry, 7-3-1, Hongo, Bunkyo, Tokyo 113-0033, Japan

20.1 Introduction

Aminoacyl-transfer RNA (tRNA) is the key molecular link in the translation of mRNA-encoded genetic information into peptides and proteins. As is now well understood, each mRNA consists of four nucleotides, designated A, U, G, and C, and a triplet of these nucleotides constitutes a codon that designates one specific amino acid. Each aminoacyl-tRNA has the relevant amino acid attached at the 3′ end, and the three nucleotides that comprise the tRNA anticodon recognize the cognate codon of an mRNA through the base-pair formation. In nature, aminoacyl-tRNAs are synthesized by protein enzymes, called aminoacyl-tRNA synthetases (ARSs), which charge specific amino acids onto specific tRNAs. There are 20 different ARSs corresponding to the 20 proteinogenic amino acids (PAAs), and these are classified into two groups, class I and II enzymes, based on their structures [1, 2]. ARSs charge amino acids at either the 2′- or 3′-OH of the 3′-end adenosine of tRNA through the formation of an ester bond. Class I ARSs utilize the 2′-OH group, whereas class II ARSs utilize the 3′-OH; however, following 2′-OH aminoacylation by class I ARSs, the ester bond eventually migrates to the 3′-OH via transesterification.

Given that the catalytic core of the ribosome, the platform for protein synthesis, consists solely of RNA, RNA enzymes likely existed prior to the emergence of protein enzymes. Therefore, aminoacylation in primitive translation systems was also likely to have been catalyzed by RNA enzymes. In order to assess the possibility that such aminoacylation ribozymes existed in the primitive world, various groups conducted proof-of-concept experiments in which ribozymes with aminoacylation activity were evolved from pools of random RNA sequences through in vitro selection. Our group was one of those involved in such research. Although our prototype ribozyme, referred to as r24, preferentially charged aromatic amino acids activated with a cyanomethyl ester (CME) leaving the group onto specific tRNAs, we aimed to obtain more versatile, rather than more specific, aminoacylation ribozymes. Thus, r24 was further evolved into Fx3, which recognizes only the 3′-end region of tRNAs

Ribozymes, First Edition. Edited by Sabine Müller, Benoît Masquida, and Wade Winkler.

and was therefore capable of utilizing diverse tRNAs. In subsequent experiments, we also developed Fx3 mutants including eFx, a more active version of Fx3, and dFx, which can use amino acid substrates bearing 3,5-dinitrobenzyl ester (DBE) leaving groups. Since dFx and eFx either do not recognize the side chain of activated amino acids or recognize very broad classes of side chains (depending on the activating group used), they can be used to charge diverse amino acids onto diverse tRNAs. Due to their flexibility with respect to both amino acid and tRNA substrates, Fx3 and its derivatives have been referred to as flexizymes. In this chapter, we introduce the history of the development of flexizymes and their substrates, which have enabled the preparation of a wide array of aminoacyl-tRNAs.

One particularly useful advantage of using flexizymes for aminoacylation is that not only PAA-tRNAs but also non-proteinogenic aminoacyl-transfer RNAs (nPAA-tRNAs) can be easily synthesized. Such misacylated nPAA-tRNAs can be used for genetic code manipulation. To date, several genetic code manipulation methodologies, including both nonsense codon suppression and genetic code reprogramming, have been developed in combination with flexizyme-mediated nPAA-tRNA preparation. For example, nPAAs including *N*-methyl-, D-, and β-amino acids as well as α-amino acids with side-chain variations have been pre-charged onto tRNAs using flexizymes and then successfully introduced into peptides. In this chapter, we also provide an overview of these applications of flexizyme technology by discussing several examples of the synthesis of peptides containing various nPAAs.

20.2 The First Ribozymes Catalyzing Acyl Transfer to RNAs

Ribozymes that catalyze acyl-transfer reactions appeared to play important roles in the primitive "RNA world" [3, 4] as well as the modern "ribonucleoprotein (RNP) world." The catalytic core of the ribosome itself is composed only of RNA, and protein synthesis is therefore catalyzed by an acyl-transferase ribozyme (ATRib) [5]. It appears almost certain that many other ATRibs would have been essential for the evolutionary transition from an RNA world to the RNP world that exists today. In particular, ribozymes catalyzing aminoacylation of tRNAs (or tRNA-like RNAs), which have now been replaced by ARSs, would seem to have been indispensable for the advent of modern protein translation systems [6].

To obtain experimental evidence supporting both the RNA world hypothesis and the idea that aminoacylation ribozymes preceded modern ARS proteins, in vitro selection campaigns to isolate ribozymes capable of catalyzing acyl-transfer reactions were carried out. A pioneering study was reported by Yarus et al. in 1995, in which artificial ribozymes that catalyzed the self-aminoacylation of their own 3′-terminal 2′/3′-OH groups using Phe-AMP and Tyr-AMP as acyl donors were identified [7]. Subsequently, Jenne and Famulok also identified a ribozyme that showed self-aminoacylation activity at an internal 2′-OH group using an *N*-biotinylated-Phe-AMP substrate [8]. Around the same time, Szostak and Lohse performed in vitro selection from an N90 random RNA pool using an aminoacylated RNA oligomer homologous to the 3′-terminal region of tRNAs to isolate ATRibs, with the aim of mimicking the function of the ribosome as a peptidyl transfer catalyst [9]. The resulting ribozyme was further minimized to yield an acyl-transferase ribozyme (originally named STE18 and later renamed ATRib) [10]. ATRib has a 3′-end single-stranded region complementary to the donor hexanucleotide; thus, it can recognize the donor and catalyze acyl transfer to its own 5′-terminal OH group to yield aminoacylated ATRib (Figure 20.1a). These pioneering studies demonstrated that many types of acyl-transfer reactions can be catalyzed by RNA [11]; however, they did not yield tRNA-like adaptor RNAs aminoacylated on their 3′-terminus. In other words, the acylated RNA products could not act as substrates for ribosomal translation.

20.3 The ATRib Variant Family: Ribozymes Catalyzing tRNA Aminoacylation via Self-Acylated Intermediates

Although the original ATRib reaction yielded a self-acylated product, subsequent experiments by Lee et al. demonstrated that the aminoacyl group on ATRib could be further transferred to a tRNA [12]. This is because the 3′-terminal region of ATRib (internal guide sequence [IGS]) is complementary to both the acyl-donor oligo RNA and some tRNAs. Thus, when the donor RNA and tRNA were incubated in the presence of ATRib, the aminoacylated ATRib generated via the first acyl-transfer reaction could interact with tRNA via the IGS and catalyze a second acyl-transfer reaction to afford an aminoacylated tRNA (Figure 20.1b).

This ATRib-catalyzed tandem acyl transfer demonstrated the potential of ATRib as a tRNA-acylating ribozyme and led to the design and construction of a new RNA library (the ATRib–N70 library), in which an N70 random sequence domain was attached to the 3′ end of the ATRib acyl-transferase domain (Figure 20.1c). To obtain new ribozymes that use a non-RNA acyl donor, the RNA library was subjected to in vitro selected *N*-biotinylated glutamine cyanomethyl ester (biotin-Gln-CME) as an acyl donor, and ATRib derivatives that self-acylated with biotin-Gln were collected using a streptavidin-agarose resin [12]. The resulting ribozyme (AD02) was an ATRib variant having a glutamine-recognition (QR) domain attached to the 3′-terminal region (Figure 20.1d). AD02 could induce a two-step acyl-transfer reaction to obtain biotin-Gln-tRNA when the ribozyme was mixed with tRNA and biotin-Gln-CME and subjected to thermocycling (denaturing at 80 °C and annealing at 25 °C). In the first step of this reaction, the QR domain recognizes both biotin-Gln-CME and the IGS of the ATRib domain, and self-acylation of the 5′-terminal OH with biotin-Gln then occurs. Subsequently, thermocycling accelerates strand exchange, resulting in the interaction of the IGS with the acceptor tRNA, and the second acyl transfer occurs to yield aminoacylated tRNA. Although the overall yield was low and sophisticated temperature control was required, AD02 represented the first example of a ribozyme catalyzing tRNA aminoacylation using a simple amino acid derivative.

Further experiments studying the critical nucleobases in AD02 allowed nonessential regions of the original 70-nucleotide (nt) sequence to be deleted, leading to a minimal QR helix–loop RNA (29 nt) that could function *in trans* [13] (Figure 20.1e). Importantly, this QR domain exhibited remarkable specificity toward the Gln side

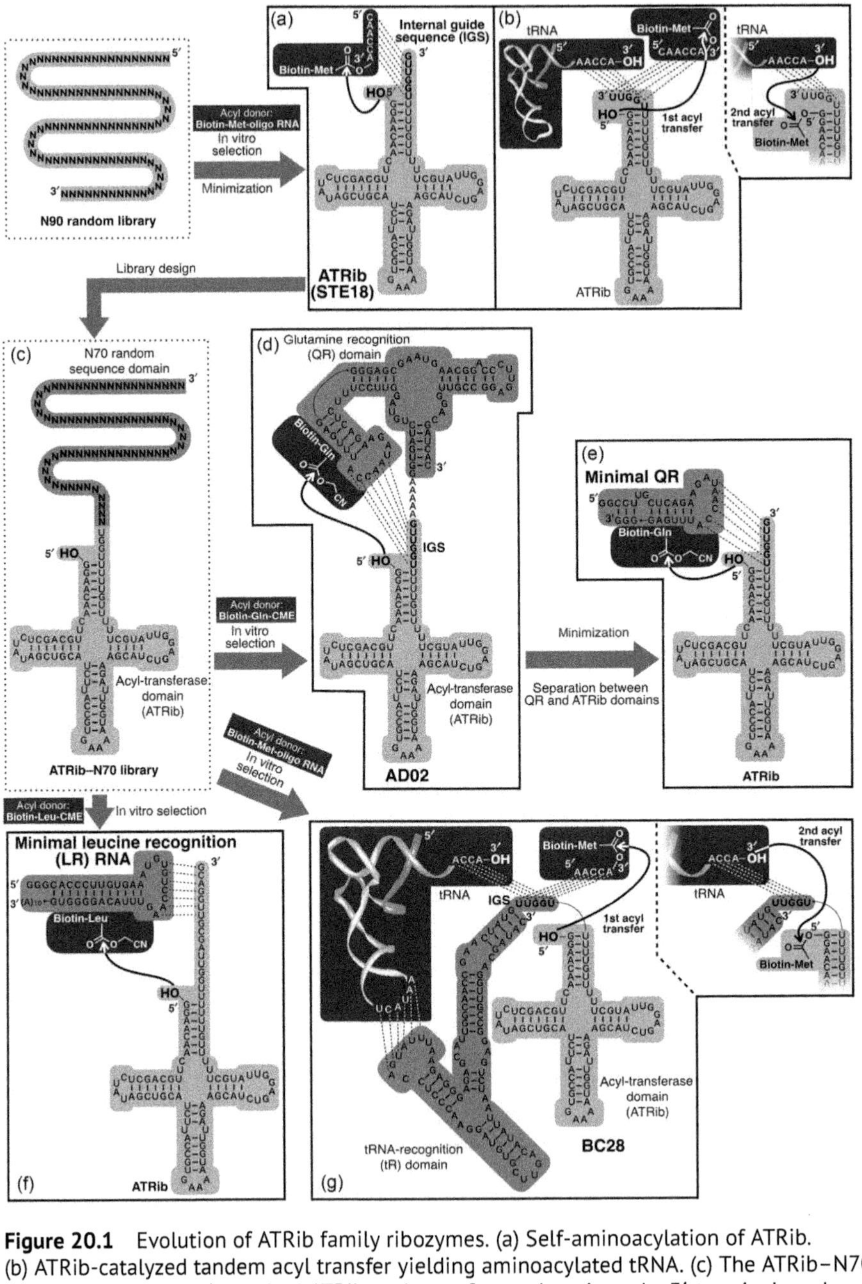

Figure 20.1 Evolution of ATRib family ribozymes. (a) Self-aminoacylation of ATRib. (b) ATRib-catalyzed tandem acyl transfer yielding aminoacylated tRNA. (c) The ATRib–N70 library containing a 5′-terminal ATRib acyl-transferase domain and a 3′-terminal random sequence domain. (d) Self-aminoacylation of AD02 with biotin-Gln. (e) Self-aminoacylation of ATRib with biotin-Gln assisted by a trans-acting minimal QR helix RNA. (f) Self-aminoacylation of ATRib with biotin-Leu assisted by a trans-acting minimal LR helix RNA. (g) BC28-catalyzed tandem acyl transfer toward a specific tRNA. The internal guide sequence (IGS) present at the 3′ end of the ATRib domain is shown in bold. Base pairs formed upon ribozyme recognition are shown as dashed lines. N: A, C, G, or U.

chain over other amino acid side chains. A similar in vitro selection campaign based on the same ATRib–N70 library using biotin-Leu-CME as the acyl donor also resulted in a helix–loop domain recognizing Leu (LR) that could act cooperatively with ATRib to yield self-acylated ATRib with biotin-Leu [14] (Figure 20.1f). These two successful examples demonstrated that the ATRib–N70 library could lead to various ATRib derivatives with directed recognition abilities.

The two ATRib-derived tRNA acylation systems described above showed high degrees of specificity for the side-chain structures of the aminoacyl donors but could accept different tRNAs in a relatively nonselective manner. To isolate an ATRib derivative with tRNA selectivity, further in vitro selection experiments were performed using biotin-Met-tRNA^{fMet} to obtain RNA sequences that transferred the biotin-Met group from the tRNA to the 5′-terminal OH group of ATRib [15]. The resulting ribozyme (BC28) retained the ATRib catalytic domain but included a new tRNA recognition (tR) domain at the 3′-terminal, including a loop region with a 5′-UUAUGA-3′ sequence. This sequence was capable of forming a loop–loop pseudoknot with the anticodon loop of tRNA^{fMet}, inducing the self-aminoacylation of the 5′-terminal OH group catalyzed by the ATRib domain. Using aminoacylated oligo RNA as an acyl donor, BC28 could selectively aminoacylated tRNA^{fMet} using IGS-guided tandem acyl transfer via the aminoacylated ATRib domain (Figure 20.1g). While BC28 was found to accept different amino acid side-chain structures, it exhibited significant selectivity for the tRNA sequence. Interestingly, the tRNA selectivity was programmable by mutating the loop sequence embedded in the tR domain to be complementary to the anticodon loop of the target tRNA. Indeed, the original BC28 and mutated BC28 achieved selective aminoacylation of the designated tRNAs in the presence of six different competing tRNAs.

A series of studies developing ATRib variants successfully demonstrated that in vitro selection of RNA libraries and rational design of selected ribozymes could afford artificial ribozymes that catalyzed tRNA aminoacylation with unique substrate specificities. However, the ATRib family of ribozymes relies on base pairing between the IGS region and two RNA substrates (acyl donor and acceptor); thus, they necessarily suffer from both substrate-induced and product-induced inhibition. The reactions also proceed by equilibrium control via the self-aminoacylated intermediates, and as such, tRNA aminoacylation always competes with the reverse reaction (in addition to spontaneous ester hydrolysis). For these reasons, the ATRib variants have generally resulted in low conversion efficiency (e.g. ~4% of tRNA was aminoacylated by AD02). Therefore, more robust tRNA aminoacylation ribozymes functioning by a different mechanism were required for downstream applications.

20.4 Prototype Flexizymes: Ribozymes Catalyzing Direct tRNA Aminoacylation

To develop a practical class of ribozymes, Saito et al. envisioned that direct aminoacylation of the 3′ end of tRNA could be a key function of ribozymes and thus designed a new RNA library (N70–tRNA library), in which an N70 random sequence domain was attached to the 5′ end of a tRNA substrate domain [16] (Figure 20.2a). This design was based on the fact that prokaryote precursor tRNAs contain variable 5′-leader sequences that are cleaved by RNase P upon maturation of tRNAs [17]. Although the biological function of the 5′-leader sequences in modern organisms is not well understood, it is plausible that they are remnants of primitive ribozymes that played a role in the self-aminoacylation of the tRNA's 3′-terminal OH group. Based on this hypothesis, the N70–tRNA library appeared to be a promising starting point to obtain experimental insight into the evolution of ARS-like ribozymes as well as to identify a new class of artificial ribozymes for aminoacylating tRNAs.

The N70–tRNA library was incubated with N-biotinylated phenylalanine cyanomethyl ester (biotin-Phe-CME) to select catalytic RNA sequences capable of attaching biotin-Phe onto the 3′-terminal OH of the tRNA domain [16]. This in vitro selection experiment resulted in a single clone containing a catalytic domain named r24. Importantly, the r24 domain was capable of directly aminoacylating the tRNA, unlike the ATRib family, and it was trans-acting in the trimolecular reaction (r24, tRNA, and amino acid substrate); thus, it acted as an ARS-like ribozyme (Figure 20.2b). Biochemical studies revealed that the site of aminoacylation was the 3′-OH group of the tRNA terminal adenosine (not the 2′-OH) and demonstrated that r24 recognized the aromatic ring of the Phe side chain [18–20]. Hence, r24 showed modest substrate selectivity for L-Phe, accepted L-Tyr and D-Phe with lower efficiency, but exhibited no activity with aliphatic amino acids. Interestingly, r24 activity was not affected by replacement of the leaving group, i.e. in addition to Phe-CME, Phe-AMP and a thioester-activated Phe could be used for the aminoacylation of the tRNA. These observations collectively indicated that neither the leaving group nor the amino group of the aminoacyl donor was involved in r24 recognition. Structural characterization by chemical probing experiments suggested six nucleotides (U_{59}–U_{62} and U_{67}–U_{68}) that constituted the essential catalytic core of r24 and were responsible for the recognition of the aromatic side chain of the substrate. The catalytic core proposed by conventional gel-based analyses was later verified by X-ray crystallographic studies of a descendant ribozyme [21]. These studies allowed for truncation of nonessential stem regions to obtain a 57-nt minimized ribozyme (r24mini) without any loss of activity [19] (Figure 20.2c).

Although the original reports postulated that a GGU motif (G_{70}–U_{72}) in r24/r24mini formed base pairs with the tRNA 3′-terminus, subsequent research suggested that the 3′-tail region of r24/r24mini invaded the tRNA acceptor stem to facilitate intensive base pairing upon RNA recognition (Figure 20.2c). This unique mechanism of tRNA recognition led to the rational design of r24mini variants, referred to as designer ribozymes [22]. One such designer ribozyme, FxfMet, retained the catalytic domain of r24mini but had a distinct 3′-tail. This 3′-tail region was designed to complement the 3′-terminal 10-nt sequence of $tRNA^{fMet}$, such that this ribozyme could recognize and exclusively aminoacylate $tRNA^{fMet}$ with biotin-Phe (Figure 20.2d). The programmability of such designer ribozymes, in which sequence complementarity dictates tRNA specificity, yielded a series of aminoacylation ribozymes discriminating their cognate tRNAs from

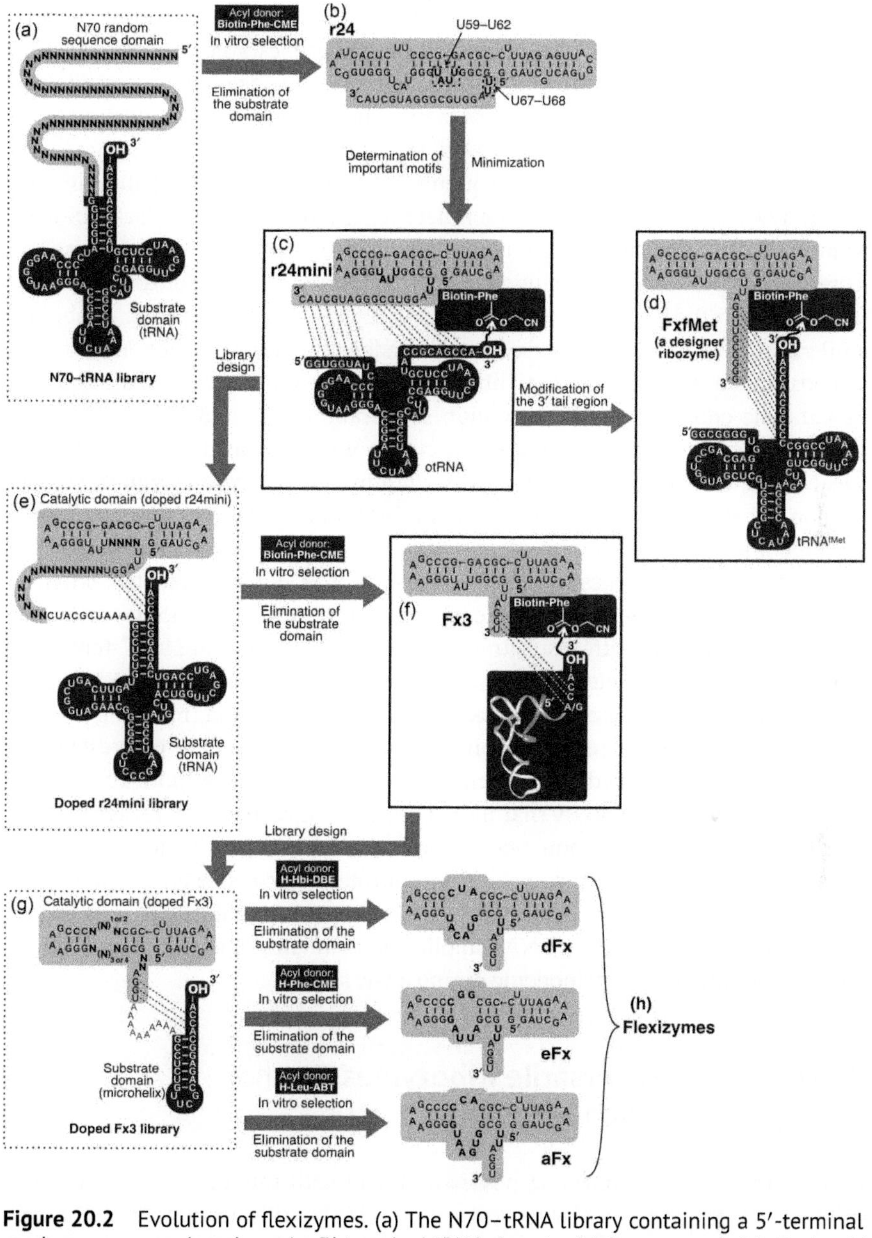

Figure 20.2 Evolution of flexizymes. (a) The N70–tRNA library containing a 5′-terminal random sequence domain and a 3′-terminal tRNA domain. (b) Sequence of r24. Nucleotides involved in the proposed catalytic core are shown in bold. (c) Aminoacylation of a specific tRNA by r24mini. (d) Aminoacylation of a specific tRNA by designer ribozymes based on programmable base pairing between the 5′-terminal region of tRNA and the 3′-terminal region of ribozymes. (e) The doped r24mini library, in which the four nucleotides adjacent to the catalytic core and the 3′-tail region were randomized. (f) Aminoacylation of various tRNAs with Phe analogs by Fx3. (g) The doped Fx3 library, in which the putative bases interacting with the aromatic side chain were randomized. (h) Flexizymes enabling aminoacylation of various tRNAs with a wide range of amino acids. Base pairs formed upon ribozyme recognition are shown as dashed lines. N: A, C, G, or U.

others by simple engineering of the 3′-tail region. This long base pairing with the tRNA substrate observed in r24/r24mini/designer ribozymes was advantageous for precise and programmable recognition of specific tRNAs; however, it inevitably led to product inhibition. Thus, these r24 family ribozymes did not, unfortunately, exhibit multiple turnovers.

The specificity for tRNA substrate sequence and amino acid side chain demonstrated by r24 family ribozymes might reflect an important feature for ARS-like ribozymes in a pre-proteinaceous world. On the other hand, aminoacylation ribozymes with low substrate specificity could provide an avenue for developing a molecular tool for the preparation of various misacylated tRNAs. In order to obtain r24mini variants with improved activity and broader substrate tolerance, Murakami et al. constructed a doped r24mini library where the four nucleotides adjacent to the catalytic core and the 3′-tail region were randomized [23] (Figure 20.2e). The four-nucleotide motif in the selected clones was completely conserved and identical to the original sequence found in r24mini, validating the importance of this region for ribozyme activity. In contrast, the randomized 3′-tail region failed to converge after the in vitro selection experiments and showed no similarity to the original sequence, suggesting that this region was not required for activity. This discovery allowed for further truncation of r24mini to yield a 45-nt ribozyme (Fx3) that recognized the tRNA substrate via three base pairs using its 3′-terminal GGU trinucleotide (Figure 20.2f). This limited interaction between Fx3 and the tRNA appeared advantageous for suppressing product inhibition, resulting in Fx3 exhibiting modest multiple turnover ability (~14 turnovers in an 8-hour reaction). Because Fx3 only recognized the 3′-terminal common sequence motif of tRNAs, it could aminoacylate any tRNA that had an A, G, or U at position 73, regardless of their body and anticodon sequences. Moreover, Fx3 accepted various aromatic amino acid residues, including some with non-proteinogenic side chains, and this flexible substrate recognition exhibited by Fx3, therefore, enabled us to prepare diverse artificial aminoacylated tRNAs, applicable to in vitro ribosomal synthesis of proteins containing non-proteinogenic amino acids [24].

20.5 Flexizymes: Versatile Ribozymes for the Preparation of Aminoacyl-tRNAs

While the prototype Fx3 ribozyme possessed broad substrate tolerance for tRNA, it exhibited significant selectivity toward amino acids bearing aromatic side chains. Thus, new ribozymes that utilize a wider variety of amino acid substrates were still required for extensive genetic code manipulation. Given that the Fx3 catalytic core favors aromatic side chains but does not recognize the leaving group, a new benzyl ester group (DBE) was designed with the aim of altering the recognition element of Fx3 from the side chain to the leaving group. For in vitro selection to isolate Fx3 variants utilizing the newly designed substrates, Murakami et al. designed a new RNA library based on Fx3 where the putative bases interacting with the aromatic side chain were randomized [25] (doped Fx3 library; Figure 20.2g).

This doped Fx3 library was subjected to three independent in vitro selection experiments using different amino acid acyl-donor substrates, identifying a series of trans-acting tRNA-acylating ribozymes, referred to as flexizymes [25] (Figure 20.2h). The in vitro selection campaign using δ-(*N*-biotinyl)amino-α-hydroxybutanoic acid-3,5-dinitrobenzyl ester (Hbi-DBE) afforded dinitro-flexizyme (dFx), which is capable of acylating tRNAs using a variety of nonaromatic amino acids. Enhanced flexizyme (eFx), isolated by in vitro selection using H-Phe-CME, showed improved tRNA aminoacylation activity compared to the original Fx3 ribozyme using CME-derivatized aromatic amino acids. Moreover, eFx also accepted aliphatic amino acids activated by 4-chlorobenzyl thioester (CBT), an approach that is appropriate for aminoacylation with β-branched bulky amino acids. Hydrophobic aminoacyl donors that are sometimes poorly soluble in water can be derivatized with 4-[(2-aminoethyl)carbamoyl]benzyl thioester (ABT) and used for aminoacylation catalyzed by amino-flexizyme (aFx) [26]. Collectively, by choosing an appropriate activating group and corresponding flexizyme, an extremely wide range of acyl donors (all 20 PAAs and diverse noncanonical acyl groups, such as non-proteinogenic α-amino acids, β-amino acids, D-amino acids, α-hydroxy acids, *N*-acyl amino acids, *N*-alkyl amino acids, simple carboxylic acids, and peptides) can be accepted by this tRNA acylation system [27] (Figure 20.3). It should be noted that these three flexizymes, similar to Fx3, only interact with the 3′-terminal common sequence motif of tRNAs, exhibiting broad substrate tolerance toward tRNAs.

Altogether, this set of three flexizymes allows for the aminoacylation of any objective tRNA with any desired acyl group, providing a highly versatile tRNA aminoacylation technology. Indeed, flexizymes have facilitated biochemical studies of proteins recognizing aminoacyl-tRNAs (e.g. editing domains of ARSs) by providing various misacylated tRNAs [28–31]. Moreover, this system has also been applied for the synthesis of artificial aminoacyl-tRNAs that can be utilized for an in vitro translation reaction with a reprogrammed genetic code, allowing for the synthesis and development of macrocyclic peptides with desirable bioactivities (see below).

20.6 Application of Flexizymes to Genetic Code Reprogramming

As mentioned above, the set of three flexizymes is a powerful tool for the synthesis of diverse nPAA-tRNAs, which are applicable for ribosomal incorporation of nPAAs into peptides. To date, various genetic code manipulation techniques that enable nPAA incorporation, such as nonsense codon suppression methods and genetic code reprogramming methods, have been developed [32–34]. Nonsense codon suppression methods utilize suppressor nPAA-tRNAs that target one specific stop codon out of the three that occur in the natural genetic code: UAG, UAA, and UGA. By comparison, genetic code reprogramming approaches assign nPAAs to one or more of the sense codons, although, since all of the 61 sense codons are already used up for introducing the 20 PAAs, the PAAs designated at the “reprogrammed” codons should be removed from the translation system to prevent the competition. In their

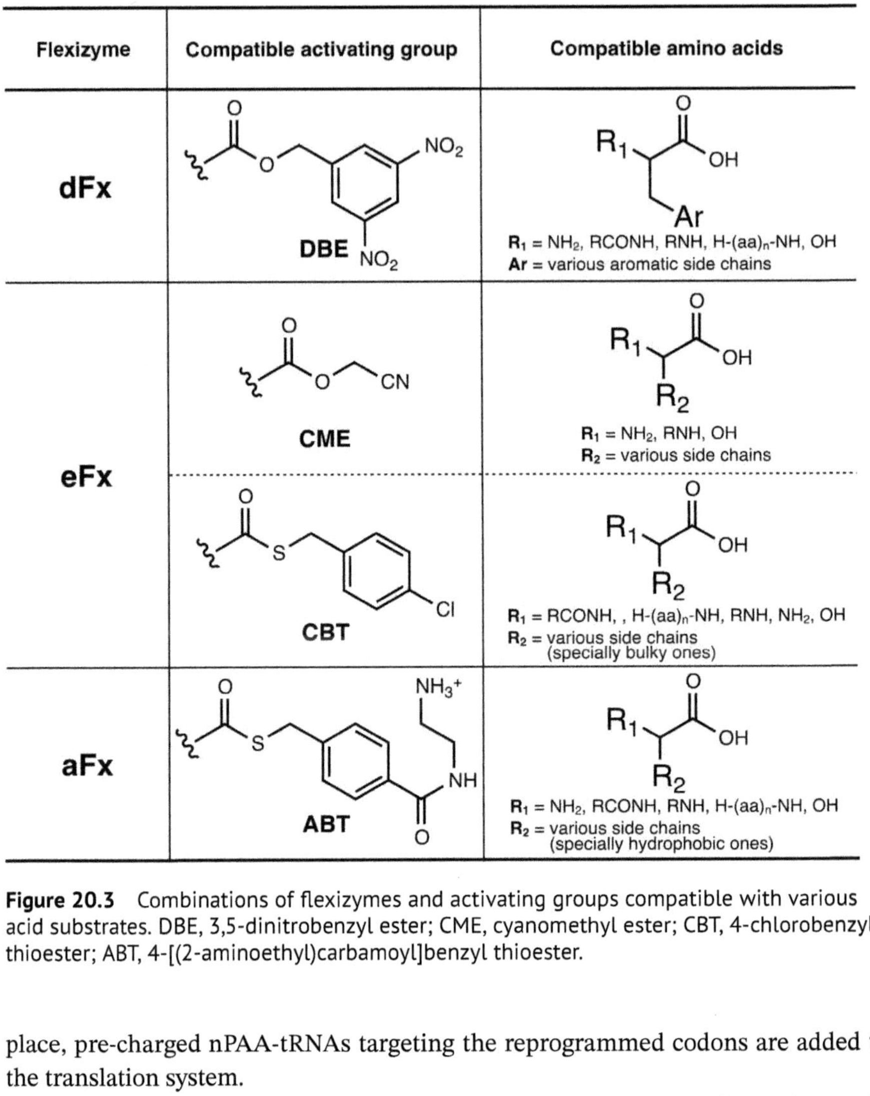

Figure 20.3 Combinations of flexizymes and activating groups compatible with various acid substrates. DBE, 3,5-dinitrobenzyl ester; CME, cyanomethyl ester; CBT, 4-chlorobenzyl thioester; ABT, 4-[(2-aminoethyl)carbamoyl]benzyl thioester.

place, pre-charged nPAA-tRNAs targeting the reprogrammed codons are added to the translation system.

Pioneering genetic code reprogramming experiments were performed as early as 1962, when Ala was introduced at the Cys codons using Ala-$tRNA^{Cys}$ prepared by chemical desulfurization of Cys-$tRNA^{Cys}$ [35]. Similarly, in 1971, phenyllactic acid (Flac) was introduced at the Phe codons to produce polyester chains using Flac-$tRNA^{Phe}$ prepared by chemical deamination of Phe-$tRNA^{Phe}$ [36]. In these experiments, Cys and Phe were virtually removed from the codon table by the chemical modification of Cys-tRNA and Phe-tRNA into Ala-tRNA and Flac-tRNA, respectively. However, as the Cys-to-Ala and Phe-to-Flac conversions were insufficient, misincorporation of Cys or Phe was observed to a significant degree. These results indicated that the complete removal of the competing PAA-tRNA from the translation system is required for accurate synthesis of the desired peptides without misincorporation.

More recently, the accuracy of nPAA incorporation during genetic code reprogramming has been improved through the use of reconstituted in vitro translation systems, such as the PURE system [37], in place of conventional S30 extract or other lysate-based translation systems. In these reconstituted systems, the ribosome and all of the other components (translation factors, ARSs, amino acids, etc.) required for translation are individually purified and recombined to construct a modular system in which unnecessary cellular components can be excluded [34, 37]. Using such systems, one can easily remove unwanted components from the reaction. For instance, by omitting a particular amino acid and/or the corresponding ARS, the codon(s) for the omitted amino acid can be "vacated," and an arbitrary nPAA can be assigned in place of the PAA by the addition of pre-charged nPAA-tRNAs. Figure 20.4 shows an example of such a reprogrammed translation system in which 16 amino acids/ARS pairs have been removed (leaving only Tyr, Lys, Met, and Asp) and three pre-charged nPAA-tRNAs have been introduced. Such a custom-made *Escherichia coli* reconstituted translation system, called flexible in vitro translation (FIT) system, allows us to delete amino acids in the cognate codons and replace them with nPAA-tRNAs prepared by the flexizymes. Using this technique, various translation substrates have been incorporated into peptides, including *N*-methyl amino acids, D-amino acids, β-amino acids, and hydroxy acids [27]. A particularly useful variation of this technique has been the initiation of translation with N^{α}-chloroacetyl amino acids at the N-terminus of the peptide, which can react with the sulfhydryl moiety of a downstream Cys to produce a non-reducible thioether bond and yield a macrocyclic scaffold [38].

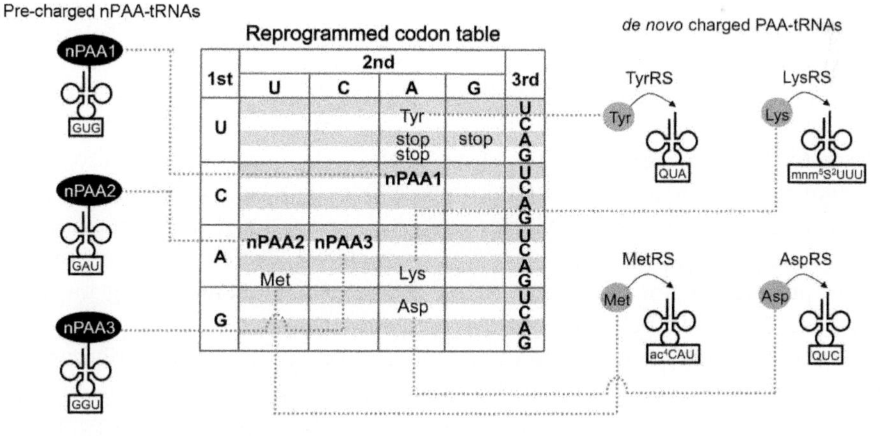

Figure 20.4 Genetic code reprogramming in a reconstituted in vitro translation system. 16 amino acid/ARS pairs (leaving only Tyr, Lys, Met, and Asp) are excluded in the reconstitution of this particular translation system. Tyr, Lys, Met, and Asp are charged by ARSs onto the corresponding tRNAs with QUA, mnm^{5}s^{2}UUU, ac^{4}CAU, and QUC anticodons, respectively, and introduced at UAY, AAR, AUG, and GAY codons, respectively. In addition, nPAA1, nPAA2, and nPAA3 are pre-charged onto tRNAs with GUG, GAU, and GGU anticodons and introduced at CAY, AUY, and ACY codons, respectively. Empty codons are not assigned to any amino acids. Q, queuosine; mnm^{5}s^{2}U, 5-methylaminomethyl-2-thiouridine; ac^{4}C, 4-acetylcytidine; Y, U or C; R, A or G.

One of the disadvantages of genetic code reprogramming is that some PAAs generally have to be omitted from the new genetic code in order to make space for nPAAs. However, because of the redundancy of the genetic code in which 61 sense codons are shared by only 20 PAAs, it is possible to create vacant codons without sacrificing any PAAs by reducing the redundancy. Iwane et al. accomplished this by artificially dividing codon "boxes" into two to create vacant codons (Figure 20.5) [39]. For instance, the valine codon box consisting of four GUN codons was divided into two, where valine (Val) and citrulline (Cit) were introduced to GUG and GUU/GUC by Val-$tRNA_{CAC}$ and Cit-$tRNA_{GAC}$, respectively. Similarly, ArgCGN and GlyGGN codon boxes were also divided into two and used for the incorporation of iodophenylalanine (^{Iodo}F) and acetyllysine (^{Ac}K) without sacrificing Arg and Gly. Using this reprogrammed genetic code containing Cit, ^{Iodo}F, and ^{Ac}K as well as the 20 PAAs, a 32-mer peptide containing 23 different amino acids (Cit, ^{Iodo}F, ^{Ac}K, and the 20 PAAs) could be successfully translated.

20.7 Development of Orthogonal tRNA/Ribosome Pairs Using Mutant Flexizymes

tRNAs have a common CCA sequence at their 3′ end, which is required for ribosome binding during translation. The Watson–Crick base pairs between the C_{74}/C_{75} of tRNA and G_{2252}/G_{2251} at the P site of 23S rRNA as well as the base pair between the C_{75} and G_{2553} at the A site are essential for peptidyl transfer activity in the ribosome (Figure 20.6c; the subscript number indicates the base number of tRNA or *E. coli* ribosome) [40–42]. For this reason, the 3′-end CCA of tRNAs is conserved among all three domains of life. As shown in Figure 20.6a, flexizymes also recognize the 3′-end C_{74}/C_{75} of tRNAs for aminoacylation. In the case of dFx, G_{45} and G_{44} form base pairs with the C_{74} and C_{75} of tRNAs. However, the introduction of compensatory mutations into the flexizymes and tRNAs can also be tolerated [43]. For instance, a dFx variant with C_{45}/C_{44} mutations can recognize a mutant tRNA with G_{74}/G_{75} mutations and efficiently charge amino acids onto such a tRNA (Figure 20.6a,b). Similarly, C_{45}/G_{44} and G_{45}/C_{44} dFx mutants can utilize G_{74}/C_{75} and C_{74}/G_{75} mutant tRNAs, respectively. Therefore, it is relatively easy to prepare mutant aminoacyl-tRNAs with 3′-end mutations using mutant flexizymes. It should be noted that such mutant aminoacyl-tRNAs cannot be recognized by wild-type ribosomes, but can be recognized by mutant ribosomes with compensatory mutations [43]. For instance, compensatory mutations of G_{75} in the tRNA and C_{2251}/C_{2553} in the 23S rRNA are tolerated and retain peptidyl transfer activity (Figure 20.6c). Importantly, mutant tRNA/ribosome pairs do not cross-react with wild-type tRNA/ribosome pairs and vice versa. Thus, if the mutant and the wild-type tRNA/ribosome pairs coexist in a translation system, the two tRNA/ribosome pairs could work independently and decode two separate genetic codes. These two independent codon tables assigned to the mutant and wild-type tRNA/ribosome pairs can be reprogrammed by genetic code reprogramming. For example, the codon table for the mutant pair can be reprogrammed such that some nPAAs are encoded through nPAA-tRNAs with 3′-end mutations, whereas all PAAs are retained in genetic code for the wild-type ribosome. Indeed, using

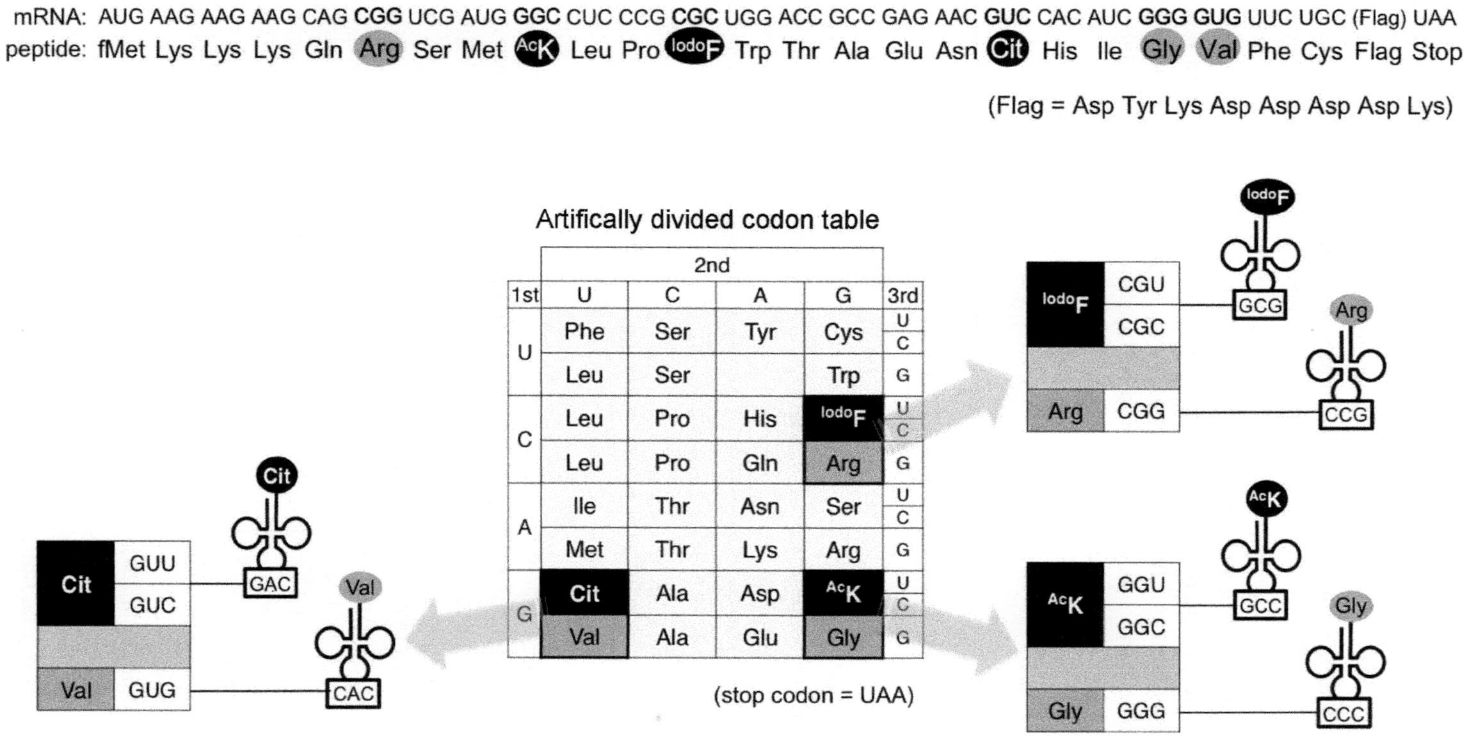

1st	U	C	A	G	3rd
U	Phe	Ser	Tyr	Cys	U
	Phe	Ser	Tyr	Cys	C
	Leu	Ser		Trp	G
C	Leu	Pro	His	IodoF	U
	Leu	Pro	His	IodoF	C
	Leu	Pro	Gln	Arg	G
A	Ile	Thr	Asn	Ser	U
	Ile	Thr	Asn	Ser	C
	Met	Thr	Lys	Arg	G
G	Cit	Ala	Asp	AcK	U
	Cit	Ala	Asp	AcK	C
	Val	Ala	Glu	Gly	G

Figure 20.5 Translation of a model peptide consisting of 23 different amino acids by means of the artificial division of codon boxes. ValGUN, ArgCGN, and GlyGGN codon boxes are divided into GUY/GUG, CGY/CGG, and GGY/GGG codons, respectively. Then, GUG, CGG, and GGG codons are assigned to Val, Arg, and Gly by ARS charged Val-tRNA, Arg-tRNA, and Gly-tRNA, whereas GUY, CGY, and GGY codons are assigned to citrulline (Cit), 4-iodophenylalanine (^{Iodo}F), and *N*-ε-acetyllysine (^{Ac}K) by means of pre-charged Cit-tRNA, ^{Iodo}F-tRNA, and ^{Ac}K-tRNA, respectively. Using this genetic code, a model peptide containing 23 different amino acids including Cit, ^{Iodo}F, and ^{Ac}K and only 20 PAAs was translated. Source: Adapted from Iwane et al. [39].

such an approach, Terasaka et al. successfully introduced azidonorvaline (Anv), Lys, and ^{Ac}Lys at the UAC, AAG, and GAC codons in the reprogrammed codon table for the mutant pair, while Tyr, Lys, and Asp were retained in the wild-type codon table. In addition, *N*-(5-FAM)-L-phenylalanine (Fph) was introduced at the initiator AUG codons of both codon tables for fluorescent labeling of peptides [43]. Then, by introducing a single mRNA (the ORF sequence is 5′-AUG AAG UAC GAC AAG UAC GAC UGA-3′) into a translation system in which both tRNA/ribosome pairs coexist, synthesis of two different peptides (Fph-Lys-Anv-^{Ac}K-Lys-Anv-^{Ac}K from the mutant tRNA/ribosome pair and Fph-Lys-Tyr-Asp-Lys-Tyr-Asp from the wild-type pair) was accomplished (Figure 20.6). As no hybrid products caused by codon cross-reading were observed, this proved that the mutant and the wild-type tRNA/ribosome pairs work orthogonally.

20.8 In Vitro Selection of Bioactive Peptides Containing nPAAs Through RaPID Display

In natural product bioactive peptides, nPAAs occur at high frequency and play important roles in biological activity. For instance, N-alkylation, D-stereochemistry, β-amino acids, unnatural side chains, and macrocyclic scaffolds are widely observed and make important contributions to membrane permeability, protease resistance, and structural rigidity [44–49]. The ability of nPAAs to enhance these characteristics has resulted in macrocyclic peptides containing nPAAs attracting attention as novel chemical class for the development of peptidic drugs. As mentioned above, using the FIT system, diverse nPAAs including *N*-methyl-, D-, and β-amino acids as well as α-amino acids with various side-chain modifications can be introduced into peptides [25, 50–56]. Regarding macrocyclic peptide synthesis by FIT, by carefully designing peptides to include self-reactive moieties (such as the combination of an N^{α}-chloroacetyl amino acid with Cys described above), diverse cyclic peptide architectures can be accessed, including backbone-to-side-chain, side-chain-to-side-chain, and backbone-to-backbone cyclization [38, 57–59].

Another advantage of FIT is that random peptide libraries containing nPAAs with extremely high diversity (more than 1×10^{12} molecules) can be easily constructed by simply using template mRNAs containing random sequences. Such random peptide libraries can be screened for affinity to an (often disease-related) target protein in a process termed Random non-standard Peptides Integrated Discovery (RaPID), which involves the combination of mRNA display selection with FIT (Figure 20.7) [60–62]. In this approach, each mRNA and its cognate peptide are covalently linked via a puromycin linker, enabling recovery of the mRNA after the pull-down of peptides that bind to an immobilized target molecule. The bound fraction can be recovered and amplified by reverse transcription and PCR, followed by transcription into mRNA, which can be used for a subsequent round of affinity selection. By repeating such selection cycles several times, it is possible to obtain ligands with exceptional binding affinity and selectivity for the target molecule.

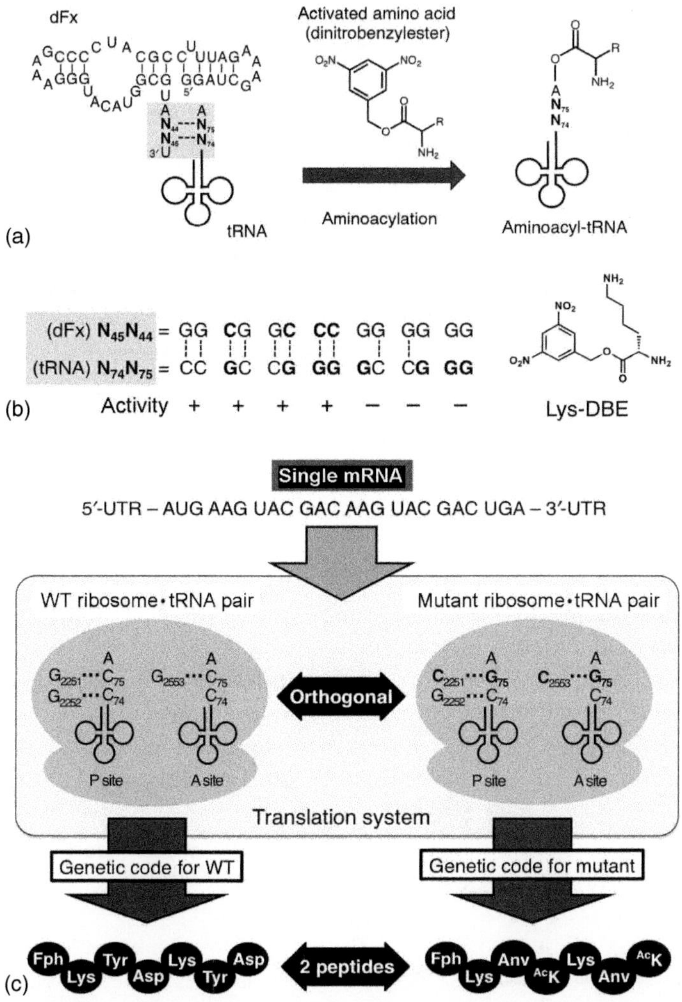

Figure 20.6 Development of an orthogonal tRNA/ribosome pair. (a) Recognition of the tRNA 3′-end region by dFx. N_{44} and N_{45} of dFx form base pairs with N_{75} and N_{74} of the tRNA. dFx charges amino acids activated by dinitrobenzylester (DBE) onto tRNAs. (b) Introduction of compensatory mutations at the 3′ end of dFx (N_{45} and N_{44}) and tRNA (N_{74} and N_{75}). Pairs of mutant dFx/tRNA variants and the activity of each to charge Lys-DBE onto a tRNA are shown. Introduction of compensatory mutations in this region is tolerated, whereas mismatched pairs lose aminoacylation activity. (c) Translation of two different peptides from a single mRNA by orthogonal tRNA/ribosome pairs. Base-pair formation between C_{75}/C_{74} of the tRNA and G_{2251}/G_{2252} of the rRNA at the P site and base-pair formation between C_{75} of the tRNA and G_{2553} of the rRNA at the A site are requisite for peptidyl transfer activity. However, a tRNA with a G_{75} mutation can be recognized by a 23S rRNA with C_{2251}/C_{2253} mutations. Importantly, mutant tRNA does not cross-react with the wild-type (WT) ribosome, and WT tRNAs are not recognized by the mutant ribosome. UAY and GAY codons are assigned to Tyr and Asp, respectively, in the genetic code for the WT ribosome/tRNAs, and to azidonorvaline (Anv) and ^{Ac}K in the genetic code for the mutant ribosome/tRNAs. *N*-(5-FAM)-L-phenylalanine (Fph) is introduced at the initiator AUG codon of both genetic codes. By translating a single mRNA, two distinct peptides can be synthesized in parallel according to the two independent genetic codes.

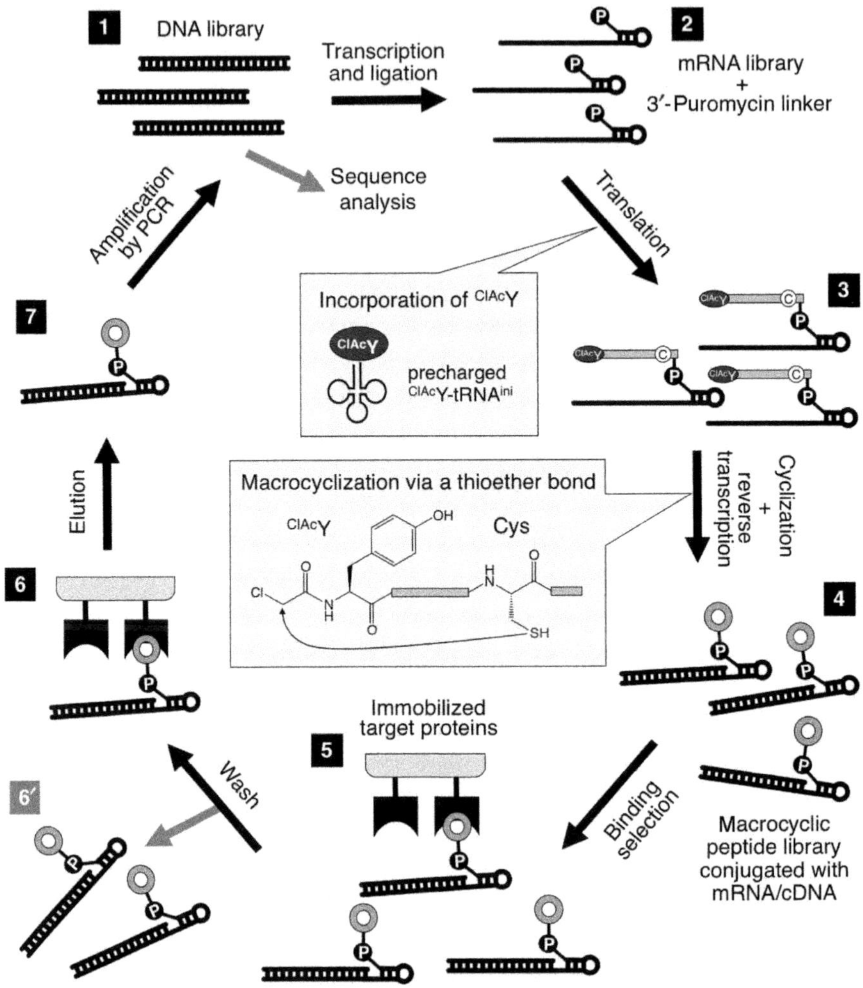

Figure 20.7 Scheme of RaPID selection using a macrocyclic peptide library. A random DNA library is transcribed into a corresponding mRNA library, followed by puromycin-linker ligation at the 3′ end. Then, random peptides with N-terminal chloroacetyltyrosine (^{ClAc}Y) are translated and covalently linked to the mRNA via the puromycin linker. After macrocyclization of the peptide via a thioether bond and reverse transcription of the mRNA to produce mRNA–cDNA hybrids, the library is mixed with immobilized target protein to selectively recover peptides with target affinity. Finally, the bound fraction is eluted and amplified by PCR, and the sequence of the DNA library is used for the next round of selection or analyzed by sequencing.

Taking advantage of the RaPID system, numerous peptide ligands have been developed to date [62–76]. As a specific example, a macrocyclic peptide library containing *N*-methyl-Phe (^{Me}Phe), *N*-methyl-Ser (^{Me}Ser), *N*-methyl-Gly (^{Me}Gly), and *N*-methyl-Ala (^{Me}Ala) was applied to the identification of peptides that inhibit the human oncoprotein E6AP, with one of the peptides obtained, CM$_{11}$-1, containing two ^{Me}Phe residues, one ^{Me}Ser residue, and one ^{Me}Gly residue, and exhibiting strong affinity for E6AP (dissociation constant, $K_D = 0.6$ nM) and the inhibition of the interaction between E6AP and the tumor suppressor P53 [62]. The rapidity with which this technique can identify novel high-affinity ligands/inhibitors to various targets of interest and the unique chemical structures of the compounds identified promise to provide a wealth of leads for potential pharmaceutical development.

20.9 tRid: A Method for Selective Removal of tRNAs from an RNA Pool

Analysis of cellular small RNAs, whose lengths are around 70–100 nt, is often complicated by the large abundance of tRNAs in the same molecular mass range. Many biologically important noncoding RNAs (e.g. pre-miRNAs and snoRNAs) are of approximately this molecular weight, and the detection of such RNAs is consequently very difficult unless they are expressed at very high levels [77]. There are various methods to analyze cellular RNA, such as northern blotting, microarray analysis, quantitative PCR, and deep sequencing. However, all of these techniques are difficult to apply to the detection of nonabundant small RNA species in the 70–100 nt range. As such, selective removal of tRNAs from cellular small RNA fractions would be beneficial for the analysis of relatively low-abundance small RNA species.

Futai et al. took advantage of flexizyme technology to develop such an approach named tRid (tRNA rid) for selective depletion of tRNAs from pooled small RNAs (Figure 20.8a) [78]. As described above, all tRNAs have a common 3′-CCA end, which can be recognized by flexizymes regardless of the body sequences of the tRNA. Thus, by using flexizyme to charge *N*-biotinylated phenylalanine (biotin-Phe) onto tRNAs indiscriminately, tRNAs can be selectively removed by streptavidin pull-down and separated from the non-tRNA small RNA fraction. Figure 20.8a shows the secondary structure of dial-Fx, specially designed for tRid, in which the 3′-end of 46th nucleotide (N_{46}) is a mixture of A, U, G, and C that pairs with all tRNA 74th discriminator base and oxidized to dialdehyde for the prevention of the ligation with a 3′-adopter RNA for reverse transcription. After the elimination of tRNA by tRid, the remaining small RNA pool was subjected to the 5′- and 3′-adaptor ligation. Sequencing of the resulting small RNA pool showed that tRid method effectively removes abundant tRNA species. We also found many unidentified novel small RNAs originating from human (HeLa) and *E. coli* cells.

20.10 Use of a Natural Small RNA Library Lacking tRNA for In Vitro Selection of a Folic Acid Aptamer: Small RNA Transcriptomic SELEX

To date, various RNA aptamer elements whose structures and functions are regulated by binding of metabolites have been found in living organisms. However,

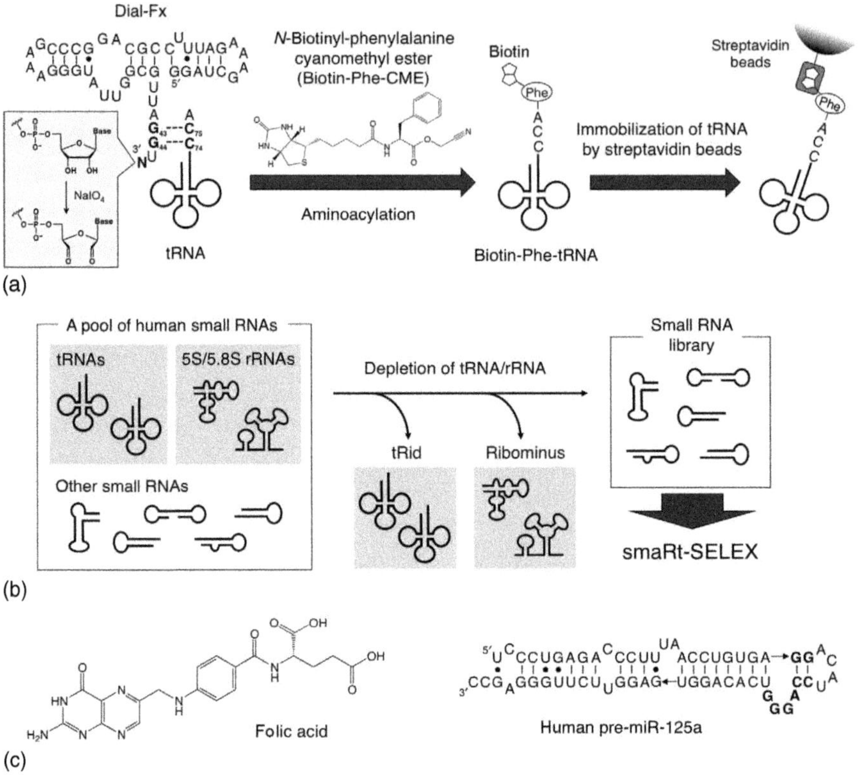

Figure 20.8 tRid, a method for selectively depleting tRNAs from an RNA pool, and its application to in vitro selection of aptamers by smaRt-SELEX. (a) Scheme of tRNA removal by charging biotin-Phe onto the 3′ end of tRNA by dial-Fx. tRNAs charged with biotin-Phe can be removed by streptavidin beads. The 46th nucleotide of dial-Fx is oxidized by sodium periodate to yield the 3′-dialdehyde. (b) Preparation of a tRNA/rRNA-depleted small RNA library for smaRt-SELEX. A pool of human small RNAs is subjected to tRNA removal by tRid and rRNA removal using oligonucleotide probes complementary to rRNAs (RiboMinus™, Thermo Fisher Scientific). (c) Structure of folic acid. (d) Structure of human pre-miR-125a as a folic acid aptamer. Folic acid binding sites are shown in bold. Source: (a) Futai et al. [78]. © 2016 Elsevier. (b) Source: Adapted from Terasaka et al. [79].

as described in the previous section, the discovery of such elements among RNAs 70–100 nt in length is complicated by the high abundance of tRNAs in this size range. Clearance of tRNAs using tRid can be used to overcome this limitation. Terasaka et al. prepared a tRNA-depleted human small RNA library by means of tRid. In addition to tRNA depletion, small rRNA fragments (5S and 5.8S rRNAs) were also eliminated using commercially available oligonucleotide probes complementary to rRNAs (RiboMinus™, Thermo Fisher Scientific) in order to further enrich low-abundance RNAs (Figure 20.8b) [79]. This tRNA/rRNA-depleted small RNA fraction was then ligated to 5′- and 3′-adaptor oligonucleotides to give a small RNA library, which could be used for the systematic evolution of ligands by exponential enrichment (SELEX) to natural aptamers (Figure 20.8c). This in vitro selection method for the discovery of naturally occurring small RNA aptamers has been referred to as small RNA transcriptomic SELEX (smaRt-SELEX).

In order to demonstrate the utility of this technique, natural aptamers for the metabolite folic acid were identified using a smaRt-SELEX approach. Folic acid immobilized on magnetic beads was used for RNA pull-down experiments starting with a tRNA/rRNA-depleted small RNA library. The recovered RNAs were amplified by RT-PCR and transcribed into a new RNA pool for the next round of selection. Iterative rounds of selection lead to the discovery that human miRNA-125a precursor (hsa-pre-miR-125a) exhibits folic acid affinity with a dissociation constant (K_D) of 2.8 μM (Figure 20.8d). This finding was confirmed by mutation analysis, which demonstrated that the bases of hsa-pre-miR-125a involved in an interaction with folic acid are the internal AGGG loop and base pairs at G_{25}/C_{32} and G_{26}/C_{31} in the stem region (Figure 20.8d, indicated by bold). Thus, these experiments demonstrated the utility of smaRt-SELEX for the identification of natural small RNA aptamers.

20.11 Summary and Perspective

In this chapter, we have introduced and discussed the history of discovery and development of ATRibs and their engineering and application as biochemical tools. Earlier research of a family of ATRibs primarily aimed at obtaining experimental evidence that supported the RNA world hypothesis. Indeed, these pioneering studies successfully proved the capacity of RNA to catalyze various biomimetic acyl-transfer reactions. However, these ATRibs could not be applied to practical tRNA aminoacylation since they exhibited very low aminoacylation efficiency and showed substrate selectivity for specific acyl donors and/or acyl acceptors, limiting their utility in biochemical applications. However, based on these earlier studies, an appropriate redesign of the RNA library where 5′ end of tRNA was covalently linked to random RNA sequences and selection of self-aminoacylating species yielded a prototype of the tRNA-acylating ribozyme. Further rational engineering approaches enabled us to devise a family of versatile tRNA acylation ribozymes, referred to as flexizymes. Importantly, later-generation flexizymes like dFx, eFx, and aFx are capable of side-chain independent aminoacylation activity and recognize the tRNA shared 3′-CCA end. Therefore, using flexizymes, diverse amino acids can be charged onto diverse tRNAs. In fact, even tRNAs with mutations in the 3′-CCA end can be used as flexizyme substrates through the use of compensatory mutations (Figure 20.6).

Because aminoacyl-tRNA is a key molecule in the translation of mRNA-encoded information into peptide sequences, in natural systems, the relationship between each codon and its cognate amino acid is strictly defined by the genetic code. However, using flexizymes, which can charge arbitrary amino acids onto any tRNA, the reprogramming of the codon–amino acid relationship is largely facilitated. Using such an approach, even nPAAs can be assigned to arbitrary codons and subsequently incorporated into peptide chains. To date, various kinds of nPAAs such as *N*-methyl-, D-, and β-amino acids as well as α-amino acids with side-chain variations have been introduced into peptides using the FIT approach, which combines the use of aminoacyl-tRNAs synthesized using flexizymes and a reconstituted in vitro translation system. Additionally, by randomizing the mRNA sequences used, it is possible to obtain very-high-diversity random peptide libraries containing nPAAs, which can be applied to in vitro selection of bioactive peptides using the RaPID system (Figure 20.7).

Although the versatility that allows any amino acid to be charged on any tRNA is the single greatest advantage of flexizymes, in some cases, this characteristic could restrict usage. For instance, mixtures of activated amino acids cannot be utilized for charging specific amino acids onto tRNAs. Likewise, charging an amino acid onto a specific tRNA from among a mix of similar tRNAs is impossible. For these reasons, current use of flexizymes involves the isolated reaction of a specific amino acid and tRNA pair, followed by the termination of the reaction, isolation of the aminoacyl-tRNA. and introduction into the downstream translation reaction. As such, coupling of aminoacylation and translation in a single reaction is impossible since various amino acids and tRNAs must be included in the translation system. This, in turn, means that the use of flexizymes in cells is highly problematic. For such aminoacylation/translation-coupled approaches to be effective, more specific flexizymes capable of recognizing specific amino acid and tRNA pairs would need to be developed. In theory, flexizymes with substrate specificity could also be used in in vivo translation systems.

Aminoacylation of tRNA by flexizymes is useful for not only ribosomal translation but also selective removal of tRNAs from mixed samples of small RNAs derived from cells. The tRid technique involves charging of biotin-Phe selectively onto the tRNAs in RNA pools, thereby allowing researchers to selectively remove tRNA species by simple streptavidin pull-down. Since tRNAs are the most abundant small RNA species, their removal greatly facilitates the detection and analysis of other nonabundant small RNAs that generally are inaccessible due to overlapping with the abundant tRNAs. Such tRNA-depleted small RNA fractions can also be used as libraries for the discovery of natural small RNA aptamers through smaRt-SELEX as evidenced by the identification of human pre-miR-125a as a folate aptamer (Figure 20.8c,d).

As discussed in this chapter, the investigation of a possible transition from RNA-dependent ancient life to RNP-dependent modern life leads to the identification of primitive acyl-transferase ribozymes. While unsuitable for practical applications, these ribozymes laid the groundwork for the discovery of later ribozymes with greater activity and broader substrate scope—flexizymes. Because of their ability to charge essentially any tRNA with any acyl group, flexizymes have proven to be exceptionally useful tools, with applications in fundamental biochemistry, genetic code reprogramming, and drug discovery, and we anticipate that additional applications will be developed in the future. Further development of flexizymes themselves may also lead to the development of variants with different characteristics such as substrate specificity, further expanding their scope for experimental application.

Acknowledgments

This work was supported by CREST of Molecular Technologies, JST to H.S., and KAKENHI grants (16H06444 and 26220204 to H.S.; 26560429 and 18H02080 to T.K.; 15K12739, 16H01131, and 17H04762 to Y.G.) from the Japan Society for the Promotion of Science.

References

1 Vasil'eva, I.A. and Moor, N.A. (2007). Interaction of aminoacyl-tRNA synthetases with tRNA: general principles and distinguishing characteristics of the high-molecular-weight substrate recognition. *Biochemistry (Mosc)* 72 (3): 247–263.
2 Beuning, P.J. and Musier-Forsyth, K. (1999). Transfer RNA recognition by aminoacyl-tRNA synthetases. *Biopolymers* 52 (1): 1–28.
3 Gilbert, W. (1986). The RNA World. *Nature* 319: 618.
4 Neveu, M., Kim, H.J., and Benner, S.A. (2013). The "strong" RNA world hypothesis: fifty years old. *Astrobiology* 13 (4): 391–403.
5 Cech, T.R. (2000). Structural biology. The ribosome is a ribozyme. *Science* 289 (5481): 878–879.
6 Hager, A.J., Pollard, J.D., and Szostak, J.W. (1996). Ribozymes: aiming at RNA replication and protein synthesis. *Chem. Biol.* 3 (9): 717–725.
7 Illangasekare, M., Sanchez, G., Nickles, T., and Yarus, M. (1995). Aminoacyl-RNA synthesis catalyzed by an RNA. *Science* 267 (5198): 643–647.
8 Jenne, A. and Famulok, M. (1998). A novel ribozyme with ester transferase activity. *Chem. Biol.* 5 (1): 23–34.
9 Lohse, P.A. and Szostak, J.W. (1996). Ribozyme-catalysed amino-acid transfer reactions. *Nature* 381 (6581): 442–444.
10 Suga, H., Lohse, P.A., and Szostak, J.W. (1998). Structural and kinetic characterization of an acyl transferase ribozyme. *J. Am. Chem. Soc.* 120 (6): 1151–1156.
11 Illangasekare, M. and Yarus, M. (1999). Aminoacyl-tRNA synthetases and self-acylating ribozymes. In: *The RNA World*, 2e, 183–196. Cold Spring Harbor Laboratory Press.
12 Lee, N., Bessho, Y., Wei, K. et al. (2000). Ribozyme-catalyzed tRNA aminoacylation. *Nat. Struct. Biol.* 7 (1): 28–33.
13 Lee, N. and Suga, H. (2001). A minihelix-loop RNA acts as a trans-aminoacylation catalyst. *RNA* 7 (7): 1043–1051.
14 Ramaswamy, K., Wei, K., and Suga, H. (2002). Minihelix-loop RNAs: minimal structures for aminoacylation catalysts. *Nucleic Acids Res.* 30 (10): 2162–2171.
15 Bessho, Y., Hodgson, D.R., and Suga, H. (2002). A tRNA aminoacylation system for non-natural amino acids based on a programmable ribozyme. *Nat. Biotechnol.* 20 (7): 723–728.

16 Saito, H., Kourouklis, D., and Suga, H. (2001). An in vitro evolved precursor tRNA with aminoacylation activity. *EMBO J.* 20 (7): 1797–1806.

17 Frank, D.N. and Pace, N.R. (1998). Ribonuclease P: unity and diversity in a tRNA processing ribozyme. *Annu. Rev. Biochem.* 67: 153–180.

18 Saito, H. and Suga, H. (2001). A ribozyme exclusively aminoacylates the 3'-hydroxyl group of the tRNA terminal adenosine. *J. Am. Chem. Soc.* 123 (29): 7178–7179.

19 Saito, H., Watanabe, K., and Suga, H. (2001). Concurrent molecular recognition of the amino acid and tRNA by a ribozyme. *RNA* 7 (12): 1867–1878.

20 Saito, H. and Suga, H. (2002). Outersphere and innersphere coordinated metal ions in an aminoacyl-tRNA synthetase ribozyme. *Nucleic Acids Res.* 30 (23): 5151–5159.

21 Xiao, H., Murakami, H., Suga, H., and Ferre-D'Amare, A.R. (2008). Structural basis of specific tRNA aminoacylation by a small in vitro selected ribozyme. *Nature* 454 (7202): 358–361.

22 Kourouklis, D., Murakami, H., and Suga, H. (2005). Programmable ribozymes for mischarging tRNA with nonnatural amino acids and their applications to translation. *Methods* 36 (3): 239–244.

23 Murakami, H., Saito, H., and Suga, H. (2003). A versatile tRNA aminoacylation catalyst based on RNA. *Chem. Biol.* 10 (7): 655–662.

24 Murakami, H., Kourouklis, D., and Suga, H. (2003). Using a solid-phase ribozyme aminoacylation system to reprogram the genetic code. *Chem. Biol.* 10 (11): 1077–1084.

25 Murakami, H., Ohta, A., Ashigai, H., and Suga, H. (2006). A highly flexible tRNA acylation method for non-natural polypeptide synthesis. *Nat. Methods* 3 (5): 357–359.

26 Niwa, N., Yamagishi, Y., Murakami, H., and Suga, H. (2009). A flexizyme that selectively charges amino acids activated by a water-friendly leaving group. *Bioorg. Med. Chem. Lett.* 19 (14): 3892–3894.

27 Goto, Y., Katoh, T., and Suga, H. (2011). Flexizymes for genetic code reprogramming. *Nat. Protoc.* 6: 779–790.

28 Das, M., Vargas-Rodriguez, O., Goto, Y. et al. (2014). Distinct tRNA recognition strategies used by a homologous family of editing domains prevent mistranslation. *Nucleic Acids Res.* 42 (6): 3943–3953.

29 Liu, Z., Vargas-Rodriguez, O., Goto, Y. et al. (2015). Homologous trans-editing factors with broad tRNA specificity prevent mistranslation caused by serine/threonine misactivation. *Proc. Natl. Acad. Sci. U.S.A.* 112 (19): 6027–6032.

30 Novoa, E.M., Vargas-Rodriguez, O., Lange, S. et al. (2015). Ancestral AlaX editing enzymes for control of genetic code fidelity are not tRNA-specific. *J. Biol. Chem.* 290 (16): 10495–10503.

31 Danhart, E.M., Bakhtina, M., Cantara, W.A. et al. (2017). Conformational and chemical selection by a trans-acting editing domain. *Proc. Natl. Acad. Sci. U.S.A.* 114 (33): E6774–E6783.

32 Noren, C., Anthony-Cahill, S., Griffith, M., and Schultz, P. (1989). A general method for site-specific incorporation of unnatural amino acids into proteins. *Science* 244: 182–188.

33 Wang, L., Brock, A., Herberich, B., and Schultz, P.G. (2001). Expanding the genetic code of *Escherichia coli*. *Science* 292 (5516): 498–500.

34 Forster, A., Tan, Z., Nalam, M. et al. (2003). Programming peptidomimetic syntheses by translating genetic codes designed *de novo*. *Proc. Natl. Acad. Sci. U.S.A.* 100 (11): 6353–6357.

35 Chapeville, F., Lipmann, F., Von Ehrenstein, G. et al. (1962). On the role of soluble ribonucleic acid in coding for amino acids. *Proc. Natl. Acad. Sci. U.S.A.* 48 (6): 1086–1092.

36 Fahnestock, S. and Rich, A. (1971). Ribosome-catalyzed polyester formation. *Science* 173 (3994): 340–343.

37 Shimizu, Y., Inoue, A., Tomari, Y. et al. (2001). Cell-free translation reconstituted with purified components. *Nat. Biotechnol.* 19 (8): 751–755.

38 Goto, Y., Ohta, A., Sako, Y. et al. (2008). Reprogramming the translation initiation for the synthesis of physiologically stable cyclic peptides. *ACS Chem. Biol.* 3 (2): 120–129.

39 Iwane, Y., Hitomi, A., Murakami, H. et al. (2016). Expanding the amino acid repertoire of ribosomal polypeptide synthesis via the artificial division of codon boxes. *Nat. Chem.* 8 (4): 317–325.

40 Moazed, D. and Noller, H.F. (1991). Sites of interaction of the CCA end of peptidyl-tRNA with 23S rRNA. *Proc. Natl. Acad. Sci. U.S.A.* 88 (9): 3725–3728.

41 Samaha, R.R., Green, R., and Noller, H.F. (1995). A base pair between tRNA and 23S rRNA in the peptidyl transferase centre of the ribosome. *Nature* 377 (6547): 309–314.

42 Kim, D.F. and Green, R. (1999). Base-pairing between 23S rRNA and tRNA in the ribosomal A site. *Mol. Cell* 4 (5): 859–864.

43 Terasaka, N., Hayashi, G., Katoh, T., and Suga, H. (2014). An orthogonal ribosome-tRNA pair via engineering of the peptidyl transferase center. *Nat. Chem. Biol.* 10 (7): 555–557.

44 Biron, E., Chatterjee, J., Ovadia, O. et al. (2008). Improving oral bioavailability of peptides by multiple N-methylation: somatostatin analogues. *Angew. Chem. Int. Ed. Engl.* 47 (14): 2595–2599.

45 Driggers, E.M., Hale, S.P., Lee, J., and Terrett, N.K. (2008). The exploration of macrocycles for drug discovery – an underexploited structural class. *Nat. Rev. Drug Discov.* 7 (7): 608–624.

46 Ovadia, O., Greenberg, S., Chatterjee, J. et al. (2011). The effect of multiple N-methylation on intestinal permeability of cyclic hexapeptides. *Mol. Pharm.* 8 (2): 479–487.

47 Beck, J.G., Chatterjee, J., Laufer, B. et al. (2012). Intestinal permeability of cyclic peptides: common key backbone motifs identified. *J. Am. Chem. Soc.* 134 (29): 12125–12133.

48 Cabrele, C., Martinek, T.A., Reiser, O., and Berlicki, Ł. (2014). Peptides containing β-amino acid patterns: challenges and successes in medicinal chemistry. *J. Med. Chem.* 57 (23): 9718–9739.
49 Ahlbach, C.L., Lexa, K.W., Bockus, A.T. et al. (2015). Beyond cyclosporine A: conformation-dependent passive membrane permeabilities of cyclic peptide natural products. *Future Med. Chem.* 7 (16): 2121–2130.
50 Goto, Y., Murakami, H., and Suga, H. (2008). Initiating translation with D-amino acids. *RNA* 14 (7): 1390–1398.
51 Kawakami, T., Murakami, H., and Suga, H. (2008). Messenger RNA-programmed incorporation of multiple N-methyl-amino acids into linear and cyclic peptides. *Chem. Biol.* 15 (1): 32–42.
52 Kawakami, T., Murakami, H., and Suga, H. (2008). Ribosomal synthesis of polypeptoids and peptoid-peptide hybrids. *J. Am. Chem. Soc.* 130 (50): 16861–16863.
53 Fujino, T., Goto, Y., Suga, H., and Murakami, H. (2013). Reevaluation of the D-amino acid compatibility with the elongation event in translation. *J. Am. Chem. Soc.* 135: 1830–1837.
54 Fujino, T., Goto, Y., Suga, H., and Murakami, H. (2016). Ribosomal synthesis of peptides with multiple β-amino acids. *J. Am. Chem. Soc.* 138 (6): 1962–1969.
55 Katoh, T., Tajima, K., and Suga, H. (2017). Consecutive elongation of D-amino acids in translation. *Cell Chem. Biol.* 24 (1): 1–9.
56 Katoh, T., Iwane, Y., and Suga, H. (2017). Logical engineering of D-arm and T-stem of tRNA that enhances D-amino acid incorporation. *Nucleic Acids Res.* 45 (22): 12601–12610.
57 Sako, Y., Morimoto, J., Murakami, H., and Suga, H. (2008). Ribosomal synthesis of bicyclic peptides via two orthogonal inter-side-chain reactions. *J. Am. Chem. Soc.* 130 (23): 7232–7234.
58 Kawakami, T., Ohta, A., Ohuchi, M. et al. (2009). Diverse backbone-cyclized peptides via codon reprogramming. *Nat. Chem. Biol.* 5 (12): 888–890.
59 Kang, T.J., Hayashi, Y., and Suga, H. (2011). Synthesis of the backbone cyclic peptide sunflower trypsin inhibitor-1 promoted by the induced peptidyl-tRNA drop-off. *Angew. Chem. Int. Ed. Eng.* 50 (9): 2159–2161.
60 Roberts, R.W. and Szostak, J.W. (1997). RNA-peptide fusions for the *in vitro* selection of peptides and proteins. *Proc. Natl. Acad. Sci. U.S.A.* 94 (23): 12297–12302.
61 Nemoto, N., Miyamoto-Sato, E., Husimi, Y., and Yanagawa, H. (1997). In vitro virus: bonding of mRNA bearing puromycin at the 3'-terminal end to the C-terminal end of its encoded protein on the ribosome in vitro. *FEBS Lett.* 414 (2): 405–408.
62 Yamagishi, Y., Shoji, I., Miyagawa, S. et al. (2011). Natural product-like macrocyclic *N*-methyl-peptide inhibitors against a ubiquitin ligase uncovered from a ribosome-expressed de novo library. *Chem. Biol.* 18: 1562–1570.
63 Hayashi, Y., Morimoto, J., and Suga, H. (2012). In vitro selection of anti-Akt2 thioether-macrocyclic peptides leading to isoform-selective inhibitors. *ACS Chem. Biol.* 7 (3): 607–613.

64 Morimoto, J., Hayashi, Y., and Suga, H. (2012). Discovery of macrocyclic peptides armed with a mechanism-based warhead: isoform-selective inhibition of human deacetylase SIRT2. *Angew. Chem. Int. Ed. Eng.* 51 (14): 3423–3427.

65 Tanaka, Y., Hipolito, C.J., Maturana, A.D. et al. (2013). Structural basis for the drug extrusion mechanism by a MATE multidrug transporter. *Nature* 496 (7444): 247–251.

66 Hipolito, C.J., Tanaka, Y., Katoh, T. et al. (2013). A macrocyclic peptide that serves as a cocrystallization ligand and inhibits the function of a MATE family transporter. *Molecules* 18 (9): 10514–10530.

67 Ito, K., Sakai, K., Suzuki, Y. et al. (2015). Artificial human Met agonists based on macrocycle scaffolds. *Nat. Commun.* 6: 6373.

68 Iwasaki, K., Goto, Y., Katoh, T. et al. (2015). A fluorescent imaging probe based on a macrocyclic scaffold that binds to cellular EpCAM. *J. Mol. Evol.* 81 (5-6): 210–217.

69 Matsunaga, Y., Bashiruddin, N.K., Kitago, Y. et al. (2016). Allosteric inhibition of a semaphorin 4D receptor plexin B1 by a high-affinity macrocyclic peptide. *Cell Chem. Biol.* 23 (11): 1341–1350.

70 Jongkees, S.A.K., Caner, S., Tysoe, C. et al. (2017). Rapid discovery of potent and selective glycosidase-inhibiting de novo peptides. *Cell Chem. Biol.* 24 (3): 381–390.

71 Yu, H., Dranchak, P., Li, Z. et al. (2017). Macrocycle peptides delineate locked-open inhibition mechanism for microorganism phosphoglycerate mutases. *Nat. Commun.* 8: 14932.

72 Kawamura, A., Munzel, M., Kojima, T. et al. (2017). Highly selective inhibition of histone demethylases by de novo macrocyclic peptides. *Nat. Commun.* 8: 14773.

73 Song, X., Lu, L.Y., Passioura, T., and Suga, H. (2017). Macrocyclic peptide inhibitors for the protein-protein interaction of Zaire Ebola virus protein 24 and karyopherin alpha 5. *Org. Biomol. Chem.* 15 (24): 5155–5160.

74 Passioura, T., Bhushan, B., Tumber, A. et al. (2018). Structure-activity studies of a macrocyclic peptide inhibitor of histone lysine demethylase 4A. *Bioorg. Med. Chem.* 26 (6): 1225–1231.

75 Nishio, K., Belle, R., Katoh, T. et al. (2018). Thioether macrocyclic peptides selected against TET1 compact catalytic domain inhibit TET1 catalytic activity. *ChemBioChem* 19 (9): 979–985.

76 Passioura, T., Watashi, K., Fukano, K. et al. (2018). De novo macrocyclic peptide inhibitors of hepatitis B virus cellular entry. *Cell Chem. Biol.* 25 (7): 906–915.

77 Palazzo, A.F. and Lee, E.S. (2015). Non-coding RNA: what is functional and what is junk? *Front. Genet.* 6: 2.

78 Futai, K., Terasaka, N., Katoh, T., and Suga, H. (2016). tRid, an enabling method to isolate previously inaccessible small RNA fractions. *Methods* 106: 105–111.

79 Terasaka, N., Futai, K., Katoh, T., and Suga, H. (2016). A human microRNA precursor binding to folic acid discovered by small RNA transcriptomic SELEX. *RNA* 22 (12): 1918–1928.

21

In Vitro Selected (Deoxy)ribozymes that Catalyze Carbon–Carbon Bond Formation

Michael Famulok

Universität Bonn, Life & Medical Sciences (LIMES) Institut, Chemical Biology and Medicinal Chemistry Unit, Gerhard-Domagk-Straße 1, 53121 Bonn, Germany

21.1 Introduction

When referring to the metabolic functions in the RNA world, Gerald Joyce stated in his seminal review "The antiquity of RNA-based evolution" [1]: "Although the central process of the RNA world was the replication of RNA genomes, some form of metabolism must have supported the process (...) The possibility of a more complex RNA-based metabolism is purely conjectural. That said, one could imagine that all of the reactions of central metabolism, now catalysed by protein enzymes, were once catalysed by ribozymes." Thus, in an RNA world, not only the activated nucleotides (nt) required for forming longer RNAs but also lipids for compartmentalization must have been available in sufficient quantities as precursors to support the continuous production and evolution of the earliest RNA/lipid-based self-replicating supramolecular systems, or protocells [2]. The conjecture that nucleotides, lipids, and cofactors arose from simpler building blocks in RNA-catalyzed metabolic reactions therefore seems quite evident. Furthermore, maintaining the complex and unstable components that constitute the early RNA world protocells likely required certain forms of energy metabolism [3, 4].

Probably, the most crucial chemical transformations that are requirements for the synthesis of all these precursors from simpler building blocks are the formation of carbon–carbon bonds and the catalysis of alkylation reactions. Whether RNA catalysts for such reactions existed in the RNA world is something that will be virtually impossible to find out based on the fossil record. The plausibility of the existence of C—C bond-forming or alkylating ribozymes was, however, experimentally tested by means of in vitro selection of ribozymes for a variety of carbon–carbon bond formations and some related alkylation reactions. The purpose of this chapter is to summarize and discuss these studies, in the light not only of in vitro selected ribozymes but also of DNAzyme-catalyzed related reactions.

Ribozymes, First Edition. Edited by Sabine Müller, Benoît Masquida, and Wade Winkler.

21.2 Diels–Alderase Ribozymes

The first examples showing that RNA molecules are capable of catalyzing a C—C bond-forming reaction were the Diels–Alder reaction ribozymes by Eaton and coworkers [5] and Seelig and Jäschke [6] groups.

For isolation, the Eaton Diels–Alderase DA22, a library of 10^{14} chemically modified RNA sequences, was employed, in which all uracil residues contained a 4-pyridylmethyl carboxamide residue at their 5-position [5]. DA22 exhibited a rate that was 800-fold higher than the non-catalyzed reaction rate, and its catalytic activity depended on the pyridyl modification in the active site, as well as on the presence of a Cu^{2+} ion. Even the position of the nitrogen atom in the active site pyridine residues is crucial for catalytic activity [7]. For a more detailed discussion of DA22 and derivatives, see Chapter 15.

The first all-RNA Diels–Alderase ribozyme was described in 1999 by Seelig and Jäschke [6]. This ribozyme belongs to the most thoroughly characterized ribozymes that were obtained by in vitro selection [8]. It catalyzes the Diels–Alder reaction between maleimide derivatives and an anthracene moiety (Figure 21.1). The ribozyme contains a preformed catalytic pocket that arranges the diene and the dienophile for the reaction to be accelerated catalytically, as revealed by mutation analyses and chemical and enzymatic structural probing [9]. The pre-organized binding pocket is a remarkable feature of this ribozyme since many other functional

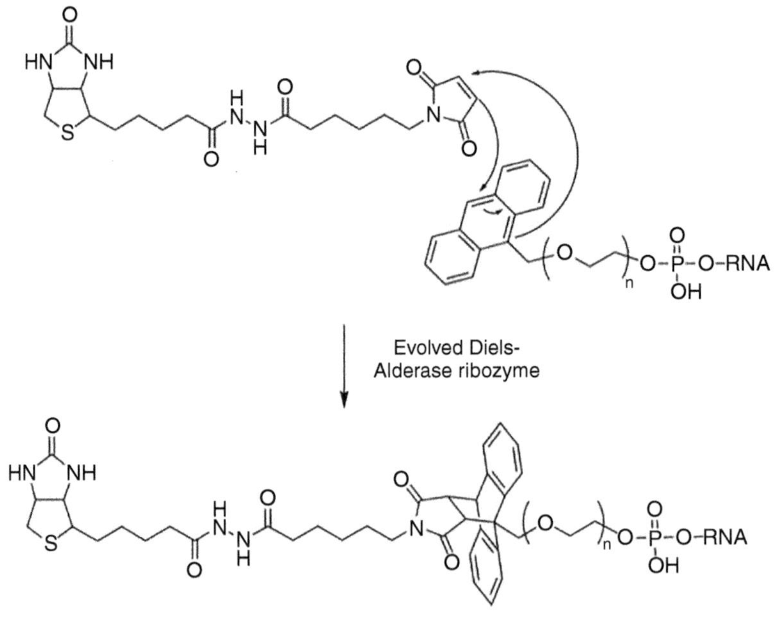

Figure 21.1 Reaction between a biotin maleimide substrate and a 5'-polyethylene glycol-anthracene-conjugated RNA Diels-Alderase ribozyme to form the Diels-Alder product.

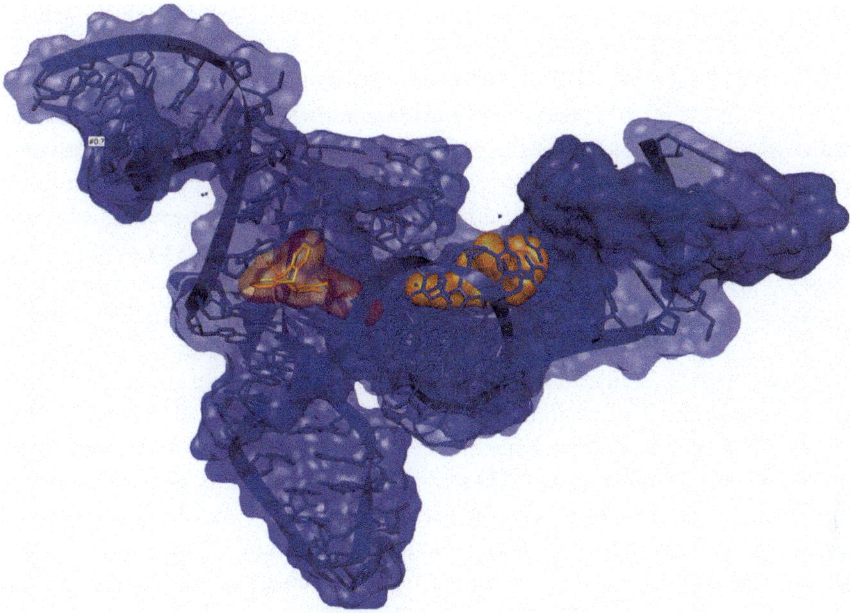

Figure 21.2 Crystal structure of selenium-modified Diels-Alder ribozyme complexed with the product of the reaction between N-pentylmaleimide and covalently attached 9-hydroxymethylanthracene. PDB: 1YLS. Source: Adapted from Serganov et al. [13].

RNAs and aptamers bind their ligands by adaptive binding mechanisms, in which substrates are recognized by induced fit binding [10]. The diversity of selected sequences contained a small conserved secondary structure motif, based on which a minimal version of the ribozyme consisting of only 49 nt was shown to catalyze the Diels–Alder reaction in a bimolecular fashion with the diene and dienophile free in solution. Moreover, the reaction occurred under multiple turnover with high enantioselectivity [11] and with precisely defined requirements on the chemical nature of the diene and dienophile substrates [12]. Deeper insight into the three-dimensional structure of the Diels–Alderase ribozyme could be gained from the crystal structural analysis of the ribozyme and its bound Diels–Alder product [13] (Figure 21.2). Not only did the crystal structure confirm the pre-organization of the substrate binding pocket and catalytic center, but also it revealed that unlike the previous Diels–Alderase DA22, no metal ion is involved in catalysis. Instead, the catalytic mechanism involves a combination of proximity, complementarity, and electronic effects.

21.3 Aldolase Ribozyme

The aldol condensation is the primary reaction for the synthesis of sugars; its reverse reaction is the primary reaction for energy metabolism in today's organisms. Therefore, the relevance of ribozymes that can catalyze aldol reactions is high in providing experimental evidence for the plausibility of the RNA world.

The first ribozyme that was shown to be able to catalyze an aldol condensation was isolated by in vitro selection in our laboratory [14]. The aldol donor was a levulinic amide, coupled to the 5′-end of every member of the RNA library consisting of about 2×10^{15} different 194 nucleotides-containing sequences (Figure 21.3). A biotinylated benzaldehyde derivative acted as the aldol acceptor [15]. All the sequences contained in the library that were able to catalyze the condensation between the aldol donor and acceptor substrates thus became biotinylated and could be separated from the inactive ones – the vast majority of sequences in the library – by streptavidin pull-down. Formally, the condensation of the ketone and the aldehyde requires the deprotonation of one of the carbon atoms neighboring the keto group in the levulinic acid. It is this formal deprotonation that makes the aldol condensation a particularly difficult reaction to be catalyzed by an RNA molecule because deprotonation at that carbon requires strongly basic conditions. From the selection, an RNA catalyst, the 11D2 ribozyme, emerged that was able to accelerate the rate of the uncatalyzed reaction 4300-fold. The activity of the 11D2 ribozyme not only strictly depended on the presence of Zn^{2+} ions but also required at least 2.6 mM Mg^{2+} ions for optimal performance. No other divalent metal ions were able to substitute Zn^{2+} ions, which were bound by 11D2 in a cooperative manner. This dependence on Zn^{2+} is remarkable insofar as some protein aldolases, the class II aldolases, also use Zn^{2+} as an essential Lewis acid cofactor for deprotonation, presumably to stabilize the enolate [16].

21.4 A DNAzyme that Catalyzes a Friedel–Crafts Reaction

The Friedel–Crafts reaction is among the most common C—C bond-forming reactions in organic synthesis. Although Friedel–Crafts alkylations play a less important role in metabolism than aldolases, there are examples of such reactions catalyzed by natural protein enzymes, such as a recently discovered small group of bacterial prenyltransferases that catalyze the C-prenylation of aromatic substrates in secondary metabolism [17]. Even in classical organic chemistry, regioselective mono-alkylation of olefins by the Friedel–Crafts reaction under aqueous conditions is not trivial to achieve.

According to this notion, it is remarkable that it could be shown that even single-stranded DNA in itself is capable of catalyzing a Friedel–Crafts reaction between an indole and an olefinic acyl imidazole substrate in the presence of Cu^{2+} [18] (Figure 21.4). By employing an in vitro selection scheme in which the partition between active versus inactive deoxyribozymes was achieved by a gel shift, a 72-nucleotide-containing DNAzyme was isolated that catalyzed the Friedel–Crafts reaction between a biotin-containing indole derivative and an acyl imidazole containing a double bond that was coupled to a short single-stranded DNA sequence, in the presence of copper nitrate. The most active among the isolated DNAzymes was M14. This DNAzyme led to 72% conversion into the Friedel–Crafts product at 50 mol%, when 2 mM of the acyl imidazole and 10 mM of the indole, neither of which tethered to the DNA, were used. Without M14

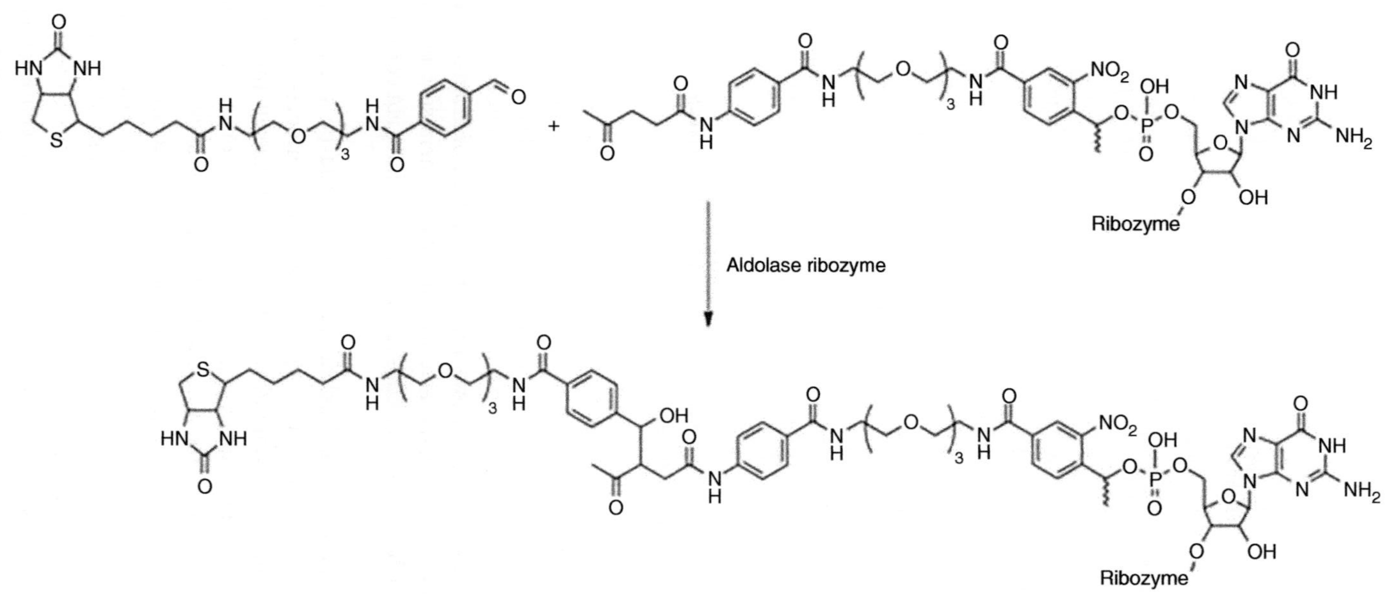

Figure 21.3 Aldol reaction between a biotinylated benzaldehyde moiety and a levulinic acid aldol donor coupled to the 5'-end of the ribozyme via a photo-cleavable linker.

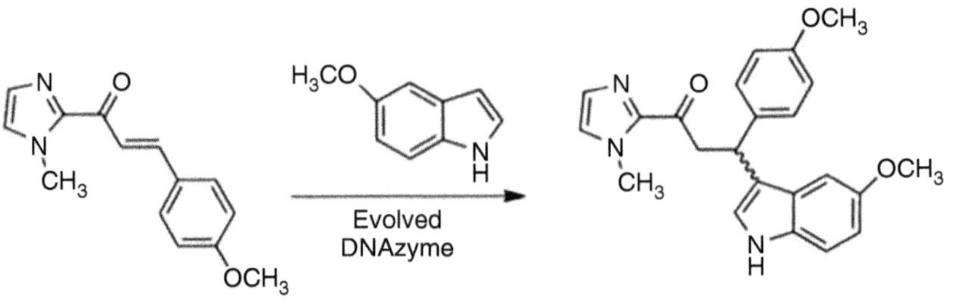

Figure 21.4 DNAzyme-catalyzed Friedel-Crafts reaction between an acyl imidazole and an indole substrate.

DNAzyme, only 7% conversion into the product was achieved under otherwise the same conditions. Still, when only 10 mol% of M14 was used, 18% conversion into the product was observed, indicating that M14 acts as a catalyst for the Friedel–Crafts reaction between the two substrates. Until now, no kinetic parameters for M14 were reported, nor is it known whether the DNAzyme leads to enantiomerically enriched or even pure product. It is also not known by which molecular mechanism the DNAzyme acts. Notwithstanding, M14 is an example showing once again that even a single-stranded DNA can be capable of catalyzing a demanding C—C bond-forming reaction as the Friedel–Crafts reaction.

21.5 Alkylating Ribozymes

Several examples of ribozymes that catalyze alkylation reactions have been described. Although some of these reactions are not C—C bond-forming ones, alkylation reactions usually involve the nucleophilic attack of a heteroatom such as sulfur or nitrogen on a positively polarized carbon atom. These reactions are also chemically demanding and show some mechanistic similarities to C—C bond-forming reactions. Alkylation reactions likely have been relevant for nucleic acid-catalyzed metabolic transformations in a hypothetical RNA world or in a primitive protocell.

The first example of an RNA catalyst for an N-alkylation reaction was the self-biotinylating alkyl transferase ribozyme BL2.8-7 [19]. In order to isolate this ribozyme, a biotin-binding RNA aptamer called BB8-5, a pseudoknot-containing RNA aptamer [20], was first selected from a pool of 5×10^{14} different RNA sequences bearing 72 randomized nucleotides flanked by a 22-mer 5′- and an 18-mer 3′-primer binding site [19]. BB8-5 then was used to create another library with a mutagenized core of the 30% mutagenized 72-mer aptamer sequence, flanked by additional 12-mer completely randomized regions 5′- and 3′ of the mutated 72-mer, plus new primer binding sites. This pool was incubated with *N*-biotinoyl-*N*′-iodoacetyl-ethylenediamine (BIE) and subjected to several cycles of in vitro selection using streptavidin agarose to separate self-biotinylated RNAs from those that did not contain a biotin residue. This selection yielded a ribozyme BL8-6

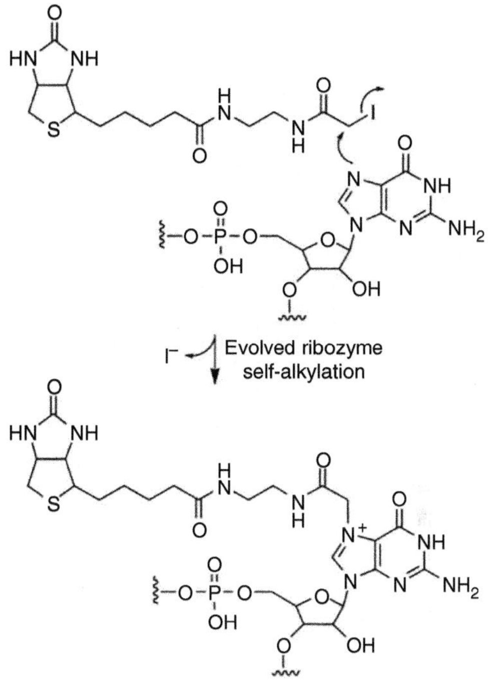

Figure 21.5 Self-alkylation reaction between a biotinylated iodoacetamide substrate and the N7-position of guanosine residue G96 of the evolved ribozyme BL8-6.

that performed the BIE-dependent self-biotinylation reaction (Figure 21.5) at a rate of 0.001 min^{-1}.

BL8-6 was then mutated with a probability of 30% substitution of non-wild-type nucleotides at each position, and the selection was repeated at increased stringency using reduced BIE concentrations and shorter incubation time. From this selection, clone BL2.8-7 emerged showing a 50-fold improved catalytic rate over BL8-6 of 0.05 min^{-1}.

The libraries from which both BL8-6 and BL2.8-7 were selected contained a nucleophilic guanosine monophosphorothioate (GMPS) group at the 5′-end of all pool members to provide a nucleophile as a potential alkylation site. Interestingly, the actual alkylation site of BL2.8-7 was not the thiol group at the 5′-end; omitting this functionality still led to self-biotinylation at the same rate. Instead, it was the N7-position of an internal guanosine residue, G96, that was alkylated.

Another example of a self-alkylating ribozyme is the in vitro selected ribozyme UV5, a 188-mer RNA sequence that catalyzes a Michael reaction between a fumaramide group and a thiol functionality [21] (Figure 21.6). The Michaelase ribozyme was selected from a pool of 2×10^{15} different RNA sequences that contained a randomized region of 142 nt. Each of the pool sequences contained a chemically modified guanosine at their 5′-end to introduce a fumaramide functionality as the Michael acceptor unit. The attachment of this functionality to the 5′-end of the RNA pool was achieved by T7 RNA polymerase transcription initiation using a synthetic guanosine derivative as the initiator nucleotide to which the fumaramide moiety was coupled via a photo-cleavable linkage [22]. The

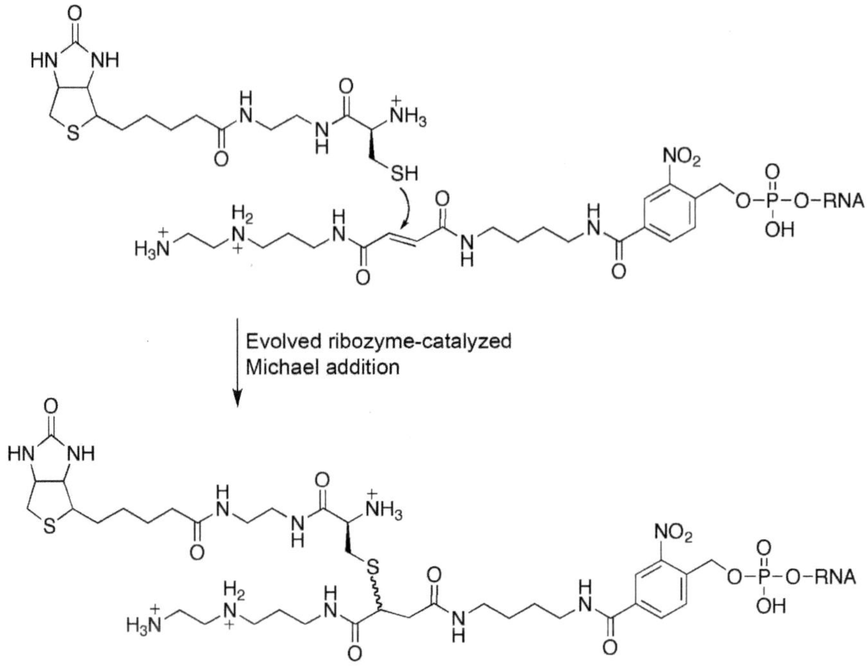

Figure 21.6 Michael reaction between a biotinylated cysteine Michael-donor substrate and the fumaramide Michael-acceptor substrate coupled to the 5′-end of the evolved ribozyme UV5.

nucleophilic Michael donor substrate was a biotinylated cysteine that was used at a concentration of 2 mM during the first 10 cycles of selection. Only those RNA molecules that catalyze the Michael addition between the cysteine thiol and the fumaramide would be covalently attached to the biotinylated cysteine substrate and thus could be separated from the unmodified ones by streptavidin chromatography. Due to the photo-cleavable linker located between the RNA and the fumaramide, RNAs bound to the streptavidin beads could be eluted by UV cleavage, reverse transcribed, polymerase chain reaction (PCR) amplified, and transcribed in the presence of the initiator guanosine. To increase stringency after 10 rounds of selection and amplification, the Michael donor substrate concentration was reduced to 400 μM in rounds 11–13, to 100 μM in rounds 14 and 15, and to 10 μM in rounds 16 and 17. Furthermore, in rounds 11–13, the incubation time was reduced from 1 hour to 5 minutes and further reduced to 1 minute in rounds 14–17. From this selection, a single family of sequences emerged that differed from one another only by a few point mutations.

The kinetic characterization of the ribozyme clone UV5 revealed a rate enhancement of 3×10^5 over the uncatalyzed reaction with a value of k_{cat}/K_M of 888 $M^{-1} \times min^{-1}$ when using the same *trans*-H fumaramide that was used during the selection. Substitution of only one hydrogen atom by a CH_3 group at either of the two possible positions of the double bond led to a 6- and 16-fold reduction of the rate enhancement, respectively, by increasing the K_M value 6- and 10-fold, respectively, suggesting that the UV5 either shows aspects of regioselectivity or that it positions the cysteine substrate near the double bonds in a way that allows the nucleophilic attack to occur at either end. That the catalyzed reaction did indeed occur at the C=C double bond of the fumaramide moiety of the Michael acceptor was verified by replacing the double bond for a C—C single bond, which completely omitted the reaction with the Michael donor substrate. Furthermore, the transfer of the biotinylated cysteine substrate onto an external 5′-fumaramide-modified RNA sequence comprising the first 20 nucleotides of the 5′-end could be achieved by using a version of UV5 that was truncated by the first 20 nt. Finally, the entire reaction could be inhibited by the addition of biotin, but not the cysteine part of the substrate, indicating that the biotin group participates in the binding and positioning of the biotinylated Michael donor substrate.

An entirely different approach for isolating self-alkylating ribozymes employed a pool of genomic RNA fragments from nine different organisms that span the three kingdoms of life [23]. As a first step, a series of electrophilic compounds was identified that were insufficiently electrophilic to react with a random RNA pool. Eight probes were identified that modified a random RNA at a level below 0.1%. These electrophilic probes were each synthesized in a biotinylated form. The biotinylated electrophiles were mixed and incubated with the pool of genomic RNA fragments

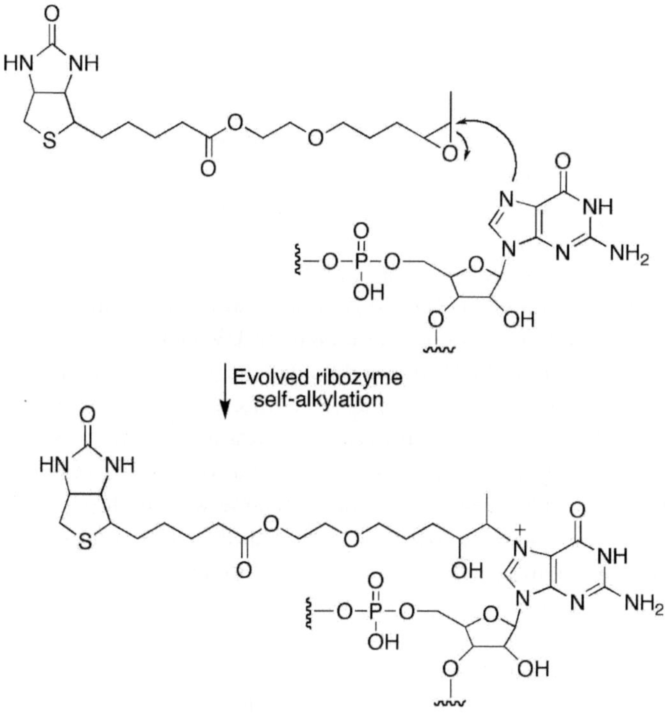

Figure 21.7 Self-alkylation reaction between a biotinylated epoxide substrate and the N7-position of guanosine residue G9 of the 42-nucleotide ribozyme found to be part of the genome of *A. pernix*.

to select self-alkylating genomic RNAs by streptavidin pull-down. As it turned out after six in vitro selection cycles, only one of the eight electrophilic probes, a disubstituted epoxide, had enriched the pool for genomic RNAs that have sufficient activity to catalyze their self-alkylation by epoxide ring opening. By RNAseq, a 63-nt fragment of the extremophile archaea *Aeropyrum pernix* emerged, among other less active genomic RNA fragments. This fragment could be further reduced to a 42-nt minimal motif that showed a catalytic efficiency that was at least 1900-fold higher than that of a random pool of 42-nt RNAs without any detectable reaction. As the site of labeling, N7 of G9 was determined (Figure 21.7).

With respect to the substrate, the modification with the biotin moiety was not required for activity, whereas shortening of the alkyl group or replacing the ester group for an amide reduced it. Interestingly, the epoxide stereochemistry was found to be an important determinant for reactivity: cis-epoxides showed a 22-fold faster reaction with the ribozyme than the corresponding trans-epoxides. Variants of the *A. pernix* 42-mer RNA that also showed appreciable self-alkylating activities were found in the mouse genome, but not in the genomes of *Escherichia coli* and *Saccharomyces cerevisiae*. Finally, the epoxide-opening ribozyme was further engineered to enrich RNAs of interest from total cellular RNA and to capture RNA-binding proteins.

21.6 Conclusion

The examples discussed here show that in vitro selected single-stranded nucleic acids are able to catalyze a variety of "demanding" carbon–carbon bond-forming reactions. Several of these ribozymes and deoxyribozymes exhibit rather moderate length, as compared with many natural functional RNAs such as riboswitches. Based on the comparison between small in vitro selected aptamer motifs and aptameric domains in natural riboswitches that are often considerably larger but bind the same ligand usually much tighter than in vitro selected aptamers, it can be extrapolated that the catalytic efficiency of carbon–carbon bond-forming ribozymes and deoxyribozymes might further improve with the length of the nucleic acid in combination with the application of Darwinian in vitro evolution. Such approaches have already led to much improved performance of artificial ribozymes, as seen, for example, in the case of the RNA ligase ribozyme. The original RNA ligase ribozyme [24] was evolved in several steps to first yield improved oligonucleotide ligases with improved catalytic performance [25] and then a variant that was able to catalyze the polymerization of mononucleotide triphosphates in a template-directed fashion [26]. The examples of ribozymes that catalyze the formation of carbon–carbon bonds or that perform alkylation reactions at heteroatoms shown in this chapter bear witness to the astonishing catalytic power of single-stranded nucleic acids. They certainly increase the plausibility of RNA-based metabolism in a – still hypothetical, but not unlikely – RNA world.

References

1 Joyce, G.F. (2002). The antiquity of RNA-based evolution. *Nature* 418 (6894): 214–221.

2 Szostak, J.W. (2017). The narrow road to the deep past: in search of the chemistry of the origin of life. *Angew. Chem. Int. Ed.* 56 (37): 11037–11043.

3 Yarus, M. (2005). Chemical biology: bring them back alive. *Nature* 438 (7064): 40.

4 Szostak, J.W., Bartel, D.P., and Luisi, P.L. (2001). Synthesizing life. *Nature* 409 (6818): 387–390.

5 Tarasow, T.M., Tarasow, S.L., and Eaton, B.E. (1997). RNA-catalysed carbon–carbon bond formation. *Nature* 389 (6646): 54–57.

6 Seelig, B. and Jäschke, A. (1999). A small catalytic RNA motif with Diels–Alderase activity. *Chem. Biol.* 6 (3): 167–176.

7 Tarasow, T.M., Tarasow, S.L., Tu, C. et al. (1999). Characteristics of an RNA Diels–Alderase active site. *J. Am. Chem. Soc.* 121 (15): 3614–3617.

8 Fiammengo, R. and Jäschke, A. (2005). Nucleic acid enzymes. *Curr. Opin. Biotechnol.* 16 (6): 614–621.

9 Keiper, S., Bebenroth, D., Seelig, B. et al. (2004). Architecture of a Diels–Alderase ribozyme with a preformed catalytic pocket. *Chem. Biol.* 11 (9): 1217–1227.

10 Hermann, T. and Patel, D.J. (2000). Adaptive recognition by nucleic acid aptamers. *Science* 287 (5454): 820–825.

11 Seelig, B., Keiper, S., Stuhlmann, F., and Jaschke, A. (2000). Enantioselective ribozyme catalysis of a bimolecular cycloaddition reaction. *Angew. Chem. Int. Ed.* 39 (24): 4576–4579.

12 Stuhlmann, F. and Jaschke, A. (2002). Characterization of an RNA active site: Interactions between a Diels–Alderase ribozyme and its substrates and products. *J. Am. Chem. Soc.* 124 (13): 3238–3244.

13 Serganov, A., Keiper, S., Malinina, L. et al. (2005). Structural basis for Diels–Alder ribozyme-catalyzed carbon–carbon bond formation. *Nat. Struct. Mol. Biol.* 12 (3): 218–224.

14 Fusz, S., Eisenfuhr, A., Srivatsan, S.G. et al. (2005). A ribozyme for the aldol reaction. *Chem. Biol.* 12 (8): 941–950.

15 Fusz, S., Srivatsan, S.G., Ackermann, D., and Famulok, M. (2008). Photocleavable initiator nucleotide substrates for an aldolase ribozyme. *J. Org. Chem.* 73 (13): 5069–5077.

16 Fessner, W.-D., Schneider, A., Held, H. et al. (1996). The mechanism of class II, metal-dependant aldolases. *Angew. Chem. Int. Ed. Engl.* 35: 2219–2221.

17 Haug-Schifferdecker, E., Arican, D., Bruckner, R., and Heide, L. (2010). A new group of aromatic prenyltransferases in fungi, catalyzing a

2,7-dihydroxynaphthalene 3-dimethylallyl-transferase reaction. *J. Biol. Chem.* 285 (22): 16487–16494.

18 Mohan, U., Burai, R., and McNaughton, B.R. (2013). In vitro evolution of a Friedel–Crafts deoxyribozyme. *Org. Biomol. Chem.* 11 (14): 2241–2244.

19 Wilson, C. and Szostak, J.W. (1995). In vitro evolution of a self-alkylating ribozyme. *Nature* 374 (6525): 777–782.

20 Wilson, C., Nix, J., and Szostak, J. (1998). Functional requirements for specific ligand recognition by a biotin-binding RNA pseudoknot. *Biochemistry* 37 (41): 14410–14419.

21 Sengle, G., Eisenführ, A., Arora, P.S. et al. (2001). Novel RNA catalysts for the Michael reaction. *Chem. Biol.* 8 (5): 459–473.

22 Eisenführ, A., Arora, P.S., Sengle, G. et al. (2003). A ribozyme with michaelase activity: synthesis of the substrate precursors. *Bioorg. Med. Chem.* 11 (2): 235–249.

23 McDonald, R.I., Guilinger, J.P., Mukherji, S. et al. (2014). Electrophilic activity-based RNA probes reveal a self-alkylating RNA for RNA labeling. *Nat. Chem. Biol.* 10 (12): 1049–1054.

24 Bartel, D.P. and Szostak, J.W. (1993). Isolation of new ribozymes from a large pool of random sequences [see comment]. *Science* 261 (5127): 1411–1418.

25 Ekland, E.H., Szostak, J.W., and Bartel, D.P. (1995). Structurally complex and highly active RNA ligases derived from random RNA sequences. *Science* 269 (5222): 364–370.

26 Ekland, E.H. and Bartel, D.P. (1996). RNA-catalysed RNA polymerization using nucleoside triphosphates. *Nature* 383 (6596): 192.

22

Nucleic Acid-Catalyzed RNA Ligation and Labeling

Mohammad Ghaem Maghami and Claudia Höbartner

Universität Würzburg, Institut für Organische Chemie, Am Hubland, 97074 Würzburg, Germany

22.1 Introduction

A large majority of biophysical approaches for studying RNA structure, folding, and dynamics as well as investigating RNA protein and RNA ligand interactions requires access to site-specifically labeled RNA. In many cases, the labels are organic fluorophores that are installed at the RNA to enable spectroscopic readout of structural changes, for example, in ensemble or single-molecule FRET experiments [1]. Other important labels contain selectively enriched isotopes, or paramagnetic nucleoside probes for NMR and EPR spectroscopy [2]. Ideally, the labels should be covalently attached and not interfere with the natural function of the RNA under investigation. Numerous methods are available that can provide the desired molecules; many of them involve combinations of multiple chemical and enzymatic steps. The most traditional method involves chemical synthesis of RNA oligonucleotides, in which modified nucleotides are introduced by solid-phase synthesis. Due to the limited length that can be achieved by repeated coupling steps and the limited availability of modified phosphoramidites, this approach is only suitable for rather short RNA targets. Longer desired molecules are then synthesized by enzymatic ligation of the modified/labeled oligonucleotides. This is often performed using T4 RNA ligase, or T4 DNA ligase in combination with a DNA splint. Other approaches involve combinations of enzymes [3], specifically optimized transcription conditions on immobilized templates for position-selective labeling of RNA [4]. In alternative ligation strategies, deoxyribozymes, also known as DNAzymes or DNA enzymes, can be used, in particular if one of the two fragments that need to be ligated can be prepared by in vitro transcription and the other fragment by solid-phase synthesis [5]. The 9DB1 DNA enzyme is most suitable for this type of DNA-catalyzed ligation of modified RNA fragments, as it can be engineered for nearly any type of ligation junction. The abovementioned techniques can be described as "divide-and-conquer" approaches for the synthesis of modified RNA, and many of these have been successfully used for many targets and resulted in important insights but still are rather laborious and often technically challenging. Therefore,

Ribozymes, First Edition. Edited by Sabine Müller, Benoît Masquida, and Wade Winkler.

post-synthetic strategies for covalent site-specific labeling of RNA have received increasing attention. A post-synthetic approach must combine recognition of a desired labeling site with the ability to form a covalent bond between the recognized target site and a functional group on the to be installed label. Since protein enzymes feature the unique combination of catalytic active sites for specific substrate recognition, several examples of enzyme-catalyzed RNA labeling approaches have recently been developed. Not only these include methyltransferase enzymes in combination with modified SAM analogs, which have been used to target unique mRNA cap structures [6], but also other nucleobase methyltransferases such as Trm1 or METTL14 have been repurposed for RNA labeling [7, 8]. In addition, C/D box snoRNPs have been engineered with custom guide RNAs, resulting in targeted alkylation of specific internal 2′-OH groups [9]. An alternative enzymatic approach involved a bacterial tRNA guanine alkyltransferase enzyme that catalyzed the exchange of an unmodified guanine nucleobase by a chemically modified preQ1 derivative, i.e. a fluorescently labeled 7-aminomethyl-7-deazaguanine analog [10]. Since this strategy is based on a tRNA-modifying enzyme, the secondary structure context of the anticodon stem-loop is necessary for the enzyme to function. Therefore, a prerequisite for this approach is the introduction of the respective sequence element into the target RNA.

Nucleic acid enzymes represent an attractive alternative to protein enzymes for site-specific labeling of RNA. A particularly interesting feature is the fact that Watson–Crick base pairing of the ribozyme binding arms can be harnessed to address the desired labeling site in the target RNA. This can be achieved with so-called "trans-acting" ribozymes or deoxyribozymes, which form an intermolecular complex with the target RNA and catalyze the reaction between two substrates without being modified themselves. Alternatively, self-labeling ribozymes can be used as tags that are attached to the RNA of interest. This chapter discusses recent examples of ribozymes to catalyze the attachment of fluorescent labels or fluorescently labeled nucleotides or oligonucleotides to an RNA of interest, either at internal or terminal positions.

22.2 Ribozymes for RNA Labeling at Internal Positions

Several strategies have been reported to develop ribozymes with the ability to label RNA molecules at internal positions. Two classes of ribozymes use electrophilic small-molecule substrates, while the twin ribozyme (see also Chapter 16) catalyzes ligation of fluorescently labeled oligonucleotides. In this section, these ribozymes will be discussed in detail. The second part of the chapter describes deoxyribozymes that can be used for similar purpose, i.e. the ligation of a fluorescently labeled oligonucleotide, or for the direct attachment of fluorescent nucleotides at an internal position.

22.2.1 Fluorescein Iodoacetamide Reactive Ribozyme

Ribozymes that catalyze a self-labeling reaction at an unknown internal position were reported by Sharma et al. in 2014 [11]. Two ribozymes were identified from an in vitro selection experiment, in which a random RNA pool containing a stretch of 116 random nucleotides was incubated with fluorescein iodoacetamide. The reaction was then quenched by adding mercaptoethanol, and the RNA pool was ethanol precipitated to remove the excess of unreacted fluorescein. Immunoprecipitation (IP)-SELEX was used to capture the active species from the bulk of inactive pool. For this purpose, the fluorescein iodoacetamide-treated pool was incubated with anti-fluorescein antibody-coated magnetic beads. Under these conditions, fluorescein-labeled RNA was immobilized onto the beads, and unreacted RNA was washed away. This immunocapture step was then followed by on-bead reverse transcription and PCR amplification to produce the double-stranded DNA template for the transcript of the next selection round. The results of the activity assay on the selection pool of various selection rounds indicate enrichment of active species as early as the 4th selection round. After 12 rounds of selection, the enriched pool was subjected to cloning and sequencing. The most active variants were denoted 1F1R and 5F1R. Both of these ribozymes were able to catalyze the self-labeling reaction in the presence of total cellular RNA and within cell lysate. Insertion of these two sequences in the 3′-UTR region of the mCherry-encoding reporter mRNA allowed site-specific labeling of this RNA in vitro. While these ribozymes prove to be useful tools for RNA labeling (Figure 22.1a), they have an extremely narrow substrate specificity range, which limits their application only to fluorescein labeling of target RNA. A significant drawback of this labeling approach is the as-yet-incomplete characterization of the in vitro selected ribozyme. The lack of information on the exact site of reaction renders any further application and engineering approaches challenging.

22.2.2 Genomically Derived Epoxide Reactive Ribozyme

Electrophilic reagents beyond iodoacetamides are promising alternatives for ribozyme-catalyzed RNA labeling, and the use of genomic RNA libraries instead of synthetic random RNA pools represents another interesting approach. In this context, a 62-nucleotide (nt)-long ribozyme derived from a genomic RNA sequence of the archaea *Aeropyrum pernix* was discovered by McDonald and coworkers also in 2014 [12]. To select for such RNA catalysts, the authors identified a total of eight electrophilic alkylation reagents with no detectable background reactivity toward RNA. An RNA pool derived from genomic fragments of nine different organisms from all three kingdoms of life was incubated with a cocktail of all the eight candidate electrophilic compounds in their biotinylated form. After five rounds of in vitro selection, the group observed signs of enriched self-alkylating activity in the RNA pool. To identify the reactive electrophilic probe, the sixth selection round was performed in the presence of each probe individually. It was identified, at this stage, that the enriched pool only reacts with the epoxide functionalized probe. After sequencing of the enriched pool and further characterization of the enriched species, the *A. pernix*-derived sequence was found as the most promising candidate since it was derived from an intact segment of the *A. pernix* genome and its activity was independent of the sequence context of the selection pool. Minimization and reselection on the partially randomized 42-nt pool based on the initial isolated core catalytic sequence led to a fivefold improvement in catalytic activity. The ribozyme

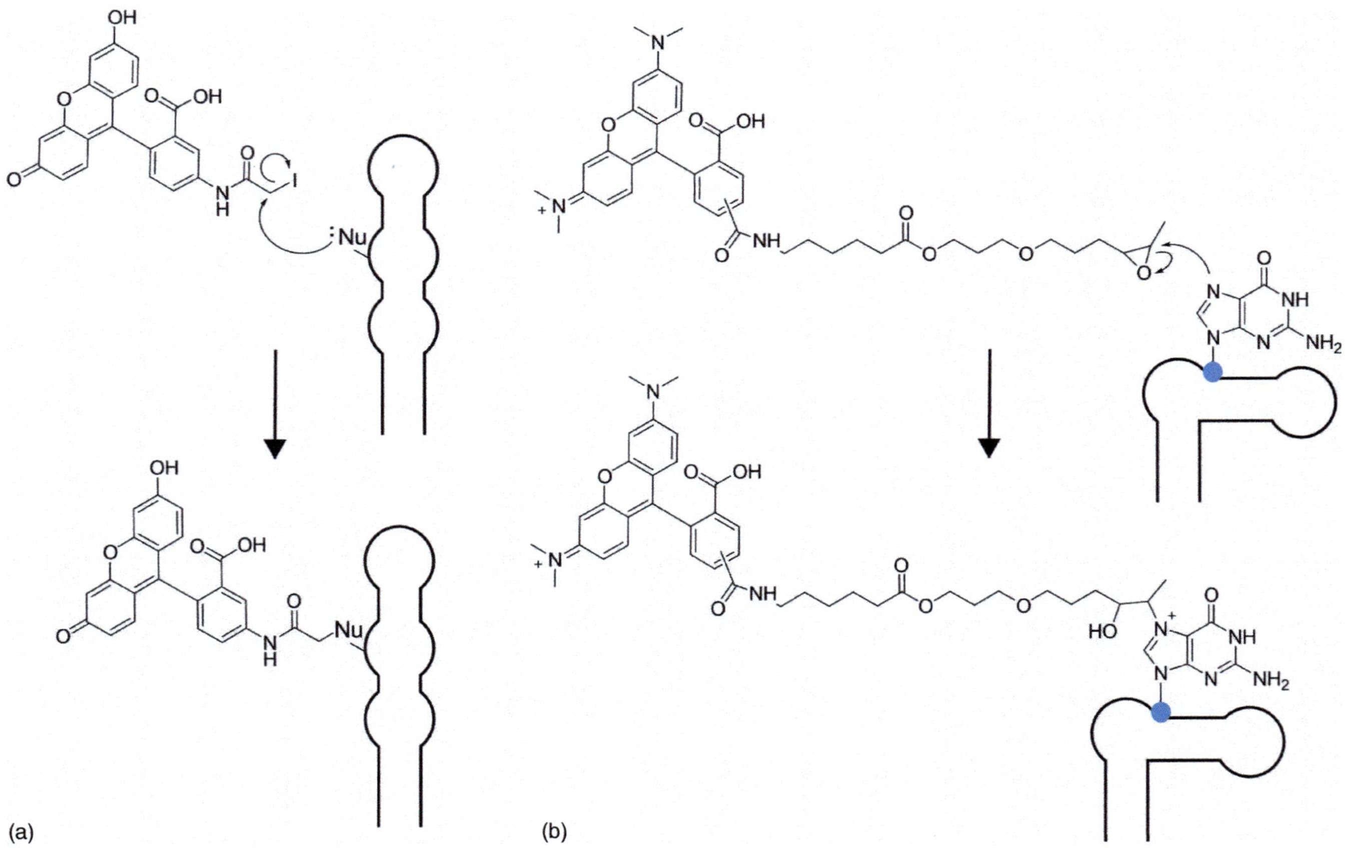

Figure 22.1 (a) Schematic representation of the iodoacetamide-reactive ribozyme labeling itself with fluorescein. (b) Fluorescent RNA labeling by guanine N7 alkylation via epoxide opening using the self-labeling ribozyme selected from *A. pernix* genomic RNA. Source: (a) Source: Adapted from Sharma et al. [11]. (b) Adapted from McDonald et al. [12].

catalyzes alkylation of a specific N7 guanine within the ribozyme sequence and thereby allows the attachment of fluorescent probes (Figure 22.1b).

McDonald and colleagues tested the reactivity of the *A. pernix* ribozyme toward a number of disubstituted epoxides derivatized with moieties other than biotin to study the alkylation substrate scope of the catalytic RNA. During these experiments, the *A. pernix* ribozyme proved to be highly flexible regarding the epoxide substrate making possible self-functionalization using a wide variety of labeling reagents and bio-orthogonal functional groups.

The ribozyme fused to the 3′-end of the human 5S rRNA demonstrated efficient self-labeling activity as an in vitro generated transcript. Labeling activity was also clearly detectable within both the extracted total cellular RNA and the cell lysate, when the fusion RNA was transcribed in vivo, via transfection of a vector encoding it, into HEK-293T cells. Insertion of the ribozyme into the Ash1 mRNA of *Saccharomyces cerevisiae* and addition of the biotinylated epoxide substrate made possible streptavidin-mediated pull-down of the Ash1 mRNA along with known interacting proteins [12].

22.2.3 Twin Ribozyme

The twin ribozyme for RNA labeling was originally designed in the research group of S. Müller [13]. The twin ribozyme represents the product of the tandem fusion of two hairpin ribozymes and facilitates cleaving out of a specific patch from the internal region of a target RNA and replacing it with a synthetic RNA fragment. Unlike the abovementioned self-alkylating ribozymes, twin ribozymes possess trans-activity and are therefore able to label target RNA sequences with minimal changes to the original sequence context.

The hairpin ribozyme (which the twin ribozyme is based on) is a small naturally occurring catalytic RNA with the ability to reversibly cleave and (re)ligate a piece of RNA. Recognition of the target sequence in this ribozyme is mediated via base complementarity. The ribozyme shows a high degree of flexibility regarding the cleavage/ligation substrate sequence; however, the presence of a G nucleotide 3′ to the cleavage/ligation site is strongly required. The cleavage reaction occurs via facilitation of an in-line attack by the 2′-OH group of the nucleotide 5′ to the cleavage site on its adjacent phosphodiester bond, resulting in the formation of a 2′,3′-cyclic phosphodiester bond on the one fragment and a free 5′-OH on the other. (Re)ligation happens via the attack of the free 5′-OH onto the 2′,3′-cyclic phosphodiester leading to the formation of a linear 3′,5′-phosphodiester bond and a free 2′-OH [13, 14].

Müller's group fused two hairpin ribozymes together, each flanked by two recognition arms complementary to the target RNA sequence (Figure 22.2). One of the arms acts also as a linker, which covalently couples the two ribozymes together. Upon annealing to the target sequence, each hairpin ribozyme cleaves its cognate cleavage site on the substrate sequence. The linker arm, which anneals to the middle cleaved fragment, is designed to have a none-paired tetraloop to destabilize the cleavage product. The RNA to be inserted will then easily displace the cleaved patch

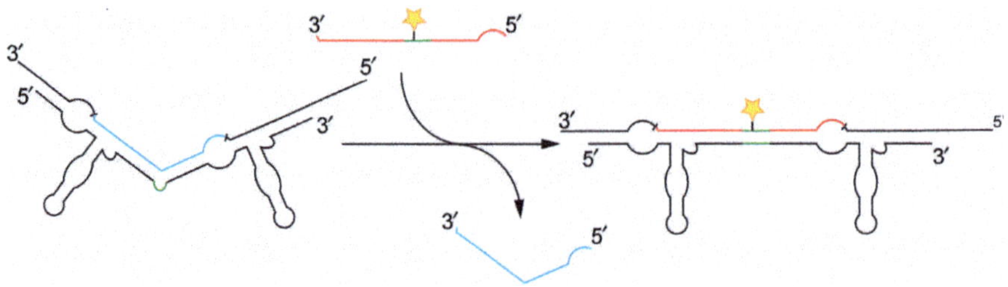

Figure 22.2 Schematic representation of RNA labeling using a twin ribozyme. A fluorescently labeled oligonucleotides is used as substrate to replace the corresponding fragment in the target RNA by two successive cleavage and ligation events. Source: Adapted from Vauleon et al. [15].

from the arm due to full sequence complementarity. The ligation product is therefore 4 nt longer than the initial substrate RNA [13].

The twin ribozyme was used to replace a 16-mer fragment in a 145-nt in vitro transcribed RNA, with a 20-nt synthetic RNA, containing an amino-modified deoxythymidine, which was derivatized with a number of fluorescent dyes (fluorescein, TAMRA, Cy5) and affinity probes (biotin). Reaction yields of around 11% and 18% were observed when the reaction was performed at 37 °C. By increasing the temperature to 55 °C, the reaction yield was improved; however, the best results were achieved at an optimal temperature of 47 °C and a yield of 53% for the fluorescein-labeled replacement oligonucleotide [15].

Interestingly, the twin ribozyme was used not only for RNA labeling but also for repair of defective transcripts, such as the correction of mutations in an mRNA to affect translation into functional proteins. This was shown in proof-of-principle work for repair of a defective EGFP transcript [16], as well as a fragment of an oncogenic mRNA that is associated with cancer cell development [17].

22.2.4 DNA as a Catalyst for Ligation of Modified RNA

As an alternative to protein enzymes T4 RNA ligase or T4 DNA ligase that join two RNA fragments via a native 3′,5′-phosphodiester bond, enzymes made entirely of DNA represent attractive tools for joining a modified synthetic RNA fragment to the remainder of the desired RNA. DNA catalysts have no natural counterparts, but in the laboratory, they are identified by in vitro selection procedures following essentially the same concept of selection and amplification from random libraries that is used for the identification of ribozymes. Several RNA-ligating deoxyribozymes have been identified by Silverman for joining RNA substrates with various combinations of functional groups [18]. Similar to the ligation reaction mediated by the twin ribozyme, deoxyribozymes can activate 5′-OH groups for ligation to 2′,3′-cyclic phosphate-terminated RNA. While most candidates preferentially resulted in the undesired 2′,5′-ligated isomer, special tricks during the selection procedure enable the in vitro evolution of DNA catalysts that result in regioselective opening of the cyclic phosphate to the desired native RNA linkages [19].

A second class of deoxyribozymes uses the 3′-OH group as nucleophile to form a new phosphodiester bond with the second RNA fragment that contains a 5′-triphosphate. The nucleophile attacks the alpha phosphate, and with the help of divalent metal ions, pyrophosphate is released, resulting in the formation of the native 3′–5′ linkage (Figure 22.3a). The most prominent and generally applicable representative of this class of DNA enzymes is called 9DB1 and was first reported in 2005 [22]. Using combinatorial mutation analyses, the catalytic core of this 9DB1 catalyst was shortened from originally 40 nt to 31 nt [23], and for each of the nucleotides in the catalytic core, the essential functional groups were identified [24]. The minimized catalyst was the first example of a DNA enzyme, for which the three-dimensional structure could be solved by X-ray crystallography (Figure 22.3a) [20]. The DNA enzyme crystallized in complex with the ligation product and the structural model showed the formation of an active site that is constructed by several long-range tertiary interactions that could not be predicted from biochemical data. Interestingly, the structure revealed an important feature of the DNA enzyme that is responsible for recognition of the ligation site, that is, the formation of Watson–Crick base pairs between two loop nucleotides and the two nucleotides at the ligation junction. This observation suggested that compensatory mutation of these base pairs should enable general application of the DNA enzyme for ligation of any 5′-triphosphorylated donor RNA, independent of the identity of the 5′-terminal nucleotide. This was indeed experimentally demonstrated, which was an additional proof that the crystal structure represented the DNA enzyme in a catalytically relevant conformation, and most importantly, this finding now allows 9DB1 derivatives to be engineered for a wide variety of ligation sites in any desired target RNA [20].

22.2.5 Site-Specific Internal Labeling of RNA with DNA Enzymes

A second class of DNA enzymes turned out to be very useful for site-specific labeling of RNA. These catalysts also resulted from in vitro selections that aimed at RNA-ligating deoxyribozymes, but instead, by activating the 3′-OH group of the terminal nucleotide for the formation of linear ligation products, these enzymes formed 2′,5′-branched RNA by using an internal adenosine 2′-OH group as nucleophile [25, 26]. The resulting DNA enzymes enable the synthesis of branched RNA and comblike DNA structures [27], which is challenging by other means. In addition, they have been used to attach structural constraints to RNA to manipulate folding of ribozymes into catalytically competent structures [28]. Most importantly, however, in the context of RNA labeling, the 10DM24 deoxyribozyme was engineered to accept mononucleotide triphosphates as substrates and attach those efficiently to the 2′-OH group of the target adenosine [29]. This strategy was expanded by demonstrating that GTP analogs with diverse bio-orthogonal 2′-functional groups could be used, and guidelines were developed for the choice of efficient target site RNA (Figure 22.3b) [21]. Interestingly, the catalytic efficiency of the 10DM24 deoxyribozyme was boosted by the addition of lanthanide cofactors, such as Tb^{3+}, which in combination with Mg^{2+} or Mn^{2+} resulted in nearly quantitative labeling yields for a wide variety of substrates [21, 30]. These

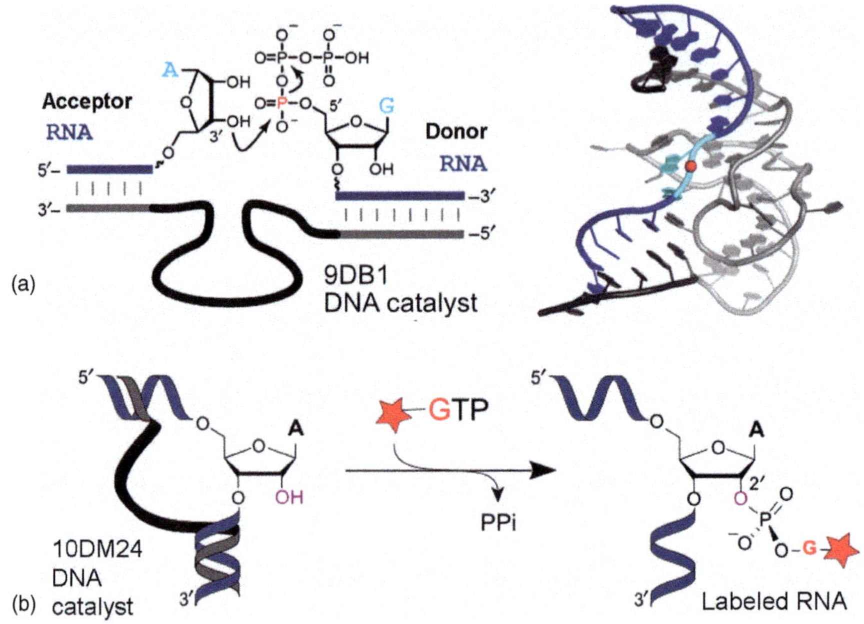

Figure 22.3 (a) DNA-catalyzed RNA ligation by the 9DB1 deoxyribozyme and structure of the DNA enzyme in complex with the ligation product. (b) 10DM24 DNA-catalyzed labeling of RNA at internal adenosine 2′-OH group with a fluorescent GTP analog. Source: (a) Ponce-Salvatierra [20]. © 2016 Springer Nature. Reproduced with permission of Springer Nature. (b) Adapted from Büttner et al. [21].

applications include fluorescent labeling of several different riboswitch RNAs followed by studies of ligand binding by fluorescence spectroscopy [21], U6 snRNA, and synthetic mRNAs for studies of RNA splicing. Interestingly, in the course of these studies in yeast cell extract, it turned out that the 2′,5′-linked label was readily removed by the natural debranching enzyme. To circumvent this problem, the 10DM24 deoxyribozyme-catalyzed labeling reaction was carried out with fluorescently labeled alpha-S-GTP analogs, resulting in stereospecific formation of phosphorothioate linkages [30].

22.3 RNA-Catalyzed Labeling of RNA at the 3′-end

While the attachment of fluorophores at internal positions is the most versatile approach to choose suitable labeling sites in a large RNA, in certain cases, the terminal labeling of RNA can also provide useful constraints. Several chemical methods have been described, including traditional periodate oxidation of the terminal ribose to a dialdehyde followed by condensation with amines, hydrazides, or hydrazones [31]. Enzymatic labeling of the 3′-end has also been achieved with terminal deoxynucleotidyltransferase or poly(A)polymerase and modified nucleotides [32, 33]. However, ribozymes have only recently been considered as potential alternatives.

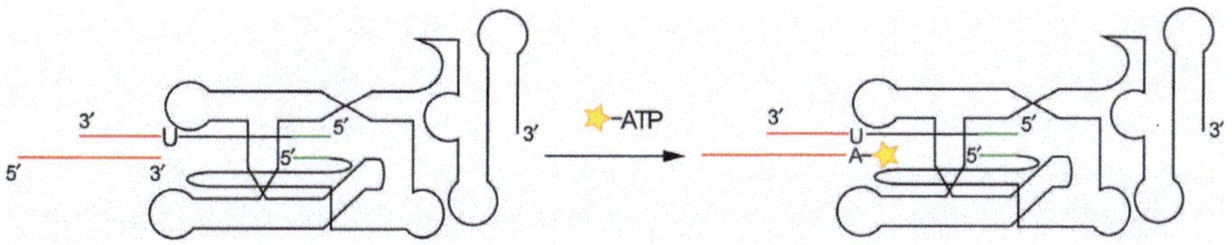

Figure 22.4 Schematic presentation of the RNA polymerase ribozyme that catalyzes 3′-terminal attachment of a fluorescent ATP analog. Source: Adapted from Samanta et al. [34].

In a recent report by Samanta et al., an RNA polymerase ribozyme's potential has been put to test with regard to labeling target RNA sequences at their 3′-end. The polymerase ribozyme codenamed 24-3 uses an RNA template to extend DNA or RNA primers via 5′ → 3′ addition of (d)NTPs [34]. The family of RNA polymerase ribozymes is derived from earlier selected ligase ribozymes, and these molecules are still intensively investigated in the context of the RNA world hypothesis and the aim to demonstrate the potential of RNA to self-replicate [35–37] (see also Chapter 13). The RNA labeling approach using these molecules is based on ribozyme-catalyzed single-nucleotide primer extension using modified NTP analogs (Figure 22.4). The RNA of interest acts as primer annealed to a template, which allows only addition of one single nucleotide in the presence of a particular NTP analog. The 5′-end of the primer has to be complementary to a short tag sequence at the 5′-terminus of the 24-3 polymerase. Samanta et al. were able to label target RNA sequences using NTP analogs carrying a wide variety of labeling moieties and bio-orthogonal functional groups either on the non-Watson–Crick face of the nucleobase or on the 2′-position of the sugar or even the alpha phosphate. The paper reports an average extension efficiency of ~61% across all the NTP analogs tested [34]. Thus, the polymerase ribozyme-mediated 3′-end labeling of target RNA molecules appears to be extremely versatile and flexible regarding the target sequence and the type of labeling required. Similar to the hairpin ribozyme above, it demonstrates a nice example of a well-studied and versatile ribozyme to be repurposed for RNA labeling.

22.4 Potential Ribozymes for RNA Labeling at the 5′-end

Other classes of ribozymes have been reported in literature with reactivity toward the 5′-triphosphate structure of RNA transcripts, and these could potentially be useful for RNA labeling as well. One group of such ribozymes contains capping ribozymes developed by M. Yarus and coworkers [38–40], and the second group of RNA catalyst is a 5′-purine nucleotide transferase developed by H. Suga and coworker [41].

The capping ribozymes are named this way due to their ability to form structures analogous to the eukaryotic mRNA 5′-cap structures. These ribozymes catalyze the attack by a terminal phosphate of small and large substrate molecules onto the 5′-triphosphate region of the target RNA, leading to the formation of a phosphoanhydride bond upon release of pyrophosphate as a leaving group (Figure 22.5a). The early reports demonstrate that the capping ribozymes are mainly specific toward the phosphoryl moiety of the ligated substrate. The authors attached a number of phosphorylated small molecules (nucleotides at various phosphorylation states and coenzymes such as NADP, FMN, thiamine, and coenzyme A) and large molecules such as tRNA to the 5′-end of the target RNA molecule [42, 43]. Phosphorylated fluorescent dyes have not been tested as substrates for these ribozymes; however, the reported scope suggests that it could be worthwhile to explore capping ribozymes for the distinct application in RNA labeling.

The 5′-purine nucleotide ribozyme named M4 catalyzes the formation of a 2′,5′-phosphodiester bond between a free purine nucleoside and the 5′-end of a triphosphorylated RNA. The ribozyme readily accepts a wide range of purine analogs including fluorescein-12-GTP; however, the reaction yields for the latter proved to be quite low [41]. Both of these ribozyme groups show trans activity; however, their flexibility regarding the substrate sequences they can modify has not been deeply investigated. Nevertheless, it is conceivable that such catalytic RNA activities can be evolved into highly efficient ribozymes for RNA labeling.

22.5 Conclusions

To conclude, nucleic acid enzymes, including ribozymes and deoxyribozymes, are highly valuable tools for the synthesis of site-specifically labeled RNA. Several of the described approaches result from repurposing or from directed evolution of other ribozyme activities, while some catalysts were specifically generated by in vitro selection for RNA labeling reactions with electrophilic substrates. The examples presented in this chapter highlight the need for further development of trans-acting RNA catalysts for site-specific labeling of RNA with small molecule substrates. Ideally, these reactions would be combined with a fluorogenic reaction, such that only successfully labeled RNA is detected, and excess of labeling reagent would not interfere with downstream analyses. Most importantly, however, new ribozymes need to be fully characterized, such that the exact target sites and structures of the resulting products are known. Engineering of nucleic acid enzymes will be much facilitated once the substrate scope on the side of the RNA target as well as the labeling agents is fully explored. A bright future is therefore expected for ribozymes and deoxyribozymes that can be designed as versatile RNA labeling agents for numerous applications in vitro and potentially also in vivo.

Acknowledgments

This work was supported by the European Research Council (ERC Consolidator Grant No. 682586). We also gratefully acknowledge support by the Max Planck Research School Molecular Biology.

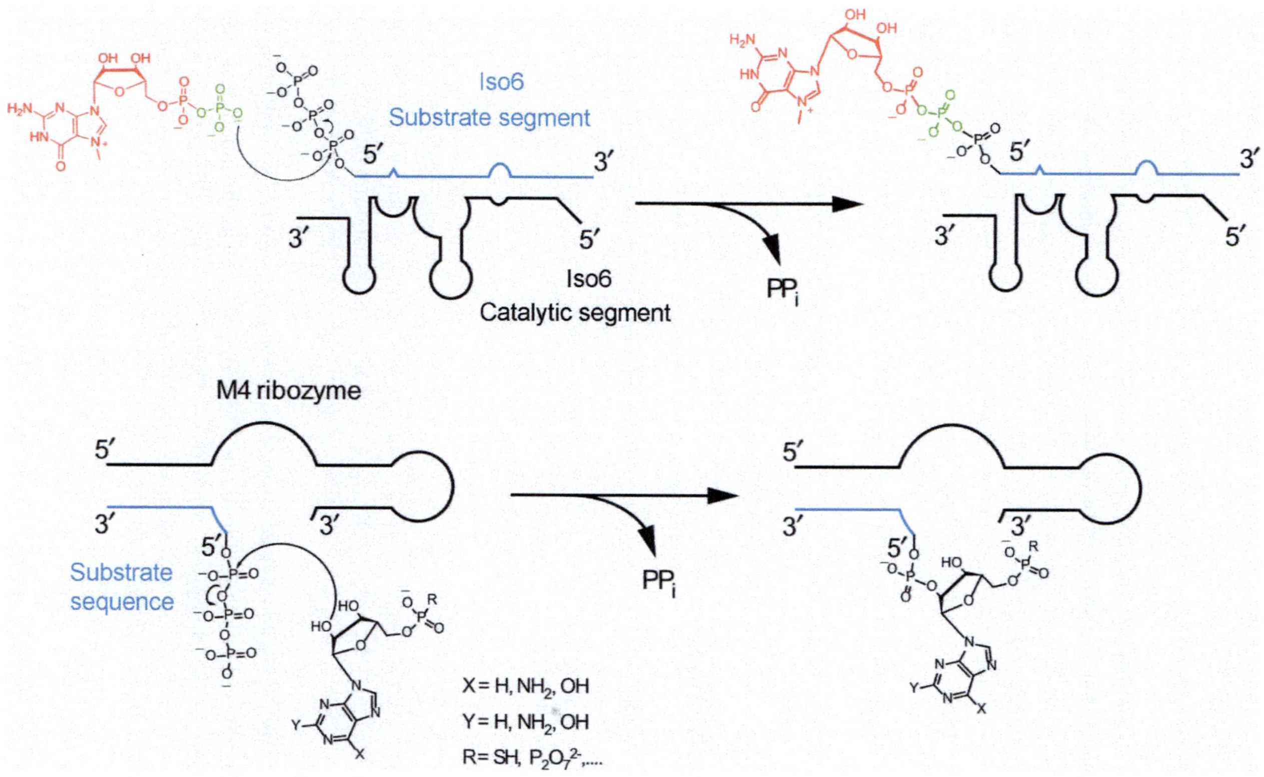

Figure 22.5 (a) Iso6 capping ribozyme exemplarily shown with m7GDP as labeling substrate. (b) Potential for 5′-labeling of RNA by ligation of a single nucleotide at the 5′-end with the M4 nucleotide transferase ribozyme. Source: (b) Adapted from Kang et al. [41].

References

1 Helm, M., Kobitski, A.Y., and Nienhaus, G.U. (2009). Single-molecule Forster resonance energy transfer studies of RNA structure, dynamics and function. *Biophys. Rev.* 1 (4): 161.

2 Marchanka, A., Kreutz, C., and Carlomagno, T. (2018). Isotope labeling for studying RNA by solid-state NMR spectroscopy. *J. Biomol. NMR* 71 (3): 151–164.

3 Keyhani, S., Goldau, T., Blumler, A. et al. (2018). Chemo-enzymatic synthesis of position-specifically modified RNA for biophysical studies including light control and NMR spectroscopy. *Angew. Chem. Int. Ed. Engl.* 57 (37): 12017–12021.

4 Liu, Y., Holmstrom, E., Yu, P. et al. (2018). Incorporation of isotopic, fluorescent, and heavy-atom-modified nucleotides into RNAs by position-selective labeling of RNA. *Nat. Protoc.* 13 (5): 987–1005.

5 Büttner, L., Seikowski, J., Wawrzyniak, K. et al. (2013). Synthesis of spin-labeled riboswitch RNAs using convertible nucleosides and DNA-catalyzed RNA ligation. *Bioorg. Med. Chem.* 21 (20): 6171–6180.

6 Muttach, F. and Rentmeister, A. (2016). A biocatalytic cascade for versatile one-pot modification of mRNA starting from methionine analogues. *Angew. Chem. Int. Ed. Engl.* 55 (5): 1917–1920.

7 Motorin, Y., Burhenne, J., Teimer, R. et al. (2011). Expanding the chemical scope of RNA:methyltransferases to site-specific alkynylation of RNA for click labeling. *Nucleic Acids Res.* 39 (5): 1943–1952.

8 Hartstock, K., Nilges, B.S., Ovcharenko, A. et al. (2018). Enzymatic or in vivo installation of propargyl groups in combination with click chemistry for the enrichment and detection of methyltransferase target sites in RNA. *Angew. Chem. Int. Ed. Engl.* 57 (21): 6342–6346.

9 Tomkuviene, M., Clouet-d'Orval, B., Cerniauskas, I. et al. (2012). Programmable sequence-specific click-labeling of RNA using archaeal box C/D RNP methyltransferases. *Nucleic Acids Res.* 40 (14): 6765–6773.

10 Alexander, S.C., Busby, K.N., Cole, C.M. et al. (2015). Site-specific covalent labeling of RNA by enzymatic transglycosylation. *J. Am. Chem. Soc.* 137 (40): 12756–12759.

11 Sharma, A.K., Plant, J.J., Rangel, A.E. et al. (2014). Fluorescent RNA labeling using self-alkylating ribozymes. *ACS Chem. Biol.* 9 (8): 1680–1684.

12 McDonald, R.I., Guilinger, J.P., Mukherji, S. et al. (2014). Electrophilic activity-based RNA probes reveal a self-alkylating RNA for RNA labeling. *Nat. Chem. Biol.* 10 (12): 1049–1054.

13 Welz, R., Bossmann, K., Klug, C. et al. (2003). Site-directed alteration of RNA sequence mediated by an engineered twin ribozyme. *Angew. Chem. Int. Ed. Engl.* 42 (21): 2424–2427.

14 Ivanov, S.A., Vauleon, S., and Muller, S. (2005). Efficient RNA ligation by reverse-joined hairpin ribozymes and engineering of twin ribozymes consisting of conventional and reverse-joined hairpin ribozyme units. *FEBS J.* 272 (17): 4464–4474.

15 Vauleon, S., Ivanov, S.A., Gwiazda, S., and Muller, S. (2005). Site-specific fluorescent and affinity labelling of RNA by using a small engineered twin ribozyme. *ChemBioChem* 6 (12): 2158–2162.

16 Balke, D., Becker, A., and Müller, S. (2016). In vitro repair of a defective EGFP transcript and translation into a functional protein. *Org. Biomol. Chem.* 14 (28): 6729–6737.

17 Balke, D., Zieten, I., Strahl, A. et al. (2014). Design and characterization of a twin ribozyme for potential repair of a deletion mutation within the oncogenic CTNNB1-DeltaS45 mRNA. *ChemMedChem* 9 (9): 2128–2137.

18 Silverman, S.K. (2009). Deoxyribozymes: selection design and serendipity in the development of DNA catalysts. *Acc. Chem. Res.* 42 (10): 1521–1531.

19 Kost, D.M., Gerdt, J.P., Pradeepkumar, P.I., and Silverman, S.K. (2008). Controlling the direction of site-selectivity and regioselectivity in RNA ligation by Zn^{2+}-dependent deoxyribozymes that use 2',3'-cyclic phosphate RNA substrates. *Org. Biomol. Chem.* 6 (23): 4391–4398.

20 Ponce-Salvatierra, A., Wawrzyniak-Turek, K., Steuerwald, U. et al. (2016). Crystal structure of a DNA catalyst. *Nature* 529 (7585): 231–234.

21 Büttner, L., Javadi-Zarnaghi, F., and Höbartner, C. (2014). Site-specific labeling of RNA at internal ribose hydroxyl groups: terbium-assisted deoxyribozymes at work. *J. Am. Chem. Soc.* 136 (22): 8131–8137.

22 Purtha, W.E., Coppins, R.L., Smalley, M.K., and Silverman, S.K. (2005). General deoxyribozyme-catalyzed synthesis of native 3'-5' RNA linkages. *J. Am. Chem. Soc.* 127 (38): 13124–13125.

23 Wachowius, F., Javadi-Zarnaghi, F., and Höbartner, C. (2010). Combinatorial mutation interference analysis reveals functional nucleotides required for DNA catalysis. *Angew. Chem. Int. Ed.* 49 (45): 8504–8508.

24 Wachowius, F. and Höbartner, C. (2011). Probing essential nucleobase functional groups in aptamers and deoxyribozymes by nucleotide analogue interference mapping of DNA. *J. Am. Chem. Soc.* 133: 14888–14891.

25 Coppins, R.L. and Silverman, S.K. (2004). A DNA enzyme that mimics the first step of RNA splicing. *Nat. Struct. Mol. Biol.* 11 (3): 270–274.

26 Zelin, E., Wang, Y., and Silverman, S.K. (2006). Adenosine is inherently favored as the branch-site RNA nucleotide in a structural context that resembles natural RNA splicing. *Biochemistry* 45 (9): 2767–2771.

27 Mui, T.P. and Silverman, S.K. (2008). Convergent and general one-step DNA-catalyzed synthesis of multiply branched DNA. *Org. Lett.* 10 (20): 4417–4420.

28 Zelin, E. and Silverman, S.K. (2009). Efficient control of group I intron ribozyme catalysis by DNA constraints. *Chem. Commun.* 45 (7): 767–769.

29 Höbartner, C. and Silverman, S.K. (2007). Engineering a selective small-molecule substrate binding site into a deoxyribozyme. *Angew. Chem. Int. Ed.* 46 (39): 7420–7424.

30 Carrocci, T.J., Lohe, L., Ashton, M.J. et al. (2017). Debranchase-resistant labeling of RNA using the 10DM24 deoxyribozyme and fluorescent modified nucleotides. *Chem. Commun.* 53 (88): 11992–11995.

31 Qiu, C., Liu, W.Y., and Xu, Y.Z. (2015). Fluorescence labeling of short RNA by oxidation at the 3'-end. *Methods Mol. Biol.* 1297: 113–120.
32 Winz, M.L., Samanta, A., Benzinger, D., and Jaschke, A. (2012). Site-specific terminal and internal labeling of RNA by poly(A) polymerase tailing and copper-catalyzed or copper-free strain-promoted click chemistry. *Nucleic Acids Res.* 40 (10): e78.
33 Winz, M.L., Linder, E.C., Andre, T. et al. (2015). Nucleotidyl transferase assisted DNA labeling with different click chemistries. *Nucleic Acids Res.* 43 (17): e110.
34 Samanta, B., Horning, D.P., and Joyce, G.F. (2018). 3'-End labeling of nucleic acids by a polymerase ribozyme. *Nucleic Acids Res.* 46 (17): e103.
35 Attwater, J., Wochner, A., and Holliger, P. (2013). In-ice evolution of RNA polymerase ribozyme activity. *Nat. Chem.* 5 (12): 1011–1018.
36 Attwater, J., Raguram, A., Morgunov, A.S. et al. (2018). Ribozyme-catalysed RNA synthesis using triplet building blocks. *Elife* 7.
37 Horning, D.P. and Joyce, G.F. (2016). Amplification of RNA by an RNA polymerase ribozyme. *Proc. Natl. Acad. Sci. U.S.A.* 113 (35): 9786–9791.
38 Huang, F. and Yarus, M. (1997). 5'-RNA self-capping from guanosine diphosphate. *Biochemistry* 36 (22): 6557–6563.
39 Huang, F., Bugg, C.W., and Yarus, M. (2000). RNA-catalyzed CoA, NAD, and FAD synthesis from phosphopantetheine, NMN, and FMN. *Biochemistry* 39 (50): 15548–15555.
40 Zaher, H.S., Watkins, R.A., and Unrau, P.J. (2006). Two independently selected capping ribozymes share similar substrate requirements. *RNA* 12 (11): 1949–1958.
41 Kang, T.J. and Suga, H. (2007). In vitro selection of a 5'-purine ribonucleotide transferase ribozyme. *Nucleic Acids Res.* 35 (12): 4186–4194.
42 Huang, F. and Yarus, M. (1997). Versatile 5' phosphoryl coupling of small and large molecules to an RNA. *Proc. Natl. Acad. Sci. U.S.A.* 94 (17): 8965–8969.
43 Jadhav, V.R. and Yarus, M. (2002). Acyl-CoAs from coenzyme ribozymes. *Biochemistry* 41 (3): 723–729.

Ribozymes

Ribozymes

Volume 2

Edited by

Sabine Müller
Benoît Masquida
Wade Winkler

WILEY-VCH

Editors

Sabine Müller
University Greifswald
Institut für Biochemie
Felix-Hausdorff-Str. 4
17489 Greifswald
Germany

Benoît Masquida
CNRS – Université de Strasbourg
UMR 7156
Génétique Mollaire Génomique
Microbiologie
4 allée Konrad Roentgen
67084 Strasbourg
France

Wade Winkler
The University of Maryland
Cell Biology & Molecular Genetics
3112 Biosciences Bldg.
MD
United States

Cover
Courtesy of Dr. Benoît Masquida

Library of Congress Card No.:
applied for

British Library Cataloguing-in-Publication Data
A catalogue record for this book is available from the British Library.

Bibliographic information published by the Deutsche Nationalbibliothek
The Deutsche Nationalbibliothek lists this publication in the Deutsche Nationalbibliografie; detailed bibliographic data are available on the Internet at <http://dnb.d-nb.de>.

Print ISBN: 978-3-527-34454-3
ePDF ISBN: 978-3-527-81455-8
ePub ISBN: 978-3-527-81453-4
oBook ISBN: 978-3-527-81452-7

Typesetting Straive, Chennai, India
Printing and Binding CPI Group (UK) Ltd, Croydon, CR0 4YY

Printed on acid-free paper

C083571_240621

Contents

Preface

Ribozymes, a neologism that appeared in 1982 in a paper of Tom Cech, triggered the blossom of a field of research on which the RNA community gathered its countless efforts over years. Thus, after the discovery of the first naturally occurring catalytic RNAs, more than 30 years ago, research in the field of ribozymes and RNA catalysis has made tremendous progress. In the 1990s, many of the catalytic RNAs known today were first identified in nature. In parallel to these discoveries, the powerful SELEX method for discovery of novel nucleic acids was invented, which allowed the development of artificial ribozymes with rather diverse functionalities. Over the years, investigation into the structure and mechanism of ribozymes led to a deep understanding of their catalytic strategies. Today, ribozymes are understood to an extent that it is possible to utilize rational design and molecular engineering to construct catalytic RNAs with pre-defined function. Yet, there is still much to be learned. Improvements in high-throughput bioinformatics approaches are still fostering the discovery of new ribozymes and novel genomic locations of known motifs in highly diverse genetic contexts for all branches of life. In spite of an apparent loss of attraction due to the discovery of the RNAi and CRISPR/cas mechanisms, which at times seemed more appealing, ribozymes still inspire activity of many research groups. Even after three decades of research following discovery of the first catalytic RNA, ribozyme research has not lost the intriguing and highly motivating flair of the first days. There are still many questions to be addressed and much is waiting to be discovered.

This book aims to survey what we have learned over the past 35 years about ribozymes and nucleic acid catalysis and to present the today state of the art. It musters over 30 chapters demonstrating this activity. From the study of artificial ribozymes developed by Darwinian selection to natural ribozymes including the large translation and splicing machineries, via ribozyme engineering, and their biochemical and biophysical studies, this book takes the reader through a thrilling journey. An enormous knowledge has been accumulated on ribozyme structure, function and mechanism. However, the wealth of data also makes apparent the missing parts, which encompass what are the cellular processes regulated by ribozymes and how this is done. For instance, only little is known about the role of ribozymes identified in the human genome, and the biological roles of the new ribozymes discovered in bacteria have just started to be studied. Ribozymes deserve a continuous look from researchers, and state of the art investigation methods need to be coupled with classical ones to unravel the still unknown.

We wish to thank all participating authors for the tremendous work. Their efforts and high-quality contributions have made this book come true. We hope we have succeeded in providing a source of information on mechanistic and structural aspects of nucleic acid catalysis, on tools and methods for characterization, engineering and application of ribozymes, as well as on the key questions, strategies and challenges in ribozyme research today.

December 2019

Sabine Müller
Benoît Masquida
Wade Winkler

Foreword

The study of ribozymes, RNA-based catalysts, and subsequently nucleic acid-based catalysis has been immensely instrumental in the push of efforts toward a deeper understanding of RNA structure and function. The field attracted many scientists trained in chemistry, biology, or computer science. In 1993, a few years after the Noble Prize in Chemistry (https://www.nobelprize.org/prizes/lists/all-nobel-prizes-in-chemistry) was awarded to Tom Cech [1] and Sidney Altman [2] for the discovery of ribozymes, the RNA Society (www.rnasociety.org) was founded and two years later the *RNA Journal* (https://rnajournal.cshlp.org) started. Ever since, the annual RNA meetings have gathered more than 1000 scientists from all over the world. Many new techniques and approaches were developed accelerating the pace of discoveries on RNA. For many years, the meetings started with a session on "RNA Catalysis." However, with the avalanche of new data, new RNAs, and new biology, the session on catalysis session has dwindled. Billions of years ago, RNA did start the chemistry of the game of life, but was overwhelmed by the initiated evolutionary processes.

The central group of the chemical and biological feats achieved by RNA is the ribose hydroxyl O2′, at the same time key actor and Achilles's heel. And a DNA phosphodiester bond is 10^4–10^5 less prone to cleavage than a RNA phosphodiester bond [3]. As Paracelsus wrote several centuries ago: "All things *are* poison *and* nothing *is* without poison; *only the* dose makes *a* thing not *a* poison".

The catalytic power of the 2′-hydroxyl group must be strongly controlled and amplified only at specific location in the RNA sequence. Catalysis generally is initiated by the formation of an anionic O2′ oxygen. In any biological system, catalysis requires accessibility, local molecular dynamics and solvent molecules (water, ions, or small ligands). A water molecule (or a hydroxide ion), or a solvated divalent ion, or an amine group from a ligand can thus capture the proton from the 2′-hydroxyl group. Afterwards, depending on the available mobility of the nucleotide, the anionic O2′ oxygen can attack its own 3′-phosphate group or another phosphodiester bond. Other types of chemical reactions have also been achieved using ribozymes as described in several chapters of this book.

Accessibility and local dynamics depend on the RNA sequence and the ensuing RNA architecture. The distinctive mark of nucleic acids is the formation of pairs between the bases of the nucleotides with the complementary Watson–Crick pairs the most frequent ones. However, such pairs form only regular helices and any complex fold or assemblies of helices require linking segments that engage in some type of non-Watson–Crick base pairs. Depending on the complexity and compactness of the RNA fold, not many bases will remain unpaired with a large number of degrees of freedom. Such regions are particularly prone to phosphodiester cleavages through hydrolysis or metal ion attack. Further, it was noticed a long time ago that, in single-stranded regions, dinucleotide steps with sequence pyrimidine-adenosine (UpA and CpA) were particularly sensitive to cleavage [4, 5]. These early observations were later thoroughly studied [6, 7]. Such cleavages are regularly observed in control lanes during gel electrophoresis. The precise molecular mechanism for these spontaneous cleavages in YpA sequences, however, has remained elusive. Recently, a surprising interpretation, based on mass spectrometry data and implying the *syn* conformation of the A and its protonation at N3 position, has been put forward [8].

In each ribozyme, cleavage occurs at a very precise dinucleotide location and generally with weak sequence dependence. The more extensively studied are the endonucleolytic, ribozymes (see Chapters 1 and 2). New data and results confirm the involvement of unexpected chemistry with anionic guanosine residues acting as a general base for capturing the O2′-hydroxyl proton [9, 10]. Interestingly, a proposal has been forwarded in which the tautomeric enol form of guanosine in which N1(G) carries an in-plane electron doublet would capture the proton from the O2′-hydroxyl group [11, 12]. Interestingly, such tautomers have been observed in functional ribosome crystals and related to the occurrence of miscoding at the first and second codon positions following the formation of tautomeric GoU pairs [13, 14]. The tautomeric ratio of G is around 1 in 10^4, a value close to the average miscoding error in bacteria [15]. The tautomeric forms trapped in a constrained environment (stacking, H-bond between the amino group of G and an anionic phosphate oxygen, minor groove contacts, etc.) are stabilized as observed in several crystal structures [12, 16]. The frequent occurrence of stacked Gs forming H-bonds through their Watson–Crick edge to phosphate anionic oxygens has been analyzed and described thoroughly [17]. It has also been shown that the tautomeric form coexists with the anionic form [18].

Overall, this book is timely and unique in its breadth of content. The book describes the great diversity of nucleic acid base catalysis beautifully. The reader is conveyed to a chemical journey extending from biologically functional ribozymes (like RNaseP or group I and II introns) to engineered and designed ribozymes as well as DNAzymes. For example, we now have three distinct structural environments with a natural 2′–5′ phosphodiester linkage: the group II ribozyme [19, 20], the spliceosome [21], and the lariat-capping ribozyme [22]. Despite the clear structural similarities between the active sites of group II ribozyme and of the spliceosome, the striking observation is how the 2′–5′ linkage promotes a highly constrained environment with nested non-Watson–Crick pairs and sugar-phosphate contacts.

Given those three-dimensional visions, one can only wonder at the structural and molecular fitness of nucleic acids, especially RNA. Here I have selected some topics that I enjoy particularly, and I apologize for not discussing many other aspects that are covered in this book. Indeed, some chapters extend to biotechnology and to ribozymes used in diagnostics and therapeutics. Finally, the major tools and techniques used for the analysis of nucleic acid structures are presented up to date. In this regard, I cannot resist citing Sydney Brenner [23, 24], Nobel Prize 2002 in Physiology or Medicine: "Progress in science depends on new techniques, new discoveries and new ideas, probably in that order."

Eric Westhof
Architecture et Réactivité de l'ARN
Université de Strasbourg
Institut de biologie moléculaire et cellulaire du CNRS
2 allée Konrad Roentgen, 67084 Strasbourg, France
August 2019

References

1 Cech, T.R., Zaug, A.J., and Grabowski, P.J. (1981). In vitro splicing of the ribosomal RNA precursor of *Tetrahymena*: involvement of a guanosine nucleotide in the excision of the intervening sequence. *Cell* 27 (3 Pt 2): 487–496.
2 Guerrier-Takada, C., Gardiner, K., Marsh, T. et al. (1983). The RNA moiety of ribonuclease P is the catalytic subunit of the enzyme. *Cell* 35 (3 Pt 2): 849–857.
3 Thompson, J.E., Kutateladze, T.G., Schuster, M.C. et al. (1995). Limits to catalysis by ribonuclease A. *Bioorg. Chem.* 23 (4): 471–481.
4 Cannistraro, V.J., Subbarao, M.N., and Kennell, D. (1986). Specific endonucleolytic cleavage sites for decay of *Escherichia coli* mRNA. *J. Mol. Biol.* 192 (2): 257–274.
5 Kierzek, R. (1992). Hydrolysis of oligoribonucleotides: influence of sequence and length. *Nucleic Acids Res.* 20 (19): 5073–5077.
6 Kaukinen, U., Lyytikainen, S., Mikkola, S., and Lonnberg, H. (2002). The reactivity of phosphodiester bonds within linear single-stranded oligoribonucleotides is strongly dependent on the base sequence. *Nucleic Acids Res.* 30 (2): 468–474.
7 Soukup, G.A. and Breaker, R.R. (1999). Relationship between internucleotide linkage geometry and the stability of RNA. *RNA* 5 (10): 1308–1325.
8 Fuchs, E., Falschlunger, C., Micura, R., and Breuker, K. (2019). The effect of adenine protonation on RNA phosphodiester backbone bond cleavage elucidated by deaza-nucleobase modifications and mass spectrometry. *Nucleic Acids Res.* 47(14): 7223–7234.
9 Wilson, T.J., Liu, Y., Li, N.S. et al. (2019). Comparison of the structures and mechanisms of the pistol and hammerhead ribozymes. *J. Am. Chem. Soc.* 141 (19): 7865–7875.

10 Bevilacqua, P.C. (2003). Mechanistic considerations for general acid-base catalysis by RNA: revisiting the mechanism of the hairpin ribozyme. *Biochemistry* 42 (8): 2259–2265.

11 Pinard, R., Hampel, K.J., Heckman, J.E. et al. (2001). Functional involvement of G8 in the hairpin ribozyme cleavage mechanism. *EMBO J.* 20 (22): 6434–6442.

12 Singh, V., Fedeles, B.I., and Essigmann, J.M. (2015). Role of tautomerism in RNA biochemistry. *RNA* 21 (1): 1–13.

13 Demeshkina, N., Jenner, L., Westhof, E. et al. (2012). A new understanding of the decoding principle on the ribosome. *Nature* 484 (7393): 256–259.

14 Rozov, A., Wolff, P., Grosjean, H. et al. (2018). Tautomeric G*U pairs within the molecular ribosomal grip and fidelity of decoding in bacteria. *Nucleic Acids Res.* 46 (14): 7425–7435.

15 Parker, J. (1989). Errors and alternatives in reading the universal genetic code. *Microbiol. Rev.* 53 (3): 273–298.

16 Westhof, E., Yusupov, M., and Yusupova, G. (2014). Recognition of Watson–Crick base pairs: constraints and limits due to geometric selection and tautomerism. *F1000Prime Rep.* 6: 19.

17 Zirbel, C.L., Sponer, J.E., Sponer, J. et al. (2009). Classification and energetics of the base-phosphate interactions in RNA. *Nucleic Acids Res.* 37 (15): 4898–4918.

18 Kimsey, I. J., Petzold, K., Sathyamoorthy, B., Stein, Z. W., and Al-Hashimi, H. M. (2015) Visualizing transient Watson-Crick-like mispairs in DNA and RNA duplexes. *Nature* 519, 315–320.

19 Toor, N., Keating, K.S., Taylor, S.D., and Pyle, A.M. (2008). Crystal structure of a self-spliced group II intron. *Science* 320 (5872): 77–82.

20 Costa, M., Walbott, H., Monachello, D. et al. (2016). Crystal structures of a group II intron lariat primed for reverse splicing. *Science* 354 (6316).

21 Wilkinson, M.E., Fica, S.M., Galej, W.P. et al. (2017). Postcatalytic spliceosome structure reveals mechanism of 3′-splice site selection. *Science* 358 (6368): 1283–1288.

22 Meyer, M., Nielsen, H., Olieric, V. et al. (2014). Speciation of a group I intron into a lariat capping ribozyme. *Proc. Natl. Acad. Sci. U.S.A.* 111 (21): 7659–7664.

23 Brenner, S. (2002). Life sentences: detective rummage investigates. *Genome Biol.* 3 (9): comment1013.1011-1013.1012.

24 Robertson, M. (1980). Biology in the 1980s, plus or minus a decade. *Nature* 285 (5764): 358–359.

Part IV

DNAzymes

23

The Chemical Repertoire of DNA Enzymes

Marcel Hollenstein

Institut Pasteur, Department of Structural Biology and Chemistry, Laboratory for Bioorganic Chemistry of Nucleic Acids, CNRS UMR3523, 28, rue du Docteur Roux, 75724 Paris Cedex 15, France

23.1 Introduction

Until the early 1980s and the discovery of catalytic ribonucleic acid (RNA) molecules by Thomas Cech [1, 2], catalysis of all biochemical transformations was believed to be chaperoned by proteinaceous enzymes. Shortly after this seminal discovery, Gerald Joyce employed a revolutionary combinatorial screening method coined SELEX (Systematic Evolution of Ligands by Exponential Enrichment) [3, 4] to isolate artificial ribozymes capable of hydrolyzing deoxyribonucleic acid (DNA) substrates [5], thus further underscoring the catalytic potential of RNA. The fundaments of the beliefs that protein enzymes controlled all biochemical catalysis were shaken anew when Ronald Breaker and Gerald Joyce isolated the first RNA-cleaving DNA-based enzyme in 1994 by application of SELEX [6]. Since the advent of the first catalytic DNA molecule, a large number of selection experiments resulted in the elaboration of a rather impressive collection of DNAzymes capable of catalyzing reactions as diverse as amide bond hydrolysis [7] and the thymine dimer photoreversion reaction [8]. Progress in the development of DNAzymes combined with their (sometimes) impressive catalytic prowess has culminated in the completion of two recent clinical trials for the treatment of skin cancer and asthma [9, 10]. However, despite these favorable traits, DNAzymes still suffer from some shortcomings, including a strong M^{2+}-cofactor dependence, which causes competition with antisense activity and severely undermines their catalytic efficiency in vivo. DNAzymes, like any other unmodified oligonucleotides, are rapidly degraded by nucleases and subjected to an efficient renal clearance. Also, unlike their RNA counterparts, no active catalytic DNA motif has been found in a living system, which probably is a direct consequence of the paucity of single-stranded DNA within cells [11].

In this chapter, the SELEX identification process as well as the catalytic scope and properties of DNAzymes are described. Moreover, the recent elucidation of crystal structures of DNAzymes constitutes the basis of a mechanistic description of these functional nucleic acids. In the last section, the inclusion of (chemical) functional groups, either by post-SELEX optimization using solid-phase

Ribozymes, First Edition. Edited by Sabine Müller, Benoît Masquida, and Wade Winkler.

synthesis or directly in the SELEX protocol via the inclusion of modified nucleoside triphosphates (dN*TPs), will be presented as a potential strategy to alleviate some of the shortcomings of DNAzymes.

23.2 Catalytic Repertoire of DNAzymes

In 1990, two seminal contributions radically changed the field of nucleic acid chemistry and introduced the possibility of conveying binding properties to RNA molecules [3, 4, 12]. These visionary experiments blended synthetic organic chemistry, which enabled the crafting of degenerate sequences by combining simultaneously all phosphoramidite building blocks during solid-phase synthesis, with the inherent capacity of amplifying nucleic acid molecules at will via the polymerase chain reaction (PCR) [13]. The first step of the combinatorial method of in vitro selection coined SELEX consists in the creation of large libraries of DNA or RNA oligonucleotides (typically around 10^{14} molecules) either by primer extension (PEX) reactions or PCR [14–16]. The fraction of the sequences stemming from these degenerate libraries that form adequate three-dimensional structures and hence enable binding to the target can be separated from the unbound species. PCR is then employed to amplify the isolated binding sequences, and the resulting enriched library is used in subsequent rounds of selection. The selection stringency (or pressure) can be altered through the rounds to favor the emergence of strong binders, for instance, by decreasing the incubation time, changing the temperature or buffer conditions, or reducing the concentration of target. At the end of the selection protocol, sequence information can be gathered by cloning and sequencing or by directly applying next generation sequencing and bioinformatics tools on the final, enriched library. The resulting binding sequences (called aptamers) are then tested individually for their binding affinity through an assessment of their equilibrium dissociation constants (K_d). Prior to the seminal contributions by the Gold and Szostak groups, another groundbreaking article by the Joyce laboratory-made use of the SELEX protocol to identify catalytic RNA molecules [5]. The selection method established for the isolation of ribozymes differs only in a single point from that described for the generation of aptamers: the RNA pool has its DNA target included in the sequence and is hence immobilized on a solid support; only the catalytically active species will be capable of freeing themselves from the solid support. As a direct consequence of these pioneering articles, the SELEX protocol could be hijacked to unravel an increasing number of catalytically active DNA molecules. However, the reader should be aware that despite important progress and improvements in the SELEX methodology, certain parameters are still dictated by empirical rules. For instance, short randomized regions (e.g. N20) in the initial libraries enable a full coverage of the sequence space but markedly reduce the diversity of structural motifs while larger degenerate sequences (≥N40) inherently convey structural flexibility but forbid the sampling of all possible combinations of molecules [13, 17]. In addition, selection experiments will also be heavily impacted by the presence of sequence bias in the starting randomized libraries, the formation of molecular parasites, the loss of sequences or over-amplification of other oligonucleotides during PCR, the presence of a negative selection step, and the initial conditions chosen for the first SELEX rounds. Clearly, the control of as many as possible of these parameters will increase the success rate of in vitro selection experiments.

In this section, the diversity of chemical transformations catalyzed by DNAzymes will be discussed to both define the scope of these biocatalysts and address their inherent limitations.

23.2.1 Hydrolytic Reactions

Most DNAzymes that have been isolated by in vitro selection so far are capable of accelerating the hydrolysis of bonds, mainly ribophosphodiesters [18]. This predominance of hydrolytic DNAzymes can be ascribed to the potential applications of such catalysts (i.e. mainly as gene-silencing agents and as biosensors) as well as to their ease of selection. In this section, the most important bond-cleaving DNAzymes will be described, along with their intrinsic properties and applications.

RNA-cleaving DNAzymes

The first-ever isolated DNAzyme cleaves a substrate containing a single embedded ribo(adenosine) nucleotide within a DNA sequence in the presence of 1 mM Pb^{2+} with a first-order rate constant (k_{obs}) of 1.4 min^{-1} [6]. The choice of the metal cofactor may seem somewhat peculiar but is directly connected to some precedents in ribozymes [19, 20] and tRNAs [21, 22]. This first DNAzyme was then rapidly followed by the selection of another RNA-cleaving DNA molecule using biologically more relevant Mg^{2+} as a cofactor that cleaved the same substrate as the Pb^{2+}-dependent DNAzyme albeit with a reduced rate constant (k_{obs} = 0.01 min^{-1}) [23]. While these early catalysts clearly demonstrated the possibility of generating DNA-based enzymes, their inherent properties limited their biological usefulness. In 1997, Santoro and Joyce selected two DNAzymes (Dz), coined 8–17 and 10–23 (Figure 23.1), which addressed these shortcomings: Dz8–17 and 10–23 are capable of hydrolyzing nearly any type of all-RNA substrates and reach kinetic perfection ($k_{cat}/K_m \sim 10^9$ M^{-1} min^{-1}) under high Mg^{2+} concentrations [28]. Particularly, DNAzyme 10–23 revealed to be capable of cleaving all possible purine–pyrimidine dinucleotide junctions (R·Y in Figure 23.1a with R = A or G; Y = U or C) under simulated physiological conditions but with AU and GU sites being the most prone to hydrolysis [29, 30]. Initially, Dz8–17 was thought to have a more limited substrate repertoire compared to Dz10–23, cleaving only GA dinucleotide junction-containing substrates [24, 28]. However, reselection experiments first allowed to lax the substrate requirements to 14 different dinucleotide junctions [31] while mutation experiments of the catalytic core then revealed Dz8–17 variants capable of hydrolyzing all 16 possible combinations [32]. Interestingly, the 8–17 catalytic motif has been observed in a number of other in vitro selections despite using different experimental conditions [24]. Indeed, incubating randomized libraries under various selection stringencies such as different metal cofactors [33–36], increased reaction temperatures [37], or shorter reaction times [38] all

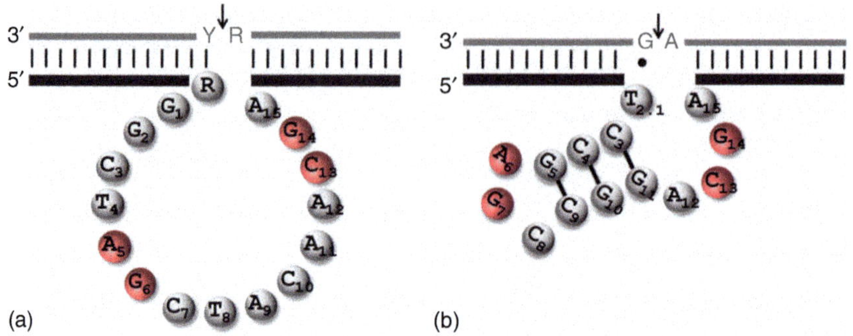

Figure 23.1 Schematic representation of the putative secondary structures of DNAzymes 10–23 (a) and 8–17 (b). Nucleotides in filled red circles are absolutely critical for catalytic activity and are strictly conserved [24–26]. Source: Adapted from Hollenstein [27].

resulted in the identification of 8–17 variants. In addition to infiltrating selection experiments, the 14–15 nt long 8–17 motif has also been found masked within larger sequence contexts that displayed no inherent catalytic activity [38, 39]. The recurrence of the 8–17 motif, a fact known as the tyranny of the small motif [40–42], hints at the possibility that this sequence is a common and efficient answer to DNA-mediated RNA cleavage. Interestingly, both Dz8–17 and 10–23 have only four strictly conserved residues in their respective catalytic cores. In Dz8–17, dA_6, dG_7, dC_{13}, and dG_{14} (Figure 23.1b) play a critical role in catalysis (see Section 23.2.3) and cannot be replaced by any other nucleotide without dramatically impairing function [25, 31, 43, 44]. In addition, nucleotide dC_8 appears to be moderately important for catalytic activity and can be substituted by other nucleotides without inflicting drastic reduction in catalytic rate constants [43], while residue $dT_{2.1}$ is important for maintaining a rG·dT wobble pair with the substrate but can be replaced by other nucleotides if the dinucleotide junction is varied [25]. All other residues constituting the catalytic core are rather variable and can be substituted or deleted without causing an important negative impact on the catalytic activity of Dz8–17. In the case of Dz10–23, dA_5, dG_6, dC_{13}, and dG_{14} (Figure 23.1a) are completely intolerant to modification, while the nucleobases on dG_2 and dT_4 appear to be critically important for the folding of the DNAzyme into its catalytic active species [25, 26, 45]. As for Dz8–17, all other residues of the catalytic core can be mutated or deleted without perturbing the activity too much. Interestingly, based on a systematic mutagenic investigation, Dz10–23 was recently proposed to be a special variation of the Dz8–17 motif [18, 25].

Due to their small size, high catalytic activity, and substrate promiscuity, both Dz8–17 and Dz10–23 have found a broad range of applications [46]. Particularly, Dz10–23 has been used as a gene-silencing agent via its capacity at selectively and specifically hydrolyzing mRNAs. In an early example, the binding arms of Dz10–23 were increased in length (to enhance target specificity) and designed to recognize the AUG translational start site at positions 816–817 in the early growth response gene 1 (*EGR1*) [47]. The Egr-1 factor binds to the promoters of numerous genes and thus acts as an important transcriptional regulator. Particularly, Egr-1 is expressed in artery walls upon injury and thus represents an important target for the development of wound repair strategies. The DNAzyme 10–23 variant coined ED5 selectively recognized and hydrolyzed a stretch of Egr-1 RNA and was capable of inhibiting smooth muscle cell regrowth following a mechanical injury in vitro [47]. Following this initial study, other reports have confirmed the usefulness of Dz10–23 at silencing genes both in vitro [48–50] and in vivo [51–53]. These efforts have recently culminated in the successful completion of the first clinical trials involving DNAzymes for the treatment of skin tumors [10] and asthma [9].

On the other hand, Dz8–17 has not been employed as intensively as Dz10–23 as a gene-silencing agent [54] but has rather been exploited as a potent biosensing device due to its sensitivity to an array of metal cations [55]. Even though Dz8–17 has initially been selected in the absence of any other metal ion than Mg^{2+}, this catalytic DNA molecule has a strong preference for Pb^{2+} as a cofactor displaying an impressive single-turnover rate constant (k_{obs}) of $5.75\,min^{-1}$ [56]. This surprising property was rapidly capitalized through the creation of a potent lead biosensor by simply appending a fluorophore-quencher system on the substrate and the enzyme strands [57]. This initial construct was then hijacked to develop other lead-sensing platforms or as a highly potent signaling system for the detection of other analytes [58–61]. In addition to lead, Dz8–17 can utilize other metal ions as cofactors for activity in the following order of preference: $Pb^{2+} \gg Zn^{2+} \gg Mn^{2+} \approx Co^{2+} > Ni^{2+} > Mg^{2+} \approx Ca^{2+} > Sr^{2+} \approx Ba^{2+}$ [56]. The development of sensing platforms based on Dz8–17 for the selective detection of other metals is hence rather difficult due to the poor selectivity for metals other than Pb^{2+} and Zn^{2+} [37, 62]. This lack of selectivity is further complicated by the inhibition caused by certain metal ions such as Tb^{3+} [63] and the insensitivity to monovalent salts alone, even at high concentrations [64, 65]. Thus, numerous in vitro selection experiments have been designed to isolate metal-specific DNAzymes for sensing purposes [55]. For instance, DNAzymes active with metal cofactors such as Hg^{2+} [66], $UO_2{}^{2+}$ [67], Ag^+ [68], Ln^{3+} [69–72], Cu^{2+} [73], and Na^+ [74] have all recently been isolated and have been or will be integrated into metal sensing systems. Other metal ions, such as Li^+ or Tl^{3+} have eluded detection by DNAzymes due to their weak interaction with DNA and chemical modifications might be necessary in these cases (*vide infra*).

Besides the isolation of metal-dependent RNA-cleaving DNAzymes for the crafting of biosensing platforms, selection experiments have also been conceived to improve the catalytic efficiency of DNAzymes, especially under physiological conditions. Moreover, other selections have been conceived to avoid the recurrence of 8–17-like motifs as well as to broaden the functional pH range and substrate scope of DNAzymes. An impressive example of a potent DNAzyme with improved catalytic properties is the Co^{2+}-dependent DNA molecule DEC22–18, which self-cleaves a substrate containing a single ribo(nucleotide) embedded within a DNA sequence with an impressive single-turnover rate constant of $\sim 10\,min^{-1}$. The self-cleaving (or cis-acting) species could be converted into the corresponding trans-cleaving species DET22–18 that displayed the highest rate constant after Dz10–23 ($k_{cat} = 7.2\,min^{-1}$) and an appreciable catalytic efficiency ($k_{cat}/K_m = 7.7 \times 10^6\,M^{-1}\,min^{-1}$) [75].

Poisoning by the tyranny of the small motif can also be eluded by using L-RNA substrates (the mirror image of the naturally occurring D-RNA) in the selection protocol. In this context, an early example yielded DNAzyme L:15–30 that cleaved a DNA substrate containing a single L-ribo(nucleotide) with an appreciable catalytic efficiency ($k_{cat}/K_m = 4.1 \times 10^5\ M^{-1}\ min^{-1}$) [76]. As expected, there is no sequence homology between L:15–30 and Dz8–17 (and 10–23). A more recent selection campaign yielded DNAzyme LRD-BT1, which efficiently hydrolyzed a substrate containing a single L-ribo(guanosine) moiety but with improved catalytic efficiency ($k_{cat}/K_m = 9.3 \times 10^6\ M^{-1}\ min^{-1}$) [77].

The substrate scope of DNAzymes can be expanded by changing, for instance, the connectivity of the RNA nucleotides within the substrate. Particularly, most natural RNAs possess a 3′–5′-phosphodiester linkage but some bacteria have been shown to have a peculiar 2′–5′-connectivity [78]. In this context, DNAzyme 2′:10–16 cleaves a substrate that contains a single 2′–5′-phosphodiester unit with high catalytic efficiency ($k_{cat}/K_m = 1.7 \times 10^7\ M^{-1}\ min^{-1}$) and 6000-fold regioselectivity over the corresponding 3′–5′-phosphodiester substrate [76]. Similarly, the Ce^{3+}-dependent Ce5 DNAzyme positively discriminates 2′–5′-linked RNA from 3′–5′-linkages and hydrolyzes these substrates with an appreciable rate constant ($k_{obs} = 0.16\ min^{-1}$) [79]. Lastly, selection experiments were carried out at low pH (3–7) in order to develop robust catalysts working under more demanding experimental conditions [18, 80]. Efficient catalysts (k_{obs} values ranging between 0.023 and 1.1 min^{-1}) [80] could be obtained regardless of the pH and were subsequently converted to trans-cleaving species [18].

All the different DNAzymes that were described in this section clearly highlight the potency of DNA to act as a catalyst, especially for hydrolysis of RNA substrates.

DNAzymes Hydrolyzing DNA Phosphodiester Linkages and Other Bonds

In the first reported in vitro selection experiment, an artificial ribozyme was evolved to hydrolyze a DNA-substrate [5]. Also, the conversion of the *Tetrahymena* ribozyme from an RNA-cleaving to a DNA-cleaving enzyme was achieved by Darwinian evolution [81]. Surprisingly, the identification of DNA-mediated DNA hydrolysis eluded scientists until the serendipitous isolation of DNAzyme 10MD5 [82]. Initially, Silverman et al. strived to select for a DNAzyme capable of hydrolyzing amide bonds and consequently used a substrate containing a tripeptide fragment embedded in a DNA strand. The resulting catalytic molecules did not hydrolyze the more labile amide linkages but rather cleaved the more robust phosphodiester moieties ($t_{1/2}$ for the uncatalyzed reactions are ~500 years for peptide bonds and range between 140 000 and 30 million years for DNA [27]) downstream of the peptidic fragment. The most active species, 10MD5, hydrolyzes all-DNA substrates with an appreciable rate constant ($k_{obs} = 0.045\ min^{-1}$), which equates to an impressive 10^{12}-fold improvement compared to the uncatalyzed reaction [82]. A reselection experiment allowed to relax both the strong pH and the metal cofactor requirements of 10MD5 since the resulting DNAzyme 9NL27 cleaved its DNA substrate in a 0.7 pH window and required only Zn^{2+} as a cofactor [83, 84]. Yet another selection experiment allowed increasing the limited substrate sequence tolerance of Dz10MD5, which only cleaved substrates containing the four-nucleotide ATG^T cleavage site to four DNAzymes capable of hydrolyzing any substrate containing an N^G dinucleotide site [85]. Interestingly, the nature of the metal cofactor could be varied from Zn^{2+} for DNAzyme 9NL27 to lanthanide ions (Ce^{3+}, Eu^{3+}, or Yb^{3+}) for DNAzymes obtained in another in vitro selection experiment without altering the catalytic efficiency (k_{obs} values ranging from 5×10^{-4} to $0.024\,min^{-1}$) [86].

Another family of DNA-cleaving DNAzymes has been identified by Breaker and coworkers, who used a circular DNA library comprising two N50 regions during the SELEX protocol, which simultaneously allowed cleavage at any position of the random-sequence domain and precluded other cleavage mechanisms (such as oxidation or depurination) from occurring [11]. The identified DNAzyme, I-R3, hydrolyzes its DNA substrate at a single location with an impressive efficiency ($k_{obs} \sim 1\,min^{-1}$ and $1.6\,min^{-1}$ at 37 °C and 45 °C, respectively), albeit in a rather narrow pH window ($\Delta pH \sim 0.3$). A bimolecular version of I-R3 was shown not only to be capable of multiple turnover but also of hydrolyzing large genomic DNA substrates. In a subsequent study, the catalytic properties of all the sequences of I-R3 containing single (i.e. 45), double (i.e. 945), and triple (i.e. 12285) mutants were inspected using deep sequencing. This analysis showed that eight out of the 15 mutated nucleotides considered in the study were required for catalysis, while five other sites appeared to be quite tolerant to single mutations and the remaining two nucleotides accept certain substitutions. A reasonable catalytic activity was retained in only 11% of the species with double mutations and further shrank to 1.1% for constructs containing triple modifications. This analysis clearly demonstrated a very narrow and defined sequence space that provides an answer to efficient DNA-mediated DNA cleavage [87].

DNA-cleaving DNAzymes are alluring tools for the development of biosensing platforms since they are more robust than their RNA-cleaving counterparts and thus lead to a lower background signal. Despite their rather recent discovery, these DNA-based catalysts have already been integrated into a number of practical sensing applications: For instance, DzI-R3 played a critical role in the sensitive detection of Zn^{2+} by a DNA supersandwich based nanopore detection system [88]. DzI-R3 was also used in conjunction with ZnO nanoparticles for the detection of Zn^{2+} [89], with TiS_2 nanosheets to develop a sensor for various biomolecules [90], with the rolling circle amplification strategy for the production of DNA size markers for gel electrophoresis applications [91], with hydrogels for the controlled release of polymers and proteins [92], with cofactor DNA strands for the construction of logic gate operations [93], and with origami structures to visualize the Zn^{2+}-dependent activity of single molecules [94]. Likewise, DNAzyme 9NL27 was engaged in a PCR-based method for the detection of Zn^{2+} [95].

Isolation of catalysts capable of hydrolyzing amide linkages or P—O bonds of amino acid side chains could result in the creation of artificial DNA-based proteases and peptidases for the selective alteration of proteins [96]. In this context, an in vitro selection experiment was carried out with DNA substrates that contained embedded ester or amide bonds to identify DNA catalysts that could promote the hydrolysis of such carbonyl linkages [97]. While DNAzymes capable of hydrolyzing ester bonds and aromatic amide (anilide) moieties could be isolated, albeit displaying modest efficiencies (k_{obs} = 0.05 and 3.5×10^{-3} min^{-1}, respectively), evolution experiments failed to yield DNA molecules capable of catalyzing the scission of aliphatic amides. This observation coincides with the results of the selection of Dz10MD5 and of early attempts at isolating amide cleaving ribozymes [98], suggesting that accelerating this reaction is a rather daunting task for unmodified nucleic acids and requires the presence of additional chemical groups (see Section 23.3.3).

The main challenge involved in the evolution of nucleic acid catalysts modifying amino acids or peptides is to devise constructs and selection protocols that compensate for the absence of Watson–Crick binding interactions between substrate and enzyme. The Silverman laboratory first developed a selection scheme that avoided direct binding of the nucleic acid scaffold with the amino acid substrate. This selection method enabled the selection of DNAzymes capable of forming nucleopeptide linkages on single amino acid [99] and tripeptidic substrates [100]. This approach was then further expanded to provide DNAzymes modifying free peptide substrates [101].

In order to expand the catalytic repertoire of DNAzymes to peptidic substrates, the Silverman laboratory devised an ingenious in vitro selection scheme (Figure 23.2) [102]. The hexapeptide substrate containing an internal phosphotyrosine moiety was affixed on a DNA anchor directly connected to the randomized library. After the dephosphorylation step, the DNA anchor was hybridized to a DNAzyme that requires a tyrosine substrate to form a nucleopeptide linkage. This DNAzyme-catalyzed reaction produced a PAGE shift, which ultimately allowed for the separation of active from inactive molecules. The application of this selection protocol enabled the identification of a DNAzyme that accelerated the hydrolysis of the monophosphoester bond located on the side chains of tyrosine and serine of peptidic substrates. More in particular, DNAzyme 14WM9 dephosphorylated tyrosine (k_{obs} = 0.19 min^{-1}) much more efficiently than serine (k_{obs} = 5.2×10^{-3} min^{-1}) and displayed multiple turnover catalysis with untethered substrates. Interestingly, substantial phosphatase activity was observed upon incubating a fragment of the protein prochlorosin ProcA2.8 equipped with the hexapeptide substrate CAAY^{P}AA and DNAzyme 14WM9. This clearly highlights the potential of nucleic acids to both bind to protein substrates and substantially accelerate the rate constant of the dephosphorylation reaction. A similar selection scheme also enabled the identification of DNAzymes capable of catalyzing the reverse reaction, i.e. DNAzymes displaying kinase activity [103].

23.2.2 DNAzymes with Ligase and Other Activities

Besides hydrolyzing various bonds (particularly ribophosphodiesters), DNAzymes are perfectly adept at performing other more complex catalytic functions. Among the various chemical transformations, the reverse reaction (ligation) has attracted considerable attention. In principle, RNA-cleaving DNAzymes should be capable of catalyzing ligation reactions since all the different steps of the kinetic schemes of these enzymes are governed by equilibria [104]. However, DNAzymes hydrolyzing

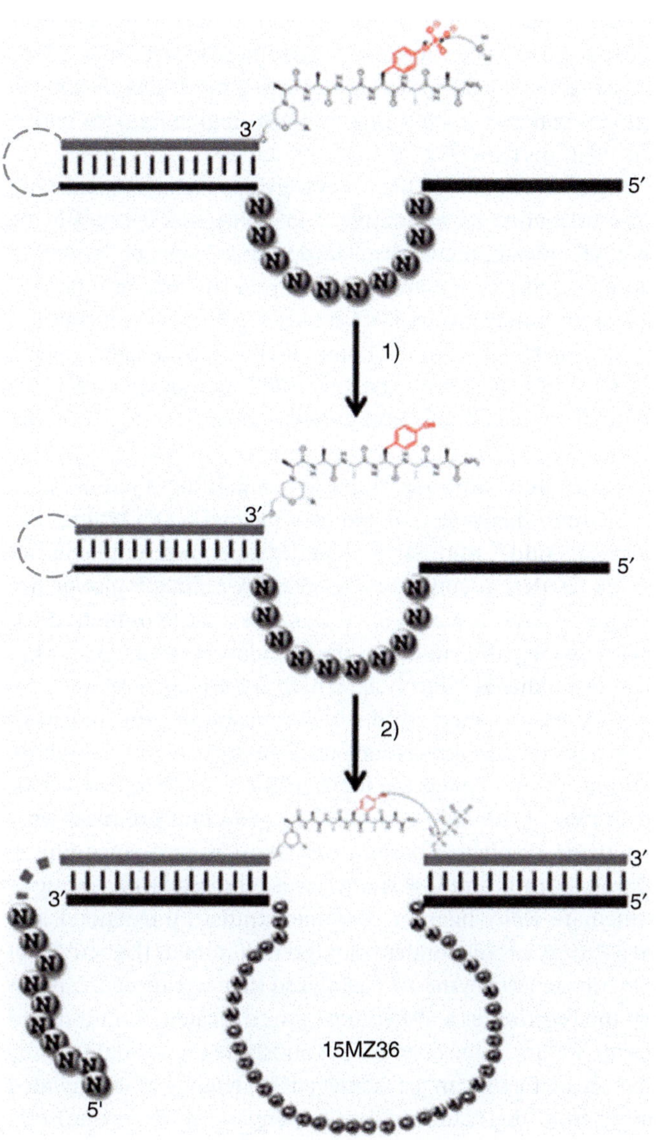

Figure 23.2 Selection scheme for the isolation of DNAzymes with phosphatase activity: (1) in the selection step, the randomized library catalyzes the removal of the phosphate moiety on tyrosine; (2) the resulting product is then incubated with a 5′-triphosphate containing RNA substrate and DNAzyme 15MZ36 [101], which accelerates the addition of the hydroxyl group of tyrosine on the triphosphate moiety. This capture step allows the separation of active from inactive species by gel electrophoresis. Source: Adapted from Chandrasekar and Silverman [102].

RNA substrates have a strong and clear preference for the cleavage over the ligation reaction: Dz10–23 hydrolyzes its RNA target with a rate constant of 0.18 min^{-1} under simulated physiological conditions while the rate of the reverse reaction is only ~4×10^{-4} min^{-1} [29]. Thus, various selection experiments were designed to identify new DNAzymes capable of ligating DNA and RNA substrates. However, the generation of DNAzymes capable of synthesizing RNA fragments with native 3′–5′-linkages is a nontrivial undertaking. Indeed, when two RNA fragments containing a 2′,3′-cyclic phosphate and a 5′-OH group were used as substrates, only DNAzymes catalyzing the formation of nonnative 2′–5′ RNA linkages could be isolated [105]. When the cyclic phosphate was replaced with a 2′-OH moiety of an internal rather than 3′-terminal nucleotide and the 5′-OH on the second fragment was substituted with a 5′-triphosphate group, DNAzymes catalyzing the formation of 2′–5′-linked branched and lariat RNAs were obtained [106]. DNAzymes capable of forming native 3′–5′-RNA linkages could be identified, but only after more stringent elements were included in the selection protocol. A first strategy coerced both RNA fragments (containing a 2′,3′-diol and a 5′-triphosphate) to base-pair to the guide arms since under these conditions, duplexes containing 3′–5′-linkages are thermodynamically favored. This selection protocol yielded a DNAzyme that efficiently ligated RNA ($k_{obs} = 8 \times 10^{-3}$ min^{-1}), albeit at the expense of a rather strict sequence requirement. This restriction could be alleviated by redirecting selection experiments toward the exclusive formation of 3′–5′-linkages instead of forcing the formation of RNA–DNA duplexes [107]. To do so, the population of each selection step is subjected to a Dz8–17-mediated hydrolysis, which selectively cleaves 3′–5′- over 2′–5′-linkages. The cleaved (shorter) products can easily be resolved from the longer 2′–5′-RNAs by gel electrophoresis and used in subsequent rounds of selection. Such a selection protocol enabled the identification of DNAzyme 9DB1, which ligates RNA with a rate of 0.04 min^{-1} in 60–70% yield and produces only 3′–5′-linkages. Moreover, 9DB1 displays a broad substrate tolerance since only a D↓RA sequence (with D = A, G, or U and R = A or G) was required [108].

DNAzymes can also mediate DNA ligation as demonstrated by Cuenoud and Szostak, who engineered a Zn^{2+}/Cu^{2+}-dependent catalyst that mimics the activity of the T4 DNA ligase. The resulting DNAzyme E47 catalyzed the formation of a phosphodiester bond through the reaction of a 5′-hydroxyl group located on one strand with a phosphorimidazolide unit on the second DNA substrate (k_{cat} = 0.07 min^{-1}) [109]. Alternatively, ligation of DNA can be achieved by feeding 5′-adenylated DNA fragments (stemming from the action of a self-adenylating DNAzyme) to a self-ligating DNA enzyme [110]. Besides ligating DNA and RNA, DNAzymes are also capable of covalently connecting the side-chains of peptides to RNA and DNA sequences (as discussed in Section 23.2.1).

The development of catalysts for organic synthesis in aqueous medium has an obvious and immediate positive impact on the environment. It concomitantly would simplify experimental processes and provide milder reaction conditions. Most approaches toward this aim focus on the generation of small molecules (mainly organocatalysts) or metal complexes [111] and nucleic acid-based catalysts would obviously be advantageous. In this context, DNAzymes capable of catalyzing the formation of C–C bonds have also been identified. A notable example is DNAzyme DAB22, which catalyzes the Diels–Alder reaction of a maleimide containing dienophile and an anthracene-based diene [112]. The Ca^{2+}-dependent DAB22 catalyzes the [4+2] cycloaddition with second-order rate constants comparable to that of a related ribozyme [113], suggesting that both types of nucleic acids are equally competent at accelerating specific chemical transformations. Another synthetically relevant reaction catalyzed by DNAzymes is the Friedel–Crafts alkylation, where C—C bonds are formed by the interaction between aromatic π nucleophiles and activated alkene or ketone electrophiles. The Cu^{2+}-dependent DNAzyme M14 catalyzed the Friedel–Crafts alkylation both as a cis- and trans-acting species rather efficiently since yields of over 70% could be obtained after 24 hours [114]. The advantage of M14 over other DNA-based catalytic systems is that no external ligand needs to be added [111]. So far, the chemical repertoire of DNAzymes in terms of C—C bond-forming reactions (and other organic reactions) is limited to the Diels–Alder and Friedel–Crafts alkylation, but precedents in ribozymes raise hope that other DNA catalysts will be identified soon.

Catalytically competent DNA molecules have also been identified for chemical transformations that markedly differ from bond forming and breaking reactions notably the peroxidation of organic substrates and the thymine dimer photoreversion. The peroxidase-mimicking DNAzymes are a well-studied class of G-quadruplex DNA-based systems that bind to the cofactor hemin and promote the oxidation of ABTS [2,2′-azinobis(3-ethylbenzthiazoline-6-sulfonic acid] to the corresponding radical cation $ABTS^{\bullet+}$ in the presence of H_2O_2. This oxidation is accompanied by a color change, and this optical signal has been hijacked for an expansive array of sensing applications and detection purposes [115, 116]. The first described DNAzyme with peroxidase activity, DzPS5.M, catalyzes the oxidation of ABTS with an activity two orders of magnitude higher than that of free hemin [117, 118]. Truncation of DzPS5.M (from 24 down to 18 nucleotides) by rational design led to DzPS2.M, which is even more active than the parent sequence for the 1-electron peroxidation reaction ($k_{cat}/K_m = 1.02\times10^4\ s^{-1}\ M^{-1}$) [119, 120]. Both DzPS5.M and PS2.M bind to hemin, which induces a structural transition from antiparallel to parallel/mixed type G-quadruplex structures leading to the catalytically competent species [121].

A related G-quadruplex based DNAzyme is capable of harnessing UV-light to reverse the formation of cyclobutane pyrimidine dimers (CPDs) [122]. CPDs are among the most common DNA lesions and arise through a UV-light induced [2+2] cycloaddition between adjacent thymidine bases. In cells, CPDs are usually removed by CPD-photolyases, which mainly use flavin adenine dinucleotide (FAD) based cofactors to reverse the reaction. Hence, Chinnapen and Sen devised a selection experiment to identify serotonin-dependent DNAzymes that could imitate the action of natural photolyases. The selection experiment yielded two distinct classes of DNAzymes that either required the tryptophan metabolite serotonin for activity or that catalyzed the photoreversion reaction in the absence of the prosthetic cofactor [8]. The most active member of the first class of DNAzymes accelerated the repair of a CPD with a high second-order rate constant ($k_{cat}/K_m = 1\times10^4\ min^{-1}\ M^{-1}$) and uses the cofactor serotonin for catalysis rather than for structural or conformational purposes [123]. On the other hand, DNAzymes stemming from the second category adopt G-quadruplex structures, which serve as light antennae as well as electron sources [124]. The most active DNAzyme of this category repairs CPD lesions with an impressive catalytic efficiency ($k_{cat}/K_m = 7.8 \times 10^6\ min^{-1}\ M^{-1}$) [8]. This highly efficient DNAzyme was converted into a catalyst that uses less damaging higher wavelength light (345 nm versus 240 nm) by substitution of a guanine nucleotide of the catalytic core with the fluorescent analog 6-methylisoxanthopterin (6MI) [125].

23.2.3 Structural and Mechanistic Considerations

Most mechanistic and structural information of DNAzymes stems from mutagenic and deletion studies along with FRET and related biochemical investigations [27]. However, two recent crystal structures have brought more detailed insight into how DNAzymes catalyze reactions. The first ever reported crystal structure is that of the RNA ligating DNAzyme 9DB1 caught in a post-catalytic state (Figure 23.3a) [126]. The catalytic core of 9DB1 is quite compact and consists of two stacks of base pairs (regions P2 and P3). A total of five paired regions can be observed (Figure 23.3a,c): two regions of 9DB1 bind to the substrate via the guide arms (P1 and P4); section P2, which contains one dG–dC Watson–Crick base pair and a dT·dG wobble pair; region P3 which contains four dG–dC Watson–Crick base pairs; and base pairing between the rA and rG of the substrate at the ligation site with two thymine nucleotides. The compact nature of the catalytic core results from not only base pairing but also extensive tertiary and stacking interactions. Importantly, the two nucleotides of the ligation junction (A_{-1} and G_1) are stacked onto each other and maintained in this position through additional stacking interactions with dA_{15} and dG_{27} (in region P2) and the aforementioned base pairs with the two dT nucleotides 29 and 30 (Figure 23.3c). These different interactions result in the formation of a duplex-like structure at the level of the ligation site. This crystal structure also gives insight into the regioselectivity (3′–5′ vs. 2′–5′) of the reaction. Indeed, the 2′-OH group of donor G_1 is maintained in an adequate geometry through hydrogen bonding with the minor groove of the dC_{12}–dG_{26} base pair (in region P2), while the formation of the P1 and P4 regions reduces the distance between the phosphate groups of A_{-1} and G_1. This combination of interactions and the involvement of the 2′-OH of A_{-1} in a hydrogen bond with the nucleobase of G_1 are believed to account for the observed regioselectivity of 9DB1. In terms of mechanism, the phosphate group of dA_{13} in region P2 lies in close proximity (i.e. 3.1 Å) from the phosphate of G_1 of the ligation site. Phosphorothioate replacement experiments on both non-bridging oxygen atoms clearly demonstrated the importance and implication of the S_p stereoisomer in the activation of the attacking 3′-OH residue (Figure 23.4a). Surprisingly, no electron density could be observed for metal ions despite the clear and important Mg^{2+}-dependence of 9DB1 [108]. Hence, the crystal structure of 9DB1 gives detailed structural information on the compact nature of the catalytic core as well as an insight into the mechanism of 9DB1, but the implication and importance of metal cofactors could not be elucidated.

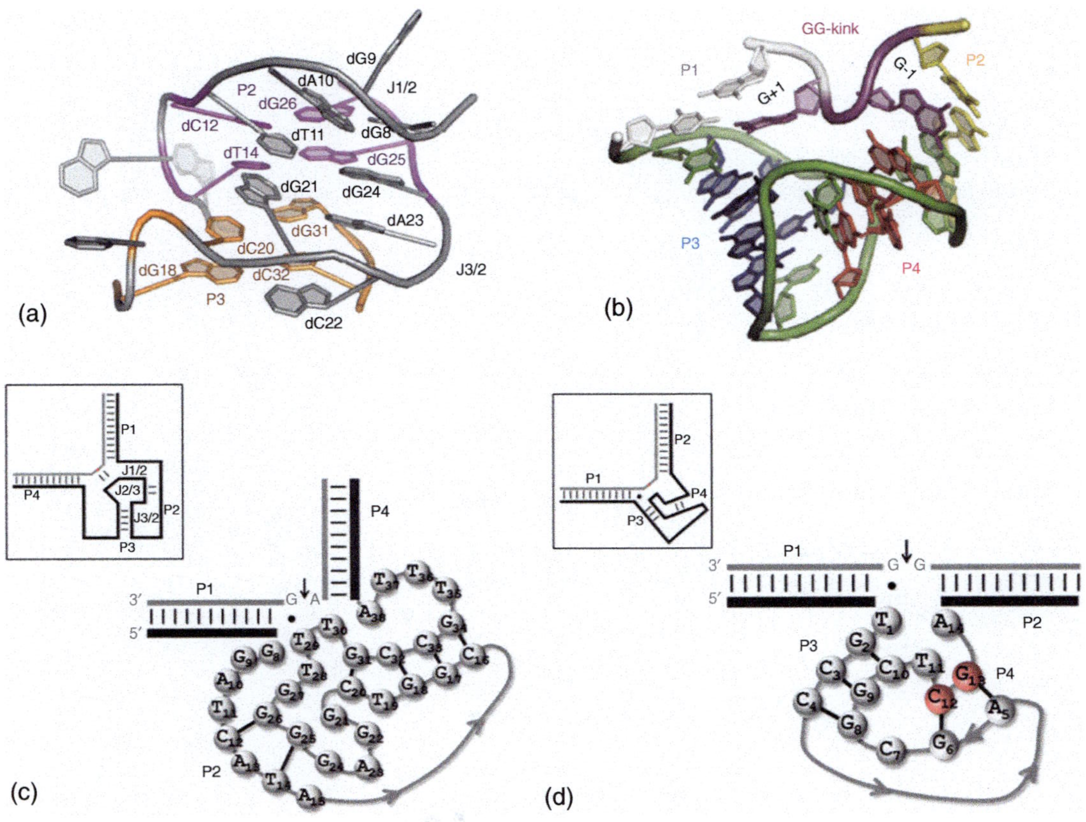

Figure 23.3 (a) Close-up view on the structure of the catalytic core of the RNA-ligating DNAzyme 9DB1. (b) Close-up view of the catalytic core of the RNA-cleaving DNAzyme 8–17. (c) and (d) Represent the sequences and observed structures of the respective DNAzymes. Source: Ponce-Salvatierra et al. [126]. © 2016 Springer Nature. Reproduced with permission of Springer Nature. (b) Liu et al. [127]. © 2017 Springer Nature. Reproduced with permission of Springer Nature.

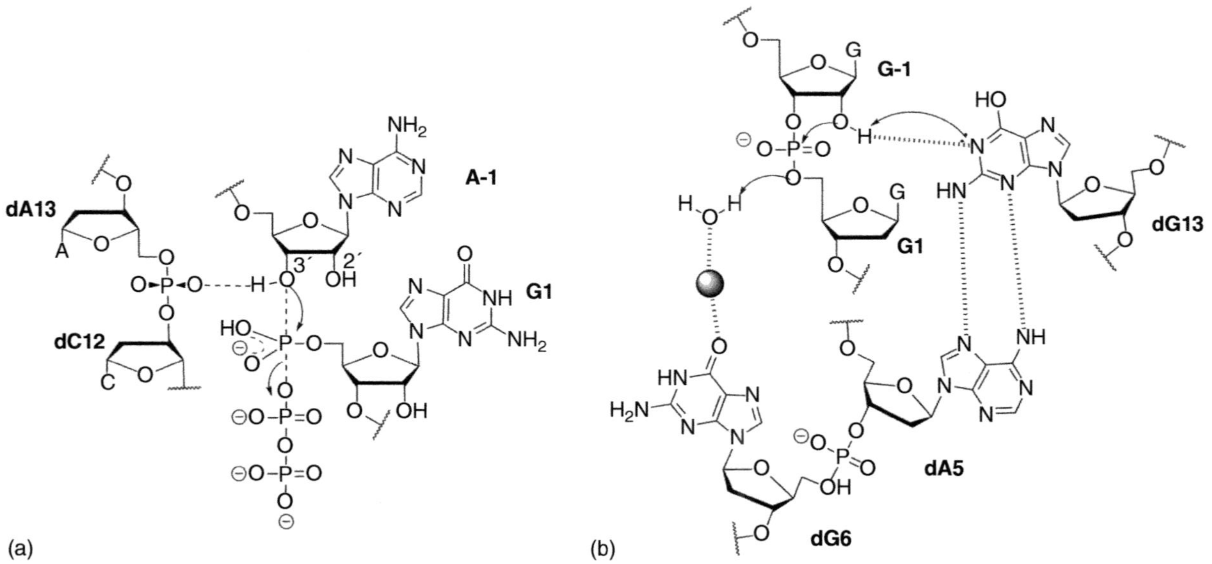

Figure 23.4 Schematic representations of the putative chemical mechanisms of (a) the RNA-ligating DNAzyme 9DB1 and (b) the RNA-cleaving DNAzyme 8–17.

The crystal structure of Dz8–17 also reveals a compact active site with four base-paired regions (Figure 23.3b). In order to elucidate the crystal structure of Dz8–17, Liu et al. equipped the substrate with a 2′OMe-G nucleotide to prevent cleavage from occurring and utilized the African swine fever virus DNA polymerase X (*Asfv*PolX) to facilitate the crystallization and molecular replacement processes. Besides capturing Dz8–17 in the presence of the cofactor Pb^{2+}, Liu et al. also obtained crystal structures of the DNAzyme-substrate and Dz8–17-native DNA complexes both in the absence of Pb^{2+}. This collection of crystal structures give a picture of the different steps in the catalytic progression of Dz8–17. As for 9DB1, four different paired regions were observed: (i) P1 and P2, which connect the substrate to the guide arms; (ii) the short duplexes P3 and P4 (three and two base pairs, respectively) that maintain the catalytic core in a pseudoknot structure (Figure 23.3d). Region P3 has been predicted previously by mutagenesis studies [28, 29, 43], while P4 that includes the critically conserved residues dC_{13} and dG_{14} was not known. As for 9DB1, the nucleotides of the cleavage junction of the substrate of Dz8–17 are maintained in a compact G–G kink through the formation of an rG_1:dT_1 wobble pair and a non-canonical rG_{-1}–dA_{14} pair. Stacking interactions are also important to maintain the configuration of P3 and P4, particularly through the interactions of the rG_1:dT_1 wobble pair with the rC_{-1}:rG_{+2} pair of the substrate (region P1) and that of the rG_{-1}–dA_{14} pair with the dG_{13}-dA_5 pair of P4. The importance of dA_{14} had been noted previously in a mutational study where the *N*3 unit was substituted by a carbon atom and either hydration or pK_a perturbations were believed to be implicated [45]. The rG_1:dT_1 wobble pair has also been predicted by mutation studies [25, 43] and photo-cross-linking assays [128]. The differences between the three crystal structures reveal that the DNAzyme does not undergo global folding upon metal binding and that metal recognition occurs through a pre-organized binding pocket as previously predicted by FRET studies [129]. In terms of mechanism, Pb^{2+} coordinates to O6 of dG_6 (involved in Watson–Crick base pairing with the critically conserved dC_{12}) and to a water molecule. This Pb^{2+}–H_2O coordination is believed to reduce the pK_a of the water molecule and hence suggests its role as a general acid to offer a proton to the O5′ of the leaving group. In addition, nucleotide dG_{13}, which has been shown to be strictly required for catalysis [25], interacts with the 2′-OH of rG_{-1}, suggesting its plausible role as a general base for the deprotonation of this unit for its subsequent in-line attack on the phosphate moiety (Figure 23.4b). Thus, this crystal structure nicely confirms previous studies which had shown the direct implications of the *N*7 of dA_5, the keto group O6 of dG_6, and the *N*1 unit of dG_{13} in the catalytic activity [24, 25, 43]. Both crystal structures will certainly help to identify positions for chemical modifications in order to improve the properties of both DNAzymes.

23.3 Chemical Modifications as Rescue and Expansion of Catalytic Activity

Since their advent in 1994, DNAzymes have evolved from a scientific curiosity to important players in the field of biocatalysis. This importance is reflected by successful clinical trials, their capacity at recognizing non-nucleosidic substrates, and the multitude of reactions described in Section 23.2. However, despite these staggering progressess, DNAzymes still suffer from a number of shortcomings, which sometimes reveal to be severe predicaments for their in vivo applications. In this section, the current limitations of DNAzymes are discussed and a possible solution to overcome some of these shortcomings will be presented, namely the inclusion of chemical modifications into the nucleosidic backbone. Particularly, chemical functionalities can be introduced either *after* the SELEX experiments by solid-phase synthesis or *during* the selection protocol via the polymerization of modified nucleoside triphosphates (dN*TPs).

23.3.1 Challenges of DNAzymes for Practical Applications

Some of the challenges that DNAzymes meet are not necessarily specific to this class of nucleic acids but also affect other types of functional and/or therapeutic oligonucleotides. Indeed, DNAzymes, as well as antisense oligonucleotides (ASO), ribozymes, aptamers, or other nucleic acids, are rather reluctant at crossing biological membranes, which results in poor cellular uptake or crossing of the blood–brain-barrier (BBB). Obviously, this limited cellular internalization imposes a severe impediment for an efficient gene silencing activity mediated by DNAzymes. Similarly, the catalytic activity of DNAzymes in vivo is often negatively impacted by changing from a rather controlled and simple buffered solvent system used during the in vitro selection protocols to very complex and crowded conditions inside cells. Also, proteins or other cellular components can bind to DNAzymes and thus restrict their binding to their intended substrates and/or reduce their catalytic potency [130]. Working under in vivo conditions also severely limits the nature and especially the concentration of the metal ions available to DNAzymes to mediate their catalytic activity. This can have profound consequences as exemplified by the case of Dz10–23: under in vitro conditions with 10–25 mM Mg^{2+} concentrations, Dz10–23 nearly reaches catalytic perfection [28]; however, under physiological conditions (i.e. $[Mg^{2+}] \sim 0.2$ mM) [131], the same DNAzyme displays rate constants and catalytic efficiencies that are reduced by several orders of magnitude [132–134]. The strong decrease of in vivo activity of DNAzymes might also be connected to a local depletion of $[Mg^{2+}]$ caused by strong binding of Mg^{2+} to free ATP [135]. This drastic drop in catalytic efficiency under physiological conditions also raises the question of whether the observed in vivo gene silencing activity of DNAzymes is really caused by a true hydrolytic event of the expected mRNA substrate or rather by a steric blockade mechanism related to that of ASO. However, this question remains controversial since conflicting reports have demonstrated that both *modus operandi* could be effective. For instance, both Dz8–17 and 10–23 were evaluated and compared for their capacity at hydrolyzing mRNA substrates under low M^{2+} concentrations. Both DNAzymes performed rather poorly under low $[Mg^{2+}]$ conditions, hinting at a possible antisense effect. On the other hand, Dz8–17 displayed a strong catalytic activity even in the presence of physiological $[Zn^{2+}]$ [134]. Intriguingly, in the course of the construction of a DNAzyme-based light-controlled gene regulation system (see Section 23.3.2), Deiters et al. observed that both the uncaged and mutated versions of Dz10–23 led to the same extent of gene silencing activity as the unmodified 10–23 control, thus clearly suggesting an antisense mechanism [136]. On the other hand, Mg^{2+}- and Zn^{2+}-dependent DNAzymes were used to modulate telomerase activity in cells via the degradation of two antagonistic transcription factors and this study clearly demonstrated that the observed effect resulted from the cleavage of the RNA substrates rather than by an antisense effect [137]. This conclusion was also comforted by modified versions of Dz10–23, which catalytically hydrolyzed integrin alpha-4 mRNA in human primary fibroblasts; an antisense effect was observed only when two critically conserved residues were mutated [138].

Another issue – inherent to all unmodified oligonucleotides – is their rapid nuclease-mediated degradation in vivo.[14] Indeed, unmodified DNA oligonucleotides generally survive less than an hour in the presence of nucleases [139–142], and their RNA counterparts are even more labile with half-lives not exceeding minutes [143, 144]. This high sensitivity to nuclease digestion is a severe impediment regardless of the in vivo application; clearly, DNAzymes need chemical modifications to survive degradation and be active under these conditions. Related to this, oligonucleotides typically have rather low molecular weights and thus are subject to rapid renal clearance since the kidneys eliminate molecules with a molecular weight <50 kDa [145]. This efficient renal clearance also needs to be overcome in order to improve the general in vivo utility of DNAzymes.

Protein enzymes can resort to a wealth of functional groups present on the side chains of the 20 natural amino acids to perform catalysis. Compared to their proteinaceous counterparts, nucleic acids catalysts have a depleted chemical arsenal consisting only of aromatic moieties, weakly nucleophilic exocyclic amines, and a free hydroxyl residue in RNA oligonucleotides. This low diversity of functional groups combined with a highly negatively charged and a rather flexible backbone structure have been thought to limit the capacity of RNA [146] and DNA [13] to catalyze reactions. While nucleic acids display a surprisingly large catalytic repertoire, certain reactions such as amide (*vide supra*) or C—C bond hydrolysis still represent arduous and demanding transformations for which no unmodified DNAzymes and ribozymes have been identified. The lack of functional group, diversity is also reflected by the fact that DNAzymes, with the exception of 8–17 and 10–23, rarely achieve catalytic perfection and that the main (positive) contribution to the catalytic efficiency essentially stems from K_m values rather than from high rate constants. Indeed, under optimal conditions (50 mM Mg^{2+}), Dz10–23 displays a k_{cat} value of 3.4 min^{-1} (and a K_m of 7.6×10^{-10} M) [28] while maximal catalytic rates of >10 s^{-1} are regularly observed for protein catalysts under in vivo conditions [147].

Consequently, the introduction of chemical modifications can alleviate some (or all) of the above-mentioned limitations.

23.3.2 Post-SELEX Modification of DNAzymes

After the selection protocol, chemical alterations can be brought to the scaffold of DNAzymes by standard solid-phase DNA synthesis using suitably modified phosphoramidite building blocks. The modifications can be introduced either at the level of the substrate-binding arms or to the catalytic core depending on the intended aim: DNAzymes can be modified in a post-SELEX manner either to improve the catalytic efficiency or to enhance the nuclease resistance and/or cell penetration capacity.

Improvement of the Catalytic Properties

Improving the catalytic efficiency through the introduction of functional groups into the scaffold of DNAzymes is an alluring yet very difficult and delicate undertaking since minute alterations of the chemical and structural environment of the catalytic core can lead to drastic reductions in function and activity. Hence, numerous mutants containing one or several modifications need to be synthesized, and their kinetic properties carefully evaluated. In an early example, amino-acid-like modifications were introduced on the nucleobases of non–critically conserved dU and dA nucleotides (on the C5 and N6 positions, respectively) of Dz10–23. The activity of the resulting mutants was then tested and this screen revealed that most modified constructs led to a decrease in catalytic efficiency compared to wild-type Dz10–23. However, two mutants containing modified dU nucleotides (**1** and **2** in Figure 23.5a) displayed rate constants that were much improved (~20–30 fold) compared to the unmodified DNAzyme and could even promote the M^{2+}-independent cleavage of the RNA substrate, albeit with low efficiency [148]. Modification of dA nucleotides with amino acid-like residues was the subject of various campaigns striving at improving the catalytic efficiency of Dz10–23 [149–152]. However, the rate enhancement observed for successful mutants (mainly based on a 7-deaza-8-aza-scaffold) remained relatively modest (~10-fold). The group of Keliang Liu followed a similar strategy but appended the modifications at the level of the N7 atom of dG nucleotides of the catalytic core of 10–23. Indeed, the inclusion of functional groups at this position of dG nucleobases was deemed to interfere less with potential π-stacking and Watson–Crick pairing interactions in which these nucleotides might be involved. A systemic study revealed that (i) removal of the N7 atom of the dG nucleobases was highly detrimental for catalytic activity; (ii) a shift of the nitrogen atom from position 7 to position 8 of the five-membered ring was tolerated for certain dG nucleotides; (iii) a combination of a shift of the nitrogen atom to position 8 of the nucleobase and the introduction of a cationic amine (**3** in Figure 23.5a) seems to be well tolerated and leads to a positive enhancement of the rate constant (~40-fold) [153].

Enhancement of Pharmacokinetic Properties

The bulk of efforts to modulate the chemical nature of the scaffold of DNAzymes mainly focused on enhancing their resistance against nuclease degradation without inducing a negative impact on their activity. To do so, the constituting nucleotides of the substrate recognition arms, and to a lesser extent, of the catalytic core of DNAzymes are replaced with modifications known to reduce the rate of degradation by nucleases (Figure 23.5b). The resulting mutants are then tested to evaluate the effect of the modifications on both the rate constants and the resistance against

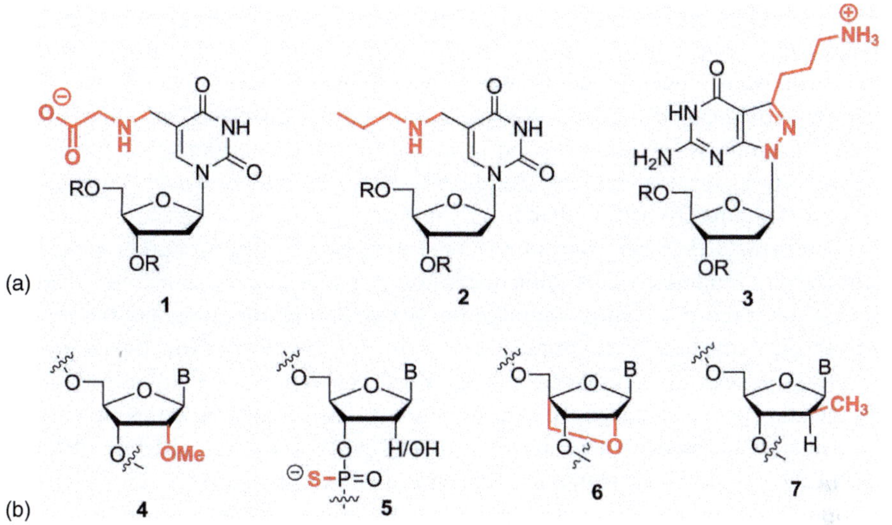

Figure 23.5 (a) Chemical structures of base modified nucleosides introduced in the scaffold of DNAzymes to enhance their catalytic activity; (b) representative sugar modifications used to improve the resistance against nuclease degradation of DNAzymes.

nucleases. In this context, a thorough screen of different modifications including 2′-methoxy-RNA (2′-OMe **4**), phosphorothioate-DNA **5**, Locked Nucleic Acid (LNA **6**), and 3′-inverted thymidine moieties enabled to identify a number of Dz10–23 constructs that displayed both an increase in catalytic activity (under multiple turnover conditions) and reduced nuclease degradation [48]. Particularly, the RNA mimics **4** and **6** appear to convey a more A-DNA-like nature to the guide arms and hence a higher substrate recognition, which in turn is believed to be responsible for the increased rate constants. However, a fine balance between high substrate binding and efficient release of the cleavage products needs to be met: an increase of both the length of the binding arms and the modification composition beyond a certain threshold might result in a slower product release and, thus, in reduced catalytic efficiency. Interestingly, 2′-OMe modifications were rather well tolerated in the catalytic core of Dz10–23, particularly at positions dC_7 and dT_8 (Figure 23.1a), suggesting that these sites are also amenable to chemical alteration. Similar modification campaigns of the 10–23 scaffold have resulted in nuclease-resistant constructs capable of hydrolyzing a number of RNA substrates including integrin alpha-4 (ITGA4) [138], insulin-like growth factor I (IGF-I) [52], miRNA in zebrafish embryos [154], hepatitis C virus (HCV) [50, 155], *Escherichia coli* 23S ribosomal RNA [156], human miRNAs hsa-miR-372 and hsa-miR-373 [157], and E6 mRNA from human papillomavirus [54, 158]. As shown in this section, most synthetic modification methods of DNAzymes to improve their nuclease resistance rely on alterations of the sugar–phosphate backbone. In a rare example involving nucleobase-modifications, Perrin et al. have incorporated guanidinium-bearing deoxyuridine nucleotides into the guide arms of Dz10–23 to enhance the stability of the resulting DNAzyme–RNA duplexes through a reduction of the global negative charge. The presence of the guanidinylated nucleobases did not preclude catalysis even though the rate constants were usually lower than for the wild-type DNAzyme [159].

An efficient gene-silencing DNAzyme needs to be capable of traveling from the site of administration directly into cells and there locate its mRNA target to accelerate its hydrolysis. At each step of this rather lengthy journey, DNAzymes might be degraded or eliminated from the body. Thus the presence of chemical modifications that help bypassing these hurdles is of high importance. The aforementioned altered sugar nucleotides are popular modifications since they enhance the resistance against degradation by circulating nucleases. On the other hand, the introduction of chemical modifications at the 5′- and/or 3′-termini of DNAzymes permit an improvement of their cell penetration capacity or to retard renal clearance. For instance, the introduction of cholesterol moieties improves the cellular delivery of short interfering RNAs (siRNAs) [160] and microRNAs (miRNAs) [161] either through a receptor-mediated mechanism or by an increased membrane permeability [162]. Similarly, a cholesterol–triethylene glycol unit was appended on the 3′-end of a modified version of Dz10–23 designed to cleave a specific RNA sequence of the epidermal growth factor receptor (EGFR). The resulting construct displayed an improved cell permeability compared to the unmodified system without affecting the allele specificity or the antiproliferative effect [163]. In this context, the appendage of high molecular mass polyethylene glycol (PEG) units to the 3′/5′ termini of therapeutic oligonucleotides, aptamers particularly, is another popular strategy to improve the circulating time by reducing renal clearance [14, 164]. So far, direct PEGylation has not been reported for DNAzymes but a polyplex delivery system consisting of transferrin-modified, PEG-stabilized polyplexes has been described for Dz10–23 [165]. This system allowed to deliver the DNAzyme to cells without the aid of a transfecting agent and prevented its rapid elimination from the body without impacting the catalytic activity [166]. Similarly, poly(propylene imine)-based dendrimers were combined with PEG-like units to facilitate the in vitro and in vivo cellular delivery of Dz10–23 [167]. Besides PEG-like units, DNAzymes have been conjugated to nanoparticulate systems to improve their cellular uptake. For instance, Dz10–23 has been adsorbed on MnO_2 nanosheets and this system was shown to present a number of salient features: (i) these nanomaterials allow the cellular delivery of DNAzymes; (ii) MnO_2 nanosheets protect DNAzymes from nuclease degradation; (iii) these sheets supply Mn^{2+} to Dz10–23, which is a more efficient cofactor than Mg^{2+}; (iv) the gene silencing activity was not negatively impacted [168]. A similar construct was recently used to monitor enzyme catalysis [169] and DNA base-excision repair [170] in live cells. DNAzymes have also been connected to gold nanoparticles, which are known to enhance cellular internalization of oligonucleotides by facilitating their endosomal escape [171]. Gold nanoparticles can easily be conjugated to nucleic acids provided 3′- or 5′-thio groups are present. When Dz10–23 was conjugated to gold nanoparticles, a very efficient silencing of the growth differentiation factor 15 was achieved in vivo, clearly demonstrating the usefulness of this approach for the cellular delivery of DNAzymes [172]. This observation was recently confirmed by the efficient in vivo silencing of the tumor necrosis factor-α by a Dz10–23-gold nanoparticle construct [173]. A related system consisting of a DNAzyme-gold nanoparticles conjugate has also been used to develop nanozymes capable of the site-selective splicing of RNA substrates [174] or detecting miRNAs in cancer cells [175]. Carbon nanotubes are another highly promising vehicle for the cellular delivery of oligonucleotides via a mechanism that depends on their functionalization and length [176, 177]. DNAzymes can also be connected to carbon nanotubes with only a small decrease in catalytic efficiency (~3-fold reduction in k_{cat}/K_m) [178]. However, this conjugation system has not been evaluated for its capacity at gene silencing in vivo as yet. The coupling of nanoparticles or PEG-like systems certainly helps for the cellular internalization of DNAzymes but does not remedy the lack of selectivity. To address this issue, DNAzymes can be conjugated to affinity ligands such as aptamers and antibodies. For instance, the anti-nucleolin aptamer AS1411 was coupled to a variant of Dz10–23 that was capable of hydrolyzing a specific stretch of the mRNA of the apoptosis protein survivin, which is overexpressed in various cancer cells. The aptamer served as a shuttle to transport the DNAzyme specifically into the desired target, i.e. retinoblastoma cell lines Y79 and WERI-Rb1, and concomitantly promoted the cellular uptake of the construct. Once inside the cytoplasm of the cancer cells, the DNAzyme catalyzed the hydrolytic degradation of survivin mRNA and suppressed the progression of retinoblastoma cancer [179]. Antibody–nucleic acid conjugates represent popular and efficient drug carrier platforms [180], even though most DNAzyme–antibody conjugates have been used for sensing purposes [116], it is foreseeable that the use of such constructs for the selective delivery of DNAzymes will be reported in the near future.

Consequently, the appendage or inclusion of sugar-modified nucleotides, cholesterol and PEG-like units, or nanoparticles onto the scaffold of DNAzymes is an alluring strategy to improve their general utility in vivo [181].

Post-SELEX Optimization for Different Applications

The post-SELEX modification using solid-phase synthesis has also been extensively used to shed light on the *modus operandi* of DNAzymes. The modifications used for mechanistic considerations range from simple mutagenic structures where the core nucleotides are substituted by all other natural nucleotides to mutants containing base and sugar modified building blocks. Mutagenic studies could pinpoint the critical residues of the catalytic cores of the RNA-cleaving DNAzymes 10–23 [26, 182] and 8–17 [43], but also of DNAzymes with photolyase [125] and peroxidase activity [183]. As mentioned previously, replacement of the constituting nucleotides of the catalytic cores of Dz10–23 and 8–17 by C3-spacers and abasic sites further enabled to refine the understanding of the mechanisms underlying these DNAzymes and to emit the hypothesis that Dz10–23 might be a variant of 8–17 [25]. Modified nucleobases have also been incorporated into the scaffolds of DNAzymes to probe the influence of more subtle factors on catalytic activity, such as minor groove interactions [45] or proton transfer within the catalytic core [44]. In this context, fluorescent nucleotides have also been included in the catalytic sites to study the folding mechanism implicated in DNAzymes. The fluorescent adenine analog 2-aminopurine (2AP) is a popular choice due to the sensitivity of its spectral features to base stacking interactions and its inherent capacity at recognizing both cytosine and thymidine [184]. Hence, the inclusion of 2AP into the dinucleotide junction of the RNA substrates of Dz8–17 and the Pb^{2+}-dependent DNAzyme GR-5 [6] revealed that both enzymes displayed a similar metal ion binding pocket, hinting at the possibility that both DNAzymes are related to each other [185]. The inclusion of a 2-AP moiety at position dA_{15} of Dz8–17 (Figure 23.1b) served as a spectroscopic probe to investigate the effect of four different metal cations (three activators: Mg^{2+}, Ca^{2+}, and Mn^{2+}; one inhibitor: Cu^{2+}) on folding and catalysis. The fluorescence of the 2-AP probe was increased in the presence of the activating metal ions due to unstacking at this location, suggesting metal ion binding at position dA_{15}; an opposite behavior was observed when inhibiting Cu^{2+} was used [186]. However, these findings were not confirmed by the crystal structure of Dz8–17. Similarly, the binding mode of monovalent metal ions was also investigated by the substitution of adenine nucleotides by 2-AP residues in the catalytic cores of the Ag^{+}-dependent DNAzyme Ag10c [187] and an aptameric sequence derived from the Na^{+}-dependent DNAzyme Ce13d [188, 189]. Fluorescent pyrenyl-modified nucleosides have also been used as surrogates of adenine nucleotides of the catalytic core of Dz10–23 to identify critically important residues [190].

The spatiotemporal control of the activity of DNAzymes can also be achieved by the inclusion of modified residues by application of the post-SELEX method. A modulation of the cleavage activity can be achieved by using an external stimulus, which can trigger the conversion from an inactive to an active form of the DNAzyme without interfering with other conditions or biological systems. In this context, light is ideal for an exogenous control since it is orthogonal to natural systems, non-invasive, and the wavelength can be adjusted to cause minimal or no damage to other biological components [191]. Hence, providing DNAzymes with photocaging moieties, i.e. light-labile temporary blocking groups, is an alluring strategy to remotely modulate their catalytic activity. The photoprotecting groups can be placed at various locations of both the nucleosidic scaffold (Figure 23.6) and the sequence of the DNAzyme. For instance, a 6-nitropiperonyloxymethylene (NPOM; **8**) was appended on position N3 of thymidine hence obliterating the formation of Watson–Crick base-pairing. When such a photocaged residue was introduced either in the guide arms or at the critically conserved position dT_4 of the catalytic core of Dz10–23, cleavage activity was completely abolished. The catalytic activity could be restored by a simple UV-light treatment (1 minutes at 365 nm), which cleanly removed all the photolabile blocking groups [136]. A photolabile unit was also affixed at the 2′-position of the single ribo(adenosine) unit of the substrate of DNAzymes 8–17 and GR5. The presence of the 2′-*O*-nitrobenzyl protecting group (**9** in Figure 23.6) abolishes cleavage activity due to the absence of a free hydroxyl group that can attack the adjacent 3′-phosphate and concomitantly enabled the intracellular delivery of the intact DNAzyme. The protecting group can be removed by irradiation at 365 nm, which restores the 2′-OH, and thus, the catalytic activity of the DNAzyme [192]. A similar strategy was used to block the catalytic core of Dz10–23 at position dT_8 with a 2′-deoxyuridylate analog equipped with phenylazo–benzoyl moieties [193]. Photolabile groups can also be appended directly on the bridging phosphate of the phosphodiester backbone. For instance, the TEEP-OH group (**10** in Figure 23.6) can be installed on phosphorothioate units of the backbone by reaction with 2-bromo-4′-hydroxyacetophenone. The TEEP-OH modifications were shown to efficiently block the activity of DNAzymes 10–23 and 8–17 and catalytically competent species could be recovered by light irradiation at 365 nm [194]. The light-controlled activity of DNAzymes can also be achieved by incorporating unnatural azo derivatives such as **11** (Figure 23.6). These azo derivatives can reversibly isomerize from a trans to the corresponding cis form, depending on the wavelength of the incident light. These units can be incorporated either into the catalytic core or binding arms [195] or into photo-trigger strands at the 5′-/3′-termini of DNAzymes 8–17 and 10–23 [196, 197].

An interesting question is whether DNA, particularly DNAzymes, can catalyze chemical transformations in organic solvents. DNA is barely soluble in any solvent, but water due to the high density of negative charge present on the phosphodiester backbone. Therefore, identifying DNA-based catalysts that are active in organic media is a non-trivial task. Certain organic solvents are compatible with DNA unto a certain specific threshold. For instance, low ethanol (EtOH) concentrations destabilize DNA duplexes while higher EtOH contents cause aggregation and eventually, precipitation of DNA [198, 199]. Consequently, DNAzymes capable of hydrolyzing [199–201] or ligating [202] RNA substrates have been shown to function in the presence of organic solvents but still depend on the presence of water. The low compatibility of DNA with hydrophobic environments can be overcome by adding surfactants that both neutralize the negative charge of the backbone and provide a hydrophobic coating [203, 204]. While the addition of cationic surfactants such as cethyltrimethylammonium bromide (CTAB) revealed to be detrimental for the catalytic activity of Dz10–23 (probably due to compaction and aggregation of the DNAzyme) [205], the addition of a cationic comb-type copolymer enhanced the rate constants of wild-type [205] and LNA-modified Dz10–23 [206]. The addition of such copolymers thus represents a valid method for coercing DNAzymes to function in organic media. As an alternative, the inclusion of hydrophobic units such as PEG on the scaffold of DNAzymes could increase their compatibility with organic solvents. Indeed, when the hemin-dependent peroxidase DNAzyme was equipped with a PEG unit, the catalytic activity could be detected only in pure MeOH [207]. PEGylated DNA-encoded libraries have also been used as a platform for the discovery of small-molecules catalyzing an aldol reaction [208], hinting at the possibility of using such a modification in SELEX for the identification of DNAzymes compatible with organic solvents.

23.3.3 Polymerization of Modified Nucleoside Triphosphates for SELEX of DNAzymes

The enzymatic (co-)polymerization of modified nucleoside triphosphates (dN*TPs) is a versatile method and a solid alternative to solid-phase synthesis for the introduction of chemical functionality into nucleic acids [209–211]. While solid-phase

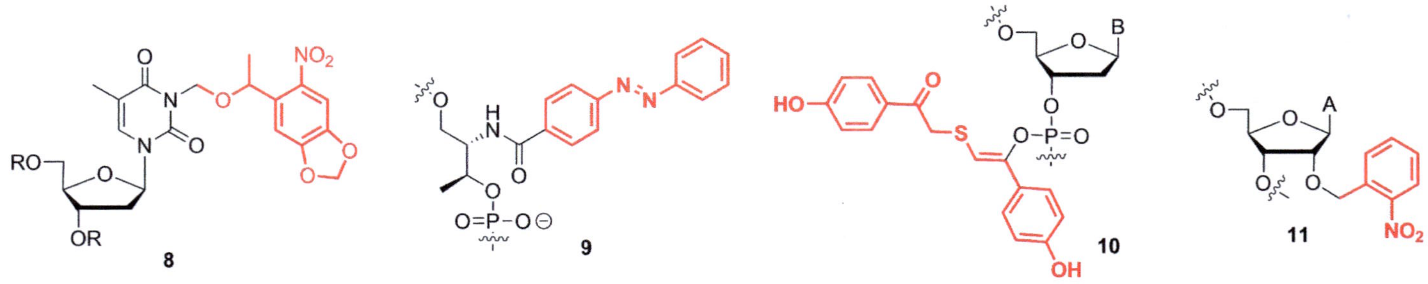

8 9 10 11

Figure 23.6 Chemical structures of photocaging groups that have been used for the spatial and temporal control of the activity of DNAzymes.

synthesis requires phosphoramidite building blocks and relies on organic chemistry reactions, the inclusion of modifications with dN*TPs only necessitates triphosphates to be compatible with a natural or engineered polymerase. However, the choice of the chemical nature and the location of the modifications on the nucleotide analogs are still dictated by empirical rules: (i) Nucleobases are usually modified at the C5-position of pyrimidines and the *N*7 of 7-deazapurines via rigid linker arms to ensure uptake by the polymerase and facilitate the synthetic routes; (ii) Even though alterations of the sugar moieties are not well tolerated, triphosphates bearing modifications at the C2′ and *O*4′ positions have been shown to be processed by polymerases; (iii) Phosphate modifications are not common and focus essentially on the substitution of a non-bridging oxygen by a sulfur atom (phosphorothioates). The resulting modified triphosphates often have excellent substrate properties and have allowed the introduction of a vast variety of functional groups, including protein-like residues [212–216], boronic acids [217, 218], carborane moieties [219], protein enzymes [220, 221], and even DNAzymes [222]. Other sites, including the C2-position of purines [223] and the *N*4 position of cytosine [224] triphosphates, have been shown to tolerate the appendage of small functional groups. Triphosphates analogs with unnatural bases have recently been proposed as a means for the facile enzymatic immobilization of DNAzymes and aptamers on solid support [225]. Also, progress in polymerase evolution has permitted to lax the substrate specificity, especially of sugar modified triphosphates [226–230].

The use of dN*TPs in selection experiments was first reported for aptamers [231] but was rapidly exported to the field of catalytic nucleic acids. Indeed, a modified dUTP equipped with an imidazole moiety (**12** in Figure 23.7) was used by Santoro and Joyce *in lieu* of its natural counterpart TTP to isolate the Zn^{2+}-dependent RNA-cleaving DNAzyme 16.2–11. Initially, the selection protocol yielded a self-cleaving species that could be converted into a trans-acting DNAzyme capable of multiple turnover. The trans-cleaving DNAzyme 16.2–11 critically depended on the presence of three modified dU units for catalysis, which were also presumably involved in base-pairing with an rA nucleotide of the substrate. Other salient structural features of Dz16.2–11 include a hairpin structure in the catalytic core formed by two modified dU–dA base pairs and a dG·dT wobble pair in the stem as well as another dG•dU wobble pair at the level of the substrate. Dz16.2–11 requires the presence of the modifications, 10 μM Zn^{2+}, and minimal Mg^{2+} (1 mM) to efficiently ($k_{cat}/K_m \sim 10^8\ M^{-1}\ min^{-1}$) cleave an all-RNA containing substrate

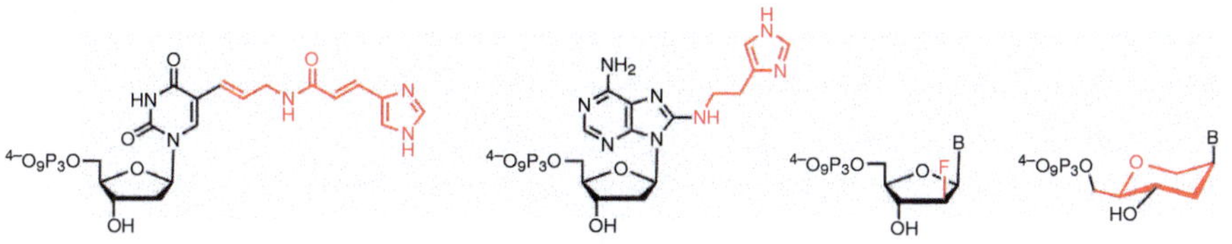

12 13 14 15

Figure 23.7 Chemical structures of dN*TPs that have been used for the isolation of modified DNAzymes.

[232]. A similar approach was used to identify artificial ribonucleases through a selection experiment with a modified triphosphate bearing a side-chain reminiscent of the amino acid tyrosine. The resulting Mg^{2+}-dependent DNAzyme self-cleaved a substrate containing a single scissile ribo(cytosine) linkage with an appreciable rate constant ($k_{obs} = 0.20\,min^{-1}$) [233].

As mentioned previously, the discovery of DNAzymes capable of operating under physiological [Mg^{2+}] or in the absence of Mg^{2+} is of high interest to improve their in vivo gene silencing efficiency. An early selection experiment carried out in the total absence of M^{2+} led to the identification of moderately active species (k_{obs} values in the $0.01\,min^{-1}$ range) [234]. Similarly, replacing all M^{2+} with spermidine [235], reducing the pH [236], or by adding a large excess of histidine (vs. Mg^{2+}) in the selection conditions led to the isolation of rather poor catalysts (k_{obs} values in the $0.01\,min^{-1}$ range) or in histidine-independent and Mg^{2+}/Ca^{2+}-dependent DNAzymes, respectively. Inspired by the M^{2+}-independent mechanism of RNase A, which involves two histidines (His12 and His119) and one lysine (Lys41) in the active site [237], Perrin et al. crafted two triphosphates equipped with side-chains mimicking these residues (i.e. histaminyl-dATP **13** [$dA^{im}TP$] in Figure 23.7 and commercially available 5-aminoallyl-dUTP [$dU^{aa}TP$]) [238]. Even though the imidazole moiety was appended at position 8 of the purine base, a position known to disfavor polymerase acceptance [239], Sequenase was capable of introducing up to three consecutive $dA^{im}MP$ units into DNA and yield fully extended products [215, 240]. Moreover, $dA^{im}TP$ was not accepted by polymerases under PCR conditions, but templates containing this nucleotide analog could efficiently be transcribed back into natural DNA under PCR conditions using the Vent (*exo*$^-$) polymerase. Thereafter, this set of modified dNTPs was used instead of the natural counterparts in a SELEX experiment, which resulted in the isolation of the M^{2+}-independent RNA-cleaving DNAzyme 9_{25}–11 [241]. This DNAzyme efficiently self-cleaved its substrate containing a single ribo(cytosine) unit embedded in a DNA sequence ($k_{obs} = 0.044\,min^{-1}$ at 37 °C [241] and $0.28\,min^{-1}$ at 13 °C [242]) in the total absence of M^{2+}. It is noteworthy mentioning that these rate constants represent 10^5–10^6-fold improvements compared to the uncatalyzed hydrolysis of RNA [243]. DNAzyme 9_{25}–11 could be converted to a corresponding trans-cleaving species (Dz9_{25}–11t) by solid-phase synthesis that exhibited a substantial catalytic efficiency ($k_{cat}/K_m = 5.3 \times 10^5\,min^{-1}\,M^{-1}$) for the M^{2+}-independent hydrolysis of the target RNA substrate [104, 244]. The presence of both base-modified nucleotides was strictly required for efficient catalysis, and the side-chains appear to be directly involved in general base and general acid catalysis rather than participating in folding or pK_a perturbation of specific nucleotides [245]. Interestingly, even minute chemical alterations of the constituting nucleotides (such as reducing the linker arm of $dA^{im}TP$ **13** by a methylene group) have a devastating effect on the outcome of selections [246] or the activity of existing DNAzymes, suggesting very compact and highly ordered catalytic cores. In addition, mechanistic investigation on the participation of the amine connecting the imidazole to the nucleobase led to the serendipitous discovery of a strategy for the light-control of the activity of DNAzymes (see Section 23.3.2). Indeed, when this amino group was substituted with a sulfur atom, the resulting thioether containing nucleotide analog underwent an unusual photo-deprotection reaction. This was exploited to develop a photocaged version of Dz8–17; the presence of a single dA unit modified with the thioether at position 8 of the nucleobase completely ablated the catalytic activity, which could be restored by irradiation at 280 nm [247]. What is more, the imidazole side chains located on dA^{im} residues are good ligands for transition metals, particularly Hg^{2+}, which strongly inhibit the catalytic activity of $Dz9_{25}$–11 [248]. This feature could potentially be exploited to sense Hg^{2+}, but efficient biosensing systems usually activate rather than inhibit the appearance of a readout signal. Therefore, a selection experiment was carried out using the same modified triphosphates ($dA^{im}TP$ **13** and $dU^{aa}TP$), a longer degenerate sequence (N40) to explore larger sequence and chemical space, and a negative selection step to exclude the participation of related metal ions such as Zn^{2+} and Cu^{2+}. The resulting modified DNAzyme 10–13 self-cleaved its RNA substrate with an appreciable rate ($k_{obs} = 0.037\ min^{-1}$) selectively and only in the presence of Hg^{2+}. Dz10–13 acts as a highly potent biosensor for mercury since concentrations as low as 100 nM can easily be detected [249]. The presence of the other side-chain, the lysine mimic on dU^{aa}, conveys another interesting property to $Dz9_{25}$–11: apurinic/apyrimidinic (AP) lyase-like activity. Indeed, when the substrate of $Dz9_{25}$–11t was equipped with an abasic site instead of the scissile ribo(cytosine) unit, the primary amines engaged in the formation of a covalent Schiff-base intermediate, which rapidly decayed, leading to strand break [250].

The isolation of $Dz9_{25}$–11 clearly demonstrated the potency of the combined use of two modified triphosphates in selection experiments. It led to some interesting applications such as AP-lyase activity and biosensing of an environmental contaminant. However, $Dz9_{25}$–11 also displayed some practical limitations: (i) maximal catalytic activity is observed at 13 °C, which is clearly not suitable for applications in the intracellular milieu; (ii) the catalytic rates are moderate ($k_{cat} = 0.03\ min^{-1}$ at 24 °C) and biphasic kinetics are observed at 13 °C under single-turnover conditions; (iii) all-RNA substrates are not recognized by the DNAzyme, which is a severe impediment for in vivo gene silencing; (iv) reselection with a longer randomized library (N40) led to the identification of 9_{25}–11 and related sequences. In order to address these limitations, a third dN*TP bearing a positively charged guanidinium residue was included in the selection experiments. This additional group was deemed to confer increased thermal stability to DNAzymes through a reduction of the negative charge repulsion [251]. Hence, the modified dUTP **16** ($dU^{ga}TP$; Figure 23.8c) was prepared by reacting $dU^{aa}TP$ with a guanidinylation reagent, and the ensuing triphosphate was shown to be a good substrate for DNA polymerases. $dU^{aa}TP$ **16** was accompanied by $dA^{im}TP$ **13**, and commercially available 5-aminoallyl-dCTP ($dC^{aa}TP$ **18** in Figure 23.8c) in two selection experiments. The first was carried out with an N20 degenerate region to ensure sampling of the entire sequence space and resulted in the identification of DNAzyme 9–86. This heavily modified DNAzyme self-cleaved a substrate containing a single scissile ribophosphodiester group with an improved efficiency ($k_{obs} = 0.13\ min^{-1}$ at 24 °C) in the absence of any prosthetic cofactor and displayed a temperature optimum at 37 °C. However, the catalytic efficiency with all-RNA substrates remained modest

(k_{obs} = 1.4×10^{-3} min^{-1}), and engineering of a trans-acting species could not be achieved [131]. The second selection experiment involved a longer N40 randomized library to explore a larger sequence space and potentially allow the formation of more complex three-dimensional structures. This resulted in the identification of DNAzyme 10–66, which self-cleaved the same substrate as 9–86 in the absence of M^{2+} with an impressive rate constant ($k_{obs} = 0.50\,min^{-1}$ at 24 °C), which increased slightly with temperature ($k_{obs} = 0.63\,min^{-1}$ at 37 °C). Interestingly, Dz10–66 could hydrolyze its substrate under conditions that do not support catalytic activity by any nucleic acid-based system. Unlike Dz9–86, Dz10–66 could be converted into a trans-cleaving species that displayed second-order rate constants in the same range as 9_{25}–11 ($k_{cat}/K_m \sim 5\times10^5\,min^{-1}\,M^{-1}$) [42]. Collectively, these selection experiments allowed raising both the temperature optimum and the catalytic efficiency compared to 9_{25}–11, which was obtained with two modifications and still maintained the possibility for multiple turnover catalysis. However, both selections failed to identify catalysts capable of efficiently hydrolyzing all-RNA targets. Thus, two additional selection experiments were carried out with the hypothesis that the formation of an A-DNA duplex between the library and the substrate could favor the identification of all-RNA cleaving catalysts. In this context, the first selection involving a substrate containing a stretch of 12 RNA nucleotides led to the identification of DNAzyme 12–91 (Figure 23.8a), which self-cleaved with an appreciable rate constant (k_{obs} = $0.06\,min^{-1}$) at a single location. Surprisingly, Dz12–91 was more proficient at catalyzing the scission of a single-embedded ribo-cytosine linkage ($k_{obs} = 0.27\,min^{-1}$) even though this substrate was not present during the selection. As in the case of Dz9–86, Dz12–91 could not be converted into a trans-cleaving catalyst. What is more, both modified DNAzymes share a strong sequence and topology homology. Indeed, of the 19 nucleotides composing the catalytic cores, only three are different and account for the ~50-fold higher cleavage efficiency with all-RNA substrates for Dz12–91 [252]. In a second selection, a longer N40 randomized region and a slightly longer RNA substrate (17 vs 12 nucleotides) were used that led to the identification of Dz7–38–32 (Figure 23.8b). This DNAzyme self-cleaved its substrate at a single location and presented biphasic kinetics with a fast and a slow phase (amplitudes of 36% and 26%, respectively) that displayed different apparent rate constants ($4.9\,min^{-1}$ and $0.06\,min^{-1}$, respectively). Dz7–38–32 could be converted into a trans-acting species that displayed a second-order rate constant of $8.2\times10^4\,min^{-1}\,M^{-1}$ (slightly lower than that of DNAzymes 9_{25}–11 and 10–66, albeit for a substrate containing 17 RNA units and not a single ribonucleotide). Interestingly, Dz7–38–32 presents no sequence homology with the related DNAzymes 10–66, 12–91, and 9–86 [253].

Another solution to M^{2+}-independent all-RNA cleavage by DNAzymes was proposed by Williams et al., who used a combination of the modified dUTP **12** (Figure 23.7) and a 7-deaza-dATP analog equipped with a propargyl amino moiety in the selection protocol. The resulting DNAzyme was capable of promoting the hydrolysis of the intended target, albeit degraded the substrate at two distinct positions. The observed rate constants at the two sites were very similar (0.06 and $0.07\,min^{-1}$). No trans-cleaving species was reported for this DNAzyme [254].

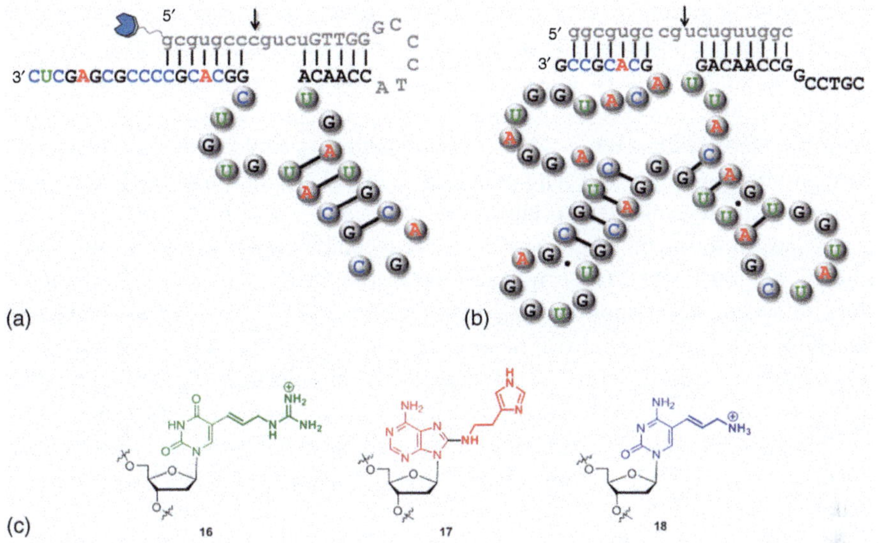

Figure 23.8 Sequences and hypothetical 2D structures of M^{2+}-independent RNA-cleaving DNAzymes obtained with dN*TPs included in SELEX: (a) Dz12–91 and (b) Dz7–38–32t. (c) Chemical structures of the modified nucleotides present in these modified DNAzymes.

Taken together, the combined use of two or three modified nucleotides equipped with amino acid-like side-chains in SELEX yielded DNAzymes capable of the M^{2+}-independent hydrolysis of RNA with multiple turnover catalysis and with important rate enhancements compared to unmodified catalysts. This clearly highlights the potential of modifications to improve the activity of DNAzymes under physiological conditions. Modifications can also bestow DNA with new reactivity, which is not accessible to unmodified nucleic acids. For instance, numerous selection experiments have been devised to identify unmodified nucleic acid catalysts capable of hydrolyzing amide bonds [216]. These selection experiments ended either in the discovery of species catalyzing the scission of phosphodiester bonds of DNA [82] or RNA [98, 255] rather than the targeted amide units or in the absence of any catalytic species. In order to address this catalytic shortcoming, Silverman et al. included modified dUTPs equipped with amine, carboxylic acid, and hydroxyl groups into selection experiments. These three selection experiments yielded different DNAzymes capable of hydrolyzing the aliphatic amide bond of the substrate with rate constants in the 10^{-3} min^{-1} range and with yields varying between 17–64% after 48 hours. Surprisingly, when one of the resulting DNAzymes was synthesized without the modified dU nucleotides, some residual catalytic activity ($k_{obs} = 5.7 \times 10^{-4}$ min^{-1}) remained [7] (Figure 23.8).

Examples of DNAzymes obtained by selection with sugar-modified nucleotides are rather scarce. An early example involved the use of 2′-fluoro-ATP, 2′-amino-CTP, and either 2′-fluoro-UTP or 2′-fluoro-5-[(*N*-imidazole-4-acetyl)propylamine]-UTP along with the unmodified GTP. The 2′-fluoro and 2′-amino-moieties were incorporated to confer enhanced nuclease resistance to the resulting ribozymes, while the imidazole-modified UTP was believed to provide a functionality that could engage in acid–base catalysis. Analysis of the clones resulting from the selection with the base-modified UTP revealed that this modification was expendable and did not contribute to the expected acid–base catalysis. In addition, most of the 2′-fluoro-nucleotides and some of the 2′-amino-cytosines could be replaced with the corresponding 2′-OMe group. The resulting 2′-OMe/2′-amino-modified ribozyme displayed an appreciable second rate constant ($k_{cat}/K_m = 10^6\ min^{-1}\ M^{-1}$) at 37 °C and, in the presence of 1 mM Mg^{2+}, showed good resistance against nuclease degradation [256]. More recently, variants of the TgoT polymerase have been crafted to accept sugar modified triphosphates such as 2′-fluoroarabino nucleic acids (FANA **14**) and 1,5-anhydrohexitol nucleic acids (HNA **15** in Figure 23.7). These polymerases revealed to be instrumental in the selection of potent aptamers against various protein targets [228]. Similarly, these polymerases served for the identification of several DNAzymes capable of cleavage of RNA substrates as well as ligating RNA and XNA (xeno nucleic acids) fragments. Of the four different backbone chemistries assessed in separate selection experiments, the FANA system yielded the most active catalyst with an observed rate constant of 0.058 min^{-1} for the hydrolysis of an all-RNA substrate. This FANAzyme could also be converted into a trans-cleaving species that displayed multiple turnover catalysis. Interestingly, some sequences obtained with the ANA (arabinonucleic acid) presented 10–23 and 8–17-like character. FANAzymes capable of ligating RNA ($k_{obs} = 2 \times 10^{-4}\ min^{-1}$) or FANA ($k_{obs} = 0.038\ min^{-1}$) sequences were also isolated [257].

Various DNAzymes structures consisting of the mirror image of natural D-DNA (the so-called Spieglmers [258]) have been constructed by replacing the constituting nucleotides of known DNAzymes by the corresponding L-building blocks [259]. So far, no Spiegelmer–DNAzymes have been selected with the corresponding L-triphosphates despite the recent crafting of polymerases capable of synthesizing L-nucleic acids [229, 260].

The generation of DNA catalysts via the inclusion of dN*TPs in the selection process is an efficient and potent method, yet one that requires both synthetic and biological steps. The identification of novel chemistries compatible with selection experiments will certainly contribute to the identification of modified DNAzymes capable of transformations not accessible to unmodified scaffolds [261].

23.4 Conclusions

Despite being the latest players in the field of functional nucleic acids, DNAzymes have grown into potent catalysts and have been integrated into numerous applications ranging from biosensing to therapeutics. DNAzymes can certainly be compared to their RNA counterparts in terms of their chemical repertoire and catalytic efficiency. As for other functional nucleic acids and therapeutic oligonucleotides, the scaffold of DNAzymes needs to be strengthened with chemical entities in order to enhance their in vivo utility. Specifically, DNAzymes have a strong dependence on metal ion cofactors, which is often incompatible with cellular environments leading to an impaired catalytic activity and potentially to gene silencing through an antisense rather than a hydrolytic mechanism. Also, DNAzymes consisting of a natural phosphodiester backbone are prone to degradation by nucleases, which is a severe predicament for in vivo applications. These (and other) hurdles can be alleviated by the inclusion of chemical modifications either by a post-SELEX approach utilizing solid-phase DNA synthesis or by including dN*TPs directly into the selection protocol (or by combining both methods). The post-SELEX optimization is particularly efficient for including modifications at the level of the binding arms and the 3′/5′-ends to improve the pharmacokinetic properties of DNAzymes.

On the other hand, the incorporation of modifications via the post-SELEX strategy often induces a perturbation of the catalytic activity and requires rather lengthy and tedious structure–activity relationship (SAR) studies. Incorporating dN*TPs into SELEX directly yields modified DNAzymes and thus obliterates the need for these SAR studies. This strategy has allowed the identification of catalysts for the hydrolysis of amide bonds, which represents one of the most difficult reactions to be catalyzed by wild-type nucleic acids. Also, this method reduces the strong metal dependence of DNAzymes by creating highly efficient M^{2+}-independent artificial ribonucleases. As for the post-SELEX optimization strategy, selecting with dN*TPs presents some disadvantages since this method is tainted by the non-negligible synthetic and biochemical efforts that are required to isolate the triphosphates and identify polymerases that accept these analogs as substrates. Nonetheless, the introduction of chemical modification into the scaffold of DNAzymes either by post-SELEX amendment, selection with dN*TPs, or a combination of both, represents a solid approach to improve the general usefulness of these DNA-based catalysts.

The quest for modified DNAzymes might receive guidance from machine-learning-based approaches, as is the case for aptamers [262] and antimicrobial peptides [263]. In addition, further progress in synthetic organic chemistry combined with novel selection protocols [264–267] will certainly help in the identification of more robust DNAzymes and modified catalysts capable of accelerating more challenging chemical transformations such as C—C bond hydrolysis or in organic media.

Acknowledgment

The author gratefully acknowledges financial support from Institut Pasteur.

References

1 Kruger, K., Grabowski, P.J., Zaug, A.J. et al. (1982). Self-splicing RNA – auto-excision and auto-cyclization of the ribosomal-RNA intervening sequence of *Tetrahymena*. *Cell* 31 (1): 147–157.
2 Cech, T.R. and Bass, B.L. (1986). Biological catalysis by RNA. *Annu. Rev. Biochem.* 55: 599–629.

3 Tuerk, C. and Gold, L. (1990). Systematic evolution of ligands by exponential enrichment: RNA ligands to bacteriophage T4 DNA polymerase. *Science* 249: 505–510.

4 Ellington, A.D. and Szostak, J.W. (1990). In vitro selection of RNA molecules that bind specific ligands. *Nature* 346: 818–822.

5 Robertson, D.L. and Joyce, G.F. (1990). Selection *in vitro* of an RNA enzyme that specifically cleaves single-stranded DNA. *Nature* 344: 467–468.

6 Breaker, R.R. and Joyce, G.F. (1994). A DNA enzyme that cleaves RNA. *Chem. Biol.* 1: 223–229.

7 Zhou, C., Avins, J.L., Klauser, P.C. et al. (2016). DNA-catalyzed amide hydrolysis. *J. Am. Chem. Soc.* 138 (7): 2106–2109.

8 Chinnapen, D.J. and Sen, D. (2004). A deoxyribozyme that harnesses light to repair thymine dimers in DNA. *Proc. Natl. Acad. Sci. U.S.A.* 101: 65–69.

9 Krug, N., Hohlfeld, J.M., Kirsten, A.M. et al. (2015). Allergen-induced asthmatic responses modified by a GATA3-specific DNAzyme. *N. Engl. J. Med.* 372 (21): 1987–1995.

10 Cho, E.A., Moloney, F.J., Cai, H. et al. (2013). Safety and tolerability of an intratumorally injected DNAzyme, Dz13, in patients with nodular basal-cell carcinoma: a phase 1 first-in-human trial (DISCOVER). *Lancet* 381 (9880): 1835–1843.

11 Gu, H., Furukawa, K., Weinberg, Z. et al. (2013). Small, highly active DNAs that hydrolyze DNA. *J. Am. Chem. Soc.* 135: 9121–9129.

12 Gold, L. (2015). SELEX: how it happened and where it will go. *J. Mol. Evol.* 81 (5–6): 140–143.

13 Perrin, D.M. (2012). Lifelike but Not Living: Selection of Synthetically Modified Bioinspired Nucleic Acids for Binding and Catalysis In *Polymer Science: A Comprehensive Reference* Lifelike but Not Living: Selection of Synthetically Modified Bioinspired Nucleic Acids for Binding and Catalysis (ed. M. Möller), 3–33. Elsevier.

14 Röthlisberger, P. and Hollenstein, M. (2018). Aptamer chemistry. *Adv. Drug Delivery Rev.* 134: 3–21.

15 Joyce, G.F. (2007). Forty years of in vitro evolution. *Angew. Chem. Int. Ed.* 46: 6420–6436.

16 Silverman, S.K. (2016). Catalyic DNA: scope, applications, and biochemistry of deoxyribozymes. *Trends Biochem. Sci.* 41 (7): 595–609.

17 Velez, T.E., Singh, J., Xiao, Y. et al. (2012). Systematic evaluation of the dependence of deoxyribozyme catalysis on random region length. *ACS Comb. Sci.* 14 (12): 680–687.

18 Liu, M., Chang, D.R., and Li, Y.F. (2017). Discovery and biosensing applications of diverse RNA-cleaving DNAzymes. *Acc. Chem. Res.* 50 (9): 2273–2283.

19 Pan, T. and Uhlenbeck, O.C. (1992). In vitro selection of RNAs that undergo autolytic cleavage with Pb^{2+}. *Biochemistry* 31 (16): 3887–3895.

20 Pan, T. and Uhlenbeck, O.C. (1992). A small metalloribozyme with a 2-step mechanism. *Nature* 358 (6387): 560–563.

21 Rubin, J.R. and Sundaralingam, M. (1983). Lead ion binding and RNA chain hydrolysis in phenylalanine transfer-RNA. *J. Biomol. Struct. Dyn.* 1 (3): 639–646.

22 Brown, R.S., Dewan, J.C., and Klug, A. (1985). Crystallographic and biochemical investigation of the lead(II)-cataylzed hydrolysis of yeast phenylalanine transfer-RNA. *Biochemistry* 24 (18): 4785–4801.

23 Breaker, R.R. and Joyce, G.F. (1995). A DNA enzyme with Mg^{2+}-dependent RNA phosphoesterase activity. *Chem. Biol.* 2: 655–660.

24 Schlosser, K. and Li, Y.F. (2010). A versatile endoribonuclease mimic made of DNA: characteristics and applications of the 8–17 RNA-cleaving DNAzyme. *ChemBioChem* 11 (7): 866–879.

25 Wang, B., Cao, L.Q., Chiuman, W. et al. (2010). Probing the function of nucleotides in the catalytic cores of the 8–17 and 10–23 DNAzymes by abasic nucleotide and C3 spacer substitutions. *Biochemistry* 49 (35): 7553–7562.

26 Zaborowska, Z., Furste, J.P., Erdmann, V.A., and Kurreck, J. (2002). Sequence requirements in the catalytic core of the "10–23" DNA enzyme. *J. Biol. Chem.* 277 (43): 40617–40622.

27 Hollenstein, M. (2015). DNA catalysis: the chemical repertoire of DNAzymes. *Molecules* 20 (11): 20777–20804.

28 Santoro, S.W. and Joyce, G.F. (1997). A general purpose RNA-cleaving DNA enzyme. *Proc. Natl. Acad. Sci. U.S.A.* 94: 4262–4266.

29 Santoro, S.W. and Joyce, G.F. (1998). Mechanism and utility of an RNA-cleaving DNA enzyme. *Biochemistry* 37 (38): 13330–13342.

30 Cairns, M.J., King, A., and Sun, L.Q. (2003). Optimisation of the 10–23 DNAzyme-substrate pairing interactions enhanced RNA cleavage activity at purine–cytosine target sites. *Nucleic Acids Res.* 31 (11): 2883–2889.

31 Cruz, R.P.G., Withers, J.B., and Li, Y.F. (2004). Dinucleotide junction cleavage versatility of 8–17 deoxyribozyme. *Chem. Biol.* 11 (1): 57–67.

32 Schlosser, K., Gu, J., Sule, L., and Li, Y.F. (2008). Sequence-function relationships provide new insight into the cleavage site selectivity of the 817 RNA-cleaving deoxyribozyme. *Nucleic Acids Res.* 36 (5): 1472–1481.

33 Faulhammer, D. and Famulok, M. (1996). The Ca^{2+} ion as a cofactor for a novel RNA-cleaving deoxyribozyme. *Angew. Chem. Int. Ed.* 35 (23–24): 2837–2841.

34 Peracchi, A. (2000). Preferential activation of the 8–17 deoxyribozyme by Ca^{2+} ions – evidence for the identity of 8–17 with the catalytic domain of the MG5 deoxyribozyme. *J. Biol. Chem.* 275 (16): 11693–11697.

35 Li, J., Zheng, W.C., Kwon, A.H., and Lu, Y. (2000). In vitro selection and characterization of a highly efficient Zn(II)-dependent RNA-cleaving deoxyribozyme. *Nucleic Acids Res.* 28 (2): 481–488.

36 Zhou, W.H., Zhang, Y.P., Ding, J.S., and Liu, J.W. (2016). In vitro selection in serum: RNA-cleaving DNAzymes for measuring Ca^{2+} and Mg^{2+}. *ACS Sens.* 1 (5): 600–606.

37 Nelson, K.E., Bruesehoff, P.J., and Lu, Y. (2005). In vitro selection of high temperature Zn^{2+}-dependent DNAzymes. *J. Mol. Evol.* 61 (2): 216–225.

38 Schlosser, K. and Li, Y.F. (2004). Tracing sequence diversity change of RNA-cleaving deoxyribozymes under increasing selection pressure during in vitro selection. *Biochemistry* 43 (30): 9695–9707.
39 Schlosser, K., Lam, J.C.F., and Li, Y.F. (2006). Characterization of long RNA-cleaving deoxyribozymes with short catalytic cores: the effect of excess sequence elements on the outcome of in vitro selection. *Nucleic Acids Res.* 34 (8): 2445–2454.
40 Joyce, G.F. (2004). Directed evolution of nucleic acid enzymes. *Annu. Rev. Biochem.* 73: 791–836.
41 Porter, E.B., Polaski, J.T., Morck, M.M., and Batey, R.T. (2017). Recurrent RNA motifs as scaffolds for genetically encodable small-molecule biosensors. *Nat. Chem. Biol.* 13 (3): 295–301.
42 Hollenstein, M., Hipolito, C.J., Lam, C.H., and Perrin, D.M. (2009). A DNAzyme with three protein-like functional Groups: enhancing catalytic efficiency of M^{2+}-independent RNA cleavage. *ChemBioChem* 10 (12): 1988–1992.
43 Peracchi, A., Bonaccio, M., and Clerici, M. (2005). A mutational analysis of the 8–17 deoxyribozyme core. *J. Mol. Biol.* 352 (4): 783–794.
44 Cepeda-Plaza, M., McGhee, C.E., and Lu, Y. (2018). Evidence of a general acid–base catalysis mechanism in the 8–17 DNAzyme. *Biochemistry* 57 (9): 1517–1522.
45 Räz, M. and Hollenstein, M. (2015). Probing the effect of minor groove interactions on the catalytic efficiency of DNAzymes 8–17 and 10–23. *Mol. BioSyst.* 11 (5): 1454–1461.
46 Kim, J., Jang, D., Park, H. et al. (2018). Functional-DNA-driven dynamic nanoconstructs for biomolecule capture and drug delivery. *Adv. Mater.* 30 (45): 1707351.
47 Santiago, F.S., Lowe, H.C., Kavurma, M.M. et al. (1999). New DNA enzyme targeting Egr-1 mRNA inhibits vascular smooth muscle proliferation and regrowth after injury. *Nat. Med.* 5: 1264–1269.
48 Schubert, S., Gül, D.C., Grunert, H.-P. et al. (2003). RNA cleaving '10–23' DNAzymes with enhanced stability and activity. *Nucleic Acids Res.* 31: 5982–5992.
49 Fahmy, R.G. and Khachigian, L.M. (2004). Locked nucleic acid modified DNA enzymes targeting early growth response-1 inhibit human vascular smooth muscle cell growth. *Nucleic Acids Res.* 32: 2281–2285.
50 Robaldo, L., Berzal-Herranz, A., Montserrat, J.M., and Iribarren, A.M. (2014). Activity of core-modified 10–23 DNAzymes against HCV. *ChemMedChem* 9 (9): 2172–2177.
51 Niewiarowska, J., Sacewicz, I., Wiktorska, M. et al. (2009). DNAzymes to mouse b1 integrin mRNA in vivo: targeting the tumor vasculature and retarding cancer growth. *Cancer Gene Ther.* 16: 713–722.
52 Fokina, A.A., Meschaninova, M.I., Durfort, T. et al. (2012). Targeting insulin-like growth factor I with 10–23 DNAzymes: 2′-*O*-methyl modifications in the catalytic core enhance mRNA cleavage. *Biochemistry* 51: 2181–2191.

53 Fokina, A.A., Stetsenko, D.A., and Francois, J.C. (2015). DNA enzymes as potential therapeutics: towards clinical application of 10–23 DNAzymes. *Expert Opin. Biol. Ther.* 15 (5): 689–711.

54 Donini, S., Clerici, M., Wengel, J. et al. (2007). The advantages of being locked – assessing the cleavage of short and long RNAs by locked nucleic acid-containing 8–17 deoxyribozymes. *J. Biol. Chem.* 282 (49): 35510–35518.

55 Zhou, W.H., Saran, R., and Liu, J.W. (2017). Metal sensing by DNA. *Chem. Rev.* 117 (12): 8272–8325.

56 Brown, A.K., Li, J., Pavot, C.M.B., and Lu, Y. (2003). A lead-dependent DNAzyme with a two-step mechanism. *Biochemistry* 42 (23): 7152–7161.

57 Li, J. and Lu, Y. (2000). A highly sensitive and selective catalytic DNA biosensor for lead ions. *J. Am. Chem. Soc.* 122 (42): 10466–10467.

58 Willner, I., Shlyahovsky, B., Zayats, M., and Willner, B. (2008). DNAzymes for sensing, nanobiotechnology and logic gate applications. *Chem. Soc. Rev.* 37: 1153–1165.

59 Wang, F., Lu, C.H., and Willner, I. (2014). From cascaded catalytic nucleic acids to enzyme-DNA nanostructures: controlling reactivity, sensing, logic operations and assembly of complex structures. *Chem. Rev.* 114 (5): 2881–2941.

60 Chang, D.R., Zakaria, S., Deng, M.M. et al. (2016). Integrating deoxyribozymes into colorimetric sensing platforms. *Sensors* 16 (12): 23.

61 McGhee, C.E., Loh, K.Y., and Lu, Y. (2017). DNAzyme sensors for detection of metal ions in the environment and imaging them in living cells. *Curr. Opin. Biotechnol.* 45: 191–201.

62 Li, L., Feng, J., Fan, Y.Y., and Tang, B. (2015). Simultaneous imaging of Zn^{2+} and Cu^{2+} in living cells based on DNAzyme modified gold nanoparticle. *Anal. Chem.* 87 (9): 4829–4835.

63 Kim, H.-K., Li, J., Nagraj, N., and Lu, Y. (2008). Probing metal binding in the 8–17 DNAzyme by Tb^{III} luminescence spectroscopy. *Chem. Eur. J.* 14: 8696–8703.

64 Bonaccio, M., Credali, A., and Peracchi, A. (2004). Kinetic and thermodynamic characterization of the RNA–cleaving 8–17 deoxyribozyme. *Nucleic Acids Res.* 32: 916–925.

65 Mazumdar, D., Nagraj, N., Kim, H.-K. et al. (2009). Activity, folding and Z-DNA formation of the 8–17 DNAzyme in the presence of monovalent ions. *J. Am. Chem. Soc.* 131: 5506–5515.

66 Liu, J. and Lu, Y. (2007). Rational design of "Turn-On" allosteric DNAzyme catalytic beacons for aqueous mercury ions withultra high sensitivity and selectivity. *Angew. Chem. Int. Ed.* 46: 7587–7590.

67 Liu, J., Brown, A.K., Meng, X. et al. (2007). A catalytic beacon sensor for uranium with parts-pertrillion sensitivity and millionfold selectivity. *Proc. Natl. Acad. Sci. U.S.A.* 104: 2056–2061.

68 Saran, R. and Liu, J.W. (2016). A silver DNAzyme. *Anal. Chem.* 88 (7): 4014–4020.

69 Huang, P.-J.J., Vazin, M., and Liu, J. (2014). In vitro selection of a new lanthanide-dependent DNAzyme for ratiometric sensing lanthanides. *Anal. Chem.* 86 (19): 9993–9999.
70 Huang, P.-J.J., Vazin, M., Matuszek, Z., and Liu, J. (2015). A new heavy lanthanide-dependent DNAzyme displaying strong metal cooperativity and unrescuable phosphorothioate effect. *Nucleic Acids Res.* 43 (1): 461–469.
71 Huang, P.-J.J., Lin, J., Cao, J. et al. (2014). Ultrasensitive DNAzyme beacon for lanthanides and metal speciation. *Anal. Chem.* 86 (3): 1816–1821.
72 Huang, P.J.J., Vazin, M., and Liu, J.W. (2016). In vitro selection of a DNAzyme cooperatively binding two lanthanide ions for RNA cleavage. *Biochemistry* 55 (17): 2518–2525.
73 Huang, P.J.J. and Liu, J.W. (2016). An ultrasensitive light-up Cu^{2+} biosensor using s new DNAzyme cleaving a phosphorothioate-modified substrate. *Anal. Chem.* 88 (6): 3341–3347.
74 Torabi, S.-F., Wu, P., McGhee, C.E. et al. (2015). In vitro selection of a sodium-specific DNAzyme and its application in intracellular sensing. *Proc. Natl. Acad. Sci. U.S.A.* 112 (19): 5903–5908.
75 Mei, S.H.J., Liu, Z.J., Brennan, J.D., and Li, Y.F. (2003). An efficient RNA-cleaving DNA enzyme that synchronizes catalysis with fluorescence signaling. *J. Am. Chem. Soc.* 125 (2): 412–420.
76 Ordoukhanian, P. and Joyce, G.F. (2002). RNA-cleaving DNA enzymes with altered regio- or enantioselectivity. *J. Am. Chem. Soc.* 124 (42): 12499–12506.
77 Tram, K., Xia, J.J., Gysbers, R., and Li, Y.F. (2015). An efficient catalytic DNA that cleaves L-RNA. *PLoS One* 10 (5): 14.
78 Greer, C.L., Javor, B., and Abelson, J. (1983). RNA ligase in bacteria – formation of a 2′,5′ linkage by an *Escherichia coli* extract. *Cell* 33 (3): 899–906.
79 Zhou, W.H., Ding, J.S., and Liu, J.W. (2016). An efficient lanthanide-dependent DNAzyme cleaving 2′–5′-linked RNA. *ChemBioChem* 17 (10): 890–894.
80 Liu, Z.J., Mei, S.H.J., Brennan, J.D., and Li, Y.F. (2003). Assemblage of signaling DNA enzymes with intriguing metal-ion specificities and pH dependences. *J. Am. Chem. Soc.* 125 (25): 7539–7545.
81 Beaudry, A.A. and Joyce, G.F. (1992). Directed evolution of an RNA enzyme. *Science* 257 (5070): 635–641.
82 Chandra, M., Sachdeva, A., and Silverman, S.K. (2009). DNA-catalyzed sequence-specific hydrolysis of DNA. *Nat. Chem. Biol.* 5: 718–720.
83 Xiao, Y., Chandra, M., and Silverman, S.K. (2010). Functional compromises among pH tolerance, site specificity, and sequence tolerance for a DNA-hyrdolyzing deoxyribozyme. *Biochemistry* 49: 9630–9637.
84 Xiao, Y., Allen, E.C., and Silverman, S.K. (2011). Merely two mutations switch a DNA-hydrolyzing deoxyribozyme from heterobimetallic (Zn^{2+}/Mn^{2+}) to monometallic (Zn^{2+}-only) behavior. *Chem. Commun.* 47 (6): 1749–1751.
85 Xiao, Y., Wehrmann, R.J., Ibrahim, N.A., and Silverman, S.K. (2012). Establishing broad generality of DNA catalysts for site-specific hydrolysis of single-stranded DNA. *Nucleic Acids Res.* 40 (4): 1778–1786.

86 Dokukin, V. and Silverman, S.K. (2012). Lanthanide ions as required cofactors for DNA catalysts. *Chem. Sci.* 3: 1707–1714.

87 Dhamodharan, V., Kobori, S., and Yokobayashi, Y. (2017). Large scale mutational and kinetic analysis of a self-hydrolyzing deoxyribozyme. *ACS Chem. Biol.* 12 (12): 2940–2945.

88 Liu, N.N., Hou, R.Z., Gao, P.C. et al. (2016). Sensitive Zn^{2+} sensor based on biofunctionalized nanopores via combination of DNAzyme and DNA supersandwich structures. *Analyst* 141 (12): 3626–3629.

89 Ma, L.Z., Liu, B.W., Huang, P.J.J. et al. (2016). DNA adsorption by ZnO nanoparticles near its solubility limit: implications for DNA fluorescence quenching and DNAzyme activity assays. *Langmuir* 32 (22): 5672–5680.

90 Li, X., Ding, X.L., Li, Y.F. et al. (2016). A TiS_2 nanosheet enhanced fluorescence polarization biosensor for ultra-sensitive detection of biomolecules. *Nanoscale* 8 (18): 9852–9860.

91 Gu, H.Z. and Breaker, R.R. (2013). Production of single-stranded DNAs by self-cleavage of rolling-circle amplification products. *Biotechniques* 54 (6): 337–343.

92 Lilienthal, S., Shpilt, Z., Wang, F. et al. (2015). Programmed DNAzyme-triggered dissolution of DNA-based. Hydrogels: means for controlled release of biocatalysts and for the activation of enzyme cascades. *ACS Appl. Mater. Interfaces* 7 (16): 8923–8931.

93 Furukawa, K. and Minakawa, N. (2014). Allosteric control of a DNA-hydrolyzing deoxyribozyme with short oligonucleotides and its application in DNA logic gates. *Org. Biomol. Chem.* 12 (21): 3344–3348.

94 Endo, M., Takeuchi, Y., Suzuki, Y. et al. (2015). Single-molecule visualization of the activity of a Zn^{2+}-dependent DNAzyme. *Angew. Chem. Int. Ed.* 54 (36): 10550–10554.

95 Xu, J.C., Sun, Y.H., Sheng, Y.J. et al. (2014). Engineering a DNA-cleaving DNAzyme and PCR into a simple sensor for zinc ion detection. *Anal. Bioanal.Chem.* 406 (13): 3025–3029.

96 Silverman, S.K. (2015). Pursuing DNA catalysts for protein modification. *Acc. Chem. Res.* 48: 1369–1379.

97 Brandsen, B.M., Hesser, A.R., Castner, M.A. et al. (2013). DNA-catalyzed hydrolysis of esters and aromatic amides. *J. Am. Chem. Soc.* 135: 16014–16017.

98 Dai, X., De Mesmaeker, A., and Joyce, G.F. (1995). Cleavage of an amide bond by a ribozyme. *Science* 267: 237–241.

99 Pradeepkumar, P.I., Höbartner, C., Baum, D.A., and Silverman, S.K. (2008). DNA-catalyzed formation of nucleopeptide linkages. *Angew. Chem. Int. Ed.* 47: 1753–1757.

100 Sachdeva, A. and Silverman, S.K. (2010). DNA-catalyzed serine side chain reactivity and selectivity. *Chem. Commun.* 46 (13): 2215–2217.

101 Wong, O.Y., Pradeepkumar, P.I., and Silverman, S.K. (2011). DNA-catalyzed covalent modification of amino acid side chains in tethered and free peptide substrates. *Biochemistry* 50: 4741–4749.

102 Chandrasekar, J. and Silverman, S.K. (2013). Catalytic DNA with phosphatase activity. *Proc. Natl. Acad. Sci. U.S.A.* 110 (14): 5315–5320.

103 Walsh, S.M., Sachdeva, A., and Silverman, S.K. (2013). DNA catalysts with tyrosine kinase activity. *J. Am. Chem. Soc.* 135 (40): 14928–14931.

104 Ting, R., Thomas, J.M., Lermer, L., and Perrin, D.M. (2004). Substrate specificity and kinetic framework of a DNAzyme with an expanded chemical repertoire: a putative RNaseA mimic that catalyzes RNA hydrolysis independent of a divalent metal cation. *Nucleic Acids Res.* 32: 6660–6672.

105 Flynn-Charlebois, A., Wang, Y.M., Prior, T.K. et al. (2003). Deoxyribozymes with 2′–5′ RNA ligase activity. *J. Am. Chem. Soc.* 125 (9): 2444–2454.

106 Wang, Y.M. and Silverman, S.K. (2003). Deoxyribozymes that synthesize branched and lariat RNA. *J. Am. Chem. Soc.* 125 (23): 6880–6881.

107 Wang, Y.M. and Silverman, S.K. (2005). Directing the outcome of deoxyribozyme selections to favor native 3′–5″ RNA ligation. *Biochemistry* 44 (8): 3017–3023.

108 Purtha, W.E., Coppins, R.L., Smalley, M.K., and Silverman, S.K. (2005). General deoxyribozyme-catalyzed synthesis of native 3′–5′ RNA linkages. *J. Am. Chem. Soc.* 127 (38): 13124–13125.

109 Cuenoud, B. and Szostak, J.W. (1995). A DNA metalloenzyme with DNA ligase activity. *Nature* 375 (6532): 611–614.

110 Sreedhara, A., Li, Y.F., and Breaker, R.R. (2004). Ligating DNA with DNA. *J. Am. Chem. Soc.* 126 (11): 3454–3460.

111 Kitanosono, T., Masuda, K., Xu, P.Y., and Kobayashi, S. (2018). Catalytic organic reactions in water toward sustainable society. *Chem. Rev.* 118 (2): 679–746.

112 Chandra, M. and Silverman, S.K. (2008). DNA and RNA can be equally efficient catalysts for carbon–carbon bond formation. *J. Am. Chem. Soc.* 130: 2936–2937.

113 Seelig, B. and Jäschke, A. (1999). A small catalytic RNA motif with Diels–Alderase activity. *Chem. Biol.* 6: 167–176.

114 Mohan, U., Burai, R., and McNaughton, B.R. (2013). In vitro evolution of a Friedel–Crafts deoxyribozyme. *Org. Biomol. Chem.* 11 (14): 2241–2244.

115 Kosman, J. and Juskowiak, B. (2011). Peroxidase-mimicking DNAzymes for biosensing applications: a review. *Anal. Chim. Acta* 707 (1): 7–17.

116 Peng, H.Y., Newbigging, A.M., Wang, Z.X. et al. (2018). DNAzyme-mediated assays for amplified detection of nucleic acids and proteins. *Anal. Chem.* 90 (1): 190–207.

117 Li, Y.F. and Sen, D. (1997). Toward an efficient DNAzyme. *Biochemistry* 36 (18): 5589–5599.

118 Travascio, P., Li, Y.F., and Sen, D. (1998). DNA-enhanced peroxidase activity of a DNA aptamer-hemin complex. *Chem. Biol.* 5 (9): 505–517.

119 Travascio, P., Bennet, A.J., Wang, D.Y., and Sen, D. (1999). A ribozyme and a catalytic DNA with peroxidase activity: active sites versus cofactor-binding sites. *Chem. Biol.* 6 (11): 779–787.

120 Poon, L.C.H., Methot, S.P., Morabi-Pazooki, W. et al. (2011). Guanine-rich RNAs and DNAs that bind heme robustly catalyze oxygen transfer reactions. *J. Am. Chem. Soc.* 133 (6): 1877–1884.

121 Nakayama, S. and Sintim, H.O. (2009). Colorimetric split G-quadruplex probes for nucleic acid sensing: improving reconstituted DNAzyme's catalytic efficiency via probe remodeling. *J. Am. Chem. Soc.* 131 (29): 10320–10333.

122 Barlev, A. and Sen, D. (2018). DNA's encounter with ultraviolet light: an instinct for self-preservation? *Acc. Chem. Res.* 51 (2): 526–533.

123 Thorne, R.E., Chinnapen, D.J.F., Sekhon, G.S., and Sen, D. (2009). A deoxyribozyme, Sero1C, uses light and serotonin to repair diverse pyrimidine dimers in DNA. *J. Mol. Biol.* 388 (1): 21–29.

124 Chinnapen, D.J.F. and Sen, D. (2007). Towards elucidation of the mechanism of UV1C, a deoxyribozyme with photolyase activity. *J. Mol. Biol.* 365 (5): 1326–1336.

125 Barlev, A. and Sen, D. (2013). Catalytic DNAs that harness violet light to repair thymine dimers in a DNA substrate. *J. Am. Chem. Soc.* 135 (7): 2596–2603.

126 Ponce-Salvatierra, A., Wawrzyniak-Turek, K., Steuerwald, U. et al. (2016). Crystal structure of a DNA catalyst. *Nature* 529 (7585): 231–234.

127 Liu, H.H., Yu, X., Chen, Y.Q. et al. (2017). Crystal structure of an RNA-cleaving DNAzyme. *Nat. Commun.* 8: 10.

128 Liu, Y. and Sen, D. (2008). A contact photo-cross-linking investigation of the ctive Site of the 8–17 deoxyribozyme. *J. Mol. Biol.* 381: 845–859.

129 Kim, H.K., Rasnik, I., Liu, J.W. et al. (2007). Dissecting metal ion-dependent folding and catalysis of a single DNAzyme. *Nat. Chem. Biol.* 3 (12): 763–768.

130 Zhou, W.H., Ding, J.S., and Liu, J.W. (2017). Theranostic DNAzymes. *Theranostics* 7 (4): 1010–1025.

131 Hollenstein, M., Hipolito, C.J., Lam, C.H., and Perrin, D.M. (2009). A self-cleaving DNA enzyme modified with amines, guanidines and imidazoles operates independently of divalent metal cations (M^{2+}). *Nucleic Acids Res.* 37 (5): 1638–1649.

132 Cieslak, M., Szymanski, J., Adamiak, R.W., and Cierniewski, C.S. (2003). Structural rearrangements of the 10–23 DNAzyme to b3 integrin subunit mRNA induced by cations and their relations to the catalytic activity. *J. Biol. Chem.* 278 (48): 47987–47996.

133 Trepanier, J., Tanner, J.E., Momparler, R.L. et al. (2006). Cleavage of intracellular hepatitis C RNA in the virus core protein coding region by deoxyribozymes. *J. Viral Hepat.* 13: 131–138.

134 Wang, F., Saran, R., and Liu, J. (2015). Tandem DNAzymes for mRNA cleavage: choice of enzyme, metal ions and the antisense effect. *Bioorg. Med. Chem. Lett.* 25: 1460–1463.

135 Victor, J., Steger, G., and Riesner, D. (2018). Inability of DNAzymes to cleave RNA in vivo is due to limited Mg concentration in cells. *Eur. Biophys. J. Biophys. Lett.* 47 (4): 333–343.

136 Young, D.D., Lively, M.O., and Deiters, A. (2010). Activation and deactivation of DNAzyme and antisense function with light for the photochemical regulation of gene expression in mammalian cells. *J. Am. Chem. Soc.* 132: 6183–6193.

137 Wang, J.S., Qin, H.S., Wang, F.M. et al. (2017). Metal-ion-activated DNAzymes used for regulation of telomerase activity in living cells. *Chem. Eur. J.* 23 (47): 11226–11229.

138 Chakrayarthy, M., Aung-Htut, M.T., Le, B.T., and Veedu, R.N. (2017). Novel chemically-modified DNAzyme targeting integrin alpha-4 RNA transcript as a potential molecule to reduce inflammation in multiple sclerosis. *Sci. Rep.* 7: 8.

139 Cazenave, C., Chevrier, M., Thuong, N.T., and Helene, C. (1987). Rate of degradation of alpha-oligodeoxynucleotides and beta-oligodeoxynucleotides in *Xenopus oocytes* – implications for anti-messenger strategies. *Nucleic Acids Res.* 15 (24): 10507–10521.

140 Shaw, J.P., Kent, K., Bird, J. et al. (1991). Modified deoxyoligonucleotides stable to exonuclease degradation in serum. *Nucleic Acids Res.* 19 (4): 747–750.

141 Kurreck, J. (2003). Antisense technologies – improvement through novel chemical modifications. *Eur. J. Biochem.* 270 (8): 1628–1644.

142 Khvorova, A. and Watts, J.K. (2017). The chemical evolution of oligonucleotide therapies of clinical utility. *Nat. Biotechnol.* 35 (3): 238–248.

143 Cummins, L.L., Owens, S.R., Risen, L.M. et al. (1995). Characterization of fully 2′-modified oligoribonucleotide heteroduplex and homoduplex hybridization and nuclease sensitivity. *Nucleic Acids Res.* 23 (11): 2019–2024.

144 Sproat, B.S., Lamond, A.I., Beijer, B. et al. (1989). Highly efficient chemical synthesis of 2′-*O*-methyloligoribonucleotides and tetrabiotinylated derivatives – novel probes that are resistant to degradation by RNA or DNA specific nucleases. *Nucleic Acids Res.* 17 (9): 3373–3386.

145 Rusconi, C.P., Roberts, J.D., Pitoc, G.A. et al. (2004). Antidote-mediated control of an anticoagulant aptamer in vivo. *Nat. Biotechnol.* 22 (11): 1423–1428.

146 Narlikar, G.J. and Herschlag, D. (1997). Mechanistic aspects of enzymatic catalysis: lessons from comparison of RNA and protein enzymes. *Annu. Rev. Biochem.* 66: 19–59.

147 Davidi, D., Noor, E., Liebermeister, W. et al. (2016). Global characterization of in vivo enzyme catalytic rates and their correspondence to in vitro k(cat) measurements. *Proc. Natl. Acad. Sci. U.S.A.* 113 (12): 3401–3406.

148 Smuga, D., Majchrzak, K., Sochacka, E., and Nawrot, B. (2010). RNA-cleaving 10–23 deoxyribozyme with a single amino acid-like functionality operates without metal ion cofactors. *New J. Chem.* 34 (5): 934–948.

149 Li, Z.W., Liu, Y., Liu, G.F. et al. (2014). Position-specific modification with imidazolyl group on 10–23 DNAzyme realized catalytic activity enhancement. *Bioorg. Med. Chem.* 22 (15): 4010–4017.

150 Zhu, J.F., Li, Z.W., Wang, Q. et al. (2016). The contribution of adenines in the catalytic core of 10–23 DNAzyme improved by the 6-amino group modifications. *Bioorg. Med. Chem. Lett.* 26 (18): 4462–4465.

151 He, J.L., Zhang, D., Wang, Q. et al. (2011). A novel strategy of chemical modification for rate enhancement of 10–23 DNAzyme: a combination of A9 position and 8-aza-7-deaza-2′-deoxyadenosine analogs. *Org. Biomol. Chem.* 9 (16): 5728–5736.

152 Wang, Q., Zhang, D., Liu, Y. et al. (2012). A structure-activity relationship study for 2′-deoxyadenosine analogs at A9 position in the catalytic core of 10–23 DNAzyme for rate enhancement. *Nucl. Acid Ther.* 22 (6): 423–427.

153 Liu, Y., Li, Z.W., Liu, G.F. et al. (2013). Breaking the conservation of guanine residues in the catalytic loop of 10–23 DNAzyme by position-specific nucleobase modifications for rate enhancement. *Chem. Commun.* 49 (44): 5037–5039.

154 Suryawanshi, H., Lalwani, M.K., Ramasamy, S. et al. (2012). Antagonism of microRNA function in zebrafish embryos by using locked nucleic acid enzymes (LNAzymes). *ChemBioChem* 13 (4): 584–589.

155 Trepanier, J.B., Tanner, J.E., and Alfieri, C. (2008). Reduction in intracellular HCV RNA and virus protein expression in human hepatoma cells following treatment with 2′-*O*-methyl-modified anti-core deoxyribozyme. *Virology* 377 (2): 339–342.

156 Vester, B., Lundberg, L.B., Sørensen, M.D. et al. (2002). LNAzymes: incorporation of LNA-type monomers into DNAzymes markedly increases RNA cleavage. *J. Am. Chem. Soc.* 124: 13682–13683.

157 Jadhav, V.M., Scaria, V., and Maiti, S. (2009). Antagomirzymes: oligonucleotide enzymes that specifically silence MicroRNA function. *Angew. Chem. Int. Ed.* 48 (14): 2557–2560.

158 Reyes-Gutierrez, P. and Alvarez-Salas, L.M. (2009). Cleavage of HPV-16 E6/E7 mRNA mediated by modified 10–23 deoxyribozymes. *Oligonucleotides* 19 (3): 233–242.

159 Lam, C.H. and Perrin, D.M. (2010). Introduction of guanidinium-modified deoxyuridine into the substrate binding regions of DNAzyme 10–23 to enhance target affinity: implications for DNAzyme design. *Bioorg. Med. Chem. Lett.* 20: 5119–5122.

160 Soutschek, J., Akinc, A., Bramlage, B. et al. (2004). Therapeutic silencing of an endogenous gene by systemic administration of modified siRNAs. *Nature* 432 (7014): 173–178.

161 Krutzfeldt, J., Rajewsky, N., Braich, R. et al. (2005). Silencing of microRNAs in vivo with antagomirs. *Nature* 438 (7068): 685–689.

162 Lorenz, C., Hadwiger, P., John, M. et al. (2004). Steroid and lipid conjugates of siRNAs to enhance cellular uptake and gene silencing in liver cells. *Bioorg. Med. Chem. Lett.* 14 (19): 4975–4977.

163 Lai, W.Y., Chen, C.Y., Yang, S.C. et al. (2014). Overcoming EGFR T790M-based tyrosine kinase inhibitor resistance with an allele-specific DNAzyme. *Mol. Ther. Nucleic Acids* 3: 9.

164 Zhou, J.H. and Rossi, J. (2017). Aptamers as targeted therapeutics: current potential and challenges. *Nat. Rev. Drug Discovery* 16 (3): 181–202.

165 Xu, Z.J., Yang, L.F., Sun, L.Q., and Cao, Y. (2012). Use of DNAzymes for cancer research and therapy. *Chin. Sci. Bull.* 57 (26): 3404–3408.

166 Pun, S.H., Tack, F., Bellocq, N.C. et al. (2004). Targeted delivery of RNA-cleaving DNA enzyme (DNAzyme) to tumor tissue by transferrin-modified, cyclodextrin-based particles. *Cancer Biol. Ther.* 3 (7): 641–650.

167 Tack, F., Bakker, A., Maes, S. et al. (2006). Modified poly(propylene imine) dendrimers as effective transfection agents for catalytic DNA enzymes (DNAzymes). *J. Drug Target.* 14 (2): 69–86.

168 Fan, H.H., Zhao, Z.L., Yan, G.B. et al. (2015). A smart DNAzyme-MnO_2 nanosystem for efficient gene silencing. *Angew. Chem. Int. Ed.* 54 (16): 4801–4805.

169 Chen, F., Bai, M., Cao, K. et al. (2017). Programming enzyme-initiated autonomous DNAzyme nanodevices in living cells. *ACS Nano* 11 (12): 11908–11914.

170 Chen, F., Bai, M., Cao, K. et al. (2017). Fabricating MnO_2 nanozymes as intracellular catalytic DNA circuit generators for versatile imaging of base-excision repair in living cells. *Adv. Funct. Mater.* 27 (45): 9.

171 Choi, C.H.J., Hao, L.L., Narayan, S.P. et al. (2013). Mechanism for the endocytosis of spherical nucleic acid nanoparticle conjugates. *Proc. Natl. Acad. Sci. U.S.A.* 110 (19): 7625–7630.

172 Yehl, K., Joshi, J.P., Greene, B.L. et al. (2012). Catalytic deoxyribozyme-modified nanoparticles for RNAi-independent gene regulation. *ACS Nano* 6 (10): 9150–9157.

173 Somasuntharam, I., Yehl, K., Carroll, S.L. et al. (2016). Knockdown of TNF-alpha by DNAzyme gold nanoparticles as an anti-inflammatory therapy for myocardial infarction. *Biomaterials* 83: 12–22.

174 Petree, J.R., Yehl, K., Galior, K. et al. (2018). Site-selective RNA splicing nanozyme: DNAzyme and RtcB conjugates on a gold nanoparticle. *ACS Chem. Biol.* 13 (1): 215–224.

175 Peng, H.Y., Li, X.F., Zhang, H.Q., and Le, X.C. (2017). A microRNA-initiated DNAzyme motor operating in living cells. *Nat. Commun.* 8: 13.

176 Lee, Y. and Geckeler, K.E. (2010). Carbon nanotubes in the biological interphase: the relevance of noncovalence. *Adv. Mater.* 22 (36): 4076–4083.

177 Villa, C.H., McDevitt, M.R., Escorcia, F.E. et al. (2008). Synthesis and biodistribution of oligonucleotide-functionalized, tumor-targetable carbon nanotubes. *Nano Lett.* 8 (12): 4221–4228.

178 Yim, T.J., Liu, J.W., Lu, Y. et al. (2005). Highly active and stable DNAzyme – carbon nanotube hybrids. *J. Am. Chem. Soc.* 127 (35): 12200–12201.

179 Subramanian, N., Kanwar, J.R., Akilandeswari, B. et al. (2015). Chimeric nucleolin aptamer with survivin DNAzyme for cancer cell targeted delivery. *Chem. Commun.* 51 (32): 6940–6943.

180 Röthlisberger, P., Gasse, C., and Hollenstein, M. (2017). Nucleic acid aptamers: emerging applications in medical imaging, nanotechnology, neurosciences, and drug delivery. *Int. J. Mol. Sci.* 18 (11): 2430.

181 Hallett, M.A., Dalal, P., Sweatman, T.W., and Pourmotabbed, T. (2013). The distribution, clearance, and safety of an anti-MMP-9 DNAzyme in normal and MMTV-PyMT transgenic mice. *Nucl. Acid Ther.* 23 (6): 379–388.

182 Zaborowska, Z., Schubert, S., Kurreck, J., and Erdmann, V.A. (2005). Deletion analysis in the catalytic region of the 10–23 DNA enzyme. *FEBS Lett.* 579: 554–558.

183 Cheng, M.P., Zhou, J., Jia, G.Q. et al. (2017). Relations between the loop transposition of DNA G-quadruplex and the catalytic function of DNAzyme. *Biochim. Biophys. Acta, Gen. Subj.* 1861 (8): 1913–1920.

184 Jean, J.M. and Hall, K.B. (2001). 2-Aminopurine fluorescence quenching and lifetimes: role of base stacking. *Proc. Natl. Acad. Sci. U.S.A.* 98 (1): 37–41.

185 Saran, R. and Liu, J.W. (2016). A comparison of two classic Pb^{2+}-dependent RNA-cleaving DNAzymes. *Inorg. Chem. Front.* 3 (4): 494–501.

186 Peracchi, A., Bonaccio, M., and Credali, A. (2017). Local conformational changes in the 8–17 deoxyribozyme core induced by activating and inactivating divalent metal ions. *Org. Biomol. Chem.* 15 (41): 8802–8809.

187 Saran, R., Yao, L., Hoang, P., and Liu, J.W. (2018). Folding of the silver aptamer in a DNAzyme probed by 2-aminopurine fluorescence. *Biochimie* 145: 145–150.

188 Zhou, W.H., Ding, J.S., and Liu, J.W. (2016). A highly specific sodium aptamer probed by 2-aminopurine for robust Na^+ sensing. *Nucleic Acids Res.* 44 (21): 10377–10385.

189 Zhou, W.H., Saran, R., Ding, J.S., and Liu, J.W. (2017). Two completely different mechanisms for highly specific Na^+ recognition by DNAzymes. *ChemBioChem* 18 (18): 1828–1835.

190 Li, Z.W., Zhu, J.F., and He, J.L. (2016). Conformational studies of 10–23 DNAzyme in solution through pyrenyl-labeled 2′-deoxyadenosine derivatives. *Org. Biomol. Chem.* 14 (41): 9846–9858.

191 Mayer, G. and Heckel, A. (2006). Biologically active molecules with a "light switch". *Angew. Chem. Int. Ed.* 45 (30): 4900–4921.

192 Hwang, K., Wu, P.W., Kim, T. et al. (2014). Photocaged DNAzymes as a general method for sensing metal ions in living cells. *Angew. Chem. Int. Ed.* 53 (50): 13798–13802.

193 Keiper, S. and Vyle, J.S. (2006). Reversible photocontrol of deoxyribozyme-catalyzed RNA cleavage under multiple-turnover conditions. *Angew. Chem. Int. Ed.* 45 (20): 3306–3309.

194 Wang, X.Y., Feng, M.L., Xiao, L. et al. (2016). Postsynthetic modification of DNA phosphodiester backbone for photocaged DNAzyme. *ACS Chem. Biol.* 11 (2): 444–451.

195 Liu, Y. and Sen, D. (2004). Light-regulated catalysis by an RNA-cleaving deoxyribozyme. *J. Mol. Biol.* 341 (4): 887–892.

196 Liang, X.G., Zhou, M.G., Kato, K., and Asanuma, H. (2013). Photoswitch nucleic acid catalytic activity by regulating topological structure with a universal supraphotoswitch. *ACS Synth. Biol.* 2 (4): 194–202.

197 Kamiya, Y., Arimura, Y., Ooi, H. et al. (2018). Development of visible-light-responsive RNA scissors based on a 10–23 DNAzyme. *ChemBioChem* 19 (12): 1305–1311.

198 Piskur, J. and Rupprecht, A. (1995). Aggregated DNA in ethanol solution. *FEBS Lett.* 375 (3): 174–178.

199 Yu, T.M., Zhou, W.H., and Liu, J.W. (2018). Ultrasensitive DNAzyme-based Ca^{2+} detection boosted by ethanol and a solvent-compatible scaffold for aptazyme design. *ChemBioChem* 19 (1): 31–36.

200 Zhou, W.H., Saran, R., Chen, Q.Y. et al. (2016). A new Na^{+}-dependent RNA-cleaving DNAzyme with over 1000-fold rate acceleration by ethanol. *ChemBioChem* 17 (2): 159–163.

201 Nakano, S., Horita, M., Kobayashi, M., and Sugimoto, N. (2017). Catalytic activities of ribozymes and DNAzymes in water and mixed aqueous media. *Catalysts* 7 (12): 14.

202 Behera, A.K., Schlund, K.J., Mason, A.J. et al. (2013). Enhanced deoxyribozyme-catalyzed RNA ligation in the presence of organic cosolvents. *Biopolymers* 99 (6): 382–391.

203 Liu, K., Zheng, L.F., Liu, Q. et al. (2014). Nucleic acid chemistry in the organic phase: from functionalized oligonucleotides to DNA side chain polymers. *J. Am. Chem. Soc.* 136 (40): 14255–14262.

204 Tanaka, K. and Okahata, Y. (1996). A DNA-lipid complex in organic media and formation of an aligned cast film. *J. Am. Chem. Soc.* 118 (44): 10679–10683.

205 Gao, J.Y., Shimada, N., and Maruyama, A. (2015). Science enhancement of deoxyribozyme activity by cationic copolymers. *Biomater. Sci.* 3 (2): 308–316.

206 Saito, K., Shimada, N., and Maruyama, A. (2016). Cooperative enhancement of deoxyribozyme activity by chemical modification and added cationic copolymer. *Sci. Technol. Adv. Mater.* 17 (1): 437–442.

207 Abe, H., Abe, N., Shibata, A. et al. (2012). Structure formation and catalytic activity of DNA dissolved in organic solvents. *Angew. Chem. Int. Ed.* 51 (26): 6475–6479.

208 Hook, K.D., Chambers, J.T., and Hili, R. (2017). A platform for high-throughput screening of DNA-encoded catalyst libraries in organic solvents. *Chem. Sci.* 8 (10): 7072–7076.

209 Hollenstein, M. (2012). Nucleoside triphosphates – building blocks for the modification of nucleic acids. *Molecules* 17 (11): 13569–13591.

210 Hocek, M. (2014). Synthesis of base-modified 2′-deoxyribonucleoside triphosphates and their use in enzymatic synthesis of modified DNA for applications in bioanalysis and chemical biology. *J. Org. Chem.* 79 (21): 9914–9921.

211 Hottin, A. and Marx, A. (2016). Structural insights into the processing of nucleobase-modified nucleotides by DNA polymerases. *Acc. Chem. Res.* 49 (3): 418–427.

212 Sakthivel, K. and Barbas, C.F. (1998). Expanding the potential of DNA for binding and catalysis: highly functionalized dUTP derivatives that are substrates for thermostable DNA polymerases. *Angew. Chem. Int. Ed.* 37 (20): 2872–2875.

213 Jäger, S. and Famulok, M. (2004). Generation and enzymatic amplification of high-density functionalized DNA double strands. *Angew. Chem. Int. Ed.* 43 (25): 3337–3340.

214 Jäger, S., Rasched, G., Kornreich-Leshem, H. et al. (2005). A versatile toolbox for variable DNA functionalization at high density. *J. Am. Chem. Soc.* 127 (43): 15071–15082.

215 Wang, Y.J., Ng, N., Liu, E.K. et al. (2017). Systematic study of constraints imposed by modified nucleoside triphosphates with protein-like side chains for use in in vitro selection. *Org. Biomol. Chem.* 15 (3): 610–618.

216 Hollenstein, M. (2013). Deoxynucleoside triphosphates bearing histamine, carboxylic acid, and hydroxyl residues – synthesis and biochemical characterization. *Org. Biomol. Chem.* 11: 5162–5172.

217 Cheng, Y., Dai, C., Peng, H. et al. (2011). Design, synthesis, and polymerase-catalyzed incorporation of click-modified boronic acid-TTP analogues. *Chem. Asian J.* 6 (10): 2747–2752.

218 Wang, K., Wang, D.Z., Ji, K.L. et al. (2015). Post-synthesis DNA modifications using a trans-cyclooctene click handle. *Org. Biomol. Chem.* 13 (3): 909–915.

219 Balintova, J., Simonova, A., Bialek-Pietras, M. et al. (2017). Carborane-linked 2′-deoxyuridine 5′-*O*-triphosphate as building block for polymerase synthesis of carborane-modified DNA. *Bioorg. Med. Chem. Lett.* 27 (21): 4786–4788.

220 Welter, M., Verga, D., and Marx, A. (2016). Sequence-specific incorporation of enzyme–nucleotide chimera by DNA polymerases. *Angew. Chem. Int. Ed.* 55 (34): 10131–10135.

221 Palluk, S., Arlow, D.H., de Rond, T. et al. (2018). De novo DNA synthesis using polymerase–nucleotide conjugates. *Nat. Biotechnol.* 36: 645.

222 Verga, D., Welter, M., Steck, A.L., and Marx, A. (2015). DNA polymerase-catalyzed incorporation of nucleotides modified with a G-quadruplex-derived DNAzyme. *Chem. Commun.* 51 (34): 7379–7381.

223 Matyasovsky, J., Perlikova, P., Malnuit, V. et al. (2016). 2-Substituted dATP derivatives as building blocks for polymerase-catalyzed synthesis of DNA modified in the minor groove. *Angew. Chem. Int. Ed.* 55 (51): 15856–15859.

224 Jakubovska, J., Tauraitė, D., Birštonas, L., and Meškys, R. (2018). N^4-Acyl-2′-deoxycytidine-5′-triphosphates for the enzymatic synthesis of modified DNA. *Nucleic Acids Res.* 46 (12): 5911–5923.

225 Rothlisberger, P., Levi-Acobas, F., Sarac, I. et al. (2017). Facile immobilization of DNA using an enzymatic his-tag mimic. *Chem. Commun.* 53 (97): 13031–13034.

226 Chen, T.J., Hongdilokkul, N., Liu, Z.X. et al. (2016). Evolution of thermophilic DNA polymerases for the recognition and amplification of C2′-modified DNA. *Nat. Chem.* 8 (6): 557–563.

227 Larsen, A.C., Dunn, M.R., Hatch, A. et al. (2016). A general strategy for expanding polymerase function by droplet microfluidics. *Nat. Commun.* 7: 9.

228 Pinheiro, V.B., Taylor, A.I., Cozens, C. et al. (2012). Synthetic genetic polymers capable of heredity and evolution. *Science* 336: 341–344.

229 Wang, Z.M., Xu, W.L., Liu, L., and Zhu, T.F. (2016). A synthetic molecular system capable of mirror-image genetic replication and transcription. *Nat. Chem.* 8 (7): 698–704.

230 Houlihan, G., Arangundy-Franklin, S., and Holliger, P. (2017). Engineering and application of polymerases for synthetic genetics. *Curr. Opin. Biotechnol.* 48: 168–179.

231 Latham, J.A., Johnson, R., and Toole, J.J. (1994). The application of a modified nucleotide in aptamer selection: novel thrombin aptamers containing -(1-pentynyl)-2'-deoxyuridine. *Nucleic Acids Res.* 22 (14): 2817–2822.

232 Santoro, S.W., Joyce, G.F., Sakthivel, K. et al. (2000). RNA cleavage by a DNA enzyme with extended chemical functionality. *J. Am. Chem. Soc.* 122 (11): 2433–2439.

233 Lam, C.H., Hipolito, C.J., Hollenstein, M., and Perrin, D.M. (2011). A divalent metal-dependent self-cleaving DNAzyme with a tyrosine side chain. *Org. Biomol. Chem.* 9 (20): 6949–6954.

234 Geyer, C.R. and Sen, D. (1997). Evidence for the metal-cofactor independence of an RNA phosphodiester-cleaving DNA enzyme. *Chem. Biol.* 4: 579–593.

235 Faulhammer, D. and Famulok, M. (1997). Characterization and divalent metal-ion dependence of in vitro selected deoxyribozymes which cleave DNA/RNA chimeric oligonucleotides. *J. Mol. Biol.* 269: 188–202.

236 Kasprowicz, A., Stokowa-Soltys, K., Jezowska-Bojczuk, M. et al. (2017). Characterization of highly efficient RNA-cleaving DNAzymes that function at acidic pH with no divalent metal-ion cofactors. *ChemistryOpen* 6 (1): 46–56.

237 Raines, R.T. (1998). Ribonuclease A. *Chem. Rev.* 98: 1045–1065.

238 Perrin, D.M., Garestier, T., and Hélène, C. (1999). Expanding the catalytic repertoire of nucleic acid catalysts: simultaneous incorporation of two modified deoxyribonucleoside triphosphates bearing ammonium and imidazolyl functionalities. *Nucleosides Nucleotides* 18: 377–391.

239 Cahova, H., Pohl, R., Bednarova, L. et al. (2008). Synthesis of 8-bromo-, 8-methyl- and 8-phenyl-dATP and their polymerase incorporation into DNA. *Org. Biomol. Chem.* 6 (20): 3657–3660.

240 Lam, C., Hipolito, C., and Perrin, D.M. (2008). Synthesis and enzymatic incorporation of modified deoxyadenosine triphosphates. *Eur. J. Org. Chem.*: 4915–4923.

241 Perrin, D.M., Garestier, T., and Hélène, C. (2001). Bridging the gap between proteins and nucleic acids: a metal-independent RNase A mimic with two protein-like functionalities. *J. Am. Chem. Soc.* 123 (8): 1556–1563.

242 Ting, R., Thomas, J.M., and Perrin, D.M. (2007). Kinetic characterization of a *cis*- and *trans*-acting M^{2+}-independent DNAzyme that depends on synthetic RNaseA-like functionality – burst-phase kinetics from the coalescence of two active DNAzyme folds. *Can. J. Chem.* 85 (4): 313–329.

243 Li, Y.F. and Breaker, R.R. (1999). Kinetics of RNA degradation by specific base catalysis of transesterification involving the 2′-hydroxyl group. *J. Am. Chem. Soc.* 121 (23): 5364–5372.

244 Lermer, L., Roupioz, Y., Ting, R., and Perrin, D.M. (2002). Toward an RNaseA mimic: a DNAzyme with imidazoles and cationic amines. *J. Am. Chem. Soc.* 124: 9960–9961.

245 Thomas, J.M., Yoon, J.-K., and Perrin, D.M. (2009). Investigation of the catalytic mechanism of a synthetic DNAzyme with protein-like functionality: an RNaseA mimic? *J. Am. Chem. Soc.* 131: 5648–5658.

246 Hipolito, C.J., Hollenstein, M., Lam, C.H., and Perrin, D.M. (2011). Protein-inspired modified DNAzymes: dramatic effects of shortening side-chain length of 8-imidazolyl modified deoxyadenosines in selecting RNaseA mimicking DNAzymes. *Org. Biomol. Chem.* 9 (7): 2266–2273.

247 Ting, R., Lermer, L., and Perrin, D.M. (2004). Triggering DNAzymes with light: a photoactive C8 thioether-linked adenosine. *J. Am. Chem. Soc.* 126 (40): 12720–12721.

248 Thomas, J.M., Ting, R., and Perrin, D.M. (2004). High affinity DNAzyme-based ligands for transition metal cations – a prototype sensor for Hg^{2+}. *Org. Biomol. Chem.* 2 (3): 307–312.

249 Hollenstein, M., Hipolito, C., Lam, C. et al. (2008). A highly selective DNAzyme sensor for mercuric ions. *Angew. Chem. Int. Ed.* 47 (23): 4346–4350.

250 May, J.P., Ting, R., Lermer, L. et al. (2004). Covalent schiff base catalysis and turnover by a DNAzyme: A M^{2+} independent AP-endonuclease mimic. *J. Am. Chem. Soc.* 126 (13): 4145–4156.

251 Deglane, G., Abes, S., Michel, T. et al. (2006). Impact of the guanidinium group on hybridization and cellular uptake of cationic oligonucleotides. *ChemBioChem* 7 (4): 684–692.

252 Hollenstein, M., Hipolito, C.J., Lam, C.H., and Perrin, D.M. (2013). Toward the combinatorial selection of chemically modified DNAzyme RNase A mimics active against all-RNA substrates. *ACS Comb. Sci.* 15 (4): 174–182.

253 Wang, Y.J., Liu, E.K., Lam, C.H., and Perrin, D.M. (2018). A densely modified M^{2+}-independent DNAzyme that cleaves RNA efficiently with multiple catalytic turnover. *Chem. Sci.* 9 (7): 1813–1821.

254 Sidorov, A.V., Grasby, J.A., and Williams, D.M. (2004). Sequence-specific cleavage of RNA in the absence of divalent metal ions by a DNAzyme incorporating imidazolyl and amino functionalities. *Nucleic Acids Res.* 32 (4): 1591–1601.

255 Joyce, G.F., Dai, X., and De Mesmaeker, A. (1996). Amide cleavage by a ribozyme: correction. *Science* 272: 18–19.

256 Zinnen, S.P., Domenico, K., Wilson, M. et al. (2002). Selection, design, and characterization of a new potentially therapeutic ribozyme. *RNA* 8 (2): 214–228.

257 Taylor, A.I., Pinheiro, V.B., Smola, M.J. et al. (2015). Catalysts from synthetic genetic polymers. *Nature* 518 (7539): 427–430.

258 Vater, A. and Klussmann, S. (2015). Turning mirror-image oligonucleotides into drugs: the evolution of Spiegelmer therapeutics. *Drug Discovery Today* 20 (1): 147–155.

259 Cui, L., Peng, R.Z., Fu, T. et al. (2016). Biostable L-DNAzyme for sensing of metal ions in biological systems. *Anal. Chem.* 88 (3): 1850–1855.

260 Pech, A., Achenbach, J., Jahnz, M. et al. (2017). A thermostable D-polymerase for mirror-image PCR. *Nucleic Acids Res.* 45 (7): 3997–4005.

261 Liu, C., Cozens, C., Jaziri, F. et al. (2018). Phosphonomethyl oligonucleotides as backbone-modified artificial genetic polymers. *J. Am. Chem. Soc.* 140 (21): 6690–6699.

262 Alipanahi, B., Delong, A., Weirauch, M.T., and Frey, B.J. (2015). Predicting the sequence specificities of DNA- and RNA-binding proteins by deep learning. *Nat. Biotechnol.* 33 (8): 831–838.

263 Yoshida, M., Hinkley, T., Tsuda, S. et al. (2018). Using evolutionary algorithms and machine learning to explore sequence space for the discovery of antimicrobial peptides. *Chem* 4 (3): 533–543.

264 Kong, D.H., Yeung, W., and Hili, R. (2017). In vitro selection of diversely functionalized aptamers. *J. Am. Chem. Soc.* 139 (40): 13977–13980.

265 Renders, M., Miller, E., Hollenstein, M., and Perrin, D. (2015). A method for selecting modified DNAzymes without the use of modified DNA as a template in PCR. *Chem. Commun.* 51 (7): 1360–1362.

266 Yu, H.Y., Zhang, S., and Chaput, J.C. (2012). Darwinian evolution of an alternative genetic system provides support for TNA as an RNA progenitor. *Nat. Chem.* 4 (3): 183–187.

267 Chen, Z., Lichtor, P.A., Berliner, A.P. et al. (2018). Evolution of sequence-defined highly functionalized nucleic acid polymers. *Nat. Chem.* 10 (4): 420–427.

24

Light-Utilizing DNAzymes

Adam Barlev and Dipankar Sen

Simon Fraser University, Department of Chemistry, 8888 University Drive, Burnaby BC V5A 1S6, Canada

24.1 Introduction

The geological record indicates that the evolution of photoenzymes that produced oxygen completely changed the composition of the earth's atmosphere at a relatively early stage of life's evolution [1]. Besides photosynthetic organisms, all clades of life show in their evolutionary history photolyase enzymes [2], which repair a common type of photoinduced DNA damage [3]. The RNA world hypothesis supposes that at an early stage of evolution, all enzymatic activities were carried out by RNA enzymes [4]. Was the RNA world confined in the dark, or were there RNA photoenzymes? We may never know with certainty, but if we observe RNA or other nucleic acid photoenzymes, then at least we can know whether or not it is a possibility.

Enzymes are defined by two overriding characteristics: binding of substrates, measured in K_M, and their ability to enhance the rate of a reaction of those substrates bound, measured as a turnover frequency [5]. Photoenzymes, which are a small but important subclass of enzymes, require the absorption of light to carry out their activity [6]. As we have learned, while proteins contemporaneously make up the majority of enzymes, both natural and artificial nucleic acid enzymes have been isolated and characterized, composed of RNA [7], DNA [8], or more exotic nucleic acid polymers [9]. A true photochemical DNAzyme (PDZ) must satisfy these same requirements as protein enzymes, reversibly binding its substrates and products, while waiting to absorb light to carry out its enhancement of reactivity, as shown in Figure 24.1. A direct measure of the efficiency of a photoenzyme is the quantum yield, which takes into account the absorbance of photons and which fraction of absorbed photons result in catalysis. Another important consideration for a photoenzyme is the energy or wavelength of the photon absorbed to trigger reactivity. Of course, to be called a DNAzyme, one must be made of DNA, which for the purposes of this discussion means an oligonucleotide with a phosphodiester backbone, and consist of mainly the four standard bases; however, also included for consideration are chemically modified DNAs with nonstandard bases. As of yet, no natural or artificial photochemical ribozymes or photochemical XNAs have been

Ribozymes, First Edition. Edited by Sabine Müller, Benoît Masquida, and Wade Winkler.

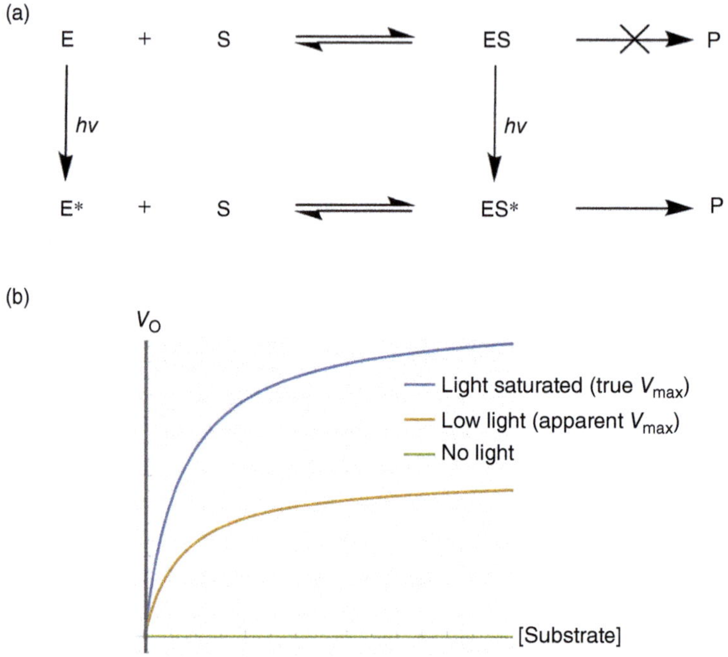

Figure 24.1 (a) A photoenzyme binds its substrates in the ground state but only performs catalysis upon absorption of light. (b) Upon saturating concentration of the substrate, an enzyme achieves a maximum rate. Similarly, a photoenzyme achieves its true maximum rate upon saturating light flux.

reported. However, there is no fundamental reason why the reactivity present in a DNA oligonucleotide would be impossible in an RNA or XNA context.

24.2 PhotoDNAzymes (PDZs)

The existing literature on photoDNAzymes is relatively new. Our own lab has developed two classes of PDZs based on their catalysis of thymine dimer repair, summarized in Figure 24.2 [10]. Indeed, the proteinaceous class of enzymes known as photolyases is exemplars of photoenzymes, capable of utilizing visible light with quantum yields near unity as a source of energy to split thymine dimers formed primarily by the absorption of UVA and UVB light by DNA and revert distorted base conformations back to their original state [11]. To carry out this photocatalysis, photolyases employ a pair of nucleotide-like cofactors to absorb light [12] and transfer an electron to its substrate dimer [13]. Inspired by photolyases, particularly, the CPD photolyase, capable of repairing cyclobutane (CPD) pyrimidine dimers formed as a result of DNA irradiation, and some previous results showing that catalytic antibodies with photolyase activity utilize tryptophan as a photosensitizer [14], we undertook an in vitro selection for a serotonin-utilizing PDZ for CPD thymine dimer repair. The substrate dimer was synthesized such that its repair led to progressive cleavage of the full-length substrate into two shorter DNA strands [15]. This selection was successful, yielding the sequence of SERO1C, a PDZ with a k_{cat} of 0.03 min^{-1} and K_M of 3 μM under the irradiation conditions [16]. SERO1C's serotonin cofactor, however, was bound much less strongly than cofactors that are in proteinaceous photolyases, with a functional K_D estimated at 73 μM. This PDZ had a maximum rate enhancement at 340 nm, the wavelength where serotonin's absorbance peaks, as expected. The in vitro selection was designed to isolate such a sequence, and it did. However, the in vitro selection was designed with a negative selection step as well, in which the library of sequences was subjected to irradiation without a supplied cofactor. From the pool of DNA sequences that could survive the negative selection round, repairing thymine dimers without a cofactor but with UV-B light, the sequence of a new DNAzyme, UV1C, was cloned [15]. This DNAzyme has been investigated intensively [17, 18], and many lines of evidence have led us to conclude that its active site contains a parallel stranded, two-layer G-quadruplex, which acts as its catalytic chromophore. Relative to double-stranded DNA, G-quadruplexes feature a tail in their absorbance spectrum extending to 305 nm [19]. This change in absorbance correlates well with its maximum rate enhancement, at 305 nm. UV1C, too, displayed multiple-turnover catalysis, with a more respectable K_M of 0.58 μM and a turnover of 4.5 min^{-1}, with its Michaelis plot shown in Figure 24.2b. By way of comparison, a catalytic antibody for thymine dimer repair showed a K_M of 6.5 μM and k_{cat} of 1.2 min^{-1}.

UV1C has been studied intensively using substrate analogs, contact cross-linking, and replacement of potential active-site guanines with the less easily oxidized base inosine [17, 18]. A variety of strategies had been attempted to improve UV1C, but the breakthrough occurred when active-site guanines were replaced with the guanine analog 6-methyl isoxanthopterin (6-MI), which redshifted the action spectrum of UV1C to the edge of the visible [20]. Interestingly, the new activity brought about by the 6MI base could also be brought to bear from a residue not in the G-quadruplex active site, thus leaving the full activity of the quadruplex-dependent pathway at shorter wavelengths and introducing a new activity throughout the UV-A region. This non-quadruplex position proved versatile for substitution, accepting even the large unusual base 7-(2,2′-bithien-5-yl)-imidazo[4,5-*b*]pyridine (Dss) [21], bringing its activity firmly into the visible range of the spectrum.

The only other true PDZ in the literature is also a catalyst for a 2+2 cycloaddition, but this time in the formation direction [22]. This PDZ uses an opening in a three-way junction to bind a small molecule capable of intermolecular 2+2 reaction in the proximity of a benzophenone base [23], which acts as a triplet photosensitizer, shown in Figure 24.2c. In addition to enhancing the rate, as compared to the uncatalyzed reaction, this DNAzyme imparts a chirality to the newly generated products. This PDZ was not analyzed for Michaelis–Menten kinetics, but the stoichiometry was such that at 20% catalyst loading, some multiple turnovers must have occurred. The results of this approach are modest improvements in rate and EE, but it is important to recognize that this PDZ was not subjected to rounds of SELEX to increase substrate binding. Indeed, to the contrary, this PDZ is probably quite promiscuous and would bind weakly to a variety of potential substrates.

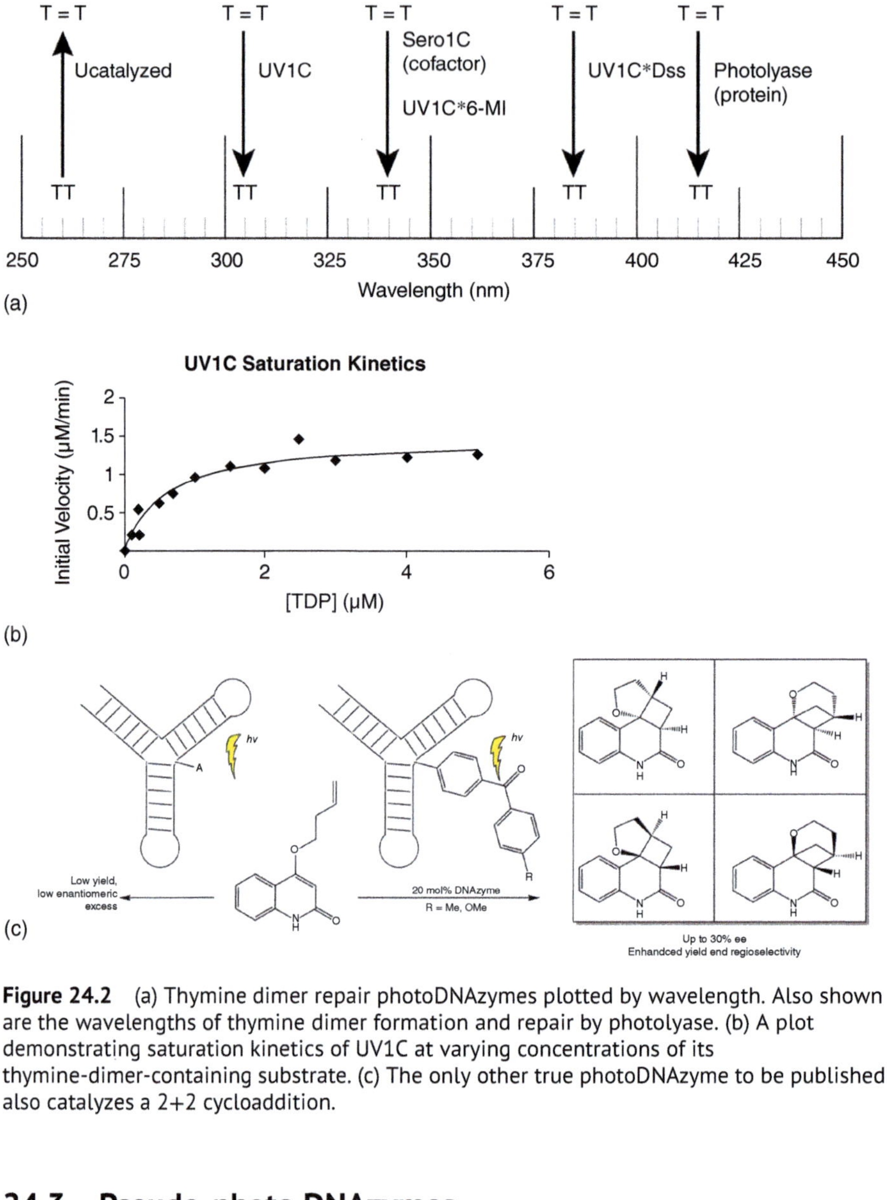

Figure 24.2 (a) Thymine dimer repair photoDNAzymes plotted by wavelength. Also shown are the wavelengths of thymine dimer formation and repair by photolyase. (b) A plot demonstrating saturation kinetics of UV1C at varying concentrations of its thymine-dimer-containing substrate. (c) The only other true photoDNAzyme to be published also catalyzes a 2+2 cycloaddition.

24.3 Pseudo-photo DNAzymes

In another category of photocatalysts lie the pseudo-photoDNAzymes. These species catalyze a photochemical reaction but are unable to turn over a substrate, because of too tight binding to their substrates and products. An example of such a system from our lab is the UV1C sequence acting upon a thymine dimer substrate in which the phosphodiester backbone is intact. We demonstrated that in fact, this system comes to photoequilibrium orders of magnitude more quickly than the uncatalyzed substrate and that the unique chemical environment in which UV1C holds its substrate leads to a shift in the photostationary state of the reaction [24]. Regardless, without the ability to turn over, this cannot be truly categorized as a PDZ.

Highly influential to our own work was the investigation of the most common oxidative damage product of guanine, 8-oxoguanine (8OG) [25], as a photocatalyst for thymine dimer repair, reported by Nguyen and Burrows [26]. In this system, 8OG acted as a chromophore with a redshifted absorption spectrum relative to that of guanine itself. In some of these experiments, an oligonucleotide containing 8OG was complementary to a thymine-dimer-containing strand, as shown in Figure 24.3a. In these cases, the 8OG strand could be coaxed to turn over by way of repeated heat denaturation and reannealing, with five repair events observed per oligonucleotide [27].

The cofactor at the active site of photolyase is a reduced flavin. Perhaps, inspired by this, when flavin was incorporated into a DNA double strand by Behrens, Carell, and coworkers, they observed irradiation-dependent repair of a nicked thymine dimer within the same oligonucleotide, schematically shown in Figure 24.3b [28]. Interestingly, the repair rates observed when the distance (in base pairs) between the flavin and the dimer was increased were only weakly affected, reiterating the long-range charge-transfer capability inherent in DNA duplexes. Being effective catalysts, these systems are only capable of a single turnover but not true enzymatic behavior.

In 2014, new evidence emerged for long-lived chargetr-ansfer states in unmodified DNA under UV irradiation [29]. Such charge transfer allows DNA, normally thought of chemically rather inert, to display unusual and unexpected kinds of reactivity under UV irradiation, with one such situation shown in Figure 24.3c. This activity is sequence-dependent thymine dimer repair, originally described by Holman, Rokita, and coworkers [30] and later investigated by Bucher, Carell, and coworkers [31]. In both cases, guanine was critical to the repair process. This is certainly an example of photocatalysis, but the catalytic sequence remains covalently attached to its substrate, so it can not strictly be classed as a PDZ. Regardless, this remarkable property of self-repair appears to be an intrinsic property of DNA as a material, and it is fascinating to speculate whether it can be expanded to a system that operates in trans. Indeed, we believe only a small fraction of the sequences that can self-repair have been identified.

24.4 Photoactive DNA Components for Future PDZ Design

Should one set out to design a new PDZ, either by selection from random sequence pools or by rational design, one should be familiar with a toolkit of parts that may be combined to give the reactivity one desires. At the level of the base, a great variety of chromophores are available for incorporation into DNA via automated oligonucleotide synthesis, such that the substituted chromophores profoundly change the absorbance spectrum of the DNA while still maintaining a DNA structure, such as shown in Figure 24.4a. One of the most commonly used such bases is 2-aminopurine (2-AP) [32]. As an adenosine analog, it extends the spectrum of DNA incorporating

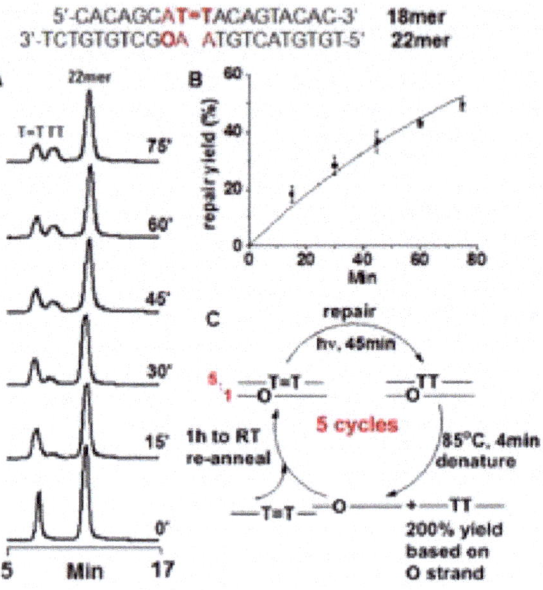

5'-CACAGCAT=TACAGTACAC-3' 18mer
3'-TCTGTGTCGOA ATGTCATGTGT-5' 22mer
A
22mer
T=T TT
75'
60'
45'
30'
15'
0'
5
Min
17
B
repair yield (%)
60
40
20
0
0
20
40
60
80
Min
C
repair
hv, 45min
T=T
TT
O
5 cycles
1h to RT
re-anneal
85°C, 4min
denature
T=T
O
TT
200% yield
based on
O strand

(a)

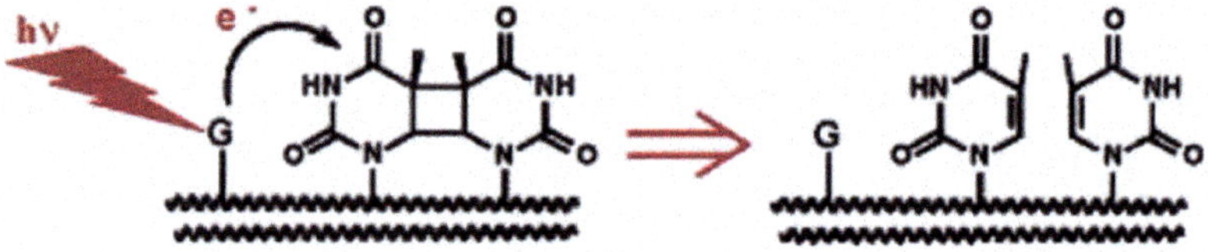

hν
e⁻
G
HN
NH
N
O
G

(b)

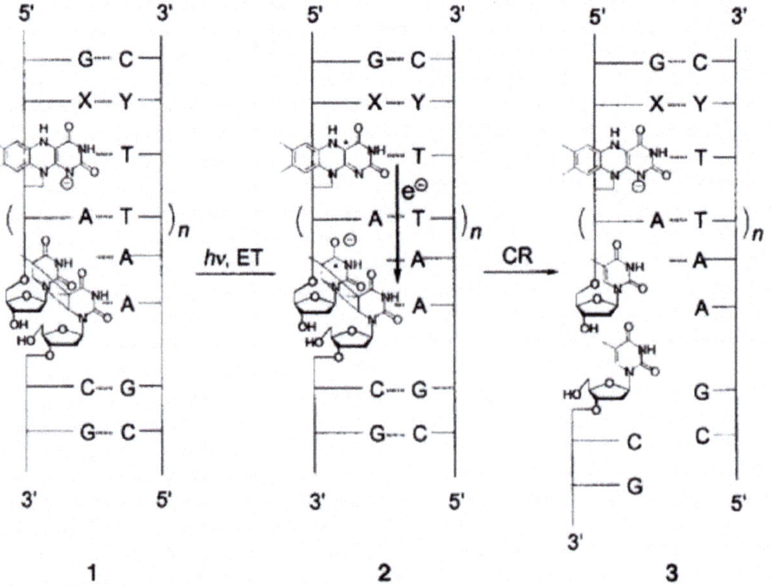

5'
3'
G C
X Y
T
A T
A
A
C G
G C
hv, ET
e⊖
CR
3'
5'
1
2
3

(c)

Figure 24.3 (a) Pseudo-photoDNAzymes are photocatalysts for a single turnover. An oligonucleotide containing 8OG is capable of repairing multiple substrates if melted and reannealed to a new substrate. Reprinted with permission from J. Am. Chem. Soc. 133: 14586–14589, Copyright (2011) American Chemical Society. (b) Guanine on its own promotes self-repair if in the same strand. Reprinted with permission from J. Am. Chem. Soc. 129: 6–7, Copyright (2007) American Chemical Society. (c) Flavin may be used to repair thymine dimers at various distances from the dimer, but in this case, repair doesn't lead to dissociation of the product to allow the association of a new substrate. Reprinted with permission from Angew. Chem. Int. Ed. 41: 1763–1766, Copyright (2002) WILEY-VCH Verlag GmbH.

it to 305 nm and beyond; however, it is noteworthy that when replacing A with 2-AP in a duplex, that duplex's stability, as measured by melting temperature, drops significantly [33]. A guanine analog we have used to some success is 6-MI [34], whose absorbance peaks at 340 nm. 6-MI is noteworthy in its ability to participate in forming G-quartets [35], and a 6MI-C base pair is known to lower the melting temperature of a duplex less than 2-AP does [36]. While these two compounds, 2-AP and 6-MI, are both more or less base-like, with hydrogen bond donors and acceptors similar to the natural nucleic acid bases, other more unusual chromophores may impart a more interesting and versatile photoreactivity.

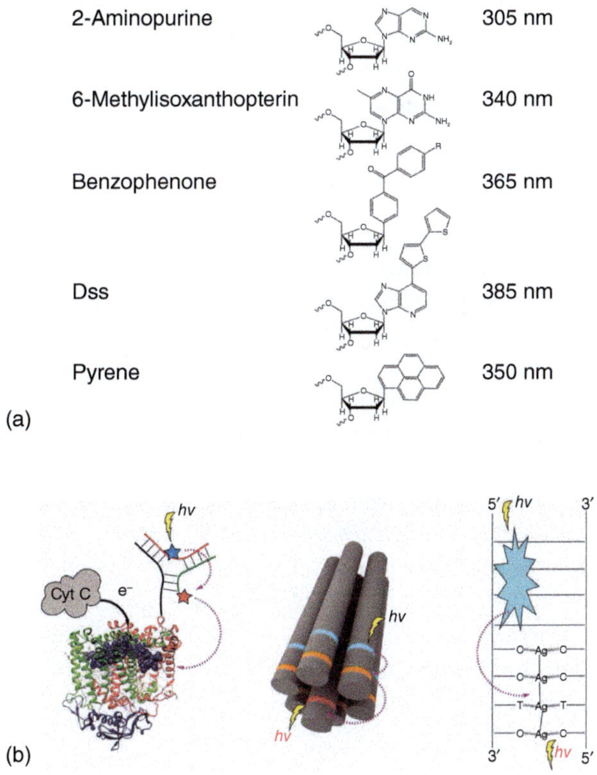

Figure 24.4 (a) The 2-Aminopurine design of photochemical DNAzymes may begin at the level of the individual 6-Methylisoxanthopterin base. A variety of wavelengths in the UV-A are available. (b) These bases may be combined by using standard duplexes, three-way junctions, and DNA origami. Beyond purely organic chromophores, inorganic nanostructures are a novel way to enhance the wavelength range of a photochemical DNAzyme. Reprinted with permission from J. Am. Chem. Soc. 136 (47): 16618–16625. Copyright (2014) American Chemical Society.

DNA, on its own, has an extremely low quantum yield for direct formation of a triplet excited state, but contact triplet–triplet transfer can be efficiently promoted by a variety of sensitizer compounds such as acetophenone, benzophenone, or methoxy benzophenone [23]. These small molecules themselves are not chiral, but when incorporated into a chiral oligonucleotide as nucleobase substitutions, these allow the transfer of chiral information to photosensitized reactions [22]. Should one be willing to depart further from the canonical, the chromophore Dss, (see above for full name), besides possessing a strong absorbance at 385 nm, is also capable of forming a base pair based exclusively on shape complementarity with its partner pyrrole-2-carbaldehyde, (Pa) [37]. To date, the Dss*Pa pair has been successfully used in an aptamer selection [38] but has only been incorporated in a PDZ system by rational insertion [20]. The existence of such a base pair should allow in vitro selection of PDZs with a significant absorbance tail extending into the visible region of the spectrum. Going even further away from standard bases, the fluorescent aromatic nucleobases developed by the Kool Lab are highly photostable, and strong stacking between adjacent modifications allows dramatic changes in their absorption spectra [39].

Flavin and flavin-like cofactors used by photolyases are obvious choices for the development of future photoDNAzymes [40], but there have been significant challenges in their use. The photophysical properties and evolutionary history of flavin and pterin cofactors have led some to speculate about their role in the RNA world, proposing that flavin and pterin cofactors comprised a system capable of using the energy from light to supply the energy to form ATP [41]. It is relatively easy to append a flavin moiety to the 5′ end of an oligonucleotide [42], but it is more synthetically challenging to incorporate a flavin in the body of a DNA strand. If what is desired is to incorporate the catalytic chemistry of flavin into a PDZ, perhaps, a non-covalent approach is possible. Indeed, flavin cofactors in protein enzymes are rarely covalently bound. RNA aptamers to flavin have already been developed [43]. Even without selection, an abasic site can act as a flavin binding site with a K_D in the micromolar range [44].

In our analysis of the reaction mechanism of the UV1C DNAzyme, it became clear that the G-quadruplex imparts a different photoreactivity than is seen in standard duplexes [45]. More and more evidence has emerged to support this supposition. The absorption spectrum of a G-quadruplex is slightly redshifted relative to that of a duplex, an observation sometimes exploited to track G-quadruplex melting via monitoring of hyperchromicity at 290 nm [19]. When excited, bases in a duplex are capable of relaxing from their excited state by proton transfer across the base pairing face [29]. The Hoogsteen base pairs formed by the guanines in a G-quadruplex, however, are incapable of such a relaxation [46]. This leads to longer excited-state lifetimes during which a photochemical reaction is possible. When an electron hole is introduced in a G-quadruplex either by a tethered photo-oxidant or pulse radiolysis, it decays at a lower rate than is observed in duplex DNA [47]. Theoretical investigations of this hole decay reiterate the role of Hoogsteen base pairing in the enhanced stability of oxidized guanines in G-quadruplex DNA relative to guanines in single-stranded and duplex DNAs [46].

Photolyases notably use two cofactors, with one specializing in harvesting light and transferring the harvested energy to another cofactor, that specializes in photocatalysis [11]. Such a strategy has not been realized yet in a PDZ, but many elements have been developed in other contexts, which could be used to imbue a PDZ with this property. A three-way junction is a common motif used in DNA nanostructures [48]. By conjugating a single strand to the photosynthetic reaction center isolated from the purple bacterium *R. sphaeroides*, Yan and coworkers were able to direct a pair of chromophores, each on their own DNA strand, to precise locations relative to the photosynthetic reaction center, as shown schematically in Figure 24.4b [49]. Without both dyes present on oligonucleotides in the junction, no photoreactivity was observed; however, when the construct was assembled as intended, energy transfer from one dye to the other dye triggered cytochrome oxidation at the reaction center. This proof-of-principle work could be easily applied to a DNAzyme system with the appropriate built-in action spectrum. In preliminary work by Yan, Liu, and coworkers, a seven-helix bundle, another common motif in DNA origami, was employed to organize several antenna chromophores to transmit their energy to a single central acceptor chromophore, mimicking a strategy seen in some photosynthetic reaction systems [50].

Nucleic acids have become a key building block in nanotechnology, both on their own in supramolecular assemblies and also in combination with inorganic components [51]. Upconverting nanoparticles absorb multiple photons in the IR and emit in the UV region of the spectrum [52]. Such a property could, in principle, be exploited to power a PDZ with a high-intensity source in the IR region. Alternatively, mismatched pyrimidine base pairs can template the formation of metal nanoclusters. These nanoclusters have been used to demonstrate the quantum coherence of UV light absorption by duplex DNA [53]. While used in this case to demonstrate the photophysical properties of duplex DNA, these inorganic clusters have an absorption in the visible and may prove a novel way to modify the action spectrum of a PDZ.

24.5 Conclusions

Nucleic acids are highly versatile substances, with many unique and complex properties. In order to be suitable genetic materials, in which capacity they do serve in living systems, they must be stable to many environmental influences, and for the most part, nucleic acids are quite photostable. In contrast to that are the known examples of photoreactivity presented in PDZs and pseudo-DNAzymes. Our understanding of the potential reactivity of PDZs and of their potential consequences for the RNA world is now just in its infancy.

References

1 Schopf, J.W. (2011). The paleobiological record of photosynthesis. *Photosynth. Res.* 107 (1): 87–101.
2 Kanai, S., Kikuno, R., Toh, H. et al. (1997). Molecular evolution of the photolyase–blue-light photoreceptor family. *J. Mol. Evol.* 45 (5): 535–548.
3 Essen, L.O. and Klar, T. (2006). Light-driven DNA repair by photolyases. *Cell. Mol. Life Sci.* 63 (11): 1266–1277.
4 Benner, S.A., Ellington, A.D., and Tauer, A. (1989). Modern metabolism as a palimpsest of the RNA world. *Proc. Natl. Acad. Sci. U.S.A.* 86 (18): 7054–7058.
5 Frey, P.A. and Hegeman, A.D. (2007). *Enzymatic Reaction Mechanisms*. Oxford University Press.
6 Begley, T.P. (1994). Photoenzymes: a novel class of biological catalysts. *Acc. Chem. Res.* 27 (12): 394–401.
7 Lorsch, J.R. and Szostak, J.W. (1996). Chance and necessity in the selection of nucleic acid catalysts. *Acc. Chem. Res.* 29 (2): 103–110.
8 Li, Y. and Sen, D. (1997). Toward an efficient DNAzyme. *Biochemistry* 36 (18): 5589–5599.
9 Pinheiro, V.B. and Holliger, P. (2014). Towards XNA nanotechnology: new materials from synthetic genetic polymers. *Trends Biotechnol.* 32 (6): 321–328.
10 Barlev, A. and Sen, D. (2018). DNA's encounter with ultraviolet light: an instinct for self-preservation? *Acc. Chem. Res.* 51 (2): 526–533.
11 Zhong, D. (2015). Electron transfer mechanisms of DNA repair by photolyase. *Annu. Rev. Phys. Chem.* 66 (1): 691–715.
12 Malhotra, K., Kim, S.T., and Sancar, A. (1994). Characterization of a medium wavelength type DNA photolyase: purification and properties of photolyase from *Bacillus firmus*. *Biochemistry* 33 (29): 8712–8718.
13 Warren, J.J., Ener, M.E., Vlcek, A. et al. (2012). Electron hopping through proteins. *Coord. Chem. Rev.* 256 (21–22): 2478–2487.
14 Cochran, A.G., Sugasawara, R., and Schultz, P.G. (1988). Photosensitized cleavage of a thymine dimer by an antibody. *J. Am. Chem. Soc.* 110 (23): 7888–7890.
15 Chinnapen, D.J.-F. and Sen, D. (2004). A deoxyribozyme that harnesses light to repair thymine dimers in DNA. *Proc. Natl. Acad. Sci. U.S.A.* 101 (1): 65–69.
16 Thorne, R.E., Chinnapen, D.J.-F., Sekhon, G.S., and Sen, D. (2009). A deoxyribozyme, Sero1C, uses light and serotonin to repair diverse pyrimidine dimers in DNA. *J. Mol. Biol.* 388 (1): 21–29.
17 Chinnapen, D.J.-F. and Sen, D. (2007). Towards elucidation of the mechanism of UV1C, a deoxyribozyme with photolyase activity. *J. Mol. Biol.* 365 (5): 1326–1336.
18 Sekhon, G.S. and Sen, D. (2009). Unusual DNA-DNA cross-links between a photolyase deoxyribozyme, UV1C, and its bound oligonucleotide substrate. *Biochemistry* 48 (27): 6335–6347.
19 Mergny, J.-L. and Lacroix, L. (2009). UV melting of G-quadruplexes. *Curr. Protoc. Nucleic Acid Chem.* 37 (1): 17.1.1–17.1.15.
20 Barlev, A. and Sen, D. (2013). Catalytic DNAs that harness violet light to repair thymine dimers in a DNA substrate. *J. Am. Chem. Soc.* 135 (7): 2596–2603.

21 Hirao, I. and Kimoto, M. (2012). Unnatural base pair systems toward the expansion of the genetic alphabet in the central dogma. *Proc. Jpn. Acad. Ser. B* 88 (7): 345–367.
22 Gaß, N., Gebhard, J., and Wagenknecht, H.-A. (2017). Photocatalysis of a [2+2] cycloaddition in aqueous solution using DNA three-way junctions as chiral photoDNAzymes. *ChemPhotoChem* 1 (2): 48–50.
23 Gaß, N. and Wagenknecht, H.-A. (2015). Synthesis of benzophenone nucleosides and their photocatalytic evaluation for [2+2] cycloaddition in aqueous media. *Eur. J. Org. Chem.* 2015 (30): 6661–6668.
24 Barlev, A., Sekhon, G.S., Bennet, A.J., and Sen, D. (2016). DNA repair by DNA: the UV1C DNAzyme catalyzes photoreactivation of cyclobutane thymine dimers in DNA more effectively than their de novo formation. *Biochemistry* 55 (43): 6010–6018.
25 Burrows, C.J. (2009). Surviving an oxygen atmosphere: DNA damage and repair. *ACS Symp. Ser. Am. Chem. Soc.* 2009: 147–156.
26 Nguyen, K.V. and Burrows, C.J. (2012). Whence flavins? Redox-active ribonucleotides link metabolism and genome repair to the RNA world. *Acc. Chem. Res.* 45 (12): 2151–2159.
27 Nguyen, K.V. and Burrows, C.J. (2011). A prebiotic role for 8-oxoguanosine as a flavin mimic in pyrimidine dimer photorepair. *J. Am. Chem. Soc.* 133 (37): 14586–14589.
28 Behrens, C., Burgdorf, L.T., Schwögler, A., and Carell, T. (2002). Weak distance dependence of excess electron transfer in DNA. *Angew. Chem. Int. Ed.* 41 (10): 1763–1766.
29 Bucher, D.B., Pilles, B.M., Carell, T., and Zinth, W. (2014). Charge separation and charge delocalization identified in long-living states of photoexcited DNA. *Proc. Natl. Acad. Sci. U.S.A.* 111 (12): 4369–4374.
30 Holman, M.R., Ito, T., and Rokita, S.E. (2007). Self-repair of thymine dimer in duplex DNA. *J. Am. Chem. Soc.* 129 (1): 6–7.
31 Bucher, D.B., Kufner, C.L., Schlueter, A. et al. (2016). UV-induced charge transfer states in DNA promote sequence selective self-repair. *J. Am. Chem. Soc.* 138 (1): 186–190.
32 Manoj, P., Mohan, H., Mittal, J.P. et al. (2007). Charge transfer from 2-aminopurine radical cation and radical anion to nucleobases: a pulse radiolysis study. *Chem. Phys.* 331 (2–3): 351–358.
33 Sholokh, M., Sharma, R., Shin, D. et al. (2015). Conquering 2-aminopurine's deficiencies: highly emissive isomorphic guanosine surrogate faithfully monitors guanosine conformation and dynamics in DNA. *J. Am. Chem. Soc.* 137 (9): 3185–3188.
34 Hawkins, M.E., Pfleiderer, W., Balis, F.M. et al. (1997). Fluorescence properties of pteridine nucleoside analogs as monomers and incorporated into oligonucleotides. *Anal. Biochem.* 244 (1): 86–95.
35 Gros, J., Rosu, F., Amrane, S. et al. (2007). Guanines are a quartet's best friend: impact of base substitutions on the kinetics and stability of tetramolecular quadruplexes. *Nucleic Acids Res.* 35 (9): 3064–3075.

36 Wojtuszewski Poulin, K., Smirnov, A.V., Hawkins, M.E. et al. (2009). Conformational heterogeneity and quasi-static self-quenching in DNA containing a fluorescent guanine analogue, 3MI or 6MI. *Biochemistry* 48 (37): 8861–8868.

37 Kimoto, M., Mitsui, T., Yokoyama, S., and Hirao, I. (2010). A unique fluorescent base analogue for the expansion of the genetic alphabet. *J. Am. Chem. Soc.* 132 (14): 4988–4989.

38 Kimoto, M., Nakamura, M., and Hirao, I. (2016). Post-ExSELEX stabilization of an unnatural-base DNA aptamer targeting VEGF165 toward pharmaceutical applications. *Nucleic Acids Res.* 44 (15): 7487–7494.

39 Xu, W., Chan, K.M., and Kool, E.T. (2017). Fluorescent nucleobases as tools for studying DNA and RNA. *Nat. Chem.* 9 (11): 1043–1055.

40 Schwögler, A. and Carell, T. (2000). Toward catalytically active oligonucleotides: synthesis of a flavin nucleotide and its incorporation into DNA. *Org. Lett.* 2 (10): 1415–1418.

41 Kritsky, M., Telegina, T., Vechtomova, Y. et al. (2010). Excited flavin and pterin coenzyme molecules in evolution. *Biochemistry (Mosc)* 75 (10): 1200–1216.

42 Kino, K., Miyazawa, H., and Sugiyama, H. (2007). User-friendly synthesis and photoirradiation of a flavin-linked Oligomer. *Genes Environ.* 29 (1): 23–28.

43 Lauhon, C.T. and Szostak, J.W. (1995). RNA aptamers that bind flavin and nicotinamide redox cofactors. *J. Am. Chem. Soc.* 117 (4): 1246–1257.

44 Sankaran, N.B., Nishizawa, S., Seino, T. et al. (2006). Abasic-site-containing oligodeoxynucleotides as aptamers for riboflavin. *Angew. Chem. Int. Ed.* 45 (10): 1563–1568.

45 Miannay, F.-A., Banyasz, A., Gustavsson, T., and Markovitsi, D. (2009). Excited states and energy transfer in G-quadruplexes. *J. Phys. Chem. C* 113 (27): 11760–11765.

46 Yang, Y., Yang, W., Su, H. et al. (2018). Mechanistic insights into the photogeneration and quenching of guanine radical cation via one-electron oxidation of G-quadruplex DNA. *Phys. Chem. Chem. Phys.* 20 (19): 13598–13606.

47 Choi, J., Park, J., Tanaka, A. et al. (2013). Hole trapping of G-quartets in a G-quadruplex. *Angew. Chem. Int. Ed.* 52 (4): 1134–1138.

48 Duckett, D.R. and Lilley, D.M. (1990). The three-way DNA junction is a Y-shaped molecule in which there is no helix-helix stacking. *EMBO J.* 9 (5): 1659–1664.

49 Dutta, P.K., Levenberg, S., Loskutov, A. et al. (2014). A DNA-directed light-harvesting/reaction center system. *J. Am. Chem. Soc.* 136 (47): 16618–16625.

50 Dutta, P.K., Varghese, R., Nangreave, J. et al. (2011). DNA-directed artificial light-harvesting antenna. *J. Am. Chem. Soc.* 133 (31): 11985–11993.

51 Pinheiro, A.V., Han, D., Shih, W.M., and Yan, H. (2011). Challenges and opportunities for structural DNA nanotechnology. *Nat. Nanotechnol.* 6: 763–772.

52 Carling, C.-J., Nourmohammadian, F., Boyer, J.-C., and Branda, N.R. (2010). Remote-control photorelease of caged compounds using near-infrared light and upconverting nanoparticles. *Angew. Chem. Int. Ed.* 49 (22): 3782–3785.

53 Volkov, I.L., Reveguk, Z.V., Serdobintsev, P.Y. et al. (2018). DNA as UV light-harvesting antenna. *Nucleic Acids Res.* 46 (7): 3543–3551.

25

Diverse Applications of DNAzymes in Computing and Nanotechnology

Matthew R. Lakin[1,2], *Darko Stefanovic*[1,2], *and Milan N. Stojanovic*[3]

[1]*University of New Mexico, Department of Computer Science, Albuquerque 1, University of New Mexico, NM 87131, USA*
[2]*University of New Mexico, Center for Biomedical Engineering, Albuquerque 1, University of New Mexico, NM 87131, USA*
[3]*Columbia University, Department of Medicine, Division of Experimental Therapeutics, 630 West 168th Street, New York, NY 10032, USA*

25.1 Introduction

Advances in molecular biology over recent decades have enabled scientists to gain unprecedented insight into the structure and function of biological systems for sensing, autoregulation, and control at the nanoscale. The field of bionanotechnology has developed in parallel with the related goal of recapitulating the function of these naturally occurring systems in synthetic analogs. The approaches taken in this endeavor have ranged from direct biomimicry to more radical re-imaginings of molecular processes that incorporate biological or biochemical components, but which operate differently from their natural counterparts. Of particular interest to us has been the development of *molecular computing* devices that process information at the nanoscale.

By molecular computing, we mean the creation of rationally designed ensembles of molecules that detect input signals, which are usually the presence or absence of other molecules (but may be more general, e.g. pH or light signals). The design of the computing ensemble is such that these input signals cause conformational shifts that in turn trigger further reactions. The goal of the molecular computing circuit designer is to design a molecule, or an ensemble of molecules, such that the triggered interactions carry out some desired computational task. By connecting different components together in a network or cascade, such systems can mimic the behavior of electronic logic circuits. The outputs from such a circuit are typically observed in the laboratory using fluorescence spectroscopy, but this is primarily a matter of convenience and a wide range of potential outputs can be linked to a molecular circuit, including the release of payload compounds (e.g. drugs, in a biomedical application) or the activation of a color change or pH change in solution. In a sense, the molecular circuit functions as a connecting fabric that can be wired in a modular fashion to link a wide range of detected inputs to a wide range of possible

Ribozymes, First Edition. Edited by Sabine Müller, Benoît Masquida, and Wade Winkler.

output channels, with potentially arbitrary computational processing happening in between.

While work on molecular computing has explored a number of different biochemical frameworks for computation, including protein-based systems [1, 2] and in vitro transcription and translation circuits [3–5], our work (and this chapter) focuses primarily on the use of DNA as a computational substrate. Because DNA interactions are sequence-specific, the intended structures, and functions can be programmed solely in terms of the complementarity of sequences; moreover, as the Watson–Crick base pairing complementarity rules are relatively straightforward, these interactions can be predicted and designed using relatively simple thermodynamic and kinetic models, for which mature software tools are available. In recent years the availability of affordable and high-quality synthetic DNA oligonucleotides [6] has greatly enhanced the capabilities of experimental groups working in this area. Indeed, a number of groups are now exploring the use of DNA as a medium for high-density archival storage of large amounts of information [7, 8]. As described above, we are interested in using DNA to implement novel information processing architectures that compute using the dynamics of interactions between molecular circuit components. The large numbers of DNA molecules present in even small reaction volumes offer great promise for massively parallel computing [9] as well as for carrying out computations that approach theoretical limits on energy consumption [10, 11].

Our research has particularly focused on the use of DNAzymes, also known as deoxyribozymes, to perform tasks at the nanoscale, specifically, computational tasks. DNAzymes are not known to occur in nature, and the known DNAzyme catalytic motifs have been isolated in in vitro evolution experiments [12–16]. We have used RNA-cleaving DNAzymes to create multi-input logic gates by grafting input-binding modules onto the DNAzyme strand, with the cleavage of fluorescently labeled reporter substrates providing an output reporting channel. This chapter reviews our work, and that of others, on using DNAzymes for computational tasks and for other tasks in nanotechnology. We discuss our work on DNAzyme-based molecular computers, which we have scaled up to build large-scale automata comprising parallel gate arrays, as well as multilayer signaling cascades. We cover our development (both experimental and theoretical) of adaptive DNAzyme-based networks that can be trained with, or learn, specific responses to input stimuli. We also review work on nanoscale molecular robots built from DNAzymes. Thus, here we present a diverse range of applications of DNAzymes in molecular computing and nanotechnology.

25.2 Loop-Based Control of DNAzyme Logic Gates

Adleman's seminal paper [9] demonstrated the potential in harnessing molecular interactions to solve computational problems. A programmed self-assembly process was adopted to construct DNA molecules whose sequences encoded candidate solutions to a small instance of the Hamiltonian path problem in graph traversal and a laboratory protocol was used to isolate the true solutions from among the candidates. Our early work in this area took an alternative approach. Rather than using individual molecules to encode solutions, we adopted a population-based approach by creating large numbers of DNA-based logic gates whose behavior could then be observed *en masse* using bulk fluorescence measurements. We were inspired in this line of thinking by a number of reviews [12, 13] on nucleic acid catalysts and aptamers, which led us to the concept of logic gates based on DNAzymes, which could be controlled using nucleic acid inputs introduced at the beginning of the experiment, and which would then proceed to compute autonomously and without any further human intervention. We also drew inspiration from contemporaneous work from the Breaker and Ellington groups on the design of allosterically regulated ribozymes, which were controlled using both small molecule and oligonucleotide effectors, and which demonstrated the implementation of multi-input logic [17–19].

We used RNA-cleaving DNAzymes, which are single strands of DNA that can catalyze the cleavage of complementary chimeric substrate molecules at a cleavage site marked by a single RNA base. Our work has focused in large part on the "E6" [15] and "8–17" [16] catalytic motifs, although numerous others exist. Figure 25.1a illustrates the basic structure of a DNAzyme strand, using the E6 structural motif as an example. The central catalytic core sequence is largely conserved (with the exception of the small central loop in the E6 motif, as we will discuss below). This domain coordinates the binding of divalent metal cation cofactors that are required for the cleavage process. The catalytic core is flanked by two variable substrate binding arms, which recognize and bind to a complementary substrate molecule and position it correctly so that cleavage can occur. Figure 25.1a shows the means by which a DNAzyme binds to a substrate molecule, cleaves the substrate at the cleavage site, and unbinds to release the two shorter product molecules back into solution. By attaching a fluorophore–quencher pair to the two ends of the substrate, cleavage can be observed due to the increase in fluorescence as the quencher is cleaved away from the fluorophore. Crucially, the DNAzyme strand functions as a catalyst in this process: it is unchanged and can hence cleave further substrates in a multiple turnover process that provides inherent signal amplification capabilities. A number of factors govern the efficiency of this process, including the catalytic efficiency of the DNAzyme itself, the buffer conditions, and the design of the substrate binding arms of the DNAzyme strand: if too short, substrates cannot bind to the DNAzyme; if too long, product release is hampered.

To design our first generation of DNAzyme-based logic gates, we identified three locations on the E6 DNAzyme motif that can serve as control modules: the two substrate binding arms and the small loop in the middle of the catalytic core. We introduced control capabilities by grafting *molecular beacons* into these regions, to serve as detection elements for nucleic acid inputs. A molecular beacon is a nucleic acid hairpin loop that can be opened by binding of a complementary nucleic acid to the loop part of the hairpin; by turning one of the substrate-binding arms of a DNAzyme into a molecular beacon such that the closing of the beacon blocked the substrate from binding to that arm, we created a DNAzyme that would only be catalytically active in the presence of the corresponding input oligonucleotide that could open the molecular beacon and enable DNAzyme-catalyzed cleavage of the substrate to proceed, as shown in Figure 25.1b. We called a DNAzyme logic gate that responds positively to the presence of input i_1 a YES i_1 gate [21]. (Here and

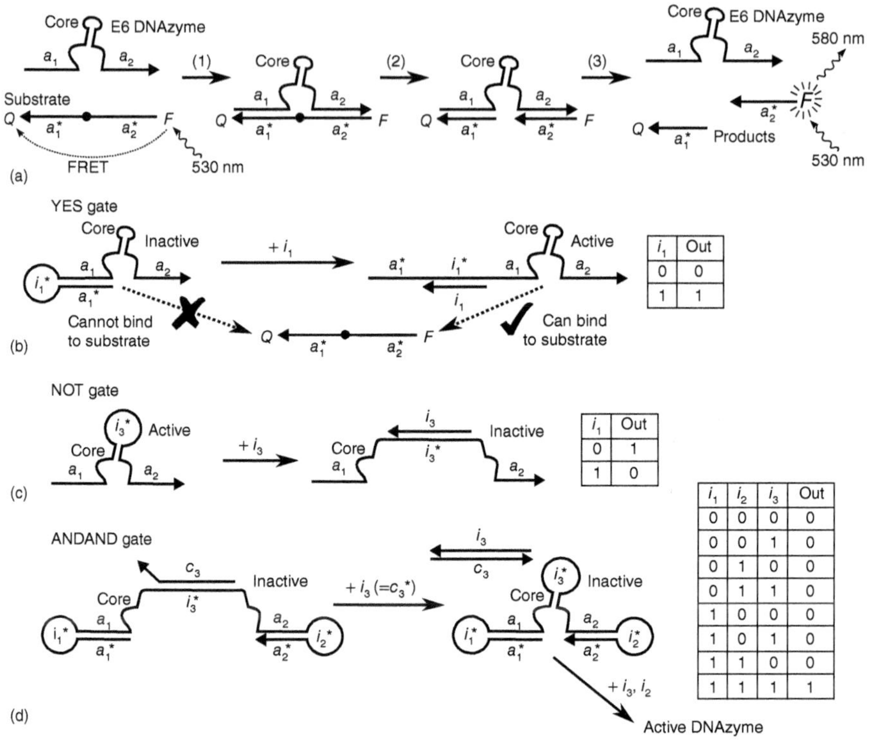

Figure 25.1 DNAzyme loop-based logic gates. (a) DNAzyme structure (here the "E6" catalytic motif) and DNAzyme-catalyzed cleavage of a substrate molecule that consists of sequences complementary to the DNAzyme's substrate binding arms (a_1^* and a_2^*), with a cleavage site in between (denoted by a small disc). The reaction is observed via increasing fluorescence when the fluorophore on one end of the substrate is cleaved away from its quencher. (b) A YES gate constructed by grafting a molecular beacon [20] input detection module onto a DNAzyme such that the closed stem of the molecular beacon blocks one of the substrate binding arms (a_1) unless it is opened by binding of the corresponding input i_1. (c) NOT gate. Extending the small loop in the catalytic core of the "E6" DNAzyme to a full-size input-binding loop allows us to implement a NOT gate. In the absence of input i_3, the catalytic core structure folds as normal and the DNAzyme is functional. However, when i_3 is added, the loop is opened, which distorts the catalytic core and prevents the DNAzyme from binding to the substrate. (d) ANDAND gate. By pre-binding the logic gate with strands complementary to the true input strands, we can invert the sense of control of any of the input-binding loops. Here, we convert an ANDANDNOT gate into an ANDAND gate by reversing the action of the i_3 input. This is achieved by pre-binding the logic gate with the c_3 strand, which binds to the input-binding loop in the catalytic core and deforms the catalytic core. Then, addition of i_3, which is complementary to c_3, serves as a third activating input, in addition to i_1 and i_2 (not shown).

henceforth, we represent an input value of 1 by the presence of the corresponding input oligonucleotide and an input value of 0 by its absence, and detect an output value of 1 by fluorescence rise due to cleavage and 0 by the lack of such a rise in detectable fluorescence.) By blocking *both* substrate binding arms with molecular beacons sensitive to different inputs, we generalized from a YES gate to an AND gate that only activates in the presence of *both* inputs and thus implements the logical formula $i_1 \wedge i_2$ [22].

The YES and AND gates only exhibit positive control of DNAzyme activity by inputs. To implement negation (that is, a DNAzyme logic gate that is active only in the *absence* of its input) we attached a molecular beacon to the small loop within the E6 catalytic core. Changing this loop does not affect the catalytic efficiency of the DNAzyme, and binding of an input strand to this loop deforms the catalytic core and thus disables the catalytic activity of the DNAzyme. Only in the absence of the complementary input can the core domain fold into its catalytically active conformation, producing a NOT gate that computes $\neg i_3$, as shown in Figure 25.1c [22]. Furthermore, we were able to implement implicit OR gates by creating multiple parallel YES gates that simply cleave the same substrate sequence. We exploited this to build large-scale parallel arrays designed compositionally in terms of their individual gates, as outlined below. We increased the complexity of the individual gates even further by attaching control loops to all three potential control sites simultaneously, to produce an ANDANDNOT gate [22, 23]. This gate computes the logic function $i_1 \wedge i_2 \wedge \neg i_3$, as it is only active in the *presence* of both positive inputs i_1 and i_2 and in the *absence* of the negative input i_3.

To develop further three-input logic gates, we employed a strategy of pre-binding gates with blocking strands complementary to certain input binding loops, so that they were initially held open as opposed to folding closed. Then, addition of the complementary input would strip off the blocking strand by a toehold-mediated strand displacement reaction [24, 25], which we describe further below. This effectively reversed the regulatory action of those loops on the DNAzyme, so a loop that was previously a positive regulator became negative, and vice versa, and allowed us to realize any three-input logic function, such as the ANDAND gate that computes the function $i_1 \wedge i_2 \wedge i_3$, shown in Figure 25.1d. From a theoretical perspective, this set of logic gates is sufficient to implement any Boolean logic formula whose disjunctive normal form (DNF) conversion contains at most three literals per clause. We then used our collection of loop-controlled DNAzyme logic gates with implicit OR connections to build a number of demonstration circuits. Early circuits that we constructed included a half-adder [26] and a full adder [23]. In these circuits, multiple outputs were reported via different substrates labeled with different fluorophore–quencher pairs.

As a larger-scale demonstration, we created DNAzyme-based *molecular automata* that could play games of strategy against human opponents. These took the form of parallel arrays of DNAzyme logic gates executing game strategies that had been encoded as suitable Boolean formulae [27]. Even for relatively simple games, these formulae tended to be large and complex, providing an ideal challenge for scaling up our DNAzyme logic arrays.

We built three generations of game-playing automata, called MAYA-I–III. The acronym stood for "Molecular Array of YES and AND gates". The first of these, MAYA-I [28] played a symmetry-pruned version of tic-tac-toe in a 3×3 section of a well plate, which we number wells 1–9 (Figure 25.2). The automaton moves first and always claims the center well (well 5): this occurs by the human adding the required

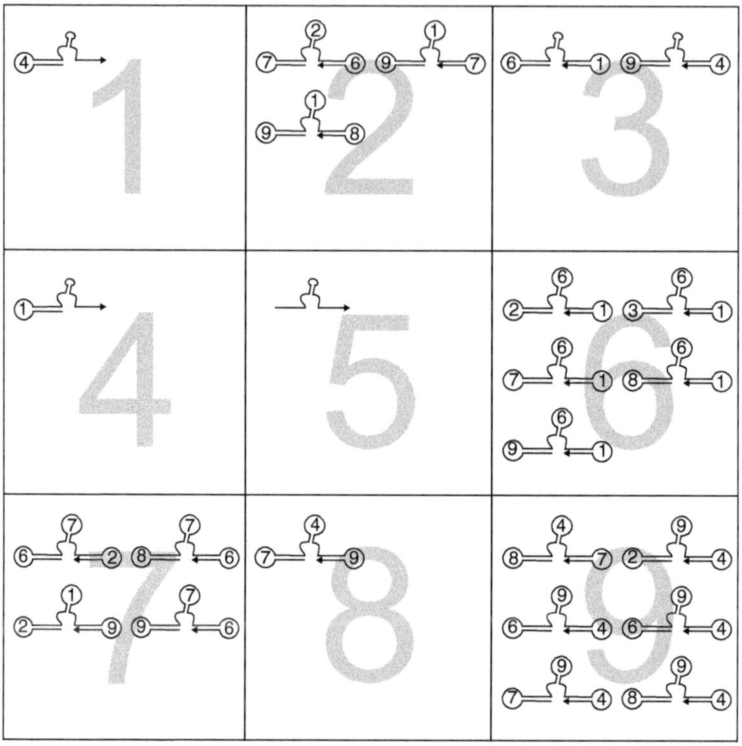

Figure 25.2 Distribution of logic gates in wells for MAYA-I, a DNAzyme-based automaton that plays a symmetry-pruned game of tic-tac-toe. The center well (5) contains a DNAzyme without any logic gate attachments, while the other wells contain logic gates. Source: Adapted from Stojanovic and Stefanovic [28].

Mg^{2+} ions to all wells; since well 5 is the only well containing a non-logic-gated DNAzyme, that DNAzyme immediately activates and produces a fluorescent signal indicating that MAYA has claimed well 5. The human moves by adding one of eight input oligonucleotides, which represent the eight remaining wells, to *all* wells in the system. These interact with the logic gates in the other wells and precisely one of these wells will light up with increased fluorescence, indicating that it is the next well claimed by the automaton. (Each well thus carries a record of all wells claimed so far.) Symmetry pruning leaves just 19 legal games, and we tested all of these [28]. The strategy encoded in the programming of the automaton is perfect, and it won 18 of the 19 possible games, with the human earning a draw only by playing perfectly.

The second-generation MAYA-II automaton [29] played an unrestricted version of tic-tac-toe, with the initial human move able to claim any of the eight remaining wells (this increased the number of legal games to 76). The number of inputs was increased from 8 to 32 by also encoding not just which wells had been claimed, but in what order. Thus the number of logic gates required also increased, from 23 for MAYA-I to a total of 128 for MAYA-II (some of these were additional gates that echoed the human's moves with a second fluorophore). From a computational standpoint,

MAYA-II is not substantially more complex than MAYA-I: the main advance was in the biochemical complexity of preparing such a large collection of logic gates. We learned that gates that work well separately may fail when placed in a parallel array with other DNAzymes and that predicting such interference and designing around would be a major challenge for the future of the molecular computing field. Indeed, in the intervening years, advances in software tools and algorithms for nucleic acid sequence design [30–33] have enabled the design of a number of large-scale molecular circuits [34–36]. Finally, our third-generation MAYA-III system [37] moved beyond fixed strategies and could be trained to play any strategy in a specially designed game. We discuss MAYA-III in detail in Section 25.4 below.

25.3 Strand Displacement Control of DNAzyme Cascades

A key limitation of the DNAzyme logic circuits described above is that they rely on implicit OR-gate connections between individual logic gates. This, and the fact that individual gates can only accommodate a maximum of three inputs, means that there exist logic formulae that cannot be implemented using that approach. Our chosen solution was to implement a mechanism for communication between DNAzyme gates, enabling them to be explicitly wired into arbitrary circuits.

Our starting point for this work was an alternative mechanism for controlling DNAzyme catalysis: the use of toehold-mediated strand displacement. Strand displacement is a powerful technique for implementing molecular computations that has previously been used to implement a range of computational frameworks including digital logic [34, 38], distributed control algorithms [36], and artificial neural networks [35]. We chose to use strand displacement for this work because it enables highly programmable reaction pathways to be engineered, an important capability for our desired application of controlling DNAzyme catalytic activity.

Our early work in this area developed a number of DNAzyme logic gates, including YES and AND gates, whose catalytic activity was controlled by the removal of inhibitor strands via toehold mediated strand displacement reactions (Figure 25.3a) [39]. The AND gates required two inputs to activate each DNAzyme simultaneously using a cooperative strand displacement reaction [41] in which each of the two input strands only displaces part of the sequestered DNAzyme strand. Rational design was used to engineer mismatched bases into strategic locations in the DNAzyme inhibitor structure to optimize the response of these gates to inputs, while simultaneously minimizing non-specific activation (leakage).

Having created DNAzyme logic gates activated by strand displacement, we then sought to connect them together to implement multilayer DNAzyme logic cascades. The challenge that we faced here was that the output of an upstream DNAzyme logic gate was the cleavage of a substrate molecule, whereas the input of a downstream gate would need to be a strand capable of activating the DNAzyme via strand displacement. Thus, we looked for a mechanism to use this upstream cleavage reaction to enable a blocked activator strand to interact with the downstream logic

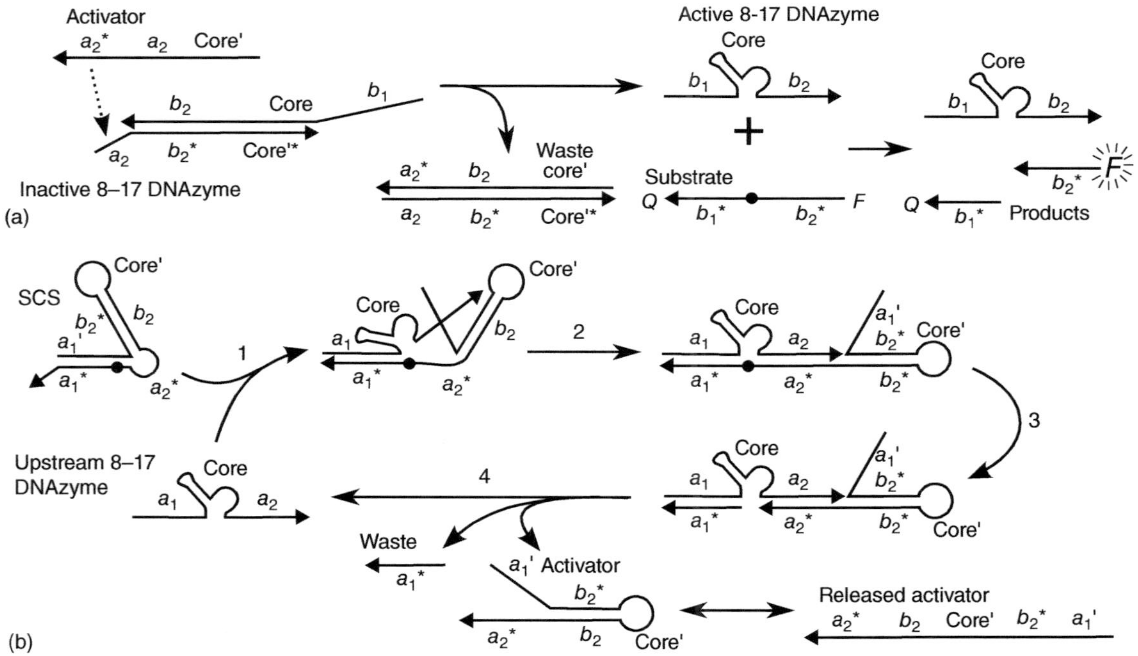

Figure 25.3 Design and application of structured chimeric substrates for multi-layer DNAzyme circuits. (a) Activation of an 8–17 DNAzyme via strand displacement triggered by an input strand, which strips the inhibitor so that the DNAzyme can fold into its active conformation and proceed to cleave substrates. (b) Mechanism of the SCS cleavage reaction. Binding of the upstream DNAzyme initially displaces the outer stem (reaction 1), opening the outer loop for the DNAzyme to bind (reaction 2). Then, the DNAzyme can cleave the SCS (reaction 3). Unbinding of the DNAzyme from the cleaved SCS recycles the DNAzyme and produces a short waste strand and a downstream activator (reaction 4). The activator can interconvert into a linear form as it is structurally weaker than the SCS. (c) Template for dengue diagnostic circuits, which compute 3-input AND functions. The diagnostic circuit for serotype DEN-k ($k \in \{1, 2, 3, 4\}$) requires the presence of two conserved sequences from the dengue genomes (which we call "DengueA" and "DengueB") and one sequence specific to the serotype of interest (which we call "DEN-k"). (d) Experimental data for all four dengue serotyping circuits, showing correct operation of all four instantiations of circuit template using all eight combinations of the two conserved sequences and the correct serotype-specific sequence. Also shown is the off-target activation in the presence of both conserved sequences and the three off-target serotype sequences in each case. Error bars represent the 95% confidence interval from three replicate experiments. Source: (a) Adapted from Brown et al. [39]; (b-d) Adapted from Brown et al. [40].

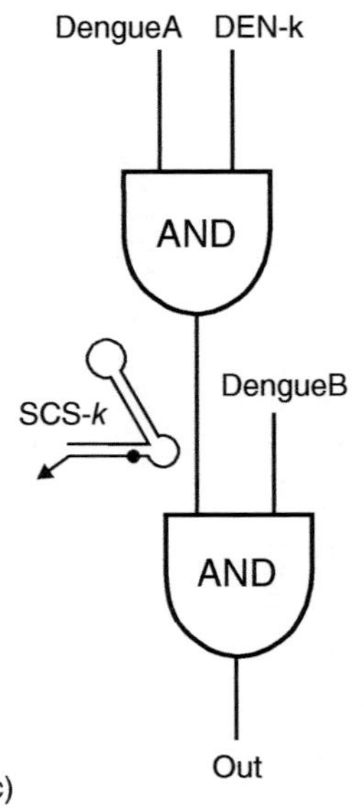

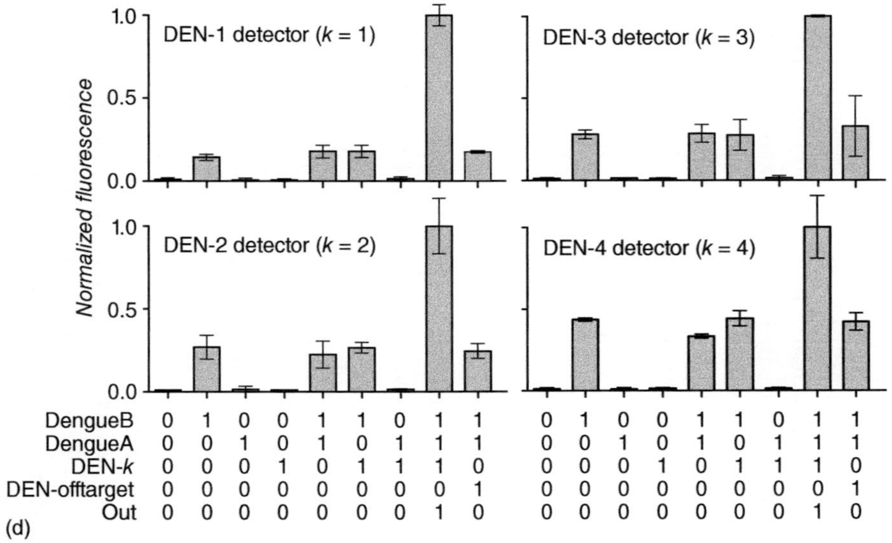

Figure 25.3 *(Continued)*

gate. We addressed this problem by developing a "structured chimeric substrate" (SCS) molecule that, in its uncleaved state, sequesters the downstream activator sequence in a relatively stable secondary structure. The key design challenge was to balance pre-cleavage stability with post-cleavage accessibility of the activator strand. We experimented with numerous designs [42], and our final design is shown in Figure 25.3b. It consists of a dual stem-loop structure, with the toehold and displacement domain for the downstream logic gate sequestered within the inner toehold and loop [40]. The interaction of the upstream DNAzyme with the folded substrate causes it to become partially linearized by strand displacement, thereby exposing the cleavage site. Cleavage of the substrate essentially removes one side of the outer stem, which weakens the secondary structure of the molecule. This makes the downstream activator available to interact with downstream gates. Thus, to summarize, we designed a substrate molecule that only releases the downstream activator strand after it is cleaved by an upstream DNAzyme. A crucial component of this design was that in using strand displacement to activate the DNAzymes, we could focus on sequestering the toehold that is required to initiate the programmed activation pathway.

We used our SCS design to implement a number of cascade circuits that could not be implemented using the parallel gate array method [40], using as our basis the compact, Zn^{2+}-dependent "8–17" DNAzyme motif [16]. We built multi-layer cascades of inactive DNAzymes wired together by SCS molecules, which are triggered by a signal activating the top layer DNAzymes, which then activate the next layer down, and so on for up to five layers [40]. We also built a multi-layer logic circuit comprising cooperative displacement AND gate DNAzymes, with wiring between the gates implemented using SCS molecules. To demonstrate practical applicability, we designed the circuit to sense synthetic oligonucleotide mimics of target sequences chosen from the genomes of the four serotypes of the dengue virus [43, 44]. We developed four circuits, one per serotype, which each detected two conserved sequences and one serotype-specific sequence and combined the results using a three-input AND circuit (Figure 25.3c). Experimental results shown in Figure 25.3d show that the logic circuits function correctly and are also serotype-specific [40]. This work demonstrates that signaling between DNAzymes can be implemented using our SCS approach, which opens up the possibility of implementing arbitrary logic functions using multilayer DNAzyme signaling cascades and logic networks.

25.4 Trainable and Adaptive DNAzyme Networks

A direction of significant interest for us is to move from molecular circuits that can execute a single, preprogrammed computation toward circuits that can adapt their behavior over time, including in response to "training" data fed to them by an experimenter. There are numerous advantages to trainable and adaptive molecular circuits. For example, designing and building molecular circuits de novo is non-trivial. Thus, building a small number of general-purpose circuits that can be trained as required to carry out a particular task could open up the field to greater use of molecular computing networks to solve practical tasks. Indeed, these networks could even be designed to autonomously observe and learn from environmental observations, without human intervention. This would be a particularly powerful capability for applications of molecular computing networks where they cannot be accessed and re-programmed manually after their initial deployment, e.g. applications in autonomous monitoring, diagnosis, and control of regulatory networks in living cells and organisms.

Our first foray in this direction was the MAYA-III molecular automaton [37], which was the successor to the MAYA-I and MAYA-II systems discussed above. Where MAYA-I and MAYA-II were preprogrammed to play a fixed strategy, the goal of MAYA-III was to develop a system that could be programmed to play any possible strategy in a simple retributive game. The game that we developed is called "tit-for-tat" and is played on a 2×2 board as follows:

(1) The human plays to claim one of the four squares.
(2) The automaton responds by claiming a square according to its programmed strategy. If that square is already taken by the human, the automaton loses.
(3) The human plays to claim another square. We assume that the human does not make the mistake of claiming a square that is already taken.
(4) The automaton responds by claiming another square according to its programmed strategy. As before, if that square is already taken by the human, the automaton loses. Otherwise, we say that the automaton wins.

An example game of tit-for-tat is illustrated in Figure 25.4a. The game was kept deliberately simple; however, there are still 81 different winning strategies. To enable strategies to be programmed in, the set of inputs to the logic array were conceptually divided into training inputs (t_j) and game-playing inputs (i_k). The MAYA-III automaton consisted of four AND gates and 12 ANDAND gates. On each case, one of the input-binding loops detected a training input, while the remainder detected gameplay inputs. Figure 25.4b,c illustrate strategies and training protocols for the tit-for-tat game using the MAYA-III automaton. Training inputs were of the form t_{mn}, where m was the move number (1 or 2) and n was the well number of the human move for which the response was being trained. The t_{mn} training input was added to the well that the automaton should claim in response to the signaled human move in the desired strategy. Gameplay inputs were of the form i_{mn}, where m was the move number (1 or 2) and n was the number of the well being claimed. This enabled the user to "train" the automaton in a process that mimics real gameplay, but which used the training inputs instead of the gameplay inputs. The result of this training step was that certain logic gates in the four wells were partially activated by their training input. Those gates would then wait until their corresponding gameplay inputs were observed before activating, and the choice of gates that were partially activated during training would determine the strategy played by the automaton in the real game, following an evolution like that illustrated in Figure 25.4d. Furthermore, we added an additional overhanging toehold on the training inputs, so that they could be removed via toehold-mediated strand displacement, thereby returning the automaton to its original, unprogrammed state (since the control loops would refold

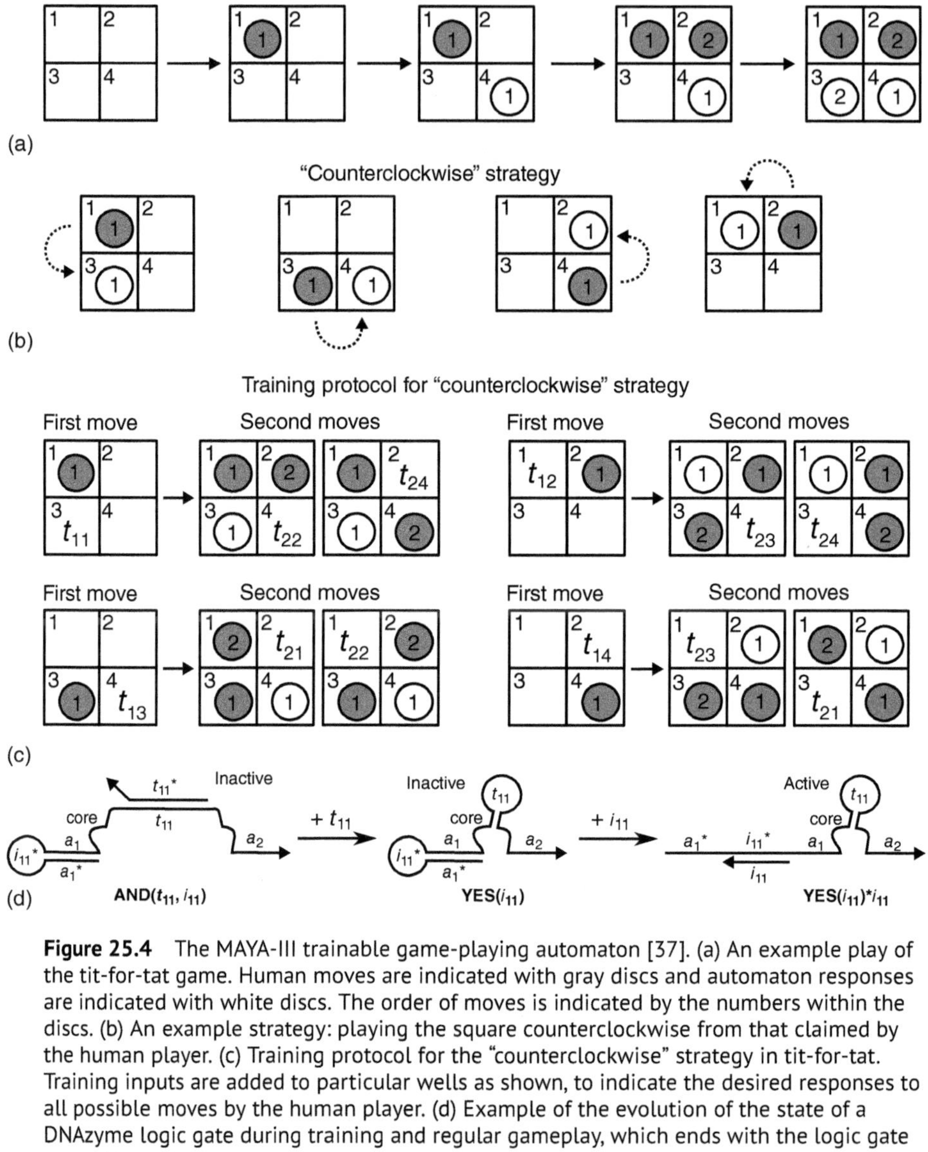

Figure 25.4 The MAYA-III trainable game-playing automaton [37]. (a) An example play of the tit-for-tat game. Human moves are indicated with gray discs and automaton responses are indicated with white discs. The order of moves is indicated by the numbers within the discs. (b) An example strategy: playing the square counterclockwise from that claimed by the human player. (c) Training protocol for the "counterclockwise" strategy in tit-for-tat. Training inputs are added to particular wells as shown, to indicate the desired responses to all possible moves by the human player. (d) Example of the evolution of the state of a DNAzyme logic gate during training and regular gameplay, which ends with the logic gate becoming active.

closed once their activating training input had been removed). Thus, the MAYA-III automaton could be programmed by example with a strategy, used in a game, and then wiped clean and re-programmed to play a different strategy.

We have also been studying the use of DNAzyme (and other) components to build adaptive molecular systems that are not only capable of rote learning by example but also of generalization from a series of training examples and through this process autonomously *learn* to carry out a particular function [45]. Our initial work in that area hypothesized a number of DNAzyme components, many of which we have subsequently realized experimentally, and used these to construct a theoretical design for a molecular network capable of learning a class of linear functions based on a series of training inputs provided by a user. The components that we assumed are as follows:

- *Conditional activation of DNAzymes by an input.* We and other researchers have demonstrated this using both loop-based [22, 28] and strand displacement-based [39, 40] activation mechanisms.
- *Release of effector strands by cleavage reactions.* Our work described above on the design of DNAzyme signaling cascades [40] showed that this can be achieved by rational design [42] of a structured substrate molecule.
- *Self-inhibition of DNAzymes.* Our previous work on loop-based control of DNAzymes demonstrated that a loop-based NOT gate can be implemented [22, 28].
- *Multi-stage cleavage reactions.* This reaction hypothesized a substrate molecule that required multiple cleavage reactions to occur in sequence in order to produce an output effector strand. As such, this would be a generalization of the single-cleavage structured substrates that we have previously developed [40, 42]. We have not realized this component experimentally; however, we note that its behavior could potentially be mimicked using multilayer DNAzyme logic cascades such as those that we have previously published [40].

The basic outline of our molecular learning algorithm is summarized in Figure 25.5a. The goal of the molecular learner is to learn a linear function of the form

$$f(x_1, \dots, x_n) = w_1 \cdot x_1 + \cdots + w_n \cdot x_n \tag{25.1}$$

where the inputs $x_1, \dots, x_n$ and the corresponding weights $w_1, \dots, w_n$ are all non-negative values in Eq. (25.1). The network is pre-initialized with certain substrate species at concentrations that encode the initial values of the weights, and then specific training species are provided whose concentrations encode the values of the first round of training inputs along with the corresponding expected output computed using the linear function f, which the system should learn. These inputs activate different DNAzymes in the circuit, whose reactions then carry out the analog molecular computation required to compute the network's current prediction based on the stored weight values and then compare that result with the expected result supplied by the user. The output of that comparison is transmitted to a feedback circuit that uses cleavage of the hypothesized multi-cleavage substrate structures described above to modify the stored weights used by the predictor. This is done by modifying the concentrations of the substrates that represent the stored weight values. Thus, subsequent additions of more training data will be processed in the context of these updated weights, and over time the system should converge to an approximately correct set of stored weight values.

Figure 25.5b shows data from a number of simulated training runs [45], which show that our hypothesized DNAzyme circuit design can indeed learn a range of different target functions (that is, weight values). This work demonstrates the potential for future applications of DNAzymes in the construction of non-trivial

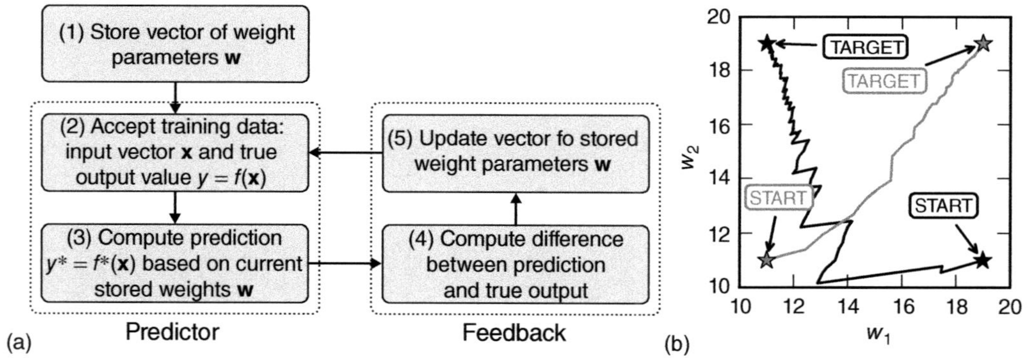

Figure 25.5 Theoretical studies of molecular learning circuits using DNAzymes. (a) Basic architecture of the molecular learning algorithm. (b) Simulated training runs, illustrating the change in two weight parameters (w_1 and w_2) from starting values toward target values, via the execution of programmed DNAzyme interactions [45].

adaptive and trainable molecular devices. More recently, we began to investigate other mechanisms for implementing learning in molecular networks, focusing in particular on toehold-mediated strand displacement [46], although that work is beyond the scope of this review. We do note that the field of molecular learning circuits is a nascent one, and we and our collaborators have performed much of the related work in this area [47–50].

25.5 DNAzyme Nanorobots

One of the most enticing applications of DNA nanotechnology is in the design and implementation of nanoscale robotic systems, dubbed "nanorobots." These systems exploit the sequence-specific nature of DNA hybridization, often in concert with enzymatic mechanisms to power the process, to turn chemical energy into mechanical work. This work seeks to engineer synthetic devices analogous to the molecular motors that play a crucial role in transporting and organizing chemical reactants in biological systems. In this section, we review contributions to this effort, focusing in particular on molecular walkers and other surface-based systems powered by DNAzyme catalysis.

Perhaps the earliest work on DNA-based molecular machines was the "molecular tweezer" system reported by Yurke et al. [51], which used DNA strand displacement reactions to reversibly switch a small DNA nanostructure between open and closed states. This work demonstrated the basic principle of converting chemical energy into mechanical work via a DNA nanostructure. Early attempts to move from this cyclical system to one that produces linear motion included the system by Yin et al. [52], which used cleavage by a restriction enzyme to power a DNA "walker" strand as it moved between anchorages protruding from a double-stranded DNA backbone, producing an irreversible "burnt bridges" motor. An early paper linking DNAzymes with molecular motors involved using the substrate binding and cleavage cycle of a 10–23 DNAzyme to actuate the molecular tweezer system described above [53, 54].

The Mao group pioneered the use of a DNAzyme that can move between anchorages that are also complementary substrates for the DNAzyme [55]. This mechanism worked by the DNAzyme cleaving its current anchorage and, in doing so, exposing one of its substrate binding arms that can then reach over and bind to the next anchorage. More recent work by Mao provided a detailed study of this mechanism for DNAzyme-based molecular walking by attaching a CdS quantum dot to the DNAzyme and using this to optically image its progress, thereby enabling detailed kinetic measurements [56]. Subsequent work has developed a range of mechanisms for DNA-based molecular walkers, including directional and reversible walking based on DNA strand displacement [57], the use of nickase enzymes to drive a molecular walker on a programmed route through a network of tracks [58, 59], the use of a DNA walker to build a programmable nanoscale "assembly line" [60], and the use of computer-controlled microfluidics to provide the fuel species needed for a molecular walker to walk considerable distances [61].

Our contribution to this area has focused primarily on a design known as a *molecular spider*. A molecular spider is a polycatalytic assembly that has a "body" attached to multiple catalytic "legs," which are just DNAzyme strands attached to the body. In practice, the body can be implemented as a streptavidin molecule, with four biotinylated DNAzyme legs attached to it via biotin–streptavidin linkages. The idea behind this design is that the body provides a physical constraint on the locations of the DNAzyme legs that, when presented with a spatial array of cleavable substrate molecules, can be used to generate motion. This is based on the fact that the interaction of the legs with the substrates irreversibly modifies them via cleavage, so that when a cleaved substrate is revisited by a leg, it is shorter, which reduces the residence time of the leg on that substrate. Its net effect is to bias [62] the motion of the walker toward self-avoidance, and we have published extensive computational simulations and analyses that study this effect [63–66]. Crucially, the catalytic activity of the DNAzyme means that the DNAzyme can interact with a large number of substrate molecules as it moves across the surface, without itself being chemically modified by the process [67]. In addition, the multiple legs reduce the likelihood of the spider "falling off" the surface, that is, of all legs unbinding from their substrates before any of the legs can rebind.

Our own experimental contribution to this area began in 2006 when we demonstrated that molecular spiders can be used to cleave multiple substrates from a nonuniform spatial substrate array [68]. In that work, a three-dimensional dextran–streptavidin matrix was used. There was no way to observe the specific behavior of individual spiders, and the progress of the cleavage reaction was monitored using surface plasmon resonance to measure the overall mass loss from the matrix assembly. In the following years, however, the rapid development of tools and techniques in structural DNA nanotechnology, in particular, the rapid rise of DNA origami [69], made it possible to conduct more detailed investigations of molecular spider behavior. This culminated in 2010 with an experiment that demonstrated the visualization of individual molecular spiders moving along a predefined track laid out on a DNA origami tile (Figure 25.6) [70]. The fact that DNA origami tiles can be designed with specific strands protruding from the ends of individual staples that form the tile was exploited to lay out the path, and atomic force microscopy (AFM) was used to visualize the progress of individual molecular spiders along the path. Statistical measurements of the ensemble behavior of molecular spiders on tiles were also obtained, using super-resolution total-internal-reflection fluorescence (TIRF) microscopy.

This work demonstrated that molecular spiders can not only explore large arrays of substrates at random but also be "programmed" to follow specific paths. This insight formed the basis of our own subsequent theoretical work, where we asked how this capability could be harnessed to carry out computation on a surface using molecular spiders [71]. We hypothesized sequences of substrate tracks laid out on a surface for multi-legged DNAzymes to walk along. We assumed that, in addition to their catalytically active "legs," each spider had an "arm" that was not used for walking but rather for carrying a DNA signal strand that represented whether that spider represented a logical "true" or "false" signal. Then, we simulated AND, OR, and NOT gates, and cascades comprising multiple such gates linked together. These

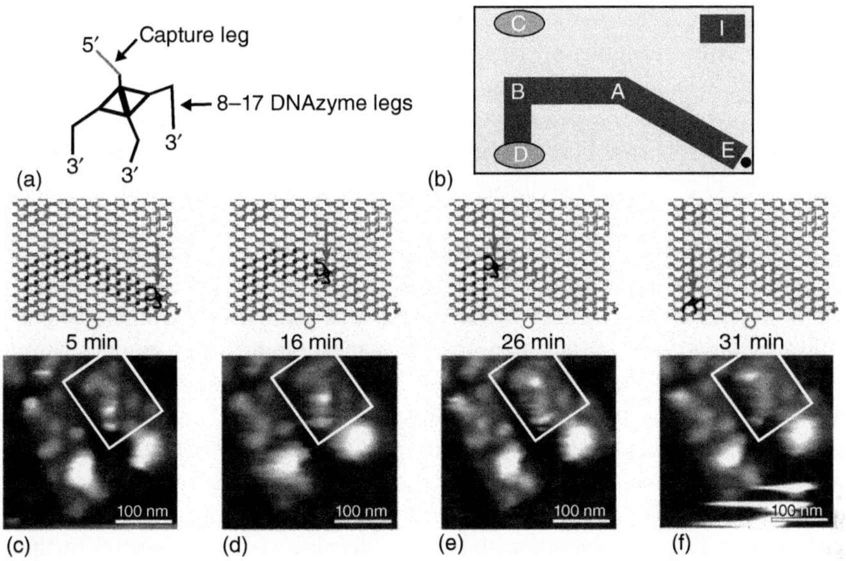

Figure 25.6 Nanoscale locomotion on patterned surfaces by molecular spiders. *(a)* A molecular spider consists of a streptavidin "body" with multiple "legs," each of which is an 8–17 DNAzyme [16]. Here, the additional leg is used to position the spider in the starting position on the DNA origami track. *(b)* The track design is patterned onto a DNA origami tile by augmenting the staple strands cleavable substrates. The spiders are programmed to follow the track "E-A-B-D." The "C" feature is an off-target destination to demonstrate that the spider follows the track to its programmed destination, and "I" is a feature to aid in identifying and orienting tiles in AFM images. *(c)–(f)* AFM images and corresponding diagrams illustrating the observed motion of a single molecular spider along the programmed track on the tile. Source: Figure initially published in [70]. Reproduced with permissioon of Springer Nature.

designs relied on a combination of track layout with additional control circuitry on the tile surface, which could prevent spiders traveling down a track depending on the value of their signal, trap spiders in a location, and modify the value of the signal carried by a spider. This work demonstrated that molecular walkers could be used to implement logical computation on a surface. A similar approach was taken by Dannenberg et al. [72] to show that a "burnt bridges" walker powered by protein nickase enzymes [58, 59] could also be used to carry out logical computations. Notably, Chatterjee et al. recently demonstrated experimentally that DNA hairpin circuits on an origami tile could be used to implement fast, robust, and scalable DNA logic circuits [73].

Researchers have explored practical applications of DNAzymes for controlled release of molecules over time. Indeed, one of us (M.N.S.) has published on the use of molecular spiders to release insulin by cleaving it away from a substrate array [74] and on how orally administrable drugs could be used to activate such a process [75]. DNAzymes have also been used to release oligonucleotide therapeutics from an extracellular matrix to inhibit the growth of cultured cancer cells [76], and as a sensor for microRNA detection in cells [77].

Finally, our more recent work has explored an alternative mechanism for implementing DNAzyme reactions on surfaces. Rather than use DNAzyme-catalyzed cleavage to move a molecular robot across the surface, instead we covalently attached DNAzymes to the headgroups of lipid molecules and embedded these in microparticle-supported lipid bilayers [78]. We chose the lipid composition to ensure that the bilayer would be fluid at room temperature, enabling the attached DNAzyme molecules to move across the surface via passive diffusion rather than an active "walking" mechanism. We also attached complementary substrate molecules to the same surface to implement a DNAzyme-catalyzed cleavage reaction. Reaction kinetics were monitored via loss of fluorescence from individual microspheres using flow cytometry. We anticipated observing multiple-turnover kinetics based on the diffusion of DNAzyme and substrate molecules on the surface; however, product inhibition prevented this. Future work to engineer a less stable complex between the surface-bound DNAzyme and the surface-bound cleaved product should enable the observation of multiple turnover in this system. However, our preliminary work in this area has demonstrated that integrating DNAzyme-based molecular computing components with lipid bilayers is a viable strategy for further research. Indeed, the integration of DNA nanostructures such as origami tiles [79], wireframe structures [80], and nanopores [81–83] with lipid bilayers is an area of much current research interest [84].

25.6 Conclusions

In conclusion, we have reviewed almost 15 years of work in our laboratories on the development of DNAzymes for applications in nanotechnology, including molecular computing and the development of molecular nanomachines. Our early research on solution-phase molecular computing systems using parallel arrays of DNAzyme logic gates [22, 28] was among the first work in this direction, and the solution-phase approach with bulk populations of logic gates is now standard in the community. Recent work has demonstrated the power of implementing molecular computing circuits in which some or all of the components are attached to surfaces [73, 85]. The advantages of localized computing circuits include faster operating speeds and enhanced scalability (due to the ability to safely reuse nucleotide sequences in different parts of a circuit). The use of molecular motors to carry out mechanical work at the nanoscale, including our own experimental and theoretical work on molecular spiders [70, 71], explores similar ideas on the interaction between programmable DNA hybridization and spatial organization.

Our subsequent work on DNAzyme cascades and multilayer logic circuits, illustrated the challenges involved in connecting substrate-cleaving DNAzymes into sequential logic circuits, due to the mismatch between the output signal (a cleavage process) and the input signal (a binding process). Inspired by protein cascades, we designed a substrate molecule such that the cleavage reaction would cause a conformational shift that exposed a downstream activator sequence [40, 42]. This substrate molecule served as an intermediary between the communicating DNAzymes, removing the need for direct interactions between DNAzymes and enabling modular circuit construction. Although our current work in this direction has focused solely on activating signals, we believe that this approach could be used to implement a range of dynamical behaviors [86] using DNAzymes, including repression, feedback cycles, and oscillators [87, 88]. We have also explored alternative methods for communication between DNAzymes, including DNAzyme ligase logic gates [89] (although these suffer from product inhibition issues due to the stability of the DNAzyme–product complex) as well as communication between DNAzymes on spatially separated particles [90].

Importantly, other groups have also studied the use of DNAzymes in computational cascades. The Kolpashchikov group has published [91] a cascade design analogous to our own, in which the cleavage of a hairpin substrate directly released a horseradish peroxidase-mimicking DNAzyme that generates a colorimetric output signal. This is an elegant approach; however, the direct interaction between DNAzymes and the fact that the downstream DNAzyme catalyzes a different reaction from the upstream one limits its scalability. The Willner group has also developed logic cascades based on DNAzymes, which used a two-strand structured substrate complex [92]: see [93] and citations therein. The Willner and Kolpashchikov groups are also leaders in the development of an alternative mechanism for controlling DNAzyme activity, whereby a multi-component DNAzyme is split into two parts that individually possess no catalytic activity. Then, triggered assembly of the DNAzyme complex switches on the catalytic activity of the DNAzyme, and this assembly process can be mediated by logic computations. There is an extensive literature on this topic [92–97], and the idea has been extended to structures derived from DNA tile architectures [98–100].

DNAzymes have been isolated that can catalyze a wide range of chemical reactions [101], including RNA ligation [102], hydrolysis of DNA [103, 104], kinase activity [105], and lysine side chain modification [106]. This work offers intriguing possibilities for DNAzymes to serve as output channels from DNA logic circuits to actuate on the environment via covalent modification of substrate molecules. Related work [107, 108] has shown that DNA reactions can be used to activate protein outputs, but the advantage of the DNAzyme approach is that the actuator can be built solely of nucleic acids, which simplifies the design and assembly processes and reduces the system cost. DNAzyme-based actuators are potentially useful for biosensing and biomedical applications, and our own work [109] and that of others has explored these possibilities. RNA-cleaving DNAzymes are of particular interest as catalytic antisense molecules, and their use has been explored for cancer therapy [110–116]. Additionally, DNAzyme logic gates have been shown to function in the cellular environment [117], as have logic gates based on DNA strand displacement [118]. Recent work on DNAzymes comprising enantiomeric L-DNA strands that are immune to degradation by intracellular processes [119] offers a way to keep DNAzymes active in the cell for extended periods of time, which is an important hurdle to overcome for practical applications. Collectively, this work offers a direct route to the development of logic-based molecular therapeutics that detect the chemical environments within individual cells, perform non-trivial information processing based on those observations, and conditionally activate a therapeutic payload, such as an antisense DNAzyme.

In the context of biodetection, the Li group has developed a powerful system for detecting specific strains of bacteria using DNAzymes that have been artificially evolved to switch on in the presence of metabolic products secreted by those bacterial strains [120–123]. Furthermore, these DNAzymes have been printed onto food packaging to serve as a warning that the product has been contaminated by pathogenic bacteria [124, 125]. This work is a highly innovative application of in vitro selection that demonstrates the power of DNAzymes not only as actuators but also as sensors. Thus, DNAzymes are already finding diverse applications in fields ranging from molecular computing and nanotechnology to biosensing and biomedical diagnostics and therapeutics.

Acknowledgments

We acknowledge our other experimental collaborators, in particular, Joanne Macdonald, Renjun Pei, Sergei Rudchenko, Steven Graves, and Carl W. Brown, III. This material is based upon work supported by the National Science Foundation under grants 1027877, 1028238, 1318833, 1422840, 1518861, and 1525553.

References

1 Katz, E. (2017). Enzyme-based logic gates and networks with output signals analyzed by various methods. *ChemPhysChem* 18 (13): 1688–1713. doi: https://doi.org/10.1002/cphc.201601402.

2 Katz, E. and Privman, V. (2010). Enzyme-based logic systems for information processing. *Chem. Soc. Rev.* 39: 1835–1857. doi: https://doi.org/10.1039/B806038J.

3 Baccouche, A., Montagne, K., Padirac, A. et al. (2014). Dynamic DNA-toolbox reaction circuits: a walkthrough. *Methods* 67 (2): 234–249. doi: https://doi.org/10.1016/j.ymeth.2014.01.015.

4 Montagne, K., Plasson, R., Sakai, Y. et al. (2011). Programming an *in vitro* DNA oscillator using a molecular networking strategy. *Mol. Syst. Biol.* 7: 466. doi: https://doi.org/10.1038/msb.2011.12.

5 Kim, J. and Winfree, E. (2011). Synthetic *in vitro* transcriptional oscillators. *Mol. Syst. Biol.* 7: 465. doi: https://doi.org/10.1038/msb.2010.119.

6 Carlson, R. (2009). The changing economics of DNA synthesis. *Nat. Biotechnol.* 27 (12): 1091–1094. doi: https://doi.org/10.1038/nbt1209-1091.

7 Goldman, N., Bertone, P., Chen, S. et al. (2013). Towards practical, high-capacity, low-maintenance information storage in synthesized DNA. *Nature* 494: 77–80. doi: https://doi.org/10.1038/nature11875.

8 Erlich, Y. and Zielinski, D. (2017). DNA fountain enables a robust and efficient storage architecture. *Science* 355 (6328): 950–954. doi: https://doi.org/10.1126/science.aaj2038.

9 Adleman, L.M. (1994). Molecular computation of solutions to combinatorial problems. *Science* 266 (5187): 1021–1024. doi: https://doi.org/10.1126/science.7973651.

10 Bennett, C.H. (1982). The thermodynamics of computation—a review. *Int. J. Theor. Phys.* 21 (12): 905–940. doi: https://doi.org/10.1007/BF02084158.

11 Qian, L., Soloveichik, D., and Winfree, E. (2011). Efficient Turing-universal computation with DNA polymers. In: *Proceedings of the 16th International Conference on DNA Computing and Molecular Programming, Lecture Notes in Computer Science*, vol. 6518 (ed. Y. Sakakibara and Y. Mi), 123–140. Springer-Verlag. doi: https://doi.org/10.1007/978-3-642-18305-8_12.

12 Osborne, S.E. and Ellington, A.D. (1997). Nucleic acid selection and the challenge of combinatorial chemistry. *Chem. Rev.* 97 (2): 349–370. doi: https://doi.org/10.1021/cr960009c.

13 Breaker, R.R. (1997). *In vitro* selection of catalytic polynucleotides. *Chem. Rev.* 97 (2): 371–390. doi: https://doi.org/10.1021/cr960008k.

14 Paul, N., Springsteen, G., and Joyce, G.F. (2006). Conversion of a ribozyme to a deoxyribozyme through in vitro evolution. *Chem. Biol.* 13: 329–338. doi: https://doi.org/10.1016/j.chembiol.2006.01.007.

15 Breaker, R.R. and Joyce, G.F. (1995). A DNA enzyme with Mg^{2+}-dependent RNA phosphoesterase activity. *Chem. Biol.* 2: 655–660. doi: https://doi.org/10.1016/1074-5521(95)90028-4.

16 Santoro, S.W. and Joyce, G.F. (1997). A general purpose RNA-cleaving DNA enzyme. *Proc. Natl. Acad. Sci. U.S.A.* 94: 4262–4266. doi: https://doi.org/10.1073/pnas.94.9.4262.

17 Robertson, M.P. and Ellington, A. (1999). *In vitro* selection of an allosteric ribozyme that transduces analytes to amplicons. *Nat. Biotechol.* 17 (1): 62–66. doi: https://doi.org/10.1038/5236.

18 Robertson, M.P. and Ellington, A. (2000). Design and optimization of effector-activated ribozyme ligases. *Nucleic Acids Res.* 28 (8): 1751–1759. doi: https://doi.org/10.1093/nar/28.8.1751.

19 Jose, A.M., Soukup, G.A., and Breaker, R.R. (2001). Cooperative binding of effectors by an allosteric ribozyme. *Nucleic Acids Res.* 29 (7): 1831–1637. doi: https://doi.org/10.1093/nar/29.7.1631.

20 Tyagi, S. and Kramer, F.R. (1996). Molecular beacons: probes that fluoresce upon hybridization. *Nat. Biotechnol.* 14 (3): 303–309. doi: https://doi.org/10.1038/nbt0396-303.

21 Stojanovic, M.N., de Prada, P., and Landry, D.W. (2001). Catalytic molecular beacons. *ChemBioChem* 2: 411–415. doi: https://doi.org/10.1002/1439-7633(20010601)2:6⟨411::AID-CBIC411⟩3.0.CO;2-I.

22 Stojanovic, M.N., Mitchell, T.E., and Stefanovic, D. (2002). Deoxyribozyme-based logic gates. *J. Am. Chem. Soc.* 124: 3555–3561. doi: https://doi.org/10.1021/ja016756v.

23 Lederman, H., Macdonald, J., Stefanovic, D., and Stojanovic, M.N. (2006). Deoxyribozyme-based three-input logic gates and construction of a molecular full adder. *Biochemistry* 45 (4): 1194–1199. doi: https://doi.org/10.1021/bi051871u.

24 Yurke, B. and Mills, A.P. Jr. (2003). Using DNA to power nanostructures. *Genet. Program. Evolv. Mach.* 4: 111–122. doi: https://doi.org/10.1023/A:1023928811651.

25 Zhang, D.Y. and Seelig, G. (2011). Dynamic DNA nanotechnology using strand-displacement reactions. *Nat. Chem.* 3: 103–113. doi: https://doi.org/10.1038/nchem.957.

26 Stojanovic, M.N. and Stefanovic, D. (2003). Deoxyribozyme-based half adder. *J. Am. Chem. Soc.* 125 (22): 6673–6676. doi: https://doi.org/10.1021/ja0296632.

27 Stefanovic, D. and Stojanovic, M.N. (2013). Computing game strategies. In: *CiE 2013: The Nature of Computation. Logic, Algorithms, Applications, Lecture Notes in Computer Science*, vol. 7921 (ed. P. Bonizzoni, V. Brattka, and B. Lowe), 383–392. Springer-Verlag. doi: https://doi.org/10.1007/978-3-642-39053-1_45.

28 Stojanovic, M.N. and Stefanovic, D. (2003). A deoxyribozyme-based molecular automaton. *Nat. Biotechnol.* 21 (9): 1069–1074. doi: https://doi.org/10.1038/nbt862.

29 Macdonald, J., Li, Y., Sutovic, M. et al. (2006). Medium scale integration of molecular logic gates in an automaton. *Nano Lett.* 6 (11): 2598–2603. doi: https://doi.org/10.1021/nl0620684.

30 Zadeh, J.N., Steenberg, C.D., Bois, J.S. et al. (2011). NUPACK: analysis and design of nucleic acid systems. *J. Comput. Chem.* 32: 170–173. doi: https://doi.org/10.1002/jcc.21596.

31 Zadeh, J.N., Wolfe, B.R., and Pierce, N.A. (2011). Nucleic acid sequence design via efficient ensemble defect optimization. *J. Comput. Chem.* 32: 439–452. doi: https://doi.org/10.1002/jcc.21633.

32 Wolfe, B.R. and Pierce, N.A. (2015). Sequence design for a test tube of interacting nucleic acid strands. *ACS Synth. Biol.* 4 (10): 1086–1100. doi: https://doi.org/10.1021/sb5002196.

33 Wolfe, B.R., Porubsky, N.J., Zadeh, J.N. et al. (2017). Constrained multistate sequence design for nucleic acid reaction pathway engineering. *J. Am. Chem. Soc.* 139 (8): 3134–3144. doi: https://doi.org/10.1021/jacs.6b12693.

34 Qian, L. and Winfree, E. (2011). Scaling up digital circuit computation with DNA strand displacement cascades. *Science* 332: 1196–1201. doi: https://doi.org/10.1126/science.1200520.

35 Qian, L., Winfree, E., and Bruck, J. (2011). Neural network computation with DNA strand displacement cascades. *Nature* 475: 368–372. doi: https://doi.org/10.1038/nature10262.

36 Chen, Y.J., Dalchau, N., Srinivas, N. et al. (2013). Programmable chemical controllers made from DNA. *Nat. Nanotechnol.* 8: 755–762. doi: https://doi.org/10.1038/nnano.2013.189.

37 Pei, R., Matamoros, E., Liu, M. et al. (2010). Training a molecular automaton to play a game. *Nat. Nanotechnol.* 5: 773–777. doi: https://doi.org/10.1038/nnano.2010.194.

38 Seelig, G., Soloveichik, D., Zhang, D.Y., and Winfree, E. (2006). Enzyme-free nucleic acid logic circuits. *Science* 314: 1585–1588. doi: https://doi.org/10.1126/science.1132493.

39 Brown, C.W. III, Lakin, M.R., Stefanovic, D., and Graves, S.W. (2014). Catalytic molecular logic devices by DNAzyme displacement. *ChemBioChem* 15 (7): 950–954. doi: https://doi.org/10.1002/cbic.201400047.

40 Brown, C.W. III, Lakin, M.R., Horwitz, E.K. et al. (2014). Signal propagation in multi-layer DNAzyme cascades using structured chimeric substrates. *Angew. Chem. Int. Ed.* 53 (28): 7183–7187. doi: https://doi.org/10.1002/anie.201402691.

41 Zhang, D.Y. (2011). Cooperative hybridization of oligonucleotides. *J. Am. Chem. Soc.* 133: 1077–1086. doi: https://doi.org/10.1021/ja109089q.

42 Lakin, M.R., Brown, C.W. III, Horwitz, E.K. et al. (2014). Biophysically inspired rational design of structured chimeric substrates for DNAzyme cascade engineering. *PLOS ONE* 9 (10): e110986. doi: https://doi.org/10.1371/journal.pone.0110986.

43 Simmons, C.P., Farrar, J.J., van Vinh Chau, N., and Wills, B. (2012). Current concepts: dengue. *N. Engl. J. Med.* 366 (15): 1423–1432. doi: https://doi.org/10.1056/NEJMra1110265.

44 Bhatt, S., Gething, P.W., Brady, O.J. et al. (2013). The global distribution and burden of dengue. *Nature* 496 (7446): 504–507. doi: https://doi.org/10.1038/nature12060.

45 Lakin, M.R., Minnich, A., Lane, T., and Stefanovic, D. (2014). Design of a biochemical circuit motif for learning linear functions. *J. R. Soc. Interface* 11 (101): 20140902. doi: https://doi.org/10.1098/rsif.2014.0902.

46 Lakin, M.R. and Stefanovic, D. (2016). Supervised learning in adaptive DNA strand displacement networks. *ACS Synth. Biol.* 5 (8): 885–897. doi: https://doi.org/10.1021/acssynbio.6b00009.

47 Banda, P., Teuscher, C., and Lakin, M.R. (2013). Online learning in a chemical perceptron. *Artif. Life* 19 (2): 195–219. doi: https://doi.org/10.1162/ARTL_a_00105.

48 Poje, J.E., Kastratovic, T., Macdonald, A.R. et al. (2014). Visual displays that directly interface and provide read-outs of molecular states via molecular graphics processing units. *Angew. Chem. Int. Ed.* 53 (35): 9222–9225. doi: https://doi.org/10.1002/anie.201402698.

49 Banda, P., Teuscher, C., and Stefanovic, D. (2014). Training an asymmetric signal perceptron through reinforcement in an artificial chemistry. *J. R. Soc. Interface* 11: 20131100. doi: https://doi.org/10.1098/rsif.2013.1100.

50 Blount, D., Banda, P., Teuscher, C., and Stefanovic, D. (2017). Feedforward chemical neural network: an in silico chemical system that learns XOR. *Artif. Life* 23 (3): 295–317. doi: https://doi.org/10.1162/ARTL_a_00233.

51 Yurke, B., Turberfield, A.J., Mills, A.P. Jr. et al. (2000). A DNA-fuelled molecular machine made of DNA. *Nature* 406: 605–608. doi: https://doi.org/10.1038/35020524.

52 Yin, P., Yan, H., Daniell, X.G. et al. (2004). A unidirectional DNA walker that moves autonomously along a track. *Angew. Chem. Int. Ed.* 43: 4906–4911. doi: https://doi.org/10.1002/anie.200460522.

53 Chen, Y., Wang, M., and Mao, C. (2004). An autonomous DNA nanomotor powered by a DNA enzyme. *Angew. Chem. Int. Ed.* 43 (27): 3554–3557. doi: https://doi.org/10.1002/anie.200453779.

54 Chen, Y. and Mao, C. (2004). Putting a brake on an autonomous DNA nanomotor. *J. Am. Chem. Soc.* 126 (28): 8626–8627. doi: https://doi.org/10.1021/ja047991r.

55 Tian, Y., He, Y., Chen, Y. et al. (2005). A DNAzyme that walks processively and autonomously along a one-dimensional track. *Angew. Chem. Int. Ed.* 44: 4355–4358. doi: https://doi.org/10.1002/anie.200500703.

56 Cha, T.G., Pan, J., Chen, H. et al. (2015). Design principles of DNA enzyme-based walkers: translocation kinetics and photoregulation. *J. Am. Chem. Soc.* 137 (29): 9429–9437. doi: https://doi.org/10.1021/jacs.5b05522.

57 Bath, J., Green, S.J., Allen, K.E., and Turberfield, A.J. (2009). Mechanism for a directional, processive, and reversible DNA motor. *Small* 5 (13): 1513–1516. doi: https://doi.org/10.1002/smll.200900078.

58 Wickham, S.F.J., Endo, M., Katsuda, Y. et al. (2011). Direct observation of stepwise movement of a synthetic molecular transporter. *Nat. Nanotechnol.* 6: 166–169. doi: https://doi.org/10.1038/nnano.2010.284.

59 Wickham, S.F.J., Bath, J., Katsuda, Y. et al. (2012). A DNA-based molecular motor that can navigate a network of tracks. *Nat. Nanotechnol.* 7: 169–173. doi: https://doi.org/10.1038/nnano.2011.253.

60 Gu, H., Chao, J., Xiao, S.J., and Seeman, N.C. (2010). A proximity-based programmable DNA nanoscale assembly line. *Nature* 465: 202–205. doi: https://doi.org/10.1038/nature09026.

61 Tomov, T.E., Tsukanov, R., Glick, Y. et al. (2017). DNA bipedal motor achieves a large number of steps due to operation using microfluidics-based interface. *ACS Nano* 11 (4): 4002–4008. doi: https://doi.org/10.1021/acsnano.7b00547.

62 Antal, T. and Krapivsky, P.L. (2007). Molecular spiders with memory. *Phys. Rev. E* 76: 021121. doi: https://doi.org/10.1103/PhysRevE.76.021121.

63 Semenov, O., Olah, M.J., and Stefanovic, D. (2011). Mechanism of diffusive transport in molecular spider models. *Phys. Rev. E* 83: 021117. doi: https://doi.org/10.1103/PhysRevE.83.021117.

64 Olah, M.J. and Stefanovic, D. (2013). Superdiffusive transport by multivalent molecular walkers moving under load. *Phys. Rev. E* 87: 062713. doi: https://doi.org/10.1103/PhysRevE.87.062713.

65 Semenov, O., Olah, M.J., and Stefanovic, D. (2013). Cooperative linear cargo transport with molecular spiders. *Nat. Comput.* 12 (2): 259–276. doi: https://doi.org/10.1007/s11047-012-9357-2.

66 Semenov, O., Mohr, D., and Stefanovic, D. (2013). First-passage time properties of molecular spiders. *Phys. Rev. E* 88: 012724. doi: https://doi.org/10.1103/PhysRevE.88.012724.

67 Stefanovic, D., Stojanovic, M.N., Olah, M.J., and Semenov, O. (2013). Catalytic molecular walkers: aspects of product release. *ECAL 2013: 12th European Conference on Artificial Life*, pp. 1134–1141. doi: https://doi.org/10.7551/978-0-262-31709-2-ch172.

68 Pei, R., Taylor, S.K., Stefanovic, D. et al. (2006). Behavior of polycatalytic assemblies in a substrate-displaying matrix. *J. Am. Chem. Soc.* 128 (39): 12693–12699. doi: https://doi.org/10.1021/ja058394n.

69 Rothemund, P.W.K. (2006). Folding DNA to create nanoscale shapes and patterns. *Nature* 440: 297–302. doi: https://doi.org/10.1038/nature04586.

70 Lund, K., Manzo, A.J., Dabby, N. et al. (2010). Molecular robots guided by prescriptive landscapes. *Nature* 465: 206–2010. doi: https://doi.org/10.1038/nature09012.

71 Mo, D., Lakin, M.R., and Stefanovic, D. (2016). Logic circuits based on molecular spider systems. *BioSystems* 146: 10–25. doi: https://doi.org/10.1016/j.biosystems.2016.03.008.

72 Dannenberg, F., Kwiatkowska, M., Thachuk, C., and Turberfield, A.J. (2015). DNA walker circuits: computational potential, design, and verification. *Nat. Comput.* 14: 195–211. doi: https://doi.org/10.1007/s11047-014-9426-9.

73 Chatterjee, G., Dalchau, N., Muscat, R.A. et al. (2017). A spatially localized architecture for fast and modular DNA computing. *Nat. Nanotechnol.* 12: 920–927. doi: https://doi.org/10.1038/nnano.2017.127.

74 Taylor, S. and Stojanovic, M.N. (2007). Is there a future for DNA-based molecular devices in diabetes management? *J. Diabetes Sci. Technol.* 1 (3): 440–444. doi: https://doi.org/10.1177/193229680700100319.

75 Taylor, S.K., Pei, R., Moon, B.C. et al. (2009). Triggered release of an active peptide conjugate from a DNA device by an orally administrable small molecule. *Angew. Chem. Int. Ed.* 48 (24): 4394–4397. doi: https://doi.org/10.1002/anie.200900499.

76 Li, F., Cha, T.G., Pan, J. et al. (2016). DNA-walker regulated cancer cell growth inhibition. *ChemBioChem* 17 (12): 1138–1141. doi: https://doi.org/10.1002/cbic.201600052.

77 Peng, H., Li, X.F., Zhang, H., and Le, X.C. (2017). A microRNA-initiated DNAzyme motor operating in living cells. *Nat. Commun.* 8: 14378. doi: https://doi.org/10.1038/ncomms14378.

78 Fabry-Wood, A., Fetrow, M.E., Brown, C.W. III et al. (2017). A microsphere-supported lipid bilayer platform for DNA reactions on a fluid surface. *ACS Appl. Mater. Interfaces* 9 (35): 30185–30195. doi: https://doi.org/10.1021/acsami.7b11046.

79 Johnson-Buck, A., Jiang, S., Yan, H., and Walter, N.G. (2014). DNA cholesterol barges as programmable membrane-exploring agents. *ACS Nano* 8 (6): 5641–5649. doi: https://doi.org/10.1021/nn500108k.

80 Perrault, S.D. and Shih, W.M. (2014). Virus-inspired membrane encapsulation of DNA nanostructures to achieve *in vivo* stability. *ACS Nano* 8 (5): 5132–5140. doi: https://doi.org/10.1021/nn5011914.

81 Burns, J.R., Stulz, E., and Howorka, S. (2013). Self-assembled DNA nanopores that span lipid bilayers. *Nano Lett.* 13 (6): 2351–2356. doi: https://doi.org/10.1021/nl304147f.

82 Burns, J.R., Göpfrich, K., Wood, J.W. et al. (2013). Lipid-bilayer-spanning DNA nanopores with a bifunctional porphyrin anchor. *Angew. Chem. Int. Ed.* 52 (46): 12069–12072. doi: https://doi.org/10.1002/anie.201305765.

83 Langecker, M., Arnaut, V., Martin, T.G. et al. (2012). Synthetic lipid membrane channels formed by designed DNA nanostructures. *Science* 338 (6109): 932–936. doi: https://doi.org/10.1126/science.1225624.

84 Langecker, M., Arnaut, V., List, J., and Simmel, F.C. (2014). DNA nanostructures interacting with lipid bilayer membranes. *Acc. Chem. Res.* 47 (6): 1807–1815. doi: https://doi.org/10.1021/ar500051r.

85 Bui, H., Shah, S., Mokhtar, R. et al. (2018). Localized DNA hybridization chain reactions on DNA origami. *ACS Nano* 12 (2): 1146–1155. doi: https://doi.org/10.1021/acsnano.7b06699.

86 Del Vecchio, D. and Murray, R.M. (2014). *Biomolecular Feedback Systems.* Princeton University Press.

87 Alon, U. (2007). *An Introduction to Systems Biology: Design Principles of Biological Circuits.* Chapman & Hall/CRC.

88 Elowitz, M.B. and Leibler, S. (2000). A synthetic oscillatory network of transcriptional regulators. *Nature* 403: 335–338. doi: https://doi.org/10.1038/35002125.

89 Stojanovic, M.N., Semova, S., Kolpashchikov, D. et al. (2005). Deoxyribozyme-based ligase logic gates and their initial circuits. *J. Am. Chem. Soc.* 127 (19): 6914–6915. doi: https://doi.org/10.1021/ja043003a.

90 Yashin, R., Rudchenko, S., and Stojanovic, M.N. (2007). Networking particles over distance using oligonucleotide-based devices. *J. Am. Chem. Soc.* 129 (50): 15581–15584. doi: https://doi.org/10.1021/ja074335t.

91 Gerasimova, Y.V., Cornett, E.M., Edwards, E. et al. (2013). Deoxyribozyme cascade for visual detection of bacterial RNA. *ChemBioChem* 14: 2087–2090. doi: https://doi.org/10.1002/cbic.201300471.

92 Elbaz, J., Lioubashevski, O., Wang, F. et al. (2010). DNA computing circuits using libraries of DNAzyme subunits. *Nat. Nanotechnol.* 5: 417–422. doi: https://doi.org/10.1038/nnano.2010.88.

93 Wang, F., Lu, C.H., and Willner, I. (2014). From cascaded catalytic nucleic acids to enzyme-DNA nanostructures: controlling reactivity, sensing, logic operations, and assembly of complex structures. *Chem. Rev.* 114: 2881–2941. doi: https://doi.org/10.1021/cr400354z.

94 Kolpashchikov, D.M. (2010). Binary probes for nucleic acid analysis. *Chem. Rev.* 110: 4709–4723. doi: https://doi.org/10.1021/cr900323b.

95 Mokany, E., Bone, S.M., Young, P.E. et al. (2010). MNAzymes, a versatile new class of nucleic acid enzymes that can function as biosensors and molecular switches. *J. Am. Chem. Soc.* 132: 1051–1059. doi: https://doi.org/10.1021/ja9076777.

96 Bone, S.M., Hasick, N.J., Lima, N.E. et al. (2014). DNA-only cascade: a universal tool for signal amplification, enhancing the detection of target analytes. *Anal. Chem.* 86 (18): 9106–9113. doi: https://doi.org/10.1021/ac501811r.

97 Gerasimova, Y.V. and Kolpashchikov, D.M. (2015). Divide and control: split design of multi-input DNA logic gates. *Chem. Commun.* 51: 870–872. doi: https://doi.org/10.1039/c4cc08241a.

98 Gerasimova, Y.V. and Kolpashchikov, D.M. (2016). Towards a DNA nanoprocessor: reusable tile-integrated DNA circuits. *Angew. Chem. Int. Ed.* 55 (35): 10244–10247. doi: https://doi.org/10.1002/anie.201603265.

99 Cox, A.J., Bengtson, H.N., Gerasimova, Y.V. et al. (2016). DNA antenna tile-associated deoxyribozyme sensor with improved sensitivity. *ChemBioChem* 17 (21): 2038–2041. doi: https://doi.org/10.1002/cbic.201600438.

100 Campbell, E.A., Peterson, E., and Kolpashchikov, D.M. (2017). Self-assembling molecular logic gates based on DNA crossover tiles. *ChemPhysChem* 18 (13): 1730–1734. doi: https://doi.org/10.1002/cphc.201700109.

101 Silverman, S.K. (2008). Catalytic DNA (deoxyribozymes) for synthetic applications—current abilities and future prospects. *Chem. Commun.* 3467–3485. doi: https://doi.org/10.1039/b807292m.

102 Flynn-Charlebois, A., Wang, Y., Prior, T.K. et al. (2003). Deoxyribozymes with 2′-5′ RNA ligase activity. *J. Am. Chem. Soc.* 125: 2444–2454. doi: https://doi.org/10.1021/ja028774y.

103 Xiao, Y., Wehrmann, R.J., Ibrahim, N.A., and Silverman, S.K. (2012). Establishing broad generality of DNA catalysts for site-specific hydrolysis of single-stranded DNA. *Nucleic Acids Res.* 40 (4): 1778–1786. doi: https://doi.org/10.1093/nar/gkr860.

104 Gu, H., Furukawa, K., Weinberg, Z. et al. (2013). Small, highly active DNAs that hydrolyze DNA. *J. Am. Chem. Soc.* 135: 9121–9129. doi: https://doi.org/10.1021/ja403585e.

105 Li, Y. and Breaker, R.R. (1999). Phosphorylating DNA with DNA. *Proc. Natl. Acad. Sci. U.S.A.* 96 (6): 2746–2751. doi: https://doi.org/10.1073/pnas.96.6.2746.

106 Brandsen, B.M., Velez, T.E., Sachdeva, A. et al. (2014). DNA-catalyzed lysine side chain modification. *Angew. Chem. Int. Ed.* 53 (34): 9045–9050. doi: https://doi.org/10.1002/anie.201404622.

107 Xiang, Y. and Lu, Y. (2011). Using personal glucose meters and functional DNA sensors to quantify a variety of analytical targets. *Nat. Chem.* 3 (9): 697–703. doi: https://doi.org/10.1038/nchem.1092.

108 Prokup, A. and Deiters, A. (2014). Interfacing synthetic DNA logic operations with protein outputs. *Angew. Chem. Int. Ed.* 53 (48): 13192–13195. doi: https://doi.org/10.1002/anie.201406892.

109 Brown, C.W. III, Lakin, M.R., Fabry-Wood, A. et al. (2015). A unified sensor architecture for isothermal detection of double-stranded DNA, oligonucleotides, and small molecules. *ChemBioChem* 16 (5): 725–730. doi: https://doi.org/10.1002/cbic.201402615.

110 Dass, C.R., Saravolac, E.G., Li, Y., and Sun, L.Q. (2002). Cellular uptake, distribution, and stability of 10–23 deoxyribozymes. *Antisense Nucleic Acid Drug Dev.* 12: 289–299. doi: https://doi.org/10.1089/108729002761381276.

111 Mitchell, A., Dass, C.R., Sun, L.Q., and Khachigian, L.M. (2004). Inhibition of human breast carcinoma proliferation, migration, chemoinvasion and solid tumour growth by DNAzymes targeting the zinc finger transcription factor EGR-1. *Nucleic Acids Res.* 32 (10): 3065–3069. doi: https://doi.org/10.1093/nar/gkh626.

112 Dass, C.R. (2004). Deoxyribozymes: cleaving a path to clinical trials. *Trends Pharmacol. Sci.* 25 (8): 395–397. doi: https://doi.org/10.1016/j.tips.2004.06.001.

113 Dass, C.R., Choong, P.F.M., and Khachigian, L.M. (2008). DNAzyme technology and cancer therapy: cleave and let die. *Mol. Cancer Ther.* 7 (2): 243–251. doi: https://doi.org/10.1158/1535-7163.MCT-07-0510.

114 Dass, C.R., Galloway, S.J., and Choong, P.F. (2010). Dz13, a c-jun DNAzyme, is a potent inducer of caspase-2 activation. *Oligonucleotides* 20 (3): 137–146. doi: https://doi.org/10.1089/oli.2009.0226.

115 Elahy, M. and Dass, C.R. (2011). Dz13: c-jun downregulation and tumour cell death. *Chem. Biol. Drug Des.* 78: 909–912. doi: https://doi.org/10.1111/j.1747-0285.2011.01166.x.

116 Kim, S.H. and Dass, C.R. (2012). Induction of caspase-2 activation by a DNA enzyme evokes tumor cell apoptosis. *DNA Cell Biol.* 31 (1). doi: https://doi.org/10.1089/dna.2011.1323.

117 Kahan-Hanum, M., Douek, Y., Adar, R., and Shapiro, E. (2013). A library of programmable DNAzymes that operate in a cellular environment. *Sci. Rep.* 3: 1535. doi: https://doi.org/10.1038/srep01535.

118 Groves, B., Chen, Y.J., Zurla, C. et al. (2016). Computing in mammalian cells with nucleic acid strand exchange. *Nat. Nanotechnol.* 11: 287–294. doi: https://doi.org/10.1038/nnano.2015.278.

119 Cui, L., Peng, R., Fu, T. et al. (2016). Biostable L-DNAzyme for sensing of metal ions in biological systems. *Anal. Chem.* 88 (3): 1850–1855. doi: https://doi.org/10.1021/acs.analchem.5b04170.

120 Tram, K., Kanda, P., Salena, B.J. et al. (2014). Translating bacterial detection by DNAzymes into a litmus test. *Angew. Chem. Int. Ed.* 53 (47): 12799–12802. doi: https://doi.org/10.1002/anie.201407021.

121 Aguirre, S.D., Ali, M.M., Kanda, P., and Li, Y. (2012). Detection of bacteria using fluorogenic DNAzymes. *J. Vis. Exp.* 63: e3961. doi: https://doi.org/10.3791/3961.

122 Ali, M.M., Aguirre, S.D., Lazim, H., and Li, Y. (2011). Fluorogenic DNAzyme probes as bacterial indicators. *Angew. Chem. Int. Ed.* 50: 3751–3754. doi: https://doi.org/10.1002/anie.201100477.

123 Huang, P.J.J., Liu, M., and Liu, J. (2013). Functional nucleic acids for detecting bacteria. *Rev. Anal. Chem.* 32 (1): 77–89. doi: https://doi.org/10.1515/revac-2012-0027.

124 Ali, M.M., Brown, C.L., Jahanshahi-Anbuhi, S. et al. (2017). A printed multicomponent paper sensor for bacterial detection. *Sci. Rep.* 7: 12335. doi: https://doi.org/10.1038/s41598-017-12549-3.

125 Yousefi, H., Ali, M.M., Su, H.M. et al. (2018). Sentinel wraps: real-time monitoring of food contamination by printing DNAzyme probes on food packaging. *ACS Nano* 12 (4): 3287–3294. doi: https://doi.org/10.1021/acsnano.7b08010.

Part V

Ribozymes/DNAzymes in Diagnostics and Therapy

26

Optimization of Antiviral Ribozymes

Alfredo Berzal-Herranz and Cristina Romero-López

Instituto de Parasitología y Biomedicina "López-Neyra" (IPBLN-CSIC), Department of Molecular Biology, PTS Granada, Av. del Conocimiento 17, Armilla, 18016 Granada, Spain

26.1 Introduction

Strategies aimed at destroying viral genomes or blocking the information they encode have long been pursued as methods of fighting viral infections. The ability of certain RNA molecules, in particular the so-called ribozymes [1], to cleave sequence-specific RNA molecules renders them excellent candidates for development as antiviral molecular tools. The first described naturally occurring ribozymes were responsible for either the self-splicing or the sequence-specific self-cleavage of the larger RNA molecules containing them – with the exception of the RNase P ribozyme that naturally catalyzes the trans-cleavage of cellular RNAs (for reviews, see [2–5]). The subsequent engineering of naturally self-cleaving RNAs made the development of ribozyme technology possible, and a great deal of effort has been invested in turning ribozymes into therapeutic agents. The development of this technology required the following: (i) the identification and isolation of the minimal domain responsible for the observed catalytic activity (from now on referred to as the "ribozyme") and the minimal oligonucleotide containing the ribozyme-cleavable phosphodiester bond (from now on referred to as the "substrate"), (ii) the demonstration that ribozymes catalyze the trans-cleavage of their cognate substrate RNA molecule [6–11] (and that engineered ribozymes behave like protein enzymes, remaining unmodified after the reaction and therefore able to perform their function time and again), and (iii) the demonstration that ribozyme specificity, which is determined by sequence complementarity with the target substrate, is modifiable without reducing catalytic activity, thus rendering any RNA molecule targetable (for a review, see [12–15]).

Traditionally, ribozymes have been classified according to their size as either large or small. Large ribozymes include the two known types of autocatalytic introns, i.e. the group I and group II introns and RNase P, an ubiquitous enzyme that participates in the biogenesis of transfer RNAs (tRNAs) [16, 17]. Although the

Ribozymes, First Edition. Edited by Sabine Müller, Benoît Masquida, and Wade Winkler.

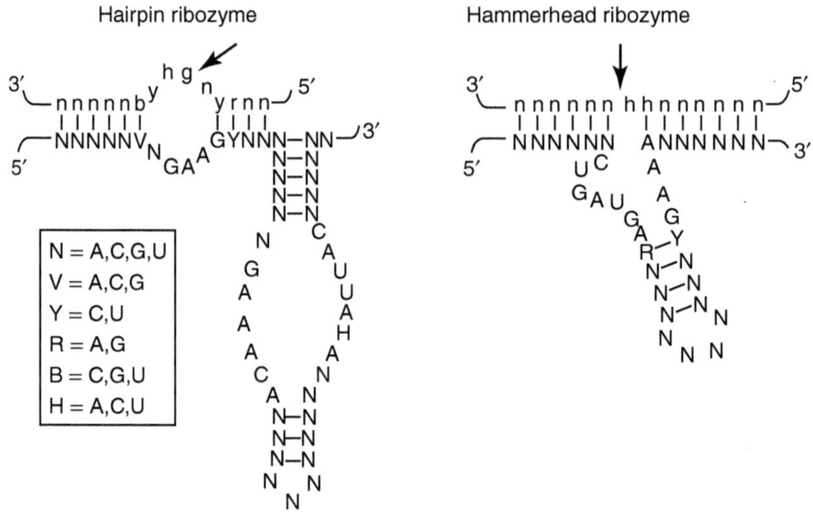

Figure 26.1 Secondary structure model and sequence requirements of the trans-cleaving hairpin and hammerhead ribozyme–substrate complexes [25, 26]. Ribozyme nucleotides are in uppercase letters. Substrate nucleotides are in lowercase letters. Arrows indicate the cleavage sites. Source: Adapted from Berzal-Herranz et al. [25].

group I and group II introns catalyze their self-splicing, they use different molecular mechanisms to achieve this [18]. The small ribozymes include the hammerhead, hairpin, hepatitis delta-type, and Varkud satellite (VS) ribozymes [7, 19–24], and it is on this class that most of the efforts to develop ribozyme-based tools has been concentrated. Each type of ribozyme is characterized by a defined and conserved structure that is required for its catalytic activity.

Although there have been attempts to develop therapeutic strategies involving most types of natural ribozyme, the hammerhead (HH) and hairpin types have been those most investigated for use as gene-inactivating agents in general and as antivirals in particular (Figure 26.1). However, a number of limitations beyond those intrinsic to the technology have jeopardized the successful use of ribozymes as effective antiviral tools. These include factors that limit the efficiency of nucleic acid-based strategies in general, such as colocalization of the ribozyme with the target RNA at an effective concentration, resistance to ribonucleases, delivery of the ribozyme to the target cells, unwanted side effects, etc. Attempts have been made to solve some of these problems, e.g. by developing ribozyme production cassettes under the control of different promoters [27] and/or coding several ribozymes with different specificities in tandem [28–30]. Other problems, however, have shown the need to improve catalytic performance. In hairpin ribozymes this was achieved by introducing sequence changes [31–33] or structural modifications [34–36]. Similar success has been reported for the consensus hammerhead ribozyme [19, 37]. Attempts have also been made to increase the short half-life of ribozymes in the cell milieu by introducing modifications, mainly in the sugar-phosphate backbone, without altering their binding or catalytic efficiency [38, 39].

Another important factor to overcome is the lack of ribozyme access to the cognate substrate sequence – a problem derived from the structure of the target RNA molecule. RNA molecules are naturally folded into more or less stable secondary/tertiary structures. It cannot be naively assumed that a substrate sequence will lie exposed, waiting for a ribozyme or oligonucleotide to recognize it and then efficiently bind to it. Rather, the structure of the target RNA molecule may hide the cleavable substrate's binding sequences. Intramolecular RNA–RNA interactions are the main factors responsible for maintaining the structure of RNA molecules, but intermolecular interactions (e.g. protein–RNA, RNA–RNA) are also commonly determinant in the final molecular conformation and thus the accessibility of the substrate sequences. Ribozymes have to compete with all the existent interactions in order to access and bind to their cleavage site. Thus, the theoretically best substrate sequences may not necessarily be available for efficient cleavage by a cognate ribozyme. This chapter summarizes the work performed at our laboratory to find ways to respond to this difficulty, i.e. optimizing ribozymes by facilitating their access to their target substrate sequences.

26.2 Antiviral Catalytic Antisense RNAs

Early in the development of ribozyme technology, researchers identified disease-causing RNA viruses, e.g. human immunodeficiency virus (HIV) and hepatitis C virus (HCV), as priority targets. Unfortunately, RNA viruses show wide genetic variability – an important hindrance in the design of efficient ribozymes. A single point mutation may alter the validity of a specific sequence as a substrate for antiviral ribozymes. Further, the regions of highest sequence conservation within viral genomic RNAs are exactly those that show the greatest degree of structural compaction, increasing the difficulty in choosing a target substrate sequence. Designing ribozymes against these viruses is challenging.

We questioned whether it was possible to use the intrinsic structural complexity of a genomic viral RNA molecule to actually facilitate the access of a designed ribozyme to its specific substrate sequence. We therefore searched for natural RNA elements or sequence motifs that could efficiently bind to RNA molecules. It is important to note that natural RNA domains specialized in promoting RNA–RNA interactions (e.g. antisense RNA-regulated systems or elements that promote retroviral genome dimerization) meet a number of structural requirements. For example, they must fold into stable stem-loops, promoting RNA–RNA interactions via loop–loop contacts known as kissing-loop interactions. This type of interaction mechanism was probably commonly used for molecule recognition in the prebiotic RNA world [40]. It is known that gene control by natural antisense RNAs relies on the efficient binding of the antisense sequence to the target molecule (rapid and specific binding), the rate-limiting step being the kissing-loop interaction between the two RNA counterparts [41]. A universally conserved YUNR sequence motif (Y = pyrimidine, U = uracil, N = any nucleotide, R = purine) has been identified in all prokaryotic antisense RNA/target pair systems; this is responsible for the

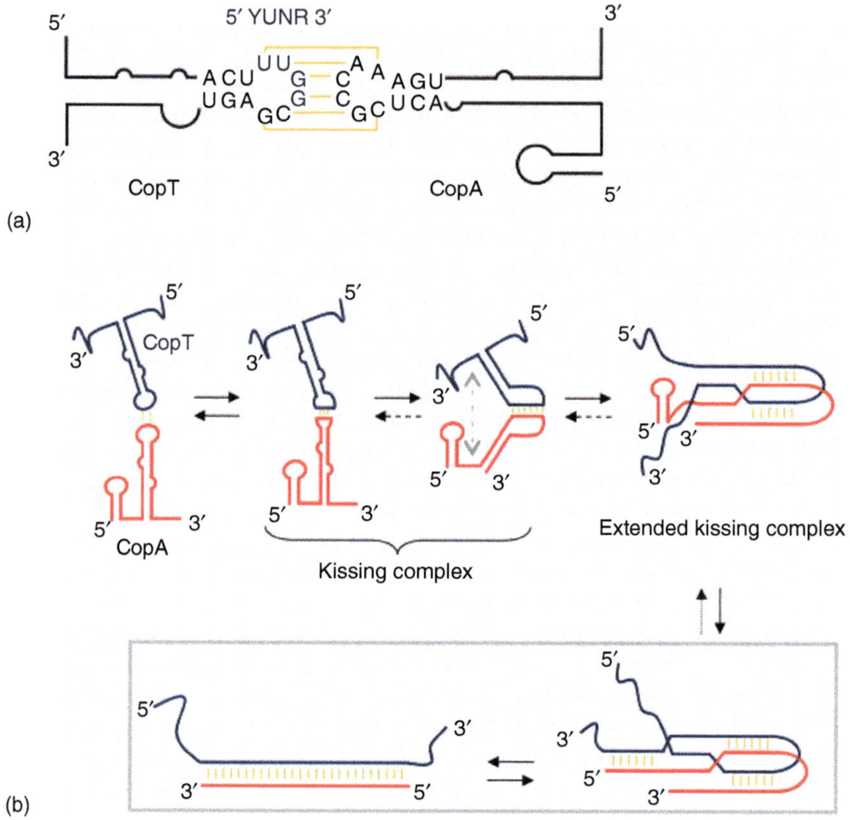

Figure 26.2 Mechanism of action of a natural antisense RNA system, illustrated by the CopA–CopT system. (a) Secondary structure model of the CopA antisense RNA and its target RNA, CopT. The sequences of nucleotides in the upper portions of the interacting stem-loop domains are shown. The YUNR motif within the interacting apical loop of CopT is shown in blue. The loop–loop interaction is indicated by yellow lines. (b) Formation pathway of the CopA–CopT complex. Dotted black arrows denote very slow reactions. Intramolecular interactions are shown by yellow lines. Source: Adapted from Kolb et al. [44].

kissing-loop interaction and therefore for the gene-regulating activity [42, 43] (Figure 26.2).

Based on this information it was hypothesized that it might be possible to optimize ribozymes by recreating the natural scenario based on kissing-loop interactions to facilitate the access of the catalytic RNAs to their cognate target substrate within viral RNA genomes. It was thought this optimization might be achieved by equipping the minimal natural ribozyme with an RNA domain specialized in promoting specific RNA–RNA binding at an existing structural domain in the viral RNA. The resulting molecules would be chimeric inhibitor RNAs named catalytic antisense RNAs, since they combine two inhibitory RNA domains: a ribozyme and an antisense domain [45–47] (Figure 26.3).

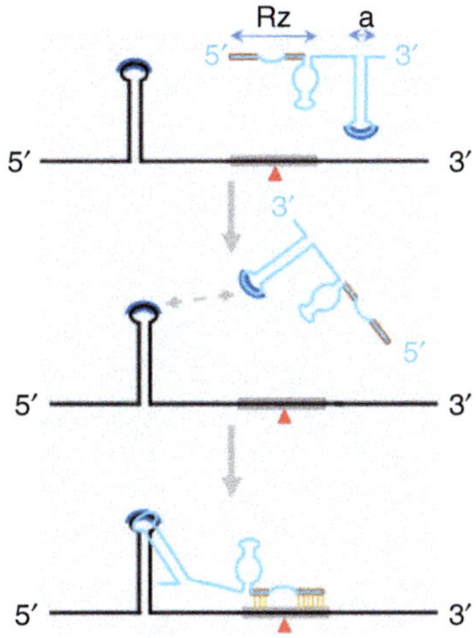

Figure 26.3 Hypothesized mode of action of catalytic antisense RNAs. The diagram shows a catalytic antisense RNA (in blue) represented by the secondary structure model of the hairpin ribozyme (Rz), linked at its 3′-end to an antisense domain (a), and folded into a stable stem-loop. The ribozyme–substrate recognition motifs are highlighted by orange boxes; the antisense nucleotides involved in the kissing-loop interaction are highlighted in dark blue. The target RNA molecule is represented in black. The anchor site in the target is shown as a stem-loop; the nucleotides of the apical loop involved in the kissing-loop interaction with the antisense domain are highlighted in dark blue. The substrate sequence of the ribozyme domain is highlighted in a gray box. The red arrowhead indicates the cleavage site.

26.2.1 HIV-1 TAR as an Anchoring Site for Optimized Ribozymes

To test the above hypothesis, we first searched for structural elements within the 5′ conserved non-translated region (5′NTR) of the HIV genome that resembled essential domains of the natural sense/antisense RNA systems (Figure 26.4). Interestingly, the highly conserved (both in terms of sequence and structure) TAR domain (trans-activation response element) folds into a stable stem closed by a 6 nt loop. This resembles stem-loop II of the well-studied natural antisense CopA RNA that supports the essential kissing-loop interaction with its target CopT molecule [48]. TAR, a 59 nt-long RNA domain encoded at position +1 with respect to the HIV transcription initiation site [49], contains a YUNR sequence motif in the apical loop as does CopT (Figure 26.4). It is the target of the Tat protein (a transcription trans-activator protein), which is essential for the viral cycle. HIV-TAR thus makes an excellent candidate target for an antisense domain.

To test the suitability of TAR as an anchor site for the hairpin and hammerhead ribozymes, a specific TAR antisense domain (which we named αTAR) was designed and covalently linked to the 3′-ribozyme end. αTAR, a 59 nt-long RNA molecule consisting of a fully complementary sequence of TAR, theoretically folds into a stable stem closed by a 6 nt-long single-stranded apical loop [46]. The cleavage efficiency of the resulting catalytic antisense RNAs was tested against two series of artificial long RNA target molecules [46]. All target molecules contained the 59 nt-long HIV-TAR domain upstream of the 14 nt-long natural substrate of the hairpin ribozyme (HP-Swt) derived from the minus strand of tobacco ring spot virus satellite RNA [(−)sTRSV] [9, 10] (Figure 26.5). The first of these series of long targets (series A) consisted of the *Escherichia coli* messenger RNA (mRNA) for the β-galactosidase α-peptide containing the 5′–3′ cassette (HIV-TAR/6 nt-long linker/HP-Swt) at five different positions in its sequence (Figure 26.5). In contrast, series B consisted of five long targets based on the mRNA of the β-galactosidase

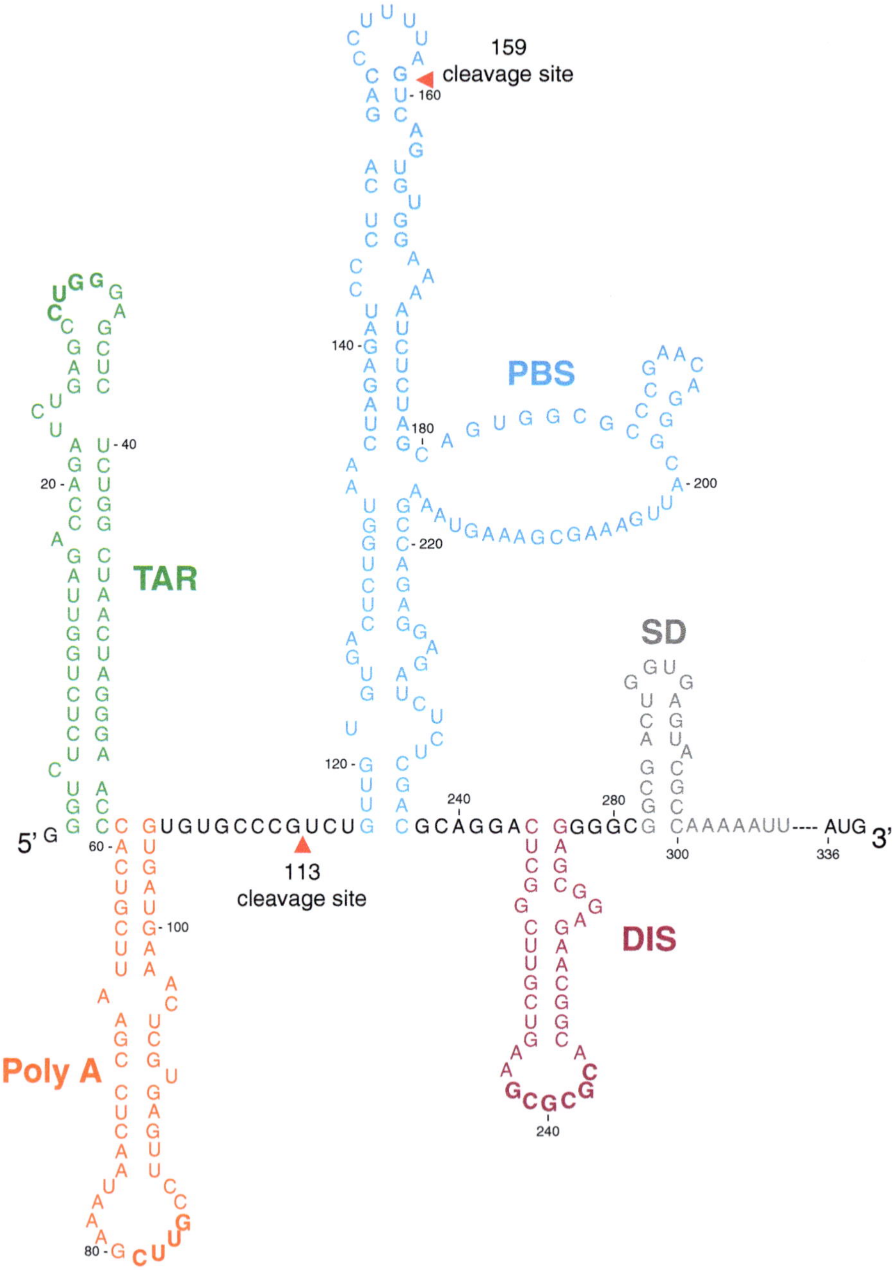

Figure 26.4 Sequence and secondary structure of the HVI-1 5′NTR. The functional domains TAR, Poly(A), primer binding site (PBS), dimerization initiation site (DIS), and splice donor (SD) signal are shown in colors. The YUNR motifs in the apical loops of TAR and PolyA, and the palindromic sequence in the apical loop of DIS (all proposed to be involved in the kissing-loop interaction with theoretical antisense domains) are shown in bold letters. Red arrowheads indicate the 113 and 159 ribozyme cleavage sites. The translation initiation codon (AUG) is indicated.

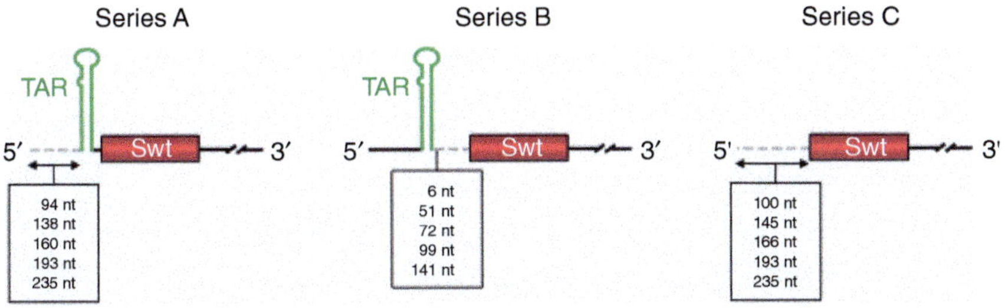

Figure 26.5 Artificial long RNA target series used to validate the usefulness of the 59 nt-long TAR RNA domain as a ribozymes anchor site. The secondary structure model of TAR is represented in green. The 14 nt-long minimal ribozyme–substrate sequence (Swt) is highlighted in a red box. Variable-length regions in constructs of the same series are indicated by a discontinuous gray line. Series C represents the control target lacking the TAR domain.

α-peptide, keeping the HP-Swt at the same positions as in series A but with the HIV-TAR element at a fixed position close to the 5′ end of the mRNA [46]. Therefore, the distance between TAR and Swt increased over the series B target molecules (Figure 26.5). Both target series were challenged by the hairpin and hammerhead ribozymes (the HP-Swt substrate is also cleavable by the hammerhead ribozyme) and the corresponding catalytic antisense molecules carrying the TAR antisense domain covalently linked to the 3′-end of the minimal catalytic domain. A third target series (series C) depleted of the HIV-TAR anchoring domain was used for cleavage efficiency analysis [46, 47]. This allowed comparisons to be made of the influence of molecular/structural contexts, i.e. the accessibility of a specific sequence and its efficient cleavage by the hairpin and hammerhead ribozymes. In addition, it allowed us to determine the usefulness of the HIV-TAR element as an anchoring site for optimized ribozymes (i.e. catalytic antisense RNAs) and simultaneously to examine how the distance, in nucleotides, between the anchoring and the cleavage sites might affect catalytic performance.

The first conclusion provided by this analysis was drawn from the observation that catalytic antisense RNAs cleaved all the TAR domain-containing target molecules more efficiently than the standard ribozymes (hairpin or hammerhead), indicating that the interaction TAR/αTAR enhances their catalytic efficiency (Table 26.1). This suggests that the ability of ribozymes to access their corresponding substrate sequence can be improved by recreating the sense/antisense mechanisms that promote a kissing-loop interaction between the ribozyme and the target molecule. Cleavage assays of target RNAs lacking the TAR domain, or containing a non-TAR antisense stem-loop domain, resulted in a fall in the K_{obs} values. Together, these results highlighted the potential of using singular RNA domains within the target molecule and embedded in a specific structural environment as anchoring elements that improve the access of ribozymes to the cleavable substrate sequence [46, 47].

The efficiencies of the catalytic antisense anti-HIV RNAs based on the hairpin (referred to as HP for the sake of simplicity in this section) and hammerhead (referred to as HH) ribozymes were further tested against a subgenomic HIV

Table 26.1 Processing of TAR-containing artificial target molecules.

Substrate	HP $(K_{obs})/(min^{-1})$	HP αTAR $(K_{obs})/(min^{-1})$	K_{rel}	HH $(K_{obs})/(min^{-1})$	HH αTAR $(K_{obs})/(min^{-1})$	K_{rel}
A1–B1	0.1307 ± 0.0125	0.2722 ± 0.0487	2.0826	0.0615 ± 0.0064	0.2747 ± 0.0130	4.4666
A2	0.2685 ± 0.0593	0.4189 ± 0.1146	1.5601	0.0364 ± 0.0120	0.0355 ± 0.0063	0.9753
B2	0.1161 ± 0.0235	0.2479 ± 0.0610	2.1352	0.0087 ± 0.0004	0.0844 ± 0.0375	9.7011
A3	0.2924 ± 0.0731	0.3623 ± 0.0895	1.2390	0.0263 ± 0.0018	0.0317 ± 0.0159	1.2053
B3	0.2028 ± 0.0388	0.2429 ± 0.0640	1.1977	0.0203 ± 0.0007	0.0702 ± 0.0102	3.4581
A4	0.2279 ± 0.0204	0.4361 ± 0.0481	1.9135	0.1003 ± 0.0096	0.1106 ± 0.0136	1.1026
B4	0.0807 ± 0.0253	0.5030 ± 0.0226	6.2330	0.0214 ± 0.0009	0.0522 ± 0.0062	2.4392
A5	0.2687 ± 0.0576	0.3813 ± 0.0958	1.4190	0.1077 ± 0.0122	0.0959 ± 0.0130	0.8904
B5	0.2666 ± 0.0347	0.3412 ± 0.0581	1.2798	0.0627 ± 0.0104	0.0790 ± 0.0055	1.2600

K_{obs} values are the mean of at least four independent experiments ± the SD. K_{rel} values correspond to the normalization of K_{obs} values to the one obtained for the hairpin (HP) or hammerhead (HH) parental ribozyme. The final extend of cleavage exceeded 90% in all cases.
Source: Puerta-Fernández et al. [46] and Puerta-Fernández et al. [47].

RNA containing the full-length 5′ non-translated region (NTR). For this, catalytic antisense sequences and ribozymes were designed against two substrate sequences within the HIV-NTR (positions 113 and 159), previously shown to be cleaved by both the HP and HH ribozymes [50, 51]. Two sets of ribozymes and their derived catalytic antisense RNAs were assayed (HP113 and HP113αTAR, HH113 and HH113αTAR, HP159 and HP159αTAR, HH159 and HH159αTAR) [47]. In vitro cleavage assays showed that the four catalytic antisense RNA molecules processed the subgenomic RNA containing the HIV-NTR more efficiently than the corresponding parental ribozyme (Figure 26.6a). Cleavage efficiencies and the improvement observed with the presence of the TAR antisense domain depended on the cleavable site and the catalytic domain type, confirming the data obtained with the artificial target RNA molecules.

The in vivo anti-HIV-1 activities of the catalytic antisense RNAs were assayed, following co-transfection of U87-CD4-CXCR4 cells with plasmid pNL4-3 (allowing a complete HIV-1 replication cycle) and the use of a plasmid construct expressing a specific inhibitory RNA of the above-described set. Antiviral activity was measured in terms of diminishing antigen p24 levels – a measurement of the viral replication level. A reduction of almost 2 orders of magnitude was achieved with HH113αTAR and HP159αTAR, whereas no significant effect was observed with the corresponding ribozymes (HH113 and HP159, respectively) or the isolated antisense domain (αTAR) [52] (Figure 26.6b). This antiviral activity cannot be explained simply by the sum of the activities of the individual ribozyme and antisense domains. All of the above results support the validity of the proposed strategy for optimizing antiviral ribozymes.

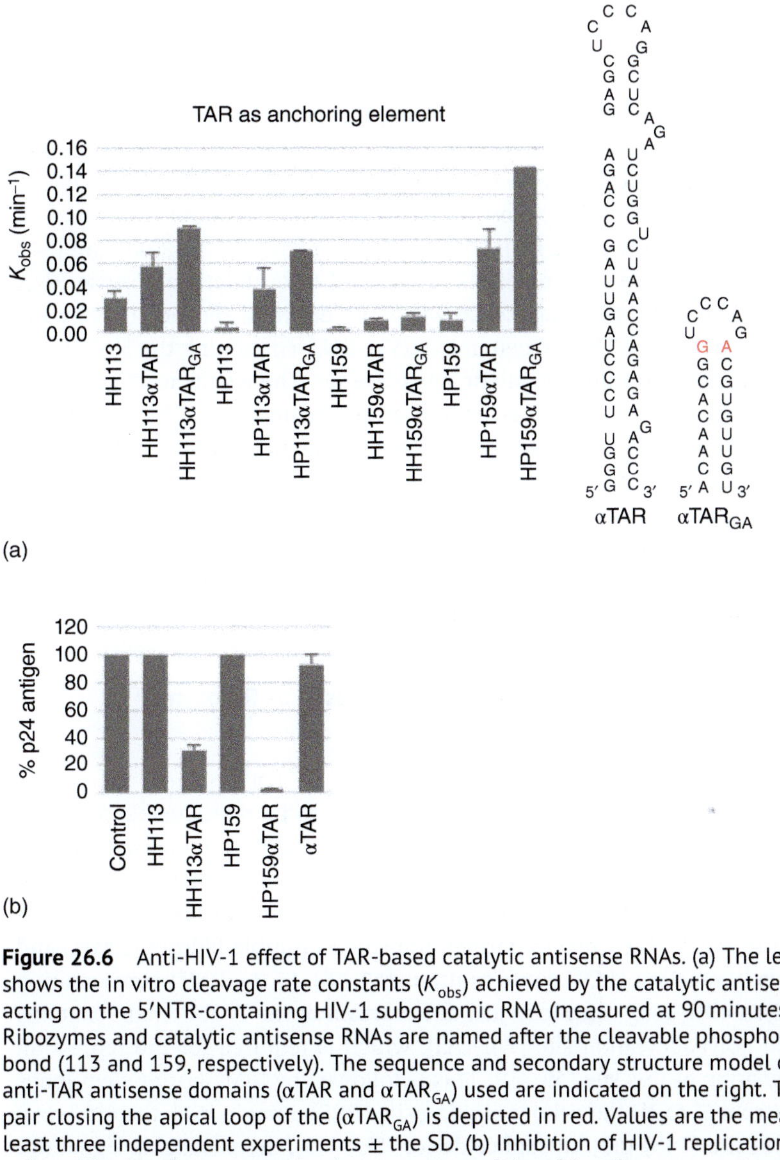

Figure 26.6 Anti-HIV-1 effect of TAR-based catalytic antisense RNAs. (a) The left bar chart shows the in vitro cleavage rate constants (K_{obs}) achieved by the catalytic antisense RNAs acting on the 5′NTR-containing HIV-1 subgenomic RNA (measured at 90 minutes). Ribozymes and catalytic antisense RNAs are named after the cleavable phosphodiester bond (113 and 159, respectively). The sequence and secondary structure model of the two anti-TAR antisense domains (αTAR and αTAR$_{GA}$) used are indicated on the right. The GA base pair closing the apical loop of the (αTAR$_{GA}$) is depicted in red. Values are the mean of at least three independent experiments ± the SD. (b) Inhibition of HIV-1 replication in cell culture. The bar chart represents the percentage of the p24 antigen as a measure of HIV-1 replication with respect to the control cells transfected with empty vector. Values were measured five days after cell transfection and are the mean of at least three independent experiments ± the SD.

Optimization of the αTAR Antisense Domain

Despite the structural features of the TAR domain, no information exists that indicates its involvement in RNA–RNA functional interactions. Therefore, there are no known RNAs that efficiently bind to the TAR domain. Efficient TAR-interacting RNA molecules can, however, be identified by performing in vitro selection. Using this approach, Toulmé et al. selected and characterized a TAR-specific RNA aptamer named $R06_{24}$ [53, 54]. Aptamers are RNA or DNA oligonucleotides that efficiently and specifically bind to a target molecule [55, 56]. $R06_{24}$ is a 24 nt-long RNA oligonucleotide that folds into a stable stem-loop structure, consisting of a 6 nt-long loop fully complementary to the apical loop of TAR, that is closed by a G·A bp, and that has an 8 bp stem with no sequence complementarity to TAR (Figure 26.6a). The presence of the G·A bp loop closure determines the structural conformation of a loop responsible for a stable loop–loop interaction with the TAR domain [54]. In an attempt to improve the anti-HIV-1 activity of the catalytic antisense RNA set described above, the αTAR antisense was replaced in all set members by the $R06_{24}$ aptamer (αTAR_{GA}). The subgenomic HIV-1 RNA containing the 5′NTR was challenged against the resulting catalytic antisense set. The results in Figure 26.6a show that the presence of the aptamer, an antisense domain selected for its ability to bind to TAR, significantly improved the cleavage rate of all the tested hairpin- and hammerhead-based catalytic antisense molecules, with up to a 16-fold improvement for the hairpin-based catalytic antisense targeting the 113 position with respect to the standard hairpin ribozyme. These results indicate that a positive correlation exists between the binding efficiency of the sense-antisense domains and the viral RNA processing efficiency of the catalytic domain.

26.2.2 HIV-1 Poly(A) and DIS Domains Can Be Used as Ribozyme Anchoring Sites

To validate our strategy of ribozyme optimization, we had to show the results for the HIV-1 TAR domain as an anchoring element to be reproducible for other structural domains in targeted viral RNAs. Along with the TAR domain, the HIV-1 5′NTR contains multiple highly conserved structural domains that play essential roles in the infectious cycle. These are embedded in two mutually exclusive conformations of the overall NTR region: termed branched multiple hairpin (BMH) and long-distance interaction (LDI) conformations [57, 58]. In the BMH conformation (Figure 26.4), a 47 nt-long stem-loop just downstream of the TAR domain contains the proximal polyadenylation site [Poly(A)], which is required for the repression of polyadenylation at this site [59, 60]. This is followed by four shorter stem-loops containing (from 5′ to 3′) the primer binding site (PBS) domain [61], the dimerization initiation site (DIS) [62], the splice donor (SD) [63], the packaging signal (ψ) [63, 64], and the translation initiation codon of Gag (reviewed in [65]). Of these, only the DIS-containing element is naturally involved in functional RNA–RNA interactions. This interaction is responsible for promoting the formation of genome homodimers – an essential viral process. The DIS structural element contains a 6 nt-long palindromic motif at the apical loop that is essential for the initial RNA–RNA interaction via a kissing-loop mechanism [66] (Figure 26.4). In addition, the Poly(A) stem-loop domain, which is involved in the polyadenylation of the viral transcripts [67], contains a YUNR motif at its apical loop (Figure 26.4), though its potential functional involvement is unknown. Both natural domains were tested as anchoring sites for anti-HIV-1 catalytic antisense RNAs targeted against

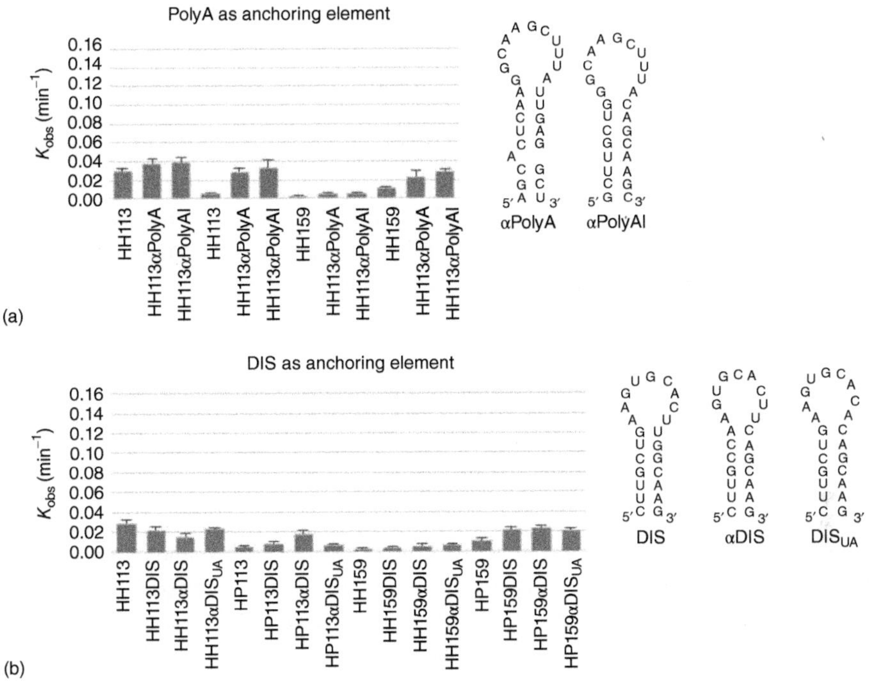

Figure 26.7 In vitro anti-HIV-1 activity of catalytic antisense RNAs. (a) Anti-HIV-1 effect of catalytic antisense RNAs using Poly(A) as the anchor site. The left bar chart shows the in vitro cleavage rate constants (K_{obs}) achieved by the catalytic antisense RNAs acting on the 5′NTR-containing HIV-1 subgenomic RNA. The sequence and secondary structure model of the two anti-Poly(A) antisense domains tested (αPolyA and αPolyAI) are indicated on the right. (b) Anti-HIV-1 effect of catalytic antisense using DIS as the anchor site. The left bar chart shows the in vitro cleavage rate constants (K_{obs}) achieved by the catalytic antisense RNAs acting on the 5′NTR-containing HIV-1 subgenomic RNA. The sequence and secondary structure model of the anti-DIS antisense domains tested (DIS, αDIS, and αDIS$_{UA}$) are indicated on the right. Values were obtained after 90 minutes and are the mean of at least three independent experiments ± the SD.

positions 113 and 159 of the HIV-1 NTR. (Note that, unlike TAR and Poly(A), the DIS element is located downstream of both the 113 and 159 ribozyme cleavage sites (Figure 26.4). For this, two different Poly(A) antisense domains were designed (Figure 26.7a): a 28 nt-long molecule fully complementary to the apical portion of the HIV-1 Poly(A) domain (named αpolyA) and a 27 nt-long molecule composed of the antisense sequence of the apical loop of the Poly(A) domain closing a short 8 bp stem of a non-related sequence (αpolyAl) (Figure 26.7a). Two sets of catalytic antisense RNAs based on the described four anti-HIV-1 ribozymes carrying either of the Poly(A) antisense sequences were constructed, and all showed significantly more efficient cleavage activity against the HIV-1 5′NTR subgenomic RNA than the standard ribozymes (Figure 26.7a).

Similarly, to test the usefulness of the HIV-1 DIS element as an anchor site, three sets of four catalytic antisense RNAs based on the described hairpin and hammerhead 113 and 159 ribozymes were constructed. They differed in the so-called antisense domain (Figure 26.7b) that consisted of either (i) the 23 nt-long apical portion of the DIS domain (since the sense sequence is the natural binder of the DIS element; named DIS), (ii) the 23 nt-long antisense sequence of the apical portion of DIS (αDIS), or (iii) the same as in (i) but containing a single point mutation, U to A, at the 5′-nucleotide of the apical loop (DIS_{UA}). Marquet et al. described the importance of the presence of a purine residue at this position for genome dimerization [68]. Even though the anchor site is located downstream of the cleavage site, all catalytic antisense RNAs – with the exception of those based on the hammerhead targeting position 113 – processed the HIV-1 NTR subgenomic RNA more efficiently than did the corresponding ribozymes (Figure 26.7b and [69]).

26.3 A General Experimental Strategy for Designing Catalytic Antisense RNAs

The results described so far confirm that the cleavage site access difficulties imposed by the structural complexity of the target molecule can be overcome by using this very same structural complexity to provide an anchor site for optimized ribozymes. However, they also indicate that the different structural RNA elements within target molecules differ in terms of their efficiency as anchor sites. Further, different antisense elements can be designed for a chosen anchor site, yielding catalytic antisense RNAs of different cleavage competence. In fact, the above results show that a perfectly matching complementary sequence need does not make for the most proficient catalytic antisense RNA. Rather, the most efficient binder makes the best antisense domain.

The methodology for designing optimized ribozymes presented up to now has all been trial and error. Not only is this a very time-consuming strategy, but it cannot guarantee that the most proficient inhibitory ribozyme-based RNA is going to be designed. Thus, answers to the following questions were pursued: Which is the most competent anchoring element within a targeted RNA molecule? And which antisense motif for a chosen anchoring element would yield the most proficient ribozyme? A strategy that would simultaneously assay all possible anchoring sites within the viral RNA molecule, plus a large variety of potential antisense elements for each of them, was thus sought. This was undertaken using a subgenomic HCV RNA containing the 5′ untranslated region (UTR) as a model.

26.3.1 Experimental Isolation of HCV IRES Catalytic Antisense RNAs

HCV has been a favorite target of ribozyme aficionados pursuing therapeutic strategies based on the use of catalytic RNAs [70]. The HCV genome is a single-stranded RNA molecule rich in structurally highly conserved RNA elements involved in a functional network of long-distance RNA–RNA interactions. These RNA elements are distributed throughout the genome but are more concentrated in the UTRs at both ends of the genome (for review, see [71, 72]). Viral translation is initiated by an internal ribosome entry site (IRES) element mostly located at the 5′UTR but also

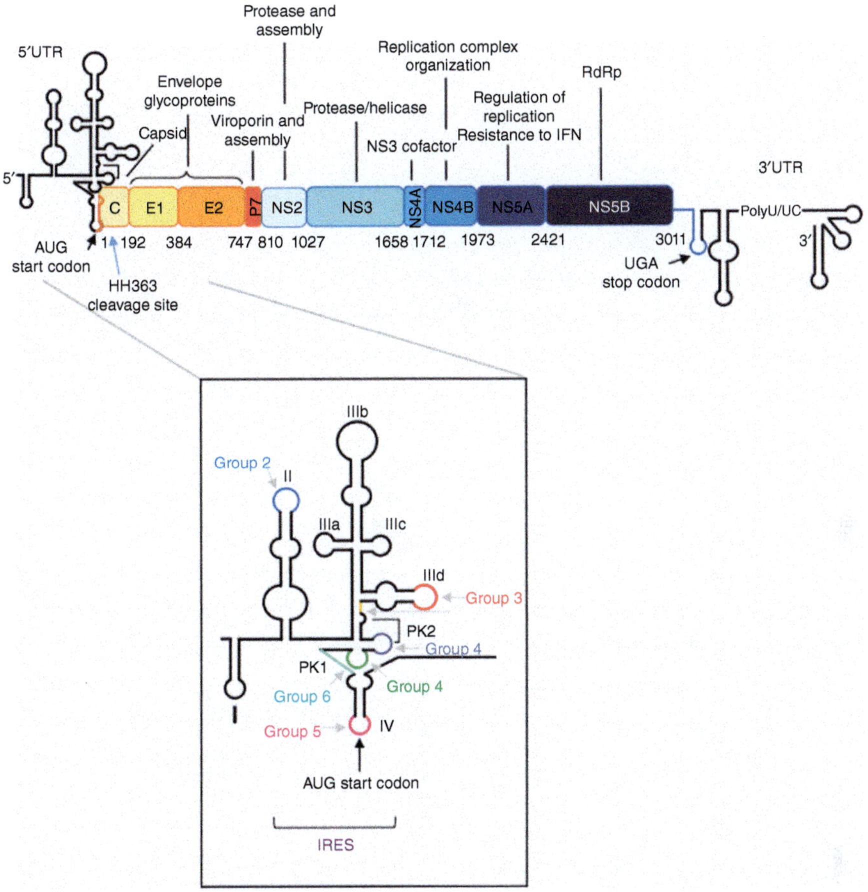

Figure 26.8 The hepatitis C virus genome is shown in the upper panel. The protein products coded by the single open reading frame (ORF) are depicted in colored boxes. The secondary structures of the UTRs flanking the ORF are indicated by solid lines. The HH363 ribozyme cleavage site is indicated. A detailed view of the genomic 5′-end containing the IRES is shown at the bottom. Different structural domains (I–IV) and the translation start codon are indicated. Colored lines indicate the positions of the complementary sequences and therefore potential binding sites for the consensus sequence motifs that define each of the identified aptamer groups (indicated using the same color code). Source: Romero-López and Berzal-Herranz [72]. © 2017 Frontiers Media S.A.

spanning around 30 nt of the core coding sequence [73, 74] (Figure 26.8). The IRES folds into a complex, compact structure containing a large number of well-defined structural elements (Figure 26.8) and therefore represents an ideal target for catalytic antisense RNAs. To meet the requirements that the strategy should simultaneously assay all possible anchoring sites within the viral RNA molecule, plus a large variety of potential antisense elements for each of them, an in vitro selection-based methodology was designed [75]. This required two sequential selection steps to be performed on a large starting population of RNA molecules composed of a fixed hammerhead ribozyme targeting position 363 within the HCV IRES and a 25 nt-long

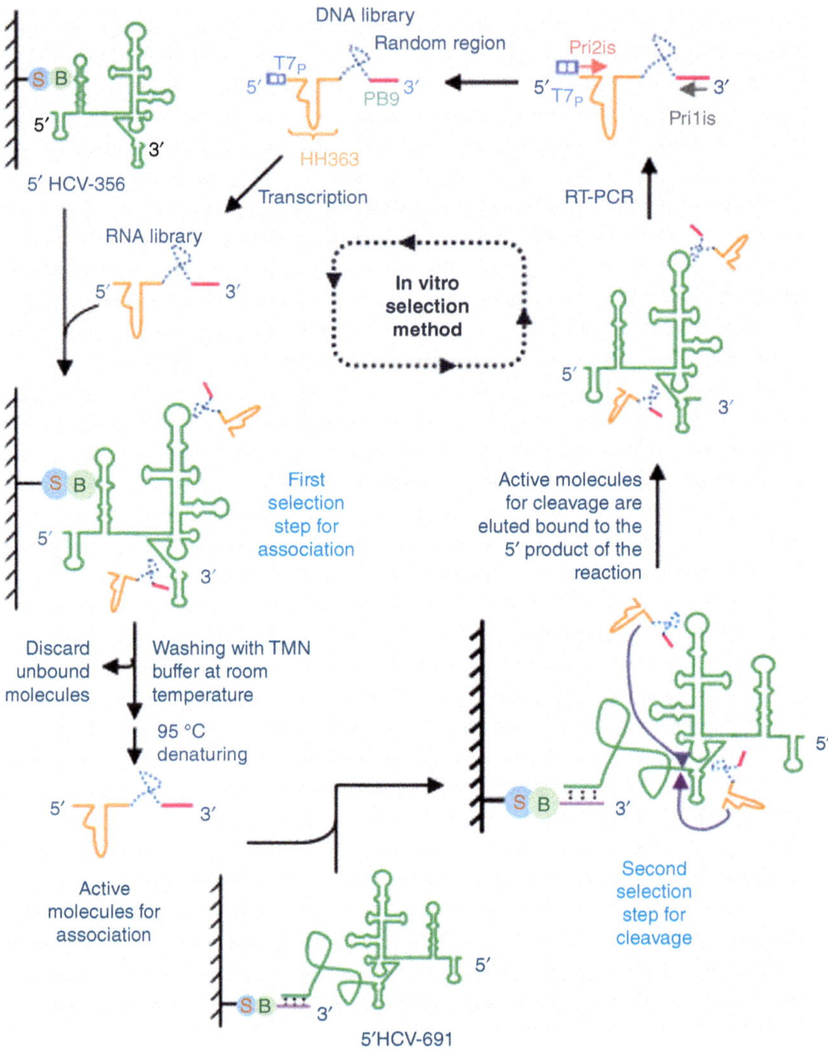

Figure 26.9 In vitro selection strategy for the identification of anti-HCV IRES catalytic antisense RNAs. The strategy designed for the isolation of catalytic antisense RNAs containing the hammerhead ribozyme HH363 (orange solid line), and an aptamer to promote binding (dark blue dotted line), for use against the HCV genomic RNA represented (solid green line). 5′HCV-356, HCV genome fragment containing only the first 356 nt, which lacks the HH363 cleavage site located at position 363 of the RNA genome. It was used during the first selection step for binding. During its synthesis, 5′HCV-356 was internally biotinylated at a theoretical ratio of one modified nucleotide per molecule. 5′HCV-691, HCV genome fragment containing the first 691 nt. It was used during the second selection step, to isolate those molecules that conserve the ability of HH363 to cleave the HCV RNA. 5′HCV-691 was trapped to the streptavidin column by a 21 nt-long 5′-end biotinylated DNA oligonucleotide sequence complementary to its 3′-end. They are excised from the column by the catalytic activity of the ribozyme domain bound to the 5′ HCV RNA cleavage product. The HCV IRES domain is identified by a schematic representation of its secondary structure. The ribozyme cleavage site is indicated by arrowheads. S, streptavidin; B, biotin ligand. Source: Romero-López et al. [75]. © 2005 Walter de Gruyter.

random sequence domain (with the binder to the anchor site being identified from this random sequence stretch). The first step selects for binding to the HCV IRES RNA fragment, and the second for the cleavage of the HCV target substrate by the catalytic domain (Figure 26.9). The latter was meant to counter select any of the selected binders that jeopardize the catalytic activity of the ribozyme to be improved. This strategy, therefore, can be considered an systematic evolution of ligands by exponential enrichment (SELEX)-like procedure [55, 56]. The iterative application of this methodology to the population of putative catalytic antisense RNAs allowed seven groups to be defined by a consensus sequence motif within the selected binding domain – which turned out to be complementary to a specific motif within the HCV IRES (Figure 26.8). For each group, binding was dependent on the consensus sequence motif but always within a structural environment forming an intrinsic feature of the complete RNA molecule. Further analysis of the inhibitory activity of the selected molecules showed this strategy allowed for good anchor sites within the IRES to be detected, without imposing any sequence or structural constraints [75–78]. Simultaneously, the strategy allowed the selection of the corresponding antisense sequence for each site from a large mix of sequences (>10^{15}). In addition, it simultaneously scanned options for the linker between the anchoring and cleavage sites, as well as other sequence motifs that might reduce the efficiency of the binder and/or the ribozyme. The IRES inhibition exerted by the different identified catalytic antisense RNAs was analyzed in in vitro translation assays. For these, a monocistronic mRNA coding for the FLuc protein, translated under the control of the HCV IRES, was incubated in rabbit reticulocyte lysates in the presence of the different inhibitory RNAs. IRES inhibition was measured via the reduction of luciferase activity. As shown in Figure 26.10, the great majority of the selected catalytic antisense RNAs reduced the activity of the IRES – indeed by as much as 90% [76].

Further characterization of specific anti-HCV catalytic antisense RNAs revealed strong inhibition (up to 60%) of the IRES activity in a human hepatoma cell line (Huh-7) and a significant reduction of HCV RNA levels (up to 70%) in a subgenomic HCV replicon system (Huh-7 NS3-3′) [76–79].

As shown for the anti-HIV catalytic antisense molecules, the inhibitory capacity of the selected anti-HCV molecules is an intrinsic feature of their entire structure that cannot be explained by the sum of the inhibitory effects exerted independently by the catalytic (ribozyme) and antisense (aptamer) domains [76].

26.4 Concluding Remarks

In vitro models have proved very useful in demonstrating that ribozymes have great potential as antiviral tools. However, attempts to use ribozymes in vivo have failed to yield acceptable therapeutic results, the consequence of a number of unforeseen or underestimated obstacles. One important obstacle is the lack of access to the target cleavable sequence – a problem arising from the intrinsic structure of RNA molecules. The work summarized here shows this can be circumvented by the addition, to the ribozyme, of an RNA domain that promotes efficient binding to a natural

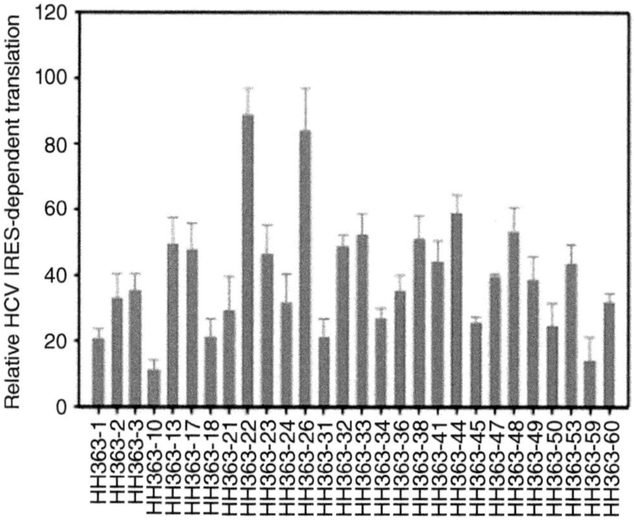

Figure 26.10 Anti-HCV IRES activity of different catalytic antisense RNAs. In vitro translation assays of the firefly luciferase gene under the control of the HCV IRES, performed in reticulocyte extracts in the presence of a catalytic antisense RNA. The bar chart shows the relative light intensity with respect to control RLuc activity translated in a cap-dependent manner and to luciferase activity in the absence of any inhibitor RNA. Values are the mean of at least three independent experiments ± the SD. Source: Romero-López et al. [76]. © 2007 Springer Nature. Reproduced with permission of Springer Nature.

structural domain within the targeted viral RNA. The latter acts as an anchor site for the ribozyme, facilitating its access to its cognate substrate sequence, thus enhancing the ribozyme's antiviral activity. Our work also involved the development of a strategy to isolate improved antiviral ribozymes. Based on SELEX methodology, this allowed us to rapidly and simultaneously assay all possible anchor sites within the target viral RNA, as well as a large variety of possible binders of each. The inefficient, costly, and time-consuming trial and error-based approach to designing optimized antiviral ribozymes is thus now superseded. The new system could be used for designing efficient ribozymes against any RNA of interest.

Acknowledgments

Work in our group is supported by the grant BFU2015-64359-P to A.B.-H.

References

1 Lilley, D.M. (2011). Mechanisms of RNA catalysis. *Philos. Trans. R. Soc. London, Ser. B* 366 (1580): 2910–2917.
2 Cech, T.R. (1990). Self-splicing of group I introns. *Annu. Rev. Biochem.* 59: 543–568.

3 Lambowitz, A.M. and Belfort, M. (1993). Introns as mobile genetic elements. *Annu. Rev. Biochem.* 62: 587–622.
4 Symons, R.H. (1992). Small catalytic RNAs. *Annu. Rev. Biochem.* 61: 641–671.
5 Altman, S., Kirsebom, L., and Talbot, S. (1993). Recent studies of ribonuclease P. *FASEB J.* 7 (1): 7–14.
6 Zaug, A.J. and Cech, T.R. (1986). The intervening sequence RNA of *Tetrahymena* is an enzyme. *Science* 231 (4737): 470–475.
7 Uhlenbeck, O.C. (1987). A small catalytic oligoribonucleotide. *Nature* 328 (6131): 596–600.
8 Haseloff, J. and Gerlach, W.L. (1988). Simple RNA enzymes with new and highly specific endoribonuclease activities. *Nature* 334 (6183): 585–591.
9 Hampel, A. and Tritz, R. (1989). RNA catalytic properties of the minimum (−)sTRSV sequence. *Biochemistry* 28 (12): 4929–4933.
10 Feldstein, P.A., Buzayan, J.M., and Bruening, G. (1989). Two sequences participating in the autolytic processing of satellite tobacco ringspot virus complementary RNA. *Gene* 82 (1): 53–61.
11 Branch, A.D. and Robertson, H.D. (1991). Efficient trans cleavage and a common structural motif for the ribozymes of the human hepatitis delta agent. *Proc. Natl. Acad. Sci. U.S.A.* 88 (22): 10163–10167.
12 Cech, T.R. (1992). Ribozyme engineering. *Curr. Opin. Struct. Biol.* 2: 605–609.
13 Symons, R.H. (1994). Ribozymes. *Curr. Opin. Struct. Biol.* 4: 322–330.
14 Tanner, N.K. (1999). Ribozymes: the characteristics and properties of catalytic RNAs. *FEMS Microbiol. Rev.* 23 (3): 257–275.
15 Puerta-Fernández, E., Romero-López, C., Barroso-delJesus, A., and Berzal-Herranz, A. (2003). Ribozymes: recent advances in the development of RNA tools. *FEMS Microbiol. Rev.* 27 (1): 75–97.
16 Altman, S. (1989). Ribonuclease P: an enzyme with a catalytic RNA subunit. *Adv. Enzymol. Relat. Areas Mol. Biol.* 62: 1–36.
17 Cech, T.R. (1993). Structure and mechanism of the large catalytic RNAs: group I and group II introns and ribonuclease P. In: *The RNA World* (eds. R.F. Gesteland and J.F. Atkins), 239–269. Cold Spring Harbor Laboratory Press.
18 Saldanha, R., Mohr, G., Belfort, M., and Lambowitz, A.M. (1993). Group I and group II introns. *FASEB J.* 7 (1): 15–24.
19 De la Peña, M., Gago, S., and Flores, R. (2003). Peripheral regions of natural hammerhead ribozymes greatly increase their self-cleavage activity. *EMBO J.* 22 (20): 5561–5570.
20 Haseloff, J. and Gerlach, W.L. (1989). Sequences required for self-catalysed cleavage of the satellite RNA of tobacco ringspot virus. *Gene* 82 (1): 43–52.
21 Burke, J.M. (1994). The hairpin ribozyme. *Nucleic Acids Mol. Biol.* 8: 105–118.
22 Fedor, M.J. (2000). Structure and function of the hairpin ribozyme. *J. Mol. Biol.* 297 (2): 269–291.
23 Asif-Ullah, M., Levesque, M., Robichaud, G., and Perreault, J.P. (2007). Development of ribozyme-based gene-inactivations; the example of the hepatitis delta virus ribozyme. *Curr. Gene Ther.* 7 (3): 205–216.
24 Lilley, D.M. (2004). The Varkud satellite ribozyme. *RNA* 10 (2): 151–158.

25 Berzal-Herranz, A., Joseph, S., Chowrira, B.M. et al. (1993). Essential nucleotide sequences and secondary structure elements of the hairpin ribozyme. *EMBO J.* 12 (6): 2567–2573.

26 Kore, A.R., Vaish, N.K., Kutzke, U., and Eckstein, F. (1998). Sequence specificity of the hammerhead ribozyme revisited; the NHH rule. *Nucleic Acids Res.* 26 (18): 4116–4120.

27 Kato, Y., Kuwabara, T., Warashina, M. et al. (2001). Relationships between the activities in vitro and in vivo of various kinds of ribozyme and their intracellular localization in mammalian cells. *J. Biol. Chem.* 276 (18): 15378–15385.

28 Shahi, S. and Banerjea, A.C. (2002). Multitarget ribozyme against the S1 genome segment of reovirus possesses novel cleavage activities and is more efficacious than its constituent mono-ribozymes. *Antiviral Res.* 55 (1): 129–140.

29 Sriram, B., Thakral, D., and Panda, S.K. (2003). Targeted cleavage of hepatitis E virus 3′ end RNA mediated by hammerhead ribozymes inhibits viral RNA replication. *Virology* 312 (2): 350–358.

30 Li, X., Kuang, E., Dai, W. et al. (2005). Efficient inhibition of hepatitis B virus replication by hammerhead ribozymes delivered by hepatitis delta virus. *Virus Res.* 114 (1-2): 126–132.

31 Berzal-Herranz, A., Joseph, S., and Burke, J.M. (1992). In vitro selection of active hairpin ribozymes by sequential RNA-catalyzed cleavage and ligation reactions. *Genes Dev.* 6 (1): 129–134.

32 Joseph, S. and Burke, J.M. (1993). Optimization of an anti-HIV hairpin ribozyme by in vitro selection. *J. Biol. Chem.* 268 (33): 24515–24518.

33 Esteban, J.A., Banerjee, A.R., and Burke, J.M. (1997). Kinetic mechanism of the hairpin ribozyme. Identification and characterization of two nonexchangeable conformations. *J. Biol. Chem* 272 (21): 13629–13639.

34 Barroso-delJesus, A., Tabler, M., and Berzal-Herranz, A. (1999). Comparative kinetic analysis of structural variants of the hairpin ribozyme reveals further potential to optimize its catalytic performance. *Antisense Nucleic Acid Drug Dev.* 9 (5): 433–440.

35 Müller, S. (2003). Engineered ribozymes as molecular tools for site-specific alteration of RNA sequence. *ChemBioChem* 4 (10): 991–997.

36 Müller, S., Appel, B., Krellenberg, T., and Petkovic, S. (2012). The many faces of the hairpin ribozyme: structural and functional variants of a small catalytic RNA. *IUBMB Life* 64 (1): 36–47.

37 De la Peña, M. and Flores, R. (2001). An extra nucleotide in the consensus catalytic core of a viroid hammerhead ribozyme: implications for the design of more efficient ribozymes. *J. Biol. Chem.* 276 (37): 34586–34593.

38 Sioud, M. (2006). Ribozymes and siRnas: from structure to preclinical applications. *Handb. Exp. Pharmacol.* 173: 223–242.

39 Burnett, J.C. and Rossi, J.J. (2012). RNA-based therapeutics: current progress and future prospects. *Chem. Biol.* 19 (1): 60–71.

40 Tomizawa, J.-I. (1993). Evolution of functional structures of RNA. In: *The RNA World* (eds. R.F. Gesteland and J.F. Atkins), 419–445. Cold Spring Harbor Laboratory Press.

41 Wagner, E.G. and Simons, R.W. (1994). Antisense RNA control in bacteria, phages, and plasmids. *Annu. Rev. Microbiol.* 48: 713–742.

42 Franch, T., Petersen, M., Wagner, E.G. et al. (1999). Antisense RNA regulation in prokaryotes: rapid RNA/RNA interaction facilitated by a general U-turn loop structure. *J. Mol. Biol.* 294 (5): 1115–1125.

43 Franch, T. and Gerdes, K. (2000). U-turns and regulatory RNAs. *Curr. Opin. Microbiol.* 3 (2): 159–164.

44 Kolb, F.A., Engdahl, H. M., and Slagter-Jager, J. G. (2000). Progression of a loop–loop complex to a four-way junction is crucial for the activity of a regulatory antisense RNA. *EMBO J.* 19 (21): 5905–5915.

45 Pérez-Ruiz, M., Sievers, D., García-López, P.A., and Berzal-Herranz, A. (1999). The antisense sequence of the HIV-1 TAR stem-loop structure covalently linked to the hairpin ribozyme enhances its catalytic activity against two artificial substrates. *Antisense Nucleic Acid Drug Dev.* 9 (1): 33–42.

46 Puerta-Fernández, E., Barroso-delJesus, A., and Berzal-Herranz, A. (2002). Anchoring hairpin ribozymes to long target RNAs by loop–loop RNA interactions. *Antisense Nucleic Acid Drug Dev.* 12 (1): 1–9.

47 Puerta-Fernández, E., Barroso-delJesus, A., Romero-López, C., and Berzal-Herranz, A. (2003). HIV-1 TAR as anchoring site for optimized catalytic RNAs. *Biol. Chem.* 384: 343–350.

48 Wagner, E.G.H. and Nordstrom, K. (1986). Structural analysis of an RNA molecule involved in replication control of plasmid R1. *Nucleic Acids Res.* 14 (6): 2523–2538.

49 Muesing, M.A., Smith, D.H., and Capon, D.J. (1987). Regulation of mRNA accumulation by a human immunodeficiency virus trans-activator protein. *Cell* 48 (4): 691–701.

50 Ojwang, J.O., Hampel, A., Looney, D.J. et al. (1992). Inhibition of human immunodeficiency virus type 1 expression by a hairpin ribozyme. *Proc. Natl. Acad. Sci. U.S.A.* 89 (22): 10802–10806.

51 Bramlage, B., Luzi, E., and Eckstein, F. (2000). HIV-1 LTR as a target for synthetic ribozyme-mediated inhibition of gene expression: site selection and inhibition in cell culture. *Nucleic Acids Res.* 28 (21): 4059–4067.

52 Puerta-Fernández, E., Barroso-delJesus, A., and Romero-López, C. (2005). Inhibition of HIV-1 replication by RNA targeted against the LTR region. *AIDS* 19 (9): 863–870.

53 Duconge, F. and Toulme, J.J. (1999). In vitro selection identifies key determinants for loop–loop interactions: RNA aptamers selective for the TAR RNA element of HIV-1. *RNA* 5 (12): 1605–1614.

54 Duconge, F., Di Primo, C., and Toulme, J.J. (2000). Is a closing "GA pair" a rule for stable loop–loop RNA complexes? *J. Biol. Chem.* 275 (28): 21287–21294.

55 Tuerk, C. and Gold, L. (1990). Systematic evolution of ligands by exponential enrichment: RNA ligands to bacteriophage T4 DNA polymerase. *Science* 249 (4968): 505–510.

56 Ellington, A.D. and Szostak, J.W. (1990). In vitro selection of RNA molecules that bind specific ligands. *Nature* 346 (6287): 818–822.

57 Huthoff, H. and Berkhout, B. (2001). Two alternating structures of the HIV-1 leader RNA. *RNA* 7 (1): 143–157.

58 Berkhout, B., Ooms, M., and Beerens, N. (2002). In vitro evidence that the untranslated leader of the HIV-1 genome is an RNA checkpoint that regulates multiple functions through conformational changes. *J. Biol. Chem.* 277 (22): 19967–19975.

59 Berkhout, B., Klaver, B., and Das, A.T. (1995). A conserved hairpin structure predicted for the poly(A) signal of human and simian immunodeficiency viruses. *Virology* 207 (1): 276–281.

60 Zarudnaya, M.I., Potyahaylo, A.L., Kolomiets, I.M., and D.M., H. (2013). Structural model of the complete poly(A) region of HIV-1 pre-mRNA. *J. Biomol. Struct. Dyn.* 31 (10): 1044–1056.

61 Sleiman, D., Goldschmidt, V., and Barraud, P. (2012). Initiation of HIV-1 reverse transcription and functional role of nucleocapsid-mediated tRNA/viral genome interactions. *Virus Res.* 169 (2): 324–339.

62 Ulyanov, N.B., Mujeeb, A., and Du, Z. (2006). NMR structure of the full-length linear dimer of stem-loop-1 RNA in the HIV-1 dimer initiation site. *J. Biol. Chem.* 281 (23): 16168–16177. Epub 12006 Apr 16166.

63 Harrison, G.P. and Lever, A.M. (1992). The human immunodeficiency virus type 1 packaging signal and major splice donor region have a conserved stable secondary structure. *J. Virol.* 66 (7): 4144–4153.

64 Clever, J.L., Miranda, D. Jr., and Parslow, T.G. (2002). RNA structure and packaging signals in the 5′ leader region of the human immunodeficiency virus type 1 genome. *J. Virol.* 76 (23): 12381–12387.

65 Berkhout, B. (1996). Structure and function of the human immunodeficiency virus leader RNA. *Prog. Nucleic Acid Res. Mol. Biol.* 54: 1–34.

66 Muriaux, D., De Rocquigny, H., Roques, B.P., and Paoletti, J. (1996). NCp7 activates HIV-1Lai RNA dimerization by converting a transient loop–loop complex into a stable dimer. *J. Biol. Chem.* 271 (52): 33686–33692.

67 Das, A.T., Klaver, B., and Berkhout, B. (1999). A hairpin structure in the R region of the human immunodeficiency virus type 1 RNA genome is instrumental in polyadenylation site selection. *J. Virol.* 73 (1): 81–91.

68 Lodmell, J.S., Ehresmann, C., Ehresmann, B., and Marquet, R. (2000). Convergence of natural and artificial evolution on an RNA loop–loop interaction: the HIV-1 dimerization initiation site. *RNA* 6 (9): 1267–1276.

69 Sánchez-Luque, F.J., Reyes-Darias, J.A., Puerta-Fernández, E., and Berzal-Herranz, A. (2010). Inhibition of HIV-1 replication and dimerization interference by dual inhibitory RNAs. *Molecules* 15 (7): 4757–4772.

70 Romero-López, C., Sánchez-Luque, F.J., and Berzal-Herranz, A. (2006). Targets and tools: recent advances in the development of anti-HCV nucleic acids. *Infect. Disord. Drug Targets* 6 (2): 121–145.

71 Romero-López, C. and Berzal-Herranz, A. (2013). Unmasking the information encoded as structural motifs of viral RNA genomes: a potential antiviral target. *Rev. Med. Virol.* 23 (6): 340–354.

72 Romero-López, C. and Berzal-Herranz, A. (2017). The 5BSL3.2 functional RNA domain connects distant regions in the hepatitis C virus genome. *Front. Microbiol.* 8: 2093.

73 Tsukiyama-Kohara, K., Iizuka, N., Kohara, M., and Nomoto, A. (1992). Internal ribosome entry site within hepatitis C virus RNA. *J. Virol.* 66 (3): 1476–1483.

74 Wang, C., Sarnow, P., and Siddiqui, A. (1993). Translation of human hepatitis C virus RNA in cultured cells is mediated by an internal ribosome-binding mechanism. *J. Virol.* 67 (6): 3338–3344.

75 Romero-López, C., Barroso-delJesus, A., Puerta-Fernández, E., and Berzal-Herranz, A. (2005). Interfering with hepatitis C virus IRES activity using RNA molecules identified by a novel in vitro selection method. *Biol. Chem.* 386 (2): 183–190.

76 Romero-López, C., Díaz-González, R., and Berzal-Herranz, A. (2007). Inhibition of hepatitis C virus internal ribosome entry site-mediated translation by an RNA targeting the conserved IIIf domain. *Cell. Mol. Life Sci.* 64 (22): 2994–3006.

77 Romero-López, C., Díaz-González, R., Barroso-delJesus, A., and Berzal-Herranz, A. (2009). Inhibition of hepatitis C virus replication and internal ribosome entry site-dependent translation by an RNA molecule. *J. Gen. Virol.* 90 (Pt 7): 1659–1669.

78 Romero-López, C., Berzal-Herranz, B., Gómez, J., and Berzal-Herranz, A. (2012). An engineered inhibitor RNA that efficiently interferes with hepatitis C virus translation and replication. *Antiviral Res.* 94 (2): 131–138.

79 Romero-López, C., Lahlali, T., Berzal-Herranz, B., and Berzal-Herranz, A. (2017). Development of optimized inhibitor RNAs allowing multisite-targeting of the HCV genome. *Molecules* 22 (5): 1–11.

27

DNAzymes as Biosensors

Lingzi Ma and Juewen Liu

University of Waterloo, Waterloo Institute for Nanotechnology, Department of Chemistry, Waterloo, ON, N2L 3G1, Canada

27.1 Introduction

Biosensors measure the concentration of target analytes based on biomolecular reactions. Representative examples include glucose meters using glucose oxidase and pregnancy test strips using antibodies. These biosensors have profoundly improved the quality of life with positive clinical impacts. Compared with traditional instrumentation-based analysis, biosensors are attractive for their lower cost, portability, and small sample volume [1]. Biosensors are composed of at least one biomolecular recognition element coupled with a signal transduction element. So far, the biosensor market is dominated by protein-based recognition elements, while nucleic acid-based sensors such as molecular beacons and DNA microarrays are mainly utilized for DNA or RNA detection. Both these sensors take advantage of the base-pairing interactions in nucleic acids for recognition [2].

For practical analytical applications performed outside cells, DNA is preferred over RNA for stability and cost considerations. Since the early 1990s, new chemical functions of DNA such as specific molecular binding (aptamers) [3, 4] and catalysis (DNAzymes) have been discovered [5]. Using such functional DNA for biosensor development has been an important direction of research [6–11]. Most of the research in this field has been focused on aptamer-based sensors since aptamers can be selected to bind essentially to any target molecules [12–14]. At the same time, DNAzymes also possess unique analytical merits. Since the theme of this book is on ribozymes and DNAzymes, we aim to describe recent developments on this front while still having a historic perspective.

DNAzymes refer to artificial DNA sequences with catalytic activities. Unlike ribozymes, DNAzymes have not yet been found in nature, and all known DNAzymes have been obtained from in vitro selection. Using DNAzymes for designing biosensors is an interesting idea, where a target analyte can either promote or inhibit the enzyme activity. So far, DNAzymes are capable of sensing a diverse range of analytes, including metal ions, small molecules, and proteins [6, 7, 15, 16].

Ribozymes, First Edition. Edited by Sabine Müller, Benoît Masquida, and Wade Winkler.

In particular, DNAzymes are excellent tools for metal ion detection while obtaining aptamers with eligible specificity for metals is quite difficult. This chapter first introduces the general mechanism of DNAzyme-catalyzed RNA cleavage reaction, and then some representative sensing DNAzymes are reviewed. Later, signaling methods such as fluorescence, colorimetry, and electrochemistry are described. Finally, application of such sensors for environmental water analysis, intracellular sensing, and bacterial cell detection is also discussed.

27.2 Advantages of DNAzyme-Based Sensors

DNAzyme-based biosensors possess a number of advantages. First, analyte-specific DNAzymes can be obtained via in vitro selection, making DNAzyme also a platform technology that can be developed to detect emerging analytes. In this regard, DNAzymes are similar to aptamers. Second, the catalytic turnover of DNAzymes may allow more sensitive detection via signal amplification, while aptamers do not have such properties on their own. DNAzymes are able to create or break chemical bonds, allowing a significant signal change, especially when the signal is distance dependent. On the contrary, aptamers rely on conformational folding for detection that sometimes results in relatively smaller signal change.

Compared with protein-based enzymes, DNAzymes are made of DNA that is less sensitive to physical environment. While DNAzymes can still be denatured upon heating, they can readily renature after cooling. Proteins are often irreversibly denatured by heating. The chemical stability of DNA against hydrolysis is around one-million-fold higher than that of RNA [17]. DNAzymes are chemically synthesized and have low batch-to-batch variation while production of antibodies often involves animals with large variation in quality. Chemical synthesis also allows convenient labeling of many functional moieties such as fluorophores, conjugation groups, and spacers. Combined with the programmability of DNA [18], DNAzymes can be applied to versatile sensor designing strategies. Finally, compared with ribozymes, DNAzymes are also more stable and less expensive.

DNAzymes also have several disadvantages. Compared with protein-based enzymes, DNAzymes often exhibit slower catalytic rates, making catalytic turnover difficult to realize in some cases. For example, cleavage rates of most RNA-cleaving DNAzymes are $\sim 1\ \mathrm{min}^{-1}$ or slower. Fortunately, from the examples reported so far, detection can often be completed within 10 minutes by measuring only initial reaction rates. In addition, DNAzymes catalyze chemical reactions, which are in general irreversible, disallowing continuous monitoring of the fluctuation of analyte concentration over a long time. Overall, the development of DNAzymes doubtlessly provides a very unique sensing platform.

27.3 General Mechanism of RNA Cleavage

While the main goal of this chapter is to introduce DNAzyme-based biosensors, it is still useful to briefly describe the general mechanism of RNA cleavage, which is

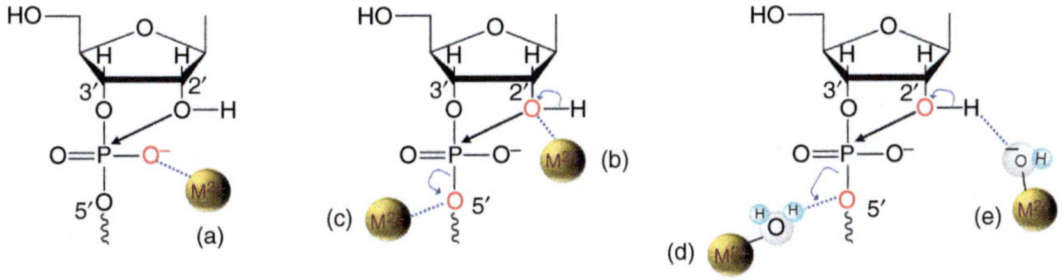

Figure 27.1 Possible catalytic functions of metal ions in the RNA cleavage reaction. Metal ions can act as (a) an electrophilic catalyst, (b, c) Lewis acids, (d) a general acid, or (e) a general base to facilitate breaking a phosphodiester bond.

the most common reaction used in biosensing. The major difference between RNA and DNA is that RNA has a 2′-OH group that can act as an internal nucleophile to attack the adjacent phosphate group (Figure 27.1a). After a penta-coordinated transition state, the leaving of the 5′-oxygen breaks a phosphodiester bond into a 2′,3′-cyclic phosphate and a 5′-OH termini. Here we emphasize on the role of metal ions since they are main targets of DNAzyme-based sensors. Apart from the structure stabilization of the negatively charged phosphate backbone, metal cofactors may participate in cleavage reaction through inter- or outer-sphere interactions. For example, a metal ion can serve as an electrophilic catalyst and coordinate with the non-bridging oxygen, making the phosphorus center more reachable for the hydroxyl group (Figure 27.1a). In an RNA cleavage reaction, a metal hydroxide (e.g. deprotonated metal-bound water) can act as a general base assisting the deprotonation of the 2′-hydroxyl group by attracting the proton (Figure 27.1d). Deprotonated 2′-hydroxyl thus becomes a much stronger nucleophile. Alternatively, a metal ion can function as a Lewis acid to coordinate directly with the 2′-oxygen (Figure 27.1b). A successful nucleophilic attack produces a penta-coordinated phosphorane bearing two negative charges. Such a highly negatively charged transition state can be neutralized by coordinating with metal ions. Meanwhile, a metal-bound water molecule is able to donate a proton to the 5′-oxygen as a general acid (Figure 27.1e). The leaving group can also be stabilized by direct coordination of a metal ion acting as Lewis acid (Figure 27.1c). Overall, metal cofactors play one or multiple roles during the RNA cleavage reaction.

Since metal ions can facilitate the reaction in many possible ways [19, 20], the exact role of metal is often unclear without high-resolution structures. So far, no nuclear magnetic resonance (NMR) structure for DNAzymes is available. Luckily, important progress has been made on X-ray crystallization. The crystal structure of an RNA-ligating DNAzyme was solved in 2016 [21]. In 2017, the structure of the 8–17 DNAzyme with Pb^{2+} cofactor was also reported, where a guanine was found to be critical for the cleavage reaction by functioning as a general base [22]. The structural observations suggested that one Pb^{2+} binds to the catalytic core and accelerates the reaction via its bound water molecule. Still, most of the metal binding information was probed indirectly using biochemical and biophysical assays.

27.4 Representative DNAzymes

Using in vitro selection, DNAzymes that can catalyze a diverse range of reactions have been isolated, ranging from the RNA/DNA cleavage, ligation, and phosphorylation to Diels–Alder reactions [23]. In the last decade, the Silverman group has pushed on this front to explore new catalytic functions of DNA [24, 25]. Selection of DNAzymes is very similar to that for ribozymes, except that no reverse transcription step is needed. To use DNAzymes as biosensors, it is important to have the reaction promoted (or inhibited) by the target analyte. This can be achieved by introducing the target in the selection step, and only the sequences that can react are selected from a large DNA library. Counterselection using competing analytes is often carried out to improve specificity, for which the active sequences are removed from the library and the inactive sequences are collected for subsequent selections [26].

For biosensing applications, DNAzymes that catalyze the RNA cleavage reaction are of great interest. This can be attributed to the fact that RNA-cleaving DNAzymes are relatively easy to select in terms of experimental protocol and they present a relatively fast rate among the different DNAzymes. In addition, using site-specific cleavage of DNA to design biosensors is quite straightforward. We briefly introduce some representative DNAzymes for biosensor development below, most of which have been widely used for biosensing.

27.4.1 DNAzymes for Pb^{2+}

A unique application of DNAzymes is to detect metal ions since selecting metal-specific aptamers has been quite challenging due to difficulties associated with immobilization of metal ions. Although some new selection strategies have been explored omitting the need for metal immobilization [27, 28], the selected aptamers are yet to be fully demonstrated good for metal detection. The first DNAzyme ever reported was isolated in the presence of Pb^{2+} by Breaker and Joyce in 1994 [5]. This DNAzyme was named GR5 (Figure 27.2a), and it cleaves a single RNA linkage embedded in a DNA substrate (so-called RNA/DNA chimeric substrate). A typical DNAzyme is composed of a conserved catalytic loop flanked by two substrate-binding domains that hybridize with substrates by base pairs. Later GR5 was found by Lu and coworkers to be highly specific for Pb^{2+}. Even nanomolar Pb^{2+} can activate the DNAzyme, while no other metal ion was found to be active, including 50 mM Mg^{2+}.

Before the report on its high Pb^{2+} specificity, the GR5 DNAzyme did not receive much attention for analytical applications. Back then, DNAzymes were believed to be useful for intracellular antiviral applications, but GR5 cleaves full RNA substrates and requires toxic Pb^{2+}, making it impractical for cleaving viral RNA inside cells.

Biosensor development was mainly drove by the 8–17 DNAzyme, which was also isolated by the Joyce and Santoro lab [35]. This DNAzyme cleaves all-RNA substrates (GR5 cleaves only a single RNA linkage) and also RNA/DNA chimeric substrates in the presence of Mg^{2+}. The sequence of its substrate-binding arms can also be adjusted to recognize different RNA substrates. Therefore, together with the

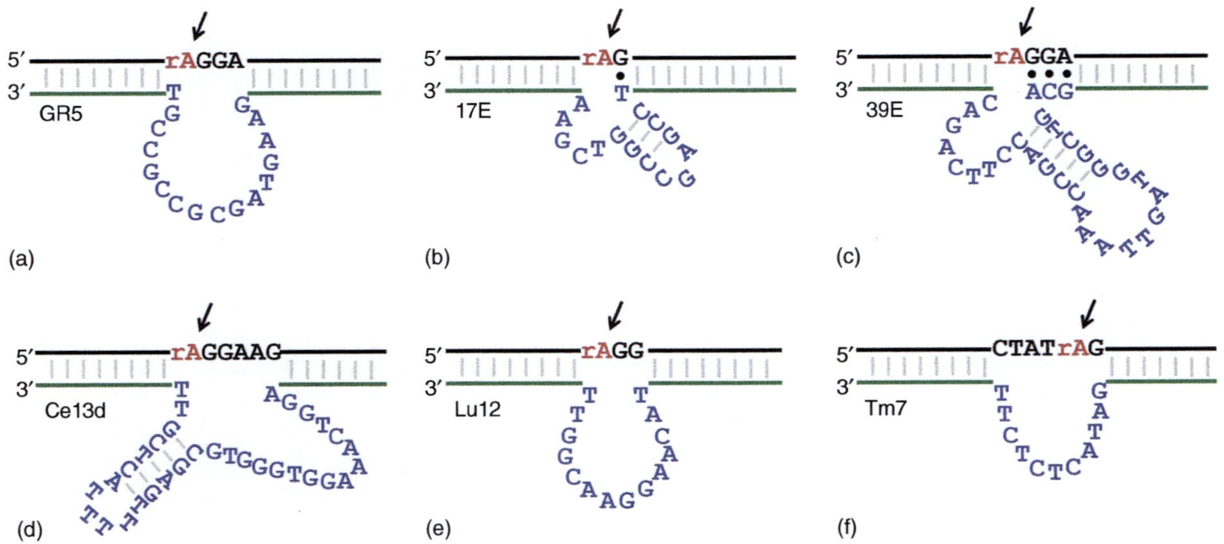

Figure 27.2 Secondary structures of two Pb^{2+}-dependent DNAzymes: (a) GR5 and (b) 17E. (c) The UO_2^{2+}-dependent DNAzyme called 39E. Secondary structures of three Ln^{3+}-dependent DNAzymes: (d) Ce13d, (e) Lu12, and (f) Tm7. Source: (a) Adapted from Zhang and Fu et al. [29]. (b) Adapted from Liu and Lu [30]. (c) Adapted from Liu and Brown et al. [31]. (d) Adapted from Huang and Lin et al. [32]. (e) Adapted from Huang and Vazin et al. [33]. (f) Adapted from Huang and Vazin et al. [34].

10–23 DNAzyme (another DNAzyme selected by the Joyce lab), they generated a lot of excitement for their potential therapeutic applications for intracellular RNA cleavage. The Lu lab selected a variant of the 8–17 DNAzyme called 17E (Figure 27.2b) by using Zn^{2+} [36], and later they found that the 17E was also highly specific for Pb^{2+} [37]. Although it is active with essentially any divalent metals at high metal concentrations such as Zn^{2+} and Cd^{2+}, millimolar Mg^{2+} and Ca^{2+} can also promote its cleavage [38, 39]. For a long time, 17E has been a model DNAzyme to develop biosensors for Pb^{2+} detection,[30, 40–42] until the revisit of the GR5 DNAzyme.

Since the initial discovery of the 17E DNAzyme in 1997, it has been reselected in many labs under different selection conditions. The Li lab has attributed this to its small catalytic motif size, high activity, and tolerance to mutation [43]. This has created a problem for in vitro selection of new DNAzymes. If the selection condition allows activity for 17E, it is highly likely to dominate the selection library.

27.4.2 DNAzymes for Lanthanides and Actinides

In 2007, the Lu lab discovered the 39E DNAzyme that was intentionally isolated in the presence of uranyl (UO_2^{2+}) (Figure 27.2c) [31]. It has excellent selectivity for UO_2^{2+} over any other metal ion by one-million-fold or more. In water, UO_2^{2+} is the most stable species of uranium, and a high concentration of uranium can cause adverse health effects to humans. In the past few years, we have isolated a series of DNAzymes that work specifically for lanthanides [44]. Lanthanides refer to a series of 15 metals with very similar properties. They can be called industrial vitamins due to their technological importance, and detecting lanthanides could be useful yet analytically very challenging. The Ce13d DNAzyme works with all the lanthanides with a similar activity (Figure 27.2d) [32], and the Lu12 DNAzyme has descending activity with increasing atomic number of lanthanides (Figure 27.2e) [33], while the Tm7 DNAzyme works only for the seven heavy lanthanides (Figure 27.2f) [34]. These are relatively hard metal ions that interact strongly with the phosphate group in nucleic acids. Thus, it is understandable that they can activate DNAzymes for RNA cleavage.

27.4.3 DNAzymes for Thiophilic Metals

Thiophilic metals are often highly toxic, and the most notable examples are mercury and cadmium. These metals are bioaccumulative and can cause irreversible damages to organs, especially in children. Therefore, biosensors for their detection are of great analytical and environmental importance [45–47]. DNAzymes specific for thiophilic metals such as Cd^{2+} have been quite challenging to select. As mentioned above, the RNA cleavage reaction often requires a metal ion to stabilize the negative charges building up in the transition state of the penta-coordinated phosphorane. However, soft thiophilic metals do not interact strongly with the phosphate group, which is a hard ligand. To solve this problem, DNA bases modified with metal ligands such as imidazole have been introduced. The Joyce lab reported a Zn^{2+}-dependent DNAzyme selected from a library containing imidazole-modified uridine (Figure 27.3a) [48]. Perrin and coworkers reported a Hg^{2+}-dependent

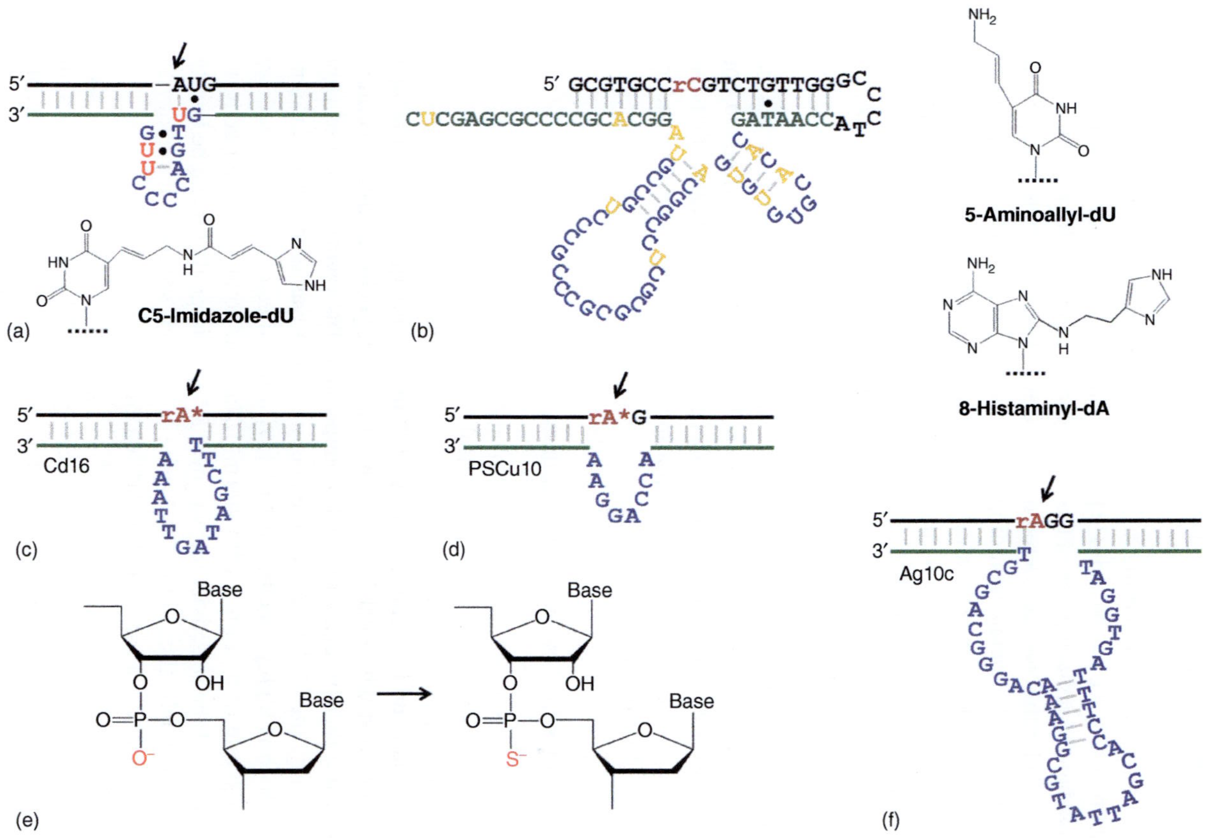

Figure 27.3 (a) The Zn^{2+}-dependent RNA-cleaving DNAzyme, where the red U indicates C5-imidazole-modified deoxyuridine. (b) The Hg^{2+}-dependent RNA-cleaving DNAzyme contains two types of modified bases (marked yellow) including 5-aminoallyl-dU and 8-histaminyl-dA. Secondary structures of (c) the Cd^{2+}-dependent DNAzyme and (d) the Cu^{2+}-dependent DNAzyme containing the PS-modified substrate (rA*). (e) Chemical structure of the PS modification at the cleavage site. (f) The Ag^{+}-specific DNAzyme named Ag10c. Source: (a) Adapted from Santoro and Joyce et al. [48]. (b) Hollenstein and Hipolito et al. [49]. Reproduced with permission of John Wiley & Sons, Inc. © 2008. (c) Adapted from Huang and Liu [50]. (d) Adapted from Huang and Liu [51]. (e) Huang and Liu [52]. © 2014 American Chemical Society. (f) Adapted from Saran and Liu [53].

DNAzyme by employing two types of modified nucleotides (Figure 27.3b) [49]. However, these modified DNAzymes attracted little analytical interest so far, likely due to the fact that these modifications are not commercially available. Even if they were available, the potential cost is very high since multiple modified bases exist in such DNAzymes, which would also discourage analytical chemists from trying them.

We solved this problem in a different way. Instead of introducing modifications to the random library, we placed a single phosphorothioate (PS) modification at the scissile phosphate by replacing one of the non-bridging oxygens with a sulfur atom (Figure 27.3e). With this, we discovered that the Ce13d DNAzyme acts as a general sensor for all thiophilic metals including Hg^{2+}, Cu^{2+}, Pb^{2+}, and Cd^{2+} [52]. Later, we introduced the same PS modification in the random library and performed new selections. This allowed us to use the natural nucleotides and normal polymerase chain reactions (PCRs) for the selection. We obtained new and highly selective DNAzymes for Cd^{2+} (Figure 27.3c) [50] and Cu^{2+} (Figure 27.3d) [51]. This success also highlighted the importance of metal binding to the scissile phosphate in the cleavage reaction.

Quite surprisingly, we also discovered a Ag^{+}-specific DNAzyme named Ag10c (Figure 27.3f) [53]. Although Ag^{+} is even more thiophilic, this DNAzyme did not contain a PS modification. Later detailed biochemical assays indicated that the Ag10c contained a silver aptamer while its phosphate stabilization was achieved by the Na^{+} or other ions in the buffer [54].

27.4.4 DNAzymes for Physiologically Abundant Metals

Four types of metal ions are abundant in biological samples: Na^{+}, K^{+}, Mg^{2+}, and Ca^{2+}. The balance of these electrolytes is critical for human health. For example, a high sodium level is often associated with hypertension and water retention. Due to the large ionic radii and low charge density, Na^{+} and K^{+} are unlikely to contribute to specific interaction with nucleic acids other than nonspecific charge screening. Nevertheless, the in vitro discovery of Na^{+}-dependent DNAzymes provided new insights. Recently, the NaA43 DNAzyme (Figure 27.4a) was reported by Lu and coworkers, which is highly specific for Na^{+} [59]. It requires only Na^{+} for activity with a cleavage rate of 0.1 min^{-1} with 400 mM Na^{+}. No other metal can activate this DNAzyme, and the selectivity for Na^{+} over K^{+} is so large that it can be hardly measured. Interestingly, the NaA43 DNAzyme has a high sequence homology to Ce13d, and both contain an aptamer for Na^{+} [60, 61].

Another Na^{+}-specific DNAzyme named EtNa (Figure 27.4b) was recently isolated in our lab. Activity study revealed that the catalysis of EtNa greatly depends on reaction solvent. EtNa is highly active toward Na^{+} in concentrated organic solvents (e.g. ethanol) [55]; however, it becomes more active and selective with Ca^{2+} in aqueous solutions [56]. The selectivity of EtNa for Ca^{2+} over Mg^{2+} reached 90-fold with 2 mM metal (Figure 27.4c), which was the highest among all reported

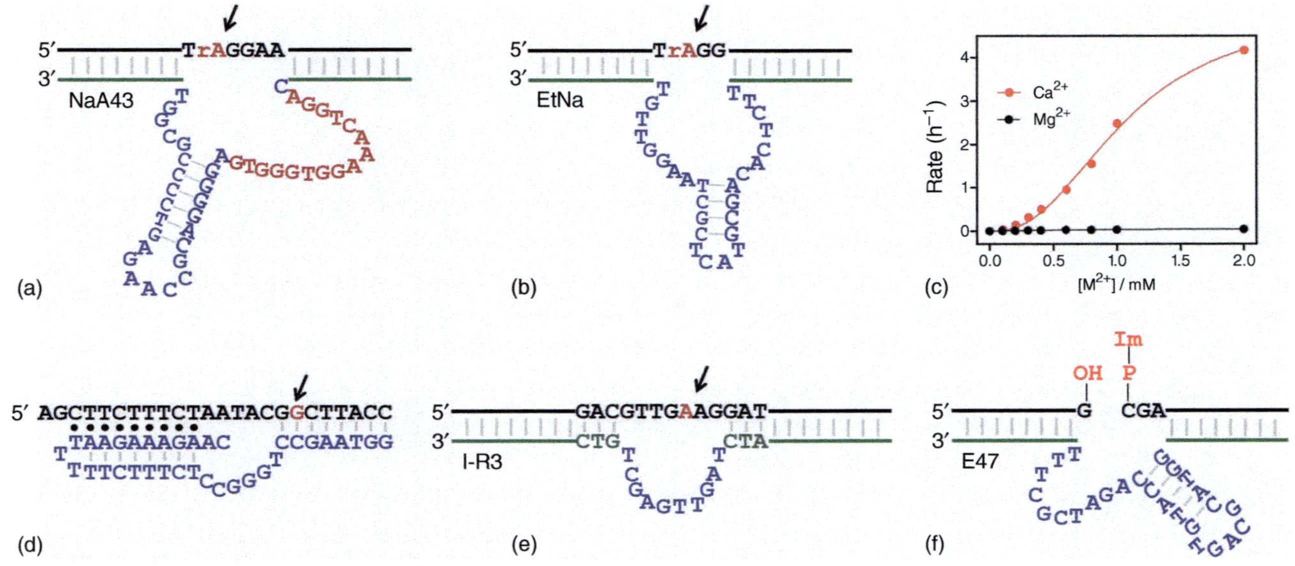

Figure 27.4 Secondary structures of two Na^+-dependent RNA-cleaving DNAzymes including (a) the NaA43 DNAzyme and (b) the EtNa DNAzyme. (c) Cleavage rates of EtNa as a function of Ca^{2+} and Mg^{2+} concentrations. (d) The Cu^{2+}-dependent DNA-cleaving DNAzyme. (e) The Zn^{2+}-dependent DNA-cleaving DNAzyme. (f) The Cu^{2+}-dependent DNA-ligating DNAzyme. Source: (b) Adapted from Zhou and Saran et al. [55]. (c) Zhou and Saran et al. [56]. Reproduced with permission of John Wiley & Sons, Inc. © 2017. (d) Adapted from Carmi and Balkhi et al. [57]. (e) Adapted from Gu and Furukawa et al. [58].

Ca^{2+}-dependent DNAzymes. The selectivity difference might be attributed to its requirement of two Ca^{2+} ions for catalysis but only one Mg^{2+} ion. Inside cells, Mg^{2+} is the most common metal cofactor for protein enzymes and ribozymes. However, all the DNAzymes selected with Mg^{2+} lack sufficient selectivity, which greatly hinders their applications for Mg^{2+} sensing.

It is interesting to note that all these RNA-cleaving DNAzymes have a very similar overall structure. The substrate strand contains a single RNA linkage and is recognized by the enzyme strand using two base-paired regions. The enzyme has a bulged loop and some also have a hairpin structure. This structure similarity has made it very easy for biosensor design. Essentially the same design can in general be applied to most of these DNAzymes.

27.4.5 Metal-Sensing DNAzymes Catalyzing Other Reactions

All of the above DNAzymes work on the RNA cleavage reaction. Aside from these, a few other types of DNAzymes have also been used for biosensor development. For example, Breaker and coworkers isolated a DNA-cleaving DNAzyme that can use Cu^{2+}in the presence of ascorbate (Figure 27.4d) [57, 62]. The reaction mechanism includes oxidative cleavage by free radicals produced by the copper species. This DNAzyme is unique since it involves DNA triplex for substrate recognition. Later, the Breaker group used Zn^{2+} to select a DNA-cleaving DNAzyme (Figure 27.4e) [58]. In this case, cleavage was hydrolytic instead of oxidative. This DNAzyme has a very narrow pH range, and likely Zn^{2+} is precipitated at high pH to form oxide that can absorb DNA and inactivate the DNAzyme [63].

In 1995, Cuenoud and Szostak isolated a DNA-ligating DNAzyme specific to Cu^{2+} (also active with a slightly higher concentration of Zn^{2+}) (Figure 27.4f) [64]. This DNAzyme however needs to use imidazole-activated substrate, which is quite unstable, thus limiting its practical applications. Many G-quadruplex (G4) DNA can bind with hemin and form DNAzyme complexes that mimic peroxidase activity (Figure 27.5a) [67]. The binding of metal ions (e.g. K^+, Na^+) stabilizes the G-quadruplex structure (Figure 27.5b) [65]. G4 DNAzyme can catalyze the oxidation of many chromogenic substrates such as 3,3′,5,5′-tetramethylbenzidine (TMB) and 2,2′-azino-bis(3-ethylbenzothiazoline-6-sulphonic acid) (ABTS) in the presence of H_2O_2 to produce color products. This DNAzyme has been widely applied in designing the signal transduction part of biosensors [68].

27.4.6 Aptazymes

Due to the chemical nature of the RNA cleavage reaction, DNAzymes are good for metal detection, while direct activation of DNAzymes using small molecules or proteins is more difficult. For certain analytes, good aptamers are available, and aptamer binding can also be converted to DNAzyme or ribozyme cleavage events by the aptazyme technology. Aptazymes contain an aptamer motif that can bind to the target and allosterically regulate the activity of the DNAzyme [11, 69]. In early works, many aptazymes were obtained by rational design to append an

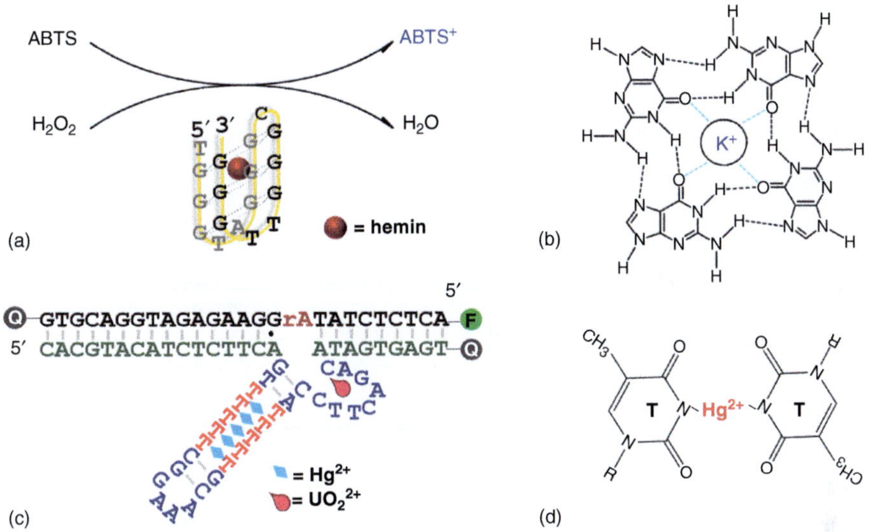

Figure 27.5 (a) G-quadruplex DNAzyme (or peroxidase-mimicking DNAzyme) that catalyzes the oxidation of ABTS by H_2O_2 into $ABTS^+$. (b) Chemical structure of a G-quartet formed by Hoogsteen hydrogen binding between four guanines that are stabilized by K^+. (c) Representative aptazyme based on the UO_2^{2+}-specific DNAzyme with T-Hg^{2+}-T base pairs for Hg^{2+} sensing. (d) Chemical structure of a T-Hg^{2+}-T base pair. Source: (b) Bhattacharyya and Mirihana et al. [65]. © 2016 Frontiers Media S.A. (c) Adapted from Liu and Lu [66].

aptamer motif to a ribozyme or DNAzyme scaffold [70–72]. A representative early example is shown in Figure 27.5c. Liu and Lu appended a few thymine–thymine mismatches to the UO_2^{2+} -specific 39E DNAzyme, and the resulting aptazyme also became highly dependent on Hg^{2+} [66]. Hg^{2+} can form T-Hg^{2+}-T base pairs, thus stabilizing the active DNAzyme structure (Figure 27.5d) [73]. Li and coworkers demonstrated a more sophisticated aptazyme for ATP, where an ATP aptamer motif (in pink) hybridizes with a DNAzyme motif (in blue), forming a hairpin structure. The binding of ATP triggers a structure change in the enzyme core and activates the DNAzyme (Figure 27.6a) [75]. Such rational design strategy has also been used to obtain aptazymes for proteins [76].

In some experiments, aptazymes can also be obtained by direct selection. This can be performed intentionally by using a known DNAzyme or ribozyme as a scaffold [77]. A randomized region is then added to a stem-loop in the enzyme to perform the selection in the presence of the target effector. Alternatively, selection can be performed from scratch with a fully random library. For example, the abovementioned Ce13d and NaA43 DNAzymes are aptazymes for Na^+, and Ag10c is an aptazyme for Ag^+. In these two examples, however, the isolation of the aptazymes was not predicted before the selection, and they were identified to be aptazymes only after careful biochemical characterization.

Apart from metal ions, aptazymes were also isolated to respond to cell-related targets. Li and coworkers have isolated a few interesting DNAzymes that are able to recognize bacterial pathogens such as *Escherichia coli* and *Clostridium*

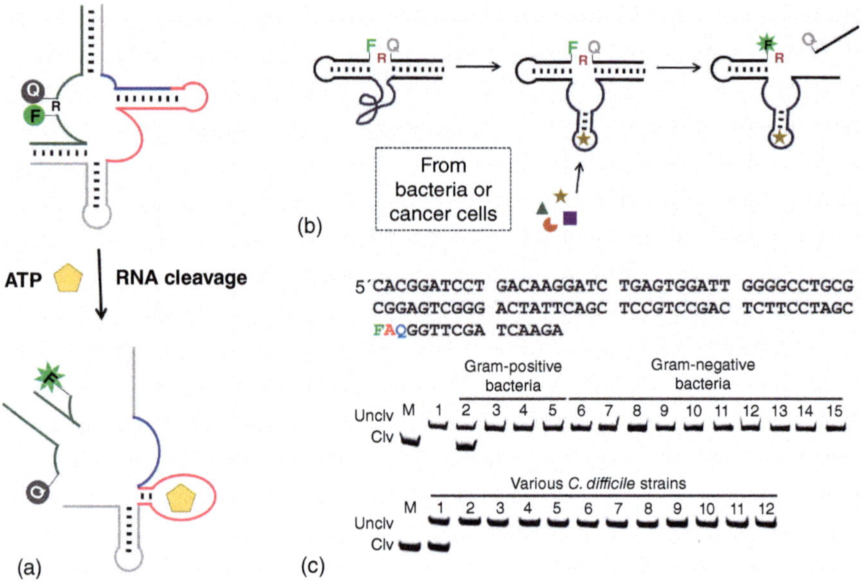

Figure 27.6 (a) The structure change of an ATP aptazyme. The ATP aptamer region is in pink color, and the catalytic loop of the DNAzyme is in blue. (b) Scheme showing the selection strategy of RNA-cleaving DNAzymes for bacteria and cancer cells. The targets used are their crude extracellular mixtures. (c) The sequence of RFD-CD1, a RNA-cleaving DNAzyme that is specific to *C. difficile*. Gel images showing the selectivity of RFD-CD1 in the presence of various types of bacteria and various strains of *C. difficile*. Source: Shen and Wu et al. [74]. Reproduced with permission of American Chemical Society.

difficile [74, 78]. In their selections, they introduced a fluorophore and a quencher simultaneously near the cleavage site so that signal can be produced right after cleavage. This design can effectively avoid the isolation of the 8–17 DNAzyme during the selection experiment. An example of their detection platform is illustrated in Figure 27.6b. As-selected DNAzymes display cleavage activity in the crude extracellular mixture of a specific bacterium and induce fluorescent signal. In the case of *C. difficile*, the selected DNAzyme, RFD-CD1, presents a high selectivity without cleavage being observed from other types of bacteria or even other strains of *C. difficile* (Figure 27.6c) [74]. Meanwhile, the target protein in this case was identified to be a transcription factor called TcdC based on a series of experiments including size screening and gene comparison. This sensing platform has also been successfully expanded to cancer cells [79]. These DNAzymes have great potential on disease diagnosis and food monitoring [80]. The main advantage of this strategy is that it allows detection without prior understanding of specific biomarkers. The specificity of these sequences can be gained naturally from the selection process. However, the aptamer motif in these cases is relatively hard to define. Detailed biochemical characterizations of these DNAzymes in terms of the cleavage mechanism have yet to be characterized. Based on the available information that the activating molecules are likely to be proteins, we believe that they can be considered as aptazymes.

27.5 DNAzyme-Based Fluorescent Sensors

After introducing the various DNAzymes, we describe available signaling methods in this section. DNA is a highly predictable molecule and DNAzymes have very similar secondary structures. Therefore, the same signaling method can be applied to different DNAzymes. To avoid redundancy, we give a representative example for each signaling methods. Due to limited space, we cannot cover all reported signaling methods, and only a few representative methods are described here. More signaling methods can be found in other review papers [6, 7, 81, 82].

27.5.1 Catalytic Beacons

Fluorescence detection has been the most popular choice due to its high sensitivity and versatility. In addition to fluorescence intensity, the peak wavelength, energy transfer, polarization, and lifetime can all be used to develop biosensors, and most of these have been utilized in DNA-based sensing. For RNA-cleaving DNAzymes, many strategies are available by labeling fluorophores. Li and Lu first reported a catalytic beacon design by labeling a fluorophore on one end of the substrate strand and a quencher on the corresponding end of the enzyme strand, resulting in quenched fluorescence in the initial state (Figure 27.7a) [37]. After cleavage, the cleaved fragment bearing the fluorophore is released due to lower melting temperature and thus induces signal enhancement. This method is quite unique to DNAzymes or ribozymes since it relies on the cleavage reaction. This catalytic beacon design is highly effective and can be used as an initial method to confirm the performance of RNA-cleaving DNAzymes as sensors.

Herein, we present one example for Ca^{2+} detection [83]. Ca^{2+} is an extremely important analyte for cellular signal transduction [84]. The EtNa DNAzyme was found to be highly active with Ca^{2+}, and we used a simple catalytic beacon design (Figure 27.7b). Usually, the kinetics of signaling is monitored for such sensors since this relies on a chemical reaction (Figure 27.7c). An advantage of monitoring kinetics is that the rate of fluorescence enhancement instead of the absolute fluorescence value can be used for quantification, allowing more reproducible calibration. The sensor has a detection limit of 17 μM Ca^{2+}. The selectivity is tested by using other metals, and only Pb^{2+} produced signal suggesting high selectivity (Figure 27.7d). Note that no signal was observed in the presence of Mg^{2+}, and this DNAzyme is the best so far for discriminating Ca^{2+} from Mg^{2+}.

The distance between the fluorophore and the quencher in Figure 27.7a can in theory vary from zero to infinite with a huge room for signal enhancement. In practice, however, signaling is mainly limited by incomplete hybridization and thus high background signals at room temperature. This was partially solved by introducing two quenchers at each end of the substrate strand serving as both inter- and intra-molecular quenchers [85]. In this case, a 10-fold increase of the signal-to-background ratio was observed when using the 17E DNAzyme as a model. With the development of nanotechnology, many studies have used nanomaterials (e.g. gold nanoparticles and graphene) as quenchers to design DNAzyme-based

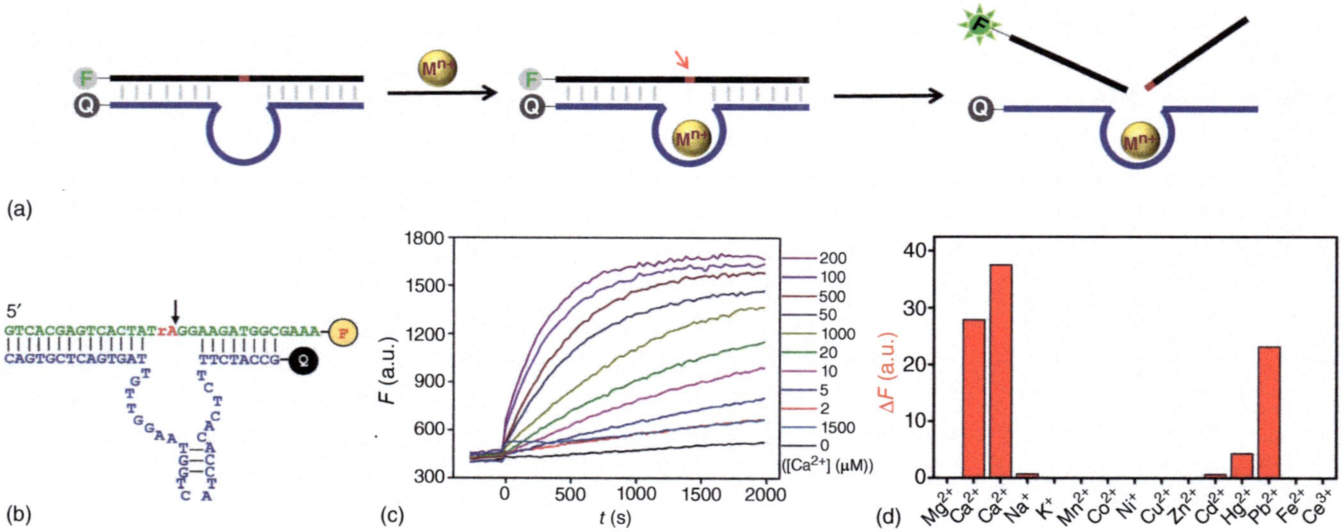

Figure 27.7 (a) A rational design of the DNAzyme-based catalytic beacon sensor. A fluorophore and a quencher were labeled at one end of the substrate strand and the enzyme strand, respectively. (b) The Ca^{2+} sensor design using the EtNa DNAzyme with a fluorophore and a quencher labels. (c) The sensitivity and (d) selectivity of Ca^{2+} detection using the EtNa DNAzyme. Source: Reprinted with permission from ref [83]. Copyright 2018 John Wiley & Sons, Inc.

sensors taking advantage of their high quenching efficiency and strong interactions with DNA [86, 87].

Apart from quenching-based methods, other fluorescent properties can also be utilized in sensor design, such as fluorescence polarization [29, 88], lifetime [89], and pyrene excimer formation [90]. For example, Wang and coworkers detected the fluorescence polarization change of tetramethylrhodamine (TMR) labeled on the substrate strand to track the activity of the GR5 DNAzyme (Figure 27.2a) [29]. An enhanced signal change was achieved by introducing TMR–guanine interactions, providing a detection limit of ~100 pM Pb^{2+}. Such methods are less dependent on fluorescence quantum yield or fluorophore concentration, and thus it is easier to build calibration curves.

27.5.2 Intracellular Sensing

An interesting application of DNAzymes is intracellular measurement of metal ions. In biology, metal ions are essential for enzymatic reactions and can regulate diverse biochemical processes [91]. Meanwhile, accumulation of certain metal ions at elevated level can be potentially toxic. As a result, quantifying and imaging metal ions are critical for understanding their functions in living cells. Recently, a few works have applied DNAzymes to address this issue by taking advantages of their high selectivity and sensitivity [92–94]. However, delivery requirement of the DNAzyme into cells makes this task challenging. The Lu lab has made significant contributions in this front. In their work, the Na^+-dependent RNA-cleaving DNAzyme, NaA43 (Figure 27.4a), was engineered into a fluorescent sensor for Na^+ detection by labeling the enzyme strand with a quencher and the substrate strand with a fluorophore and a quencher (Figure 27.8a) [59]. The sensor presents a high selectivity for Na^+ over competing metal ions with a detection limit of 135 μM. To achieve Na^+ imaging in living cells, α-helical cationic polypeptide was used to deliver the sensor into the cell cytosol with high delivery efficiency. Furthermore, photocaging strategy was used to prevent cleavage during the transportation [95]. As shown in Figure 27.8a, the 2′-hydroxyl group was caged with a photolabile *o*-nitrobenzyl group. After the delivery, the cleavage activity of the sensor was resumed by irradiating 365 nm light. The influx of extracellular Na^+ into cells induced the cleavage reaction and fluorescence increase within 30 minutes (Figure 27.8b). Their following work further achieved the endogenous Na^+ detection by amplifying the signal using catalytic hairpin assembly (CHA) [96].

27.5.3 Internally Labeled DNAzymes

The above strategies employ a fluorophore/quencher pair that is labeled far away from the cleavage site and do not significantly interfere with the DNAzyme activity. However, a high background signal is often observed in these cases due to the incomplete hybridization. As a result, the signal enhancement of 8–17 DNAzyme is normally less than 10-fold when using these methods [37, 85]. Labeling a fluorophore/quencher pair near the cleavage site can be quite effective since it eliminates the need for releasing the cleaved substrate fragments (Figure 27.9a). Thus,

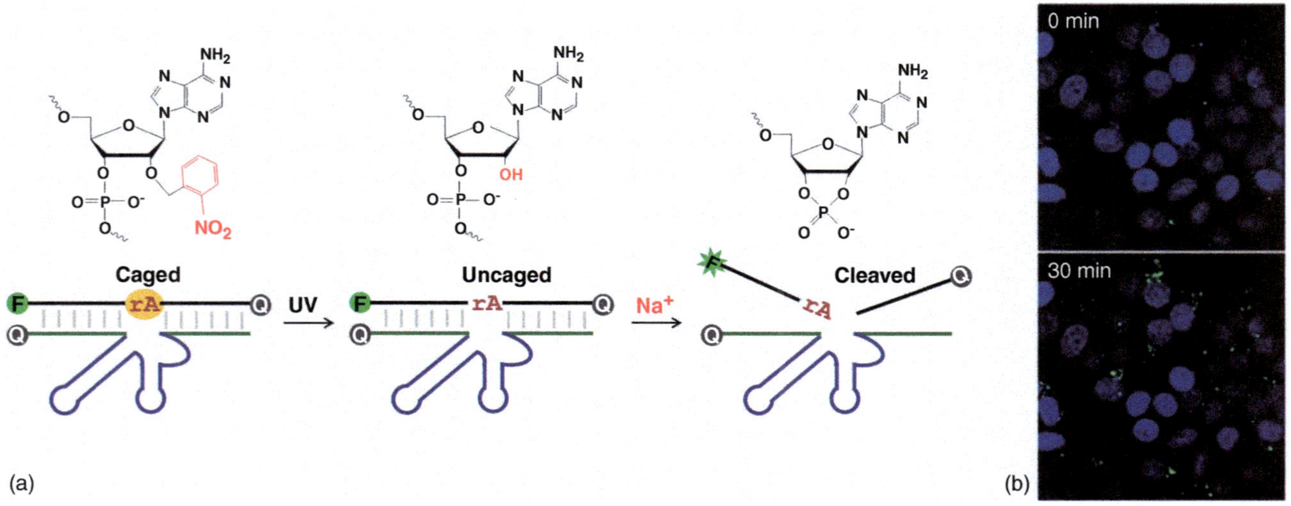

Figure 27.8 (a) The NaA43 DNAzyme-based intracellular sensing of Na^+ based on the photocaging strategy. (b) Confocal microscopy images of HeLa cells transfected with caged NaA43 sensors. (Scale bar: 20 μm.) Source: (a) Adapted from Torabi and Wu et al. [59].

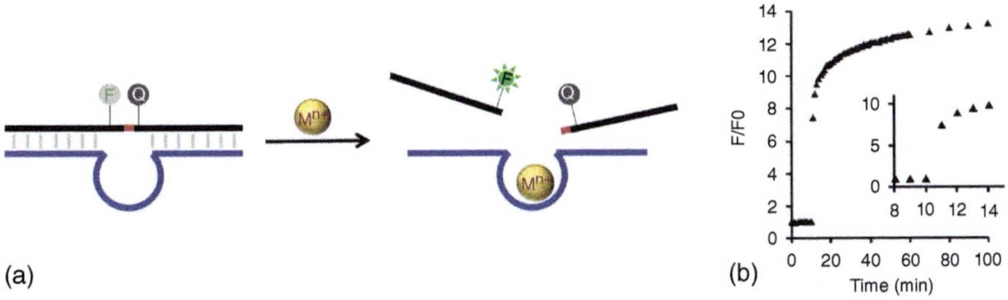

Figure 27.9 (a) Sensing design of an internally labeled RNA-cleaving DNAzyme. A fluorophore/quencher pair was attached near the cleavage site. (b) The fluorescence enhancement of the internally labeled 8–17 DNAzyme as a function of time. Source: (b) Mei and Liu et al. [97]. © 2003 American Chemical Society.

fluorescence increase can be observed right after the cleavage step with a low background signal. Li and Chiuman studied different attaching positions based on 8–17 DNAzyme and achieved 15- to 85-fold signal enhancement [98]. Nevertheless, the post-modification may induce steric hindrance near the cleavage site, resulting in suppressed catalytic activity.

Li and coworkers solved this problem by performing in vitro selection with such labels already present in the library [97]. Based on their distance optimization, a fluorophore/quencher pair should be attached to two adjacent nucleotides of the cleavage site considering signal generation. For instance, the first DNAzyme selected by this method exhibits a rate constant of ~7 min^{-1} and a high fluorescence enhancement as shown in Figure 27.9b. As mentioned previously, this strategy has been successfully applied in selecting DNAzymes that are specific for bacterial pathogens and cancer cells. They selected a DNAzyme probe that is able to detect breast cancer cells, MDA-MB-231, with great specificity and sensitivity [79]. The selected DNAzyme can induce fluorescent signal as a function of protein concentration in cancer cell lysates with a detection limit of 0.5 $\mu g\,ml^{-1}$. Besides, this fluorogenic probe is highly selective toward breast cancer cells over normal cells and other tumor cells. However, as discussed previously, it is challenging to precisely identify the target protein species in the cell lysates.

27.5.4 Folding-Based Detection

In the above designs, at least two labels are required to effectively induce fluorescence signal for detection. In certain special cases, DNAzymes with a single label are also able to detect metal ions based on local folding. Fluorescent nucleobases have been widely used as probes to study nucleic acids [99]. Their fluorescence is often sensitive to the local environment that can be utilized to probe DNA conformational change. As base analogs, they have less interference on DNA structure. Therefore, folding of DNAzymes by metal ion binding could be detected by fluorescent nucleobases and generate sufficient signal change. Our group reported the 2-aminopurine (2AP)-probed RNA-cleaving DNAzymes for Na^+ sensing [100]. The enzyme loop of the Ce13d DNAzyme has been found to contain a conserved

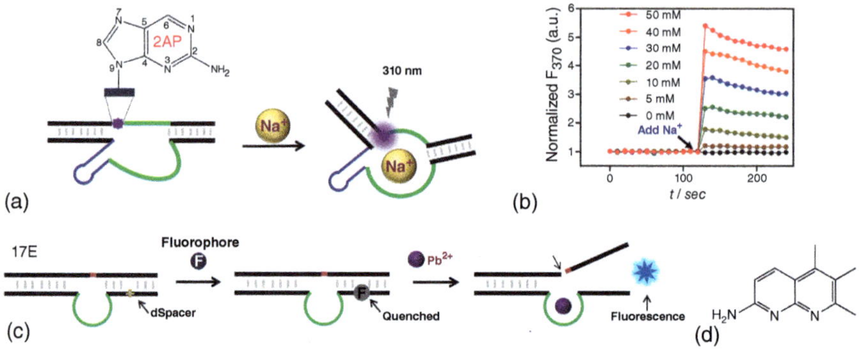

Figure 27.10 (a) The folding-based sensing of Na^+ using 2AP-labeled Ce13d DNAzyme and the chemical structure of 2-aminopurine. (b) The fluorescence kinetics of the folding-based sensor as response to different Na^+ concentrations. Source: Reprinted with permission from Ref. [100]. Copyright 2016 Oxford University Press. (c) A label-free detection of Pb^{2+} using an abasic site-containing 17E DNAzyme. (d) The chemical structure of ATMND.

Na^+ aptamer [61]. In this sensor, we labeled a deoxy-2AP at the cleavage site of the Ce13d DNAzyme (Figure 27.10a). 2AP is a fluorescent adenine analog whose emission intensity strongly depends on the local base stacking environment. Therefore, fluorescence intensity change can be observed upon Na^+ binding, indicating a conformational change of the enzyme loop. As shown in Figure 27.10b, the fluorescence increased with increasing concentrations of Na^+. The sensor was further improved by rational mutations with a detection limit of 0.4 mM Na^+. Compared with other fluorescent sensors, such a folding-based sensor is insensitive to ionic strength that is advantageous for Na^+ detection.

In addition to these covalent fluorophore-labeling strategies, label-free DNAzyme sensors have also been developed relying on DNA staining dyes. Such dyes are fluorescent when binding to the duplex regions in DNA, and after cleavage, the content of duplex regions is usually decreased [101]. In an interesting example, Lu and coworkers designed an abasic site-containing DNAzyme for label-free detection of Pb^{2+} [102]. Lead is a dangerous contamination in soil and water that may threaten life and health. In this work, the Pb^{2+}-dependent 17E DNAzyme was modified with dSpacer in its duplex regions (Figure 27.10c). A fluorophore called 2-amino-5,6,7-trimethyl-1,8-naphthyridine (ATMND) (Figure 27.10d) can bind to the dSpacer with its fluorescence being quenched. With the addition of Pb^{2+}, the substrate was cleaved and released the fluorophore, giving rise to the fluorescence signal. The sensor shows a detection limit of 4 nM Pb^{2+} under optimized condition with high selectivity.

27.6 Colorimetric Sensors Based on DNAzymes

For fluorescent sensors, a fluorimeter is required for measurement of the signal. Colorimetric detection is attractive because it may potentially eliminate the need for instruments and can be directly measured using the naked eye. Since DNA does not absorb light in the visible region, a color label is needed for colorimetric sensing. The extinction coefficient of typical organic dyes is on the order of $10^5\ M^{-1}\ cm^{-1}$, meaning that at least 1 μM of the dye (and thus DNA) is needed for visual detection. Such a high concentration of DNA would decrease the sensitivity of the sensor. Fortunately, the development of nanoscience in the past few decades has produced new colorimetric labels using metallic nanoparticles such as gold. In addition, taking advantage of the programmability of DNA, DNAzymes can be coupled with other functional materials or components to induce color change. Several colorimetric sensing strategies are reviewed in this section.

27.6.1 Using DNA-Functionalized Gold Nanoparticles

Gold nanoparticles (AuNPs) possess excellent optical properties for colorimetric biosensor design. First, AuNPs are extremely bright with extinction coefficients higher than 10^8 for 13 nm and exceeding 10^{10} for 50 nm particles [103]. These values are three to 5 orders of magnitude higher than most organic dyes. Therefore, AuNPs can be observed by the naked eye at low nM and even pM concentrations, allowing highly sensitive detection. For example, most pregnancy test strips use AuNPs of ~50 nm as a color label. In addition, AuNPs have distance-dependent color: disperse AuNPs are red, while aggregated particles are purple or blue due to the coupling of their surface plasmon.

Attaching DNA to AuNPs has been performed since 1996, and it is now a quite mature technology [104, 105]. While the early work in the field focused on DNA detection [103], after introducing aptamers and DNAzymes, these AuNPs can also be used to detect many other analytes [7]. In this section, we introduce a few representative examples. Lu and Liu developed a Pb^{2+} sensor by using the 17E DNAzyme (Figure 27.2b) to control the disassembly of AuNPs [30]. AuNPs were modified with two types of linker DNAs that can hybridize with the substrate (Figure 27.11a). DNA-functionalized AuNPs were first assembled into a tail-to-tail manner to produce a blue color aggregate. The presence of Pb^{2+} cleaved the substrate and triggered the disassembly of AuNPs, leading to a color change to red. The Pb^{2+}detection by such a "light-up" sensor is rapid (within 5 minutes) at room temperature with high sensitivity and selectivity. Interestingly, when the AuNPs were arranged in a head-to-tail manner, the DNAzyme was inactivated likely due to the steric hindrance from the AuNPs (Figure 27.11b).

27.6.2 Label-Free Detection

The above example requires conjugation of thiolated DNA to AuNPs. While this allows formation of highly controllable structures, the cost of DNA is quite high. Apart from using DNA-labeled AuNPs, label-free colorimetric sensing has also been achieved. Citrate-coated AuNPs tend to aggregate out of the solution in the presence of even a low concentration of salt such as 50 mM NaCl. Nonstructured and short single-stranded DNA (ssDNA) oligonucleotides can be easily absorbed

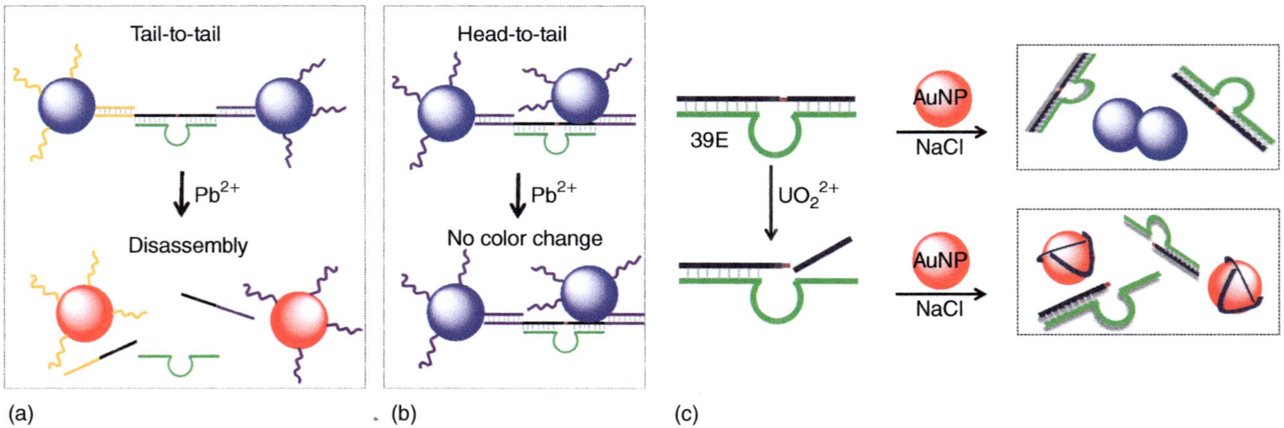

Figure 27.11 A colorimetric Pb^{2+} sensor by using the 17E DNAzyme to direct the assembly of AuNPs. (a) In the tail-to-tail manner, the cleavage of the DNAzyme in the presence of Pb^{2+} leads to the disassembly of the aggregates. (b) In the head-to-tail manner, Pb^{2+} fails to induce cleavage of the DNAzyme due to the steric hindrance. (c) A colorimetric uranium sensor using the 39E DNAzyme and label-free gold nanoparticles.

on AuNPs and prevent them from aggregation. On the contrary, the adsorption of double-stranded DNA (dsDNA) on AuNPs is slower due to its shielded bases. This discovery was first reported by Rothberg and Li, and they used it to monitor PCR progress [106]. Taking advantage of this, Lu and coworkers proposed a colorimetric uranium sensor using DNAzymes and label-free gold nanoparticles [107]. Uranyl (UO_2^{2+}) salts exposed in the environment might accumulate in human body and cause damage to the kidneys, liver, lungs, and brain. The 39E DNAzyme (Figure 27.2c) is UO_2^{2+} specific and can cleave substrate strands as shown in Figure 27.11c. The released substrate fragments can then bind to AuNPs and restrain the aggregation induced by salt. Without UO_2^{2+}, intact DNAzyme complex fails to bind to AuNPs, resulting in a color change from red to blue after adding the same amount of salt. By measuring the plasmon resonance peak shift of AuNPs, they obtained a detection limit of 1 nM after 6 minutes of reaction time. Compared with the labeled method, the authors concluded that the label-free sensor is more sensitive, cost-effective, and time saving. On the other hand, however, such label-free methods are more susceptible to interference. For example, proteins and other molecules in real environmental or biological samples may also adsorb on AuNPs to increase their colloidal stability. At the same time, the sample may also contain polyvalent metals that can induce AuNP aggregation. These may complicate analysis and produce false results.

27.6.3 Hydrogel-Assisted Colorimetric Detection

AuNPs can also be used as simple colorimetric indicators without direct contacting with DNAzymes. A rising interest is applying soft materials such as hydrogels and liposomes to improve the performance of DNA-based sensors [108–110]. In general, these materials are biocompatible, cost-effective, easy for storage, and excellent DNA carriers. DNAzymes have been incorporated with hydrogels for colorimetric metal sensing [111, 112]. For example, Huang and Wu et al. combined DNAzyme-assisted hydrogel and AuNPs to perform colorimetric detection of Ln^{3+} [113]. Figure 27.12a described the working principle. The polyacrylamide chains were grafted with short DNA strands (pA1 and pA2) that are complementary to two ends of substrate strands. The Ln^{3+}-dependent DNAzyme, Ce13d (Figure 27.2d), serves as functional regions and hybridizes with substrate strands. Without Ln^{3+}, the DNA hybridization directs the formation of three-dimensional hydrogel and confines the AuNPs. The addition of Ln^{3+} cleaves the substrate strands and breaks the hydrogel crosslinks, leading to an increased AuNP concentration in the supernatant. The concentration-dependent color change (Figure 27.12c) allows the colorimetric detection of Ln^{3+} with a detection limit of 20 nM (Ce^{3+}). The detection is sensitive only to Ln^{3+} mainly due to the specificity of the Ce13d DNAzyme (Figure 27.12b). However, a relatively long operation time is needed for this detection (~2 hours).

27.6.4 Coupled with G4 DNAzyme

Another mechanism of producing color change is to use an enzyme that can convert colorless substrates into colored products. A representative example is

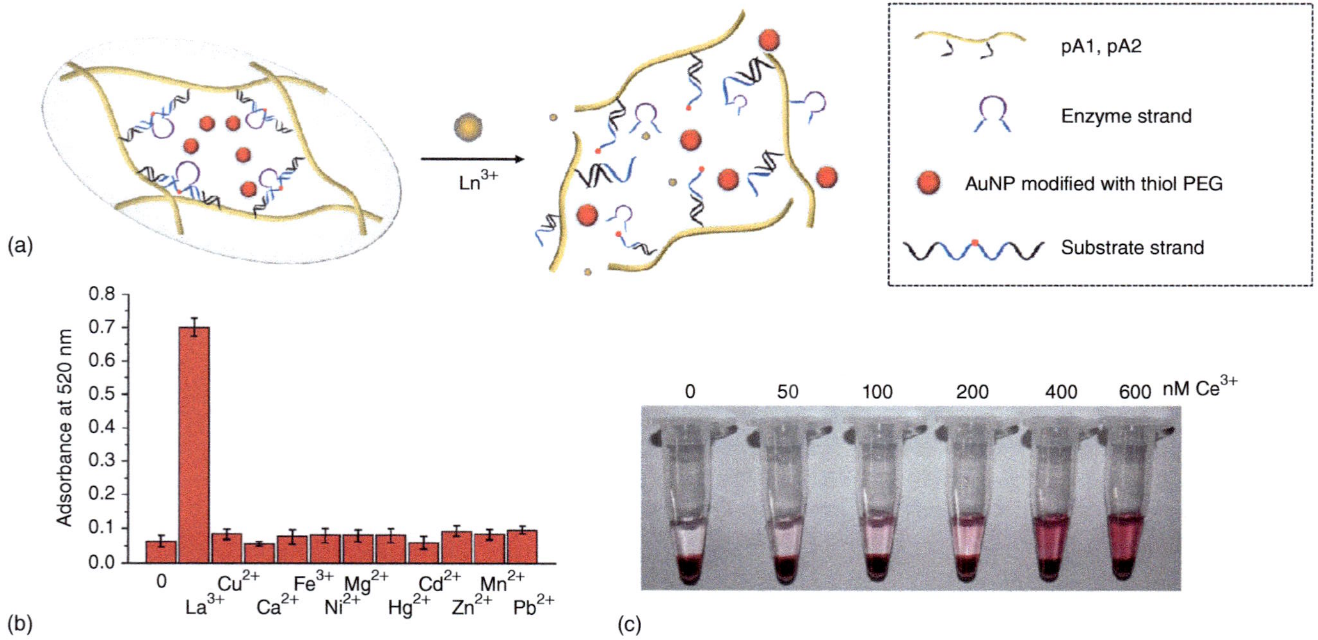

Figure 27.12 (a) The working principle of the target-responsive DNAzyme hydrogel for colorimetric detection of Ln^{3+}. (b) The hydrogel responds to Ln^{3+} but not to other metal ions. (c) The visual color change of the supernatants in the presence of various Ce^{3+} concentrations. Source: Huang and Wu et al. [113]. Reproduced with permission of Springer Nature.

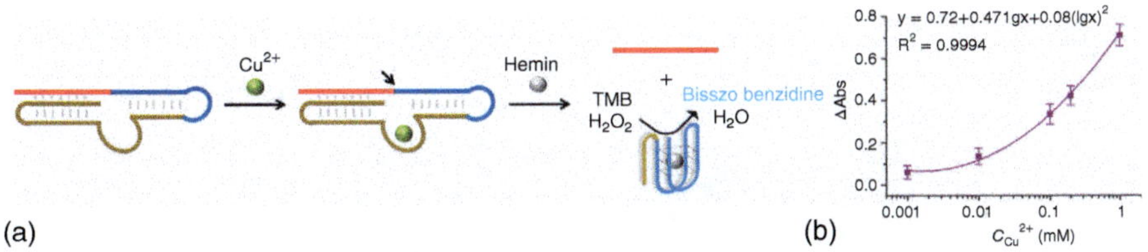

Figure 27.13 (a) Colorimetric sensing of Cu^{2+} using the dual-DNAzyme system. The G4 DNAzyme domain is marked in blue color that can bind to hemin after the cleavage reaction. (b) The nonlinear relationship between the absorbance change of the dual-DNAzyme system and Cu^{2+} concentration. Source: Yin and Ye et al. [114]. © 2009 American Chemical Society.

enzyme-linked immunosorbent assay (ELISA), where a horseradish peroxidase (HRP)-labeled antibody is used to generate color change by using chromogenic substrates such as TMB and ABTS in the presence of H_2O_2. As introduced previously, G4 DNA/hemin complex can mimic peroxidase-like activity and has been widely used as labels (Figure 27.5a). Not surprisingly, many work combined RNA-cleaving DNAzymes and peroxidase-mimicking G4 DNAzymes for metal ion sensing [114, 115]. For example, Yin and Ye et al. reported colorimetric sensing of Cu^{2+} using a dual-DNAzyme system [114]. As shown in Figure 27.13a, G4 DNAzyme sequence (blue colored) was embedded in the binding region of the Cu^{2+}-dependent DNAzyme (Figure 27.4d). In the presence of Cu^{2+}, the complex performed self-cleavage, allowing the G4 domain to interact with hemin. The G4 DNAzyme then self-assembled and catalyzed the oxidation of TMB into blue-colored bisszo benzidine. The absorbance change at 450 nm is nonlinearly related to the Cu^{2+} concentration in the range of 0.001–1.0 mM (Figure 27.13b).

27.7 Electrochemical Sensors and Other Sensors

The above optical sensing approaches are mainly performed in bulk solutions. DNAzyme-based sensing has also been performed on planar surface using electrochemical signaling. In general, the signal generation always relies on the conformational change of the DNAzyme induced by the cleavage reaction. For example, Xiao and Rowe et al. utilized the 8–17 DNAzyme for electrochemical sensing of Pb^{2+} by covalently labeling a methylene blue dye at the enzyme strand (Figure 27.14a) [41]. The cleavage of the substrate triggers the conformational change, allowing methylene blue to transfer electrons to the electrode surface.

In addition to using covalently labeled DNAzymes, label-free strategies were also reported. For example, $Ru\left(NH_3\right)_6^{3+}$ was used as electrochemical signal transducer (Figure 27.14c) [116]. The Pb^{2+}-dependent DNAzyme (8–17) was immobilized on a gold surface without any other modifications. $Ru\left(NH_3\right)_6^{3+}$ can easily bind to negatively charged phosphate backbone by electrostatic interactions. With the addition of Pb^{2+}, the substrate was cleaved and escaped from the electrode surface. Therefore, the amount of $Ru\left(NH_3\right)_6^{3+}$ near the Au electrode decreased, resulting

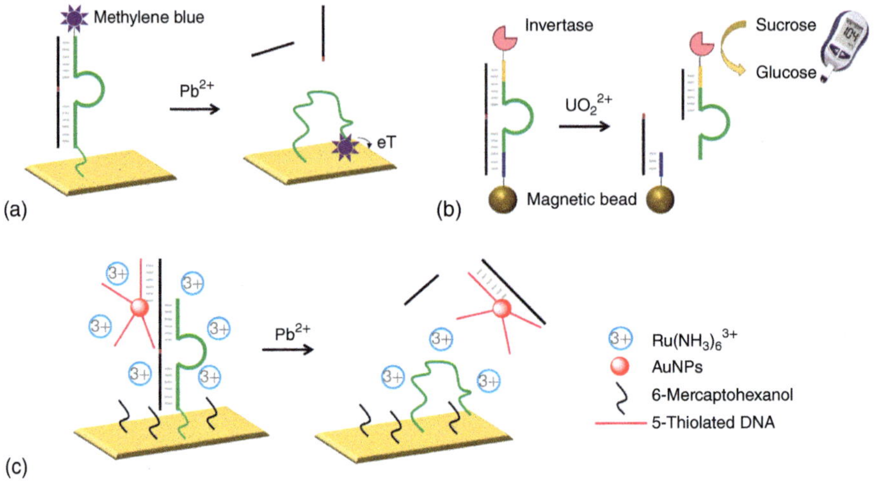

Figure 27.14 (a) Methylene-blue-modified DNAzyme generates electrochemical signals for Pb^{2+} detection. (b) Invertase-modified DNAzyme generates glucose for indirect sensing of UO_2^{2+}. (c) Electrochemical sensor using 8–17 DNAzyme and $Ru(NH_3)_6{}^{3+}$for Pb^{2+} detection. Source: (a) Adapted from Xiao and Rowe et al. [41]. (c) Adapted from Shen and Chen et al. [116].

in a lower signal. Remarkably, DNA-modified AuNPs can hybridize with the cleaved substrate and transport more DNAs away from surface. Herein, amplified electrochemical signals were obtained, enabling an ultrasensitive detection (limit of detection (LOD) of ~1 nM).

Besides, electrochemical detection is attractive because it is less affected by sample matrix and can work in serum. A remarkable work was reported by the Lu lab in which they incorporated invertase with DNAzymes to quantify metal ions through the detection of glucose [117]. As illustrated in Figure 27.14b, the 39E DNAzymes (Figure 27.2c) were attached to magnetic beads for future separation. In the presence of UO_2^{2+}, the invertase-labeled DNAzyme is released and catalyzes sucrose to glucose, which can be easily measured by a glucose meter. Interestingly, this platform can be used for various analytical targets (e.g. cocaine, adenosine) when applying corresponding aptamers.

27.8 DNAzyme Sensors Coupled with Signal Amplification Mechanisms

Signal amplification in biosensors refers to coupling enzymes to the biorecognition element so that each target analyte can produce more than one signaling molecule due to the catalytic turnover of the enzyme. Therefore, an improved sensitivity can be expected. The best example is probably the enzyme labels into second antibody in ELISA assays, where each enzyme label (usually HRP) can oxidize thousands of substrate molecules to produce colorimetric signal. Likewise, increasing efforts have been put in developing signal amplifying methods for the DNAzyme-based sensors [118, 119].

It is notable that DNAzymes themselves are already capable of signal amplification since each metal ion can in principle cleave multiple DNAzyme molecules. Using typical fluorescent metal ion detection as an example (Figure 27.7a), each cleavage reaction can light up one fluorophore. While each metal ion can in principle cleave more than one DNAzyme, few DNAzymes have taken advantage of this type of direct signal amplification. The reason is that most DNAzymes are quite slow, especially at low metal concentrations (e.g. below 1 μM), when signal amplification is needed. At low metal concentrations, most DNAzymes are below 0.1 min^{-1} (very few can reach ~1 min^{-1} such as GR5 and 17E with Pb^{2+}). With such a slow rate, to have more than one cleavage event for each metal, 10 minutes or longer is often needed, and a long sensing time is analytically undesirable. At even higher metal concentrations, the metal is even more than the DNA concentration, and thus there is little chance for each metal to cleave more than one DNAzyme on average. Lu and coworkers demonstrated a few catalytic turnovers in their UO_2^{2+}-specific DNAzyme at very low metal concentrations [31]. Overall, relying on DNAzyme alone does not have a general practical analytical value for metal detection.

To overcome this limitation, Zhang and Wang et al. combined molecular beacons with catalytic beacons to enable multiple enzymatic turnovers [120]. Molecular beacons are ssDNA that can form a stem-loop structure (Figure 27.15a). The background fluorescence can be efficiently quenched and is independent of the DNAzyme/substrate ratios. Herein, DNAzymes are allowed to catalyze more than one substrate, resulting in higher fluorescent signal at same concentration of metal ion. In the 17E DNAzyme model, this sensing platform displayed a detection limit of 600 pM Pb^{2+} that is much more sensitive than the DNAzyme-based catalytic beacons. Remarkably, the sensing with a DNAzyme/substrate ratio of 2 : 3 required only 10 minutes for completion.

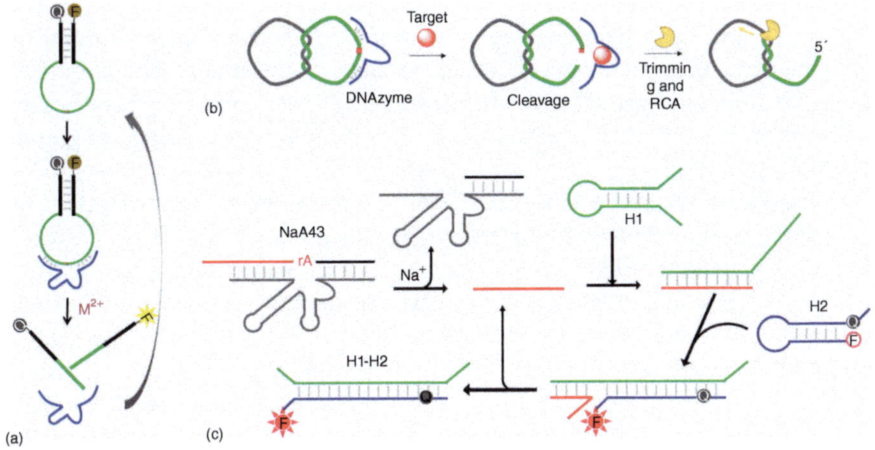

Figure 27.15 (a) Amplified sensing platform combining molecular beacons with catalytic beacons to achieve multiple enzymatic turnovers. (b) Coupling DNAzyme with RCA for sensitive detection of *E. coli*. (c) The CHA-amplified DNAzyme sensor for endogenous Na^+ imaging. Source: Adapted from Liu and Zhang et al. [121].

To have a real impact, other faster enzymes are needed, and sometimes the target molecule is no longer metal ions. The key here is to have one cleavage event to be converted to multiple fluorescence signals. As for DNAzyme-based sensing, enzymatic amplification of DNA is a useful approach for signal amplification. A representative strategy combined DNAzyme activity with the quantitative polymerase chain reaction (QPCR) [122]. The DNAzyme sequence was carefully designed allowing uncleaved substrates to be amplified but not cleaved ones. Therefore, the cleavage efficiency of the 17E DNAzyme can be quantified by QPCR. This strategy was able to detect Pb^{2+} with a dynamic range from 10 nM to 5 μM and a detection limit of 1 nM. However, a disadvantage of this approach is that it requires long processing time and thermal control. Isothermal DNA amplification process such as rolling cycle amplification (RCA) has also been applied for signal amplifying [121, 123, 124]. Recently, Li and coworkers coupled an *E. coli*-specific DNAzyme with RCA for *E. coli* detection (Figure 27.15b) [121]. The design of two interlocked ssDNA rings prevents the template strand (in gray) from being used for RCA. The binding of the target cleaves the substrate that serves as a primer to initiate the RCA. The RCA product can then be detected by a duplex-binding dye, 3,3′-diethylthiadicarbocyanine. This *E. coli* sensor exhibits a detection limit of as low as 10 cells ml^{-1}.

Another strategy incorporates the RNA-cleaving DNAzymes into enzyme-free catalytic reactions for amplified detection [125, 126]. Compared with enzyme-based methods, they are liberated from protein enzymes. For instance, many works have applied hybridization chain reaction (HCR) and CHA to detect target DNA at pM to fM level. The Willner group has made great contributions in this field. For example, they reported the use of Mg^{2+}-dependent DNAzymes in HCR as shown in Figure 27.16 [127]. The DNAzyme motif is partially embedded in two hairpin structures, H1 and H2, with its cleavage activity being caged. The target DNA can unlock a hairpin structure (H3) and initiate the cross-linking of the two hairpins, H1 and H2. Therefore, the growth of DNA polymer exposes multiple DNAzyme structures and induces amplified cleavage reactions. The amplified sensing platform was applied to detect an oncogene, BRCA1, presenting a detection limit of 10 fM. However, this method also suffers from long processing time up to 10 hours.

Recently, Wu and Fan et al. reported a CHA reaction designed to improve the sensitivity of the Na^+-dependent DNAzyme [96]. In the presence of Na^+, the cleavage substrate fragment (red colored) initiates the CHA reaction by opening the hairpin DNA, H1 (Figure 27.15c). The opened H1 was designed to further unlock H2 and to recover the fluorescence. Since the duplex formed between H1 and H2 is stronger, the cleavage substrate fragment is displaced and released. As a result, multiple CHA reactions can be induced with only one cleavage product, resulting in an increased signal at low Na^+ concentrations. With an improved sensitivity of 14 μM, the CHA-amplified DNAzyme sensor has been used to image the endogenous Na^+ in living cells.

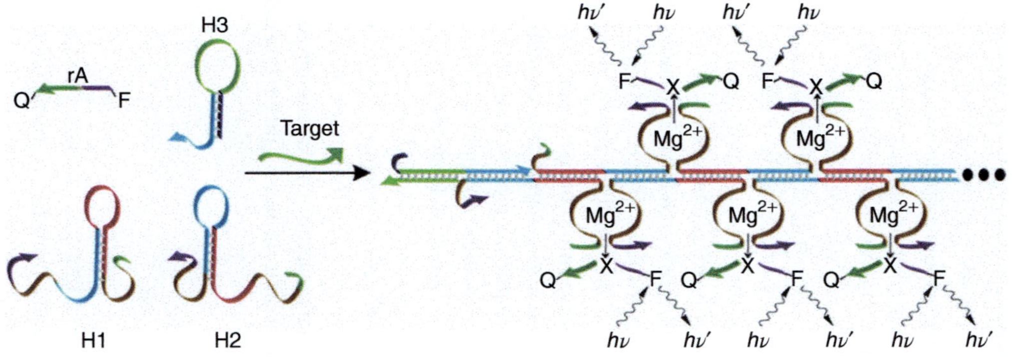

Figure 27.16 A sensing strategy using Mg^{2+}-dependent DNAzymes in HCR for amplified detection of target DNA. Source: Wang and Elbaz et al. [127]. © 2011 American Chemical Society.

27.9 Conclusions

In summary, significant progress has been made in the last decade in selecting new DNAzymes and also in developing more sensitive signaling methods. In particular, the inherent sensitivity and selectivity of RNA-cleaving DNAzymes for specific metal ions have been utilized for metal ion detection. Most metal-sensing DNAzymes can reach low nM and even pM detection limit with excellent selectivity. Most of the current work has been performed in clean buffers with a few real samples tested for proof-of-concept purpose. Future work will likely to focus on testing such sensors in real samples to understand sample matrix effect. For example, environmental water samples contain high concentrations of dissolved organic matters such as humic acid, and they may chelate metals to decrease the concentration of free metal ions. For biological samples such as serum and cells, their high protein content may make it difficult to detect metals at physiologically relevant levels other than for Na^+, K^+, Mg^{2+}, and Ca^{2+}. DNA does not have a high metal binding affinity, and it cannot compete with protein for metal binding, especially for transition metals. To increase metal binding affinity, modified DNA might be used for developing DNAzymes for this purpose. Another research front is to integrate DNAzyme-based sensors with convenient assay platforms. Currently, most measurements require handling of small volume of liquids and multiple operation steps such as dilution, annealing, mixing, and addition. Integration of these operations in a user-friendly, cost-effective, and portable device is also of significant challenge.

On the fundamental side, their sophisticated structure and catalytic mechanism still remain undiscovered to a great extent. Compared with protein enzymes and ribozymes, limited progress has been made in decoding the structure of DNAzymes. Just in 2017, Liu and Yu et al. studied and revealed the crystal structure of the 17E DNAzyme that was selected in 1997 [22]. Therefore, future efforts could be put in exploring their metal binding mechanism and their roles in DNAzyme catalysis. This in turn might help create a rational design of DNAzyme-based sensors by knowing the important nucleotides, upon which probes might be labeled to achieve a large extent signal change.

Acknowledgment

We thank Dr. Sona Jain for proofreading this chapter. Funding for the Liu lab that funded a part of this work is mainly from the Natural Sciences and Engineering Research Council of Canada (NSERC).

References

1 Turner, A.P.F. (2000). Biosensors-sense and sensitivity. *Science* 290 (5495): 1315–1317.

2 Wang, K., Tang, Z. et al. (2009). Molecular engineering of DNA: molecular beacons. *Angew. Chem. Int. Ed.* 48 (5): 856–870.

3 Ellington, A.D. and Szostak, J.W. (1990). In vitro selection of RNA molecules that bind specific ligands. *Nature* 346 (6287): 818–822.
4 Tuerk, C. and Gold, L. (1990). Systematic evolution of ligands by exponential enrichment: RNA ligands to bacteriophage T4 DNA polymerase. *Science* 249 (4968): 505–510.
5 Breaker, R.R. and Joyce, G.F. (1994). A DNA enzyme that cleaves RNA. *Chem. Biol.* 1 (4): 223–229.
6 Zhou, W., Saran, R., and Liu, J. (2017). Metal sensing by DNA. *Chem. Rev.* 117 (12): 8272–8325.
7 Liu, J., Cao, Z., and Lu, Y. (2009). Functional nucleic acid sensors. *Chem. Rev.* 109 (5): 1948–1998.
8 Cho, E.J., Lee, J.W., and Ellington, A.D. (2009). Applications of aptamers as sensors. *Annu. Rev. Anal. Chem.* 2: 241–264.
9 Tan, W., Donovan, M.J., and Jiang, J. (2013). Aptamers from cell-based selection for bioanalytical applications. *Chem. Rev.* 113 (4): 2842–2862.
10 Schlosser, K. and Li, Y. (2010). A versatile endoribonuclease mimic made of DNA: characteristics and applications of the 8-17 RNA-cleaving DNAzyme. *ChemBioChem* 11 (7): 866–879.
11 Famulok, M., Hartig, J.S., and Mayer, G. (2007). Functional aptamers and aptazymes in biotechnology, diagnostics, and therapy. *Chem. Rev.* 107 (9): 3715–3743.
12 Navani, N.K. and Li, Y.F. (2006). Nucleic acid aptamers and enzymes as sensors. *Curr. Opin. Chem. Biol.* 10 (3): 272–281.
13 Song, S., Wang, L., Li, J. et al. (2008). Aptamer-based biosensors. *TrAC, Trends Anal. Chem.* 27 (2): 108–117.
14 Zhou, W., Huang, P.J., Ding, J., and Liu, J. (2014). Aptamer-based biosensors for biomedical diagnostics. *Analyst* 139 (11): 2627–2640.
15 Zhang, X.B., Kong, R.M., and Lu, Y. (2011). Metal ion sensors based on DNAzymes and related DNA molecules. *Annu. Rev. Anal. Chem.* 4 (1): 105–128.
16 Hwang, K., Hosseinzadeh, P., and Lu, Y. (2016). Biochemical and biophysical understanding of metal ion selectivity of DNAzymes. *Inorg. Chim. Acta* 452: 12–24.
17 Li, Y. and Breaker, R.R. (1999). Kinetics of RNA degradation by specific base catalysis of transesterification involving the 2′-hydroxyl group. *J. Am. Chem. Soc.* 121 (23): 5364–5372.
18 Jones, M.R., Seeman, N.C., and Mirkin, C.A. (2015). Programmable materials and the nature of the DNA bond. *Science* 347 (6224): 1260901.
19 Ward, W.L., Plakos, K., and DeRose, V.J. (2014). Nucleic acid catalysis: metals, nucleobases, and other cofactors. *Chem. Rev.* 114 (8): 4318–4342.
20 Sigel, R.K. and Pyle, A.M. (2007). Alternative roles for metal ions in enzyme catalysis and the implications for ribozyme chemistry. *Chem. Rev.* 107 (1): 97–113.
21 Ponce-Salvatierra, A., Wawrzyniak-Turek, K., Steuerwald, U. et al. (2016). Crystal structure of a DNA catalyst. *Nature* 529 (7585): 231–234.
22 Liu, H., Yu, X. et al. (2017). Crystal structure of an RNA-cleaving DNAzyme. *Nat. Commun.* 8 (1): 2006.

23 Joyce, G.F. (2004). Directed evolution of nucleic acid enzymes. *Annu. Rev. Biochem.* 73: 791–836.

24 Silverman, S.K. (2016). Catalytic DNA: scope, applications, and biochemistry of deoxyribozymes. *Trends Biochem. Sci* 41 (7): 595–609.

25 Silverman, S.K. (2009). Deoxyribozymes: selection design and serendipity in the development of DNA catalysts. *Acc. Chem. Res.* 42 (10): 1521–1531.

26 Bruesehoff, P.J., Li, J., Augustine, A.J. 3rd, and Lu, Y. (2002). Improving metal ion specificity during in vitro selection of catalytic DNA. *Comb. Chem. High Throughput Screening* 5 (4): 327–335.

27 Rajendran, M. and Ellington, A.D. (2008). Selection of fluorescent aptamer beacons that light up in the presence of zinc. *Anal. Bioanal.Chem.* 390 (4): 1067–1075.

28 Qu, H., Csordas, A.T. et al. (2016). Rapid and label-free strategy to isolate aptamers for metal ions. *ACS Nano* 10 (8): 7558–7565.

29 Zhang, D., Fu, R., Zhao, Q. et al. (2015). Nanoparticles-free fluorescence anisotropy amplification assay for detection of RNA nucleotide-cleaving DNAzyme activity. *Anal. Chem.* 87 (9): 4903–4909.

30 Liu, J. and Lu, Y. (2005). Stimuli-responsive disassembly of nanoparticle aggregates for light-up colorimetric sensing. *J. Am. Chem. Soc.* 127 (36): 12677–12683.

31 Liu, J., Brown, A.K. et al. (2007). A catalytic beacon sensor for uranium with parts-per-trillion sensitivity and millionfold selectivity. *Proc. Natl. Acad. Sci. U.S.A.* 104 (7): 2056–2061.

32 Huang, P.J., Lin, J., Cao, J. et al. (2014). Ultrasensitive DNAzyme beacon for lanthanides and metal speciation. *Anal. Chem.* 86 (3): 1816–1821.

33 Huang, P.J., Vazin, M., and Liu, J. (2014). In vitro selection of a new lanthanide-dependent DNAzyme for ratiometric sensing lanthanides. *Anal. Chem.* 86 (19): 9993–9999.

34 Huang, P.J., Vazin, M., Matuszek, Z., and Liu, J. (2015). A new heavy lanthanide-dependent DNAzyme displaying strong metal cooperativity and unrescuable phosphorothioate effect. *Nucleic Acids Res.* 43 (1): 461–469.

35 Santoro, S.W. and Joyce, G.F. (1997). A general purpose RNA-cleaving DNA enzyme. *Proc. Natl. Acad. Sci. U.S.A.* 94 (9): 4262–4266.

36 Li, J., Zheng, W., Kwon, A.H., and Lu, Y. (2000). In vitro selection and characterization of a highly efficient Zn(II)-dependent RNA-cleaving deoxyribozyme. *Nucleic Acids Res.* 28 (2): 481–488.

37 Li, J. and Lu, Y. (2000). A highly sensitive and selective catalytic DNA biosensor for lead ions. *J. Am. Chem. Soc.* 122 (42): 10466–10467.

38 Brown, A.K., Li, J., Pavot, C.M., and Lu, Y. (2003). A lead-dependent DNAzyme with a two-step mechanism. *Biochemistry* 42 (23): 7152–7161.

39 Kasprowicz, A., Stokowa-Sołtys, K., Wrzesiński, J. et al. (2015). In vitro selection of deoxyribozymes active with Cd^{2+} ions resulting in variants of DNAzyme 8–17. *Dalton Trans.* 44 (17): 8138–8149.

40 Wang, H., Kim, Y. et al. (2009). Engineering a unimolecular DNA-catalytic probe for single lead ion monitoring. *J. Am. Chem. Soc.* 131 (23): 8221–8226.

41 Xiao, Y., Rowe, A.A., and Plaxco, K.W. (2007). Electrochemical detection of parts-per-billion lead via an electrode-bound DNAzyme assembly. *J. Am. Chem. Soc.* 129 (2): 262–263.

42 Liu, J. and Lu, Y. (2003). A colorimetric lead biosensor using DNAzyme-directed assembly of gold nanoparticles. *J. Am. Chem. Soc.* 125 (22): 6642–6643.

43 Cruz, R.P., Withers, J.B., and Li, Y. (2004). Dinucleotide junction cleavage versatility of 8–17 deoxyribozyme. *Chem. Biol.* 11 (1): 57–67.

44 Huang, P.J.J., Vazin, M., Lin, J.J. et al. (2016). Distinction of individual lanthanide ions with a DNAzyme beacon array. *ACS Sensors* 1 (6): 732–738.

45 Lin, Y.W., Huang, C.C., and Chang, H.T. (2011). Gold nanoparticle probes for the detection of mercury, lead and copper ions. *Analyst* 136 (5): 863–871.

46 Kim, H.N., Ren, W.X., Kim, J.S., and Yoon, J. (2012). Fluorescent and colorimetric sensors for detection of lead, cadmium, and mercury ions. *Chem. Soc. Rev.* 41 (8): 3210–3244.

47 Que, E.L., Domaille, D.W., and Chang, C.J. (2008). Metals in neurobiology: probing their chemistry and biology with molecular imaging. *Chem. Rev.* 108 (5): 1517–1549.

48 Santoro, S.W., Joyce, G.F., Sakthivel, K. et al. (2000). RNA cleavage by a DNA enzyme with extended chemical functionality. *J. Am. Chem. Soc.* 122 (11): 2433–2439.

49 Hollenstein, M., Hipolito, C., Lam, C. et al. (2008). A highly selective DNAzyme sensor for mercuric ions. *Angew. Chem. Int. Ed.* 47 (23): 4346–4350.

50 Huang, P.J. and Liu, J. (2015). Rational evolution of Cd^{2+}-specific DNAzymes with phosphorothioate modified cleavage junction and Cd^{2+} sensing. *Nucleic Acids Res.* 43 (12): 6125–6133.

51 Huang, P.J. and Liu, J. (2016). An Ultrasensitive Light-up Cu^{2+} biosensor using a new DNAzyme cleaving a phosphorothioate-modified substrate. *Anal. Chem.* 88 (6): 3341–3347.

52 Huang, P.J. and Liu, J. (2014). Sensing parts-per-trillion Cd^{2+}, Hg^{2+}, and Pb^{2+} collectively and individually using phosphorothioate DNAzymes. *Anal. Chem.* 86 (12): 5999–6005.

53 Saran, R. and Liu, J. (2016). A silver DNAzyme. *Anal. Chem.* 88 (7): 4014–4020.

54 Saran, R., Kleinke, K., Zhou, W. et al. (2017). A silver-specific DNAzyme with a new silver aptamer and salt-promoted activity. *Biochemistry* 56 (14): 1955–1962.

55 Zhou, W., Saran, R., Chen, Q. et al. (2016). A new Na^{+}-dependent RNA-cleaving DNAzyme with over 1000-fold rate acceleration by ethanol. *ChemBioChem* 17 (2): 159–163.

56 Zhou, W., Saran, R., Huang, P.J. et al. (2017). An exceptionally selective DNA cooperatively binding two Ca^{2+} ions. *ChemBioChem* 18 (6): 518–522.

57 Carmi, N., Balkhi, S.R., and Breaker, R.R. (1998). Cleaving DNA with DNA. *Proc. Natl. Acad. Sci. U.S.A.* 95 (5): 2233–2237.

58 Gu, H., Furukawa, K., Weinberg, Z. et al. (2013). Small, highly active DNAs that hydrolyze DNA. *J. Am. Chem. Soc.* 135 (24): 9121–9129.

59 Torabi, S.F., Wu, P. et al. (2015). In vitro selection of a sodium-specific DNAzyme and its application in intracellular sensing. *Proc. Natl. Acad. Sci. U.S.A.* 112 (19): 5903–5908.

60 Torabi, S.F. and Lu, Y. (2015). Identification of the same Na^{+}-specific DNAzyme motif from two in vitro selections under different conditions. *J. Mol. Evol.* 81 (5–6): 225–234.

61 Zhou, W., Zhang, Y., Huang, P.J. et al. (2016). A DNAzyme requiring two different metal ions at two distinct sites. *Nucleic Acids Res.* 44 (1): 354–363.

62 Carmi, N., Shultz, L.A., and Breaker, R.R. (1996). In vitro selection of self-cleaving DNAs. *Chem. Biol.* 3 (12): 1039–1046.

63 Ma, L., Liu, B., Huang, P.J. et al. (2016). DNA adsorption by ZnO nanoparticles near its solubility limit: implications for DNA fluorescence quenching and DNAzyme activity assays. *Langmuir* 32 (22): 5672–5680.

64 Cuenoud, B. and Szostak, J.W. (1995). A DNA metalloenzyme with DNA ligase activity. *Nature* 375 (6532): 611–614.

65 Bhattacharyya, D., Mirihana Arachchilage, G., and Basu, S. (2016). Metal cations in G-quadruplex folding and stability. *Front. Chem.* 4: 38.

66 Liu, J. and Lu, Y. (2007). Rational design of "turn-on" allosteric DNAzyme catalytic beacons for aqueous mercury ions with ultrahigh sensitivity and selectivity. *Angew. Chem. Int. Ed.* 46 (40): 7587–7590.

67 Li, Y., Geyer, C.R., and Sen, D. (1996). Recognition of anionic porphyrins by DNA aptamers. *Biochemistry* 35 (21): 6911–6922.

68 Kosman, J. and Juskowiak, B. (2011). Peroxidase-mimicking DNAzymes for biosensing applications: a review. *Anal. Chim. Acta* 707 (1-2): 7–17.

69 Knudsen, S.M. and Ellington, A.D. (2006). Aptazymes: alloesteric ribozymes and deoxyribozymes as biosensors. In: *Aptamer Handbook* (ed. S. Klussmann), 290–310. Weinheim: Wiley-VCH.

70 Zivarts, M., Liu, Y., and Breaker, R.R. (2005). Engineered allosteric ribozymes that respond to specific divalent metal ions. *Nucleic Acids Res.* 33 (2): 622–631.

71 Breaker, R.R. (2002). Engineered allosteric ribozymes as biosensor components. *Curr. Opin. Biotechnol.* 13 (1): 31–39.

72 Soukup, G.A. and Breaker, R.R. (1999). Engineering precision RNA molecular switches. *Proc. Natl. Acad. Sci. U.S.A.* 96 (7): 3584–3589.

73 Kondo, J., Yamada, T. et al. (2014). Crystal structure of metallo DNA duplex containing consecutive Watson–Crick-like T-Hg(II)-T base pairs. *Angew. Chem. Int. Ed.* 53 (9): 2385–2388.

74 Shen, Z., Wu, Z. et al. (2016). A catalytic DNA activated by a specific strain of bacterial pathogen. *Angew. Chem. Int. Ed.* 55 (7): 2431–2434.

75 Shen, Y., Chiuman, W., Brennan, J.D., and Li, Y. (2006). Catalysis and rational engineering of trans-acting pH6DZ1, an RNA-cleaving and fluorescence-signaling deoxyribozyme with a four-way junction structure. *ChemBioChem* 7 (9): 1343–1348.

76 Hartig, J.S., Najafi-Shoushtari, S.H. et al. (2002). Protein-dependent ribozymes report molecular interactions in real time. *Nat. Biotechnol.* 20 (7): 717–722.

77 Koizumi, M., Soukup, G.A., Kerr, J.N.Q., and Breaker, R.R. (1999). Allosteric selection of ribozymes that respond to the second messengers cGMP and cAMP. *Nat. Struct. Biol.* 6 (11): 1062–1071.

78 Ali, M.M., Aguirre, S.D., Lazim, H., and Li, Y. (2011). Fluorogenic DNAzyme probes as bacterial indicators. *Angew. Chem. Int. Ed.* 50 (16): 3751–3754.

79 He, S., Qu, L., Shen, Z. et al. (2015). Highly specific recognition of breast tumors by an RNA-cleaving fluorogenic DNAzyme probe. *Anal. Chem.* 87 (1): 569–577.

80 Yousefi, H., Ali, M.M., Su, H.-M. et al. (2018). Sentinel wraps: real-time monitoring of food contamination by printing DNAzyme probes on food packaging. *ACS Nano*. 12(4): 3287–3294.

81 Wang, H., Yang, R., Yang, L., and Tan, W. (2009). Nucleic acid conjugated nanomaterials for enhanced molecular recognition. *ACS Nano* 3 (9): 2451–2460.

82 Song, S., Qin, Y., He, Y. et al. (2010). Functional nanoprobes for ultrasensitive detection of biomolecules. *Chem. Soc. Rev.* 39 (11): 4234–4243.

83 Yu, T., Zhou, W., and Liu, J. (2018). Ultrasensitive DNAzyme-based Ca^{2+} detection boosted by ethanol and a solvent-compatible scaffold for aptazyme design. *ChemBioChem* 19 (1): 31–36.

84 Lew, V.L., Tsien, R.Y., Miner, C., and Bookchin, R.M. (1982). Physiological intracellular calcium concentration level and pump-leak turnover in intact red cells measured using an incorporated calcium chelator. *Nature* 298 (5873): 478–481.

85 Liu, J. and Lu, Y. (2003). Improving fluorescent DNAzyme biosensors by combining inter- and intramolecular quenchers. *Anal. Chem.* 75 (23): 6666–6672.

86 Liu, M., Zhao, H., Chen, S. et al. (2011). A "turn-on" fluorescent copper biosensor based on DNA cleavage-dependent graphene-quenched DNAzyme. *Biosens. Bioelectron.* 26 (10): 4111–4116.

87 Kim, J.H., Han, S.H., and Chung, B.H. (2011). Improving Pb^{2+} detection using DNAzyme-based fluorescence sensors by pairing fluorescence donors with gold nanoparticles. *Biosens. Bioelectron.* 26 (5): 2125–2129.

88 Amaral, N.B., Zuliani, S., Guieu, V. et al. (2014). Catalytic DNA-based fluorescence polarization chiral sensing platform for L-histidine detection at trace level. *Anal. Bioanal.Chem.* 406 (4): 1173–1179.

89 Kim, H.K., Li, J., Nagraj, N., and Lu, Y. (2008). Probing metal binding in the 8–17 DNAzyme by TbIII luminescence spectroscopy. *Chem. Eur. J* 14 (28): 8696–8703.

90 Nagatoishi, S., Nojima, T., Juskowiak, B., and Takenaka, S. (2005). A pyrene-labeled G-quadruplex oligonucleotide as a fluorescent probe for potassium ion detection in biological applications. *Angew. Chem. Int. Ed.* 117 (32): 5195–5198.

91 Simkiss, K. (1979). Metal ions in cells. *Endeavour* 3 (1): 2–6.

92 Wu, P., Hwang, K., Lan, T., and Lu, Y. (2013). A DNAzyme-gold nanoparticle probe for uranyl ion in living cells. *J. Am. Chem. Soc.* 135 (14): 5254–5257.

93 Zhou, W., Liang, W., Li, D. et al. (2016). Dual-color encoded DNAzyme nanostructures for multiplexed detection of intracellular metal ions in living cells. *Biosens. Bioelectron.* 85: 573–579.

94 Wang, W., Satyavolu, N.S.R. et al. (2017). Near-infrared photothermally activated DNAzyme-gold nanoshells for imaging metal ions in living cells. *Angew. Chem. Int. Ed.* 56 (24): 6798–6802.

95 Hwang, K., Wu, P., Kim, T. et al. (2014). Photocaged DNAzymes as a general method for sensing metal ions in living cells. *Angew. Chem. Int. Ed.* 53 (50): 13798–13802.

96 Wu, Z., Fan, H. et al. (2017). Imaging endogenous metal ions in living cells using a DNAzyme-catalytic hairpin assembly probe. *Angew. Chem. Int. Ed.* 56 (30): 8721–8725.

97 Mei, S.H., Liu, Z., Brennan, J.D., and Li, Y. (2003). An efficient RNA-cleaving DNA enzyme that synchronizes catalysis with fluorescence signaling. *J. Am. Chem. Soc.* 125 (2): 412–420.

98 Chiuman, W. and Li, Y. (2006). Efficient signaling platforms built from a small catalytic DNA and doubly labeled fluorogenic substrates. *Nucleic Acids Res.* 35 (2): 401–405.

99 Xu, W., Chan, K.M., and Kool, E.T. (2017). Fluorescent nucleobases as tools for studying DNA and RNA. *Nat. Chem.* 9 (11): 1043–1055.

100 Zhou, W., Ding, J., and Liu, J. (2016). A highly specific sodium aptamer probed by 2-aminopurine for robust Na^+ sensing. *Nucleic Acids Res.* 44: 354–363.

101 Ferrari, D. and Peracchi, A. (2002). A continuous kinetic assay for RNA-cleaving deoxyribozymes, exploiting ethidium bromide as an extrinsic fluorescent probe. *Nucleic Acids Res.* 30 (20): e112–e112.

102 Xiang, Y., Tong, A., and Lu, Y. (2009). Abasic site-containing DNAzyme and aptamer for label-free fluorescent detection of Pb^{2+} and adenosine with high sensitivity, selectivity, and tunable dynamic range. *J. Am. Chem. Soc.* 131 (42): 15352–15357.

103 Rosi, N.L. and Mirkin, C.A. (2005). Nanostructures in biodiagnostics. *Chem. Rev.* 105 (4): 1547–1562.

104 Mirkin, C.A., Letsinger, R.L., Mucic, R.C., and Storhoff, J.J. (1996). A DNA-based method for rationally assembling nanoparticles into macroscopic materials. *Nature* 382 (6592): 607–609.

105 Alivisatos, A.P., Johnsson, K.P. et al. (1996). Organization of nanocrystal molecules' using DNA. *Nature* 382 (6592): 609.

106 Li, H. and Rothberg, L. (2004). Colorimetric detection of DNA sequences based on electrostatic interactions with unmodified gold nanoparticles. *Proc. Natl. Acad. Sci. U.S.A.* 101 (39): 14036–14039.

107 Lee, J.H., Wang, Z.D., Liu, J.W., and Lu, Y. (2008). Highly sensitive and selective colorimetric sensors for uranyl (UO_2^{2+}): development and comparison of labeled and label-free DNAzyme-gold nanoparticle systems. *J. Am. Chem. Soc.* 130 (43): 14217–14226.

108 Li, J., Mo, L., Lu, C.H. et al. (2016). Functional nucleic acid-based hydrogels for bioanalytical and biomedical applications. *Chem. Soc. Rev.* 45 (5): 1410–1431.

109 Liu, J.W. (2011). Oligonucleotide-functionalized hydrogels as stimuli responsive materials and biosensors. *Soft Matter* 7 (15): 6757–6767.

110 Lee, J., Kim, H.J., and Kim, J. (2008). Polydiacetylene liposome arrays for selective potassium detection. *J. Am. Chem. Soc.* 130 (15): 5010–5011.

111 YunáZhang, W. and JamesáYang, C. (2011). DNAzyme crosslinked hydrogel: a new platform for visual detection of metal ions. *Chem. Commun.* 47 (33): 9312–9314.

112 Huang, Y., Ma, Y., Chen, Y. et al. (2014). Target-responsive DNAzyme cross-linked hydrogel for visual quantitative detection of lead. *Anal. Chem.* 86 (22): 11434–11439.

113 Huang, Y., Wu, X. et al. (2017). Target-responsive DNAzyme hydrogel for portable colorimetric detection of lanthanide(III) ions. *Science China-Chemistry* 60 (2): 293–298.

114 Yin, B.C., Ye, B.C., Tan, W. et al. (2009). An allosteric dual-DNAzyme unimolecular probe for colorimetric detection of copper(II). *J. Am. Chem. Soc.* 131 (41): 14624–14625.

115 Elbaz, J., Shlyahovsky, B., and Willner, I. (2008). A DNAzyme cascade for the amplified detection of Pb^{2+} ions or L-histidine. *Chem. Commun.* 13: 1569–1571.

116 Shen, L., Chen, Z. et al. (2008). Electrochemical DNAzyme sensor for lead based on amplification of DNA-Au bio-bar codes. *Anal. Chem.* 80 (16): 6323–6328.

117 Xiang, Y. and Lu, Y. (2011). Using personal glucose meters and functional DNA sensors to quantify a variety of analytical targets. *Nat. Chem.* 3 (9): 697–703.

118 Peng, H., Newbigging, A.M., Wang, Z. et al. (2018). DNAzyme-mediated assays for amplified detection of nucleic acids and proteins. *Anal. Chem.* 90 (1): 190–207.

119 Liu, M., Chang, D., and Li, Y. (2017). Discovery and biosensing applications of diverse RNA-cleaving DNAzymes. *Acc. Chem. Res.* 50 (9): 2273–2283.

120 Zhang, X.B., Wang, Z., Xing, H. et al. (2010). Catalytic and molecular beacons for amplified detection of metal ions and organic molecules with high sensitivity. *Anal. Chem.* 82 (12): 5005–5011.

121 Liu, M., Zhang, Q. et al. (2016). Programming a topologically constrained DNA nanostructure into a sensor. *Nat. Commun.* 7: 12074.

122 Wang, F., Wu, Z., Lu, Y., Wang, J. et al. (2010). A label-free DNAzyme sensor for lead(II) detection by quantitative polymerase chain reaction. *Anal. Biochem.* 405 (2): 168–173.

123 Ali, M.M. and Li, Y. (2009). Colorimetric sensing by using allosteric-DNAzyme-coupled rolling circle amplification and a peptide nucleic acid–organic dye probe. *Angew. Chem. Int. Ed.* 121 (19): 3564–3567.

124 Liu, M., Zhang, Q., Chang, D. et al. (2017). A DNAzyme feedback amplification strategy for biosensing. *Angew. Chem. Int. Ed.* 56 (22): 6142–6146.

125 Lu, C.H., Wang, F.A., and Willner, I. (2012). Zn^{2+}-ligation DNAzyme-driven enzymatic and nonenzymatic cascades for the amplified detection of DNA. *J. Am. Chem. Soc.* 134 (25): 10651–10658.

126 Liu, S., Cheng, C., Gong, H., and Wang, L. (2015). Programmable Mg^{2+}-dependent DNAzyme switch by the catalytic hairpin DNA assembly for dual-signal amplification toward homogeneous analysis of protein and DNA. *Chem. Commun.* 51 (34): 7364–7367.

127 Wang, F., Elbaz, J., Orbach, R. et al. (2011). Amplified analysis of DNA by the autonomous assembly of polymers consisting of DNAzyme wires. *J. Am. Chem. Soc.* 133 (43): 17149–17151.

28

Compartmentalization-Based Technologies for In Vitro Selection and Evolution of Ribozymes and Light-Up RNA Aptamers

Farah Bouhedda and Michael Ryckelynck

Université de Strasbourg, CNRS, Architecture et Réactivité de l'ARN, UPR 9002, Institut de Biologie Moléculaire et Cellulaire, 2, allée Konrad Roentgen, F-67084 Strasbourg, France

28.1 Introduction

In the 1980s, ribonucleic acid molecules able to catalyze chemical reactions, also called ribozymes [1], have been discovered in cells. First evidence of this catalytic capacity came with the finding that RNAs forming so-called self-splicing introns can excise themselves out of larger RNAs without the assistance of cell cofactor [2]. Since then, a large number of other self-splicing and self-cleaving RNAs (e.g., hammerhead or hairpin ribozymes) have been brought to light [3–7]. Besides these self-modifying activities, the capacity of RNA to work as a true recycling enzyme was also established first by the discovery that the 5′ end transfer RNA (tRNA) maturating activity of the ribonuclease P was actually carried by the RNA moiety of the ribonucleoprotein [8]. The list of RNA-based recycling enzymes was later further enriched by the discovery that several additional activities central to cell life (e.g. peptidyl transferase and splicing) are also catalyzed by the RNA component (23S/28S RNA in prokaryotes/eukaryotes and small nuclear RNA [snRNA], respectively) of the ribonucleoprotein complexes (respectively the ribosome and the spliceosome [9]).

Altogether, the abovementioned natural ribozymes bring strong support to the RNA world hypothesis [10], which postulates that, between the prebiotic chemical world and the contemporary DNA/RNA/protein one, primitive systems may have existed in which both genetic information storage and catalytic activity would have been supported by RNA (or a polymer alike) molecules. Yet, being able to resurrect such a primitive system, and so validate the theory, requires the demonstration that RNA is indeed able to catalyze a much wider palette of activities (e.g. alkylation, polymerization, isomerization, carbon–carbon additions, etc.) in order to sustain replication and metabolism. The quest for these ancient catalysts partly motivated the development of in vitro selection and evolution technologies starting from early 1990s (see below) and led to the identification of artificial RNAs endowed with a variety of functions. Not only were these newly isolated ribozymes important on

Ribozymes, First Edition. Edited by Sabine Müller, Benoît Masquida, and Wade Winkler.

a fundamental point of view, but also some of them were found to have important applications, for instance, in bioengineering or biotechnology [11–14]. As in nature, in vitro selection and evolution processes require the phenotype to stay associated with its encoding genotype, and when searching for catalysts, the way this genotype/phenotype link is established and maintained may directly impact the performances of the output RNA. In this chapter, we will first give a rapid overview of technologies allowing for selecting ribozymes based on their self-modification prior to introducing compartmentalization-based technologies and discussing how the latter may represent a significant advance in the development of ribozymes as well as RNAs endowed with other functions, with a special emphasis on light-up aptamers.

28.2 Selection of Self-Modifying Ribozymes

The proper maintenance of the genotype/phenotype link is crucial when selecting/evolving biological molecules. Fortunately, when working with nucleic acids, the polymer (DNA or RNA) can be both the support of the genetic information and its effector. Therefore, experiments can be designed such that RNAs having the activity of interest primarily modify themselves by adding a mark making them selectable (Figure 28.1a). Then, using the proper partition method, libraries can be specifically and exponentially enriched in molecules displaying the target activity. Such a strategy of systematic evolution of ligands by exponential enrichment (SELEX) [15, 16] was initially devised for the isolation of aptamers able to specifically recognize a target (of size ranging from ions to a whole cell [17–19]) starting from a large pool of RNAs ($\sim 10^{15}$ molecules). RNAs that can specifically interact with the target can then be partitioned from the bulk using, for instance, beads substituted with the target molecule. Upon stringent washes aiming at eliminating unbound RNAs and weak binders, the bound molecules of interest are eluted, recovered, and amplified by reverse transcriptase (RT)-polymerase chain reaction (PCR). Finally, the genes are in vitro transcribed to regenerate material used to prime a new round of selection [16, 20]. SELEX technology was also adapted to the identification of catalysts, and it allowed, for instance, the discovery of RNAs endowed with activities as diverse as aminoacyl-tRNA synthetase [14], peptidyl transferase [21], pyrimidine synthetase [22], or Diels–Alderase [23] (more comprehensive reviews on the topic can be found in [24–26]).

The most impressive example of SELEX-derived ribozyme, which is also perhaps one of the most relevant to the RNA world theory, was the isolation by Bartel and Szostak of an RNA ligase ribozyme that can catalyze the formation of a 3′,5′-phosphodiester bond [27]. This ligase was isolated from a starting pool of 10^{15} different RNA molecules randomized over a 220 nucleotide-long region and carrying a 5′ triphosphate. Using a clever selection scheme (Figure 28.2), the molecules were challenged to ligate the 3′ end of a substrate oligonucleotide to their own triphosphorylated 5′ ends. To enrich libraries in potent catalysts, the mixture was then flown through an affinity resin made of beads substituted in oligonucleotides complementary to the ligated one. Upon a washing step, captured molecules were

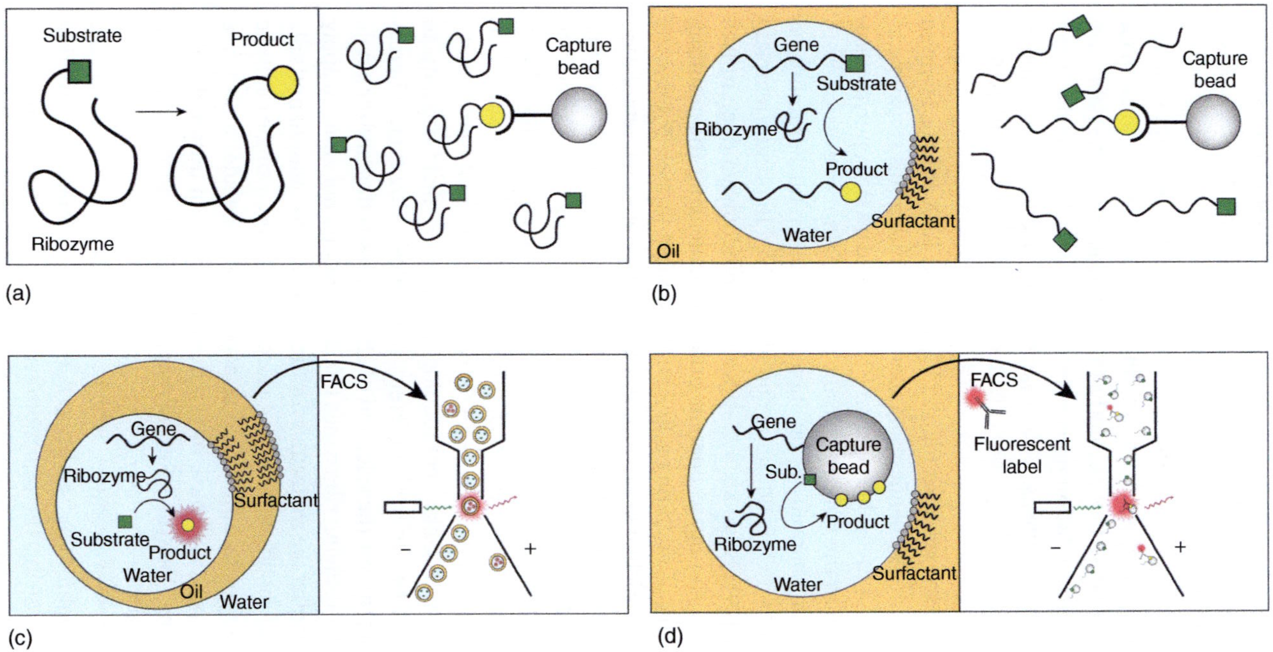

Figure 28.1 Establishment of genotype/phenotype link. (a) Strategy based on RNA self-modification. Active ribozymes modify themselves by converting a covalently attached substrate (green square) into product (yellow circle). Product-displaying RNAs are then partitioned from the bulk using, for instance, a resin substituted with capture molecules. (b) In vitro compartmentalization (IVC) using modified genes. Genes attached to a substrate of the target activity are encapsulated (with a dilution adjusted to have at most one gene per droplet) into water-in-oil (W/O) droplets stabilized by a surfactant together with an expression mixture. Upon transcription, genes encoding active ribozymes are expected to have their grafted substrate converted into product allowing for partitioning them from the bulk. (c) Double emulsion-based IVC. Genes are first individualized together with an expression machinery and a fluorogenic substrate of the reaction into W/O droplets stabilized by a surfactant. These droplets are then re-emulsified to form water-in-oil-in-water (W/O/W) droplets that can be analyzed and handled with a FACS. (d) Microbead display. Genes are individualized onto microbeads prior to encapsulating them into W/O droplets together with an expression mixture. Active ribozymes then convert substrate (Sub.) into product displayed at the surface of the beads. Finally, reaction products are fluorescently labeled, and the beads are FACS sorted.

eluted and reverse-transcribed, and the resulting cDNAs were PCR amplified. The PCR was performed in two regimes: first using selective primers enriching the library in active molecules and then using nonselective primers allowing for adding a transcription promoter. Iterating such a cycle while increasing the selection pressure (e.g., gradually shortening the incubation time) allowed for progressively enriching the library in efficient catalysts. Indeed, performing four rounds of this in vitro selection allowed for enriching the libraries in a first generation of ligase ribozymes likely to be suboptimal when considering the very large sequence space to explore (4^{220} different variants). Consequently, the performances of the pool were next optimized by an in vitro evolution process in which rounds of selection were interspersed by rounds of mutagenesis. This strategy finally led to the isolation of the class I ligase, a 186 nucleotide-long ribozyme that can accelerate 7 million times the ligation reaction.

Interestingly, the RNA ligase was also found to extend an RNA primer by up to four nucleotides in a templated-dependent manner [28], pointing this ribozyme as being a good starting point for the isolation of an RNA-dependent RNA polymerase, an activity that can be seen as the holy grail in the search of ribozymes relevant to the RNA world. The extension capacity of the molecule was then further improved by equipping the ribozyme with an additional domain involved in primer/template (P/T) duplex recognition [29]. To do so, a 76-nucleotide randomized domain was appended to the 3′ end of the ligase domain prior to challenging the resulting ribozymes for their capacity of extending a P/T duplex grafted to their 5′ ends. As before, libraries were first enriched in potent catalysts (using here a gel retardation partition approach) prior to refining their performance through an in vitro process. This led to the isolation of the R18 RNA-dependent RNA polymerase, a ribozyme able to elongate *in trans* and in a template-dependent manner an RNA primer by up to 14 nucleotides [29]. Yet, this molecule was found unable to synthesize more than one RNA helical turn, a limitation attributable to the selection format (see below).

The main strength of SELEX-like methodologies relies on their capacity to handle very large libraries (up to 10^{15} different molecules). However, whereas these approaches are powerful for the selection of efficient cis-acting and/or single turnover catalysts (e.g., self-cleaving ribozymes [30, 31]), they do not allow to apply selection pressures challenging the molecules for their capacity to catalyze multiple turnovers. Indeed, in a self-modification-based selection strategy, a single catalytic event is often sufficient to make a molecule selectable, and the addition of free competitive substrate is expected to have only a limited impact on intramolecular reaction efficiency. Selecting ribozymes for their single turnover capacity may also lead to the isolation of catalysts displaying a very high affinity for their substrate and to some extent to reaction products, which, in some cases, can render the enzyme prone to product inhibition (see below). Moreover, the need for tethering the substrate to the RNA during the selection/evolution process may restrict its accessibility to the active site and, doing so, limit the identification of catalysts able to optimally process their substrates, as this was likely the case during the isolation of the R18 RNA-dependent RNA polymerase [29]. These different limitations of the SELEX-like technologies stimulated the development and the use of in vitro

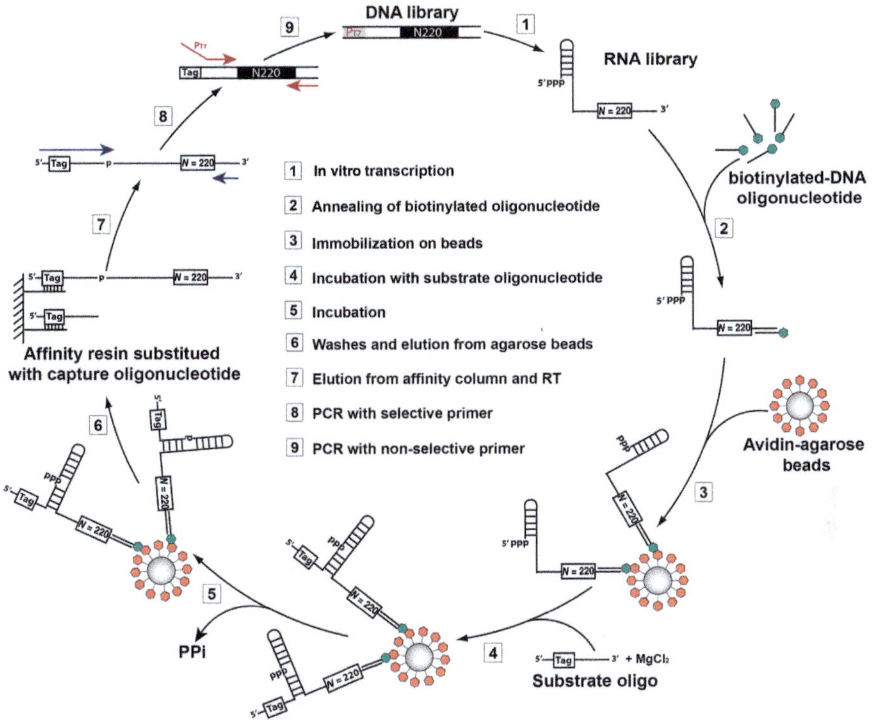

Figure 28.2 Isolation of an RNA ligase ribozyme by a self-modification strategy. A starting DNA library made of 1.6×10^{15} different molecules randomized over a region of 220 nucleotides was in vitro transcribed (step 1). To preserve their solubility, resulting RNAs were then annealed to biotinylated oligonucleotides (step 2) and immobilized onto avidin-agarose beads (step 3). Substrate oligonucleotide was then added to the mixture together with $MgCl_2$ (step 4). Active ribozymes catalyzed the ligation between oligonucleotide 3′ end and ribozyme 5′ end together with the release of an inorganic pyrophosphate (PP_i) molecule (step 5). Self-modified RNAs (i.e. active ribozymes) were then captured on a resin substituted with an oligonucleotide complementary to a tag sequence present in the 5′ part of the substrate oligonucleotide (step 6). Upon elution (step 7), the fraction of active molecules was next further enriched by a RT-PCR using primers selective of the ligation product (step 8) prior to amplifying the enriched pool using nonselective primers bringing the T7 RNA polymerase promoter (step 9) and, doing so, producing a gene library ready to enter a new round of selection.

compartmentalization (IVC) methodologies that are better suited for the selection of enzymes characterized by both high reaction acceleration rates and efficient recycling.

28.2.1 Ribozyme Discovery Using In Vitro Compartmentalization

Besides the direct attachment of a substrate to the catalyst, the genotype/phenotype link can be established by confining the catalyst with its substrate into small compartments in which the edge acts as a barrier that physically maintains the genotype with its phenotype (Figure 28.1b–d). Inspired by Nature, Tawfik and Griffiths introduced in the late 1990s the concept of IVC in which reactions are carried out into water-in-oil (W/O) droplets [32]. Typically, a gene library is diluted into an in vitro expression mixture supplemented in catalyst substrate just prior to being dispersed in millions of W/O droplets, each one containing at most one gene. The addition of a surfactant in the oil phase allows for stabilizing the microcompartments and, doing so, preserving the genotype/phenotype linkage integrity [33]. In this scheme, upon emulsification (by stirring or extrusion), genes are expressed within the droplets in which substrate will be turned over into product *in trans* all the better that the enzyme has an elevated catalytic constant (k_{cat}).

Selection by Gene Modification

In the original IVC format, the selection operates at the level of the gene that is either the substrate of the reaction [34] or physically attached to a substrate molecule [32, 35–38]. Upon expression, should the gene expression product display the searched activity, the substrate will be converted into a selectable product (Figure 28.1b). As a consequence, upon DNA recovery, those genes encoding active enzymes can be partitioned away from the rest of the library using approaches such as affinity purification [37], electrophoretic separation [38], or PCR amplification of nuclease-resistant DNAs [36]. Interestingly, when using IVC, the selection pressure can easily be tuned by adjusting free substrate concentration in the reaction mixture. Therefore, to make the gene selectable, the encoded enzyme has to be both highly efficient in catalyzing the reaction and getting recycled.

Initially, IVC was mainly applied to the evolution of protein enzymes such as methyltransferases [32, 35], endonucleases [39], or DNA polymerases [34, 40, 41]. Later, Griffiths and coworkers also applied this technology to the selection of ribozymes and demonstrated its robustness by isolating RNAs able to efficiently catalyze *in trans* the Diels–Alder cycloaddition between a 1,3-diene and an alkene dienophile [37]. To do so, the authors devised an IVC procedure (Figure 28.3a) in which each gene contained in the starting library of 10^{11} molecules was conjugated to an anthracene via a polyethylene glycol (PEG) linker prior to being diluted into an in vitro transcription medium. The mixture was then dispersed into micrometer-scale W/O droplets using a stirring process. After in vitro transcription occurred, droplets were supplemented in Mg^{2+} and biotinylated maleimide substrate. Finally, upon the last incubation, those droplets containing active Diels–Alderase ribozymes also contained genes biotinylated via the formation of a cycloadduct between anthracene and biotinylated maleimide, making these genes isolable using streptavidin-coated magnetic beads. Performing a total of nine rounds of IVC selection allowed the isolation of very active ribozymes with catalytic rates closely approaching the maximal theoretical catalytic rate of the reaction [37]. Using the same strategy of gene modification, Zaher and Unrau isolated B6.61, an improved version of the R18 RNA-dependent RNA polymerase ribozyme endowed with superior extension capacity (up to 20 nucleotides polymerized) as well as an improved fidelity [38]. Based on what precedes, IVC proved to be a powerful way of selecting efficient catalysts (proteins and RNAs), but library preparation is made complex by the need to label genes with a substrate, a limitation overcome in the later version of IVC presented hereafter.

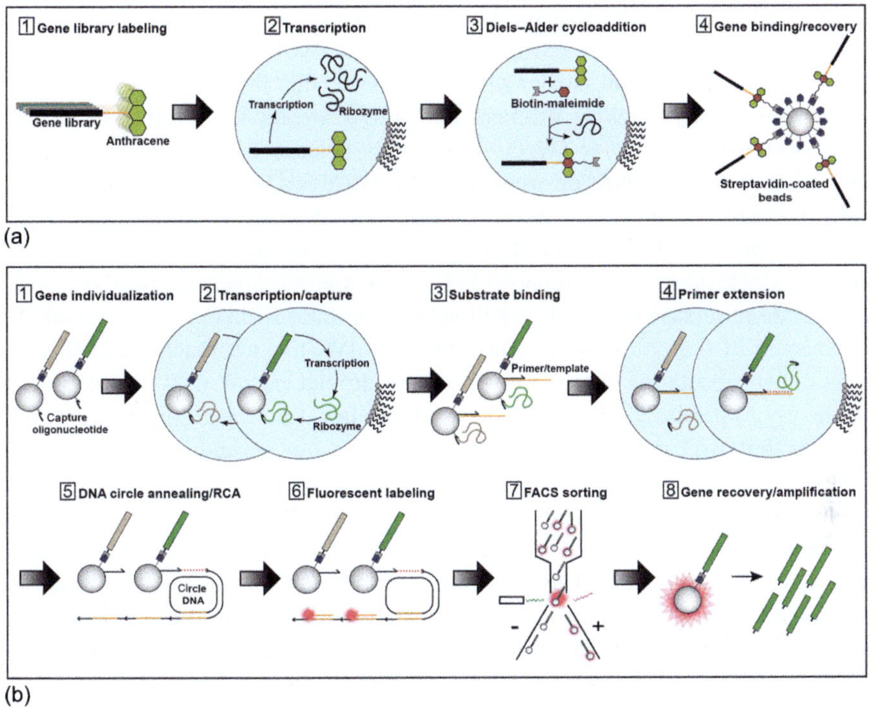

Figure 28.3 Isolation of ribozymes using in vitro compartmentalization (IVC). (a) Selection of Diels–Alderase ribozymes using IVC as described by [37]. Genes from a library were conjugated with an anthracene via a PEG linker (step 1) prior to being individualized into water-in-oil droplets. Genes were first transcribed in the droplets (step 2) before Mg^{2+} and biotin-maleimide were added by diffusion (step 3). In droplets containing active Diels–Alderase ribozymes, the formation of the cycloadduct by reaction of biotinylated maleimide with the anthracene carried by the gene led to gene biotinylation, making the gene selectable on beads substituted with streptavidin (step 4). (b) Selection of RNA-dependent RNA polymerase ribozymes using IVC as described by [42]. Biotinylated ribozyme-coding genes were first individualized onto streptavidin-coated magnetic beads together with a hairpin oligonucleotide complementary to RNA 5′end (step 1). The beads were then mixed with an expression mixture and encapsulated into water-in-oil droplets in which DNA was transcribed into RNAs that were captured on the beads (step 2). After breaking of emulsion, beads were further decorated with primer/template duplexes (step 3). The resulting beads were then re-emulsified into a new set of water-in-oil droplets in which RNAs were released from the beads by disulfide bridge reduction and active ribozymes were free to extend the primer by copying the template of the duplex (step 4). Upon reaction, beads were collected and the presence of extension products revealed by a rolling circle amplification (RCA, step 5) followed by the fluorescent labeling of the amplification products using fluorescently-labeled probes (step 6). Finally, beads were FACS sorted (step 7) and genes of interest recovered by PCR amplification (step 8).

Screening Catalyst-Coding Gene Libraries Using Double Emulsions

A first alternative to gene modification consists in using a fluorogenic substrate (i.e. a molecule that is poorly fluorescent in its substrate form but that become fluorescent upon the action of an enzyme that converts it into product) of the target activity. Using a fluorogenic substrate makes the identification of droplets/genes of interest straightforward through fluorescence measurement. Yet, sorting them from the bulk remains challenging when using W/O emulsion since laboratory devices such as fluorescence-activated cell sorter (FACS) cannot easily handle objects suspended in an oil carrier phase. However, this limitation can be overcome by re-emulsifying W/O droplets into a FACS-compatible aqueous phase to form water-in-oil-in-water (W/O/W) droplets (Figure 28.1c). Therefore, the combination of double emulsion-based IVC and FACS is a very useful tool for very high-throughput screening of catalyst-coding gene libraries. Even though it has never been applied to RNA, this technology proved its worth by evolving enzymes such as β-galactosidase [43], proteases [44], DNA methyltransferase [45], or even dihydrofolate reductase [46]. Yet, it still suffers limitations linked to the elevated polydispersity of the emulsion common to every IVC-like approach (see below), the risk of encapsulating more than one W/O droplet per W/O/W droplet and the difficulty to precisely modify droplet content after it has been formed, a set of limitations that can partly be addressed by jointly using beads and IVC.

Screening Catalyst-Coding Gene Libraries Using Microbead Display
Genotype can also be linked to its phenotype by capturing both genes and reaction products at the surface of micrometer beads (Figure 28.1d). In this scenario, W/O droplets are transiently used to express genes and perform the activity assay in highly parallel manner while preserving information confinement. As pioneered by Griffiths and coworkers [47], microbead display consists in individualizing the genes of a libraries onto streptavidin-conjugated beads (carrying at most one gene molecule) displaying also a large copy number of a capture molecule (e.g. antibody or oligonucleotide). Each bead is then encapsulated into a W/O droplet together with an expression mixture and optionally with the substrate of the reaction. Upon synthesis, gene expression products (protein or RNA) are captured at bead surface (Figure 28.3b). Beads are then recovered and re-emulsified into a new set of W/O droplets together with a substrate of the enzyme either attached to the bead [42] or free in the droplet and captured on demand (e.g. upon a light activation step, [48]). Upon reaction, droplets are recovered and fluorescently labeled using a short-lived reactive fluorescent reporter [47], a secondary fluorescently labeled antibody [48] or a DNA amplification approach (e.g. rolling circle amplification) in the presence of fluorescent oligonucleotides [42]. At the end of the process, each bead carries both a genotype (i.e. the gene) and a measure of its phenotype (i.e. a fluorescent signal proportional to enzyme activity) that can be analyzed in high-throughput regime by FACS. Moreover, concentrating the fluorescence at bead surface allows for increasing the detection sensitivity.

Microbead display allowed to isolate not only proteins for their binding capacities [47] or for their enzymatic properties [48] but also ribozymes. Ellington and collaborators [49] adapted the IVC microbead display procedure to the selection of trans-acting ligase ribozymes. To do so, they prepared a doped mutant library of the b1-207t, the trans-acting form of the Bartel Class I ligase [50], and individualized the genes at the surface of the beads together with a large number of a substrate oligonucleotide (the first substrate of the ribozyme). Then, beads were isolated into W/O droplets together with an in vitro transcription mixture and a large number of a second fluorescently labeled oligonucleotides (the second substrate of the ribozyme). Upon gene transcription, ribozymes catalyzed the ligation of both oligonucleotides and, doing so, labeled the beads with the fluorescent ligation product proportionally to ribozyme activity. Finally, FACS sorting the beads allowed the authors to isolate, in only four rounds of selection, a variant only five mutations away from the parental ribozyme and with comparable kinetics. An even more appealing exemplification of microbead display came with the isolation by Holliger and coworkers [42] of a significantly improved variant of the R18 RNA polymerase [29]. Indeed, whereas the original R18 ribozyme was able to generate a short 14 nucleotide-long RNA fragment (see above), the use of microbead display (Figure 28.3b) allowed the identification of the tC19Z ribozyme, an RNA-dependent RNA polymerase capable of synthesizing up to 95 nucleotide-long RNAs endowed with a catalytic activity [42], a polymerase highly relevant to the RNA world. Finally, more recently, microbead display also proved to be efficient at selecting high affinity [51], structure switching [52], and light-up RNA aptamers [53].

Emulsion-Free In Vitro Compartmentalization

Even though emulsion-based IVC is the most widely used method, compartmentalization can be also achieved without the use of oil, for example, in ice as nicely showed by Holliger and coworkers [54]. Indeed, when an aqueous solution is cooled down below its freezing point, a biphasic system forms, whereby solutes are excluded from the growing ice crystals and are concentrated in an interstitial liquid: the eutectic phase. The effectiveness of this approach was demonstrated by the isolation of a new mutant of the R18 RNA polymerase, the so-called RNA polymerase Y, able to synthesize an RNA as long as itself [55]. Discovering such a processive ribozyme represents a great advance in the field by showing that, in principle, a self-replicating ribozyme could have existed while the life was developing in an RNA world. Moreover, isolating new ribozymes in ice not only allows to expand the number of synthetic molecules available but also provides strong support to the RNA world theory showing that ice, widespread on the early Earth, can provide a favorable environment for the evolution of catalytic RNAs.

Since its introduction two decades ago, IVC demonstrated its suitability for the isolation of efficient catalysts. Yet, even though the technology became more and more versatile and accurate over its different development stages, it still suffers from two main limitations. First, the high degree of polydispersity (a quantification of droplet-to-droplet size variation) of the produced emulsions directly challenges the extent to which IVC can be quantitative since droplet size variance may be significantly higher than the biological one. Said differently, the same gene would have different phenotypes in different sized droplets making it challenging to distinguish small activity differences. Second, it is difficult to modify droplet content after their formation in a controlled way without damaging them. Indeed, even though it is possible to deliver hydrophobic substrates, or ligands, through the oil phase inside the water droplet [37] or even to deliver water-soluble components thanks to nanodroplets, or swollen micelles [56], none of these techniques allows a precise control over droplet content. Moreover, even though in microbead display the beads can be recovered from one emulsion and encapsulated into a new one, the method accuracy stays limited by the poor control over droplet size dispersity. Nevertheless, these limitations can be overcome by transposing IVC to a microfluidic format offering an exquisite control over droplet production and manipulation.

28.2.2 Microfluidic-Assisted In Vitro Compartmentalization

A microfluidic device is usually made of a network of micrometer-scale channels imprinted into a polymer bound onto a glass slide. When circulating into such small channels, liquids change their flow properties (e.g., they adopt a laminar flow characterized by a low mixing efficiency), making it possible to finely control them. For instance, focusing together an aqueous phase and two oil streams (supplemented with a surfactant molecule to stabilize the formed emulsion) coming from orthogonal channels and forcing the resulting co-flow to pass through a small orifice allows to generate W/O droplets of highly controlled and reproducible (less than 3% of polydispersity) picoliter volumes [57, 58]. Moreover, by using dedicated microfluidic modules, it is possible to individually act on these droplets after their formation. Indeed, using droplet fusion device [59] or a picoinjector [60] allows to add reagents into each droplet at rates of several thousand events per second without challenging droplet integrity. Furthermore, droplets can also be split [61], they can be collected and incubated off-chip, and their fluorescence can be analyzed and used to sort droplets accordingly [62]. Altogether, these different modules allow to perform every basic operations required to transpose IVC to a microfluidic format (microfluidic-assisted IVC or μIVC).

Selection of Artificial Ribozymes Using μIVC

μIVC has been applied and adapted for the discovery of several new and improved protein enzymes [63]. Moreover, the application of this technology to the evolution of RNAs has been recently pioneered by our group [64]. Combining the use of four different microfluidic chips allowed devising an in vitro ultrahigh-throughput pipeline (Figure 28.4) able to analyze libraries made of million mutants in highly quantitative manner and in a single day. More precisely, genes contained in a library are first diluted into a PCR amplification mixture prior to being individualized into small (2.5 pl) droplets (Figure 28.4a) collected and thermocycled to amplify each DNA molecule several tens of thousands of times. Such amplification is of prime importance as it allows both to reduce the incubation time during the later steps and it also strongly reduces the biological variance inherent to expressing a single gene into droplets [65]. Upon thermocycling, droplets are reinjected into a droplet fusion device where they are synchronized and fused one-to-one with larger (16 pl) droplets containing an in vitro transcription mixture (Figure 28.4b). An off-chip incubation step of the collected droplets allows gene transcription to occur. Next, droplets are reinjected into a picoinjection device (Figure 28.4c) in which a controlled volume of ribozyme substrate is injected into each droplet prior to allow the reaction to proceed. Finally, droplets are reinjected into a last device in which their fluorescence is individually analyzed and used to sort them accordingly (Figure 28.4d). We used this pipeline to improve the catalytic properties of the trans-acting form of the X-motif ribozyme [30, 66], a nuclease ribozyme displaying an extremely high activity in its cis-acting form [31, 67] but suffering from a significant product inhibition *in trans* [64], a feature anticipated for ribozymes isolated using a SELEX strategy and displaying strong affinity for their substrate (see above). Performing nine rounds of screening interspersed by rounds of mutagenesis allowed us to identify two X-motif mutants (iXm1 and iXm2) that no longer suffered from product inhibition and were able to catalyze the *in trans* cleavage of substrate RNA under multiple-turnover conditions with a 28-fold improved catalytic constant ($k_{cat} = 0.48\ min^{-1}$) compared with the starting X-motif molecule ($k_{cat} = 0.017\ min^{-1}$).

Extending the Application Scope of μIVC to the Light-Up RNA Aptamers
Since the μIVC pipeline introduced above uses a fluorescent readout, its application can be extended to any RNA whose function leads to a fluorescent signal as this is typically the case with light-up RNA aptamers, a class of RNAs that specifically interact with small fluorogenic dyes (also called fluorogens) to form a fluorescent complex [68]. These modules (fluorogen/aptamer couples) are functional homologues of fluorescent proteins, and their use may lead to significant breakthroughs in the field of imaging provided the complexes they form are bright and stable and offer a contrast high enough to make possible the prolonged visualization of biologically relevant RNAs.

A first promising light-up aptamer was an RNA called Spinach [69], a functional mimic of the green fluorescent protein (GFP) isolated by SELEX for its capacity to bind the 3,5-difluoro-4-hydroxybenzylidene imidazolinone (DFHBI, a fluorogen that mimics the fluorophore naturally found in the GFP). Yet, this aptamer was rapidly found to be suboptimal [70]. Using a simplified version of our μIVC pipeline and thanks to the exquisite control offered over reaction conditions (e.g., warming and complete replacement of potassium ions by sodium), we developed iSpinach [71], an optimized (largely salt-insensitive, brighter, and more thermostable) Spinach-derived aptamer well suited for in vitro applications and recently found to be adapted for imaging RNA in living cells [72]. Even though promising, the Spinach/DFHBI module (as well as its numerous derivatives) still faces two main limitations: the low affinity between the fluorogen and the aptamer and the very low photostability of the complex (half-life of around one second under constant illumination [73]). In light of these limitations, the Mango RNA aptamer that interacts with and activates Thiazole Orange 1-Biotin (TO1-Biotin) fluorescence [74] is much more attractive as it forms with its fluorogen a complex displaying an affinity (nM K_D) several orders of magnitude higher than Spinach does. Yet, the original Mango/TO1-Biotin complex is limited by its modest brightness. Nevertheless, using μIVC, we recently isolated three new Mango-like aptamers (Mango II, III, and IV), all of them forming more fluorescent complexes and one of them (Mango II) displaying an affinity for TO1-Biotin even 10 times higher than the original Mango molecule [73]. These markedly improved properties allowed us to image small noncoding RNA (i.e., 5S ribosomal RNA [rRNA], U6 RNA, and the box C/D scaRNA mgU2-47) both in fixed and living mammalian cells.

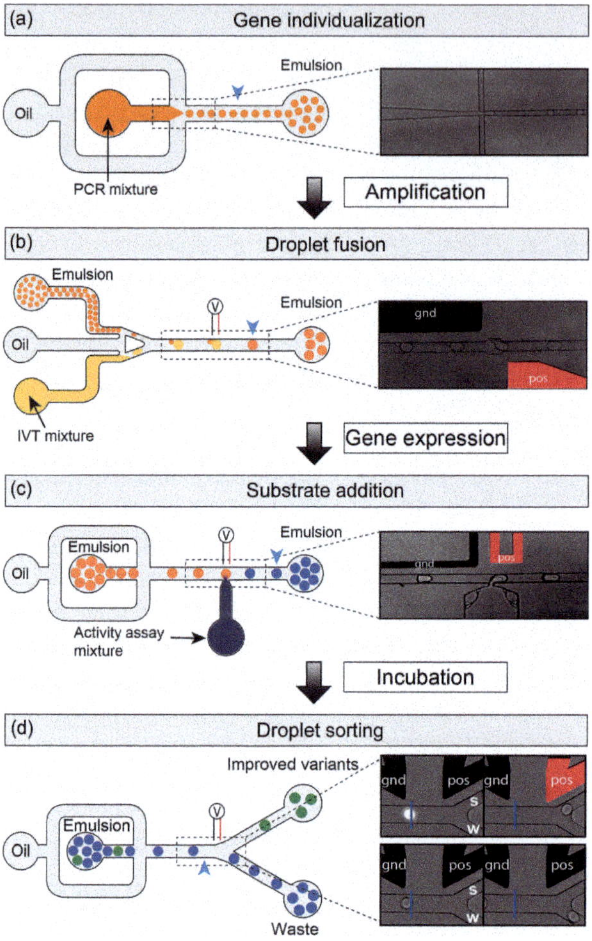

Figure 28.4 Droplet-based microfluidic screening strategy. (a) A gene library is diluted into a PCR mixture prior to dispersing the resulting aqueous phase (dark orange) into small 2.5 pl droplets carried by a fluorinated oil phase (gray) supplemented with a fluorosurfactant. The emulsion is then collected and thermocycled to allow the DNA to be PCR amplified. (b) Upon thermocycling, PCR droplets (dark orange) are reinjected into a fusion device and paired with 16 pl droplets generated on-chip and containing an in vitro transcription mixture (light orange). Then, pairs of droplets are electrocoalesced by passing between a pair of electrodes. (c) Upon incubation, droplets are reinjected into a picoinjection device in which each droplet receive a controlled volume (i.e. 4 pl) of fluorogenic substrate solution (dark blue) prior to being collected and incubated off-chip. (d) Finally, the emulsion is reinjected into an analysis and sorting device where droplets are spaced by an oil stream (gray) and their fluorescence is analyzed. Droplets in which an active ribozyme converted the fluorogenic substrate into fluorescent product (green droplets) are sorted from the bulk by applying a transient electric field. Indeed, applying such a field (schematized by red electrodes) allows deflecting droplets of interest (green droplets) into the "sort" channel ("s" on the micrograph), whereas the others flow toward the "waste" channel ("w" on the micrograph). A micrograph of the key place of each device is shown on the right. Moreover, for each device, the place where the fluorescence measurement is made is indicated by a blue arrow. Blue lines show the location of the laser lines used to scan the droplets during sorting. Finally, for each device, ground (gnd) and energized (pos) electrodes are shown, respectively, in black and red (when the electrode is energized). Further information can be found in [64]. Source: Ryckelynck et al. [64]. © 2015 RNA Society.

28.3 Conclusions

While decades ago RNA was perceived as a labile intermediate in gene expression with a secondary role in cell life, current knowledge point that, in contrast, RNA is actually at the center of regulatory networks. Indeed, RNA can be endowed with a variety of functions such as guiding protein complex through base pairing to other nucleic acids, specific recognition of small molecules via aptamer domains, and even catalyzing chemical reactions via ribozyme domains. Therefore, being able to manipulate, engineer existing ribozymes, or even create new ones represent important milestones for fields such as synthetic biology or even for creating minimal artificial cells relevant to the RNA world. To this end, important methodological developments allowed to propose more and more efficient technologies able to handle large gene libraries to search for optimized RNAs. In vitro selection of self-modifying ribozymes was the first set of technologies used to reach these goals. Yet, even though a variety of ribozymes were identified by these approaches, many of them are still suboptimal and represent perhaps more of a good starting point than a final product. This is well exemplified by the case of the RNA polymerase ribozyme for which the story started with the isolation of an RNA ligase [27] that was progressively engineered until reaching the synthetic capacity of producing an RNA as long as itself [55, 75]. Identifying such an efficient catalyst requires the use of technologies selecting ribozymes for their actual capacity to catalyze a large number of turnover and to be efficiently recycled rather than for their potential to do so. In this view, compartmentalization-based screening technologies (by compartmentalizing reactions either in W/O droplets or in ice) such as IVC represent an important advance in the field by selecting catalysts for every key enzymatic property (i.e. substrate recognition, affinity and specificity, efficient product release, elevated acceleration rate, and turnover). Yet, IVC suffers from difficulties to control the size of the compartments as well as modify them after they have been formed. However, transposing IVC to a microfluidic format now allows to overcome these last limitations by both generating highly monodisperse emulsions and precisely acting (fuse, split, analyze, or sort) on them and on demand [63]. Interestingly, this ultrahigh-throughput μIVC technology is applicable not only to ribozymes but also to any other class of RNA provided a fluorescent readout can be set up. Among future applications of this technology, one can easily imagine the handling and the evolution of entire RNA-based networks involving several RNA catalysts working together in the same droplet and acting therefore as the minimal metabolism relevant to the one that would have existed in the RNA world. Moreover, the very high throughput and the exquisite control over reaction conditions will also make possible the development of RNA-based tools with applications in biosensing, imaging, cell engineering, and reprogramming just as a few examples.

Acknowledgments

This work has been published under the framework of the LabExNetRNA (ANR-10-LABX-0036_NETRNA) and benefits from a funding from the state managed by the French National Research Agency as part of the Investments for the Future program. It also received the financial support of the Agence Nationale de la Recherche (ANR-16-CE11-0010-01, project BrightRiboProbes) and was supported by the Université de Strasbourg and the Centre National de la Recherche Scientifique.

References

1 Kruger, K., Grabowski, P.J., Zaug, A.J. et al. (1982). Self-splicing RNA: autoexcision and autocyclization of the ribosomal RNA intervening sequence of *Tetrahymena*. *Cell* 31 (1): 147–157.
2 Cech, T.R., Zaug, A.J., and Grabowski, P.J. (1981). In vitro splicing of the ribosomal RNA precursor of *Tetrahymena*: involvement of a guanosine nucleotide in the excision of the intervening sequence. *Cell* 27 (3 Pt 2): 487–496.
3 Zaug, A.J. and Cech, T.R. (1992). The intervening sequence RNA of *Tetrahymena* is an enzyme. 1986. *Biotechnology* 24: 57–62.
4 Jacquier, A. and Rosbash, M. (1986). Efficient trans-splicing of a yeast mitochondrial RNA group II intron implicates a strong 5′ exon-intron interaction. *Science* 234 (4780): 1099–1104.
5 Uhlenbeck, O.C. (1987). A small catalytic oligoribonucleotide. *Nature* 328 (6131): 596–600.
6 Haseloff, J. and Gerlach, W.L. (1988). Simple RNA enzymes with new and highly specific endoribonuclease activities. *Nature* 334 (6183): 585–591.
7 Hampel, A. and Tritz, R. (1989). RNA catalytic properties of the minimum (−)sTRSV sequence. *Biochemistry* 28 (12): 4929–4933.
8 Guerrier-Takada, C., Gardiner, K., Marsh, T. et al. (1983). The RNA moiety of ribonuclease P is the catalytic subunit of the enzyme. *Cell* 35 (3 Pt 2): 849–857.
9 Shi, Y. (2017). The spliceosome: a protein-directed metalloribozyme. *J. Mol. Biol.* 429 (17): 2640–2653.
10 Gilbert, W. (1986). Origin of life: the RNA world. *Nature* 319: 618.
11 Lieber, A., He, C.Y., Polyak, S.J. et al. (1996). Elimination of hepatitis C virus RNA in infected human hepatocytes by adenovirus-mediated expression of ribozymes. *J. Virol.* 70 (12): 8782–8791.
12 de Feyter, R. and Li, P. (2000). Technology evaluation: HIV ribozyme gene therapy, Gene Shears Pty Ltd. *Curr. Opin. Mol. Ther.* 2 (3): 332–335.
13 Khan, A.U. (2006). Ribozyme: a clinical tool. *Clin. Chim. Acta* 367 (1–2): 20–27.
14 Morimoto, J., Hayashi, Y., Iwasaki, K., and Suga, H. (2011). Flexizymes: their evolutionary history and the origin of catalytic function. *Acc. Chem. Res.* 44 (12): 1359–1368.
15 Ellington, A.D. and Szostak, J.W. (1990). In vitro selection of RNA molecules that bind specific ligands. *Nature* 346 (6287): 818–822.
16 Tuerk, C. and Gold, L. (1990). Systematic evolution of ligands by exponential enrichment: RNA ligands to bacteriophage T4 DNA polymerase. *Science* 249 (4968): 505–510.

17 Gold, L., Polisky, B., Uhlenbeck, O., and Yarus, M. (1995). Diversity of oligonucleotide functions. *Annu. Rev. Biochem.* 64: 763–797.

18 Wilson, D.S. and Szostak, J.W. (1999). In vitro selection of functional nucleic acids. *Annu. Rev. Biochem.* 68: 611–647.

19 Kaur, H. (2018). Recent developments in cell-SELEX technology for aptamer selection. *Biochim. Biophys. Acta, Gen. Subj.* 1862 (10): 2323–2329.

20 Darmostuk, M., Rimpelova, S., Gbelcova, H., and Ruml, T. (2015). Current approaches in SELEX: an update to aptamer selection technology. *Biotechnol. Adv.* 33 (6 Pt 2): 1141–1161.

21 Zhang, B. and Cech, T.R. (1998). Peptidyl-transferase ribozymes: trans reactions, structural characterization and ribosomal RNA-like features. *Chem. Biol.* 5 (10): 539–553.

22 Unrau, P.J. and Bartel, D.P. (1998). RNA-catalysed nucleotide synthesis. *Nature* 395 (6699): 260–263.

23 Seelig, B. and Jaschke, A. (1999). A small catalytic RNA motif with Diels–Alderase activity. *Chem. Biol.* 6 (3): 167–176.

24 Bartel, D.P. and Unrau, P.J. (1999). Constructing an RNA world. *Trends Cell Biol.* 9 (12): M9–M13.

25 Jaschke, A. (2001). Artificial ribozymes and deoxyribozymes. *Curr. Opin. Struct. Biol.* 11 (3): 321–326.

26 Muller, U.F. (2006). Re-creating an RNA world. *Cell. Mol. Life Sci.* 63 (11): 1278–1293.

27 Bartel, D.P. and Szostak, J.W. (1993). Isolation of new ribozymes from a large pool of random sequences. *Science* 261 (5127): 1411–1418.

28 Ekland, E.H. and Bartel, D.P. (1996). RNA-catalysed RNA polymerization using nucleoside triphosphates. *Nature* 382 (6589): 373–376.

29 Johnston, W.K., Unrau, P.J., Lawrence, M.S. et al. (2001). RNA-catalyzed RNA polymerization: accurate and general RNA-templated primer extension. *Science* 292 (5520): 1319–1325.

30 Tang, J. and Breaker, R.R. (2000). Structural diversity of self-cleaving ribozymes. *Proc. Natl. Acad. Sci. U.S.A.* 97 (11): 5784–5789.

31 Breaker, R.R., Emilsson, G.M., Lazarev, D. et al. (2003). A common speed limit for RNA-cleaving ribozymes and deoxyribozymes. *RNA* 9 (8): 949–957.

32 Tawfik, D.S. and Griffiths, A.D. (1998). Man-made cell-like compartments for molecular evolution. *Nat. Biotechnol.* 16 (7): 652–656.

33 Miller, O.J., Bernath, K., Agresti, J.J. et al. (2006). Directed evolution by in vitro compartmentalization. *Nat. Methods* 3 (7): 561–570.

34 Ghadessy, F.J., Ong, J.L., and Holliger, P. (2001). Directed evolution of polymerase function by compartmentalized self-replication. *Proc. Natl. Acad. Sci. U.S.A.* 98 (8): 4552–4557.

35 Lee, Y.F., Tawfik, D.S., and Griffiths, A.D. (2002). Investigating the target recognition of DNA cytosine-5 methyltransferase HhaI by library selection using in vitro compartmentalisation. *Nucleic Acids Res.* 30 (22): 4937–4944.

36 Cohen, H.M., Tawfik, D.S., and Griffiths, A.D. (2004). Altering the sequence specificity of *Hae*III methyltransferase by directed evolution using in vitro compartmentalization. *Protein Eng. Des. Sel.* 17 (1): 3–11.

37 Agresti, J.J., Kelly, B.T., Jaschke, A., and Griffiths, A.D. (2005). Selection of ribozymes that catalyse multiple-turnover Diels–Alder cycloadditions by using in vitro compartmentalization. *Proc. Natl. Acad. Sci. U.S.A.* 102 (45): 16170–16175.

38 Zaher, H.S. and Unrau, P.J. (2007). Selection of an improved RNA polymerase ribozyme with superior extension and fidelity. *RNA* 13 (7): 1017–1026.

39 Doi, N., Kumadaki, S., Oishi, Y. et al. (2004). In vitro selection of restriction endonucleases by in vitro compartmentalization. *Nucleic Acids Res.* 32 (12): e95.

40 Ghadessy, F.J., Ramsay, N., Boudsocq, F. et al. (2004). Generic expansion of the substrate spectrum of a DNA polymerase by directed evolution. *Nat. Biotechnol.* 22 (6): 755–759.

41 Ong, J.L., Loakes, D., Jaroslawski, S. et al. (2006). Directed evolution of DNA polymerase, RNA polymerase and reverse transcriptase activity in a single polypeptide. *J. Mol. Biol.* 361 (3): 537–550.

42 Wochner, A., Attwater, J., Coulson, A., and Holliger, P. (2011). Ribozyme-catalyzed transcription of an active ribozyme. *Science* 332 (6026): 209–212.

43 Mastrobattista, E., Taly, V., Chanudet, E. et al. (2005). High-throughput screening of enzyme libraries: in vitro evolution of a beta-galactosidase by fluorescence-activated sorting of double emulsions. *Chem. Biol.* 12 (12): 1291–1300.

44 Tu, R., Martinez, R., Prodanovic, R. et al. (2011). A flow cytometry-based screening system for directed evolution of proteases. *J. Biomol. Screening* 16 (3): 285–294.

45 Bernath, K., Hai, M., Mastrobattista, E. et al. (2004). In vitro compartmentalization by double emulsions: sorting and gene enrichment by fluorescence activated cell sorting. *Anal. Biochem.* 325 (1): 151–157.

46 Aharoni, A., Amitai, G., Bernath, K. et al. (2005). High-throughput screening of enzyme libraries: thiolactonases evolved by fluorescence-activated sorting of single cells in emulsion compartments. *Chem. Biol.* 12 (12): 1281–1289.

47 Sepp, A., Tawfik, D.S., and Griffiths, A.D. (2002). Microbead display by in vitro compartmentalisation: selection for binding using flow cytometry. *FEBS Lett.* 532 (3): 455–458.

48 Griffiths, A.D. and Tawfik, D.S. (2003). Directed evolution of an extremely fast phosphotriesterase by in vitro compartmentalization. *EMBO J.* 22 (1): 24–35.

49 Levy, M., Griswold, K.E., and Ellington, A.D. (2005). Direct selection of trans-acting ligase ribozymes by in vitro compartmentalization. *RNA* 11 (10): 1555–1562.

50 Bergman, N.H., Johnston, W.K., and Bartel, D.P. (2000). Kinetic framework for ligation by an efficient RNA ligase ribozyme. *Biochemistry* 39 (11): 3115–3123.

51 Wang, J., Gong, Q., Maheshwari, N. et al. (2014). Particle display: a quantitative screening method for generating high-affinity aptamers. *Angew. Chem. Int. Ed.* 53 (19): 4796–4801.

52 Qu, H., Csordas, A.T., Wang, J. et al. (2016). Rapid and label-free strategy to isolate aptamers for metal ions. *ACS Nano* 10 (8): 7558–7565.

53 Gotrik, M., Sekhon, G., Saurabh, S. et al. (2018). Direct selection of fluorescence-enhancing RNA aptamers. *J. Am. Chem. Soc.* 140 (10): 3583–3591.

54 Attwater, J., Wochner, A., Pinheiro, V.B. et al. (2010). Ice as a protocellular medium for RNA replication. *Nat. Commun.* 1: 76.

55 Attwater, J., Wochner, A., and Holliger, P. (2013). In-ice evolution of RNA polymerase ribozyme activity. *Nat. Chem.* 5 (12): 1011–1018.

56 Bernath, K., Magdassi, S., and Tawfik, D.S. (2005). Directed evolution of protein inhibitors of DNA-nucleases by in vitro compartmentalization (IVC) and nano-droplet delivery. *J. Mol. Biol.* 345 (5): 1015–1026.

57 Anna, S.L., Bontoux, N., and Stone, H.A. (2003). Formation of dispersions using "flow focusing" in microchannels. *Appl. Phys. Lett.* 82 (3): 364–366.

58 Zhu, P. and Wang, L. (2016). Passive and active droplet generation with microfluidics: a review. *Lab Chip* 17 (1): 34–75.

59 Chabert, M., Dorfman, K.D., and Viovy, J.L. (2005). Droplet fusion by alternating current (AC) field electrocoalescence in microchannels. *Electrophoresis* 26 (19): 3706–3715.

60 Abate, A.R., Hung, T., Mary, P. et al. (2010). High-throughput injection with microfluidics using picoinjectors. *Proc. Natl. Acad. Sci. U.S.A.* 107 (45): 19163–19166.

61 Link, D.R., Anna, S.L., Weitz, D.A., and Stone, H.A. (2004). Geometrically mediated breakup of drops in microfluidic devices. *Phys. Rev. Lett.* 92 (5): 054503-1 – 054503-4.

62 Baret, J.C., Miller, O.J., Taly, V. et al. (2009). Fluorescence-activated droplet sorting (FADS): efficient microfluidic cell sorting based on enzymatic activity. *Lab Chip* 9 (13): 1850–1858.

63 Autour, A. and Ryckelynck, M. (2017). Ultrahigh-throughput improvement and discovery of enzymes using droplet-based microfluidic screening. *Micromachines* 8 (4): 128.

64 Ryckelynck, M., Baudrey, S., Rick, C. et al. (2015). Using droplet-based microfluidics to improve the catalytic properties of RNA under multiple-turnover conditions. *RNA* 21 (3): 458–469.

65 Woronoff, G., Ryckelynck, M., Wessel, J. et al. (2015). Activity-fed translation (AFT) assay: a new high-throughput screening strategy for enzymes in droplets. *ChemBioChem* 16 (9): 1343–1349.

66 Lazarev, D., Puskarz, I., and Breaker, R.R. (2003). Substrate specificity and reaction kinetics of an X-motif ribozyme. *RNA* 9 (6): 688–697.

67 Emilsson, G.M., Nakamura, S., Roth, A., and Breaker, R.R. (2003). Ribozyme speed limits. *RNA* 9 (8): 907–918.

68 Bouhedda, F., Autour, A., and Ryckelynck, M. (2018). Light-up RNA aptamers and their cognate fluorogens: from their development to their applications. *Int. J. Mol. Sci.* 19 (1): 19010044-1 - 19010044-21.

69 Paige, J.S., Wu, K.Y., and Jaffrey, S.R. (2011). RNA mimics of green fluorescent protein. *Science* 333 (6042): 642–646.

70 Strack, R.L., Disney, M.D., and Jaffrey, S.R. (2013). A superfolding Spinach2 reveals the dynamic nature of trinucleotide repeat-containing RNA. *Nat. Methods* 10 (12): 1219–1224.

71 Autour, A., Westhof, E., and Ryckelynck, M. (2016). iSpinach: a fluorogenic RNA aptamer optimized for in vitro applications. *Nucleic Acids Res.* 44 (6): 2491–2500.

72 Wang, Z., Luo, Y., Xie, X. et al. (2018). In situ spatial complementation of aptamer-mediated recognition enables live-cell imaging of native RNA transcripts in real time. *Angew. Chem. Int. Ed.* 57 (4): 972–976.

73 Autour, A., Jeng, S.C., Cawte, A.D. et al. (2018). Fluorogenic RNA Mango aptamers for imaging small non-coding RNAs in mammalian cells. *Nat. Commun.* 9 (1): 656.

74 Dolgosheina, E.V., Jeng, S.C., Panchapakesan, S.S. et al. (2014). RNA mango aptamer-fluorophore: a bright, high-affinity complex for RNA labeling and tracking. *ACS Chem. Biol* 9 (10): 2412 – 2420.

75 Attwater, J., Raguram, A., Morgunov, A.S. et al. (2018). Ribozyme-catalysed RNA synthesis using triplet building blocks. *eLife*: 7: e35255-1 – e35255-25.

Part VI

Tools and Methods to Study Ribozymes

29

Elucidation of Ribozyme Mechanisms at the Example of the Pistol Ribozyme

Christoph Falschlunger, Josef Leiter, and Ronald Micura

University of Innsbruck, Department of Organic Chemistry, Innrain 80/82, 6020 Innsbruck, Austria

29.1 Introduction

Comparative genomic analysis revealed the identity of four novel classes of self-cleaving ribozymes termed twister, twister sister, pistol, and hatchet, thus expanding the sequence space of catalytically active RNAs [1, 2]. One of them, the pistol ribozyme (Figure 29.1a), is of particular interest because its strand scission rate is among the fastest of known nucleolytic RNAs [3]. In this sense, applications capitalizing on high speed cleavage combined with small molecule binding aptamers have been presented recently [4, 5]. They concern the engineering of chemically regulated self-cleaving ribozymes (commonly referred to as aptazymes) that are an emerging promising class of genetic devices that allow dynamic control of gene expression [6, 7].

29.2 Structural Aspects – Overall Fold and Cleavage Site Architecture

To date, two three-dimensional structures of the pistol ribozyme have been solved by X-ray crystallography and deposited on the protein data bank exhibiting a resolution of 2.70 Å for *env25* pistol [8] and 2.97 Å for *env27* pistol [9], respectively (Figure 29.1b,c) (PDB code 5K7C and 5KTJ). Both structures represent a precatalytic conformation that has been trapped based on mutants carrying a hydrogen atom instead of the native 2′OH nucleophile at the cleavage site. Moreover, the two structures are in very good agreement with respect to the overall topology and the cleavage site alignments [10, 11].

The pistol RNA fold is characterized by the formation of a six-base-pair pseudoknot involving complementary loop segments between the hairpin and the internal loop of the consensus secondary structure motif (Figure 29.1). The pseudoknot double helix is coaxially stacked between stems P1 and a small segment involving

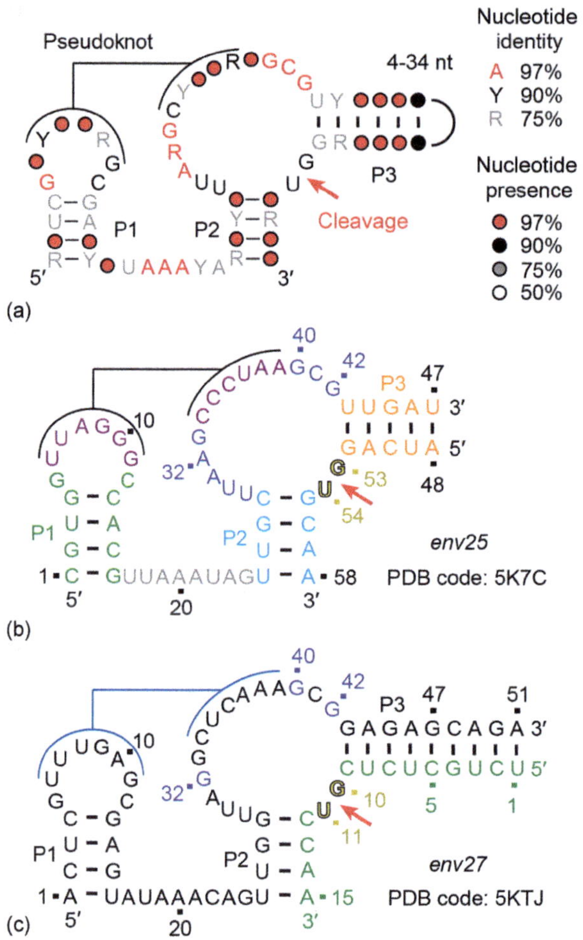

Figure 29.1 Pistol ribozymes. (a) Consensus RNA sequence and secondary structure model for pistol ribozymes based on the alignment of 500 unique representatives found in DNA sequence databases composed of complete bacterial genomes, microbial metagenomic DNA, and bacteriophage DNA. Source: Adapted from Refs. [1, 3]. The arrowhead designates the ribozyme-mediated cleavage site. (b) Color-coded representation of the secondary structure of the *env25* pistol ribozyme highlighting stems and pseudoknots. (c) Color-coded representation of the secondary structure of the *env27* pistol ribozyme highlighting the bimolecular assembly of the ribozyme (black) and substrate (green with cleavage site in yellow) strands.

noncanonical pairing interactions, followed by stem P3 (Figure 29.2a). Furthermore, an A-minor motif interaction between the single-stranded sequence (linking P1 to P2) and P1 is characteristic and dictates the orientation of stem P2 (Figure 29.2b). Thus, P2 becomes positioned next to the pseudoknot stem and thereby forces the substrate strand into an extended conformation with a splayed apart G53–U54 cleavage site whose scissile phosphate is directed toward the core of the ribozyme fold (numbering throughout the text follows the *env25* pistol ribozyme in [8], Figure 29.1b). Thereby, the nucleobases of G53 and U54 are tightly held in place by intercalation of their nucleobases and H-bond interactions, with G53 stacked on top of P3 and U54 stacked on top of P2 (Figure 29.2). The three nucleosides that come closest to the scissile phosphate between G53 and U54 are highly conserved; they include a purine nucleoside (A32 in *env25*; G32 in *env27*) and a guanosine (G40) and, more distant, another guanosine (G42) that interconnects the two former via H-bond networks (Figure 29.2a). The thus formed cleft A32(G32)–G42–G40 appears prefolded and ideal to receive and accommodate the cleavage site of the substrate

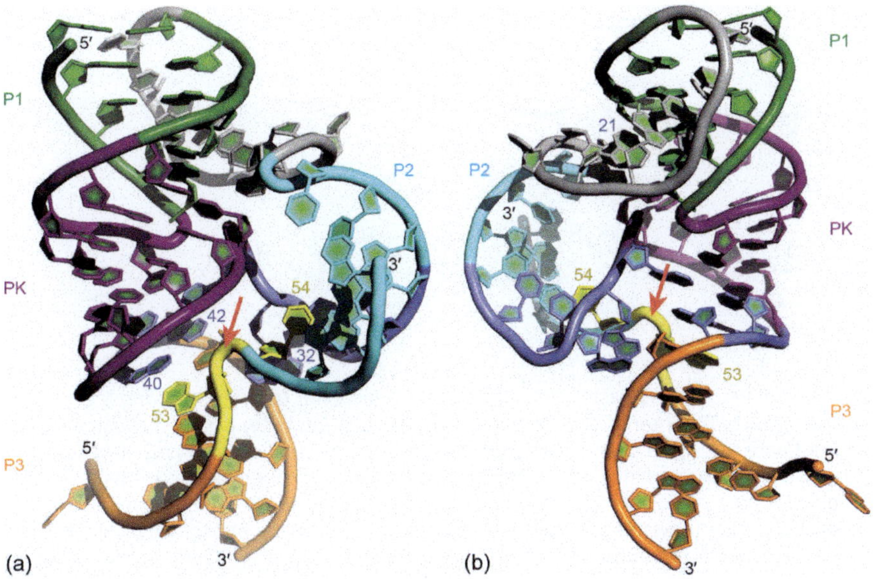

Figure 29.2 Three-dimensional fold of the *env25* pistol ribozyme [8] in cartoon presentation and color coded as in Figure 29.1b, with the G53–U54 cleavage site highlighted by a red arrow. (a) Front view. (b) Back view. PDB code 5K7C. Source: (a) Reproduced with permission from Ren et al. [8]. Copyright 2016, Springer Nature.

strand. We note that not the nucleoside identity of the cleavage site, rather its two-nucleoside length between P3 and P2, is strictly conserved (Figure 29.1a) [1, 3].

29.3 Cleavage Mechanism and Catalysis

Self-cleaving ribozymes can apply up to four major strategies to enhance strand scission rates with up to a 10 million-fold over spontaneous RNA degradation (Figure 29.3) [3, 13–15]. These strategies lower the energy barrier of the transition state by arranging an in-line alignment of the 2′O nucleophile with the to-be-cleaved P—O5′ bond (α-factor), by stabilizing the enhanced negative charge at the non-bridging oxygens in the transition state (β-factor), by activating the 2′O nucleophile by proton abstraction (γ-factor), and/or by donating a proton to the emerging negative charge at 5′O leaving group (δ-factor).

29.3.1 Role of the Conserved Guanosine-40

The structure of the *env25* pistol ribozyme reveals that the two highly conserved nucleotides G40 and A32 are located very close to the scissile phosphate (Figure 29.4a) [8]. Hence, they are prime candidates for their involvement in the chemical mechanism. In particular, G40 is in excellent position to serve as base in general acid–base catalysis: first, its N1 is in 2.8 Å distance to the scissile

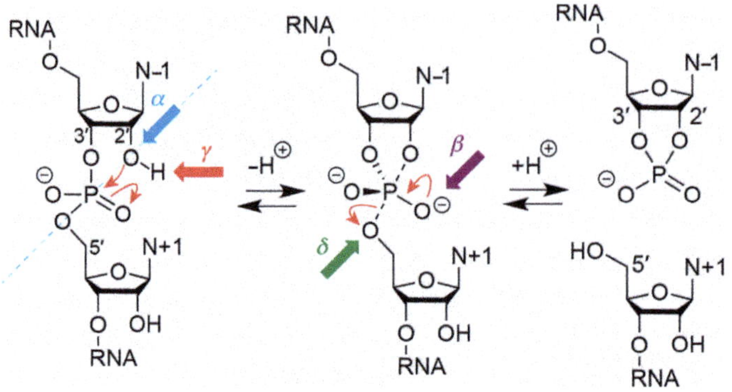

Figure 29.3 RNA phosphodiester cleavage by phosphodiester transfer involving the 2′-hydroxyl group. The internucleotide linkage ("scissile" phosphate) passes through a pentacoordinate transition state that results in two cleavage products carrying either a 2′,3′-cyclic phosphate terminus or a 5′-hydroxyl terminus. The four catalytic strategies that can impact on the reaction are as follows: *α*, in-line nucleophilic attack, S_N2-type (blue); *β*, neutralization of the (developing) negative charge on non-bridging phosphate oxygens (purple); *γ*, deprotonation of the 2′-hydroxyl group (red); and *δ*, neutralization of negative charge on the 5′-oxygen atom by protonation (green). Source: Reproduced with permission from Neuner et al. [16]. Copyright 2017, John Wiley & Sons.

phosphate attacking 2′O atom (modeled onto dG53), and second, the estimated 2′O to P–O5′ angle τ at the cleavage site is 167°, which is close to its ideal angle (180°) for in-line alignment. Interestingly, a hydrated magnesium ion is observed to be outer sphere coordinated to the O6 of G40 with an O6-to-Mg^{2+} distance of 3.9 Å. The positively charged metal ion may contribute to shift the pK_a value of the G40 N1-H toward neutrality so as to trigger ionization or tautomerization. The second available structure (*env27*) is also supportive for such a scenario: at a similar position to the magnesium ion in *env25*, cobalt hexamine with a quadratic bipyramidal coordination sphere of the ammonia molecules provides N to O6 distances of about 3.4 Å [9].

Furthermore, we point out that the *env27* structure implies a direct involvement of G40 in stabilizing the transition state (β-factor) because of the observed bifurcated hydrogen bonding pattern apparent between G40 N1-H and N2-H with one of the non-bridging oxygens of the scissile phosphate. Note that the orientation of the phosphate in the catalytic cleft is slightly distinct between the two reported structures [8, 9]. For *env27*, the estimated 2′O to P–O5′ angle τ at the cleavage site is only 126° (compared with 167° for *env25*), and this conformation accounts for the Watson–Crick face of G40 to approach the non-bridging oxygen of the scissile phosphate more closely (Figure 29.5).

29.3.2 Role of the Conserved Purine Nucleoside-32

In the *env25* pistol ribozyme structure, A32 is located near the 5′O leaving group of U54. The closest functional groups (N3 and 2′OH of A32) are positioned in 5 Å distance, and therefore, their direct participation in catalysis (i.e. protonation

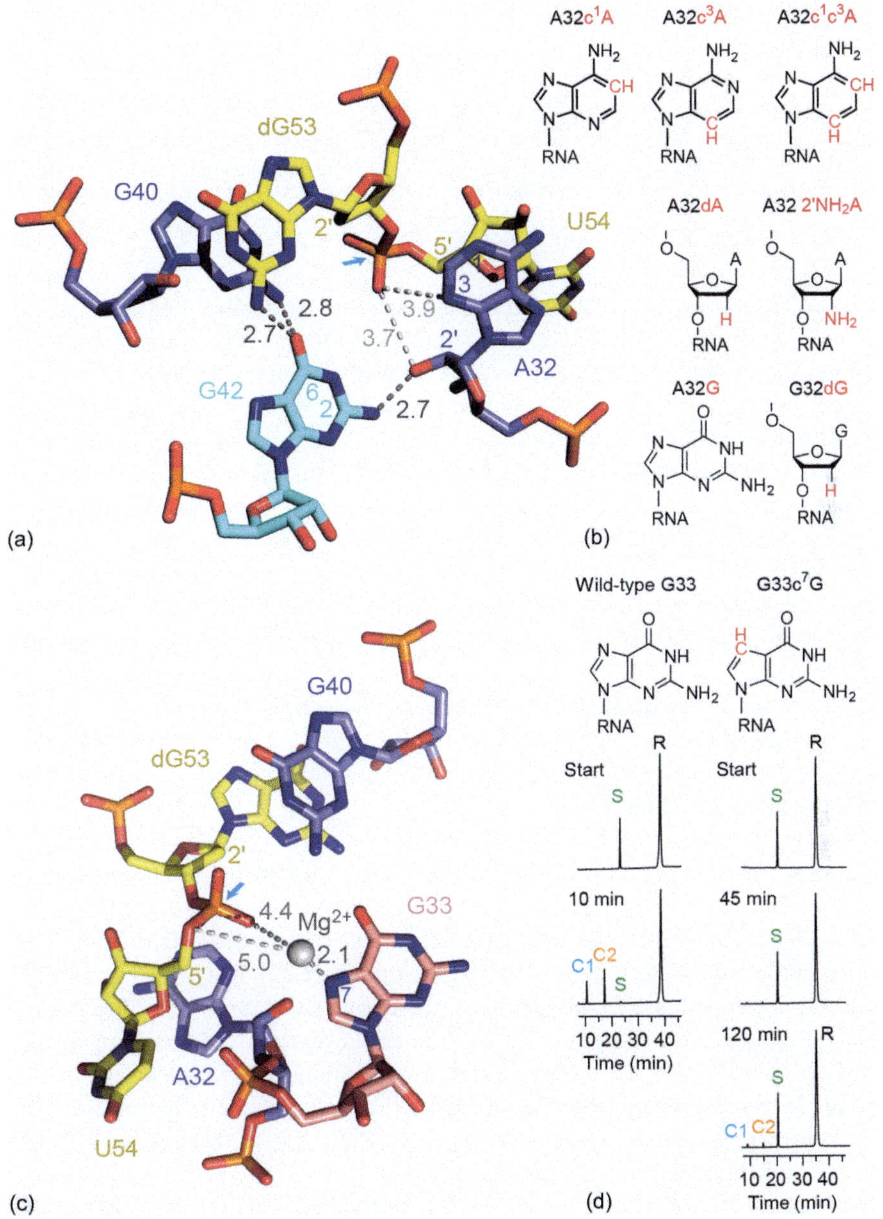

Figure 29.4 Active site of the *env25* pistol ribozyme (PDB code 5K7C) [8]. (a) Interactions of A32 and G42 with the cleavage site (yellow); numbers in gray indicate distances in Å; the cyan arrow indicates the scissile phosphate. (b) Chemical structures of modified nucleosides that were analyzed for their impact on phosphodiester cleavage activity. (c) Interactions of G33 and its N7 inner-sphere coordinated, hydrated Mg^{2+} with the cleavage site (yellow); numbers in gray indicate distances in Å; the cyan arrow indicates the scissile phosphate. (d) Cleavage of the c^7G33-modified pistol ribozyme is significantly impaired (c. 1000-fold); analysis of the cleavage reaction: HPLC traces at different time points are depicted, cleavage product C1 is 5′-UGCAA, and cleavage product C2 is 5′-AUCAGG-2′,3′-cyclic phosphate (see also RNA sequences in Figure 29.1b). Source: Reproduced with permission from Neuner et al. [16]. Copyright 2017, John Wiley & Sons.

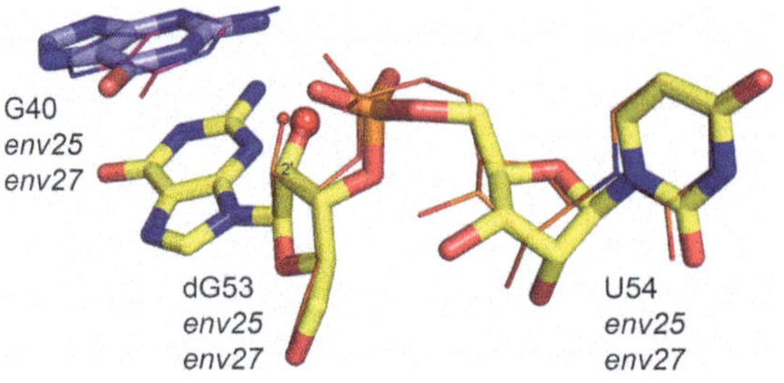

Figure 29.5 Overlay of cleavage site G53–U54 (yellow) and G40 of the X-ray structures PDB code 5KTJ (*env27* thin lines) [9] and PDB code 5K7C (*env25* sticks) [8]. Note that the 2′-OH group has been modeled onto dG53; the C2′ is labeled with 2′.

of the 5′O) would require additional conformational rearrangements along the reaction coordinate to attain distances that allow for direct hydrogen transfer. To draw conclusions with respect to the chemical mechanism of phosphodiester cleavage based on "static" conformational snapshots of X-ray structures solely is critical, and therefore, functional assays are inevitable to obtain a comprehensive and reliable picture on how RNA catalyzes the reaction. To this end, atom-specific mutagenesis is a powerful tool to evaluate the impact of single functional groups on ribozyme activity [12, 17]. Pistol ribozymes equipped with the modification 1-deazaadenosine (c^1A), 3-deazaadenosine (c^3A), or 1,3-dideazaadenosine ($c^{13}A$) in position 32 (Figure 29.4b) all exhibited cleavage rates that were as fast as the wild type or even slightly expedited [16]. Thus, participation of the (protonated) nucleobase-32 as acid in general acid–base catalysis is unlikely. This conclusion was further supported by nuclear magnetic resonance (NMR) spectroscopic studies of a ^{13}C2-A32 labeled pistol ribozyme. A pK_a value of 4.7 was determined for A32, slightly shifted compared with 3.7 in the context of short single-stranded fragments under the same conditions [8].

Although the 2′OH of A32 is rather close to the non-bridging oxygen of the scissile phosphate (3.9 Å) and to the 5′O leaving group (5 Å), the tightest interaction of this OH group is its entanglement in a H-bond (2.8 Å) with the N2 group of G42 (Figure 29.4a) [8]. This H-bond may lower the pK_a of A32 2′OH that potentially supports proton donation to the 5′O leaving group. Interestingly, a pistol mutant carrying A32 2′NH_2 (Figure 29.4b) instead of the native 2′OH group exhibited strand scission rates that were compromised at pH values lower than 7.0. According to previous reports in the literature, ribose 2′NH_2 modifications in RNA are protonated at neutral pH to a significant extent (pK_a 6.2) [18]. Clearly, protonation of A32 2′NH_2 would interfere with H-bond formation to the H-donor of G42 N2 and weaken the formation of the structurally important scaffold of A32–G42–G40 (Figure 29.4a). Another possible mechanistic route involves the stabilization of the transition state via H-bonding of A32 2′OH to the non-bridging oxygen of the scissile phosphate, thereby contributing to rate enhancement.

It is very likely that the A32 2′OH contributes in combined manner via the multiple pathways addressed above. Undoubtedly, its role is vital and its importance clearly reflected in the observation that 2′-deoxy A32 pistol mutants cleaved much slower. We also point out that in the *env27* pistol ribozyme, guanosine instead of adenosine resides in position 32; the corresponding 2′-deoxy G32 mutant experienced the same decrease in cleavage activity as observed for the corresponding

A32/dA32 juxtaposition in the *env25* ribozyme (Figure 29.4b) [16] and is therefore supportive to the proposed roles of the A32 2′OH.

29.3.3 Role of the Conserved Guanosine-33

Both structures (*env25* and *env27*) display a hydrated Mg^{2+} ion near the 5′O leaving group (c. 5.5 Å) that is held in place through inner-sphere coordination to N7 of G33 (2.1 and 2.3 Å) (Figure 29.4c). The significance of this hydrated Mg^{2+} ion was revealed by the corresponding 7-deazaguanosine (c^7G)–33 pistol mutant (Figure 29.4d). Compared with all above discussed mutants, the c^7G33 modification – which implies the deletion of the metal binding site – caused the largest drop (about 1000-fold) in cleavage rate, namely, to an observed rate k_{obs} of about 4×10^{-3} min^{-1} [16]. The hydrated Mg^{2+} ion is likely involved in proton transfer from an acidified Mg^{2+}-bound water molecule to the 5′O leaving group. Furthermore, it can also not be excluded that the hydrated Mg^{2+} assists in the stabilization of the transition state through charge compensation and/or hydrogen bonding.

Preliminary studies have been reported on the role of metal ions for the reactivity of pistol ribozymes utilizing phosphorothioate replacements instead of the scissile phosphate [3]. The phosphorothioates were applied as diastereomeric mixture, with both R_P and S_P chirality at the phosphor center. The observation was that the sulfur-containing RNAs cleaved only a fraction of 50% and cleavage could not be restored in the presence of thiophilic manganese ions. This finding was interpreted that direct (inner-sphere) interaction of a metal ion with a non-bridging oxygen of the scissile phosphate (which would contribute to stabilize the emerging negative charge in the transition state) is unlikely [3]. This finding, however, could also indicate an intense interaction of the G40 Watson–Crick face with the scissile phosphate in the transition state, as a non-bridging sulfur at the scissile position would significantly decrease hydrogen bonding strength. To improve our understanding of the observed thio effects, structure solutions of a transition state analog (e.g. in form of vanadate complexes as had been solved for hairpin and hammerhead ribozymes) would help significantly.

29.4 Mechanistic Proposal for the Pistol Ribozyme

Collectively, our current mechanistic understanding of the pistol ribozyme is based on a series of structural and functional studies [1, 3, 8, 9, 11, 16]. They shed light on the structural role of the 2′-OH of the purine nucleoside-32 and the catalytic role of the hydrated Mg^{2+} coordinated to G33 of the pistol ribozyme. The suggested mechanism for pistol ribozyme phosphodiester cleavage involves a combination of catalytic strategies (Figure 29.6). In the precatalytic state, the attacking 2′OH of G53 is orientated nearly ideal for in-line attack at the scissile phosphate (α-factor). Furthermore, G40 is positioned at an ideal distance to activate the attacking 2′OH of G53 (γ-factor). With respect to protonation of the 5′O leaving group (δ-factor), we

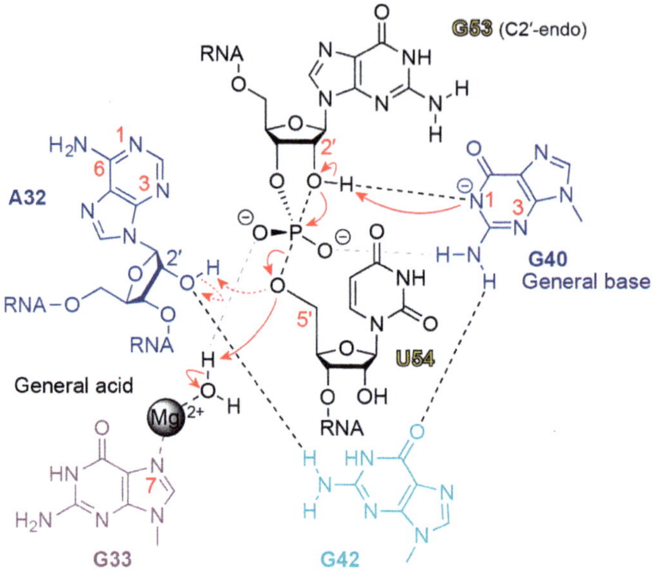

Figure 29.6 Current understanding of the chemical mechanism for phosphodiester cleavage in pistol ribozymes [16]. G40 assists in proton abstraction of the G53 2′OH nucleophile that attacks (red arrows) the phosphor atom of the to-be-cleaved phosphodiester, which is drawn in a pentavalent transition state. Release of the 5′O leaving group is supported by proton transfer from a hydrated Mg^{2+} ion, itself inner sphere coordinated to N7 of G33. The primary role of A32 is to structure the cleft of A32–G42–G40 by hydrogen bonding (and stacking to A30; not shown) for the accommodation of the cleavage site. The dashed red arrows indicate a less efficient, alternative path of proton transfer for leaving group stabilization. Source: Reproduced with permission from Neuner et al. [16]. Copyright 2017, John Wiley & Sons.

exclude H^+ donation from a protonated A32. The A32 2′OH, which is structurally tightly engaged to stabilize the active site architecture, may fulfill this role; however, it is more likely that H^+ donation to the 5′O leaving group originates from one of the water molecules of the hydrated Mg^{2+} cation that is inner sphere coordinated to the N7 of G33. We note that this Mg^{2+} ion allows water-mediated hydrogen bond formation to the non-bridging oxygen of the scissile phosphate or may become repositioned transiently for direct inner-sphere coordination. Such an arrangement could contribute to the stabilization of the transition state (β-factor). One of the available crystal structures (*env27*) also implies a direct involvement of G40 in stabilizing the transition state (β-factor) because of the observed bifurcated hydrogen bonding pattern between the G40 nucleobase with one of the non-bridging oxygens of the scissile phosphate [9].

We finally note that a first interpretation of the three-dimensional pistol structures implied A32 as an ideal candidate for general acid–base catalysis, likely involved in leaving group stabilization (δ-factor). For several other ribozyme classes, such as hairpin or the twister ribozymes, active site adenines had been indeed identified to act as general acid [19–24]. Only the thorough combination of structural studies with atomic mutagenesis and NMR spectroscopic experiments (that both required multistep syntheses of deaza-nucleobase and $^{13}C/^{15}N$ labeled nucleoside building blocks for RNA solid-phase synthesis [16, 25–29]) lead to a clearer picture of the precise roles of active site nucleosides in the chemical mechanism.

29.5 Outlook

Further important insights into the mechanism of pistol ribozymes are awaited from crystal structures of transition state analogs (e.g. vanadate mimics [19, 30–35]) and post-cleavage states. Likewise, exploring the conformational dynamics of nucleobases involved in catalysis would contribute to a deeper mechanistic understanding. The expected micro- to millisecond timescale of nucleoside dynamics in the active site and of nucleosides at tertiary interactions should be accessible by using selective ^{13}C labeling patterns and relaxation dispersion NMR experiments [27, 36–39]. In this way, populations of nucleoside conformations (concerning e.g. ribose pucker, glycosidic bonds, etc.) that are crucial for approaching the transition state of phosphodiester cleavage could be revealed. Additionally, using selective ^{15}N labeling patterns, nucleobase tautomers and/or ionic forms potentially involved in catalysis could be made visible by modern NMR spectroscopic techniques. This has been exemplified recently in a related context, namely, for wobble base pair formation of the frequently encountered tRNA modification uridine-5-oxyacetic acid (cmo^{5}U) [36].

References

1 Weinberg, Z., Kim, P.B., Chen, T.H. et al. (2015). New classes of self-cleaving ribozymes revealed by comparative genomics analysis. *Nat. Chem. Biol.* 11 (8): 606–610.

2 Roth, A., Weinberg, Z., Chen, A.G.Y. et al. (2014). A widespread self-cleaving ribozyme class is revealed by bioinformatics. *Nat. Chem. Biol.* 10 (1): 56–60.

3 Harris, K.A., Lünse, C.E., Li, S. et al. (2015). Biochemical analysis of pistol self-cleaving ribozymes. *RNA* 21 (11): 1852–1858.

4 Kobori, S., Takahashi, K., and Yokobayashi, Y. (2017). Deep sequencing analysis of aptazyme variants based on a pistol ribozyme. *ACS Synth. Biol.* 6 (7): 1283–1288.

5 Nomura, Y., Chien, H.-C., and Yokobayashi, Y. (2017). Direct screening for ribozyme activity in mammalian cells. *Chem. Commun. (Cambridge, England)* 53 (93): 12540–12543.

6 Machtel, P., Bąkowska-Żywicka, K., and Żywicki, M. (2016). Emerging applications of riboswitches – from antibacterial targets to molecular tools. *J. Appl. Genet.* 57 (4): 531–541.

Crystal structures of pistol ribozyme transition state analog and product have been solved recently [40]. These structures further contribute to an in-depth understanding of the catalytic mechanism.

7 Berens, C., Groher, F., and Suess, B. (2015). RNA aptamers as genetic control devices: the potential of riboswitches as synthetic elements for regulating gene expression. *Biotechnol. J.* 10 (2): 246–257.

8 Ren, A., Vušurović, N., Gebetsberger, J. et al. (2016). Pistol ribozyme adopts a pseudoknot fold facilitating site-specific in-line cleavage. *Nat. Chem. Biol.* 12 (9): 702–708.

9 Nguyen, L.A., Wang, J., and Steitz, T.A. (2017). Crystal structure of Pistol, a class of self-cleaving ribozyme. *Proc. Natl. Acad. Sci. U.S.A.* 114 (5): 1021–1026.

10 Ren, A., Micura, R., and Patel, D.J. (2017). Structure-based mechanistic insights into catalysis by small self-cleaving ribozymes. *Curr. Opin. Chem. Biol.* 41: 71–83.

11 Gasser, C., Gebetsberger, J., Gebetsberger, M., and Micura, R. (2018). SHAPE probing pictures Mg^{2+}-dependent folding of small self-cleaving ribozymes. *Nucleic Acids Res.* 46 (14): 6983–6995.

12 Lang, K., Erlacher, M., Wilson, D.N. et al. (2008). The role of 23S ribosomal RNA residue A2451 in peptide bond synthesis revealed by atomic mutagenesis. *Chem. Biol.* 15 (5): 485–492.

13 Breaker, R.R., Emilsson, G.M., Lazarev, D. et al. (2003). A common speed limit for RNA-cleaving ribozymes and deoxyribozymes. *RNA* 9 (8): 949–957.

14 Emilsson, G.M., Nakamura, S., Roth, A., and Breaker, R.R. (2003). Ribozyme speed limits. *RNA* 9 (8): 907–918.

15 Kellerman, D.L., York, D.M., Piccirilli, J.A., and Harris, M.E. (2014). Altered (transition) states: mechanisms of solution and enzyme catalyzed RNA 2′-*O*-transphosphorylation. *Curr. Opin. Chem. Biol.* 21: 96–102.

16 Neuner, S., Falschlunger, C., Fuchs, E. et al. (2017). Atom-specific mutagenesis reveals structural and catalytic roles for an active-site adenosine and hydrated Mg^{2+} in pistol ribozymes. *Angew. Chem. Int. Ed.* 56 (50): 15954–15958.

17 Shan, S.o., Yoshida, A., Sun, S. et al. (1999). Three metal ions at the active site of the *Tetrahymena* group I ribozyme. *Proc. Natl. Acad. Sci. U.S.A.* 96 (22): 12299–12304.

18 Aurup, H., Tuschl, T., Benseler, F. et al. (1994). Oligonucleotide duplexes containing 2′-amino-2′-deoxycytidines: thermal stability and chemical reactivity. *Nucleic Acids Res.* 22 (1): 20–24.

19 Rupert, P.B., Massey, A.P., Sigurdsson, S.T., and Ferré-D'Amaré, A.R. (2002). Transition state stabilization by a catalytic RNA. *Science* 298 (5597): 1421–1424.

20 Kath-Schorr, S., Wilson, T.J., Li, N.-S. et al. (2012). General acid–base catalysis mediated by nucleobases in the hairpin ribozyme. *J. Am. Chem. Soc.* 134 (40): 16717–16724.

21 Hieronymus, R., Godehard, S.P., Balke, D., and Müller, S. (2016). Hairpin ribozyme mediated RNA recombination. *Chem. Commun. (Cambridge, England)* 52 (23): 4365–4368.

22 Wilson, T.J., Liu, Y., Domnick, C. et al. (2016). The novel chemical mechanism of the twister ribozyme. *J. Am. Chem. Soc.* 138 (19): 6151–6162.

23 Košutić, M., Neuner, S., Ren, A. et al. (2015). A mini-twister variant and impact of residues/cations on the phosphodiester cleavage of this ribozyme class. *Angew. Chem. Int. Ed.* 54 (50): 15128–15133.

24 Gebetsberger, J. and Micura, R. (2017). Unwinding the twister ribozyme: from structure to mechanism. *Wiley Interdiscip. Rev.: RNA* 8 (3): e1402–e1402.

25 Erlacher, M.D., Lang, K., Wotzel, B. et al. (2006). Efficient ribosomal peptidyl transfer critically relies on the presence of the ribose 2′-OH at A2451 of 23S rRNA. *J. Am. Chem. Soc.* 128 (13): 4453–4459.

26 Mairhofer, E., Fuchs, E., and Micura, R. (2016). Facile synthesis of a 3-deazaadenosine phosphoramidite for RNA solid-phase synthesis. *Beilstein J. Org. Chem.* 12 (1): 2556–2562.

27 Juen, M.A., Wunderlich, C.H., Nußbaumer, F. et al. (2016). Excited states of nucleic acids probed by proton relaxation dispersion NMR spectroscopy. *Angew. Chem. Int. Ed.* 55 (39): 12008–12012.

28 Strebitzer, E., Nußbaumer, F., Kremser, J. et al. (2018). Studying sparsely populated conformational states in RNA combining chemical synthesis and solution NMR spectroscopy. *Methods* 148: 39–47.

29 Neuner, S., Santner, T., Kreutz, C., and Micura, R. (2015). The "speedy" synthesis of atom-specific ^{15}N imino/amido-labeled RNA. *Chemi. Eur. J.* 21 (33): 11634–11643.

30 Salter, J., Krucinska, J., Alam, S. et al. (2006). Water in the active site of an all-RNA hairpin ribozyme and effects of Gua8 base variants on the geometry of phosphoryl transfer. *Biochemistry* 45 (3): 686–700.

31 Torelli, A.T., Krucinska, J., and Wedekind, J.E. (2007). A comparison of vanadate to a 2′-5′ linkage at the active site of a small ribozyme suggests a role for water in transition-state stabilization. *RNA* 13 (7): 1052–1070.

32 Mir, A. and Golden, B.L. (2016). Two active site divalent ions in the crystal structure of the hammerhead ribozyme bound to a transition state analogue. *Biochemistry* 55 (4): 633–636.

33 Mir, A., Chen, J., Robinson, K. et al. (2015). Two divalent metal ions and conformational changes play roles in the hammerhead ribozyme cleavage reaction. *Biochemistry* 54 (41): 6369–6381.

34 Geraldes, C.F. and Castro, M.M. (1989). Interaction of vanadate with monosaccharides and nucleosides: a multinuclear NMR study. *J. Inorg. Biochem.* 35 (2): 79–93.

35 Davies, D.R. and Hol, W.G.J. (2004). The power of vanadate in crystallographic investigations of phosphoryl transfer enzymes. *FEBS Lett.* 577 (3): 315–321.

36 Strebitzer, E., Rangadurai, A., Plangger, R. et al. (2018). 5-Oxyacetic acid modification destabilizes double helical stem structures and favors anionic Watson–Crick like cmo^5U-G base pairs. *Chem. Eur. J.* 24 (71): 18903–18906.

37 Chen, B., Longhini, A.P., Nußbaumer, F. et al. (2018). CCR5 RNA pseudoknots: residue and site-specific labeling correlate internal motions with microRNA binding. *Chem. Eur. J.* 24 (21): 5462–5468.

38 Xue, Y., Gracia, B., Herschlag, D. et al. (2016). Visualizing the formation of an RNA folding intermediate through a fast highly modular secondary structure switch. *Nat. Commun.* 7 (1): ncomms11768.

39 Kimsey, I.J., Petzold, K., Sathyamoorthy, B. et al. (2015). Visualizing transient Watson–Crick-like mispairs in DNA and RNA duplexes. *Nature* 519 (7543): 315–320.

40 Teplova, M., Falschlunger, C., Krasheninina, O. et al. (2020). Crucial roles of two hydrated Mg^{2+} ions in reaction catalysis of the pistol ribozyme. *Angew. Chem. Int. Ed.* 59 (7): 2837–2843.

30

Strategies for Crystallization of Natural Ribozymes

Benoît Masquida[1], Diana Sibrikova[1], and Maria Costa[2]

[1] *CNRS – Université de Strasbourg, UMR 7156, Génétique Moléculaire Génomique Microbiologie, 4 allée, Konrad Roentgen, 67084 Strasbourg, France*
[2] *Université Paris-Saclay, CEA, CNRS, Institute for Integrative Biology of the Cell (I2BC), 91198, Gif-sur-Yvette, France*

30.1 Introduction

The group of ribozymes acting like turnover enzymes is up to now restricted to the ribonuclease P [1], the ribosome [2], and the spliceosome [3]. Other natural ribozymes are endonucleolytic RNAs, catalyzing the reversible cleavage of a phosphodiester bond within their own chain. This group contains the hammerhead [4], hairpin [5], hepatitis delta virus (HdV) [6], and Varkud satellite (VS) [7] ribozymes, as well as group I [8, 9] and group II introns [9]. More recently discovered ribozymes include twister [10], twister sister, pistol, and hatchet [11], as well as GlmS [12], which is also a riboswitch. When the cleavage site intervenes at the boundaries of the ribozyme, like in the case of the HdV ribozyme, the structure of the initial or pre-catalytic RNA is expected to be quite similar to the 3′ cleavage product. Yet, most frequently, the cleavage site intervenes at an internal position. In the course of the reaction, the rate of pre-cleavage form decreases in the solution until the catalytic equilibrium is reached. As fragments of the ribozyme itself, the RNA cleavage products may not be able to fold properly since cleavage affects the higher-order structure of the RNA by splitting structural domains. Therefore, a solution of active ribozymes ends up as a mix of different characteristic RNA precursors and products, to which shorter species resulting from the degradation of unstable cleavage products should be added.

Crystallization of biomolecules requires chemical and conformational homogeneity [13]. RNA incubation in crystallization droplets usually lasts a couple of days or weeks at temperatures ranging from 15 to 37 °C, at pH 6– 8, in the presence of monovalent and divalent and/or trivalent ions, like magnesium salts or coordinated transition metals such as cobalt, osmium, or iridium that help in solving the phase problem using anomalous diffraction or scattering methods. Lanthanides can also be used for the same purpose. Apart from crystallization agents such as ammonium sulfate, oily alcohols (MPD, 2-Methyl-2,4-pentanediol), or polyethylene glycols (PEG) of

Ribozymes, First Edition. Edited by Sabine Müller, Benoît Masquida, and Wade Winkler.

different molecular weights, various additives such as polyamines (spermine, spermidine) can be used. Though not optimal, these conditions allow for some ribozyme activity, meaning that pre- and post-cleavage forms of the ribozyme may coexist in the sample before the crystallization occurs. Preventing ribozymes to catalyze their own reaction is thus a mandatory aspect of ribozyme structural biology that requires designing unreactive constructs yet still able to fold into the catalytically relevant conformation.

Endonucleolytic ribozymes catalyze RNA cleavage through a phosphodiester transesterification pathway based on the exchange of two phosphodiester bonds. The conservation of energy storage in this reaction allows for its reversibility. The reaction consists of a nucleophilic attack promoted by the 2′-hydroxyl group adjacent to the reactive phosphate group and generates 2′,3′-cyclic phosphate and 5′-hydroxyl product termini. This 2′-OH group needs to be activated to become an efficient nucleophile to attack the 3′-adjacent phosphate group. However, the reaction relies on additional chemical and structural requirements. Going stepwise from activation to product formation, the catalytic pathway consists of successive chemical and conformational steps. Structural biologists need to find strategies to block the ribozymes under close scrutiny at a single step in order to isolate conformers corresponding to a well-defined state along the catalytic pathway. As a general rule, success stems from good knowledge of the ribozyme catalytic behavior deduced from biochemical studies prior to the design of the RNA constructs, which will be assayed for crystallization. Several strategies can be employed to prevent the reaction to occur. This chapter reviews these strategies and illustrates them with representative examples of natural small ribozymes and self-splicing introns.

30.2 Strategies to Inactivate the Nucleophile

Endonucleolytic ribozymes exhibit different chemical properties (exhaustively summarized in [14]) to become efficient catalysts. In brief, the phosphodiester transesterification reaction consists of a bimolecular nucleophilic substitution (S_N2). The S_N2-like reaction is characterized by the formation of the new bond, generating a pentavalent phosphorous intermediate, followed by cleavage of the leaving group, a mechanism that energetically culminates with the formation of a phosphorane transition state. This situation leads to chirality inversion if the atom targeted by the nucleophile is asymmetric. The transesterification mechanism assumes that the nucleophilic hydroxyl group is activated by deprotonation following the action of a general base. The structure of the ribozyme provides an architecture that favors the placement of the nucleophile in line with the phosphorus and leaving 5′-oxygen atoms. The alignment of these three atoms is required for the formation of the pentavalent phosphorane transition state, which adopts a trigonal bipyramidal geometry. The transition state needs also to be stabilized structurally by the ribozyme architecture, notably through an increased number of H-bonds. Finally, the leaving oxyanion should be stabilized by an acid, which gives a proton to stabilize the conjugate base, the 5′-hydroxyl. This catalytic mechanism is very much similar to the one operated by the RNase A enzyme, in which all the abovementioned atoms are aligned to perform the S_N2-like reaction [15]. All these steps add on individually to result in the final catalytic rate. Conversely, disturbance of any of these steps decreases the overall reaction rate. In principle, the modification of the chemical groups involved in any of these events can help decrease the overall catalytic rate sufficiently to render the ribozyme inactive enough over the incubation time necessary for crystallization. The most straightforward strategy is to remove or replace the nucleophilic oxygen atom by molecular engineering. Another strategy consists of mutating the residues that will either activate the nucleophile or stabilize the leaving group. Finally, the scissile phosphate group can be omitted by the synthesis of split oligonucleotides or by circular permutation (CP).

Another critical aspect regarding ribozymes crystallization is related to the integrity of their architecture. The 5′ and 3′ boundaries of ribozymes are not always easy to define. Moreover, loops can contain additional domains dispensable for the ribozyme catalytic architecture. Usually, efforts are made to reduce the length of the ribozyme constructs while retaining significant catalytic activity. This strategy can lead to the removal of regions that are somewhat dispensable for the catalysis but critical for the architecture. Nevertheless, some domains may also confer to the ribozymes too much flexibility, which may prevent folding as a single conformation. In these cases, adequate domain replacement monitored by conformational analysis can help improve the situation. Illustrations of these situations will be given below.

30.3 When the Cleavage Site is at the Edge of the Ribozyme

The location of the cleavage site at the 5′-boundary of the ribozyme constitutes the most ideal situation since the 3′-RNA product corresponds to the entire ribozyme domain and the other one to a dispensable short fragment. The cases of the hepatitis delta (HdV, [16]) and of the GlmS [17, 18] ribozymes well illustrate this situation, where the ribozyme domains are located downstream from the cleavage site. The cleavage product corresponding to the ribozyme core resulting from an in vitro cleavage assay can be purified. Yet, the study of the catalytic mechanism requires the presence of the scissile nucleotide. The latter can be added by the hybridization of a ribozyme domain made from an oligonucleotide mimicking the unreacted substrate by bearing a specific chemical modification inhibiting catalysis.

When the structure of the HdV ribozyme [16, 19] was solved (Figure 30.1), the only RNA structures that were known apart from a variety of helical structures were of transfer RNAs (tRNAs) [21, 22], of two versions of the hammerhead ribozymes [23, 24], and of the P4–P6 domain from the *Tetrahymena* intron [25]. In this context, the main goal was to expand the RNA structural repertoire by solving the structure of a new RNA architecture. When embedded in a viroid, the ribozyme allows to cleave the multimeric RNA copy resulting from the viroid RNA replication through a rolling-circle mechanism into individual genomes and simultaneously to ligate their ends to form circles [26]. In spite of the absence of

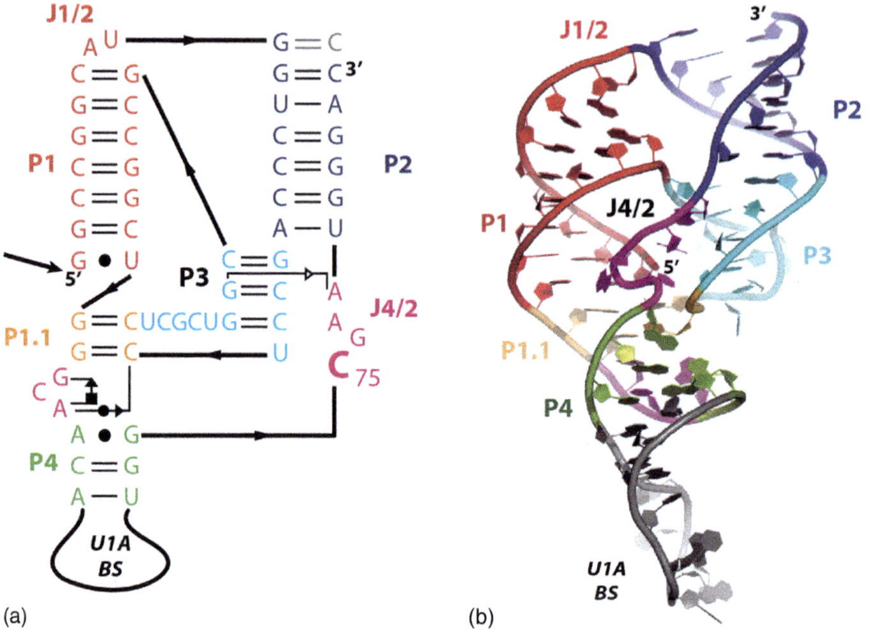

Figure 30.1 The folded core of the HdV ribozyme corresponds to the 3′-cleavage product. panel (a) presents the secondary structure of the ribozyme and panel (b) the corresponding crystal structure. The arrow indicates the position of the cleavage site upstream from the 5′-end. The structural elements are color-coded identically on the secondary and three-dimensional (3D) structure picture. The 3D structure has been derived from PDB files 1drz. The tertiary interactions are depicted using the Leontis–Westhof nomenclature. Source: Leontis and Westhof [20].

sequence requirements and of need for catalytic ions, the 5′-hydroxyl cleavage product of the ribozyme was purified to sort out the misfolded intermediates and meet the chemical and conformational homogeneity of the sample required for crystallization. Among the structural features unraveled by the first crystal structure of the HdV ribozyme [16], a pseudoknot (P1.1) formed by two consecutive G=C base pairs was observed between residues from loop L3 and the junction of stems P1 and P4 (J1/4), in addition to the one already known from comparative sequence analysis that leads to the formation of the P2 stem. P1.1 is actually instrumental for positioning the first residue of the ribozyme 3′-cleavage product, G+1, which is involved in a wobble pair with U37. Residue C75 has been identified as the residue responsible for the rate-limiting proton transfer step of the catalytic reaction. C75 should either abstract the proton of the 2′-hydroxyl of U-1 [27, 28] or stabilize the 5′-oxyanion of the leaving group [29]. In agreement with these data, the crystal structure shows that the N3 Watson–Crick group of C75 establishes a hydrogen bond with the 5′-hydroxyl group of G1. However, this property would imply that the pK_a of C75(N3) was around ~7 instead of 4.2 as observed for the free cytosine. This hypothesis was addressed by Nuclear Magnetic Resonance NMR, and only a relatively weak increase of the pK_a was observed in the context of the HdV cleavage product [30]. Further crystallographic studies were carried out to investigate the structure of the pre-cleavage form of the ribozyme. Catalysis was prevented in these constructs following strategies that will be presented below.

The GlmS RNA is a ribozyme and also a riboswitch that binds glucosamine-6-phosphate (GlcN6P), a bacterial and fungal cell wall precursor [31]. The cellular concentration of GlcN6P negatively controls the RNA amount of *glms*, the gene producing the GlcN6P synthase [12] (Figure 30.2a). While several crystal structures of the aptamer domain of this riboswitch have been determined, the folding energetic landscape between the GlcN6P bound and unbound conformations of the riboswitch has only been investigated recently [32]. The first structural study from the *Thermoanaerobacter tengcongensis* GlmS constructs gathered crystal structures with or without the scissile bond residue allowing comparison between the conformations of the pre- and post-cleavage forms [17, 33]. Yet, the binding of glucose-6-phosphate (Glc6P), an inhibitor of the GlmS riboswitch, but not of GlcN6P, could be observed. Only models of the catalytic mechanism could be deduced from these structures. Just after reporting the crystal structure of the *T. tengcongensis* GlmS riboswitch, another research group published the crystal structure of the *Bacillus anthracis* GlmS riboswitch in the presence of the ligand (Figure 30.2b). The direct involvement of the amino group of GlcN6P to protonate the leaving 5′-oxyanion during catalysis could be demonstrated [18].

This set of crystal structures shows that when the cleavage site intervenes at the 5′-end, the addition of a couple of 5′-nucleotides does not have major effects on the overall ribozyme architecture. However, the addition of 5′-nucleotides restores the cleavage site, requiring the inhibition of the nucleophile activation. Strategies used to prevent nucleophile activation are described below.

30.4 Removal or Neutralization of the Catalytic 2′-Hydroxyl Group

With a lesser knowledge of the catalytic mechanism, meaning when the nucleophile position has been identified, but not much other informations are available, the easiest strategy is either to remove the 2′-hydroxyl group or to modify it by methylation. Removal may lead to a loss of information by releasing some of the constraints on the active nucleotide, while methylation preserves the 2′-*O*-atom, which potentially confers the opportunity to observe its role in catalysis.

30.4.1 The Hammerhead Ribozyme

The hammerhead ribozyme was the first catalytic RNA, the structure of which was solved by crystallography [23]. The ~50-nt hammerhead ribozyme was found in the 1980s in the satellite RNA of the avocado sunblotch viroid [4]. More recently, genome surveys have revealed that the hammerhead motif is actually widespread

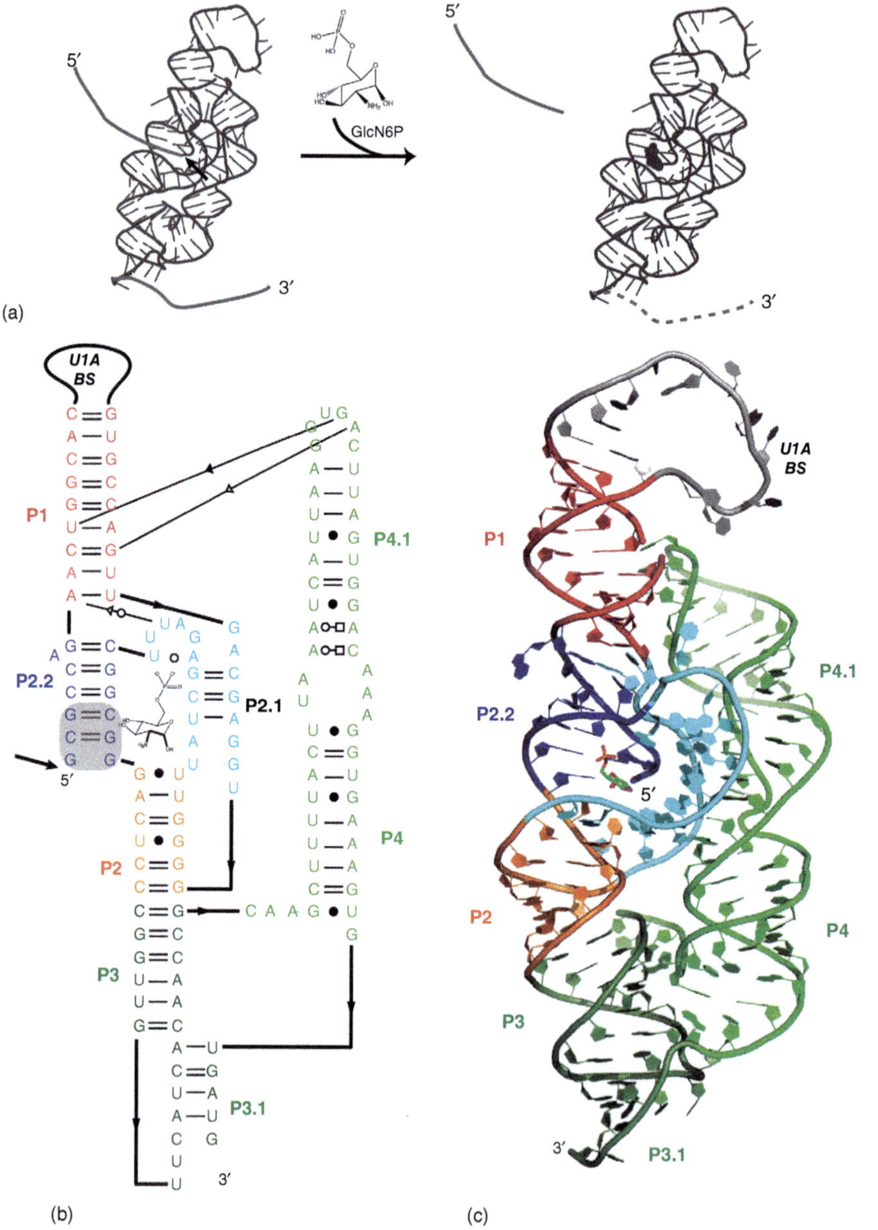

Figure 30.2 The GlmS riboswitch is also a ribozyme. (a) The GlmS riboswitch does not change drastically its conformation upon binding of the GlcN6P. Rather, it undergoes cleavage of its 5′-sequence, which leads to degradation of the mRNA (sketched as a dashed line). (b) The secondary structure of the *Bacillus anthracis* riboswitch is depicted with the region where GlcN6P binds framed in a gray square. (c) The 3D structure of the riboswitch (PDB: 3g9c) is represented on this panel with the same color code as in (b).

in genomes, from bacteria to humans [34]. This fairly short RNA allowed for defining a trans-acting construct based on the hybridization of a ribozyme domain and a substrate strand [35, 36] that acts like a true enzyme/substrate complex.

In the first hammerhead crystal structure, a minimal ribozyme domain was produced by in vitro transcription, whereas the substrate strand was chemically synthesized as a DNA strand. This strategy allowed for the systematic replacement of all 2′-hydroxyl groups of the substrate strand by hydrogen atoms and consequently prevented cleavage at the scissile phosphate. Since the RNA strand geometry cannot fold like B-DNA due to the presence of the 2′-hydroxyl groups, the DNA strand was forced to adopt the A-form [23]. In the structure of the all-RNA hammerhead ribozyme published shortly afterward, a two-strand strategy to reconstitute the minimal ribozyme was again chosen. In this case, the nucleotide bearing the proton of the nucleophilic 2′-hydroxyl group was changed to a 2′-*O*-methyl group to block catalysis [24]. In order to impair catalysis during crystallization, neutralization of the nucleophile by chemical modification is preferred over the depletion of the attacking 2′-hydroxyl group. This is because the former strategy is more likely to provide relevant insight into the catalytic mechanism. However, introduction of such chemical modifications relies on solid-phase chemical synthesis (see [37, 38] for review) of short RNA oligonucleotides carrying the desired modified nucleotides. Indeed, the all-RNA hammerhead ribozyme showed the same conformation as the DNA–RNA hybrid construct. Yet, in both crystal structures, the conformation of the active site was not in agreement with the biochemical and enzymatic data previously obtained for the ribozyme. Also, in these structures, the conformation at the cleavage site did not show the expected alignment of the three atoms involved in the transesterification reaction (Figure 30.3a), namely, the 2′-hydroxyl groups of the catalytic residue on one side and the phosphorus atom and the 5′-hydroxyl groups of the contiguous 3′-residue on the other side. Subsequently, intensive research was conducted using the minimal RNA construct to try to observe its active conformation by freeze-trapping an active version of the ribozyme at low pH in the presence of magnesium ions [39]. But again, the conformation trapped in the crystal was still not fully supportive of an in-line S_N2-like mechanism for the self-cleavage reaction.

The answer actually came 10 years later when a kinetic and thermodynamic characterization of the crystal structure of a longer hammerhead ribozyme derived from *Schistosoma mansoni* was performed (Figure 30.3c). This hammerhead ribozyme version comprises additional critical tertiary contacts between peripheral segments that improve catalysis by 1000-fold at sub-millimolar magnesium concentration [40, 41]. For crystallization purposes, a two-strand construct was also used to reconstitute the ribozyme, and the nucleophilic 2′-hydroxyl group was again neutralized by methylation. The resulting crystal structure revealed that the active site of this longer ribozyme forms a three-way junction. Stem II is closed by an apical loop, which interacts with the internal loop embedded in stem I, itself formed between the ribozyme and the substrate strand [42]. This peripheral tertiary interaction induces a major conformational change in the distal three-way junction that forces the in-line positioning of the tree atomic groups involved in the catalytic

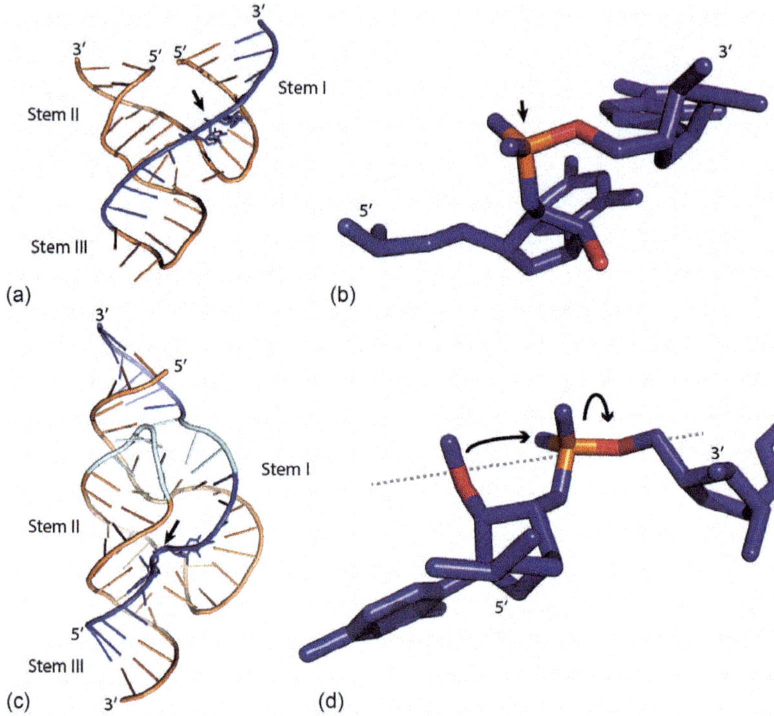

Figure 30.3 This figure displays the structures of the minimal (PDB: 1mme) (a) and the extended (PDB: 3zd5) (c) versions of the hammerhead ribozyme. (b, d) How the residues around the scissile bond are organized according to the absence or presence of the tertiary interaction ((b) cyan) between the loop of stem II and the internal loop of stem I, respectively. It is straightforward from (d) that the three atoms involved in the S_N2-like reaction are ideally aligned, with respect to (b).

reaction. The right panel of Figure 30.3 shows the very different positions adopted by the dinucleotide around the scissile bond. Taken together, these observations indicate that the historical hammerhead construct that was used as a model system for a long time despite its poor activity had in fact been too much shortened to keep the relevant pre-catalytic conformation [43]. In conclusion, designing a streamline version of a ribozyme for crystallization purposes should rely, first, on careful sequence analysis of the available ribozyme variants and, second, on careful biochemical and kinetic analysis of each construct in order to select the molecule that more closely mimics the natural full-length version.

30.4.2 Group I Intron

The ~300-nt group I ribozymes are able to self-splice along with a two-step trans-esterification mechanism. First, exogenic guanosine (exoG) binds a pocket in the catalytic core to play the role of the nucleophile. During the first step, exoG trans-esterifies the bond between the 5′-exon and the intron and remains attached to the 5′-end of the intron. Following a profound conformational change leading to the

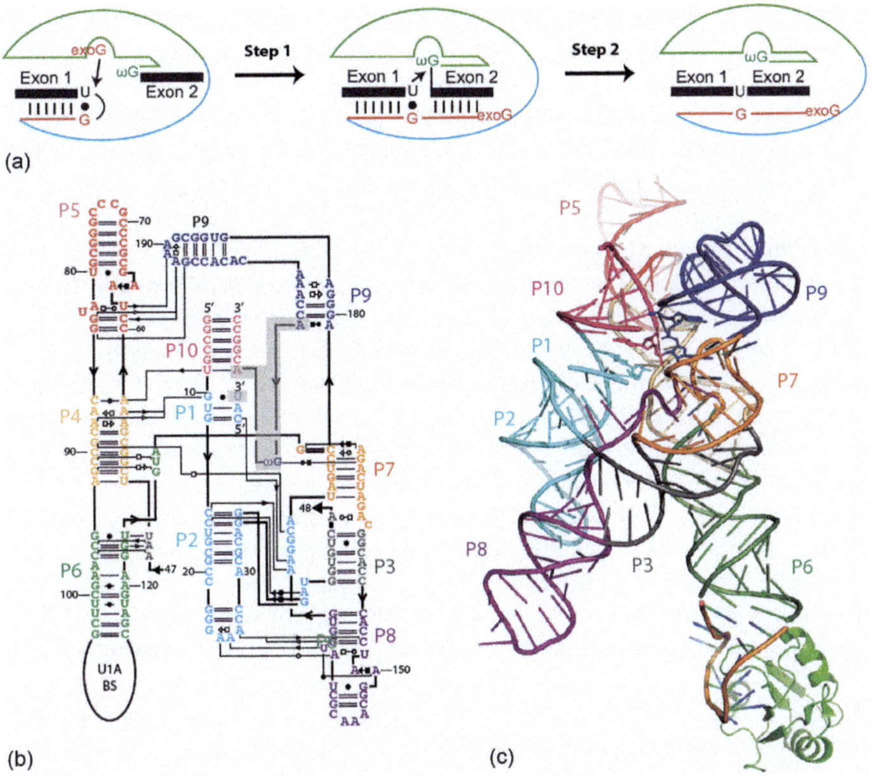

Figure 30.4 (a) Splicing pathway mediated by the group I introns in general (see text for details). (b) Secondary structure depicting the tertiary interactions observed in the 3D structure (c). The residues that have been modified as deoxynucleotides are framed in gray on the secondary structure diagram and represented in sticks mode on the 3D panel, showing how they are close in space. The U1A protein and the RNA binding site are visible in the lower right corner. These modified nucleotides were integrated into two distinct oligonucleotides. The first one constitutes a short 5′-exon forming P1 (5′-CAU-3′). The second one spans from residue 190 up to the 3′-end of P10. The conformation corresponds to the pre-cleavage state prior to the second catalytic step (see (a)).

replacement of exoG by the ultimate G residue (ωG) of the intron in its pocket, the 3′-residue of the 5′-exon cleaves the bond between the intron and the 3′-exon (Figure 30.4a). This pathway results in exon ligation and intron release. Capturing intermediates along this complex pathway requires the inactivation of well-chosen residues (Figure 30.4b,c).

The self-splicing group I intron from the pre-tRNA^{Ile} of bacterium *Azoarcus* [44, 45] was crystallized bound to its exons in a conformation preceding the second catalytic step of splicing. This intron/exon complex was stabilized by four specific 2′-deoxy substitutions: two at intron positions A205 and ωG, one at the end of the 5′-exon, and one at the beginning of the 3′-exon. Collectively, these modifications slow down exon ligation by a million-fold and allow for crystal nucleation and growth over ∼2 weeks. The crystallization construct was obtained by the hybridization of a transcript corresponding to the majority of the intron with an RNA–DNA oligonucleotide spanning the 3′-section of the intron and the beginning of the 3′-exon and a trinucleotide mimicking a short 5′-exon ending with a dT residue. This three-piece strategy greatly facilitated the introduction of the substitutions since all of the 2′-deoxynucleotides were inserted during chemical synthesis of the RNA–DNA chimeric oligonucleotides (Figure 30.4). In addition, a U1A binding site was introduced in the L6 loop of the intron to optimize crystallization by promoting crystal packing interactions between molecules in solution [46]. Note that U1A contains selenomethionine residues used to solve the phase problem without performing heavy-atom soaking or co-crystallization.

The crystallized intron/exons complex revealed the structural basis for 5′- and 3′-splice site selection and substrate alignment used by group I intron ribozymes. However, positioning of the metal ions into the active site was found not to be fully consistent with available biochemical data presumably because of the 2′-deoxy substitution at the terminal intron guanosine ωG [44]. Indeed, only subsequent crystallization of an all-ribose version of the same construct, containing the catalytically essential ωG O2′ ligand, allowed the correct positioning of the metal ions in the active site and provided crystallographic evidence for the "two-metal-ion" mechanism of group I intron splicing [47]. In this case, the same 5′-exon 5'-CAT-3' trimer was used to reduce sufficiently the exon ligation reaction and permit crystal nucleation. Again, this example nicely illustrates that crystallization of complete ribozymes in their active state is a challenging task: modification of catalytically important positions is necessary to block or reduce the activity of the ribozyme; still, such changes should not alter the catalytically competent structure of the active site.

Other structures of group I introns have been solved by X-ray crystallography [48, 49]. Yet, their conformations did not represent any active state and thus did not require the introduction of any chemical modification. In the structure of the *Tetrahymena* ribozyme [48], the substrate helix was not included in the construct. In the ribozyme of the *orf142* from the phage Twort [49], the ribozyme core was crystallized with a tetrameric RNA mimicking the substrate. However, since this pseudo-substrate does not correspond to any catalytic step, proper docking onto the ribozyme core could not be observed.

30.4.3 Group II Intron

Group II introns, which are not evolutionarily related to group I introns, constitute the second class of self-splicing ribozymes known to date. They are abundant in bacteria and bacteria-derived organelles of fungi, plants, and algae. Group II introns are large catalytic RNAs of ~600 nucleotides which, in association with the reverse transcriptase they encode, can also behave as mobile retrotransposable elements. Despite their bacterial origin, it is now widely believed that group II introns played a determinant part in eukaryotic evolution as the ancestors of the nuclear premessenger introns and their splicing machinery (the spliceosome) [50]. This evolutionary hypothesis is grounded on the extensive structural and mechanistic similarities shared by the two splicing systems. Both group II intron and nuclear pre-messenger

RNA splicing proceed through two sequential transesterification reactions and generate a typical intron form called the "lariat" intron. Self-splicing of group II introns is initiated by the 2′-hydroxyl group of highly conserved adenosine, lying unpaired in domain VI (bpA in Figure 30.5a), which promotes a nucleophilic attack at the 5′-splice site. This first splicing step results in the formation of a specific branched intron intermediate, the "lariat" intermediate, in which the "branch point" adenosine (bpA) becomes covalently linked to the first intron nucleotide, a conserved guanosine, through a 2′,5′-phosphodiester bond. It follows the second transesterification reaction, during which the 3′-OH of the cleaved 5′-exon attacks the 3′-splice site, resulting in the ligation of the flanking exons and the release of the intron "lariat." Because transesterification reactions are chemically reversible, excised group II lariats are then able to invade specific RNA or DNA targets by catalyzing the splicing reactions in the reverse order and orientation: this "reverse splicing" pathway plays an essential role in the biological activities of group II introns, since reverse splicing into DNA triggers the genomic mobility of the intron.

Recent X-ray crystal structures of an excised group II intron lariat, alone or bound to its 5′-exon substrate (Figure 30.5), uncovered the structural basis for reverse splicing by group II introns [51]. These structures revealed for the first time that the 2′(A)-5′(G) branch plays a direct and crucial role in organizing the active site for the second step of splicing. Remarkably, the structures also show that after the completion of the splicing reaction, the last nucleotide of the freed intron lariat remains firmly positioned in the catalytic center with its 3′-OH activated by a highly coordinated metal ion (M1 in Figure 30.5c) and poised for catalysis of the reverse splicing reaction. In the absence of its 5′- and 3′-exon substrates, the excised group II lariat is a stable RNA; therefore, the purified lariat sample could be crystallized without any modification. Formation of a stable lariat/5′-exon complex, however, required co-crystallization of the lariat with an unreactive analog of the 5′-exon. This analog consisted of a 7-mer RNA carrying a 3′-deoxyribose at its 3′-end (Figure 30.5a,c). Removal of the terminal 3′-hydroxyl group of the 5′-exon was necessary to block RNA lariat "debranching", a reaction that is mechanistically analogous to the reverse of the first step of splicing and, accordingly, leads to the production of linear intron molecules covalently attached to the 5′-exon. Comparison of the lariat structures with and without this substrate revealed that binding of the 5′-exon drives a local conformational rearrangement of domain V that extends into the active site and contributes to the coordination of a second catalytic metal ion. Interestingly, native crystals (grown in magnesium-containing solutions) lacked electron density corresponding to this second metal ion, most probably due to the absence of the critical 3′-OH group of the 5′-exon. Evidence for this second catalytic metal ion was nevertheless obtained through crystal soaking in ytterbium – an anomalous scatterer lanthanide that mimics highly coordinated magnesium ions but binds with higher affinity. Indeed, ytterbium soakings revealed the presence of two metal ions in the reaction center (Y1 and Y2 in Figure 30.5c), one of which perfectly superimposes with metal ion M1 already identified in the native crystals. Altogether, the two sets of crystals allowed the direct observation of all but three metal-ion coordinations that build up the "two metal-ion" transition state for the reverse splicing reaction

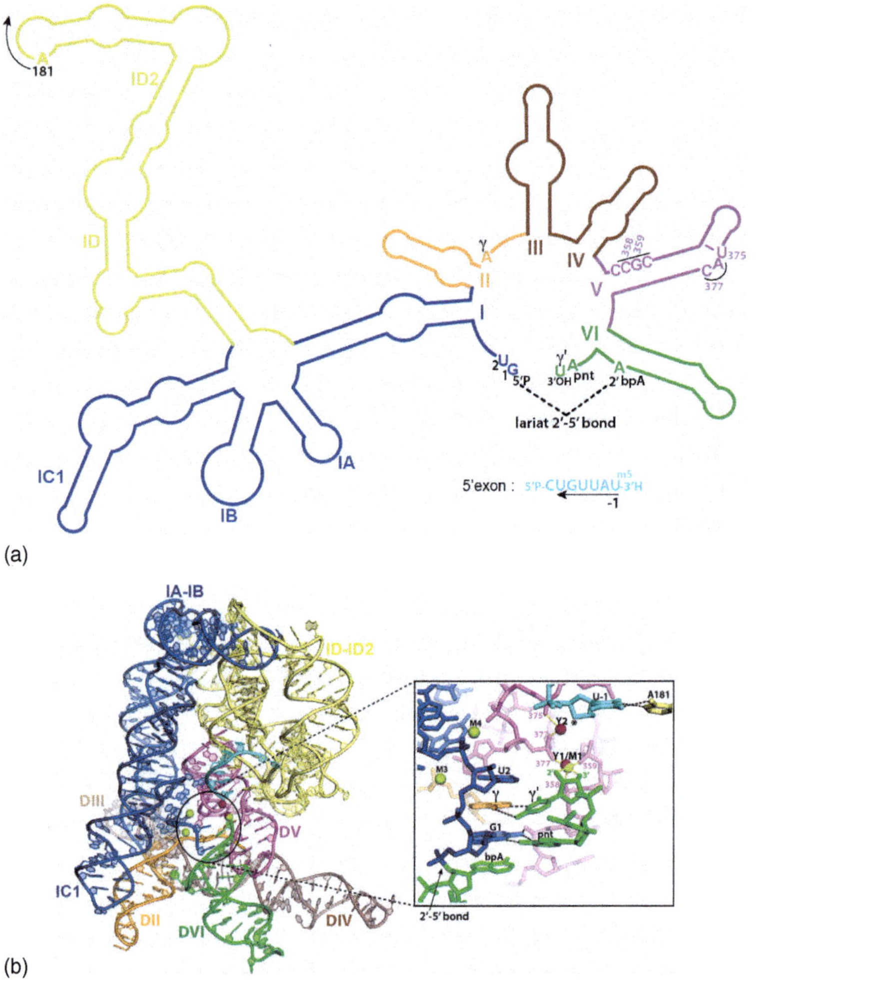

Figure 30.5 Structure of a group II intron lariat primed for reverse splicing. (a) Secondary structure outline, colored by domains, of the crystallized group II intron lariat. The intron crystallized is an engineered version of the *Oceanobacillus iheyensis* group II ribozyme [51]. Only the active site nucleotides or those involved in binding of the catalytic metal ions are shown. The RNA oligonucleotide corresponding to the unreactive analog of the 5′-exon used for co-crystallization is shown in cyan with the 3′-terminal nucleotide ^{m5}U3′H standing for a 5-methyluridine with a 3′-deoxyribose (the m5 base modification was imposed by the chemical synthesis of the 7-mer RNA and did not interfere with the base pairing of the oligonucleotide to the intron lariat). The arrow spanning a section of the terminal loop in subdomain ID2 indicates the 5′-exon binding site, which is complementary to the sequence of the 5′-exon. (b) Crystal structure of the group II intron lariat bound to its 5′-exon RNA at 3.5 Å resolution (PDB entry 5j02). Coloring of the structure is according to (a). (c) Close-up view of the active site showing the network of hydrogen bonds (dashed black lines) and stacking interactions that involve the 2′,5′-branch nucleotides, the intron boundaries, and the conserved γ nucleotide. The architecture of the active site directly promotes positioning of the terminal ribose of the lariat into the reaction center in a configuration poised for catalysis of the reverse splicing reaction. The catalytic metal ions (Y1/M1 and Y2) identified in the reaction center establish inner-sphere coordinations (dashed yellow lines) with the 2′- and 3′-hydroxyl groups of the last intron nucleotide (γ′) and the non-bridging phosphate oxygens of highly conserved nucleotides of domain V (these nucleotides are numbered in (a)). The asterisk indicates that the terminal ribose of the 5′-exon analog lacks a 3′-hydroxyl group.

(Figure 30.5c; [51]). Importantly, the transition state model derived from this crystallographic work is identical to the one prevailing for group I intron self-splicing (see section 30.4.2; [47]).

30.4.4 The Hairpin Ribozyme

The hairpin ribozyme provides another example where the precise definition of the domains important for folding was critical to solving a catalytically relevant structure. Like the hammerhead and HdV ribozymes, the hairpin ribozyme is part of the satellite RNA of a virus (tobacco ringspot virus) and chops the multimeric RNA resulting from a rolling-circle amplification [5]. The hairpin ribozyme was named after the shape of the secondary structure of the minimal fragment initially studied (Figure 30.6). This construct consists of two helical domains (A and B), each incorporating an internal loop [52]. In the initially characterized hairpin ribozyme construct, a kink between domains A and B promotes docking of the internal loops so that catalysis occurs. However, structural studies in solution have shown that the linker between the domains is very flexible [53, 54]. Closer analysis of the natural structural context indicated that the ribozyme actually folded from a four-way junction, bringing into play two additional helices that stack on top of domains A and B in order to form an RNA holiday-like junction [55]. These junctions, like the related DNA holiday junctions instrumental in genomic recombination events [56], form an "X" with each branch corresponding to a helix.

The 5′-helices of domains A and B are tethered so that a kink between them results in the docking of internal loops A and B. In a minimal construct presenting only domains A and B, this situation results in high conformational heterogeneity due to constant competition between helical stacking at the edge of the two domains and docking of the internal loops. Förster resonance energy transfer (FRET) experiments indicated that upon folding, the natural four-way junction was forming a dominant conformation forcing the two domains required for catalysis to dock productively. The four-way junction context thus increases dramatically the rate of interaction between domains A and B as compared to the original and much more flexible hairpin minimal construct [53]. This four-way junction hairpin ribozyme RNA to which a U1A protein binding site was added to improve crystallization [46] was finally solved by X-ray crystallography [57]. Like in the case of the all-RNA hammerhead ribozyme, the nucleophilic 2′-hydroxyl group from the substrate strand was neutralized by methylation. The structure showed close to perfect alignment between the three atoms involved in the S_N2-like reaction as well as pointed to a network of hydrogen bonds stabilizing the pre-catalytic state. In particular, the scissile G residue is pulled out from domain A by forming a trans-Watson–Crick base pair with an unpaired C of domain B, which orients its backbone favorably for the S_N2-like reaction (Figure 30.6d). It is worth to note that the structure of an all-RNA substrate hairpin ribozyme could be solved despite its catalytic activity. In this case, the constraints from the crystal packing together with the occurrence of the ligation reaction seem to prevent product strand dissociation [58].

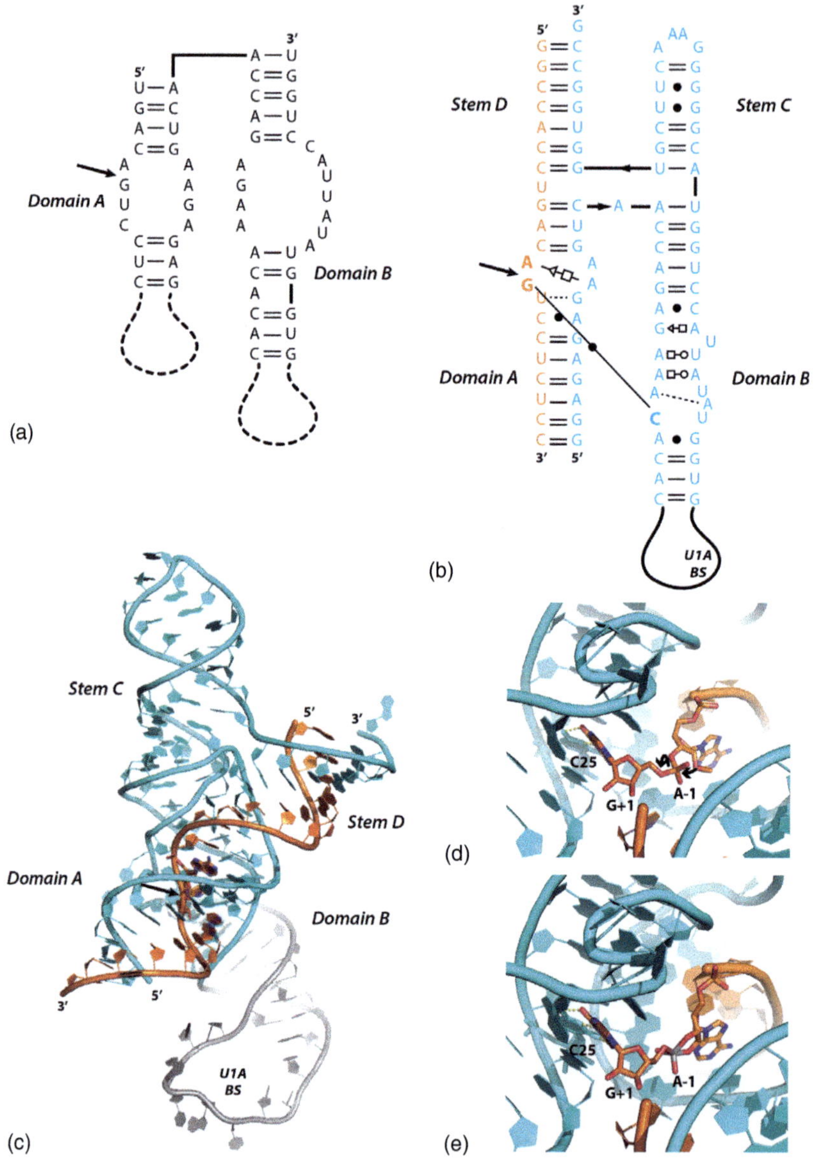

Figure 30.6 The hairpin ribozyme can be active as a short or an extended version, provided that the internal loops of domains A and B establish specific interactions. In the short version (a), domains A and B are tethered by a flexible linker, whereas in the natural context, a four-way junction is formed (b). The black arrow points to the cleavage site. (c) The 3D structure corresponding to (b) (PDB: 1m5k). The orange substrate strand bound in trans in the RNA construct extends through domain A and stem D. The topology of the four-way junction is clearly visible on (c). (d) A close-up of the scissile dinucleotide (orange) embedded in the ribozyme core (blue). The curved arrows show how the electrons are supposedly transferred during catalysis from the O2′-hydroxyl group of A9 to the O5′ of G10. (e) The vanadate complex (PDB: 1m5o) mimicking the transition state in a similar orientation shows that G1 still interacts within loop B with residue C25 (blue sticks). The presence of vanadate perturbs significantly the pucker of the A-1 ribose moiety.

In order to go beyond the structure of the ground state and to elucidate the catalytic mechanism, another modification strategy of the ribozyme was undertaken to study the structure of the transition state. This strategy relies on the coordination of both the 2′- and 3′-hydroxyl groups of the residue bearing the nucleophile and of the 5′-hydroxyl group of the downstream nucleotide by a vanadate ion. Vanadate coordination fairly well mimics the phosphorane intermediate generated during the cleavage reaction [58]. To allow binding of the vanadate ion, the substrate was split into two fragments resembling the reaction products with free 2′-, 3′-, and 5′-hydroxyl groups. This strategy leads to the removal of the scissile phosphate, which is replaced by the vanadate ion. The crystal structure reveals how the transition state is stabilized throughout increasing the number of hydrogen bonds formed with the two nucleotides in the vicinity of the vanadate (Figure 30.6e). Noteworthy, the structure of the ribozyme with a split substrate displaying a 2′,3′-cyclic phosphate at the cleavage site could also be solved to complete the snapshots necessary to describe the entire catalytic process [58]. The structure of the transition state of the hammerhead ribozyme was also studied using vanadate ion binding to the cleavage site [59, 60]. The catalytic site undertakes a substantial reshaping under these conditions and allows for describing more accurately the role of the residues and ions potentially involved in the reaction, notably G12, the general base, and the two magnesium ions close to the scissile bond. The vanadate structure also suggests that a water molecule bound to a divalent cation could act as the general acid that stabilizes the 5′ oxyanion.

30.5 Removal of the Scissile Phosphodiester Bond Using Circular Permutation

Success of the split substrate strategy relies on the efficient hybridization of the RNA strands to their target as in the case of the hairpin ribozyme. However, this prerequisite is not always met since annealing efficiency depends on the length and concentration of three different RNA species, each one potentially engaged in individual folding events that might perturb the formation of a well-folded trimer. Moreover, if the substrate binds to a region with an intricate fold, hybridization itself may be inefficient. An example that illustrates this situation is given with the lariat-capping (LC) ribozyme [61], which is the topic of an entire chapter in the present book (Chapter 5). In brief, this ribozyme catalyzes a transesterification reaction called "branching" that is initiated by the nucleophilic attack of a 2′-hydroxyl group from a U residue on a C residue residing two nucleotides upstream. This reaction leads to the formation of a short 3-nt lariat in which the U residue is covalently attached to the C residue through a 2′,5′-phosphodiester bond [62]. From a chemical point of view, the lariat is identical to those generated by group II introns and the spliceosome during the first step of splicing [3], except that the lariat loops involved in splicing span hundreds of nucleotides instead. In the case of the LC ribozyme, the 2′,5′-branch-point structure is well buried into the ribozyme core, where each nucleotide from the lariat weaves specific stacking and hydrogen bond interactions with specific core residues. These noncanonical interactions could not be deduced from comparative sequence analysis since the lariat sequence is fully conserved among all LC ribozyme homologues [63]. Most importantly, the sequence stretch 3′ from the cleavage site directly contributes to building up the three-way junction of the regulatory domain, which works by forming a receptor for a residue within a loop of the ribozyme core, instrumental for structuring the catalytic site. Initial efforts to crystallize the ribozyme using synthetic oligonucleotides were unsuccessful because the hybrids could not properly fold.

To overcome this problem, a circular permutation strategy was implemented to swap the extremities of the ribozyme to the residues around the scissile phosphodiester, while the natural ends located in the stem were tethered by a UUCG tetraloop [64] (Figure 30.7). Without any starting G residues, the 5′-CAU sequence resulting from this permutation did not follow at all the sequence requirements of the enzyme used to produce the RNA in vitro, the T7 RNA polymerase. For production purposes, a 5′-hammerhead ribozyme domain was thus inserted upstream from the LC ribozyme. Since the hammerhead was introduced to increase the transcription yield of the LC ribozyme, the optimal sequence transcribed by the T7 RNA polymerase was introduced right downstream from the T7 promoter [64]. The version of the ribozyme based solely on the three-way junction proved to be inefficient at cleaving, certainly because the LC ribozyme could fold faster, preventing the hammerhead ribozyme to adopt a structure prone to cleave. Consequently, the hammerhead was optimized for fast-folding by preserving the tertiary contacts observed in the *Schistosoma* ribozyme [41]. In addition, an HdV ribozyme was added at the 3′ end of the LC ribozyme to guarantee that the construct would not present any additional residue after its terminal G, ωG.

The structure of the circularly permutated LC ribozyme allowed to uncover all the interactions that the three lariat residues mediate with the other nucleotides from the core. It is worth noting that a water molecule interacts with the 2′-hydroxyl group of U232 and the 5′ hydroxyl of C230 and could be considered reminiscent of the scissile phosphate group that is available in the natural LC ribozyme but missing in the circularly permutated form due to the construct design (Figure 30.7). Nevertheless, ωG, which is important for catalytic activity, could not be observed due to structural disorder. This indicates that the conformation captured in the crystal corresponds to a snapshot of the LC ribozyme right after catalysis and before the release of the RNA products.

Interestingly, circular permutations intervene naturally in RNA sequences. Domain shuffling can lead to RNA circular permutations, notably in back-splicing events resulting in the appearance of circular RNAs originating from pre-mRNAs [65, 66]. As far as the active RNA structure is maintained, the position of the 5′- and 3′-ends can be located in different regions. Along with the same idea, loops closing helices can adopt any length from 3 nt to several kilobases as far as the few nucleotides contributing to tertiary interactions are preserved, if any. Such examples of natural circularly permutated RNAs have been described for the signal recognition particle (SRP) [67] and for the hammerhead and HdV-like ribozymes [68, 69]. Circular permutations can also be engineered to incorporate chemically modified

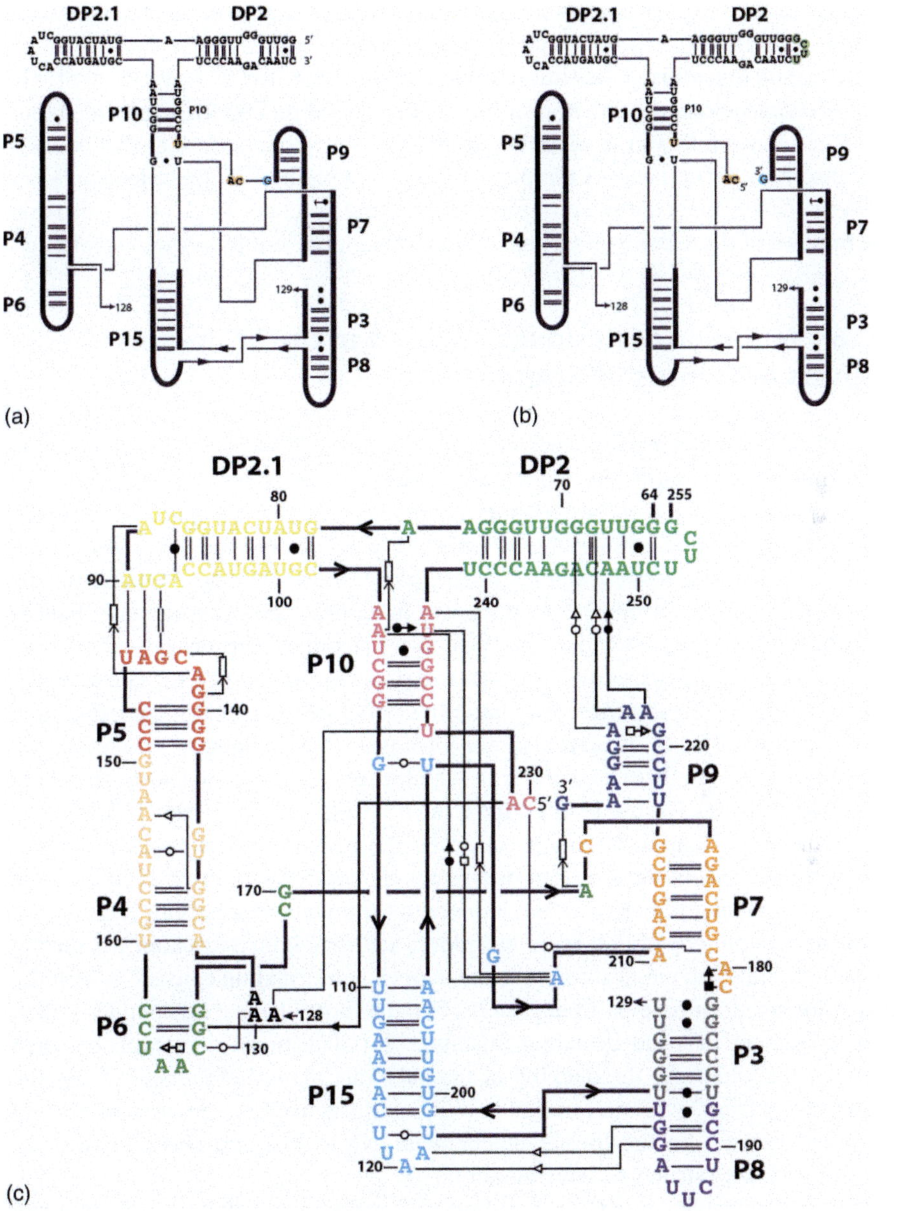

Figure 30.7 Organization of the LC ribozyme constructs. (a) Representation of the secondary structure of the wild-type LC ribozyme. Both the 5′ and 3′ ends originate from DP2. The 3′-G residue from the ribozyme (ωG, blue circle) is tethered to the three lariat-forming nucleotides (orange circles). In the circularly permutated (CP) construct (b), the 5′- and 3′-ends are placed at the lariat C and ribozyme ωG, respectively. The continuity of the RNA chain is preserved by closing DP2 using a UUCG loop (green circles). The secondary structure and tertiary interactions deduced from the crystal structure of the CP construct are represented on (c).

residues further used in biochemical studies [70–73]. Examples involving ribozymes include the RNase P where circular permutations were engineered to insert photo-agents to perform cross-linking studies [71] or to study folding kinetics [74].

In principle, circular permutation can be engineered on any RNA, provided that the 5′-sequence follows the requirement of the RNA polymerase that will be used to transcribe it from the DNA template. CP is thus a method of choice to express an RNA as a single molecule and probe its structure and function, based on the idea that the removal of the scissile phosphate group does not prevent homogenous folding.

30.6 Mutation of Residues Involved in the Acido-Basic Aspects of Transesterification

The 154-nt VS ribozyme is a fairly large ribozyme within the family of nucleolytic ribozymes. It is found in the mitochondria of a specific isolate of the *Neurospora* fungus [7]. Inactivation of the VS ribozyme for crystallization purpose was achieved following an original strategy based on modifying the residues involved in the proton transfer necessary for stabilizing the catalytic intermediates [75]. In brief, its structure is organized around three three-way junctions encompassing seven helices (numbered 1–7). From each of those three-way junctions protrudes one stem–loop, each one contacting a specific region of the ribozyme (Figure 30.8). The stacking between stems from the three-way junctions generates a helical region spanning four helices (1–4). The structure shows a feature so far unique to this ribozyme. The ribozyme crystallizes as a dimer in which the substrate stem (numbered 1) of one ribozyme docks in the second ribozyme in a catalytically active conformation [75]. This conformation is characterized by the formation of a kissing complex between the loops of helix 1 from one ribozyme and the loop from helix 5 of the second. Interestingly, these two 7-nt loops adopt similar conformations reminiscent of the anticodon loop of tRNAs. In each loop, a U-turn intervenes after the second nucleotide allowing the next four ones to adopt a conformation consistent with the formation of a four base-pair helix. In tRNAs, the fourth helical nucleotide is often hyper-modified to avoid decoding mistakes [76]. The trans-interaction between loops 1 and 5 leads to the docking of the substrate along helix 2 and places the internal loop from helix 6 in position to build up the active site. The observed dimer is in agreement with the former biochemical, NMR data and small-angle X-ray scattering (SAXS) structure determination [77].

The strategy implemented by the authors was motivated by obtaining the RNA as a single strand to be able to purify it under native conditions since the dimer formed in the course of the transcription reaction. The structures of two mutants were solved, each one strongly affecting proton transfer in the catalytic site. The G638A structure was solved by single-wavelength anomalous diffraction (SAD), and the A756G structure was solved by the molecular replacement method using the G638A structure as a search model. Residues important for catalysis locate in stems 1 and 6. In helix 1, G638 acts as a general base to abstract the proton of the 2′-hydroxyl group of the cross-strand stacked guanine. Helix 6 harbors A756, the general acid that stabilizes

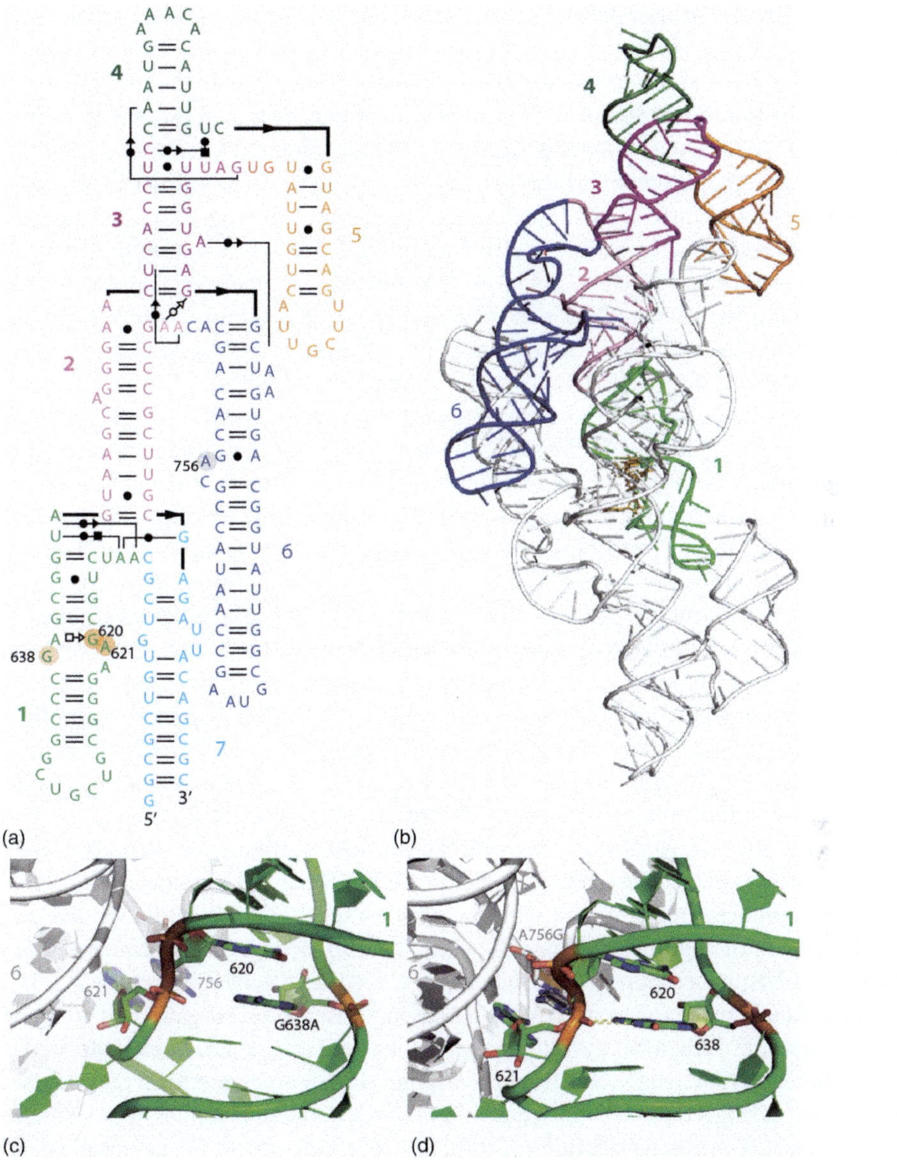

Figure 30.8 The secondary structure of the crystallized construct of the VS ribozyme (a) How the three three-way junctions in the VS ribozyme lead to coaxial stacking of helices 1–4 (bold-face numbers). The residues around the scissile phosphate (G620–A621) and the general base (G638) and acid (A756) are framed in a colored circle (orange for the RNA containing the scissile phosphate and gray for the RNA interacting in trans). (b) 3D structure of the VS ribozyme dimer. The molecule corresponding to (a) is represented using the same color code. The ribozyme interacting in trans is gray colored. (c) The close-up of the catalytic site in the G638A mutant crystal structure (PDB: 4r4v) shows how the residues around the scissile phosphate of A621 are splayed apart and the cross-strand stacking between residues 638 and 620. (d) These features are maintained in the A756G mutant crystal structure (PDB: 4r4p). However, the presence of two G residues weaves a set of hydrogen bonds to the scissile phosphate that is not observed for the G638A mutant.

the leaving oxyanion after the formation of the phosphorane transition state. The two X-ray structure models of the mutant forms G638A and A756G of the VS ribozyme unravel how the two residues contribute to both the structure and the catalysis.

In the G638A structure, the O2′, phosphorus, and O5′ atoms of the scissile bond between G620 and A621 are not perfectly in line, showing that the mutation significantly affects the activation of the reaction (Figure 30.8c). The extrusion of A621 from its helix and further stabilization through A-minor binding to a G=C pair in helix 2 indicate how the in-line positioning of the three atoms taking part in the S_N2-like reaction may be easily achieved. However, A638, which stacks on the face of G620 points its Watson–Crick edge toward the phosphate group of A621. This situation places N1 of A638 in close enough proximity of the O2′-group from G620 and supports the notion that G638 is the general base involved in the activation of the reaction. In the G638A crystal structure, the general acid A756 also points toward the phosphate group of A621. Noteworthy, a water molecule is clamped by hydrogen bonds between N1 of A756 and O2P of A621. The position of this water molecule could mimic the O5′-ligand of the phosphorane intermediate that would be resolved by proton transfer from A756, which would be protonated on its N1 atom (Figure 30.8).

In the A756G structure model (Figure 30.8d), the three groups involved in the S_N2-like reaction are placed in-line much better than in the G638A VS ribozyme RNA. Similar to the G638A crystal structure, A621 is extruded from its helix to make an A-minor contact within the shallow groove of helix 2. The distance between the O2′-nucleophile and the N1 group of G638 is increased by the C3′-endo pucker of the G620 ribose. This group would actually reside in a closer distance, if the ribose were in a C2′-endo conformation. The offset of the O2′ leaves the N1 and N2 groups of G638 free to form hydrogen bonds with the O1P and O2P atoms of A621. The O2P and O5′-groups of A621 are also in contact with the N1 and N2 groups from G756. All these contacts are consistent with the notion that the A756G mutation affects the transfer of the proton stabilizing the leaving group through a set of contacts that locks the phosphate group, greatly increasing the required activation energy.

The catalytic mechanism of the HdV ribozyme was also investigated following a mutational approach. After the publication of the initial crystal structure of the product form of the ribozyme, a construct corresponding to a precursor form of the HdV ribozyme bearing a C75-to-U mutation and two additional nucleotides at the 5′-end was studied. Ten crystal structures of this construct were solved in the presence of various divalent cations including magnesium, calcium, strontium, barium, manganese, cobalt, copper, and iridium in order to identify both the two players of the acid/base reaction [78]. The C75-to-U mutation prevents the deprotonation of the N3 position, thus inactivating the ribozyme. The crystal structures show the presence of an ion in the vicinity of the nucleophile. The model, which could be derived from these complementary studies, is that the cation acts as the general base, abstracting the proton from the 2′-hydroxyl nucleophile U-1, and C75 acts as the general acid, stabilizing the 5′-oxyanion of G1. Another crystal structure containing three consecutive deoxy residues at the 5′-end of the substrate strand was also published [79]. The original crystal structures of the product form [16] and of the C75U mutated construct [78] were subject to refinement using the ERRASER program suite in order to rule out structural ambiguities that led to a better interpretation of the catalytic mechanism [80].

30.7 Recently Discovered Ribozymes

In 2015, a set of new ribozymes was identified by comparative genomics analysis [10, 11]. These ribozymes were named after the features of their secondary structures. Twister is "twisted" by a double pseudoknot, like twister sister (TS). Pistol and Hatchett look alike these objects. These ribozymes constitute additional families of endonucleolytic ribozymes, which also yield 5′-hydroxyl and 2′,3′-cyclic phosphodiester ends. With the benefit of the acquired knowledge on the crystallization of ribozymes, it took only a few years to get their crystal structures. This paragraph is dedicated to those ribozymes and to the strategies that were used to crystallize them.

30.7.1 The Twister Ribozyme

The twister ribozyme [10] is widespread within bacteria, fungi, plants, and animals. It seems to be important in genetic control, although such a function has not been demonstrated yet. The ~50-nt minimal twister ribozyme secondary structure is composed of three helices of about four base pairs separated by two internal loops. The crystal structure of the *Oryza sativa* twister ribozyme illustrates how the ribozyme folds [81]. Basically, the second internal loop adopts a sharp kink allowing stems 2 and 3 to end up side by side in a parallel orientation stabilized by the formation of a first pseudoknot (T1) between the apical loop and the 3′-strand of the first internal loop (L1). This conformation is further stabilized by a second pseudoknot (T2) resulting from the interaction of the 3′-nucleotides from the apical loop with the 5′-strand of the second internal loop (L2). The 5′-strand of the first internal loop contains a UA step where the cleavage occurs. T1 is instrumental in positioning the scissile adenosine residue (Figure 30.9a,b).

In the construct crystallized by the group of David Lilley [81], catalysis was prevented by the incorporation of a 2′-deoxyuridine (dU) in an RNA entirely chemically synthesized. A crystal packing contact involving the dU perturbs its orientation and is invoked to prevent the observation of the S_N2-characteristic O2′-P-O5′ in-line orientation. The likeliness of the mechanism of an acid/base-catalyzed reaction was demonstrated by a combination of molecular modeling studies and pH-dependent catalytic measurements of mutated ribozymes. These studies point to an acid-base cleavage mechanism, like most nucleolytic ribozymes, with G45 as a general base, since a G45-to-A mutation shifts the pK_a of the reaction. The position of A29 in the crystal structure could not explain thoroughly the acid pK_a observed in catalytic measurement experiments, leaving open the question of its involvement. Two other crystal structures of twister ribozymes from *O. sativa* and from an environmental sample without the identification of the organism of origin (*env*) were solved but

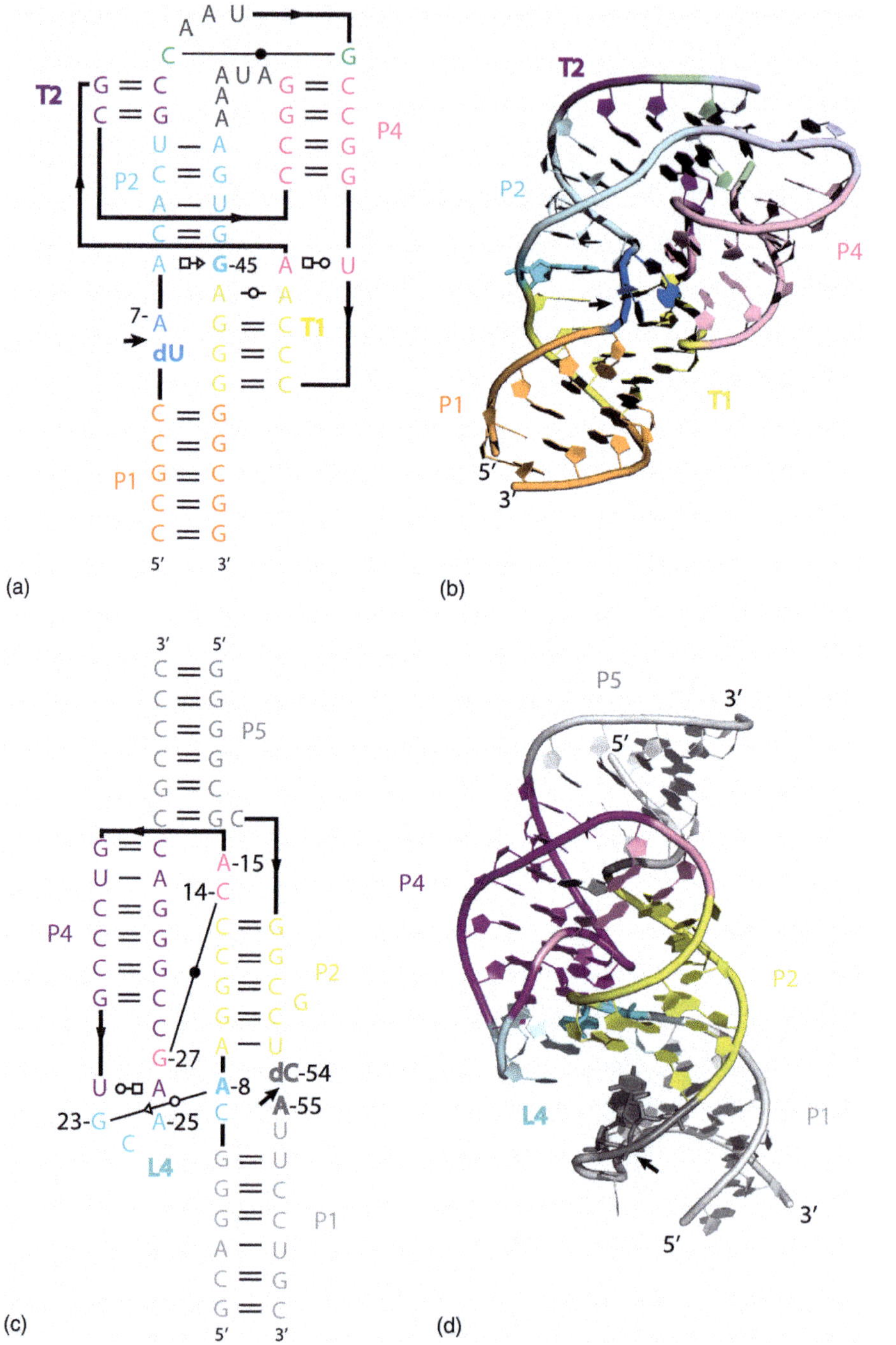

Figure 30.9 The twister (a, b) and twister sister (TS) (c, d) ribozymes. In spite of similarities at the secondary structure level, these two ribozymes adopt distinct architectures as can be seen from the secondary structures deduced from the 3D structures. The cleavage sites are indicated by short black arrows on all panels. The structure of the twister ribozyme from *Oryza sativa* is represented (a, b) from PDB file 4oji. The structure of the twister sister ribozyme prepared from PDB file 5t5a is represented on (c, d).

at lower resolutions and with geometries not permitting to decipher the catalytic mechanism [82].

An additional crystal structure from the group of Patel describing the structure of another environmental twister ribozyme (*env22*) displays a conformation where a modeled O2′ could be properly aligned with the -P-O5′ scissile bond [83]. This construct was formed by the hybridization of two RNA fragments of 19 and 37 nucleotides obtained by chemical synthesis. The 19-mer contained a dU residue at position 5 to prevent phosphotransfer at residue 6. In this structure, dU5 does not contribute to packing, which rationalizes its better orientation. The structure also confirms the involvement of G48 (equivalent to *O. sativa* G45) in the acid/base reaction through the interaction of its N1 group with a non-bridging oxygen atom of the scissile phosphate. The structure also presents a hexa-coordinated magnesium ion, which seems to be involved in catalysis.

Following up, the structure of a mini-twister ribozyme derived from the *env22* construct was solved, in which stem P1 was depleted, since in-solution studies indicated that it was dispensable for activity [84]. This construct was inactivated by 2′-*O*-methylation of U5. The conformation of the scissile dinucleotide is very close to the one observed in the structure containing a dU5 [83], although the atoms involved in the S_N2-like mechanism are more off-line than in the dU5 substituted ribozyme, a situation apparently due to the bulkiness of the methyl group that influences the pucker of the ribose ring. Nevertheless, in-depth biochemical studies demonstrated the catalytic role of the magnesium ion coordinating the scissile phosphate as well as of A6. Even though more studies are necessary to precisely identify the role of each factor, the situation is now that twister seems to adopt a mixed strategy based on metal ion catalysis on the one hand and on nucleotide-driven acid/base catalysis on the other hand [84].

30.7.2 The Twister Sister Ribozyme

The twister sister (TS) ribozyme [11] was named after secondary structure features that look alike those of the twister ribozyme. However, its 3D structure [85] departs significantly from twister's, notably by a cleavage site on the 3′-side of the first helix, which forms a distinctive catalytic pocket based on a metal ion with a putative role as a general base activating the nucleophile during catalysis. The 3D structure of the TS ribozyme is organized around two helical stacks weaving tertiary interactions resulting from the 180° rotation of one helical segment with respect to the other, leading to contacts between the apical loop (L4) and the first internal loop (L1). This twist forms a three-way junction locked by docking of a couple of unpaired nucleotides in specific receptors through stacking interactions and hydrogen bonding (Figure 30.9c,d). Like in twister ribozyme, L4 is responsible for interactions with the two internal loops, L1 and L2. It provides a receptor for A8, which donates its sugar edge to the sugar edges of G23 and gets stacked in between A25 and A26. G27, the ultimate nucleotide of the apical loop, is stretched by the backbone rotamers to reach the most distal region to base-pair with C14, providing helical continuity between P2 and A15 from the three-way junction. In L1, C54, and A55, which comprise the scissile bond, adopt a quasi-helical conformation and appear somehow disconnected from the ribozyme core, except by interaction with a magnesium ion having the O2 atom of C54 in its first shell of hydration. The helical conformation of the nucleotides at the scissile bond is not compatible with the S_N2-like mechanism and indicates that the structure [85] corresponds to a pre-catalytic state. The RNA strand design strategy was based, like in the case of twister, on the depletion of the nucleophilic 2′-hydroxyl group by chemical synthesis of two distinct strands. A 40-mer ribozyme incorporated two 5-bromocytidines, in order to solve the phase problem by anomalous diffraction methods, and a 22-mer substrate incorporated the 2′-deoxycytidine.

In another structure of a TS ribozyme solved by the group of Patel [86], the architecture is based on a four-way junction and not on a three-way junction like in the above example. The structure was solved by heavy-atom anomalous diffraction methods using iridium hexammine. The four-way junction as well as an extended L1 internal loop forms a significantly different organization of the overall structure and around the catalytic site. Firstly, the additional stem P3 extends the helical stack formed by P1 and P2 generating a four-way junction where the angle between the two helical stacks (P1–P2–P3 and P4–P5) is close to 90°. The expanded L1 provides an opportunity for a set of noncanonical base pairs, which also chelate several magnesium ions, in close vicinity of the catalytic site. Despite the fact that a similar strategy was applied to deplete the nucleophile by chemical synthesis of a ribozyme and of a substrate strand incorporating the required modification, the two nucleotides around the scissile bond do not adopt a helical conformation. While L4 serves as a receptor for residue A7 (equivalent to A8 from 5t5a), G35 (equivalent to G27 from 5t5a) interacts with a magnesium ion, which itself is connected to the backbone of the 5′-strand of L1. The intricate set of interactions between L1 and L4 residues seems to promote specific contacts within the purine-rich L1 that lead to the extrusion of the 2′-deoxycytidine residue (C62) in a conformation that could favor the S_N2-like reaction, since A63 is clamped within the helix by means of hydrogen bonds and stacking interactions blocked by magnesium ions. Surprisingly, no ion is observed bridging the A residues from L4 with the groove of L1 as observed in the structure from the group of Lilley [85].

In spite of the application of the same strategy to inactivate the nucleophile by depletion of the 2′-hydroxyl group, these two structures present very different features. The interpretation of the catalytic mechanism needs to be reinforced in one case (5t5a) with specific biochemical studies, whereas in the other case (5y85), the interpretation is more straightforward directly from the structure. The sequences of the ribozymes themselves seem to be responsible for these observations and point to the difficulty to yield exploitable structural data to shed light on catalytic mechanisms.

30.7.3 The Pistol Ribozyme

Crystal structures of the pistol ribozyme were solved quickly after its discovery and the first functional studies [87, 88]. The *env25* ribozyme [87] was made up of two chemically synthesized annealed strands, an 11-mer substrate strand, and a 47-mer ribozyme core. The *env27* ribozyme was made up of a 15-nt substrate, while the core was obtained by in vitro transcription [88]. In both cases, cleavage was prevented by the incorporation of a deoxy residue in place of G53 (*env25*) or G10 (*env27*). The secondary structure is composed of a short hairpin followed by a longer one. The latter is interrupted by a very asymmetric internal loop, the longest strand of which forms a half-helical turn pseudoknot with the loop of the former hairpin. In the 3D structure, the pseudoknot is clamped between stems P1 and P3, making an extended helical core. The strand connecting P1 to P2 elegantly follows the shallow groove of P1 and sets P2 off the ribozyme core. Since the final structures showed very similar features, we describe details from the *env25* structure. This arrangement, together with the constraints due to the formation of the P2 helix, produces the sharp kink necessary to splay apart the nucleotides around the scissile phosphate group of C54 (Figure 30.10).

Initial biochemical analysis [89], NMR, and fluorescence spectroscopy data obtained on the crystallized forms of the ribozymes point to the N3 group of A32 to be responsible for the stabilization of the 5′-oxyanion leaving group, while G40 would be able to activate the nucleophile. The dG modification does not prevent a magnesium ion to be close to the scissile phosphate. However, this ion seems to play a structural role rather than a catalytic one, since it is not directly coordinated to any RNA atom.

30.8 Conclusions and Perspectives

All the examples reported in this review show how structural models offer a powerful heuristic framework to design and interpret the results of biochemical and mutational studies toward a better understanding of the catalytic mechanisms of ribozymes. This means that even though significant knowledge is required to inactivate the ribozyme to crystallize, it is from the structures obtained that the conclusive experiments may be designed. Efficient and diverse strategies for controlling ribozyme catalysis in the context of crystallization studies is thus a valuable addition to the biochemist's toolbox. In these strategies, the solid phase chemical synthesis of oligonucleotides holds a central role.

Technical progresses in chemical and enzymatic synthesis, in heavy-atom derivatization of RNA, in software both for RNA folding predictions and crystal structure refinement as well as data collection and treatment, and in detector sensitivity allows for a faster and more efficient method for solving RNA structures. The main bottleneck remains crystallization, since the stability of an RNA molecule does depend not only on its architecture but also on its folding process, as well as on the crystal packing. In the future, X-ray free-electron laser (XFEL) facilities may help to release the pressure on obtaining large crystals, as in a recent study, in which distinct forms of a riboswitch could be observed simultaneously, [90] and thus become a more routine technique.

To summarize, the modifications that require lesser knowledge on catalysis are the deoxy substitution of the nucleophile or its methylation. Another strategy consists

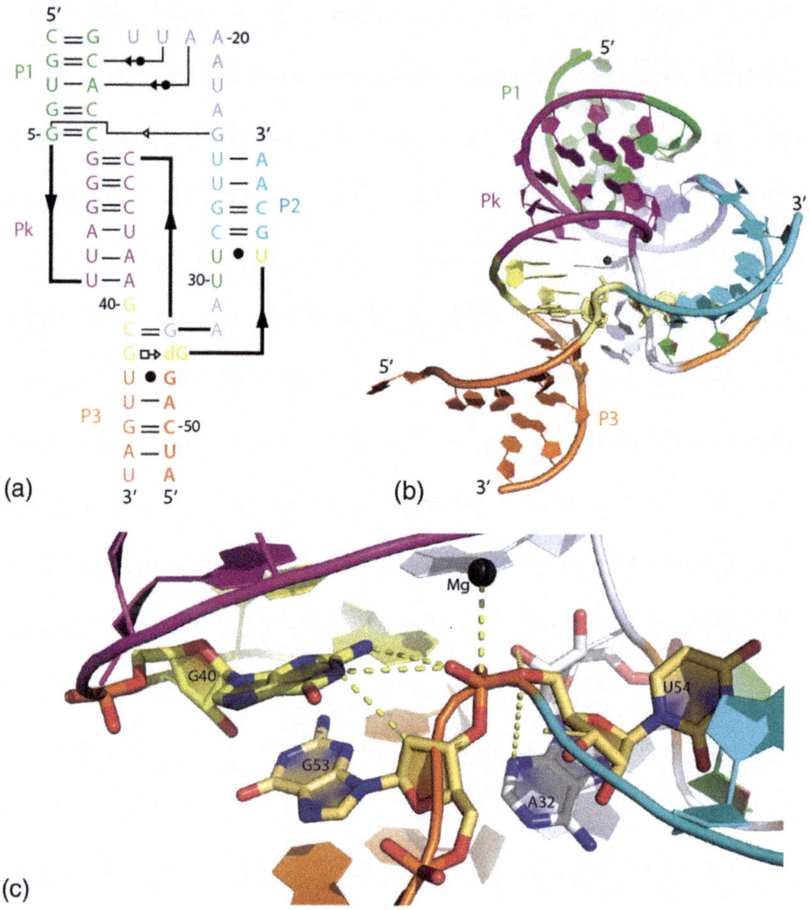

Figure 30.10 Structural aspects of the pistol ribozyme (PDB: 5k7e). (a) The secondary structure shows the topology of the RNA, emphasizing how the pseudoknot Pk is made from the interactions of the two loops. (b) The overall 3D structure shows how P1, Pk, and P2 stack on each other to form an extended helical core on the side, of which P2 docks and induces the kink at the cleavage site. (c) A close-up of the catalytic site shows the kink at the scissile phosphate between G53 and U54. Distances between atomic groups that could be important during the catalytic reaction are indicated by yellow dashed lines (G40(N1)-U54(O1P) 4.0 Å; G40(N2)-U54(O1P) 4.1 Å; G40(N1)-G53(C2′) 3.5 Å; Mg-U54(P) 4.1 Å; A32(O2′)-U54(O5′) 5.1 Å; and A32(N3)-U54(O5′) 5.4 Å).

of using a split substrate and results in the loss of the scissile phosphate. All these modifications perturb the catalytic site organization, and structural interpretations should be made with the support of biochemical data obtained on the unmodified ribozyme form. The construction of a circularly permutated ribozyme is always possible and mimics the split substrate strategy. In this case, in vitro transcription can be used to produce a single-strand RNA of any length opening up the possibility to purify the RNA of interest under native conditions directly from the transcription mix. Finally, studying the transition state structural mimic using a vanadate ion constitutes a fourth strategy that can be added on top of the split substrate or circular permutation. When nucleotide candidates involved in nucleophile activation or leaving group stabilization are identified, it is also possible to design mutated constructs with very poor catalytic properties.

Acknowledgments

BM is supported by CNRS and by the LabEx MitoCross (LABEX ANR-11-LABX-0057_MITOCROSS). DS is supported by an IdEx doctoral grant obtained from the University of Strasbourg (BGPC 17/No ARC 26349). MC acknowledges funding from the BIG LidEx program.

References

1 Guerrier-Takada, C., Gardiner, K., Marsh, T. et al. (1983). The RNA moiety of ribonuclease P is the catalytic subunit of the enzyme. *Cell* 35: 849–857.

2 Cech, T.R. (2000). The ribosome is a ribozyme. *Science* 289 (5481): 878–879. https://doi.org/10.1126/science.1289.5481.1878.

3 Galej, W.P., Toor, N., Newman, A.J., and Nagai, K. (2018). Molecular mechanism and evolution of nuclear pre-mRNA and group II intron splicing: insights from cryo-electron microscopy structures. *Chem. Rev.*118 (8), 4156–4176.

4 Hutchins, C.J., Rathjen, P.D., Forster, A.C., and Symons, R.H. (1986). Self-cleavage of plus and minus RNA transcripts of avocado sunblotch viroid. *Nucleic Acids Res.* 14 (9): 3629–3640.

5 Buzayan, J.M., Gerlach, W.L., and Bruening, G. (1986). Non-enzymatic cleavage and ligation of RNAs complementary to a plant virus satellite RNA. *Nature* 323: 349–353.

6 Sharmeen, L., Kuo, M.Y.-P., Dinter-Gottlieb, G., and Taylor, J. (1988). Antigenomic RNA of human hepatitis delta virus can undergo self-cleavage. *J. Virol.* 62: 2674–2679.

7 Saville, B.J. and Collins, R.A. (1990). A site-specific self-cleavage reaction performed by a novel RNA in *Neurospora* mitochondria. *Cell* 61: 685–696.

8 Cech, T.R., Zaug, A.J., and Grabowski, P.J. (1981). In vitro splicing of the ribosomal RNA precursor of *Tetrahymena*: involvement of a guanosine nucleotide in the excision of the intervening sequence. *Cell* 27: 487–496.

9 Michel, F., Jaquier, A., and Dujon, B. (1982). Comparison of fungal mitochondrial introns reveals extensive homologies in RNA secondary structure. *Biochimie* 64: 867–881.

10 Roth, A. Weinberg, Z., Chen, A. G., et al. (2014). A widespread self-cleaving ribozyme class is revealed by bioinformatics. *Nat. Chem. Biol.* 10 (1): 56–60.

11 Weinberg, Z. Kim, P. B., Chen, T. H., et al. (2015). New classes of self-cleaving ribozymes revealed by comparative genomics analysis. *Nat. Chem. Biol.* 11 (8): 606–610.

12 Winkler, W.C., Nahvi, A., Roth, A. et al. (2004). Control of gene expression by a natural metabolite-responsive ribozyme. *Nature* 428 (6980): 281–286.

13 Kondo, J., Sauter, C., and Masquida, B. (2014). RNA crystallization. In: *Handbook of RNA biochemistry* (eds. R.K. Hartmann, A. Bindereif, A. Schön and E. Westhof), 481–498. Wiley VCH Verlag Gmbh & Co.

14 Lilley, D.M.J. (2017). How RNA acts as a nuclease: some mechanistic comparisons in the nucleolytic ribozymes. *Biochem. Soc. Trans.* 45 (3): 683–691.

15 Raines, R.T. (1998). Ribonuclease A. *Chem. Rev.* 98 (3): 1045–1066.

16 Ferré-D'Amaré, A.R., Zhou, K.H., and Doudna, J.A. (1998). Crystal structure of a hepatitis delta virus ribozyme. *Nature* 395 (6702): 567–574.

17 Klein, D.J. and Ferre-D'Amare, A.R. (2006). Structural basis of glmS ribozyme activation by glucosamine-6-phosphate. *Science* 313 (5794): 1752–1756.

18 Cochrane, J.C., Lipchock, S.V., and Strobel, S.A. (2007). Structural investigation of the GlmS ribozyme bound to Its catalytic cofactor. *Chem. Biol.* 14 (1): 97–105.

19 Ferre-D'Amare, A.R. and Doudna, J.A. (2000). Crystallization and structure determination of a hepatitis delta virus ribozyme: use of the RNA-binding protein U1A as a crystallization module. *J. Mol. Biol.* 295 (3): 541–556.

20 Leontis, N.B. and Westhof, E. (2001). Geometric nomenclature and classification of RNA base pairs. *RNA* 7 (4): 499–512.

21 Sussman, J.L., Holbrook, S.R., Warrant, R.W. et al. (1978). Crystal structure of yeast phenylalanine tRNA. I. Crystallographic refinement. *J. Mol. Biol.* 123: 607–630.

22 Westhof, E., Dumas, P., and Moras, D. (1985). Crystallographic refinement of yeast aspartic acid transfer RNA. *J. Mol. Biol.* 184: 119–145.

23 Pley, H.W., Flaherty, K.M., and McKay, D.B. (1994). Three-dimensional structure of a hammerhead ribozyme, *Nature.* 372 (3 Nov): 68–74.

24 Scott, W.G., Finch, J.T., and Klug, A. (1995). The crystal structure of an all-RNA hammerhead ribozyme : a proposed mechanism for RNA catalytic cleavage. *Cell* 81: 991–1002.

25 Cate, J.H. Gooding, A. R., Podell, E et al. (1996). Crystal structure of a group I ribozyme domain: principles of RNA packing. *Science* 273: 1678–1684.

26 Branch, A.D. and Robertson, H.D. (1984). A replication cycle for viroids and other small infectious RNA's. *Science* 223 (4635): 450–455.

27 Tanner, M.A., Anderson, E.M., Gutell, R.R., and Cech, T.R. (1997). Mutagenesis and comparative sequence analysis of a base triple joining the two domains of group I ribozymes. *RNA* 3 (9): 1037–1051.

28 Shih, I.H. and Been, M.D. (2001). Involvement of a cytosine side chain in proton transfer in the rate-determining step of ribozyme self-cleavage. *Proc. Natl. Acad. Sci. U.S.A.* 98 (4): 1489–1494.

29 Nakano, S., Proctor, D.J., and Bevilacqua, P.C. (2001). Mechanistic characterization of the HdV genomic ribozyme: assessing the catalytic and structural contributions of divalent metal ions within a multichannel reaction mechanism. *Biochemistry* 40 (40): 12022–12038.

30 Luptak, A., Ferre-D'Amare, A.R., Zhou, K. et al. (2001). Direct pK_a measurement of the active-site cytosine in a genomic hepatitis delta virus ribozyme. *J. Am. Chem. Soc.* 123 (35): 8447–8452.

31 Milewski, S. (2002). Glucosamine-6-phosphate synthase – the multi-facets enzyme. *Biochim. Biophys. Acta* 1597 (2): 173–192.

32 Savinov, A. and Block, S.M. (2018). Self-cleavage of the glmS ribozyme core is controlled by a fragile folding element. *Proc. Natl. Acad. Sci. U.S.A.* 115 (47), 11976–11981.

33 Klein, D.J. and Ferre-D'Amare, A.R. (2009). Crystallization of the glmS ribozyme-riboswitch. *Methods Mol. Biol.* 540: 129–139.

34 Jimenez, R.M., Delwart, E., and Luptak, A. (2011). Structure-based search reveals hammerhead ribozymes in the human microbiome. *J. Biol. Chem.* 286 (10): 7737–7743.

35 Fedor, M.J. and Uhlenbeck, O.C. (1990). Substrate sequence effects on "hammerhead" RNA catalytic efficiency. *Proc. Natl. Acad. Sci. U.S.A.* 87 (5): 1668–1672.

36 Ruffner, D.E., Dahm, S.C., and Uhlenbeck, O.C. (1989). Studies on the hammerhead RNA self-cleaving domain. *Gene* 82 (1): 31–41.

37 Caruthers, M.H. (2011). A brief review of DNA and RNA chemical synthesis. *Biochem. Soc. Trans.* 39 (2): 575–580.

38 Hartsel, S.A., Kitchen, D.E., Scaringe, S.A., and Marshall, W.S. (2005). RNA oligonucleotide synthesis via 5′-silyl-2′-orthoester chemistry. *Methods Mol. Biol.* 288: 33–50.

39 Scott, W.G., Murray, J.B., Arnold, J.R.P. et al. (1996). Capturing the structure of a catalytic RNA intermediate: the hammerhead ribozyme. *Science* 274: 2065–2069.

40 Khvorova, A., Lescoute, A., Westhof, E., and Jayasena, S.D. (2003). Sequence elements outside the hammerhead ribozyme catalytic core enable intracellular activity. *Nat. Struct. Biol.* 10 (9): 708–712.

41 Canny, M.D. Jucker, F. M., Kellogg, E. et al. (2004). Fast cleavage kinetics of a natural hammerhead ribozyme. *J. Am. Chem. Soc.* 126 (35): 10848–10849.

42 Martick, M. and Scott, W.G. (2006). Tertiary contacts distant from the active site prime a ribozyme for catalysis. *Cell* 126 (2): 309–320.

43 Nelson, J.A. and Uhlenbeck, O.C. (2006). When to believe what you see. *Mol. Cell* 23 (4): 447–450.

44 Adams, P.L., Stahley, M.R., Kosek, A.B. et al. (2004). Crystal structure of a self-splicing group I intron with both exons. *Nature* 430 (6995): 45–50.

45 Adams, P.L. Stahley, M. R., Kosek, A. B et al. (2004). Crystal structure of a group I intron splicing intermediate. *RNA* 10 (12): 1867–1887.

46 Ferre-D'Amare, A.R. (2010). Use of the spliceosomal protein U1A to facilitate crystallization and structure determination of complex RNAs. *Methods.* 52 (2), 159–167.

47 Stahley, M.R. and Strobel, S.A. (2005). Structural evidence for a two-metal-ion mechanism of group I intron splicing. *Science* 309 (5740): 1587–1590.

48 Guo, F., Gooding, A.R., and Cech, T.R. (2004). Structure of the *Tetrahymena* ribozyme; base triple sandwich and metal ion at the active site. *Mol. Cell* 16 (3): 351–362.

49 Golden, B.L., Kim, H., and Chase, E. (2005). Crystal structure of a phage Twort group I ribozyme-product complex. *Nat. Struct. Mol. Biol.* 12 (1): 82–89.
50 Lambowitz, A.M. and Belfort, M. (2015). Mobile bacterial group II introns at the crux of eukaryotic evolution. *Microbiol. Spectr.* 3 (1) MDNA3-0050-2014. doi:10.1128/microbiolspec.MDNA3-0050-2014
51 Costa, M., Walbott, H., Monachello, D. et al. (2016). Crystal structures of a group II intron lariat primed for reverse splicing. *Science* 354 (6316). aaf9258
52 Hampel, A., Tritz, R., Hicks, M., and Cruz, P. (1990). 'Hairpin' catalytic RNA model : evidence for helices and sequence requirement for substrate RNA. *Nucleic Acids Res.* 18 (2): 299–304.
53 Murchie, A.I.H., Thomson, J.B., Walter, F., and Lilley, D.M. (1998). Folding of the hairpin in its natural conformation achieves close physical proximity of the loops. *Mol. Cell* 1 (6): 873–881.
54 Walter, N.G., Burke, J.M., and Millar, D.P. (1999). Stability of hairpin ribozyme tertiary structure is governed by the interdomain junction. *Nat. Struct. Biol.* 6 (6): 544–549.
55 McKinney, S.A. Tan, E., Wilson, T. J. et al. (2004). Single-molecule studies of DNA and RNA four-way junctions. *Biochem. Soc. Trans.* 32 (Pt 1): 41–45.
56 Lilley, D.M. and Norman, D.G. (1999). The Holliday junction is finally seen with crystal clarity. *Nat. Struct. Biol.* 6 (10): 897–899.
57 Rupert, P.B. and Ferre-D'Amare, A.R. (2001). Crystal structure of a hairpin ribozyme-inhibitor complex with implications for catalysis. *Nature* 410 (6830): 780–786.
58 Rupert, P.B., Massey, A.P., Sigurdsson, S.T., and Ferre-D'Amare, A.R. (2002). Transition state stabilization by a catalytic RNA. *Science* 298 (5597): 1421–1424.
59 Mir, A. Chen, J., Robinson, K., et al. (2015). Two divalent metal ions and conformational changes play roles in the hammerhead ribozyme cleavage reaction. *Biochemistry* 54 (41): 6369–6381.
60 Mir, A. and Golden, B.L. (2016). Two active site divalent ions in the crystal structure of the hammerhead ribozyme bound to a transition state analogue. *Biochemistry* 55 (4): 633–636.
61 Meyer, M. Nielsen, H., Olieric, V., et al. (2014). Speciation of a group I intron into a lariat capping ribozyme. *Proc. Natl. Acad. Sci. U.S.A.* 111 (21): 7659–7664.
62 Nielsen, H., Westhof, E., and Johansen, S. (2005). An mRNA is capped by a 2′, 5′ lariat catalyzed by a group I-like ribozyme. *Science* 309 (5740): 1584–1587.
63 Tang, Y., Nielsen, H., Masquida, B. et al. (2014). Molecular characterization of a new member of the lariat capping twin-ribozyme introns. *Mobile DNA* 5: 25.
64 Meyer, M. and Masquida, B. (2014). Cis-Acting 5′ hammerhead ribozyme optimization for in vitro transcription of highly structured RNAs. *Methods Mol. Biol.* 1086: 21–40.
65 Li, X., Yang, L., and Chen, L.L. (2018). The biogenesis, functions, and challenges of circular RNAs. *Mol. Cell* 71 (3): 428–442.
66 Franz, A. Rabien, A., Stephan, C. et al. (2018). Circular RNAs: a new class of biomarkers as a rising interest in laboratory medicine. *Clin. Chem. Lab. Med.*56 (12), 1992-2003.

67 Plagens, A., Daume, M., Wiegel, J., and Randau, L. (2015). Circularization restores signal recognition particle RNA functionality in *Thermoproteus*. *eLife* 4. DOI: 10.7554/eLife.11623

68 Hammann, C., Luptak, A., Perreault, J., and de la Pena, M. (2012). The ubiquitous hammerhead ribozyme. *RNA* 18 (5): 871–885.

69 Webb, C.H. and Luptak, A. (2011). HdV-like self-cleaving ribozymes. *RNA Biol.* 8 (5). 719-727

70 Gott, J.M., Pan, T., LeCuyer, K.A., and Uhlenbeck, O.C. (1993). Using circular permutation analysis to redefine the R17 coat protein binding site. *Biochemistry* 32 (49): 13399–13404.

71 Harris, M.E. and Christian, E.L. (1999). Use of circular permutation and end modification to position photoaffinity probes for analysis of RNA structure. *Methods* 18 (1): 51–59.

72 Ohuchi, S.J., Sagawa, F., Sakamoto, T., and Inoue, T. (2015). Altering the orientation of a fused protein to the RNA-binding ribosomal protein L7Ae and its derivatives through circular permutation. *Biochem. Biophys. Res. Commun.* 466 (3): 388–392.

73 Truong, J., Hsieh, Y.F., Truong, L. et al. (2018). Designing fluorescent biosensors using circular permutations of riboswitches. *Methods*. 143, 102–109.

74 Pan, T., Fang, X., and Sosnick, T. (1999). Pathway modulation, circular permutation and rapid RNA folding under kinetic control. *J. Mol. Biol.* 286 (3): 721–731.

75 Suslov, N.B. DasGupta, S., Huang, H., et al. (2015). Crystal structure of the Varkud satellite ribozyme. *Nat. Chem. Biol.* 11 (11): 840–846.

76 Fernandez-Millan, P. Schelcher, C., Chihade, J. et al. (2016). Transfer RNA: from pioneering crystallographic studies to contemporary tRNA biology. *Arch. Biochem. Biophys.* 602: 95–105.

77 Wilson, T.J. and Lilley, D.M. (2011). Do the hairpin and VS ribozymes share a common catalytic mechanism based on general acid–base catalysis? A critical assessment of available experimental data. *RNA* 17 (2): 213–221.

78 Ke, A., Zhou, K., Ding, F. et al. (2004). A conformational switch controls hepatitis delta virus ribozyme catalysis. *Nature* 429 (6988): 201–205.

79 Chen, J.H. Yajima, R., Chadalavada, D. M., et al. (2010). A 1.9 A crystal structure of the HdV ribozyme precleavage suggests both Lewis acid and general acid mechanisms contribute to phosphodiester cleavage. *Biochemistry* 49 (31): 6508–6518.

80 Kapral, G.J. Jain, S., Noeske, J. et al. (2014). New tools provide a second look at HdV ribozyme structure, dynamics and cleavage. *Nucleic Acids Res.* 42 (20): 12833–12846.

81 Liu, Y., Wilson, T.J., McPhee, S.A., and Lilley, D.M. (2014). Crystal structure and mechanistic investigation of the twister ribozyme. *Nat. Chem. Biol.* 10 (9): 739–744.

82 Eiler, D., Wang, J., and Steitz, T.A. (2014). Structural basis for the fast self-cleavage reaction catalyzed by the twister ribozyme. *Proc. Natl. Acad. Sci. U.S.A.* 111 (36): 13028–13033.

83 Ren, A. Kosutic, M., Rajashankar, K. R., et al. (2014). In-line alignment and Mg^{2+} coordination at the cleavage site of the env22 twister ribozyme. *Nat. Commun.* 5: 5534.

84 Kosutic, M. Neuner, S., Ren, A. et al. (2015). A mini-twister variant and impact of residues/cations on the phosphodiester cleavage of this ribozyme class. *Angew. Chem. Int. Ed.* 54 (50): 15128–15133.

85 Liu, Y., Wilson, T.J., and Lilley, D.M.J. (2017). The structure of a nucleolytic ribozyme that employs a catalytic metal ion. *Nat. Chem. Biol.* 13 (5): 508–513.

86 Zheng, L. Mairhofer, E., Teplova, M. et al. (2017). Structure-based insights into self-cleavage by a four-way junctional twister-sister ribozyme. *Nat. Commun.* 8 (1): 1180.

87 Ren, A. Vušurović, N., Gebetsberger, J. et al. (2016). Pistol ribozyme adopts a pseudoknot fold facilitating site-specific in-line cleavage. *Nat. Chem. Biol.* 12: 702.

88 Nguyen, L.A., Wang, J., and Steitz, T.A. (2017). Crystal structure of Pistol, a class of self-cleaving ribozyme. *Proc. Natl. Acad. Sci. U.S.A.* 114 (5): 1021–1026.

89 Harris, K.A., Lunse, C.E., Li, S. et al. (2015). Biochemical analysis of pistol self-cleaving ribozymes. *RNA* 21 (11): 1852–1858.

90 Stagno, J.R. Liu, Y., Bhandari, Y. R., et al. (2017). Structures of riboswitch RNA reaction states by mix-and-inject XFEL serial crystallography. *Nature* 541 (7636): 242–246.

31

NMR Spectroscopic Investigation of Ribozymes

Bozana Knezic, Oliver Binas, Albrecht Eduard Völklein, and Harald Schwalbe

Goethe University, Center for Biomolecular Magnetic Resonance (BMRZ), Institute for Organic Chemistry and Chemical Biology, Max von Laue Straße 7, 60438 Frankfurt am Main, Germany

31.1 Introduction

Nuclear magnetic resonance (NMR) spectroscopy is a highly versatile physical method to study structure and function of proteins, oligonucleotides, and their complexes at atomic resolution. With the advent of labeling RNA with NMR-active stable isotopes ^{13}C and ^{15}N, NMR is able to characterize biologically relevant RNAs and RNA–ligand and RNA–protein complexes and determine key aspects not only to further our understanding in structural terms but also to understand enzyme reactivity.

We outline the power of NMR spectroscopy to investigate structure, dynamics, and function of catalytically active RNA ribozymes. We will introduce NMR-based methods and also cover practical aspects of NMR in the first part of the chapter. In the second part, we will provide a summary of NMR studies of the ribozyme classes shown in Figure 31.1.

31.2 Methods and Preparation

NMR experiments of RNAs [1, 2] can be conducted in solution but recently also in the solid state. Solid-state NMR emerged as a new technique for the structural characterization of RNA just a few years ago, introducing several new aspects to NMR of RNA. While liquid-state NMR is limited in sample size due to unbeneficial relaxation properties of huge systems, resolution in ssNMR is independent of sample size that only affects signal intensity. Resolution in ssNMR is instead dependent on sample preparation. Typical freeze-drying methods like lyophilization yield samples, featuring insufficient resolution for structure determination due to structural heterogeneity [2]. With new advances like micro-crystallization [3], it is possible to reach line widths of 0.3 ppm, with solution NMR still being far superior. An advantage of

Ribozymes, First Edition. Edited by Sabine Müller, Benoît Masquida, and Wade Winkler.

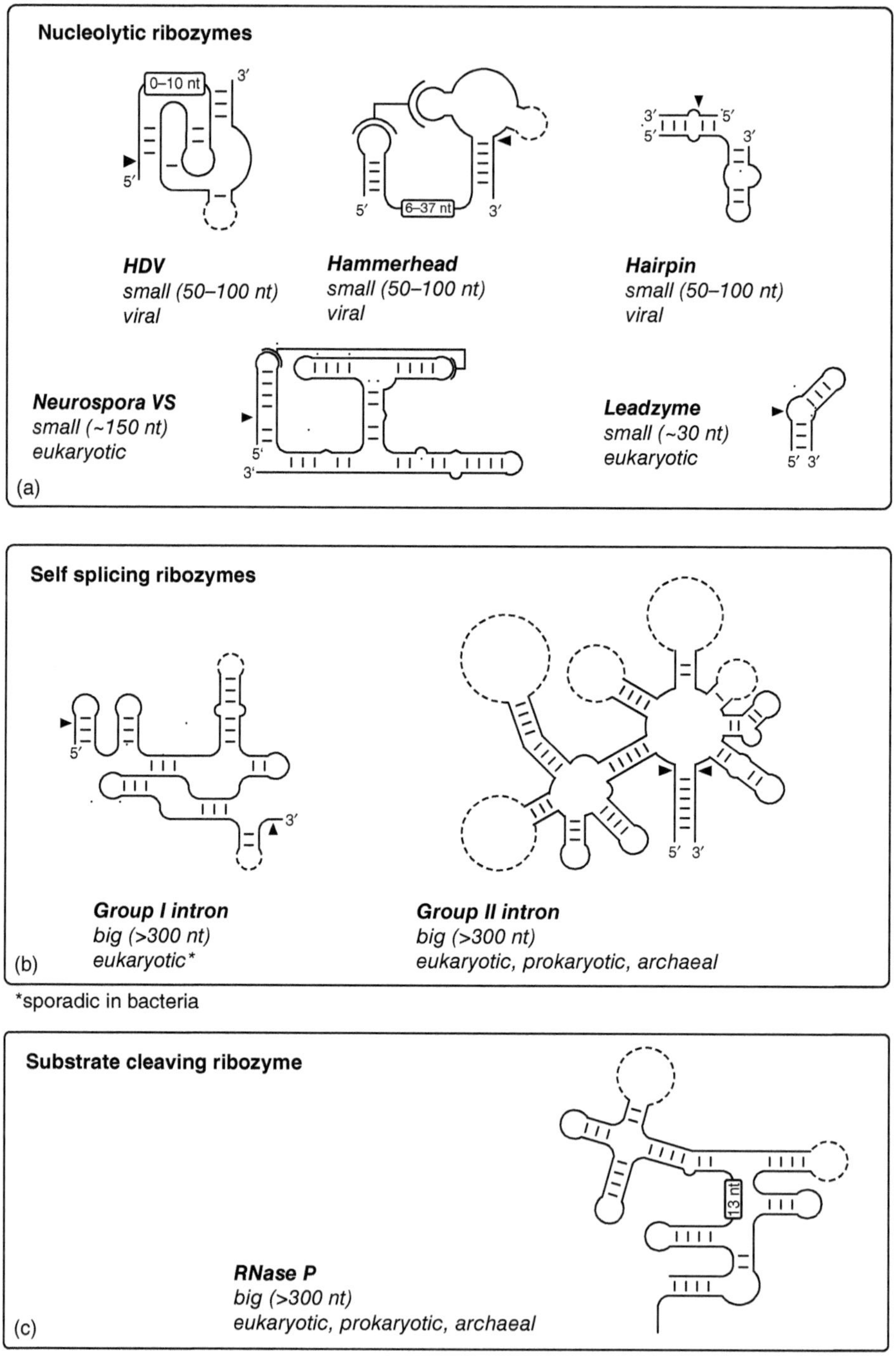

Figure 31.1 Different classes of ribozymes are shown. Ribozymes can be classified into three different types, namely, ribozymes with nucleolytic activity (a), with self-splicing activity (b), and with substrate-cleaving activity (c). Occurrence in different kingdoms of life (prokaryotes, eukaryotes, and archaea) is pointed out below in each ribozyme. Nucleotides are abbreviated as nt. Base-paired regions are depicted schematically. Small loops are represented by non-dashed lines. Loops of variable size are represented by dashed lines.

investigating RNA with ssNMR is the correlation of heteronuclei with dipolar interactions, which is not possible in solution NMR due to molecular tumbling [4]. As RNA is sparse of protons, this provides an interesting benefit over solution NMR. In 2015, Carlomagno and coworkers showed that it is possible to determine a structure of a 26-nt RNA with ssNMR [5]. Since ssNMR on RNA is an emerging technique, not many studies exist featuring RNA or ribozymes. Accordingly we will focus on NMR studies on ribozymes in solution in this review.

Depending on the kind of information, solution NMR experiments of ribozyme require between 50 μM and 0.5 mM sample in a volume of 0.3 ml. For an initial screening with simple experiments, quantities of 2–5 nmoles can be sufficient, while more sophisticated experiments require over 150 nmoles of substance. Experiments can be conducted on unlabeled samples, but labeling of RNAs with ^{13}C and ^{15}N is typically performed.

An NMR setup for studies on biomacromolecules consists of an NMR spectrometer with a helium-cooled superconducting magnet with field strength between 14 and 21 T, resulting in a ^{1}H frequency in the range of 400–950 MHz. A console with the required radiofrequency (rf) electronics including frequency generator and power amplifier for 4–6 channels allowing to apply rf pulse on the NMR-active nuclei ^{1}H, ^{2}H, ^{13}C, ^{15}N, ^{19}F, and ^{31}P is required. A multichannel probe, containing the RNA sample in an NMR tube of 0.3–0.5 ml volume, and the circuit for rf application at any desired temperature are required as well. In order to increase the signal-to-noise ratio for different NMR experiments, the receiving coil can be cooled, and different setups for detected nuclei have been implemented.

There are different ways to prepare isotope-labeled RNAs. Enzymatic in vitro transcription of RNA (Figure 31.2a) is probably the most straightforward and cost-efficient way to generate RNA samples and requires – besides the DNA template and common laboratory chemicals – only RNA polymerase and the respective ribonucleoside triphosphates (rNTPs). The DNA template contains the necessary promoter sequence combined with the target DNA sequence. In common practice, DNA oligonucleotides are produced via PCR or linearized DNA plasmids. Isotope-labeled rNTPs, which serve as raw material, are commercially available at reasonable prices [6, 7]. RNA polymerases that provide the required throughput are as well commercially available, but we typically prepare it in-house [8–11]. Purification of the transcribed target RNA involves desalting, followed by HPLC or preparative PAGE purification. Advantages of the method include the high yields of RNA at reasonable cost and the acceptable preparative effort.

Helmling et al. established a fast and high-throughput compatible native method [12]. This method involves centrifugal devices to remove NTPs and salts right after the in vitro transcription or ligation process. Furthermore, in the case of using NMR buffer as washing material, the purified sample may already be concentrated and prepared for the transfer to an NMR tube. Residual enzymes and DNA cannot be removed by this technique. On the other hand, these residues are generally present in such low concentrations or are non-labeled with NMR-active nuclei ^{13}C or ^{15}N; they do not disturb the structure and function of the target ribozyme nor the relevant NMR signals. Unfortunately all these methods do not allow the incorporation

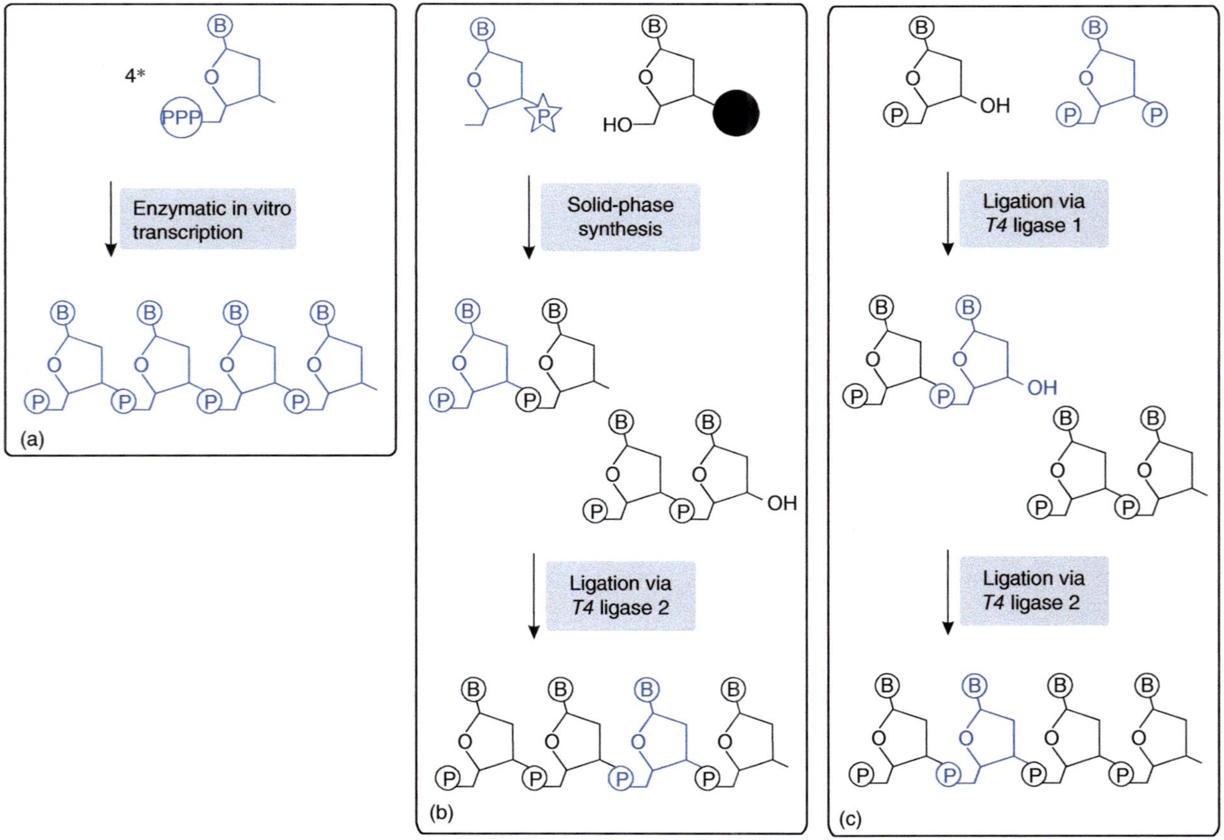

Figure 31.2 Outline of three different RNA labeling methods. Labeled nucleotides are highlighted in blue. (a) In vitro transcription of RNA with labeled rNTPs. (b) Solid-phase synthesis of RNA with side-specific incorporation of one labeled nucleotide phosphoramidite and a subsequent ligation via T4 RNA ligase 2. The black dot marks the solid phase, e.g. CPG, which binds the 3′-end of the synthesized RNA. (c) Chemo-enzymatic ligation of RNA via T4 RNA ligase 1 and 2 with side-specific incorporation of a nucleotide bisphosphate.

of specifically labeled sites in a straightforward manner, limiting the production to uniformly or nucleotide-specific labeled samples in most cases.

31.2.1 Labeling of Particular RNA Regions

For detailed NMR investigations exploiting the full potential of this biophysical technique, the ribozyme or particular regions of a ribozyme have to be labeled specifically.

Solid-phase synthesis is a well-known and established method for the site-specific labeling of oligonucleotides. Synthesis proceeds from the 3′-end to the 5′-end, and the 3′-end is bound to a solid phase. The standard procedure is a fast and automated cycle: deprotection of the 5′-end is followed by coupling with the 3′-end of a nucleoside phosphoramidite and capping of the phosphate [13]. At this point, a modified nucleoside phosphoramidite can be implemented at the 5′-end for modifications at a specific position of a long ribozyme; it is possible to ligate another RNA sequence to the 5′-end of the modified RNA. The *T4* RNA ligase 2 is applicable in this case and accomplishes ligation. The procedure is shown in Figure 31.2b. Yet, the length of the target ribozyme is limited in terms of quality and quantity needed for an NMR sample [14, 15]. Side products accumulate during solid-phase synthesis with increasing oligonucleotide length. Specific side reactions including 5′-2′ coupling are more likely to happen in longer RNAs [13]. Further, with every synthetic step, parts of the molecule have to be protected, and the protection group has to be eliminated, leading to lower synthetic yield. The purification of the target ribozyme is another limiting step toward preparation of mg quantities required for an NMR sample. While preparative PAGE purification offers no limitation regarding the target ribozyme length, HPLC purification can only be done for a maximum length of 60 nt in average while still maintaining nucleotide resolution [16]. In conclusion, solid-phase synthesis becomes increasingly expensive for longer RNAs and is practically and routinely used only for short constructs. For the preparation of RNAs longer than ~80 nt, the chemo-enzymatic ligation is more suitable to overcome the solid-phase synthesis limitations [17].

The chemo-enzymatic procedure with *T4* RNA ligases 1 and 2, outlined in Figure 31.2c, allows the site-specific labeling of RNAs with no limitation. For this purpose, the target RNA sequence is split at certain positions according to labeling requirements. Oligoribonucleotides with an OH group at C3′ at the 3′-terminal nucleotide are ligated with isotopically labeled or modified nucleoside 3′,5′-bisphosphates using *T4* RNA ligase 1 [18, 19]. In a follow-up step, the ribozyme with implemented modification is dephosphorylated by RNA phosphatase, e.g. rSAP. The *T4* RNA ligase 2 catalyzes the formation of a phosphodiester bond between the 3′-terminal hydroxyl from the labeled oligoribonucleotide and a 5′-terminal phosphate of the residual ribozyme sequence [20]. This ligation step requires support from a DNA splint, which brings the two oligoribonucleotides in close proximity. Subsequently the side specifically labeled ribozyme has to be purified to separate the individual ligation components. HPLC is the fastest and the most common strategy to purify the product ribozyme. However, preparative PAGE purification is often applied for longer ribozymes [21]. The limitation of this approach is given by the type of labeling, in particular, if nonnatural nucleosides have been incorporated.

31.2.2 Photolabile Caging of RNAs

The photocaging strategy is based on the incorporation of one or more nucleotides, which have a photolabile group at functional groups of the nucleotide [22, 23]. This photolabile group can prevent the nucleotide from formation of a Watson–Crick base pair with another nucleotide. By irradiation with UV light of a certain wavelength, the modification can be eliminated, regenerating the original and inhibited function of the ribozyme (Figure 31.3). In this case, the eliminated photolabile group must not interfere with the structure or function of the original ribozyme [24, 25]. This strategy is widely used in cases where the ribozyme is able to assume more than one secondary structure to trap one of these structures. Photolabile groups are named photolabile cages too, as they "cage" a particular secondary structure of a target ribozyme. Therefore, it is possible to screen through the conformational space of different ribozyme structures.

One of the widely used photolabile cages is the NPE – *ortho*-nitrophenyl ethyl – group [26, 27]. This photolabile cage is bound to the base-pair site of each nucleotide. It protects the nucleotide from pairing with the analogous nucleotide and therefore inhibits one potential secondary structure of the ribozyme. For the analysis of such a ribozyme, an NMR sample is prepared, containing the photolabile NPE group. That kind of sample has to be protected from light, in order to preserve the photolabile NPE group from detachment. The examination is realized in a special NMR tube with a hollow plunger, which can be connected to a laser, while it is plugged into an NMR spectrometer. Through this connection, real-time NMR (RT-NMR) experiments become possible [28]. A laser pulse of 365 nm is sufficient to remove the NPE group from the ribozyme, and as a result, the caged secondary structure of the ribozyme will immediately recover. Observed NMR spectra will reveal potential rearrangements as signal differences, especially in the imino region.

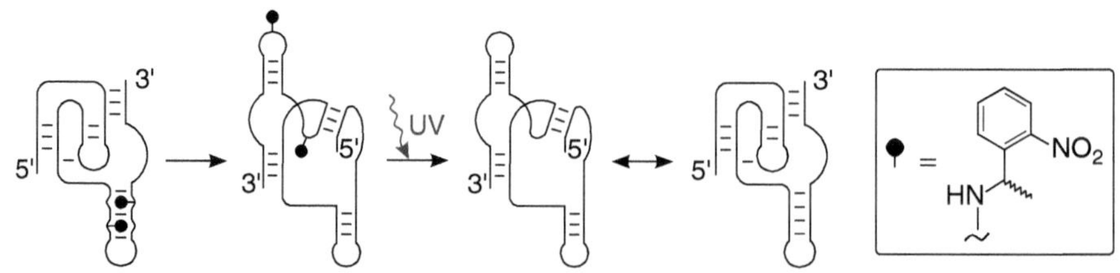

Figure 31.3 Illustration of the effect caused by the incorporation of a photolabile caging group in the target RNA (HDV ribozyme). The caged RNA prefers a secondary structure (ligase conformation), where the cage does not disturb any hydrogen bonds. After irradiation with UV light, the photolabile cage is cleaved off, and the RNA secondary structure (HDV conformation) is recovered. The NPE group is shown on the right.

31.2.3 Initial Screening of RNA Constructs by NMR

A typical NMR investigation of an RNA sample starts with mapping of the conformational space the RNA adopts in solution. In the case of ribozymes, for instance, multiple folding topologies or stages of RNA processing can be present. Simple ^{1}H-1D NMR spectra provide an overview of the topological coherence of the analyzed RNA.

The most relevant resonances in this approach are the ^{1}H resonances of the imino protons in uracil and guanine nucleobases. In the absence of hydrogen bonds or other means to protect from solvent exchange, the exchange rate with solvent water molecules exceeds the acquisition time regime of an NMR measurement [29], thus rendering these protons invisible to NMR. Imino protons involved in a stable base pair on the other hand resonate in the region between 10 and 15 ppm and can be detected as their exchange speed is substantially slower [30]. These imino protons are referred to as protected from exchange. Furthermore, exchange protected imino protons in a Watson–Crick-type base pair resonate in the region between 12 and 15 ppm while noncanonically base-paired imino protons might exhibit a different chemical shift. These characteristics paired with the comparatively high dispersion of their chemical shifts make imino protons ideal probes for the analysis of base-pair stability and conformational homogeneity of the RNA in general.

Aromatic protons of all bases resonate between 6 and approximately 8.5 ppm. Since their exchange with the solvent is neglectable (but for the H8 of purine nucleotides), aromatic protons are visible for all bases and typically feature lower line width. Therefore these resonances are better suited for quantification than imino proton resonances. Multiple peaks of a single well-separated aromatic proton resonance hint to the presence of heterogeneities in the sample. Therefore, peak integration can yield reliable results as possible differences in exchange are not to be expected. Unfortunately, due to the high number of aromatic protons in a longer RNA and their low chemical shift dispersion, the region is often too crowded to be properly analyzed in a 1D experiment.

The short measurement time of only a few minutes qualifies 1D-NMR for high-throughput screening approaches or screenings of a single sample under multiple conditions.

Probing of general base-pair stabilities can be accomplished by measuring at different temperatures while monitoring the imino proton region. Destabilization of base pairs is instantly visible by broadening and subsequent diminishing of imino proton peaks. 1D measurements at different temperatures can serve as a valuable basis to establish experimental conditions for further characterization [31].

Furthermore, perturbation of chemical shifts can probe the effects of interaction partners such as small-molecular-weight ligands or proteins on an RNA sample. Upon addition of such an interaction partner, the strength of the chemical shift perturbation gives information about the location of the binding event in the case of assigned resonances [32]. Additionally, the spectral characteristics allow the classification of the interaction into a strong intermediate or weak exchange regime. In a titration process, thermodynamic parameters such as the dissociation constant can be determined from such data.

Especially in the process of functional explorations of ribozymes, the interaction with metal ions is a highly investigated topic. As magnesium ions play a catalytic role in numerous ribozymatic reactions, Mg^{2+} binding is extensively studied [33–35]. Even though magnesium has an NMR-active isotope with a natural abundance of 10%, NMR of Mg^{2+} is seldom measured. Alternatively, it is reasonable to spike samples with Mn^{2+} [36] or Cd^{2+} [37] ions to substitute for Mg^{2+}. As the addition of Mn^{2+} leads to strong paramagnetic relaxation effects, a structural placement of the binding location is possible. Cd^{2+} ions on the other hand lead to prominent chemical shift perturbation in ^{31}P-1D spectra, which will be discussed in more detail later on.

With 1D-NMR techniques, even kinetic traces can be determined in the case of reasonably slow reactions while the ongoing process is simultaneously monitored on an atomic level [26].

The high number of possible applications make 1D-NMR one of the best suited tools for the characterization and screening of RNA and ribozymes in particular while keeping the experimental effort on an attainable level.

31.3 Resonance Assignment

In order to determine secondary structure of a ribozyme, NOESY experiments are very useful [38]. Nuclear Overhauser effect and exchange spectroscopy (NOESY) experiments are 2D experiments, which are based on the nuclear Overhauser effect arising between protons of a target ribozyme. To investigate the base-pairing pattern of a target ribozyme, exchangeable protons in the imino region (15–10 ppm) have to be examined. In this case, the observation of proton resonances in the direct and indirect dimension results in a 2D NMR spectrum with diagonal and cross signals. Diagonal peaks arise for every G or U residue. Cross signals on the other hand display the NOE contacts between neighboring paired Gs and Us. Figure 31.4 shows imino cross and diagonal peaks in the top panel.

In general, a so-called imino walk can determine elements of the secondary structure by displaying paired bases, due to spatial proximity. In all remaining cases, paired bases indicated by diagonal signals in this kind of 2D spectrum do not have any related paired base in their close surrounding (<5 Å). To differentiate between the bases, it is essential to have a look at the positions and intensities of the respective signals. G:U base pairs, for instance, are usually located in the region from 12 to 10 ppm and exhibit an intense NOE cross peak. With the information on the imino region, it is possible to examine NOE contacts to the aromatic H2 protons as well as the amino protons of cytosine residues. As especially the cytosine amino protons lie in close proximity to the aromatic protons and aromatic protons in the stem are typically within a distance below 5 Å to each other, an assignment of these is possible from the NOESY alone. The aromatic assignment can be obtained as well from NOE to the intraresidual H1′ protons and their NOE to the aromatic proton of the neighboring base. Further, NOE contacts from the aromatic spin system to the sugar protons of a target ribozyme enable these assignments in principle as well. Unfortunately, the peak dispersion for sugar protons other than H1′ is poor. As a

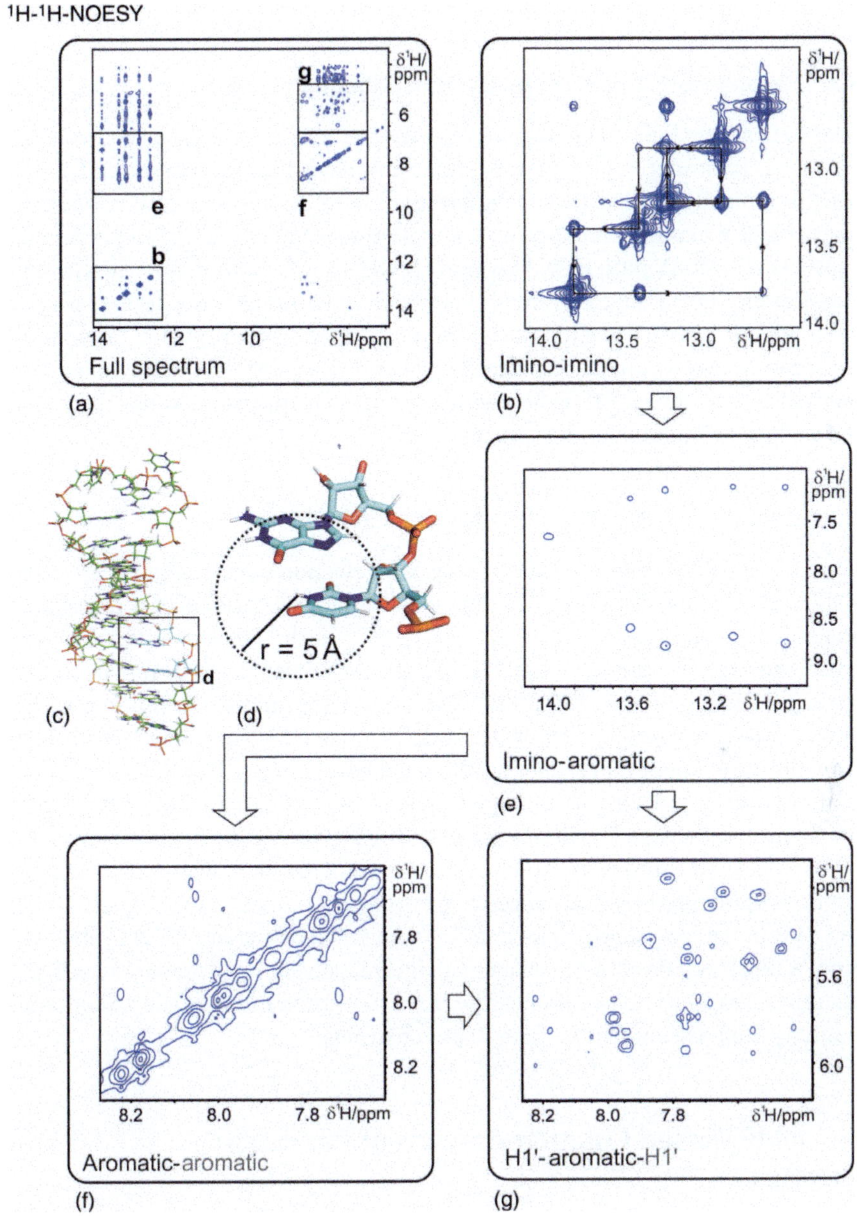

Figure 31.4 Graphical representation of further assignment of an RNA, utilizing 3D heteronuclear NMR experiments. The magnetization transfer is depicted below the spectra, with INEPT transfers shown in blue and TOCSY transfers shown in red. Arrows depict how the information from one spectrum is conferred to the other spectra. Spectra are measured on a 21-nt RNA at 600 MHz.

consequence, this assignment becomes more difficult with larger ribozymes, due to more crowded spectra.

31.3.1 Resonance Assignment by Uniform Labeling

In order to facilitate the proton assignment, heteronuclear NMR experiments have to be applied, assuming ^{15}N- and/or ^{13}C-labeled ribozyme samples (Figure 31.5).

These isotopically labeled RNA samples allow a multitude of experiments that are inaccessible at natural abundance. Most commonly, for a resonance assignment, HCP [39], HCP-TOCSY [40], HCN [41], and HCCH-TOCSY [42] are applied to monitor connectivity of H3′ and H5′ sugar protons of neighboring bases, H1′ of neighboring bases, intra-base aromatic to H1′ correlation, and assignment of sugar protons based on H1′ resonances, respectively. In contrast to a NOESY assignment, the assignment with the help of 3D experiments is performed utilizing coupling between nuclei, leading to unambiguous assignments.

31.3.2 Resonance Assignment by Selective Labeling

Selective labeling is a method, which allows the detection of one single nucleotide by NMR spectroscopy. Assuming a single ^{13}C^{15}N-labeled residue in a ribozyme, an unambiguous assignment of proton resonances of the labeled residue can be achieved, running ^{13}C or ^{15}N X-edited ^{1}H-1D experiments. Assignment of the heteronuclei takes place via ^{1}HX-HSQC. This allows in principle the full resonance assignment of a ribozyme if a selectively labeled sample for each residue is available. The isotopic enrichment for specific residues can be as low as 1% [43].

An alternative approach, which targets specifically the assignment of the sugar region, is given by sparse labeling within the sugar. Bases with sugar moieties, in which only specific carbons are labeled, can be prepared chemically or enzymatically, while the required NTPs are accessible by in vivo production in *Escherichia coli* [44]. Common labeling schemes include C3′ and C5′ labeling or C1′, C2′, and C4′ labeling. While this type of isotopic labeling requires considerable preparative effort, the resulting samples display significantly more resolution in the sugar regions and can simplify their assignment in huge RNA constructs.

31.4 NMR-Based Characterization of Particular Ribozymes

31.4.1 Hepatitis Delta Virus (HDV)

The human hepatitis delta virus ribozyme is known for its ability to cleave itself, yielding a 5′ hydroxyl residue and a 2′,3′-cyclic phosphate group [45]. Some early studies found great similarity between the genomic and antigenomic HDV secondary structures [45, 46]. Kuo et al. [47] and Perrotta et al. [48, 49] showed that it is possible to reduce the genomic HDV ribozyme sequence to a minimum,

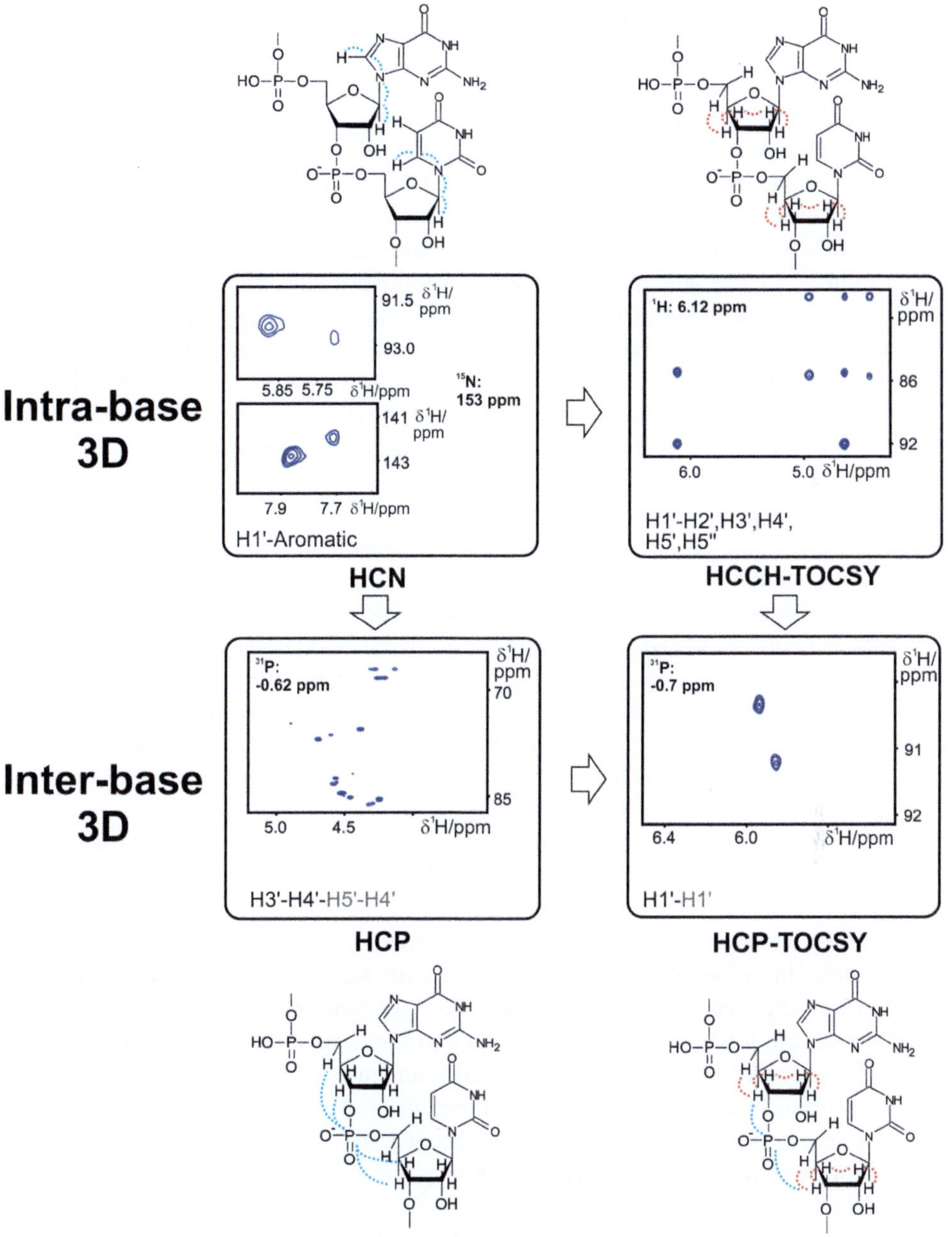

Figure 31.5 Graphical representation of the assignment of an RNA, utilizing ^{1}H-^{1}H-NOESY and a 3D structure calculated from NOE data. In the case of inter-base connectivities, the neighboring base is written in gray below the respective spectrum. (a) Relevant region of a NOESY of RNA to facilitate assignment of imino, aromatic, and H1′ protons. Zooms of the regions annotated are shown in (b), (e), (f), and (g). (c) Three-dimensional structure calculated from NOE data. (d) Zoom of two adjacent bases, outlining the range of the NOE. The spectrum of a 21-nt RNA was measured at 700 MHz.

which is still able to cleave itself. Further investigation of Perrotta et al. [48, 49] and Rosenstein et al [46] based on an antigenomic minimal HDV sequence identified a pseudoknot structure as key structural motif required for proper cleavage.

In order to have further insight into the secondary and tertiary structures of HDV ribozymes, NMR can be applied. Lee et al. performed NMR experiments on a 74-mer HDV ribozyme [50]. Melting of the ribozyme was measured via ^{1}H-1D experiments recorded at temperatures between 12 and 77 °C. The proton resonances of non-Watson–Crick-bound loop regions – shifted between 9 and 12 ppm – became less sharp at higher temperatures. This effect was attributed to faster solvent exchange. On the other hand, proton resonances located at 12–15 ppm did not change at different temperatures indicating higher stability for base pairs in the stems. Further, Lee et al. were able to observe a signal from an A-U imino proton at 14.28 ppm, which split into two proton resonances at 12–16 °C indicating the existence of two conformations in slow exchange. The spectra are shown in Figure 31.6.

For the assignment of the imino proton resonances, ^{1}H-^{1}H-NOESY spectra were analyzed. These 2D spectra exhibit NOE contacts between almost all imino and amino protons. The NOE cross peaks are located at 6.7–8.2 ppm and are in general only visible for G-C base pairs. Two of the imino signals at 13.47 and 14.28 ppm, however, display one NOE contact to the aromatic H2 proton between 7 and 8 ppm referring to the presence of A-U base pairs. Moreover, tetraloop formation has been observed due to three imino resonances belonging to adjacent bases as suggested by inter-base cross peaks. In conclusion, Lee et al. demonstrated that the 74-nt-long HDV ribozyme assumes multiple conformations, while its major conformer is in agreement with the proposed pseudoknot structure.

The role of metal ions (Mg^{2+} and Na^{+}) in the cleavage reaction of HDV ribozymes was investigated by Ganguly et al. [51] by MD simulations. To independently cross-validate the theoretical data, NMR spectroscopy was applied. In order to show, whether cleavage is preceded by metal-mediated protonation of the 2′-hydroxyl, 3′-adenosine monophosphate was titrated with NaOH and $Ca(OH)_2$, and ^{1}H-1D NMR spectra were acquired. Samples were prepared with 10% D_2O, as a 100% D_2O sample tends to feature an N-linked proton pK_a shift of about 0.4–0.6 units [52, 53]. The chemical shift of H1′ proton resonances, which are far apart from the water resonance, was displayed as a function of the pH. In conclusion, divalent ions seemed to have a greater effect on the cleavage activity, involving a precluded protonation of the 2′-hydroxyl at the cleavage site.

Other high-resolution NMR research was done by Kolk et al. [54]. They used an isolated hairpin sequence from the antigenomic HDV ribozyme to examine the pseudoknot secondary structure element. They chose a sequence containing 7 nt, which is important for cleavage activity. This sequence was extended to a 19-nt-long hairpin loop. ^{1}H-^{1}H NOESY and ^{1}H-^{1}H TOCSY NMR spectra were recorded, and most of the proton resonances were assigned. The NMR data were used to calculate a more precise loop conformation by MD simulations.

The crystal structure of HDV ribozyme published by Ferre-D Amare et al. [55] was approved by Tanaka et al. by screening HDV sequences with ^{13}C^{15}N-labeled

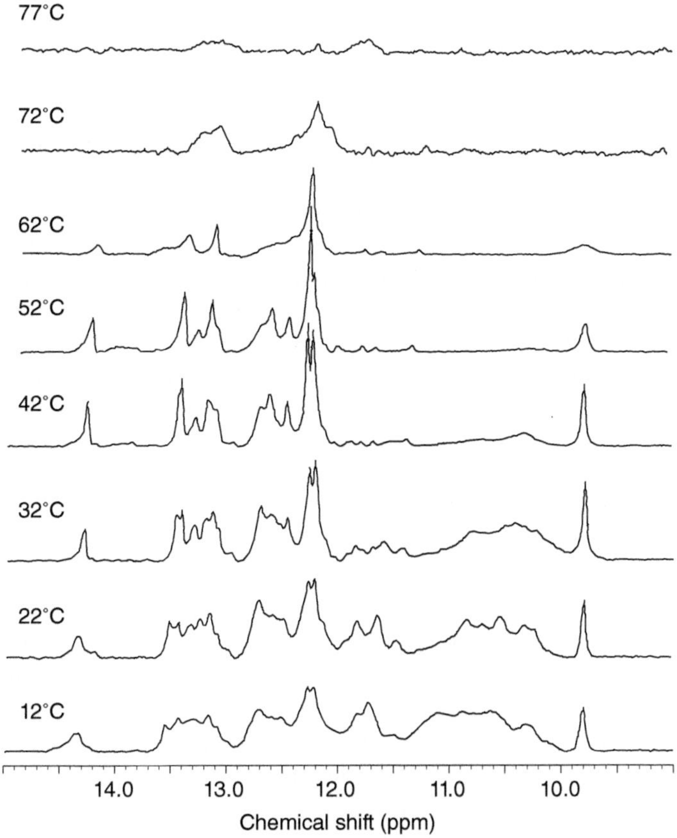

Figure 31.6 ^{1}H-1D NMR spectra of the HDV ribozyme measured and published by Lee et al. [50]. The spectra were recorded at different temperatures in order to investigate the melting of the ribozyme. The split signal of an A-U imino proton at 14.28 ppm appearing at temperatures between 12 and 22 °C points toward two conformations. The original picture was scanned from a written issue and vectorized to improve the quality with permission of the original publisher. Source: Lee et al. [50]. Reproduced with permission from Federation of European Biochemical Societies. © 1993 Federation of European Biochemical Societies.

Gs, designed in respect of the pseudoknot model [56–58]. For NMR analysis, the ribozyme sequences were split into three separate ribozyme parts to facilitate the proton assignment. The imino proton resonances from ^{1}H-^{1}H-NOESY and ^{1}H-^{15}N-HSQC experiments were partially assigned, pointing out the similarity with the crystal structure.

As a starting point, the imino proton resonances of a G*U base pair were assigned first, since the resonances are shifted to the range of 10–12 ppm and exhibit a strong NOE cross signal. The other NOE cross signals were assigned via a walk through the imino region of the NOESY spectrum. Further, NMR samples containing a substrate sequence were prepared. NOESY and HSQC spectra of these full-length constructs did not show any major changes, except the imino proton resonances of the substrate. In order to determine temperature and Mg^{2+} dependencies, ^{1}H-1D spectra were recorded at different conditions. The findings hinted a catalytic effect of Mg^{2+} during cleavage reactions. Another HDV ribozyme-based publication from Andrej Luptak et al. considers pK_a determination via ^{13}C NMR measurements [59]. They focused on a C suspected to be involved in the cleavage reaction and examined the ^{13}C chemical shifts of this carbon atom, which allows the calculation of the pK_a value.

In order to understand the effect of mutational evolution, Schultes et al. have analyzed the interconversion of a shortened HDV ribozyme (HDVP) and a synthetic class lll ligase (LigP), as both catalyze fundamentally different reactions, do not have any related structural elements, and do not share any evolutionary history [60]. They connected both prototypes by creating a neutral network, which is defined as a set of sequences with various genotypes but a specific phenotype. Starting with one prototype sequence, 86 neutral mutations were introduced to generate the other prototype sequence, whereby each mutation step passes on the biologically important activity of the respective prototype. As a result, they show an intersecting sequence (INT) in the middle of this network, a ribozyme featuring catalytic activities of both prototypes.

At this point, NMR spectroscopy can be very helpful to screen through conformational space to understand the correlation between sequence, structure, and function of the neutral network. The screening through conformational space can be achieved by the preparation of NMR samples of particular constructs exhibiting strong change in structure and/or structural dynamics. Hence, we choose five ligase constructs and four HDV constructs from the neutral network, situated between the intersecting sequence and the particular prototype [61]. Figure 31.7 shows the imino proton region of the ^{1}H-1D NMR spectra of the selected constructs.

The ^{1}H-1D NMR spectra of the constructs INT, Lig1, and HDV1 show very heterogeneous but similar imino proton resonances, without alteration of chemical shifts although the catalytic activity increases strongly with the incorporation of mutations into the sequence of INT. This indicates the existence of multiple secondary structures for one sequence and shows that the increase in catalytic activity is caused by the incorporation of mutations but not by a secondary structure rearrangement. Moreover, the homogeneity enhances with the incorporation of mutations converging in the direction of the respective prototype, suggesting an improvement of secondary structure stability. In order to further investigate the structure of INT, it was forced into just one structure by caging two nucleotides with a photolabile group and detecting the refolding process by RT-NMR spectroscopy [62].

The investigation of nucleotide polymerization, an activity, which is related to what was named later on ribozymes, was first studied by R. Naylor and P. T. Gilham in 1966 [63]. Further, in vitro evolution-based screenings could show the existence of autocatalytic systems of RNA template-directed condensation of two RNA substrates [64]. An efficient self-replicating ribozyme R3C was found by Natasha Paul and Gerald F. Joyce [65, 66], leading to the discussion of the origin of life with reference to the "RNA world" hypothesis.

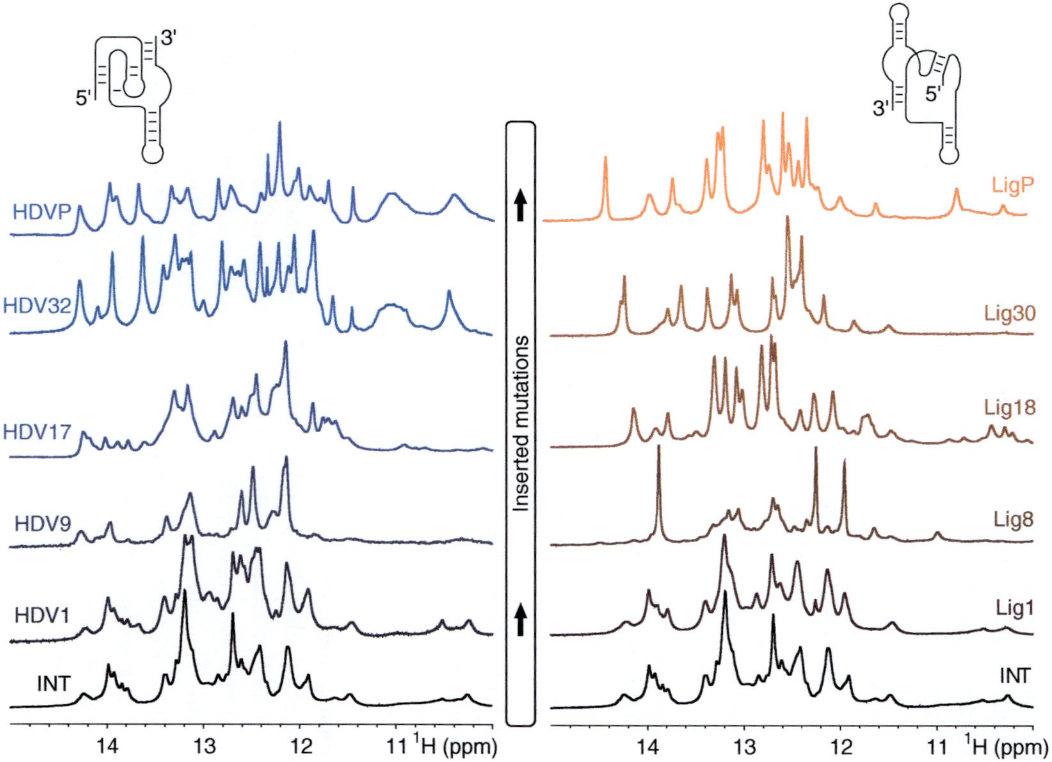

Figure 31.7 The imino proton regions of 1H-1D NMR spectra of different ribozyme sequences are shown [61]. The sequences were obtained from a neutral network of a HDV ribozyme and a class III ligase [60]. The spectra were acquired via "jump-and-return" experiments using an 800 MHz Bruker spectrometer. The samples were set up to a concentration of 100–400 μM in phosphate buffer. The secondary structures and the direction of mutation insertion are shown. Source: Adapted from Knezic [61].

31.4.2 Hammerhead Ribozyme

The hammerhead ribozyme is a self-cleaving ribozyme with a structural motif consisting of three stems and a highly conserved catalytic core. The ribozyme was first discovered in plant pathogens and viral satellites as a part of a rolling circle replication mechanism [67]. Catalytic activity is achieved by Mg^{2+}-mediated formation of an in-line conformation around the scissile phosphodiester bond and subsequent cleavage reaction [68]. In several studies, NMR was employed to investigate structural and functional aspects of the hammerhead ribozyme.

After the determination of the crystal structure [69, 70], additional structural characterization was carried out in solution, utilizing NMR. In-depth characterization of the catalytic core took place by ^{1}H-^{1}H-NOESY, ^{1}H-1D, and ^{13}C-HSQC spectroscopy, hinting at structural changes, leading to an active conformation, different to the crystal structure. These studies were carried out in the absence and presence of Mg^{2+} ions, and structural differences between cleaved and uncleaved constructs were assessed [71]. Further studies were carried out determining the structure in solution and the formation of the active conformation also involving mutation studies, relying mostly on ^{1}H-^{1}H-NOESY data [72–74]. The global structure of the hammerhead ribozyme was determined by measurement of residual dipolar coupling (RDC), showing fundamental differences from the crystal structure [75]. In addition, an NMR structure of a widespread loop–loop interaction motif could be determined in 2009 using ^{1}H-^{1}H-NOESY data and various 3D methods on $^{13}C^{15}N$-labeled samples to determine structural parameters [76].

NMR-aided functional investigation of the hammerhead ribozyme centers mostly on the characterization of the active state of the ribozyme and the interstate transitions. As the catalytically active state involves binding of a divalent cation, the determination of metal binding sites and the thermodynamics and kinetics of such a binding are of high interest. In 1997, Herschlag et al. could show that phosphorothioate modification on the scissile phosphate of the ribozyme had deleterious effects on magnesium binding. The affinity could be restored by spiking with small amounts of Cd^{2+} [77]. As this modification induces a considerable chemical shift in the ^{31}P signal of the modified phosphate, it enables unambiguous assignment of its resonance and therefore ^{31}P monitoring of ribozyme activity on a specific ^{31}P-1D peak. Subsequent studies allowed the localization of metal binding sites in the catalytic core by either ^{31}P CSP analysis [78] or exploitation of the deleterious effect of manganese spiking to the bound phosphates signal due to paramagnetic relaxation [79]. While the phosphorothioate method led to significant results in the characterization of metal binding, possible structural alterations introduced by the modification could not be ruled out [80]. In following studies, the method was often avoided in favor of less invasive strategies. Lilley et al. were able to determine a folding model upon metal interaction by ^{19}F modification of uridine residues in and next to the catalytically important CUGA sequence. The resulting ^{19}F resonances were monitored in 1D transient studies and revealed a two-step folding model [81]. ^{15}N-detected metal binding assays for the identification of metal binding sites were introduced, using selectively and site-specifically ^{15}N-labeled RNA and exploiting the large CSP introduced by metal binding [82, 83]. Extensive NMR studies in the late 2000s helped the in-depth characterization of the catalytic process of the hammerhead ribozyme. Uncoupling the metal binding event from the cleavage reaction by incorporation of a photocleavable protecting group, both processes could be monitored separately in atomic resolution [84].

31.4.3 Hairpin Ribozyme

During the investigation of plant satellite RNA in the late 1980s, the hairpin ribozyme was first identified in negative-strand satellite RNA of the tobacco ringspot virus [85]. The short ribozyme catalyzes cleavage and joining reactions releasing a substrate with a 5′-hydroxyl group, leaving the ribozyme with a 2′,3′-cyclic phosphate upon cleavage. It consists of four helical stems and two unpaired loops with the scissile phosphate located in one of the loops. Nucleobases in the loop regions generally termed loop A and B are highly conserved throughout species [86].

Structural investigation of the hairpin ribozyme by NMR spectroscopy mainly took place in the late 1990s. For the data collection preceding the calculation of the solution structure of loop A, region-selective ^{13}C labeling was employed to facilitate the assignment of aromatic protons and to allow the application of ^{13}C-filtered NOESY experiments. Along with the data collected from conventional NOESY and ^{13}C-NOESY-HMQC experiments, the structure could be determined [87]. Further structural studies naturally focused on the solution structure of the loop B that was determined few years later. In the according study, 3D-HCCH-TOCSY along with the typical 2D-NOESY and 2D-TOCSY spectra were used for assignment, while the NOE data were gathered from 2D-NOESY and ^{13}C-3D-NOESY-HMQC spectra [88].

NMR spectroscopy was further utilized in studies characterizing the function of the hairpin ribozyme. Determination of the metal binding sites took place by chemical shift mapping, paramagnetic line broadening, and intermolecular NOE. Corresponding binding sites could only be identified in loop B, though no structural change was indicated upon binding of Mn^{2+}, Mg^{2+}, or $Co(NH3)_6{}^{3+}$, leading to the assumption that metal ions do not facilitate the catalytic activity but help in maintaining a correct tertiary structural fold [36]. Monitoring the ^{13}C chemical shift perturbation upon pH change, pK_a values significantly different from the normally observed values could be determined for loop adenine residues. It was concluded that protonation of adenine plays a role in catalytic activity of the ribozyme [89].

The hairpin ribozyme served as a model system in later NMR studies, employing atom- and site-selective labeling of the loop A domain proving that six ^{15}N residues can be labeled and unambiguously distinguished by indirect ^{13}C tagging [90]. In a later study, the adenine-binding mutant of the hairpin ribozyme (ADHR1) [91] and its catalytic mechanism was investigated. Kinetic experiments and monitoring of ^{15}N-HSQC, ^{13}C-HSQC, and ^{31}P-1D spectra under various conditions revealed a significantly higher reaction barrier than in the wild-type ribozyme with dependence on the adenine cofactor [92].

31.4.4 *Neurospora* VS Ribozyme

In the class of nucleolytic ribozymes also consisting of HDV, hammerhead, and hairpin ribozymes, the *Neurospora* Varkud satellite ribozyme is the largest with a length around 150 nucleotides. It catalyzes ligation or cleavage by transesterification on a specific loop site of a substrate and can act in cis or in trans. It is composed of six helical structures with two three-way junctions [93]. As a crystal structure of the ribozyme could only be determined recently [94], structural investigation was carried out via NMR and started in 2000 with the structure determination of the substrate stem-loop. Distance restraints for the structure calculation were determined from ^{1}H-^{1}H-NOESY, while COSY- and TOCSY-type experiments yielded the assignment [95]. Subsequent NMR spectroscopic studies of the *Neurospora* VS ribozyme were mostly carried out in the Legault lab and focused on the determination of the global structure. An NMR structure of a mutated stem-loop I, which mimics the active cleavage site of the ribozyme, was determined with the internal loop featuring a G-A sheared base pair and an A-G-A base triplet [96]. The structure of stem-loop V was solved in the presence of Mg^{2+} ions, with the binding sites determined by paramagnetic relaxation enhancement of a Mn^{2+} spiked sample monitored by ^{13}C-HSQC experiments. Four distinct binding sites could be determined in the loop region [97]. In 2008, kinetics and substrate binding of the stem-loop V were investigated. By combining magnesium-dependent ^{31}P and ^{15}N7 chemical shift perturbations with native PAGE data for several mutants, it became apparent that the loop residues tolerate a multitude of mutations retaining binding of the substrate while deletion of U700 substantially reduces activity [98]. Further structural studies addressed the A730 loop located in helix VI [99] that is involved in a loop–loop interaction not unlike the hairpin ribozyme [88] as well as the substrate recognition site [100] and the III–IV–V junction. Distance and angular restraints were determined from ^{1}H-^{1}H-NOESY data, while magnesium binding sites in the III-IV-V junction structure were determined by Mn^{2+}-induced paramagnetic relaxation enhancement [101]. Shortly afterward, the structure of the II–III–VI junction was determined as well, showing remarkable structural diversity in the junction core including an A·G base pair and minor groove base triplets. For the unambiguous resonance assignment in the junction, an A-specific H(NC)-TOCSY-(C)H experiment and 3D HCCH-TOCSY and HCCH-COSY experiments were utilized. Summarizing all the structural data determined for the separate elements, a model of the full structure of the 150-nucleotide ribozyme could be determined by NMR [102].

31.4.5 Leadzyme

Selection experiments opting to find RNA cleavage reactions catalyzed by Pb^{2+} ions yielded several ribozymes first described by Pan et al. [103]. Later the same year they derived a 30-nucleotide ribozyme from a suggested alternative secondary structure of one of the cleavage active variants [104]. It features an asymmetric six-nucleotide internal loop and is rapidly cleaved in the presence of Pb^{2+}. This specific structure, one of many lead ribozymes, is referred to as leadzyme.

Legault et al. used heteronuclear NMR experiments to investigate the leadzyme. They observed a significant upfield shift for the C2 resonance of adenine 25 (A25) [105]. The C2 resonance shifts 8 ppm with a pH range from 4.4 to 8.3, and a pK_a of 6.5 was determined. A25 is in close proximity to the site of cleavage.

A solution structure of the leadzyme was determined by Hoogstraten et al., and a model of a lead-binding bulged guanine base critical for ribozyme function was proposed [106]. A feature of the structure is a protonated AH^+-C base pair between the cleavage site and the regular A-form RNA. Due to the involvement in a base pair, it is unlikely that the protonated A25 participates in a "histidine-like" catalytic mechanism. The structure also showed a conformation that requires rearrangement to reach the trigonal bipyrimidal transition state proposed for ribozyme reaction. The chemical shift analyses showed that Pb^{2+}, Mg^{2+}, or Na^+ addition does not result in large structural rearrangement. Complementary studies by Katahira et al. also found little change in the imino spectrum upon addition of lead ions [107]. An analysis of the relaxation rate along the B1 field ($T1\rho$) of the purine C8 showed diverse dynamics for the residues of the internal loop [106]. The longer $T1\rho$ of A8 compared with helical purine regions indicated a sub-nanosecond fluctuation. In contrast, G7 and G24 fluctuate on a micro- to millisecond timescale. These results were also found in structural probing assays using chemical modification. The low solvent accessibility of G24 could be explained through base stacking with G23 and A25.

31.4.6 Ribonuclease P

Ribonuclease P is a large multi-turnover ribozyme that cleaves 5′ ends of pre-tRNA in vivo facilitating its maturation. It consists of a protein domain and an RNA domain that are both essential in vivo while the RNA alone is proven to be catalytically active in vitro and is therefore considered a ribozyme with a protein cofactor [108]. The large RNA domain comprises 18 helical stems with the catalytic center located around helix P4 [109]. NMR-assisted structural investigation of RNase P began in 1996 with the structure determination of the 31-nucleotide sequence around the conserved substrate binding GGU motif. To facilitate the resonance assignment prior to the structure calculation, cytidines were partially deuterated in this study, simplifying the spectra especially in the sugar regions [110]. Structural research of RNase P moved on not much later with the determination of the NMR structure of the RNase P protein from data obtained in ^{13}C- and ^{15}N-correlated 3D-NOESY experiments [111]. Investigation of the RNA proceeded with the structure of helix P4 that is located in the catalytic core next to the binding site of the pre-tRNA. Metal binding sites were determined by adding $Co(NH_3)_6{}^{3+}$ ions that gave rise to ^1H-^1H-NOESY cross peaks to adjacent protons [112]. Another study could show that the pre-tRNA as well binds Mg^{2+} ions even prior to RNase P binding [113]. For the highly resolved NMR structure of helix P5.1 from bacterial RNase P type II, the structure calculation was carried out with distance restraints from ^1H-^1H-NOESY. Dihedral restraints from ^1H-^{31}P-hetero TOCSY, HCP, and HMQC-TOCSY as well as RDC data were implemented for refinement, resulting in a high-resolution structure [114].

Further research focused mainly on aspects of ribozyme function by employing a variety of methods to assess metal binding and dynamics. The highly conserved and catalytically important C and U residues, forming a bulge in the center of the P4 helix, could be identified as a pivot point for structural rearrangement upon Mg^{2+} binding. RDC and ^{15}N relaxation data were measured in the presence and absence of Mg^{2+}, revealing remarkable flexibility around these specific residues while mapping the global conformation of the helix [115]. In the protein domain, the RNA binding surface was determined flexible in contrast to the protein being relatively rigid in general, as shown by ^{15}N relaxation data of the backbone [116]. A three-step folding pathway was proposed by monitoring CSP in ^{15}N-HSQC spectra a few years later. Metal binding sites of the protein could be localized using paramagnetic relaxation enhancement by adding $[Cr(CN)_6]^{3-}$ [117]. Metal binding of the RNA domain was reviewed in 2010 when EXAFS spectroscopy data shed new light on the metal binding sites of the P4 helix, characterizing the binding of a Zn^{2+} ion in the catalytic core. Complexation of a Mg^{2+} ion between two conserved G residues as found by PRE of $Co(NH_3)_6{}^{3+}$ is described in great detail showing a six-coordinated geometry [118].

Other studies concerned the stability of the pre-tRNA substrate stem whose opening is considered a rate-limiting step in substrate binding. Kinetics of pre-tRNA constructs with several modifications were determined while noninvasiveness was checked via ^{1}H-1D NMR of the imino region [119]. NMR proves to be particularly useful when hard evidence of structural change is needed, as shown in a binding assay of the whole RNase P complex [120].

31.4.7 Group I Intron

Group I introns have the ability to self-splice. Therefore, the diverse transcripts known for this ribozyme include a guanosine binding site [121, 122]. This binding site is essential for the self-splicing process, which is composed of two transesterification steps requiring a bound guanosine [123]. Due to the large size of the group I intron, the different regions have been studied independently by NMR spectroscopy. A 20-nt-long RNA with a UUGC tetraloop was studied by Frederic H.-T. Allain et al. [124]. This tetraloop construct has the properties of the P1 helix of the group I intron, which is located at the cleavage site. Using the assignment data of ^{1}H-1D, ^{1}H-^{1}H-NOESY, ^{13}C-correlated 3D-NOESY, ^{1}H-^{13}C-^{31}P HCP, and H-COSY-H,C-HMQC spectra, it was possible to simulate a solution structure of P1 by MD simulations. The stability of P1 was investigated by Joon-Hwa Lee et al. [125]. NH_3 and $HPO_4{}^{2-}$ have been used as base catalysts for the base-pair opening. The effect of those was observed by ^{1}H-1D and ^{1}H-^{1}H-NOESY experiments, and based on the imino proton exchange rates, the equilibrium constants have been determined. Mayumi Amano investigated a 17-nt-long RNA from the 2.1 loop region of group I intron by NMR spectroscopy [126]. With the help of NOE contacts from ^{1}H-1D and ^{1}H-^{1}H-NOESY spectra, imino proton, H1′, and sugar proton assignments were accomplished. Further, Mayumi Amano studied a 20-nt RNA from the P9.1 region of group I intron [127]. Therefore, imino proton resonances from ^{1}H-1D spectra were assigned. Another investigation of both hairpin structures, the 17 mer and 20 mer, showed low-range interactions between those [128]. A 25-nt-long model RNA from the P4–P6 region of group I intron was analyzed by ^{1}H-^{1}H-NOESY, COSY, and natural-abundance ^{1}H-^{13}C HMQC experiments [129]. Data from assigned NOE contacts were converted to constraints for MD simulations of the structure. Michael Chastain et al. concluded that a triple helix is formed in the minor groove of the ribozyme. One year later, Michael Chastain and Ignacio Tinoco, Jr., extended this study [130]. Other MD simulations of a slightly larger RNA model of P4–P6 were based on the resonance data from ^{1}H-1D, ^{1}H-^{1}H-NOESY, COSY, TOCSY, and x-filtered NOESY spectra [131]. The largest model construct of P4–P6 was investigated by Eric C. Johnson et al. [132]. The NOESY and WaterLOGSY spectra were received including a screening through a ligand library. The resonance changes of free ligands due to faster or slower tumbling were recorded, showing which library compound is bound to the ribozyme. One specific part of the P4–P6 domain is the P5 helix. One of the first studies was done by Jeffrey A. McDowell et al. [133]. They chose a duplex motive similar to P5 and determined the 3D solution structure of it by MD simulations. The data needed for these calculations were extracted from ^{1}H-1D, ^{1}H-^{1}H-NOESY, DQF-COSY, and ^{1}H-^{31}P-HETCOR spectra. Furthermore, the solution structure of a 25-nt-long model RNA including a 5-nt-sized bulge loop was defined, based on 1D proton spectra and ^{1}H-^{1}H-NOESY, COSY, ^{1}H-^{13}C-HMQC, TOCSY, and homonuclear 3D-TOCSY-NOESY spectra [134]. Another NMR-based solution structure study incorporated $Co(NH_3)_6{}^{3+}$ and $Mg(H_2O)_6{}^{2+}$, in order to test secondary structure dependent metal ion binding [135]. For the analysis of the P5abc region of group I intron, this ribozyme domain was truncated, yielding a 56-nt-long RNA (tP5abc) [136]. NOE signal assignment was done for ^{1}H-1D and ^{1}H-^{1}H-NOESY spectra. An additional titration with Mg^{2+} did not cause any secondary structure rearrangement. In a different study, $Co(NH_3)_6{}^{3+}$ was used to determine ion binding sites [137]. Therefore ^{1}H-1D, ^{1}H-^{1}H-NOESY, ^{1}H-^{15}N-HMQC, and HNN-COSY spectra have been recorded, and the extracted data was used for MD simulations. Minxue Zheng et al. investigated secondary structure changes, caused by tertiary structure formation, applying ^{1}H-1D, ^{1}H-^{1}H-NOESY, DQF-COSY (double-quantum-filtered), TOCSY, fast HSQC, INEPT, and HMQC experiments and restrained MD simulation [138]. The native structure of tP5abc was analyzed by Hashim M. Al-Hashimi's group [139]. The combination of ^{1}H-1D, ^{1}H-^{1}H-NOESY, ^{1}H-^{15}N-HSQC, and ZZ-exchange NMR experiments and SHAPE data gave rise to the hypothesis that the P5c region independently forms a native structure [140]. Furthermore, Brant Gracia et al. point out the great ability of single base pairs to stabilize secondary structures [141]. They applied SOFAST-HMQC experiments on a P5abc construct combined with the intron core and later on incorporated a single mutation. Satoru Watanabe et al. investigated the guanosine binding site [142]. The imino proton resonances from ^{1}H-1D and ^{1}H-^{1}H-NOESY spectra for a 31-nt-long model RNA of P7, P9.0, and terminal 3′G residue have been assigned. Additional studies on the P7–P9 domain were done by Aya Kitamura et al. including ^{1}H-^{13}C-HSQC, ^{1}H-^{15}N-HSQC, ^{1}H-^{1}H-NOESY, and COSY experiments and subsequent structure calculations, in order to determine the solution structure [143, 144].

31.4.8 Group II Intron

The group II intron is a self-splicing ribozyme differentiated into six domains [145]. Lan Zhang et al. [146] have achieved the crystal structure of the domains 5 and 6. To further investigate the secondary structure of this ribozyme, the solution structures of both domains have been characterized successfully by NMR spectroscopy. Assignment was achieved by analysis of 1H,1H-NOESY with the aid of 3D-HCCH-TOCSY and HNN-COSY experiments. The extracted NOE restraints led to a defined structure of domain 6 with an RMSD of 4.2 Å. The high RMSD is attributed to the overall flexibility in the bulge region, leading to a global conformational heterogeneity [147]. The structure of domain 5 showed flexibility in the loop region that was consistent with the broad peaks observed in the spectra. RDC analysis was therefore only applied for stem residues [148].

31.5 Closing Remarks

This overview points out that some ribozymes are already well analyzed by NMR spectroscopy. This method provided a good insight into the diverse structural elements of the ribozymes shown in Figure 31.1. Further, it was possible to characterize kinetic and dynamic aspects of the catalytically active molecules, due to a multitude of experiments and applicable conditions. It should be further noted that the benefit of NMR spectroscopy, in contrast to crystallography, is the capability of mimicking natural conditions.

References

1 Fürtig, B., Richter, C., Wöhnert, J., and Schwalbe, H. (2003). NMR spectroscopy of RNA. *ChemBioChem* 4: 936–962.

2 Cherepanov, A.V., Glaubitz, C., and Schwalbe, H. (2010). High-resolution studies of uniformly ^{13}C,^{15}N-labeled RNA by solid-state NMR spectroscopy. *Angew. Chem. Int. Ed.* 49: 4747–4750.

3 Huang, W., Bardaro, M.F., Varani, G., and Drobny, G.P. (2012). Preparation of RNA samples with narrow line widths for solid state NMR investigations. *J. Magn. Reson.* 223: 51–54.

4 Marchanka, A. and Carlomagno, T. (2014). Solid-state NMR and RNA structure: a new partnership? In: *eMagRes*, 119–128. Chichester, UK: Wiley.

5 Marchanka, A., Simon, B., Althoff-Ospelt, G., and Carlomagno, T. (2015). RNA structure determination by solid-state NMR spectroscopy. *Nat. Commun.* 6: 1–7.

6 Batey, R.T., Puglisi, J.D., Williamson, J.R. et al. (2007). Preparation of isotopically labeled ribonucleotides for multidimensional NMR spectroscopy of RNA. *Nucleic Acids Res.* 20: 4515–4523.

7 Quant, S. et al. (1994). Chemical synthesis of 13C-labelled monomers for the solid-phase and template controlled enzymatic synthesis of DNA and RNA oligomers. *Tetrahedron Lett.* 35: 6649–6651.

8 Milligan, J.F., Groebe, D.R., Witherell, G.W., and Uhlenbeck, O.C. (1987). Oligoribonucleotide synthesis using T7 RNA polymerase and synthetic DNA templates. *Nucleic Acids Res.* 15: 8783–8798.

9 Chamberlin, M. and Ring, J. (1973). Characterization of T7-specific ribonucleic acid polymerase. II. Inhibitors of the enzyme and their application to the study of the enzymatic reaction. *J. Biol. Chem.* 248: 2245–2250.

10 Guillerez, J., Lopez, P.J., Proux, F. et al. (2005). A mutation in T7 RNA polymerase that facilitates promoter clearance. *Proc. Natl. Acad. Sci. U.S.A* 102: 5958–5963.

11 Losick, R. (1972). In vitro transcription. *Annu. Rev. Biochem.* 41: 409–446.

12 Helmling, C. et al. (2015). Rapid NMR screening of RNA secondary structure and binding. *J. Biomol. NMR* 63: 67–76.

13 Marshall, W.S. and Kaiser, R.J. (2004). Recent advances in the high-speed solid phase synthesis of RNA. *Curr. Opin. Chem. Biol.* 8: 222–229.

14 Walker, G.C., Uhlenbeck, O.C., Bedows, E., and Gumport, R.I. (1975). T4-induced RNA ligase joins single-stranded oligoribonucleotides. *Proc. Natl. Acad. Sci. U.S.A* 72: 122–126.

15 Silber, R., Malathi, V.G., and Hurwitz, J. (1972). *Proc. Natl. Acad Sci. U. S. A* 69: 3009–3013.

16 Huang, Z., Jayaseelan, S., Hebert, J. et al. (2013). Single-nucleotide resolution of RNAs up to 59 nucleotides by high-performance liquid chromatography. *Anal. Biochem.* 435 (1): 35–43.

17 Smalley, M.K. and Silverman, S.K. (2004). Site-specific fluorescent labeling of large RNAs with pyrene. *Curr. Protocol Nucl Acid Chem.* 19: 11.11.11–11.11.12.

18 England, T.E. and Uhlenbeck, O.C. (1978). 3′-Terminal labelling of RNA with T4 RNA ligase [25]. *Nature* 275: 560–561.

19 Beaucage, S.L. and Caruthers, M.H. (1981). Deoxynucleoside phosphoramidites – a new class of key intermediates for deoxypolynucleotide synthesis. *Tetrahedron Lett.* 22: 1859–1862.

20 England, T.E. and Uhlenbeck, O.C. (1978). Enzymic oligoribonucleotide synthesis with T4 RNA ligase. *Biochemistry* 17: 2069–2076.

21 Hengesbach, M. et al. (2008). RNA intramolecular dynamics by single-molecule FRET. In: *Current Protocols in Nucleic Acid Chemistry*, 11.12.11–11.12.22. Hoboken, NJ: Wiley.

22 Keyhani, S., Goldau, T., Blümler, A. et al. (2018). Chemo-enzymatic synthesis of position-specifically modified RNA for biophysical studies including light control and NMR spectroscopy. *Angew. Chem. Int. Ed.* 57: 12017–12021.

23 Barltrop, J.A. and Schofield, P. (1965). Organic photochemistry. Part II. Some photosensitive protecting groups. *J. Chem. Soc.*: 4758.

24 Kaplan, J.H., Forbush, B., and Hoffman, J.F. (1978). Rapid photolytic release of adenosine 5′-triphosphate from a protected analog: utilization by the sodium: potassium pump of human red blood cell ghosts. *Biochemistry* 17: 1929–1935.

25 Pelliccioli, A.P. and Wirz, J. (2002). Photoremovable protecting groups: reaction mechanisms and applications. *Photochem. Photobiol. Sci.* 1: 441–458.

26 Wenter, P., Fürtig, B., Hainard, A. et al. (2005). Kinetics of photoinduced RNA refolding by real-time NMR spectroscopy. *Angew. Chem. Int. Ed.* 44: 2600–2603.

27 Wenter, P., Fürtig, B., Hainard, A. et al. (2006). A caged uridine for the selective preparation of an RNA fold and determination of its refolding kinetics by real-time NMR. *ChemBioChem* 7: 417–420.

28 Fürtig, B. et al. (2007). Time-resolved NMR studies of RNA folding. *Biopolymers* 86: 360–383.

29 Snoussi, K. and Leroy, J.-L. (2001). Imino proton exchange and base-pair kinetics in RNA duplexes. *Biochemistry* 40: 8898–8904.

30 Steinert, H.S., Rinnenthal, J., and Schwalbe, H. (2012). Individual basepair stability of DNA and RNA studied by NMR-detected Solvent exchange. *Biophys. J.* 102: 2564–2574.

31 Nikolova, E.N. and Al-Hashimi, H.M. (2010). Thermodynamics of RNA melting, one base pair at a time. *RNA* 16: 1687–1691.

32 Ferner, J. et al. (2009). Structures of HIV TAR RNA-ligand complexes reveal higher binding Stoichiometries. *ChemBioChem* 10: 1490–1494.

33 Orita, M. et al. (1996). Magnesium-mediated conversion of an inactive form of a hammerhead ribozyme to an active complex with its Substrate. *J. Biol. Chem.* 271: 9447–9454.

34 Cate, J.H., Hanna, R.L., and Doudna, J.A. (1997). A magnesium ion core at the heart of a ribozyme domain. *Nat. Struct. Biol.* 4: 553–558.

35 Hampel, A. and Cowan, J.A. (1997). A unique mechanism for RNA catalysis: the role of metal cofactors in hairpin ribozyme cleavage. *Chemistry & Biology* 4: 513–517.

36 Butcher, S.E., Allain, F.H.T., and Feigon, J. (2000). Determination of metal ion binding sites within the hairpin ribozyme domains by NMR. *Biochemistry* 39: 2174–2182.

37 Bonneau, E. and Legault, P. (2014). NMR localization of divalent cations at the active site of the neurospora vs ribozyme provides insights into RNA–metal-ion interactions. *Biochemistry* 53: 579–590.

38 Kumar, A., Wagner, G., Ernst, R.R., and Wuethrich, K. (1981). Buildup rates of the nuclear Overhauser effect measured by two-dimensional proton magnetic resonance spectroscopy: implications for studies of protein conformation. *J. Am. Chem. Soc.* 103: 3654–3658.

39 Marino, J.P. et al. (1994). Three-dimensional triple-resonance 1H, 13C, 31P experiment: sequential through-bond correlation of ribose protons and intervening phosphorus along the RNA oligonucleotide backbone. *J. Am. Chem. Soc.* 116: 6472–6473.

40 Marino, J.P. et al. (1995). Sequential correlation of anomeric ribose protons and intervening phosphorus in RNA oligonucleotides by a 1H,13C,31P triple resonance experiment: HCP-CCH-TOCSY. *J. Biomol. NMR* 5: 87–92.

41 Sklenář, V., Peterson, R.D., Rejante, M.R., and Feigon, J. (1993). Two-and three-dimensional HCN experiments for correlating base and sugar resonances in 15N, 13C-labeled RNA oligonucleotides. *J. Biomol. NMR* 3: 721–727.

42 Kay, L.E., Xu, G.Y., Singer, A.U. et al. (1993). A gradient-enhanced HCCH-TOCSY experiment for recording side-chain 1H and 13C correlations in H_2O samples of proteins. *J. Magn. Reson., Ser. B* 101: 333–337.
43 Phan, A.T. and Patel, D.J. (2002). A site-specific low-enrichment 15N, 13C isotope-labeling approach to unambiguous NMR spectral assignments in nucleic acids. *J. Am. Chem. Soc.* 124: 1160–1161.
44 Thakur, C.S., Sama, J.N., Jackson, M.E. et al. (2010). Selective 13C labeling of nucleotides for large RNA NMR spectroscopy using an *E. coli* strain disabled in the TCA cycle. *J. Biomol. NMR* 48: 179–192.
45 Sharmeen, L., Kuo, M.Y., Dinter-Gottlieb, G., and Taylor, J. (1988). Antigenomic RNA of human hepatitis delta virus can undergo self-cleavage. *J. Virol.* 62: 2674–2679.
46 Rosenstein, S.P. and Been, M.D. (1991). Evidence that genomic and antigenomic RNA self-cleaving elements from hepatitis delta virus have similar secondary structures. *Nucleic Acids Res.* 19: 5409–5416.
47 Kuo, M.Y., Sharmeen, L., Dinter-Gottlieb, G., and Taylor, J. (1988). Characterization of self-cleaving RNA sequences on the genome and antigenome of human hepatitis delta virus. *J. Virol.* 62: 4439–4444.
48 Perrotta, A.T. and Been, M.D. (1991). A pseudoknot-like structure required for efficient self-cleavage of hepatitis delta virus RNA. *Nature* 350: 434–436.
49 Perrotta, A.T. and Been, M.D. (1991). The self-cleaving domain from the genomic RNA of hepatitis delta virus: sequence requirements and the effects of denaturant. *Nucleic Acids Res.* 19: 1979–1979.
50 Lee, B.-S., Wu, H.-N., and Huang, T.-H. (1993). The catalytic domain of human hepatitis delta virus RNA. *FEBS Lett.* 324: 296–300.
51 Ganguly, A., Thaplyal, P., Rosta, E. et al. (2014). Quantum mechanical/molecular mechanical free energy simulations of the self-cleavage reaction in the hepatitis delta virus ribozyme. *J. Am. Chem. Soc.* 136: 1483–1496.
52 Nakano, S., Chadalavada, D.M., and Bevilacqua, P.C. (2000). General acid-base catalysis in the mechanism of a hepatitis delta virus ribozyme. *Science* 287: 1493–1497.
53 Schowen, K.B. and Schowen, R.L. (1982). Solvent isotope effects of enzyme systems. *Methods Enzymol.* 87: 551–606.
54 Kolk, M.H., Heus, H.A., and Hilbers, C.W. (1997). The structure of the isolated, central hairpin of the HDV antigenomic ribozyme: novel structural features and similarity of the loop in the ribozyme and free in solution. *EMBO J.* 16: 3685–3692.
55 Ferré-D'Amaré, A.R., Zhou, K., and Doudna, J.A. (1998). Crystal structure of a hepatitis delta virus ribozyme. *Nature* 395: 567–574.
56 Tanaka, Y. et al. (2002). Imino proton NMR analysis of HDV ribozymes: nested double pseudoknot structure and Mg2+ ion-binding site close to the catalytic core in solution. *Nucleic Acids Res.* 30: 766–774.
57 Tanaka, Y. et al. (1999). Design and NMR analysis of HDV ribozymes for structural investigation. *Nucleic Acid Symp. Ser.* 17: 221–222.

58 Tanner, N.K. et al. (1994). A three-dimensional model of hepatitis delta virus ribozyme based on biochemical and mutational analyses. *Curr. Biol.* 4: 488–498.

59 Lupták, A., Ferré-D'Amaré, A.R., Zhou, K. et al. (2001). Direct pK_a measurement of the active-site cytosine in a genomic hepatitis delta virus ribozyme. *J. Am. Chem. Soc.* 123: 8447–8452.

60 Schultes, E.A. and Bartel, D.P. (2000). One sequence, two ribozymes: implications for the emergence of new ribozyme folds. *Science* 289: 448–452.

61 Knezic, B. (2016) *Untersuchung der Sequenzabhängigkeit der RNA-Struktur.*

62 Keyhani, S. (2017) *Unpublished data.*

63 Naylor, R. and Gilham, P.T. (1966). Studies on some interactions and reactions of oligonucleotides in aqueous solution. *Biochemistry* 5: 2722–2728.

64 Zielinski, W.S. and Orgel, L.E. (1987). Autocatalytic synthesis of a tetranucleotide analogue. *Nature* 327: 346–347.

65 Paul, N. and Joyce, G.F. (2002). A self-replicating ligase ribozyme. *Proc. Natl. Acad. Sci. U.S.A* 99: 12733–12740.

66 Lincoln, T.A. and Joyce, G.F. (2009). Self-sustained replication of an RNA enzyme. *Science* 323: 1229–1232.

67 Prody, G.A., Bakos, J.T., Buzayan, J.M. et al. (1986). Autolytic processing of dimeric plant virus satellite RNA. *Science* 231: 1577–1580.

68 Murray, J.B. et al. (1998). The structural basis of hammerhead ribozyme self-cleavage. *Cell* 92: 665–673.

69 Pley, H.W., Flaherty, K.M., and McKay, D.B. (1994). Three-dimensional structure of a hammerhead ribozyme. *Nature* 372: 68–74.

70 Scott, W.G., Finch, J.T., and Klug, A. (1995). The crystal structure of an AII-RNA hammerhead ribozyme: a proposed mechanism for RNA catalytic cleavage. *Cell* 81: 991–1002.

71 Simorre, J.-P., Legault, P., Hangar, A.B. et al. (1997). A conformational change in the catalytic core of the hammerhead ribozyme upon cleavage of an RNA substrate. *Biochemistry* 36: 518–525.

72 Simorre, J.-P. et al. (1998). Structural variation induced by different nucleotides at the cleavage site of the hammerhead ribozyme. *Biochemistry* 37: 4034–4044.

73 Sakamoto, T. et al. (1997). Hairpin structure of an RNA 28-mer, which contains a sequence of the enzyme component of a hammerhead ribozyme system: evidence for tandem G: a Pairs that are not of side-by-side type. *J. Biochem.* 122: 556–562.

74 Ojha, R.P. et al. (1997). Structure of an Anti-HIV-1 hammerhead ribozyme complex with a 17-mer DNA substrate analog of HIV-1 gag RNA and a mechanism for the cleavage reaction: 750 MHz NMR and computer experiments. *J. Biomol. Struct. Dyn.* 15: 185–215.

75 Bondensgaard, K., Mollova, E.T., and Pardi, A. (2002). The global conformation of the hammerhead ribozyme determined using residual dipolar couplings. *Biochemistry* 41: 11532–11542.

76 Dufour, D., de la Peña, M., Gago, S. et al. (2009). Structure–function analysis of the ribozymes of chrysanthemum chlorotic mottle viroid: a loop–loop interaction motif conserved in most natural hammerheads. *Nucleic Acids Res.* 37: 368–381.

77 Peracchi, A., Beigelman, L., Scott, E.C. et al. (1997). Involvement of a specific metal ion in the transition of the hammerhead ribozyme to its catalytic conformation. *J. Biol. Chem.* 272: 26822–26826.

78 Maderia, M., Hunsicker, L.M., and DeRose, V.J. (2000). Metal–phosphate interactions in the hammerhead ribozyme observed by 31 P NMR and phosphorothioate substitutions. *Biochemistry* 39: 12113–12120.

79 Hansen, M.R. et al. (1999). Identification and characterization of a novel high affinity metal- binding site in the hammerhead ribozyme. *Rna* 5: 1099–1104.

80 Suzumura, K.-i., Takagi, Y., Orita, M., and Taira, K. (2004). NMR-based reappraisal of the coordination of a metal ion at the Pro-R p oxygen of the A9/G10.1 site in a hammerhead ribozyme. *J. Am. Chem. Soc.* 126: 15504–15511.

81 Hammann, C., Norman, D.G., and Lilley, D.M.J. (2001). Dissection of the ion-induced folding of the hammerhead ribozyme using 19F NMR. *Proc. Natl. Acad. Sci. U.S.A* 98: 5503–5508.

82 Wang, G., Gaffney, B.L., and Jones, R.A. (2004). Differential binding of Mg2+ , Zn2+ , and Cd2+ at two sites in a hammerhead ribozyme Motif, determined by 15 N NMR. *J. Am. Chem. Soc.* 126: 8908–8909.

83 Tanaka, Y. et al. (2002). Identification of the metal ion binding site on an RNA Motif from hammerhead ribozymes uising 15 N NMR spectroscopy. *J. Am. Chem. Soc.* 124: 4595–4601.

84 Fürtig, B. et al. (2008). NMR-spectroscopic characterisation of phosphodiester bond cleavage catalyzed by the minimal hammerhead ribozyme. *RNA Biology* 5: 41–48.

85 Hampel, A. and Tritz, R. (1989). RNA catalytic properties of the minimum (-)sTRSV sequence. *Biochemistry* 28 (12): 4929–4933.

86 Fedor, M.J. (1999). Tertiary structure stabilization promotes Hairpin ribozyme ligation. *Biochemistry* 38: 11040–11050.

87 Cai, Z. and Tinoco, I. (1996). Solution structure of loop A from the hairpin ribozyme from tobacco ringspot virus satellite. *Biochemistry* 35: 6026–6036.

88 Butcher, S.E., Allain, F.H.T., and Feigon, J. (1999). Solution structure of the loop B domain from the hairpin ribozyme. *Nat. Struct. Biol.* 6: 212–216.

89 Ravindranathan, S., Butcher, S.E., and Feigon, J. (2000). Adenine protonation in domain B of the hairpin ribozyme. *Biochemistry* 39: 16026–16032.

90 Shallop, A.J., Gaffney, B.L., and Jones, R.A. (2004). Use of both direct and indirect 13C tags for probing nitrogen interactions in Hairpin ribozyme models by 15N NMRI. *Nucleosides Nucleotides Nucleic Acids* 23: 273–280.

91 Meli, M., Vergne, J., and Maurel, M.-C. (2003). In vitro selection of adenine-dependent Hairpin ribozymes. *J. Biol. Chem.* 278: 9835–9842.

92 Buck, J. et al. (2009). NMR spectroscopic characterization of the adenine-dependent Hairpin ribozyme. *ChemBioChem* 10: 2100–2110.

93 Lilley, D.M.J. (2004). The Varkud satellite ribozyme. *RNA* 10: 151–158.

94 Suslov, N.B. et al. (2015). Crystal structure of the Varkud satellite ribozyme. *Nat. Chem. Biol.* 11: 840–846.

95 Michiels, P.J.A., Schouten, C.H.J., Hilbers, C.W., and Heus, H.A. (2000). Structure of the ribozyme substrate hairpin of Neurospora VS RNA: a close look at the cleavage site. *RNA* 6: 1821–1832.

96 Hoffmann, B. et al. (2003). NMR structure of the active conformation of the Varkud satellite ribozyme cleavage site. *Proc. Natl. Acad. Sci. U.S.A* 100: 7003–7008.

97 Campbell, D.O., Bouchard, P., Desjardins, G., and Legault, P. (2006). NMR structure of Varkud satellite ribozyme stem–loop V in the presence of magnesium ions and localization of metal-binding sites. *Biochemistry* 45: 10591–10605.

98 Bouchard, P. et al. (2008). Role of SLV in SLI substrate recognition by the Neurospora VS ribozyme. *RNA* 14: 736–748.

99 Desjardins, G., Bonneau, E., Girard, N. et al. (2011). NMR structure of the A730 loop of the Neurospora VS ribozyme: insights into the formation of the active site. *Nucleic Acids Res.* 39: 4427–4437.

100 Bouchard, P. and Legault, P. (2014). Structural insights into substrate recognition by the neurospora varkud satellite ribozyme: importance of u-turns at the kissing-loop junction. *Biochemistry* 53: 258–269.

101 Bonneau, E. and Legault, P. (2014). Nuclear magnetic resonance structure of the III–IV–V three-way junction from the Varkud satellite ribozyme and identification of magnesium-binding sites using paramagnetic relaxation enhancement. *Biochemistry* 53: 6264–6275.

102 Bonneau, E., Girard, N., Lemieux, S., and Legault, P. (2015). The NMR structure of the II–III–VI three-way junction from the Neurospora VS ribozyme reveals a critical tertiary interaction and provides new insights into the global ribozyme structure. *RNA* 21: 1621–1632.

103 Pan, T. and Uhlenbeck, O.C. (1992). In vitro Selection of RNAs that undergo autolytic cleavage with Pb2+. *Biochemistry* 31: 3887–3895.

104 Pan, T. and Uhlenbeck, O.C. (1992). A small metalloribozyme with a two-step mechanism. *Nature* 358: 560–563.

105 Legault, P. and Pardi, A. (1994). In situ probing of adenine protonation in RNA by 13C NMR. *J. Am. Chem. Soc.* 116: 8390–8391.

106 Legault, P., Hoogstraten, C.G., Metlitzky, E., and Pardi, A. (1998). Order, dynamics and metal-binding in the lead-dependent ribozyme 1. *J. Mol. Biol.* 284: 325–335.

107 Katahira, M. et al. (1996). NMR studies of a lead ribozyme and its non-cleavable analogue. *Nucleosides Nucleotides* 15: 489–503.

108 Guerrier-Takada, C., Gardiner, K., Marsh, T. et al. (1983). The RNA moiety of ribonuclease P is the catalytic subunit of the enzyme. *Cell* 35: 849–857.

109 Marquez, S.M., Chen, J.L., Evans, D., and Pace, N.R. (2006). Structure and function of eukaryotic ribonuclease P RNA. *Molecular Cell* 24: 445–456.

110 Glemarec, C. et al. (1996). The NMR structure of 31mer RNA domain of Escherichia coli RNase P RNA using its non-uniformly deuterium labelled counterpart [the 'NMR-window' concept]. *Nucleic Acids Res.* 24: 2022–2035.

111 Spitzfaden, C. et al. (2000). The structure of ribonuclease P protein from Staphylococcus aureus reveals a unique binding site for single-stranded RNA. *J. Mol. Biol.* 295: 105–115.

112 Schmitz, M. and Tinoco, I. (2000). J, Solution structure and metal-ion binding of the P4 element from bacterial RNase P RNA. *Rna* 6: 1212–1225.

113 Zuleeg, T., Hartmann, R.K., Kreutzer, R., and Limmer, S. (2001). NMR spectroscopic evidence for Mn2+(Mg2+) binding to a precursor-tRNA microhelix near the potential RNase P cleavage site. *J. Mol. Biol.* 305: 181–189.

114 Leeper, T.C. et al. (2002). Structure of the UGAGAU hexaloop that braces Bacillus RNase P for action. *Nat. Struct. Biol.* 9: 397–403.

115 Getz, M.M., Andrews, A.J., Fierke, C.A., and Al-Hashimi, H.M. (2007). Structural plasticity and Mg2+ binding properties of RNase P P4 from combined analysis of NMR residual dipolar couplings and motionally decoupled spin relaxation. *Rna* 13: 251–266.

116 Henkels, C.H., Chang, Y.-C., Chamberlin, S.I., and Oas, T.G. (2007). Dynamics of backbone conformational heterogeneity in *Bacillus subtilis* ribonuclease P protein. *Biochemistry* 46: 15062–15075.

117 Chang, Y.-C., Franch, W.R., and Oas, T.G. (2010). Probing the folding intermediate of *Bacillus subtilis* RNase P protein by nuclear magnetic resonance. *Biochemistry* 49: 9428–9437.

118 Koutmou, K.S. et al. (2010). NMR and XAS reveal an inner-sphere metal binding site in the P4 helix of the metallo-ribozyme ribonuclease P. *Proc. Natl. Acad. Sci. U.S.A* 107: 2479–2484.

119 Cuzic, S., Heidemann, K.A., Wöhnert, J., and Hartmann, R.K. (2008). *Escherichia coli* RNase P RNA: substrate ribose modifications at G+1, but not nucleotide - 1/+73 base pairing, affect the transition state for cleavage chemistry. *J. Mol. Biol.* 379: 1–8.

120 Chen, W.Y. et al. (2011). Cooperative RNP assembly: complementary rescue of structural defects by protein and RNA subunits of archaeal RNase P. *J. Mol. Biol.* 411: 368–383.

121 Reinhold-Hurek, B. and Shub, D.A. (1992). Self-splicing introns in tRNA genes of widely divergent bacteria. *Nature* 357: 173–176.

122 Michel, F. and Westhof, E. (1990). Modelling of the three-dimensional architecture of group I catalytic introns based on comparative sequence analysis. *J. Mol. Biol.* 216: 585–610.

123 Cech, T.R., Herschlag, D., Piccirilli, J.A., and Pyle, A.M. (1992). RNA catalysis of a group I ribozyme. *J. Biolumin. Chemilumin.* 267: 17479–17482.

124 Allain, F.H.T. and Varani, G. (1995). Structure of the P1 helix from Group I self-splicing Introns. *J. Mol. Biol.* 250: 333–353.

125 Lee, J.-H. and Pardi, A. (2007). Thermodynamics and kinetics for base-pair opening in the P1 duplex of the Tetrahymena group I ribozyme. *Nucleic Acids Res.* 35: 2965–2974.

126 Amano, M. (2001). NMR solution structure of the 2.1 loop of Tetrahymena group I intron. *ucleic Acids Res.* 1: 169–170.

127 Amano, M. (2000). NMR solution structure of the L 9.1a region of Tetrahymena group I intron. *Nucleic Acids Symp. Ser.* 44: 281–282.

128 Amano, M. (2003). The new loop-loop interactions between the peripheral domains and three-dimensional model of Tetrahymena group I intron. *Nucleic Acids Symp. Ser.* 3: 173–174.

129 Chastain, M. and Tinoco, I. (1992). A base-triple structural domain in RNA. *Biochemistry* 31: 12733–12741.

130 Chastain, M. and Tinoco, I. (1993). Nucleoside triples from the Group I Intron. *Biochemistry* 32: 14220–14228.

131 Nowakowski, J. and Tinoco, I. (1996). Conformation of an RNA molecule that models the P4/P6 junction from group I introns. *Biochemistry* 35: 2577–2585.

132 Johnson, E.C., Feher, V.A., Peng, J.W. et al. (2003). Application of NMR SHAPES Screening to an RNA Target. *J. Am. Chem. Soc.* 125: 15724–15725.

133 McDowell, J.A. and Turner, D.H. (1996). Investigation of the structural basis for thermodynamic stabilities of tandem GU mismatches: Solution structure of (rGAGGUCUC)2 by two-dimensional NMR and simulated annealing. *Biochemistry* 35: 14077–14089.

134 Luebke, K.J., Landry, S.M., and Tinoco, I. (1997). Solution conformation of a five-nucleotide RNA bulge loop from a group I intron. *Biochemistry* 36: 10246–10255.

135 Colmenarejo, G. and Tinoco, I. (1999). Structure and thermodynamics of metal binding in the P5 helix of a group I intron ribozyme 1 1Edited by P. E. Wright. *J. Mol. Biol.* 290: 119–135.

136 Silverman, S.K., Zheng, M., Wu, M. et al. (1999). Quantifying the energetic interplay of RNA tertiary and secondary structure interactions. *Rna* 5: 1665–1674.

137 Rüdisser, S. and Tinoco, I. (2000). Solution structure of cobalt(III)hexammine complexed to the GAAA tetraloop, and metal-ion binding to G·A mismatches. *J. Mol. Biol.* 295: 1211–1223.

138 Zheng, M., Wu, M., and Tinoco, I. (2001). Formation of a GNRA tetraloop in P5abc can disrupt an interdomain interaction in the Tetrahymena group I ribozyme. *Proc. Natl. Acad. Sci. U.S.A* 98: 3695–3700.

139 Xue, Y., Gracia, B., Herschlag, D. et al. (2016). Visualizing the formation of an RNA folding intermediate through a fast highly modular secondary structure switch. *Nat. Commun.* 7: ncomms11768.

140 Koculi, E., Cho, S.S., Desai, R. et al. (2012). Folding path of P5abc RNA involves direct coupling of secondary and tertiary structures. *Nucleic Acids Res.* 40: 8011–8020.

141 Gracia, B. et al. (2016). RNA Structural Modules Control the Rate and Pathway of RNA Folding and Assembly. *J. Mol. Biol.* 428: 3972–3985.

142 Watanabe, S. et al. (1996). An RNA fragment consisting of the P7 and P9.0 stems and the 3′-terminal guanosine of the Tetrahymena group I intron. *Nucleic Acids Res.* 24: 1337–1344.

143 Kitamura, A. et al. (2012). The guanosine binding mechanism of the Tetrahymena group I intron. *Nucleic Acids Symp. Ser.* 42: 191–192.

144 Kitamura, A. et al. (2002). Solution structure of an RNA fragment with the P7/P9.0 region and the 3′-terminal guanosine of the Tetrahymena group I intron. *Rna* 8: 440–451.

145 Lehmann, K. and Schmidt, U. (2003). Group II introns: structure and catalytic versatility of large natural ribozymes. *Crit. Rev. Biochem. Mol. Biol.* 38: 249–303.

146 Zhang, L. and Doudna, J.A. (2002). Structural insights into Group II intron catalysis and branch-site selection. *Science* 295: 2084–2088.

147 Sigel, R.K.O. et al. (2004). Solution structure of domain 5 of a group II intron ribozyme reveals a new RNA motif. *Nat. Struct. Mol. Biol.* 11: 187–192.

148 Erat, M.C., Zerbe, O., Fox, T., and Sigel, R.K.O. (2007). Solution structure of domain 6 from a self-splicing group II intron ribozyme: A Mg2+ binding site is located close to the stacked branch adenosine. *ChemBioChem* 8: 306–314.

32

Studying Ribozymes with Electron Paramagnetic Resonance Spectroscopy

Olav Schiemann

University of Bonn, Institute of Physical and Theoretical Chemistry, Wegelerstr. 12, Bonn 53115, Germany

32.1 Introduction

During the last decades, various ribozymes have been discovered [1, 2]. In order to understand the principles underlying the mechanisms of their function, biochemical, biophysical, and computational methods are needed that enable one to follow conformational changes and dynamics on the scale of atoms and domains, to measure parameters of the microenvironment around the cleavage/ligation site, to monitor the formation and breaking of hydrogen bonds, to resolve changes in the ligand sphere of metal ions involved, and to unravel the time scales of the individual steps in the catalytic mechanism. On the biochemical site, experiments in which the catalytic activity of a ribozyme is monitored in dependence of site-specific modifications on nucleotides have been very successful in unraveling the influence of metal ions and pK values of bases. An example for this is the decrease in self-cleavage activity upon phosphorothioate substitution in the active site and the recovery of the activity when exchanging Mg^{2+} for Mn^{2+} [3, 4]. Methods that provide global structures with atomic resolution are X-ray crystallography [5, 6] and nuclear magnetic resonance (NMR) spectroscopy [7, 8]. The advantage of the latter is that it is performed in solution and that it provides information about the dynamics of the ribozyme. However, NMR is restricted to ribozyme sizes below ~100 kDa. In contrast, cryo-EM [9, 10] does not need crystals and has no size restriction, but data are acquired in the frozen state. Beyond these methods with atomistic resolution, fluorescence resonance energy transfer (FRET) provides the dynamics of domains on the nanometer and single molecule scale [11, 12]. In addition, it yields the timescale of conformational changes, which is complementary to the one of NMR, microseconds to milliseconds for FRET [13] versus seconds to minutes for NMR [14]. Small angle X-ray scattering (SAXS) [15, 16] is an interesting method for obtaining the overall shape of a ribozyme, and native mass spectrometry, e.g. in the form of LILBID [17, 18], can help to identify ribozyme–protein complexes and to unravel their constituents.

Complementary to all these methods, electron paramagnetic resonance (EPR) or electron spin resonance (ESR) spectroscopy [19–21] is a method that has no size

Ribozymes, First Edition. Edited by Sabine Müller, Benoît Masquida, and Wade Winkler.

restriction, can be performed in solution or even whole cells, provides information about microenvironments, and yields atomistic resolution on a local scale and coarse-grained global structures. However, it requires the presence of unpaired electrons, which can be introduced by paramagnetic for diamagnetic metal ion exchange, e.g. Mn^{2+} for Mg^{2+}, or by site-specific attachment of small organic spin labels, e.g. nitroxides. In the following, the basics of EPR are briefly explained followed by a section on site-directed spin labeling of RNA, and then several examples are summarized related to the application of EPR spectroscopy to ribozymes.

32.2 EPR Methods

32.2.1 Magnetic Interactions

Below, a brief introduction is given into the interactions of an unpaired electron within an applied magnetic field and with its surrounding. A more detailed introduction into EPR can be found here [19–21].

An unpaired electron has a spin $S = ½$ and associated with this a magnetic spin quantum number m_s, which can have the values $\pm½$. Within an applied magnetic field B_0, the $m_s = -½$ state is lower in energy than the $m_s = +½$ state, with the energy difference ΔE between them given by (Figure 32.1a)

$$\Delta E = g\mu_B B_0 = h\nu \tag{32.1}$$

where μ_B is the Bohr magneton and g is the g-factor of the unpaired electron. According to the Boltzmann equation, the population of the $m_s = -½$ state is larger than that of the energetically higher lying state $m_s = +½$. Using microwave radiation and according to the selection rule $\Delta m_s = \pm 1$, the spin can be flipped from $m_s = -½$ to $m_s = +½$, if the microwave frequency ν matches the energy difference ΔE (resonance condition). This gives rise to one line whose position on the magnetic field axis B_0 is dependent on the g-factor. Often EPR spectra are recorded from frozen samples, because they yield better signal intensities and the electron spin relaxation times are slowed down, enabling pulsed EPR measurements. As a consequence, the g-factor is now described by a g-tensor with three principle components g_{xx}, g_{yy}, and g_{zz}. If all three g-values are identical, the spectrum is called isotropic; if two are identical ($g_\perp$ and $g_\parallel$), the system is axial; and if all three are different, it is orthorhombic (Figure 32.2). The g-tensor adopts for unpaired electrons in ions or molecules largely different values in the range and thus contains information about the type of radical and its structure.

In addition to the interaction between the unpaired electron and B_0, the unpaired electron also interacts with nuclei, having a nuclear spin $I > 0$ (Figure 32.1b). This leads to a splitting of the EPR line in dependence of the number and type of nuclei as well as the strength of their interaction with the unpaired electron. This interaction is called hyperfine coupling whose magnitude is given by the hyperfine coupling constant A (Figure 32.1b). In the frozen state also A is described by a tensor with the three principal values A_{xx}, A_{yy}, and A_{zz}. The magnitude of the hyperfine coupling

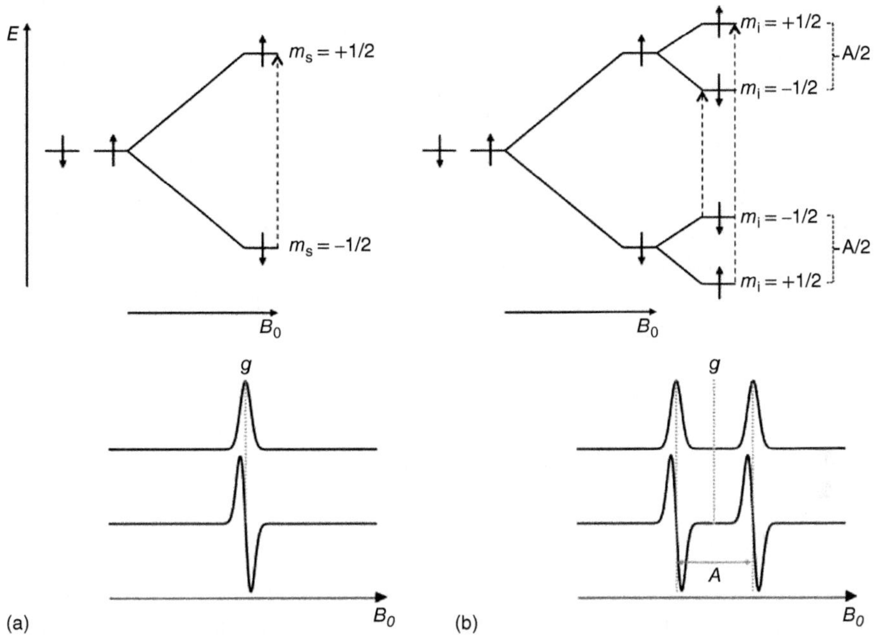

Figure 32.1 The energy levels and corresponding EPR spectra of an unpaired electron in (a) an applied magnetic field B_0 and (b) strongly coupled to a nucleus with $I = ½$, e.g. ^1H. The energy separation between the m_I states in one m_s state is half the hyperfine coupling constant A. The selection rules for the EPR transitions (dashed arrows) are $\Delta m_s = \pm 1$ and $\Delta m_I = 0$. The EPR spectra are shown in absorption, as recorded in pulsed EPR experiments, and in first derivative mode, as recorded in cw EPR experiments.

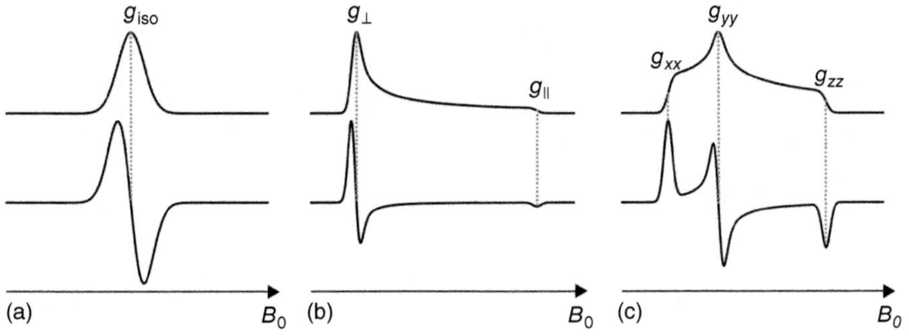

Figure 32.2 EPR spectra for (a) an isotropic g-tensor with g_{iso}, (b) an axial g-tensor with $g_\perp$ and $g_\parallel$, and (c) an orthorhombic g-tensor with g_{xx}, g_{yy}, and g_{zz}.

is given by the amount of spin density on the nucleus, its magnetic moment, and its distance and the orientation of this distance vector with respect to B_0. For nuclei with $I > 1/2$ also, the quadrupole coupling has to be taken into account.

If more than one unpaired electron is present in a molecule, they can also couple with each other, giving rise to the dipolar coupling constant D. This coupling constant depends with $1/r^3$ on the distance r between the unpaired electrons and the angle between this distance vector and B_0. If several unpaired electrons are centered on one nucleus, e.g. in the case of high-spin metal ions like Mn^{2+}, the zero-field splitting constants D and E occur, which do also contain structural information.

The aim of EPR spectroscopy is to identify, separate, and quantify all these interactions. In order to achieve this, one uses multifrequency and pulsed EPR methods in combination with site-directed spin and isotope labeling. In the last step, the obtained EPR parameters are translated into structural or dynamics information, which is usually achieved with the help of computational methods like molecular dynamics (MD) [22] simulations or density functional theory (DFT) [23].

32.2.2 Multifrequency cw EPR [21]

The most common EPR spectra are continuous wave (cw) EPR spectra recorded at X-band frequencies (9 GHz). Usually this provides access to large hyperfine coupling constants and largely anisotropic g-tensors. Like in NMR spectroscopy, going to higher magnetic fields/frequencies yields better signal intensities and better resolved spectra (Figure 32.3).

In addition, recording spectra at different principal microwave frequencies facilitates to separate the magnetic field-dependent Zeeman interaction from the other contributions. Thus, multifrequency cw EPR spectroscopy is well suited to obtain

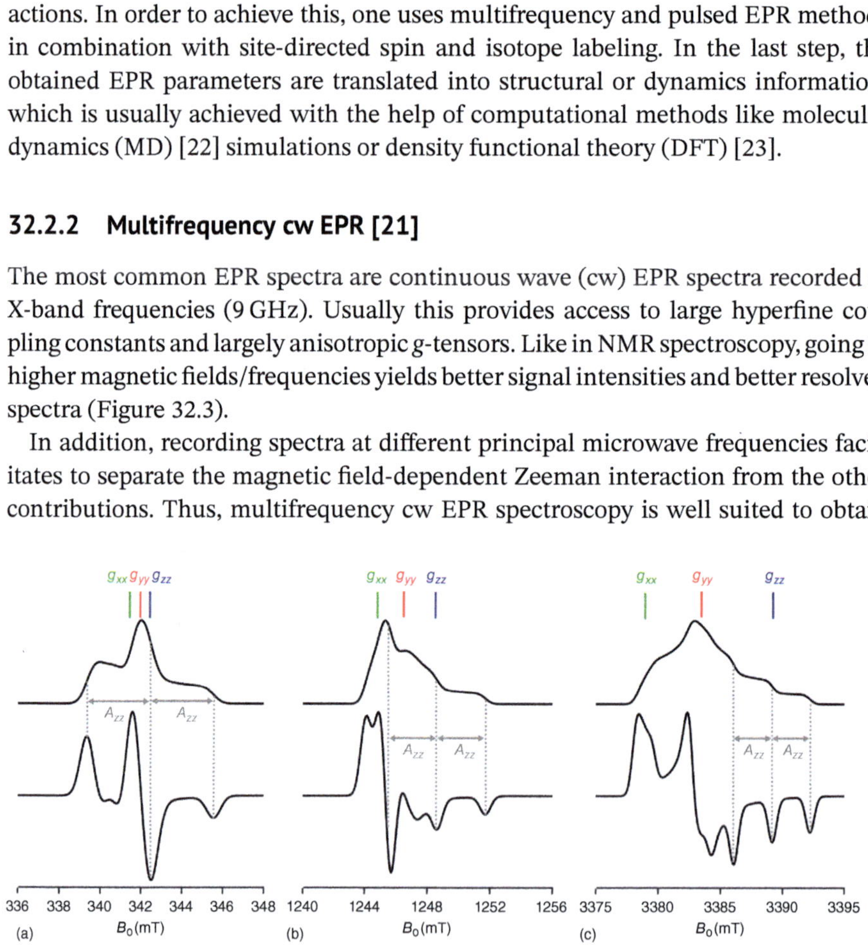

Figure 32.3 EPR spectra of a nitroxide in the frozen state at (a) X-band (9 GHz), (b) Q-band (33 GHz), and (c) W-band (95 GHz) with the $A(^{14}N)$- and g-tensor indicated (top, spectrum in absorption as obtained in typical field swept pulsed EPR experiments; bottom, the same spectrum in first derivative as obtained in typical cw EPR experiments). The hyperfine coupling to the ^{14}N nucleus yields three line, because $I(^{14}N) = 1$. The A_{zz} component of the ^{14}N-hyperfine coupling tensor is the largest and resolved, whereas the A_{xx} and A_{yy} are small and usually hidden under the line width.

the g-tensor and large hyperfine coupling constants. Smaller hyperfine coupling constants can be resolved with pulsed hyperfine spectroscopy (see section 32.3.2). The obtained g- and A-tensors are then translated into structural information by means of computational methods, most commonly by DFT calculations [23]. The usual approach is such that DFT methods are used to calculate the tensors for a set of proposed structures, the calculated tensors are then compared with the experimental ones, and the best fitting one is assumed to be the true structure. The sensitivity of the g- and A-tensors on the structure around an unpaired electron can, for example, nicely be seen in their changes in dependence of the number of hydrogen bonds on a nitroxide [24]. In addition, the g- and A-tensors give access to the microenvironment around a nitroxide spin label, yielding, for example, the local pH value [25]. Beyond the g- and A-tensors, cw EPR spectra also contain information about the mobility/dynamics. Prominent examples for this are the spectra of nitroxide spin labels [26]. These spectra change their shape and width in dependence of the rotational correlation time τ_c of the nitroxide (Figure 32.4). The rotational correlation time depends on the overall tumbling of the biomolecule and the rotational freedom of the nitroxide at the site of its attachment. It thus provides access to information about the local and overall dynamics of, e.g. ribozyme domains. However, in order to disentangle both contributions, a multifrequency approach should be used [26, 27].

32.2.3 Pulsed Hyperfine Spectroscopy

Whereas cw EPR is well suited to resolve the g-tensor and large hyperfine coupling constants, pulsed hyperfine spectroscopy [28] is used to unravel large and small hyperfine coupling constants as well as quadrupole parameters. The term hyperfine spectroscopy subsumes a large range of pulsed EPR methods aiming at selecting from the cw EPR spectrum, in which all interactions are present but mostly hidden under the line width, the wanted hyperfine coupling constants. Most prominent examples for such methods are electron spin echo envelope modulation (ESEEM) [29], hyperfine sublevel correlation experiment (HYSCORE) [30], electron–electron double resonance (ELDOR) electron-detected NMR (EDNMR) [31], and electron nuclear double resonance (ENDOR) [32]. For the respective pulse sequences, see Figure 32.5. With respect to ribozymes these methods are important for, e.g. determining the structure of metal ion binding sites. Important nuclei with a magnetic moment are in this respect ^{1}H (I = ½), ^{14}N (I = 1), and ^{31}P (I = ½) and after isotope substitution ^{2}H (I = 1), ^{15}N (I = ½), and ^{17}O (I = 5/2) [33–35].

32.2.4 Pulsed EPR Dipolar Spectroscopy (PDS)

Whereas the abovementioned methods yield structural information on an atomistic but local scale, pulsed EPR dipolar spectroscopy (PDS) [36, 37] provides access to the arrangement of ribozyme domains with respect to each other and how they change conformations during function. The idea of all PDS methods is to measure the dipolar coupling ν_{dd}, which depends, according to Eq. (32.2), on the distance r between both spin centers and the angle θ between the distance vector r and the z-axis of B_0.

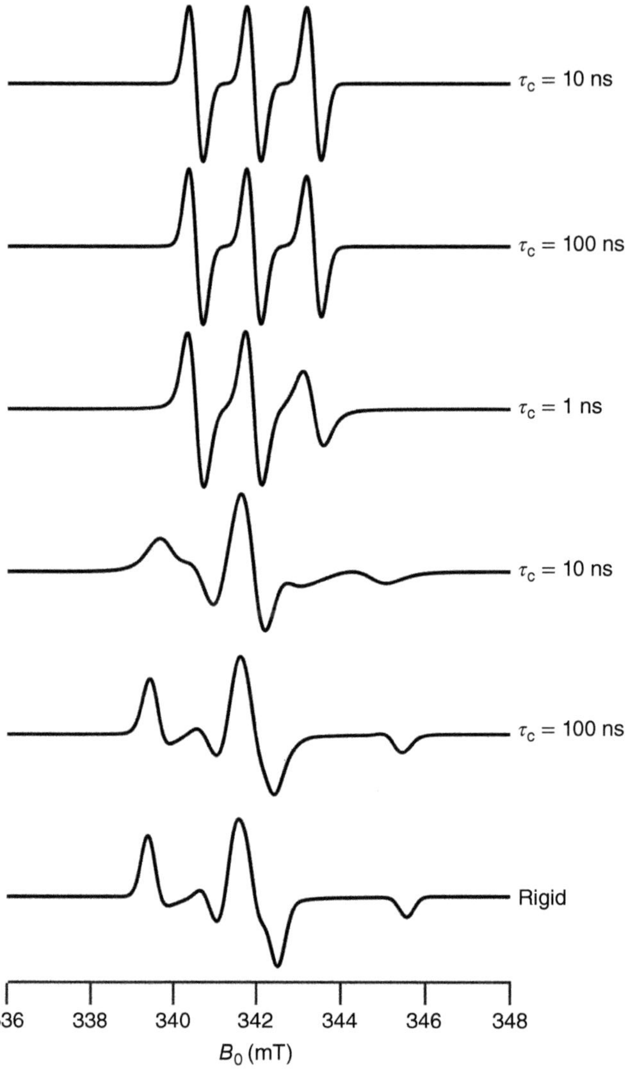

Figure 32.4 Dependence of cw X-band EPR nitroxide spectra on the rotational correlation time τ_c. For small τ_c, e.g. for a nitroxide in the liquid state at room temperature, the nitroxide spectrum is isotropic with three lines of equal intensity. If the sample is frozen, the spectrum is anisotropic and considerably broader. In between, the width and shape of the spectrum gradually changes with τ_c.

In Eq. (32.2), D is the dipolar coupling constant defined by Eq. (32.3). In Eq. (32.3), μ_B is the Bohr's magneton, g_A and g_B are the g-values of electrons A and B, μ_0 is the permeability in vacuum, and h is the Planck constant:

$$\nu_{dd} = D\left(1 - 3\cos^2\theta\right) \tag{32.2}$$

$$D = \frac{\mu_B^2 g_A g_B \mu_0}{4\pi h}\frac{1}{r^3} \tag{32.3}$$

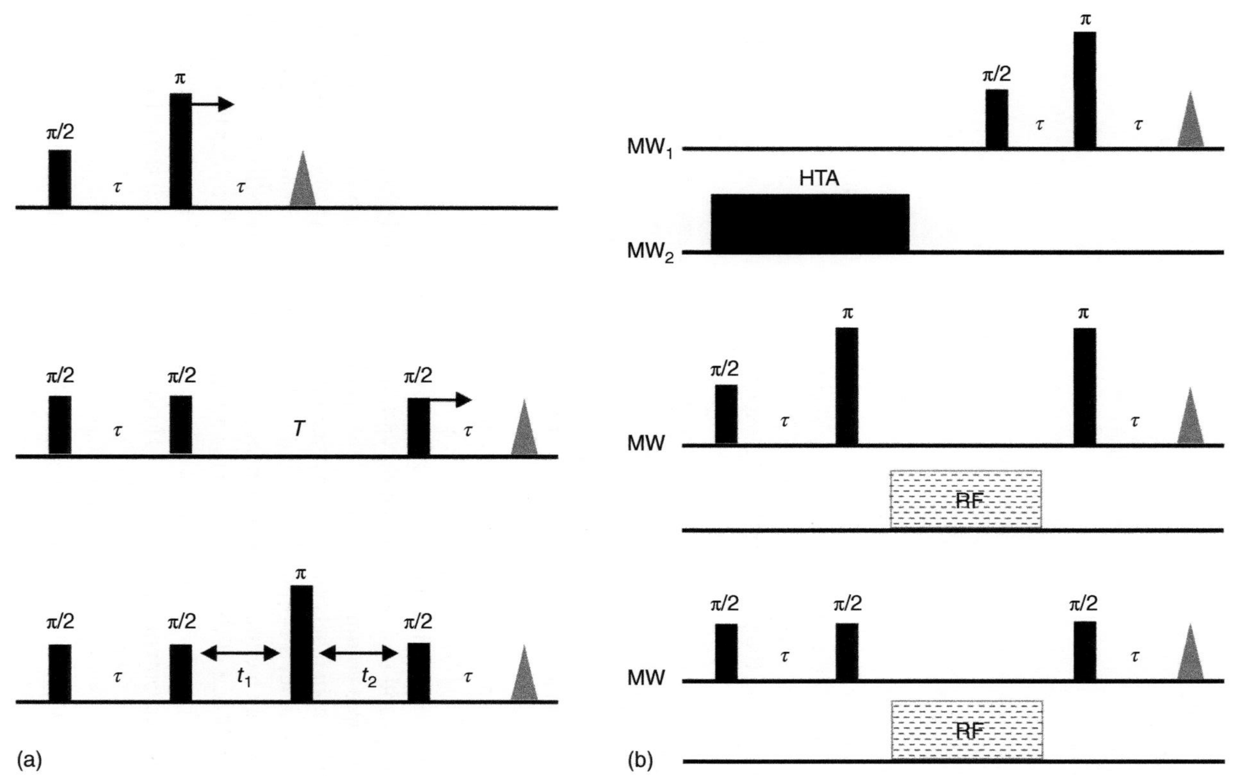

Figure 32.5 Pulse sequences for commonly used pulsed hyperfine experiments. (a) 2-Pulse SEEM (top), 3-pulse ESEEM (middle), and HYSCORE (bottom) and b) EDNMR (top), Davies ENDOR (middle), and Mims ENDOR (bottom). In the EDNMR, the high turning angle (HTA) pulse is applied at a different microwave frequency MW_2 than the Hahn-echo detection sequence (MW_1). In both ENDOR experiments, the nuclei transitions are driven by a radiofrequency (RF) pulse whose frequency is swept while the pulse delays are kept constant. The inter-pulse delays are denoted as τ, t, and T.

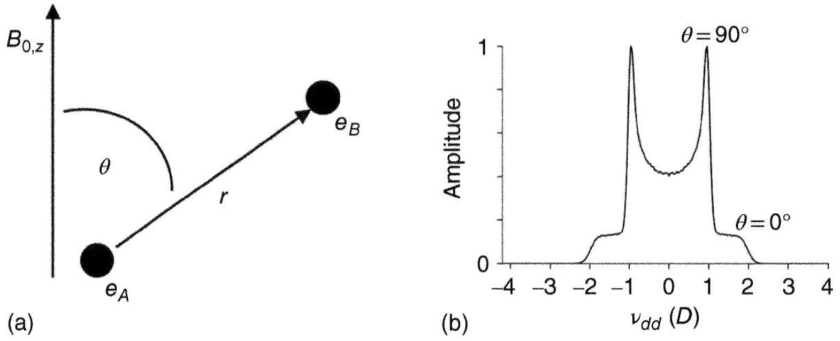

Figure 32.6 Scheme of (a) the electron A–electron B arrangement within B_0 and (b) the Pake pattern. The frequency of the dipolar coupling ν_{dd} at the peak corresponding to $\theta = 90°$ is the dipolar coupling constant D in Eq. (32.3), which is directly related to the distance r.

If both spin centers are randomly oriented with respect to r, the full Pake pattern will be obtained in a PDS measurement from which the distance r can be easily read of (Figure 32.6). If the orientation of the spin centers with respect to r is fixed to the same angle in each molecule, then the PDS measurement becomes orientation selective, considering the pulses in the PDS experiment excite only a fraction of the EPR spectrum [38, 39]. In this case, the PDS measurement is repeated at various positions on the EPR spectrum, yielding different parts of the Pake pattern. Further analysis yields not only r but also the orientation of the spin centers and thus of, e.g. the domains with respect to each other. In a simpler analysis approach, all fractions of the Pake pattern are added up, thereby minimizing the orientation selection, and then the distance r can be extracted but the angular information is lost.

The pulse sequences of the most relevant PDS methods are shown in Figure 32.7. The most prominent PDS method is pulsed electron–electron double resonance (PELDOR or DEER) [37, 40–42] usually used in combination with site-directed spin labeling of ribozymes with nitroxides [43]. Nowadays, PELDOR experiments are performed at Q-band and in combination with a 150 W microwave amplifier [44]. In this case, sample volumes of 80 μl at a concentration of 25 μmolar in spins are routinely used. If spin labels with narrow spectral width like trityl labels [45–50] are used, single-frequency PDS methods like double quantum coherence (DQC) [51] or single-frequency technique for refocusing dipolar couplings (SIFTER) [52] are useful [53, 54]. If one of the spin centers is a fast relaxing metal center like Mn^{2+} [55–57], Cu^{2+} [58], or Fe^{3+} [59], a pulse sequence called relaxation-induced dipolar Modulation enhancement (RIDME) [60, 61] can be used, in which the spin of the metal center is not flipped by a pulse but by its own relaxation, while the slowly relaxing spin label is observed. Using shaped pulses instead of the usually applied rectangular pulses can lead to increased sensitivity and distance limits [61].

Conversion of PDS Data into Distance Distributions

In each of the PDS experiments, the echo signal intensity is monitored in dependence of an incremented time interval between pulses. This yields a time trace in which the intensity of the echo is plotted against the time of this time

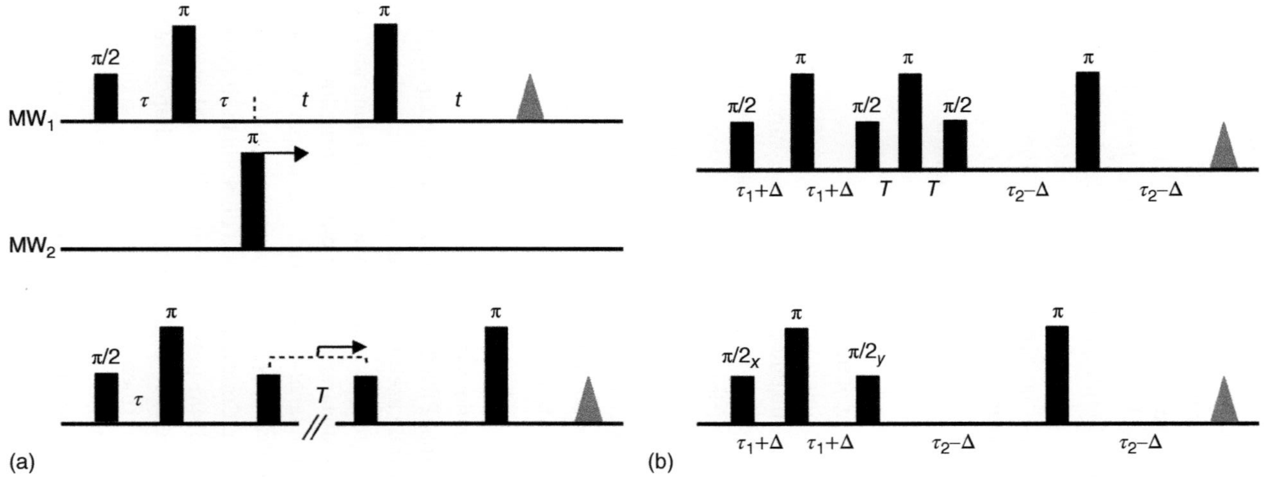

Figure 32.7 Pulse sequences of the most often used PDS methods: (a) 4-pulse PELDOR (top), 5-pulse RIDME (bottom), (b) DQC (top), and SIFTER (bottom). The inter-pulse delays are denoted as τ_1, τ_2, t, and T.

interval (Figure 32.8). Such a time trace shows an oscillation whose frequency encodes the dipolar coupling D between the spin centers within the same molecule and thus the distance between them. In addition, also the random, intermolecular distances between spin centers on different molecules contribute to the time trace. In the PELDOR experiment, the latter interaction leads to an exponential decay [62], which can be removed from the time trace by fitting the last part of the time trace where all oscillation is damped. In the other PDS experiments, the background decay is governed by additional effects and is less well defined. Having extracted the wanted intramolecular contribution, the time trace can be converted into a distance distribution using different programs, the most famous one being DeerAnalysis [63]. In the conversion from the time to distance domain, one should take the following considerations into account [62–66]:

- The oscillations should be visible in order to facilitate a reliable background correction.
- The time traces should have a signal-to-noise ratio that enables a reliable background correction and fitting of the time trace.
- The upper distance limit is determined by the length of the time trace; the longer the time trace, the larger the distances that can be measured. In other words, a time trace with a length of 1 μs will not enable the measurement of a 7 nm distance.
- The error of the distance distribution needs to be determined. Within DeerAnalysis it needs at least to be validated with the validation tool, and minor peaks should be interpreted with great care.
- The width of the distance distribution is influenced by the magnitude of the alpha value chosen for the regularization; the choice of it should rely on one of the criteria implemented in DeerAnalysis.

Translation of Distance Distributions into Structures

One should keep in mind that PDS measures the dipolar coupling between the spin centers. For example, in the case of nitroxides, the spin is localized on the N–O group. However, the N–O group is attached via a rather long and flexible linker to the biomolecule. Thus, the distance does not only become longer but is also influenced by the conformers the label–linker structure adopts. Consequently, in order to translate the observed distances into ribozyme structures, the label conformers have to be taken into account. Commonly this is done via in silico labeling programs like mtsslWizard [67, 68], MMM [69], or Rosetta [70]. In these approaches, structural models of ribozymes are used as input parameters, then the labels are attached to the models, the label conformers are calculated, and the distances between the conformer clouds are determined (Figure 32.9). The calculation of the label conformers is done in different ways and depends on the program chosen. The ribozyme model that yields distance distributions, which fit the measured distance distributions the best, is then regarded as the true structure. Alternatively, the arrangement of domains with respect to each other can also be unraveled via docking software, e.g. implemented in mtsslWizard [68]. In general, the in silico labeling and generation of label conformers works well, if the label extents into the solution and does not

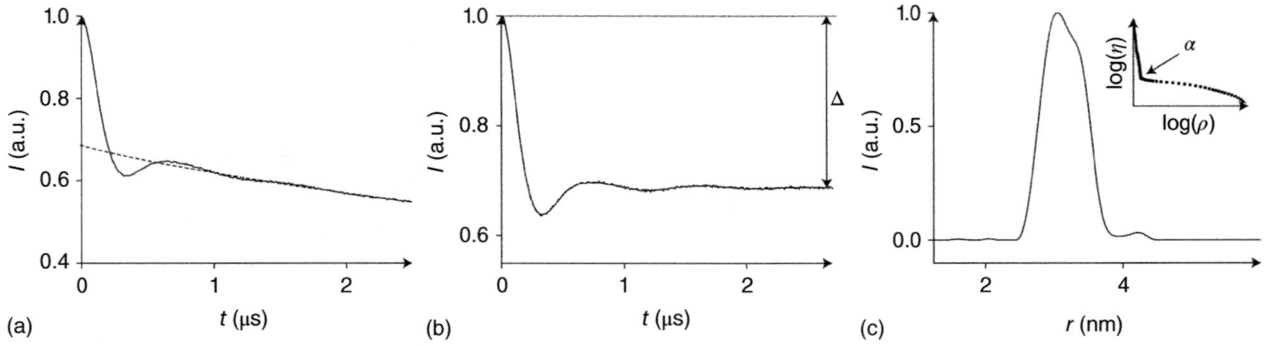

Figure 32.8 Conversion of the PELDOR time trace into a distance distribution. (a) Experimental time trace (solid line) overlaid with the intermolecular decay fitted to the last third of the time trace. (b) The background- and zero-time-corrected time traces with modulation depth Δ indicated. (c) The distance distributions with the L-curve for the α-values shown as inset. One can use the intersection of both lines as a criterion for choosing the α-value. In this inset η is the smoothness of the distribution and ρ quantifies the goodness of the fit.

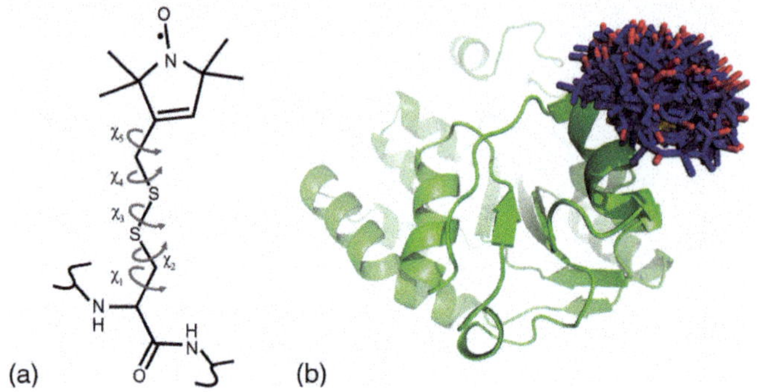

Figure 32.9 Scheme showing (a) the rotatable bonds for the nitroxide MTSSL used for protein labeling and (b) the corresponding conformer cloud of MTSSL bound to a biomolecule generated with mtsslWizard.

interact strongly with the protein. If the label interacts with the ribozyme, the in silico programs are usually not good in reproducing this [71]. A better approach in this case might be a full MD simulation taking also local ribozyme dynamics and rearrangements into account as shown for model systems and the T4 lysozyme [72–74].

32.3 Site-Directed Spin Labeling

During the last two decades, several techniques have been established for binding spin labels to RNA in a site-specific manner. In the following, a brief overview is given; more detailed and complete reviews on this topic can be found here [75–77].

The labeling strategies can be grouped into those where the label is attached during the synthesis of the RNA and those where the labeling step comes after RNA synthesis. In the former case, the label is covalently bound to the phosphoramidite, which is then bound to the evolving RNA strand at the desired position during the automated solid-support RNA synthesis. In the latter case, a usually commercially available and modifiable phosphoramidite is incorporated at the desired position into the RNA strand during RNA strand synthesis, and the labeling step is done after the RNA synthesis is completed. Overall, synthesis techniques exist to attach the label site selectively at the phosphate, the sugar, or the base moiety. The wide range of labeling methods and sites enable to tailor the labeling to the scientific question and to the adequate EPR method and to circumvent structural/functional perturbations. In any case it has to be checked whether the RNA is structurally and/or functionally perturbed. This can be done, for example, via functional assays, UV/vis-based melting studies, and CD spectroscopy.

32.3.1 Labeling During RNA Synthesis

In 2007 Barhate et al. [78] reported on the incorporation of a phosphoramidite into DNA in which a nitroxide is rigidly attached to a cytidine. This spin label has been named **Ç** (Figure 32.10). In 2012, Höbartner et al. [79] extended this approach to RNA phosphoramidites and presented the incorporation of the rigid label **Çm** into several RNA secondary structures (Figure 32.10). Such rigid labels enable detailed studies on the oligonucleotide dynamics at both the local scale in the liquid state using cw EPR [80] and the nanometer scale using orientation selective PELDOR [81].

A disadvantage of the labeled phosphoramidites is that reagents used during the RNA synthesis can reduce the nitroxide group. A possible solution was presented by Weinrich et al. [82]: A phosphoramidite of cytidine was labeled at the exocyclic amine group with a nitroxide (see section 32.3.2) [83] whose NO group was protected via alkylation with a light-sensitive group, **1** (Figure 32.10). The protection group can be cleaved photochemically after the RNA synthesis, leaving after subsequent oxidation by oxygen (air) a TEMPO-based nitroxide bound at the 4-amino group of uridine. With this labeling strategy, PELDOR measurements were successfully carried out [82].

An additional limitation of the phosphoramidite strategy is the significant effort needed to synthesize the labeled phosphoramidites. An approach that alleviates this issue is based on Sonogashira C–C cross-coupling of the alkyne-functionalized nitroxide **TPA** (Figure 32.10) to an iodine-modified base during the RNA synthesis [84, 85]. This strategy enables labeling uridine, cytidine, and adenine, although the cross-coupling step is tedious. In all three cases mentioned above, the label is bound to the bases of the RNA.

32.3.2 Postsynthetic Labeling of RNA

In the postsynthetic approach, the label is covalently bound to the RNA after the RNA strand synthesis. In order to bind the label site-specifically, a phosphoramidite with a unique functional group at the phosphate, the base, or the sugar moiety is incorporated at the desired site into the RNA strand. The modification is chosen such that a corresponding functional group on the label reacts selectively and with high efficiency with this modification. Examples for this are shown in Figure 32.11 and are discussed below.

With respect to labeling the phosphate group, Qin's lab introduced a nucleotide independent strategy in which a phosphorothioate is introduced during the RNA synthesis at the desired site and then the thiol-specific spin label 3-iodomethyl-1-oxy-2,2,5,5-tetramethylpyrroline is attached at this site, leading to **3**. The linker formed is very short with only one methylene group. However, different diastereomers are formed at the phosphate, and its negative charge is lost [86, 87]. In a second step, the strategy was extended to attach a nitroxide via two bonds to two neighboring phosphorothioates, **4**, constraining the conformational degree of freedom even further, which is great for orientation selective PELDOR measurements and dynamics studies at room temperature [88].

Labeling the ribose sugar was achieved by Saha et al. who synthesized a isothiocyanato-functionalized tetraethyl isoindoline nitroxide and attached it covalently to the 2′-amino group of RNA duplexes, **5** [89]. Compared to tetramethyl

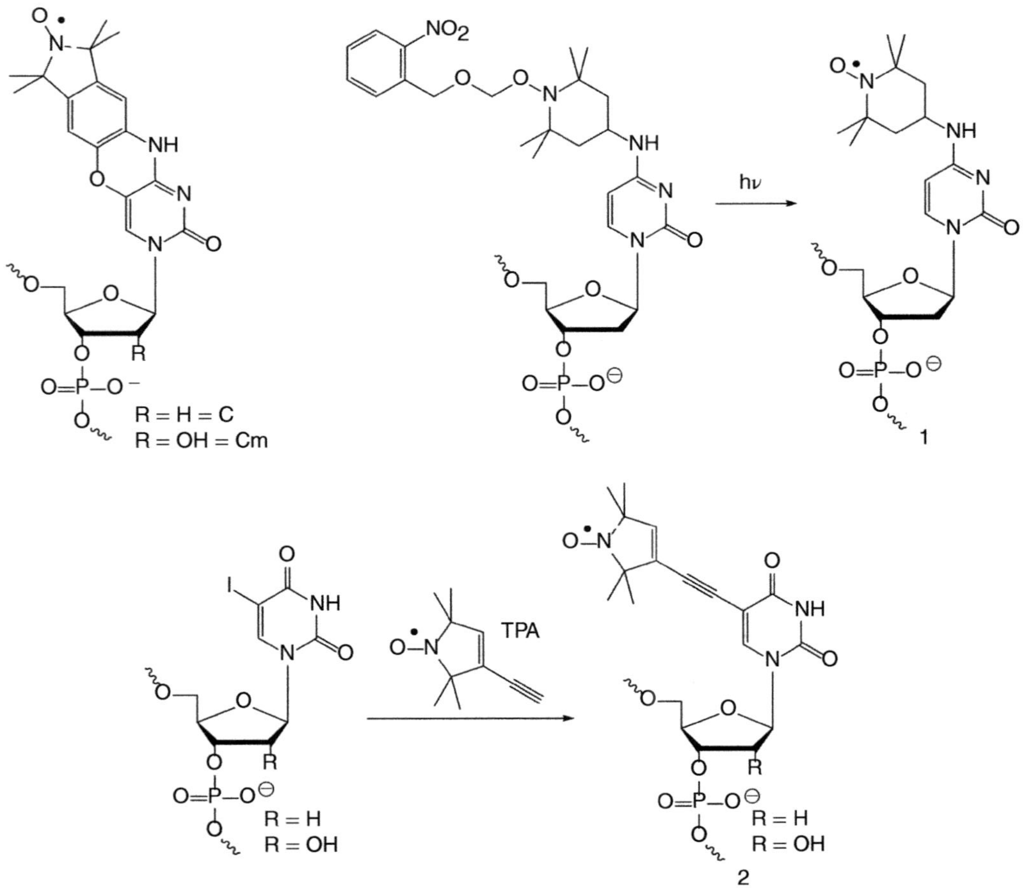

Figure 32.10 Spin labels for RNA labeling during synthesis.

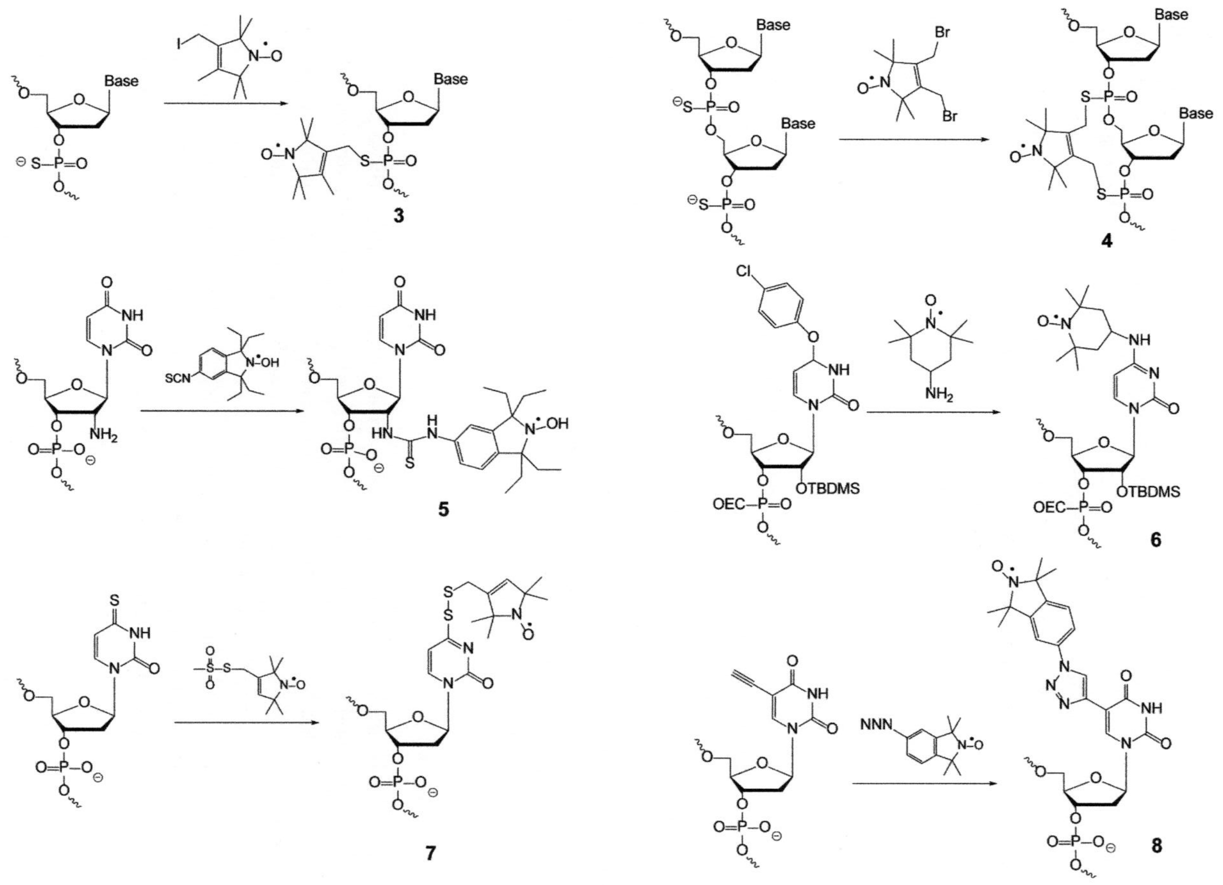

Figure 32.11 Spin labeling of RNA postsynthetically.

nitroxides, the tetraethyl spin label showed increased stability under reducing conditions, making it a promising spin label for *in-cell* measurements [89–91]. cw EPR measurements in the liquid state indicated that the dynamics of the label as such is strongly reduced, giving access to the dynamics of the RNA [89].

For the postsynthetic, site-selective labeling of bases, several strategies have been established. Höbartner bound nitroxides to the exocyclic amino groups in cytidine **6**, adenine, and guanine, where the nitroxide ring is directly attached to the amino group [83]. PELDOR-based distance measurements on RNA duplexes spin labeled in this way were achieved up to distances of 8.1 nm [92].

Thio-modified RNA bases, e.g. 4-thiouridine or 2-thiocytidine, can also be used for site-selective reactions. Importantly, they are naturally present in t-RNAs, or commercially available thio-modified phosphoramidites can be incorporated during the solid-phase synthesis. These can then be reacted with, e.g. iodoacetamide-functionalized [93, 94] or methanethiosulfonate-functionalized nitroxide spin labels (MTSSL) [95], **7**.

Using copper(I)-catalyzed "click" chemistry, Kerzhner et al. established an easy to use, robust, and high-yield method by which an azide-functionalized nitroxide is coupled in solution to the base 4-ethynyl-2′-deoxy-urdinine, **8** [96]. Strand cleavage by Cu(I) can be avoided by controlling the reaction length, by complexing Cu(I) and quenching free Cu(I) after the reaction. The phosphoramidite of 4-ethynyl-2′-deoxy-urdinine is commercially available, and the spin label can be easily synthesized. In addition, the hydrogen bonding pattern of the bases is not disrupted. Interestingly, the label is rather mobile at room temperature in the liquid state but locks into a rigid conformer when lowering the temperature. This gives rise to strongly orientation-selective PELDOR spectra, which is advantageous for dynamics studies based on PELDOR [96].

32.3.3 Labeling Long RNAs

Most of the approaches for labeling RNAs of lengths that exceed those amenable by automated solid-phase synthesis rely on ligating shorter RNA strands, which have been labeled by one of the methods described above. The ligation steps are performed by using a T4-ligase in combination with DNA splints (Figure 32.12a). The first example for this is the double labeling of the 72-nucleotide-long noncoding RNA RsmZ, which enabled a detailed NMR/PELDOR study on its complex formation with proteins (see section 32.4.5) [97, 98]. Here, shorter RNA strands were labeled at 4-thiouridines with an iodoacetamide-functionalized nitroxide [93, 94] and then ligated to different unlabeled RNA fragments using T4-DNA ligase and the corresponding DNA splints. A T4-RNA ligase in combination with DNA splints was used to build the doubly spin labeled 59-nucleotide-long full-length TAR RNA. The labeling was done with protected TEMPO, which could be released upon light excitation, as described above [82]. This enabled PELDOR studies on the TAR domain motions in dependence of arginine amide [82]. Esquiaqui et al. [94] employed a T4-DNA ligase in combination with a DNA splint to spin label the 232-nucleotide-long glycine riboswitch. In that way, they were able to

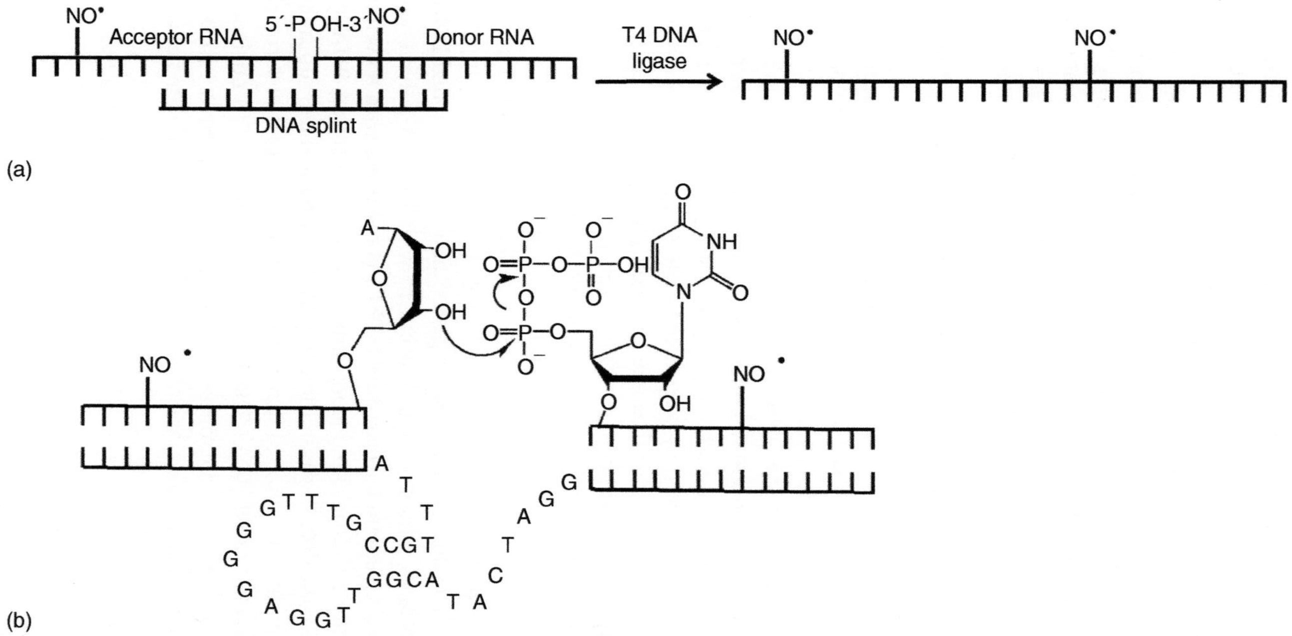

Figure 32.12 Obtaining long, labeled RNAs via ligation using (a) a T4-ligase and a DNA splint and (b) a deoxyribozyme.

follow RNA backbone dynamics of the glycine riboswitch in dependence of Mg^{2+}, K^+, and glycine using cw EPR measurements. By employing a T4-DNA ligase in combination with a DNA splint, two RNA strands labeled could be ligated via the "click" reaction to yield the 81-nucleotide-long TPP riboswitch [99]. In this ligation protocol, reduction of the nitroxide label was not observed and enabled PELDOR experiments on TPP conformations in solution.

Höbartner et al. employed instead of a T4-ligase a deoxyribozyme that catalyzes the formation of the 3′–5′ phosphodiester bond between the two spin-labeled RNA strands in the presence of divalent metal ions (Figure 32.12b) [100, 101]. The deoxyribozyme is commercially available, and the ligation protocol does not reduce the label. In this way they were able to doubly spin label the 53-nucleotide-long SAM-III and the 118-nucleotide-long SAM-I riboswitches. However, the deoxyribozyme has rather stringent sequence requirements for the RNA ends.

Fedin, Karpova, and Bagryanskaya labs achieved labeling of long RNAs in a different way (Figure 32.13a) [102]. They use a 4-[*N*-(2-chloroethyl-*N*-methyl)amino]-benzyl-phosphoroamide attached to the 5′ end of a short DNA sequence, which is complementary to the RNA sequence where the spin label is to be attached to. The DNA strand binds at this RNA site and alkylates the first 3′-RNA base after the DNA–RNA duplex forming a DNA–RNA cross-link. The cross-link is broken by hydrolysis, leaving an amino group connected via an aliphatic linker to the RNA base. This amino group can then be labeled with *N*-hydroxysuccinimide esters of nitroxides. Using this method they were able to site-specifically label bases within the 330-nucleotide-long internal ribosome entry site (IRES) of the hepatitis C virus RNA and to study its dynamics with cw EPR and PELDOR [104].

Kath–Schorr lab [103] recently introduced an interesting posttranscriptional method that enables labeling of very long RNAs and may open up even labeling long RNAs within cells (Figure 32.13b). Their approach is based on the nonnatural base pair dTPT3–dNaM introduced by Romesberg [105]. The nonnatural nucleotide dNaM, depicted as dX in Figure 32.13b, is placed in a single-stranded DNA region at the desired position, which is then transcribed with T7 RNA polymerase in the presence of the complementary base TPT3. TPT3 is modified with a cyclopropene group (**Y** in Figure 32.13b). This incorporates **Y** into the forming RNA strand at the position where the dNaM was in the DNA. The cyclopropene group is then coupled to a tetrazine-functionalized nitroxide Tet^{NO} via an inverse electron demand Diels–Alder cycloaddition (iEDDA). The approach is applied to a 18-nucleotide-long duplex RNA, which reveals that rather narrow and well-defined PELDOR time traces and distance distributions are obtained although the linker between base and N–O group is rather long and flexible. MD simulations indicate that the label bends back onto the RNA forming contacts with the GC base pairs at the end of the duplex [103].

32.3.4 Non-covalent Labeling

All labeling strategies mentioned above require the covalent binding of the label to the RNA strand. An alternative is non-covalent spin labeling [106]. Here, a

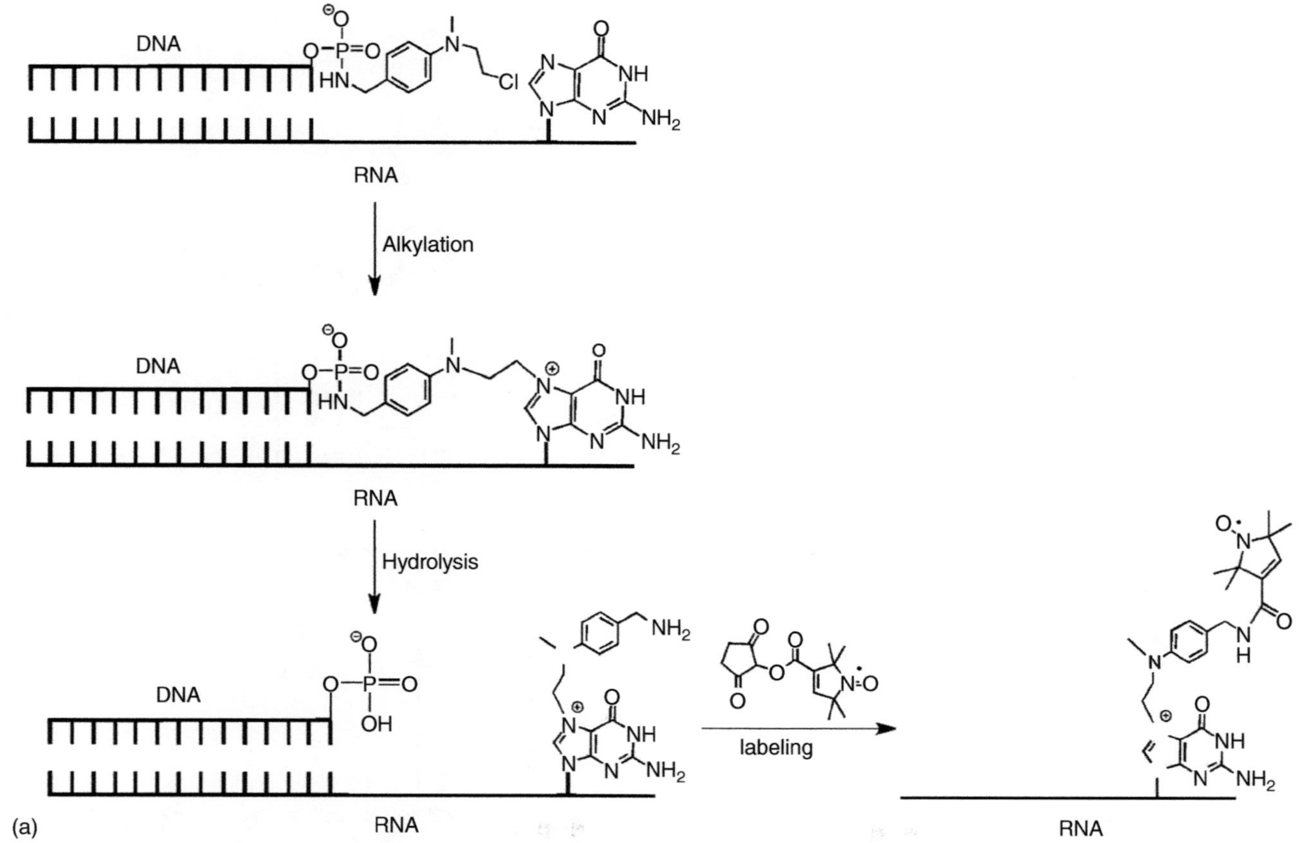

Figure 32.13 Obtaining long, labeled RNAs via (a) sequence recognition, alkylation [102], and labeling and (b) by posttranscriptional labeling [103]. Source: Adapted from [103].

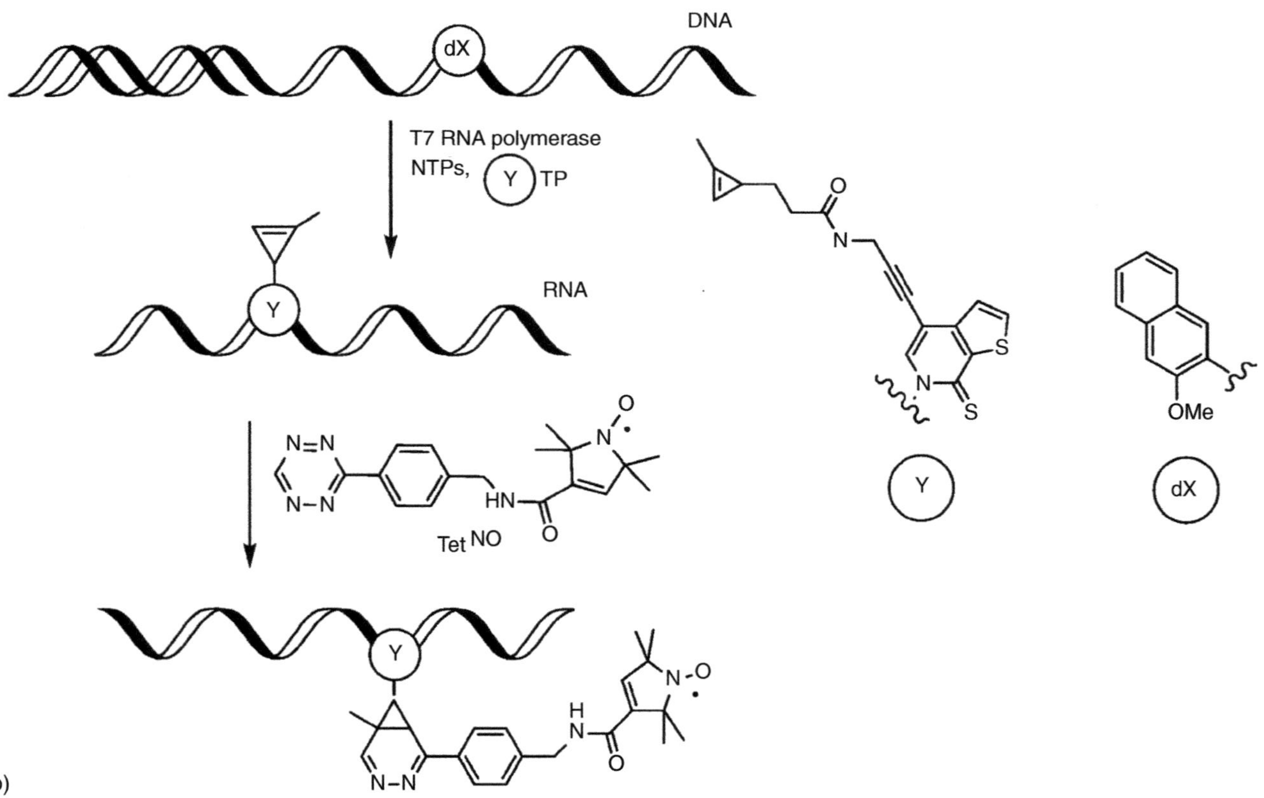

(b)

Figure 32.13 (*Continued*)

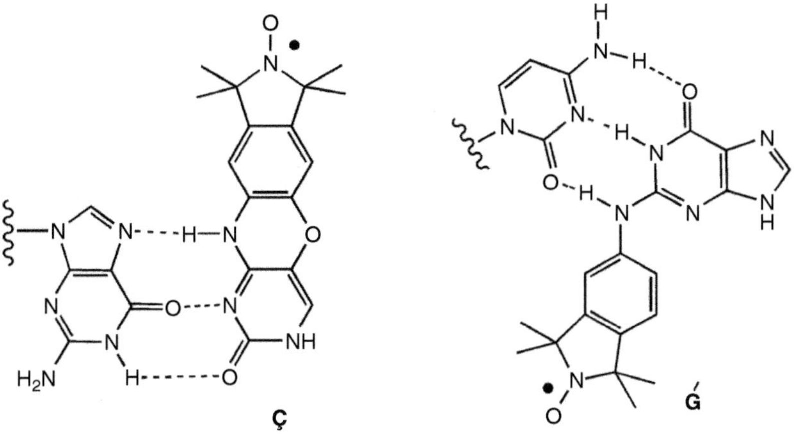

Figure 32.14 Non-covalent labeling with **ç** and **Ġ**.

nucleobase is labeled with a nitroxide, which then binds via stacking into an abasic site and hydrogen bonds to the opposing, complementary base. Site specificity is achieved by placing the abasic site at the desired position. For DNA, the lab of Sigurdsson introduced a non-covalent analog of the **Ç** label, **ç** (Figure 32.14) [107], and Reginsson et al. [108] used it in combination PELDOR to study a DNA–protein complex.

In 2016, Kamble et al. synthesized an isoindoline nitroxide derivate of the nucleobase guanine (Figure 32.14), called **Ġ**. The label is prepared in one step from readily available starting materials [109]. cw EPR measurements show the successful incorporation into an abasic site of an RNA duplex [110]. Moreover, strong orientation dependence could be observed through pulsed EPR measurements on duplex RNA [109]. A drawback of the non-covalent labeling approach is that the achieved binding constants of the labeled nucleobases do not lead to complete binding. Thus, the EPR spectra are a superposition of free and bound label.

32.3.5 Beyond Labeling with Nitroxides

All labeling strategies summarized above make use of nitroxides as the spin-bearing group. During the last years other paramagnetic centers were introduced that are tailored with their chemical and EPR properties to the EPR spectroscopic method. For example, triarylmethyl-based radicals [45–50] have raised significant interest because they give rise to one EPR line, only, which increases the signal-to-noise in EPR measurements and enables the use of single-frequency PDS methods [53, 54]. Furthermore, their long phase memory times in the liquid state enable PDS measurements at room temperature if the tumbling of the biomolecule is restrained [49]. In this regard, Shevlev et al. [50] reported a strategy for labeling DNA with a trityl group **9** (Figure 32.15) and performed PDS measurements at 298 K on DNA duplexes [111]. In addition, trityl labels show increased stability under reducing condition as compared with most nitroxides [90], which enabled in cell RIDME measurements

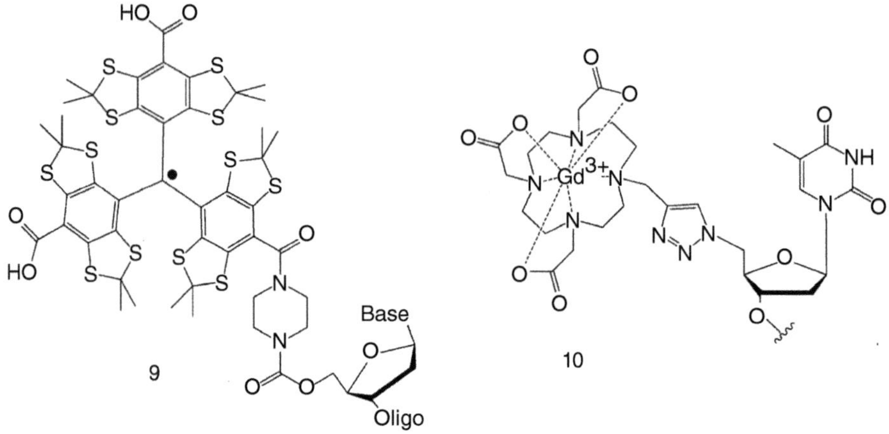

9

10

Figure 32.15 A trityl- **9** and Gd(III)-based **10** spin label for oligonucleotides.

between a trityl and a Fe(III) ion in a cytochrome [48]. In this context, also the labeling with Gd(III) labels [112] has to be mentioned. This made remarkable advancements during the last years and has been shown to be also very well-suited for *in cell* PDS measurements [113]. With respect to oligonucleotides, the Gd^{3+} label **10** (Figure 32.15) has been coupled to azide-modified 5′ ends of DNAs. However, neither the trityl nor the Gd(III) labels have been applied to RNA, yet.

Excitingly, Qin's lab reported recently the labeling of oligonucleotides with NV centers, which allowed EPR measurements on single DNA strands [114]. Also, metal ion binding sites in ribozymes can be labeled by an exchange of a diamagnetic metal ion for a paramagnetic one. Most prominent is here the exchange of Mn(II) for Mg(II) [115], which has been used in combination with pulsed hyperfine spectroscopy and cw EPR to structurally characterize these binding sites and to address the question where the binding site is localized [116].

32.4 Examples for Applications of EPR to Ribozymes

Metal ions are important for the folding and catalysis of ribozymes. Here, examples will be provided in which the number of binding sites, their affinity, local structure, and position within the ribozyme fold have been determined. In addition, examples will be highlighted where EPR enabled monitoring conformational changes in ribozymes and other functional RNAs on the nanometer scale.

32.4.1 Hammerhead Ribozymes

One of the best-characterized ribozymes to date is the hammerhead ribozyme (HHRz) (Figure 32.16), which is a self-cleaving motif of the mRNA of the *tobacco ringspot* virus. In the early days, the available crystal structures of the minimal construct (mHHRz) and biochemical data did not lead to a clear picture of the

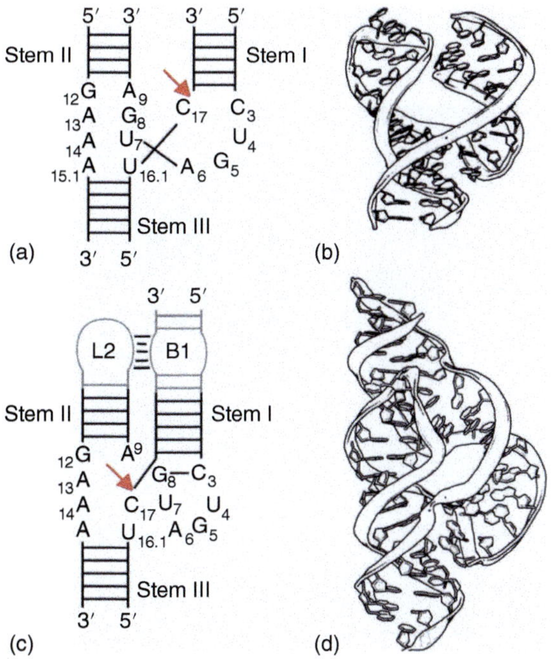

Figure 32.16 Secondary and tertiary arrangements of the m- (a, b) and tsHHRz (c, d). Source: Taken from [117].

cleavage mechanism, especially with respect to as to whether metal(II) ions are directly involved in the cleavage step [4, 118–120].

Mn^{2+} Binding Site in the HHRz

The lab of DeRose was the first to address this question with EPR spectroscopic methods [121]. They substituted the naturally used Mg^{2+} by paramagnetic Mn^{2+}. Using cw EPR, they titrated Mn^{2+} to the mHHRz and monitored the cw X-band EPR signal intensity of free Mn^{2+} in dependence of the amount of Mn^{2+} added. Since Mn^{2+} bound to the mHHRz does not give an EPR spectrum at room temperature in solution, due to fast relaxation, they could calculate the amount of bound and free Mn^{2+} and constructed from this binding isotherms. Based on these data they found four high-affinity binding sites with a K_d of 4 μM and further low-affinity binding sites. Increasing the monovalent ion concentration from 100 mM NaCl to 1 M NaCl reduced the number of high-affinity sites to a single one with a K_d of 10 μM. They proved that the mHHRz is still active upon Mn^{2+} for Mg^{2+} exchange and found that the maximum activity is only reached if the high-affinity sites are occupied [121].

Then in a second step, they used ESEEM to identify the ligand sphere of the single high-affinity Mn^{2+} binding site under high-salt conditions. By comparison of fingerprint ESEEM spectra of Mn^{2+} bound to the mHHRz and GMP in combination with ^{15}N isotope labeling, they identified a guanine N7 nitrogen as being directly bound to Mn^{2+} [122]. In subsequent $^{1/2}H$-, $^{14/15}N$-, and ^{31}P-ENDOR studies, they identified the water molecules and a phosphate group in the first coordination sphere. Combining the data they suggested the binding site structure in Figure 32.17c [123, 124].

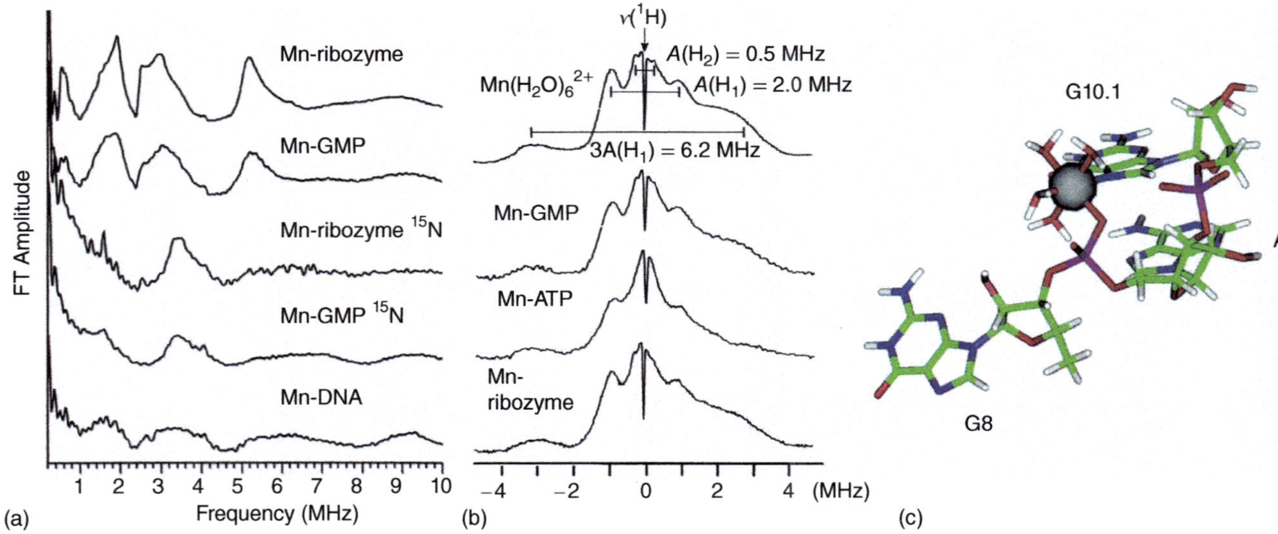

Figure 32.17 (a) 3-Pulse ESEEM spectra of Mn^{2+} bound to different constructs. (b) Q-band ^1H-ENDOR spectra of various Mn^{2+} constructs. . (c) The structure of the single high-affinity Mn^{2+} binding site in the mHHRz. Source: (a) Taken from [122]. (b) Taken from [123]. (c) Taken from [124].

In 2003, Schiemann et al. [125] elucidate whether the single high-affinity Mn^{2+} binding site observed in solution by EPR is the same one as found in the crystal structure at G10.1. In their work, they could confirm the number and affinities of the Mn^{2+} binding sites determined before by DeRose. Additionally, they demonstrated that binding of neomycin B leads to the release of Mn^{2+} including the release from the high-affinity sites. Then they used 3-pulse ESEEM and HYSCORE to obtain the ^{14}N-hyperfine coupling tensor and the ^{14}N-quadrupole coupling parameters. These parameters were then compared with the ^{14}N-EPR parameters obtained from DFT calculations of different model structures including the one from the Mn^{2+} binding site in the X-ray structure. The best agreement was obtained for the structure from the crystal structure, which led to the conclusion that it is localized at G10.1 and that its first coordination sphere is the same as in the crystal. Based on this, it was concluded that this ion site is not directly involved in catalysis because it is too far away from the cleavage site.

At about this time Khvorova et al. [126] found that including the naturally occurring loop at stem I leads to an HHRz construct (tsHHRz) that is catalytically much more active than the mHHRz. This loop was proposed to interact with the loop at stem II, locking both stems [127]. This raised the question as to whether this also leads to a change in the high-affinity binding sites. Kisseleva et al. [128] approached this question by comparing the binding of Mn^{2+} with the mHHRz and the tsHHRz. cw EPR titration demonstrated that the tsHHRz has already a single high-affinity binding site with a K_d of $\leq$10 nM at low monovalent salt concentrations (100 mM), at least 2 orders of magnitude lower than observed for the mHHRz. This parallels the observation that the tsHHRz is catalytic active at physiologically salt concentration, whereas the mHHRz is not. Interestingly, the binding competition experiments with neomycin B showed no release of the high-affinity Mn^{2+} for the tsHHRz. In order to answer the question whether this high-affinity binding site is closer to the cleavage site and maybe is involved in catalysis, 3-pulse ESEEM and HYSCORE fingerprint spectra were recorded and compared with the ones from the mHHRz [128]. Unchanged spectra led to the conclusion that also in the tsHHRz, the high-affinity site is located at G10.1. The differences in the K_d's were attributed to the larger conformational freedom of the mHHRz. In addition, a *W*-band ^{31}P-ENDOR study on the m- and tsHHRz revealed a strong isotropic phosphorus hyperfine coupling of about 8 MHz for the Mn^{2+} binding site in both ribozymes. DFT calculations on the structure of the G10.1 binding site from the crystal did yield the same strong coupling constant, underpinning that the site is the same in both constructs [129]. The functional role of the high-affinity Mn^{2+} was further investigated with stopped-flow kinetics experiments [130]. The cleavage rate of the tsHHRz was found to be 20-fold faster than that for mHHRz under the same conditions and that Mn^{2+} leads to a faster cleavage rate than for Mg^{2+}. Interestingly, the investigated Mn^{2+} concentrations needed for the ribozyme activity observed were several folds higher than the dissociation constant previously found with EPR spectroscopy. Summarizing all findings, the high-affinity Mn^{2+} ions are structural ions necessary for the correct folding but are not directly involved in the cleavage reaction. These findings were in agreement with the X-ray structure of the tsHHRz [117, 131] (Figure 32.18).

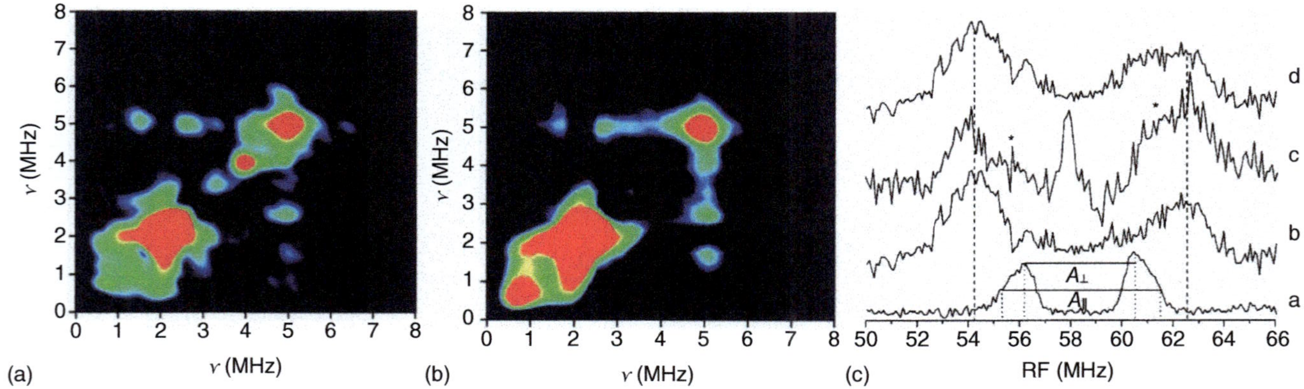

Figure 32.18 (a) HYSCORE spectrum of the high-affinity Mn^{2+} binding site in the mHHRz and (b) in the tsHHRz. (c) W-band ENDOR of Mn^{2+} in a, phosphate buffer; b, the tsHHRz; c, the mHHRz; d, the sum of a and b. Source: (a,b) Taken from [128]. (c) Taken from [129].

Folding of the HHRz

The Mg^{2+} dependence of the loop–loop interaction in the tsHHRz as part of the folding and for the catalysis itself was investigated with cw EPR by Kim et al. [132]. A nitroxide spin label was attached at three different sites in the substrate strand via 2′-amino labeling at the sugar, and the dynamics of the spin label was monitored as a function of Mg^{2+} concentration by cw EPR spectroscopy. Increasing the Mg^{2+} concentration causes broadening of the cw EPR spectra at position U1.12 and C1.9, whereas it induces a distinctive splitting at position U1.16. Site U1.16 was assigned to monitor the docking of stems I and II. The docking occurs at low Mg^{2+} concentration $[Mg^{2+}]_{1/2dock} = 0.7$ mM at 0.1 M NaCl, whereas higher Mg^{2+} concentrations are needed for the cleavage reaction to occur. PELDOR measurements on the tsHHRz by Kim et al. (Figure 32.19) [133] supported the findings made above by cw EPR. A pair of nitroxide spin labels was attached to 2′-amino groups at a uridine and a cytidine in the loops of stems I and II. A change of the distance between the labeled positions was observed upon increasing the Mg^{2+} concentration, in agreement with the Mg^{2+}-dependent interaction between both loops.

The internal changes in the structure and/or dynamics of the mHHRz upon folding were studied by Edwards et al. [134]. For this study, a tempo derivate was used to label the 2′-position of a uridine in the catalytic core, without disrupting catalytic activity. The folding of the mHHRz was followed in dependence of temperature, metal ion identity, ionic strength, and cleavage inhibitors by monitoring the changes in the rotational correlation time of the label. This resulted in the conclusion that two folding pathways are operative for the mHHRz [134].

32.4.2 Diels–Alder Ribozyme

In 1999, the lab of Jäschke discovered the Diels–Alder ribozyme, which is an artificial ribozyme catalyzing the Diels–Alder reaction [135]. A crystal structure of this ribozyme was reported in 2005 [136], revealing eight bound Mg^{2+} ions. Two of them were suggested to be due to the crystal packing, and one showed only week outer-sphere contacts (Figure 32.20a). Using cw EPR spectroscopy,

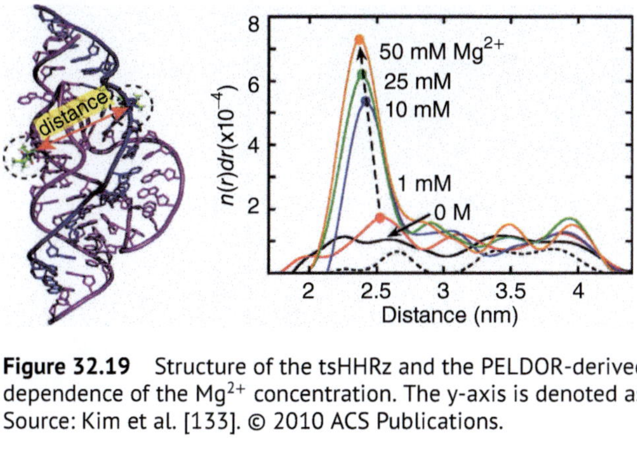

Figure 32.19 Structure of the tsHHRz and the PELDOR-derived distance distributions in dependence of the Mg^{2+} concentration. The y-axis is denoted as the probability density. Source: Kim et al. [133]. © 2010 ACS Publications.

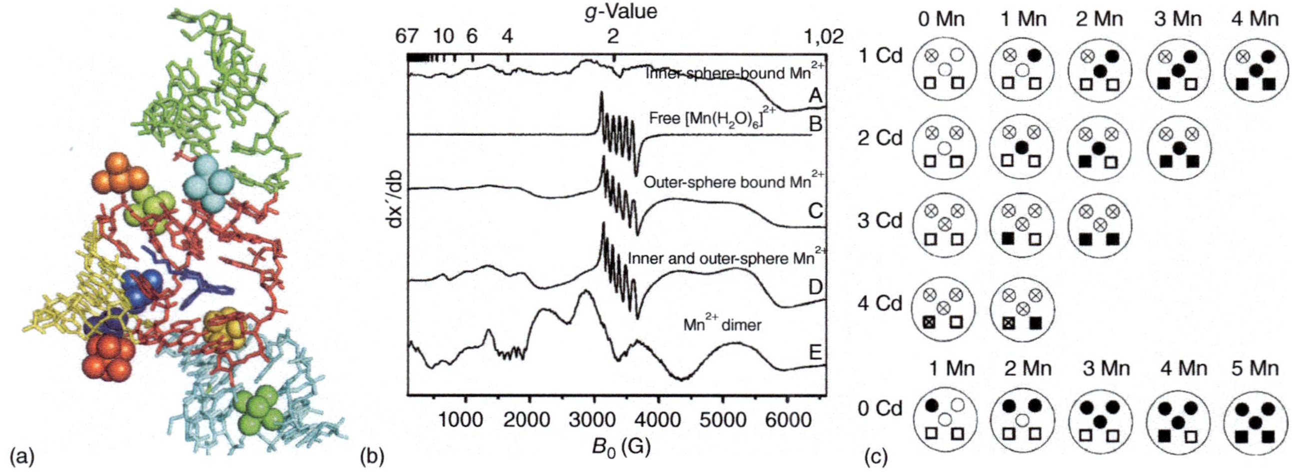

Figure 32.20 (a) Structure of the Diels–Alder ribozyme. (b) cw X-band EPR spectra of the different binding sites selected with (c) different ratios of Mn^{2+}/Cd^{2+}/ribozyme. Source: Kisseleva et al. [137].

Kisseleva et al. [137] determined five high-affinity Mn^{2+} binding sites in solution and several low-affinity ones, fitting to the X-ray structure. They also found that Cd^{2+} has a higher affinity for these binding sites than Mn^{2+} and can replace the latter. Thus, competition experiments were performed at different equivalents of Mn^{2+} and Cd^{2+}, and cw X-band EPR spectra were recorded at cryogenic temperatures. This showed that the five binding sites had differentiable affinities and could be occupied sequentially. Since the EPR spectra changed according to the occupied site, it was possible to characterize them. Two sites could be identified as binding Mn^{2+} in an outer-sphere manner, three could be identified as binding it in an inner-sphere manner, and two Mn^{2+} ions are in close contact with a Mn–Mn distance of 6 Å. These structural findings were in agreement with the crystal structure (Figure 32.20b,c).

32.4.3 Group I Intron

Qin and Herschlag labs studied the docking of the P1 domain onto the *Tetrahymena* group I intron (Figure 32.21) [138]. The P1 domain is a duplex region formed between the exon junction substrate strand and the internal guide sequence (IGS). This duplex domain is believed to first form an open complex and then to fold back onto the ribozyme, forming a closed complex. They monitored the nanosecond dynamics of the P1 domain via cw EPR and found that the single-stranded J1/2 junction modulates the P1 dynamics in the open state but less than in the closed state of the complex. In a later study, the rigid spin label **Ç** is used to unravel the length and sequence dependence of the J1/2 junction on the nanosecond dynamics of P1 [139].

32.4.4 Ribosome

The largest ribozyme complex studied so far with EPR spectroscopy is the ribosome. The lab of Bagryanskaya spin-labeled an mRNA fragment at the 5′-end and at an internal adenine base at C8, which positions the spin labels in the P- and A-sites

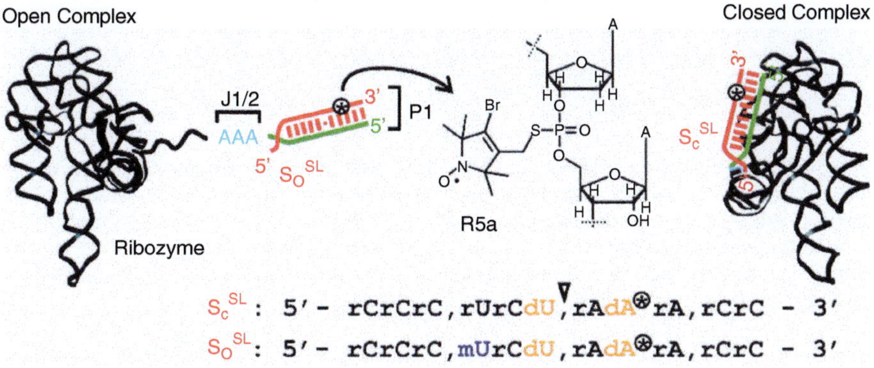

Figure 32.21 Spin labeling of the *Tetrahymena* group I intron and sketches of the open and closed complex forms. Source: Grant et al. [138]. © 2009 American Chemical Society.

of the ribosome [140]. They then measure the distances between both labels on the mRNA with PELDOR in dependence of the tRNA loading of the ribosome. They found that binding of neither the 60S subunit nor the tRNA at the A site changes the arrangement of the mRNA at the codon–anticodon interface [140]. In a later PELDOR study [141], they follow structural changes of a doubly spin labeled 11-mer mRNA model upon binding to human 80S ribosomes. They find that stable complexes are formed with the single-stranded mRNA, the tRNAs, and the ribosome and that tRNAs bound at the A- and E-sites have an influence on the stability of these complexes. Surprisingly, they also see the formation of a complex without tRNAs, in which the mRNA maybe bound to the ribosomal protein uS3, which positions the mRNA away from the ribosomal mRNA channel. In addition, they find that the ribosome shifts the equilibrium of single strand versus duplex mRNA toward the single-stranded mRNA.

In contrast to mRNAs, Goldfarb lab localized a Mn^{2+} binding site within the hairpin loop HP92 of the 23S rRNA by measuring Mn^{2+}–nitroxide distances via W-band PELDOR (Figure 32.22) [142]. The spin labels were attached to 2′-amino groups of

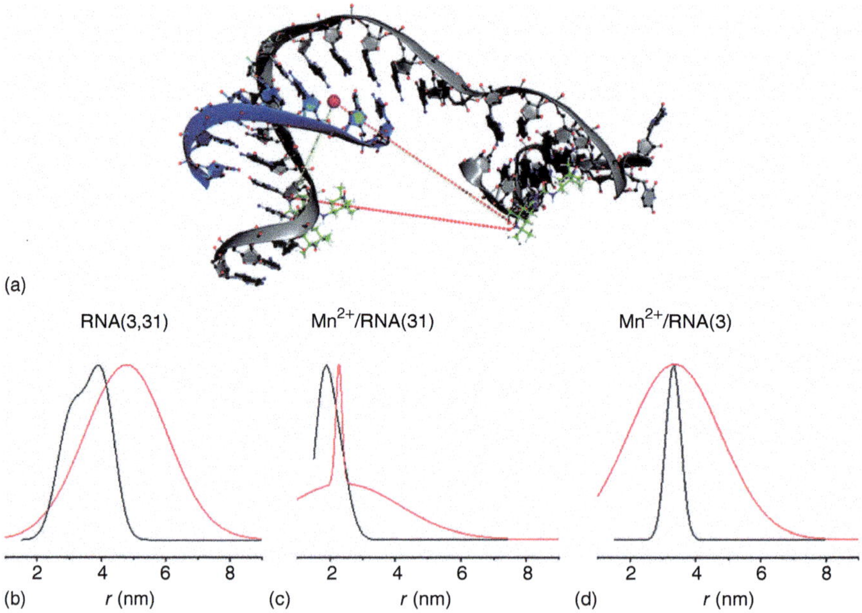

Figure 32.22 (a) Structure of HP92 with Mn^{2+} indicated as red sphere and the labels at position 3 (left) and 31 (right) as green sticks. The red lines are the spin–spin distances. The distance distributions for (b) the two labels, (c) Mn^{2+} and the label at position 31, and (d) Mn^{2+} and the label at position 3. The red lines are the experimental, and the black ones the modeled distributions. Source: Kaminker et al. [142]. © 2015 Royal Society of Chemistry. Reproduced with permission of Royal Society of Chemistry.

uridines at two different sites. They interpret their data such that the Mn^{2+} is bound in an outer-sphere manner in the minor groove of the stem and close to the loop.

32.4.5 Non-ribozyme RNAs

Riboswitches

In the case of riboswitches, several studies have used site-directed spin labeling of the riboswitch, following its conformational changes upon ligand binding by PELDOR. Such an approach revealed an equilibrium between two conformers for the synthetic tetracycline riboswitch in the apo state. Ligand binding shifts this equilibrium toward one of these conformers [143]. In a follow-up study, this picture is refined taking Mg^{2+} concentrations into account and using the **Çm** label [144]. For the preQ riboswitch, an equilibrium between two conformations is also seen for the apo state, with one conformation dominating the equilibrium, whereas binding of the ligand preQ leads to the domination of the other conformer in the equilibrium (Figure 32.23) [99]. For the neomycin riboswitch, PELDOR data reveal a pre-organized riboswitch that does not change its structure upon binding of the neomycin ligand [145]. Also for the cocaine aptamer, only minor distance changes are observed. However, the conformational flexibility is reduced upon ligand binding [146].

Oligonucleotide–Protein Complexes

Recently, Qin's lab reported on nucleic acid-dependent conformational changes in the CRISPR/Cas9 system [147]. Single spin labels were attached to two different cysteines in the spyCas9 protein, and the changes in mobility were monitored upon adding different nucleic acids. The data are interpreted as indicating large-scale domain rearrangements in spyCas9 upon RNA binding. Similarly, Grohmann et al. studied the effect of RNA binding on the RNA polymerase subunits F/E by FRET and PELDOR [148]. The protein was also labeled, with either nitroxides or fluorophores. Both FRET and PELDOR indicate that the protein conformation does not change upon RNA binding, suggesting that it acts as a rigid guiding rail for the emerging RNA. Interestingly, PELDOR-derived distance distributions fit nicely to the crystal structure, whereas FRET-derived ones do not. Related to this, a paper from 2016 reports conformational changes in the helicase PcrA upon DNA and ADP/AMPPNP binding, whereas XPD shows no conformational changes (Figure 32.24) [149].

Last but not least, a joint NMR/PELDOR approach enabled a detailed study of how the noncoding RNA RsmZ sequesters, stores, and releases the protein dimers RsmE [97, 98, 150]. It was found that the RsmE dimers are loaded onto RsmZ sequentially, specifically, and cooperatively. For the formed 70 kDa complex, two different structures could be resolved. In this study, the RNA was twofold spin-labeled, and the PELDOR-derived distance distributions provided important long-range constraints for docking the subunits of the complex and for identifying the presence of two complex structures.

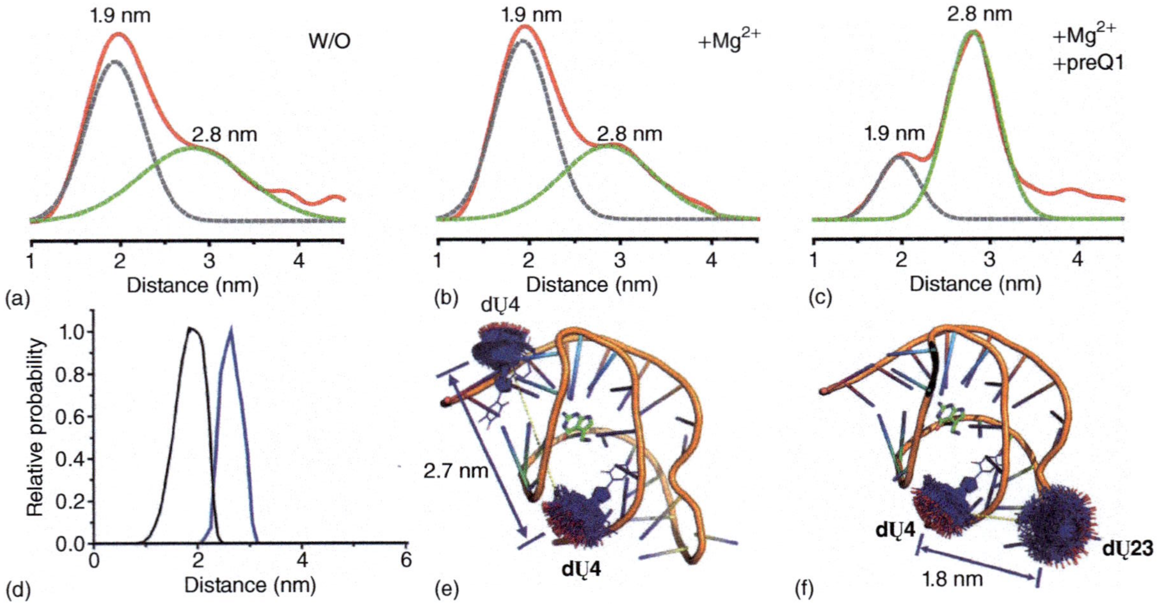

Figure 32.23 Top: The derived distance distributions (red lines) for the doubly labeled preQ1 riboswitch dU4-dU32 (a) in the apo form, (b) in the presence of Mg^{2+}, and (c) with Mg^{2+} and preQ1. Each distance distribution was deconvoluted by two Gaussian components (gray and green dashed lines). (d) MtsslWizard distance distributions for preQ1-unbound (black) and preQ1-bound states (blue). Next are shown the crystal structures of the preQ1 riboswitch in the preQ1-unbound (e, PDB 3Q51) and preQ1-bound states (f, PDB 3Q50), where the spin labels have been attached by means of the program mtsslWizard. Source: Kerzhner et al. [99]. © 2018 American Chemical Society.

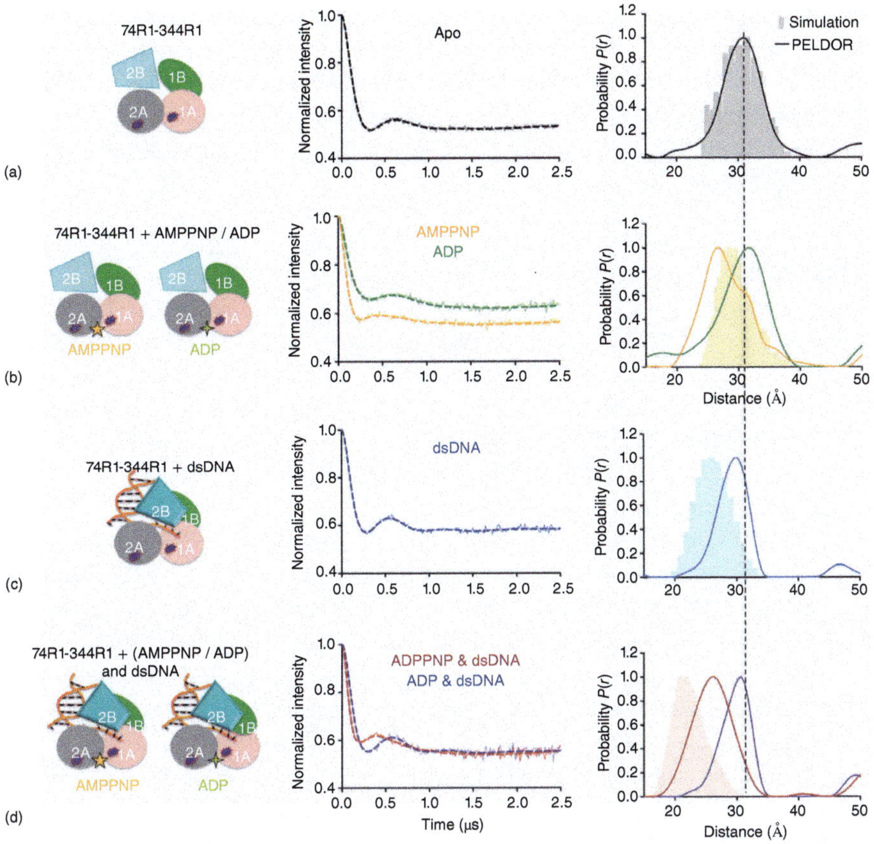

Figure 32.24 PELDOR measurements on (a) PcrA with labels at different sites and in the presence of (b) ADP or AMPPNP, (c) DNA, and (d) ADP + DNA and AMPPNP + DNA. Source: Constantinescu-Aruxandei et al. [149]. © 2016 Oxford University Press.

32.5 Conclusion

EPR spectroscopy in combination with site-directed spin labeling or metal ion exchange is established as a versatile method to gain information about structures and dynamics of ribozymes and RNA systems in general. RNA protein complexes as large as the ribosome have been studied with EPR, providing valuable information. On a local scale, EPR yields atomistic resolution of Mn^{2+} binding sites in ribozymes, and the binding sites can be counted and localized and their affinity determined. Long-range EPR-based distance measurements are precise, disentangle domain arrangements and show how they change during function or binding of effectors. Line-shape analysis of cw EPR spectra and orientation selective PELDOR measurements offer access to detailed information about RNA dynamics. Whereas cw EPR-based line-shape analysis can be performed in the liquid state at room temperature, PELDOR and pulsed hyperfine methods are applied in the frozen state. Excitingly, first PELDOR measurements on trityl-labeled oligonucleotides at room temperature have been reported [49, 50], and several examples appeared in the literature, where oligonucleotides or proteins have been studied with EPR within whole cells [48, 91]. Thus, one can hope that it will be possible to study ribozymes with EPR in the near future within their truly natural environment and without size restriction.

References

1 Hartig, J.S. (ed.) (2016). *Ribozymes: Methods and Protocols*. New York: Humana Press.

2 Lilley, D.M.J. and Eckstein, F. (eds.) (2008). *Ribozymes and RNA Catalysis*, 1. Cambridge: Royal Society of Chemistry.

3 Hud, N.V. (ed.) (2009). *Nucleic Acid-Metal Ion Interactions*, 1–433. Cambridge: Royal Society of Chemistry.

4 Eckstein, F. and Lilley, D.M.J. (eds.) (1997). *Catalytic RNA*. Berlin: Springer.

5 Westhof, E. (2015). Twenty years of RNA crystallography. *RNA* 21 (4): 486–487.

6 Khatter, H., Myasnikov, A.G., Natchiar, S.K., and Klaholz, B.P. (2015). Structure of the human 80S ribosome. *Nature* 520 (7549): 640–645.

7 Barnwal, R.P., Yang, F., and Varani, G. (2017). Applications of NMR to structure determination of RNAs large and small. *Arch. Biochem. Biophys.* 628: 42–56.

8 Fürtig, B., Buck, J., Richter, C., and Schwalbe, H. (2012). Functional dynamics of RNA ribozymes studied by NMR spectroscopy. In: *Ribozymes. Methods and Protocols*, vol. 848 (ed. J.S. Hartig), 185–199. New York: Humana Press.

9 Crowther, R.A.e. (2016). *The Resolution Revolution: Recent Advances in CryoEM*, 1ste, 1. Amsterdam: Academic Press Online-Resource (xx, 445 Seiten).

10 Wu, J., Niu, S., Tan, M. et al. (2018). Cryo-EM structure of the human ribonuclease P holoenzyme. *Cell* 175 (5): 1393–1404.

11 Lakowicz, J.R. (2006). *Principles of fluorescence spectroscopy*. New York: Springer.

12 Stephenson, J.D., Kenyon, J.C., Symmons, M.F., and Lever, A.M. (2016). Characterizing 3D RNA structure by single molecule FRET. *Methods* 103: 57–67.

13 Oikawa, H., Takahashi, T., Kamonprasertsuk, S., and Takahashi, S. (2018). Microsecond resolved single-molecule FRET time series measurements based on the line confocal optical system combined with hybrid photodetectors. *Phys. Chem. Chem. Phys.* 20 (5): 3277–3285.

14 Fürtig, B., Buck, J., Manoharan, V. et al. (2007). Time-resolved NMR studies of RNA folding. *Biopolymers* 86 (5-6): 360–383.

15 Li, J., Jiao, A., Chen, S. et al. (2018). Application of the small-angle X-ray scattering technique for structural analysis studies: a review. *J. Mol. Struct.* 1165: 391–400.

16 Chen, Y. and Pollack, L. (2016). SAXS studies of RNA: structures, dynamics, and interactions with partners. *Wiley Interdiscip. Rev. RNA* 7 (4): 512–526.

17 Peetz, O., Hellwig, N., Henrich, E. et al. (2019). LILBID and nESI: Different Native Mass Spectrometry Techniques as Tools in Structural Biology. *J. Am. Soc. Mass. Spectrom.* 30 (1): 181–191.

18 Morgner, N., Barth, H.D., Brutschy, B. et al. (2008). Binding sites of the viral RNA element TAR and of TAR mutants for various peptide ligands, probed with LILBID: a new laser mass spectrometry. *J. Am. Soc. Mass. Spectrom.* 19 (11): 1600–1611.

19 Goldfarb, D. and Stoll, S.e. (2018). *Handbook of EPR Spectroscopy: Fundamentals and Methods*. Hoboken: Wiley, p 1 online resource.

20 Schweiger, A. and Jeschke, G. (2001). *Principles of Pulse Electron Paramagnetic Resonance*. Oxford: Oxford University Press.

21 Misra, S.K.e. (2011). *Multifrequency Electron Paramagnetic Resonance: Theory and Applications*, XXXIII. Weinheim: Wiley-VCH, 1022 str.

22 Dror, R.O., Dirks, R.M., Grossman, J.P. et al. (2012). Biomolecular simulation: a computational microscope for molecular biology. *Annu. Rev. Biophys.* 41: 429–452.

23 Kaupp, M., Bühl, M., and Malkin, V.G. (eds.) (2004). *Calculation of NMR and EPR Parameters*. Weinheim: Wiley-VCH.

24 Gast, P., Herbonnet, R.T., Klare, J. et al. (2014). Hydrogen bonding of nitroxide spin labels in membrane proteins. *Phys. Chem. Chem. Phys.* 16 (30): 15910–15916.

25 Kirilyuk, I.A., Bobko, A.A., Grigor'ev, I.A., and Khramtsov, V.V. (2004). Synthesis of the tetraethyl substituted pH-sensitive nitroxides of imidazole series with enhanced stability towards reduction. *Org. Biomol. Chem.* 2 (7): 1025–1030.

26 Nesmelov, Y.E. and Thomas, D.D. (2010). Protein structural dynamics revealed by site-directed spin labeling and multifrequency EPR. *Biophys. Rev.* 2 (2): 91–99.

27 Zhang, Z., Fleissner, M.R., Tipikin, D.S. et al. (2010). Multifrequency electron spin resonance study of the dynamics of spin labeled T4 lysozyme. *J. Phys. Chem. B* 114 (16): 5503–5521.

28 Berliner, L. and Hanson, G.e. (2009). Applications to metalloenzymes and metals in medicine. In: *High Resolution EPR*. New York: Springer, p 1 online resource.

29 Dikanov, S.A. and Tsvetkov, Y.D. (1992). *Electron spin echo envelope modulation (ESEEM) spectroscopy*. Boca Raton: CRC Press.

30 Van Doorslaer, S. (2017). Hyperfine spectroscopy: ESEEM. *eMagRes* 6 (1): 51–69.

31 Cox, N., Nalepa, A., Lubitz, W., and Savitsky, A. (2017). ELDOR-detected NMR: a general and robust method for electron–nuclear hyperfine spectroscopy? *J. Magn. Reson.* 280: 63–78.

32 Van Doorslaer, S. and Vinck, E. (2007). The strength of EPR and ENDOR techniques in revealing structure–function relationships in metalloproteins. *Phys. Chem. Chem. Phys.* 9 (33): 4620–4638.

33 Hoffman, B.M., DeRose, V.J., Doan, P.E. et al. (1993). Metalloenzyme active-site structure and function through multifrequency CW and Pulsed ENDOR.

In: *EMR of Paramagnetic Molecules* (eds. L.J. Berliner and J. Reuben), 151–218. Boston: Springer.

34 DeRose, V.J. and Hoffman, B.M. (1995). Protein structure and mechanism studied by electron nuclear double resonance spectroscopy. *Methods Enzymol.* 246: 554–589.

35 Britt, R.D. (1996). Electron spin echo methods in photosynthesis research. In: *Biophysical Techniques in Photosynthesis* (eds. J. Amesz and A.J. Hoff), 235–253. Dordrecht: Springer.

36 Timmel, C.R. and Jeffrey, H. (2014). *Structural Information from Spin-Labels and Intrinsic Paramagnetic Centres in the Biosciences*. Berlin: Springer.

37 Schiemann, O. and Prisner, T.F. (2007). Long-range distance determinations in biomacromolecules by EPR spectroscopy. *Q. Rev. Biophys.* 40 (1): 1–53.

38 Denysenkov, V.P., Prisner, T.F., Stubbe, J., and Bennati, M. (2006). High-field pulsed electron-electron double resonance spectroscopy to determine the orientation of the tyrosyl radicals in ribonucleotide reductase. *Proc. Natl. Acad. Sci. USA* 103 (36): 13386–13390.

39 Schiemann, O., Cekan, P., Margraf, D. et al. (2009). Relative orientation of rigid nitroxides by PELDOR: beyond distance measurements in nucleic acids. *Angew. Chem. Int. Ed.* 48 (18): 3292–3295.

40 Tsvetkov, Y.D., Bowman, M.K., and Grishin, Y.A. (2019). Nanoscale distance measurement in the biological, materials and chemical sciences. In: *Pulsed Electron–Electron Double Resonance*. Cham: Springer.

41 Jeschke, G. (2016). Dipolar spectroscopy – double-resonance methods. *eMagRes* 5 (3): 1459–1475.

42 Ward, R. and Schiemann, O. (2013). Structural information from oligonucleotides. In: *Structural Information from Spin-Labels and Intrinsic Paramagnetic Centres in the Biosciences, Structure and Bonding*, vol. 152 (eds. C.R. Timmel and J.R. Harmer), 249–281. Berlin: Springer.

43 Likhtenshteĭn, G.I., Yamauchi, J., Nakatsuji, S. et al. (2008). *Nitroxides: Applications in Chemistry, Biomedicine, and Materials Science*. Weinheim: Wiley-VCH.

44 Polyhach, Y., Bordignon, E., Tschaggelar, R. et al. (2012). High sensitivity and versatility of the DEER experiment on nitroxide radical pairs at Q-band frequencies. *Phys. Chem. Chem. Phys.* 14 (30): 10762–10773.

45 Reginsson, G.W., Kunjir, N.C., Sigurdsson, S.T., and Schiemann, O. (2012). Trityl radicals: spin labels for nanometer-distance measurements. *Chem. Eur. J.* 18 (43): 13580–13584.

46 Jassoy, J.J., Meyer, A., Spicher, S. et al. (2018). Synthesis of nanometer sized bis- and tris-trityl model compounds with different extent of spin–spin coupling. *Molecules* 23 (3): 682–700.

47 Fleck, N., Hett, T., Brode, J. et al. (2019). C–C cross-coupling reactions of trityl radicals: spin density delocalization, exchange coupling, and a spin label. *J. Organomet. Chem.* 84 (6): 3293–3303.

48 Jassoy, J.J., Berndhäuser, A., Duthie, F. et al. (2017). Versatile trityl spin labels for nanometer distance measurements on biomolecules in vitro and within cells. *Angew. Chem. Int. Ed.* 56 (1): 177–181.

49 Yang, Z., Liu, Y., Borbat, P. et al. (2012). Pulsed ESR dipolar spectroscopy for distance measurements in immobilized spin labeled proteins in liquid solution. *J. Am. Chem. Soc.* 134 (24): 9950–9952.

50 Shevelev, G.Y., Krumkacheva, O.A., Lomzov, A.A. et al. (2014). Physiological-temperature distance measurement in nucleic acid using triarylmethyl-based spin labels and pulsed dipolar EPR spectroscopy. *J. Am. Chem. Soc.* 136 (28): 9874–9877.

51 Borbat, P.P. and Freed, J.H. (2013). Pulse dipolar electron spin resonance: distance measurements. In: *Structural Information from Spin-Labels and Intrinsic Paramagnetic Centres in the Biosciences, Structure and Bonding*, vol. 152 (eds. C.R. Timmel and J.R. Harmer), 1–82. Berlin: Springer.

52 Jeschke, G., Pannier, M., Godt, A., and Spiess, H.W. (2000). Dipolar spectroscopy and spin alignment in electron paramagnetic resonance. *Chem. Phys. Lett.* 331 (2): 243–252.

53 Kunjir, N.C., Reginsson, G.W., Schiemann, O., and Sigurdsson, S.T. (2013). Measurements of short distances between trityl spin labels with CW EPR, DQC and PELDOR. *Phys. Chem. Chem. Phys.* 15 (45): 19673–19685.

54 Meyer, A., Jassoy, J.J., Spicher, S. et al. (2018). Performance of PELDOR, RIDME, SIFTER, and DQC in measuring distances in trityl based bi- and triradicals: exchange coupling, pseudosecular coupling and multi-spin effects. *Phys. Chem. Chem. Phys.* 20 (20): 13858–13869.

55 Meyer, A. and Schiemann, O. (2016). PELDOR and RIDME measurements on a high-spin manganese(II) bisnitroxide model complex. *J. Phys. Chem. A* 120 (20): 3463–3472.

56 Keller, K., Mertens, V., Qi, M. et al. (2017). Computing distance distributions from dispolar evolution data with overtones: RIDME spectroscopy with Gd(III)-based spin labels. *Phys. Chem. Chem. Phys.* 19 (27): 17856–17876.

57 Akhmetzyanov, D., Ching, H.Y., Denysenkov, V. et al. (2016). RIDME spectroscopy on high-spin Mn^{2+} centers. *Phys. Chem. Chem. Phys.* 18 (44): 30857–30866.

58 Meyer, A., Abdullin, D., Schnakenburg, G., and Schiemann, O. (2016). Single and double nitroxide labeled bis(terpyridine)-copper(II): influence of orientation selectivity and multispin effects on PELDOR and RIDME. *Phys. Chem. Chem. Phys.* 18 (13): 9262–9271.

59 Abdullin, D., Duthie, F., Meyer, A. et al. (2015). Comparison of PELDOR and RIsDME for distance measurements between nitroxides and low-spin Fe(III) ions. *J. Phys. Chem. B* 119 (43): 13534–13542.

60 Milikisyants, S., Scarpelli, F., Finiguerra, M.G. et al. (2009). A pulsed EPR method to determine distances between paramagnetic centers with strong spectral anisotropy and radicals: the dead-time free RIDME sequence. *J. Magn. Reson.* 201 (1): 48–56.

61 Spindler, P.E., Schöps, P., Kallies, W. et al. (2017). Perspectives of shaped pulses for EPR spectroscopy. *J. Magn. Reson.* 280: 30–45.

62 Jeschke, G. and Polyhach, Y. (2007). Distance measurements on spin-labelled biomacromolecules by pulsed electron paramagnetic resonance. *Phys. Chem. Chem. Phys.* 9 (16): 1895–1910.

63 Jeschke, G., Chechik, V., Ionita, P. et al. (2006). DeerAnalysis 2006 – a comprehensive software package for analyzing pulsed ELDOR data. *Appl. Magn. Reson.* 30 (3): 473–498.

64 Ackermann, K. and Bode, B.E. (2018). Pulse EPR distance measurements to study multimers and multimerisation. *Mol. Phys.* 116 (12): 1513–1521.

65 Baber, J.L., Louis, J.M., and Clore, G.M. (2015). Dependence of distance distributions derived from double electron–electron resonance pulsed EPR spectroscopy on pulse-sequence time. *Angew. Chem. Int. Ed.* 54 (18): 5336–5339.

66 Edwards, T.H. and Stoll, S. (2018). Optimal Tikhonov regularization for DEER spectroscopy. *J. Magn. Reson.* 288: 58–68.

67 Hagelueken, G., Abdullin, D., and Schiemann, O. (2015). mtsslSuite: Probing biomolecular conformation by spin-labeling studies. *Methods Enzymol.* 563: 595–622.

68 Hagelueken, G., Abdullin, D., Ward, R., and Schiemann, O. (2013). mtsslSuite: In silico spin labelling, trilateration and distance-constrained rigid body docking in PyMOL. *Mol. Phys.* 111 (18-19): 2757–2766.

69 Polyhach, Y., Bordignon, E., and Jeschke, G. (2011). Rotamer libraries of spin labelled cysteines for protein studies. *Phys. Chem. Chem. Phys.* 13 (6): 2356–2366.

70 Hirst, S.J., Alexander, N., McHaourab, H.S., and Meiler, J. (2011). RosettaEPR: an integrated tool for protein structure determination from sparse EPR data. *J. Struct. Biol.* 173 (3): 506–514.

71 Abdullin, D., Hagelueken, G., and Schiemann, O. (2016). Determination of nitroxide spin label conformations via PELDOR and X-ray crystallography. *Phys. Chem. Chem. Phys.* 18 (15): 10428–10437.

72 Schiemann, O., Piton, N., Mu, Y. et al. (2004). A PELDOR-based nanometer distance ruler for oligonucleotides. *J. Am. Chem. Soc.* 126 (18): 5722–5729.

73 Islam, S.M., Stein, R.A., McHaourab, H.S., and Roux, B. (2013). Structural refinement from restrained-ensemble simulations based on EPR/DEER data: application to T4 lysozyme. *J. Phys. Chem. B* 117 (17): 4740–4754.

74 Abdullin, D., Fleck, N., Klein, C. et al. (2019). Synthesis of μ_2-oxo-bridged iron(III) tetraphenylporphyrin-spacer-nitroxide dimers and their structural and dynamics characterization by using EPR and MD simulations. *Chem. Eur. J.* 25 (10): 2586–2596.

75 Qin, P.Z. and Warncke, K. (eds.), (2015), Methods Enzymol; 563. *Investigations of Biological Systems by Using Spin Labels, Spin Probes, and Intrinsic Metal Ions, Part A*, 1e, 1. Cambridge: Academic Press Online-Resource XX, 684 Seiten.

76 Qin, P.Z. and Warncke, K. (eds.), (2015), Methods Enzymol; Vol. 564. *Investigations of Biological Systems by Using Spin Labels, Spin Probes, and Intrinsic Metal Ions, Part B*, 1e, 613. Cambridge: Academic Press.

77 Shelke, S.A. and Sigurdsson, S.T. (2013). Site-directed nitroxide spin labeling of biopolymers. In: *Structural Information from Spin-Labels and Intrinsic Paramagnetic Centres in the Biosciences* (eds. C.R. Timmel and J.R. Harmer), 121–162. Berlin: Springer.

78 Barhate, N., Cekan, P., Massey, A.P., and Sigurdsson, S.T. (2007). A nucleoside that contains a rigid nitroxide spin label: a fluorophore in disguise. *Angew. Chem. Int. Ed.* 46 (15): 2655–2658.

79 Höbartner, C., Sicoli, G., Wachowius, F. et al. (2012). Synthesis and characterization of RNA containing a rigid and nonperturbing cytidine-derived spin label. *J. Organomet. Chem.* 77 (17): 7749–7754.

80 Cekan, P., Smith, A.L., Barhate, N. et al. (2008). Rigid spin-labeled nucleoside C: a nonperturbing EPR probe of nucleic acid conformation. *Nucleic Acids Res.* 36 (18): 5946–5954.

81 Marko, A., Denysenkov, V., Margraf, D. et al. (2011). Conformational flexibility of DNA. *J. Am. Chem. Soc.* 133 (34): 13375–13379.

82 Weinrich, T., Jaumann, E.A., Scheffer, U. et al. (2018). A cytidine phosphoramidite with protected nitroxide spin label: synthesis of a full-length TAR RNA and investigation by in-line probing and EPR spectroscopy. *Chem. Eur. J.* 24 (23): 6202–6207.

83 Sicoli, G., Wachowius, F., Bennati, M., and Höbartner, C. (2010). Probing secondary structures of spin-labeled RNA by pulsed EPR spectroscopy. *Angew. Chem. Int. Ed.* 49 (36): 6443–6447.

84 Piton, N., Mu, Y., Stock, G. et al. (2007). Base-specific spin-labeling of RNA for structure determination. *Nucleic Acids Res.* 35 (9): 3128–3143.

85 Schiemann, O., Piton, N., Plackmeyer, J. et al. (2007). Spin labeling of oligonucleotides with the nitroxide TPA and use of PELDOR, a pulse EPR method, to measure intramolecular distances. *Nat. Protoc.* 2 (4): 904–923.

86 Cai, Q., Kusnetzow, A.K., Hubbell, W.L. et al. (2006). Site-directed spin labeling measurements of nanometer distances in nucleic acids using a sequence-independent nitroxide probe. *Nucleic Acids Res.* 34 (17): 4722–4730.

87 Qin, P.Z., Haworth, I.S., Cai, Q. et al. (2007). Measuring nanometer distances in nucleic acids using a sequence-independent nitroxide probe. *Nat. Protoc.* 2 (10): 2354–2365.

88 Nguyen, P.H., Popova, A.M., Hideg, K., and Qin, P.Z. (2015). A nucleotide-independent cyclic nitroxide label for monitoring segmental motions in nucleic acids. *BMC Biophys.* 8: 6.

89 Saha, S., Jagtap, A.P., and Sigurdsson, S.T. (2015). Site-directed spin labeling of 2′-amino groups in RNA with isoindoline nitroxides that are resistant to reduction. *Chem. Commun.* 51 (66): 13142–13145.

90 Jagtap, A.P., Krstic, I., Kunjir, N.C. et al. (2015). Sterically shielded spin labels for in-cell EPR spectroscopy: analysis of stability in reducing environment. *Free Radical Res.* 49 (1): 78–85.

91 Krstic, I., Hänsel, R., Romainczyk, O. et al. (2011). Long-range distance measurements on nucleic acids in cells by pulsed EPR spectroscopy. *Angew. Chem. Int. Ed.* 50 (22): 5070–5074.

92 Halbmair, K., Seikowski, J., Tkach, I. et al. (2016). High-resolution measurement of long-range distances in RNA: pulse EPR spectroscopy with TEMPO-labeled nucleotides. *Chem. Sci.* 7 (5): 3172–3180.

93 Sprinzl, M., Krämer, E., and Stehlik, D. (1974). On the structure of phenylalanine tRNA from yeast. *Eur. J. Biochem.* 49 (3): 595–605.

94 Esquiaqui, J.M., Sherman, E.M., Ye, J.D., and Fanucci, G.E. (2016). Conformational flexibility and dynamics of the internal loop and helical regions of the Kink–Turn motif in the glycine riboswitch by site-directed spin-labeling. *Biochemistry* 55 (31): 4295–4305.

95 Qin, P.Z., Hideg, K., Feigon, J., and Hubbell, W.L. (2003). Monitoring RNA base structure and dynamics using site-directed spin labeling. *Biochemistry* 42 (22): 6772–6783.

96 Kerzhner, M., Abdullin, D., Wiecek, J. et al. (2016). Post-synthetic spin-labeling of RNA through click chemistry for PELDOR measurements. *Chem. Eur. J.* 22 (34): 12113–12121.

97 Duss, O., Yulikov, M., Jeschke, G., and Allain, F.H. (2014). EPR-aided approach for solution structure determination of large RNAs or protein-RNA complexes. *Nat. Commun.* 5: 3669.

98 Duss, O., Michel, E., Yulikov, M. et al. (2014). Structural basis of the non-coding RNA RsmZ acting as a protein sponge. *Nature* 509 (7502): 588–592.

99 Kerzhner, M., Matsuoka, H., Wübben, C. et al. (2018). High-yield spin labeling of long RNAs for electron paramagnetic resonance spectroscopy. *Biochemistry* 57 (20): 2923–2931.

100 Büttner, L., Seikowski, J., Wawrzyniak, K. et al. (2013). Synthesis of spin-labeled riboswitch RNAs using convertible nucleosides and DNA-catalyzed RNA ligation. *Bioorg. Med. Chem.* 21 (20): 6171–6180.

101 Wawrzyniak-Turek, K. and Höbartner, C. (2014). Deoxyribozyme-mediated ligation for incorporating EPR spin labels and reporter groups into RNA. *Methods Enzymol.* 549: 85–104.

102 Babaylova, E.S., Ivanov, A.V., Malygin, A.A. et al. (2014). A versatile approach for site-directed spin labeling and structural EPR studies of RNAs. *Org. Biomol. Chem.* 12 (19): 3129–3136.

103 Domnick, C., Hagelueken, G., Eggert, F. et al. (2019). Posttranscriptional spin labeling of RNA by tetrazine-based cycloaddition. *Org. Biomol. Chem.* 17 (7): 1805–1808.

104 Babaylova, E.S., Malygin, A.A., Lomzov, A.A. et al. (2016). Complementary-addressed site-directed spin labeling of long natural RNAs. *Nucleic Acids Res.* 44 (16): 7935–7943.

105 Li, L., Degardin, M., Lavergne, T. et al. (2014). Natural-like replication of an unnatural base pair for the expansion of the genetic alphabet and biotechnology applications. *J. Am. Chem. Soc.* 136 (3): 826–829.

106 Belmont, P., Chapelle, C., Demeunynck, M. et al. (1998). Introduction of a nitroxide group on position 2 of 9-phenoxyacridine: easy access to spin labelled DNA-binding conjugates. *Bioorg. Med. Chem. Lett.* 8 (6): 669–674.

107 Shelke, S.A. and Sigurdsson, S.T. (2010). Noncovalent and site-directed spin labeling of nucleic acids. *Angew. Chem. Int. Ed.* 49 (43): 7984–7986.

108 Reginsson, G.W., Shelke, S.A., Rouillon, C. et al. (2013). Protein-induced changes in DNA structure and dynamics observed with noncovalent site-directed spin labeling and PELDOR. *Nucleic Acids Res.* 41 (1): e11.

109 Kamble, N.R., Gränz, M., Prisner, T.F., and Sigurdsson, S.T. (2016). Noncovalent and site-directed spin labeling of duplex RNA. *Chem. Commun.* 52 (100): 14442–14445.

110 Shelke, S.A., Sandholt, G.B., and Sigurdsson, S.T. (2014). Nitroxide-labeled pyrimidines for non-covalent spin-labeling of abasic sites in DNA and RNA duplexes. *Org. Biomol. Chem.* 12 (37): 7366–7374.

111 Shevelev, G.Y., Gulyak, E.L., Lomzov, A.A. et al. (2018). A versatile approach to attachment of triarylmethyl labels to DNA for nanoscale structural EPR studies at physiological temperatures. *J. Phys. Chem. B* 122 (1): 137–143.

112 Mahawaththa, M.C., Lee, M.D., Giannoulis, A. et al. (2018). Small neutral Gd(III) tags for distance measurements in proteins by double electron–electron resonance experiments. *Phys. Chem. Chem. Phys.* 20 (36): 23535–23545.

113 Yang, Y., Yang, F., Li, X.Y. et al. (2019). In-cell EPR distance measurements on ubiquitin labeled with a rigid PyMTA-Gd(III) tag. *J. Phys. Chem. B* 123 (5): 1050–1059.

114 Shi, F., Kong, F., Zhao, P. et al. (2018). Single-DNA electron spin resonance spectroscopy in aqueous solutions. *Nat. Methods* 15 (9): 697–699.

115 Feig, A.L. (2000). The use of manganese as a probe for elucidating the role of magnesium ions in ribozymes. *Metal Ions in Biol. Sys.* 37: 157–182.

116 Abdullin, D., Florin, N., Hagelueken, G., and Schiemann, O. (2015). EPR-based approach for the localization of paramagnetic metal ions in biomolecules. *Angew. Chem. Int. Ed.* 54 (6): 1827–1831.

117 Scott, W.G., Horan, L.H., and Martick, M. (2013). The hammerhead ribozyme: structure, catalysis, and gene regulation. *Prog. Mol. Biol. Transl. Sci.* 120: 1–23.

118 Pley, H.W., Flaherty, K.M., and McKay, D.B. (1994). Three-dimensional structure of a hammerhead ribozyme. *Nature* 372 (6501): 68–74.

119 Scott, W.G., Finch, J.T., and Klug, A. (1995). The crystal structure of an all-RNA hammerhead ribozyme: a proposed mechanism for RNA catalytic cleavage. *Cell* 81 (7): 991–1002.

120 Scott, W.G., Murray, J.B., Arnold, J.R. et al. (1996). Capturing the structure of a catalytic RNA intermediate: the hammerhead ribozyme. *Science* 274 (5295): 2065–2069.

121 Horton, T.E., Clardy, D.R., and DeRose, V.J. (1998). Electron paramagnetic resonance spectroscopic measurement of Mn^{2+} binding affinities to the hammerhead ribozyme and correlation with cleavage activity. *Biochemistry* 37 (51): 18094–18101.

122 Morrissey, S.R., Horton, T.E., Grant, C.V. et al. (1999). Mn^{2+}–nitrogen interactions in RNA probed by electron spin–echo envelope modulation spectroscopy: application to the hammerhead ribozyme. *J. Am. Chem. Soc.* 121 (39): 9215–9218.

123 Morrissey, S.R., Horton, T.E., and DeRose, V.J. (2000). Mn^{2+} sites in the hammerhead ribozyme investigated by EPR and continuous-wave Q-band ENDOR spectroscopies. *J. Am. Chem. Soc.* 122 (14): 3473–3481.

124 Vogt, M., Lahiri, S., Hoogstraten, C.G. et al. (2006). Coordination environment of a site-bound metal ion in the hammerhead ribozyme determined by ^{15}N and ^{2}H ESEEM spectroscopy. *J. Am. Chem. Soc.* 128 (51): 16764–16770.

125 Schiemann, O., Fritscher, J., Kisseleva, N. et al. (2003). Structural investigation of a high-affinity Mn^{II} binding site in the hammerhead ribozyme by EPR spectroscopy and DFT calculations. Effects of neomycin B on metal-ion binding. *Chembiochem: A European J. Chem. Biol.* 4 (10): 1057–1065.

126 Khvorova, A., Lescoute, A., Westhof, E., and Jayasena, S.D. (2003). Sequence elements outside the hammerhead ribozyme catalytic core enable intracellular activity. *Nat. Struct. Biol.* 10 (9): 708–712.

127 Penedo, J.C., Wilson, T.J., Jayasena, S.D. et al. (2004). Folding of the natural hammerhead ribozyme is enhanced by interaction of auxiliary elements. *RNA* 10 (5): 880–888.

128 Kisseleva, N., Khvorova, A., Westhof, E., and Schiemann, O. (2005). Binding of manganese(II) to a tertiary stabilized hammerhead ribozyme as studied by electron paramagnetic resonance spectroscopy. *RNA* 11 (1): 1–6.

129 Schiemann, O., Carmieli, R., and Goldfarb, D. (2007). W-band ^{31}P-ENDOR on the high-affinity Mn^{2+} binding site in the minimal and tertiary stabilized hammerhead ribozymes. *Appl. Magn. Reson.* 31 (3–4): 543–552.

130 Kisseleva, N., Khvorova, A., Westhof, E. et al. (2008). The different role of high-affinity and low-affinity metal ions in cleavage by a tertiary stabilized cis hammerhead ribozyme from tobacco ringspot virus. *Oligonucleotides* 18 (2): 101–110.

131 Martick, M. and Scott, W.G. (2006). Tertiary contacts distant from the active site prime a ribozyme for catalysis. *Cell* 126 (2): 309–320.

132 Kim, N.K., Murali, A., and DeRose, V.J. (2005). Separate metal requirements for loop interactions and catalysis in the extended hammerhead ribozyme. *J. Am. Chem. Soc.* 127 (41): 14134–14135.

133 Kim, N.K., Bowman, M.K., and DeRose, V.J. (2010). Precise mapping of RNA tertiary structure via nanometer distance measurements with double electron–electron resonance spectroscopy. *J. Am. Chem. Soc.* 132 (26): 8882–8884.

134 Edwards, T.E. and Sigurdsson, S.T. (2005). EPR spectroscopic analysis of U7 hammerhead ribozyme dynamics during metal ion induced folding. *Biochemistry* 44 (38): 12870–12878.

135 Seelig, B. and Jäschke, A. (1999). A small catalytic RNA motif with Diels–Alderase activity. *Chemistry & Biology* 6 (3): 167–176.

136 Serganov, A., Keiper, S., Malinina, L. et al. (2005). Structural basis for Diels–Alder ribozyme-catalyzed carbon–carbon bond formation. *Nat. Struct. Mol. Biol.* 12 (3): 218–224.

137 Kisseleva, N., Kraut, S., Jäschke, A., and Schiemann, O. (2007). Characterizing multiple metal ion binding sites within a ribozyme by cadmium-induced EPR silencing. *HFSPJ* 1 (2): 127–136.

138 Grant, G.P., Boyd, N., Herschlag, D., and Qin, P.Z. (2009). Motions of the substrate recognition duplex in a group I intron assessed by site-directed spin labeling. *J. Am. Chem. Soc.* 131 (9): 3136–3137.

139 Nguyen, P., Shi, X., Sigurdsson, S.T. et al. (2013). A single-stranded junction modulates nanosecond motional ordering of the substrate recognition duplex of a group I ribozyme. *Chembiochem : A European Journal of Chemical Biology* 14 (14): 1720–1723.

140 Malygin, A.A., Graifer, D.M., Meschaninova, M.I. et al. (2015). Doubly spin-labeled RNA as an EPR reporter for studying multicomponent supramolecular assemblies. *Biophys. J.* 109 (12): 2637–2643.

141 Malygin, A.A., Graifer, D.M., Meschaninova, M.I. et al. (2018). Structural rearrangements in mRNA upon its binding to human 80S ribosomes revealed by EPR spectroscopy. *Nucleic Acids Res.* 46 (2): 897–904.

142 Kaminker, I., Bye, M., Mendelman, N. et al. (2015). Distance measurements between manganese(II) and nitroxide spin-labels by DEER determine a binding site of Mn^{2+} in the HP92 loop of ribosomal RNA. *Phys. Chem. Chem. Phys.* 17 (23): 15098–15102.

143 Wunnicke, D., Strohbach, D., Weigand, J.E. et al. (2011). Ligand-induced conformational capture of a synthetic tetracycline riboswitch revealed by pulse EPR. *RNA* 17 (1): 182–188.

144 Hetzke, T., Vogel, M., Gophane, D.B. et al. (2019). Influence of Mg^{2+} on the conformational flexibility of a tetracycline aptamer. *RNA* 25 (1): 158–167.

145 Krstic, I., Frolow, O., Sezer, D. et al. (2010). PELDOR spectroscopy reveals preorganization of the neomycin-responsive riboswitch tertiary structure. *J. Am. Chem. Soc.* 132 (5): 1454–1455.

146 Grytz, C.M., Marko, A., Cekan, P. et al. (2016). Flexibility and conformation of the cocaine aptamer studied by PELDOR. *Phys. Chem. Chem. Phys.* 18 (4): 2993–3002.

147 Vazquez Reyes, C., Tangprasertchai, N.S., Yogesha, S.D. et al. (2017). Nucleic acid-dependent conformational changes in CRISPR-Cas9 revealed by site-directed spin labeling. *Cell Biochem. Biophys.* 75 (2): 203–210.

148 Grohmann, D., Klose, D., Klare, J.P. et al. (2010). RNA-binding to archaeal RNA polymerase subunits F/E: a DEER and FRET study. *J. Am. Chem. Soc.* 132 (17): 5954–5955.

149 Constantinescu-Aruxandei, D., Petrovic-Stojanovska, B., Schiemann, O. et al. (2016). Taking a molecular motor for a spin: helicase mechanism studied by spin labeling and PELDOR. *Nucleic Acids Res.* 44 (2): 954–968.

150 Duss, O., Yulikov, M., Allain, F.H.T., and Jeschke, G. (2015). Combining NMR and EPR to determine structures of large RNAs and protein-RNA complexes in solution. *Methods Enzymol.* 558: 279–331.

33

Computational Modeling Methods for 3D Structure Prediction of Ribozymes

Pritha Ghosh[1], Chandran Nithin[1], Astha Joshi[1], Filip Stefaniak[1], Tomasz K. Wirecki[1], and Janusz M. Bujnicki[1,2]

[1] *International Institute of Molecular and Cell Biology in Warsaw, Laboratory of Bioinformatics and Protein Engineering, ul. Ks. Trojdena 4, PL-02-109 Warsaw, Poland*
[2] *Adam Mickiewicz University, Faculty of Biology, Institute of Molecular Biology and Biotechnology, Bioinformatics Laboratory, ul. Umultowska 89, PL-61-614 Poznan, Poland*

33.1 Introduction

The information necessary for RNA molecules to fold into the three-dimensional (3D) structure is contained within the primary sequence, and the folding of RNA happens instantaneously inside the cells [1]. In spite of this, accurate prediction of 3D structures of RNA molecules exceeding 50 nucleotides (nt) in length is a major challenge [2]. Moreover, advances in high-throughput nucleic acid sequencing techniques have led to an uneven growth rate of RNA sequence information (https://www.ncbi.nlm.nih.gov/refseq/statistics) as compared with that of RNA 3D structures (https://www.rcsb.org/stats/growth/growth-rna). This may be partly attributed to the difficulties associated with the experimental determination of RNA 3D structures, which is currently more challenging than protein structure determination [3–5]. To address this problem, computational methods have been either adapted from the existing analogous tools for proteins or developed de novo specifically for RNA. These methods not only provide new insights into the biology of RNA molecules for which there are no experimentally solved structures but also complement our understanding of RNAs for which there are existing complete or partial structures. The success of such methods depends on various factors and will be discussed later in this chapter.

RNA molecules play important roles in the metabolic activities, cell signaling, and regulation of gene expression [6]. While some of these functions are performed by RNA molecules as independent units, a vast majority of these activities are performed in complex with proteins, metal ions, and/or other ligand molecules [7–17]. Further, the discovery of “ribozymes” or catalytic RNAs in the early 1980s by Thomas R. Cech [18, 19] and Sidney Altman [20] has demonstrated that RNA molecules can also catalyze chemical reactions almost as efficiently as the long-

Ribozymes, First Edition. Edited by Sabine Müller, Benoît Masquida, and Wade Winkler.

known protein enzymes [21]. This discovery has also challenged one of the long-standing dogmas in biochemistry: "all enzymes are proteins."

Naturally occurring ribozyme families are conserved in the evolution. Within families, they may vary in size, primary sequence, and overall shape, but the catalytic site is organized around a family-specific structural framework (common core), similarly to protein enzymes [22–25]. The catalytic core may be adorned by variable peripheral elements [25–30]. Divalent cations are important for the functioning of many ribozymes either to provide structural stability and facilitate active site formation or to directly participate in catalysis [31].

The relative dearth of experimentally solved structures of ribozymes stimulated the application of computational methods for structure prediction to build 3D structural models. The experimentally solved and computationally modeled 3D structures for the naturally occurring ribozyme classes are listed in Table 33.1. Below, we discuss various approaches for the computational modeling of 3D structures of ribozymes, and we illustrate them with examples.

33.2 Computational Modeling Approaches

Contemporary RNA 3D structure prediction approaches have been influenced by developments in the field of protein 3D structure prediction, which can be categorized into two general classes: "template-based modeling" and "template-free modeling" [104]. The RNA molecule for which the 3D model needs to be built is often referred to as the "target" molecule. A template structure is generally an experimentally determined structure of evolutionarily related RNA molecule. The strategy of choice is dependent on the data available: due to the limited availability of templates, the template-based modeling is possible only for some classes of RNA molecules. In this section, we focused on template-based and template-free modeling approaches, as well as a combination of both (Figure 33.1). Each of these methods is illustrated with case studies from literature in this section.

33.2.1 Template-Based Modeling Approach

Template-based modeling of macromolecular 3D structures exploits the observation that despite the accumulation of divergent mutations, evolutionarily related or homologous macromolecules adopt the same architecture [106]. Structural modeling based on this principle was first developed for proteins and later adapted for modeling of RNAs and is popularly referred to as "comparative modeling," "homology modeling," or "template-based modeling" [107]. This approach heavily relies on macromolecular structure databases. The first step in the modeling process is to identify a "template" structure. Following the selection of the best template, the sequence of the target RNA molecule must be aligned to the template to determine the correspondence between them. Models generated by this method are highly accurate, and the level of accuracy is dependent on the degree of homology between the target and the template. However, the major limitation of the template-based modeling method is that it can accurately predict 3D structures of RNA, only if a homologous experimentally solved structure exists that can be provided as a template.

Table 33.1 List of experimentally solved and computationally modeled structures for the naturally occurring ribozyme classes.

Name of ribozyme	Experimentally solved structure(s)		Computational model(s)
	PDB ID	Reference(s)	Reference(s)
AspLC	Not yet available	—	[32]
glmS	2GCS, 2GCV, 2H0X, 2H0W, 2HO6, 2HO7	[33]	[34]
Group I intron	1L8V, 1U6B, 1TUT, 1X8W, 1Y0Q, 6BJX, 6D8L, 6D8M, 6D8N, 6D8O	[35–40]	[23, 41, 42]
Group II intron	4E8Q, 4R0D, 4Y1N, 4Y1O, 5G2Y, 5J01, 5J02, 6EZ0	[43–48]	[49]
Hairpin ribozyme	1M5K, 1M5O, 1M5P, 1M5V, 1X9C, 1X9K, 1ZFT, 1ZFV, 1ZFX, 2BCY, 2BCZ, 2D2K, 2D2L, 2FGP, 2NPY, 2NPZ, 2OUE, 3B58, 3B5A, 3B5F, 3B5S, 3B91, 3BBI, 3BBK, 3BBM, 3GS1, 3GS5, 3GS8, 4G6P, 4G6R, 4G6S	[50–55]	[56]
Hammerhead ribozyme	1MME, 1NYI, 1Q29, 1RMN, 3ZD5, 5DH6, 299D, 2QUS, 2RPK, 1HMH, 299D 2OEU, 300D, 301D, 359D, 379D, 3ZD5, 3ZP8, 3ZD4, 5DI2, 5DI4, 5DQK, 5DH6, 5DH7, 5DH8, 5EAQ, 5EAO	[57–70]	[66]
Hatchet ribozyme	6JQ5, 6JQ6	[71]	—
Hepatitis delta virus (HDV)	1DRZ, 1SJ3, 1SJ4, 1SJF, 1VBX, 1VBY, 1VBZ, 1VC0, 1VC5, 1VC6, 2OIH, 2OJ3, 3NKB, 4PRF	[72–76]	[77]

(Continued)

Table 33.1 (Continued)

Name of ribozyme	Experimentally solved structure(s)		Computational model(s)
	PDB ID	Reference(s)	Reference(s)
Lariat capping ribozyme	4P8Z, 4P9R, 4P95, 6GYV, 6G7Z	[78]	[79, 80]
Pistol ribozyme	5KTJ, 5K7C, 5K7D, 5K7E, 6R47	[81–83]	—
RNase P	1F6X, 1F6Z, 1F78, 1F79, 1F7F, 1F7G, 1F7H, 1F7I, 1JOX, 1JP0, 1U9S, 1XSG, 1XSH, 1XST, 1XSU 2A2E, 2CD1, 2CD3, 2CD5, 2CD6, 3DHS	[84–90]	—
RNA polymerase ribozyme	3HHN, 3IVK	[91]	—
Twister ribozyme	4OJI, 4QJH, 4QJD, 4RGE, 4RGF, 5DUN	[92–95]	—
Twister-sister ribozyme	5T5A, 5Y85, 5Y87	[96]	[97]
VS ribozyme	1HWQ, 1OW9, 1TBK, 1TJZ, 4R4P, 4R4V	[98–102]	[103]

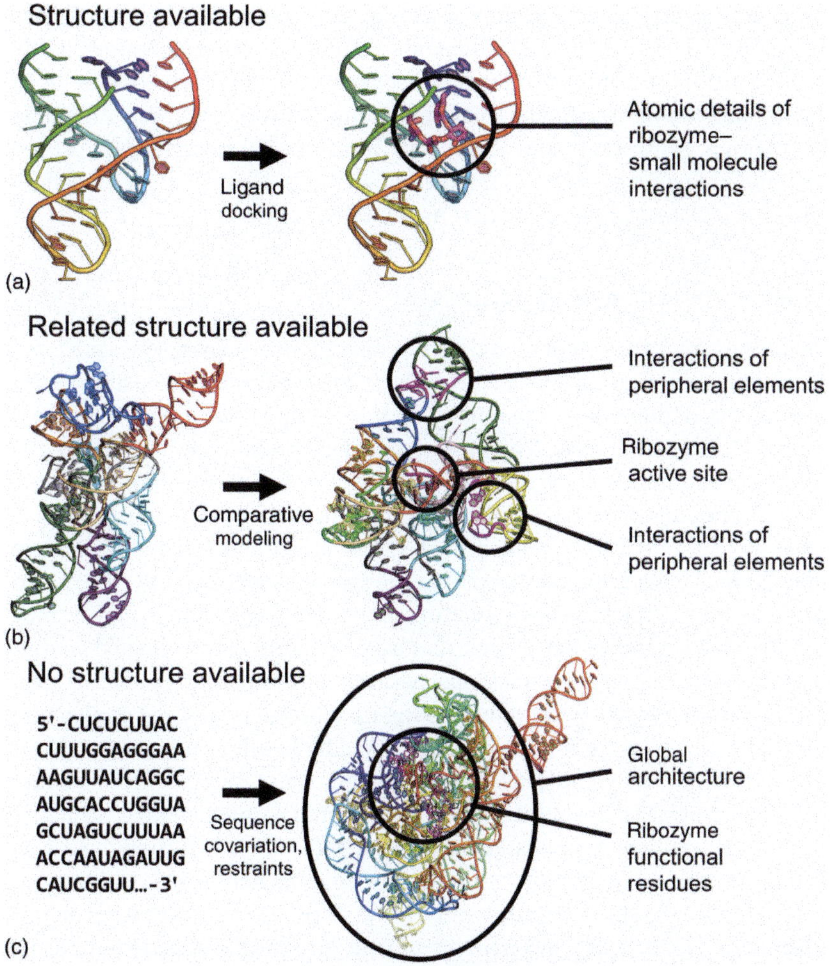

Figure 33.1 An overview of applications of modeling approaches depending on the availability of experimentally determined ribozyme structures. (a) The binding mode of neomycin to the active site of the hammerhead ribozyme was modeled by macromolecular docking using the available crystal structure. (b) A comparative model of the lariat capping ribozyme based on the *Azoarcus* group I intron crystal structure [79] can be used to predict interactions necessary for the structure and function of the ribozyme. (c) A model of *Tetrahymena thermophila* group I intron constructed using restraints based on sequence covariation and stereochemical constraints [22]. This model predicts the global fold of the RNA molecule and the arrangement of functional residues in the ribozyme. Source: (a) Adapted from Hermann and Westhof [105].

A number of template-based methods are available for modeling of RNA 3D structures, which can be applied to ribozymes. MacroMolecular Builder (previously RNABuilder), from the Altman group [108], models a target sequence by using interatomic distances and internal coordinates (with distance and torsion angles) from the aligned regions of the template structure. ModeRNA was developed in our laboratory [109, 110]. According to the target–template alignment, it copies the coordinates of conserved residues from the template structure and inserts variable regions from a database of fragments. ModeRNA's special feature is the ability to model RNA structures with posttranscriptionally modified residues. Both MMB and ModeRNA were tested on modeling group I intron structures. Assemble from the Westhof group [111] and RNA2D3D from the Shapiro group [112] are semiautomatic interactive RNA 3D modeling methods. Assemble was used to model the *Allovahlkampfia* LC ribozyme (AspLC) structure [32].

33.2.2 Template-Free Modeling Approach

The template-free modeling approach does not directly utilize complete structures of particular macromolecules. Instead, it relies on general knowledge, such as the fundamental laws of physics and chemistry and/or on statistical information from the database of known structures. This type of modeling methods is computationally more expensive and hence much slower than template-based methods. There are various approaches to reduce the associated computational cost, and one of them is to coarse-grain the system by treating groups of atoms as single interaction centers or "pseudo-atoms." This drastically improves the speed of calculations [113–115] at the cost of accuracy of the modeled structures. Template-free methods may or may not faithfully predict a native-like RNA structure with a precisely estimated energy.

Many template-free modeling methods perform a simulation of folding and thereby offer valuable insights not only into the final folded structure but also into the important aspects of the folding process. Vfold3D [116] and iFoldRNA [117] are based on molecular dynamics (MD) and attempt to predict the RNA folding process based on physical principles. They can attempt to predict RNA structure based on sequence alone. NAST is also based on MD, but it is heavily based on additional information and employs restraints to guide the folding [115]. NAST has been used to model the *Azoarcus* ribozyme [118]. SimRNA is a method developed in our laboratory [119, 120], inspired by protein structure prediction methods such as CABS [121] and REFINER [122]. SimRNA performs prediction of RNA 3D structures using a Monte Carlo simulation. It can use sequence information alone, but it can be supplemented by various restraints. All these methods use coarse-grained representations to speed up calculations.

RNAComposer, from the Adamiak group [123, 124], is an example of a method that assembles a complete RNA structure from a number of smaller fragments derived from existing structures, based on secondary structure (ss) restraints. It was used in the RNA-Puzzles Rounds II [79] and III [103] to accurately predict the structures of the lariat capping ribozyme (problem 5) and the Varkud satellite (VS) ribozyme (puzzle 7), respectively. This tool was also used for the modeling of hammerhead ribozyme-like motifs identified in archaeal genomes [125]. Fragment assembly of RNA with full-atom refinement (FARFAR) in the Rosetta framework, from the Das group [79, 126], is a de novo modeling method that builds nucleotide-resolution models of small motifs in RNA, as well as complex RNA folds. It was used in the RNA-Puzzles Rounds II [79] and III [103] to accurately predict the structures of the lariat capping ribozyme (problem 5) and the VS ribozyme (puzzle 7), respectively. The MC-Fold/MC-Sym pipeline [127] also predicts RNA structure by assembling local structural motifs, which are however generated de novo. This method was employed in the modeling of catalytic core of hammerhead ribozyme [128].

33.2.3 Combination of Modeling Approaches

The difficulties associated with the 3D structural modeling of ribozymes, especially for molecules with complex structures, can be best solved by using a modeling protocol that involves both the template-based and template-free modeling methods described above. Such a combined modeling protocol has been initially developed for 3D structure prediction of proteins [129]. Later, RNA 3D modeling exercises performed in our laboratory and in other groups have also indicated that the combined protocol yields more accurate results as compared with each of the methods when used individually [130]. Of course, the basic condition is the availability of the template structure or a related RNA. At first, template-based (homology-guided) modeling is used to determine the 3D structure of the ribozyme's conserved core. This step also generates approximate (not necessarily correct) conformations of peripheral elements, which are absent from the reference structure (template). Following this, template-free modeling is used to predict the structures of these peripheral elements and their interactions with each other, while the conserved core (modeled in the previous step based on homology to a reference structure) is kept constrained. The latter ensures that the conformational search space is reduced, leading to the decrease in the computational time required for the folding simulations. Additionally, spatial restraints derived from the experimental data can be used to improve the predictions of long-range interactions and validate the structural predictions.

In the case of availability of multiple templates (structures with good sequence identity and coverage, with two or more regions of the target sequence) for modeling, each of the different regions of the target sequence may be individually modeled based on their corresponding "best" templates. Following this, structural restraints may be obtained from each of these homology-guided models and provided as inputs to the template-free folding simulation. The final model is expected to retain the same spatial arrangement of tertiary motifs, as in the individual "partial" models. A model built in this manner is expected to be more accurate than the one built based on a single template structure.

As an example, our group used ModeRNA [107] in combination with SimRNA [119] and QRNAS [131] in the RNA-Puzzles Rounds II [79] and III [103] to accurately predict the structures of the lariat capping ribozyme (problem 5) and the VS ribozyme (puzzle 7), respectively.

33.2.4 Modeling of RNA Interactions with Ligands

RNA molecules are known to be regulated by small-molecule ligands. Many RNAs are also targets of small-molecule drugs; for example, the bacterial ribosome is one of the major antibiotic targets [16], and viral RNAs are targeted by inhibitors [132].

Hence, it is important to identify small molecules that can bind to RNA and model the 3D structure of the RNA–ligand complex. Small-molecule ligands are docked on the RNA receptor, and the docked poses are scored to identify the preferred binding mode. The scoring functions for ranking the docked poses can be knowledge-based or derived from first principles [133], as in RNA folding or RNA–protein docking. Though docking is the usual method of choice for modeling of RNA–ligand interactions, other methods like MD simulations [97] and ab initio folding methods can also be used to predict 3D structures of RNA–ligand complexes.

33.3 Case Studies

In this section, we have reviewed examples of ribozyme structures that have been modeled using various computational methods. The case studies have been discussed in the order of precision: from cases where a related structure was known that could be used as a template to cases where very little was known about the target structure. It should be noted that the choice of modeling methods is usually based on the following considerations: availability of experimentally solved homologous structures, biological questions to be addressed, and the scope of the computational methods in addressing mechanistic details of these questions.

33.3.1 Hairpin Ribozyme

The hairpin ribozyme is one of the smallest catalytic RNAs that has been found in viroid, virusoid, and satellite RNAs in plant viruses [134]. It folds into a structure composed of two domains, A and B. The 3D structure model of the hairpin ribozyme was built in 1997 by the Gait group [56], following a two-step modeling approach. The first step involved the modeling of the individual domains and loops. In the second step, guided docking of domains A and B was performed using data derived from cross-linking experiments.

33.3.2 Lariat Capping Ribozyme

The lariat capping ribozyme has evolved from a group I intron ancestor and developed specific architectural features; hence it has been classified as a separate family of ribozymes [78]. In 2008, molecular modeling of this ribozyme was performed by Nielsen, Westhof, and Masquida groups [80], following the protocol described by the Westhof group [135]. In this method, ss was deduced based on a sequence alignment. In the subsequent step, the ss was partitioned into multiple modules, for each of which 3D models were built. Following this, an iterative assembly of these models was performed to obtain a complete 3D structure. At each iteration, the models were refined and validated using experimental restraints. The modeling of the 3D structure of this ribozyme was also one of the "problems" presented at Round II of the RNA-Puzzles competition [79].

33.3.3 Group I Intron

Since their discovery in 1982, group I introns have been well studied across different kingdoms of life [136]. Biochemical and molecular biology studies have described the mechanism of group I intron functions, long before the first crystal structure was solved in 1996 [137–140]. Modeling approaches guided by a bioinformatics sequence covariation analysis [22] and experimental data such as chemical and enzymatic probing [137] or systematic nucleotide substitution [141] helped in gaining a structural insight on the functional mechanism of regulation of group I introns [142, 143]. In 1987, the first 3D model of the catalytic center of *Tetrahymena thermophila* group I intron was proposed by Kim and Cech [144]. The modeling approach was based on deriving the RNA structure rules from the crystal structure of the $tRNA^{Phe}$ [145]. Additional constraints were also imposed based on the results of comparative sequence analysis [146, 147] and the experimental chemical and enzymatic data for the accessibility of each nucleotide [138, 139].

A few years later, the Westhof group modeled the entire core of group I introns present in *T. thermophila* (TtLSU) and *Saccharomyces cerevisiae* (ScLSU) ribosomal RNA (rRNA) precursor [22]. The long-range tertiary contacts were incorporated in the modeling process, based on the covariation data derived by aligning 87 distant group I intron sequences. The detailed protocol followed for modeling was described by the Westhof group [135]. Later, they used a similar modeling approach to build 3D models of two full-length group I introns, namely, TtLSU and intron present in cytochrome b gene of *Saccharomyces douglas* (SdCob.1) [23]. The modeling approach was extended by incorporating the tertiary interactions identified by phylogenetic or experimental studies. A similar approach was later employed by the Westhof group to model *S. cerevisiae* mitochondrial intron (bI5) [41] and *Azoarcus* intron [42].

33.3.4 Varkud Satellite (VS) Ribozyme

The VS ribozyme is the largest nucleolytic ribozyme found in natural isolates of *Neurospora* [148]. The 3D structure was modeled by our group in a template-free manner with ss restraints, as a part of the RNA-Puzzles Round III [103]. Based on hints from the problem description, manual docking of modeled monomeric structures was performed to propose the structure of the dimer.

To date, only a few examples of modeling interactions of ribozymes with small molecules and ions have been described. In general, ion interactions were modeled either with MD or using 3D reference site interaction model (3D-RISM) [149], while for small-molecule ligands, molecular docking in combination with MD (for ligands) was used.

33.3.5 Twister-Sister (TS) Ribozyme

The twister-sister (TS) ribozyme [150] is a member of a group of diverse catalytic RNA species called nucleolytic ribozymes, which process replication intermediates, transcripts, and control gene expression by cleaving a specific phosphodiester linkage within the RNA [151, 152]. The crystal structure of this ribozyme (PDB ID: 5T5A) was solved by the Lilley group in 2017 [96]. It points to multiple binding sites for divalent metal ions, which stabilize tertiary contacts in the crystal. Following the experimental determination of the TS ribozyme structure, 3D-RISM [149] was used by the York group [97] to predict the occupation sites for cations both in a crystal packing environment and in a solution. Though the predictions for the former had been found to be consistent with that of the experimental data, it revealed differences in the predicted cation binding sites among the crystal and the solution systems. Hence, the authors went ahead to model the functionally active state of the TS ribozyme using MD simulations and molecular solvation theory-based analysis. Independent 2 ns MD simulations with Mg^{2+} at the alternative predicted site lead to spontaneous rearrangement of the active site residues without affecting the experimentally inferred global fold.

33.3.6 Hammerhead Ribozyme

The Westhof group used MD simulations to prove the existence of the μ-bridging between two Mg^{2+} in the experimentally solved hammerhead ribozyme structure and a hydroxide anion [105]. According to simulations' results, this magnesium cluster, located on the deep groove side, provides a hydroxyl anion to activate 2′-hydroxyl nucleophile after a conformational change in the RNA. The outcome of this research greatly facilitated the understanding of the mechanism of hammerhead ribozyme cleavage. Hermann and Westhof also modeled interactions of aminoglycoside antibiotics with the hammerhead ribozyme [105]. First, initial models were generated by performing molecular docking of the ligand to the rigid crystal structure of the RNA. Additional restraints were applied on the ammonium groups of the antibiotics in order to position them in the sites occupied by Mg^{2+} in the experimentally solved structure, which had been removed before docking. To assess the stability of the complex, generated models were subjected to MD simulations, and the time-dependent hydrogen bonding patterns in the RNA–ligand complex were evaluated.

33.3.7 Hepatitis Delta Virus (HDV) Ribozyme

Cations and water molecules also play a crucial role in self-cleavage of the hepatitis delta virus (HDV) ribozyme as they are involved in both folding and cleavage reaction catalysis. The Šponer group used 200 ns of explicit-solvent MD simulations to predict binding sites and stability of monovalent and divalent cations in HDV ribozyme substrate and product forms [153]. Obtained models are in agreement with a proposed concerted mechanism of the catalysis reaction. This system was also studied by the York group who used 3D-RISM in combination with MD simulations to assess the role of the metal ion in the catalysis process [154].

33.3.8 Ligand Docking

The Schroeder and Westhof groups used manual docking of neomycin B into the model of the catalytic core of the group I intron [155] with the aid of FRODO application [156]. Ligand conformers were derived by MD simulations using the program AMBER. Again, as modeling results suggested, ammonium groups of the antibiotic act as a replacement for the magnesium ions present in the unbound structure.

33.4 Future Perspectives

The study of biological systems is restricted due to the limited availability of experimentally solved 3D structures [157]. Computational modeling plays an important role in providing key insights to questions, which cannot be addressed by experimental methods alone. It must be emphasized that the type of the system to be modeled (e.g. the size of the RNA molecule) and the availability of the starting data (e.g. the availability of a related structure and/or experimental data that can be used as restraints) strongly determine the applicability of computational modeling methods. It is not recommended to run any modeling method as a "black box," as different methods are applicable to answer different questions.

In this chapter, we have discussed different methods for the modeling of ribozyme 3D structures supported by examples from literature illustrating their applications. "Pure" template-based modeling methods have very limited application in this scenario – they can be used only for modeling ribozymes that are related to other RNA molecules for which structures have been already determined. Hitherto, macromolecular modeling in combination with experimental restraints has proved to be a powerful tool in predicting the 3D structures with better accuracy. Consequently, combining the template-based modeling with template-free folding that uses experimentally derived restraints can be employed to model structures for many ribozymes. In cases where the related experimentally determined structure is available and the resulting model is expected to have very high accuracy, detailed modeling may be performed in the presence of ions, water molecules, ligands, or proteins to identify the interactions and modes of ribozyme actions. However, it should be noted that in the absence of sufficient data, modeling of details may not lead to meaningful predictions.

Currently, one of the serious limitations of the prediction of ribozyme–ligand complexes is the lack of advanced bioinformatics tools that take the flexibility of interacting partners into consideration. The development of such methodology is a major goal for future research.

Acknowledgments

We thank Elżbieta Purta, Katarzyna Merdas, Smita Priyadarshini Pilla, Rohit Suratekar, and Amal Thomas for critical reading of the manuscript and valuable comments. The research on RNA structure and interactions in the Bujnicki laboratory was funded primarily by the Foundation for Polish Science (FNP, grants TEAM/2009-4/2 and TEAM/2016-3/18) and by the European Research Council (ERC, StG grant RNA+P = 123D), and it is currently supported by the Polish

National Science Center (NCN, grant MAESTRO 2017/26/A/NZ1/01083). J.M.B. was also supported by the "Ideas for Poland" fellowship from the FNP. P.G. and F.S. were supported by the Foundation for Polish Science grant TEAM/2016-3/18. C.N. was supported by IIMCB statutory funds. A.J. was supported by Polish National Science Centre (NCN, grant 2017/24/T/NZ1/00360). T.K.W. was supported by Polish National Science Centre (NCN, grant OPUS 2017/25/B/NZ2/01294).

References

1 Ferré-D'Amaré, A.R. and Doudna, J.A. (1999). RNA folds: insights from recent crystal structures. *Annu. Rev. Biophys. Biomol. Struct.* 28: 57–73.

2 Wang, J., Mao, K., Zhao, Y. et al. (2017). Optimization of RNA 3D structure prediction using evolutionary restraints of nucleotide–nucleotide interactions from direct coupling analysis. *Nucleic Acids Res.* 45 (11): 6299–6309.

3 Doudna, J.A. (2000). Structural genomics of RNA. *Nat. Struct. Biol.* 7 (Suppl): 954–956.

4 Ke, A. and Doudna, J.A. (2004). Crystallization of RNA and RNA–protein complexes. *Methods.* 34 (3): 408–414.

5 Scott, L.G. and Hennig, M. (2008). RNA structure determination by NMR. *Methods Mol. Biol.* 452: 29–61.

6 Morris, K.V. and Mattick, J.S. (2014). The rise of regulatory RNA. *Nat. Rev. Genet.* 15 (6): 423–437.

7 Nithin, C., Ghosh, P., and Bujnicki, J.M. (2018). Bioinsformatics tools and benchmarks for computational docking and 3D structure prediction of RNA–protein complexes. *Genes.* 9 (9): 432.

8 Lunde, B.M., Moore, C., and Varani, G. (2007). RNA-binding proteins: modular design for efficient function. *Nat. Rev. Mol. Cell Biol.* 8 (6): 479–490.

9 Nagai, K. (1996). RNA – protein complexes. *Curr. Opin. Struct. Biol.* 6 (1): 53–61.

10 Butcher, S.E., Allain, F.H., and Feigon, J. (2000). Determination of metal ion binding sites within the hairpin ribozyme domains by NMR. *Biochemistry* 39 (9): 2174–2182.

11 Cusack, S. (1999). RNA–protein complexes. *Curr. Opin. Struct. Biol.* 9 (1): 66–73.

12 Hermann, T. (2003). Chemical and functional diversity of small molecule ligands for RNA. *Biopolymers.* 70 (1): 4–18.

13 Thomas, J.R. and Hergenrother, P.J. (2008). Targeting RNA with small molecules. *Chem. Rev.* 108 (4): 1171–1224.

14 Nithin, C., Mukherjee, S., and Bahadur, R.P. (2019). A structure-based model for the prediction of protein–RNA binding affinity. *RNA.* 25 (12): 1628–1645.

15 Ghosh, P., Mathew, O.K., and Sowdhamini, R.J.B.b. (2016). RStrucFam: a web server to associate structure and cognate RNA for RNA-binding proteins from sequence information. *BMC Bioinformatics.* 17 (1): 411.

16 Wilson, D.N. (2014). Ribosome-targeting antibiotics and mechanisms of bacterial resistance. *Nat. Rev. Microbiol.* 12 (1): 35–48.

17 Nithin, C., Mukherjee, S., and Bahadur, R.P. (2017). A non-redundant protein-RNA docking benchmark version 2.0. *Proteins* 85 (2): 256–267.

18 Zaug, A.J. and Cech, T.R. (1980). In vitro splicing of the ribosomal RNA precursor in nuclei of *Tetrahymena*. *Cell* 19 (2): 331–338.

19 Cech, T.R., Zaug, A.J., and Grabowski, P.J. (1981). In vitro splicing of the ribosomal RNA precursor of *Tetrahymena*: involvement of a guanosine nucleotide in the excision of the intervening sequence. *Cell* 27 (3 Pt 2): 487–496.

20 Guerrier-Takada, C., Gardiner, K., Marsh, T. et al. (1983). The RNA moiety of ribonuclease P is the catalytic subunit of the enzyme. *Cell* 35 (3): 849–857.

21 Emilsson, G.M., Nakamura, S., Roth, A., and Breaker, R.R. (2003). Ribozyme speed limits. *RNA* 9 (8): 907–918.

22 Michel, F. and Westhof, E. (1990). Modelling of the three-dimensional architecture of group I catalytic introns based on comparative sequence analysis. *J. Mol. Biol.* 216 (3): 585–610.

23 Lehnert, V., Jaeger, L., Michel, F., and Westhof, E. (1996). New loop-loop tertiary interactions in self-splicing introns of subgroup IC and ID: a complete 3D model of the *Tetrahymena thermophila* ribozyme. *Chem. Biol.* 3 (12): 993–1009.

24 Doherty, E.A. and Doudna, J.A. (2001). Ribozyme structures and mechanisms. *Annu. Rev. Biophys. Biomol. Struct.* 30: 457–475.

25 Serganov, A. and Patel, D.J. (2007). Ribozymes, riboswitches and beyond: regulation of gene expression without proteins. *Nat. Rev. Genet.* 8 (10): 776–790.

26 Cech, T.R. (1989). Conserved sequences and structures of group I introns: building an active site for RNA catalysis – a review. In: (eds. M. Belfort, D.A. Shub) *RNA: Catalysis, Splicing, Evolution*, 191–203. Amsterdam: Elsevier.

27 Cech, T.R. (1990). Self-splicing of group I introns. *Annu. Rev. Biochem.* 59: 543–568.

28 Zhao, C. and Pyle, A.M. (2017). Structural insights into the mechanism of group II intron splicing. *Trends Biochem. Sci.* 42 (6): 470–482.

29 Kazantsev, A.V., Rambo, R.P., Karimpour, S. et al. (2011). Solution structure of RNase P RNA. *RNA* 17 (6): 1159–1171.

30 Mitra, S., Laederach, A., Golden, B.L. et al. (2011). RNA molecules with conserved catalytic cores but variable peripheries fold along unique energetically optimized pathways. *RNA* 17 (8): 1589–1603.

31 Hanna, R. and Doudna, J.A. (2000). Metal ions in ribozyme folding and catalysis. *Curr. Opin. Chem. Biol.* 4 (2): 166–170.

32 Tang, Y., Nielsen, H., Masquida, B. et al. (2014). Molecular characterization of a new member of the lariat capping twin-ribozyme introns. *Mobile DNA* 5: 25.

33 Klein, D.J. and Ferré-D'Amaré, A.R. (2006). Structural basis of glmS ribozyme activation by glucosamine-6-phosphate. *Science* 313 (5794): 1752–1756.

34 McCown, P.J., Roth, A., and Breaker, R.R. (2011). An expanded collection and refined consensus model of glmS ribozymes. *RNA* 17 (4): 728–736.

35 Battle, D.J. and Doudna, J.A. (2002). Specificity of RNA–RNA helix recognition. *Proc. Natl. Acad. Sci. U.S.A.* 99 (18): 11676–11681.

36 Adams, P.L., Stahley, M.R., Kosek, A.B. et al. (2004). Crystal structure of a self-splicing group I intron with both exons. *Nature* 430 (6995): 45–50.

37 Znosko, B.M., Kennedy, S.D., Wille, P.C. et al. (2004). Structural features and thermodynamics of the J4/5 loop from the *Candida albicans* and *Candida dubliniensis* group I introns. *Biochemistry* 43 (50): 15822–15837.

38 Guo, F., Gooding, A.R., and Cech, T.R. (2004). Structure of the *Tetrahymena* ribozyme: base triple sandwich and metal ion at the active site. *Mol. Cell* 16 (3): 351–362.

39 Golden, B.L., Kim, H., and Chase, E. (2005). Crystal structure of a phage Twort group I ribozyme–product complex. *Nat. Struct. Mol. Biol.* 12(1): 82–9.

40 Shoffner, G.M., Wang, R., Podell, E. et al. (2018). In crystallo selection to establish new RNA crystal contacts. *Structure* 26 (9): 1275–1283.e1273.

41 Jaeger, L., Westhof, E., and Michel, F. (1991). Function of P11, a tertiary base pairing in self-splicing introns of subgroup IA. *J. Mol. Biol.* 221 (4): 1153–1164.

42 Rangan, P., Masquida, B., Westhof, E., and Woodson, S.A. (2003). Assembly of core helices and rapid tertiary folding of a small bacterial group I ribozyme. *Proc. Natl. Acad. Sci. U.S.A.* 100 (4): 1574–1579.

43 Marcia, M. and Pyle, A.M. (2012). Visualizing group II intron catalysis through the stages of splicing. *Cell* 151 (3): 497–507.

44 Robart, A.R., Chan, R.T., Peters, J.K. et al. (2014). Crystal structure of a eukaryotic group II intron lariat. *Nature* 514 (7521): 193–197.

45 Zhao, C., Rajashankar, K.R., Marcia, M., and Pyle, A.M. (2015). Crystal structure of group II intron domain 1 reveals a template for RNA assembly. *Nat. Chem. Biol.* 11 (12): 967–972.

46 Qu, G., Kaushal, P.S., Wang, J. et al. (2016). Structure of a group II intron in complex with its reverse transcriptase. *Nat. Struct. Mol. Biol.* 23 (6): 549–557.

47 Erat, M.C., Besic, E., Oberhuber, M. et al. (2018). Specific phosphorothioate substitution within domain 6 of a group II intron ribozyme leads to changes in local structure and metal ion binding. *J. Biol. Inorg. Chem.* 23 (1): 167–177.

48 Costa, M., Walbott, H., Monachello, D. et al. (2016). Crystal structures of a group II intron lariat primed for reverse splicing. *Science* 354 (6316).

49 Dai, L., Chai, D., Gu, S.-Q. et al. (2008). A three-dimensional model of a group II intron RNA and its interaction with the intron-encoded reverse transcriptase. *Mol. Cell* 30 (4): 472–485.

50 Rupert, P.B., Massey, A.P., Sigurdsson, S.T., and Ferre-D'Amare, A.R. (2002) Transition state stabilization by a catalytic RNA. *Science.* 298(5597): 1421–4.

51 Alam, S., Grum-Tokars, V., Krucinska, J. et al. (2005). Conformational heterogeneity at position U37 of an all-RNA hairpin ribozyme with implications for metal binding and the catalytic structure of the S-turn. *Biochemistry* 44 (44): 14396–14408.

52 Salter, J., Krucinska, J., Alam, S. et al. (2006). Water in the active site of an all-RNA hairpin ribozyme and effects of Gua8 base variants on the geometry of phosphoryl transfer. *Biochemistry* 45 (3): 686–700.

53 MacElrevey, C., Spitale, R.C., Krucinska, J., and Wedekind, J.E. (2007). A posteriori design of crystal contacts to improve the X-ray diffraction properties of a small RNA enzyme. *Acta Crystallogr., Sect. D: Biol. Crystallogr.* 63 (Pt 7): 812–825.

54 Spitale, R.C., Volpini, R., Heller, M.G. et al. (2009). Identification of an imino group indispensable for cleavage by a small ribozyme. *J. Am. Chem. Soc.* 131 (17): 6093–6095.

55 Liberman, J.A., Guo, M., Jenkins, J.L. et al. (2012). A transition-state interaction shifts nucleobase ionization toward neutrality to facilitate small ribozyme catalysis. *J. Am. Chem. Soc.* 134 (41): 16933–16936.

56 Earnshaw, D.J., Masquida, B.t., MuÈller, S. et al. (1997). Inter-domain cross-linking and molecular modelling of the hairpin ribozyme. *J. Mol. Biol.* 274 (2): 197–212.

57 Feig, A.L., Scott, W.G., and Uhlenbeck, O.C. (1998). Inhibition of the hammerhead ribozyme cleavage reaction by site-specific binding of Tb. *Science* 279 (5347): 81–84.

58 Scott, W.G., Finch, J.T., and Klug, A. (1995). The crystal structure of an all-RNA hammerhead ribozyme: a proposed mechanism for RNA catalytic cleavage. *Cell* 81 (7): 991–1002.

59 Anderson, M., Schultz, E.P., Martick, M., and Scott, W.G. (2013). Active-site monovalent cations revealed in a 1.55-Å-resolution hammerhead ribozyme structure. *J. Mol. Biol.* 425 (20): 3790–3798.

60 Schultz, E.P., Vasquez, E.E., and Scott, W.G. (2014). Structural and catalytic effects of an invariant purine substitution in the hammerhead ribozyme: implications for the mechanism of acid-base catalysis. *Acta Crystallogr., Sect. D: Biol. Crystallogr.* 70 (Pt 9): 2256–2263.

61 Martick, M. and Scott, W.G. (2006). Tertiary contacts distant from the active site prime a ribozyme for catalysis. *Cell* 126 (2): 309–320.

62 Chi, Y.-I., Martick, M., Lares, M. et al. (2008). Capturing hammerhead ribozyme structures in action by modulating general base catalysis. *PLoS Biol.* 6 (9): e234.

63 Martick, M., Lee, T.-S., York, D.M., and Scott, W.G. (2008). Solvent structure and hammerhead ribozyme catalysis. *Chem. Biol.* 15 (4): 332–342.

64 Dunham, C.M., Murray, J.B., and Scott, W.G. (2003). A helical twist-induced conformational switch activates cleavage in the hammerhead ribozyme. *J. Mol. Biol.* 332 (2): 327–336.

65 Murray, J.B., Terwey, D.P., Maloney, L. et al. (1998). The structural basis of hammerhead ribozyme self-cleavage. *Cell* 92 (5): 665–673.

66 Tuschl, T., Gohlke, C., Jovin, T.M. et al. (1994). A three-dimensional model for the hammerhead ribozyme based on fluorescence measurements. *Science* 266 (5186): 785–789.

67 Dufour, D., de la Peña, M., Gago, S. et al. (2009). Structure-function analysis of the ribozymes of chrysanthemum chlorotic mottle viroid: a loop-loop interaction motif conserved in most natural hammerheads. *Nucleic Acids Res.* 37 (2): 368–381.

68 Mir, A., Chen, J., Robinson, K. et al. (2015). Two divalent metal ions and conformational changes play roles in the hammerhead ribozyme cleavage reaction. *Biochemistry* 54 (41): 6369–6381.

69 Scott, W.G., Murray, J.B., Arnold, J.R. et al. (1996). Capturing the structure of a catalytic RNA intermediate: the hammerhead ribozyme. *Science* 274 (5295): 2065–2069.

70 Mir, A. and Golden, B.L. (2016). Two active site divalent ions in the crystal structure of the hammerhead ribozyme bound to a transition state analogue. *Biochemistry* 55 (4): 633–636.

71 Zheng, L., Falschlunger, C., Huang, K. et al. (2019). Hatchet ribozyme structure and implications for cleavage mechanism.*Proc Natl Acad Sci USA*. 116 (22): 10783–10791.

72 Ferré-D'Amaré, A.R., Zhou, K., and Doudna, J.A. (1998). Crystal structure of a hepatitis delta virus ribozyme. *Nature* 395 (6702): 567–574.

73 Ke, A., Zhou, K., Ding, F. et al. (2004). A conformational switch controls hepatitis delta virus ribozyme catalysis. *Nature* 429 (6988): 201–205.

74 Kapral, G.J., Jain, S., Noeske, J. et al. (2014). New tools provide a second look at HDV ribozyme structure, dynamics and cleavage. *Nucleic Acids Res.* 42 (20): 12833–12846.

75 Ke, A., Ding, F., Batchelor, J.D., and Doudna, J.A. (2007). Structural roles of monovalent cations in the HDV ribozyme. *Structure* 15 (3): 281–287.

76 Chen, J.-H., Yajima, R., Chadalavada, D.M. et al. (2010). A 1.9 Å crystal structure of the HDV ribozyme precleavage suggests both lewis acid and general acid mechanisms contribute to phosphodiester cleavage. *Biochemistry* 49 (31): 6508–6518.

77 Tanner, N.K., Schaff, S., Thill, G. et al. (1994). A three-dimensional model of hepatitis delta virus ribozyme based on biochemical and mutational analyses. *Curr. Biol.* 4 (6): 488–498.

78 Meyer, M., Nielsen, H., Oliéric, V. et al. (2014). Speciation of a group I intron into a lariat capping ribozyme. *Proc. Natl. Acad. Sci. U.S.A.* 111 (21): 7659–7664.

79 Miao, Z., Adamiak, R.W., Blanchet, M.-F. et al. (2015). RNA-Puzzles Round II: assessment of RNA structure prediction programs applied to three large RNA structures. *RNA* 21 (6): 1066–1084.

80 Beckert, B., Nielsen, H., Einvik, C. et al. (2008). Molecular modelling of the GIR1 branching ribozyme gives new insight into evolution of structurally related ribozymes. *EMBO J.* 27 (4): 667–678.

81 Ren, A., Vušurović, N., Gebetsberger, J. et al. (2016). Pistol ribozyme adopts a pseudoknot fold facilitating site-specific in-line cleavage. *Nat. Chem. Biol.* 12 (9): 702–708.

82 Nguyen, L.A., Wang, J., and Steitz, T.A. (2016) Crystal structure of Pistol, a class of self-cleaving ribozyme. *Proc Natl Acad Sci U S A*. 114(5): 1021–1026.

83 Wilson, T.J., Liu, Y., Li, N.-S. et al. (2019). Comparison of the structures and mechanisms of the Pistol and Hammerhead ribozymes. *J. Am. Chem. Soc.* 141 (19): 7865–7875.

84 Torres-Larios, A., Swinger, K.K., Krasilnikov, A.S. et al. (2005). Crystal structure of the RNA component of bacterial ribonuclease P. *Nature* 437 (7058): 584–587.

85 Kazantsev, A.V., Krivenko, A.A., and Pace, N.R. (2009). Mapping metal-binding sites in the catalytic domain of bacterial RNase P RNA. *RNA* 15 (2): 266–276.

86 Schmitz, M. (2004). Change of RNase P RNA function by single base mutation correlates with perturbation of metal ion binding in P4 as determined by NMR spectroscopy. *Nucleic Acids Res.* 32 (21): 6358–6366.

87 Krasilnikov, A.S., Xiao, Y., Pan, T., and Mondragón, A. (2004). Basis for structural diversity in homologous RNAs. *Science* 306 (5693): 104–107.

88 Leeper, T.C., Martin, M.B., Kim, H. et al. (2002). Structure of the UGAGAU hexaloop that braces *Bacillus* RNase P for action. *Nat. Struct. Biol.* 9 (5): 397–403.

89 Schmitz, M. and Tinoco, I. Jr. (2000). Solution structure and metal-ion binding of the P4 element from bacterial RNase P RNA. *RNA* 6 (9): 1212–1225.

90 Schmitz, M., Tinoco Jr., I. (2000) Solution structure and metal-ion binding of the P4 element from bacterial RNase P RNA. *RNA* 6(9): 1212–1225.

91 Shechner, D.M., Grant, R.A., Bagby, S.C. et al. (2009). Crystal structure of the catalytic core of an RNA-polymerase ribozyme. *Science* 326 (5957): 1271–1275.

92 Liu, Y., Wilson, T.J., McPhee, S.A., and Lilley, D.M.J. (2014). Crystal structure and mechanistic investigation of the twister ribozyme. *Nat. Chem. Biol.* 10 (9): 739–744.

93 Eiler, D., Wang, J., and Steitz, T.A. (2014). Structural basis for the fast self-cleavage reaction catalyzed by the twister ribozyme. *Proc. Natl. Acad. Sci. U.S.A.* 111 (36): 13028–13033.

94 Ren, A., Košutić, M., Rajashankar, K.R. et al. (2014). In-line alignment and Mg^{2+} coordination at the cleavage site of the env22 twister ribozyme. *Nat. Commun.* 5: 5534.

95 Košutić, M., Neuner, S., Ren, A. et al. (2015). A mini-twister variant and impact of residues/cations on the phosphodiester cleavage of this ribozyme class. *Angew. Chem. Int. Ed.* 54 (50): 15128–15133.

96 Liu, Y., Wilson, T.J., and Lilley, D.M.J. (2017). The structure of a nucleolytic ribozyme that employs a catalytic metal ion. *Nat. Chem. Biol.* 13 (5): 508–513.

97 Gaines, C.S. and York, D.M. (2017). Model for the functional active state of the TS ribozyme from molecular simulation. *Angew. Chem. Int. Ed.* 56 (43): 13392–13395.

98 Suslov, N.B., DasGupta, S., Huang, H. et al. (2015). Crystal structure of the Varkud satellite ribozyme. *Nat. Chem. Biol.* 11 (11): 840–846.

99 Hoffmann, B., Mitchell, G.T., Gendron, P. et al. (2003). NMR structure of the active conformation of the Varkud satellite ribozyme cleavage site. *Proc. Natl. Acad. Sci. U.S.A.* 100 (12): 7003–7008.

100 Campbell, D.O. and Legault, P. (2005). Nuclear magnetic resonance structure of the Varkud satellite ribozyme stem-loop V RNA and magnesium-ion binding from chemical-shift mapping. *Biochemistry* 44 (11): 4157–4170.

101 Flinders, J. and Dieckmann, T. (2004). The solution structure of the VS ribozyme active site loop reveals a dynamic “hot-spot”. *J. Mol. Biol.* 341 (4): 935–949.

102 Flinders, J. and Dieckmann, T. (2001). A ph controlled conformational switch in the cleavage site of the VS ribozyme substrate RNA1. *J. Mol. Biol.* 308 (4): 665–679.

103 Miao, Z., Adamiak, R.W., Antczak, M. et al. (2017). RNA-Puzzles Round III: 3D RNA structure prediction of five riboswitches and one ribozyme. *RNA* 23 (5): 655–672.

104 Magnus, M., Matelska, D., Lach, G. et al. (2014). Computational modeling of RNA 3D structures, with the aid of experimental restraints. *RNA Biol.* 11 (5): 522–536.

105 Hermann, T. and Westhof, E. (1998). Aminoglycoside binding to the hammerhead ribozyme: a general model for the interaction of cationic antibiotics with RNA. *J. Mol. Biol.* 276 (5): 903–912.

106 Chothia, C. and Lesk, A.M. (1986). The relation between the divergence of sequence and structure in proteins. *EMBO J.* 5(4): 823–6.

107 Rother, K., Rother, M., Boniecki, M. et al. (2011). RNA and protein 3D structure modeling: similarities and differences. *J. Mol. Model.* 17 (9): 2325–2336.

108 Flores, S.C., Wan, Y., Russell, R., and Altman, R.B. (2010). Predicting RNA structure by multiple template homology modeling. *Pac. Symp. Biocomput.*: 216–227.

109 Rother, M., Milanowska, K., Puton, T. et al. (2011). ModeRNA server: an online tool for modeling RNA 3D structures. *Bioinformatics* 27 (17): 2441–2442.

110 Rother, M., Rother, K., Puton, T., and Bujnicki, J.M. (2011). ModeRNA: a tool for comparative modeling of RNA 3D structure. *Nucleic Acids Res.* 39 (10): 4007–4022.

111 Jossinet, F., Ludwig, T.E., and Westhof, E. (2010). Assemble: an interactive graphical tool to analyze and build RNA architectures at the 2D and 3D levels. *Bioinformatics* 26 (16): 2057–2059.

112 Martinez, H.M., Maizel, J.V. Jr., and Shapiro, B.A. (2008). RNA2D3D: a program for generating, viewing, and comparing 3-dimensional models of RNA. *J. Biomol. Struct. Dyn.* 25 (6): 669–683.

113 Tozzini, V. (2010). Multiscale modeling of proteins. *Acc. Chem. Res.* 43 (2): 220–230.

114 Cragnolini, T., Laurin, Y., Derreumaux, P., and Pasquali, S. (2015). Coarse-grained HiRE-RNA model for ab Initio RNA folding beyond simple molecules, including noncanonical and multiple base pairings. *J. Chem. Theory Comput.* 11 (7): 3510–3522.

115 Jonikas, M.A., Radmer, R.J., Laederach, A. et al. (2009). Coarse-grained modeling of large RNA molecules with knowledge-based potentials and structural filters. *RNA* 15 (2): 189–199.

116 Xu, X., Zhao, P., and Chen, S.J. (2014). Vfold: a web server for RNA structure and folding thermodynamics prediction. *PLoS One* 9 (9): e107504.

117 Sharma, S., Ding, F., and Dokholyan, N.V. (2008). iFoldRNA: three-dimensional RNA structure prediction and folding. *Bioinformatics* 24 (17): 1951–1952.

118 Flores, S.C., Jonikas, M., Bruns, C. et al. (2012). Methods for building and refining 3D models of RNA. In: *RNA 3D Structure Analysis and Prediction* (eds. N. Leontis and E. Westhof), 143–166. Berlin, Heidelberg: Springer.

119 Boniecki, M.J., Lach, G., Dawson, W.K. et al. (2016). SimRNA: a coarse-grained method for RNA folding simulations and 3D structure prediction. *Nucleic Acids Res.* 44 (7): e63.

120 Magnus, M., Boniecki, M.J., Dawson, W., and Bujnicki, J.M.J.N.a.r. (2016). SimRNAweb: a web server for RNA 3D structure modeling with optional restraints. *Nucleic Acids Res.* 44 (W1): W315–W319.

121 Kolinski, A. (2004). Protein modeling and structure prediction with a reduced representation. *Acta Biochim. Pol.* 51 (2): 349–371.

122 Boniecki, M., Rotkiewicz, P., Skolnick, J., and Kolinski, A. (2003). Protein fragment reconstruction using various modeling techniques. *J. Comput.-Aided Mol. Des.* 17 (11): 725–738.

123 Popenda, M., Szachniuk, M., Antczak, M. et al. (2012). Automated 3D structure composition for large RNAs. *Nucleic Acids Res.* 40 (14): e112.

124 Purzycka, K.J., Popenda, M., Szachniuk, M. et al. (2015). Chapter one – Automated 3D RNA structure prediction using the RNAComposer method for riboswitches1. In: *Methods in Enzymology*, vol. 553 (eds. S.-J. Chen and D.H. Burke-Aguero), 3–34. Academic Press.

125 Gupta, A. and Swati, D. (2017). Hammerhead ribozymes in archaeal genomes: a computational hunt. *Interdiscip. Sci.* 9 (2): 192–204.

126 Das, R. and Baker, D. (2007). Automated *de novo* prediction of native-like RNA tertiary structures. *Proc. Natl. Acad. Sci. U.S.A.* 104 (37): 14664–14669.

127 Parisien, M. and Major, F. (2008). The MC-Fold and MC-Sym pipeline infers RNA structure from sequence data. *Nature* 452 (7183): 51–55.

128 Pinard, R., Lambert, D., Walter, N.G. et al. (1999). Structural basis for the guanosine requirement of the hairpin ribozyme. *Biochemistry* 38 (49): 16035–16039.

129 Koliński, A. and Bujnicki, J.M. (2005). Generalized protein structure prediction based on combination of fold-recognition with de novo folding and evaluation of models. *Proteins* 61 (Suppl 7): 84–90.

130 Piatkowski, P., Kasprzak, J.M., Kumar, D. et al. (2016). RNA 3D structure modeling by combination of template-based method ModeRNA, template-free folding with SimRNA, and refinement with QRNAS. In: *RNA Structure Determination: Methods and Protocols* (eds. D.H. Turner and D.H. Mathews), 217–235. New York, NY: Springer.

131 Stasiewicz, J., Mukherjee, S., Nithin, C., and Bujnicki, J.M. (2019). QRNAS: software tool for refinement of nucleic acid structures. *BMC Struct. Biol.* 19 (1): 5.

132 Thomas, H. (2016). Small molecules targeting viral RNA. *Wiley Interdiscip. Rev.: RNA* 7 (6): 726–743.

133 Stefaniak, F., Chudyk, E.I., Bodkin, M. et al. (2015). Modeling of ribonucleic acid-ligand interactions. *Wiley Interdiscip. Rev.: Comput. Mol. Sci.* 5 (6): 425–439.

134 Burke, J.M. (1996). Hairpin ribozyme: current status and future prospects. *Biochem. Soc. Trans.* 24 (3): 608–615.

135 Masquida, B. and Westhof, E. (2005). Modeling the architecture of structured RNAs within a modular and hierarchical framework. In: (eds. Roland K. Hartmann, Albrecht Bindereif, Astrid Schön, Eric Westhof) *Handbook of RNA Biochemistry*, 536–545.

136 Nielsen, H. and Johansen, S.D. (2009). Group I introns: moving in new directions. *RNA Biol.* 6 (4): 375–383.

137 Inoue, T. and Cech, T.R. (1985). Secondary structure of the circular form of the *Tetrahymena* rRNA intervening sequence: a technique for RNA structure analysis using chemical probes and reverse transcriptase. *Proc. Natl. Acad. Sci. U.S.A.* 82 (3): 648–652.

138 Tanner, N.K. and Cech, T.R. (1985). Self-catalyzed cyclization of the intervening sequence RNA of *Tetrahymena*: inhibition by methidiumpropyl-EDTA and localization of the major dye binding sites. *Nucleic Acids Res.* 13 (21): 7759–7779.

139 Cech, T.R., Tanner, N.K., Tinoco, I. Jr. et al. (1983). Secondary structure of the *Tetrahymena* ribosomal RNA intervening sequence: structural homology with fungal mitochondrial intervening sequences. *Proc. Natl. Acad. Sci. U.S.A.* 80 (13): 3903–3907.

140 Cate, J.H., Gooding, A.R., Podell, E. et al. (1996). Crystal structure of a group I ribozyme domain: principles of RNA packing. *Science* 273 (5282): 1678–1685.

141 Couture, S., Ellington, A.D., Gerber, A.S. et al. (1990). Mutational analysis of conserved nucleotides in a self-splicing group I intron. *J. Mol. Biol.* 215 (3): 345–358.

142 Winter, A.J., Groot Koerkamp, M.J., and Tabak, H.F. (1990). The mechanism of group I self-splicing: an internal guide sequence can be provided in trans. *EMBO J.* 9 (6): 1923–1928.

143 Michel, F., Hanna, M., Green, R. et al. (1989). The guanosine binding site of the *Tetrahymena* ribozyme. *Nature* 342 (6248): 391–395.

144 Kim, S.H. and Cech, T.R. (1987). Three-dimensional model of the active site of the self-splicing rRNA precursor of *Tetrahymena*. *Proc. Natl. Acad. Sci. U.S.A.* 84 (24): 8788–8792.

145 Levitt, M. (1969). Detailed molecular model for transfer ribonucleic acid. *Nature* 224 (5221): 759–763.

146 Fox, G.E. and Woese, C.R. (1975). 5S RNA secondary structure. *Nature* 256 (5517): 505–507.

147 Noller, H.F. and Woese, C.R. (1981). Secondary structure of 16S ribosomal RNA. *Science* 212 (4493): 403–411.

148 Lilley, D.M. (2004). The Varkud satellite ribozyme. *RNA* 10 (2): 151–158.

149 Luchko, T., Gusarov, S., Roe, D.R. et al. (2010). Three-dimensional molecular theory of solvation coupled with molecular dynamics in Amber. *J. Chem. Theory Comput.* 6 (3): 607–624.

150 Weinberg, Z., Kim, P.B., Chen, T.H. et al. (2015). New classes of self-cleaving ribozymes revealed by comparative genomics analysis. *Nat. Chem. Biol.* 11 (8): 606–610.

151 Lilley, D.M.J. and Eckstein, F. (2008). *Ribozymes and RNA Catalysis*. Royal Society of Chemistry.

152 Wilson, T.J. and Lilley, D.M.J. (2009). Biochemistry. The evolution of ribozyme chemistry. *Science* 323 (5920): 1436–1438.

153 Krasovska, M.V., Sefcikova, J., Réblová, K. et al. (2006). Cations and hydration in catalytic RNA: molecular dynamics of the hepatitis delta virus ribozyme. *Biophys. J.* 91 (2): 626–638.

154 Radak, B.K., Lee, T.-S., Harris, M.E., and York, D.M. (2015). Assessment of metal-assisted nucleophile activation in the hepatitis delta virus ribozyme from molecular simulation and 3D-RISM. *RNA* 21 (9): 1566–1577.

155 Hoch, I., Berens, C., Westhof, E., and Schroeder, R. (1998). Antibiotic inhibition of RNA catalysis: neomycin B binds to the catalytic core of the td group I intron displacing essential metal ions[1], [1]Edited by M. Yaniv. *J. Mol. Biol.* 282 (3): 557–569.

156 Jones, T.A. (1978). A graphics model building and refinement system for macromolecules. *J. Appl. Crystallogr.* 11 (4): 268–272.

157 Ponce-Salvatierra, A., Astha, J., Merdas, K. et al. (2019). Computational modeling of RNA 3D structure based on experimental data. 39 (2): BSR20180430.

Index

Ribozymes, First Edition. Edited by Sabine Müller, Benoît Masquida, and Wade Winkler.

d

e

h

i

k

l